More Than Just a Textbook

Internet

Step 1 Connect to Chemistry
glencoe.com

Step 2 Connect to online resources by using  codes and directly access the chapter you want.

CCAA7244c1

Enter this code with the appropriate chapter number.

Check out the following features on your **Online Learning Center:**

Study Tools

Online Student Edition
Interactive Tutors
Personal Tutors
Vocabulary PuzzleMaker
Multilingual Science Glossary
Chapter Tests
Standardized Test Practice

Concepts In Motion
- Interactive Tables
- Interactive Figures

Study to Go
- Section Self-Check Quizzes
- e-flash cards

Extensions

Prescreened Web Links
Periodic Table Links
Career Links

WebQuest Projects
Science Fair Ideas

For Teachers

Teacher Bulletin Board
Teaching Today

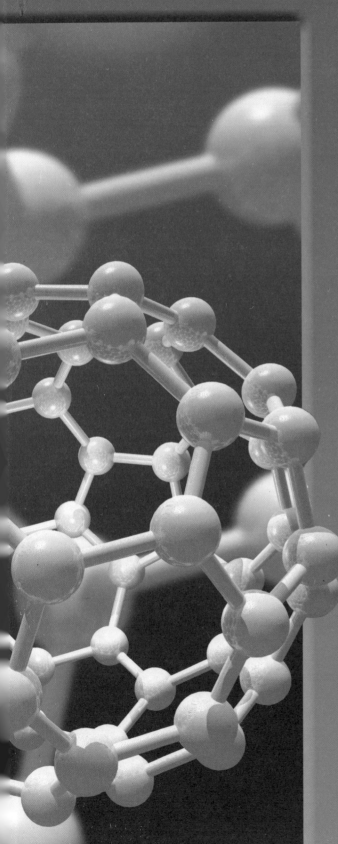

Safety Symbols

These safety symbols are used in laboratory and investigations in this book to indicate possible hazards. Learn the meaning of each symbol and refer to this page often. *Remember to wash your hands thoroughly after completing lab procedures.*

SAFETY SYMBOLS	HAZARD	EXAMPLES	PRECAUTION	REMEDY
DISPOSAL	Special disposal procedures need to be followed.	certain chemicals, living organisms	Do not dispose of these materials in the sink or trash can.	Dispose of wastes as directed by your teacher.
BIOLOGICAL	Organisms or other biological materials that might be harmful to humans	bacteria, fungi, blood, unpreserved tissues, plant materials	Avoid skin contact with these materials. Wear mask or gloves.	Notify your teacher if you suspect contact with material. Wash hands thoroughly.
EXTREME TEMPERATURE	Objects that can burn skin by being too cold or too hot	boiling liquids, hot plates, dry ice, liquid nitrogen	Use proper protection when handling.	Go to your teacher for first aid.
SHARP OBJECT	Use of tools or glassware that can easily puncture or slice skin	razor blades, pins, scalpels, pointed tools, dissecting probes, broken glass	Practice common-sense behavior and follow guidelines for use of the tool.	Go to your teacher for first aid.
FUME	Possible danger to respiratory tract from fumes	ammonia, acetone, nail polish remover, heated sulfur, moth balls	Make sure there is good ventilation. Never smell fumes directly. Wear a mask.	Leave foul area and notify your teacher immediately.
ELECTRICAL	Possible danger from electrical shock or burn	improper grounding, liquid spills, short circuits, exposed wires	Double-check setup with teacher. Check condition of wires and apparatus.	Do not attempt to fix electrical problems. Notify your teacher immediately.
IRRITANT	Substances that can irritate the skin or mucous membranes of the respiratory tract	pollen, moth balls, steel wool, fiberglass, potassium permanganate	Wear dust mask and gloves. Practice extra care when handling these materials.	Go to your teacher for first aid.
CHEMICAL	Chemicals that can react with and destroy tissue and other materials	bleaches such as hydrogen peroxide; acids such as sulfuric acid, hydrochloric acid; bases such as ammonia, sodium hydroxide	Wear goggles, gloves, and an apron.	Immediately flush the affected area with water and notify your teacher.
TOXIC	Substance may be poisonous if touched, inhaled, or swallowed.	mercury, many metal compounds, iodine, poinsettia plant parts	Follow your teacher's instructions.	Always wash hands thoroughly after use. Go to your teacher for first aid.
FLAMMABLE	Open flame may ignite flammable chemicals, loose clothing, or hair.	alcohol, kerosene, potassium permanganate, hair, clothing	Avoid open flames and heat when using flammable chemicals.	Notify your teacher immediately. Use fire safety equipment if applicable.
OPEN FLAME	Open flame in use, may cause fire.	hair, clothing, paper, synthetic materials	Tie back hair and loose clothing. Follow teacher's instructions on lighting and extinguishing flames.	Always wash hands thoroughly after use. Go to your teacher for first aid.

 Eye Safety Proper eye protection should be worn at all times by anyone performing or observing science activities.

 Clothing Protection This symbol appears when substances could stain or burn clothing.

 Animal Safety This symbol appears when safety of animals and students must be ensured.

 Radioactivity This symbol appears when radioactive materials are used.

 Handwashing After the lab, wash hands with soap and water before removing goggles.

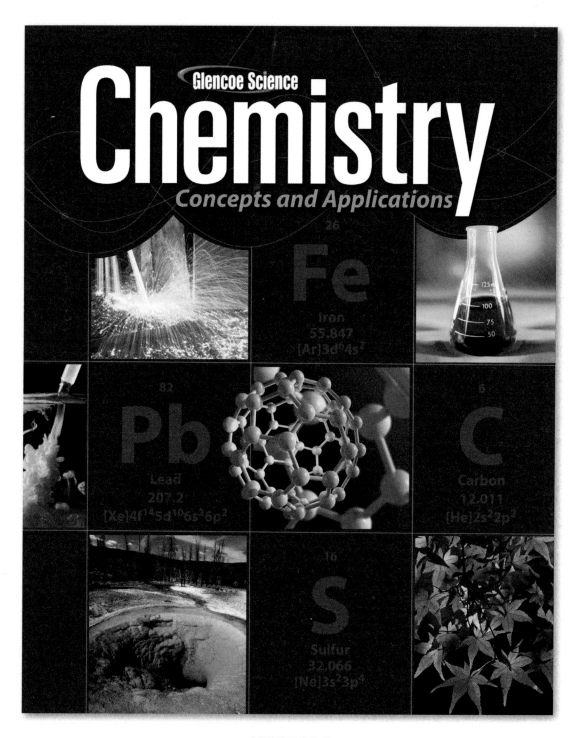

AUTHORS
John S. Phillips • Victor S. Strozak • Cheryl Wistrom • Dinah Zike

About the Cover:
Chemistry is all around you!

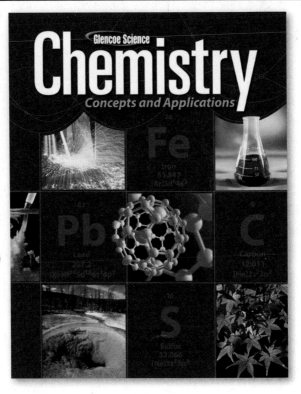

Molten iron is poured at a temperature of 1538°C.

Mixing lead(II) nitrate and potassium iodide produces the bright yellow solid lead(II) iodide.

Yellow sulfur deposits rim a hot spring in Yellowstone National Park.

Copper(II) sulfate forms a clear, blue solution in water.

Carbon atoms form many shapes, including the molecules nicknamed buckyballs.

Anthocyanins give grapes a purple color and Japanese maple leaves a red color in autumn.

The **McGraw·Hill** *Companies*

 Glencoe

Copyright © 2009 The McGraw-Hill Companies, Inc. All rights reserved. No part of this publication may be reproduced or distributed in any form or by any means, or stored in a database or retrieval system, without prior written consent of the McGraw-Hill Companies, Inc., including, but not limited to, network storage or transmission, or broadcast for distance learning.

Send all inquiries to:
Glencoe/McGraw-Hill
8787 Orion Place
Columbus, OH 43240-4027

ISBN: 978-0-07-880723-7
MHID: 0-07-880723-9

Printed in the United States of America.

2 3 4 5 6 7 8 9 10 079/043 12 11 10 09 08

Contents in Brief

Student Guide

Reading for Information xviii
Scavenger Hunt xxi

Chapters

1. Chemistry: The Science of Matter 2
2. Matter Is Made of Atoms 48
3. Introduction to the Periodic Table 82
4. Formation of Compounds 116
5. Types of Compounds 150
6. Chemical Equations and Reactions 186
7. Completing the Model of the Atom 226
8. Periodic Properties of the Elements 254
9. Chemical Bonding 298
10. The Kinetic Theory of Matter 336
11. Behavior of Gases 368
12. Chemical Quantities 402
13. Water and Its Solutions 434
14. Acids, Bases, and pH 478
15. Acids and Bases React 514
16. Oxidation-Reduction Reactions 552
17. Electrochemistry 582
18. Organic Chemistry 620
19. The Chemistry of Life 664
20. Chemical Reactions and Energy 702
21. Nuclear Chemistry 738

Student Resources 778

Chemistry Skill Handbook 783
Supplemental Practice 807
Safety Handbook 844
Reference Tables 846
Solutions to Problems 858
Try At Home Labs 868
Glossary/Glosario 879
Index 897

About the Authors

John S. Phillips is the Director of Educational Technology at Forest Ridge School in Bellevue, Washington. He coordinates the school's 1:1 laptop program, and supports teachers with the intergration of laptops into their curriculum. He has taught chemistry at the high school and college levels for almost thirty years. Dr. Phillips has coordinated and led programs and workshops for teachers from kindergarten through college that encourage and support creative science teaching. He earned a BA in chemistry at Western Maryland College and a PhD in chemistry from Purdue University. He is a member of the American Chemical Society, National Science Teachers Association, and Sigma Xi.

Victor S. Strozak is a science educator with 45 years teaching and administrative experience at both the high school and college levels. He holds a BS degree in chemistry from St. John's University, an MS in chemistry from New York University, and a PhD in Science Education from New York University. Dr. Strozak taught chemistry and mathematics for six years at Xaverian High School in Brooklyn, New York and then moved on to New York City College of Technology, where he spent the next 31 years as a Professor of Chemistry, Dean of Science and Mathematics, and director of numerous science education projects. Dr. Strozak is currently the Senior Research Associate in Science Education at the Center for Advanced Study in Education at the Graduate Center of the City University of New York (CUNY). He is the Co-PI and project manager for a university-wide NSF Graduate Teaching Fellows in K-12 Education project in which doctoral science students will collaborate with science teachers in New York City high schools to incorporate authentic research experiences into the high school science curriculum.

Cheryl Wistrom is an associate professor of chemistry at Saint Joseph's College in Rensselaer, Indiana where she has been honored with both the Science Division and college faculty teaching awards. She has taught chemistry, biology, and science education courses at the college level since 1990 and is also a licensed pharmacist who works in a hospital pharmacy. She earned her BS degree in biochemistry at Northern Michigan University, a BS in pharmacy at Purdue University, and her MS and PhD in biological chemistry at the University of Michigan. She has published several research papers involving senescence in human fibroblast cells. Dr. Wistrom is the director of the Little Einstein Science Camp, an annual day camp for elementary students.

Dinah Zike is an international curriculum consultant and inventor who has developed educational products and three-dimensional, interactive graphic organizers for over 30 years. As president and founder of Dinah-Might Adventures, L.P., Dinah is the author of over 100 award-winning educational publications, including The Big Book of Science. Ms. Zike has a BS and an MS in educational curriculum and instruction from Texas A & M University. Dinah Zike's Foldables are an exclusive feature of McGraw-Hill textbooks.

Reviewers and Consultants

High School Reviewers

Each teacher reviewed selected chapters of *Chemistry: Concepts and Applications* and provided feedback and suggestions regarding the effectiveness of the instruction.

Jon L. Allan, MS
University High School
Spokane, WA

William Allen, MEd
Stevens Point Area Senior
 High School
Stevens Point, WI

Eddie Anderson
Oak Ridge High School
Oak Ridge, TN

Lawrence Bacci
Rochester Adams High School
Rochester, MI

Susan H. Brierley
Garfield High School
Seattle, WA

Michael Chan, PhD
Director of Science
Rochester City School District
Rochester, NY

Lauren Clare
Charlotte High School
Punta Gorda, FL

Robert A. Cooper, MEd
Pennsbury High School
Fairless Hills, PA

Sharon Doerr
Oswego High School
Oswego, NY

Jeffrey L. Engel, MEd, EdS
Madison County High School
Danielsville, GA

Richard A. Garst
Ironwood High School
Glendale, AZ

Jo Marie Hansen
Twin Falls High School
Twin Falls, ID

Cynthia Harrison, MSA
Parkway South High School
Manchester, MO

Vince Howard, MEd
Kentridge High School
Kent, WA

Stephen Hudson
Mission High School
San Francisco, CA

Israel E. Iyoke
Skyline High School
Dallas, TX

Michael Krein, MS
Coordinator of Chemistry
Stamford High School
Stamford, CT

David J. Lee
Franklin D. Roosevelt High
 School
Dallas, TX

**Sister John Ann Proach,
OSF, MA, MS**
Science Curriculum
 Chairperson
Archdiocese of Philadelphia
Bishop McDevitt High School
Wyncote, PA

Eva M. Rambo, MAT
Bloomington South High
 School
Bloomington, IN

Nancy Schulman, MS
Manalapan High School
Manalapan, NJ

Durgha Shanmugan, MD
Hillcrest High School
Dallas, TX

Tim Watts, MEd
Assistant Principal
Warren County Middle
 School
Front Royal, VA

Jason E. Wirth
Marysville Schools
Marysville, OH

Content Consultants

Content consultants each reviewed selected chapters of *Chemistry: Concepts and Applications* for content accuracy and clarity.

Larry B. Anderson, PhD
Associate Professor
The Ohio State University
Columbus, OH

Ildiko V. Boer, MA
Assistant Professor
County College of Morris
Randolph, NJ

Marcia C. Bonneau, MS
Lecturer
State University of New York
Cortland, NY

James H. Burness, PhD
Associate Professor
Penn State University
York, PA

Larry Cai
Graduate Teaching Associate
The Ohio State University
Columbus, OH

Sheila Cancella, PhD
Department Chair, Science &
 Engineering
Raritan Valley Community
 College
Somerville, NJ

James Cordray, MS
Berwyn, IL

Jeff Hoyle, PhD
Associate Professor
Nova Scotia Agricultural
 College
Truro, Nova Scotia
Canada

Teresa Anne McCowen, MS
Senior Lecturer
Butler University
Indianapolis, IN

Lorraine Rellick, PhD
Assistant Professor
Capital University
Columbus, OH

Marie C. Sherman, MS
Chemistry Teacher
Ursuline Academy
St. Louis, MO

Charles M. Wynn, PhD
Chemistry Professor
Eastern Connecticut State
 University
Willimantic, CT

Contributing Writers and Consultants

Contributing Writers
Contributing writers helped develop chapter elements, features, labs, and handbooks.

Helen Frensch, MA
Santa Barbara, CA

Nicholas Hainen, MA
Former Chemistry Teacher
Worthington High School
Worthington, OH

Zoe A. Godby Lightfoot, MS
Former Chemistry Teacher
Carbondale Community High School
Marion, IL

Mark V. Lorson, PhD
Chemistry Teacher
Jonathan Alder High School
Plain City, OH

Robert Roth, MS
Pittsburgh, PA

Richard G. Smith, MAT
Chemistry Teacher
Bexley High School
Bexley, OH

Patricia West
Oakland, CA

Safety Consultant
The Safety Consultant reviewed labs and lab materials for safety and implementation.

Kenneth R. Roy, PhD
Director of Environmental
  Health and Safety
Glastonbury Public Schools
Glastonbury, CT

Contents

Your book is divided into chapters that are organized around Themes, Big Ideas, and Main Ideas of chemistry.

THEMES are overarching concepts used throughout the entire book that help you tie what you learn together. They help you see the connections among major ideas and concepts.

BIG Idea appears in each chapter and help you focus on topics within the themes. The Big Ideas are broken down even further into Main Ideas.

MAIN Idea draws you into more specific details about chemistry. All the Main Ideas of a chapter add up to the chapter's Big Idea.

THEMES
Energy
Macro to Submicroscopic
Conservation
Systems and Interactions
Equilibrium and Change

BIG Idea
One per chapter

MAIN Idea
One per section

Student Guide
Reading for Information xviii
Scavenger Hunt. xxi

Chapter 1
Chemistry: The Science of Matter 2
1.1 The Puzzle of Matter 4
1.2 Properties and Changes of Matter. 32

Chapter 2
Matter Is Made of Atoms. 48
2.1 Atoms and Their Structures 50
2.2 Electrons in Atoms. 67

Chapter 3
Introduction to the Periodic Table. 82
3.1 Development of the Periodic Table 84
3.2 Using the Periodic Table. 93

Chapter 4
Formation of Compounds 116
4.1 The Variety of Compounds. 118
4.2 How Elements Form Compounds. 128

Chapter 5
Types of Compounds 150
5.1 Ionic Compounds. 152
5.2 Covalent Compounds 168

Contents

Chapter 6
Chemical Equations and Reactions 186
6.1 Chemical Equations. 188
6.2 Types of Reactions. 200
6.3 Nature of Reactions. 208

Chapter 7
Completing the Model of the Atom 226
7.1 Present-Day Atomic Theory 228
7.2 The Periodic Table and Atomic Structure 241

Chapter 8
Periodic Properties of the Elements 254
8.1 Main Group Elements 256
8.2 Transition Elements 280

Chapter 9
Chemical Bonding. 298
9.1 Bonding of Atoms 300
9.2 Molecular Shape and Polarity 313

Chapter 10
The Kinetic Theory of Matter . . 336
10.1 Physical Behavior of Matter 338
10.2 Energy and Changes of State 346

Chapter 11
Behavior of Gases. 368
11.1 Gas Pressure 370
11.2 The Gas Laws 380

Chapter 12
Chemical Quantities 402
12.1 Counting Particles of Matter 404
12.2 Using Moles 413

Chapter 13
Water and Its Solutions 434
13.1 Uniquely Water 436
13.2 Solutions and Their Properties. 451

Chapter 14
Acids, Bases, and pH 478
14.1 Acids and Bases 480
14.2 Strengths of Acids and Bases. 497

Chapter 15
Acids and Bases React 514
15.1 Acid and Base Reactions 516
15.2 Applications of Acid-Base Reactions. 531

Chapter 16
Oxidation-Reduction Reactions 552
16.1 The Nature of Oxidation-Reduction Reactions. 554
16.2 Applications of Oxidation-Reduction Reactions. 563

Chapter 17
Electrochemistry. 582
17.1 Voltaic Cells: Electricity from Chemistry 584
17.2 Electrolysis: Chemistry from Electricity 600

Contents

Chapter 18
Organic Chemistry 620
- 18.1 Hydrocarbons 622
- 18.2 Substituted Hydrocarbons 640
- 18.3 Plastics and Other Polymers 647

Chapter 19
The Chemistry of Life 664
- 19.1 Molecules of Life 666
- 19.2 Reactions of Life 689

Chapter 20
Chemical Reactions and Energy 702
- 20.1 Energy Changes in Chemical Reactions 704
- 20.2 Measuring Energy Changes 715
- 20.3 Photosynthesis 729

Chapter 21
Nuclear Chemistry 738
- 21.1 Types of Radioactivity 740
- 21.2 Nuclear Reactions and Energy 756
- 21.3 Nuclear Tools 763

Student Resources

Student Resources 778

Chemistry Skill Handbook 783
- Measurement in Science 783
- Relating SI, Metric, and English Measurements 786
- Making and Interpreting Measurements ... 789
- Expressing the Accuracy of Measurements .. 790
- Expressing Quantities with Scientific Notation 792
- Computations with the Calculator 796
- Using Dimensional Analysis 798
- Organizing Information 800
- Ratios, Fractions, and Percents 805
- Operations Involving Fractions 806

Supplemental Practice 808

Safety Handbook 844
- Safety Guidelines in the Chemistry Laboratory 844
- First Aid in the Laboratory 844
- Safety Symbols 845

Reference Tables 846
- D-1 The Modern Periodic Table 846
- D-2 Color Key 848
- D-3 Symbols and Abbreviations 848
- D-4 Alphabetical Table of the Elements ... 849
- D-5 Properties of Elements 850
- D-6 Electron Configurations of the Elements 853
- D-7 Useful Physical Constants 855
- D-8 Names and Charges of Polyatomic Ions 855
- D-9 Solubility Guidelines 856
- D-10 Solubility Product Constants 856
- D-11 Acid-Base Indicators 857

Solutions to Problems 858

Try At Home Labs 868

Glossary/Glosario 879

Index 897

Labs

LAUNCH Lab — Begin each chapter with a hands-on introduction.

Chapter

1. Why is the mass different? 3
2. What's inside? 49
3. Versatile Materials 83
4. Observe Evidence of Change 117
5. Elements, Compounds, and Mixtures . 151
6. Observe a Chemical Reaction 187
7. Observe Electric Charge 227
8. Periodic Properties 255
9. Oil and Vinegar Dressing 299
10. Temperature and Mixing 337

Chapter

11. Volume and Temperature of a Gas ... 369
12. How much is a mole? 403
13. Solution Formation 435
14. Testing Household Products 479
15. What is the role of a buffer? 515
16. Observe a Redox Reaction 553
17. A Lemon Battery 583
18. Model Simple Hydrocarbons 621
19. Testing for Simple Sugars 665
20. Increasing the Rate of Reaction 703
21. The Penetrating Power of Radiation... 739

Labs

MiniLab
Practice scientific methods and hone your lab skills with these quick activities.

Section		
1.1	Observe Mixing	21
1.2	Paper Chromatography of Inks	22
1.3	Create an Alloy	25
1.4	Analyze Cereal	28
1.5	Synthesize a Polymer	38
2.1	Model Isotopes	61
2.2	Line Emission Spectra	75
3.1	Predict Properties	87
3.2	Trends in Reactivity	96
4.1	Iron Versus Rust	120
4.2	The Formation of Ionic Compounds	133
5.1	Chemical Weather Predictor	164
5.2	Chemical Bonds in Bone	169
6.1	Energy Change	194
6.2	A Simple Exchange	203
6.3	Starch-Iodine Clock Reaction	218
7.1	Flame Tests	232
7.2	Model Electrons in Atoms	244
8.1	Trends in Atomic Radii	260
8.2	The Ion Charges of a Transition Element	283
9.1	Coffee Filter Chromatography	310
9.2	Model Molecules	323
10.1	Diffusion Rates	341
10.2	Vaporization Rates	355

Section		
11.1	Mass and Volume of a Gas	373
11.2	How Straws Work	384
12.1	Counting by Mass	408
12.2	Determine the Amount of Reactant	418
13.1	Surface Tension	443
13.2	Hard and Soft Water	452
14.1	Reactions of Acids	482
14.2	Antacids	504
15.1	Acidic, Basic, or Neutral?	518
15.2	Buffers	533
16.1	Corrosion of Iron	557
16.2	Test for Alcohol	568
17.1	Lemon Voltage	586
17.2	Electrolysis	602
18.1	Test for Unsaturated Oil	630
18.2	A Synthetic Aroma	645
18.3	Absorbent Polymers	652
19.1	Extract DNA	687
19.2	Fermentation by Yeast	696
20.1	Heat In, Heat Out	708
20.2	Dissolving—Exothermic or Endothermic?	722
21.1	Model a Chain Reaction	758
21.2	Test for Radon	770

Labs

CHEMLAB

Apply the skills you developed in Launch Labs and MiniLabs in these chapter-culminating, hands-on labs.

Chapter

- **1.1** Observe a Candle **10**
- **1.2** Kitchen Chemicals **18**
- **1.3** The Composition of Pennies **36**
- **2** Conservation of Mass **54**
- **3** The Periodic Table of the Elements **98**
- **4** The Formation and Decomposition of Zinc Iodide **134**
- **5** Ionic or Covalent Compounds **170**
- **6** Explore Chemical Changes **204**
- **7** Metals, Reaction Capacities, and Valence Electrons **234**
- **8** Reactions and Ion Charges of the Alkaline Earth Elements **266**
- **9** Separating Candy Colors **326**

Chapter

- **10** Molecules and Energy **360**
- **11** Boyle's Law **386**
- **12** Analyze a Mixture **422**
- **13** Identify Solutions **456**
- **14** Household Acids and Bases **506**
- **15** Titration of Vinegar **544**
- **16** Oxidation and Reduction Reactions . . **560**
- **17** Oxidation-Reduction and Electrochemical Cells **592**
- **18** Identify Polymers **649**
- **19** Catalytic Decomposition **674**
- **20** Energy Content of Food **720**
- **21** Model Radioactive Decay **748**

ChemLabs

Everyday Chemistry

Discover chemistry in everyday experiences.

Chapter		
1	You Are What You Eat	17
2	Fireworks—Getting a Bang Out of Color	74
3	Metallic Money	108
4	Elemental Good Health	126
5	Hard Water	158
6	Whitening Whites	192
	Stove in a Sleeve	219
7	Colors of Gems	246
8	The Chemistry of Matches	273
9	Jiggling Molecules	318
10	Freeze Drying	351
11	Popping Corn	395

Chapter		
12	Air Bags	417
13	Soaps and Detergents	455
	Antifreeze	466
14	Balancing pH in Cosmetics	505
15	Hiccups	534
16	Lightning-Produced Fertilizer	571
17	Manufacturing a Hit CD	611
18	Chemistry and Permanent Waves	655
19	Clues to Sweetness	680
	Fake Fats and Designer Fats	684
20	Catalytic Converters	711
21	Radon—An Invisible Killer	771

Real-World Chemistry Features

Chemistry and Society
Examine chemistry in the news on complex issues in society.

Chapter 1	Natural Versus Synthetic Chemicals	29
Chapter 2	Recycling Glass	58
Chapter 4	The Rain Forest Pharmacy	144
Chapter 13	Water Treatment	447
Chapter 14	Atmospheric Pollution	495
Chapter 15	Artificial Blood	537
Chapter 18	Recycling Plastics	657

Chemistry and Technology
Discover recent advancements that have influenced chemistry.

Chapter 3	Metals That Untwist	106
Chapter 5	Carbon Allotropes: From Soot to Diamonds	174
Chapter 6	Mining the Air	214
Chapter 7	Hi-Tech Microscopes	236
Chapter 8	Carbon and Alloy Steels	286
Chapter 9	Chromatography	324
Chapter 10	Fractionation of Air	352
Chapter 11	Health Under Pressure	388
Chapter 12	Improving Percent Yield in Chemical Synthesis	424
Chapter 13	Versatile Colloids	470
Chapter 14	Manufacturing Sulfuric Acid	485
Chapter 16	Forensic Blood Detection	574
Chapter 17	Copper Ore to Wire	606
Chapter 20	Alternative Energy Sources	724
Chapter 21	Archaeological Radiochemistry	750

How It Works
Examine how chemistry helps make familiar things work.

Chapter 5	Cement	166
Chapter 6	Light Sticks	195
Chapter 8	Inert Gases in Lightbulbs	282
Chapter 10	Pressure Cookers	357
Chapter 11	Tire-Pressure Gauge	375
Chapter 13	A Portable Reverse Osmosis Unit	468
Chapter 15	Taste	519
Chapter 15	pH Indicators	543
Chapter 16	Breathalyser Test	569
Chapter 17	The Pacemaker: Helping a Broken Heart	595
Chapter 17	Nicad Rechargeable Batteries	597
Chapter 17	Hydrogen-Oxygen Fuel Cell	598
Chapter 20	Hot and Cold Packs	706
Chapter 21	Smoke Detectors	744

In the Field
Investigate a day in the life of people working in the field of chemistry.

Chapter 1	Forensic Scientist	12
Chapter 6	Plant-Care Specialist	210
Chapter 9	Chemist	316
Chapter 13	Wastewater Operator	448
Chapter 14	Cosmetic Bench Chemist	490
Chapter 17	Metal Plater	612
Chapter 18	Pharmacist	634
Chapter 19	Biochemist	676

Cross-Curricular Connections

It may not have occurred to you that chemistry is an integral part of all your courses, not just the sciences. Learn in these features how chemistry is connected to literature, art, and history, as well as to the sciences of physics, biology, health, and Earth science.

Literature
Jules Verne and His Icebergs............26
The Language of a Chemist.............95

Art
China's Porcelain......................161
Glass Sculptures......................344
Asante Brass Weights..................411
Art Forger van Meegerend—
Villain or Hero?......................754

History
Politics and Chemistry—Elemental
Differences...........................56
Hydrogen's Ill-Fated Lifts.............140
Lead Poisoning in Rome................270
Linus Pauling: An Advocate of
Knowledge and Peace..................305

Physics
Auroroa Borealis.......................71
Niels Bohr—Atomic Physicist and
Humanitarian.........................230
Rocket Booster Engines................566

Biology
Air in Space..........................201
Fluorides and Tooth Decay.............278
Measurement of Blood Gases...........487
Vision and Vitamin A..................632
Function of Hemoglobin................694
A Biological Mystery Solved
with Tracers.........................766

Earth Science
Weather Balloons.....................383
Cave Formation......................525
Bacterial Refining of Ores............723

Concepts in Motion

Interactive Tables Check your understanding by viewing interactive versions of some of the tables in your text.

Chapter

- **1** Table 1.3 Some Common Alloys 23
- **2** Table 2.1 Particles of an Atom 65
- **3** Table 3.5 Properties of Metals and Nonmetals 103
- **5** Table 5.5 Naming Prefixes 165
  Table 5.7 Names of Common Acids and Bases 180
- **7** Table 7.3 The Electron Configurations of the Noble Gases 244
- **8** Carbon and Alloy Steels 286
- **9** Table 9.1 Physical Properties of Ionic and Covalent Compounds 300

Chapter

- **13** Table 13.1 The Uniqueness of Water 436
- **14** Table 14.1 Common Industrial Acids and Bases 481
  Table 14.3 Common Strong Acids and Bases 498
- **18** Table 18.2 The First Ten Alkanes 624
- **19** Table 19.1 Biological Macromolecules 667
- **20** Table 20.1 Predicting Whether a Reaction is Spontaneous 714
- **21** Table 21.1 Comparison of Chemical and Nuclear Reactions 742

Concepts In Motion

Interactive Figures Enhance and enrich your knowledge of chemistry concepts through animations of visuals.

Chapter

2
- Figure 2.4 Nitrogen Cycle 53
- Figure 2.20 Balmer Series 73
- Figure 2.21 Electron Transitions 73

4
- Figure 4.17 Sodium Chloride Ionic Bond ... 131
- Figure 4.24 Strong, Weak, and Nonelectrolytes 142

6
- Figure 6.12 Precipitate Formation 206
- Figure 6.22 Limiting Reactants 218

7
- Figure 7.2 Structure of the Atom 229

8
- Figure 8.3 Atomic Radii 257

9
- Figure 9.2 Electronegativity 302
- Figure 9.8 Bond Types 308
- Figure 9.18 Molecular Shapes 321

Chapter

11 Figure 11.7 The Gas Laws 380
12 Figure 12.6 Molar Mass 407
13 Figure 13.15 Dissolution of Compounds ... 451
- Figure 13.28 Osmosis 467

15 Figure 15.20 Neutralization Reactions 541
16 Figure 16.4 Redox Reactions 557
17 Figure 17.6 Voltaic Cell 590
18 Figure 18.5 Isomers of Pentane 628
19 Figure 19.23 Structure of DNA 687
20 Figure 20.9 Calorimetry 715
21 Figure 21.13 Nuclear Chain Reaction 757
- Figure 21.14 Critical Mass 758
- Figure 21.15 Nuclear Power 759

Reading for Information

When you read *Chemistry: Concepts and Applications*, you need to read for information. Science is nonfiction writing; it describes real-life events, people, ideas, and technology. Here are some tools that *Chemistry: Concepts and Applications* has to help you read.

Before You Read

By reading the **BIG Idea**, **MAIN Idea**, and **Launch Lab** prior to reading the chapter or section, you will get a preview of the coming material.

> The **BIG Idea** describes what you will learn in the chapter. The **MAIN Idea** within a chapter supports the Big Idea of the chapter. Each section of the chapter has a Main Idea that describes the focus of the section.

Source: Chapter 12, p. 402

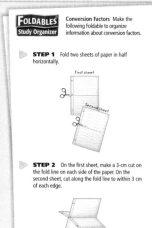

Source: Chapter 12, p. 403

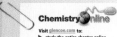

Each chapter starts with a hands-on introduction to the material being covered. Read and perform the **Launch Lab** to discover new concepts and tie some concepts to what you have previously learned.

OTHER WAYS TO PREVIEW

- Read the chapter title to find out what the topic will be.
- Skim the photos, illustrations, captions, graphs, and tables.
- Look for key terms that are boldfaced and highlighted.
- Create an outline using section titles and heads.

Reading for Information

As You Read

Within each section you will find a tool to deepen your understanding and tools to check your understanding.

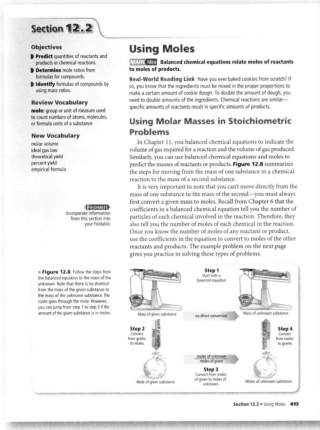

Source: Section 12.2, p. 413

> The **Real-World Reading Link** describes how the section's content may relate to you.

Source: Section 12.2, p. 414

> **Example Problems** take you step-by-step to solve problems in chemistry. Reinforce the skills you've learned by working through the **Practice Problems.**

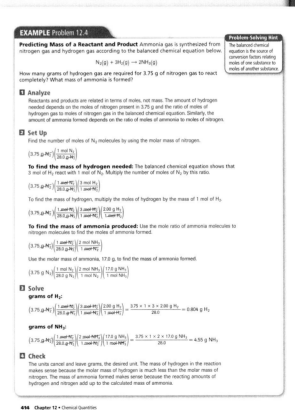

OTHER READING SKILLS

- Ask yourself what is the **BIG Idea**?
  What is the **MAIN Idea**?
- Relate the information in *Chemistry: Concepts and Applications* to other areas you have studied.
- Predict events or outcomes by using clues and information that you already know.
- Change your predictions as you read and gather new information.

Reading for Information

After You Read

Follow up your reading with a summary and assessment of the material to evaluate if you understood the text.

Each section concludes with an assessment. The assessment contains a summary and questions. The summary reviews the section's key concepts while the questions test your understanding.

Source: Chapter 12, p. 428

At the end of each chapter you will find a Study Guide. The chapter's vocabulary terms as well as key concepts are listed here. Use this guide for review and to check your comprehension.

Source: Chapter 12, p. 429

OTHER WAYS TO REVIEW

- State the **BIG Idea**.
- Relate the **MAIN Idea** to the **BIG Idea**.
- Use your own words to explain what you read.
- Apply this information in other school subjects or at home.
- Identify sources you could use to find out more information about this topic.

Scavenger Hunt

Chemistry: Concepts and Applications contains a wealth of information. Complete this fun activity so you will know where to look to learn as much as you can.

As you complete this scavenger hunt, either alone, with your teacher, or with others, you will quickly learn how *Chemistry: Concepts and Applications* is organized and how to get the most out of your reading and study time.

1. How many chapters are in this book?

2. On what page does the glossary begin? What glossary is online?

3. Where can you find a listing of Laboratory Safety Symbols?

4. If you want to find all the Launch Labs, MiniLabs, and ChemLabs, where in the front do you look?

5. How can you quickly find the pages that have information about plant-care specialist?

6. What is the name of the table that summarizes the Key Concepts of a chapter?

7. Where can you find reference tables? What are the page numbers?

8. On what page can you find the **BIG Idea** for Chapter 1? On what pages can you find a **MAIN Idea** for Chapter 2?

9. Where can you find information on hydrogen?

10. Name four activities that are found at **Chemistry Online**.

11. What study tool shown at the beginning of a chapter can you make from notebook paper?

12. Where do you go to view the **Concepts In Motion**?

Scavenger Hunt   xxi

CHAPTER 1
Chemistry: The Science of Matter

BIG Idea Everything is made of matter.

1.1 The Puzzle of Matter
MAIN Idea Most everyday matter occurs as mixtures—combinations of two or more substances.

1.2 Properties and Changes of Matter
MAIN Idea Matter can undergo physical and chemical changes.

ChemFacts

- Many of the processes that occur around you are the result of chemistry in action.
- The atmosphere of Earth is primarily made up of four different chemicals.
- Chemists study chemical reactions that go on all around you.
- Buildings, cars, and concrete are all mixtures of different chemicals.

Start-Up Activities

LAUNCH Lab

Why is the mass different?

Matter is anything that has mass and takes up space. The three most common forms matter can take on Earth are solids, liquids, and gases. How do the masses of these three states of matter compare?

Materials
- balloons (3)
- table salt
- balance
- funnel
- graduated cylinder
- water
- scissors
- string

Procedure
1. Read and complete the lab safety form.
2. Measure and record the mass of a balloon.
3. Insert the narrow end of a funnel into the opening of the balloon and fill it with water without stretching the bulb of the balloon. Tie off the balloon's end, measure and record the mass of the balloon and water.
4. Repeat steps 1 and 2 using salt. The size of the salt-filled balloon should be approximately the same size as the water-filled balloon.
5. Repeat steps 1 and 2 using air. Blow just enough air into the balloon so that it is approximately the same size as the balloons filled with water and salt.

Analysis
1. **Calculate** the masses of the water, the salt, and the air.
2. **Compare** the masses of the solid, the liquid, and the gas.

Inquiry How can three objects with the same volume have different masses?

Chemistry Online

Visit glencoe.com to:
▶ study the entire chapter online
▶ explore Concepts In Motion
▶ take Self-Check Quizzes
▶ use Personal Tutors
▶ access Web Links for more information, projects, and activities
▶ find the Try at Home Lab, Comparing Frozen Liquids

FOLDABLES Study Organizer

Properties and Changes Make a Foldable to help you organize your study of chemical and physical changes and the properties of matter.

▶ **STEP 1** Fold up the bottom of a horizontal sheet of paper about 5 cm as shown.

▶ **STEP 2** Fold the paper in half.

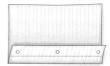

▶ **STEP 3** Unfold once and staple to make two pockets. Label the pockets *Chemical* and *Physical*.

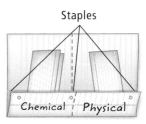

FOLDABLES Use this Foldable with Section 1.2. As you read the section, use index cards or quarter-sheets of paper to summarize what you learn about the properties and changes of matter. Insert these into the appropriate pockets of your Foldable.

Chapter 1 • Chemistry: The Science of Matter **3**

Section 1.1

Objectives

- **Classify** matter according to its composition.
- **Distinguish** among elements, compounds, homogeneous mixtures, and heterogeneous mixtures.
- **Relate** the properties of matter to structure.

New Vocabulary

chemistry
matter
mass
property
scientific model
qualitative
quantitative
substance
mixture
physical change
physical property
solution
alloy
solute
solvent
aqueous solution
element
compound
formula

The Puzzle of Matter

MAIN Idea Most everyday matter occurs as mixtures—combinations of two or more substances.

Real-World Reading Link If you enjoy working puzzles, you might like puzzles that have a lot of pieces—thousands or even tens of thousands. No single piece gives much information about the completed puzzle. But have you ever worked a puzzle without looking at the picture on the box?

A Picture of Matter

When you begin a study of chemistry, you start to work on a challenging puzzle—the puzzle of matter. Every chunk of matter is a puzzle piece. Your puzzle box is the universe, and the box contains many different kinds of pieces. Your job, like that of a puzzle solver, is to figure out how to connect all the different pieces.

Composition and Behavior **Chemistry** is the science that investigates and explains the structure and properties of matter. **Matter** is anything that takes up space and has mass. Matter is all around you: the metal and plastic of a telephone, the paper and ink of a book, the glass and liquid of a bottle of soda, the air you breathe, and the materials that make up your body.

Mass is the measure of the amount of matter that an object contains. What isn't matter? The heat and light from a lamp are not matter; neither are thoughts, ideas, radio waves, or magnetic fields. **Figure 1.1** compares the masses of two chunks of matter. On pages 785–791 of the Skill Handbook in the back of this book, you'll find a description of the International System of Units (SI) and the units used to measure mass and other quantities.

■ **Figure 1.1** Mass is a measure of the amount of matter in an object.

The water in the container has a mass of 1 kg. The kilogram is the unit of mass in SI.

The bus has a mass of about 14,000 kg (15 tons).

The structure of matter refers to its composition—what matter is made of—as well as to how matter is organized. The **properties** of matter describe the characteristics and behavior of matter, including the changes that matter undergoes. **Figure 1.2** compares some different kinds of matter in terms of composition and behavior.

FIGURE 1.2

Composition and Behavior of Matter

The composition of a sample of matter—both the elements it contains and how those elements are arranged—affects its behavior.

Salt and water have different compositions, so it isn't surprising that they have different properties. Salt is made of the elements sodium and chlorine, while water is made of hydrogen and oxygen. You couldn't wash your hair with salt, just as you wouldn't sprinkle water on your popcorn.

Aspirin and sucrose (table sugar) are both composed of carbon, hydrogen, and oxygen, but you wouldn't use aspirin to sweeten your cereal or take a spoonful of sugar for a headache. Even though aspirin and sugar contain the same elements, differences in their structures determine their individual behaviors.

Aspartame and saccharin are substances with different compositions but similar tastes. Saccharin is made of carbon, hydrogen, nitrogen, oxygen, sodium, and sulfur. Aspartame also contains carbon, hydrogen, nitrogen, and oxygen, but it contains no sodium or sulfur. The way in which the components of aspartame and saccharin are organized must be a major factor in causing both of them to have a sweet taste.

Section 1.1 • The Puzzle of Matter 5

Iron is strong, yet can be flattened and stretched.

Iron turns to liquid at a high temperature.

Iron is attracted by a magnet.

Iron conducts electricity.

■ **Figure 1.3** Many properties of iron are easy to observe.

You can determine some properties of matter just by examining or manipulating it. What color is it? Is it a solid, a liquid, or a gas? If it's solid, is it soft or hard? Does it burn? Does it dissolve in water? Does something happen when you mix it with another kind of matter? You determine all of these properties by examining and manipulating matter, as shown in **Figure 1.3.**

Although you can find out a lot about matter just by looking at it and doing simple tests, you usually can't tell what something is made of only by looking at it. Measurements usually must be made or chemical changes observed. **Figure 1.4** shows, by a simple experiment, that sugar is composed of carbon, hydrogen, and oxygen. Most matter does not reveal its composition so easily.

■ **Figure 1.4** When concentrated sulfuric acid is added to sucrose, an interesting reaction occurs.
Observe *What are some signs that a change is occuring?*

The sugar breaks down and forms water (composed of hydrogen and oxygen).

The water is released as steam.

Black carbon is left behind.

6 Chapter 1 • Chemistry: The Science of Matter

■ **Figure 1.5** When you look at a skyscraper, you see its size and shape. These properties are the result of the building's structure, which is hidden from view under the exterior skin of the building.

The macroscopic level of matter Observations of the composition and the behavior of matter are based on a macroscopic view. Matter that is large enough to be seen is called macroscopic, so all of your observations in chemistry and everywhere else start from this perspective. The macroscopic world is the one you touch, feel, smell, taste, and see. The properties of iron shown in **Figure 1.3** are seen from a macroscopic perspective. But if you want to describe and understand the structure of iron, you must use a different perspective—one that allows you to see what can't be seen. What you can see of the skyscraper shown in **Figure 1.5** is similar to a macroscopic perspective of matter. You cannot see the organization of the steel beams and bolts that hold the building together, or the system of pipes, wires, and ventilation ducts that thread through the building.

In the same way, the appearance and the properties of a piece of matter are the result of its structure. Although you may get hints of the actual structure from a macroscopic view, you must go to a submicroscopic perspective to understand how the hidden structure of matter influences its behavior.

The submicroscopic level of matter The submicroscopic view gives you a glimpse into the world of atoms. It is a world so small that you cannot see it even with the most powerful light microscope, hence the term submicroscopic. You probably learned in earlier science courses that matter is made up of atoms. Compared with the macroscopic world, atoms are so small that if the period at the end of this sentence were made of carbon atoms, it would be composed of more than 100,000,000,000,000,000,000 (100 quintillion) carbon atoms. If you could count all of those atoms at a rate of three per second, it would take you a trillion years to finish. Fortunately, in chemistry, you will spend your time becoming acquainted with atoms rather than counting them.

■ **Figure 1.6** Although a scanning tunneling microscope (STM) provides a glimpse of the submicroscopic world, it does not truly produce pictures of atoms, at least not as we think of pictures in the macroscopic world.

To make an image with an STM, a probe like this one moves up and down in response to the position of atoms. A computer converts the probe's motion into a bumpy-looking image.

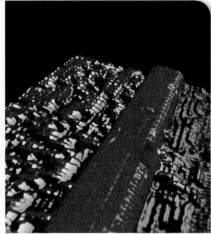
This STM image shows a nanowire in red. The nanowire is 10 atoms wide.

Although atoms cannot be seen unaided, the scanning tunneling microscope (STM) is capable of producing images on a computer screen that show the locations of individual atoms, as shown in **Figure 1.6.** The STM provides a visible perspective of the submicroscopic world; it can even be used to move individual atoms around on a surface.

Using Models in Chemistry

In your study of chemistry, you will use both macroscopic and submicroscopic perspectives. For example, sucrose and aspirin are both composed of carbon, hydrogen, and oxygen atoms, but they have different behaviors and functions. These differences must come about because of differences in the number of atoms and the microscopic arrangement of their atoms. **Figure 1.7** shows models that reveal these submicroscopic differences.

■ **Figure 1.7** The different submicroscopic arrangements of the atoms in aspirin (left) and sucrose (right) cause the differences in their behavior. Don't worry about understanding the complete meaning of these structures. Each ball represents an atom, and each bar between atoms represents a chemical connection between the atoms.
Determine *How many of each element are in each molecule?*

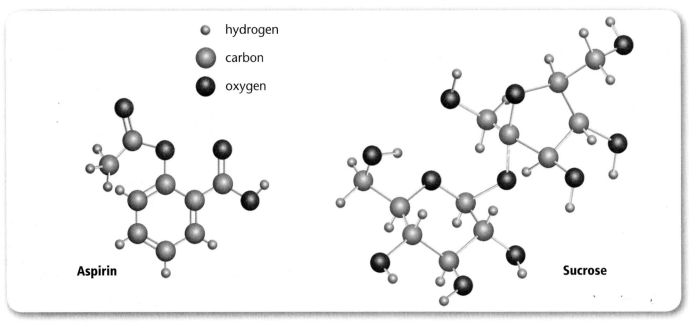

Different types of models The drawings in **Figure 1.7** are tools that allow you to study what you cannot actually see—in this case, the arrangements of atoms. These drawings represent one type of model used in science. Sometimes, a model is something you can see and manipulate. These are the kinds of models you are already familiar with—model cars and airplanes or perhaps an architect's scale model of a proposed building. **Figure 1.8** shows that models are important tools in many fields. Models are used, tested, and revised constantly through new experiments. A model of the submicroscopic structure of a piece of matter must be able to explain the observed macroscopic behavior of that matter and predict behavior that has yet to be observed.

Scientists use many types of models to represent things that are hard to visualize. Chemists also use several different types of models to represent matter, as you will soon learn. The model of aspirin (acetylsalicylic acid) in **Figure 1.8** shows an example of a scientific model. A **scientific model** is a thinking device that helps you understand and explain macroscopic observations. Scientific models are built on investigation and experimentation. Data from many experiments are collected and a visual, verbal, or mathematical model is created. In Greece, a model of matter based on atoms was discussed about 2500 years ago, but this model was not a scientific model because it was never supported by experiments. It was the nineteenth century before a scientific model of matter was proposed. This atomic model was developed and verified by experiments. It has withstood 200 years of prediction and experimentation with only slight modifications.

■ **Figure 1.8** Models help you see and understand structure, whether you're creating a new jet or studying the composition of aspirin. Chemists use computers to create models similar to the model of aspirin below in the search for new drugs to treat disease.
Compare and contrast the model of aspirin shown in Figure 1.7 with the model shown in Figure 1.8.

A model of a prospective new jet is made and studied before an airplane is actually built.

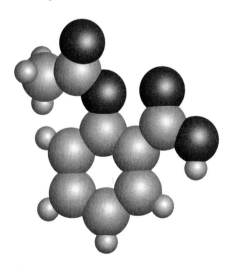

This model is another, more informative way to represent the arrangement of atoms in the compound aspirin.

CHEMLAB 1

OBSERVE A CANDLE

Background
You have seen candles burn, perhaps on a birthday cake. But you probably have never considered a burning candle from a chemist's point of view. Michael Faraday, a nineteenth century chemist, found much to observe as a candle burns. He wrote a book and gave talks on the subject. In this ChemLab, you will investigate a burning candle and the products of combustion.

Question
What are the requirements for and characteristics of a candle flame? What are the products of the combustion of the candle?

Objectives
- **Observe** a candle flame and perform several tests.
- **Interpret** observations and the results of the tests.

Preparation

Materials
large birthday candles
matches
shallow metal dish
25 mL of limewater solution
250-mL beaker
500-mL Erlenmeyer flask
solid rubber stopper to fit the flask
wire gauze square
tongs

Safety Precautions
WARNING: *Keep all combustible materials, including clothing, away from the match and candle flames. Do not allow the limewater to splash into your eyes. If it does, immediately rinse your eyes for 15 minutes and notify your teacher.*

Procedure

1. Read and complete the lab safety form.
2. Light a candle and allow a drop or two of liquid wax to fall into the center of the metal dish. Press the candle upright onto the melted wax before it can solidify. If the candle burns too low during the following procedures, repeat this step with a new candle.
3. Observe the flame of the burning candle for a few minutes. Try to observe what is burning and where the burning takes place. Observe the different regions of the flame. Make at least eight observations, and record them in a data table like the one shown.
4. Light a second candle and hold the flame about 2–4 cm to the side of the first candle flame. Gently blow out the first candle flame, then quickly move the flame of the second candle into the smoke from the first flame. Record your observations.
5. Relight the standing candle. With tongs, hold the wire gauze over the flame, perpendicular to the candle. Slowly lower the gauze onto the flame. Do not allow the gauze to touch the candle wax. If the flame goes out, quickly move the wire gauze off to the side. Record your observations.
6. Fill the 250-mL beaker with cold tap water, dry the outside of the beaker, and hold it about 3–5 cm above the candle flame. Record your observations.
7. Pour tap water into the pan or dish to a depth of about 1 cm.
8. Quickly lower an Erlenmeyer flask over the candle so that the mouth of the flask is below the surface of the water. Allow the flask to remain in place for approximately one minute. Record your observations.

9. Lift the flask out of the water, turn it upright, and add about 25 mL of limewater. Stopper the flask and swirl the solution for approximately one minute. Record your observations. If the solution becomes cloudy or chalky, calcium carbonate was formed, indicating the presence of carbon dioxide in the flask.

Analyze and Conclude

1. **Describe** the different types of changes that took place.
2. **Infer** Do your results in step 4 indicate that the candle wax burns as a solid, a liquid, or a vapor? Explain.
3. **Interpret** One requirement for combustion is the presence of fuel. Interpret your results from steps 5 and 8 to determine the other requirements.
4. **Explain** Based upon your analysis of the observations from steps 6 and 9, what are two products of the combustion of the candle?

Apply and Assess

1. **Explain** Sir Humphry Davy invented a safety lamp for miners in which a flame was surrounded by a wire gauze cylinder. Can you explain the reason why the lamp was constructed in this way?

INQUIRY EXTENSION

What change in water level occurred in step 8? Propose an explanation for this change.

Data and Observations

Data Table	
Procedure step	Observations
3	
4	
5	
6	
8	
9	

In the Field

Meet Dr. John Thornton
Forensic Scientist

A poem by William Blake contains these words: "To see a world in a grain of sand and a heaven in a wildflower." In Dr. Thornton's work, that favorite line of poetry is literally true. To discover what Dr. Thornton means, read this interview, in which he shares his thoughts about the world of physical evidence.

On the Job

Q: Dr. Thornton, could you tell us what you do at your lab, Forensic Analytical Specialties?

A: We analyze physical evidence from the scene of a crime—hair, fiber, body fluids, bullets, paint, soil, glass, shoe impressions, fingerprints, drugs, and plant material. Any of this tangible, physical evidence can associate a person with a crime scene. That's half of it, the "whodunit"; the other half is the "howdunit."

Q: Can you give some specific examples of each part of your analysis?

A: Fingerprints are the classic means of establishing that a person was at the scene of a crime. We develop fingerprints with chemicals: ninhydrin for prints on paper, and cyanoacrylate, which is actually superglue, on other items. As for the "how," here's an example of a case. A person claims he shot a person while defending himself against strangulation. If there's a gunpowder pattern on the victim's clothing, that would tend to support the suspect's story. On the other hand, if there's no powder, indicating that the person was several feet away, that story doesn't hold up.

Q: It must be almost impossible to avoid leaving some kind of evidence behind, right?

A: Yes. Frequently, the evidence that is most incriminating is so small that the perpetrator is oblivious to it, such as grains of pollen, sand, or tiny diatoms. Those would be a signature of a particular geographic location.

Q: Has your work ever helped locate a missing person?

A: A few years ago, when a young woman was missing, a shoe was found near the freeway. Inside the shoe was about a thimbleful of fibers. I went through her sock drawer and found socks that matched every single type of fiber that was in the shoe. That clue indicated the direction of her disappearance, and we later found her.

Early Influences

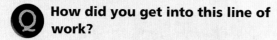

Q How did you get into this line of work?

A When I was in the seventh or eighth grade, I was just browsing through the local public library and stumbled upon a book called *Crime Investigation,* by Paul Kirk, who was probably the foremost forensic scientist at the time. From that point on, I knew what I wanted to do. What I didn't know was that I would later enroll in Kirk's university class. Even more improbably, I eventually took his place on the faculty after he retired.

Q Were you a kid who liked things like decoder rings and puzzles?

A No, I was just a farm boy in the Central Valley. Everything there was flat, and I guess I was looking for some other horizons. I did that by reading.

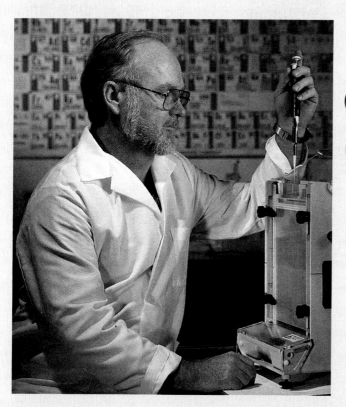

Personal Insights

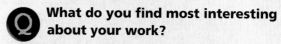

Q What do you find most interesting about your work?

A Forensic science is really a study of how the world is put together. My work involves chemistry, physics, botany, and geology, so it's impossible to get bored.

Q How do you go about tackling the problems posed by a set of clues at a crime scene?

A At any crime scene, you want to use physical evidence to tell a story. It's like looking at a tapestry from the back side: you can vaguely see that over here there's a unicorn, and over here there's a tree, and this might be a fence. It's difficult to get the clarity. I generally just sit down and think about things for a while. I let ideas wash over me from different directions. Then I play what-if: What if this happened at the scene, then what might have happened next? I try all of these ideas on for size, then I try to winnow the scientific wheat from the chaff.

Q What kind of satisfaction do you derive from your work?

A I feel that the quality of justice is enhanced by a full and perhaps contentious airing of all the relevant issues and the physical evidence that we provide.

CAREER CONNECTION

Forensic chemists are assisted by people in these lines of work:

Crime Lab Technologist (Collects and analyzes physical evidence) BA from college with crime laboratory program

Fingerprint Classifier (Files and matches fingerprints) High school followed by study at police schools

Private Investigator (Collects facts pertaining to a crime) High school and detective-training program

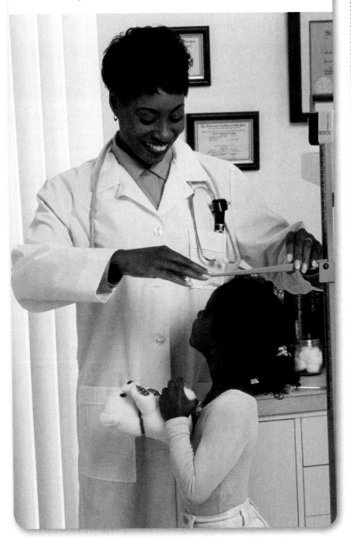

■ **Figure 1.9** In a medical examination, many of the doctor's observations are quantitative. Measured quantities, such as height and weight, can be compared with those of previous examinations to see growth.

Classifying Matter

Examples of matter range from grains of rice to the stars, from a drop of water to the rocks of the Grand Canyon, from a potato chip to a computer chip. Matter is all the stuff around us; it is all part of the same puzzle. You know that all these bits of matter connect in some way, just like the pieces of a jigsaw puzzle. But, there are so many different sizes and kinds of matter. How can you begin to make sense of the puzzle of matter?

Just as you do when you work a jigsaw puzzle, you first classify the pieces before trying to make connections. You might find pieces that share common properties, such as a flat edge or the same color, and place them in separate piles. By placing the pieces in these separate, smaller piles, it is easier to see how they might connect to form the completed picture.

Classification by composition A powerful way to classify matter is by its composition. This is the broadest type of classification. When you examine an unknown piece of matter, you first ask, "What is it made of?" For example, sucrose is composed of the elements carbon, hydrogen, and oxygen. This kind of description is a qualitative expression of composition. A **qualitative** observation is one that can be made without measurement.

After a qualitative analysis, the next question that you might ask is how much of each of the elements is present. For sucrose, the answer to that question is that 100.0 g of sucrose contains 42.1 g of carbon, 51.4 g of oxygen, and 6.5 g of hydrogen. This is a quantitative expression of composition. A **quantitative** observation is one that uses measurement. You make quantitative measurements every day when you answer such questions as, "What's the temperature?" "How long was the touchdown pass?" "How much do you weigh?" **Figure 1.9** shows some quantitative measurements being made.

Pure substance or a mixture? Another general way to classify matter by composition is in terms of purity. There are only two categories. A sample of matter is either pure—made up of only one kind of matter—or it is a mixture of different kinds of matter. What does it mean to say that matter is pure?

■ **Figure 1.10** Products such as orange juice are often advertised as pure. From a chemist's point of view, they are not pure but instead are complex mixtures containing many different substances.

Pure substances The word *pure* is often used to describe common things, as in **Figure 1.10.** In chemistry, pure means that every bit of the matter being examined is the same substance. A **substance** is matter with the same fixed composition and properties. Any sample of pure matter is a substance.

If the sugar in a bag from the supermarket is a substance, then it is pure sucrose. Every bit of matter in the bag must have the same properties and the same fixed composition as every other bit. Now, consider a bag of high-quality, dry, white sand. White sand is the common name for a substance called silicon dioxide. It is white and crystalline like sugar, and, when the sand is examined, every particle has the same fixed composition (53.2 percent oxygen, 46.8 percent silicon). Both sand and sucrose are substances, but they have different properties and compositions.

Mixtures Suppose you mix pure sand with sucrose. You might not be able to see any difference, but if you add this mixture to your tea, as shown in **Figure 1.11,** it's a different story. The sweet taste of the sugar is still there, but so is the grittiness of the sand. Every particle is not the same, so the properties are not the same throughout. Some parts taste sweet, while others are gritty and tasteless. The composition is also not fixed, but instead depends on how much sand, sugar, and tea are mixed into the water.

■ **Figure 1.11** Pure sugar is different from a mixture of sugar and sand. Each component of a mixture retains its behavior. Sugar dissolves and sweetens the tea, but sand is insoluble. It settles to the bottom of the cup.

Section 1.1 • The Puzzle of Matter

Most of the matter you encounter every day is a mixture. A **mixture** is a combination of two or more substances in which the basic identity of each substance is not changed. In the sugar and sand mixture, the sand does not influence the properties of the sugar, and the sugar does not influence the properties of the sand. They are in contact with each other, but they do not interact with each other. **Figure 1.12** shows some common mixtures.

FIGURE 1.12

Mixtures

Unlike pure substances, mixtures can have varying compositions. For example, two samples of seawater might contain different amounts of salt.

Sand, pulverized stone, minerals, salts, and substances from decayed plants and animals make up the rich mixture called soil.

This seascape shows two mixtures: seawater, a mixture of water, salts, and other substances; and air, a mixture of gases including nitrogen, oxygen, carbon dioxide, and water vapor.

Blood is a complex mixture of many substances including water, proteins, glucose, fats, amino acids, and carbon dioxide.

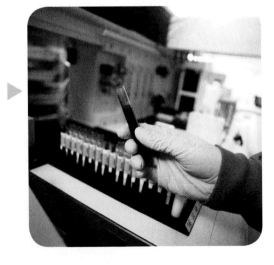

Solder is an alloy of tin and lead used by plumbers and electricians to seal a joint or connect one piece of metal to another.

Everyday Chemistry

You Are What You Eat

The morning is half over and you need some energy, so you eat a piece of sour-apple candy. You have just consumed four chemicals—the sugar fructose for sweet taste and energy, citric acid for tartness, methyl butanoate for apple flavor, and red dye #2 for color. *Chemical* is just another name for "substance." Whenever you eat or drink, you consume chemicals that your body needs for energy, growth, and repair.

What chemicals are found in your body? The percentages of elements that make up the human body are shown in **Figure 1** below. The elements found in your body are not free elements; they are in the form of compounds. For example, 50–65 percent of your body is water, a compound of hydrogen and oxygen. Carbon is the most abundant element found in all of the major molecules in living organisms, such as DNA, proteins, and carbohydrates. Therefore, it is not surprising that most of a human body's mass is made of oxygen, carbon, and hydrogen.

Your body is a complex chemical factory. It checks to see if the right amount of each compound is present, decomposes the food you eat, and uses the new substances formed to make compounds needed for growth and repair. It breaks down other food components to obtain the energy it needs.

Figure 2 Breakfast foods are mixtures of many different compounds.

The chemistry of food You might start the day with a breakfast of melon, eggs, whole-wheat toast, milk, and tea, or you might start the day off with a breakfast of fruit, pastries, coffee, and orange juice, as shown in **Figure 2**. All of these foods are mixtures of many different compounds. If you decide to sweeten your tea with sugar, you will be adding a single compound—sucrose—to a mixture of water, caffeine, tannin, butyl alcohol, isoamyl alcohol, phenyl ethyl alcohol, benzyl alcohol, geraniol, hexyl alcohol, and essential oils.

If you eat scrambled eggs, you are consuming a mixture of water, ovalbumin, coalbumin, ovomucoid, mucia, globulins, amino acids, lipovitellin, livetin, cholesterol, lecithin, lipids (fats), fatty acids, butyric acid, acetic acid, lutein, zeaxanthine, and vitamin A with a little sodium chloride (salt) sprinkled in. Doesn't that sound mouthwatering?

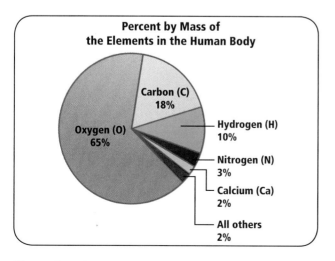

Figure 1 The human body is composed of many different elements.

Percent by Mass of the Elements in the Human Body
- Oxygen (O) 65%
- Carbon (C) 18%
- Hydrogen (H) 10%
- Nitrogen (N) 3%
- Calcium (Ca) 2%
- All others 2%

Explore Further

1. **Analyze** What color is burned food? What element is usually found in food that is burned? What element do you think most foods have in common?

2. **Interpret** Find out how organic and inorganic compounds differ. Which type makes up most of the human body? Why is the other type important to life?

CHEMLAB 2

SMALL SCALE

KITCHEN CHEMICALS

Background

The chemical and physical properties of a substance make up a sort of fingerprint that characterizes the substance. In this ChemLab, you will test four unknown solids using three different liquids. The unknowns are common materials that you'd probably find in your kitchen. The results of your tests will give you the information you need to unravel the compositions of mixtures of two solids and three solids.

Questions

How can you identify a substance by comparing its properties with those of known substances?

Objectives

- **Observe** the chemical reactions and the physical changes of four common kitchen materials with three test reagents.
- **Compare and interpret** the reactions of the test reagents with five two-solid and three-solid mixtures of the common kitchen materials.
- **Infer** the composition of each of five unknown mixtures by comparing their reactions with those of the known materials.

Preparation

Materials

96-well microplate
9 test tubes
spatulas
3 thin-stemmed pipettes
masking tape
marking pen

Safety Precautions

WARNING: *Do not touch or taste any of the solids or liquids, even though you may believe you know their identities.*

Procedure

1. Read and complete the lab safety form.
2. Label four test tubes A, B, C, and D. Label five test tubes 1 through 5.
3. In the four lettered test tubes, place about 1 g of each of the labeled samples supplied by your teacher. These are the common kitchen materials.
4. In the numbered test tubes, place about 1 g of each of the numbered samples supplied by your teacher. These are the unknown mixtures. If you use the same spatula for each material, rinse and dry it before dipping into the next solid to avoid contaminating one material with another.
5. Label three long columns of wells on the microplate I, II, and III. Label nine rows of wells on the microplate A, B, C, and D, and 1–5 as shown in the photo below.

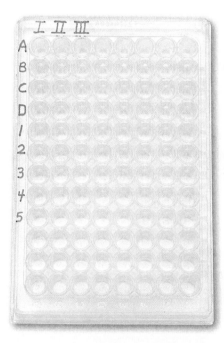

6. Place the microplate on a sheet of white paper.
7. Add a small amount of each material to the row of three wells that has the appropriate letter or number.

8. Observe and record the texture of each of the nine materials in a data table like the one shown.

9. Label the three pipettes I, II, and III. From the containers of reagent liquids supplied by your teacher, draw into the bulb of each pipette the liquid corresponding to the label number on that pipette.

10. Add 3 drops of liquid I to each of the nine materials in column I.

11. Observe any changes that take place, and record them in the data table.

12. Repeat steps 10 and 11 using liquid II and then liquid III.

Analyze and Conclude

1. **Interpret Data** What properties and reactions characterize each of the four kitchen solids?

2. **Draw Conclusions** Can you positively identify the solids that are contained in any of the five mixtures? If so, identify the solids and explain your conclusions.

3. **Infer** If you are unable to conclusively identify the solids in any of the mixtures, what are their likely identities? Explain.

Apply and Assess

1. **Infer** Two of the four original solids, baking powder and baking soda, are often used in making baked goods. What characteristic probably makes them useful in baking? Which solids display this characteristic?

2. **Explain** Baking powder is a mixture of two or more compounds, and it reacts with water or any other liquid that contains water. Baking soda is a single compound that reacts with acidic solutions but not with water. Which of the solids do you think is baking powder? Explain.

INQUIRY EXTENSION

One of the solids is an organic compound you might have learned about in a biology course. It produces a characteristic color when combined with iodine. Which solid gave this reaction? What is the identity of this compound? Design an experiment to determine whether this compound is present in other foods.

Data and Observations

Data Table					
Solid	Color	Texture	Reaction with liquid I	Reaction with liquid II	Reaction with liquid III
A					
B					
C					
D					
1					
2					
3					
4					

The separation of mixtures into substances One characteristic of a mixture is that it can be separated into its components by physical processes. The word physical tells you that the process does not change the identity of a substance. How could you separate a mixture of sand and sugar into pure sand and pure sucrose? The simplest physical means would be to look at it with a microscope and separate the bits of sugar and sand with tweezers. This would require a lot of time and patience. You are right if you think that there must be an easier way.

Separating mixtures by using physical changes is one easier way. A **physical change** is a change in matter that does not involve a change in the identity of the substance. Examples of physical changes include boiling, freezing, melting, evaporating, dissolving, and crystallizing. Separation of a mixture by physical changes takes advantage of the different physical properties of the mixed substances. **Physical properties** are characteristics of a sample of matter that can be observed or measured without any change in its identity. Examples of the physical properties of a sample of matter include its solubility, melting point, boiling point, color, density, electrical conductivity, and physical state (solid, liquid, or gas).

Look at **Figure 1.13** to see one way to separate a mixture of sugar and sand using differences in the physical properties of the two substances. The separation is possible because of the difference in solubility of sugar and sand. The first two steps shown in **Figure 1.13** involve the separation of the sand—which does not dissolve in the water—from the mixture. To separate the dissolved sugar from the mixture, the water is evaporated. Step 4 of **Figure 1.13** shows that only the sugar remains in the beaker.

■ **Figure 1.13** Because sugar and sand have some different physical properties, a separation can be made by using a series of physical changes.

Step 1 Water is added and the mixture is stirred. The sugar dissolves in the water, but the sand does not.

Step 2 The mixture is passed through a filter that traps the sand but allows the sugar solution to pass through.

Step 3 The sugar solution is heated to evaporate the water.

Step 4 When all of the water has evaporated, pure sugar remains in the beaker.

MiniLab 1.1

Observe Mixing

50 mL + 50 mL = ? Observation of some mixtures, such as water and alcohol, can provide clues to the structure of matter.

Procedure

1. Read and complete the lab safety form.
2. Fill a **100-mL graduated cylinder** to approximately the 50.0-mL mark with **water** that has been tinted with **food coloring.** Insert a **thermometer** and read the temperature of the water. Remove the thermometer. Read and record the exact volume of the water.
3. Tilt the graduated cylinder over almost as far as you can without spilling water and add **ethanol** very slowly so that it does not mix with the water. Raise the cylinder slowly as you add more alcohol.
4. Add the final alcohol with a **dropper** until the level is exactly 100.0 mL. Assume the ethanol and water have the same temperature.
5. Use a **stirring rod** to mix the contents of the cylinder as rapidly as possible. Immediately insert a thermometer. Read and record the temperature of the mixture.
6. Remove the thermometer and stirring rod, taking care that all liquid drips back into the cylinder. Read and record the volume of the mixture to the nearest 0.1 mL.

Analysis

1. **Infer** Was heat absorbed or released as the liquids mixed? How do you know?
2. **Describe** what happened to the volume when the liquids mixed. Suggest a way to explain for your observation.

Types of mixtures Sometimes, when you look at a sample of matter, it's easy to tell that the sample is a mixture. This kind of mixture is called a heterogeneous mixture. The prefix *hetero* means "different." A heterogeneous mixture is one that does not have a uniform composition and in which the individual substances remain distinct. The components of the mixture exist as distinct regions, often called phases. In other words, you can see the different substances in the mixture. Recall the mixture of sand and water. It is a heterogenous mixture. Orange juice and a piece of granite, as shown in **Figure 1.14,** are examples of heterogeneous mixtures.

VOCABULARY

WORD ORIGIN

Heterogeneous
comes from the Greek words *hetero*, which means *different*, and *genea*, which means *source*

■ **Figure 1.14** If you look at a granite rock, you can see areas of different color that indicate that the rock is composed of crystals of different substances.

The separation of sand and sugar shown in **Figure 1.13** took advantage of a difference in the physical properties of the two substances. Sugar dissolves in water, but sand does not. When sugar dissolves in water, the two pure substances, sugar and water, combine physically to form a mixture that has a constant composition throughout. This means that no matter where you sample the mixture, you find the same combination of sugar and water. Even with a powerful light microscope, you could not pick out a bit of pure sugar or a drop of pure water. This type of mixture is called a homogeneous mixture. The prefix *homo* means "the same."

MiniLab 1.2

Paper Chromatography of Inks

How can you separate the dyes in ink? The inks in marking pens are often mixtures of dyes of several basic colors, called pigments. In this MiniLab, you will use the technique of chromatography to analyze the ink from several pens.

Procedure

1. Read and complete the lab safety form.
2. Obtain a **clear plastic cup** that is at least 6 cm high. From a **coffee filter,** use **scissors** to cut a strip of paper about 2.5 cm wide and about 2.5 cm longer than the height of the cup.
3. Place the paper strip in the cup so that the bottom of the strip just rests on the bottom of the cup.
4. Push a **pencil** through the top of the paper in such a way that when the pencil rests on the top of the cup, the paper is suspended with its lower edge just touching the bottom.
5. Prepare several strips in the same way, one strip for each type of marker ink you will analyze.
6. Using one **water-soluble ink marker** for each strip, draw a narrow, horizontal line across the strip about 2 cm up from the bottom. If possible, include one black or brown marker.
7. Add about 1 cm of **water** to the cup and suspend the first strip in the water. The marker line must be above the water level when the strip is suspended in the cup.

8. Cover the top of the cup loosely with clear plastic wrap to reduce evaporation. Observe and record the effect on the marker line as the water moves up the paper.
9. Remove the strip from the cup or beaker when the water level has risen to just below the pencil. Lay the strip on a paper towel to dry.
10. Repeat procedures 6 through 9 with each of your marker strips.

Analysis

1. **Examine** Capillary action is the movement of a liquid upward through small pores that exist in some materials. Did you note any evidence of capillary action in this MiniLab?
2. **Analyze** Does your evidence indicate that any of the marker inks were composed of more than one pigment?
3. **Identify** Which ink colors contained the greatest number of pigments?

Table 1.1 Some Common Alloys

Name of Alloy	Percent Composition by Mass	Uses
Stainless steel	73–79% iron (Fe) 14–18% chromium (Cr) 7–9% nickel (Ni)	kitchen utensils, knives, corrosion-resistant applications
Bronze	70–95% copper (Cu) 1–25% zinc (Zn) 1–18% tin (Sn)	statues, castings
Brass	50–80% copper (Cu) 20–50% zinc (Zn)	plating, ornaments
Sterling silver	92.5% silver (Ag) 7.5% copper (Cu)	jewelry, tableware
14-karat gold	58% gold (Au) 14–28% silver (Ag) 14–28% copper (Cu)	jewelry
18-karat white gold	75% gold (Au) 12.5% silver (Ag) 12.5% copper (Cu)	jewelry
Solder (electronic)	63% tin (Sn) 37% lead (Pb)	electrical connections

Homogeneous mixtures are the same throughout. Another name for a homogeneous mixture is **solution**. Even though solutions might appear to be one pure substance, their compositions can vary. For example, you could make your tea very sweet by dissolving a lot of sugar in it or less sweet by dissolving only a little bit of sugar.

When you hear the word solution, something dissolved in water probably comes to mind. But liquid solutions do not have to contain water. Gasoline is a liquid solution of several substances, but it contains no water. Some solutions are gases. Air, for example, is a homogeneous mixture of several gases. Some solutions are solid. **Alloys** are solid solutions that contain different metals and sometimes nonmetallic substances. Steel, for example, is a general term for a range of homogeneous mixtures of iron and substances such as carbon, chromium, manganese, nickel, and molybdenum. Because pure gold is soft and bends easily, most gold jewelry is not made from pure gold, but rather from an alloy of gold with silver and copper. **Table 1.1** shows some common alloys and their compositions.

When you dissolve sugar in water, sugar is the **solute**—the substance being dissolved. The substance that dissolves the solute, in this case water, is the **solvent**. When the solvent is water, the solution is called an **aqueous solution**. Many of the solutions you encounter are aqueous solutions, for example, soft drinks, tea, contact-lens cleaner, and other clear cleaning liquids. In addition, most of the processes of life occur in aqueous solutions.

> **VOCABULARY**
> **WORD ORIGIN**
> **Homogeneous**
> comes from the Greek *homo*, which means *alike*, and *genea*, which means *source*

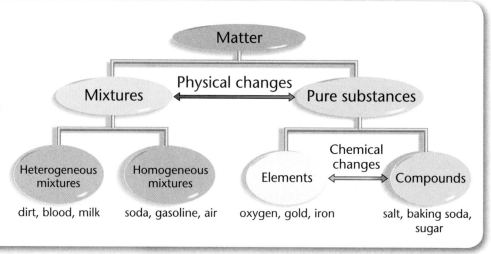

■ **Figure 1.15** The chart shows that mixtures can be either heterogeneous or homogeneous and can often be separated into pure substances by physical changes. Pure matter can be elements or compounds.

Examine *Where on this chart would you classify aluminum? Pizza?*

Substances: Pure Matter

Although the chemistry of the real world is mostly the chemistry of mixtures, all matter is composed of substances. Dig a hole, buy something at a grocery store, pick an apple from a tree, or take a deep breath. The stuff you dig, buy, pick, or inhale is a mixture. However, the behavior of mixtures is based on the composition, structure, and behavior of the pure substances that compose them. **Figure 1.15** summarizes the classification of matter from a chemical point of view.

Elements: the building blocks If you classify an unknown piece of matter as a pure substance, it means that the matter is made of only one substance. But there are two types of pure substances illustrated in **Figure 1.15**. One type of pure substance can be broken down into simpler substances. This type of substance is called a compound. Another type of substance cannot be broken down into simpler substances. Such a substance is called an element. **Elements** are the simplest form of matter. **Figure 1.16** shows two examples of elements.

■ **Figure 1.16** Both gold and diamonds are examples of elements. Diamonds are one form of pure carbon.

24 Chapter 1 • Chemistry: The Science of Matter

MiniLab 1.3

Create an Alloy

What happens when a penny and zinc react? The alchemists of old tried unsuccessfully to convert common metals to gold. Alchemists were not early chemists, but their practical knowledge about elements and compounds contributed to the work of the earliest true chemists. Like the alchemists, you will not turn copper into gold, but by allowing the copper in a penny to react with zinc under certain conditions, you can create an interesting alloy of the two metals.

Procedure

1. Read and complete the lab safety form.
2. Clean a **pre-1982 penny** with **steel wool** or a **pencil eraser**.
3. Place 1 g of **granular zinc** in an **evaporating dish**. Add 20 mL of **1M zinc chloride solution ($ZnCl_2$)**. Use **tongs** to place the penny in the dish, and put the evaporating dish on a **hot plate**.
4. Heat the mixture until it just starts to boil. This should take about two minutes. Carefully stir the mixture with the tongs and turn the penny. Continue to heat and stir gently until the penny becomes covered with zinc and appears gray in color. This usually takes less than a minute.
5. Use the tongs to remove the penny from the liquid. Rinse the penny in a beaker of cold tap water, then pat it dry with a **paper towel.**
6. Using tongs to hold the penny, gently heat it in the cooler, outer portion of a **Bunsen burner flame** until it changes color. Record your observations.
7. Continue heating gently for two or three seconds longer, then immediately immerse the penny in a fresh **beaker** of cold water.
8. Keep the penny immersed in the beaker of water for about one minute. After the penny has cooled for about a minute, remove it from the water and pat it dry. Record your final observations.

Analysis

1. **Evaluate** What evidence indicates that you created an alloy of copper and zinc? Explain.
2. **Determine** What is the probable identity of this alloy?
3. **Infer** What do you think you would see if you cut the penny in two and examined the cut edge with a microscope?

Millions of substances are known to chemists, but only 117 are elements. These 117 elements combine with each other to form all the millions of known compounds. That's why elements are often referred to as the building blocks of matter. All the substances of the universe are elements, compounds formed from elements, or mixtures of elements and compounds.

Of the 117 known elements, only about 92 occur naturally on Earth. Copper, oxygen, and gold are examples of naturally occurring elements. The remainder are synthesized, usually in barely detectable amounts, in high-energy nuclear experiments. Less than half of the 92 naturally occurring elements are abundant enough to play a significant role in the chemistry of everyday stuff.

The fact that most matter is composed of a relatively small number of building blocks simplifies the puzzle of matter. However, the observation that such a small number of pieces creates such a variety of compounds means that elements must connect to each other in countless ways.

Literature Connection

Jules Verne and His Icebergs

Imagine you are underwater in a submarine in an Antarctic iceberg field. While sleeping in your bunk, you are awakened by a violent shock and thrown into the middle of your cabin. Surveying the situation, you realize that the submarine is on its side. This is what happens to Professor Pierre Aronnax when he is aboard Captain Nemo's *Nautilus* in Jules Verne's science fiction novel *Twenty Thousand Leagues Under the Sea*.

How does Jules Verne explain the accident? In Verne's book, published in 1869, Captain Nemo correctly explains that a mountain of floating ice, an iceberg like that shown in **Figure 1,** has turned over. When an iceberg is undermined at its base by warmer water or repeated shocks, its center of gravity rises, and the whole thing turns bottom up. As the bottom of the iceberg turned upward, it struck the *Nautilus,* slid under its hull, and raised it onto an ice bed, where the ship lay on its side.

What concepts are involved? Density and the center of gravity of an iceberg, such as the one shown in **Figure 1,** account for the accident. Water is the only substance known that expands when it freezes. This is due to water molecules lining up less efficiently, in regard to volume, when water is frozen. An iceberg floats because the density of frozen freshwater is about 0.9 g/mL. This is less than the density of the salt water surrounding it, which is about 1.025 g/mL. The density of freshwater is about 1.0 g/mL, which is why an ice cube will also float in a glass of water. However, because the density of frozen freshwater, or ice, is not that much less than freshwater or salty water, ice floats with most of its mass below the water's surface. On average, 75 to 87.5 percent of an iceberg floats under the surface of the water.

An object's center of gravity is the point where all the weight of the object seems to be located.

Figure 1 Icebergs are mountains of floating ice.

The higher an object's center of gravity is above its support, the less stable the object is. If something happens to the bottom of an iceberg such that the center of gravity shifts above the waterline, the iceberg turns over.

An iceberg's center of gravity changes when large pieces of ice break off the berg. This can be caused by water freezing in cracks, the vibration of the waves, or thunderous vibrations resulting from the breaking of other bergs. When an iceberg becomes unstable, small movements can cause it to topple over.

Captain Nemo understood the concepts of density and center of gravity. He used these concepts to guide his decisions when determining the best route of escape for the *Nautilus* and its crew when they became trapped in an ice field in Antarctica after the accident with the iceberg. The concept of density is extremely important when navigating a submarine in the ocean, especially in an ocean with icebergs.

Connection to Chemistry

1. **Acquire Information** Find out what differences exist between icebergs in the northern hemisphere and in the southern hemisphere.
2. **Apply** If an iceberg were floating in freshwater instead of seawater, would more or less of its mass be above the water's surface? Explain.

■ **Figure 1.17** This photograph shows a monument that depicts an early version of the periodic table in St. Petersburg, Russia. Even though the alphabet used for Russian is much different than that used for English, it is still easy to recognize the symbols for the chemical elements. The symbols for the chemical elements are a universal language through which chemists can communicate worldwide.

Organizing the elements Your classroom might have a large chart labeled *The Periodic Table of the Elements* hanging on the wall. This table is the tool you will use most in your chemistry course. A similar table is printed on pages 92–93 of this book, as well as inside the back cover. The periodic table organizes elements in a way that provides a wealth of chemical information—much more than you will notice right now.

All of the named elements on the periodic table are represented by a one- or two-letter symbol. As **Figure 1.17** indicates, these chemical symbols are a universal shorthand that is used to make chemistry communication understandable around the world. Just as it is easier and quicker for you to write USA instead of *United States of America*, it is easier to write Al instead of the word *aluminum*. As you can see, the symbol for aluminum is taken directly from the element's name, but some elements have symbols that don't correspond to their English names. Their symbols usually correspond to their names in Latin. Some examples are shown in **Table 1.2**.

Table 1.2 Some Historic Chemical Symbols

Element	Symbol	Origin	Language
Antimony	Sb	stibium	Latin
Copper	Cu	cuprum	Latin
Gold	Au	aurum	Latin
Iron	Fe	ferrum	Latin
Lead	Pb	plumbum	Latin
Potassium	K	kalium	Latin
Silver	Ag	argentum	Latin
Sodium	Na	natrium	Latin
Tin	Sn	stannum	Latin
Tungsten	W	wolfram	German

MiniLab 1.4

Analyze Cereal

What's this stuff in my cereal? Many breakfast cereals have additives that increase their nutritional values. In this MiniLab, you will test a common cereal for the presence of one of these additives.

Procedure

1. Read and complete the lab safety form.
2. Tape a small, strong **magnet** to a **pencil** at the eraser end.
3. Place a sample of dry, fortified, cold **cereal** in a **plastic bag**.
4. Thoroughly crush the cereal with a **rolling pin** or other heavy object.
5. Pour the crushed cereal into a **beaker** and cover it with **water**.
6. Stir the cereal/water mixture for about ten minutes with your pencil-magnet stirrer. Stir slowly and easily for the last minute.
7. Remove the magnet from the cereal and examine it carefully. Record your observations.

Analysis

1. **Determine** The substance attracted to your magnet is a common element. What is it?
2. **Infer** Why do you think that this element is added to the cereal?

■ **Figure 1.18** The elements that make up a compound are chemically combined to form a new substance with a unique set of properties. Silver bromide has a unique set of physical and chemical properties and a fixed composition of 57.45 percent silver and 42.55 percent bromine.

Compounds You've learned that a compound is a pure substance that can be broken down into elements. A more complete definition is that a **compound** is a chemical combination of two or more different elements joined together in a fixed proportion. For example, water is always 11.2 percent hydrogen and 88.8 percent oxygen by mass. It doesn't matter whether a sample comes from a faucet, an iceberg, a river, or a rain puddle. Once you isolate the water from other substances in the sample, you will find that every sample of water is made up of exactly the same proportion of hydrogen and oxygen. Every compound has its own fixed composition, and that composition results in a unique set of chemical and physical properties. The properties of the compound are different from the properties of the elements that compose the compound, as shown in **Figure 1.18**.

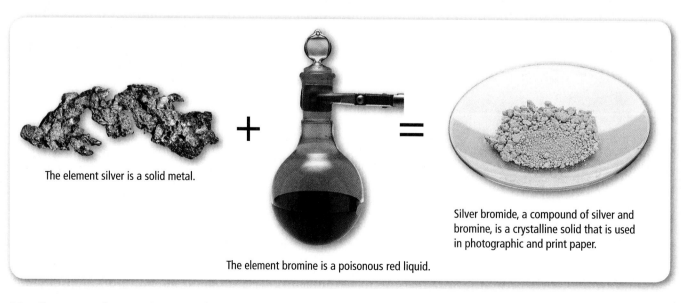

The element silver is a solid metal.

The element bromine is a poisonous red liquid.

Silver bromide, a compound of silver and bromine, is a crystalline solid that is used in photographic and print paper.

Chemistry & Society

Natural Versus Synthetic Chemicals

If you look carefully at the vitamin display in a drugstore, you'll see some bottles marked *all natural*. Most of these vitamins have been extracted from plant or animal sources. Are natural vitamins, drugs, and other substances better than the purified or synthetic ones produced by drug companies? Maybe this will help you decide.

Aspirin: A common synthetic drug To get rid of a headache, would you rather drink a cup of willow bark tea or take two aspirin? The active ingredients in both remedies have similar chemical structures, and both are effective against headaches. However, the chemical from the willow bark, salicylic acid, has several harmful side effects including stomach pain. So does aspirin, but it is a more effective painkiller and can be taken in lower doses. Also, willow bark contains many other chemicals. After years of research, scientists made aspirin in the laboratory from salicylic acid and acetic anhydride. It contains only one active ingredient, acetylsalicylic acid. Not only is aspirin a pure substance, but also the difference in its structure eliminates the serious stomach pain caused by salicylic acid.

Development of new drugs What happens when a chemical discovered in nature is shown to be a potential treatment or cure for a disease? Scientists use the following procedures to make safe, effective drugs:

(1) isolate and purify the drug,
(2) determine its composition and structure,
(3) search for a way to make it synthetically,
(4) look for a less expensive, easy way to produce it in large quantities, and
(5) try changing the structure and composition of the original compound to improve on nature's model.

Taxol: A cancer drug Scientists discovered that taxol, a chemical found in the bark of the Pacific yew tree, as shown in **Figure 1**, reduced the size of ovarian and breast cancer tumors in 30 percent of patients.

Figure 1 The bark of the Pacific yew tree contains a cancer fighting chemical.

But scientists were concerned that demand for the drug might wipe out the population of yew trees, so they started to look for new sources of the drug. Chemists Andrea and Donald Stierle found a taxol-producing fungus growing on one yew. Other scientists discovered that needles from the European yew contain a chemical similar to taxol.

Chemists experimented to figure out taxol's structure. In 1994, they succeeded in producing pure taxol in the lab. You might be wondering which taxol is better—the purified natural one or the synthetic one. Actually, the chemical structure and purity of both is identical. However, being able to make taxol synthetically is a real advantage; drug companies may be able to produce it more cheaply, and they can work on modifying its structure to make it more effective.

ANALYZE the Issue

1. **Acquire Information** Find out why scientists were concerned that the use of taxol could endanger the Pacific yew and whether this is still a concern.
2. **Research** In what ways are the structures of salicylic acid and acetylsalicylic acid (aspirin) alike? In what ways are they different?
3. **Debate** the pros and cons of using natural herbal drugs, purified natural drugs, and synthetic drugs.

Table 1.3 Some Common Compounds

Compound Name	Formula	Uses
Acetaminophen	$C_8H_9NO_2$	pain reliever
Acetic acid	CH_3OOOH	tart ingredient in vinegar
Ammonia	NH_3	fertilizer, household cleaner when dissolved in water
Ascorbic acid	$C_6H_8O_6$	vitamin C
Aspartame	$C_{14}H_{18}N_2O_5$	artificial sweetener
Acetylsalicylic acid	$C_9H_8O_4$	pain reliever
Baking soda	$NaHCO_3$	cooking
Butane	C_4H_{10}	lighter fuel
Caffeine	$C_8H_{10}N_4O_2$	stimulant in coffee, tea, some soft drinks
Calcium carbonate	$CaCO_3$	antacid
Carbon dioxide	CO_2	carbonating agent in soda
Ethanol	C_2H_5OH	disinfectant, alcoholic beverages
Ethylene glycol	$C_2H_6O_2$	antifreeze
Hydrochloric acid	HCl	called muriatic acid, cleans mortar from brick
Magnesium hydroxide	$Mg(OH)_2$	antacid
Methane	CH_4	natural gas, fuel
Phosphoric acid	H_3PO_4	flavoring in soda
Potassium tartrate	$K_2C_4H_4O_6$	cream of tartar, cooking
Propane	C_3H_8	fuel for cooking
Sodium hydroxide	$NaOH$	drain cleaner
Sucrose	$C_{12}H_{22}O_{11}$	sweetener
Sulfuric acid	H_2SO_4	battery acid
Water	H_2O	washing, cooking, cleaning

More than 10 million compounds are known and the number keeps growing. Some common compounds are listed in **Table 1.3**. New compounds are discovered and isolated from natural chemical sources such as plants and colonies of bacteria. Compounds are also synthesized in laboratories where they are tested for a variety of uses ranging from medicine to manufacturing.

Because the supply of useful chemicals from natural sources is often limited, chemists work to synthesize these compounds in the laboratory. The effort to synthesize taxol, an anticancer compound found in the bark of the Pacific yew tree, is an example of how nature often provides the lead in compound synthesis. If taxol can be made in the laboratory, then chemical engineers will try to find a way to produce it on an industrial scale.

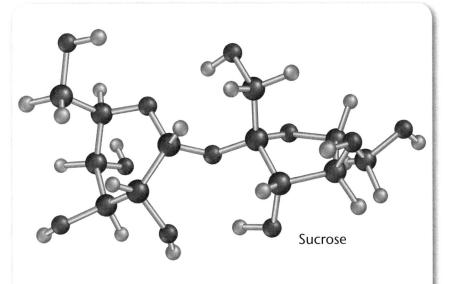

■ **Figure 1.19** Common table sugar is a compound that chemists call sucrose. In the model of sucrose above, gray spheres represent carbon atoms, blue represent hydrogen, and red represent oxygen.
Verify that the model contains 12 carbon atoms, 22 hydrogen atoms, and 11 oxygen atoms.

Formulas of compounds The second column in **Table 1.3** gives the chemical formulas for the compounds listed. A chemical **formula** is a combination of the chemical symbols that show what elements make up a compound and the number of atoms of each element. For example, the chemical formula of the compound sucrose is $C_{12}H_{22}O_{11}$. The formula tells you, in a compact way, that sucrose contains carbon, hydrogen, and oxygen. It also tells you that the smallest unit of sucrose, a molecule, contains 12 carbon atoms, 22 hydrogen atoms, and 11 oxygen atoms. **Figure 1.19** shows a submicroscopic view of sucrose.

Formulas provide a shorthand way of describing a submicroscopic view of a compound. In later chapters, you'll learn more about compounds and why elements combine to form compounds. You'll also learn how to determine the formulas for many compounds.

Section 1.1 Assessment

Section Summary

- Chemists study matter.
- Macroscopic observations reflect the submicroscopic structure of matter.
- Mixtures are heterogeneous or homogeneous (solutions).
- Substances are classified as elements or compounds. Elements are the building blocks of all matter.

1. **MAIN Idea** What three characteristics of matter do chemists study?
2. **Compare and contrast** a mixture and a pure substance.
3. **Contrast** How does a compound differ from a mixture?
4. **Apply** The element oxygen is a gas that makes up about 20 percent of Earth's atmosphere. Oxygen also is the most abundant element in Earth's crust, yet Earth's crust is not a gas. Explain this apparent conflict.
5. **Support** Matter may be subdivided into elements, compounds, heterogeneous mixtures, and homogeneous mixtures. Describe one material found in a household that belongs in each category. Support your answer.

Section 1.2

Objectives

◗ **Distinguish** between physical and chemical properties.
◗ **Contrast** chemical and physical changes.
◗ **Apply** the law of conservation of matter to chemical changes.

Review Vocabulary

matter: anything that takes up space and volume

New Vocabulary

volatile
density
chemical property
chemical change
chemical reaction
law of conservation of mass
energy
exothermic
endothermic

Properties and Changes of Matter

MAIN Idea Matter can undergo physical and chemical changes.

Real-World Reading Link When the refuse truck pulls up at the landfill to empty its cargo, you know that it is unloading all the throwaways of modern life. But each item of trash had its own journey from raw material to product to trash. And each might have further use through recycling. It all depends on the properties of the matter making up the trash.

Physical Properties

Physical properties are those that don't involve changes in composition. Many physical properties are qualitative descriptions of matter, such as *the solution is blue; the solid is hard;* or *the liquid boils at a low temperature*. Other physical properties are quantitative, which means that they can be measured with an instrument. Examples include *an ice cube melts at 0°C; iron has a density of 7.86 g/mL;* or *a mass of 35.7 g of sodium chloride dissolves in 100 mL of water*.

Figure 1.20 illustrates one physical property of sodium chloride (table salt). When table salt is added to water, a physical change occurs in that you can no longer see the salt. If you were to taste the salt water, however, you would know that the salt is still there. When the water evaporates, salt crystals will be left behind.

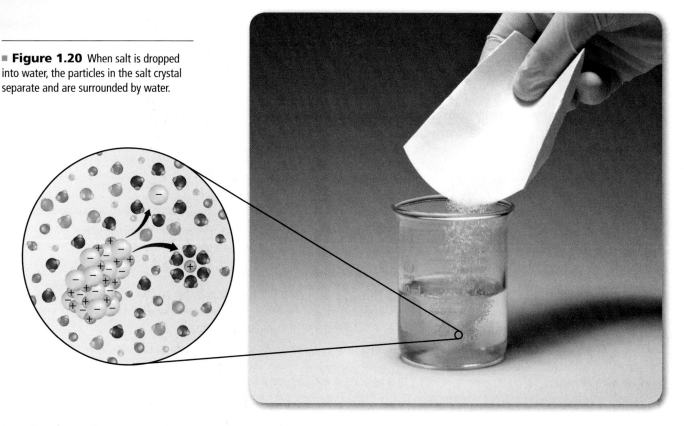

■ **Figure 1.20** When salt is dropped into water, the particles in the salt crystal separate and are surrounded by water.

32 Chapter 1 • Chemistry: The Science of Matter

■ **Figure 1.21** Because freezing and melting occur at the same temperature, both phases of water, solid and liquid, can be present when the temperature is exactly 0°C.

States of matter Most matter on Earth exists in one of three physical states: solid, liquid, or gas. A fourth state of matter, called plasma, is less familiar. You will read about it in Chapter 10. The physical state of a substance depends on its temperature. You know that if you put liquid water into a freezer, it changes to solid water (ice), and if you heat liquid water to 100°C in a teakettle, it boils and becomes gaseous water (steam). But the physical state of a substance usually means its state at room temperature—about 20°C–25°C. At room temperature, the physical state of water is liquid, salt is a solid, and oxygen is a gas.

Change of state A physical property, closely related to the physical state of a substance, is the temperature at which a substance changes from one state to another. Water, for example, freezes (and melts) at 0°C. Table salt (NaCl) melts (and freezes) at a much higher temperature, 804°C, and oxygen (O_2) freezes (and melts) at a much lower temperature, −218°C. The melting point and the freezing point of a substance are the same temperature, as you can see in **Figure 1.21,** which shows both water and ice at 0°C. Water boils at 100°C, but it also condenses from a gas to a liquid at the same temperature. Therefore, boiling point and condensation point are also the same temperature for each substance.

Changes in state are examples of physical changes because there is no change in the identity of the substance. Ice can melt back to liquid water, and steam will condense on a cool surface to liquid water. Some substances are described as **volatile**, which means that they change to a gas easily at room temperature. Alcohol and gasoline are more volatile than water. The substance naphthalene, used as mothballs, is an example of a solid substance that is volatile. You can readily smell alcohol, gasoline, and mothballs when open containers of these substances are present in a room because the liquid or solid has changed to gas and the molecules are present in the air.

◀ **FOLDABLES**
Incorporate information from this section into your Foldable.

TRY AT HOME 🏠 LAB
See page 868 for **Comparing Frozen Liquids**

Section 1.2 • Properties and Changes of Matter 33

■ **Figure 1.22** Density compares masses of equal volumes. The boxes have the same volume. You could think of density in these photos as mass/box or g/box. The usual units are g/mL.

Infer *Would the volume of the rocks have to increase or decrease in order to match the mass of the plastic foam pictured?*

Plastic foam balls that fit into the box have much less mass than the stones that fill the box.

Matter in stone is packed much more densely than it is in plastic foam.

Density Density is another physical property of matter. Consider two identical boxes, one filled with plastic foam balls and the other filled with stones. If you lifted each box, as shown in **Figure 1.22,** you might say that the box of plastic foam balls is light and that the box of stones is heavy. The plastic foam balls occupy a certain amount of space or volume (the box), but they have little mass because of the particular structure of plastic foam. The stones occupy the same volume (the box), but they have a larger mass because of the particular structure of stone. The structure of stone packs much more mass into a given volume than the structure of plastic foam does. When you compare the mass of the box of plastic foam with the mass of an identical box of stones, you are observing the different densities of the two materials. **Density** is the amount of matter (mass) contained in a unit of volume. Plastic foam has a low density, or a small mass per unit of volume. Stones have a large density, or a large mass per unit of volume.

In science, the density of solids and liquids is usually measured in units of grams (mass) per milliliter (volume), or g/mL. **Table 1.4** gives the densities of some common materials.

To find the density of a chunk of matter, it is necessary to measure both its mass and its volume. One technique for measuring these quantities is shown in **Figure 1.23.** It could be used for any object that is heavier than water and does not dissolve in water.

Table 1.4 Densities of Some Common Materials

Material	Density, g/mL
Water (4.0°C)	1.000
Ice (0°C)	0.917
Helium (25°C)	0.000164
Air (25°C)	0.00119
Aluminum	2.70
Lead	11.34
Gold	19.31
Cork	0.22–0.26
Sugar	1.59
Balsa wood	0.12

FIGURE 1.23

Determining Density

Here's one way to determine the density of a solid such as lead, using displacement.

1. **Start with a known amount of water in a graduated cylinder.**
   Pour water into the graduated cylinder. Read the exact volume, 30.0 mL, at the bottom of the meniscus.

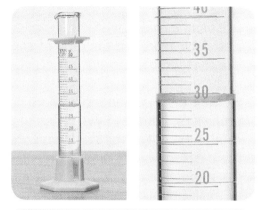

2. **Measure the mass of the cylinder with the water.**
   The mass of the cylinder with the water is 106.82 g.

3. **Carefully add a mass of lead to the graduated cylinder. Remeasure the mass of the cylinder with the water and the lead (155.83 g).**
   Subtract the mass of the water from the mass of water + lead to determine the mass of the lead.
   The mass of lead = 155.83 g − 106.82 g = 49.01 g

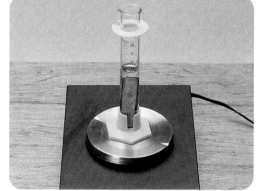

4. **Measure the volume of all the material in the cylinder (34.5 mL).**
   Subtract the volume of water from the volume of water + lead to determine the volume of lead by displacement.
   The volume of lead = 34.5 mL − 30.0 mL = 4.5 mL.
   The density of lead equals the mass of lead divided by the volume of lead.
   Density = $\dfrac{49.01 \text{ g}}{4.5 \text{ mL}}$ = 11 g/mL

Calculate *If 9.72 g of an unknown metal displaces 3.6 mL of water, what is its density?*

CHEMLAB 3

THE COMPOSITION OF PENNIES

Background
U.S. pennies have been composed of copper and zinc since 1959, but the ratio of copper to zinc has changed over the years because of increases in the price of copper. Copper and zinc are both metallic elements and they share many physical properties, but they have different densities. Pure copper has a density of 9.0 g/mL, while pure zinc has a density of 7.1 g/mL. By measuring the density of pennies from different years, it's possible to track changes in the composition of the penny.

Question
What are the approximate compositions of pennies having various mint dates?

Objectives
- **Measure** mass and volume and determine the density of pennies.
- **Interpret** class data to determine approximately when the composition of pennies changed.

Safety Precautions

Preparation

Materials
5 pennies of varying mint dates
balance sensitive to 0.01 g
50-mL graduated cylinder having 1-mL graduations

Procedure

1. Read and complete the lab safety form.
2. Record the mint date of each penny in a table like the one shown.
3. Determine the mass of each penny and record the mass to the nearest 0.01 g.
4. Half-fill the graduated cylinder with tap water. Read and record the volume to the nearest 0.1 mL.
5. Carefully add the five pennies to the cylinder so that no water splashes out. Jiggle the cylinder to dislodge any trapped air bubbles. Read and record the total volume of the water and the five pennies.

Data and Observations

Data Table

Volume of water (mL)	
Volume of water + 5 pennies (mL)	
Volume of 5 pennies (mL)	
Average volume of a penny (mL)	

Data Table

Mint Date	Mass (g)	Density (g/mL)

Analyze and Conclude

1. **Calculate** Subtract the initial volume of water from the volume of the water plus pennies to calculate the volume of the five pennies. Divide the volume by 5 to calculate the average volume of a penny. Record this volume in your data table.
2. **Calculate** the density of each penny by dividing its mass by the average volume. Post your results so they are available to the other members of your class.
3. **Observe and Infer** Does your group data show any pattern that relates the densities of the pennies to the mint dates?
4. **Classify** Classify the pennies of each year that your group examined as mostly copper (8.96 g/mL) or mostly zinc (7.13 g/mL).
5. **Graph** Design a graph to indicate the average density of the pennies for each year that your group examined.
6. **Infer** Look at all the data from your class. In what year do you think the composition of the penny changed? Use data to support your conclusion.

Apply and Assess

1. **Explain** how you might determine the identity of an irregularly shaped solid that is soluble in water.
2. **Research** In 1943, all pennies issued by the U.S. mint were struck from zinc-plated steel. One of these pennies is shown in the photo. Research to learn why steel pennies were struck in 1943. What was the purpose of the zinc coating?
3. **Infer** Why would increases in the price of copper cause the mint to change the composition of pennies?

INQUIRY EXTENSION

What factors do you think might lead to error in your density measurements? Which of these factors could not be corrected by improved technique?

FACT of the Matter
Weight is different from mass. The weight of an object is a measure of the force of gravity on the object. Scientists generally say "weigh an object" when they really mean "measure the mass of an object on a balance."

Chemical Properties and Changes

Physical properties alone are not enough to describe a substance. For a complete description, you need to know about another set of properties called chemical properties. **Chemical properties** are those that can be observed only when there is a change in the composition of the substance. Chemical properties describe the ability of a substance to react with other substances or to decompose. A chemical property of iron, for example, is that it rusts at room temperature. Rusting is a chemical reaction in which iron combines with oxygen to form a new substance, iron oxide. Aluminum reacts with oxygen too, but the compound formed, aluminum oxide, coats the aluminum and protects it from further oxidation. Platinum does not react with oxygen at room temperature. Lack of reactivity is also a chemical property.

Have you ever noticed that hydrogen peroxide (H_2O_2) solution always comes in brown bottles? It is packaged this way because in bright light, hydrogen peroxide breaks down into water and oxygen gas. The instability of a substance—its tendency to break down into different substances—is another chemical property. **Figure 1.24** illustrates other chemical properties of some substances. A chemical property always relates to a **chemical change**, the change of one or more substances into other substances. Another term for chemical change is **chemical reaction**. Terms such as *decompose, explode, rust, oxidize, corrode, tarnish, ferment, burn,* or *rot* generally refer to chemical reactions.

MiniLab 1.5

Synthesize a Polymer

How can you make slime? In this MiniLab, you will work with a polymer, polyvinyl alcohol. A polymer is a large molecule that consists of a chain of smaller repeating units. By allowing polyvinyl alcohol to react with a borax solution, you will cross-link polymer molecules to form a gel not unlike the commercial green stuff you might know. You can then investigate some of the properties of the gel.

Procedure
1. Read and complete the lab safety form.
2. Place about 20 mL of **polyvinyl alcohol solution** in a **plastic cup**.
3. Use a **craft stick** to stir the solution vigorously as you add about 3 mL of **borax solution**. Add a drop of **food coloring** if you want colored slime.
4. Continue to stir the solutions together until they form a gel.
5. Remove the gel from the cup, shape it with your hands, and perform several tests on it. Does it flow? Does it stretch or break? Can it be flattened? Record your observations.
6. Put your product in a self-sealing plastic bag, and dispose of it according to your teacher's instructions.

Analysis
1. **Infer** What effect does cross-linking the polymer have on its properties? Can you explain this effect?
2. **Classify** Name some polymers that occur in or are used to produce common materials.

FIGURE 1.24

Chemical Properties and Changes

The chemical properties of a substance describe how that substance changes to one or more new substances. It may change by reacting with another substance or by breaking down into simpler substances.

Bromine reacts with many elements, especially metals. For example, bromine reacts vigorously with aluminum, releasing a large amount of energy in the forms of light and heat. Aluminum bromide is formed by this reaction.

Iron combines with oxygen to form iron oxide. The reaction shown is the same reaction as the rusting of iron, but it occurs much faster in the pure oxygen in the flask.

The stable compound water can be decomposed into hydrogen gas (left tube) and oxygen gas (right tube) by passing an electric current through it.

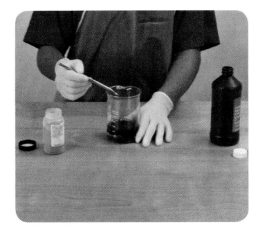

Hydrogen peroxide is another compound of hydrogen and oxygen. When the compound manganese dioxide is added to hydrogen peroxide, the peroxide breaks down rapidly into water and oxygen gas.

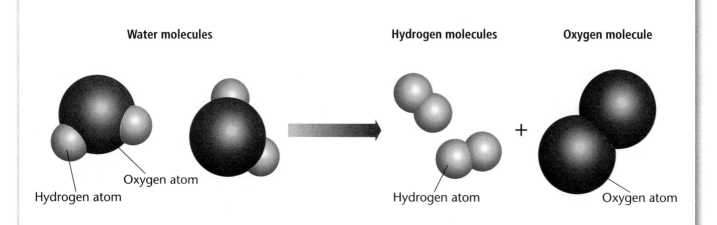

■ **Figure 1.25** Two water molecules contain two oxygen atoms and four hydrogen atoms. When they decompose, they form one molecule of oxygen containing two oxygen atoms and two molecules of hydrogen containing four hydrogen atoms. Because all matter consists of atoms and the number of atoms is the same before and after the chemical change, you can say that matter is conserved.

■ **Figure 1.26** Ammonium nitrate is a white powder capable of releasing large amounts of energy when it breaks down.

Atoms and chemical change All matter is made of atoms, and any chemical change involves only a rearrangement of the atoms. Consider the chemical change shown in **Figure 1.25,** in which water is broken down into hydrogen gas and oxygen gas. The reaction involves only hydrogen atoms and oxygen atoms. Whatever atoms are contained in the water that decomposes are found in the hydrogen and oxygen molecules that are formed. **Figure 1.25** uses models of molecules of water, oxygen, and hydrogen to show that atoms do not just appear. Atoms do not just disappear. This is an example of the **law of conservation of mass,** which says that in a chemical change, matter is neither created nor destroyed. It would be equally correct to call this the law of conservation of matter.

Chemical change and energy All chemical changes also involve some sort of energy change. Energy is either absorbed or released as the chemical change takes place. **Energy** is the capacity to do work. Work is done whenever something is moved. A carpenter does work whenever he or she picks up a hammer and drives a nail. The objects being moved can also be atoms and molecules. This is the kind of work that is encountered in chemistry.

Many reactions release energy. For example, burning wood is a chemical change in which cellulose and other substances in the wood combine with oxygen from the air to produce mainly carbon dioxide and water. Energy is also produced and released in the form of heat and light. Chemical reactions that give off heat energy are called **exothermic** reactions. A dramatic example is the combustion of ammonium nitrate (NH_4NO_3), shown in **Figure 1.26.**

Some chemical changes absorb energy. Chemical reactions that absorb heat energy are called **endothermic** reactions. When baking soda ($NaHCO_3$) is mixed into some kinds of cookie dough and the dough is baked in a hot oven, the baking soda absorbs energy and breaks down into carbon dioxide, water, and sodium carbonate (Na_2CO_3). The gases—carbon dioxide and water—puff up the cookies. **Figure 1.27** shows another endothermic reaction.

Photosynthesis The most important endothermic process on Earth is probably photosynthesis. Photosynthesis is a series of chemical reactions that absorb light energy from the Sun and produce sugars from carbon dioxide and water. Oxygen is given off as a product of the reactions. Green plants, shown in **Figure 1.28,** algae, and many kinds of bacteria carry out photosynthesis. When you eat sugars and starches, you are eating the substances formed by endothermic photosynthesis reactions. Your cells break down these substances in an exothermic process that supplies you with energy.

■ **Figure 1.27** When the two compounds ammonium thiocyanate and barium hydroxide octahydrate are mixed, a reaction occurs in which heat is absorbed from the surroundings. The flask becomes so cold that if a film of water is on the bottom of the flask, it will freeze the flask to a wooden block.

Summarize *Where is the energy transferred in an exothermic reaction and in an endothermic reaction?*

■ **Figure 1.28** When you eat vegetables from a garden such as this, you are consuming substances that plants have synthesized using energy from sunlight.

Section 1.2 • Properties and Changes of Matter

Figure 1.29 The rust provides an important clue to the composition of this old ship: it is made of an iron-based metal. As you continue to study chemistry, you will learn to infer more and more from the properties of matter you observe.

Connecting Ideas

Chemistry makes connections among the composition, structure, and behavior of matter. As you study chemistry, you will see how knowing the structure of an atom of an element enables you to predict the chemical behavior of that element. You'll learn how the state of a substance at room temperature provides clues to the way its atoms are arranged. You'll find out why salt dissolves in water, how batteries work, why compounds containing carbon are important to life, and how a nuclear reactor works. As illustrated in **Figure 1.29,** chemical and physical properties are important clues to the structure and behavior of matter.

SUPPLEMENTAL PRACTICE
For more practice with chemical properties and changes, see Supplemental Practice, page 807.

Section 1.2 Assessment

Section Summary

- Every substance has a unique set of physical and chemical properties.
- The density of a sample of matter is the amount of matter (mass in grams) in a unit volume (usually a milliliter).
- Chemical changes—also called chemical reactions—involve substances forming different substances.
- In a chemical reaction, atoms are never created or destroyed.
- Physical and chemical changes absorb or release energy.

6. **MAIN Idea Identify** each property below as either a chemical or a physical property of the substance.
   a) Aluminum bends easily.
   b) Copper sulfate dissolves in water.
   c) Magnesium burns in air.
   d) Gold jewelry is unaffected by perspiration.
   e) Baking soda is a white powder.
   f) Fluorine is a highly reactive element.

7. **Explain** What are the three common states of matter?

8. **Apply** A friend tells you that a newspaper is completely gone because it was burned up in a fire. Use the law of conservation of mass to write an explanation telling your friend what really happened to the newspaper.

9. **Infer** On cool evenings, a campfire keeps you warm. Is the burning of logs an exothermic or endothermic reaction?

CHAPTER 1 Study Guide

BIG Idea Everything is made of matter.

Section 1.1 The Puzzle of Matter

MAIN Idea Most everyday matter occurs as mixtures—combinations of two or more substances.

Vocabulary
- alloy (p. 23)
- aqueous solution (p. 23)
- chemistry (p. 4)
- compound (p. 28)
- element (p. 24)
- formula (p. 31)
- mass (p. 4)
- matter (p. 4)
- mixture (p. 16)
- physical change (p. 20)
- physical property (p. 20)
- property (p. 4)
- qualitative (p. 14)
- quantitative (p. 14)
- scientific model (p. 9)
- solute (p. 23)
- solution (p. 22)
- solvent (p. 23)
- substance (p. 15)

Key Concepts
- Chemists study matter.
- Macroscopic observations reflect the submicroscopic structure of matter.
- Mixtures are heterogeneous or homogeneous (solutions).
- Substances are classified as elements or compounds. Elements are the building blocks of all matter.

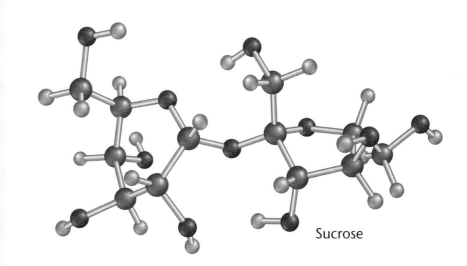

Sucrose

Section 1.2 Properties and Changes of Matter

MAIN Idea Matter can undergo physical and chemical changes.

Vocabulary
- chemical change (p. 38)
- chemical property (p. 38)
- chemical reaction (p. 38)
- density (p. 34)
- endothermic (p. 41)
- energy (p. 40)
- exothermic (p. 40)
- law of conservation of mass (p. 40)
- volatile (p. 33)

Key Concepts
- Every substance has a unique set of physical and chemical properties.
- The density of a sample of matter is the amount of matter (mass in grams) in a unit volume (usually a milliliter).
- Chemical changes—also called chemical reactions—involve substances forming different substances.
- In a chemical reaction, atoms are never created or destroyed.
- Physical and chemical changes absorb or release energy.

Chapter 1 Assessment

Understand Concepts

10. What is chemistry?

11. List the symbols for the following elements: iron, sodium, antimony, tungsten. Explain why these symbols do not correspond to the English names of the elements.

12. If you know the melting point of a pure substance, what does it tell you about its boiling point? Its freezing point?

13. Which compound or compounds listed in **Table 1.3** contain the element sodium? The element chlorine?

14. What information does the chemical formula for a compound provide about the submicroscopic structure of the compound?

15. What is mass?

16. If you say that sulfur is yellow, you are stating a property of sulfur. What is meant by property? Is color a physical or a chemical property?

17. List three properties of iron.

18. What is the solvent in an aqueous solution of salt? Why are aqueous solutions important?

19. Could two objects with the same volume have different masses? Which, if either, would contain more matter?

20. What is meant by pure matter?

21. What is energy? What part does it play in chemistry?

22. Distinguish between exothermic and endothermic reactions, and give an example of each.

23. How is a qualitative observation different from a quantitative observation? Give an example of each.

24. Explain how a mixture is different from a compound.

25. What is an element? A compound? Give an example of each.

26. Sucrose ($C_{12}H_{22}O_{11}$), is 51.5 percent oxygen by mass and only 6.4 percent hydrogen by mass, yet there are twice as many hydrogen atoms as oxygen atoms in the formula of sucrose. How can this be?

27. Classify these as chemical changes or physical changes.
    a) Water boils.
    b) A match burns.
    c) Sugar dissolves in tea.
    d) Sodium reacts with water.
    e) Ice cream melts.

28. Classify these as heterogeneous or homogeneous mixtures.
    a) salt water
    b) vegetable soup
    c) 14-k gold
    d) concrete

29. What is meant by density? Is there any way that a bag of plastic foam balls could be heavier than a bag of stones?

30. Gold freezes at 1064°C. What is the melting point of gold?

31. Mercury freezes at −38.9°C; nitrogen boils at −195.8°C. How can a boiling point be lower than a freezing point?

32. Use **Table D.5,** page 850. Over what temperature range is iron (Fe) a liquid? How does this liquid temperature range compare with that of krypton (Kr)?

33. Is the diagram in **Figure 1.30** a molecule of nitroglycerin or a model of nitroglycerin? Explain.

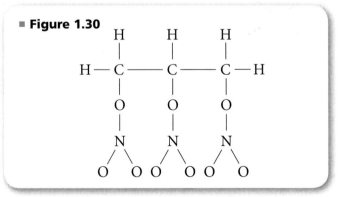

■ Figure 1.30

34. The chemical formula of water is H_2O. The chemical formula of hydrogen peroxide is H_2O_2. Which compound has the highest percentage of hydrogen by mass?

Chapter 1 Assessment

Apply Concepts

35. **Everyday Chemistry** Is the overall processing of food in your body an exothermic or an endothermic process?

36. An ice cube floats in your glass of water, just as an iceberg floats in the ocean. What conclusion can you draw about the density of liquid water and the density of solid water?

37. **Chemistry and Society** A drug with a well-known brand name may also be made by another company and sold at a lower price. Both drugs are specified to have the same chemical formula and are approved by the Food and Drug Administration. Would it be safer to use the well-known brand? Explain.

38. A doorknob is coated with the alloy brass. What is an alloy? Give an example of another alloy.

39. A recipe for salad dressing tells you to mix together vinegar, oil, herbs, salt, pepper, and chopped garlic. Describe the completed salad dressing in terms of substances and mixtures, both heterogeneous and homogeneous. What solutes or solvents are in the salad dressing?

40. Oxygen makes up more than 46 percent of Earth's crust and 61 percent of the human body. It also constitutes 21 percent of Earth's gaseous atmosphere. Explain the difference between the oxygen in the atmosphere and the oxygen in Earth's crust and the human body.

41. Some iron filings are mixed with table salt. Describe two ways you could separate the filings from the salt.

42. Ethanol melts at −114.1°C and boils at 78.5°C. What is the physical state of ethanol at room temperature?

43. Using the information in question 42, what are the freezing points and the condensation points of ethanol?

44. Is the melting of candle wax an exothermic or endothermic change?

Think Critically

Observe and Infer

45. **ChemLab 1** From macroscopic observations of a burning candle, what conclusions can you draw about events occurring on the submicroscopic level?

Observe and Infer

46. **ChemLab 2** Why is starch not used in baking for the purpose of making cakes fluffy?

Measure in SI

47. **ChemLab 3** Suppose you have a sample of an unknown mineral with a mass of 86 g. You place the sample in a graduated cylinder filled with water to the 55-mL mark. The sample sinks to the bottom of the cylinder and the water level rises to the 71-mL mark. What is the density of the sample?

Compare and Contrast

48. **MiniLab 1** Is the mixing of ethanol and water an exothermic or an endothermic process? Is it a chemical or a physical change?

Design an Experiment

49. **MiniLab 2** Marker inks of different colors, when separated by chromatography, might be found to contain some pigments of the same color. For example, a black marker and a blue marker might both contain a blue pigment. Design an experiment to show whether or not two pigments of the same color, found in different markers, are likely to be the same pigment.

Apply Concepts

50. **MiniLab 3** The appearance of a penny made of zinc and copper is different after heating the penny in zinc chloride solution. Is the new coating an element, a compound, or a mixture? Explain.

Chapter 1 Assessment

Observe and Infer

51. MiniLab 4 Are the chemicals in your food elements or compounds? Explain.

Observe and Infer

52. MiniLab 5 Is a polymer an element or a compound? Explain.

53. Two solid compounds burn in the presence of oxygen. Can you conclude that they are the same compound?

Interpret Chemical Structures

54. What does the formula C_2H_6O tell you about the substance it represents? Be as specific as possible.

Cumulative Review

In Chapters 2–21, this heading will be followed by review questions about skills and ideas you have learned in earlier chapters.

Skill Review

Table 1.5 Chemical and Physical Changes

Observation	Is it chemical or physical?	Explanation
1. Brewing tea or coffee		
2. Heating a pot of water		
3. Decomposing food scraps		
4. A seed germinating		
5. Dissolving sugar in water		
6. Vinegar and oil not mixing		
7. A termite eating wood and producing methane gas		
8. The density of salt water increasing as salt is added		

55. Make and Use a Table Complete **Table 1.5** by identifying each change or process as physical or chemical and justify your choice. When your table is complete, look at the two groups of processes. Within each group, list similarities shared by the processes. Are there differences within each group?

WRITING in Chemistry

56. Food Ingredients Look at labels on food products in your home or in the grocery store. Choose one item and list its ingredients. Describe it as a mixture or pure substance. Choose one ingredient and look it up in a chemical handbook. What is its formula? What are some of its physical and chemical properties? Find out why this substance is an ingredient in the item you chose. Find a similar item, made by a different company, and determine whether the two items have the same ingredients. Summarize your findings in an advertisement for the product.

Problem Solving

57. A miner found a nugget that had a gold color. He realized that it could be precious gold metal or iron pyrite (FeS_2), which is a compound of iron and sulfur called fool's gold. The nugget had a mass of 16.5 g and displaced 3.3 mL of water. From this information, determine whether the miner had found gold.

58. Air is a mixture of primarily nitrogen (78 percent) and oxygen (21 percent), with trace amounts of argon, neon, helium, and krypton. These pure elements have many applications. They are separated and purified by cooling air in a process known as fractionation. As air cools, the different elements liquefy based on their boiling points. In what order do the elements liquefy? Use **Table D.4.**

Cumulative
Standardized Test Practice

1. Matter is defined as anything that
   a) exists in nature.
   b) is solid to the touch.
   c) is found in the universe.
   d) has mass and takes up space.

2. Which of the following has no mass?
   a) air
   b) light
   c) atoms
   d) water

3. A model is best defined as
   a) a thinking device based on macroscopic observations.
   b) a thinking device based on microscopic observations.
   c) a thinking device based on constant experimentation.
   d) a thinking device based on a consensus of scientists.

4. When a spoonful of sugar is added to hot tea, the sugar is called a(n)
   a) solution.
   b) alloy.
   c) solvent.
   d) solute.

5. The chemical formula of chalk is $CaCO_3$. Identify the elements and calculate the number of atoms of each element in the smallest unit of chalk.
   a) 1 calcium atom; 3 cobalt atoms
   b) 1 calcium atom; 1 carbon atom; 3 oxygen atoms
   c) 1 calcium atom; 1 chlorine atom; 3 oxygen atoms
   d) 1 calcium atom; 3 carbon atoms; 3 oxygen atoms

6. Which of the following substances is volatile?
   a) perfume
   b) water
   c) coal
   d) sugar

Elements	Density (g/mL)	Boiling Point (°C)	Melting Point (°C)
Argon	1.78	−186	−189
Bromine	3.12	59	−7
Gallium	5.91	2403	30
Sodium	0.97	883	98
Tungsten	19.35	5660	3410

Use information from the data table above to answer Questions 7–9.

7. Which element is a liquid at room temperature?
   a) argon
   b) bromine
   c) gallium
   d) sodium

8. Which element has the fewest atoms per cubic centimeter?
   a) argon
   b) bromine
   c) gallium
   d) sodium

9. Which element will exist as a gas in any location on Earth?
   a) argon
   b) bromine
   c) gallium
   d) sodium

10. The law of conservation of mass states that
    a) matter cannot be created or destroyed.
    b) matter can be created but not destroyed.
    c) matter can be destroyed but not created.
    d) matter is always being created and destroyed.

NEED EXTRA HELP?

If You Missed Question...	1	2	3	4	5	6	7	8	9	10
Review Section...	1.1	1.1	1.1	1.1	1.1	1.2	1.2	1.2	1.2	1.2

CHAPTER 2: Matter Is Made of Atoms

BIG Idea Atoms are the fundamental building blocks of matter.

2.1 Atoms and Their Structures
MAIN Idea An atom is made of a nucleus containing protons and neutrons; electrons move around the nucleus.

2.2 Electrons in Atoms
MAIN Idea Each element has a unique arrangement of electrons.

ChemFacts

- The word *atom* originally was used to denote a particle that could not be broken down into smaller pieces.
- Atoms are so small that special techniques are used to visualize them.
- Metal atoms with a positive charge (shown in blue) are surrounded by a sea of negatively charged electrons (shown in red).

Start-Up Activities

LAUNCH Lab

What's inside?

Much of the fun is trying to figure out what's inside a wrapped present before you open it. Chemists had similar experiences trying to determine the structure of the atom. How good are your skills of observation and deduction?

Materials
- a wrapped box from your instructor

Procedure
1. Obtain a wrapped box from your instructor.
2. Using as many observation methods as you can, and without unwrapping or opening the box, try to figure out what the object inside the box is.
3. Record the observations you make throughout this discovery process.

Analysis
1. **Explain** How were you able to determine things such as size, shape, number, and composition of the object in the box?
2. **Describe** What senses did you use to make your observations?
3. **Infer** What additional methods of observation could you use to help you determine what might be in the box?

Inquiry Compare your methods of observation with another group's. Did the other group use methods that your group did not consider? Would using these methods change what you think is in the box?

Chemistry Online

Visit glencoe.com to:
- study the entire chapter online
- explore Concepts In Motion
- take Self-Check Quizzes
- use Personal Tutors
- access Web Links for more information, projects, and activities
- find the Try at Home Lab, Comparing Atom Sizes

FOLDABLES Study Organizer

The Atom Make the following Foldable to help you organize your study of the structure of the atom.

▶ **STEP 1** Fold a sheet of paper in half lengthwise. Make the back edge about 2 cm longer than the front edge.

▶ **STEP 2** Fold into thirds.

▶ **STEP 3** Unfold and cut along one fold line to make one small tab and one large tab.

▶ **STEP 4** Label as shown.

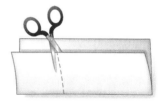

FOLDABLES Use this Foldable with Sections 2.1 and 2.2. As you read the section, record information about the atom and its parts.

Chapter 2 • Matter Is Made of Atoms **49**

Section 2.1

Objectives
- **Relate** historic experiments to the development of the modern model of the atom.
- **Illustrate** the modern model of an atom.
- **Interpret** the information available in an element block of the periodic table.

Review Vocabulary
energy: the capacity to do work

New Vocabulary
atom
atomic theory
law of definite proportions
hypothesis
experiment
theory
scientific method
scientific law
electron
proton
isotope
neutron
nucleus
atomic number
mass number
atomic mass unit

Atoms and Their Structures

MAIN Idea An atom is made of a nucleus containing protons and neutrons; electrons move around the nucleus.

Real-World Reading Link A football team might try out different plays in order to develop the best-possible game plan. As they see the results of their plans, coaches make adjustments to refine the team's tactics. Similarly, scientists over the last 200 years have developed different models of the atom, refining their models as they collected new data.

Early Ideas About Matter

The current model of the composition of matter is based on hundreds of years of work that began when observers realized that different kinds of matter exist and that these kinds of matter have different properties and undergo different changes. About 2500 years ago, the Greek philosophers thought about the nature of matter and its composition. They proposed that matter was a combination of four fundamental elements—air, earth, fire, and water, as shown in **Figure 2.1.** These Greek philosophers also argued the question of whether matter could be divided endlessly into smaller and smaller pieces or whether there was an ultimate smallest particle of matter that could not be divided any further. The Greek philosophers were keen observers of nature but, unlike modern scientists, couldn't test their ideas with experiments.

■ **Figure 2.1** Many Greek philosophers thought that matter was composed of four elements: earth, air, water, and fire. They also associated properties with each element. The pairing of opposite properties, such as hot and cold, and wet and dry, mirrored the symmetry they observed in nature. These early ideas were based solely on observations, not on scientific experiments.

Democritus (460–370 B.C.) was a philosopher who proposed that the world is made of empty space and tiny particles called atoms. Democritus thought that **atoms** are the smallest particles of matter and that different types of atoms exist for every type of matter. In fact, this is the definition scientists use today. The idea that matter is made of fundamental particles called atoms is known as the **atomic theory** of matter.

Democritus's ideas were met with criticism from other philosophers, especially from Aristotle, 384–322 B.C. He rejected the notion of atoms because it did not agree with his own ideas about matter. He did not believe that empty space could exist. Because Aristotle was one of the most influential philosophers of his time, Democritus's atomic theory was eventually rejected.

Modern Atomic Theory

In 1782, French Antoine Lavoisier (1743–1794) made measurements of chemical reactions in sealed containers. He observed that the mass of reactants in a container before a reaction was equal to the mass of the products after the reaction. For example, in a sealed container, 2.0 g of hydrogen and 16.0 g of oxygen react to form 18.0 g of water. Lavoisier concluded that when a chemical reaction occurs, matter is neither created nor destroyed but only changed. Lavoisier's conclusion became known as the law of conservation of matter. This is another name for the law of conservation of mass that you learned in Chapter 1.

Figure 2.2 illustrates the law of conservation of matter. In the reaction shown, sodium carbonate, a solid that is used in making glass, reacts with hydrochloric acid. One of the products of this reaction is a gas, carbon dioxide. If this reaction proceeded without the balloon over the mouth of the flask, you might think that mass was not conserved because the reading on the balance would get lower as carbon dioxide gas escaped. However, with the balloon in place, you can clearly see that the mass of the products of the reaction is identical to the mass of the reactants.

VOCABULARY
WORD ORIGIN
Atom
comes from the Greek word *atomos*, which means *indivisible*

■ **Figure 2.2** When sodium carbonate and hydrochloric acid react in a flask, carbon dioxide gas is one of the products. A balloon placed over the mouth of the flask prevents this gas from escaping.
Describe *How can you tell that matter was not destroyed in this reaction?*

Pyrite

Marcasite

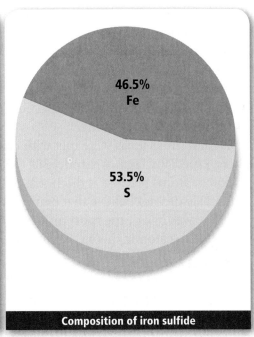
46.5% Fe
53.5% S
Composition of iron sulfide

■ **Figure 2.3** Iron sulfide (FeS$_2$) looks something like gold. It fooled many people and got the name *fool's gold* as a result. Pyrite and marcasite are both iron sulfide, but they are different minerals because their crystal structures are different. The composition of iron sulfide minerals is 46.5 percent iron and 53.5 percent sulfur by mass. These proportions are the same for every sample of iron sulfide mineral.

Proust's Contribution In 1799, another French chemist, Joseph Proust, observed that the composition of water is always 11.2 percent hydrogen and 88.8 percent oxygen by mass. Proust studied many other compounds and observed that the elements that composed the compounds were always in a certain proportion by mass. This principle is now referred to as the **law of definite proportions**. **Figure 2.3** illustrates this law.

Dalton's Atomic Theory John Dalton (1766–1844), an English schoolteacher and chemist, studied the results of experiments by Lavoisier, Proust, and many other scientists. He realized that an atomic theory of matter must explain the experimental evidence. For example, if matter were composed of indivisible atoms, then a chemical reaction would only rearrange those atoms, and no atoms would form or disappear. This idea would explain the law of conservation of mass. Also, if each element consisted of atoms of a specific type and mass, then a compound would always consist of a certain combination of atoms that never varied for that compound. Thus, Dalton's theory explained the law of definite proportions, as well.

Dalton proposed his atomic theory of matter in 1803. Although his theory has been modified slightly to accommodate new discoveries, Dalton's theory was so insightful that it has remained essentially intact. The following statements are the main points of Dalton's atomic theory.

1. All matter is made of atoms.

2. Atoms are indestructible and cannot be divided into smaller particles. (Atoms are indivisible.)

3. All atoms of one element are exactly alike, but they are different from atoms of other elements.

Dalton's atomic theory gave chemists a model of the particle nature of matter. However, it also raised new questions. If all elements are made of atoms, why are there so many different elements? What makes one atom different from another atom? Experiments in the late nineteenth century began to suggest that atoms are made of even smaller particles. Present-day chemistry explains the properties and behavior of substances in terms of three of these smaller particles. You will learn more about each of these particles in Section 2.2.

Atomic theory, conservation of matter, and recycling Where do atoms go when waste is incinerated or buried in a landfill? As you have just learned, atoms are never created or destroyed in everyday chemical processes. When waste is burned or buried in a landfill, the atoms of the waste may combine with oxygen or other substances to form new compounds. In natural processes, atoms are recycled.

Figure 2.4 shows how elemental nitrogen (N_2) from the atmosphere is converted into compounds that are used on Earth and then returns to the atmosphere. Lightning, bacteria, industrial processes, and even lichens on tree branches convert nitrogen from the atmosphere into different compounds. Elemental nitrogen (N_2) in the atmosphere, ammonia (NH_3) in some animal wastes, and nitrates (NO_3^-) in the soil are examples of substances in the environment that contain nitrogen atoms. These compounds enter the plant and animal food chain. When plant and animal wastes decompose, bacteria in the soil produce free nitrogen through a process called denitrification. This free nitrogen returns to the atmosphere.

In recent years, small towns, large cities, and entire states have discovered the benefits of recycling paper, plastic, aluminum, and glass. Labels on many supermarket bags, cardboard boxes, greeting cards, and other paper products say "Made from recycled paper." Scrap aluminum is easily recycled and made into new aluminum cans or other aluminum products. Have you noticed how newly paved roadways sparkle? The sparkle is a result of the addition of recycled glass to the paving material. Even ground-up tires can be added to asphalt for paving. By reusing the atoms in manufactured materials, we are imitating nature and conserving natural resources.

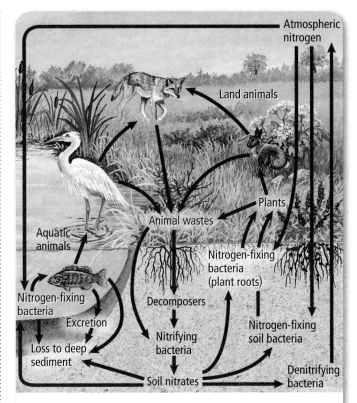

■ **Figure 2.4** Nitrogen is used and reused as it is cycled continuously through the environment. At each stage of the diagram, nitrogen might be converted into different forms, but atoms are not created or destroyed.

Concepts In Motion

Interactive Figure To see an animation of the nitrogen cycle, visit glencoe.com.

Section 2.1 • Atoms and Their Structures   53

CHEMLAB

CONSERVATION OF MASS

Background
Hydrochloric acid reacts with metals such as zinc, magnesium, and aluminum. A colorless gas escapes, and the metal seems to disappear. What happens to the atoms of the metals in these reactions? Are they destroyed? If not, where do they go?

Question
What happens to the atoms of a metal when they react with an acid?

Objectives
- **Infer** what happens to atoms during a chemical change.
- **Compare** results to the law of conservation of mass.

Preparation

Materials
granular zinc
1M HCl
125-mL Erlenmeyer flask
10-mL graduated cylinder
hot plate
balance
spatula
oven mitt

Safety Precautions

WARNING: *Use care when handling hydrochloric acid and hot objects. Be careful not to inhale the fumes from the flask.*

Procedure

1. Read and complete the lab safety form.
2. Measure the mass of a clean, dry, 125-mL Erlenmeyer flask to the nearest 0.01 g. Record its mass in a data table.
3. Obtain a sample of granular zinc, shown below, and add it to your flask. Measure the mass of the flask with the zinc in it. Record the total mass in the data table.

4. Subtract to find the mass of zinc. The mass of zinc must be between 0.20 g and 0.28 g. If you have too much zinc, remove some with a clean spoon or spatula until the mass is in the range described.
5. Add 10 mL of 1M hydrochloric acid (HCl) to the flask. Swirl the contents and look for signs of a chemical change. Record your observations in the data table.
6. Set the flask on a hot plate as shown in the photo below.

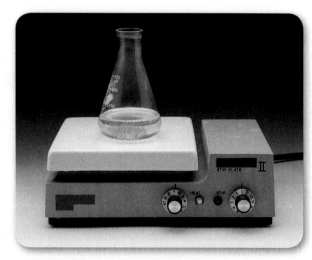

7. Set the hot plate on low to heat the flask gently. The liquid in the flask should not boil. Again look for signs that a chemical change is taking place, and record your observations in the data table.

8. Eventually, all the metallic zinc will disappear. It is important to heat the flask slowly. When the liquid in the flask is almost gone, a white solid will appear. Stop heating as soon as you see this solid. Use an oven mitt to remove the flask from the hot plate. The heat retained by the flask should be enough to completely evaporate the remaining liquid.

9. Let the flask cool. When all the liquid has evaporated, measure the mass of the cool, dry flask and its contents. Record this mass in the data table.

Analyze and Conclude

1. **Calculate** the mass of zinc chloride produced in the reaction.
2. **Summarize** What happened to the zinc?
3. **Interpret** How did the mass of the product, zinc chloride, compare with the mass of the zinc?
4. **Explain** How can you account for the difference in mass?

5. **Infer** Why were the flask and its contents heated?

Apply and Assess

1. **Calculate** Chemists have determined that zinc chloride is 48 percent zinc. Use this information to compute the mass of zinc in your product. How does this mass compare with the mass of zinc you started with?
2. **Error Analysis** If the difference in the question above is greater than 0.04 g, how can you account for it?
3. **Summarize** How does this experiment support the law of conservation of mass?

INQUIRY EXTENSION
Analyze How could you modify the procedure so that the law of conservation of matter is better demonstrated?

Data and Observations

Data Table	
Mass of empty Erlenmeyer flask	
Mass of Erlenmeyer flask with zinc sample	
Mass of zinc sample	
Mass of Erlenmeyer flask with reaction product (zinc chloride, $ZnCl_2$)	
Mass of zinc chloride produced	
What did you see when hydrochloric acid was first added to the flask?	
What did you see when you began to heat the flask?	
What did you see when all the liquid had evaporated?	

Chapter 2 • ChemLab

History Connection

Politics and Chemistry—Elemental Differences

In the time of Antoine Lavoisier (1743–1794), many scientists were still trying to explain matter as combinations of the elements air, earth, fire, and water. Lavoisier's work changed the way chemistry was done, and today he is recognized as the first modern chemist. However, like other scientists of the eighteenth century, Lavoisier could not earn a living solely as a chemist, so he also collected taxes for the king.

Remaking chemistry Lavoisier set out to reorganize chemistry. Lavoisier's habit of carefully weighing reactants and products in experiments led him to discover that the mass of materials before a chemical change equals the mass of the products after it, which is the basis of the law of conservation of matter. He also discovered that combustion is the result of reaction with oxygen.

International recognition Many scientists held Lavoisier in high esteem. Benjamin Franklin made a point to observe experiments by Lavoisier when he was in France soliciting support for the cause of the American Revolution. Lavoisier's experiments were also followed closely by Thomas Jefferson.

In 1774, the British chemist Joseph Priestley, shown in **Figure 1,** discussed one particular experiment with Lavoisier. Priestley explained that after heating "calx of mercury" (which we know today as mercury(II) oxide), metallic mercury remained and a gas was given off. When he placed a candle in the gas, it burned more brightly. He also found that if a mouse is placed in a closed jar with the gas, the mouse can breathe it and live. Priestley's gas was oxygen, but because he believed in an older theory of matter called the phlogiston theory, Priestley did not recognize it.

Figure 1 Joseph Priestley's experiments contributed to Lavoisier's conclusions.

Lavoisier repeated Priestley's experiment and came to the conclusion that air is not a simple substance but a mixture of two different gases. One of these gases, oxygen, supports combustion and promotes breathing. Lavoisier gave oxygen its name.

Political price Lavoisier belonged to the professional class, along with many of the leaders of the French Revolution. In spite of his class and the high regard for Lavoisier in the scientific world, his connection with tax collecting made him a target of suspicion. Lavoisier was arrested and condemned to death in a trial that lasted less than a day. An appeal to spare his life was denied when a judge stated that France did not need scientists or chemists. That same day, he was guillotined and his body was thrown into a common grave. A friend responded to Lavoisier's death by stating that it took only an instant to cut off his head, but it may take 100 years before France produces another like it.

Connection to Chemistry

1. **Analyze** Why was Lavoisier's role in the discovery of oxygen important even though he merely repeated Priestley's experiment?
2. **Apply** Lavoisier showed that a person uses more oxygen when working than when resting. Explain the reasoning behind Lavoisier's findings.

Hypotheses, Theories, and Laws

The first step to solving a problem, such as what makes up matter, is observation. Scientists use their senses to observe the behavior of matter at the macroscopic level. They then form a **hypothesis**, which is a testable prediction to explain their observations. For example, Lavoisier thought that matter was indestructible. He formed a hypothesis that all the matter present before a chemical change would still be there after the change.

To find out whether a hypothesis is supported, it must be tested by repeated experiments and investigations. An **experiment** is an investigation with a control designed to test a hypothesis. Scientists accept hypotheses that are supported by experiments and investigations and reject hypotheses that can't stand up to testing. Lavoisier performed numerous careful experiments using different chemical changes to support his hypothesis.

A body of knowledge builds based on hypotheses that are supported or rejected. Scientists develop theories to organize their knowledge. In nonscientific speech and writing, the word *theory* is often used to mean an unsupported notion about something. However, in the scientific sense, a **theory** is an explanation based on many observations and supported by the results of many investigations. For example, Dalton's atomic theory was based on observations of matter performed again and again by many scientists. As scientists gather more information, a theory may have to be revised or replaced with another theory. **Figure 2.5** summarizes a systematic approach used by scientists to answer questions and solve problems called a **scientific method**. Scientists describe their scientific methods when they publish their results. This allows scientists to verify the work of others.

A **scientific law** is a fact of nature that is observed so often that it becomes accepted as truth. That the Sun rises in the east each morning is a law of nature. A law can generally be used to make predictions but does not explain why something happens. In fact, theories explain laws. One part of Dalton's atomic theory explains why the law of conservation of matter is true.

■ **Figure 2.5** Scientists make observations that lead to hypotheses. A hypothesis must be tested by multiple investigations. If the results do not support the hypothesis, they become the observations that lead to a new hypothesis. A hypothesis that is refined and supported becomes a theory that explains a fact or phenomenon in nature.

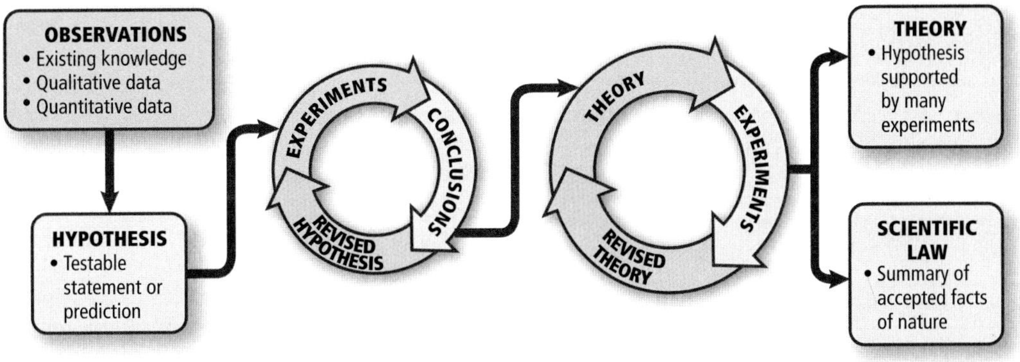

Chemistry & Society

Recycling Glass

Imagine that you arrive home thirsty from a hot day at school. You reach into the refrigerator and pull out an ice-cold glass bottle of apple juice. In seconds, the empty bottle is all that remains. What do you do with the glass bottle? Unless the bottle is recycled, it is thrown away as trash to be deposited in a landfill.

Problems of recycling glass Glass manufacturers call recycled crushed glass that can be reused cullet shown in **Figure 1**. There is no difficulty in collecting colored or uncolored container glass for recycling. Problems arise in producing the quality of cullet required by the industries that will use it.

Solutions in recycling methods Because of the color it adds, colored cullet can not be mixed with uncolored cullet for use in the glass container industry. Because the glass container industry uses 90 percent of the cullet produced, recycling centers usually sort glass by color. The sorted glass is dropped down separate chutes and processed by separate glass crushers. The end products are delivered separately to glass plants, which pay more for color-sorted cullet.

Figure 2 High quality cullet is used by the glass container industry.

Finding uses for recycled glass There are two types of cullet. High-quality cullet is color-sorted and contaminant-free. Low-quality cullet is not sorted by color and might include contaminants, such as plastics, metals, and ceramics. High-quality cullet is primarily used in the glass container industry as illustrated in **Figure 2**. Some other uses for high-quality cullet include abrasives, highway aggregate in asphalt, bead manufacturing, and fiberglass. Some uses of low-quality cullet are fiberglass insulation, roadbed aggregate used under asphalt, highway reflector beads, and decorative tile.

One other use of cullet is in restoring beach sand that has eroded. Erosion along some of Florida's beaches is being controlled by mixing finely-crushed cullet with beach sand to patch areas where the sand is being removed by natural processes faster than it is being replaced.

ANALYZE the Issue

1. **Acquire Information** Research how new glass is manufactured. Learn whether cullet is routinely used in most glass manufacturing. Does the use of cullet reduce the cost of manufacturing glass? Does the use of cullet reduce the energy requirement for producing glass?

2. **Infer** In France, a wine bottle is reused eight times before it is recycled. In the United States, it is rare for a company to reuse wine bottles. If the wine companies in the United States started to reuse wine bottles, what marketing and distribution problems might need to be solved to begin a business to wash wine bottles?

Figure 1 Cullet can be reused in many industries.

The Discovery of Atomic Structure

Dalton's atomic theory was almost correct. Dalton had assumed that atoms are the ultimate particles of matter and can't be broken up into smaller particles and that all atoms of the same element are identical. However, his theory had to be modified as new discoveries were made in the late nineteenth and early twentieth centuries. Today, we know that atoms are made of smaller particles and that atoms of the same element can be nearly, but not exactly, the same. In this section, you'll examine the discoveries that led to the modern atomic theory.

The electron Because of Dalton's atomic theory, most scientists in the 1800s believed that the atom was like a tiny solid ball that could not be broken up into parts. In 1897, a British physicist, J.J. Thomson, discovered that this solid-ball model was not accurate.

Thomson's experiments used a vacuum tube such as that shown in **Figure 2.6**. A vacuum tube has had all gases pumped out of it. At each end of the tube is a metal piece called an electrode, which is connected through the glass to a metal terminal outside the tube. These electrodes become electrically charged when they are connected to a high-voltage electrical source. When the electrodes are charged, rays travel in the tube from the negative electrode, which is the cathode, to the positive electrode, the anode. Because these rays originate at the cathode, they are called cathode rays.

Thomson found that the rays bent toward a positively charged plate and away from a negatively charged plate. He knew that objects with like charges repel each other and objects with unlike charges attract each other. Thomson concluded that cathode rays are made up of invisible, negatively charged particles referred to as **electrons**. These electrons had to come from the matter (atoms) of the negative electrode.

FOLDABLES
Incorporate information from this section into your Foldable.

■ **Figure 2.6** When a high voltage is applied to a cathode-ray tube, cathode rays form a beam that produces a green glow on the fluorescent plate. The pole of a magnet bends the cathode rays in a direction at a right angle to the direction of the field.

Determine *What would happen if the magnet were reversed?*

FACT of the Matter

Thomson did not give the electron its name. That was done in 1874 by an Irish scientist, George J. Stoney, who first calculated its charge even though he could not prove it existed.

From Thomson's experiments, scientists concluded that atoms were not just neutral spheres, but somehow were composed of electrically charged particles. In other words, atoms were composed of smaller particles, referred to as subatomic particles. Further experiments showed that an electron has a mass equal to 1/1837 the mass of a hydrogen atom, the lightest atom.

Reason should tell you that there must be a lot more to the atom than electrons. Matter is not negatively charged, so atoms can't be negatively charged either. If atoms contained extremely light, negatively charged particles, then they must also contain positively charged particles—probably with a much greater mass than electrons. Scientists immediately worked to discover such particles.

Protons and neutrons In 1886, scientists discovered that a cathode-ray tube emitted rays not only from the cathode but also from the positively charged anode. These rays travel in a direction opposite to that of cathode rays. Like cathode rays, they are deflected by electrical and magnetic fields, but in directions opposite to the way cathode rays are deflected. Thomson was able to show that these rays had a positive electrical charge. Years later, scientists determined that the rays were composed of positively charged subatomic particles called **protons**. The amount of charge on an electron and on a proton is equal but opposite, but the mass of a proton is much greater than the mass of an electron. The mass of a proton was found to be only slightly less than the mass of a hydrogen atom.

At this point, it seemed that atoms were made up of equal numbers of electrons and protons. However, in 1910, Thomson discovered that neon consisted of atoms of two different masses, as shown in **Figure 2.7**. Atoms of an element that are chemically alike but differ in mass are called **isotopes** of the element. Today, chemists know that neon consists of three naturally occurring isotopes.

Because of the discovery of isotopes, scientists hypothesized that atoms contained still a third type of particle that explained these differences in mass. Calculations showed that such a particle should have a mass equal to that of a proton but no electrical charge. The existence of this neutral particle, called a **neutron**, was confirmed in the early 1930s.

■ **Figure 2.7** These diagrams represent the two isotopes of neon that Thomson observed. Both nuclei have ten protons, but the one on the left has ten neutrons and the one on the right has twelve neutrons.

Determine How many protons and neutrons are in each atom?

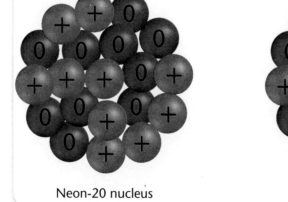

Neon-20 nucleus   Neon-22 nucleus

MiniLab 2.1

Model Isotopes

How can pennies be used to model isotopes? Many elements have several naturally occurring isotopes. Isotopes are atoms of the same element that are identical in every property except mass. When chemists refer to the mass of an atom of one of these elements, they really mean the average of the atomic masses of the isotopes of that element. You can use pennies to represent isotopes.

Procedure

1. Read and complete the lab safety form.
2. Obtain a **bag of pennies.**
3. Sort the pennies by date. Group the pre-1982 pennies and the post-1982 pennies.
4. Determine the mass of ten pennies from each group. Record the mass to the nearest 0.01 g. Divide each total mass by ten to get the average mass for one penny in that group.
5. Count the number of pennies in each group.
6. Using the data, determine the total mass of all of the pre-1982 pennies. In the same way, calculate the total mass of the post-1982 pennies.

Analysis

1. **Analyze** What is the mass of all the pennies?
2. **Calculate** Use the mass of all pennies and the total number of pennies to calculate the average mass of a penny. How does this mass compare with the average masses for each group?
3. **Infer** Why were you directed to determine the mass of ten pennies of a group and divide the mass by ten to get the mass of one penny? Why not just determine the mass of one penny from each group?

Rutherford's gold foil experiment While all of these subatomic particles were being discovered, scientists were also trying to determine how the particles of an atom were arranged. After the discovery of the electron, scientists pictured an atom as tiny particles of negative charge embedded in a ball of positive charge. You could compare this early atomic model to a ball of chocolate-chip cookie dough (except with submicroscopic chocolate chips). Almost at the same time, a Japanese physicist, Hantaro Nagaoka, proposed a different model in which electrons orbited a central, positively charged nucleus—something like Saturn and its rings. Both models are shown in **Figure 2.8.**

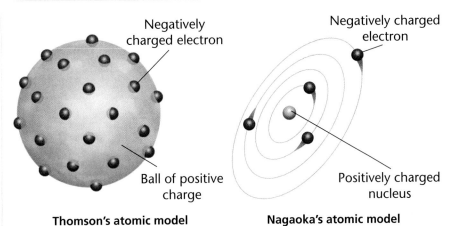

■ **Figure 2.8** Thomson pictured the atom as consisting of electrons embedded in a ball of positive charge. Nagaoka's model resembled a planet with moons orbiting in a flat plane. He believed the core had a positive charge, and the negative electrons were in orbit around it.

Personal Tutor For an online tutorial on the gold foil experiment, visit glencoe.com.

In 1909, a team of scientists led by Ernest Rutherford in England carried out the first of several important experiments that revealed an arrangement far different from the cookie-dough model of the atom. Rutherford's experimental setup is shown in **Figure 2.9.**

The experimenters set up a lead-shielded box containing radioactive polonium, which emitted a beam of positively charged subatomic particles through a small hole. Today, we know that the particles of the beam consisted of clusters containing two protons and two neutrons and are called alpha particles. The sheet of gold foil was surrounded by a screen coated with zinc sulfide, which glows when struck by the positively charged particles of the beam.

■ **Figure 2.9** Gold is a metal that can be pounded into a foil only a few atoms thick. Rutherford's group took advantage of this property in their experiment. If the cookie-dough model were true, Rutherford and his team would have expected to see the alpha particles pass straight through the foil.

1. Rutherford's lead-shielded box contained radioactive polonium. As polonium decays, it emits helium nuclei, which consist of two protons and two neutrons. These nuclei are called alpha particles. Because they have no electrons, they are positively charged.

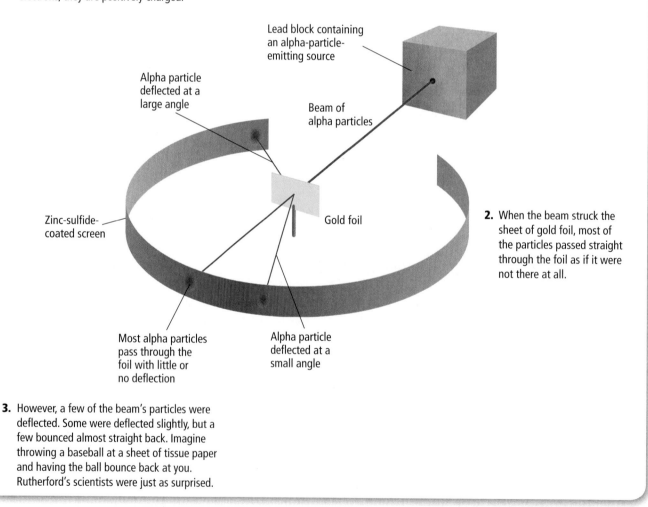

2. When the beam struck the sheet of gold foil, most of the particles passed straight through the foil as if it were not there at all.

3. However, a few of the beam's particles were deflected. Some were deflected slightly, but a few bounced almost straight back. Imagine throwing a baseball at a sheet of tissue paper and having the ball bounce back at you. Rutherford's scientists were just as surprised.

The nuclear model of the atom To explain the results of the experiment, Rutherford's team proposed a new model of the atom. Because most of the particles passed through the foil, they concluded that the atom is nearly all empty space. Because so few particles were deflected, they proposed that the atom has a small, dense, positively charged central core called a **nucleus**. The new model of the atom as pictured by Rutherford's group in 1911 and the modern nuclear model are shown in **Figure 2.10.**

Imagine how this model affected the view of matter when it was announced in 1911. When people looked at a strong steel beam or block of stone, they were asked to believe that this heavy, solid matter was practically all empty space. Even so, this atomic model has proven to be reasonably accurate.

As an example of the emptiness of an atom, consider the simplest atom, hydrogen, which consists of an electron and a nucleus of one proton. If hydrogen's proton were increased to the size of a golf ball, the electron would be about a mile away and the atom would be about two miles in diameter. To get an idea of how small an atom is, consider that there are roughly 6,500,000,000,000,000,000,000 (6.5 sextillion, or 6.5×10^{21}) atoms in a drop of water. If you need more help in expressing very large and very small quantities using scientific notation, study page 794 in the *Skill Handbook*.

> **TRY AT HOME 🏠 LAB**
> See page 868 for **Comparing Atom Sizes.**

■ **Figure 2.10** Rutherford's experiments confirmed a model of the atom with a dense, positively charged nucleus.
Infer *Why are the alpha particles deflected?*

Rutherford's Model

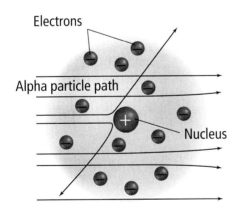

Remember that alpha particles are positively charged. If the nucleus were negatively charged, the particles would attract each other. However, in Rutherford's experiment, some of the alpha particles were deflected. This suggests that the nucleus is small, dense, and positively charged.

Modern Model

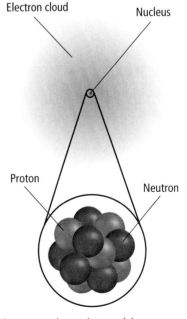

In the present-day nuclear model, atoms are composed of a nucleus containing protons and neutrons. A cloud of electrons surrounds the nucleus.

SUPPLEMENTAL PRACTICE

For more practice with atomic numbers, see Supplemental Practice, page 808.

Atomic Numbers

Look at the periodic table hanging on the wall of your classroom or the one inside the back cover of your textbook. Notice that the elements are numbered from 1 for hydrogen, 2 for helium, through 8 for oxygen, and on to numbers above 100 for the newest elements created in the laboratory. At first glance, it may seem that these numbers are just a way of counting elements, but the numbers mean much more than that. Each number is the atomic number of that atom. The **atomic number** of an element is the number of protons in the nucleus of an atom of that element. It is the number of protons that determines the identity of an element, as well as many of its chemical and physical properties, as you will see later in the course.

Because atoms have no overall electric charge, an atom must have as many electrons as it has protons in its nucleus. Therefore, the atomic number of an element also tells the number of electrons in a neutral atom of that element.

Isotopes Neutrons are neutral particles also found in the nucleus of all atoms except the simplest isotope of hydrogen. The mass of a neutron is almost the same as the mass of a proton. The sum of the protons and neutrons in the nucleus is the **mass number** of that particular atom. Isotopes of atoms of an element have different mass numbers because they have different numbers of neutrons. An isotope is identified by writing the name or symbol of the atom followed by its mass number. Recall that Thomson discovered that naturally occurring neon is a mixture of isotopes. The three neon isotopes are Ne-20 (90.5 percent), Ne-21 (0.2 percent), and Ne-22 (9.3 percent). All neon isotopes have ten protons and ten electrons. Ne-20 has ten neutrons, Ne-21 has 11 neutrons, and Ne-22 has 12 neutrons. Look at the isotopes of potassium in **Figure 2.11**.

■ **Figure 2.11** Potassium has three naturally occuring isotopes: potassium-39, potassium-40, and potassium-41.
List *the number of protons, neutrons, and electrons in each potassium isotope.*

	Potassium-39	Potassium-40	Potassium-41
Protons	19	19	19
Neutrons	20	21	22
Electrons	19	19	19

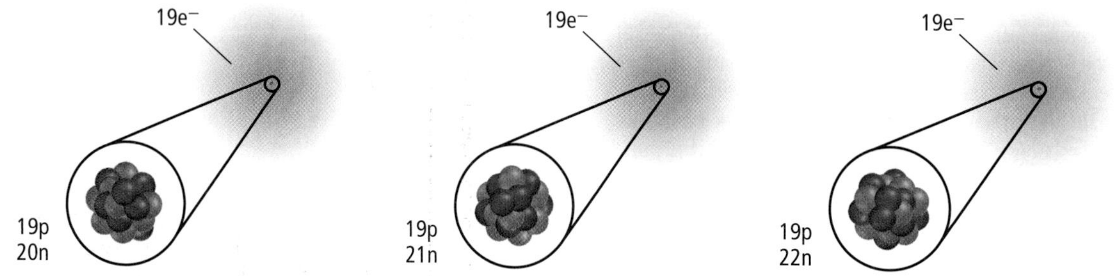

64 Chapter 2 • Matter Is Made of Atoms

Table 2.1 Particles of an Atom

Particle	Symbol	Charge	Mass Number	Mass in grams	Mass in u
Proton	p	1+	1	1.67×10^{-24}	1.01
Neutron	n	0	1	1.67×10^{-24}	1.01
Electron	e^-	1^-	0	9.11×10^{-28}	0.00055

Atomic Mass

Recall that there are 6.5×10^{21} atoms in a drop of water. Considering the size of this number, you can understand that atoms have extremely small masses. To state the mass of an atom in grams, you must deal with small numbers. **Table 2.1** summarizes the properties of protons, neutrons, and electrons, and shows you their masses in grams. Notice the shorthand symbols for the particles. You will see these symbols elsewhere in this book.

As you can see, you must use some small numbers to state the mass of an atom in grams. In order to have a simpler way of comparing the masses of individual atoms, chemists have devised a different unit of mass called an **atomic mass unit**, which is given the symbol u. An atom of the carbon-12 isotope contains six protons and six neutrons and has a mass number of 12. Chemists have defined the carbon-12 atom as having a mass of 12 atomic mass units. Therefore, 1 u = 1/12 the mass of a carbon-12 atom. As you can see in **Table 2.1**, 1 u is approximately the mass of a single proton or neutron.

Refer again to the periodic table. Each box on the table contains several pieces of information about an element, as shown in **Figure 2.12**. The number at the bottom of each box is the average atomic mass of that element. This number is the weighted average mass of all the naturally occurring isotopes of that element. Scientists use an instrument called the mass spectrometer to determine the number, mass, and abundance of isotopes of an element.

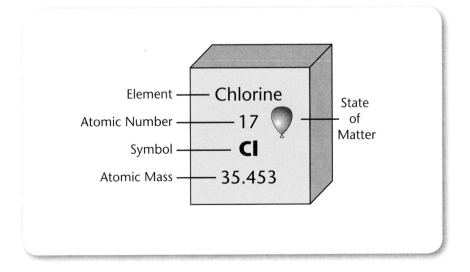

■ **Figure 2.12** Each block of the periodic table shows the atomic number of the element, its symbol and name, its physical state at room temperature, and the average mass of its atoms.

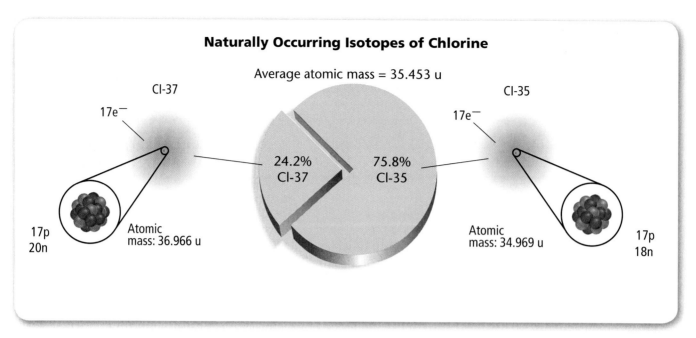

Figure 2.13 To calculate the weighted average atomic mass, you need to calculate the mass contribution of the naturally occurring isotopes. For chlorine, you need to consider two isotopes, Cl–35 and Cl–37.

Today, scientists have measured the mass and abundance of the isotopes of all but the most unstable elements. These data have been used to calculate the average atomic masses of most elements. The isotopes of chlorine are illustrated in **Figure 2.13.** As you review the figure, imagine that you have 1000 atoms of chlorine. On average, 758 of those atoms (rounded to the nearest whole atom) have masses of 34.969 u, for a total mass of 758 × 34.969 u = 26,507 u. Similarly, 242 atoms have masses of 36.966 u, resulting in a total mass of 8946 u. The mass for all 1000 atoms is 26,507 u + 8946 u = 35,453 u. This means that the average mass of one chlorine atom is about 35.453 u.

Section 2.1 Assessment

Section Summary

- Scientists make hypotheses based on observations.
- Dalton's atomic theory states that matter is made of indestructible atoms.
- Experiments in the late nineteenth and early twentieth centuries revealed that the mass of an atom is concentrated in a tiny nucleus.
- The number of protons in an atom's nucleus is called the atomic number and equals the number of electrons in the atom.
- Atoms of the same element always have the same number of protons and electrons.

1. **MAIN Idea** **Diagram** the structure of a typical atom. Identify where each subatomic particle is located.

2. **Conclude** What conclusions about the nature of matter did scientists draw from the law of definite proportions?

3. **Explain** How did scientists know that cathode rays had a negative electric charge?

4. **Contrast** How do isotopes of an element differ from one another?

5. **List** Review **Figure 2.12** on page 65. Then, refer to the periodic table at the back of the book and list all the information given in the table for the element bromine.

6. **Infer** Why did Rutherford conclude that an atom's nucleus has a positive charge instead of a negative charge? Summarize the conclusions that Rutherford's team made about the structure of an atom.

7. **Evaluate** The isotope of carbon that is used to date prehistoric fossils contains six protons and eight neutrons. What is the atomic number of this isotope? How many electrons does it have? What is its mass number?

Section 2.2

Objectives
- **Relate** the electron to modern atomic theory.
- **Compare** electron energy levels in an atom.
- **Illustrate** valence electrons by Lewis electron dot structures.

Review Vocabulary
atom: smallest particle of a given type of matter

New Vocabulary
electromagnetic spectrum
emission spectrum
energy level
electron cloud
valence electron
Lewis dot diagram

Electrons in Atoms

MAIN Idea Each element has a unique arrangement of electrons.

Real-World Reading Link Imagine climbing a ladder and trying to stand between the rungs. Unless you could stand on air, it would not work. When atoms are in various energy states, electrons behave in much the same way as a person climbing up and down the rungs of a ladder.

Electrons in Motion

Recall from Section 2.1, the atom is mostly empty space, except that the space isn't entirely empty. This space is occupied by the atom's electrons. Now, look more closely at how scientists have learned about the motion and arrangement of electrons.

Electron motion and energy Considering that electrons are negative and that an atom's nucleus contains positively charged protons, why aren't electrons pulled into the nucleus and held there? Scientists in the early twentieth century wondered the same thing.

Niels Bohr (1885–1962), a Danish scientist who worked with Rutherford, proposed that electrons must have enough energy to keep them in constant motion around the nucleus. He compared the motion of electrons to the motion of planets orbiting the Sun. Although the planets are attracted to the Sun by gravitational force, they move with enough energy to remain in stable orbits around the Sun. In the same way, when we launch satellites, rockets give them enough energy of motion so that the satellites stay in orbit around Earth, as shown in **Figure 2.14.** Electrons have energy of motion that enables them to overcome the attraction of the positive nucleus and keeps them moving around the nucleus. Bohr's view of the atom, proposed in 1913, was called the planetary model.

■ **Figure 2.14** This satellite, as it travels around Earth, is subject to balanced forces. If its energy of motion is increased, its speed increases, and it moves to an orbit farther away from Earth.

■ **Figure 2.15** A satellite can orbit Earth at almost any altitude, depending on the amount of energy used to launch it. Electrons, however, can occupy orbits of only certain energies.

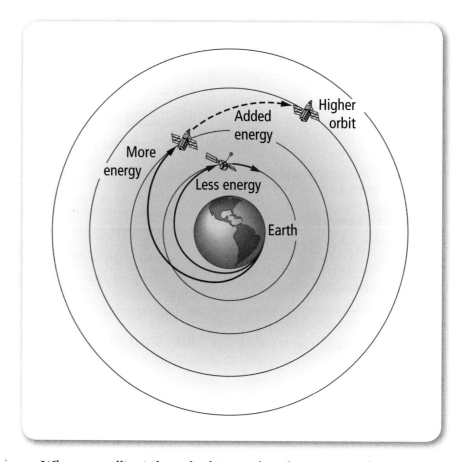

When a satellite is launched into orbit, the amount of energy determines how high above Earth it will orbit. Given a little more energy, the satellite will go into a slightly higher orbit; with a little less energy, it will have a lower orbit as shown in **Figure 2.15.** However, it seemed that electrons did not behave the same way. Instead, experiments showed that electrons occupied orbits of only certain amounts of energy. A model of the atom had to explain these observations.

The Electromagnetic Spectrum

To boost a satellite into a higher orbit requires energy from a rocket. One way to increase the energy of an electron is to supply energy in the form of high-voltage electricity. Another way is to supply electromagnetic radiation, also called radiant energy. Electromagnetic radiation travels in the form of waves that have both electric and magnetic properties. These electromagnetic waves can travel through empty space. For instance, radiant energy from the Sun travels to Earth every day. Electromagnetic waves travel through a vaccuum at the speed of light, which is approximately 300 million meters per second.

Two properties of waves are frequency and wavelength. The number of vibrations per second is the frequency of the wave. The scientific unit for frequency is the hertz (Hz). Wavelength is the distance between corresponding points on two consecutive waves. A low frequency results in a long wavelength, and a high frequency results in a shorter wavelength.

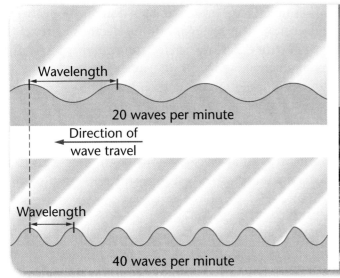

Waves transfer energy If you've ever seen water waves breaking on a shoreline or heard objects in a room vibrate from the effects of loud sound waves, you already know that waves transfer energy from one place to another. Electromagnetic waves have many of the same characteristics as other waves, shown in **Figure 2.16.** However, unlike sound waves and water waves, electromagnetic waves can travel through empty space.

Electromagnetic radiation includes radio waves that carry broadcasts to your radio and microwave radiation used to heat food in a microwave oven, radiant energy used to toast bread, and the most familiar form, visible light. All of these forms of radiant energy are parts of the whole range of electromagnetic radiation called the **electromagnetic spectrum**. A portion of this spectrum is shown in **Figure 2.17.**

■ **Figure 2.16** Waves transfer energy. Homes built above a seashore that is continually pounded by ocean waves can become endangered by erosion resulting from this transfer of energy. In the photo above, a concrete wall has been built to protect the homes you see.

■ **Figure 2.17** White light is a mixture of all colors of visible light. Whenever white light passes through a prism or a diffraction grating, it is broken into a range of colors called the visible light spectrum. When sunlight passes through raindrops, it is broken into the colors of the rainbow.

Section 2.2 • Electrons in Atoms **69**

Visible light makes up only a very narrow portion of the electromagnetic spectrum. A more complete look at the electromagnetic spectrum is shown in **Figure 2.18.** The electromagnetic spectrum includes all forms of electromagnetic radiation, with the only differences in the types of radiation being their frequencies and wavelengths. Note that higher-frequency electromagnetic waves have higher energy than lower-frequency waves. Thus, a radio wave with a wavelength of 30,000 m—almost 19 miles—has much lower energy than a gamma ray with a wavelength of 3×10^{-14} m. This is an important fact to remember when you study the relationship of light to atomic structure.

■ **Figure 2.18** All forms of electromagnetic energy interact with matter, and the ability of these different waves to penetrate matter is a measure of the energy of the waves. Note that the spectrum does not cut off at the limits shown below. There are longer radio waves and shorter gamma rays.

A. Radio waves have the lowest frequencies on the electromagnetic spectrum. In the AM radio band, frequencies range from 550 kHz (kilohertz) to 1700 kHz and wavelengths from about 200 m to 600 m.

B. Microwaves are low-frequency, low-energy waves that are used for communications and cooking.

C. Infrared waves have less energy than visible light. Infrared radiation is given off by the human body and most other warm objects. We experience infrared rays as the radiant heat you feel near a fire or an electric heater.

D. Visible light waves are the part of the electromagnetic spectrum to which our eyes are sensitive. Our eyes and brain interpret different frequencies as different colors.

E. Ultraviolet waves are slightly more energetic than visible light waves. Ultraviolet radiation is the part of sunlight that causes sunburn. Ozone in Earth's upper atmosphere absorbs most of the Sun's ultraviolet radiation.

F. X-rays have lower frequencies than gamma rays have but are still considered to be high-energy rays. X-rays pass through soft body tissue but are stopped by harder tissue such as bone.

G. Gamma rays have the highest frequencies and the shortest wavelengths. Because gamma rays are the most energetic rays in the electromagnetic spectrum, they can pass through most substances.

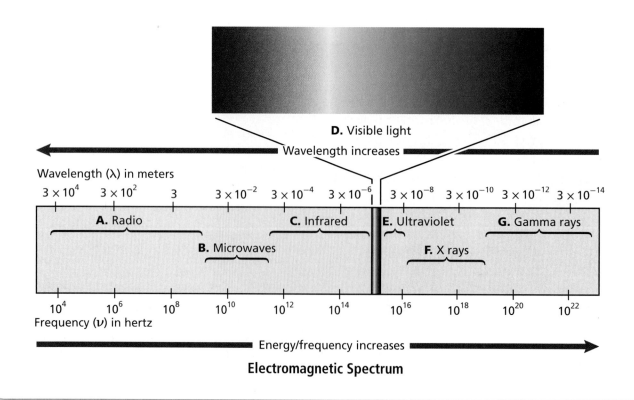

Electromagnetic Spectrum

Physics Connection

Aurora Borealis

In real life, you might never see colored lights such as those brightening the night sky in **Figure 1.** You are looking at the aurora borealis—a fantastic light show seen only in high northern latitudes. An aurora occurs from 100 to 1000 km above Earth. The lights were once thought to be reflections from the polar ice fields, but now scientists know that auroras are the most visible effect of the Sun's activity on Earth's atmosphere.

Cause of auroras An aurora is attributed to solar wind, which is a continuous flow of electrons and protons from the Sun. These high-energy, electrically charged particles flow like a current through twisted bundles of magnetic fields that connect Earth's upper atmosphere to the Sun. The particles collide with oxygen and nitrogen in the ionosphere and transfer energy to them. The energy causes electrons in these atoms to jump to higher energy levels. When the electrons return to lower energy levels, they release the absorbed energy as light.

Characteristics of the aurora When the frequencies of radiant energy released by the oxygen and nitrogen are in the visible range, they can be seen as an aurora. Atomic oxygen, releasing energy at altitudes from 100 to 200 km, is the source of the greenish-white light common in auroras. High-altitude atomic oxygen, releasing energy above 200 km, can emit a dark-red light produced during strong magnetic storms.

Nitrogen atoms lower in the ionosphere, below 100 km, produce a red light when they are struck by electrons. This is the faint red light that is often seen along the bottom edge of auroras. High in the atmosphere, above 200 km, nitrogen atoms can emit blue and violet light.

Figure 1 Aurora borealis as seen in polar latitudes.

The aurora is most frequently seen in polar latitudes because the high-energy protons and electrons move along Earth's magnetic field lines. Because these lines emerge from Earth near the magnetic poles, it is there that the particles interact with oxygen and nitrogen to produce a fantastic display of light. Auroras also may be seen in extreme southern latitudes. These displays are the aurora australis.

The aurora borealis is named after the Roman goddess of the dawn, Aurora, and the Greek name for the north wind, Boreas. The aurora australis is named after the same Roman goddess, Aurora, and the Latin word for "of the South," which is *australis*.

Connection to Chemistry

1. **Apply** How does the aurora borealis relate to the structure of an atom?
2. **Infer** What characteristic of an aurora indicates that it is caused by solar winds rather than a reflection of polar ice?

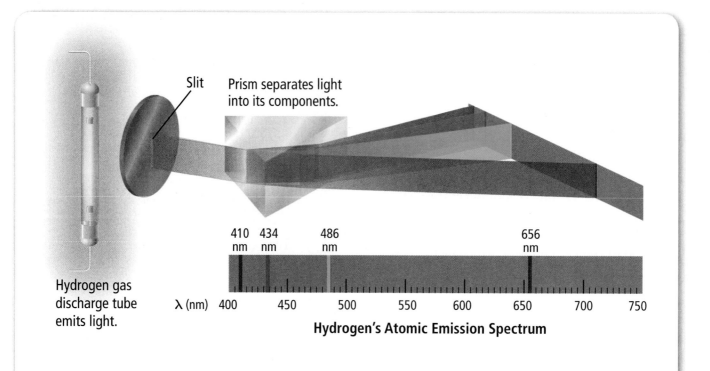

■ **Figure 2.19** The purple light emitted by hydrogen can be separated into its different components using a prism. Hydrogen has an atomic emission spectrum that contains four lines of different wavelengths.
Determine *which line has the highest energy.*

Electrons and Light

What does the electromagnetic spectrum have to do with electrons? It's all related to the energy of motion of the electron and the radiant energy. Scientists passed a high-voltage electric current through hydrogen, which absorbed some of that energy. These hydrogen atoms gave off the absorbed energy in the form of light, as shown in **Figure 2.19**. Passing that light through a prism revealed that the light consisted of just a few specific frequencies—not a whole range of colors as with white light. The spectrum of light released from excited atoms of an element is called the **emission spectrum** of that element. Each element has a different emission spectrum.

Evidence for energy levels
What could explain the fact that the emission spectrum of hydrogen consists of just a few lines? Bohr theorized that electrons absorbed energy and moved to higher energy states. Then, these excited electrons gave off that energy as light when they fell back to a lower energy state. But, why were only certain frequencies of light given off? To answer this question, Bohr suggested that electrons could have only certain amounts of energy. When they absorb energy, electrons absorb only the amount needed to move to a specific higher energy state. Then, when the electrons fall back to a lower energy state, they emit only certain amounts of energy, resulting in specific colors of light. This relationship is illustrated in **Figure 2.20**.

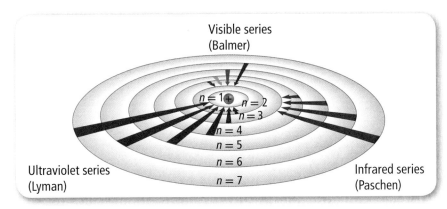

■ **Figure 2.20** This diagram shows how scientists interpret the emission spectrum. When an electron drops from a higher-energy orbit to a lower-energy orbit, a photon is emitted. The ultraviolet, visible, and infrared series correspond to electrons dropping to $n = 1$, $n = 2$, and $n = 3$, respectively. Notice that drops of electrons to the lowest level emit ultraviolet frequencies, and shorter drops of electrons to the third level emit infrared frequencies.

Concepts In Motion

Interactive Figure To see an animation of the Balmer Series, visit glencoe.com.

Because electrons can have only certain amounts of energy, Bohr reasoned they can move around the nucleus only at distances that correspond to those amounts of energy. These regions of space in which electrons can move about the nucleus of an atom are called **energy levels**. Energy levels in an atom are like rungs on a ladder. When you climb up or down a ladder, you must step on a rung. You can't step between the rungs.

The same principle applies to the movement of electrons between energy levels in an atom. Like your feet on a ladder, electrons can't hover between energy levels. Electrons must absorb just certain amounts of energy to move to higher levels. The amount is determined by the energy difference between the levels. When electrons drop back to lower levels, they give off the difference in energy in the form of light. You can compare the movement of electrons between energy levels to climbing up and down that ladder, as shown in **Figure 2.21**.

Concepts In Motion

Interactive Figure To see an animation of the electron transitions, visit glencoe.com.

■ **Figure 2.21** Only certain energy levels are allowed. The energy levels are similar to the rungs of a ladder. The four visible lines correspond to electrons dropping from a higher n to the orbit $n = 2$. As n increases, the hydrogen atom's energy levels are closer to each other.

Compare *the rungs on a ladder with the energy levels in an atom.*

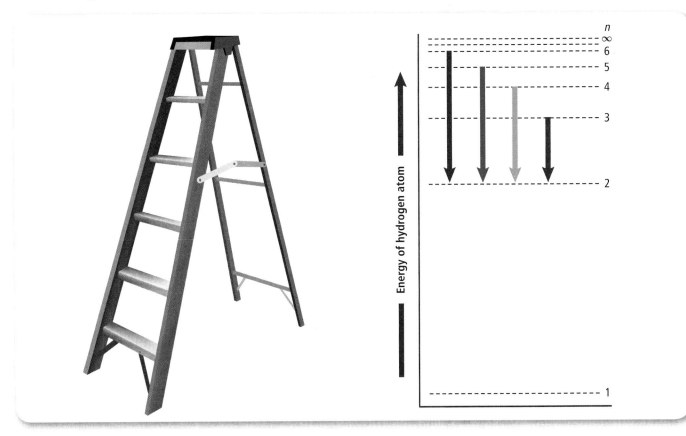

Section 2.2 • Electrons in Atoms

Everyday Chemistry

Fireworks—Getting a Bang out of Color

When you see a colorful display of fireworks, such as shown in **Figure 1,** you probably don't think about chemistry. However, creators of fireworks displays have to keep chemistry in mind. Packing what will provide "the rocket's red glare" for the 4th of July display is done by hand. Employees learn to follow proper precautions to avoid fires and explosions.

Chemistry of fireworks Typical fireworks contain an oxidizer, a fuel, a binder, and a color producer. The oxidizer is the main component, making up 38–64 percent of the material. A common oxidizer is potassium perchlorate ($KClO_4$). The presence of chlorine in the oxidizer adds brightness to the colors by producing light-emitting chloride salts that make each color flame sparkle. When it oxidizes a fuel such as aluminum or sulfur, it produces an exothermic reaction with noise and flashes. Aluminum or magnesium makes the flash dazzling blue-white. The loud noise comes from the rapidly expanding gases. Other fuels not only raise the temperature but also bind the loosely assembled materials together.

Flame Color	Color-Producing Salts

Red	strontium salts, lithium salts
	  strontium carbonate, $SrCO_3$ = bright red
	  lithium carbonate, Li_2CO_3 = red

Orange	calcium salts
	  calcium chloride, $CaCl_2$
	  calcium sulfate, $CaSO_4 \cdot H_2O$

Yellow	sodium salts
	  sodium nitrate, $NaNO_3$
	  cryolite, Na_3AlF_6

Green	barium salts barium chloride, $BaCl$ = bright green

Blue	copper salts
	  copper acetoarsenite, $C_4H_6As_6Cu_4O_{16}$ = blue
	  copper (I) chloride, $CuCl$ = turquoise blue

Purple	mixture of strontium salts (red) and copper salts (blue)

Producing color A metallic salt, such as those shown in the table, is added to produce a specific color-emission spectrum. Care must be taken in selecting the ingredients so that the oxidizer is not able to react with the metallic salt during storage and cause an explosion.

Explore Further

1. **Apply** Based on your knowledge of the electromagnetic spectrum, which of the salts or compounds in the table has an emission spectrum with the shortest wavelength?

2. **Infer** Oxidizers in fireworks are chemicals that break down rapidly to release oxygen, which then burns the fuel. Why is it necessary to provide an oxidizer in the mixture rather than relying on oxygen in the air?

Figure 1 Colorful fireworks are used at special events.

MiniLab 2.2

Line Emission Spectra

How can you observe and compare emission spectra of white light and several elements? Emission spectra of elements result from electron transitions within atoms. They provide information about the arrangements of electrons in the atoms.

Procedure

1. Read and complete the lab safety form.
2. Observe the emitted light from an **incandescent bulb** through a **diffraction grating** as you hold it close to your eye. Hold the grating by the cardboard edge and avoid touching the transparent material that encloses the diffraction grating. Record your observations.
3. Next, observe the light produced by the **spectrum tube** containing hydrogen gas and record your observations. It may be necessary for you to move to within a meter of the spectrum tube in order to effectively observe the emission spectrum. **WARNING:** *The spectrum tube operates at a high voltage. Under no circumstances should you touch the spectrum tube or any part of the transformer.*
4. Repeat step 2 with other **spectrum tubes** as your teacher designates.

Analysis

1. **Explain** Why do only certain colors appear in the emission spectra of the elements?
2. **Infer** If each hydrogen atom contains only one electron, how are several emission spectral lines possible?
3. **Interpret** Why do other elements emit many more spectral lines than hydrogen atoms?

The Electron Cloud Model

As a result of continuing research throughout the twentieth century, scientists today realize that energy levels are not neat, planetlike orbits around the nucleus of an atom. Instead, they are spherical regions of space around the nucleus in which electrons are most likely to be found. Examine **Figure 2.22**. Electrons themselves take up little space but travel rapidly through the space surrounding the nucleus. These spherical regions where electrons travel may be depicted as clouds around the nucleus. The space around the nucleus of an atom where the atom's electrons are most likely to be found is called the **electron cloud**.

FOLDABLES
Incorporate information from this section into your Foldable.

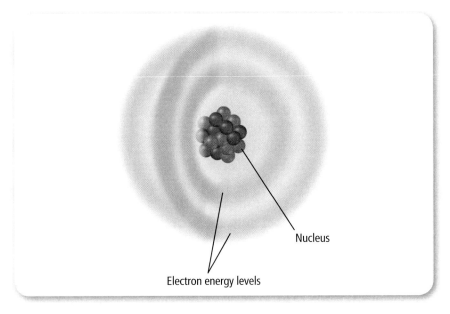

■ **Figure 2.22** In the electron cloud model of an atom, energy levels are concentric spherical regions of space around the nucleus. The darker areas represent regions where electrons in that level are most likely to be found. Electrons are less likely to be found in the lighter regions of each level.

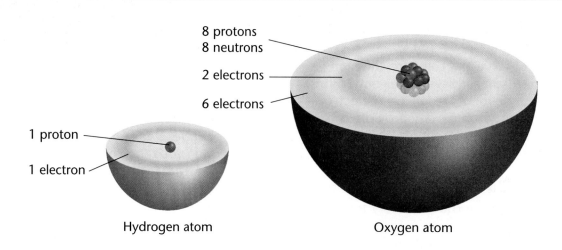

■ **Figure 2.23** A hydrogen atom has only one electron. It's in the first energy level. An oxygen atom has eight electrons. Two of these fill the first energy level, and the remaining six are in the second energy level.
Identify *An atom of helium has two electrons. In which energy level would you place these electrons?*

Electrons in energy levels How are electrons arranged in energy levels? Each energy level can hold a limited number of electrons. The lowest energy level is the smallest and the closest to the nucleus. This first energy level holds a maximum of two electrons. The second energy level is larger because it is farther away from the nucleus. It holds a maximum of eight electrons. The third energy level is larger still and holds a maximum of 18 electrons.

Figure 2.23 shows the energy levels and electron clouds in hydrogen and oxygen atoms. As you review art of atoms and molecules throughout this book, keep in mind that the identity of atoms can always be determined by their color. For example, the outer shells of oxygen atoms will always be red, and hydrogen, blue. Refer to **Table D.1** on page 846 for complete color conventions.

Valence electrons You will learn more about the details of electron arrangement in Chapter 7. For now, it's important to learn about the electrons in the outermost energy level of an atom. The electrons in the outermost energy level are called **valence electrons**. As you can see in **Figure 2.23,** hydrogen has one valence electron and oxygen has six valence electrons. You can also use the periodic table as a tool to predict the number of valence electrons in any atom in groups 1, 2, 13, 14, 15, 16, 17, and 18. All atoms in group 1, like hydrogen, have one valence electron. Likewise, atoms in group 2 have two valence electrons. Atoms in groups 13–18 have three through eight valence electrons, respectively.

Why do you need to know how to determine the number of outer-level electrons that are in an atom? Recall that at the beginning of this section, it was stated that when atoms come near each other, it is the electrons that interact. In fact, it is the valence electrons that interact. Therefore, many of the chemical and physical properties of an element are directly related to the number and arrangement of valence electrons.

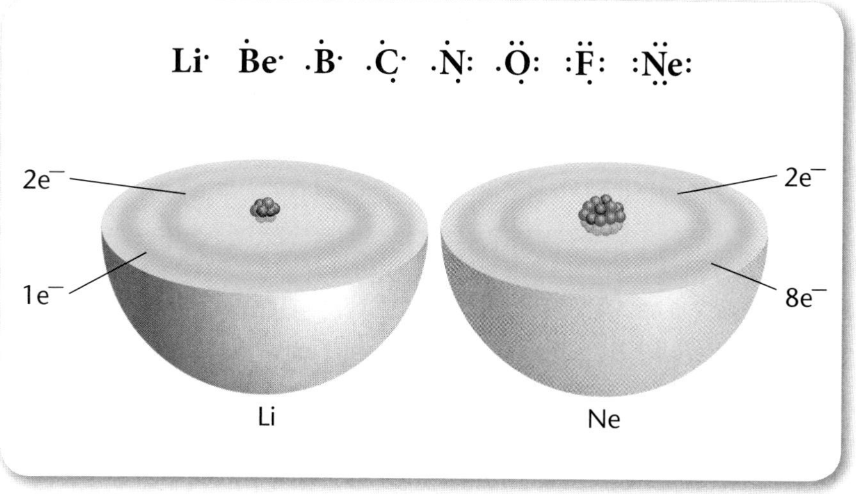

Li· ·Be· ·Ḃ· ·Ċ· ·Ṅ· ·Ö· :F̈: :N̈e:

Figure 2.24 The Lewis dot diagrams for these elements illustrate how valence electrons change from element to element across a row on the periodic table.

Lewis dot diagrams Because valence electrons are so important to the behavior of an atom, it is useful to represent them with symbols. A **Lewis dot diagram** illustrates valence electrons as dots (or other small symbols) around the chemical symbol of an element. Each dot represents one valence electron. In the dot diagram, the element's symbol represents the core of the atom—the nucleus plus all the inner electrons. The Lewis dot diagrams for several elements are shown in **Figure 2.24**.

Connecting Ideas

A Lewis dot diagram is a convenient, shorthand method to represent an element and its valence electrons. You have used the periodic table already as a source of information about the symbols, the names, the atomic numbers, and the average atomic masses of elements. In Chapter 3, you will learn that the arrangement of elements on the periodic table yields even more information about the electronic structures of atoms and how those structures can help you to predict many of the properties of elements.

SUPPLEMENTAL PRACTICE

For more practice with drawing Lewis dot diagrams, see Supplemental Practice, page 808.

Section 2.2 Assessment

Section Summary

- Electrons move around an atom's nucleus in specific energy levels.
- Energy levels are spherical regions in which electrons are likely to be found.
- The greater the energy of the level, the farther from the nucleus the level is located.
- Electrons can absorb energy and move to a higher energy level.
- Lewis dot diagrams can be used to represent the valence electrons in a given atom.

8. **MAIN Idea** **Describe** For each element, describe how many electrons are in each energy level and then draw the Lewis dot diagram for each atom.
   a) argon; 18 electrons
   b) magnesium; 12 electrons
   c) nitrogen; 7 electrons
   d) aluminum; 13 electrons
9. **Infer** What change occurs within an atoms when it emits light?
10. **Explain** How does the modern electron cloud model of the atom differ from Bohr's original planetary model of the atom?
11. **Describe** how scientists concluded that electrons occupy specific energy levels.
12. **Apply** Give an everyday example of how you can tell that light waves transfer energy.

CHAPTER 2 Study Guide

BIG Idea Atoms are the fundamental building blocks of matter.

2.1 Atoms and Their Structures

MAIN Idea An atom is made of a nucleus containing protons and neutrons; electrons move around the nucleus.

Vocabulary
- atom (p. 51)
- atomic mass unit (p. 65)
- atomic number (p. 64)
- atomic theory (p. 51)
- electron (p. 59)
- experiment (p. 57)
- hypothesis (p. 57)
- isotope (p. 60)
- law of definite proportions (p. 52)
- mass number (p. 64)
- neutron (p. 60)
- nucleus (p. 63)
- proton (p. 60)
- scientific law (p. 57)
- scientific method (p. 57)
- theory (p. 57)

Key Concepts
- Scientists make hypotheses based on observations.
- Dalton's atomic theory states that matter is made of indestructible atoms.
- Experiments in the late nineteenth and early twentieth centuries revealed that the mass of an atom is concentrated in a tiny nucleus.
- The number of protons in an atom's nucleus is called the atomic number and equals the number of electrons in the atom.
- Atoms of the same element always have the same number of protons and electrons.

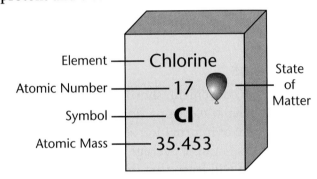

2.2 Electrons in Atoms

MAIN Idea Each element has a unique arrangement of electrons.

Vocabulary
- electromagnetic spectrum (p. 69)
- electron cloud (p. 75)
- emission spectrum (p. 72)
- energy level (p. 73)
- Lewis dot diagram (p. 77)
- valence electron (p. 76)

Key Concepts
- Electrons move around an atom's nucleus in specific energy levels.
- Energy levels are spherical regions in which electrons are likely to be found.
- The greater the energy of the level, the farther from the nucleus the level is located.
- Electrons can absorb energy and move to a higher energy level.
- Lewis dot diagrams can be used to represent the valence electrons in a given atom.

Chapter 2 Assessment

Understand Concepts

13. How does Dalton's atomic theory explain the law of conservation of mass?

14. Why is it necessary to perform repeated investigations in order to support a hypothesis?

15. What is the atomic number of calcium? What does that number tell you about an atom of calcium?

16. If the nucleus of an atom contains 12 protons, how many electrons are there in a neutral atom? Explain.

17. An atom of sodium has 11 protons, 11 electrons, and 12 neutrons. What is its atomic number? What is its mass number?

18. **Figure 2.25** shows a circle graph of the abundance of the two kinds of silver atoms found in nature. The more abundant isotope has an atomic mass of a little less than 107, but the average atomic mass of silver on the periodic table is about 107.9. Explain why it is higher.

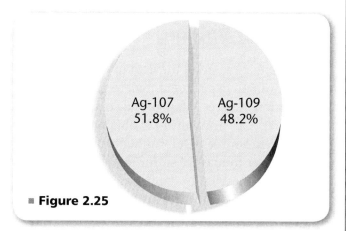

■ **Figure 2.25**

19. Describe the relationship among frequency, wavelength, and energy of electromagnetic waves.

20. Draw the Lewis dot diagram for silicon, which contains 14 electrons.

Apply Concepts

History Connection

21. Lavoisier's discovery that air was not a substance but rather a mixture of nitrogen and oxygen is described as "history-making." Why do you think this description is accurate?

Physics Connection

22. How does solar wind differ from electromagnetic radiation from the Sun?

Everyday Chemistry

23. Rocket boosters used to launch the space shuttle must carry both fuel and an oxidizer. Why would an oxidizer be needed in such rockets?

Chemistry and Society

24. Even though the law of conservation of mass applies to both, why is waste glass a more serious problem than leaves that have fallen from trees or food waste?

Think Critically

Relate Cause and Effect

25. How was Bohr's model of the atom consistent with the results of Rutherford's gold foil experiment?

Observe and Infer

26. How did scientists account for the fact that the emission spectrum of hydrogen is not continuous but consists of only a few lines of certain colors?

Compare and Contrast

27. What is the relationship between an atom that has 12 protons, 12 neutrons, and 12 electrons and one that has 12 protons, 13 neutrons, and 12 electrons?

Chapter 2 Assessment

Interpret Data

28. MiniLab 1 Suppose you obtained the data in **Table 2.2** below for MiniLab 1. Complete the table and determine the average penny mass of the pennies in the mixture.

Table 2.2	Type of Penny	
	Pre-1982	Post-1982
Mass of 10	30.81 g	25.33 g
Mass of 1		
Number of pennies in mixture	34	55
Mass of each type		
Average penny mass		

Apply Concepts

29. MiniLab 2 Which of the lines of the visible emission spectrum of hydrogen represents the greatest energy drop?

Observe and Infer

30. ChemLab Why do you think the amount of zinc that you could use in the experiment was limited to 0.28 g?

Make Comparisons

31. An electromagnetic wave has a frequency of 10^{21} Hz. Use **Figure 2.18** on page 70 to determine which type of electromagnetic wave has this frequency. Compare the wavelength and the energy of this wave with those of a second wave with a frequency of 10^{17} Hz.

Cumulative Review

32. Distinguish among a mixture, a solution, and a compound. (*Chapter 1*)

33. Suppose you are given two identical bottles filled completely with clear liquids. Describe a way you could determine whether the liquids have different densities. You cannot open the bottles. (*Chapter 1*)

Skill Review

34. Relate Cause and Effect A chemistry student decided to weigh a bag of microwave popcorn before and after popping it. Following the directions on the bag, she opened it immediately after microwaving it to let the steam escape and then weighed it. She noticed that the mass of the unpopped popcorn was 0.5 g more than the mass of the popped popcorn. Was matter destroyed when the popcorn popped? Explain your answer.

WRITING in Chemistry

35. Phlogiston theory Research the phlogiston theory. How did people who believed the phlogiston theory explain chemical changes such as burning, the oxidation of metals, and the smelting of ores? Write a paragraph in which you attempt to answer the question "What was phlogiston?" using modern scientific terms.

Problem Solving

36. A chemist recorded the data in **Table 2.3** below in an experiment to determine the composition of three samples of a compound of copper (Cu) and sulfur (S) obtained from three different sources. Determine the ratio of mass of copper to mass of sulfur for each sample. How do these ratios compare? What law of chemistry does this experiment illustrate? Does the result mean that atoms of copper and sulfur occur in this compound in the same numerical ratio? Explain your answer.

Table 2.3	Summary of Results		
Sample	Mass of sample	Mass of Cu	Mass of S
1	5.02 g	3.35 g	1.67 g
2	10.05 g	6.71 g	3.34 g
3	99.6 g	66.4 g	33.2 g

Cumulative
Standardized Test Practice

1. Why would Greek philosophers not be considered scientists?
   a) They proposed inaccurate theories about matter and the structure of the universe.
   b) They did not have modern scientific instruments at their disposal.
   c) They did not test their hypotheses with extensive experimentation.
   d) Many of their observations were based on myths and superstitions.

2. Positively charged, subatomic particles located in the nucleus of an atom are called
   a) electrons.   c) neutrons.
   b) isotopes.   d) protons.

3. Why did Rutherford conclude from his gold foil experiment that an atom is mostly comprised of empty space?
   a) The positively charged particles shot into the foil were deflected by the nuclei of the gold atoms.
   b) The positively charged particles shot into the foil were attracted to electrons of the gold atoms.
   c) Most of the particles shot through the gold foil passed straight through the material.
   d) The radioactive particles shot into the gold foil caused the gold atoms to give off their own radiation.

4. What will determine the distance between an orbiting electron and the nucleus of an atom?
   a) the amount of energy in the electron
   b) the mass of the electron
   c) the energy level holding the electron
   d) the electromagnetic frequency of the electron

Chemist's Notes About Statements Made During a Food Chemistry Lecture
A. The total number of atoms in food and drink taken into the body must equal the total number of atoms stored or expelled by the body.
B. A packaged food cake can sit on a shelf for years without growing mold.
C. Based on experimental evidence gathered on large numbers of test subjects, it is believed that the consumption of excessive amounts of soft drinks increases the chance of kidney disease.
D. In a study group of 34 people, the average person lost 4 pounds during a weeklong diet of whole grains, fresh fruit, and fresh vegetables.

Use the table above to answer questions 5–7.

5. Which statement is a hypothesis?
   a) A   c) C
   b) B   d) D

6. Which statement is a theory?
   a) A   c) C
   b) B   d) D

7. Which statement is a scientific law?
   a) A   c) C
   b) B   d) D

8. Which of the following is NOT a mixture?
   a) orange juice
   b) liquid soap
   c) salt
   d) air

NEED EXTRA HELP?

If You Missed Question . . .	1	2	3	4	5	6	7	8
Review Section . . .	2.1	2.1	2.1	2.2	2.1	2.1	2.1	1.1

CHAPTER 3
Introduction to the Periodic Table

BIG Idea Periodic trends in the properties of atoms allow us to predict physical and chemical properties.

3.1 Development of the Periodic Table
MAIN Idea The periodic table evolved over time as scientists discovered more useful ways to compare and organize the elements.

3.2 Using the Periodic Table
MAIN Idea Elements are organized in the periodic table according to their electron configurations.

ChemFacts

- Like these bowling shoes, the periodic table is organized vertically and horizontally.
- There are 117 elements in the current periodic table. Only 92 of them occur naturally.
- More than 100 years ago, chemists began searching for a way to organize information about the elements.

Start-Up Activities

LAUNCH Lab

Versatile Materials

A variety of processes can be used to shape metals into different forms. Because of their physical properties, metals are used in a wide range of applications.

Materials
- tape
- samples of copper
- light socket with bulb, wires, and battery

Procedure
1. Read and complete the lab safety form.
2. Observe the different types of copper metal that your teacher gives you. Write down as many observations as you can about each of the copper samples.
3. Try gently bending each copper sample (do not break the samples). Record your observations.
4. Connect each copper sample to the circuit. Record your observations.

Analysis
1. **Compare** What properties of copper are similar in all of the samples?
2. **Contrast** How do the samples of copper differ?
3. **List** several common applications of copper. What properties make metals such as copper so versatile?

Inquiry Would your results be the same with a piece of lead or aluminum? Design an experiment to test your hypothesis.

Chemistry Online

Visit glencoe.com to:
- study the entire chapter online
- explore **Concepts In Motion**
- take Self-Check Quizzes
- use Personal Tutors
- access Web Links for more information, projects, and activities
- find the Try at Home Lab, Element Hunt

FOLDABLES Study Organizer

Classifying Elements Make the following Foldable to organize information about the classes of the elements.

▶ **STEP 1** Fold a sheet of paper into thirds lengthwise.

▶ **STEP 2** Unfold and draw vertical lines along the folds. Draw three horizontal lines to divide the paper into four rows.

▶ **STEP 3** Label the columns as follows: *Classification, Representative Elements,* and *Uses.* Label the rows *Metals, Nonmetals,* and *Metalloids.*

FOLDABLES Use this Foldable with Section 3.2. As you read the section, list the elements according to their class and identify some uses.

Chapter 3 • Introduction to the Periodic Table

Section 3.1

Objectives
- **Summarize** the steps in the historical development of the periodic table.
- **Predict** similarities in properties of the elements by using the periodic table.

Review Vocabulary
electron cloud: space around the nucleus of an atom where the atom's electrons are found

New Vocabulary
periodicity
periodic law

■ **Figure 3.1** The metals copper, silver, and gold are referred to as the coinage metals because these elements are used in making coins.
Infer Why were copper, silver, and gold grouped together?

Development of the Periodic Table

MAIN Idea The periodic table evolved over time as scientists discovered more useful ways to compare and organize the elements.

Real-Word Reading Link The seasons change from spring to summer to autumn to winter. Farmers know they can plant crops in spring and harvest them in summer or autumn. Similarly, early chemists looked for regularly repeating trends in the properties and behavior of the elements.

The Search for a Periodic Table

By 1860, scientists had already discovered 60 elements and determined their atomic masses. They noticed that some elements had similar properties. They gave each group of similar elements a name. Copper, silver, and gold shown in **Figure 3.1** were called the coinage metals; lithium, sodium, and potassium were known as the alkali metals; chlorine, bromine, and iodine were called the halogens. Chemists also saw differences among the groups of elements and between individual elements. They wanted to organize the elements into a system that would show similarities while acknowledging differences. It was logical to use atomic mass as the basis for these early attempts.

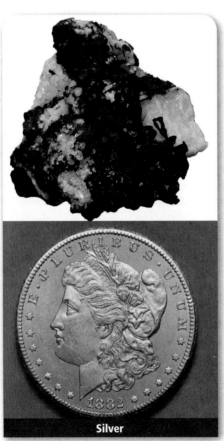

Copper

Silver

Gold

84 Chapter 3 • Introduction to the Periodic Table

Table 3.1 The Halogen Triad

Element	Atomic mass (u)	Density (g/mL)	Melting point (°C)	Boiling point (°C)
Chlorine	35.453	0.00321	−101	−34
Bromine	79.904	3.12	−7	59
Iodine	126.904	4.93	114	185

Döbereiner's triads In 1829, the German chemist J.W. Döbereiner classified some elements into groups of three, which he called triads. The elements in a triad had similar chemical properties, and their physical properties varied in an orderly way according to their atomic masses. **Table 3.1** shows the atomic mass, density, melting point, and boiling point for each of three elements in the halogen triad—chlorine, bromine, and iodine. **Figure 3.2** shows these three elements at room temperature.

Table 3.1 shows that the atomic mass of the three elements increases from 35.453 u to 79.904 u to 126.904 u from chlorine to bromine to iodine. More importantly, the atomic mass of bromine—the middle element—is 79.904 u, about the average of the atomic masses of chlorine and iodine:

$$\frac{35.453 \text{ u} + 126.904 \text{ u}}{2} = 81.179 \text{ u}$$

The fact that the atomic mass of the middle element lies about midway between the other two members of the triad is an important characteristic of triads. The table also shows that density, melting point, and boiling point all increase as atomic mass increases. The values for bromine are between those of chlorine and iodine. Iodine, with the highest atomic mass, has the highest density, melting point, and boiling point.

■ **Figure 3.2** As atomic mass increases, the state of the elements in this triad changes. At room temperature, chlorine (left) is a gas, bromine (center) is a liquid, and iodine (right) is a solid. Colors change from greenish yellow to reddish orange to violet.

Table 3.2	Metal Triad			
Element	Atomic mass (u)	Density (g/mL)	Melting point (°C)	Boiling point (°C)
Calcium	40.078	1.55	842	1500
Strontium	87.62	2.60	777	1412
Barium	137.327	3.62	727	1845

The triad in **Table 3.2** shows a relationship among the densities of three metals that is true for many triads. The density of strontium (2.60 g/mL) is near the average of the densities of calcium (1.55 g/mL) and barium (3.62 g/mL):

$$\frac{1.55 \text{ g/mL} + 3.62 \text{ g/mL}}{2} = 2.58 \text{ g/mL}$$

Density increases with increasing atomic mass, as it does in the chlorine, bromine, and iodine triad.

The average of the atomic masses of calcium and barium is about 88.703 u, which is close to 88.720 u, the actual atomic mass of strontium. Melting points for calcium, strontium, and barium also show a trend. However, the pattern of the boiling points in this triad is irregular. The irregular pattern of boiling points is typical of triads involving metals.

Döbereiner's triads were useful because they grouped elements with similar properties and revealed an orderly pattern in some of their physical and chemical properties. The concept of triads suggested that the properties of an element are related to its atomic mass.

Mendeleev's Periodic Table Russian chemist Dmitri Mendeleev was a professor of chemistry at the University of St. Petersburg when he developed a periodic table of the elements, shown in **Figure 3.3**. Mendeleev was studying the properties of the elements and realized that the chemical and the physical properties of the elements repeated in an orderly way when he organized the elements according to increasing atomic mass. For example, beryllium resembled magnesium, and boron resembled aluminum. Patterns of repeated properties began to appear. Mendeleev recognized these patterns as a way to identify new elements and predict their properties.

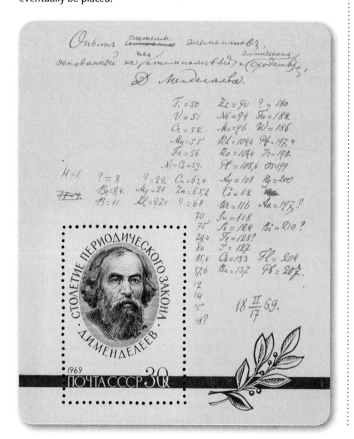

■ **Figure 3.3** Elements in horizontal rows of Mendeleev's first table displayed similar properties. Mendeleev wrote question marks in the table in places where unknown elements would eventually be placed.

Mendeleev's first table In 1869, Mendeleev published a table of the elements organized by increasing atomic mass. He listed the elements in a vertical column starting with the lightest. When he reached an element that had properties similar to another element already in the column, he began a second column. In this way, elements with similar properties were placed in horizontal rows. Notice the question mark at atomic mass 180 and its position next to Zr = 90. Mendeleev's unknown element at 180 proved to be hafnium, which was finally discovered in 1923. As illustrated by **Figure 3.4,** the chemical and physical properties of zirconium and hafnium are so similar that the two elements always occur together in nature and are difficult to separate.

Mendeleev's table was widely accepted because it was, to date, the clearest and most consistent arrangement of elements. Mendeleev left blank spaces in the table. Undiscovered elements would eventually occupy these spaces. By noting trends in the properties of known elements, Mendeleev was able to predict the properties of the yet-to-be-discovered elements scandium, gallium, and germanium.

■ **Figure 3.4** This zircon crystal contains zirconium (Zr = 90 on Mendeleev's table). It also contains hafnium, Mendeleev's unknown element (Hf = 180).

MiniLab 3.1

Predict Properties

What are the properties of unknown elements? When Mendeleev arranged the elements according to their atomic masses, some elements didn't fit. He resolved this problem by predicting the existence and properties of elements that were unknown at the time.

Procedure

1. Unknown element A is in group 14 below silicon and above tin. Unknown element B is in group 16 below sulfur and above tellurium.
2. The following information is given for the surrounding elements and in this sequence: symbol, density in g/mL, melting point in K, atomic radius in pm. Si, 2.4, 1680, 118; As, 5.72, 1087, 121; Sn, 7.3, 505, 141; Ga, 5.89, 303, 134; S, 2.03, 392, 103; Br, 3.1, 266, 119; Te, 6.24, 723, 138.
3. Refer to the locations of these elements on the portion of the periodic table above, and average the values of the four surrounding elements to predict the densities, melting points, and atomic radii of elements A and B.

Group 13	Group 14	Group 15	Group 16	Group 17
	Si		S	
Ga	Element A	As	Element B	Br
	Sn		Te	

Analysis

1. **Determine** What are the identities of elements A and B?
2. **Analyze** How do your predicted values for the three properties of element A compare with the actual values? Look up the actual values in a chemical handbook or obtain them from your teacher.
3. **Compare** How do your predicted values for the three properties of element B compare with the actual values?
4. **Explain** If your predicted values are fairly close to the actual values, how do you explain the approximate correlation?

Table 3.3 Mendeleev's Table of 1871

Group	I	II	III	IV	V	VI	VII	VIII
Formula of oxide	R_2O	RO	R_2O_3	RO_2	R_2O_5	RO_3	R_2O_7	RO_4
	H							
	Li	Be	B	C	N	O	F	
	Na	Mg	Al	Si	P	S	Cl	
	K	Ca	eka-	Ti	V	Cr	Mn	Fe, Co, Ni
	Cu	Zn	eka-	eka-	As	Se	Br	
	Rb	Sr	Yt	Zr	Nb	Mo	—	Ru, Rh, Pd
	Ag	Cd	In	Sn	Sb	Te	I	
	Cs	Ba	Di	Ce	—	—	—	
	—	—	—	—	—	—	—	
			Er	La	Ta	W	—	Os, Ir, Pt
	Au	Hg	Tl	Pb	Bi	—	—	
	—	—	—	Th	—	U	—	

Mendeleev's updated table Mendeleev later developed an improved version of his table with the elements arranged in horizontal rows. This arrangement, shown in **Table 3.3,** was the forerunner of today's periodic table. Patterns of changing properties repeated for the elements across horizontal rows. Elements in vertical columns showed similar properties.

An analogy can be made to the pattern of changes in the monthly calendar. Across the horizontal rows of a calendar are the days of the week—each changing from Sunday to Saturday, and then repeating the next week. The same days of the week fall in the vertical columns. In general, similar activities occur on the same day of the week througout the month. For instance, you might have soccer games every Saturday and music lessons on Thursday afternoons.

Periodicity Mendeleev's insight was a significant contribution to the development of chemistry. He showed that the properties of the elements repeat in an orderly way from row to row of the table. This repeated pattern is an example of periodicity in the properties of elements. **Periodicity** is the tendency to recur at regular intervals—like the appearance of Halley's comet every 76 years or the return of the full moon every 28 days.

One of the tests of a scientific theory is the ability to use it to make successful predictions. Mendeleev correctly predicted the properties of several undiscovered elements. In order to group elements with similar properties in the same columns, Mendeleev had to leave some blank spaces in his table. He suggested that these spaces represented undiscovered elements.

Mendeleev predicts elements **Table 3.3** shows Mendeleev's table in which there are two spaces for unknown elements to the right of zinc (Zn). He called these spaces eka-aluminum and eka-silicon, respectively. Based on their locations, Mendeleev predicted several of the properties of these undiscovered elements. Both elements were discovered during his lifetime. French chemists discovered eka-aluminum in 1875 and named it gallium (Ga), as shown in **Figure 3.5**. Eka-silicon was discovered in Germany in 1886 and was named germanium (Ge). The striking resemblance between Mendeleev's predicted properties and the actual properties of gallium and germanium, shown in **Table 3.4,** was one of the factors that led chemists to accept his theory of periodicity among the elements and his organization of the elements into a periodic table.

Mendeleev was so confident of the periodicity of the elements that he placed some elements in groups with others of similar properties even though arranging them strictly by atomic mass would have resulted in a different arrangement. An example is tellurium. The accepted atomic mass of tellurium (Te) was 128, so it should have been placed after iodine (I), which had an accepted atomic mass of 127. But the properties of tellurium logically placed it in the group with oxygen (O) and sulfur (S), ahead of iodine, and the properties of iodine matched those of chlorine (Cl) and bromine (Br). Mendeleev placed tellurium with oxygen and sulfur and assumed that the atomic mass of 128 was incorrect.

Moseley Mendeleev's placement of tellurium turned out to be correct even though his assumption about its atomic mass was not. The inconsistency between the positions of the elements and their atomic masses was solved in 1913 by English chemist Henry Moseley. Moseley rearranged the elements by increasing atomic number rather than by Mendeleev's order of increasing atomic mass. Moseley's insight resulted in the structure of the modern periodic table in **Figure 3.6** on pages 90–91.

■ **Figure 3.5** Eka-aluminum was discovered in 1875 and named gallium. Gallium's melting point is so low that the metal melts from the heat of the human hand.

SUPPLEMENTAL PRACTICE

For practice with the Periodic Table, see Supplemental Practice, page 809.

Table 3.4 Properties of Germanium

Property	Predicted (1869)	Actual (1886)
Atomic mass	72 u	72.61 u
Color	Dark gray	Gray-white
Density	5.5 g/mL	5.32 g/mL
Melting point	Very high	937°C
Formula of oxide	EsO_2*	GeO_2
Density of oxide	4.7 g/mL	4.70 g/mL
Oxide solubility in HCl	Slightly dissolved by HCl	Not dissolved by HCl
Formula of chloride	$EsCl_4$*	$GeCl_4$

* Es stands for eka-silicon

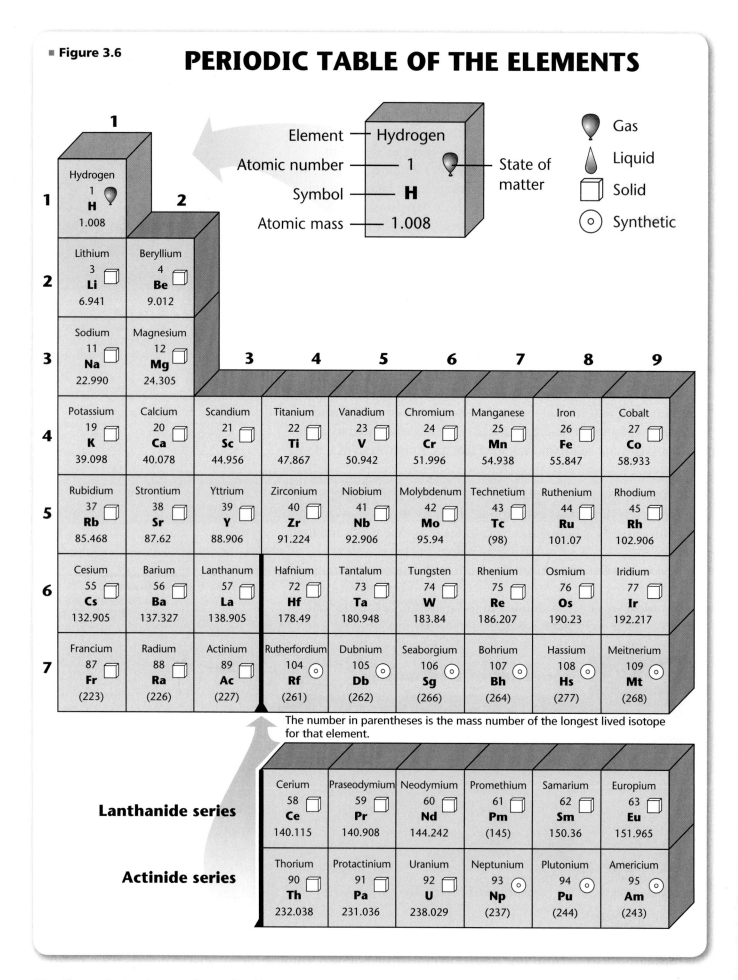

■ Figure 3.6

Legend: Metal, Metalloid, Nonmetal, Recently observed

13	14	15	16	17	18
					Helium 2 **He** 4.003
Boron 5 **B** 10.811	Carbon 6 **C** 12.011	Nitrogen 7 **N** 14.007	Oxygen 8 **O** 15.999	Fluorine 9 **F** 18.998	Neon 10 **Ne** 20.180
Aluminum 13 **Al** 26.982	Silicon 14 **Si** 28.086	Phosphorus 15 **P** 30.974	Sulfur 16 **S** 32.066	Chlorine 17 **Cl** 35.453	Argon 18 **Ar** 39.948

10	11	12	13	14	15	16	17	18
Nickel 28 **Ni** 58.693	Copper 29 **Cu** 63.546	Zinc 30 **Zn** 65.39	Gallium 31 **Ga** 69.723	Germanium 32 **Ge** 72.61	Arsenic 33 **As** 74.922	Selenium 34 **Se** 78.96	Bromine 35 **Br** 79.904	Krypton 36 **Kr** 83.80
Palladium 46 **Pd** 106.42	Silver 47 **Ag** 107.868	Cadmium 48 **Cd** 112.411	Indium 49 **In** 114.82	Tin 50 **Sn** 118.710	Antimony 51 **Sb** 121.757	Tellurium 52 **Te** 127.60	Iodine 53 **I** 126.904	Xenon 54 **Xe** 131.290
Platinum 78 **Pt** 195.08	Gold 79 **Au** 196.967	Mercury 80 **Hg** 200.59	Thallium 81 **Tl** 204.383	Lead 82 **Pb** 207.2	Bismuth 83 **Bi** 208.980	Polonium 84 **Po** 208.982	Astatine 85 **At** 209.987	Radon 86 **Rn** 222.018
Darmstadtium 110 **Ds** (281)	Roentgenium 111 **Rg** (272)	Ununbium *112 **Uub** (285)	Ununtrium *113 **Uut** (284)	Ununquadium *114 **Uuq** (289)	Ununpentium *115 **Uup** (288)	Ununhexium *116 **Uuh** (291)		Ununoctium *118 **Uuo** (294)

*The names and symbols for elements 112, 113, 114, 115, 116, and 118 are temporary. Final names will be selected when the elements' discoveries are verified.

Gadolinium 64 **Gd** 157.25	Terbium 65 **Tb** 158.925	Dysprosium 66 **Dy** 162.50	Holmium 67 **Ho** 164.930	Erbium 68 **Er** 167.259	Thulium 69 **Tm** 168.934	Ytterbium 70 **Yb** 173.04	Lutetium 71 **Lu** 174.967
Curium 96 **Cm** (247)	Berkelium 97 **Bk** (247)	Californium 98 **Cf** (251)	Einsteinium 99 **Es** (252)	Fermium 100 **Fm** (257)	Mendelevium 101 **Md** (258)	Nobelium 102 **No** (259)	Lawrencium 103 **Lr** (262)

Concepts In Motion
Interactive Figure To see an animation of the periodic table, visit glencoe.com.

The Modern Periodic Table

Döbereiner and Mendeleev both observed similarities and differences in the properties of elements and tried to relate them to atomic mass. Look again at the modern periodic table shown on pages 90 and 91 and notice that, as in Mendeleev's table, elements with similar chemical properties appear in the same group. For example, tellurium is in the same group as oxygen and sulfur, where Mendeleev placed it.

Mendeleev based his periodic table on 60 or so elements. At present, elements up to atomic number 118 have been discovered or synthesized. Many of the elements now known as the transition elements, lanthanides, and actinides were unknown in 1869 but today occupy the center of the table. The noble gases, such as the neon in the sign shown in **Figure 3.7**, were also unknown in Mendeleev's time, but now fill column 18 of the table.

Periodic law Due to the modern arrangement of increasing atomic number, there are several places in the modern table where an element of higher atomic mass comes before one of lower atomic mass. Atomic number increases by one as you move from element to element across a row. Each row (except the first) begins with a metal and ends with a noble gas. In between, the properties of the elements change in an orderly progression from left to right. The pattern in properties repeats after column 18. This regular cycle illustrates periodicity in the properties of the elements. The statement that the physical and chemical properties of the elements repeat in a regular pattern when they are arranged in order of increasing atomic number is known as the **periodic law**.

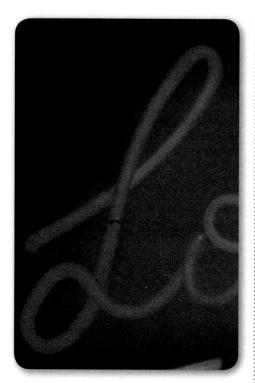

■ **Figure 3.7** Noble gases, like neon, are used to illuminate glass tubes to make neon signs.

Section 3.1 Assessment

Section Summary

- In his periodic table, Mendeleev organized the elements according to increasing atomic mass.

- Mendeleev placed elements with similar properties into groups. He recognized that patterns were useful as a way to identify unknown elements on the table.

- The modern periodic law states that the physical and chemical properties of the elements repeat in a regular pattern when they are arranged in order of increasing atomic number.

1. **MAIN Idea Contrast** How does the modern periodic law differ from Mendeleev's periodic law?

2. **Explain** What are two factors that contributed to the widespread acceptance of Mendeleev's periodic table?

3. **Compare** Which of the Döbereiner triads shown are still listed in the same column of the modern periodic table?

Triad 1	Triad 2	Triad 3
Li	Mn	S
Na	Cr	Se
K	Fe	Te

4. **Interpret Data** Use the periodic table to separate these 12 elements into six pairs of elements having similar properties.
Ca, K, Ga, P, Si, Rb, B, Sr, Sn, Cl, Bi, Br

Section 3.2

Objectives
- **Relate** an element's valence electron structure to its position in the periodic table.
- **Use** the periodic table to classify an element as a metal, nonmetal, or metalloid.
- **Compare** the properties of metals, nonmetals, and metalloids.

Review Vocabulary
periodicity: the tendency to recur at regular intervals

New Vocabulary
period
group
noble gas
metal
transition element
lanthanide
actinide
nonmetal
metalloid
semiconductor

Using the Periodic Table

MAIN Idea Elements are organized in the periodic table according to their electron configurations.

Real-World Reading Link Have you ever tried to look up information listed in tables in the sports section or the financial section of the newspaper? Or have you used a train or bus schedule to plan a trip? Some tables use many abbreviations and symbols. If you're not familiar with the codes, it's difficult to obtain any useful information, but once you figure out the symbols, it isn't difficult at all.

Relationship of the Periodic Table to Atomic Structure

Chemists invariably have a copy of the periodic table on the walls of their offices and laboratories, like that shown in **Figure 3.8.** It's a ready reference to a wealth of information about the elements, and it helps them think about their work, make predictions, and plan experiments based on those predictions. When you learn how to use the periodic table, you'll also find yourself referring to the periodic table to help you organize all the information you're learning about the elements.

In the modern periodic table, elements are arranged according to atomic number. Recall from Chapter 2 that the atomic number of an atom represents the number of electrons it has. The lineup starts with hydrogen, which has one electron. Helium comes next in the first horizontal row because helium has two electrons.

■ **Figure 3.8** Once you are familiar with the setup and symbols used in the periodic table, you will be able to obtain information about an element just by looking at its position in the table.

VOCABULARY

WORD ORIGIN

Periodic
comes from the Greek *periodus*, which means *period of time*

Periods and Groups

Notice on the periodic table that lithium, which has three electrons, starts a new horizontal row, called a **period**, in the table. Why does this happen? Why does the first period have only two elements? You read in Chapter 2 that electrons in atoms occupy discrete energy levels. The first energy level in an atom can only hold two electrons. The third electron in lithium must be at a higher energy level.

Lithium starts a new period at the far left in the table and becomes the first element in a group. A **group**, sometimes called a family, consists of the elements in a vertical column. Groups are numbered from left to right. Lithium is the first element in group 1 and in period 2.

Elements with atomic numbers 4 through 10 follow lithium and fill the second period. Each has one more electron than the element that preceded it. Neon, with atomic number 10, is at the end of the period. Eight electrons are added in period 2 from lithium to neon, so eight electrons must be the number that can occupy the second energy level. The next element, sodium with an atomic number of 11, begins period 3. Sodium's 11th electron is in the third energy level. The third period repeats the pattern of the second period.

Atomic structure of elements within a period The first period is complete with two elements, hydrogen and helium. Hydrogen has one electron in its outermost energy level, so it has one valence electron. Can you see that helium must have two valence electrons? Every period after the first starts with a group 1 element. These elements have one electron at a higher energy level than the noble gas of the preceding period. Therefore, group 1 elements have one valence electron. As you move from one element to the next across periods 2 and 3, the number of valence electrons increases by one. Group 18 elements have the maximum number of eight valence electrons in their outermost energy level. Group 18 elements are called the **noble gases**. The noble gases, with a full outer energy level, are generally unreactive.

The period number of an element is the same as the number of its outermost energy level. The valence electrons of an element in the second period, for example, are in the second energy level. A period 3 element such as aluminum (Al), shown in **Figure 3.9,** has its valence electrons in the third energy level.

■ **Figure 3.9** This sculpture is made of aluminum, the period 3 element in group 13.
Determine *How many electrons are in each energy level of aluminum?*

Literature Connection

The Language of a Chemist

... I have enjoyed looking at the world from unusual angles, inverting, so to speak, the instrumentation; examining matters of technique with the eye of a literary man, and literature with the eye of a technician.

In these words, Primo Levi, shown in **Figure 1,** describes the central paradox of his life. Born in Turin, Italy, in 1919, Primo Levi was educated and trained as a chemist. In 1944, he was arrested as a member of the Italian Anti-Fascist resistance movement. Levi confessed to being Jewish, and he was deported to a concentration camp at Auschwitz, Poland.

Primo Levi's knowledge of chemistry played a pivotal role in keeping his body as well as his spirit intact, as described in his memoirs *Survival in Auschwitz* and *The Reawakening*. With his knowledge of chemistry, he was selected to work in a factory connected to the concentration camp where synthetic rubber was manufactured. As a factory worker, Levi was able to avoid hard labor in the freezing temperatures outdoors. After the war, Levi continued to write until his death in Turin in April 1987.

Chemist and writer In an essay, "The Language of Chemists (I)," Levi reflects on the many ways in which chemists represent reality. He traces the history of benzene from an ancient resin to the discovery of its structural formula. As he does, Levi makes the reader aware of how chemists use both words and symbols to describe a material.

Elements of a life Each essay in his memoir *The Periodic Table* carries the name of an element. Some elements are related to autobiographical events; others inspire the author to reflect on human nature and the natural world through fictional stories. Throughout the book, the properties and descriptions of the elements—inert, volatile, poisonous, comparable, essential—often reflect the properties of life itself. The first chapter, "Argon," recounts how Levi's family is resistant to change, just like the inert argon gas.

Figure 1 Primo Levi was a chemist and a writer.

"Hydrogen" is a recollection of his days as a chemistry student. "Lead" is a fictional story about a family that makes its living mining lead. Ironically, they all die young from an illness caused by the poisonous metal they mined to make a living.

In "Nickel," Levi explains that he had to take jobs under false names so his employers would not know he was Jewish; similar to how nickel is often mistaken for other metals. In "Carbon," Levi finds an element that unites all living things. He follows an atom of carbon on its journey through leaves, milk, blood, and finally muscle, where the author ends his tale. In 2006, *The Periodic Table* was voted the best science book ever written by London's Royal Institution.

Connection to Chemistry

1. **Interpret** Explain this quote by Levi: *"Chemistry is the art of separating, weighing and distinguishing: these are useful exercises also for the person who sets out to describe events or give body to his own imagination."*

2. **Infer** Why do you think Levi chose the title *The Periodic Table*?

MiniLab 3.2
Trends in Reactivity

How can you compare the reactivities of elements? You've discovered that trends in the physical and chemical properties of elements occur both horizontally and vertically on the periodic table. In this MiniLab, you'll compare the reactivities of two elements in group 2 (magnesium and calcium) and three elements in group 17 (chlorine, bromine, and iodine).

Procedure

1. Read and complete the lab safety form.
2. To compare the reactivities of magnesium and calcium, use **forceps** to drop a **small piece of each element** into **two small beakers** containing **water** to a depth of about 1 cm. Observe how rapidly each element reacts with water to produce bubbles of hydrogen.
3. Pour 1 mL of **NaBr solution** into a **small test tube,** and add three drops of **chlorine water.** Stir the solution with the tip of a **microtip pipette.** Add 1 mL of **lighter fluid.** Draw the entire mixture into the microtip pipette. Squeeze the pipette bulb, expelling the mixture back into the test tube. Repeat this step several times so that the liquids are thoroughly mixed. **WARNING:** *Use care when handling chlorine water.*
4. Draw the liquids back into the pipette, invert it, and cover the tip with a cap made by cutting the bulb of another microtip pipette.
5. Set the inverted pipette in a small beaker, and allow time for the upper layer to separate from the lower layer. If no color is evident in the upper layer, expel the liquids back into the test tube, and add five more drops of chlorine water. Then repeat the remainder of steps 3 and 4 until a color is detected in the upper layer.
6. Using another test tube and microtip pipette, repeat steps 3, 4, and 5 with 1 mL of an **NaI solution,** chlorine water, and lighter fluid.

Analysis

1. **Determine** which of the two metallic elements, magnesium or calcium, is the more reactive? Are the more reactive metals located toward the top or the bottom of the periodic table?
2. **Interpret** your results from steps 3 and 4. If chlorine is more reactive than bromine, it will produce an orange color in the upper layer. If chlorine is more reactive than iodine, it will produce a violet color characteristic of iodine in the upper layer. Given this information, how do the reactivities of chlorine, bromine, and iodine compare? Are the more reactive nonmetallic elements toward the top or the bottom of the periodic table?

■ **Figure 3.10** The number of valence electrons is the same for all members of a group.

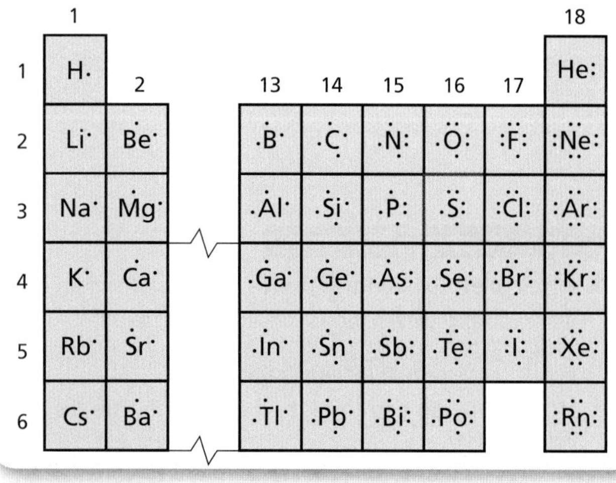

	1	2		13	14	15	16	17	18
1	H·								He:
2	Li·	Be·		·B·	·C·	·N:	·O:	:F:	:Ne:
3	Na·	Mg·		·Al·	·Si·	·P:	·S:	:Cl:	:Ar:
4	K·	Ca·		·Ga·	·Ge·	·As:	·Se:	:Br:	:Kr:
5	Rb·	Sr·		·In·	·Sn·	·Sb:	·Te:	:I:	:Xe:
6	Cs·	Ba·		·Tl·	·Pb·	·Bi:	·Po:		:Rn:

Atomic structure of elements within a group The number of valence electrons changes from one to eight as you move from left to right across a period; when you reach group 18, the pattern repeats. For the main group elements, the group number is related to the number of valence electrons. The main group elements are those in groups 1, 2, 13, 14, 15, 16, 17, and 18. For elements in groups 1 and 2, the group number equals the number of valence electrons. For elements in groups 13, 14, 15, 16, 17, and 18, the second digit in the group number is equal to the number of valence electrons. **Figure 3.10** illustrates the Lewis dot diagrams for the main group elements. Use **Figure 3.10** to verify the relationship between group number and valence electrons.

Valence electrons and chemical properties Because elements in the same group have the same number of valence electrons, they have similar properties. Sodium is in group 1 because it has one valence electron. Because the other elements in group 1 also have one valence electron, they have similar chemical properties.

Chlorine is in group 17 and has seven valence electrons. All the other elements in group 17 also have seven valence electrons and, as a result, they have similar chemical properties. Throughout the periodic table, elements in the same group have similar chemical properties because they have the same number of valence electrons.

Because the periodic table relates group and period numbers to valence electrons, it's useful in predicting atomic structure and, therefore, chemical properties. For example, oxygen, in group 16 and period 2, has six valence electrons (the same as the second digit in the group number), and these electrons are in the second energy level (because oxygen is in the second period). Oxygen has the same number of valence electrons as all the other elements in group 16 and, therefore, similar chemical properties. **Figure 3.11** shows representations of the distribution of electrons in energy levels in the first three elements of group 16.

Common names for some groups Four groups have commonly used names: the alkali metals in group 1, the alkaline earth metals in group 2, the halogens in group 17, and the noble gases in group 18. The word halogen is from the Greek words for "salt former" so named because the compounds that halogens form with metals are saltlike. The elements in group 18 are called noble gases because they are much less reactive than most of the other elements.

Personal Tutor For an online tutorial on valence electrons, visit glencoe.com.

■ **Figure 3.11** Oxygen, sulfur, and selenium have six electrons in their outermost energy levels as predicted by their group number. Note in the diagram of selenium that the third energy level can hold 18e⁻. You will learn more about energy levels in Chapter 7.

Develop *What might a representation of tellurium, another group 16 element, look like?*

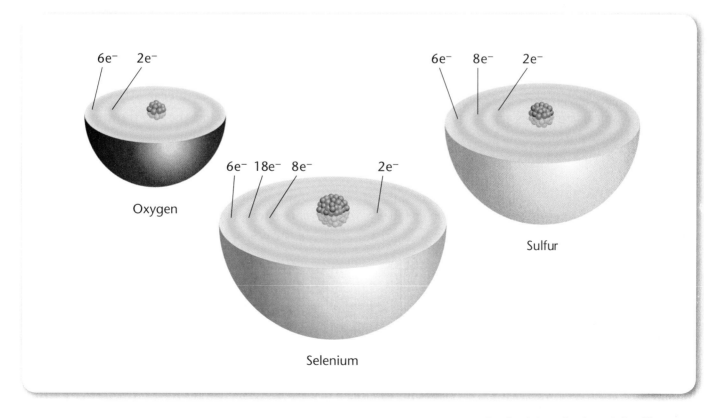

CHEMLAB

THE PERIODIC TABLE OF THE ELEMENTS

Background
In the mid-1800s, scientists found that when all of the known elements were arranged according to their properties and increasing atomic mass, the elements produced a predictable pattern. In this lab, you will investigate some representative elements in several of the vertical groups, or families, of the periodic table and classify the elements as metals, nonmetals, or metalloids. In general, metals are solids at room temperature. They have a metallic luster and are malleable. Metals conduct electricity, and many react with acids. On the other hand, nonmetals can be either solid, liquid, or gas at room temperature. If a nonmetal is a solid, it's likely to be brittle rather than malleable. Nonmetals do not conduct electricity and do not react with acids. Metalloids combine some of the characteristics of both metals and nonmetals.

Your data will allow you to classify some elements as metals, nonmetals, or metalloids and to determine general trends in metallic and nonmetallic characteristics within the periodic table.

Question
What is the pattern of metallic and nonmetallic properties of the elements in the periodic table?

Objectives
- **Observe** the properties of samples of the elements, including metals, nonmetals, and metalloids.
- **Classify** the elements as metals, nonmetals, or metalloids.
- **Analyze** your results to discover trends in the properties of the elements in the periodic table.

Preparation

Materials
stoppered test tubes containing small samples of carbon, nitrogen, oxygen, magnesium, aluminum, silicon, red phosphorus, sulfur, chlorine, calcium, selenium, tin, iodine, and lead
plastic dishes containing samples of carbon, magnesium, aluminum, silicon, sulfur, and tin
micro-conductivity apparatus
1*M* HCl
test tubes (6)
test-tube rack
10-mL graduated cylinder
spatula
small hammer
glass marking pencil

Safety Precautions
WARNING: *Be cautious when using 1M HCl. If any of the acid touches your skin or eyes, immediately rinse with water and notify your teacher. Never test chemicals by tasting.*

Procedure

1. Read and complete the lab safety form.
2. Prepare a table like the one shown for your data and observations.
3. Observe and record the appearance of each of the elements. Your description should include physical state, color, and any other observable characteristics such as luster. Don't open any of the test tubes.
4. Remove a small sample of each of the six elements in the dishes. Place the samples on a hard surface designated by your teacher. Gently tap each of the elements with a small hammer. A malleable element is one that flattens when tapped. It is brittle if it shatters when tapped. Record your observations in the data table.

5. Test the conductivity of each of the six elements in the dishes by touching the electrodes of the micro-conductivity apparatus to a piece of the element. If the bulb lights, you have evidence of conductivity. Record your observations in the data table. Wash the electrodes with water, and dry between testing each element.

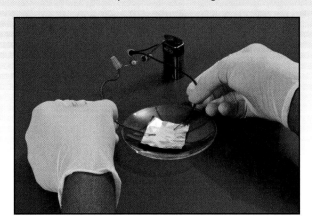

6. Use a graduated cylinder to measure 5 mL of water into each of the six test tubes.
7. Label each test tube with the symbol of one of the elements.
8. Using a spatula, put a small sample of each of the six elements (a 1-cm length of ribbon or 0.1–0.2 g of solid) into a test tube labeled with the symbol of the element.
9. Add approximately 5 mL of $1M$ HCl to each of the test tubes and observe the elements for at least 1 min. Evidence of reaction is the formation of bubbles of hydrogen on the element. Record your observations in the data table.

Analyze and Conclude

1. **Interpret data** Which elements displayed the general characteristics of metals?
2. **Interpret data** Which elements displayed the general characteristics of nonmetals?
3. **Interpret data** Which elements displayed a mixture of metallic and nonmetallic characteristics?

Apply and Assess

1. **Construct** an abbreviated periodic table with seven 1-inch squares across and five squares down. Label the squares across the top from left to right as groups 1 and 2 and 13–17. Label the squares down the side as periods 2–6. Write the appropriate atomic number and element symbol in each of the squares. Based upon your answers to the Analyze and Conclude questions, write your classification of metal, nonmetal, or metalloid for each of the elements you observed and/or tested.
2. **Conclude** Do the metallic characteristics of the elements across a period seem to increase from left to right or from right to left?
3. **Conclude** Do the metallic characteristics of the elements in a group seem to increase from top to bottom or from bottom to top?
4. **Illustrate** The metalloids indicate the approximate border between metals and nonmetals on the periodic table. Based upon your observations, draw a dark line along this border.

INQUIRY EXTENSION

Investigate Were there any element samples that did not fit into one of the three categories? Explain. What additional investigations could you conduct to learn even more about these elements' characteristics?

Data and Observations

Data Table				
Element	Appearance	Malleable or Brittle	Electrical Conductivity	Reaction with HCl

Physical States and Classes of the Elements

Other practical information can be obtained from the periodic table. The arrangement of the table helps you determine the physical state of an element; whether the element is synthetic or natural; and whether the element is a metal, a nonmetal, or a metalloid.

Physical States of the Elements The periodic table on pages 90 and 91 shows the states of the elements at room temperature and normal atmospheric pressure. Most of the elements are solid. Only two elements are liquids, and the gaseous elements, except for hydrogen, are located in the upper-right corner of the table.

Some elements are not found in nature but are produced artificially in particle accelerators like the one shown in **Figure 3.12.** These are known as synthetic elements. The synthetic elements, made by means of nuclear reactions, are marked on the periodic table. They include technetium, element 43, and all the elements after uranium, element 92. Although small amounts of neptunium and plutonium, elements 93 and 94, have been found in uranium ores, it is likely that they are the products of nuclear bombardment by radiation from uranium atoms.

Classifying Elements The color coding in the periodic table identifies which elements are metals (blue), nonmetals (yellow), and metalloids (green). The majority of the elements are metals. They occupy the entire left side and center of the periodic table. Nonmetals occupy the upper-right-hand corner. Metalloids are located along the boundary between metals and nonmetals. Each of these classes has characteristic chemical and physical properties, so by knowing whether an element is a metal, a nonmetal, or a metalloid, you can make predictions about its behavior.

> **FOLDABLES**
> Incorporate information from this section into your Foldable.

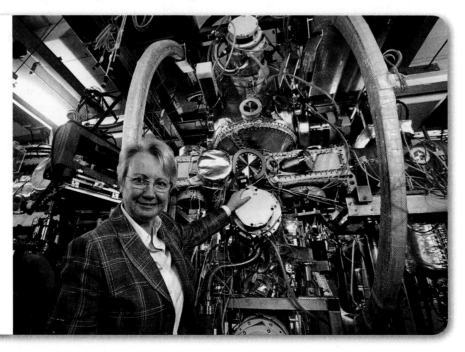

■ **Figure 3.12** The particle accelerator UNILAC in Darmstadt, Germany creates synthetic, heavy elements through a series of chemical reactions.

■ **Figure 3.13** Alloys of metals can be custom designed to meet a wide range of uses, such as prosthetic legs (left) and scaffolding for skyscrapers (right).

Discover *Give examples of other alloys and their usages.*

Metals Metals are almost everywhere. They make up many of the things you use every day—cars and bikes, jewelry, coins, electrical wires, household appliances, and computers. Because of their strength and durability, they're used in buildings and bridges, and for limb replacement as shown in **Figure 3.13. Metals** are elements that have luster, conduct heat and electricity, and usually bend without breaking. With the exception of tin, lead, and bismuth, metals have one, two, or three valence electrons. All metals except mercury are solids at room temperature; in fact, most have extremely high melting points.

The periodic table shows that most of the metals (coded blue) are not main group elements. A large number are located in groups 3–12. Notice the elements in the fourth period beginning with scandium (Sc), atomic number 21, and ending with zinc (Zn), atomic number 30. These ten elements mark the first appearance of elements in groups 3–12. From the fourth period to the bottom of the table, each period has elements in these groups.

Transition elements The elements in groups 3–12 of the periodic table are called the **transition elements**. All transition elements are metals. Many are common, including chromium (Cr), iron (Fe), nickel (Ni), copper (Cu), zinc (Zn), silver (Ag), and gold (Au). Some are less common metals but still important, such as titanium (Ti), manganese (Mn), and platinum (Pt). Some period 7 transition elements are synthetic and radioactive.

While the chemistry of the main group metals is highly predictable, that of the transition elements is less so. The unpredictability in the behavior and properties of the transition metals is due to the more complicated atomic structure of these elements.

FACT of the Matter

A fascinating feature of chromium chemistry is the many colorful compounds it produces, such as bright yellow potassium chromate and brilliant orange potassium dichromate. These and other colored compounds are responsible for the element's name. Chromium comes from the Greek word khroma, which means "color." Trace amounts of chromium in otherwise-colorless mineral crystals produce the brilliant colors of rubies and emeralds.

Section 3.2 • Using the Periodic Table

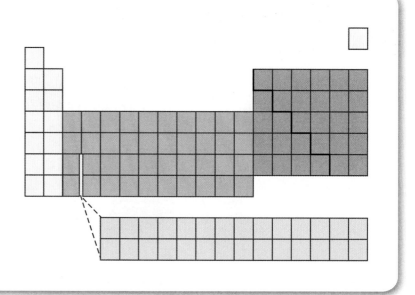

Figure 3.14 The elements separated from the rest of the table (green) have atomic numbers 58–71 and 90–103. If this section were placed in the table, the table would become too wide to be useful.

Inner transition elements In the periodic table, two series of elements, atomic numbers 58–71 and 90–103, are placed below the main body of the table. These elements are separated from the main table because placing them in their proper position would make the table very wide. **Figure 3.14** shows the position of these elements in green—separated out from the rest of the table. The elements in these two series are known as the inner transition elements. Many of these elements were unknown in Mendeleev's time, but he did know of some of them and suspected that more would be discovered.

The first series of inner transition elements is called the **lanthanides** because they follow element number 57, lanthanum. The lanthanides consist of the 14 elements from number 58 (cerium, Ce) to number 71 (lutetium, Lu). Because their natural abundance on Earth is less than 0.01 percent, the lanthanides are sometimes called the rare earth elements. All of the lanthanides have similar properties.

The second series of inner transition elements, the **actinides**, have atomic numbers ranging from 90 (thorium, Th) to 103 (lawrencium, Lr). All of the actinides are radioactive, and none beyond uranium (92) occur in nature. Like the transition elements, the chemistry of the lanthanides and actinides is unpredictable because of their complex atomic structures. What could be happening at the subatomic level to explain the properties of the inner transition elements? In Chapter 7, you'll study an expanded theory of the atom to answer this question.

Nonmetals Although the majority of the elements in the periodic table are metals, many nonmetals are abundant in nature. The nonmetals oxygen and nitrogen make up 99 percent of Earth's atmosphere. Carbon, another nonmetal, is found in more compounds than all the other elements combined. The many compounds of carbon, nitrogen, and oxygen are important in a wide variety of applications like the one shown in **Figure 3.15.**

Figure 3.15 A rain forest illustrates the interactions among carbon, nitrogen, and oxygen, which are important to the life cycle.

Table 3.5 Properties of Metals and Nonmetals

Metals	Nonmetals
Bright metallic luster	Non-lustrous, various colors
Solids are easily deformed	Solids may be hard or soft, usually brittle
Good conductors of heat and electricity	Poor conductors of heat and electricity
Loosely held valence electrons	Tightly held valence electrons

TRY AT HOME LAB
See page 869 for **Element Hunt**.

Most **nonmetals** don't conduct electricity, are much poorer conductors of heat than metals, and are brittle when solid. Many are gases at room temperature; those that are solids lack the luster of the metals. Their melting points tend to be lower than those of metals. With the exception of carbon, nonmetals have five, six, seven, or eight valence electrons. **Table 3.5** summarizes the properties of metals and nonmetals.

Metalloids Metalloids have some chemical and physical properties of metals and other properties of nonmetals. In the periodic table, the metalloids lie along the border between metals and nonmetals. Silicon (Si) is probably the most well-known metalloid. Some metalloids such as silicon, germanium (Ge), and arsenic (As) are semiconductors. A **semiconductor** is an element that does not conduct electricity as well as a metal, but does conduct slightly better than a nonmetal. The ability of a semiconductor to conduct an electric current can be increased by adding a small amount of certain other elements. **Figure 3.16** illustrates a silicon semiconductor, which made the computer revolution possible.

Atomic structure of metals, metalloids, and nonmetals The differences in the properties of the three classes of elements occur because of the different ways the electrons are arranged in the atoms. The number and arrangement of valence electrons and the tightness with which the valence electrons are held in an atom are important factors in determining the behavior of an element. In general, the valence electrons in a metal are loosely bound to the positive nucleus. They are free to move in the solid metal and are easily lost. This freedom of motion accounts for the ability of metals to conduct electricity. On the other hand, the valence electrons in atoms of nonmetals and metalloids are tightly held and are not easily lost. When undergoing chemical reactions, metals tend to lose valence electrons, whereas nonmetals tend to share electrons or gain electrons from other atoms.

VOCABULARY
WORD ORIGIN
Semiconductor
comes from the Latin *conductus*, which means *to escort or guide*

■ **Figure 3.16** Silicon is an ideal semiconductor used to make computer chips.

FIGURE 3.17

General Properties of Metals, Nonmetals, and Metalloids

Most properties of metals, nonmetals, and metalloids are determined by their valence electron configurations. The number of valence electrons that a metal has varies with its position in the periodic table. Valence electrons in metal atoms tend to be loosely held. Nonmetals have four or more tightly held electrons, and metalloids have three to seven valence electrons.

1. Familiar Metals

Polished silverware and copper jewelry are valued for their beautiful metallic luster. The ability of copper wire to conduct electricity makes it useful in electrical circuits. Some metals can be molded into objects, like this bronze figurine. Bronze is an alloy of two metals—copper and tin.

Bronze bull

Rock-crushing machine

Many transition elements are important as structural materials. Iron is made into steel by mixing it with carbon. Sometimes, other metals are added to produce special properties. Iron mixed with manganese produces a steel hard enough to be used to make the jaws on rock-crushing machines. The combination of iron with vanadium produces a tough alloy used, among other things, to make the crankshafts in automobile engines.

Exposure to air and moisture causes the iron in steel to rust. Plating a steel surface with chromium protects it from corrosion. The appearance of some consumer products is enhanced by plating with bright, shiny chromium.

Most metals are solids at room temperature; mercury is the only metal that is a liquid at room temperature. Mercury is poisonous and should never be handled.

Chromium

Mercury

104 Chapter 3 • Introduction to the Periodic Table

2. Some Lanthanides and Actinides

Compounds of europium and ytterbium are used in the picture tubes of color televisions. Neodymium is used in some high-power lasers.

High-power laser

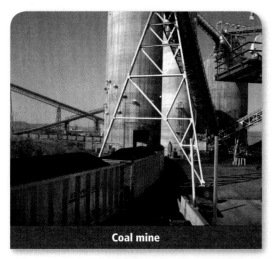
Coal mine

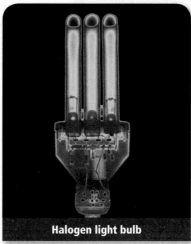
Halogen light bulb

3. Carbon and Some Other Nonmetals

Coal is nearly pure carbon. It is mined, often from strip mines, and burned as a fuel. Natural gas and oil are also carbon-rich fuels. Even though their appearances and physical properties are different, graphite and diamond are both naturally occurring forms of carbon. Bromine and iodine are used in high-intensity halogen lamps. Liquid nitrogen is used to maintain low temperatures; it can freeze the moisture in air, seen here as a white cloud.

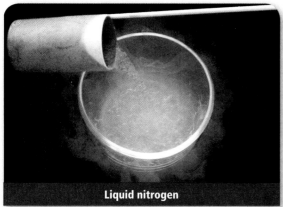
Liquid nitrogen

4. Metalloids

Silicon looks like a metal, but it's brittle and doesn't conduct heat and electricity well. Its melting point, 1410°C, is close to that of many metals. Elemental silicon (left) is melted, formed into a single crystal of pure silicon, and purified (back center). The crystal is sliced into thin wafers (right), and these are used to produce electronic devices (front center).

Silicon

Section 3.2 • Using the Periodic Table

CHEMISTRY AND TECHNOLOGY

Metals That Untwist

Those coated-wire ties for closing plastic bags are a great invention. The wire is easy to twist and, once twisted, stays that way. It would be very surprising if a twisted tie began to untwist spontaneously. However, some wires, such as those comprising the eye glasses in **Figure 1**, do spontaneously untwist.

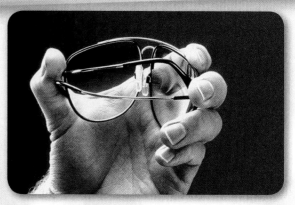

■ **Figure 1** Eyeglasses made with shape-memory metals

Shape-Memory Metals
Some alloys have a remarkable property. They revert to their previous shapes when heated or when the stress that caused their shapes is removed. These alloys are called shape-memory alloys. A piece of shape-memory alloy wire that is bent and then heated by a small electric current will revert to its original shape. Not only do shape-memory alloys help your eyeglasses fit longer, they exert gentle, continuous pressure on your teeth, reducing the number of trips to the orthodontist to have your braces tightened.

Different Solid Phases
Melting is the transition of a material from a solid to a liquid. Transitions from one phase to another can also take place within a solid. A solid can have two phases if it has two possible crystalline structures. It is the ability to undergo these changes in crystalline structure that gives shape-memory metals their properties. For example, an alloy having equal amounts of two metals may have one possible crystalline structure that is called the austenite phase. If the austenite phase is cooled under controlled conditions, the material takes on the martensite phase. As shown in **Figure 2,** the new crystal structure doesn't change the positions of the atoms throughout the material. However, the new internal organization gives the alloy new properties.

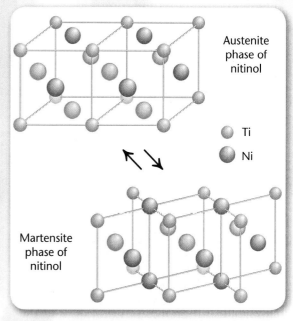

■ **Figure 2** Austenite and martensite phases of nitinol

Nitinol

Nitinol is an alloy of nickel and titanium that has the austenite phase structure. Both the nickel and titanium atoms are arranged in cubes. As you can see in **Figure 2,** each nickel atom is at the center of a cube of titanium atoms, and a titanium atom is at the center of each cube of nickel atoms. The diagram in **Figure 3** illustrates what happens if nitinol is shaped into a straight wire (a), heated (b), and cooled past its transition temperature (c)—it takes on the martensite phase. Notice that its external shape doesn't change; it's still straight (d). But its structure in (d) allows it to be bent by an external stress in (e). Now, if the wire is heated, the stress is released and it reverts to its initial shape in the martensite phase (f).

Robotic Arm

Robotic arms, like that shown in **Figure 4,** have been developed that contain shape-memory alloy wires that act as muscles and can move the fingers in a robotic hand. The hand movements are so precise that people who are both blind and deaf can feel and interpret the hand's movements as it signs to them in American Sign Language. To do this, an optical character scanner reads texts and converts the characters into input signals, which then drive the fingers to form appropriate symbols.

Clot Captor

The device called a clot captor is used to capture blood clots in the venae cavae (main veins) before they reach the lungs. Folded inside its sheath, the clot captor is only 3 mm in diameter. But once inserted into the vein, the wire unfolds to a diameter of 28 mm. The clot captor is made of nitinol wire that has a transition temperature just below body temperature. The wire is folded and placed in a sheath. As the sheath is inserted into the vein, the wire is bathed in a cold saline solution. Slowly, the sheath is withdrawn. As the wire warms to body temperature, it unfolds into its opened, umbrella-like shape.

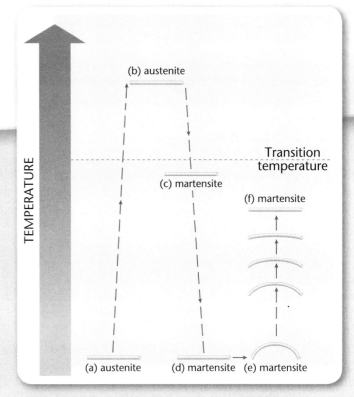

■ **Figure 3** Austenite and martensite phases are related to temperature

■ **Figure 4** Robotic Arm

Discuss the Technology

1. **Infer** Twisted nitinol-wire eyeglass frames unbend spontaneously at room temperature. Is the transition temperature of the frames above or below room temperature?
2. **Apply** Design a simple lever that could be raised and lowered smoothly using nitinol wires.

Everyday Chemistry

Metallic Money

The concept that a certain amount of a metallic substance could stand as a measure of goods and services dates back to the ancient Greeks. They were the first to use metal coins as measures of wealth.

Coinage metals Historically, copper, silver, and gold were logical choices for use in coins. These metals are not abundant in Earth's crust. Most metals occur combined in compounds, but copper, silver, and gold are native elements that commonly reside close to Earth's surface, making them relatively easy to mine. People revere these pure metals because of their beauty and rarity. Their properties allow them to be easily shaped, stamped, and marked for value.

Figure 1 World coins

Contemporary coinage metals Rarity makes metals more expensive. Gold and silver have become so expensive that they have vanished from most of the coins in the world shown in **Figure 1.** The United States eliminated gold from its coins in 1934. Silver was eliminated in 1971. A Kennedy half-dollar minted in 1972 has a coating of 75 percent copper and 25 percent nickel on a pure copper core. A Kennedy half-dollar minted in 1970 has a plating of 80 percent silver and 20 percent copper on a core of 21 percent silver and 79 percent copper.

Dimes and quarters minted after 1964 are made of 75 percent copper and 25 percent nickel coated on a platen of 100 percent copper. If you hold a dime or a quarter on edge, you can see the silvery-colored coating sandwiching the copper.

Figure 2 United States one-cent coins

The United States one-cent coin, shown in **Figure 2,** is often called a penny. However, the U.S. Mint's official name for the coin is a cent. Cents were made entirely of copper from 1793–1837. After 1837, the composition of the cent has varied mostly between bronze (95 percent copper and 5 percent tin and zinc), brass (95 percent copper and 5 zinc), and a composition of 87.5 percent copper and 12.5 percent nickel. From 1982 to present, the cent has been made of a 97.5 percent zinc platen and 2.5 percent copper plating. Due to the increasing price of copper and zinc, the U.S. Mint now makes the cent at a loss; it costs more to make the cent than the cent's actual value.

Explore Further

1. **Identify** one chemical and one physical property of copper, silver, and gold that made them the ideal coinage metals.
2. **Infer** What metals make up today's U.S. coinage metals?
3. **Acquire Information** What are some other uses of copper, silver, and gold?

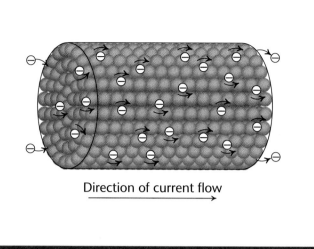

Mobile Electrons | **Fixed Electrons**

■ **Figure 3.18**
Left: In a conductor such as a copper wire, valence electrons are free to move to produce an electric current.
Right: Valence electrons in silicon are localized between neighboring silicon atoms. These electrons hold the atoms together in the crystal and are fixed. Therefore, no electrons are available to carry an electric current.

Semiconductors and Their Uses

Your television, computer, handheld electronic games, and calculator are electronic devices that depend on silicon semiconductors. All have miniature electrical circuits that use silicon's properties as a semiconductor. You learned that metals generally are good conductors of electricity, nonmetals are poor conductors, and semiconductors fall in between the two extremes, but how do semiconductors work?

Electrons and electricity An electric current is a flow of electrons. Most metals conduct an electric current because their valence electrons are not held tightly by the positive nucleus and are free to move. A copper wire is an example of a good conductor of electric current. **Figure 3.18a** illustrates the flow of electrons in copper.

At room temperature, pure silicon is not a good conductor of electricity. Silicon has four valence electrons, but they are held tightly between neighboring atoms in the crystal structure. You can see this clearly when you look at the structure of silicon in **Figure 3.18b**.

Electrical conduction by a semiconductor The electrical conductivity of a semiconductor such as silicon can be increased by a process known as doping. Doping is the addition of a small amount of another element to a crystal of a semiconductor. If a small amount of phosphorus, which has five valence electrons, is added to a crystal of silicon having only four valence electrons, each phosphorus atom provides an extra electron to the crystal structure. These extra electrons are free to move throughout the crystal to form an electric current. A phosphorus-doped silicon crystal is shown in **Figure 3.19**. A semiconductor such as phosphorus-doped silicon is called an *n*-type semiconductor because extra electrons (negatively charged) are present in the crystal structure.

■ **Figure 3.19** In phosphorus-doped silicon, the extra electrons from the phosphorus atoms are not needed to hold the crystal together. They are free to move and carry an electric current.

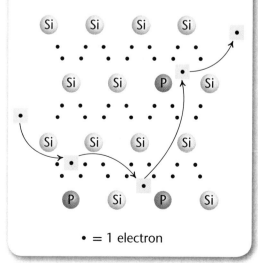

Section 3.2 • Using the Periodic Table

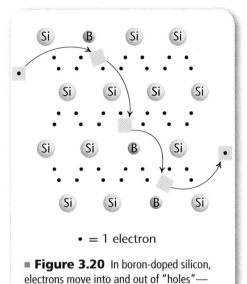

■ **Figure 3.20** In boron-doped silicon, electrons move into and out of "holes"—spaces that are lacking an electron. This movement is an electric current.

Silicon can also be doped with an element such as boron that has three valence electrons. Boron has fewer valence electrons than silicon, so when boron is added to a crystal of silicon, a shortage of electrons results. "Holes" are created in the silicon crystal structure. These are locations where there should be an electron but there is not because boron has one less electron than silicon. The movement of electrons into and out of these holes produces an electric current. Boron-doped silicon is an example of a *p*-type semiconductor because the holes act as if they are positive charges moving throughout the crystal. **Figure 3.20** illustrates boron-doped silicon.

Diodes Many semiconductor devices are made by combining *n*- and *p*-type semiconductors to form a diode. This combination of semiconductors permits electrical current to flow in only one direction, from the negative terminal to the positive terminal.

Transistors like those shown in **Figure 3.21** are the key components in electrical circuits in devices such as computers, calculators, hearing aids, and televisions. They are used to increase the strength of, or amplify, electrical signals. Transistors are exceedingly small, and their compact size and efficient operation have allowed the miniaturization of many electronic devices, such as laptop computers, heart pacemakers, and hearing aids. Transistors may be constructed by placing a *p*-type semiconductor between two *n*-type semiconductors, called an *npn*-junction, or by placing an *n*-type semiconductor between two *p*-type semiconductors, called a *pnp*-junction.

■ **Figure 3.21** Transistors (center) are used to increase the strength of electrical signals in such devices as television remote controls, telephones and cell phones.

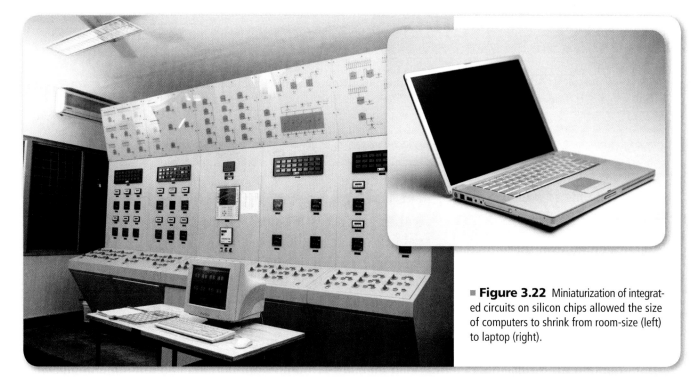

■ **Figure 3.22** Miniaturization of integrated circuits on silicon chips allowed the size of computers to shrink from room-size (left) to laptop (right).

Transistors, diodes, and other semiconductor devices are incorporated onto thin slices of silicon to form integrated circuits. Integrated circuits may contain hundreds of thousands of devices on a slice of silicon called a chip. The small size of a chip—only a few millimeters in width—has made possible the amazing growth of computer technology, as shown in **Figure 3.22.**

Connecting Ideas

You already know that elements combine chemically to form compounds. The ways that elements combine depend entirely on their valence electrons. With your knowledge of the periodic table, you will be able to explain why elements combine and predict what compounds they will form.

Section 3.2 Assessment

Section Summary

▶ Atomic structure and the number of valence electrons can be related to an element's position on the periodic table.

▶ Elements are classified as metals, nonmetals, or metalloids.

▶ The number of valence electrons and how tightly they are held determine the chemical properties of an element.

▶ The electrical conductivity of semiconductors can be increased by adding small amounts of other elements.

5. **MAIN Idea Explain** How does the arrangement of elements in periods relate to electron configuration?

6. **Illustrate** Draw a periodic table and indicate where metals are usually found on the periodic table. Where are nonmetals found? Metalloids?

7. **Contrast** What are the major differences in the physical properties of metals, nonmetals, and metalloids?

8. **Conclude** What can you conclude from the periodic table about the properties of the element barium?

9. **Infer** Germanium has the same type of structure and semiconducting properties as silicon. What type of semiconductor would you expect arsenic-doped germanium to be?

CHAPTER 3 Study Guide

BIG Idea Periodic trends in the properties of atoms allow us to predict physical and chemical properties.

Section 3.1 Development of the Periodic Table

MAIN Idea The periodic table evolved over time as scientists discovered more useful ways to compare and organize the elements.

Vocabulary
- periodicity (p. 88)
- periodic law (p. 92)

Key Concepts
- In his periodic table, Mendeleev organized the elements according to increasing atomic mass.
- Mendeleev placed elements with similar properties into groups. He recognized that patterns were useful as a way to identify unknown elements on the table.
- The modern periodic law states that the physical and chemical properties of the elements repeat in a regular pattern when they are arranged in order of increasing atomic number.

Section 3.2 Using the Periodic Table

MAIN Idea Elements are organized in the periodic table according to their electron configurations.

Vocabulary
- actinide (p. 102)
- group (p. 94)
- lanthanide (p. 102)
- metal (p. 101)
- metalloid (p. 103)
- noble gas (p. 94)
- nonmetal (p. 102)
- period (p. 94)
- semiconductor (p. 103)
- transition element (p. 101)

Key Concepts
- Atomic structure and the number of valence electrons can be related to an element's position on the periodic table.
- Elements are classified as metals, nonmetals, or metalloids.
- The number of valence electrons and how tightly they are held determine the chemical properties of an element.
- The conductivity of semiconductors can be increased by adding small amounts of other elements.

Chapter 3 Assessment

Understand Concepts

10. Describe element number 18 in terms of its period and group number, family name, and closest neighboring elements.

11. What is the group number of each of the following families of elements? Write the symbols for the elements in each family.
    a) alkalai metals   c) alkaline earth metals
    b) halogens   d) noble gases

12. How many valence electrons are in an atom of each of the following elements?
    a) Ne   d) Sr   g) Sn
    b) Br   e) Na   h) In
    c) S   f) As

13. Classify each of the elements in question 12 as a metal, nonmetal, or metalloid.

14. Draw the Lewis dot diagram for each of these elements. What is the group number of each element?
    a) Cl   e) Kr
    b) Mg   f) Cs
    c) C   g) O
    d) Bi   h) P

15. An incomplete set of atomic mass, density, and melting point data is given in **Table 3.6** for three elements in a triad. Following the patterns of Döbereiner's triads, predict a likely number for each missing value.

Table 3.6	Triad Data		
Element	Atomic mass (u)	Density (g/mL)	Melting point (K)
K	39.098		336
Rb		1.53	313
Cs	132.905	1.87	

16. The modern periodic law states that properties of the elements are arranged in a predictable pattern. What is this pattern?

Apply Concepts

17. Which of these elements have similar chemical properties?
    a) Be   d) F
    b) Sr   e) Ar
    c) Cs   f) I

18. Americium (Am) is an actinide that is used in smoke detectors. What property of actinides make this element useful in a smoke detector?

19. **Chemistry and Technology** How is the transformation from one crystalline phase to another different from melting or boiling?

20. **Everyday Chemistry** What properties of nickel and zinc make them good substitutes for copper and silver in coins?

21. **Literature Connection** What two fields did Primo Levi combine in his life's work?

Think Critically

Use a Table

22. **MiniLab 1** Mendeleev predicted the existence of eka-boron, which was unknown in his day. Eka-boron was located between calcium and titanium on the periodic table. What is the name of eka-boron now?

Relate Cause and Effect

23. **ChemLab** Explain how the metallic properties of the elements change as you move across a period from left to right. Why do the properties change in this manner?

Make Predictions

24. **MiniLab 2** Of the alkali metals lithium, potassium, and cesium, predict which is the most reactive and which is the least reactive.

25. The formulas of the chlorides of lithium, beryllium, boron, and carbon are LiCl, $BeCl_2$, BCl_3, and CCl_4 respectively. Use the periodic table to predict the formulas of the chloride of potassium, magnesium, aluminum, and silicon.

Chapter 3 Assessment

Table 3.7 — Atomic Mass and Density of Select Elements

Element	Helium	Neon	Argon	Krypton
Atomic mass (u)	4.00	20.2	39.9	83.8
Density (g/mL)	0.179	0.901	1.78	3.74

Interpret Data

26. MiniLab 2 Graph density versus atomic mass for the elements in **Table 3.7** above. Describe the relationship between atomic mass and density for the elements. Use your periodic table to locate the elements. Based on these data, what can you say about the trend in density as you move down a column of elements?

Cumulative Review

27. How do the elements differ from most of the matter you see around you? (*Chapter 1*)

28. Which element has the highest boiling point: mercury, nitrogen, or sodium? (*Chapter 1*)

29. A neutral atom of argon has 22 neutrons and 18 electrons. What are the mass number and atomic number of argon? (*Chapter 2*)

30. Argon's 18 electrons are arranged in energy levels. How many energy levels are needed to accommodate argon's electrons and how many electrons are in each energy level? (*Chapter 2*)

31. Describe the arrangement of electrons in an isotope of argon that has 21 neutrons and 18 electrons. (*Chapter 2*)

32. What would you expect to see in the emission spectrum of argon to give evidence of the existence of energy levels? (*Chapter 2*)

33. An atom with a mass number of 196 has 40 fewer protons than neutrons. Find the atomic number and the identity of this atom. (*Chapter 2*)

34. Why is it impossible to change copper (group 11, period 4) into gold (group 11, period 6) as the alchemists tried to do? (*Chapter 2*)

Skill Review

35. Look at the periodic table on pages 90 and 91. Find pairs of elements that would reverse order if the table were arranged according to increasing atomic mass as Mendeleev's periodic table was.

36. Write the symbol of the element that has valence electrons that fit each description.
 a) two electrons in the third energy level
 b) seven electrons in the fourth energy level
 c) four electrons in the sixth energy level
 d) eight electrons in the fifth energy level
 e) one electron in the first energy level
 f) six electrons in the second energy level

WRITING in Chemistry

37. Mendeleev made predictions about germanium and several other elements. Three of these were gallium (Ga), scandium (Sc), and polonium (Po). Write a news story describing the accuracy of his predictions.

Problem Solving

38. The density of aluminum is 2.7 g/mL and that of iron is 7.9 g/mL. Assume that manufacturers of soda cans could use the same volume of each metal to make soda cans. Compare the mass of an aluminum soda can with that of a comparable can made from iron. Explain your answer.

39. The chemical formula for zinc sulfide is ZnS. Use the periodic table to predict the formulas of the following similar compounds.
 a) zinc oxide   c) mercury sulfide
 b) cadmium oxide   d) zinc selenide

Cumulative Standardized Test Practice

Halogen Triad Table				
Element Symbol	Atomic Mass (u)	Density (g/mL)	Melting Point (°C)	Boiling Point (°C)
Cl	35.5	0.00321	−101	−34
Br	79.9	3.12	−7	59
I	127	4.93	114	185

Use the Halogen Triad Table above to answer Questions 1 and 2.

1. In which state does chlorine exist at room temperature?
   a) gas
   b) plasma
   c) liquid
   d) solid

2. As the atomic mass of a halogen increases, the point at which it turns from a liquid to a gas
   a) decreases.
   b) increases.
   c) fluctuates.
   d) remains constant.

3. Why was Mendeleev's periodic table a powerful tool for the science of chemistry in the nineteenth century?
   a) Mendeleev's periodic table organized element data into columns and rows.
   b) Mendeleev's periodic table allowed chemists to measure the densities of elements.
   c) Mendeleev's table allowed chemists to measure boiling and melting points of elements.
   d) Mendeleev's table allowed him to predict the properties of undiscovered elements.

4. A substance is said to be solid if
   a) it is hard and rigid.
   b) it can be compressed into a smaller volume.
   c) it takes the shape of its container.
   d) its matter particles are close together.

5. It can be predicted that element 118 would have properties similar to a(n)
   a) alkali earth metal.
   b) halogen.
   c) metalloid.
   d) noble gas.

6. Which of the following statements is true about the number of valence electrons and the Lewis dot diagram of elements in the same group?
   a) Elements in the same group have the same number of valence electrons but a different Lewis dot diagram.
   b) Elements in the same group have the same number of valence electrons and the same Lewis dot diagram.
   c) Elements in the same group have a different number of valence electrons but the same Lewis dot diagram.
   d) Elements in the same group have a different number of valence electrons and a different Lewis dot diagram.

7. A semiconductor is
   a) a substance that conducts electricity well.
   b) a substance that conducts electricity poorly.
   c) a substance that conducts electricity better than metals but not as well as nonmetals.
   d) a substance that conducts electricity better than nonmetals but not as well as metals.

8. What is the definition of an atomic mass unit?
   a) 1/12 of the mass of a carbon-12 atom
   b) a small unit used to measure subatomic particle masses
   c) the mass of one proton or one neutron
   d) the mass of one electron

NEED EXTRA HELP?								
If You Missed Question . . .	1	2	3	4	5	6	7	8
Review Section . . .	3.1	3.1	3.1	1.2	2.1	3.2	3.2	3.2

CHAPTER 4 Formation of Compounds

BIG Idea Most elements can form compounds.

4.1 The Variety of Compounds
MAIN Idea The properties of compounds differ from the properties of the elements that form the compounds.

4.2 How Elements Form Compounds
MAIN Idea Compounds form when electrons in atoms rearrange to achieve a stable configuration.

ChemFacts

- Chemical reactions beneath the surface of Mono Lake in California form tufa spires.
- Calcium ions from underground springs combine with carbonate ions in the lake water to produce the mineral in these rocks.
- Under water, the height of tufa towers can increase by as much as 2 cm per day.

Start-Up Activities

LAUNCH Lab

Observe Evidence of Change

When substances go through a chemical reaction to form new compounds, the new substances have different physical properties than the original reactants. What evidence shows a chemical reaction has occurred?

Materials
- 500-mL beaker
- 25-mL graduated cylinder
- water
- red food coloring
- laundry bleach
- stirring rod

Procedure
WARNING: *Perform this investigation in a well-ventilated area or under a hood. Do not breathe the fumes. Bleach can damage skin or clothing. Immediately notify your teacher of any spills.*

1. Read and complete the lab safety form.
2. Pour 300 mL of water into a 500-mL beaker.
3. Add three drops of red food coloring and stir the mixture until the water turns uniformly red.
4. Obtain 15 mL of bleach from your teacher and pour the bleach into the beaker.
5. Stir the mixture and record your observations.

Analysis
1. **Describe** any changes in the solution that you observed.
2. **Infer** Based on what you observed, did a new substance form? Explain.

Inquiry Does bleach cause a color change when you combine it with all compounds? How could you test for this?

Compound Formation Make the following Foldable to organize information about compound formation at the atomic level.

▶ **STEP 1** Fold a sheet of paper in half lengthwise.

▶ **STEP 2** Fold the top down about 2 cm.

▶ **STEP 3** Unfold and draw lines along all folds. Label the columns as follows: *Ionic Compounds* and *Covalent Compounds*.

FOLDABLES Use this Foldable with Section 4.2. As you read the section, record information about ionic and covalent compounds in the appropriate columns on your Foldable.

Visit **glencoe.com** to:
▶ study the entire chapter online
▶ explore **Concepts In Motion**
▶ take Self-Check Quizzes
▶ use Personal Tutors
▶ access Web Links for more information, projects, and activities
▶ find the Try at Home Lab, Mixing Ionic and Covalent Liquids

Chapter 4 • Formation of Compounds

Section 4.1

Objectives

- **Distinguish** the properties of compounds from those of the elements of which they are composed.
- **Compare and contrast** the properties of sodium chloride, water, and carbon dioxide.

Review Vocabulary

chemical property: a property that can be observed when there is a change in the composition of a substance

The Variety of Compounds

MAIN Idea The properties of compounds differ from the properties of the elements that form the compounds.

Real-World Reading Link Just as no two human fingerprints are the same, no two substances have exactly the same combination of chemical and physical properties. Like the detective who uses fingerprints to identify a suspect, chemists examine the properties of substances as clues about microscopic structure and compound formation.

Table Salt

What is the most popular food additive? In most kitchens, the answer is salt. It is used in cooking and at the table to enhance the flavor of food. Chemists refer to table salt as sodium chloride. The chemical name tells you what elements make up the compound; sodium chloride contains the elements sodium and chlorine.

Even though it's possible to make sodium chloride from its elements in the laboratory, salt is so abundant on Earth that it is used to manufacture the elements sodium and chlorine. Sodium chloride occurs naturally in large, solid, underground deposits throughout the world and is dissolved in the world's oceans. Salt can be obtained by mining these solid deposits and by the evaporation of seawater, as shown in **Figure 4.1**. In either case—mining or evaporation—the solid obtained contains substances other than sodium chloride. This raw solid is refined until it is almost entirely sodium chloride before it reaches your table. Thus, the terms table salt and sodium chloride are often interpreted to mean the same thing.

■ **Figure 4.1** Whether obtained from a mine or from the sea, sodium chloride always has the same chemical composition.

Underground salt mining accounts for about 90 percent of the world's salt production. These salt deposits formed when ancient seas evaporated millions of years ago.

Salt can also be harvested from the sea. As seawater evaporates, salt is left behind.

■ **Figure 4.2** Salt is the most common food seasoning. Also, salt lowers the melting point of ice to about 9.4°C (15°F). If the air temperature is below 9.4°C, the salt won't do much good when spread on the road.

Besides preserving food and enhancing food's flavor, sodium chloride plays crucial roles in living things—including the use of sodium ions to send signals through the nervous system. If you live in an area that gets snow and ice in the winter, salt is sometimes used to melt ice on roads. **Figure 4.2** illustrates salt's role in nutrition and road salt stored in a salt barn.

Physical properties of salt You already know some of the physical properties of table salt. It is a white solid at room temperature. If you look at table salt under a magnifier, you'll notice that the grains of table salt are little crystals shaped like cubes. These crystals are hard, but when you press down on them with the back of a spoon, the crystals shatter. This shattering shows that the crystals are brittle. If you heat sodium chloride to a temperature of about 800°C, it melts and forms liquid salt. Solid sodium chloride does not conduct electricity, but melted sodium chloride does. Salt also dissolves easily in water. The resulting solution is an excellent conductor of electricity, as shown in **Figure 4.3.**

■ **Figure 4.3** In order to light the bulb, electric current must flow between the two electrodes.

As you can see, no current flows through the dry salt crystals. Neither does pure water alone conduct electricity.

However, when the salt is dissolved in water, the solution conducts electricity and the bulb lights.

Section 4.1 • The Variety of Compounds **119**

MiniLab 4.1

Iron Versus Rust

How can you use attraction to a magnet to determine if rust is a different substance than iron? One property of iron that you are probably familiar with is that it is attracted to a magnet. This property can help you compare iron to the rust that forms on iron objects as they weather and wear. To observe this comparison, all you need is a magnet.

Procedure

1. Read and complete the lab safety form.
2. Obtain a small wad of **fresh steel wool** in one **small paper cup** and another small wad of **rusty steel wool** in another small paper cup.
3. Obtain a **3"× 5" index card** and a **magnet** wrapped in a **plastic bag.**
4. Test the fresh steel wool with the magnet. Record your observations.
5. Hold the rusty steel wool over the white card, and lightly rub the rusty steel wool between your thumb and forefinger. Some fine rust powder should fall onto the white card.
6. Next, hold the card up and slowly move the magnet under the card. Record your observations.

Analysis

1. **Describe** What effect did the magnet have on the fresh steel wool?
2. **Explain** What did you observe when the magnet was moved under the card with the rust powder?
3. **Determine** Was the material on the card a pure substance? How does your experimental evidence support your answer?
4. **Infer** What evidence do you have that the rust is a different substance from iron?

Chemical properties of salt Recall from Chapter 1 that a chemical property can best be observed when there is a change in a substance's composition. Salt does not react readily with other substances. Crystals of salt, shown in **Figure 4.4,** could sit in a salt shaker for hundreds or even thousands of years and still remain salt. It does not have to be handled in any special way or be stored in a special container. Compounds with these chemical properties are referred to as stable or unreactive. Even though it is easiest to observe a chemical property when substances change composition, salt's lack of reactivity is an important property that provides clues about its submicroscopic structure.

■ **Figure 4.4** Crystals of table salt are unreactive and do not need to be handled in a special way to keep them from reacting.

You can get more clues about salt's submicroscopic structure by answering the question: How do the properties of salt compare with the properties of sodium and chlorine? The properties of sodium and chlorine are easier to observe as both of these elements tend to react readily.

Properties of sodium Sodium is a shiny, silvery-white, soft, solid element, as you can see in **Figure 4.5a.** From its location on the left side of the periodic table, you know that it is a metallic element. Sodium melts to form a liquid when it is heated above 98°C. Sodium must be stored under oil because it reacts with oxygen and water vapor in the air. In fact, it is one of the most reactive of the common elements. When a piece of sodium is dropped into water, it reacts so violently that it catches fire and sometimes explodes. Because of its high reactivity, the free element sodium is never found in the environment. Instead, sodium is always found combined with other elements.

Properties of chlorine The element chlorine, shown in **Figure 4.5b,** is a pale green, poisonous gas with a choking odor. Because chlorine kills living cells and is slightly soluble in water, it is an excellent disinfectant for water supplies and swimming pools. You can tell that chlorine is a nonmetal by its position in the upper-right portion of the periodic table. Chlorine gas must be cooled to −34°C before it turns to a liquid. Like sodium, it is among the most reactive of the elements and must be handled with extreme care.

Chlorine is needed for many industrial processes, such as the manufacture of bleaches and plastics. Because of its industrial importance, large quantities of chlorine must be transported in railroad tank cars, tanker trucks, and river barges. If a train that is carrying chlorine derails, entire communities are evacuated until the danger of a chlorine leak passes.

Sodium and chlorine react The reaction between sodium and chlorine, shown in **Figure 4.5c,** proceeds vigorously once it has begun. Yet, when sodium and chlorine react to form sodium chloride, two dangerous elements combine to form a stable, safe substance that we consume every day. What could be happening in such a change? You will find the details in Section 4.2, but first look at two other common compounds whose properties are different than those of sodium chloride.

■ **Figure 4.5** The properties of sodium chloride are vastly different from those of elemental sodium and chlorine.

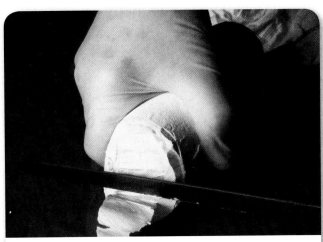

a. Sodium is a metal, but it is soft enough to be cut with a knife. Where it has been freshly cut, sodium has a silvery luster that is typical of many metals.

b. Chlorine is a greenish, poisonous gas at room temperature. If you've ever used liquid chlorine bleach or gone to a public swimming pool, you've probably smelled compounds containing chlorine.

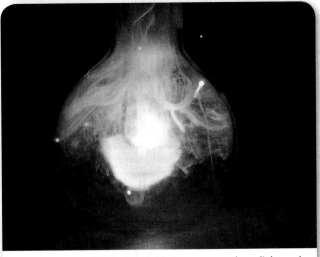

c. The reaction of sodium and chlorine generates heat, light, and produces a white, crystalline solid—sodium chloride.

Section 4.1 • The Variety of Compounds

Table 4.1	Composition of Inhaled and Exhaled Air	
Substance	Percent in inhaled air	Percent in exhaled air
Nitrogen	78	75
Oxygen	21	16
Argon	0.9	0.9
Carbon dioxide	0.03	4
Water vapor	Variable (0 to 4)	Greater than inhaled air

Carbon Dioxide

Carbon dioxide is a colorless gas. Take a deep breath and hold it for a few seconds. What you have inhaled is air, a colorless mixture of nitrogen and oxygen gases with small amounts of argon, water vapor, and carbon dioxide. Now, exhale. The mixture of gases that you exhale contains more than 100 times the amount of carbon dioxide that was in the air that you inhaled, as you can see in **Table 4.1**. In contrast, the quantity of oxygen is reduced by five percent. While the air was in your lungs, chemical and physical processes in your body exchanged some of the oxygen for carbon dioxide.

Carbon dioxide is an important chemical link between the plant and animal world. Green plants and other plantlike organisms take in carbon dioxide and give off oxygen during photosynthesis. Both plants and animals, including humans, use oxygen and give off carbon dioxide during cellular respiration.

Physical properties of carbon dioxide Carbon dioxide, like sodium chloride, is a compound, but its properties differ from those of sodium chloride. For example, salt is a solid at room temperature, but carbon dioxide is a colorless, odorless, and tasteless gas. When carbon dioxide is cooled below −80°C, the gas changes directly to white, solid carbon dioxide without first becoming a liquid. Because the solid form of carbon dioxide does not melt to a liquid, it is called dry ice, shown in **Figure 4.6**.

■ **Figure 4.6** Solid carbon dioxide is called dry ice. It is often used to ship perishable items, such as the meat shown below. The dry ice in the flask at right is immersed in water and is producing bubbles of carbon dioxide gas. The white vapors you see are condensed water vapor carried along with the cold carbon dioxide gas.
Infer *How does the density of carbon dioxide compare to the density of air?*

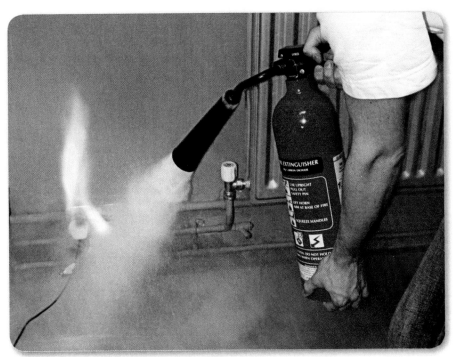

■ **Figure 4.7** A fire needs oxygen to burn. Carbon dioxide does not support burning. In fact, it is often used to put out fires. Fire extinguishers like this one are filled with compressed carbon dioxide.

Synthesize *How are both the chemical and physical properties of carbon dioxide helpful in fighting this fire?*

Carbon dioxide is soluble in water, as anyone who has ever opened a carbonated beverage knows. A water solution of carbon dioxide is a weak conductor of electricity. You can make carbon dioxide from its elements by burning carbon in air. Coal and charcoal are mostly carbon.

Chemical properties of carbon dioxide Because carbon dioxide is denser than air, it displaces air and deprives a fire of a supply of oxygen. Like sodium chloride, carbon dioxide is relatively stable. Carbon dioxide is used in some types of fire extinguishers because it does not support burning, as shown in **Figure 4.7.**

Photosynthesis is probably the most significant chemical reaction of carbon dioxide. In photosynthesis, plants use energy from the Sun to combine carbon dioxide and water chemically to make simple sugars. Plants use these sugars as raw materials to make many other kinds of compounds, from cellulose in wood and cotton to oils such as corn oil and olive oil. Photosynthesis is only one part of a natural cycle of chemical reactions known as the carbon cycle.

Properties of carbon As with sodium chloride, the properties of carbon dioxide differ from the properties of its elements. Carbon is a nonmetal and is fairly unreactive at room temperature. However, at higher temperatures, it reacts with many other elements. Charcoal is approximately 90 percent carbon. As anyone with a charcoal grill knows, carbon burns and is an excellent source of heat, as shown in **Figure 4.8.** Carbon forms a huge variety of compounds. In fact, the majority of compounds that make up living things contain carbon. Carbon compounds are so significant that an entire branch of chemistry, called organic chemistry, is dedicated to their study.

■ **Figure 4.8** When you burn charcoal to cook food, carbon in the charcoal combines with oxygen in the air to produce carbon dioxide. When elements combine to form a substance that is more stable, the reaction often gives off energy in the form of heat.

Section 4.1 • The Variety of Compounds

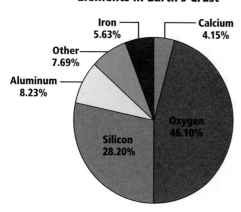

Oxygen makes up about 46 percent of Earth's crust. Nearly all of this oxygen occurs in compounds with other elements. Silicon is the next most abundant element in Earth's crust. You might not be surprised to learn that much of the oxygen is tied up in silicon dioxide, commonly known as sand.

The human body is composed of many different elements. Oxygen, carbon, hydrogen, and nitrogen are the most abundant elements in your body. The trace elements, which make up less than 2% of the body's mass, are critical for the body to live and grow.

■ **Figure 4.9** Oxygen is abundant, both in Earth's crust and in the human body.

Properties of oxygen Oxygen is a colorless, odorless, and tasteless gas that makes up about 21 percent of the air you breathe. When materials such as paper or wood burn in air, they react with oxygen, which is why people commonly say that oxygen supports burning.

Oxygen gas becomes a liquid when it is cooled to −183°C, and it is slightly soluble in water. In fact, the gills of fishes absorb dissolved oxygen from their water environments. Oxygen is more reactive than carbon and combines with many other elements. Rusting is a prime example of oxygen's reactivity. The element iron combines with oxygen from air. Many of the compounds that make up Earth's crust contain oxygen. Oxygen is the most abundant element in Earth's crust, as illustrated in **Figure 4.9.**

Water

Water is the third familiar compound about which you will investigate the submicroscopic structure of compounds. The formal chemical name of water is dihydrogen monoxide, but nobody calls it that. Water covers approximately 70 percent of Earth's surface and also makes up about 70 percent of the mass of the average human body. Oxygen makes up much more of the mass of a water molecule than hydrogen, which is why oxygen accounts for a large percentage of a person's mass, as illustrated in **Figure 4.9.**

Physical properties of water Water is excellent at dissolving other substances. It is often called the universal solvent in recognition of this valuable property. Water plays a vital role in the transport of dissolved materials, whether the aqueous solution is flowing down a river; or through your blood vessels.

■ **Figure 4.10** The iceberg is an example of water in its solid state—ice. In the ocean, water is in the liquid state. The atmosphere contains water vapor. Clouds form when water vapor condenses into small droplets of liquid water.

Water is the only one of the three compounds that occurs in Earth's environment in all three common states of matter, as shown in **Figure 4.10.** At sea level, liquid water boils into gaseous water (steam) at 100°C and freezes to solid water (ice) at 0°C. Because the density of ice is lower than the density of water, icebergs float on the ocean. Pure water does not conduct electricity in any of its states.

Chemical properties of water Water is a stable compound; it doesn't break down under normal conditions and does not react with many other substances. Perhaps the most interesting property of water is its ability to act as a medium in which chemical reactions occur. Nearly all of the chemical reactions in the human body and many important reactions on Earth occur in an aqueous solution. Recall from Chapter 1 that an aqueous solution is a homogeneous mixture in which water is the solvent. Without water, reactions between the solutes could not occur or would occur extremely slowly.

In addition, water and carbon dioxide are the starting materials for photosynthesis, the process that makes most life on Earth possible. Now, compare the properties of water with those of its component elements, hydrogen and oxygen.

Properties of hydrogen The properties of oxygen were described for you on page 124. Like oxygen, the element hydrogen is a colorless, odorless, tasteless gas. Hydrogen is the lightest and most abundant element in the universe and is usually classified as a nonmetal. Hydrogen is a reactive element. Because of its reactivity, it is seldom found as a free element on Earth. Instead, it is present in a variety of compounds, particularly water. Hydrogen reacts vigorously with many elements, including oxygen, as shown in **Figure 4.11.** This reaction forms water. The temperature of hydrogen gas must be lowered to a frigid −253°C before it turns to liquid. Hydrogen does not conduct electricity and is only slightly soluble in water.

■ **Figure 4.11** This welder is using a torch that burns hydrogen in oxygen. The heat produced is so intense that the torch can even be used underwater. The product of the reaction is water.

Section 4.1 • The Variety of Compounds

Everyday Chemistry

Elemental Good Health

Do you know that 60 elements are commonly found in the human body? Scientists think most of them play some role in life processes. The elements currently known to be essential are listed in the table below.

Elements Essential for Life	
Dietary Minerals	Non-Dietary Minerals
Ca, P, K, S, Cl, Na, Mg, Fe, F, Zn, Si, Cu, B, V, Se, Mn, I, Mo, Cr, Co	O, C, H, N

Hydrogen, carbon, oxygen, and nitrogen make up almost 96 percent of the mass of the human body. Dietary minerals, also essential for life processes, are derived from Earth's crust. Plants take up the minerals from soil, and you eat the plants.

Different roles Because calcium compounds make the hard parts of bones and teeth, calcium is needed for their growth and maintenance. Calcium is also needed for muscle contraction, regulating the heartbeat, and for the activation of enzymes. Along with calcium, phosphorus is critical for bone and teeth formation as well as for regulating the heartbeat. Phosphorus also aids in synthesizing proteins for the growth, maintenance, and repair of all tissues and cells. Fluorine helps in the formation and maintenance of teeth and may prevent osteoporosis, which is the disintegration of bone. Iron is an important element because it is the active part of the blood's hemoglobin molecule, which carries oxygen to the cells.

Aiding iron are copper and cobalt. Copper is essential to hemoglobin formation, while cobalt is essential to red blood cell formation. You might not know that magnesium is necessary for nerve and muscle function, as is potassium. Zinc and selenium are necessary for the cell division and growth and for the functioning of the immune system.

Different amounts Maintaining the proper level of each dietary mineral in your body is important for health. Nutritionists have established amounts of these elements that you should have in your daily diet. The amounts are described as the Dietary Reference Intake (DRI). For example, the DRI for calcium for teenagers is 1300 milligrams, while that of iodine is 150 micrograms (0.000150 g). You may think an amount of 150 micrograms can't be too important, but it is crucial to the function of your thyroid gland, which helps control your metabolism and growth. If you use iodized salt, which contains a little potassium iodide, you probably are getting the proper DRI of iodine. Likewise, eating a well-balanced diet of the five food groups and oils will maintain the proper levels of all the dietary minerals that are elemental to good health. The five food groups, some of which are featured in **Figure 1,** consist of grains, vegetables, fruits, oils, milk, meat and beans.

Figure 1 A variety of foods are needed for a well balanced diet

Explore Further

1. **Infer** Why might cooking foods in boiling water reduce their mineral content?
2. **Acquire Information** Why might consuming more than the Dietary Reference Intake of minerals be harmful?

■ **Figure 4.12** Here is another example of two elements reacting to form a compound whose properties are different from either element. Heated steel wool (iron) reacts with chlorine gas to form iron(III) chloride, the brown cloud you see in the flask.

Using Clues to Make a Case

Figure 4.12 shows another example of a chemical reaction—the reaction of iron with chlorine gas to form iron(III) chloride. The brown gas forms from the reaction of a green gas and a solid metal. Once again, atoms of elements reacted to form a substance that is quite different from the original atoms. If atoms of elements always combined in the same way, it's likely that all compounds would be similar. However, you've just studied compounds that have greatly differing properties. On the submicroscopic level, this indicates that atoms must combine in different ways to form different kinds of products. With the knowledge of the structure of atoms that you gained in Chapters 2 and 3, you can now examine the different ways that atoms combine.

SUPPLEMENTAL PRACTICE

For practice classifying elements and compounds, see Supplemental Practice, page 810.

Section 4.1 Assessment

Summary

- When compounds form, they have properties that differ greatly from the properties of the elements of which they are made.
- The properties of compounds depend on what happens to their constituent atoms when the compounds form.
- Macroscopic properties provide clues about what happens on the submicroscopic level.

1. **MAIN Idea** **Contrast** the properties of a compound with the elements from which it is composed. Use water as an example.

2. **Classify** the following substances as elements or compounds.
   a) table salt
   b) water
   c) sulfur
   d) chlorine gas
   e) carbon dioxide gas
   f) dry ice

3. **Examine** Give an example of a clue that indicates that atoms can combine chemically with each other in more than one way.

4. **Compare and Contrast** Using sodium chloride and carbon dioxide as examples, what can you say about the chemical reactivity of compounds compared to the elements of which they are composed?

5. **Reference** Use the periodic table to find the elements that make up the compounds discussed in this section. Which elements are metals? Which elements are nonmetals? Compare the properties of the three compounds in terms of whether their component elements are two nonmetals or a metal and a nonmetal.

Section 4.2

Objectives

- **Model** two types of compound formation, ionic and covalent, at the submicroscopic level.
- **Demonstrate** how and why atoms achieve chemical stability by bonding.
- **Compare,** using examples, the effect of covalent and ionic bonding on the physical properties of compounds.

Review Vocabulary

valence electron: an electron in the outermost energy level of an atom

New Vocabulary

octet rule
noble gas configuration
ion
ionic compound
ionic bond
crystal
covalent bond
covalent compound
molecule
electrolyte
interparticle force

How Elements Form Compounds

MAIN Idea Compounds form when electrons in atoms rearrange to achieve a stable configuration.

Real-World Reading Link As a child, you might have played with plastic building blocks that connected only in certain ways. If so, you probably noticed that the shape of the object you built depended on the limited ways the blocks interconnected. Building compounds out of atoms works in a similar way.

Atoms Collide

When substances react, particles of the substances must collide, as shown in **Figure 4.13**. It is what happens during that collision that determines what kind of product forms. How does the reaction of sodium and chlorine atoms to form salt differ from the reaction of hydrogen and oxygen atoms to form water?

When atoms collide with each other, what really comes into contact? Recall from Chapter 2 that the nucleus is tiny compared to the size of the atom's electron cloud. Also, the nucleus of an atom is buried deep in the center of the electron cloud. Therefore, it is highly unlikely that atomic nuclei would ever collide during a chemical reaction. In fact, reactions between atoms involve only their electron clouds.

When you studied the periodic table in Chapter 3, you learned that the properties of elements repeat because the pattern of the valence electrons repeats in each period. It is this arrangement of valence electrons of an atom that is primarily responsible for the atom's chemical properties. You will not be surprised, then, to learn that it is the valence electrons of colliding atoms that interact. But what kinds of interactions between valence electrons are possible? For some additional clues to what takes place when atoms combine to form compounds, look at a group of elements with unusual properties—the noble gases.

■ **Figure 4.13** Just as a stick must hit the *piñata* hard enough to break it open, particles in chemical reactions must collide with a sufficient amount of energy for a reaction to occur.

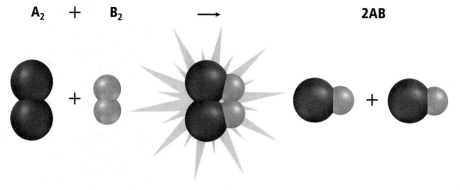

$A_2 + B_2 \longrightarrow 2AB$

Incandescent lightbulbs are filled with noble gases, usually argon and krypton, to protect the filament. The tungsten filament gets so hot that it will react with all but the most inert elements.

Neon produces a bright orange color, argon produces blue, and helium gives off yellow-white. Lighting designers obtain different colors by adding mercury or other substances to the gas mixture. Sometimes, colored glass is used.

Chemical Stability

The elements of group 18 are notorious for their almost complete lack of chemical reactivity. In fact, they have some practical uses just because they are not reactive, as you can see in **Figure 4.14.** These elements are used in eye-catching light displays often referred to as neon lights. Each gives off a different color of light when a high-voltage electric current passes through. Despite the fact that all of these elements occur naturally in the environment, none of these elements has ever been found naturally in the environment in a compound.

This group of unreactive elements used to be called the inert gases because chemists thought these elements could never react to form compounds. However, in the 1960s, chemists were able to react fluorine, under conditions of high temperature and pressure, with krypton and with xenon to form compounds. Since that time, a few additional compounds of xenon and krypton have been produced. Still, no one has been successful at synthesizing compounds of helium, neon, and argon. Now that chemists know that these elements aren't completely inert, they are called noble gases.

The octet rule Noble gases are unlike any other group of elements on the periodic table because of their extreme stability. The lack of reactivity of noble gases indicates that atoms of these elements must be stable. As you know, the elements of a vertical group on the periodic table have similar arrangements of valence electrons. Each noble gas has eight valence electrons, except for helium, which has two. **Figure 4.15** illustrates the valence electrons of the noble gases. Because electron arrangement determines chemical properties, the electron arrangements of the noble gases must be the cause of their lack of reactivity with other elements.

■ **Figure 4.14** Noble gases are used in both incandescent and neon lights.

■ **Figure 4.15** Each element in group 18 has a stable arrangement of valence electrons—two for helium and eight for the rest of the group.

Section 4.2 • How Elements Form Compounds

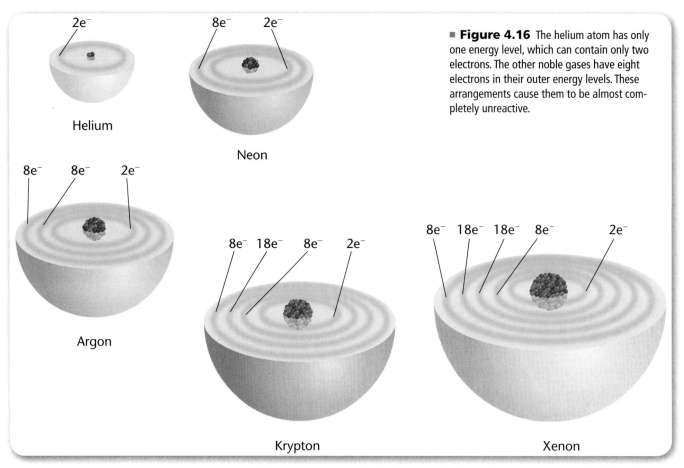

Figure 4.16 The helium atom has only one energy level, which can contain only two electrons. The other noble gases have eight electrons in their outer energy levels. These arrangements cause them to be almost completely unreactive.

Atoms combine because they become more stable by doing so. What does the electron arrangement of noble gases have to do with the way other elements react? Compare the electron arrangements of the noble gases in **Figure 4.16.** The modern model of how atoms react to form compounds is based on the fact that the stability of a noble gas results from the arrangement of its valence electrons. This model of chemical stability is called the octet rule. The **octet rule** says that an atom becomes stable by having eight electrons in its outer energy levels (or two electrons in the case of some of the smallest atoms). In other words, elements become stable by achieving the same configuration of valence electrons as one of the noble gases, a **noble gas configuration**.

Ways to Achieve Stability

If atoms collide with enough energy, their outer electrons may rearrange to achieve a stable octet of valence electrons—a noble gas configuration—and the atoms will form a compound. Remember that electrons are particles of matter, so the total number of electrons cannot change during chemical reactions. Next, think about how valence electrons might rearrange among colliding atoms so that each atom has a stable octet. There are only two possibilities to consider. The first is a transfer of valence electrons between atoms. The second possibility is a sharing of valence electrons between atoms. The reactions discussed in Section 4.1 provide good examples of both of these possibilities.

FACT of the Matter

Radon (Rn) is the last member of the noble gas group. It is a radioactive element that is formed by the radioactive decay of radium. Although it is found naturally on Earth, its presence is fleeting because it decays rapidly to other elements.

Electrons can be transferred You know that when sodium and chlorine are mixed, a reaction occurs and the compound sodium chloride forms. **Figure 4.5** showed the macroscopic view of this reaction. What can be happening at the submicroscopic level? Begin by picturing a collision between a sodium atom (Na) and a chlorine atom (Cl). Locate these atoms on the periodic table. Sodium is in group 1, so it has one valence electron. Chlorine is in group 17 and has seven valence electrons.

In Chapter 2, you have used Lewis dot diagrams to represent an atom and its valence electrons. You will now be able to use these models to show what happens when atoms combine. The Lewis dot diagrams of sodium and chlorine atoms are shown below.

$$\text{Na}\cdot \qquad \cdot\ddot{\underset{\cdot\cdot}{\text{Cl}}}:$$

How can the valence electrons of atoms rearrange to give each atom a stable configuration of valence electrons? If the one valence electron of sodium is transferred to the chlorine atom, chlorine becomes stable with an octet of electrons. Because the chlorine atom now has an extra electron, it has a negative charge. Also, because sodium lost an electron, it now has an unbalanced proton in the nucleus and therefore has a positive charge.

$$\text{Na}\cdot\,\curvearrowright\,\cdot\ddot{\underset{\cdot\cdot}{\text{Cl}}}: \rightarrow [\text{Na}]^+ + [:\ddot{\underset{\cdot\cdot}{\text{Cl}}}:]^-$$

It's easy to see chlorine has achieved a stable octet of electrons, but how does sodium become stable by losing an electron? Look at the position of sodium on the periodic table. By losing its lone valence electron, sodium will have the outer electron arrangement of neon. Sodium's stable octet consists of the eight electrons in the energy level below the level of the lost electron. **Figure 4.17** summarizes the reaction between sodium and chlorine.

> **FACT of the Matter**
> Take a deep breath. A noble gas, argon (Ar), composes approximately 1 percent of the breath of air you just took.

■ **Figure 4.17** To form ions, sodium loses an electron and chlorine gains an electron.
Explain how the transfer of the electron makes both atoms more stable.

Concepts In Motion
Interactive Figure To see an animation of sodium chloride ionic bond formation, visit glencoe.com.

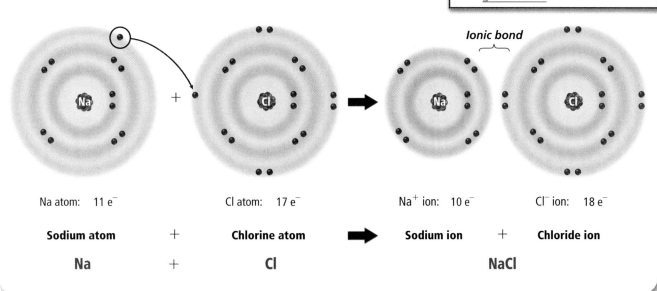

Na atom: 11 e⁻   Cl atom: 17 e⁻   Na⁺ ion: 10 e⁻   Cl⁻ ion: 18 e⁻

Sodium atom + **Chlorine atom** → **Sodium ion** + **Chloride ion**

Na + **Cl** → **NaCl**

Table 4.2 Reaction of Sodium and Chlorine

	Sodium atom	Chlorine atom	Sodium ion	Chloride ion
Formula	Na	Cl	Na$^+$	Cl$^-$
Number of protons	11	17	11	17
Number of electrons	11	17	10	18
Number of outermost electrons	1	7	8	8

FOLDABLES Incorporate information from this section into your Foldable.

Now that each atom has an octet of outermost electrons, they are no longer neutral atoms; they are charged particles called ions. An **ion** is an atom or group of combined atoms that has a charge because of the loss or gain of electrons. Ions always form when valence electrons rearrange by electron transfer between atoms. A compound that is composed of ions is called an **ionic compound**. Note that only the arrangement of electrons has changed. Nothing about the atom's nucleus has changed. This result is clear when you compare the atoms and ions in **Table 4.2.**

Ionic attraction Objects with opposite charges attract each other. Once they have formed, the positive sodium ion and negative chloride ion are strongly attracted to each other. The strong attractive force between ions of opposite charge is called an **ionic bond**. The force of the ionic bond holds ions together in an ionic compound.

Even the smallest visible grain of salt contains several quintillion sodium and chloride ions. Every positively charged sodium ion attracts all nearby negatively charged chloride ions and every negatively charged chloride ion attracts all nearby positively charged sodium ions. Therefore, these ions do not arrange themselves into isolated pairs. Instead, the ions organize themselves into a cube-shaped arrangement, as shown in **Figure 4.18.** This well-organized structure is a crystal. A **crystal** is a regular, repeating arrangement of atoms, ions, or molecules.

■ **Figure 4.18** The transfer of a single electron changes a reactive metal, sodium, and a poisonous gas, chlorine, into the stable and safe compound sodium chloride. The structure of a sodium chloride crystal is highly ordered. When viewed with a scanning electron microscope, the cubic shape of sodium chloride crystals is visible.

MiniLab 4.2

The Formation of Ionic Compounds

What other atoms give up and gain electrons to form ions? A sodium atom reacts by losing an electron to form a sodium 1+ ion. A chlorine atom gains one electron thus forming a chloride 1− ion. In this MiniLab, you will consider other combinations of atoms.

Procedure

1. Read and complete the lab safety form.
2. Cut three paper disks about 7 cm in diameter to represent the elements Li, S, Mg, O, Ca, N, Al, and I. Use a **different color of paper** for each element. Write the symbol of each element on the appropriate disks.
3. Select atoms of lithium and sulfur, and lay the circles side-by-side on a piece of **corrugated cardboard.**
4. Using **thumbtacks** of one color for lithium and another color for sulfur, place one tack for each valence electron on the disks, spacing the tacks evenly around the perimeters.
5. Transfer tacks from the metallic atoms to the nonmetallic atoms so that both elements achieve noble gas electron arrangements. Add more atoms if needed.
6. Once you have created a stable compound, write the ion symbols and charges and the formula and name for the resulting compound on the cardboard.
7. Repeat steps 3–6 for the remaining combinations of atoms.

Analysis

1. **Explain** Why did you have to use more than one atom in some cases? Why couldn't you just take more electrons from a metal atom or add extras to a nonmetal atom?
2. **Identify** the noble gases that have the same electron structures as each ion produced.

Results of ionic attraction How does ionic bonding affect the macroscopic properties of a substance? The cubic shape of a salt crystal is a result of the cubic arrangement of sodium ions and chloride ions. Given the strong attractive force between sodium and chloride ions and the degree of organization among them, it's not surprising that sodium chloride is a solid at room temperature.

Melting ionic compounds All particles of matter are in constant motion. Raising the temperature of matter causes its particles to move faster. In order for a solid to melt, its temperature must be raised until the motion of the particles overcomes the attractive forces and the crystal organization breaks down. Breaking the strong crystal structure of sodium chloride requires a lot of energy, which is the reason that sodium chloride must be heated to more than 800°C before it melts.

Shattering ionic crystals When you apply pressure to salt crystals, they don't bend or flatten out. When enough force is applied, the salt crystals suddenly shatter. This hardness and brittleness provide macroscopic evidence for the strength and rigidity of the submicroscopic structure of the salt crystal. Trying to break an ionic crystal is like trying to break a well-laid brick wall, like that shown in **Figure 4.19.** It takes a great deal of force to break the structure.

■ **Figure 4.19** The crystal structure of an ionic compound is similar to a well-laid brick wall. When enough force is applied, it shatters at the lines of the structure.

CHEMLAB

THE FORMATION AND DECOMPOSITION OF ZINC IODIDE

Background
Compounds are chemical combinations of elements. Many chemical reactions of elements to form compounds are spectacular but must be run under special laboratory conditions because they are dangerous. If elements react or spontaneously (without outside intervention or energy input once the reaction has begun) form compounds, that is a good indication that the compound is more stable than the free elements. To break a stable compound into its component elements, energy must be put into the compound.

Electricity is often used as the energy source to break down compounds. The process of decomposing a compound into its component elements by electricity is called electrolysis. **Figure 1.24** on page 39 illustrated a device used in the electrolysis of water, which decomposes water into hydrogen and oxygen gas. In this lab, you will use a similar, simplified device—shown in the photo on the opposite page—in the electrolysis of a zinc iodide solution.

Question
Can a compound be synthesized from its elements and then decomposed back into its original elements?

Objectives
- **Compare** a compound with its component elements.
- **Observe** and monitor a chemical reaction.
- **Observe** the decomposition of the compound back to its elements.

Preparation

Materials
10 × 150-mm test tube
test-tube rack
test-tube holder
100-mL beaker
spatula
plastic stirring rod
zinc
iodine crystals
distilled water
9-V battery with terminal clip and leads
two 20-cm insulated copper wires stripped at least 1 cm on each end

Safety Precautions

WARNING: *Iodine crystals are toxic and can stain the skin. Use care when using solid iodine. The reaction of zinc and iodine releases heat. Always use the test-tube holder to handle the reaction test tube.*

Procedure

1. Read and complete the lab safety form.
2. Obtain a test tube and a small beaker. Place the test tube upright in a test-tube rack.
3. Carefully add approximately 1 g of zinc dust and about 10 mL of distilled water to the test tube.
4. Carefully add about 1 g of iodine to the test tube. Record your observations in a table like the one shown.
5. Stir the contents of the test tube thoroughly with a plastic stirring rod until there is no more evidence of a reaction. Record any observations of physical or chemical changes.
6. Allow the reaction mixture to settle. Using a test-tube holder, carefully pick up the test tube and pour off the solution phase into the small beaker.
7. Add water to the beaker to bring the volume up to about 25 mL.
8. Obtain a 9-V battery with wire leads and two pieces of copper wire. Attach the copper wires to the wire leads from the battery. Make sure that the wires are not touching each other.

134 Chapter 4 • Formation of Compounds

9. Dip the wires into the solution, as shown in the photo above. Observe what takes place and record your observations.
10. After two minutes, remove the wires from the solution and examine the wires. Again, record your observations.

Analyze and Conclude

1. **Observe and Infer** What evidence was there that chemical reactions occurred?
2. **Compare and Contrast** How did you know if a reaction was complete?
3. **Infer** What term is used to describe a reaction in which heat is produced? How can you account for the heat produced in the reaction?
4. **Draw Conclusions** Why do you think the reaction between zinc and iodine stopped?
5. **Check Your Hypothesis** What evidence do you have that the compound was decomposed by electrolysis?

Apply and Assess

1. **Determine** What role did the water play in this reaction?
2. **Analyze** Do you think zinc iodide is an ionic or covalent compound? What evidence do you have to support your conclusion?
3. The formula of zinc iodide is ZnI_2. Use Lewis dot diagrams to show how it forms from its elements.

INQUIRY EXTENSION
How would this experiment have been different if a solution of sodium chloride had been used?

Data and Observations

Data Table	
Step	Observations
4. Addition of iodine to zinc	
5. Reaction of iodine and zinc	
9. Electrolysis of solution	
10. Examination of wires	

Table 4.3 — Formulas of Some Common Compounds

Formula	H_2O	$C_9H_8O_4$	$C_{12}H_{22}O_{11}$
Common Name	Water	Aspirin	Sucrose
Example			

Representing compounds with formulas Rather than writing out the name *sodium chloride* every time you refer to it, you can use a much simpler system. You can write its chemical formula, NaCl. The formula of a compound tells what elements make up the compound and how many atoms of each element are present in one unit of the compound. When sodium and chlorine atoms react, they form ions in a one-to-one ratio. As a second example, water is written as H_2O. This formula means that water consists of two hydrogen atoms combined with one oxygen atom.

The example of water brings up a second way that atoms can combine to achieve a stable outer level of electrons. **Table 4.3** shows three common compounds that form in this way.

Electrons can be shared In Section 4.1, you learned that the reaction of hydrogen and oxygen forms water. What happens when hydrogen and oxygen atoms collide? Hydrogen has only one valence electron. Oxygen, a group 16 element, has six valence electrons. Could these atoms achieve the stable electron configuration of a noble gas by transfer of electrons? If the oxygen atom could pick up two more valence electrons, it would have a stable octet—the noble gas configuration of neon.

What about hydrogen? Could a hydrogen lose its single valence electron? Your first inclination might be to treat hydrogen just like sodium, but be careful. If hydrogen loses an electron, it is left with no electrons, and that isn't the electron structure of a noble gas. Maybe hydrogen could gain an electron so that its electron arrangement is like that of helium. But, both atoms cannot gain electrons.

Colliding atoms transfer electrons only when one atom has a stronger attraction for valence electrons than has the other atom. In the case of sodium and chlorine, chlorine attracts sodium's valence electron strongly, whereas sodium holds its electron weakly. Therefore, the electron moves from sodium to chlorine, forming positive and negative ions in the process. You will learn more about this process and the factors that influence it in Chapter 9. In the case of hydrogen and oxygen, neither atom attracts electrons strongly enough to take electrons from the other atom. When atoms collide with enough energy to react, but neither attracts electrons strongly enough to take electrons from the other atom, the atoms combine by sharing valence electrons.

To understand how water forms, start by looking at the Lewis dot diagram of hydrogen (H) and oxygen (O).

$$H\cdot \qquad \cdot\ddot{\underset{..}{O}}:$$

Hydrogen requires one more electron to have the same electron arrangement as helium, while oxygen requires two more electrons to have neon's arrangement. Hydrogen and oxygen can share one electron from each atom. This sharing is shown by placing two dots representing electrons between the atoms.

$$H:\ddot{\underset{..}{O}}:$$

This arrangement makes hydrogen stable by giving it two valence electrons, but it leaves oxygen with only seven valence electrons. Oxygen's octet can be completed by sharing an electron with another hydrogen atom. (This explains why water has the formula H_2O.)

$$H:\ddot{\underset{..}{O}}: \quad + \quad H\cdot \quad \rightarrow \quad H:\ddot{\underset{..}{\underset{H}{O}}}:$$

Just as in the case of the formation of sodium chloride by ionic bonding, all the parts present before the reaction are still there after the reaction. What has changed in the combining of hydrogen and oxygen atoms? The valence electrons have rearranged and are now shared between the oxygen and the hydrogens. This is what always happens in a chemical reaction; electrons rearrange. **Figure 4.20** summarizes the reaction between hydrogen and oxygen.

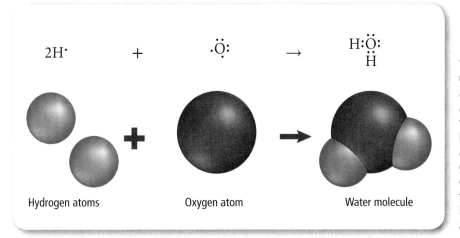

■ **Figure 4.20** The stability of the atoms in a water molecule results from a cooperative arrangement in which the eight valence electrons (six from oxygen and one each from two hydrogens) are distributed among the three atoms. By sharing an electron pair with the oxygen, each hydrogen claims two electrons in its outer level. The oxygen, by sharing two electrons with two hydrogens, claims a stable octet in its outer level. By this method, each atom achieves a stable noble gas configuration.

Iron(III) chloride is a typical ionic compound. It is crystalline at room temperature, melts at a high temperature (300°C), and dissolves in water. Iron(III) chloride, also called ferric chloride, can be used to treat sewage and drinking water and to etch copper.

Ethanol is a typical covalent compound. It is a liquid at room temperature but evaporates readily into the air. Ethanol boils at 78°C and freezes at −114°C. Unlike many covalent compounds, ethanol dissolves in water. In fact, ethanol rubbing alcohol is a mixture of ethanol and water.

■ **Figure 4.21** Ionic bonds and covalent bonds form in different ways. As a result, ionic compounds and covalent compounds often have different properties.

FOLDABLES
Incorporate information from this section into your Foldable.

Electron sharing produces molecules The attraction of two atoms for a shared pair of electrons is called a **covalent bond**. Notice that in a covalent bond, atoms share electrons and neither atom has an ionic charge. A compound whose atoms are held together by covalent bonds is a **covalent compound**.

Water is one example of a covalent compound. Although water is made up of hydrogen and oxygen atoms, these have combined into water molecules, each having two hydrogen atoms bonded to one oxygen atom. A **molecule** is an uncharged group of two or more atoms held together by covalent bonds. Sometimes chemists refer to covalent compounds as molecular compounds. The terms mean the same thing. **Figure 4.21** compares an ionic compound to a covalent compound.

More than two electrons can be shared When charcoal burns, carbon atoms collide with oxygen to form carbon dioxide (CO_2). Carbon (C) is in group 14 and has four valence electrons. Oxygen (O) is in group 16 and has six valence electrons. How can these three atoms—two atoms of oxygen and one atom of carbon—combine in such a way that all three have a stable configuration?

Carbon and oxygen are like hydrogen and oxygen when it comes to sharing or transferring electrons. Neither atom is able to attract electrons away from the other atom. In fact, two nonmetallic elements usually achieve stability by sharing electrons to form a covalent compound. On the other hand, if the reacting atoms are a metal and a nonmetal, they are much more likely to transfer electrons and form an ionic compound. You will learn more about the nature of chemical bonds and how to distinguish ionic and covalent bonds in Chapter 9.

Consider the reaction of carbon and oxygen to form carbon dioxide. Look at the Lewis dot diagrams for the participating atoms.

$$:\ddot{O}\cdot \quad \cdot\dot{C}\cdot \quad \cdot\ddot{O}:$$

You need to arrange the 16 valence electrons from these three atoms to produce a molecule in which all three atoms have a stable configuration. You know that at least one bond must exist between the carbon and each oxygen. This fact makes a good starting point. Have each oxygen share an electron with carbon as in the following dot structures.

$$:\ddot{O}:\dot{C}:\ddot{O}:$$

This arrangement gives carbon six electrons and each oxygen seven, but still no atom has an octet. What else can be done? There's no law in chemistry that says atoms must bond by sharing only one pair of electrons. What happens if they share two pairs? You now have double covalent bonds, as shown in this Lewis dot diagram.

$$:\ddot{O}::C::\ddot{O}:$$

Count all the electrons around each atom, including the shared electrons. Each atom has a stable octet. By sharing electrons, the three atoms achieve a more stable arrangement than they had as three separate atoms. The properties of carbon dioxide are a result of the unique properties of carbon dioxide molecules, not the properties of carbon or oxygen atoms.

In addition to double bonds such as those in carbon dioxide, atoms can form triple bonds. **Figure 4.22** illustrates the triple bond in nitrogen gas (N_2) along with the bonds in hydrogen gas and oxygen gas. The N_2 triple bond is extremely stable. Nitrogen gas makes up almost 80 percent of our air, but due to the stability of this triple bond, it requires high temperatures to react with other chemicals in our environment.

Chemistry Online

Personal Tutor For an online tutorial on multiple covalent bonds, visit glencoe.com.

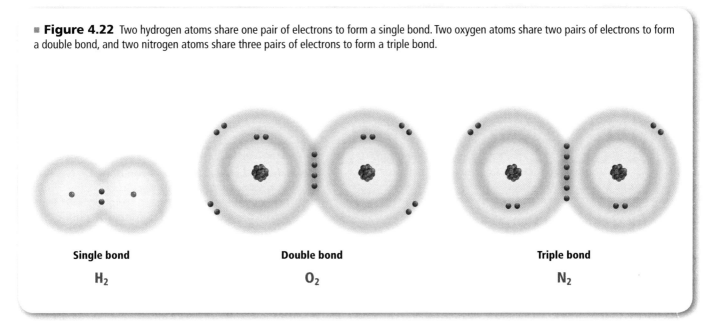

■ **Figure 4.22** Two hydrogen atoms share one pair of electrons to form a single bond. Two oxygen atoms share two pairs of electrons to form a double bond, and two nitrogen atoms share three pairs of electrons to form a triple bond.

Single bond
H_2

Double bond
O_2

Triple bond
N_2

Section 4.2 • How Elements Form Compounds

History Connection

Hydrogen's Ill-Fated Lifts

Imagine the surprise of the citizens of Paris on that morning of December 1, 1783. There were Jacques Charles and his assistant gliding above their rooftops in a basket suspended from a large balloon. Charles and his assistant filled the balloon with hydrogen and became the first humans to ride in such a lighter-than-air vehicle. By World War I, hydrogen-filled balloons were used to carry military personnel aloft to observe troop movements.

The Hindenburg In 1936, Germany launched the *Hindenburg*, a rigid airship originally designed to use helium for lift. Helium is slightly less buoyant than hydrogen. It is a noble gas and does not react with anything, whereas hydrogen burns explosively in air. However, the Germans had to continue to use hydrogen in the airship because they had no source of helium. In the following year, the Hindenburg carried more than 1300 passengers in trans-atlantic flights.

While attempting to dock in Lakehurst, New Jersey, on May 1, 1937, the airship burned rapidly when the hydrogen ignited. Thirty-six people were killed. The accident shown in **Figure 1** was the final chapter in the use of hydrogen in lighter-than-air vehicles.

Figure 1 *Hindenburg* accident

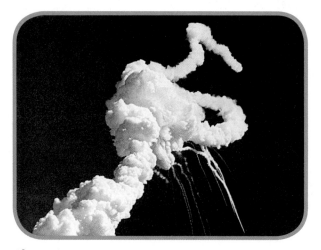

Figure 2 *Challenger* accident

Once ignited, this reaction occurs spontaneously. However, the reaction can be controlled, such as in the main engine of the space shuttle.

Space shuttle fuel The main engine of the space shuttle booster uses liquid hydrogen and oxygen as fuel. These materials are stored in separate sections of a huge external fuel tank attached beneath the shuttle. The energy released during the reaction thrusts the shuttle into orbit.

During the launch of the space shuttle *Challenger* on January 28, 1986, the reaction became uncontrolled. One minute and 13 seconds after takeoff, the external tank and shuttle exploded, killing the shuttle's seven crew members. The accident shown in **Figure 2** was caused by defects in the design of O-rings that joined sections of the solid-fuel booster engines, which are attached to the sides of the shuttle.

The chemical reaction The reaction that destroyed the *Hindenburg* was the burning of hydrogen gas.

$$2H_2(g) + O_2(g) \rightarrow 2H_2O(g)$$

Connection to Chemistry

1. **Infer** Much research has been done on developing hydrogen-fueled cars and trucks. Why would such a car be almost entirely nonpolluting? What factors do you think might affect the public's acceptance of such a vehicle?
2. **Hypothesize** What do you think is the reason that so much energy is produced when hydrogen reacts with oxygen? Recall what happens when these atoms bond.

How do ionic and covalent compounds compare?

Now you can relate the submicroscopic models of the formation of sodium chloride, water, and carbon dioxide to their macroscopic properties mentioned in Section 4.1. When elements combine, they form either ions or molecules. No other possibilities exist. The substances change dramatically, whether they change from sodium atoms to sodium ions or hydrogen and oxygen atoms to water molecules. This change explains why compounds have different properties from the elements that make them up.

Explaining the properties of ionic compounds The physical properties of ionic compounds are directly related to the way that the ions are held together. Ionic compounds are composed of well-organized, tightly bound ions. These ions form a strong, three-dimensional crystal structure. This model of the submicroscopic structure explains the general observation that ionic compounds are crystalline solids at room temperature. Just like sodium chloride, these solids are generally hard, rough, and brittle.

Ionic compounds usually have to be heated to high temperatures in order to melt them because the attractions between ions of opposite charge are strong. It takes a lot of energy to break the well-organized network of bound ions. Another property that characterizes most ionic compounds includes solubility in water. Two typical ionic compounds are shown in **Figure 4.23.** Compare their appearance with the properties described above.

■ **Figure 4.23** Here are two typical ionic compounds. Note that they are both solid at room temperature and soluble in water. Not all ionic solids are soluble in water, though.

Copper(II) sulfate ($CuSO_4$) is sometimes used to treat the growth of algae in swimming pools and in water-treatment plants.

Sodium hydrogen carbonate ($NaHCO_3$), commonly known as baking soda, is also called bicarbonate of soda. It is used to make baked foods rise.

■ **Figure 4.24** A sodium chloride solution conducts electricity well because it is an electrolyte. Sucrose, a covalent compound, does not conduct electricity in solution because it is not an electrolyte. Covalent compounds do not contribute ions to solution.

Concepts In Motion

Interactive Figure To see an animation of strong, weak, and non-electrolytes, visit glencoe.com.

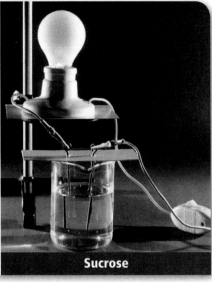

Sodium chloride | Sucrose

■ **Figure 4.25** Ionic compounds experience stronger interparticle forces than covalent compounds.

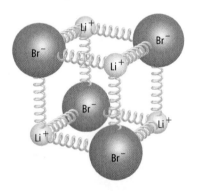

In an ionic compound such as lithium bromide (LiBr), interparticle forces are strong because of the attraction between ions of opposite charges.

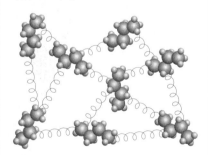

Butane (C_4H_{10}) molecules have no electrical charge, so the attraction between them is weak. At room temperature, butane is a gas unless held at pressure.

Another physical property of ionic compounds is the tendency to dissolve in water. When they dissolve in water, the solution conducts electricity. Ionic compounds also conduct electricity in the liquid (melted) state. Any compound that conducts electricity when dissolved in water or melted is an **electrolyte**. Therefore, ionic compounds are electrolytes.

In order for an ionic compound to conduct electricity, the ions must be free to move. Ionic compounds in the solid state do not conduct electricity because the ions are locked into position. Dissolving the compound in water frees the ions. Ionic compounds also become good conductors when they melt. This is evidence that melting frees the ions to move. **Figure 4.24** compares ionic and covalent solids dissolved in water. When sugar, a covalent compound, dissolves, it contributes no ions to the solution. Therefore, the solution does not conduct electricity. Sugar is not an electrolyte.

Explaining the properties of covalent compounds As with ionic compounds, the submicroscopic model of the formation of covalent compounds explains many of the properties of these compounds. In particular, you can use this model to explain why typical covalent compounds such as water and carbon dioxide have properties so different from ionic compounds.

In order to explain these differences, consider the submicroscopic organization of covalent compounds. Covalent compounds are composed of molecules. As you learned in this chapter, the atoms that compose molecules are held together by strong forces, called covalent bonds, that make the molecule a stable unit. The molecules themselves have no ionic charge, so the attractive forces between molecules are relatively weak.

The forces between particles that make up a substance are called **interparticle forces**. When these forces are between molecules, they are called intermolecular forces. These forces are illustrated in **Figure 4.25**. It is the great difference in the strength of interparticle forces in covalent compounds compared to those of ionic compounds that explains many of the differences in their physical properties.

142 Chapter 4 • Formation of Compounds

Whereas all ionic compounds are solids at room temperature, many covalent compounds are liquids or gases at room temperature. Note, however, that many covalent compounds—sugar, for example—will form crystals if there is enough attractive force between molecules. In Chapter 9, you will learn why some molecules attract each other. Many of the covalent compounds that are solids at room temperature will melt at low temperatures. Examples are sugar and the compounds that make up candle wax and fat. Compare the properties of some covalent substances in **Figure 4.26**. Covalent compounds do not conduct electricity in the pure state. Many covalent compounds, such as those in gasoline and vegetable oil, do not dissolve in water, although others such as sugar will dissolve. The solubility of covalent compounds in water varies, but in general, covalent compounds are usually less soluble in water than ionic compounds. Additionally, covalent compounds do not conduct electricity even when dissolved in water. What accounts for these differences?

TRY AT HOME 🏠 LAB

See page 869 for **Mixing Ionic and Covalent Liquids**

■ **Figure 4.26** Covalent compounds are composed of molecules in which atoms are bonded by electron sharing. Because of weak interparticle forces between molecules, covalent compounds tend to be gases or liquids at room temperature. They tend to be insoluble in water, although many are extremely soluble.

Table sugar ($C_{12}H_{22}O_{11}$) is called sucrose. It is an example of a covalent compound that is a crystalline solid soluble in water.

Candle wax is a mixture of covalent compounds. Because the molecules are larger and heavier, they form solids but melt at low temperatures.

Oil is a mixture of covalent compounds. Spilled oil does not dissolve in water, but instead floats on the water in thin layers.

In places where natural gas is not available, many people use propane (C_3H_8) to heat their homes and cook food. It is delivered to homes and businesses in pressurized tanks.

Chemistry & Society

The Rain Forest Pharmacy

Long ago, Samoan healers dispensed a tea brewed from the bark of a native rain forest tree, *Homalanthus nutans*, to help victims of a viral disease called yellow fever. Researchers have since identified the bark's active ingredient, prostratin. Prostratin is now being investigated for possible use as a drug for other viral diseases.

Using plants and products from plants for medicinal purposes has a long history. Such common drugs as aspirin, codeine, and quinine were originally derived from plants. However, only about 0.5 percent of all plant species have been studied intensively for chemical makeup and medicinal benefit. Because more than 50 percent of the world's estimated 250,000 to 422,000 species of flowering plants live in rain forests, researchers are now combing these areas of dense vegetation for new substances to fight diseases.

Screening plants One method of collecting plants for drug research, such as the plant shown in **Figure 1,** is to select many different species of plants from one locale and test them for possible medicinal uses. Because of the great diversity of plants, researchers hope to find an even greater diversity of chemical substances. Agencies such as the National Cancer Institute routinely use this screening method. Another method of screening plants for research is the phylogenetic survey, in which researchers study close relatives of plants already known to produce beneficial medicinal substances. Researchers hypothesize that similarities in the evolution of these plants may have produced similar properties in the substances in these plants.

Ethnobotanical approach Interest is growing in another method of screening plants for drug research. This method is based on the work of ethnobotanists. Ethnobotanists study the medicinal uses of plants by native peoples. The discovery of prostratin is an example of the ethnobotanical approach to screening plants for potential use in drug research.

■ **Figure 1** A Matses Indian shaman scrapes bark from a medicinal plant in the Amazon basin in Peru.

The future Isolating and testing substances from plants could take years. Within that time, rain forests might be significantly altered. Most rain forests are found in developing countries where the residents want to work and grow enough food to be self-sufficient. Because of deforestation to obtain lumber as well as space for agriculture and grazing, rain forest environments are disappearing. Some scientists are setting up preserves within the rain forests to maintain the forests' biodiversity. Others are attempting to grow disappearing plant species in rain forest nurseries. Scientists are also teaching the local residents that the rain forest has more economic value left intact than it does as cut timber and cleared land. If left intact, the rain forest can continuously regrow nuts, fruits, oil-producing plants, and medicinal plants.

ANALYZE the Issue

1. **Infer** Why might drug researchers investigate a plant that has few pests as a potentially useful plant?
2. **Acquire Information** Research present day uses of rain forests. How might research into the pharmaceutical value of rain forest substances have an impact on the present and future uses of rain forests?

Interparticle forces make the difference Many of the differences between the properties of ionic and covalent compounds result from their interparticle forces. Interparticle forces are the key to determining the state of matter of a substance at room temperature. You already know that ions are held rigidly in the solid state by the strong forces between them. Because there are much weaker interparticle forces among covalent molecules, they are held together less tightly. Thus, they are more likely to be gases or liquids at room temperature.

Because there are no ions in covalent compounds, you do not expect them to be electrical conductors. Ionic compounds tend to be soluble in water while molecular compounds do not. This difference is also explained by interparticle forces. Ions are attracted by water molecules, but many covalent molecules are not and, therefore, do not dissolve. Solubility in water and the nature of water solutions is a major topic in chemistry. You will learn more about solutions in Chapter 13.

Connecting Ideas

Now that you know that two major types of compounds exist, you may wonder where else you come into contact with these compounds on a day-to-day basis. In Chapter 5, you'll look at more examples of ionic and covalent compounds, including compounds more complex than the ones used as examples in this chapter. You'll also learn the important practical skill of naming and writing the formulas of compounds, as well as how to identify a few special categories of compounds such as acids, bases, and organic compounds.

SUPPLEMENTAL PRACTICE

For more practice classifying elements and compounds, see Supplemental Practice, page 810.

Section 4.2 Assessment

Summary

- Atoms become stable by reacting to achieve the valence electron structure of a noble gas.
- To achieve stability, some atoms transfer electrons to another atom, thus forming ions of opposite charge. The ions attract each other and form an ionic compound.
- Some atoms share electrons to form covalent compounds.
- Two atoms can share more than one pair of electrons.
- In ionic compounds, the interparticle forces are attractions between ions of opposite electric charge. They are much stronger than the interparticle forces between molecules of covalent compounds.

6. **MAIN Idea Diagram** two ways in which atoms become stable by combining with each other. Compare and contrast the compounds that result from two kinds of combinations.

7. **Explain** why sodium chloride is a neutral compound even though it is made up of ions that have positive and negative charges.

8. **Infer** Why do you think that sodium chloride has to be heated to 800°C before melting, but candle wax will start to melt at 50°C?

9. **Explain** why ionic compounds conduct an electric current in solution, but covalent compounds do not.

10. **Make Predictions** Look at the following diagram and explain how you can determine whether the compound being formed is ionic or covalent. Do you think it is more likely to be a solid or a gas at room temperature? Explain.

    $Ca: + \cdot \ddot{Br}: + \cdot \ddot{Br}: \longrightarrow [Ca]^{2+} + [:\ddot{Br}:]^- + [:\ddot{Br}:]^-$

11. **Apply** Potassium metal will react with sulfur to form an ionic compound. Use the periodic table to determine the number of valence electrons for each element. Draw a Lewis dot diagram to show how they would combine to form ions. How would you write the formula for the resulting compound?

CHAPTER 4 Study Guide

Download quizzes, key terms, and flash cards from glencoe.com.

BIG Idea Most elements can form compounds.

Section 4.1 The Variety of Compounds

MAIN Idea The properties of compounds differ from the properties of the elements that form the compounds.

Key Concepts
- When compounds form, they have properties that differ greatly from the properties of the elements of which they are made.
- The properties of compounds depend on what happens to their constituent atoms when the compounds form.
- Macroscopic properties provide clues about what happens on the submicroscopic level.

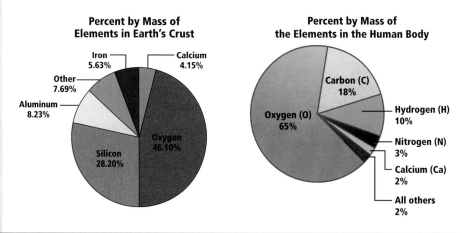

Section 4.2 How Elements Form Compounds

MAIN Idea Compounds form when electrons in atoms rearrange to achieve a stable configuration.

Vocabulary
- covalent bond (p. 138)
- covalent compound (p. 138)
- crystal (p. 132)
- electrolyte (p. 142)
- interparticle force (p. 142)
- ion (p. 132)
- ionic bond (p. 132)
- ionic compound (p. 132)
- molecule (p. 138)
- noble gas configuration (p. 130)
- octet rule (p. 130)

Key Concepts
- Atoms become stable by reacting to achieve the valence electron structure of a noble gas.
- To achieve stability, some atoms transfer electrons to another atom, thus forming ions of opposite charge. The ions attract each other and form an ionic compound.
- Some atoms share electrons to form covalent compounds.
- Two atoms can share more than one pair of electrons.
- In ionic compounds, the interparticle forces are attractions between ions of opposite electric charge. They are much stronger than the interparticle forces between molecules of covalent compounds.

Chapter 4 Assessment

Understand Concepts

12. If you tried to breathe pure carbon dioxide, you would suffocate. Why, then, is carbon dioxide essential to all life on Earth?

13. The compounds sodium chloride, water, and carbon dioxide were discussed in Section 4.1. For each of the following descriptions, tell which compound fits best.
    a) makes up 70 percent of Earth's surface
    b) formed from a metal and a nonmental
    c) a dense gas at room temperature
    d) can commonly be found in all three states of matter

14. Describe two processes by which elements can combine to form stable compounds. Name the type of bonding that results from each process.

15. Sodium reacts with fluorine to form sodium fluoride (NaF), an additive in toothpaste that prevents tooth decay. Use the format shown in **Table 4.2** to analyze this reaction.

16. What are three general properties of ionic compounds? How are these properties related to the structure of the compounds?

17. What are three general properties of covalent compounds? How are these properties related to the structure of the compounds?

18. Why would you never expect to find Na_2Cl as a stable compound?

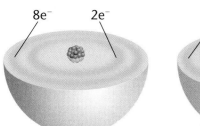 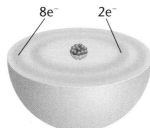

■ **Figure 4.27**

19. **Figure 4.27** shows a sodium atom and a neon atom. How is a sodium ion different from a sodium atom? From a neon atom? How are a sodium ion and a neon atom similar?

20. The term *isoelectronic* is used to describe atoms and ions that have the same number of electrons. Which of the following are isoelectronic? Na^+, Ca^{2+}, Ne, K, O^{2-}, P^{3-}

Apply Concepts

21. An unknown compound dissolves in water, but the solution does not conduct electricity. Is the compound is more likely to be an ionic or a covalent compound? Explain.

22. A bag of pretzels lists the sodium content of the pretzels. Considering that sodium reacts violently with water, why don't you explode when you eat these sodium-containing pretzels?

23. Hydrazine is a compound with the formula N_2H_4. What kind of compound is hydrazine? Describe the formation of hydrazine from nitrogen and hydrogen atoms.

24. What does it mean to say that a chemical reaction is a rearrangement of matter?

25. When sulfur reacts with metals, it often forms ionic compounds. Draw a Lewis dot diagram of a sulfur atom. Then, draw the Lewis dot diagram of the ion it will form. Name an element that has the same valence electron structure as a sulfur ion.

Everyday Chemistry

26. Carbon monoxide gas (CO) binds strongly to the iron atom of hemoglobin in blood. How does this action cause harm to the body?

History Connection

27. Do you think a rocket powered by hydrogen and oxygen causes a great deal of atmospheric pollution? Explain.

Chemistry and Society

28. List two economic factors that contribute to the loss of rain forests.

Chapter 4 Assessment

Think Critically
Interpret Chemical Structures

■ Figure 4.28

29. **MiniLab 2** Which of these compounds could the tack model in **Figure 4.28** below represent: magnesium chloride, potassium sulfide, calcium oxide, or aluminum bromide?

Observe and Infer

30. **MiniLab 1** Brass is an alloy of copper and zinc. Neither metal is magnetic. People who buy brass antiques at auctions, shows, or shops often carry a small magnet with them. What do you think they learn with this magnet?

Error Analysis

31. **ChemLab** In the electrolysis of zinc iodide, what limits your ability to recover all of the original zinc and iodine?

Interpret Data

32. **ChemLab** In the synthesis of zinc iodide, the zinc was added in excess. What does *in excess* mean? What two experimental observations did you make to confirm this?

Cumulative Review

33. When carbon in charcoal burns in air to form carbon dioxide, is the process endothermic or exothermic? How do you know? *(Chapter 1)*

34. How many energy levels are occupied by electrons in a calcium atom? A calcium ion? A bromine atom? A bromide ion? *(Chapter 2)*

35. What arrangement of valence electrons do carbon and other group 14 elements have in common? *(Chapter 3)*

Skill Review

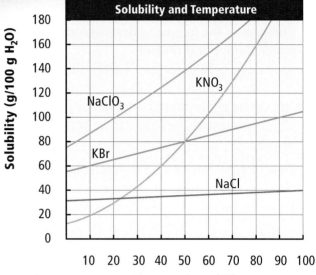

■ Figure 4.29

36. **Interpret a Graph** Look at the solubility graph in **Figure 4.29** and answer the following questions.
    a) Which of the compounds on the graph are ionic compounds? How do you know?
    b) How many grams of KBr will dissolve in 100 g of water at 60°C?
    c) People think that salt is very soluble in water. Is that thinking accurate? Defend your answer.

WRITING in Chemistry

37. Write a newspaper article about the discovery of the noble gases. Find out who discovered each of them, how and when they were discovered, and where they are found on Earth.

Problem Solving

38. Hydrogen and chlorine react to form hydrogen chloride (HCl). Hydrogen chloride is a gas at room temperature; it becomes a liquid if it is cooled to −85°C. On the basis of this evidence, do you think that hydrogen chloride is ionic or covalent? Explain.

Cumulative
Standardized Test Practice

1. What is NOT a physical property of table sugar?
   a) white, crystalline solid
   b) breaks down into carbon and water vapor when heated
   c) tastes sweet
   d) dissolves in water

2. Why does oxygen make up such a large percentage of Earth's crust?
   a) Large quantities of oxygen are trapped in soil and rock deep beneath Earth's surface.
   b) Large quantities of water, which is made of oxygen, are trapped beneath Earth's surface.
   c) Low temperatures far beneath Earth's surface freeze atmospheric oxygen into solid form.
   d) Oxygen is highly reactive and bonds with other substances forming solid compounds.

3. The octet rule states
   a) atoms become less stable with eight electrons in their outer energy level.
   b) atoms become more stable with eight electrons in their outer energy level.
   c) atoms will change their configurations and become a noble gas.
   d) atoms will become more reactive and chemically bond with noble gases.

4. A compound composed of electrically charged atoms is called a(n)
   a) octet compound.
   b) chemically stable compound.
   c) ionic compound.
   d) covalent compound.

Water Density	
State of Water	Density (g/mL)
Solid	0.917
Liquid	1.000
Gaseous (25°C)	0.008

5. According to the table above, how do the density differences of water in different states ensure the survival of aquatic life in Canada?
   a) In the winter, ice floats on lakes and ponds, insulating aquatic creatures from cold temperatures.
   b) In the winter, ice sinks to the bottom of lakes and ponds, causing organisms to hibernate.
   c) In the summer, water vapor sinks to the bottom of ponds and lakes, carrying needed oxygen.
   d) In the winter, water vapor sinks to the bottom of ponds and lakes, carrying needed oxygen.

6. Periodic law states that elements show a
   a) repetition of their physical properties when arranged by increasing atomic radius.
   b) repetition of their chemical properties when arranged by increasing atomic mass.
   c) periodic repetition of their properties when arranged by increasing atomic number.
   d) periodic repetition of their properties when arranged by increasing atomic mass.

7. An atom has no net electric charge because
   a) its subatomic particles carry no electrical charge.
   b) the positively charged protons cancel out the negatively charged neutrons.
   c) the positively charged neutrons cancel out the negatively charged electrons.
   d) the positively charged protons cancel out the negatively charged electrons.

NEED EXTRA HELP?							
If You Missed Question . . .	1	2	3	4	5	6	7
Review Section . . .	1.1	4.1	4.2	4.2	4.1	3.1	2.2

CHAPTER 5
Types of Compounds

BIG Idea There are two types of compounds: ionic and covalent.

5.1 Ionic Compounds
MAIN Idea Atoms in ionic compounds are held together by the attraction of oppositely charged ions.

5.2 Covalent Compounds
MAIN Idea Atoms in covalent compounds are held together by shared electrons.

ChemFacts

- Ninety-seven percent of the world's water resides in the ocean.
- The oceans teem with chemical compounds that contribute to the biological diversity and productivity of this precious resource.
- Coral reefs are composed of the compound calcium carbonate. Ions dissolved in seawater combine to form a variety of salt compounds including sodium chloride. Dissolved gases like carbon dioxide are also present in small concentrations.

Start-Up Activities

LAUNCH Lab

Elements, compounds, and mixtures

How are elements, compounds, and mixtures different?

Materials

Plastic freezer bag containing the following labeled items:
- copper wire
- chalk (calcium carbonate)
- small package of salt
- piece of granite
- sugar water in a vial
- small piece of wax

Procedure

1. Read and complete the lab safety form.
2. Construct a data table, and use it to record your observations.
3. Obtain a bag of objects. Identify and describe each object. Classify it as an element, compound, heterogeneous mixture, or homogeneous mixture.

Analysis

1. **Summarize** your observations. What are some differences among elements, compounds, and mixtures?
2. **Determine** if you know the name of a substance, how can you find out whether or not it is an element?

Inquiry How could you test a substance to determine its classification as an element, compound, or mixture?

Chemistry Online

Visit glencoe.com to:
- study the entire chapter online
- explore **Concepts In Motion**
- take Self-Check Quizzes
- use Personal Tutors
- access Web Links for more information, projects, and activities
- find the Try at Home Lab, Iron Ink

 **Types of Compounds** Make the following Foldable to compare ionic compounds and covalent compounds.

▶ **STEP 1** Fold a sheet of paper in half from top to bottom and then in half from side to side.

▶ **STEP 2** Unfold the paper once. Cut along the fold of the top half to make two tabs.

▶ **STEP 3** Label the tabs *Ionic Compounds* and *Covalent Compounds*.

FOLDABLES Use this Foldable with Sections 5.1 and 5.2. As you read these sections, record information about ionic compounds and covalent compounds under the appropriate tabs on your Foldable.

Chapter 5 • Types of Compounds

Section 5.1

Objectives

▶ **Apply** ionic charge to writing formulas for ionic compounds.
▶ **Apply** formulas to name ionic compounds.
▶ **Interpret** the information in a chemical formula.

Review Vocabulary

ion: an atom or group of combined atoms that has a charge because of the loss or gain of electrons

New Vocabulary

binary compound
formula unit
oxidation number
polyatomic ion
hydrate
hygroscopic
deliquescent
anhydrous

FOLDABLES Incorporate information from this section into your Foldable.

Ionic Compounds

MAIN Idea Atoms in ionic compounds are held together by the attraction of oppositely charged ions.

Real-World Reading Link Seawater is an aqueous solution that contains many dissolved substances, mainly dissolved sodium chloride. Another ionic compound found dissolved in seawater is magnesium chloride. Some common ionic compounds used in everyday life include potassium chloride, a salt substitute used by people avoiding sodium for health reasons; potassium iodide, added to table salt to prevent iodine deficiency; and sodium fluoride, added to many toothpastes to strengthen tooth enamel.

Binary Ionic Compounds

Recall from Chapter 4 that the submicroscopic structure of ionic compounds helps explain why they share certain macroscopic properties such as high melting points, brittleness, and the ability to conduct electricity when molten or when dissolved in water. What is it about the structure of these compounds that contributes to these properties? The answer resides in the ions that they are made from.

You have learned that ionic compounds are composed of oppositely charged ions held together strongly in well-organized units. Because of this structure, they usually are hard solids at room temperature and are difficult to melt. When they melt or dissolve in water, their three-dimensional structure breaks apart, and the ions are released from the structure. These charged ions are now free to move and conduct an electric current.

Look at the structure of magnesium oxide in **Figure 5.1**. The structure of magnesium oxide is a repeating pattern of magnesium ions (Mg^{2+}) and oxide ions (O^{2-}). Each Mg^{2+} ion is surrounded by six O^{2-} ions, which in turn, are each surrounded by six Mg^{2+} ions.

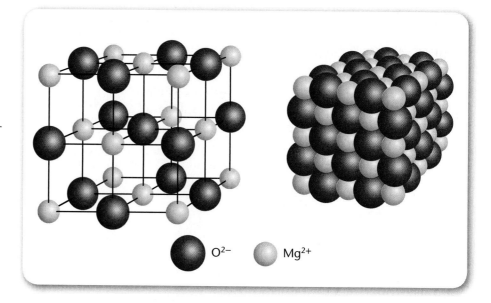

■ **Figure 5.1** The structure of magnesium oxide (MgO) is neutral because the number of negative charges equals the number of positive charges. The observable properties of ionic compounds such as MgO are due to the strong forces of attraction between these charges.

Assess How many O^{2-} ions surround each Mg^{2+} ion? How many Mg^{2+} ions surround each O^{2-}?

152 Chapter 5 • Types of Compounds

■ **Figure 5.2** These lead oxides— black lead(IV) oxide (PbO_2), yellow lead(II) oxide (PbO), and orange lead(III) oxide (Pb_2O_3) are all binary compounds. Though the formulas contain different numbers of atoms, each compound is composed only of lead and oxygen.

Chemical formulas are a crucial component of the language of chemistry, used to communicate information about substances. As a first step in studying this new language, you will learn how to name and write formulas for ionic compounds.

Sodium chloride (NaCl) contains only sodium and chlorine, and potassium iodide (KI) contains only potassium and iodine. Each is an example of a **binary compound,** which is defined as a compound that contains only two elements. Binary ionic compounds can contain more than one ion of each element, as in calcium fluoride (CaF_2), but they are not composed of more than two different elements. **Figure 5.2** shows several lead oxides, or examples of binary compounds.

Naming binary ionic compounds To name a binary ionic compound, first write the name of the positively charged ion, usually a metal, and then add the name of the nonmetal or negatively charged ion. Modify the negatively charged ion to end in *-ide*. For example, in **Figures 5.1** and **5.2** you saw magnesium oxide and three different lead oxides. The compound formed from potassium and chlorine is called potassium chloride.

Formulas for binary ionic compounds You are already familiar with one formula for an ionic compound—NaCl. Sodium chloride contains sodium ions that have a 1+ charge and chloride ions that have a 1− charge. You have learned that compounds are electrically neutral. This means that the sum of the charges in an ionic compound must always equal zero. Thus, one Na^+ balances one Cl^- in NaCl. When you write a formula, you add subscripts to the symbols for the ions until the algebraic sum of the ions' charges is zero. The smallest subscript to both ions that results in a net charge of zero is 1. However, no subscript needs to be written because it is inferred that only one ion or atom of an element is present if there is no subscript. The formula NaCl indicates that sodium chloride contains both sodium and chloride ions, that there is one sodium ion present for every chloride ion in the compound, and that the compound will therefore have no overall charge.

VOCABULARY
WORD ORIGIN
Binary
comes from the Latin word *bini*, which means *two by two*

Section 5.1 • Ionic Compounds

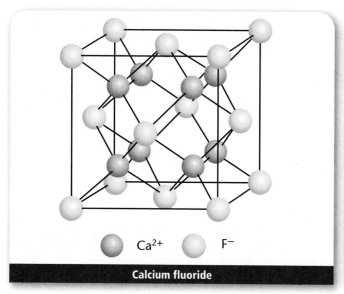

Ca²⁺ F⁻
Calcium fluoride
Fluorite

■ **Figure 5.3** When calcium fluoride forms from Ca and F, the two valence electrons from calcium are transferred to two fluorine atoms, leaving each Ca with a 2+ charge and each F with a 1− charge. Calcium fluoride occurs naturally as the mineral fluorite.
Examine *How many Ca^{2+} ions and F^- ions does one formula unit of CaF_2 contain?*

If more than one ion of a given element is present in a compound, the subscript indicates how many ions are present. The mineral known as fluorite is calcium fluoride, which has the formula CaF_2. This formula indicates that there is one calcium ion for every two fluoride ions in the compound. In an ionic compound, a formula represents the smallest ratio of atoms or ions in the compound.

Formula units versus molecules In a covalent compound, the smallest unit of the compound is a molecule, so a formula represents a single molecule of a compound. However, ionic compounds do not form molecules. Their structures are repeating patterns of ions. Should the formula of calcium fluoride be written as CaF_2, Ca_2F_4, or even Ca_3F_6? A properly written formula has the simplest possible ratio of the ions present. This simplest ratio of ions is called a **formula unit**.

Figure 5.3 shows the structure of calcium fluoride. Each formula unit of calcium fluoride consists of one calcium ion and two fluoride ions. Each of these three ions has a stable octet configuration of electrons, and the formula unit has no overall charge. Although the sum of the ionic charges in both CaF_2 and Ca_2F_4 is zero, only CaF_2 is a correct formula. One formula unit of calcium fluoride has the formula CaF_2.

Predicting charges on ions You have studied ionic compounds in which sodium becomes a positive ion with a single positive charge and calcium becomes a positive ion with two positive charges. Examine the periodic table to see if there is a way to predict the charge that different elements will have when they become ions. Which elements will lose electrons and which will gain electrons?

The noble gases each have eight electrons in their outer-energy levels. Metals have few outer-level electrons so they tend to lose them and become positive ions. Sodium must lose just one electron, becoming an Na^+ ion. Calcium must lose two electrons, becoming a Ca^{2+} ion. Most nonmetals, on the other hand, have outer-energy levels that contain four to seven electrons, so they tend to gain electrons and become negative ions. Trace the gain and loss of electrons in the example shown in **Figure 5.4**.

FACT of the Matter

Rain falling over or near an ocean is often salty because bits of salt are picked up by the wind. Crops raised in some coastal areas—such as the famous artichoke fields south of San Francisco—often have a distinctive and highly sought-after flavor because of the salty rainfall.

154 Chapter 5 • Types of Compounds

Group number and ion charge For the elements in the main groups of the periodic table—groups 1, 2, and 13–18—group numbers can be used to predict charges. Because all elements in a given group have the same number of electrons in their outer-energy levels, they must lose or gain the same number of electrons to achieve a stable noble-gas electron configuration. Metals lose electrons and nonmetals gain electrons when they form ions. The charge on the ion is known as the **oxidation number** of the atom. The oxidation numbers for many elements in the main groups are arranged by group number in **Table 5.1**. Oxidation numbers for elements in groups 3–12, the transition elements, cannot be predicted by group number.

Using oxidation number to write formulas
Alumina is the common name for aluminum oxide. It is used to produce aluminum metal, to make sandpaper and other abrasives, and to separate mixtures of chemicals by a technique called chromatography. Aluminum is in group 13, so it loses its three outer electrons to become an Al^{3+} ion. Oxygen is in group 16 and has six valence electrons, so it gains two electrons to become an O^{2-} ion.

If aluminum gives up three electrons, but oxygen only takes two, one of aluminum's electrons has not been taken up. Because all the electrons must be accounted for, more than one oxygen atom must be involved in the reaction. But, oxygen cannot gain only one electron, so a second aluminum atom must be present to contribute a second electron to oxygen. In all, two Al^{3+} ions must combine with three O^{2-} ions to form Al_2O_3. Remember that the charges in the formula for aluminum oxide must add up to zero.

$$Ca: \rightarrow Ca^{2+} \qquad \ddot{O}: \rightarrow :\ddot{O}:^{2-}$$

$$Ca^{2+} + :\ddot{O}:^{2-} \rightarrow CaO$$

■ **Figure 5.4** The compound commonly called lime is calcium oxide. It is used to make steel and cement and is added to acidic lakes and soil to neutralize the effects of acidity. Calcium is a metal that loses two electrons to become a Ca^{2+} ion; oxygen is a nonmetal that must gain two electrons to achieve the stable octet of the noble gas neon, so it becomes an O^{2-} ion. Because a formula unit must be neutral, one Ca^{2+} ion can combine with only one O^{2-} ion. The formula for calcium oxide is CaO.

$$\dot{Al}\cdot + \dot{Al}\cdot + :\ddot{O}\cdot + :\ddot{O}\cdot + :\ddot{O}\cdot \rightarrow$$
$$Al^{3+} + Al^{3+} + :\ddot{O}:^{2-} + :\ddot{O}:^{2-} + :\ddot{O}:^{2-}$$

Table 5.1	Ionic Charges of Representative Elements					
Group Number	Oxidation Number	Examples	Group Number	Oxidation Number	Examples	
Metals			Nonmetals			
1	1+	Li^+, Na^+, K^+	15	3–	N^{3-}, P^{3-}	
2	2+	Mg^{2+}, Ca^{2+}	16	2–	O^{2-}, S^{2-}	
13	3+	B^{3+}, Al^{3+}	17	1–	F^-, Cl^-, Br^-, I^-	

EXAMPLE Problem 5.1

Writing a Simple Formula Write the formula for an ionic compound containing sodium and sulfur.

1 Analyze
- Sodium is in group 1, so it has an oxidation number of 1+. Sulfur is in group 16 and has an oxidation number of 2−.

2 Set up
- Write the symbols for sodium and sulfur ions in formula form, placing the positive ion first: Na^+S^{2-}

3 Solve
- The formula as written has one positive charge and two negative charges. One more positive charge is needed to balance the 2− charge. Add a second sodium ion and place the subscript 2 after the symbol for sodium. The correct formula is then written as Na_2S.

4 Check
- Check to be sure that the charges of the ions are correct and that the overall charge of the formula is zero.
- $2(1+) + (2-) = 0$ The formula as written is correct.

PRACTICE Problems

Solutions to Problems Page 858

1. Write the formula for each compound.
   - a) lithium oxide
   - b) calcium bromide
   - c) sodium oxide
   - d) aluminum sulfide

2. Write the formula for the compound formed from each pair of elements.
   - a) barium and oxygen
   - b) strontium and iodine
   - c) lithium and chlorine
   - d) radium and chlorine

SUPPLEMENTAL PRACTICE

For more practice writing the formulas of ionic compounds, see Supplemental Practice, page 810.

Compounds containing polyatomic ions The ions you have studied so far have contained only one element. However, some ions contain more than one element. An ion that has two or more different elements is called a **polyatomic ion.** In a polyatomic ion, a group of atoms is covalently bonded together when the atoms share electrons. Although the individual atoms have no charge, the group as a whole has an overall charge. **Figure 5.5** shows models of three common polyatomic ions.

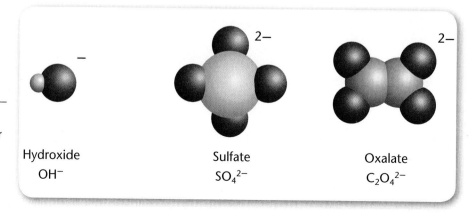

■ **Figure 5.5** Polyatomic ions are composed of more than one atom bonded together covalently. The charge is carried by the whole ion, not any single atom within the ion. Polyatomic ions form ionic bonds with other ions to produce ionic compounds.

Hydroxide OH^-

Sulfate SO_4^{2-}

Oxalate $C_2O_4^{2-}$

Ionic compounds may contain positive metal ions bonded to negative polyatomic ions, such as in NaOH; positive polyatomic ions bonded to negative nonmetal ions, such as in NH$_4$I; or positive polyatomic ions bonded to negative polyatomic ions, such as in NH$_4$NO$_3$. To write the formula for an ionic compound containing one or more polyatomic ions, simply treat the polyatomic ion as if it were a single-element ion by keeping it together as a unit. Remember that the sum of the positive and negative charges must equal zero.

Multiples of a polyatomic ion in a formula can be indicated by placing the entire polyatomic ion, without the charge, in parentheses. Write a subscript outside the parentheses to show the number of polyatomic ions in the compound. Never change the subscripts within the polyatomic ion. To do so would change the composition of the ion. The formula for the compound that contains one magnesium ion and two nitrate ions is Mg(NO$_3$)$_2$.

Naming compounds of polyatomic ions To name a compound containing a polyatomic ion, follow the same rules as used in naming binary compounds. Name the positive ion first, followed by the negative ion. Do not change the ending of the negative polyatomic ion name. The name of the compound composed of calcium and the carbonate ion is calcium carbonate.

What is the formula for calcium carbonate? Calcium is in group 2, so its ion has a 2+ charge. The carbonate ion has a 2– charge, as shown in **Table 5.2.** To form a neutral compound, one Ca^{2+} ion must combine with one CO$_3^{2-}$ ion to produce CaCO$_3$.

FOLDABLES
Incorporate information from this section into your Foldable.

Personal Tutor For an online tutorial on naming ionic compounds, visit glencoe.com.

Table 5.2 Common Polyatomic Ions

Ion	Name	Ion	Name
NH$_4^+$	ammonium	IO$_4^-$	periodate
NO$_2^-$	nitrite	C$_2$H$_3$O$_2^-$	acetate
NO$_3^-$	nitrate	H$_2$PO$_4^-$	dihydrogen phosphate
OH$^-$	hydroxide	CO$_3^{2-}$	carbonate
CN$^-$	cyanide	SO$_3^{2-}$	sulfite
MnO$_4^-$	permanganate	SO$_4^{2-}$	sulfate
HCO$_3^-$	hydrogen carbonate	S$_2$O$_3^{2-}$	thiosulfate
ClO$^-$	hypochlorite	O$_2^{2-}$	peroxide
ClO$_2^-$	chlorite	CrO$_4^{2-}$	chromate
ClO$_3^-$	chlorate	Cr$_2$O$_7^{2-}$	dichromate
ClO$_4^-$	perchlorate	HPO$_4^-$	hydrogen phosphate
BrO$_3^-$	bromate	PO$_4^{3-}$	phosphate
IO$_3^-$	iodate	AsO$_4^{3-}$	arsenate

Everyday Chemistry

Hard Water

The term *hard water* doesn't describe water's physical state. Hard water is any water containing a large amount of dissolved minerals, resulting in a large amount of calcium ions and magnesium ions. As groundwater travels through soil and rock, it dissolves small amounts of calcium and magnesium, along with other minerals. The more calcium and magnesium that is dissolved in the groundwater, the harder the water is. Water hardness is measured in milligrams per liter (mg/L) or parts per million (ppm). Hard water is considered to be any water that has at least 150 mg/L or ppm of dissolved minerals.

Soap Scum Hard water is easy to identify because it interferes with almost every aspect of cleaning, including laundering, dishwashing, and bathing. A common complaint by people using hard water is that soaps and detergents do not lather well. One of the compounds in soap that helps produce lather is sodium stearate ($NaC_{18}H_{35}O_2$), which dissolves in water.

In hard water, calcium ions react with the stearate ions to form calcium stearate ($Ca(C_{18}H_{35}O_2)_2$). This material is insoluble and forms soap scum, as shown in **Figure 1**. Soap scum is often seen as a ring around sinks or tubs. Soap scum on hair makes it dull, lifeless, and sticky. Doing laundry in hard water makes clothes hard, scratchy, and dull. Washing dishes in hard water causes a soap scum film to develop on dishes and spots to form when dry. When water softeners remove calcium and magnesium ions from the water, lather results and no additional soap scum forms.

Scales Hard water causes more problems than just soap that doesn't lather and soap scum. When hard water is heated, the calcium and magnesium minerals dissolved in the water form solid scales on the inside of water pipes.

Figure 1 Soap scum deposits on the sink and is difficult to remove because it is insoluble in water.

The solid that forms can clog and ruin the pipes. The scales also act as insulation, requiring more energy to heat the water. Hard water also causes solid scales to coat the insides of tea and coffee pots. Many industries monitor water hardness constantly to avoid costly breakdowns in boilers, cooling towers, and other equipment.

Ion Exchangers A common way of softening water—that is, reducing the number of calcium and magnesium ions—is by an ion exchanger. The ion exchanger usually contains a material, called a resin, made up of very small polystyrene (plastic) beads that are coated with sodium ions. As hard water passes through the resin, a calcium ion or magnesium ion in the water is exchanged for two sodium ions. Thus, the water leaving the ion exchanger has fewer calcium and magnesium ions, but many more sodium ions. Because sodium does not precipitate as a solid in pipes or cause soap not to lather, problems associated with hard water are eliminated.

Explore Further

1. **Interpret** What is the charge of the stearate ion?
2. **Think Critically** Why do two sodium ions replace one calcium or one magnesium ion in an ion exchanger?
3. **Acquire Information** Why are detergents more effective than soaps in hard water?

EXAMPLE Problem 5.2

Writing a Formula Containing a Polyatomic Ion Write the formula for the compound that contains lithium and carbonate ions.

1 Analyze
- Lithium is in group 1, so its ion has a 1+ charge. According to **Table 5.2**, the carbonate ion has a 2− charge, and its structure is CO_3^{2-}.

2 Set up
- Write the symbols for lithium carbonate in formula form.

$Li^+CO_3^{2-}$

3 Solve
- Determine the correct ratio of lithium ions to carbonate ions by examining their charges. The sum of the positive and negative charges does not equal zero. Two lithium ions are needed to balance the carbonate ion. Because you cannot change the charges of the ions, you must add a subscript of 2 to Li^+. The correct formula for lithium carbonate is Li_2CO_3.

4 Check
- Check to be sure that the overall charge of the formula is zero.

$2(1+) + (2-) = 0$ The formula as written is correct.

EXAMPLE Problem 5.3

Writing a More Complex Formula Write the formula for the compound that contains aluminum and sulfate ions.

1 Analyze
- Aluminum is in group 13 and has an oxidation number of 3+. According to **Table 5.2**, the sulfate ion has a 2− charge.

2 Set up
- Write the symbols for aluminum sulfate in formula form.

$Al^{3+}SO_4^{2-}$

3 Solve
- Determine the correct ratio of aluminum ions to sulfate ions by examining their charges. The sum of the positive and negative charges does not equal zero. To achieve neutrality, you must find the least common multiple of 3 and 2. The least common multiple is 6. How many Al^{3+} ions will be needed to make a charge of 6+, and how many SO_4^{2-} ions will be needed to make a charge of 6−? It will be necessary to have two Al^{3+} ions in the compound to balance three SO_4^{2-} ions. You should add a subscript of 2 to the aluminum ion and a subscript of 3 to the sulfate ion. The entire polyatomic ion must be placed in parentheses to indicate that three sulfate ions are present. Thus, the correct formula for aluminum sulfate is $Al_2(SO_4)_3$.

4 Check
- Check to be sure that the overall charge of the formula is zero.

$2(3+) + 3(2-) = 0$ The formula as written is correct.

SUPPLEMENTAL PRACTICE

For more practice writing formulas containing polyatomic ions, see Supplemental Practice, page 811.

PRACTICE Problems

Solutions to Problems Page 858

3. Write the formula for the compound made from each set of ions.
   a) ammonium and sulfite ions
   b) calcium and monohydrogen phosphate ions
   c) ammonium and dichromate ions
   d) barium and nitrate ions
4. Write the formula for each compound.
   a) sodium phosphate
   b) magnesium hydroxide
   c) ammonium phosphate
   d) potassium dichromate

Compounds of Transition Elements

In Chapter 3, you learned that the elements known as transition elements are located in groups 3–12 in the periodic table. Transition elements form positive ions just as other metals do, but most transition elements can form more than one type of positive ion. In other words, transition elements can have more than one oxidation number. For example, copper can form both Cu^+ and Cu^{2+} ions, and iron can form both Fe^{2+} and Fe^{3+} ions. **Figure 5.6** shows the two compounds that iron forms with the sulfate ion. Zinc and silver are two exceptions to the variability of other transition elements; each forms one type of ion. The zinc ion is Zn^{2+} and the silver ion is Ag^+.

■ **Figure 5.6** Iron forms both Fe^{2+} and Fe^{3+} ions, each of which can combine with the sulfate ion.

Iron(II) sulfate ($FeSO_4$), is a blue-green crystalline substance (above) that is used in fertilizer (below) and as a food supplement.

Iron(III) sulfate ($Fe_2(SO_4)_3$), (above) is a yellow crystalline substance that is used as a coagulant in water-purification and sewage-treatment plants (below).

Art Connection

China's Porcelain

See how proudly the dragon, an ancient symbol of Chinese culture, prances around the vase in **Figure 1.** Proud it should be, because this vase represents one of the greatest achievements of Chinese technology and art—glazed porcelain.

Clay, glaze, and fire By the third to sixth century A.D., the Chinese had invented glazed porcelain. They found that if a clay vessel, such as a bowl, is covered with a transparent glaze and then heated to a high temperature, a translucent ceramic material forms. This material is glazed porcelain.

Unlike a fired-clay vessel, which remains slightly porous and opaque, the translucent vessel is sealed by a glasslike covering. By changing the chemical composition of the glaze, Chinese artisans were able to change the quality and color of the glaze. For example, when they added materials that reacted with each other to form tiny gas bubbles in the glaze, the porcelain appeared brighter because the surface of the bubbles reflected light.

Figure 2 Transition elements add color to glazed porcelain.

Colored glazes One of the most important steps in glazing pottery was the addition of materials to the glaze to produce colored porcelains such as the vase in **Figure 2.** These materials were solutions of transition element ions, such as iron, manganese, chromium, cobalt, copper, and titanium.

During the firing of the glaze, these metals formed oxides. Because the metal ions in the oxides reflected only certain wavelengths of light, the glazes colored the porcelain. By varying the concentration and charges of the metal ions in the glaze, the Chinese were able to produce subtly colored porcelains. For example, cobalt produces a blue glaze, chromium a pink or green glaze depending on charge, and manganese a purple glaze. These beautiful colors have remained vivid over the course of thousands of years, and the techniques are still being used today.

Figure 1 The dragon art remains brilliant after hundreds of years due to porcelain glaze.

Connection to Chemistry

1. **Apply** Why are porcelain dishes superior to wooden dishes?
2. **Think Critically** What properties of metallic compounds make them useful as colored glazes?

Table 5.3 — Compounds of Copper and Chlorine

Copper Ion	Chloride Ion(s)	Formula	Name
Cu^+	Cl^-	$CuCl$	copper(I) chloride
Cu^{2+}	$2Cl^-$	$CuCl_2$	copper(II) chloride

Naming the compounds of transition elements Chemists must have a way to distinguish between the names of compounds formed from different ions of a transition element. They achieve this by using a Roman numeral to indicate the oxidation number of a transition element ion. This Roman numeral is placed in parentheses after the name of the element. No additional naming system is needed for zinc and silver compounds because they occur with only one oxidation state. **Table 5.3** shows the naming of the two different ionic compounds formed when chloride ions combine with each of the two copper ions. Copper occurs with oxidation states of 1+ and 2+. When a Cu^+ ion combines with a Cl^- ion, copper(I) chloride results. When a Cu^{2+} ion combines with a Cl^- ion, two chloride ions are required to balance the charge ($CuCl_2$), and copper(II) chloride is produced.

The chemical names of common transition element ions are shown in **Table 5.4.** When you solve Practice Problems 5 and 6, you will become familiar with these names. The different ions of a transition element often form compounds of different colors. For example, **Table 5.4** shows the three most common chromium ions—Cr^{2+}, Cr^{3+}, and Cr^{6+}. The oxides of those ions are CrO, Cr_2O_3, and CrO_3. CrO is black, Cr_2O_3 is green, and CrO_3 is red. Determine the oxidation number for chromium in each of these compounds.

Table 5.4 — Names of Common Ions of Selected Transition Elements

Element	Ion	Chemical Name	Element	Ion	Chemical Name
Chromium	Cr^{2+}	chromium(II)	Iron	Fe^{2+}	iron(II)
	Cr^{3+}	chromium(III)		Fe^{3+}	iron(III)
	Cr^{6+}	chromium(VI)	Manganese	Mn^{2+}	manganese(II)
Cobalt	Co^{2+}	cobalt(II)		Mn^{3+}	manganese(III)
	Co^{3+}	cobalt(III)		Mn^{7+}	manganese(VII)
Copper	Cu^+	copper(I)	Mercury	Hg^+	mercury(I)
	Cu^{2+}	copper(II)		Hg^{2+}	mercury(II)
Gold	Au^+	gold(I)	Nickel	Ni^{2+}	nickel(II)
	Au^{3+}	gold(III)		Ni^{3+}	nickel(III)
				Ni^{4+}	nickel(IV)

■ **Figure 5.7** Iron disulfide (FeS$_2$) is commonly called fool's gold. In this ionic compound, sulfur exists in the unusual form S$_2^{2-}$, and iron has an oxidation number of 2+.

Formulas of the compounds of transition elements

Suppose you wanted to write the formula for a compound containing a transition element. Look back at Sample Problem 1, where you learned to write the formula for a compound containing sodium and sulfur. How would you write the formula if it were iron(II) rather than sodium that combined with sulfur? Iron(II) has an oxidation number of 2+, and its ion can be written as Fe^{2+}. You know that the sulfide ion has a charge of 2− and can be written as S$_2^{2-}$. The charges balance in this case, and the formula for iron(II) sulfide is written as FeS$_2$. Iron(II) sulfide is known as the mineral pyrite. As shown in **Figure 5.7,** you can see why pyrite is sometimes called fool's gold.

You can write the formula for iron(III) sulfide in the same way. Just follow the steps in Sample Problem 3. The correct formula for iron(III) sulfide is Fe$_2$S$_3$. The Roman numeral refers to the oxidation number of the iron, not to how many ions are in the formula.

How can you name a compound containing a transition element if you are given the formula? Determining the charge of the transition element ion gives the clue needed to name the compound. In the formula Cr(NO$_3$)$_3$, you must determine the charge of the chromium ion in order to name the compound. Look first at the negative ion. Knowing that the nitrate ion has a charge of 1− and that there are three nitrate ions with a total charge of 3−, you can see that the chromium ion must have a charge of 3+ to maintain neutrality. Thus, this compound is named chromium(III) nitrate.

TRY AT HOME 🏠 LAB

See page 870 for **Iron Ink**

PRACTICE Problems
Solutions to Problems Page 858

5. Write the formula for the compound made from each pair of ions.
   a) copper(I) and sulfite
   b) tin(IV) and fluoride
   c) gold(III) and cyanide
   d) lead(II) and sulfide
6. Write the names of the compounds.
   a) Pb(NO$_3$)$_2$
   b) Mn$_2$O$_3$
   c) Ni(C$_2$H$_3$O$_2$)$_2$
   d) HgF$_2$

SUPPLEMENTAL PRACTICE

For more practice with the formulas of the compounds of transition elements, see Supplemental Practice, page 810.

Section 5.1 • Ionic Compounds

■ **Figure 5.8** Sodium hydroxide has a strong attraction for water molecules. Sodium hydroxide will absorb water molecules from the surrounding air and begin to dissolve. Eventually, it will absorb enough water to dissolve completely.

FOLDABLES Incorporate information from this section into your Foldable.

Hydrates

Many ionic compounds are prepared by crystallization from a water solution, and water molecules become a part of the crystal. A compound in which there is a specific ratio of water to ionic compound is called a **hydrate.** In a hydrate, the water molecules are chemically bonded to the ionic compound.

Hygroscopic substances Does your chemistry instructor often remind students to make sure that the lids on jars of chemicals are tightly closed? There is a good reason for sealing the jars tightly; some ionic compounds can easily absorb water molecules from water vapor in the air. **Figure 5.8** shows what happens as sodium hydroxide (NaOH) absorbs moisture from the air.

MiniLab 5.1

Chemical Weather Predictor

How can you make a weather predictor? Adding water to anhydrous cobalt(II) chloride to form a hydrate changes the color of the compound. If you find that the color of the compound changes with the weather, cobalt(II) chloride can serve as a weather predictor.

Procedure

1. Read and complete the lab safety form.
2. Place 5 mL of 95 percent **ethanol** in a **small beaker.**
3. Use a **spatula** to add a small amount of **cobalt(II) chloride** to the beaker. Stir until the compound dissolves.
4. Dip a **cotton swab** into the pink solution, and use it to write the chemical formula of cobalt(II) chloride on a piece of white paper.
5. Dry the paper by holding it over a hot plate set on low or by putting it in a sunny location. What color is the formula now?
6. Keep your weather predictor in a convenient location, and check its color each morning and afternoon. Keep a three-week log of the time, the current weather, and the color of the treated paper.

Analysis

1. **Determine** the formula of cobalt(II) chloride.
2. **Analyze** The hydrate of cobalt(II) chloride has six water molecules bonded to it. What is its formula?
3. **Infer** From your observations, are you able to conclude that the cobalt(II) chloride test paper is a reliable weather predictor? Justify your answer.

Compounds that easily absorb water from air are called **hygroscopic** substances, and one example is sodium carbonate (Na_2CO_3). Substances that are so hygroscopic that they take up enough water from the air to dissolve completely and form a liquid solution are called **deliquescent.** Compounds that form hydrates are often used as drying agents, also called desiccants, because they absorb so much water from the air when they become hydrated.

Formulas for hydrates To write the formula for a hydrate, write the formula for the compound and then place a dot followed by the number of water molecules per formula unit of the compound. The dot in the formula represents a ratio of compound formula units to water molecules. For example, $CaSO_4 \cdot 2H_2O$ is the formula for a hydrate of calcium sulfate that contains two molecules of water for each formula unit of calcium sulfate. This hydrate is used to make portland cement and plaster of paris. To name hydrates, write the regular name for the compound followed by the word *hydrate,* to which a prefix has been added to indicate the number of water molecules present. Use **Table 5.5** to find the correct prefix to use. The name of the compound with the formula $CaSO_4 \cdot 2H_2O$ is calcium sulfate dihydrate.

Heating hydrates can drive off the water. This results in the formation of an **anhydrous** compound—one in which all of the water has been removed. In some cases, an anhydrous compound may have a different color from that of its hydrate, as shown in **Figure 5.9**.

Table 5.5 Naming Prefixes

Molecules of Water	Prefix
1	mono-
2	di-
3	tri-
4	tetra-
5	penta-
6	hexa-
7	hepta-
8	octa-
9	nona-
10	deca-

Concepts In Motion

Interactive Table Explore naming prefixes at glencoe.com.

■ **Figure 5.9** Copper(II) sulfate pentahydrate ($CuSO_4 \cdot 5H_2O$) is used as a fungicide in water reservoirs and on some crops, such as grapes. It is also commonly used by hobbyists to grow crystals.

Copper(II) sulfate pentahydrate ($CuSO_4 \cdot 5H_2O$) is the bright blue hydrate of copper(II) sulfate ($CuSO_4$).

As $CuSO_4 \cdot 5H_2O$ is heated, the water is driven off.

Eventually, all the water is driven off, leaving white anhydrous $CuSO_4$ behind.

How It Works

Cement

People have been using binding materials for thousands of years. The rocks in the Egyptian pyramids shown in **Figure 1** are held together by a mixture of sand and the mineral gypsum, which is a compound of calcium sulfate dihydrate. When this dihydrate is heated, water evaporates, forming a compound with one water molecule per two calcium sulfate formula units. Today, we know this binding material as plaster of paris. Cement is another binding material.

■ **Figure 1** Egyptian pyramid

1 Cement is made from a mixture of limestone and clay. The most important minerals in clay are the combinations of aluminum, oxygen, and silicon known as aluminum silicates.

2 Before this limestone-clay mixture can be used, it must be heated. Heating drives off carbon dioxide and forms new ionic compounds. This new mixture of calcium silicates, calcium aluminates, and calcium aluminum ferrates forms in clumps called clinker.

3 The clinker is ground and mixed with small amounts of calcium sulfate. This mixture is called portland cement, as shown in **Figure 2**.

■ **Figure 2** Portland cement being poured into a concrete mixer

4 Portland cement is the basic ingredient in concrete, mortar, and grout. The most common use for portland cement is in the production of concrete, which is used to construct roads and buildings.

5 Concrete is a mixture of material consisting of sand and gravel, portland cement, and water. When water is mixed with the sand, gravel, and portland cement, silicate compounds hydrate and form gelatinous materials called gels.

6 The hardening process takes several days. During this time, some water is removed from the gels that formed around the sand and gravel, and calcium hydroxide absorbs carbon dioxide from the air to reform calcium carbonates. Fibers that form from the cement materials interlock and strengthen the concrete.

Think Critically

1. **Determine** What is the formula for calcium sulfate dihydrate which makes up gypsum?
2. **Analyze** Tricalcium aluminate also becomes hydrated during solidification of cement to form $Ca_3Al_2O_6 \cdot 6H_2O$. What is the name of this hydrate?

Interpreting Formulas

You have learned how to write a formula to represent a formula unit of an ionic compound. Sometimes, it might be necessary to represent more than one formula unit of a compound. To do this, place a coefficient before the formula. Two formula units of NaCl are represented by 2NaCl, three formula units by 3NaCl, and so on.

A formula summarizes how many atoms of each element are present. Each formula unit of sodium chloride contains one sodium ion and one chloride ion, therefore, three formula units would be the equivalent of three sodium ions and three chloride ions. How many oxygen atoms are present in $3HNO_3$? Each formula unit contains three oxygen atoms. Because there are three formula units, a total of nine atoms of oxygen are present.

As another example, consider how many atoms of hydrogen are in one formula unit of ammonium sulfate. The formula for ammonium sulfate, a food additive common in bread as shown in **Figure 5.10,** is $(NH_4)_2SO_4$. Each ammonium ion contains four atoms of hydrogen. Because two ammonium ions are present, there are eight atoms of hydrogen in a formula unit of ammonium sulfate. How many hydrogen atoms are in $3(NH_4)_2SO_4$? To find out, simply multiply the eight hydrogen atoms in one formula unit by three formula units: 24 hydrogen atoms are present.

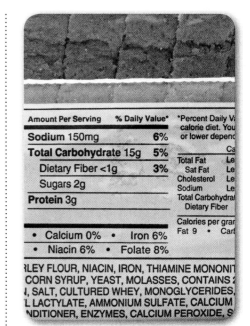

■ **Figure 5.10** Ammonium sulfate is often added to baked goods as a dough conditioner. Dough conditioners help ensure that the bread will rise evenly, especially in cases where large volumes of dough are being mixed.

Section 5.1 Assessment

Section Summary

▶ The position of an element in the periodic table indicates what charge its ions will have.

▶ Binary ionic compounds are named by naming the metal element and then the nonmetal element, with its ending changed to *-ide*.

▶ Subscripts in formulas indicate how many atoms of each element are present in a compound.

▶ Polyatomic ions can combine with ions of opposite charge to form ionic compounds.

▶ Most transition elements can form two or more positively charged ions. The oxidation number of the transition element is indicated by a Roman numeral in parentheses.

▶ Hydrates are ionic compounds bonded to water molecules.

7. **MAIN Idea** **Explain** why ionic compounds cannot conduct electricity when they are in the solid state.

8. **Write** formulas for each of the following ionic compounds.
   a) manganese(II) carbonate
   b) barium iodide dihydrate
   c) aluminum oxide
   d) magnesium sulfite
   e) ammonium nitrate
   f) sodium cyanide

9. **Name** the ionic compound represented by each formula.
   a) Na_2SO_4
   b) CaF_2
   c) $MgBr_2 \cdot 6H_2O$
   d) Na_2CO_3
   e) $KMnO_4$
   f) $Ni(OH)_2$
   g) $NaC_2H_3O_2$

10. **Interpret** What information does the formula $3Ni(HCO_3)_2$ tell you about the number of atoms of each element that are present?

11. **Infer** Calcium nitrate $(Ca(NO_3)_2)$, an ionic compound used as a fertilizer, dissolves in water. What ions are released from the solid as $Ca(NO_3)_2$ dissolves? How many of each type of ion are released for every formula unit that dissolves?

12. **Examine** the ingredient label on a tube of toothpaste. Write formulas for as many of the chemical names listed as you can. Identify whether each ingredient is an ionic compound or not.

Section 5.2

Objectives
- **Compare** the properties of covalent and ionic substances.
- **Distinguish** among allotropes of an element.
- **Apply** formulas to name covalent compounds.

Review Vocabulary
anhydrous: a compound in which all of the water has been removed, usually by heating

New Vocabulary
distillation
molecular element
allotrope
organic compound
inorganic compound
hydrocarbon

FOLDABLES
Incorporate information from this section into your Foldable.

Covalent Compounds

MAIN Idea Atoms in covalent compounds are held together by shared electrons.

Real-World Reading Link What compounds can you name that are liquids or gases at room temperature? Water, carbon dioxide, and ammonia are just a few examples. Because most ionic compounds are solids at room temperature, the odds are pretty good that any compounds you thought of are members of the other major class of compounds described in Chapter 4—the covalent compounds.

Properties of Covalent Compounds

You know that ionic compounds share many properties. The properties of a covalent compound—a substance that has atoms held together by covalent rather than ionic bonds—are more variable than the properties of ionic compounds. Some covalent compounds, such as polyethylene plastic and the fats in butter, are soft; rubber is elastic; and diamond and quartz are both hard.

Comparing ionic and covalent compounds Although covalent compounds have different properties, some generalities can be made to distinguish them from ionic compounds. Covalent compounds usually have lower melting points, and most are not as hard as ionic compounds, as shown in **Figure 5.11.** In addition, most covalent compounds are less soluble in water than ionic compounds and they are not electrolytes.

The properties of most ionic and covalent compounds vary enough that their differences can be used to classify and separate them from one another. The separation of water from salt by distillation is one example that makes use of these property differences. **Distillation** is the method of separating substances in a mixture by evaporation of a liquid and subsequent condensation of its vapor. As you learned, solar stills make use of this method. A simple lab-distillation apparatus is shown in **Figure 5.12.**

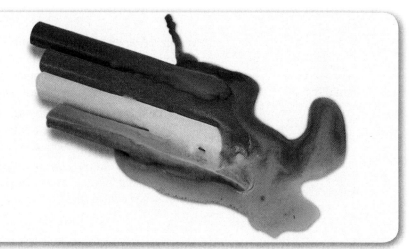

■ **Figure 5.11** Crayons are made of covalent compounds. They are soft and are insoluble in water. If you have ever left crayons out in the Sun, you know that they also have a low melting point.

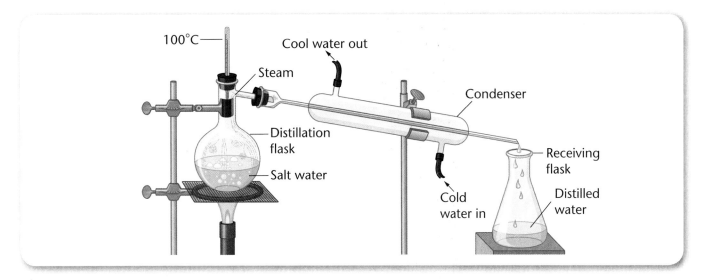

■ **Figure 5.12** A soluble ionic compound such as NaCl can be separated from water using a distillation apparatus like this one. As the salt water is boiled in the distillation flask, the water turns to steam and the salt is left behind. The steam passes through a water-cooled condenser, where it condenses into pure, distilled water. The distilled water is collected in the receiving flask.

How does the submicroscopic structure of covalent compounds contribute to their macroscopic properties? Because there are no ions, strong networks held together by the attractions of opposite charges do not occur. The interparticle forces between molecules are weak and easy to break. These weak forces explain the softness and low melting points of most covalent compounds. Most covalent compounds are not electrolytes because they do not easily form ions.

MiniLab 5.2

Chemical Bonds in Bone

How can you determine the type of chemical bonds found in bone? Calcium is an important part of some compounds in bones and eggshells. Vinegar contains acetic acid, which reacts with these calcium compounds to form calcium acetate.

Procedure

1. Read and complete the lab safety form.
2. Pick most of the meat from a small **chicken bone**.
3. Place the bone in a **beaker**, cover it with **vinegar**, and cover the beaker with a **watch glass**.
4. Label the beaker with your name and leave it for two days in the area indicated by your teacher.
5. Use **forceps** to remove the bone from the beaker, and blot it on a **paper towel**. Examine the bone for changes.
6. Replace the bone in the vinegar and let it soak for two more days. Repeat step 5.
7. Straighten a **paper clip.** Hold the clip with forceps, dip it into the vinegar solution, and hold it in the blue flame of a **Bunsen burner.** Calcium in the vinegar will produce orange-red flame test results.

Analysis

1. **Describe** the change in the properties of the bone after two days and after four days.
2. **Infer** what was the probable source of any calcium you might have observed in the flame test?
3. **Conclude** What effect do ionic calcium compounds have on the properties of bone? Do their properties in bone seem to correlate with those of typical ionic compounds?
4. **Explain** how the properties of the bone, after soaking, reflect the presence of covalent compounds.

CHEMLAB

IONIC OR COVALENT COMPOUNDS

Background
You cannot tell for sure whether a compound is ionic or covalent simply by looking at a sample of it because both types of compounds can look similar. However, simple tests can be done to classify compounds by type because each type has a set of characteristic properties shared by most members. Ionic compounds are usually hard, brittle, and water-soluble; have high melting points; and can conduct electricity when dissolved in water. Covalent compounds can be soft, hard, or flexible; are usually less water-soluble; have lower melting points; and cannot conduct electricity when dissolved in water.

Question
How can you identify ionic compounds and covalent compounds by their physical properties?

Objectives
- **Examine** the properties of several common substances.
- **Interpret** the property data to classify each substance as ionic or covalent.

Preparation

Materials
glass microscope slide
grease pencil or crayon
hot plate
spatula
4 small beakers (50- or 100-mL)
stirring rod
balance
conductivity tester
graduated cylinder, small
thermometer (must read up to 150°C)
1- to 2-g samples of any 4 of the following:
   salt substitute (KCl), fructose, aspirin, paraffin, urea, table salt, table sugar, Epsom salt

Safety Precautions
WARNING: *Use care when handling hot objects.*

Procedure

1. Read and complete the lab safety form.
2. Use a grease pencil or crayon to draw lines dividing a glass slide into four parts. Label the parts A, B, C, and D.
3. Make a data table similar to the one shown in Data and Observations.
4. Use a spatula to place about one-tenth (about 0.1 to 0.2 g) of the first of your four substances on section A of the slide.
5. Repeat step 4 with your other three substances on sections B, C, and D. Be sure to use a clean spatula for each sample. Record in your data table which substance was put on each section.
6. Place the slide on a hot plate. Turn the heat setting to medium and begin to heat the slide.
7. Gently hold a thermometer so that the bulb just rests on the slide. Be careful not to disturb your compounds.

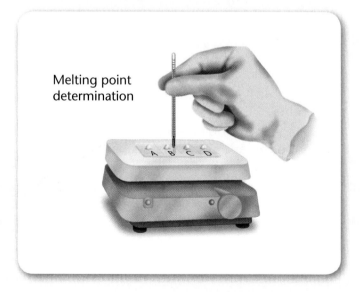

Melting point determination

170 Chapter 5 • Types of Compounds

8. Continue heating until the temperature reaches 135°C. Observe each section on the slide, and record which substances have melted. Turn off the hot plate.
9. Label four beakers with the names of your four substances.
10. Measure the mass of equal amounts of the four substances (1–2 g of each), and place the samples in their labeled beakers.
11. Add 10 mL of distilled water to each beaker.
12. Stir each substance, using a clean stirring rod for each sample. Note on your table whether or not the sample dissolved completely.
13. Test each substance for the presence of electrolytes by using a conductivity tester. Record whether or not each acts as a conductor.

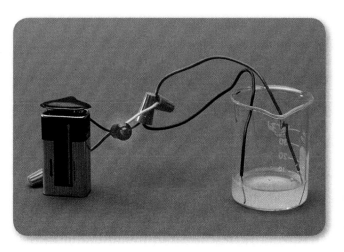

Analyze and Conclude

1. **Interpret Observations** What happened to the forces between the particles when a substance melted?
2. **Compare and Contrast** Did all compounds melt at the same temperature?
3. **Classify** Complete your data table by classifying each of the substances you tested as ionic or covalent compounds based on your observations.

Apply and Assess

1. **Contrast** What are the differences in properties between ionic and covalent compounds?
2. **Compare** How did the melting points of the ionic compounds and the covalent compounds compare? What factors affect melting point?
3. **Explain** The solutions of some covalent compounds are good conductors of electricity. Explain how this can be true when ions are required to conduct electricity.

INQUIRY EXTENSION
Infer Consider a mixture of sand, salt, and water. Design an experiment that makes use of the differences in properties of these materials to separate them.

Data and Observations

Data Table

Substance	Did it melt?	Did it dissolve in water?	Did the solution conduct electricity?	Classification
A				
B				
C				
D				

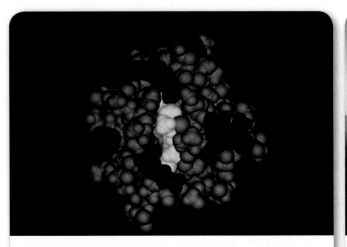
Cytochrome c contains many thousands of atoms of carbon, hydrogen, oxygen, nitrogen, and sulfur linked together by covalent bonds.

Cytochrome c is found in all living cells that derive energy by breaking down food molecules in the presence of oxygen. It is found in especially large quantities in hard-working muscle tissue.

■ **Figure 5.13** Covalent substances can be as simple as two atoms linked together, as in elemental hydrogen (H_2) and iodine (I_2). These simple molecules contrast greatly in size with the large, complex molecules, such as cytochrome c, that make up living things.

FOLDABLES
Incorporate information from this section into your Foldable.

Molecular Elements

Molecules vary greatly in size. They can contain from just two atoms to thousands or millions of atoms, as **Figure 5.13** shows. Most elements usually occur naturally in a combined form with another element; that is, they occur as compounds. However, in some cases, two or more atoms of the same element can bond together to form a molecule. A molecule that forms when atoms of the same element bond together is called a **molecular element.** Note that molecular elements are not compounds—they contain atoms of only one element. Why do atoms of these elements bond so readily to identical atoms? When they bond together, each atom achieves the stability of a noble-gas electron configuration.

Diatomic molecules Seven nonmetal elements are found naturally as molecular elements of two identical atoms. The elements whose natural state is diatomic are hydrogen (H_2), nitrogen (N_2), oxygen (O_2), fluorine (F_2), chlorine (Cl_2), bromine (Br_2), and iodine (I_2). These molecules are referred to as diatomic molecules. All except bromine and iodine are gases at room temperature; Br_2 is a liquid, and I_2 is a solid.

What can you learn by examining the structures of the diatomic molecules? Lewis dot diagrams offer clues. As an example, the chlorine atom has seven valence electrons and needs one more to achieve the configuration of the noble gas argon. If two chlorine atoms combine, they share a single pair of electrons, and each atom attains a stable octet configuration.

$$:\ddot{Cl}\cdot + \cdot\ddot{Cl}: \rightarrow :\ddot{Cl}:\ddot{Cl}:$$

Hydrogen, fluorine, bromine, and iodine molecules also are formed by the sharing of a single pair of electrons. Two oxygen atoms share two pairs of electrons to form O_2, and two nitrogen atoms share three pairs of electrons to form N_2. **Figure 5.14** shows the double bond in O_2 and the triple bond in N_2.

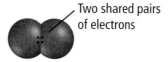

■ **Figure 5.14** Multiple covalent bonds form when two atoms share more than one pair of electrons. **a.** Two oxygen atoms form a double bond. **b.** A triple bond forms between two nitrogen atoms.

Allotropes Molecules of a single element that differ in crystalline or molecular structure are called **allotropes.** The properties of allotropes are usually different even though they contain the same element. This is because structure can be more important than composition in determining the properties of molecules.

Allotropes of phosphorus Phosphorus has three common allotropes: white, red, and black. All are formed from P_4 molecules that are joined in different ways, giving each allotrope a unique structure and set of properties. **Figure 5.15** compares white and red allotropes of phosphorus.

Allotropes of carbon Carbon occurs as several important allotropes with different properties. Diamond is a crystal in which the atoms of carbon are held rigidly in place in a three-dimensional network. In graphite, the carbon atoms are held together closely in flat layers that can slide over each other. This property makes graphite soft and greasy-feeling and useful as a dry lubricant in locks.

Another set of carbon allotropes, the fullerenes, consist of carbon atom clusters. These molecules are unusually stable and are an exciting area of research for chemists because of their potential use as superconductors. The Chemistry & Technology feature on pages 174–176 provides an in-depth look at carbon's allotropes.

VOCABULARY
WORD ORIGIN
Allotrope
comes from the Greek *allos,* meaning *other,* and *tropos,* meaning *way*

■ **Figure 5.15** White phosphorus and red phosphorus are the most common allotropes of phosphorus. White phosphorus will ignite spontaneously in air, whereas the red form won't ignite unless it comes in contact with a flame. For these reasons, white phosphorus must be stored under water, and red phosphorus is used in the strike pad of safety matches.

Section 5.2 • Covalent Compounds

CHEMISTRY AND TECHNOLOGY

Carbon Allotropes: From Soot to Diamonds

Carbon is the most versatile element in forming allotropes. Atoms of carbon can take on an incredible number of arrangements, each varying from the other and each forming a different allotrope. With all their diversity, these substances have one thing in common: they are made up solely of covalently bonded carbon atoms.

■ **Figure 1** Graphite and its structure

Graphite
The most familiar form of carbon is graphite shown in **Figure 1.** Mixed with a little clay and formed into a rod, it becomes the lead in a pencil. Look at the structure of graphite in **Figure 1.** As you can see, the carbon atoms are linked to each other in a continuous sheet of hexagons (six-sided figures). Note that each carbon atom connects three different hexagons. It's clear that the structure of graphite is well organized. The arrays of hexagons are arranged in layers that are loosely held together. The looseness between layers is why graphite is useful in pencils. As you write, the surface of the paper pulls off the loosely held layers of carbon atoms.

Carbon Blacks
Carbon blacks make up most of the soot that collects in chimneys and becomes a fire hazard. They are formed by the incomplete burning of hydrocarbon compounds, as shown in **Figure 2.** Each microscopic chunk of a carbon black is made of millions of jumbled chunks of layered carbon atoms, as shown in **Figure 2.** However, the layers lack the organization of graphite, giving carbon black its haphazard structure. Carbon blacks are used in the production of printing inks and rubber products.

■ **Figure 2** Carbon black forming and its structure

Diamond

Another allotrope of elemental carbon is diamond, as shown in **Figure 3**. Diamond is the hardest natural substance. It's often used on the tips of cutting tools and drills. Can the structure of diamond explain its hardness? Look at the model of diamond in **Figure 3**. Every carbon atom is attached to four other carbon atoms which, in turn, are each attached to four more carbon atoms. Diamond is one of the most organized of all substances. In fact, every diamond is one huge molecule of carbon atoms. This organization of covalently bonded carbons throughout diamond accounts for its hardness. If you tried to write with a diamond, you'd only tear your paper because layers of carbon atoms do not slip off as they do in graphite. The organization of carbon atoms into diamond occurs under extreme pressure and temperature, often at depths of 200 km and over a long period of time.

Charcoals

Charcoals—the kind you draw with or cook with—are another type of poorly organized carbon molecules. Charcoals are produced from the burning of organic matter. If you look closely at a chunk of charcoal, you can see that it is extremely porous. All of these pores, pock marks, and holes give charcoal a large surface area. Some charcoal, called activated charcoal, has as much as 1000 m^2 of surface area per gram. This property makes activated charcoal useful in filtering water, as shown in **Figure 4**. Molecules, atoms, and ions responsible for unwanted odors and tastes in water are attracted to and held by the surface of the activated charcoal as water passes through it, as shown in this water-filtering pitcher.

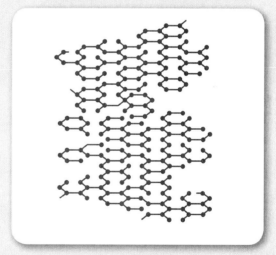

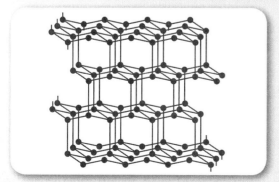

■ **Figure 3** Diamond and its structure

■ **Figure 4** Charcoal and its structure

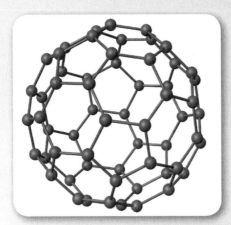

■ **Figure 5** Buckminsterfullerene model and a geodesic dome

Fullerenes

Figure 5 shows a model of buckminsterfullerene, C_{60}, which was named after engineer and architect Buckminster Fuller, who invented the geodesic dome shown in **Figure 5**. Both the dome and the molecule are unusually stable. The molecule is one of a group of highly organized allotropes of carbon called fullerenes. Buckminsterfullerene, sometimes called buckyballs, was discovered in soot in 1985. Its soccer-ball shape was confirmed in 1991. Since then, other naturally occurring and artificially produced fullerenes have been identified. Fullerenes have even-numbered molecular formulas such as C_{70} and C_{78}. The molecules of some fullerenes are hollow spheres, whereas molecules of others are hollow tubes. The cagelike structures of fullerenes are very flexible. After crashing into steel plates at speeds of 7000 m/s (about 16,000 mph), C_{60} molecules rebound with their original shapes intact.

Linear Acetylenic Carbon

This threadlike allotrope of carbon, shown in **Figure 6,** is organized into long spirals of bonded carbon atoms. Each spiral contains 300 to 500 carbon atoms. It's produced by using a laser to zap a graphite rod in a glass container filled with argon gas. The allotrope splatters on the glass walls and is then removed. Because they conduct electricity, these carbon filaments may have uses in microelectronics. Some linear acetylenic carbons may eventually form fullerenes, or carbon black.

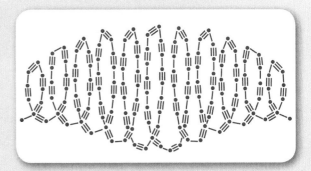

■ **Figure 6** Linear acetylenic carbon model

Discuss the Technology

1. **Apply** From their structures, predict how buckminsterfullerene, diamond, and graphite rank in increasing order of mass density. Explain.

2. **Think Critically** How would you describe a molecule of buckminsterfullerene? A molecule of linear acetylenic carbon? A molecule of diamond?

3. **Acquire Information** What might be some possible uses of fullerenes and linear acetylenic carbons?

■ **Figure 5.16** In this water purifier, ozone is bubbled through water in a tank that is 3.048 m tall and 0.0508 m in circumference. The ozone reacts with large biological molecules, killing microorganisms.

Allotropes of oxygen The oxygen that we breathe is diatomic oxygen, O_2. Although this form of oxygen is the most common in our atmosphere, oxygen also exists as O_3—ozone. The structure of ozone consists of three atoms of oxygen rather than the two atoms in diatomic oxygen.

$$\ddot{O}::\ddot{O}:\ddot{O}:$$

Like diatomic oxygen, ozone occurs naturally. You may have smelled the sharp odor of ozone during an electrical storm as it formed in the atmosphere by the action of lightning. Lightning and ultraviolet light can both provide the energy to convert diatomic oxygen to ozone. Small amounts of ozone also are formed in TV sets or computer monitors when an electrical discharge passes through the oxygen in the air. Have you ever smelled it when sitting close to the screen? Because ozone is harmful to living things, it is advisable that you not sit close to your TV set or computer.

Although ozone formed near the surface of Earth is an undesirable component of smog, ozone also has many uses, such as the water purifier shown in **Figure 5.16.** The layer of ozone found high in Earth's atmosphere is helpful because it shields living things from harmful ultraviolet radiation from the sun.

Formulas and Names of Covalent Compounds

Covalent compounds make up a large group; millions of covalent compounds are already known, and scientists are likely to discover or create many others. How can you possibly begin to study so many compounds? Before you can study their structures, properties, and potential uses, you should be able to name the compounds and write their chemical formulas. Fortunately, chemists have devised a naming system for covalent compounds that is based on a much smaller number of rules than there are compounds.

■ **Figure 5.17** The compound represented by the formula CS_2 is named carbon disulfide because two sulfur atoms are bonded to one carbon.

■ **Figure 5.18** Nitric acid and copper react to produce a dense cloud of brown nitrogen dioxide gas. Nitrogen dioxide is an air pollutant found in smog.

Naming binary inorganic compounds Substances are either organic or inorganic. Compounds that contain carbon, with a few exceptions, are classified as **organic compounds.** Compounds that do not contain carbon are called **inorganic compounds.** How are inorganic compounds held together? If inorganic compounds contain only two nonmetal elements, they are bonded covalently and are referred to as binary compounds.

The suffix *-ide* To name these compounds, write out the name of the first nonmetal and follow it by the name of the second nonmetal with its ending changed to *-ide*. How do you know which element to write first? You write the element that is farther to the left in the periodic table first, with the exceptions of a few compounds that contain hydrogen. If both elements are in the same group, name the element that is closer to the bottom of the periodic table first. For example, sulfur dioxide is a compound containing sulfur and oxygen. The sulfur is named first because it is closer to the bottom of the periodic table than oxygen is.

Indicating the number of atoms Because nonmetal atoms can share different numbers of electron pairs, several different compounds can be formed from the same two nonmetal elements. To name the compound correctly, add a prefix to the name of each element to indicate how many atoms of each element are present in the compound. The same prefixes that were used to indicate the number of water molecules in hydrates are used here. For example, CS_2, shown in **Figure 5.17,** is named carbon disulfide. Refer to **Table 5.5** to review these prefixes.

A few other rules are helpful when naming covalent compounds. If only one atom of the *first* element is listed, the prefix *mono* is usually omitted. Also, if the vowel combinations *o-o* or *a-o* appear next to each other in the name, the first of the pair is omitted to simplify pronunciation. Thus, mononitrogen monooxide, NO, becomes nitrogen monoxide.

Now you are ready to practice naming covalent compounds like the brown gas, nitrogen dioxide (NO_2), shown in **Figure 5.18.** Several different molecules can be formed when different numbers of nitrogen and oxygen atoms combine. Look at their formulas in the first column of **Table 5.6,** and try to name them without looking at the names listed in the second column.

Table 5.6	Formulas and Names of Some Molecular Compounds
Formula	Name
NO	nitrogen monoxide
NO_2	nitrogen dioxide
N_2O	dinitrogen monoxide
N_2O_5	dinitrogen pentoxide

Writing the formula from the name Consider two compounds that contain carbon and oxygen. The carbon contained in wood is converted to carbon dioxide when wood burns completely. The formula for this product is CO_2. If the carbon in wood burns incompletely, the highly toxic gas carbon monoxide is produced. What is the formula for carbon monoxide? To write the formula of a covalent compound for which you are given the name, first write the symbols of each element in the order given in the name. Then add the appropriate subscript after each element that has two or more atoms present. Remember that the prefixes in the name tell how many atoms of each element are present. If an element has no prefix, it is understood that there is only one atom of that element. Therefore, the formula for carbon monoxide is CO.

For more practice, the compound sulfur hexafluoride contains the elements sulfur and fluorine. There is only one sulfur atom; thus, the symbol S does not require a subscript. The prefix *hexa-* tells you that there are six fluorine atoms in the compound, so the subscript 6 must be added to the F. The formula for sulfur hexafluoride is SF_6. Follow the rules for writing a formula for a covalent compound as you examine the formula shown in **Figure 5.19**.

N_2O_3

dinitrogen trioxide

■ **Figure 5.19** The formula for dinitrogen trioxide is written N_2O_3. Analyze the name of this compound to determine how its formula is written.

PRACTICE Problems   Solutions to Problems Page 858

13. Name the following covalent compounds.
    a) S_2Cl_2
    b) CS_2
    c) SO_3
    d) P_4O_{10}
14. Write the formulas for each of the following covalent compounds.
    a) carbon tetrachloride
    b) iodine heptafluoride
    c) dinitrogen monoxide
    d) sulfur dioxide

SUPPLEMENTAL PRACTICE

For more practice with the formulas of covalent compounds, see Supplemental Practice, page 812.

Common names A few inorganic covalent compounds have common names that all scientists use in place of formal names. Two of these compounds are water and ammonia. The chemical name for water is dihydrogen monoxide because each molecule contains two hydrogen atoms and one oxygen atom. If you wanted to get a glass of water at a restaurant, would you ask for dihydrogen monoxide? Probably not, at least not if you were really thirsty. Most people would not understand you because you used a name that even chemists never use for water. Although the formal names of both ionic and covalent compounds are simple to write once you learn the rules for the language of chemistry, there are good reasons for sometimes using common names. Which name you use will depend on your audience.

Concepts In Motion

Interactive Table Explore names of common acids and bases at glencoe.com.

Table 5.7	Names of Common Acids and Bases
Formula	Name
Acids	
HCl	hydrochloric acid
H_2SO_4	sulfuric acid
H_3PO_4	phosphoric acid
HNO_3	nitric acid
$HC_2H_3O_2$	acetic acid (an organic compound)
Bases	
NaOH	sodium hydroxide
KOH	potassium hydroxide
NH_3	ammonia

Common acids and bases Acids and bases are additional examples of inorganic compounds that are sometimes known by common rather than formal names. A few names of common acids and bases that you will frequently use in chemistry laboratory experiments are listed in **Table 5.7.** They often do not follow the rules you have been learning, but they will soon become so familiar that their formulas and names will be easy to remember.

Naming organic compounds You have learned that most compounds that contain carbon are organic compounds. Organic compounds make up the largest class of covalent compounds known. This occurs because carbon is able to bond to other carbon atoms in rings and chains of many sizes.

Even the most complex organic compound are defined as a **hydrocarbon,** an organic compound that contains only the elements hydrogen and carbon. Hydrocarbons occur naturally in fossil fuels such as coal, natural gas, and petroleum and are used mainly as fuels and within the raw materials used for making other organic compounds.

A carbon atom can form four covalent bonds. In the simplest hydrocarbon, methane, a single carbon is bonded to four hydrogen atoms. Methane is the main component of the natural gas that you burn when you light a Bunsen burner. The next simplest hydrocarbon, ethane, is formed when two carbon atoms bond to each other as well as to three hydrogen atoms each. The formulas and names of the first ten hydrocarbon chains are shown in **Table 5.8.** Note that the names of hydrocarbons are derived from the number of carbon atoms in the molecules. Do you recognize any of these hydrocarbons? What is propane used for? **Figure 5.20** illustrates its structure and one common use.

Table 5.8	Hydrocarbons
Formula	Name
CH_4	methane
C_2H_6	ethane
C_3H_8	propane
C_4H_{10}	butane
C_5H_{12}	pentane
C_6H_{14}	hexane
C_7H_{16}	heptane
C_8H_{18}	octane
C_9H_{20}	nonane
$C_{10}H_{22}$	decane

Methane

Propane

■ **Figure 5.20** The structures of methane and propane are shown here. Methane is the main component in natural gas, and propane is used in gas grills. Many other hydrocarbons are also used for fuel.
Determine How many carbon atoms does methane contain? Propane?

Connecting Ideas

Formulas represent the known composition of real substances, but just because a formula can be written doesn't mean the compound actually exists. For example, you could easily write the formula HeP_2, but no such compound has ever been isolated. Compounds containing the noble gases helium, neon, and argon have never been found. In the next chapter, you will study the chemical changes that elements and compounds undergo as they react to form new substances. You will also learn how to represent these changes in the language of chemistry.

Section 5.2 Assessment

Section Summary

- Covalent compounds generally have low melting points, low water solubility, and little or no ability to act as electrolytes.
- Some elements exist in different structural forms called allotropes.
- Binary covalent compounds are named by writing the two elements in the order they are found in the formula, changing the ending of the second element to -ide.
- Greek prefixes indicate how many atoms of each element are present.
- Hydrocarbons are common compounds composed of hydrogen and carbon.

15. **MAIN Idea** **Explain**, in terms of electron arrangement, why carbon typically forms four bonds.

16. **Write** the name of the molecular compound represented by each formula.
    a) BF_3
    b) PBr_5
    c) C_2H_6
    d) IF_7
    e) NO
    f) SiO_2

17. **Explain** what allotropes are and give two examples.

18. **Apply** Write the formula for each of the following covalent compounds.
    a) carbon monoxide
    b) phosphorus pentachloride
    c) sulfur hexafluoride
    d) dinitrogen pentoxide
    e) iodine trichloride
    f) heptane

19. **Evaluate** A tank of a substance delivered to a factory is labeled C_4H_{10}. What is the name of the substance in the tank? What is it likely used for?

CHAPTER 5 Study Guide

Download quizzes, key terms, and flash cards from glencoe.com.

BIG Idea There are two types of compounds: ionic and covalent.

Section 5.1 Ionic Compounds

MAIN Idea Atoms in ionic compounds are held together by the attraction of oppositely charged ions.

Vocabulary
- anhydrous (p. 165)
- binary compound (p. 153)
- deliquescent (p. 165)
- formula unit (p. 154)
- hydrate (p. 164)
- hygroscopic (p. 165)
- oxidation number (p. 155)
- polyatomic ion (p. 156)

- The position of an element in the periodic table indicates what charge its ions will have.
- Binary ionic compounds are named by first naming the metal element and then the nonmetal element, with its ending changed to -ide.
- Subscripts in formulas indicate how many atoms of each element are present in a compound.
- Polyatomic ions can combine with ions of opposite charge to form ionic compounds.
- Most transition elements can form two or more positively charged ions. The oxidation number of the transition element is indicated by a Roman numeral in parentheses.
- Hydrates are ionic compounds bonded to water molecules.

Section 5.2 Covalent Compounds

MAIN Idea Atoms in covalent compounds are held together by shared electrons.

Vocabulary
- allotrope (p. 173)
- distillation (p. 168)
- hydrocarbon (p. 180)
- inorganic compound (p. 178)
- molecular element (p. 172)
- organic compound (p. 178)

- Covalent compounds generally have low melting points, low water solubility, and little or no ability to act as electrolytes.
- Some elements exist in different structural forms called allotropes.
- Binary covalent compounds are named by writing the two elements in the order they are found in the formula, changing the ending of the second element to -ide.
- Greek prefixes indicate how many atoms of each element are present.
- Hydrocarbons are common compounds composed of hydrogen and carbon.

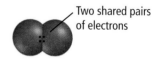
Two shared pairs of electrons

(a) :Ö· + ·Ö: → :O=O:

(b) :N· + ·N: → ·N≡N·

Three shared pairs of electrons

Chapter 5 Assessment

Understand Concepts

20. Which substances are ionic and which are covalent?
    a) magnesium sulfate
    b) hexane
    c) carbon monoxide
    d) ozone
    e) cesium chloride
    f) cobalt(II) chloride

21. Write the formula for the binary ionic compound that forms from each pair of elements.
    a) manganese(III) and iodine
    b) calcium and oxygen
    c) aluminum and fluorine

22. Write the name for each compound containing polyatomic ions.
    a) $Ca(C_2H_3O_2)_2$
    b) $NaOH$
    c) $(NH_4)_2SO_3 \cdot H_2O$
    d) $MgSO_4$
    e) $NaNO_2$
    f) $Ca(OH)_2$

23. Make a table comparing the properties of ionic and covalent compounds.

24. The metals in the following compounds can have various oxidation numbers. Predict the charge on each metal ion, and write the name for each compound.
    a) $FeCl_3$
    b) CuF_2
    c) $AuBr_3$
    d) $SnBr_4$
    e) FeS
    f) $Pb(C_2H_3O_2)_2$

25. How can water that contains dissolved ionic compounds be separated from the ionic compounds?

26. Name the covalent compound that is represented by each of the following formulas.
    a) NO
    b) IBr
    c) N_2O_4
    d) CO
    e) SiO_2
    f) ClF_3

27. What happens to the composition of a hydrate when it is heated?

Apply Concepts

28. Predict the effect of increasing acidity of rain on the rate of formation of limestone caves.

29. How could you determine quantitatively whether the ionic compound table salt or the covalent compound table sugar is more soluble in water?

30. Explain why most elements do not occur naturally in their pure states.

Everyday Chemistry

31. What is hard water and how is it treated?

Art Connection

32. Suppose an artisan wanted to coat a clay vessel with a faint pink glaze. What material should he or she add to a transparent glaze to achieve this result?

Chemistry and Technology

33. Use the structural organization of graphite to explain why it is a good lubricant.

Think Critically

Design an Experiment

34. **MiniLab 1** Design an experiment to determine the minimum amount of water required to change the color of the anhydrous cobalt compound weather predictor.

Make Predictions

35. **ChemLab** Would you expect a warm or a cool saturated solution of potassium nitrate (KNO_3) in water to be a better electrolyte? Explain.

Apply Concepts

36. Mercury(I) is unusual in that it often forms an ion that links with another mercury(I) ion. Thus, two mercury(I) ions are linked together in a single unit. What is the charge on this double ion? Write the formula for the compound that this ion forms with chlorine.

Chapter 5 Assessment

Use a Table

37. **Table 5.9** lists melting points for a number of ionic compounds. Do the melting points of the sodium and potassium compounds increase or decrease as you move down group 17? What does this suggest about the strength of the ionic bonds between these metals and the group 17 nonmetals?

Table 5.9 — Melting Points of Several Compounds

Compound	Melting Point (°C)
NaCl	804
NaI	651
KCl	773
KBr	730
NaF	993
KI	680
NaBr	755

Cumulative Review

38. How are physical changes different from chemical changes? (*Chapter 1*)

39. How does the atomic number compare with the number of electrons in a neutral atom? (*Chapter 2*)

40. How can the periodic table be used to determine the number of valence electrons in an element? (*Chapter 3*)

WRITING in Chemistry

41. Write a set of descriptions comparing the structures of a soccer ball, a geodesic dome, and buckminsterfullerene. Are their similarities a coincidence?

Skill Review

42. **Make and Use Graphs** Use the data in **Table 5.10** to construct a graph of melting point versus number of carbons and a graph of water solubility versus number of carbons. What is the relationship between chain length (number of carbon atoms) and melting point? How are chain length and water solubility related?

Table 5.10 — Number of Carbons Compared to Melting Point and Water Solubility

Number of Carbon Atoms	Melting Point (°C)	Water Solubility (g per 100 mL)
1 (methane)	−183	0.0024
2 (ethane)	−172	0.0059
3 (propane)	−188	0.012
4 (butane)	−138	0.037
5 (pentane)	−130	0.036
6 (hexane)	−95	0.0138
7 (heptane)	−91	0.0052
8 (octane)	−57	0.0015
9 (nonane)	−54	insoluble
10 (decane)	−30	insoluble

Problem Solving

43. Write the formulas for phosphorus trioxide and phosphorus pentoxide.
  a) What percent of the atoms in phosphorus trioxide are phosphorus? What percent are oxygen?
  b) What percent of the atoms in phosphorus pentoxide are phosphorus? What percent are oxygen?

Cumulative Standardized Test Practice

1. Barium chloride ($BaCl_2$) is a binary compound because it contains
   a) two elements.
   b) two ions.
   c) two oxidized elements.
   d) two bonds.

2. An oxidation number is
   a) the number of electrons an atom will lose.
   b) the number of electrons an atom will gain.
   c) the overall charge of an atom.
   d) the overall charge of an ion.

3. Why is the gas hydrogen rarely found in elemental form on Earth?
   a) Hydrogen is a rare element.
   b) Hydrogen deposits are difficult to access.
   c) Hydrogen easily forms compounds.
   d) Hydrogen is an unreactive atmospheric gas.

4. What is the correct chemical formula for the ionic compound formed by the calcium ion (Ca^{2+}) and the acetate ion ($C_2H_3O_2^-$)?
   a) $CaC_2H_3O_2$
   b) $CaC_4H_6O_8$
   c) $(Ca)_2C_2H_3O_2$
   d) $Ca(C_2H_3O_2)_2$

5. The volume of an atom is made up mostly of
   a) protons.
   b) neutrons.
   c) electrons.
   d) empty space.

6. Which is NOT a quantitative measurement of a pencil?
   a) length
   b) mass
   c) color
   d) diameter

7. Copper(II) sulfate has the chemical formula
   a) $CuSO_4$
   b) Cu_2SO_4
   c) $Cu_2(SO_4)_2$
   d) CuS_2O_8

Use the table below to answer questions 8–9

Formula and Names of Common Compounds Containing Nitrogen		
Formula	Molecular Compound Name	Common Name
?	Nitrogen monoxide	Nitrogen monoxide
NH_3	?	Ammonia
?	Dinitrogen tetrahydride	Hydrazine
N_2O	?	Nitrous oxide (Laughing gas)
NO_2	?	Nitrogen dioxide

8. What is the molecular compound name for laughing gas?
   a) mononitrogen dioxide
   b) nitrogen dioxide
   c) dinitrogen monoxide
   d) dinitrogen oxide

9. What is the molecular formula of hydrazine?
   a) N_4H_2
   b) N_2H_4
   c) $N_2(OH)_4$
   d) $N_4(OH)_2$

10. Elements in the same group of the periodic table have the same
    a) number of valence electrons.
    b) physical properties.
    c) number of protons.
    d) electron configuration.

NEED EXTRA HELP?

If You Missed Question . . .	1	2	3	4	5	6	7	8	9	10
Review Section . . .	5.1	5.1	4.1	5.1	2.1	1.1	5.1	5.2	5.2	3.2

CHAPTER 6: Chemical Equations and Reactions

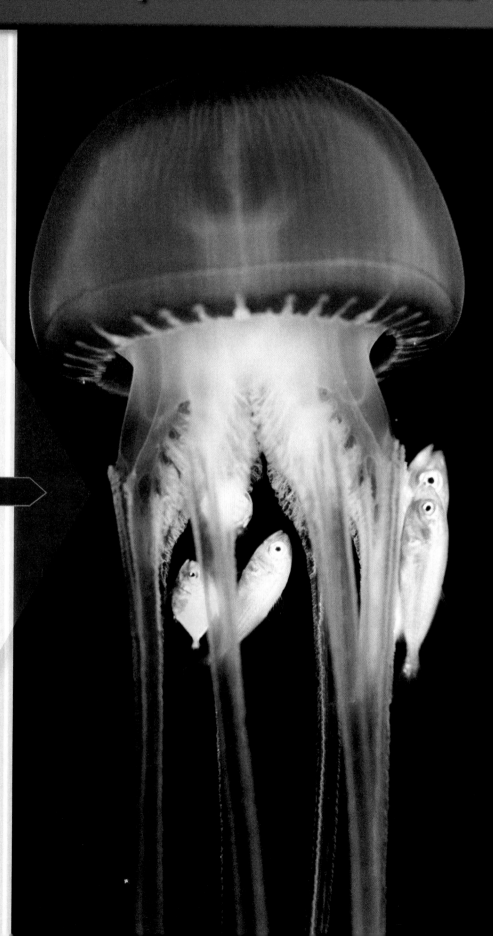

BIG Idea Millions of chemical reactions in and around you absorb or release energy as they transform reactants into products.

6.1 Chemical Equations
MAIN Idea Balanced chemical equations represent chemical reactions.

6.2 Types of Reactions
MAIN Idea There are five types of chemical reactions: synthesis, decomposition, single displacement, double displacement, and combustion reactions.

6.3 Nature of Reactions
MAIN Idea External factors modify the direction and rate of chemical reactions.

ChemFacts

- Bioluminescence is the production of light by living things.
- Functions of bioluminescence include attracting mates or prey, communication, camouflage, and defense.
- Bioluminescence is much more common in the sea than on the land.
- According to one estimate, roughly half of the known types of jellyfish are bioluminescent. Almost all of the deep-sea species of jellyfish are bioluminescent.

Start-Up Activities

LAUNCH Lab

Observe a Chemical Reaction

What evidence can you observe that a reaction takes place?

Safety Precautions

Materials

- 10-mL graduated cylinder
- 100-mL beaker
- stirring rod
- 0.01M potassium permanganate ($KMnO_4$)
- 0.01M sodium hydrogen sulfite ($NaHSO_3$)

Procedure

1. Read and complete the lab safety form.
2. Measure 5.0 mL of 0.01M potassium permanganate solution ($KMnO_4$) and pour it into a 100-mL beaker.
3. Add 5.0 mL of 0.01M sodium hydrogen sulfite solution ($NaHSO_3$) to the potassium permanganate solution while stirring. Record your observations.
4. Slowly add additional 5.0-mL portions of the $NaHSO_3$ solution until the $KMnO_4$ solution turns colorless. Record your observations.
5. Record the total volume of the $NaHSO_3$ solution you used to cause the beaker's contents to become colorless.

Analysis

1. **Infer** What evidence do you have that a reaction occurred?

Inquiry Would anything more have happened if you continued to add $NaHSO_3$ solution to the beaker? Explain.

FOLDABLES Study Organizer

Chemical Reactions Make the following Foldable to help you organize information about how chemical reactions are classified.

▶ **STEP 1** Fold a sheet of paper lengthwise, keeping the margin visible on the left side.

▶ **STEP 2** Cut the top flap into five tabs.

▶ **STEP 3** Label as follows: *Chemical Reactions, Synthesis, Decomposition, Single Displacement, Double Displacement,* and *Combustion.*

FOLDABLES Use this Foldable with Section 6.2. As you read the section, summarize each type of chemical reaction and provide examples.

Visit glencoe.com to:
▶ study the entire chapter online
▶ explore **concepts in Motion**
▶ take Self-Check Quizzes
▶ use Personal Tutors
▶ access Web Links for more information, projects, and activities
▶ find the Try at Home Lab, Preventing a Chemical Reaction

Chapter 6 • Chemical Equations and Reactions **187**

Section 6.1

Objectives
- **Relate** chemical changes and macroscopic properties.
- **Demonstrate** how chemical equations describe chemical reactions.
- **Illustrate** how to balance chemical reactions by changing coefficients.

Review Vocabulary
energy: the capacity to do work

New Vocabulary
reactant
product
coefficient

Chemical Equations

MAIN Idea Balanced chemical equations represent chemical reactions.

Real-World Reading Link What do you remember about last summer's Fourth of July? Amazing bursts of color from fireworks shot over a lake? Mouthwatering aromas coming from a barbecue grill? Do the processes that result in those colors and smells have anything in common?

Chemical Reactions

You read in Chapter 1 that substances undergo both physical and chemical changes. A physical change does not change the substance itself, but a chemical change does. Do the chemicals making up fireworks, charcoal, and barbecued food undergo chemical changes? As indicated in **Figure 6.1**, they do.

When a substance undergoes a chemical change, it takes part in a chemical reaction. After it reacts, it no longer has the same chemical identity. While it might seem amazing that a substance can undergo a change and become part of a different substance, chemical reactions occur around you all the time. Chemical reactions can be used to heat a home, power a car, manufacture fabrics for clothing, make medicines, and produce paints and dyes in your favorite colors. Reactions also provide energy for walking, running, working, and thinking.

Many important clues indicate when chemical reactions occur. None of them alone proves that such a change occurs because some physical changes, such as boiling, involve one or more of these signs. Examine the images in **Figure 6.2** to see what clues to look for.

■ **Figure 6.1** When substances undergo chemical changes, observable differences usually occur. The brilliant colors of these fireworks, the aroma of cooking food, and the light and heat from burning charcoal all come from chemical changes.

FIGURE 6.2

Signs of a Chemical Reaction

If you know what signs to look for, you can determine whether or not a chemical reaction has taken place.

Color changes often accompany chemical changes. If you place brownish-red iodine solution on a freshly cut potato, it reacts with white starch to produce a blue compound.

Precipitation of a solid from a solution can result from a chemical change. For example, when a solution of sodium fluoride (NaF) is mixed with a solution of calcium chloride ($CaCl_2$), solid calcium fluoride (CaF_2) precipitates out of solution.

Energy changes occur during all chemical changes. Heat or light can be absorbed or released during a chemical reaction. For example, oxygen rapidly reacts with fuels such as charcoal and wood during burning.

Odor changes can indicate that a substance has undergone a chemical change. When bread is baked, delicious new aromas result from chemical changes that take place while the bread is in the oven.

Gas release sometimes occurs as a result of a chemical change. The gas that fills an automobile's air bag results from a chemical reaction involving sodium azide (NaN_3).

Section 6.1 • Chemical Equations

Compounds in wood react with oxygen when they burn to form water and carbon dioxide.

Energy is needed by our bodies to perform daily activities. This energy is provided when glucose combines with oxygen in cells, also forming carbon dioxide and water.

■ **Figure 6.3** The substances produced when wood burns are the same as those produced by the reaction that provides energy for your body.
Identify *the reactants and the products in these reactions.*

VOCABULARY
WORD ORIGIN
Product
comes from the Latin word *productum* meaning *something produced*

Chemical Equations

In order to completely understand a chemical reaction, you must be able to describe any changes that take place. Part of that description involves recognizing what substances react and what substances form. A substance that undergoes a reaction is called a **reactant**. When reactants undergo a chemical change, each new substance formed is called a **product**. For example, a familiar chemical reaction involves the reaction between iron and oxygen (the reactants) that produces rust, which is iron(III) oxide (the product). The simplest reactions involve a single reactant or a single product, but some reactions involve many reactants and many products. **Figure 6.3** illustrates two familiar reactions that involve oxygen. In the first case, oxygen combines with molecules in wood in a reaction that releases energy in the form of heat and light. In the second, oxygen combines with glucose in a reaction that releases energy for use by your body.

The description Several possible observations help determine when a chemical reaction has taken place. But these observations don't completely describe what happens between reactants to form products. Have you ever seen what happens when baking soda and vinegar are mixed together? They react quickly, as you can tell by the bubbles that seem to explode out of the mixture, as shown in **Figure 6.4.** In describing this reaction, you could say that baking soda and vinegar turn into bubbles. But does that completely explain what happens? What are the bubbles made of? Do all of the atoms in vinegar and baking soda form bubbles? The reaction involves more than what can be determined by observation alone. Just as you can write a sentence to tell others what happened on your way to school today, chemists represent the changes taking place in a reaction by writing equations.

■ **Figure 6.4** Vinegar (acetic acid) and baking soda react vigorously, forming a bubbly product. This reaction is used in fire extinguishers because the bubbles produced contain carbon dioxide, which is effective in putting out fires. A word equation describing this reaction can be written as follows:
vinegar + baking soda ⟶
    sodium acetate + water + carbon dioxide

Word equations The simplest way to represent a reaction is by using words to describe all the reactants and products, with an arrow placed between them to represent change, as shown in **Figure 6.4** and its caption. As you can see in this word equation, reactants are placed to the left of the arrow, and products are placed to the right. Plus signs are used to separate reactants and also to separate products.

Vinegar and baking soda are common names. The compound in vinegar that is involved in the reaction is acetic acid, and baking soda is sodium hydrogen carbonate. These scientific names can also be used in a word equation.

acetic acid + sodium hydrogen carbonate ⟶
    sodium acetate + water + carbon dioxide

Chemical equations Word equations describe reactants and products, but they are long, awkward, and do not show the formulas of the substances involved. Word equations can be converted into chemical equations by substituting chemical formulas for the names of compounds and elements. Recall from Chapter 5 that these formulas can be written by using the oxidation numbers of the elements and the charges of the polyatomic ions. For example, the equation for the reaction of vinegar and baking soda can be written using the chemical formulas of the reactants and products.

$$HC_2H_3O_2 + NaHCO_3 \rightarrow NaC_2H_3O_2 + H_2O + CO_2$$
acetic acid   sodium   sodium   water   carbon
              hydrogen   acetate   dioxide
              carbonate

By examining a chemical equation, you can determine exactly what elements make up the substances that react and form.

Everyday Chemistry

Whitening Whites

Why might playing soccer on a muddy soccer field be worrisome? The white socks shown in **Figure 1** might never be white again! However, most of the evidence of the muddy game will be removed with a 3–6 percent aqueous solution of sodium hypochlorite.

Household bleach Perhaps the most popular type of household bleach is an aqueous solution of sodium hypochlorite (NaClO). Sodium hypochlorite is made by reacting chlorine gas (Cl_2) with an aqueous solution of sodium hydroxide (NaOH), as shown in the following equation.

$$Cl_2(g) + 2NaOH(aq) \rightarrow$$
$$NaClO(aq) + NaCl(aq) + H_2O(l)$$

In an aqueous solution, NaClO does not exist as a complete unit, but as sodium ions (Na^+), and hypochlorite ions (ClO^-). The ingredient responsible for the bleaching action in this type of bleach is the hypochlorite ion. Many other solid and liquid bleaches contain hydrogen peroxide (H_2O_2), instead of sodium hypochlorite. In these bleaches, the active substance in solution is the perhydroxyl ion (HOO^-). What do the ClO^- and HOO^- ions have in common?

Bleaching reactions As you can see, each of these polyatomic ions carries a single negative charge. If each could react with a hydrogen ion, the following reactions would happen.

$$ClO^- + H^+ \rightarrow HCl + O$$

$$HOO^- + H^+ \rightarrow H_2O + O$$

Because the compounds HCl and H_2O are more stable than the hypochlorite and perhydroxyl ions, these reactions occur. The reactions result in a bleaching action because the released oxygen reacts with the substances that cause the stain. The molecules of the compounds that cause color in a stain are structured in a way that gives them the physical property of producing color.

Figure 1 Bleach will not remove the stains. It will make the stains colorless and the socks look clean.

In the reaction between these compounds and atomic oxygen, the compound or compounds formed have different structures. These structures do not have physical properties that produce color. So a bleach does not remove a stain by removing the substance that caused the stain. A bleach bleaches by rendering a colored compound colorless so you just can't see it.

History of household bleach Sodium hypochlorite, the active ingredient in bleach, was discovered by the French chemist Berthollet in 1787. Its ability to whiten clothing was quickly discovered. By the end of the nineteenth century, sodium hypochlorite was used as a disinfectant after it was discovered that bleach was very effective at killing disease-causing bacteria. Besides being used to remove the evidence of stains on clothing, household bleach is also used today in homes, schools, hospitals, swimming pools, drinking water supplies, and for disinfecting hard surfaces and surgical instruments.

Explore Further

1. **Apply** Liquid bleaches containing sodium hypochlorite are often sold in opaque, plastic containers because sunlight causes the compound to decompose to produce oxygen gas and sodium chloride. Write the balanced chemical equation for this reaction.

2. **Infer** Why do you think bleaches containing sodium hypochlorite tend to damage finer fabrics more than bleaches containing hydrogen peroxide?

Physical state It might also be important to know the physical state of each reactant and product. How can we indicate that the CO_2 formed during this reaction is a gas and that the water is a liquid? Symbols in parentheses are put after formulas to indicate the state of the substance. Solids, liquids, gases, and water (aqueous) solutions are indicated by the symbols (s), (l), (g), and (aq). The following equation shows these symbols added to the equation for the reaction of vinegar and baking soda.

$$HC_2H_3O_2(aq) + NaHCO_3(s) \rightarrow NaC_2H_3O_2(aq) + H_2O(l) + CO_2(g)$$

Now the equation tells us that mixing an aqueous solution of acetic acid (vinegar) with solid sodium hydrogen carbonate (baking soda) results in the formation of an aqueous solution of sodium acetate, liquid water, and carbon dioxide gas. If you had examined this equation before you mixed vinegar and baking soda, you could have predicted that bubbles would form.

Energy and chemical equations Noticeable amounts of energy are often released or absorbed during a chemical reaction. Some reactions absorb energy. If energy is absorbed, the reaction is known as an endothermic reaction. **Figure 6.5** shows an example of an endothermic reaction.

For a reaction that absorbs energy, the word energy is sometimes written along with the reactants in the chemical equation. For example, the equation for the reaction in which water breaks down into hydrogen and oxygen gases shows that energy must be added to the reaction.

$$2H_2O(l) + energy \rightarrow 2H_2(g) + O_2(g)$$

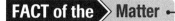

Noble gases are relatively unreactive, but they are not totally inert. The first compound containing a noble gas was synthesized by Neil Bartlett in 1962. He made xenon hexafluoride using the following reaction.

$$Xe(g) + 3F_2(g) \rightarrow XeF_6(s)$$

Stable compounds of krypton and radon have also been synthesized.

■ **Figure 6.5** The reaction of ammonium chloride and barium hydroxide octahydrate is endothermic. If these reactants are mixed at room temperature, which is about 20°C, the temperature in the mixture drops as energy is absorbed by the reaction.

Reactions that occur in a Bunsen burner, a gas grill, or mechanisms used to power an automobile release energy. As you will recall from Chapter 1, reactions that release heat energy are called exothermic reactions. When writing a chemical equation for a reaction that produces energy, the word energy is sometimes written along with the products. For example, the equation for the reaction that occurs when you light methane (CH_4) in a Bunsen burner shows that energy is released. Most of this energy is in the form of heat.

$$CH_4(g) + 2O_2(g) \rightarrow CO_2(g) + 2H_2O(g) + energy$$

The word energy is not always written in an equation. It is used only if it is important to know whether energy is released or absorbed. For the burning of methane, energy would be written in the equation because the release of heat is an important part of burning a fuel. Energy would also be written in the equation that describes the reaction when water is broken down into hydrogen and oxygen because the reaction would not occur without the addition of energy. In many reactions, such as the formation of rust, energy might be released or absorbed but it is not included in the equation because it is not important to know about the energy involved in that particular reaction. In other chemical reactions, such as the one shown in the How It Works feature on the next page, energy is released mainly in the form of light rather than heat.

MiniLab 6.1

Energy Change

How can energy changes be observed? All chemical reactions involve an energy change. This change might be so slight that it can be detected only with sensitive instruments, or it may be quite noticeable.

Procedure
1. Read and complete the lab safety form.
2. Place 25 g of **iron powder** and 1 g of **NaCl** in a **resealable plastic bag**.
3. Add 30 g of **vermiculite** to the bag, seal the zipper, and shake the bag to mix the contents.
4. Add 5 mL of **water** to the bag, reseal the zipper, and gently squeeze and shake the contents to mix them.
5. Hold the bag between your hands and note any changes in temperature.

Analysis
1. **Classify** What did you observe? What type of reaction produces this kind of change?
2. **Apply** What practical application for this reaction does the photo above suggest?

HOW IT WORKS

Light Sticks

The light stick was developed in the early 1960s by research and development firms and by the military, but was not patented until 1976. Light given off by a light stick is energy released from a chemical reaction. Reactions in which light is given off are called chemiluminescent reactions. The light from a light stick is only temporary; when the reactants are used up, light is no longer produced.

Interactions among various combinations of chemicals produce different colors of light. The most common colors are yellow and green because these are the easiest colors to produce in light sticks. The most difficult colors to produce are red and purple. Light sticks provide light when no electricity is available or when light is needed underwater. Light sticks are also a source of entertainment at parties, fairs, concerts, and other nighttime activities.

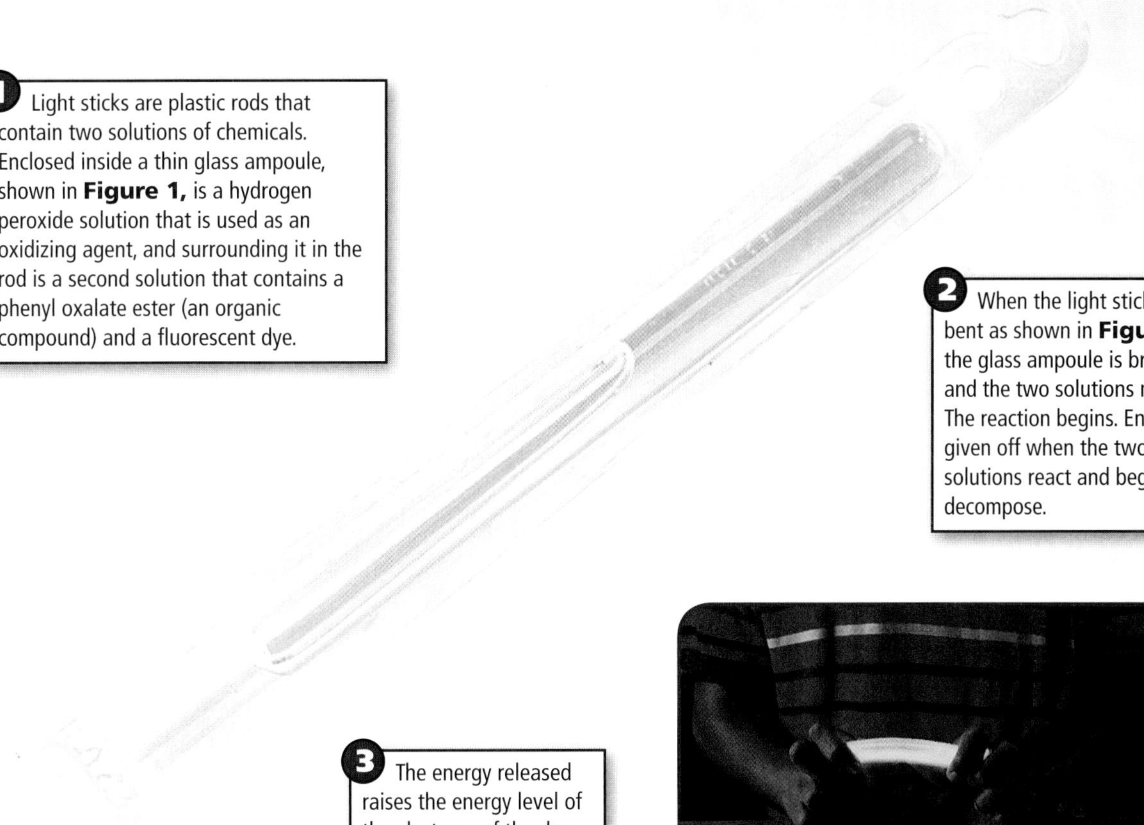

1 Light sticks are plastic rods that contain two solutions of chemicals. Enclosed inside a thin glass ampoule, shown in **Figure 1,** is a hydrogen peroxide solution that is used as an oxidizing agent, and surrounding it in the rod is a second solution that contains a phenyl oxalate ester (an organic compound) and a fluorescent dye.

2 When the light stick is bent as shown in **Figure 2,** the glass ampoule is broken and the two solutions mix. The reaction begins. Energy is given off when the two solutions react and begin to decompose.

3 The energy released raises the energy level of the electrons of the dye molecules.

■ **Figure 1** Light stick construction

4 When the excited electrons drop back down to their original energy levels, the extra energy is given off as light. This light is sometimes referred to as cool light because no noticeable heat is given off in the reaction.

■ **Figure 2** The reaction in the light stick begins when the two solutions mix.

Think Critically

1. **Explain** why the chemical reaction in the light stick is not exothermic, even though it produces energy.
2. **Infer** What are some advantages of light sticks over conventional light sources?

■ **Figure 6.6** One of the experiments that Lavoisier used to discover the law of conservation of mass was the decomposition of mercury(II) oxide (HgO). When HgO is heated, it breaks apart to form mercury metal (Hg) and oxygen gas (O_2). Lavoisier weighed the amount of HgO that decomposed and found it to be the same as the total weight of Hg and O_2 produced.

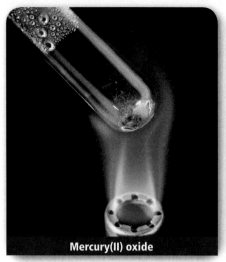
Mercury(II) oxide

Liquid mercury and oxygen

Balanced Chemical Equations

What do you think happens to the atoms in reactants when they are converted into products? Some products, such as the carbon dioxide produced from baking powder when a cake is baked, seem to disappear. What really happens to them?

The law of conservation of mass Recall from Chapter 2 that you can test a reaction to determine whether the same amount of matter is contained in the products and the reactants. That type of experiment was first carried out by the French scientist Antoine Lavoisier (1743–1794). His results indicated that the mass of the products is always the same as the mass of the reactants that react to form them, as indicated by **Figure 6.6.** The law of conservation of mass summarizes these findings: matter is neither created nor destroyed during a chemical reaction.

Conservation of atoms Remember that atoms don't change in a chemical reaction; they are just rearranged. The number and kinds of atoms present in the reactants of a chemical reaction are the same as those present in the products. You can think of conservation of mass in terms of conservation of atoms. For a chemical equation to accurately represent a reaction, the same numbers of each kind of atom must be on the left side of the arrow as are on the right side. If an equation demonstrates the conservation of atoms, it is said to be balanced.

How can you count atoms in an equation? The easiest way to learn is to practice—first with a simple reaction and then with some that are more complex. For example, consider the equation that represents breaking down carbonic acid (H_2CO_3) into water (H_2O) and carbon dioxide (CO_2).

$$H_2CO_3(aq) \rightarrow H_2O(l) + CO_2(g)$$

Because a subscript after the symbol for an element represents how many atoms of that element are found in a compound, you can see that there are two hydrogen, one carbon, and three oxygen atoms on each side of the arrow. All of the atoms in the reactants are the same as those found in the products.

Examine the equation for the formation of sodium carbonate (Na_2CO_3) and water from the reaction between sodium hydroxide (NaOH) and carbon dioxide (CO_2).

$$NaOH(aq) + CO_2(g) \rightarrow Na_2CO_3(s) + H_2O(l)$$

Do both sides of the equation have the same number of each type of atom? No. One carbon atom is on each side of the arrow, but the sodium, oxygen, and hydrogen atoms are not balanced. The equation, as written, does not truly represent the reaction because it does not show conservation of atoms.

Coefficients To indicate more than one unit taking part or being formed in a reaction, a number called a **coefficient** is placed in front of it to indicate how many units are involved. Place a coefficient of 2 in front of the sodium hydroxide formula in the previous equation.

$$2NaOH(aq) + CO_2(g) \rightarrow Na_2CO_3(s) + H_2O(l)$$

Is the equation balanced now? Two sodium atoms are on each side. How many oxygen atoms are on each side? You should be able to find four on each side. How about hydrogen atoms? Now two are on each side. Because one carbon atom is still on each side, the entire equation is balanced; it now represents what happens when sodium hydroxide and carbon dioxide react.

The balanced equation tells us that when sodium hydroxide and carbon dioxide react, two units of sodium hydroxide react with each molecule of carbon dioxide to form one unit of sodium carbonate and one molecule of water. **Figure 6.7** shows you another example of a balanced equation.

Why can't subscripts be changed when balancing an equation? Changing a subscript changes the identity of that substance. Look at the equation in Example Problem 1. Changing the subscript of the oxygen in water to 2 changes water, H_2O, to hydrogen peroxide, H_2O_2, a different compound. Changing a coefficient simply means that you are changing the amount of that substance compared to the other substances in the reaction. Changing the coefficient of water to 2 means there are two molecules of water, $2H_2O$. The identity of water stays the same.

■ **Figure 6.7** When coal burns, carbon (C) reacts with oxygen (O_2) to form carbon dioxide (CO_2). Examine the balanced equation for this reaction.
Determine *If a piece of coal contains ten billion C atoms, how many molecules of O_2 will it react with? How many molecules of CO_2 will be formed?*

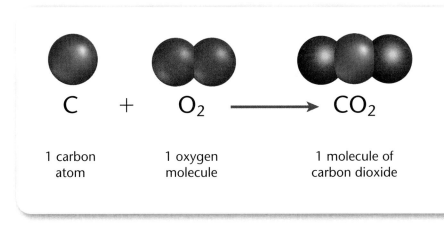

C + O_2 → CO_2

1 carbon atom | 1 oxygen molecule | 1 molecule of carbon dioxide

EXAMPLE Problem 6.1

Writing a Simple Equation Write word and chemical equations for the reaction of hydrogen and oxygen gases to form gaseous water and release energy. This reaction powers the main stage of the space shuttle.

> **Problem-Solving Hint**
> Be sure to change only coefficients, not subscripts, when balancing equations.

1 Analyze
- To write a word equation for the reaction, write the names of the reactants, draw an arrow, then write the name of any product. If there is more than one reactant or product, plus signs should separate them.

$$\text{hydrogen} + \text{oxygen} \longrightarrow \text{water} + \text{energy}$$

2 Set Up
- To write the chemical equation, use chemical formulas to replace the names of the reactants and products in the word equation you wrote. Then add symbols to represent the physical state of each compound. Remember that hydrogen and oxygen occur as diatomic gases.

$$H_2(g) + O_2(g) \longrightarrow H_2O(g) + \text{energy}$$

3 Solve
- To balance the atoms on each side of the arrow, count the number of atoms of each type on each side of the arrow. On the left are two hydrogen atoms and two oxygen atoms. On the right are also two hydrogen atoms but only one oxygen atom. Change the coefficient of the water to 2 so that the number of oxygen atoms will be balanced. Because that puts four hydrogen atoms on the right side of the arrow, the coefficient of hydrogen gas also must be changed to 2.

$$2H_2(g) + O_2(g) \longrightarrow 2H_2O(g) + \text{energy}$$

4 Check
Make sure there are equal numbers of each element on each side of the equation.

EXAMPLE Problem 6.2

Writing an Equation Write word and chemical equations for the reaction that takes place when an aqueous solution of magnesium chloride is added to a silver nitrate solution. Aqueous magnesium nitrate and solid silver chloride form.

1 Analyze
- To write a word equation for the reaction, write the names of the reactants, draw an arrow, then write the name of any product. If there is more than one reactant or product, plus signs should separate them.

$$\text{magnesium chloride} + \text{silver nitrate} \longrightarrow \text{magnesium nitrate} + \text{silver chloride}$$

2 Set Up
- To write the chemical equation, use chemical formulas to replace the names of the reactants and products in the word equation you wrote. Remember to use the oxidation number of an element and the charge on a polyatomic ion to write a correct formula. Then add symbols to represent the physical state of each compound.

$$MgCl_2(aq) + AgNO_3(aq) \longrightarrow Mg(NO_3)_2(aq) + AgCl(s)$$

3 Solve
- To balance the atoms on each side of the arrow, count the number of atoms of each type on each side of the arrow. On the left are one magnesium atom and two chlorine atoms. On the right, there is also one magnesium atom but only one chlorine atom. Change the coefficient of the AgCl to 2 so that the number of chlorine atoms will be balanced. Because that puts two silver atoms on the right side of the arrow, the coefficient of AgNO₃ also must be changed to 2. This balances the oxygen and nitrogen atoms, as well.

$$MgCl_2(aq) + 2AgNO_3(aq) \rightarrow Mg(NO_3)_2(aq) + 2AgCl(s)$$

4 Check
- Make sure there are equal numbers of each element on each side of the equation.

SUPPLEMENTAL PRACTICE

For more practice writing balanced chemical equations, see Supplemental Practice, page 814

PRACTICE Problems
Solutions to Problems Page 858

Write word equations and chemical equations for the following reactions.

1. Magnesium metal and water combine to form solid magnesium hydroxide and hydrogen gas.
2. An aqueous solution of hydrogen peroxide (dihydrogen dioxide) and solid lead(II) sulfide combine to form solid lead(II) sulfate and liquid water.
3. When energy is added to solid manganese(II) sulfate heptahydrate crystals, they break down to form liquid water and solid manganese(II) sulfate monohydrate.
4. Solid potassium reacts with liquid water to produce aqueous potassium hydroxide and hydrogen gas.

Section 6.1 Assessment

Summary
- Chemical equations—used to represent reactions—are written using symbols and formulas for elements and compounds.
- Equations are balanced by changing coefficients.
- Chemical equations can tell you how elements and compounds change during a reaction and whether a reaction is endothermic or exothermic.
- A balanced chemical equation reflects the law of conservation of mass.

5. **MAIN Idea** **Explain** Why is it important to balance a chemical equation?
6. **Apply** Write balanced chemical equations for the reactions described.
   a) sodium metal + chlorine gas → sodium chloride crystals
   b) propane gas + oxygen → carbon dioxide + water vapor + energy
   c) zinc metal + hydrochloric acid → zinc chloride solution + hydrogen
7. **Explain** How can you tell whether a chemical reaction has taken place?
8. **Calculate** Use the law of conservation of mass to determine the following:
   a) The number of grams of CO_2 that form from 4.00 g C and 10.67 g O_2.

   $$C + O_2 \rightarrow CO_2$$

   b) The number of grams of water formed if 7.75 g H_2CO_3 forms 5.50 g CO_2.

   $$H_2CO_3 \rightarrow H_2O + CO_2$$

9. **Write** In the catalytic converter of a car, a reaction occurs when nitrogen monoxide gas (NO) reacts with hydrogen gas. Ammonia gas and water vapor are formed. Write a balanced equation for this reaction.

Self-Check Quiz glencoe.com

Section 6.2

Objectives

- **Distinguish** among the five major types of chemical reactions.
- **Classify** a reaction as belonging to one of five major types.

Review Vocabulary

reactant: a substance that undergoes a reaction

New Vocabulary

synthesis
decomposition
single displacement
double displacement
combustion

Types of Reactions

MAIN Idea There are five types of chemical reactions: synthesis, decomposition, single displacement, double displacement, and combustion reactions.

Real-World Reading Link Chemistry has a lot in common with cooking. If you combine the right amounts of the right ingredients in the correct order under certain conditions, you will get the right food product. What do you have to do to become a good cook? A lot of study and probably even more practice are necessary. The same is true of chemistry.

Why Reactions Are Classified

Why are reactions grouped into classes? Think about why you might need to classify plants and animals. For example, classifying the cats in **Figure 6.8** into the informal categories *pet* versus *wild animal* allows you to make predictions about where each will be found and how you should behave if you are near one. Organizing your information about cats in this way helps you begin to understand them and gives you a framework to make predictions about any unfamiliar types of cats that you encounter. Similarly, classifying chemical reactions will make it easier to remember familiar reactions and to make predictions about unfamiliar reactions.

■ **Figure 6.8** Quickly classifying an animal can help you decide how to behave around it. For example, it might be safe to pet a friendly cat, but you wouldn't want to approach a dangerous wild animal closely. Similarly, classifying reactions can help you understand the reaction and make predictions about it.

Biology Connection

Air in Space

The concentration of carbon dioxide in Earth's atmosphere is regulated through a complex interplay of human, biological, and geological mechanisms. In a space vehicle, these mechanisms aren't available. If not controlled, carbon dioxide from the respiration of astronauts would become toxic to them. So how is air quality maintained within a space vehicle?

A 66-m³ atmosphere On board the space shuttle, the crew must have an atmosphere similar to Earth. Odors and contaminating gases must be removed, carbon dioxide must be regulated, and a warm environment needs to be maintained. The volume of the crew compartment of a space shuttle orbiter is about 66 m³. The air is maintained at a pressure of 100.3 kPa, which is similar to atmospheric pressure at sea level.

In space, the composition of the air in a crew compartment, shown in **Figure 1**, is maintained at 79 percent nitrogen and 21 percent oxygen, which is almost identical to that of Earth's atmosphere. Oxygen is carried in the orbiter as a liquid stored in two cryogenic tanks in the mid-fuselage. The gaseous oxygen from the tanks passes through pressurizing and heating nozzles and moves into the crew compartment. A five-member crew will normally consume about 4 kg of oxygen every day. Nitrogen is supplied from two systems, each made of two storage tanks, also in the mid-fuselage of the orbiter. The atmosphere of the compartment is recycled about every seven minutes.

Filtering the air During recycling, odors are removed by filters containing activated charcoal granules, which absorb from the air the chemical substances that cause the odor. Carbon dioxide gas (CO_2) is removed from the air by reacting it with solid lithium hydroxide (LiOH).

$$CO_2(g) + 2LiOH(s) \rightarrow Li_2CO_3(s) + H_2O(g)$$

The lithium hydroxide is stored in canisters that are changed every 12 hours. The used canisters are then stored for disposal when the orbiter returns to Earth.

Air Temperature Outer space is extremely cold. Therefore, you might think it would be difficult to heat the orbiter. However, the electronic equipment onboard the orbiter generates so much heat that one problem is finding ways to get rid of the excess heat. Another problem caused by the extreme cold in outer space is temperatures varying greatly in different parts of the orbiter. Therefore, the temperature control system has to perform two major functions: 1) distribute heat where it is needed on the orbiter so that vital systems do not freeze and 2) get rid of the excess heat. The temperature of the crew compartment is controlled by the cabin heat exchanger. Cool water is circulated around the crew cabin to remove excess heat. The heat is then transferred to other orbiter systems, which ultimately radiate excess heat to outer space.

Figure 1 The air in a crew compartment is made similar to Earth's atmosphere.

Connection to Chemistry

1. **Apply** Name the products of the reaction between carbon dioxide and lithium hydroxide.
2. **Research** How does the chemical removal of carbon dioxide from the atmosphere of the orbiter compare with the geochemical removal of carbon dioxide from Earth's atmosphere?

■ **Figure 6.9** When iron rusts, iron metal (Fe) and oxygen gas (O_2) combine to form iron(III) oxide (Fe_2O_3). The balanced equation for this synthesis reaction is as follows:
$$4Fe(s) + 3O_2(g) \longrightarrow 2Fe_2O_3(s)$$
State *How many reactants are there in this reaction? How many products?*

FOLDABLES
Incorporate information from this section into your Foldable.

Major classes of reactions

Just as there are thousands of species of animals, there are many different types of chemical reactions. Five types are common. If you can classify a reaction into one of the five major categories by recognizing patterns that occur, you already know a lot about the reaction.

Synthesis reactions In one type of reaction, two substances—either elements or compounds—combine to form a compound. Whenever two or more substances combine to form a single product, the reaction is called a **synthesis** reaction.

An example of a synthesis reaction involving elements as reactants is shown in **Figure 6.9.** A synthesis reaction also occurs when two compounds combine, such as when rainwater combines with carbon dioxide in the air to form carbonic acid, or when an element and a compound combine, as when carbon monoxide combines with oxygen to form carbon dioxide.

Decomposition reactions A **decomposition** reaction is one in which a compound breaks down into two or more simpler substances. The compound may break down into individual elements, such as when mercury(II) oxide decomposes into mercury and oxygen. The products may be an element and a compound, such as when hydrogen peroxide decomposes into water and oxygen, or the compound may break down into simpler compounds, as shown in **Figure 6.10.**

■ **Figure 6.10** Hydrogen peroxide slowly decomposes, but other compounds decompose quickly and violently. Compounds that decompose explosively can be used in demolitions, such as TNT (trinitrotoluene).

Iron displaces copper ions in solution

Chlorine displaces bromine

Single-displacement reactions A **single-displacement** reaction is one in which one element takes the place of another in a compound. **Figure 6.11** shows the single-displacement reaction between solid iron and an aqueous solution of copper(II) sulfate. **Figure 6.11** also shows the single-displacement reaction between chlorine gas and an aqueous solution of sodium bromide. As you can see from these photos, the reacting element can replace the first part of a compound, or it can replace the last part of a compound. You will sometimes hear the term *single-replacement reaction* instead of the term *single-displacement reaction*. These two terms mean exactly the same thing.

■ **Figure 6.11** If an iron (Fe) nail is placed into an aqueous solution of copper(II) sulfate ($CuSO_4$), the iron displaces the copper ions in solution, and copper metal (Cu) forms on the nail.
$Fe(s) + CuSO_4(aq) \rightarrow FeSO_4(aq) + Cu(s)$
When the chlorine gas (Cl_2) in the flask on the left is bubbled through an aqueous solution of sodium bromide (NaBr), the chlorine replaces the bromine in the compound. The reddish-brown bromine (Br_2) can be seen in the solution.
$Cl_2(g) + 2NaBr(aq) \rightarrow 2NaCl(aq) + Br_2(l)$

MiniLab 6.2

A Simple Exchange

Single-displacement reactions do not automatically occur just because an element and a compound are mixed together. Whether or not such a reaction occurs depends on how reactive an element is compared to the element it is to displace.

Procedure
1. Read and complete the lab safety form.
2. Half-fill a **test tube** with **0.1M AgNO₃ solution.** Clean a piece of **copper wire** or **copper foil** with **steel wool.**
3. Drop the copper foil or copper wire into the solution. Place the test tube in a **test tube rack.**
4. Keep the test tube absolutely still. Observe what happens over a half-hour period of time.

Analysis
1. **Observe** What changes do you observe in the wire? In the solution?
2. **Conclude** Does copper displace silver in silver nitrate? Does silver displace copper in copper(II) nitrate? How do you know?
3. **Apply** Write a balanced equation for the reaction.

CHEMLAB

EXPLORE CHEMICAL CHANGES

Background
Most reactions can be classified into five major types. As you carry out this investigation, you'll observe examples of each of these types. In doing so, you will also learn to recognize many of the physical changes that accompany reactions.

Question
What are some of the physical changes that indicate that a reaction has occurred?

Objective
- **Observe** physical changes that take place during chemical reactions.
- **Compare** changes that take place during different types of chemical reactions.

Preparation

Materials
125-mL flasks (4)
balance
hot plate
watch glass
spatula
stirring rod
lab burner
file
new penny
250-mL flask
ice
tongs
100-mL graduated cylinder
large test tube and one-hole stopper with glass tube and rubber tubing attached
ring stand
test-tube clamp
$0.1M$ $CuSO_4$
granular copper (Cu)
powdered sulfur (S)
calcium carbonate ($CaCO_3$)
finely ground saturated $Ca(OH)_2$ solution (limewater)
$6M$ HCl
$0.5M$ Na_2CO_3
$0.5M$ $CuCl_2$

Safety Precautions
WARNING: *Use care when handling hot objects. Dispose of the reaction mixture and products as instructed by your teacher.*

Procedure

1. Read and complete the lab safety form.
2. For each of the following reactions, record in a data table all changes that you observe.

Synthesis Reaction

1. Place 50 mL $0.1M$ $CuSO_4$ in a 125-mL flask.
2. Place 1.6 g granular copper and 0.8 g powdered sulfur on a watch glass and mix together thoroughly with a spatula.
3. Heat the flask on a hot plate set at high until the solution begins to boil.
4. Stir the Cu/S mixture into the boiling $CuSO_4$ solution.
5. Continue boiling until a black solid forms.

Decomposition Reaction

1. Place 100 mL of saturated $Ca(OH)_2$ solution (limewater) in the 250-mL flask.
2. Add finely ground $CaCO_3$ to a large test tube until it is one-fourth full. Stopper the tube with the stopper/glass tube/rubber tubing assembly, and clamp the tube to the ring stand.
3. Light a laboratory burner, and begin to heat the test tube. Submerge the end of the rubber tubing into the limewater so that any gas produced in the tube will bubble through the limewater.
4. Continue heating the $CaCO_3$ until you observe a change in the limewater. The presence of CO_2 causes limewater to become cloudy.

Single-Displacement Reaction
1. Place 30 mL 6M HCl in a 125-mL flask.
2. Using a file, cut six 0.2-cm notches evenly spaced around the perimeter of a new penny.
3. Place the penny in the flask of acid and leave it in a fume hood overnight.

Double-Displacement Reaction
1. Add 25 mL 0.5M Na_2CO_3 and 25 mL 0.5M $CuCl_2$ to a 125-mL flask.
2. Swirl the flask gently until you observe the formation of a precipitate.

Combustion Reaction
1. Light a laboratory burner and adjust the air and gas supplies until the flame is blue. Observe what happens.
2. Using tongs, hold a flask or beaker with ice in it about 10 cm over the flame for approximately one minute. Move the flask away from the flame and observe the bottom of the flask.

Data and Observations

Data Table	
Reaction	Observations
Synthesis	
Decomposition	
Single displacement	
Double displacement	
Combustion	

Analyze and Conclude
1. **Infer** Which observations noted during each of the reactions indicated that a reaction had occurred?
2. **Compare and Contrast** What did all of the reactions have in common?
3. **Determine** Write the name and formula of the
   a) black solid formed in the synthesis reaction.
   b) gaseous product of the decomposition reaction.
   c) solid product of the decomposition reaction.
   d) pale blue precipitate in the double-displacement reaction.
   e) liquid product of the combustion reaction.
4. **Explain** how the penny changed during the single-displacement reaction. What would happen if a pre-1983 penny, which is solid copper, were used?
5. **Recognize** Is energy a reactant or product of the combustion reaction?

Apply and Assess
1. **Error Analysis** Were there any physical changes that often occur during a reaction that you did not observe while doing this ChemLab? If so, what were they?
2. **Apply** Write balanced chemical equations for all of the reactions carried out.

INQUIRY EXTENSION
Analyze and Explain During the single-displacement reaction, did all of the pennies tested by your class react in the same way? Explain why some pennies might react differently than others.

■ **Figure 6.12** When clear aqueous solutions of lead(II) nitrate ($Pb(NO_3)_2$) and potassium iodide (KI) are mixed, a double-displacement reaction takes place and a yellow solid appears in the mixture. This solid is lead(II) iodide (PbI_2). It precipitates out because it is insoluble in water, unlike the two reactants and the other product.
$Pb(NO_3)_2(aq) + 2KI(aq) \rightarrow PbI_2(s) + 2KNO_3(aq)$

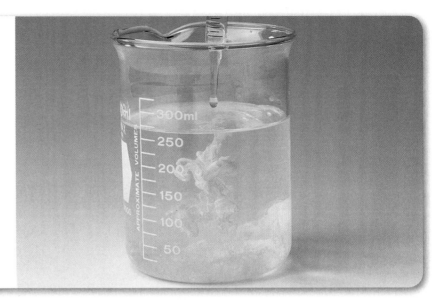

Concepts In Motion

Interactive Figure To see an animation of a precipitate forming, visit glencoe.com.

Double-displacement reactions A **double-displacement** reaction is one in which the positive ions of two ionic compounds are interchanged. For a double-displacement reaction to take place, at least one of the products must be a precipitate or water. An example of a double-displacement reaction is shown in **Figure 6.12.** As with single-displacement reactions, you might sometimes hear *double-displacement reactions* called *double-replacement reactions*. Again, the terms are interchangeable.

Combustion reactions The fifth common type of reaction is a combustion reaction. A **combustion** reaction is one in which a substance rapidly combines with oxygen to form one or more oxides, as occurs in the cars shown in **Figure 6.13.**

Although there are exceptions, many thousands of specific reactions fit into these five classes. Also, sometimes reactions fit more than one class. For example, burning coal according to the following equation is both a combustion reaction and a synthesis reaction.

$$C + O_2 \rightarrow CO_2$$

Table 6.1 summarizes important information about each of these reaction types.

VOCABULARY

WORD ORIGIN

Combustion
comes from the Latin *comburere* meaning *to burn up*

■ **Figure 6.13** Octane (C_8H_{18}) burns in a combustion reaction, forming carbon dioxide and water. Octane is one of the primary components of gasoline, making this one of the most common everyday combustion reactions.
$2C_8H_{18} + 25O_2 \rightarrow 16CO_2 + 18H_2O$
Infer *Is this reaction exothermic or endothermic? How can you tell?*

Table 6.1 Types of Reactions

Reaction Type	General Equation
Synthesis	element/compound + element/compound → compound Examples: $2Na(s) + Cl_2(g) \rightarrow 2NaCl(s)$ $CaO(s) + SiO_2(l) \rightarrow CaSiO_3(l)$
Decomposition	compound → two or more elements/compounds Examples: $PCl_5(s) \rightarrow PCl_3(s) + Cl_2(g)$ $2Ag_2O(s) \rightarrow 4Ag(s) + O_2(g)$
Single Displacement	*element a + compound bc → element b + compound ac Example: $2Al(s) + Fe_2O_3(s) \rightarrow 2Fe(s) + Al_2O_3(s)$ element d + compound bc → element c + compound bd Example: $Cl_2(aq) + 2KBr(aq) \rightarrow 2KCl(aq) + Br_2(aq)$
Double Displacement	compound ac + compound bd → compound ad + compound bc Examples: $PbCl_2(s) + Li_2SO_4(aq) \rightarrow PbSO_4(s) + 2LiCl(aq)$ $BaCl_2(aq) + H_2SO_4(aq) \rightarrow 2HCl(aq) + BaSO_4(s)$
Combustion	element/compound + oxygen → oxide(s) Examples: $CH_4(g) + 2O_2(g) \rightarrow CO_2(g) + 2H_2O(g)$ $C_6H_{12}O_6(s) + 6O_2(g) \rightarrow 6CO_2(g) + 6H_2O(l)$

*The letters a, b, c, and d each represent different elements or parts of compounds. For example, in compound ac, a represents the positive part of the compound, and c represents the negative part.

SUPPLEMENTAL PRACTICE
For more practice classifying reactions, see Supplemental Practice, page 815.

Section 6.2 Assessment

Summary

- Although thousands of individual chemical reactions are known, most can be classified into five major types that are based on patterns of behavior of reactants and products.
- The five general classes of reactions are synthesis, decomposition, single displacement, double displacement, and combustion.
- Sometimes classes of reactions overlap. For example, some combustion reactions are also synthesis reactions.

10. **MAIN Idea** **Explain** why classifying reactions can be useful.

11. **Diagram** Using different symbols to represent different atoms, draw pictures to represent an example of each of the following kinds of reactions.
    a) synthesis
    b) decomposition
    c) single displacement
    d) double displacement
    e) combustion

12. **Classify** each of the following reactions.
    a) $N_2O_4(g) \rightarrow 2NO_2(g)$
    b) $2Fe(s) + O_2(g) \rightarrow 2FeO(s)$
    c) $2Al(s) + 3Cl_2(g) \rightarrow 2AlCl_3(s)$
    d) $BaCl_2(aq) + Na_2SO_4(aq) \rightarrow BaSO_4(s) + 2NaCl(aq)$
    e) $Mg(s) + CuSO_4(aq) \rightarrow Cu(s) + MgSO_4(aq)$

13. **Infer** When a candle burns, wax undergoes a combustion reaction. Would a candle burn longer in the open or when covered with an inverted glass jar? Explain.

14. **Compare** When a fungus breaks down the wood in a fallen tree, the biological process is called decomposition. What does that process have in common with the chemical decomposition reaction type?

Section 6.3

Objectives
- **Demonstrate** factors that influence the direction of a reaction.
- **Classify** factors that influence the rate of a reaction.

Review Vocabulary
synthesis: reaction in which two or more substances combine to form a single product

New Vocabulary
equilibrium
dynamic equilibrium
Le Châtelier's principle
soluble
insoluble
activation energy
concentration
limiting reactant
catalyst
enzyme
inhibitor

Nature of Reactions

MAIN Idea External factors modify the direction and rate of chemical reactions.

Real-World Reading Link You know that changes constantly occur, but some changes are not permanent. For example, you can freeze liquid water into ice, but then that ice melts and becomes liquid water again. In other words, the freezing process is reversed. Can chemical reactions be reversed? Can the product of a reaction become a reactant?

Reversible Reactions

Many reactions can change direction. These reactions are called reversible reactions. The reactions that occur in a car's battery, shown in **Figure 6.14,** are reversible. When you start a car, chemical reactions in the battery supply the energy that you need. This process discharges the battery. When the car is running, energy from moving engine parts drives the chemical reactions in the reverse direction, recharging the battery.

Not all chemical changes are reversible. Food is digeseted, paint hardens, and fuel burns. These chemical changes result in new products, and the reactions are said to go to completion because at least one of the reactants is completely used up and the reaction stops. The reactions can't be reversed. But what happens when a reaction reverses?

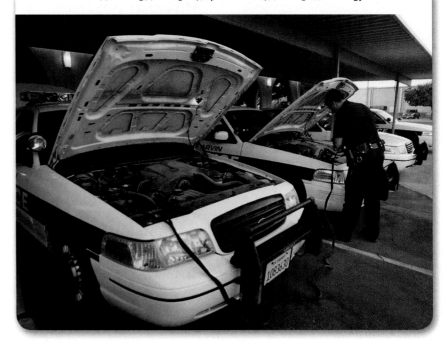

■ **Figure 6.14** When an automobile battery releases energy when the car isn't running, the reaction below moves to the right. As the car runs, mechanical energy from moving engine parts drives the reaction in the reverse direction.

$$Pb(s) + PbO_2(s) + 2H_2SO_4(aq) \rightleftharpoons 2PbSO_4(s) + 2H_2O(l) + energy$$

■ **Figure 6.15** First thing in the morning, this train is empty. At each stop, people get on, and the number of passengers on the train increases. During rush hour, the number of passengers on the train will not be changing much because the number getting off the train at each stop will be about equal to the number getting on.

Picture what happens in a subway station similar to the one shown in **Figure 6.15.** When the door of a train opens for the first time that day, passengers rush in, streaming across the platform and through the doors. No passengers get out because this is the first train of the day. When the train stops at the next station, more passengers rush in. A few probably leave the train at this point, as well. With each successive stop, more passengers will be getting on, and others will also be getting off. To describe a similar situation in chemical reactions in which a reaction automatically reverses and there is no net (overall) change, we use the term *equilibrium*.

Equilibrium When no net change occurs in the amount of reactants and products, a system is said to be in **equilibrium**. In most situations, chemical reactions exist in equilibrium when products and reactants form at the same rate. Such a system, in which opposite actions are taking place at the same rate, is said to be in **dynamic equilibrium**.

Reactions at equilibrium have reactants and products changing places, much like the passengers getting on and off a subway train. In equilibrium, reactants are never fully used up because they are constantly being formed from products. Eventually, reactants and products form at the same rate.

One example of a reversible reaction that reaches equilibrium occurs when lime (CaO), which is used to make soils less acidic, and carbon dioxide (CO_2) is formed by decomposing limestone ($CaCO_3$).

$$CaCO_3(s) \rightleftharpoons CaO(s) + CO_2(g)$$

Notice that the single arrow in the equation has been replaced with a double arrow. Because an arrow shows what direction the reaction is going, the double arrow indicates that the reaction can go in either direction. In this case, $CaCO_3$ decomposes into CaO and CO_2. But, as those products form, CO_2 and CaO combine to form $CaCO_3$. The rate, or speed, of each reaction can be determined by how quickly a reactant disappears. Eventually, the rates of the two reactions are equal, and an equilibrium exists.

In the Field

Meet Caroline Sutliff
Plant-Care Specialist

An old gardening adage goes like this: A weed is just a flower that is growing where you don't want it. Landscape gardener Caroline Sutliff works to eradicate those hardy, invasive plants, such as *Taraxacum officinale* (better known as dandelions). In this interview, she tells about how she wages her weed battles with garden tools and chemicals.

On the Job

Q: Ms. Sutliff, what do you do on the job?

A: I take care of flowers, shrubs, and trees. The company for which I work maintains the grounds of business properties, such as real estate offices and banks, as well as private homes. My job involves working with chemicals such as herbicides, fungicides, and pesticides.

Q: Are those chemicals hazardous?

A: They can be if you don't treat them properly. I have a spraying license that I earned through the Agricultural Service Extension, so I have learned all the safety precautions. Mixing the chemicals to get them ready for use is more hazardous than actually spraying them. For mixing, I wear a plastic-coated jacket and pants, heavy boots, rubber gloves, and a face mask.

Q: Do you and your coworkers know what to do in case of a mishap?

A: We carry eyewash and first-aid kits in all of our trucks. We're trained to react quickly if we get a chemical on our clothes. In that case, we must remove the affected clothing and wash up quickly. When some chemicals come in contact with your skin, they are absorbed into the body and stored in fat cells. The cumulative effect could be harmful. Because I'm very careful and follow directions exactly, I haven't had any problems.

Q: Why are chemicals necessary in your work?

A: On large properties, it's a matter of economy and time—two hours of hand weeding and cultivating versus a few minutes of spraying. However, chemical use is just part of our program. Before we spray chemicals, we cultivate, do some hand weeding, and fertilize and aerate beds of plants. Healthy plants are more likely to have a fighting chance against weeds and insects.

Q Do you use the information you learned in high school chemistry classes?

A Yes. It's important to be familiar with the properties of chemicals—to know what can be safely mixed and what can't. The container used for mixing is important chemically, too. For example, stainless steel containers can erode and change the makeup of chemicals so they won't have the same effects. Heavy plastic is used instead.

Q How do you see the field of plant care changing in the years ahead?

A I think people will become even more aware of the value of plants, particularly trees. Trees can do so much to improve a neighborhood. They clean the air, shade the sidewalks, and beautify an area. Here in Iowa City, an organization called Project Green helps plant trees in public places. I hope to get involved with that group as a volunteer.

Early Influences

Q As a child, were you interested in plants?

A I wasn't interested in planting things because I wanted immediate results. Plants don't work that way. Now I have more patience and love to tend to the flowers around my own home.

Q Did you plan to enter the plant-care field later on?

A No. My intentions were to get a degree in psychology and then go to law school. While I was looking for a job as a paralegal, I began working in plant care and discovered I love working outside. Now I can't imagine being inside all day at a desk job.

Personal Insights

Q Would you recommend that students investigate a plant-care job like yours?

A Only if they like hard, physical work. My workdays last 10 or 12 hours, and I'm out in all kinds of weather. I certainly don't need to go to a gym to stay in shape! It's satisfying for me to see that more women are entering this field.

CAREER CONNECTION

The following careers are also associated with plant care.

Horticulturist Master's degree, research, and fieldwork

Landscape Architect Bachelor's degree, often followed by a licensing examination

Soil Conservation Technician Two-year college program

Chapter 6 • In The Field **211**

Reaching equilibrium Saying that a reaction has reached equilibrium does not mean that equal amounts of reactants and products are present. Equilibrium just means that no net change is taking place—the amounts of reactants and products are not changing. Often, a reaction at equilibrium contains differing amounts of reactants and products.

Consider the following reaction in which phosphorus pentachloride (PCl_5) decomposes into phosphorus trichloride (PCl_3) and chlorine (Cl_2).

$$PCl_5(g) \rightleftharpoons PCl_3(g) + Cl_2(g)$$

Actual measurements of the amounts of reactant and products show that the system is in equilibrium. But the measurements also show that there is more PCl_5 than PCl_3 and Cl_2 present, even though the rates of the reactions are equal. The reversible reaction will favor the direction that produces the most stable products, which are those that are least likely to change. In this example, PCl_5 is less likely to decompose than the two products, PCl_3 and Cl_2, are likely to combine. Reversible reactions at equilibrium are in a balance that favors stability.

Le Châtelier's principle If a reaction reaches equilibrium, how can you obtain large quantities of a product? Won't the product constantly become a reactant? Keep in mind that reactions in equilibrium are stable. French scientist Henri Louis Le Châtelier proposed in 1884 that disturbing an equilibrium will make a system readjust to reduce the disturbance and regain equilibrium. This principle regarding changes in equilibrium is called Le Châtelier's principle. In other words, **Le Châtelier's principle** states that if a stress is applied to a system at equilibrium, the system shifts in the direction that relieves the stress.

For example, imagine a dog drinking from a water dish with a reservoir attached, as shown in **Figure 6.16**. Before drinking, the water level in the reservoir rests at a stable equilibrium position. As the dog drinks, the equilibrium is stressed and the water level in the reservoir falls until it establishes a new equilibrium.

Chemical engineers can apply this tendency of reactions to stay at equilibrium to find ways to increase the yield of products in a reaction. If products are removed from a reaction at equilibrium, more reactants will go on to form products so that balance is regained. If this removal continues, most of the reactants can be converted into products. For example, recall the reaction in which limestone ($CaCO_3$) decomposes into lime (CaO) and carbon dioxide (CO_2).

$$CaCO_3(s) \rightleftharpoons CaO(s) + CO_2(g)$$

If the carbon dioxide is removed as it is produced, the reaction will favor the formation of more carbon dioxide to reestablish equilibrium. Thus, the reaction will shift in the direction that also produces more lime as a product.

■ **Figure 6.16** Before the dog started to drink, the water in this container reached a stable equilibrium level. As the dog drinks, this level is stressed and water comes from the bottle into the bowl to reestablish a new equilibrium position.

When one product is a gas and other products and reactants are not gases, as in the previous example and in **Figure 6.17,** you can see how the gas can be removed from the reaction. How can products that are not isolated gases be removed from a reaction? They usually can't be picked out manually because of the mix of reactants and products.

A product that does not dissolve in water can be removed if all other products and the reactants dissolve in water. A compound is **soluble** in a liquid if it dissolves in it; it is **insoluble** if it does not. An insoluble product will form a solid precipitate that sinks to the bottom of a liquid solution, as shown in **Figure 6.18.** The precipitate cannot react easily because it is somewhat isolated from the other substances in the reaction. For reactions that take place in solution, forming precipitates removes products and favors the forward reaction.

Adding more reactants has basically the same effect as removing products. For the reaction to reestablish equilibrium amounts of reactants and products, more products must be made if more reactants are added.

Effects of energy Adding or removing energy, usually in the form of heat, can also influence the direction of a reaction. Because energy is a part of any reaction, it can be thought of as a reactant or a product. Just as adding more reactants to a reaction pushes it to the right, so does adding more energy to an endothermic reaction. For example, the equation for a reaction that produces aluminum metal from bauxite, an aluminum ore, shows that energy must be added for products to form.

$$3C + 2Al_2O_3(s) + energy \rightleftharpoons 4Al(l) + 3CO_2(g)$$

If more energy is added, the reaction goes to the right, forming more aluminum and carbon dioxide.

For an exothermic reaction, adding more energy pushes the reaction to the left. In the Haber process for producing ammonia (NH_3) from hydrogen (H_2) and nitrogen (N_2), for example, energy is produced.

$$3H_2(g) + N_2(g) \rightleftharpoons 2NH_3(g) + energy$$

Adding energy favors the formation of nitrogen and hydrogen. Thus, temperature must be carefully controlled in the Haber process so that the desired product, ammonia, is produced in large quantities.

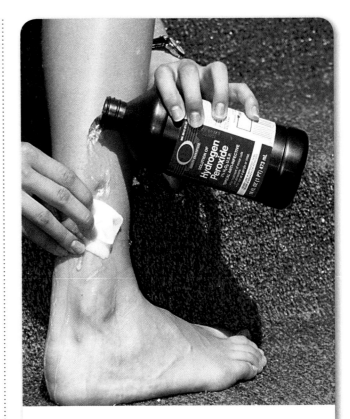

■ **Figure 6.17** When hydrogen peroxide (H_2O_2), is poured onto a wound, the hydrogen peroxide decomposes to form water and oxygen. The gaseous oxygen bubbles away. The removal of oxygen drives the reaction forward.
$$2H_2O_2(aq) \rightarrow 2H_2O(l) + O_2(g)$$

■ **Figure 6.18** When potassium hydroxide is added to an aqueous solution of calcium chloride, calcium hydroxide and potassium chloride form. Calcium hydroxide is nearly insoluble in water, so it precipitates out as a solid.

CHEMISTRY AND TECHNOLOGY

Mining the Air

Diatomic nitrogen makes up about 78 percent of Earth's atmosphere. Nitrogen is an essential element for life, yet only a few organisms can use atmospheric nitrogen directly. A few species of soil bacteria can produce ammonia, NH_3, from atmospheric nitrogen. Other species of bacteria can then convert the ammonia into nitrite and nitrate ions, which can be absorbed and used by plants.

The Haber Process

Not all usable nitrogen compounds are produced naturally from atmospheric nitrogen. Ammonia, shown in storage containers in **Figure 1,** can also be synthesized. The process of synthesizing large amounts of ammonia from nitrogen and hydrogen gases, still carried on today in plants such as the one in **Figure 2,** was invented by Fritz Haber, a German research chemist. The process was first demonstrated in 1909.

In the Haber process, the reactants are hydrogen gas (H_2) and nitrogen gas (N_2). The products are ammonia gas (NH_3) and heat. The process produces high yields of ammonia by manipulating three factors that influence the reaction—pressure, temperature, and catalytic action.

$$3H_2(g) + N_2(g) \rightarrow 2NH_3(g) + \text{heat}$$

■ **Figure 1** Ammonia storage tanks

■ **Figure 2** Ammonia production plant

Pressure

In the ammonia synthesis reaction, three molecules of H_2 and one molecule of N_2 produce two molecules of product NH_3. According to Le Châtelier's principle, if pressure on the reaction or the system is increased, the forward reaction will speed up to lessen the stress because two molecules exert less pressure than four molecules. Increased pressure will also cause the reactants to collide more often, thus increasing the reaction rate. Haber's apparatus used a total pressure of 2×10^5 kPa, which was the highest pressure he could achieve in his laboratory.

Temperature

Two factors influence the temperature at which the process is carried out. A low temperature favors the forward reaction because this reduces the stress of the heat generated by the reaction. But high temperature increases the rate of reaction because of more collisions between the reactants. Haber carried out the process at a temperature of about 600°C.

Catalyst

A catalyst is used to decrease the activation energy and thus increase the rate at which equilibrium is reached. Haber used osmium and uranium as catalysts in his apparatus.

The Process

Haber's process, detailed in **Figure 3**, incorporated several operations that increased the yield of ammonia. The reactant gases entering the chamber were warmed by heat produced by the reaction. The reactant-product mixture was allowed to cool slowly after reacting over the catalyst. Ammonia was removed from the process by liquefaction, while unreacted nitrogen and hydrogen were recycled into the process.

The Haber process gained the attention of BASF, a leading German chemical manufacturer. Carl Bosch, a German industrial chemist employed by BASF, and his colleagues designed, constructed, and tested new reaction chambers; improved pressurizing pumps; and used inexpensive catalysts. By 1913, Bosch had built an operating ammonia plant in Oppau, Germany. Throughout World War I, the Oppau plant and a later-built plant supplied Germany with ammonia for the production of explosives. Today, the Haber process, also called the Haber-Bosch process, is the main source for producing commercial ammonia. Much of this ammonia is used in the production of agricultural fertilizers being applied to a field in **Figure 4**.

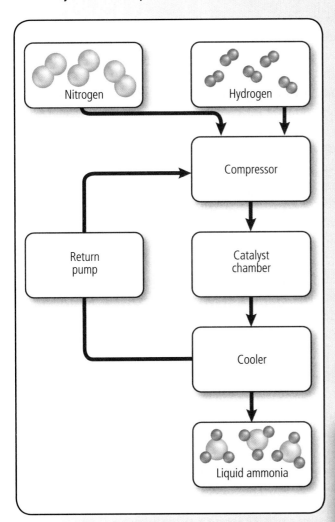

■ **Figure 3** Haber's process

■ **Figure 4** Ammonia fertilizer being applied to a field

Discuss the Technology

1. **Apply** How would slow cooling of the reactant-product mixture affect the equilibrium of the reaction?

2. **Acquire Information** How is ammonia used in the production of many agricultural fertilizers?

■ **Figure 6.19** The speedometer of a racing car might register the car's speed in miles per hour or in kilometers per hour. For the sprinter, you could divide the distance of the sprint in meters by the time in seconds in order to get the sprinter's speed in meters per second. The rate of a chemical reaction is similar, but involves how quickly reactants form or products disappear.

Activation Energy

At equilibrium, reactants and products are being formed and broken down at the same rate. How fast is this rate? If you pour a solution of potassium iodide into a solution of lead nitrate, a yellow precipitate of lead iodide seems to form instantly. If you strike a match, you can see that it takes a little time before the match burns completely. However, other reactions proceed very slowly. To understand why reactions occur at different rates, look at what happens when reactions take place.

For a reaction to occur between two substances, particles of those substances must collide. Not only do they have to collide, they have to hit each other with enough force to cause a change to take place. The amount of energy the particles must have when they collide with each other is called the **activation energy** of the reaction. A high activation energy means that few collisions have enough energy to react, and the reaction tends to be slow. On the other hand, a low activation energy means more collisions have enough energy to cause a reaction, and the reaction is faster.

Consider the highly exothermic reaction between hydrogen and oxygen. This reaction produces enough energy to power the main stage of the space shuttle. However, hydrogen and oxygen molecules can exist in the same container for years without reacting. A spark provides the activation energy needed to start the reaction. Once started, the reaction produces enough energy to keep itself going. Activation energy differs from reaction to reaction. Some reactions have such high activation energies that the reactions do not take place under normal conditions.

Rate of Reaction

Look at two examples of finding rates in **Figure 6.19.** In both cases, the rate is the distance covered per unit time. Similar to the way in which you can determine the speed of a race car or of a runner, you can determine the rate of a chemical reaction. To determine how fast a reaction is taking place, measure how quickly one of the reactants disappears or how quickly one of the products appears. Either measurement will give an amount of substance changed per unit time, which is the rate of reaction.

The rate of a reaction is important to a chemical engineer designing a process to get a good yield of product. The faster the rate, the more product that can be made in a fixed amount of time. Rate of reaction is also important to food processors who hope to slow down reactions that cause food to spoil. Can the rate of a reaction be changed? Four major factors affect this rate.

Effect of temperature One factor that affects rate of reaction is temperature. Most reactions go faster at higher temperatures. Baking a cake speeds up the reactions that change the liquid batter into a spongy product. Lowering the temperature slows down most reactions, as described in **Figure 6.20**. Photographic film and batteries stay useful longer if they are kept cool because the lower temperature slows the reactions that can ruin these products.

Effect of concentration Changing the amounts of reactants present can also alter reaction rate. The amount of substance present in a certain volume is called the **concentration** of the substance. Raising the concentration of a reactant will speed up a reaction because there are more particles per volume. More particles result in more collisions, so the reaction rate increases, as illustrated by **Figure 6.21**.

In most cases, concentration is increased by adding more reactants. If a fire is burning slowly, fanning the flames increases the amount of oxygen available and the fire burns faster. If a gas is involved, concentration may be increased by increasing pressure. Increasing pressure does not increase the number of particles, but it brings the particles closer together so collisions are more frequent. The Haber process, for example, uses high pressures to increase the rate of reaction of hydrogen and nitrogen to form ammonia.

Lowering concentration decreases the rate of reaction. Many valuable historic documents are stored in sealed cases with most of the air removed to decrease the number of particles that might react with the paper.

■ **Figure 6.20** Removing heat slows reactions. That's why freezing food helps keep it from spoiling as quickly.

■ **Figure 6.21** Adding more bumper boats increases the chance that a collision will occur. Taking some away decreases the odds of two boats meeting. When more particles are added to a reaction mixture, the chance that they will collide and react is increased.

MiniLab 6.3

Starch-Iodine Clock Reaction

How soon does an observable change occur? In a clock reaction, an observable change indicates at some point in time after two or more different reactants are mixed that a reaction has taken place. Altering any factors that change reaction rate will change how quickly the observable change is seen.

Procedure

1. Read and complete the lab safety form.
2. Use a **wax pencil** to label **five large test tubes** with the numbers 1 through 5, and place them in a **test-tube rack.**
3. Place 10 mL of **starch-containing solution** in each of the five test tubes.
4. Place tube 4 in an **ice bath** and tube 5 in a **water bath** at 35°C for at least ten minutes.
5. Add the following amounts of **iodine-producing solution** to the indicated test tubes, stir carefully with a clean **stirring rod or a wire stirrer,** and record the time it takes for the blue color to appear.
   a) 10 mL to tube 1
   b) 5 mL to tube 2
   c) 20 mL to tube 3
   d) 10 mL to tube 4
   e) 10 mL to tube 5
6. Summarize your results in a table.

Analysis

1. **Conclude** What effect does changing the amount of a reactant have on the rate of a reaction? Why?
2. **Summarize** What effect does lowering the temperature have on the rate of a reaction? Why?

The limiting reactant Keep in mind that it doesn't matter how much the concentration of one reactant is increased if another reactant is depleted. Sometimes, when two reactants are present in a reaction, more of one than the other is available for reacting. The one that there is not enough of is called the **limiting reactant**. When it is completely used up, the reaction must stop.

Figure 6.22 illustrates the concept of limiting reactants. From the available tools, four complete sets consisting of a pair of pliers, a hammer, and two screwdrivers can be assembled. The remaining tools—a pair of pliers and two screwdrivers—do not make up a complete set. Similarly, chemical reactions are limited when one of the reactants is used up.

■ **Figure 6.22** Each tool set must have one hammer, so only four sets can be assembled.
Determine *What is the limiting reactant in this example?*

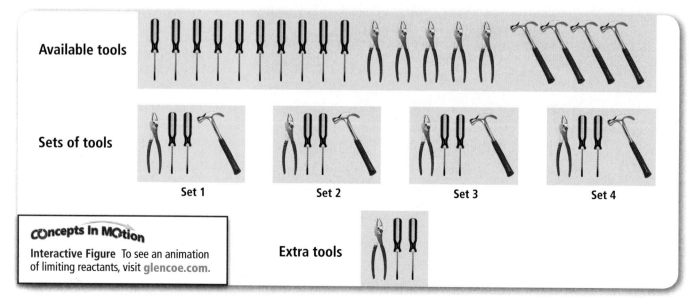

Concepts In Motion
Interactive Figure To see an animation of limiting reactants, visit glencoe.com.

Everyday Chemistry

Stove in a Sleeve

That Napoleon stated "An army marches on its stomach," can be debated. However, few argue about the meaning of the statement. One of the necessities of military operations is supplying troops with meals, also called *rations*. In the past, front-line troops often ate cold rations. Heating food by fire took too long, and the resulting smoke might have drawn enemy fire. What was needed was a portable, smokeless, self-contained method of quickly heating rations with readily available materials. The flameless ration heater was developed to meet these needs.

Flameless ration heaters The flameless ration heater (FRH) is used to warm an individual portion of rations called a ready-to-eat meal, or MRE (*M*eal, *R*eady-to-*E*at). It uses the oxidation of magnesium to generate heat. A heat-releasing chemical reaction occurs when magnesium and water are combined that forms magnesium hydroxide, $Mg(OH)_2$, and hydrogen gas. But pure magnesium can't be used in the heaters. Magnesium reacts with oxygen in the air to form a coating of magnesium oxide (MgO) that prevents the magnesium from reacting with water.

A magnesium-iron alloy, composed of 95 percent magnesium and five percent iron, and a small amount of sodium chloride provide the desired effects. These materials are mixed with a powdered plastic to form a porous pad. Then the heat-producing reaction is started by adding water to the pad.

Hot MREs MREs replaced C-rations, or combat rations, in the early 1980s. C-rations consisted of six cans: an entrée, cheese, crackers, candy, dessert, and an accessory pack. C-rations did not have much of a variety of entrées, unlike the 24 varieties of entrées in MREs today. Examples of some entrées found currently in MREs are beef ravioli, chicken with noodles, pot roast with vegetables, beef stew, chili and macaroni, and pork ribs. There are also vegetarian MREs, which include entrées of cheese and vegetable omelet, vegetable lasagna, and cheese tortellini. All of the items included in an MRE come in a pouch inside a cardboard sleeve. An MRE consists of an entrée, crackers, cheese, peanut butter or jelly, dessert, candy, powdered drink mix, instant coffee, a beverage bag, an accessory pack, and a flameless ration heater.

The MRE entrée pouch slips into a small, plastic bag containing the magnesium-iron alloy pad. After the water is added, the bag is placed inside the sleeve in which the MRE was packed. The reaction transfers heat to the MRE pouch. The cardboard sleeve is an insulator to prevent heat loss. The reaction produces enough heat to raise the temperature of an 8-ounce ration serving by 60°C in about 12 minutes and keep the ration warm for about an hour. The heater and MRE can fit inside the pocket of a uniform, so soldiers, such as those in **Figure 1,** can carry them easily. A hot meal-to-go might just be the morale booster a field soldier needs.

Figure 1 MREs provide quick and easy hot meals to field soldiers.

Explore Further

1. **Classify** Is the chemical reaction in the heater classified as an endothermic or an exothermic reaction?

2. **Apply** What other applications might flameless heating have?

TRY AT HOME LAB

See page 870 for **Preventing a Chemical Reaction**

Catalysts Another way to change the rate of a reaction is to add or take away a catalyst. A **catalyst** is a substance that speeds up the rate of a reaction without being permanently changed or used up itself. Even though they increase the rate of reactions, catalysts do not change the position of equilibrium. Therefore, they do not affect how much product you can get from a reaction—just how fast a given amount of product will form.

How does a catalyst speed up a chemical reaction? You know that in chemical reactions, chemical bonds are broken and new ones are formed. The energy needed to break these bonds is the activation energy of the reaction. A catalyst speeds up the reaction by lowering the activation energy. This process can be compared to kicking a football over a goalpost. A player can kick the football over the goalpost at its regulation height, but if the height of the goalpost is lowered, the ball can be kicked over it using less energy.

Enzymes—biological catalysts Many different compounds are able to act as catalysts. The most powerful catalysts are those found in nature. They are needed to speed up the reactions necessary for a cell to function efficiently. These biological catalysts are called **enzymes**. Enzymes help your body use food for fuel, build up your bones and muscles, and store extra energy as fat. Enzymes are involved in almost every process in a cell. For example, proteases are enzymes that break down proteins, as shown in **Figure 6.23**. These enzymes occur naturally in cells to help with recycling proteins so their parts can be used over and over. Proteases are also used in many common products, including contact lens-cleaning solution and meat tenderizer.

The importance of enzymes to our health can be seen when someone lacks a gene for a particular enzyme. For example, lactose intolerance results when an enzyme that breaks down lactose, the sugar in dairy products, is not produced in a person's digestive system. When lactose is not broken down, it accumulates in the intestine, causing bloating and diarrhea.

■ **Figure 6.23** Gelatin made with fresh pineapple will not set properly, whereas gelatin made with canned pineapple will. Fresh pineapple contains active protease enzymes that break down protein molecules in the gelatin. Canned pineapple has been heated. Enzymes are heat-sensitive, so the proteases in the canned fruit are not active.

Inhibitors Adding a catalyst speeds up a reaction. Would you ever want to slow down a reaction? Reactions that have undesirable products sometimes have to be slowed down. Food undergoes reactions that cause it to spoil. Medications decompose, destroying or limiting their effectiveness. Can any substances slow down these reactions?

A substance that slows a reaction is called an **inhibitor**. Just as catalysts don't make reactions occur, inhibitors don't completely stop reactions. An inhibitor is placed in bottles of hydrogen peroxide to prevent it from decomposing too quickly into water and oxygen. If no inhibitor were added, the shelf life of hydrogen peroxide products would be much shorter because the molecules would decompose at a faster rate. Inhibitors are also added to many food products, including the dog food shown in **Figure 6.24,** to slow spoilage.

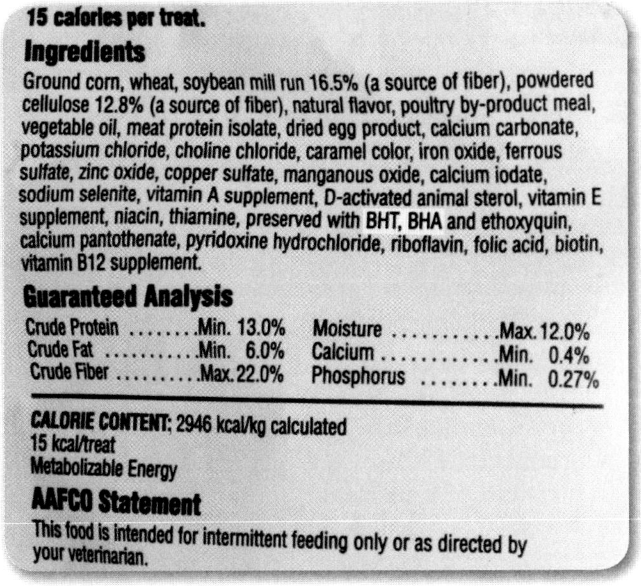

Connecting Ideas

Now you are familiar with the types of chemical reactions that occur around you. You know the importance of equilibrium and are aware of what factors can affect the rate of reaction. But why do some substances react with each other while others don't? In the next chapter, you will find out how the structure of an atom affects the way it reacts with other atoms.

■ **Figure 6.24** This dog food contains BHA and BHT. Both of these chemicals inhibit some of the reactions that cause food to spoil.

Section 6.3 Assessment

Section Summary

- Reversible reactions are those in which the products can react to reform the reactants.
- Equilibrium occurs when forward and reverse reactions take place at the same rate.
- At equilibrium, there is no net change in the amounts of products and reactants.
- According to Le Châtelier's principle, if a stress is applied to a system at equilibrium, the system shifts in the direction that relieves the stress.

15. **MAIN Idea** List and describe four factors that can influence the rate of a reaction.

16. **Explain** whether an exothermic reaction that is at equilibrium will shift to the left or to the right to readjust after each of the following procedures is followed.
    a) Products are removed.
    b) Reactants are added.
    c) Heat is added.
    d) Heat is removed.

17. **Explain** the difference between an inhibitor and a catalyst.

18. **Analyze** Hydrogen gas can be produced by the reaction of magnesium and hydrochloric acid, as shown by this equation.
    $$Mg(s) + 2HCl(aq) \rightarrow MgCl_2(aq) + H_2(g)$$
    In a particular reaction, 6 billion molecules of HCl were mixed with 1 billion atoms of Mg.
    a) Which reactant is limiting?
    b) How many molecules of H_2 are formed when the reaction is complete?

19. **Infer** Catalytic converters on cars use the metals rhodium and platinum as catalysts to convert potentially dangerous exhaust gases to carbon dioxide, nitrogen, and water. Why don't cars need to have the rhodium and platinum replaced after they are used?

CHAPTER 6 Study Guide

Download quizzes, key terms, and flash cards from glencoe.com.

BIG Idea Millions of chemical reactions in and around you absorb or release energy as they transform reactants into products.

Section 6.1 Chemical Equations

MAIN Idea Balanced chemical equations represent chemical reactions.

- coefficient (p. 197)
- product (p. 190)
- reactant (p. 190)

Key Concepts

- Chemical equations—used to represent reactions—are written using symbols and formulas for elements and compounds.
- Equations are balanced by changing coefficients.
- Chemical equations can tell you how elements and compounds change during a reaction and whether a reaction is endothermic or exothermic.
- A balanced chemical equation reflects the law of conservation of mass.

Section 6.2 Types of Reactions

MAIN Idea There are five types of chemical reactions: synthesis, decomposition, single displcacement, double displacement, and combustion reactions.

- combustion (p. 206)
- decomposition (p. 202)
- double displacement (p. 206)
- single displacement (p. 203)
- synthesis (p. 202)

Key Concepts

- Although thousands of individual chemical reactions are known, most can be classified into five major types that are based on patterns of behavior of reactants and products.
- The five general classes of reactions are synthesis, decomposition, single displacement, double displacement, and combustion.
- Sometimes classes of reactions overlap. For example, some combustion reactions are also synthesis reactions.

Section 6.3 Nature of Reactions

MAIN Idea External factors modify the direction and rate of chemical reactions.

- activation energy (p. 216)
- catalyst (p. 220)
- concentration (p. 217)
- dynamic equilibrium (p. 209)
- enzyme (p. 220)
- equilibrium (p. 209)
- inhibitor (p. 221)
- insoluble (p. 213)
- Le Châtelier's principle (p. 212)
- limiting reactant (p. 218)
- soluble (p. 213)

Key Concepts

- Reversible reactions are those in which the products can react to reform the reactants.
- Equilibrium occurs when forward and reverse reactions take place at the same rate.
- At equilibrium, there is no net change in the amounts of products and reactants.
- According to Le Châtelier's principle, if a stress is applied to a system at equilibrium, the system shifts in the direction that relieves the stress.

Vocabulary PuzzleMaker glencoe.com

Chapter 6 Assessment

Understand Concepts

20. List one piece of evidence you expect to see that indicates that a chemical reaction is taking place in each of the following situations.
    a) A slice of bread gets stuck in the toaster and burns.
    b) An unused carton of milk is left on a kitchen counter for two weeks.
    c) Wood burns in a campfire.
    d) A light stick is bent to activate it.

21. Use the equation to answer the questions.

    $$2Sr(s) + O_2(g) \rightarrow 2SrO(s)$$

    a) What is the physical state of strontium?
    b) What is the coefficient of strontium oxide?
    c) What is the subscript of oxygen?
    d) How many reactants take part in this reaction?

22. Use word equations to describe the chemical equations given.
    a) $AgNO_3(aq) + NaBr(aq) \rightarrow AgBr(s) + NaNO_3(aq)$
    b) $C_5H_{12}(l) + 8O_2(g) \rightarrow 5CO_2(g) + 6H_2O(g)$
    c) $CoCO_3(s) + energy \rightarrow CoO(s) + CO_2(g)$
    d) $BaCO_3(s) + C(s) + H_2O(g) \rightarrow 2CO(g) + Ba(OH)_2(s)$

23. Explain why subscripts should never be changed when balancing an equation.

24. Balance the equations.
    a) $Fe(s) + O_2(g) \rightarrow Fe_3O_4(s)$
    b) $NH_4NO_3(s) \rightarrow N_2O(g) + H_2O(g)$
    c) $COCl_2(g) + H_2O(l) \rightarrow HCl(aq) + CO_2(g)$
    d) $Sn(s) + NaOH(aq) \rightarrow Na_2SnO_2(aq) + H_2(g)$

Apply Concepts

25. One pollutant produced in automobile engines is nitrogen dioxide (NO_2). It changes to form nitrogen oxide (NO) and oxygen atoms when exposed to sunlight. Of what reaction type is this an example?

26. If you place a piece of a saltine cracker on your tongue for a few minutes, it begins to taste sweet. Do you think a chemical or physical change leads to this taste change? Why?

Everyday Chemistry

27. Explain why combining different cleaning products, such as those that contain ammonia and bleach, sometimes has deadly consequences. How could product manufacturers make the average person aware of hazards like this?

Chemistry and Technology

28. Why are high temperature and pressure needed for the Haber process?

Everyday Chemistry

29. What purposes might there be for a unit similar to an MRE that uses an endothermic reaction?

How It Works

30. When do you think it might be important that light sticks come in different colors?

Biology Connection

31. Why is it necessary that the gases breathed on the space shuttle are about the same composition as air on Earth instead of pure oxygen?

Think Critically

Apply Concepts

32. **ChemLab** Why can't you find a reaction in the ChemLab that can be classified as two different reaction types?

Form a Hypothesis

33. **MiniLab 1** Does the reaction taking place in MiniLab 1 go to completion, or does it reach an equilibrium? How do you know?

Chapter 6 Assessment

Make Predictions

34. MiniLab 3 Predict the effect that each of the following on the speed of the starch-iodine clock reaction.
  a) adding an inhibitor to the reaction
  b) adding a catalyst to the reaction
  c) raising the temperature of the reaction

Cumulative Review

35. Compare the number of electrons in the outer energy levels of metals, nonmetals, and metalloids. *(Chapter 3)*

36. Write a balanced equation for the formation of sodium chloride from sodium metal and chlorine gas. *(Chapter 4)*

37. Compare properties of ionic and covalent compounds. *(Chapter 5)*

Skill Review

38. Relate Cause and Effect Use what you have learned about factors that influence rates of chemical reactions to explain each of the following statements.
  a) Colors in the fabric of curtains in windows exposed to direct sunlight often fade.
  b) Meat is preserved longer when stored in a freezer rather than a refrigerator.
  c) Taking one or two aspirin will not harm most people, but taking a whole bottle at once can be fatal.
  d) Butylated hydroxyanisole (BHA) is an antioxidant often added to food, paints, plastics, and other products as a preservative.

39. Model Use toothpicks and Styrofoam balls of different colors to balance the following equations.
  a) $C_2O(g) + H_2O(l) \rightarrow HClO(aq)$
  b) $Fe_2O(s) + CO(g) \rightarrow Fe(s) + CO_2(g)$
  c) $H_2(g) + N_2(g) \rightarrow NH_3(g)$
  d) $ZnO(s) + HCl(aq) \rightarrow ZnCl_2(aq) + H_2O(l)$

40. Interpret Graphs Figure 6.25 shows the concentrations of compounds A and B as they take part in a reversible reaction.
  a) Identify the reactant and product.
  b) How long did it take for the reaction to reach equilibrium?
  c) Explain how the graph will change if more product is added one minute after the reaction reaches equilibrium.

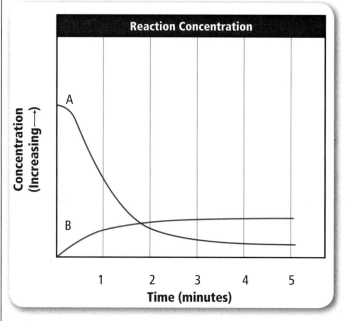

■ Figure 6.25

WRITING in Chemistry

41. Write a short report describing five chemical reactions that occur in your neighborhood every day. Describe the changes you observe as a result of these reactions.

Problem Solving

42. Balance the equations and classify each reaction as one of the five general types.
  a) $Al(s) + H_2SO_4(aq) \rightarrow Al_2(SO_4)_3(aq) + H_2(g)$
  b) $CS_2(l) + O_2(g) \rightarrow CO_2(g) + SO_2(g)$
  c) $H_2SO_4(aq) + NaCN(s) \rightarrow HCN(g) + Na_2SO_4(aq)$

Cumulative
Standardized Test Practice

1. Which is an example of a polyatomic ion?
   a) CO_2
   b) Mg^{2+}
   c) MnO^{4-}
   d) NaCl

2. Which chemical equation includes the formation of water?
   a) $2H_2O \rightarrow 2H_2 + O_2$
   b) $HC_2H_3O_2 + NaHCO_3 \rightarrow NaC_2H_3O_2 + H_2O + CO_2$
   c) $H_2 + O_2 \rightarrow H_2O_2$
   d) $H_2O\ (aq) \rightarrow H_2O\ (s)$

3. A system reaches chemical equilibrium when
   a) no new product is formed by the forward reaction.
   b) the reverse reaction no longer occurs in the system.
   c) the concentration of reactants in the system equals the concentration of products.
   d) the rate at which the forward reaction occurs equal the rate of the reverse reaction.

4. How do the properties of the compound magnesium oxide (MgO) compare with the properties of the elements magnesium and oxygen?
   a) The compound has completely different properties from the two elements.
   b) The compound has the same properties as the two elements.
   c) The compound has similar properties as the two elements.
   d) The properties of compounds and elements cannot be compared.

Free Energy of Chemical Reactions

Chemical Reaction	Free Energy (kJ/mol)
a) $CH_4(g) + 2O_2(g) \rightarrow CO_2(g) + 2H_2O(l)$	-890
b) $2H_2(g) + O_2(g) \rightarrow 2H_2O(g)$	-458
c) $2H_2(g) + O_2(g) \rightarrow 2H_2O(l)$	-572
d) $C_2H_4(g) + H_2(g) \rightarrow C_2H_6(g)$	+137
e) $6CO_2 + 6H_2O \rightarrow C_6H_{12}O_6 + 6O_2 + 6H_2O$	+470

Use the table above to answer Question 6.

5. Which chemical reactions are endothermic?
   a) a, b, and c
   b) d and e
   c) all of the reactions
   d) none of the reactions

6. Isotopes differ in
   a) the number of electrons in their atoms.
   b) the number of neutrons in their atoms.
   c) the number of protons in their atoms.
   d) the amount of empty space in their atoms.

7. Which element is a metalloid?
   a) aluminum   c) arsenic
   b) argon   d) calcium

8. What type of reaction is described by this equation?
   $Cs(s) + H_2O(l) \rightarrow CsOH(aq) + H_2(g)$
   a) synthesis
   b) decomposition
   c) single displacement
   d) double displacement

NEED EXTRA HELP?

If You Missed Question...	1	2	3	4	5	6	7	8
Review Section...	5.1	6.1	6.3	4.1	6.1	2.1	3.2	6.2

CHAPTER 7
Completing the Model of the Atom

BIG Idea Valence electrons determine the properties of elements and their positions in the periodic table.

7.1 Present-Day Atomic Theory
MAIN Idea Electrons are organized in energy levels and sublevels.

7.2 The Periodic Table and Atomic Structure
MAIN Idea A predictable pattern can be used to determine electron arrangement in an atom.

ChemFacts

- Fireworks were first used to frighten away evil spirits in the twelfth century.
- In the United States, fireworks were used to celebrate the first Independence Day in 1777.
- The colors displayed by fireworks depend on the chemical composition of the display shell.

Start-Up Activities

LAUNCH Lab

Observe Electric Charge

Electric charge plays an important role in atomic structure and throughout chemistry. How can you observe the behavior of electric charge using common objects?

Materials
- metric ruler
- hole punch
- plastic comb
- paper
- 10-cm-long piece of clear plastic tape (4)

Procedure

1. Read and complete the lab safety form.
2. Cut out small round pieces of paper using the hole punch and spread them out on a table. Run a plastic comb through your hair. Bring the comb close to the pieces of paper. Record your observations.
3. Fold a 1-cm-long portion of each piece of tape back on itself to form a handle. Stick two pieces of tape firmly to your desktop. Quickly pull both pieces of tape off of the desktop and bring them close together so that their non-sticky sides face each other. Record your observations.
4. Firmly stick one of the remaining pieces of tape to your desktop. Firmly stick the last piece of tape on top of the first. Quickly pull the pieces of tape as one from the desktop and then pull them apart. Bring the two tape pieces close together so that their non-sticky sides face each other. Record your observations.

Analysis

1. **Interpret** your observations using your knowledge of electric charge.
2. **Determine** which charges on the objects you used are similar and which ones are different.
3. **Explain** how you can tell.

Inquiry How can you relate the different charges you have observed to the structure of matter?

FOLDABLES Study Organizer

Energy Levels Make the following Foldable to organize information about energy levels and sublevels.

> **STEP 1** Fold a sheet of paper in half lengthwise.

> **STEP 2** Fold the page in half and then in half again.

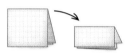

> **STEP 3** Unfold and cut along the fold lines of the top flap to make four tabs.

> **STEP 4** Label the tabs as follows: s, p, d, and f.

FOLDABLES Use this Foldable with Sections 7.1 and 7.2. As you read these sections, summarize information about energy levels and sublevels under the appropriate tab.

Visit glencoe.com to:
▶ study the entire chapter online
▶ explore **Concepts In Motion**
▶ take Self-Check Quizzes
▶ use Personal Tutors
▶ access Web Links for more information, projects, and activities
▶ find the Try at Home Lab, Comparing Orbital Sizes

Chapter 7 • Completing the Model of the Atom **227**

Section 7.1

Objectives
- **Relate** emission spectra to the electron configurations of atoms.
- **Examine** energy sublevels and orbitals within the atom.

Review Vocabulary
electromagnetic spectrum: the entire range of electromagnetic radiation

New Vocabulary
sublevel
aufbau principle
Heisenberg uncertainty principle
orbital
electron configuration

Present-Day Atomic Theory

MAIN Idea Electrons are organized in energy levels and sublevels.

Real-World Reading Link When documents are organized in file cabinets, they are arranged by topic in file folders and placed in cabinet drawers. In a similar way, electrons are organized in sublevels and levels.

Expanding the Model of the Atom

Recall from Chapter 2 that John Dalton proposed an atomic theory based on the law of conservation of matter in 1803. In his view, atoms were the smallest particles of matter. Then, in 1897, J.J. Thomson discovered electrons. The existence of electrons meant that the atom was made up of even smaller particles. These particles include protons and neutrons, as well as electrons.

At first, it wasn't clear how these subatomic particles were arranged. Scientists thought they were simply mixed together like the ingredients in cookie dough. But in 1909, Ernest Rutherford performed an experiment in which he aimed atomic particles at a thin sheet of gold foil. He found that most particles went right through the foil, but some were deflected.

These results suggested that most of the atom is empty space and that almost all of the mass of the atom is contained in a tiny nucleus. Rutherford proposed a nuclear model of the atom in which protons and neutrons make up the nucleus, and electrons move around in the space outside the nucleus.

In 1913, the Danish physicist Niels Bohr suggested that electrons revolve around the nucleus much like planets revolve around the Sun. Bohr's model was consistent with the emission spectrum produced by the hydrogen atom, but the model couldn't be used to explain more complicated atoms. **Figure 7.1** illustrates the development of the atomic theory up to Bohr's model.

■ **Figure 7.1** Atomic theory has changed over the past 2000. Experimental evidence reveals the complex nature of the submicroscopic world. Because electrons determine an element's chemical properties, chemists need an atomic model that describes the arrangement of electrons.
Summarize the development of the atomic model from Dalton to Bohr.

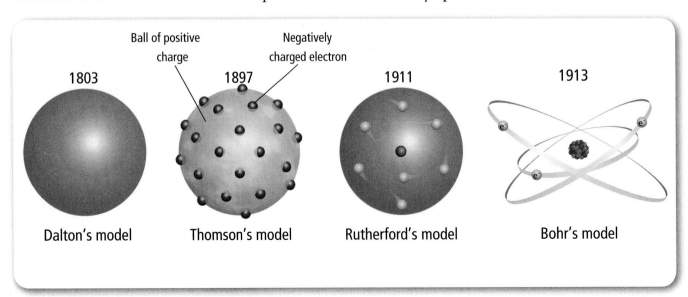

228 Chapter 7 • Completing the Model of the Atom

By 1935, the current model of the atom had evolved. This model explains electron behavior by interpreting the emission spectra of all the elements. It pictures energy levels as regions of space where there is a high probability of finding electrons. Before going on to the modern atomic theory, take another look at what you already know about atoms and electrons.

The present-day model In the present-day model of the atom, shown in **Figure 7.2**, neutrons and protons form a nucleus at the center of the atom. Negatively-charged electrons are distributed in the space around the nucleus. The electrons with the most energy are farthest from the nucleus in the outermost energy level. Recall from Chapter 2 that evidence for the existence of energy levels came from the interpretation of the emission spectra of atoms. It's important to know about energy levels in atoms because it helps to explain how atoms create chemical bonds and result in the formation of ionic or covalent compounds.

Valence electrons and the periodic table The periodic table reflects each element's electron arrangement. In Chapter 3, you learned that the number of valence electrons is equal to the group number for elements in groups 1 and 2 and is equal to the second digit of the group number for elements in groups 13 through 18. The period number represents the outermost energy level in which valence electrons are found. For example, lithium—in group 1, period 2—has one valence electron in the second energy level; sulfur—in group 16, period 3—has six valence electrons in the third level.

You can use the periodic table to determine a complete energy level diagram for any element. The atomic number of sulfur is 16. This means that sulfur has a total of 16 electrons. Sulfur is in group 16 and period 3, so it has six valence electrons in the third energy level. The remaining ten electrons must be distributed over the first two energy levels. **Figure 7.3** illustrates energy levels in sulfur.

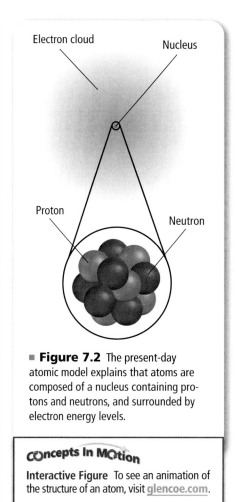

■ **Figure 7.2** The present-day atomic model explains that atoms are composed of a nucleus containing protons and neutrons, and surrounded by electron energy levels.

Concepts In Motion

Interactive Figure To see an animation of the structure of an atom, visit glencoe.com.

■ **Figure 7.3** Energy levels represent regions of space where electrons are distributed around the nucleus of an atom. Electrons located farthest away from the nucleus have the greatest amount of energy potential.

Relate the energies of the electrons in levels 1, 2, and 3 to the placement of the electrons in the model of the sulfur atom on the right.

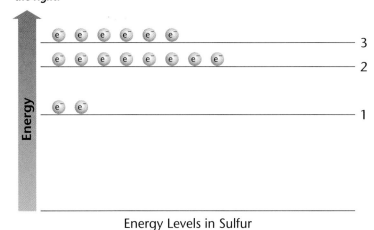
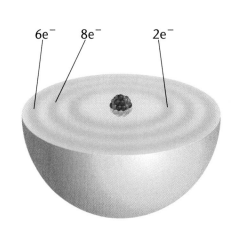

Physics Connection

Niels Bohr—Atomic Physicist and Humanitarian

Scientists are sometimes portrayed as strange and aloof. In fact, just like other people, they work, they interact with others, and they deal with everyday problems. They may differ from other people only in the way they conceptualize and experiment in areas of science. But they are like other people in that they may have strong ethical beliefs and the courage to fight for those beliefs. Niels Bohr's life exemplified these characteristics.

Bohr's atomic model Bohr, shown in **Figure 1,** had developed his atomic model by age 28. Nine years later, in 1922, he received the Nobel Prize in Physics for this work. The basic ideas of Bohr's model were that electrons move around the atom's nucleus in circular paths called orbits. These orbits are definite distances from the nucleus and represent energy levels that determine the energies of the electrons. Those electrons orbiting closest to the nucleus have the lowest energy; those farthest from the nucleus have the highest energy. If electrons absorb energy, they move to higher energy levels. When they drop down to lower energy levels, they release energy.

Theories change Parts of Bohr's theory are still accepted today, whereas other parts are outdated. Electrons are thought to move around the nucleus, but not in definite paths. The existence of energy levels is correct, but now we know with greater assurance that there are regions of probability of finding electrons. An electron cannot be expected to be in an exact place.

Atomic bomb In 1939, Bohr attended a scientific conference in the United States, where he reported that Lise Meitner and Otto Hahn had discovered how to split uranium atoms through a process called fission. This dramatic discovery laid the groundwork for the development of the atomic bomb.

Figure 1 Niels Bohr is known for his work on atomic structure and radiation.

Bohr escaped from the Nazis In 1940, the Nazis invaded and occupied Niels Bohr's homeland, Denmark. Bohr was opposed to the Nazi regime, but he continued his position as director of the Copenhagen Institute for Theoretical Physics until 1943. When he learned that the Nazis planned to arrest him and force him to work in Germany on an atomic project, he and his family fled to Sweden under frightening conditions. In 1943, Bohr came to the United States and worked with scientists from all over the world on the Manhattan Project.

Connection to Chemistry

1. **Interpret** Bohr explained that the electrons in the outermost energy level determine the chemical properties of an element. What did he mean?

2. **Think Critically** How important to chemistry is the physicist's work on atomic structure?

3. **Acquire Information** Research information on the Manhattan Project. Find out about the scientists involved, their nationalities, ages, genders, and their fields of expertise.

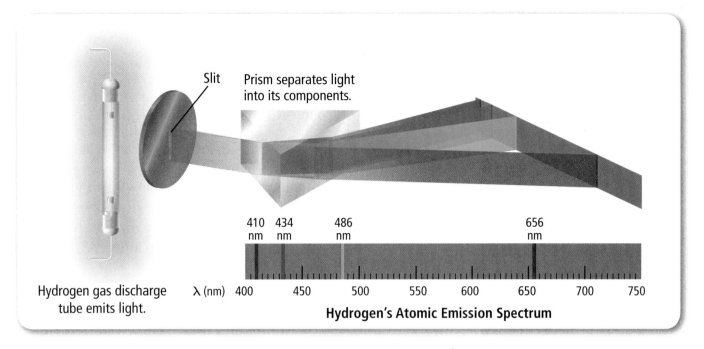

Figure 7.4 The purple light emitted by hydrogen can be separated into its different components using a prism. Hydrogen produces an emission spectrum of four different colors of visible light.

Emission spectrum of hydrogen Recall electromagnetic radiation from Chapter 2. It can be described as energy having a range of frequencies and wavelengths. The energy of the radiation directly affects the frequency of the waves. The greater the energy, the higher the frequency of a wave, and the greater the energy, the shorter the wavelength. These relationships are used to calculate the exact amount of energy released in electromagnetic radiation by the electrons in atoms. Light is a form of electromagnetic radiation.

In the emission spectrum of hydrogen, shown in **Figure 7.4,** you can see the four different colors of visible light released by hydrogen as its electron moves from higher energy levels to a lower energy level. In Chapter 2, the analogy of energy levels as rungs on a ladder suggested that electrons can move from one energy level to another, but they can't lie between energy levels. By absorbing a specific amount of energy, an electron can jump to a higher energy level. Then, when it falls back to the lower energy level, the electron releases the same amount of energy in the form of radiation with a definite frequency. **Figure 7.5** illustrates how electron transitions between energy levels are related to the amount of energy produced.

Vocabulary
Word origin

Spectrum
comes from the Latin word *spectrum*, meaning *appearance, specter*

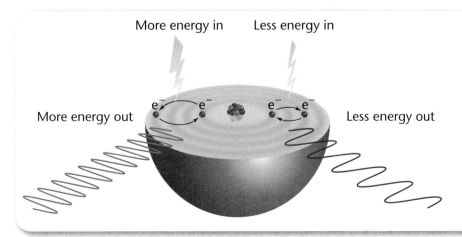

Figure 7.5 The number of energy levels to which an electron can jump depends on the amount of energy it absorbs. When an electron falls back to its original level, it emits energy in the form of light. The wavelength and color of the light depend on how much the energy of the electron decreases. The greater the energy given off, the more toward the violet end of the spectrum the color will be. Less energy produced results in a shift toward the red end of the spectrum.

MiniLab 7.1

Flame Tests

How do flame colors vary for different elements? The vivid colors of a fireworks display result from electrons in metal atoms moving from higher energy levels to lower energy levels due to the addition of heat energy. Similarly, when compounds containing metals are heated, a technique called a flame test can be used to identify unknown metals.

Procedure

1. Read and complete the lab safety form.
2. Obtain from your teacher six **wooden splints** that are labeled and have been soaked in **saturated solutions of lithium chloride, sodium chloride, potassium chloride, calcium chloride, strontium chloride,** and **barium chloride.**
3. Light a laboratory **burner** or a **propane torch** held in a metal holder. Adjust the burner to give the hottest blue flame possible.
4. In turn, hold the soaked end of each of the splints in the flame for a short time. Observe and record the color of each flame. Extinguish the splint as soon as the flame is no longer colored.
5. Obtain a splint that has been soaked in an **unknown solution.** Perform the flame test and identify the unknown metallic element.

Analysis

1. **Describe** the colors produced by each of the metallic elements.
2. **Identify** your unknown element. Justify your choice.
3. **Explain** how you might test an unknown crystalline substance to determine whether it is table salt. A taste test is never recommended for an unknown compound.

Multi-electron atoms The emission spectrum of hydrogen has four characteristic lines that identify it. Similarly, each element has a characteristic set of spectral lines. This means that the energy levels within the atom must also be characteristic of each element. But when scientists first investigated multi-electron atoms, they found that their spectra were far more complex than would be anticipated by the simple set of energy levels predicted for hydrogen. **Figure 7.6** shows spectra for three elements—hydrogen (H), mercury (Hg), and neon (Ne).

■ **Figure 7.6** The big gaps between spectral lines indicate that electrons are moving between energy levels that have a large difference in energy. The groups of fine lines indicate that electrons are moving between energy levels that are close in energy.

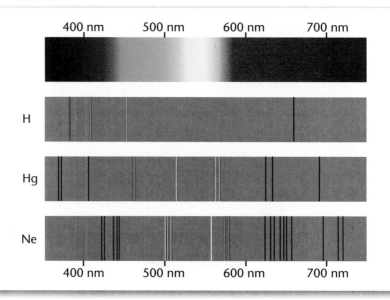

232 Chapter 7 • Completing the Model of the Atom

Energy sublevels Notice that the spectra for mercury and neon in **Figure 7.6** have many more lines than the spectrum of hydrogen. Some lines are grouped close together, and there are big gaps between these groups of lines. The big gaps correspond to the energy released when an electron jumps from one energy level to another. The interpretation of the closely spaced lines is that they represent the movement of electrons from levels that are not very different in energy. This suggests that **sublevels**—divisions within a level—exist within each energy level. If electrons are distributed over one or more sublevels within an energy level, then these electrons would have only slightly different energies. The energy sublevels are named s, p, d, and f.

Each energy level has a specific number of sublevels, which is the same as the number of the energy level for the first four periods on the periodic table. For example, the first energy level has one sublevel. It's called the 1s sublevel. The second energy level has two sublevels, the 2s and 2p sublevels. The third energy level has three sublevels: the 3s, 3p, and 3d sublevels; and the fourth energy level has four sublevels: the 4s, 4p, 4d, and 4f sublevels. Within a given energy level, the energies of the sublevels, from lowest to highest, are s, p, d, and f.

The **aufbau principle** states that each electron occupies the lowest energy sublevel available. Therefore, you need to learn the sequence of atomic orbitals. This sequence, known as the aufbau diagram, is shown in **Figure 7.7**. Notice how the sublevels within an energy level are close together. This explains the groups of fine lines in an element's emission spectrum. For example, you might expect three spectral lines with slightly different frequencies because electrons fall from the 3s, 3p, and 3d sublevels to the 2s sublevel. Because each of these electrons initially has slightly different energy within the third energy level, each emits slightly different radiation.

FOLDABLES
Incorporate information from this section into your Foldable.

■ **Figure 7.7** This diagram shows the energy of each sublevel relative to the energy of other sublevels.
Apply Which sublevel has the greater energy, 4d or 5p?

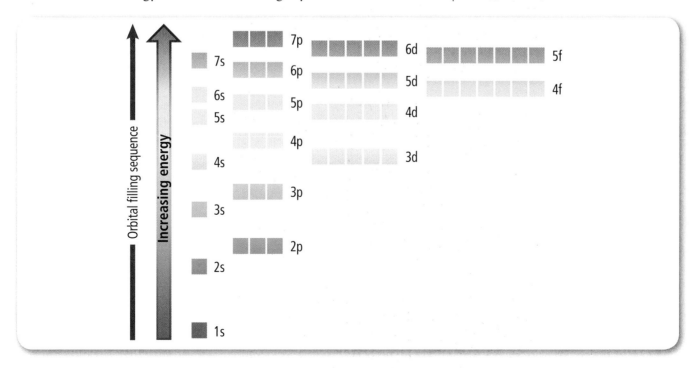

Section 7.1 • Present-Day Atomic Theory 233

CHEMLAB

METALS, REACTION CAPACITIES, AND VALENCE ELECTRONS

Background
Many metals react with acids, producing hydrogen gas. If a metal reacts with an acid in this manner, the amount of hydrogen produced is related to the number of valence electrons in the atoms of the metal. In this ChemLab, you will react equal numbers of atoms of magnesium and aluminum with hydrochloric acid and compare the amount of hydrogen gas generated by the two metals. Each reaction proceeds only as long as there are metal atoms to react.

Question
How do the reactions of hydrochloric acid with magnesium and aluminum compare, and how are these capacities related to the valence electrons in the atoms of the two elements?

Objectives
Compare the reactions using magnesium and aluminum.

Interpret the results of the investigation in terms of the numbers of valence electrons in the atoms of the two elements.

Preparation

Materials
thin-stemmed micropipettes (2)
50-mL graduated cylinder
water trough or plastic basin
1M hydrochloric acid (HCl)
3M hydrochloric acid (HCl)
magnesium ribbon
aluminum foil
transparent, waterproof tape
plastic wrap
forceps (2)

Safety Precautions
WARNING: The bulb of the micropipette may become hot during the reaction. Hold the stem portion of the pipette with forceps.

Procedure
1. Read and complete the lab safety form.
2. Cut small slits in two micropipettes as shown on the dotted line below. Obtain a 0.020-g sample of magnesium ribbon from your teacher and insert the magnesium into the bulb of one pipette. Obtain a 0.022-g sample of aluminum foil and insert it into the bulb of the other pipette. Seal the slits with transparent, waterproof tape and label the two pipettes *magnesium (Mg) foil* and *aluminum (Al) foil*.

3. Fill the water trough or plastic basin nearly full of water.
4. Fill the 50-mL graduated cylinder with water and cover the top with plastic wrap. Hold the wrap tightly so that no water can escape and invert the cylinder, placing the top beneath the surface of the water in the trough. Remove the plastic wrap. No appreciable amount of air should be in the cylinder. If this is not the case, repeat the procedure. The inverted cylinder may be clamped in place or held by hand.

5. Using the pipette containing the magnesium, squeeze most of the air from the pipette bulb and draw 3 mL of $1M$ hydrochloric acid into the pipette.

   WARNING: *Handle the hydrochloric acid solution with care. It is harmful to eyes, skin, and clothing. If any acid contacts your skin or eyes or any spillage occurs, rinse immediately with water and notify your teacher.*

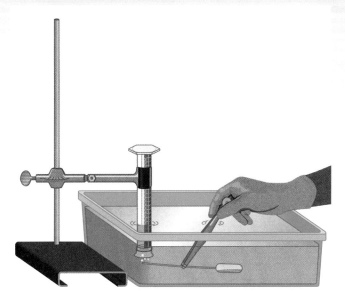

6. Hold the pipette with forceps as shown, and quickly immerse it in the water in the trough so that the pipette tip is inside the open end of the graduated cylinder.
7. Collect the hydrogen gas in the graduated cylinder. Allow the reaction to proceed until the magnesium ribbon is completely gone.
8. While the graduated cylinder is still in place, read the volume of hydrogen gas produced and record it in a table like the one shown.
9. Remove the tape from the pipette and rinse the pipette with water.
10. Repeat steps 5–9 using the micropipette containing aluminum and $3M$ HCl. Read and record the volume of hydrogen gas produced.

Data and Observations

Data Table

Volume of Hydrogen	
Hydrogen from magnesium (mL)	
Hydrogen from aluminum (mL)	

Analyze and Conclude

1. **Interpret Data** The volume of the hydrogen gas you collected is proportional to the number of molecules of hydrogen produced in the two reactions. Which element, aluminum or magnesium, produced more hydrogen molecules?
2. **Draw Conclusions** Which element has the greater ability to react per atom? Use your volume data to express the relative reaction capacities of the two elements as a ratio of small, whole numbers.
3. **Relate Concepts** In this investigation, the atoms of both metals react by losing electrons to form positive ions. Relate the ratio of reaction capacities to the number of valence electrons each element has.

Apply and Assess

Summarize Write the balanced equations for the two chemical reactions you performed.

INQUIRY EXTENSION

Predict Had you performed this investigation with sodium, using approximately the same number of atoms as you used of magnesium and aluminum, what volume of hydrogen gas would have been produced? **WARNING:** *Because sodium reacts explosively with water, it cannot be used safely in this investigation.*

CHEMISTRY AND TECHNOLOGY

Hi-Tech Microscopes

If you were a chemistry student in the 1960s or earlier, you would have been told that no one could see an atom. Now, with the aid of computers and revolutionary microscopes, it's possible to generate two- and three-dimensional images of atoms. It's even possible to move atoms around and observe their electron clouds. What kinds of instruments can accomplish feats that were not even thought of a decade or two ago? They include three kinds of microscopes: scanning probe, scanning tunneling, and atomic force.

Scanning Probe Microscope (SPM)

SPMs use probes to sense surfaces and produce three-dimensional pictures of a substance's exterior, as shown in **Figure 1.** The precise arrangement of atoms can be viewed in three dimensions. This is accomplished by measuring changes in the current as the probe passes over a surface. SPMs can pick up atoms one at a time,

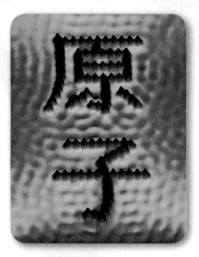

Figure 2 Atoms arranged to form Japanese kanji characters

move them around, and form letters, such as these Japanese kanji characters in **Figure 2.** Accomplishments like this suggest that in the future, all the information in the Library of Congress could be stored on an 20.3 cm silicon wafer. Scientists may be able to use SPMs to learn why two surfaces stick together or how to shrink chip circuits to make computers work faster. They may even be able to put a thousand SPM tips on a 2-cm^2 silicon chip.

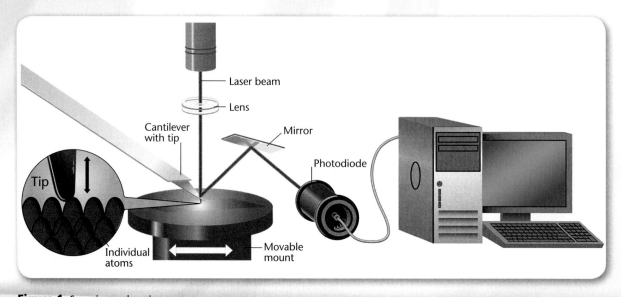

Figure 1 Scanning probe microscope

Scanning Tunneling Microscope (STM)

The scanning tunneling microscope also provides new technologies for chemists and physicists. The red areas in **Figure 3** show the valence electrons of metal atoms that are free to move about in a metallic crystal. On the surface of the crystal, they can move only in two dimensions and behave like waves. Two imperfections on the surface of the crystal cause the electrons to produce concentric wave patterns.

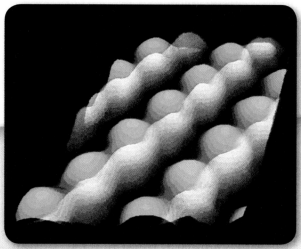

Figure 4 Electron clouds of gallium and arsenic

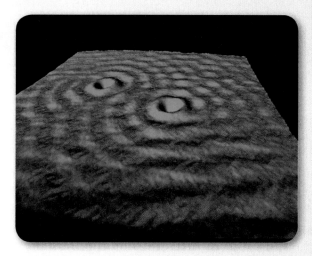

Figure 3 Electrons produced wave-like patterns.

Atomic Force Microscope (AFM)

The AFM, invented in 1985, uses repulsive forces between atoms on the probe's tip and those on the sample's surface to form images on the computer screen. An advantage of the AFM is that it does not have to work in a vacuum, so its samples require no special preparation. AFMs can provide images of molecules in living tissue and peel off membranes of living cells one layer at a time. **Figure 4** shows electron clouds of gallium and arsenic.

Another current STM technology uses hydrogen to push aside surface atoms of semiconductors to reveal the atomic structure underneath. Using this method, atoms can be removed one layer at a time.

Recently, the STM has succeeded in forming images of electron clouds in atoms and molecules. The electron clouds of molecules can be calculated. An STM color-enhanced image of gallium arsenide is shown in **Figure 4**. The electron clouds of gallium appear blue. The electron clouds of arsenic appear red. Chemists have figured out the structure of atoms and molecules using indirect evidence from chemical reactions, physical experiments, and mathematical calculations. Now, they can view images of atoms while they experiment with them and base their work on both direct and indirect evidence.

Discuss the Technology

1. **Compare and Contrast** Some scanning tunneling microscopes can remove layers of atoms. How does this capability compare with what the scanning probe microscope can do?
2. **Acquire Information** Find information on the Near-Field Scanning Optical Microscope.
3. **Infer** Compare the diagrams of the molecules with the computer-generated images of the molecules produced by the STM. How are the conclusions about the molecular structure obtained using conventional methods supported by the photos?

Table 7.1 — Distribution of Electrons in the First Four Energy Levels

Energy Level	Electrons in Level	Sublevel	Electrons in Sublevel
1	2	1s	2
2	8	2s	2
2	8	2p	6
3	18	3s	2
3	18	3p	6
3	18	3d	10
4	32	4s	2
4	32	4p	6
4	32	4d	10
4	32	4f	14

Electron Distribution in Energy Levels

Each sublevel is occupied by a specific number of electrons. An s sublevel has a maximum of two electrons. A p sublevel can have six electrons, a d sublevel can have ten electrons, and an f sublevel can have 14 electrons. **Table 7.1** shows how electrons are distributed in the sublevels of the first four energy levels.

Notice that the first energy level has one sublevel, the 1s sublevel. The maximum number of electrons in an s sublevel is two, so the first energy level is filled when it reaches two electrons. The second energy level has two sublevels, the 2s and 2p sublevels. The maximum number of electrons in a p sublevel is six, so the second energy level is filled when eight electrons are present—two in the 2s sublevel and six in the 2p sublevel. Look at the third and fourth energy levels and you can see that the third energy level can hold ten more electrons than the second because there is a d sublevel. The fourth can hold 14 more electrons than the third because there is an f sublevel.

Orbitals In the 1920s, Werner Heisenberg determined that it's impossible to measure accurately both the position and velocity of an electron at the same time. This principle is known as the **Heisenberg uncertainty principle**. In 1932, Heisenberg was awarded the Nobel Prize in Physics for this discovery because it led to the development of the electron cloud model to describe electrons in atoms.

Probability and the electron cloud model We can estimate the probability of finding an electron using a mathematical model developed by Erwin Schrödinger. Pretend that you can photograph the single electron that is attracted to the nucleus of a hydrogen atom. Once every second, you click the shutter on your camera. After a few hundred seconds, you develop the film and create a scatter diagram illustrating the position of the electron over time.

VOCABULARY
WORD ORIGIN
Orbital
comes from the Latin word *orbita* meaning *wheel, track, course, circuit*

■ **Figure 7.8**

Most of the time, hydrogen's electron is within the fuzzy cloud in the two-dimensional drawing.

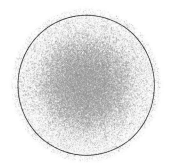

A circle, with the nucleus at the center and enclosing 95 percent of the cloud, defines an orbital in two dimensions.

The spherical model represents hydrogen's 1s orbital in three dimensions.

Sometimes, the electron is close to the nucleus. Other times, it's far away. Most of the time, it's in a small region of space that looks like a cloud. This electron cloud doesn't have a sharp boundary; its edges are fuzzy. If you ask someone to look at the picture shown in **Figure 7.8** and to tell you where the electron is, he or she could say only that it's probably somewhere in the cloud. But this cloud of electron probability can be useful. If you draw a line around the outer edge enclosing about 95 percent of the cloud, within the region enclosed by the sphere, you can expect to find the electron about 95 percent of the time. The space in which there is this high probability of finding the electron is called an **orbital**.

Electrons in orbitals Orbitals are regions of space located around the nucleus of an atom, each having the energy of the sublevel of which it is a part. Orbitals can have different sizes and shapes. There are four types of orbitals that accommodate the electrons for all the atoms of the known elements. Two simple rules apply to these four types. First, an orbital can hold a maximum of two electrons. Second, an orbital has the same name as its sublevel. There is only one s orbital with, at most, 2 electrons. A p sublevel has, at most, 6 electrons, and, thus, there are three p orbitals, each with 2 electrons. **Figure 7.9** shows the shapes of s and p orbitals.

FOLDABLES
Incorporate information from this section into your Foldable.

■ **Figure 7.9** Orbitals have characteristic shapes that depend on the number of electrons in the energy sublevels.

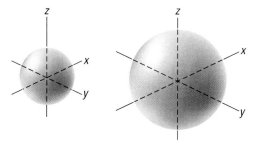

1s orbital   2s orbital

All s orbitals are spherical, and their size increases with increasing principal quantum number.

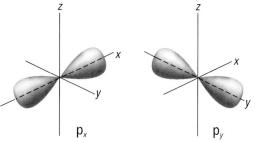

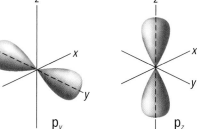

The three p orbitals are dumbshell-shaped and are oriented along the three perpendicular x, y, and z axes.

FACT of the Matter

When an electron in an atom is provided with more and more energy, it jumps to higher and higher energy levels. The electron moves farther and farther from the nucleus. The end result is that the electron is lost, and what remains is an ion.

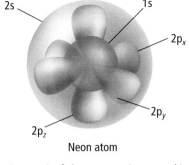

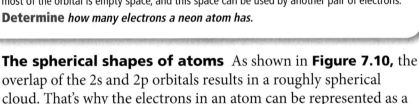
Neon atom

■ **Figure 7.10** Imagine a pair of electrons moving around in an orbital. At any instant, most of the orbital is empty space, and this space can be used by another pair of electrons.
Determine how many electrons a neon atom has.

The spherical shapes of atoms As shown in **Figure 7.10,** the overlap of the 2s and 2p orbitals results in a roughly spherical cloud. That's why the electrons in an atom can be represented as a series of fuzzy, concentric spheres. In the next section, you'll learn how the periodic table can be used to predict the electron configurations of the atoms.

Electron configurations In any atom, electrons are distributed into sublevels and orbitals in the way that creates the most stable arrangement; that is, the one with lowest energy. This most stable arrangement of electrons in sublevels and orbitals is called an **electron configuration.**

Electrons are organized in orbitals and sublevels beginning with the innermost sublevels and continuing to the outermost. At any sublevel, electrons fill the s orbital first, then the p. For example, the first energy level holds two electrons. These electrons pair up in the 1s orbital. The second energy level has four orbitals and can hold eight electrons. The first two electrons pair up in the 2s orbital, and the remaining six pair up in the three 2p orbitals.

SUPPLEMENTAL PRACTICE
For more practice with electron configuration and energy, see Supplemental Practice on p. 817.

Section 7.1 Assessment

Section Summary

▶ The position of an element in the periodic table reveals the number of valence electrons the element has.

▶ The outermost valence electrons determine the properties of an element.

▶ Electrons are found only in levels of fixed energy in an atom.

▶ Energy levels have sublevels. Each sublevel can hold a specific number of electrons.

▶ Sublevels can be subdivided into s, p, d, and f orbitals. Sublevels hold 2, 6, 10, and 14 electrons respectively.

1. **MAIN Idea Predict** How many s orbitals can the third energy level have? How many p orbitals? How many d orbitals?

2. **Predict** How many electrons can be held by each sublevel of the fourth energy level?

3. **Describe** the shape of a p orbital. How do p orbitals at the same energy level differ from one another?

4. **Apply Concepts** What are two ways in which the elements with atomic numbers 12 and 15 are different?

5. **Explain** Sodium vapor contained in a bulb or tube emits a brilliant yellow light when connected to a high-voltage source. Why do sodium atoms produce this color of light?

Section 7.2

Objectives
- **Distinguish** between the s, p, d, and f blocks on the periodic table and relate them to an element's electron configuration.
- **Predict** the electron configurations of elements using the periodic table.

Review Vocabulary
orbital: space in which there is a high probability of finding an electron

New Vocabulary
inner transition element

FOLDABLES Incorporate information from this section into your Foldable.

■ **Figure 7.11** The periodic table is divided into four blocks-s, p, d, and f.
Analyze *What is the relationship between the maximum number of electrons an energy sublevel can hold and the size of that block on the diagram?*

The Periodic Table and Atomic Structure

MAIN Idea A predictable pattern can be used to determine electron arrangement in an atom.

Real-World Reading Link Suppose you were about to play your first game of Chinese checkers. How would you know where to place the marbles on the game board to start the game? Without the rules of the game or maybe a diagram of the starting game board, you wouldn't know what to do.

Patterns of Atomic Structure

Electrons occupy energy levels by filling the lowest level first followed by higher energy levels in numerical order. Valence electrons of the main group elements occupy the s and p sublevels of the outermost energy level. The position of any element on the periodic table shows which sublevels—s, p, d, and f—the valence electrons occupy.

Orbitals and the periodic table The shape of the modern periodic table is a direct result of the order in which electrons fill energy sublevels and their orbitals. The periodic table in **Figure 7.11** is divided into blocks that show the sublevels occupied by the electrons of the atoms. Notice that groups 1 and 2 (the active metals) have valence electrons in s orbitals, and groups 13–18 (metals, metalloids, and nonmetals) have valence electrons in both s and p orbitals. Therefore, all the main group elements have their valence electrons in s or p sublevels. Groups 1 and 2 are designated as the s block of the periodic table, and groups 13–18 are designated as the p block. Note that groups 3–12 are designated as the d block and that each row of this block, except for that in period 7, has ten elements. The block beneath the table is the f block, and each row in this region contains 14 elements.

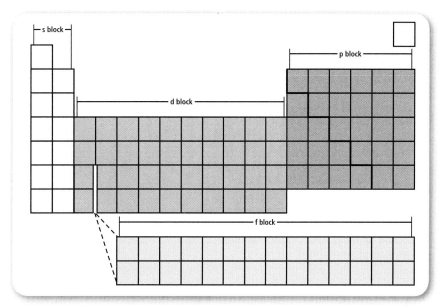

Building Electron Configurations

Chemical properties repeat when elements are arranged on the periodic table by atomic number because electron configurations repeat in a certain pattern. Refer back to **Figure 7.7** and the Aufbau principle. Hydrogen has a single electron in the first energy level. Its electron configuration is $1s^1$. This is standard notation for electron configurations. The number 1 refers to the energy level, the letter s refers to the sublevel, and the superscript refers to the number of electrons in the sublevel. Helium has two electrons in the 1s sublevel. Its electron configuration is $1s^2$. Helium has a completely filled first energy level. When the first energy level is filled, additional electrons must go into the second energy level. Electrons enter the sublevel that will give the atom the most stable configuration, that which has the lowest energy.

Second period configurations Lithium begins the second period. Its first two electrons fill the first energy level, so the third electron occupies the second level. Lithium's electron configuration is $1s^2 2s^1$. Beryllium has two electrons in the 2s sublevel, so its electron configuration is $1s^2 2s^2$. As you continue to move across the second period, electrons begin to enter the orbitals of the 2p sublevel. Each successive element has one more electron in the 2p orbitals. Carbon, for example, has four electrons in the second energy level. Two of these are in the 2s sublevel and two are in the 2p sublevels. The electron configuration for carbon is $1s^2 2s^2 2p^2$. At element number 10, neon, the p sublevel is filled with six electrons. The electron configuration for neon is $1s^2 2s^2 2p^6$. Neon has eight valence electrons; two are in an s sublevel and six are in p sublevels. The diagram in **Figure 7.12** can be used to help write the electron configuration for any element.

■ **Figure 7.12** An aufbau diagram helps to illustrate the predictable pattern for filling electron sublevels. Use the following steps and the aufbau diagram to write electron configurations.

1. Determine the number of electrons in one atom of the element for which you are writing the electron configuration. The number of electrons in a neutral atom equals the element's atomic number.

2. Starting with 1s, write the electron configuration by following the diagonal arrows from the top of the sublevel diagram to the bottom. When you complete one line of arrows, move to the right, to the beginning of the next line of arrows. As you proceed, add superscripts indicating the number of electrons in each sublevel. Continue only until you have sufficient atomic sublevels to account for the total number of electrons in one atom of the element.

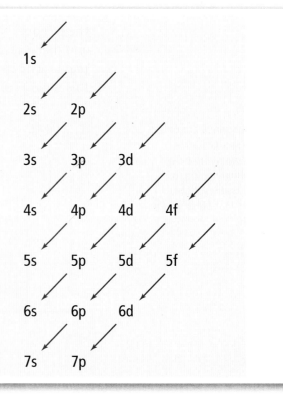

Third period configurations Sodium, atomic number 11, begins the third period and has a single 3s electron beyond the configuration of neon. Sodium's electron configuration is $1s^2 2s^2 2p^6 3s^1$. If you compare this with the electron configuration of lithium, $1s^2 2s^1$, it's easy to see why sodium and lithium have similar chemical properties. Each has a single electron in the valence level.

Noble gas notation Notice that neon's configuration has an inner core of electrons that is identical to the electron configuration in helium ($1s^2$). This insight simplifies the way electron configurations are written. Neon's electron configuration can be abbreviated [He]$2s^2 2p^6$. Using noble gas notation, neon's electron configuration is represented by an inner core of electrons from the noble gas in the preceding row (He), followed by the orbitals filled in the subsequent period. The abbreviated electron configuration for sodium is [Ne]$3s^1$, where the neon core represents sodium's ten inner electrons. Its similarity to the [He]$2s^1$ configuration of lithium shows clearly that these group 1 elements have the same number of valence electrons in the same type of sublevel.

Table 7.2 shows electron configurations for all the elements in the second and third periods. Notice that elements in the same group have similar configurations. This is important because it shows that the periodic trends in properties, observed in the periodic table, are really the result of repeating patterns of electron configuration.

Concepts in MOtion

Interactive Table Explore electron configurations and orbital diagrams at glencoe.com.

Table 7.2 Electron Configurations of Second and Third Period Elements

Second Period Elements	Configuration	Third Period Elements	Configuration
Lithium	[He] $2s^1$	Sodium	[Ne]$3s^1$
Beryllium	[He] $2s^2$	Magnesium	[Ne]$3s^2$
Boron	[He] $2s^2 2p^1$	Aluminum	[Ne]$3s^2 3p^1$
Carbon	[He] $2s^2 2p^2$	Silicon	[Ne]$3s^2 3p^2$
Nitrogen	[He] $2s^2 2p^3$	Phosphorus	[Ne]$3s^2 3p^3$
Oxygen	[He] $2s^2 2p^4$	Sulfur	[Ne]$3s^2 3p^4$
Fluorine	[He] $2s^2 2p^5$	Chlorine	[Ne]$3s^2 3p^5$
Neon	[He] $2s^2 2p^6$	Argon	[Ne]$3s^2 3p^6$

MiniLab 7.2

Model Electrons in Atoms

How can you model probability distribution? Modern atomic theory cannot tell you exactly where the electrons in atoms are. However, it does define regions in space called orbitals, where there is a 95 percent probability of finding an electron. The lowest energy orbital in any atom is called the 1s orbital. In this MiniLab, you will simulate the probability distribution of the 1s orbital by noting the distribution of impacts or hits around a central target point.

Procedure

1. Read and complete the lab safety form.
2. Obtain two pieces of **blank, white, 8½" × 11" paper** and draw a small but visible mark in the center of each of the papers. Hold the papers together toward a light and align the center marks exactly.
3. Around the center dot of one of the papers, which you will call the target paper, use a **compass** to draw concentric circles having radii of 1 cm, 3 cm, 5 cm, 7 cm, and 9 cm. Number the areas of the target 1, 2, 3, 4, and 5, starting with number 1 at the center.
4. Lay the target paper face up on top of a piece of **poster board**.
5. Cover the target paper with a piece of **carbon paper**, carbon side down. Then place the second piece of white paper on top with the center mark facing up. Use **tape** to fasten the three layers of paper in place on the poster board and to secure the poster board to the floor.
6. Stand over the target paper and drop a **dart** 100 times from chest height, attempting to hit the center mark.
7. Remove the tape from the papers. Separate the white papers and the carbon paper. Tabulate and record the number of hits in each area of the target paper.

Analysis

1. **Make an Analogy** How many hits did you record in each of the target areas? What does each hit represent in the model of the atom?
2. **Graph** Make a graph plotting the number of hits on the vertical axis and the target area on the horizontal axis.
3. **Analyze** Which of the target areas has the highest probability of a hit? Relate your findings to the model of the atom.

Noble gas configurations Each period ends with a noble gas, which shows that the noble gases have filled energy levels and stable electron configurations. The electron configurations for all the noble gases are shown in **Table 7.3**. All except helium have eight valence electrons. However, helium's two electrons fill its outermost energy level and are thus a stable configuration. These stable electron configurations explain the lack of reactivity of the noble gases. Noble gases don't need to form chemical bonds to acquire stability.

Table 7.3 The Electron Configurations of the Noble Gases

Noble Gas	Electron Configuration	Noble Gas	Electron Configuration
Helium	$1s^2$	Krypton	$[Ar]\,4s^2 3d^{10} 4p^6$
Neon	$[He]\,2s^2 2p^6$	Xenon	$[Kr]\,5s^2 4d^{10} 5p^6$
Argon	$[Ne]\,3s^2 3p^6$	Radon	$[Xe]\,6s^2 4f^{14} 5d^{10} 6p^6$

Fourth period configuration You might expect that after the 3p orbitals are filled in argon, the next electron would occupy a 3d orbital, but this is not the case. Potassium follows argon and begins the fourth period. Its configuration is [Ar]4s^1. Compare potassium's configuration to the configurations of lithium and sodium in **Table 7.2,** and recall that potassium is chemically similar to the group 1 elements. Experimental evidence indicates that the 4s and 3d sublevels are close in energy, with the 4s sublevel having a slightly lower energy. Thus, the 4s sublevel fills first because that order produces an atom with lower energy. The next element after potassium is calcium. Calcium completes the filling of the 4s sublevel. It has the electron configuration [Ar]4s^2.

Transition Elements Notice in the fourth period that calcium is followed by a group of ten elements beginning with scandium and ending with zinc. These are transition elements. Now the 3d sublevel begins to fill, producing configurations with the lowest possible energy. Ten electrons are added across the row and fill the 3d sublevel. Just as the s block at each energy level adds two electrons and fills one s orbital, and the p block at each energy level adds six electrons and fills three p orbitals, so the d block adds ten electrons. There are five d orbitals. The f block adds 14 electrons, and they are accommodated in seven orbitals. The electron configuration of scandium, the first transition element, is [Ar]4s^{2}3d^1. Zinc, the last of the series, has the configuration [Ar]4s^{2}3d^{10}. **Figure 7.13** shows the configurations of the 3d transition elements chromium and copper. The six elements following the 3d elements, gallium to krypton, fill the 4p orbitals and complete the fourth period.

■ **Figure 7.13** Ten electrons are added across the d block, filling the d sublevel. Notice that chromium and copper have only one electron in the 4s orbital. Such unpredictable exceptions show that the energies of the 4s and 3d sublevels are close.

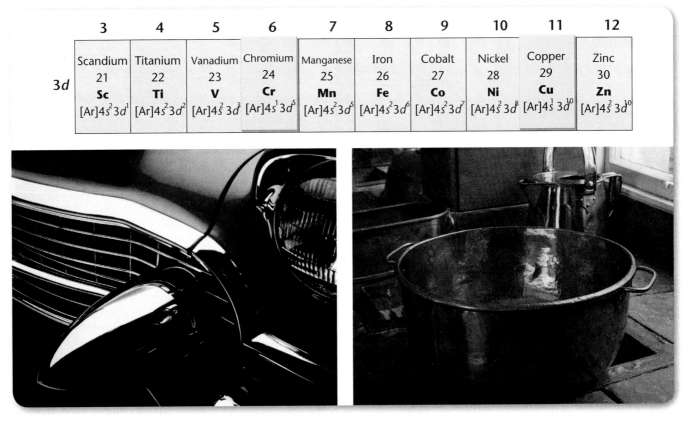

	3	4	5	6	7	8	9	10	11	12
3d	Scandium 21 **Sc** [Ar]4s^{2}3d^1	Titanium 22 **Ti** [Ar]4s^{2}3d^2	Vanadium 23 **V** [Ar]4s^{2}3d^3	Chromium 24 **Cr** [Ar]4s^{1}3d^5	Manganese 25 **Mn** [Ar]4s^{2}3d^5	Iron 26 **Fe** [Ar]4s^{2}3d^6	Cobalt 27 **Co** [Ar]4s^{2}3d^7	Nickel 28 **Ni** [Ar]4s^{2}3d^8	Copper 29 **Cu** [Ar]4s^{1}3d^{10}	Zinc 30 **Zn** [Ar]4s^{2}3d^{10}

Everyday Chemistry

Colors of Gems

Have you ever wondered what produces the gorgeous colors in a stained-glass window or in the rubies, emeralds, or sapphires mounted on a ring? Compounds of transition elements are responsible for creating the entire spectrum of colors.

Transition elements color gems and glass Transition elements have many important uses, but one that is often overlooked is their role in giving colors to gemstones and glass. Although not all compounds of transition elements are colored, most inorganic colored compounds contain a transition element such as chromium, iron, cobalt, copper, manganese, nickel, cadmium, titanium, gold, or vanadium. The color of a compound is determined by three factors:
1. the identity of the metal
2. its oxidation number
3. the negative ion combined with it

Amethyst (purple), citrine (yellow-brown), and rose quartz (pink) are quartz crystals with transition element impurities scattered throughout. Blue sapphires are composed of aluminum oxide (Al_2O_3) with the impurities iron(II) oxide (FeO) and titanium(IV) oxide (TiO_2). If trace amounts of chromium(III) oxide (Cr_2O_3) are present in the Al_2O_3, the resulting gem is a red ruby. A second kind of gemstone is one composed entirely of a colored compound. Most are transition element compounds, such as rose-red rhodochrosite ($MnCO_3$), black-grey hematite (Fe_2O_3), or green malachite ($CuCO_3 \cdot Cu(OH)_2$).

Figure 1 Quartz Crystals

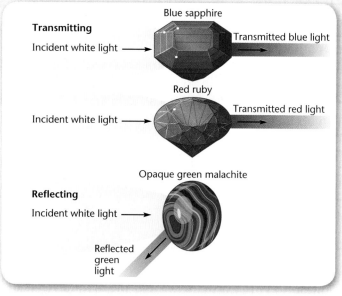

Figure 2 Gems are the color of the light they transmit.

Impurities give gemstones their color Crystals have fascinating properties. The clear, colorless quartz crystal shown in **Figure 1** is pure silicon dioxide (SiO_2). But a crystal that is colorless in its pure form may exist as a variety of colored gemstones when tiny amounts of transition element compounds are present.

How metal ions interact with light to produce color Why does the presence of Cr_2O_3 in Al_2O_3 make a ruby red? The Cr^{3+} ion absorbs yellow-green colors from white light striking the ruby, and the remaining red-blue light is transmitted, resulting in a deep red color as shown in **Figure 2**. This same process occurs in all gems. Trace impurities absorb certain colors of light from white light striking or passing through the stone. The remaining colors of light that are reflected or transmitted produce the color of the gem.

246 Chapter 7 • Everyday Chemistry

Adding transition elements to molten glass for color Glass is colored by often adding transition element compounds to the glass while it is molten. This is true for stained glass, glass used in glass blowing, as shown in **Figure 3,** and glass in the form of ceramic glazes. The table below shows some transition elements and some metals and nonmetals that produce color in glass.

Figure 3 Glass blowing

Elements that Produce Color in Glass			
Element	Oxidation State	Color	Intensity
Transition Metals			
Cobalt	oxidized/reduced	blue	very strong
Iron	oxidized	green	strong
	reduced	blue	medium
Copper	oxidized	blue-green	strong
	reduced	colorless	
Chromium	very oxidized	yellow-green	
	reduced	emerald green	very strong
Manganese	oxidized	purple	very strong
	reduced	colorless	
Uranium	oxidized	yellow, fluorescent	weak
Titanium		yellow-brown	weak
Metals			
Gold		bluish red	strong
Copper		red	strong
Silver		yellow	strong
Nonmetals			
Sulfur	very reduced	reddish amber	weak
Selenium	slightly oxidized	pink	weak

Explore Further

1. **Apply** Explain why iron(III) sulfate is yellow, iron(II) thiocyanate is green, and iron(III) thiocyanate is red.
2. **Acquire Information** Research what impurities give amethyst, rose quartz, and citrine their colors.

Multiple oxidation states Like most metals, the transition elements lose electrons to attain a more stable electron configuration. Most have multiple oxidation numbers because their s and d orbitals are so close in energy that electrons can be lost from both orbitals. For example, cobalt (atomic number 27) forms two fluorides—one with the formula CoF_2 and another with the formula CoF_3. In the first case, cobalt gives up two electrons to fluorine. In the second case, cobalt gives up three electrons.

The rusting of the transition element iron shows how iron can have more than one oxidation number. In the process of rusting, as illustrated in **Figure 7.14,** iron reacts with oxygen in the presence of water to form the compound FeO. In FeO, iron's oxidation number is 2 + because it has lost its two 4s electrons. FeO continues to react with oxygen and water to form the familiar orange-brown compound called rust, Fe_2O_3. In this oxide, iron's oxidation number is 3 + because it has lost two 4s electrons and one 3d electron. Rust (Fe_2O_3) forms when two iron atoms give up a total of six electrons to three oxygen atoms.

■ **Figure 7.14** Once beautiful and sturdy, mighty iron can turn into a mass of rust as a result of the action of air and water.

Infer Which properties of iron make it suitable for building trains? Which properties are not suitable for trains?

Inner transition elements The two rows beneath the main body of the periodic table are the lanthanides (atomic numbers 57 to 70) and the actinides (atomic numbers 89 to 102). These two series are called **inner transition elements** because their last electron occupies a 5f orbital in the seventh period. As with the d-level transition elements, the energies of sublevels in the inner transition elements are so close that electrons can move back and forth between them. This results in variable oxidation numbers, but the most common oxidation number for all of these elements is 3+.

The Sizes of Orbitals

Hydrogen and the group 1 elements each have a single valence electron in an s orbital. Hydrogen's configuration is $1s^1$; the valence electron configuration of lithium is $2s^1$. For sodium, it's $3s^1$; for potassium, $4s^1$. Continuing down the column, rubidium, cesium, and francium have valence configurations of $5s^1$, $6s^1$, and $7s^1$, respectively. How do these s orbitals differ from one another? As you move down the column, the energy of the outermost sublevel increases. The higher the energy, the farther the outermost electrons are from the nucleus. The s orbitals, occupied by the valence electrons of hydrogen and the group 1 elements, are described as spheres around the nucleus. As the valence electron gets farther from the nucleus, the s orbital it occupies gets larger and larger. **Figure 7.15** shows the relative sizes of the 1s, 2s, and 3s orbitals. Notice that in the 3-D cross section of overlapping 1s, 2s, and 3s orbitals, the size of the atom increases with respect to the element's sublevel on the periodic table. Elements within the same group share similar valence structures but do not have the same number of energy levels and do not yield the same amount of energy.

TRY AT HOME 🏠 LAB

See p. 871 for Comparing Orbital Sizes

FOLDABLES
Incorporate information from this section into your Foldable.

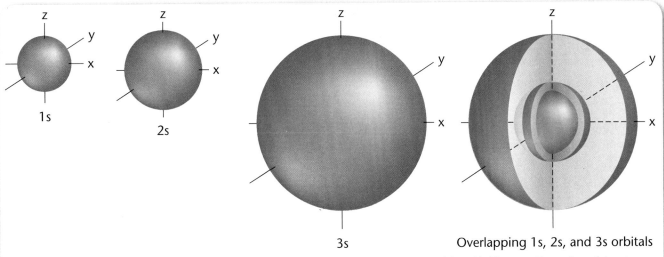

■ **Figure 7.15** As the number of the outermost energy level increases, the size and energy of the orbital increase. The nucleus of the atom is at the intersection of the coordinate axes. The model on the right shows the overlap of the 1s, 2s, and 3s sublevels. This model could represent sodium.

Connect Ideas

An important skill that you learned in this chapter is how to use the periodic table to write electron configurations. It should be clear to you now that the organization of the table arises from the electron configurations of the elements. With this added insight, you are ready to learn in Chapter 8 about trends in properties and patterns of behavior of the elements. Knowing electron configurations and periodic trends will help you organize what may seem to be a vast amount of information.

SUPPLEMENTAL PRACTICE

For more practice with electron configuration, see Supplemental Practice, page 818.

Section 7.2 Assessment

Section Summary

▶ The organization of the periodic table reflects the electron configurations of the elements.

▶ The active metals occupy the s block of the periodic table. Metals, metalloids, and nonmetals fill the p block.

▶ Within a period of the periodic table, the number of valence electrons for main group elements increases from one to eight.

▶ The transition elements, groups 3–12, occupy the d block of the periodic table. These elements can have valence electrons in both s and d sublevels.

▶ The lanthanides and actinides, called the inner transition elements, occupy the f block of the periodic table. Their valence electrons are in s and f sublevels.

6. **MAIN Idea Apply** Use the periodic table to help you write electron configurations for the following atoms. Use the appropriate noble gas abbreviations.

   a) Ca
   b) Mg
   c) Si
   d) Cl
   e) Ne

7. **Identify** the elements that have the following electron configurations:

   a) $1s^2 2s^2$
   b) $1s^2$
   c) $1s^2 2s^2 2p^5$
   d) $1s^2 2s^2 2p^2$
   e) $[Ne]3s^2 3p^4$
   f) $[Ar]4s^1$

8. **Compare** a filled and an unfilled orbital.

9. **Apply Concepts** What is the lowest energy level that can have an s orbital? What block of the periodic table is designated as the s block? Are the elements in this block mostly metals, metalloids, or nonmetals?

10. **Explain** Why do the fourth and fifth periods of the periodic table contain 18 elements rather than eight elements as in the second and third periods?

11. **Write** the electron configuration for gold (Au) using the aufbau diagram on page 242. How many electrons does this atom have? Write its electron configuration based on the noble gas xenon (Xe).

CHAPTER 7 Study Guide

Download quizzes, key terms, and flash cards from glencoe.com.

BIG Idea Valence electrons determine the properties of elements and their positions in the periodic table.

Section 7.1 Present-Day Atomic Theory

MAIN Idea Electrons are organized in energy levels and sublevels.

Vocabulary
- aufbau principle (p. 233)
- electron configuration (p. 240)
- Heisenberg uncertainty principle (p. 238)
- orbital (p. 239)
- sublevel (p. 233)

Key Concepts
- The position of an element in the periodic table reveals the number of valence electrons the element has.
- The outermost valence electrons determine the properties of an element.
- Electrons are found only in levels of fixed energy in an atom.
- Energy levels have sublevels. Each sublevel can hold a specific number of electrons.
- Sublevels can be subdivided into s, p, d, and f orbitals. Sublevels hold 2, 6, 10, and 14 electrons respectively.

Section 7.2 The Periodic Table and Atomic Structure

MAIN Idea A predictable pattern can be used to determine electron arrangement in an atom.

Vocabulary
- inner transition element (p. 248)

Key Concepts
- The organization of the periodic table reflects the electron configurations of the elements.
- The active metals occupy the s block of the periodic table. Metals, metalloids, and nonmetals fill the p block.
- Within a period of the periodic table, the number of valence electrons for main group elements increases from one to eight.
- The transition elements, groups 3–12, occupy the d block of the periodic table. These elements can have valence electrons in both s and d sublevels.
- The lanthanides and actinides, called the inner transition elements, occupy the f block of the periodic table. Their valence electrons are in s and f sublevels.

Chapter 7 Assessment

Understand Concepts

12. Explain what an electron cloud is.
13. Describe the shapes of the s and p orbitals.
14. What is an octet of electrons? What is the significance of an octet?
15. How many electrons fill an orbital?
16. How many *d* orbitals can be in an energy level? What is the maximum number of *d* electrons in an energy level? What is the lowest energy level that can have *d* orbitals?
17. What is the maximum number of electrons in each of the first four energy levels?
18. How many p orbitals can an energy level have? What is the lowest energy level that can have p orbitals?
19. What does each of the following symbols represent: 2s, 4d, 3p, 5f ?
20. Complete **Table 7.4**.

Table 7.4 Number of Electrons Per Sublevel

Sublevel	Maximum Number of Electrons
2s	
3p	
4d	
4f	

21. Why does the first period contain only two elements?
22. Indicate which elements are inner transition elements in **Table 7.5**.

Table 7.5 Inner Transition Elements

	yes	no
V		
Er		
Hg		
Cl		
Po		
Cm		

23. Where is the f region of the periodic table? Name the two series of elements that occupy the f region.
24. What are valence electrons? Where are they located in an atom? Why are they important?
25. Where are the valence electrons represented in this electron configuration of sulfur, $[Ne]3s^23p^4$?

Apply Concepts

Physics Connection

26. What significance did Bohr's model of the atom have in the development of the modern theory of the atom?

Everyday Chemistry

27. Why do the transition metals impart color to gemstones such as emeralds and rubies?

Chemistry and Technology

28. Explain why the development of the scanning probe, the scanning tunneling, and the atomic force microscopes is important to modern chemists.
29. Identify the elements that have the following electron configurations in **Table 7.6**.

Table 7.6 Electron Configurations

Electron Configuration	Elements
$[Kr]5s^24d^3$	
$1s^22s^22p^1$	
$[Xe]6s^2$	
$[Ar]4s^{13}d^{10}$	

30. How is the light given off by fireworks similar to an element's emission spectrum?
31. Sodium and oxygen combine to form sodium oxide, which has the formula Na_2O. Use the periodic table to predict the formulas of the oxides of potassium, rubidium, and cesium. What periodic property of the elements are you using?

Chapter 7 Assessment

Think Critically

Observe and Infer

32. **MiniLab 2** Why are electrons best described as electron clouds?

33. **ChemLab** In an experiment to find out the reaction capacities of magnesium and aluminum, different amounts of the two metals were used, but the amount of hydrochloric acid was kept the same. Would it be possible to obtain correct results in this way? Explain.

Interpret Data

Compare and Contrast

34. Refer to the periodic table and compare the similarities and differences in the electron configurations of the following pairs of elements: F and Cl, O and F, Cl and Ar.

Cumulative Review

35. For elements that form positive ions, how does the charge on the ion relate to the number of valence electrons the element has? *(Chapter 2)*

36. Describe John Dalton's model of the atom, and compare and contrast it with the present-day atomic model. *(Chapter 2)*

37. How many atoms of each element are present in five formula units of calcium permanganate? *(Chapter 5)*

Skill Review

38. **Use a Graph** Use **Figure 7.16** to determine what happens to frequency when energy is doubled.

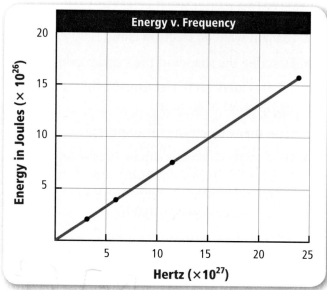

■ **Figure 7.16**

WRITING in Chemistry

39. The word laser is an acronym for light amplification by stimulated emission of radiation. Lasers have a number of applications besides light shows. Research how lasers are produced, what substances are used in lasers, and the ways laser light is used. Write a paper describing your findings.

Problem Solving

40. Use **Figure 7.17** below to list the types of radiation in order of increasing wavelength: microwaves, ultraviolet radiation from the sun, X-rays used by dentists and doctors, the red-light in a calculator display, and gamma rays. Which type of radiation has the highest energy? The lowest energy?

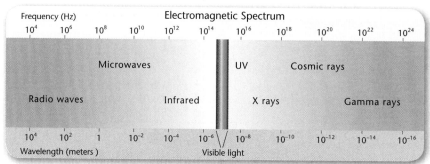

■ **Figure 7.17**

252 Chapter 7 • Completing the Model of the Atom

Cumulative
Standardized Test Practice

1. The present-day model of the atom replaced
   a) Dalton's model.
   b) Thomson's model.
   c) Rutherford's model.
   d) Bohr's model.

2. When an electron falls back to its original energy level, it
   a) emits energy.
   b) absorbs energy.
   c) transfers energy.
   d) stores energy.

Use the periodic table and the table below to answer Questions 3 to 5.

Electron Configurations for Selected Transition Metals			
Element	Symbol	Atomic Number	Electron Configuration
Vanadium	V	23	$[Ar]4s^23d^3$
Yttrium	Y	39	$[Kr]5s^24d^1$
			$[Xe]6s^24f^{14}5d^6$
Scandium	Sc	21	$[Ar]4s^23d^1$
Cadmium	Cd	48	

3. Using noble-gas notation, what is the electron configuration of Cd?
   a) $[Kr]4d^{10}4f^2$
   b) $[Ar]4s^23d^{10}$
   c) $[Kr]5s^24d^{10}$
   d) $[Xe]5s^24d^{10}$

4. Which element is a nonmetal?
   a) tin
   b) tungsten
   c) argon
   d) arsenic

5. What is the element that has the electron configuration $[Xe]6s^24f^{14}5d^6$?
   a) La
   b) Ti
   c) W
   d) Os

6. What is the complete electron configuration of a scandium atom?
   a) $1s^22s^22p^63s^23p^64s^23d^1$
   b) $1s^22s^22p^73s^23p^74s^23d^1$
   c) $1s^22s^22p^53s^23p^54s^23d^1$
   d) $1s^22s^12p^73s^13p^74s^23d^1$

7. Which of the following sublevels has the greatest amount of energy?
   a) s
   b) p
   c) d
   d) f

8. When an iron nail is placed in water, a chemical reaction causes the nail to rust. What is (are) the reactant(s) in this reaction?
   a) iron
   b) water
   c) iron and rust
   d) iron and water

9. Which statement is NOT true about allotropes?
   a) Allotropes contain only one element.
   b) Allotropes have different oxidation numbers.
   c) The properties of allotropes are different.
   d) Allotropes have different molecular structures.

10. Which would be the result if nitrogen gas replaced noble gases inside incandescent light bulbs?
    a) The tungsten filament in the bulb would burn with a dimmer light.
    b) The tungsten filament in the bulb would combust.
    c) Chemical reactions would make the light bulb last longer.
    d) Chemical reactions would reduce the lifespan of the bulb.

NEED EXTRA HELP?										
If You Missed Question...	1	2	3	4	5	6	7	8	9	10
Review Section...	7.1	7.1	7.2	3.2	7.2	7.2	7.1	6.1	5.2	3.2

CHAPTER 8
Periodic Properties of the Elements

BIG Idea Trends among elements in the periodic table include their sizes and their abilities to lose or gain electrons.

8.1 Main Group Elements
MAIN Idea For the main group elements, both metallic character and atomic radius decrease from left to right across a period.

8.2 Transition Elements
MAIN Idea The properties of the transition elements result from electrons filling d orbitals; for the inner transition elements, electrons fill f orbitals.

ChemFacts

- Paints are classified by their color, brightness, and intensity.
- Elements are classified by the numbers of their protons and electrons.
- There are 115 known elements in the universe.
- About 90 percent of the atoms in the universe are hydrogen.

Start-Up Activities

LAUNCH Lab

Periodic Properties

Is melting point a periodic property?

Materials
- notebook paper
- graphing calculator or graph paper
- **Table D.5** on pages 850–852

Procedure
1. Create a data table to record the melting points of the period 4 elements (elements 19–36). Your table should contain the atomic numbers, elements' names, and melting points.
2. Use the periodic table to find the elements' names. Use **Table D.5** on pages 850–852 to find the melting points, in °C, of these elements. Record the melting points in your data table.
3. Using either a graphing calculator or graph paper, graph the melting points. Place the atomic number on the *x*-axis and melting point on the *y*-axis. Be sure to label the axes and to give the graph a title.

Analysis
1. **Describe** the trends you see in the graph that you created.
2. **Infer** Does melting point appear to be a periodic property? Explain your answer.

Inquiry Are boiling point and density periodic properties? Make a prediction. Then, test your prediction.

Visit glencoe.com to:
- study the entire chapter online
- explore **Concepts In Motion**
- take Self-Check Quizzes
- use Personal Tutors
- access Web Links for more information, projects, and activities
- find the Try at Home Lab, The Solubility of Iodine

FOLDABLES Study Organizer

Periodic Trends Make the following Foldable to help you organize your study of trends on the periodic table.

▶ **STEP 1** Layer five sheets of paper about 2 cm apart vertically.

▶ **STEP 2** Fold up the bottom edges of the sheets to form ten layered tabs. Staple along the fold.

▶ **STEP 3** Label the top flap *Periodic Table Groups*. Label the tabs in order with the numbers and names of the main group element families. Label the last tab *Transition elements*.

1. Alkali metals
2. Alkaline earth metals
Group 13
Group 14
Group 15
Group 16
17. Halogens
18. Noble gases
Transition elements

FOLDABLES Use this Foldable with Sections 8.1 and 8.2. As you read these sections, record information about electron configurations, chemical behavior, and the uses of elements.

Chapter 8 • Periodic Properties of the Elements **255**

Section 8.1

Objectives
- **Relate** the position of main group elements on the periodic table to their electron configurations.
- **Predict** the chemical behaviors of the main group elements.
- **Relate** chemical behavior to electron configuration and atomic size.

Review Vocabulary
inner transition element: one of the elements in the actinide or lanthanide series

New Vocabulary
alkali metal
alkaline earth metal
halogen

Main Group Elements

MAIN Idea For the main group elements, both metallic character and atomic radius decrease from left to right across a period.

Real-World Reading Link You have a chemical stockroom in your own home. Under the kitchen sink, you might find dishwasher detergent, steel-wool soap pads, and ammonia window cleaner. In your pantry, you might find vinegar, baking powder, and baking soda. These products contain many simple compounds, such as sodium chloride and sodium hydroxide.

Patterns of Behavior of Main Group Elements

Recall from Chapter 7 that elements in the same group (a vertical column) in the periodic table have the same number of valence electrons, and therefore, have similar properties. But elements in a period (a horizontal row) have properties that differ from one another. This is because the number of valence electrons increases from one to eight as you move from left to right in any row in the periodic table, except the first. As a result, the character of the elements changes. **Figure 8.1** illustrates the main group elements of the periodic table. Each period begins with two or more metallic elements, which are followed by one or two metalloids. The metalloids are followed by nonmetallic elements, and every period ends with a noble gas.

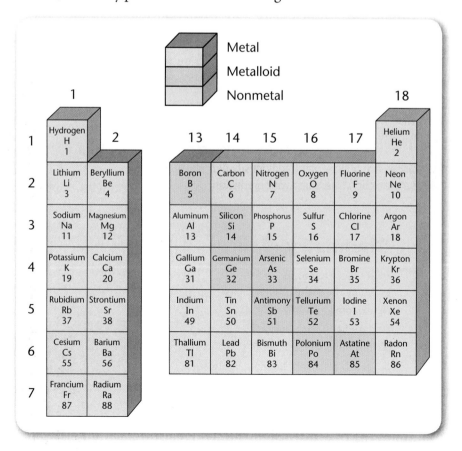

■ **Figure 8.1** The pattern of metal-metalloid-nonmetal - noble gas is typical for the main group elements in each period. Remember that the most active metals, groups 1 and 2, are in the s block of the periodic table. The metalloids, nonmetals, and less active metals fall within the p block of the periodic table.

State the name of the metal that begins period 3. Which nonmetal ends period 4?

256 Chapter 8 • Periodic Properties of the Elements

Patterns in atomic size Recall from Chapter 7 that the size of an atom increases as you move down a column because the electrons are found in energy levels farther away from the nucleus. How does atomic radius change across a period? Consider Period 2. You might expect the size of the second-period atoms to increase across the period from lithium (Li) on the left to fluorine (F) on the right because the number of electrons increases. The opposite is true, however. The lithium atom, with only three electrons, is actually larger than the fluorine atom, which has nine. This trend is the same for all of the periods. **Figure 8.2** and **Figure 8.3** show the periodic trends in atomic size.

Effect of increasing nuclear charge To understand why atomic radii decrease across a period, it helps to recall what determines the radius of an atom. An atomic radius is defined as one-half the distance between the nuclei of two identical atoms that are bonded together. This radius is related to the number of electrons in the valence level and the number of protons in the nucleus of the atom. Compare a lithium atom to a beryllium atom. There are three protons in the nucleus of a lithium atom, so there is an attractive force of 3+ acting on lithium's lone valence electron. In contrast, the nucleus of a beryllium atom has four protons, contributing to an overall attractive force of 4+. Beryllium's two valence electrons are therefore pulled closer to the nucleus. The stronger the attractive force, the greater the pull of electrons toward the nucleus of an atom.

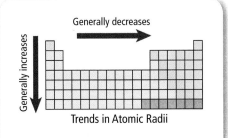

■ **Figure 8.2** Atomic radii generally decrease from left to right in a period and generally increase down a group.

Concepts In Motion

Interactive Figure To see an animation of the trends in atomic radii, visit glencoe.com.

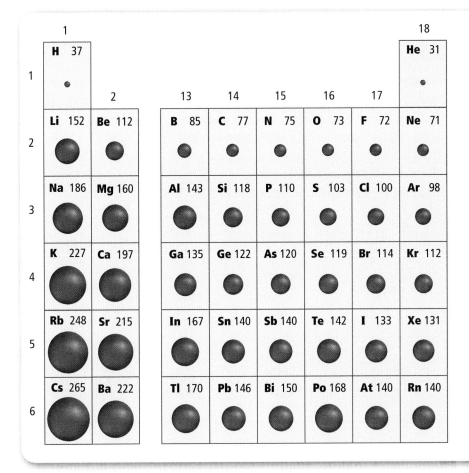

■ **Figure 8.3** The atomic radii of main group elements, given in picometers (10^{-12} m), vary as you move from left to right across a period and down a group.

Apply *Why do atomic radii increase as you move down a group?*

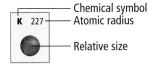

Section 8.1 • Main Group Elements **257**

■ **Figure 8.4** The cesium ion (Cs⁺) is larger than the sodium ion (Na⁺) so it is possible for eight chloride ions (Cl⁻) to fit around a single cesium ion in the cesium chloride crystal lattice. The smaller sodium ion can accommodate only six chloride ions in the sodium chloride structure.

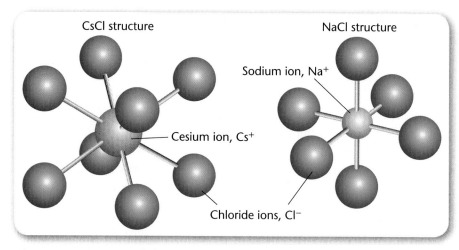

Ionic size Atomic size plays an important role in the chemical reactivity of an element. Ionic size is equally as important in determining the structure of solid ionic compounds and how ions behave in solution. **Figure 8.4** shows how the structure of two ionic compounds differ due to the size of their positive ions. How does the size of an atom change when it becomes an ion?

Positive ions When metallic atoms lose one or more electrons to become positive ions, they acquire the configuration of the noble gas in the preceding period. This means that the outermost electrons of the ion are in a lower energy level than the valence electrons of the neutral atom. The electrons that are not lost experience a greater attraction to the nucleus and pull together in a tighter bundle with a smaller radius. The result is that all positive ions have smaller radii than their corresponding neutral atoms. **Figure 8.5** compares positive ions of lithium and sodium.

Negative ions When a neutral atom gains electrons to become a negative ion, the atom acquires the electron configuration of the noble gas at the end of its period. But the nuclear charge doesn't increase with added electrons. Fluorine has a nuclear charge of 9+, but it must hold ten electrons to become a fluoride ion (F⁻). The result is that all the electrons are held less tightly, and the radius of the ion is larger than the neutral atom, as shown in **Figure 8.5**.

■ **Figure 8.5** Lithium (Li) and sodium (Na) each lose one valence electron. The ions that form are smaller than the neutral atoms because the remaining electrons reside in lower energy levels and are therefore attracted more strongly to the nuclei. Fluorine (F) and chlorine (Cl) become negative ions by adding one valence electron. When an electron is added, the charge on the nucleus is not strong enough to hold the increased number of electrons as closely as it holds the electrons in the neutral atom.

Summarize Will an ion formed by oxygen be larger or smaller than the neutral atom? Why?

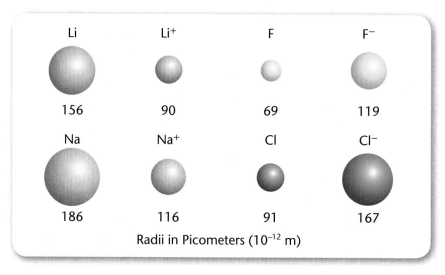

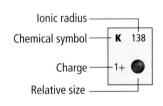

Figure 8.6 Notice that the negative ions in groups 15–17 are all much larger than the positive ions of the same period. Because like charges repel, adding electrons causes negative ions to be larger due to increased repulsion in the electron cloud. Moving from group 15 to group 17, each ion has one less electron and one more positive charge in the nucleus, causing a trend in radius size similar to that of the positive ions.

Explain *why the ionic radii increase for both positive and negative ions as you move down a group.*

Patterns in ionic radii In **Figure 8.5**, you can see that the sodium ion is larger than the lithium ion. This trend in increasing ionic size continues as you move down the periodic table in group 1, as shown in **Figure 8.6**. Within a period, ions of atoms with 1+, 2+, and 3+ charges (groups 1, 2, and 13) decrease in size from left to right. Although the ions have the same electron configuration, nuclear charge increases from left to right, resulting in a stronger attraction of electrons to the nucleus and consequently a smaller size. Notice the same trend for the negative ions in groups 15, 16, and 17. Ionic radii decrease because nuclear charge increases. **Figure 8.6** also shows how the sizes of both positive and negative ions change across a period.

Patterns in chemical reactivity in period 2 You've already noticed that the character of the period 2 elements changes from metal to metalloid to nonmetal to noble gas as you move across the period. How are the electron configurations of these elements related to the tendency of the metals to lose electrons, the nonmetals to share or gain electrons, and the noble gases to be unreactive?

Lithium is the most active second-period metal because it can attain the noble-gas configuration of helium by losing a single electron. If lithium loses one electron from its 2s sublevel, its electron configuration changes from $1s^2 2s^1$ to $1s^2$. The resulting lithium ion has a 1+ charge and the same electron configuration as a helium atom. Although this is not an octet, it is a noble-gas configuration. Elements tend to react in ways that allow them to achieve the stable configuration of the nearest noble gas.

Section 8.1 • Main Group Elements

Beryllium, the next element in the second period, must lose a pair of 2s electrons to acquire the helium configuration. It's harder to lose two electrons than it is to lose one, so beryllium is slightly less reactive than lithium. Nevertheless, beryllium does react by losing both of its 2s electrons and forming a 2+ ion with the helium electron configuration.

If this pattern continued, you might expect boron to lose three electrons to attain the helium configuration. Sometimes boron reacts by losing electrons, but more often it reacts by sharing electrons. Boron is the only metalloid in the period, which means boron sometimes behaves like a metal and loses electrons like its neighboring metals, lithium and beryllium. When it loses electrons, boron achieves the noble-gas configuration of helium. More often, though, boron acts like a nonmetal and shares electrons. Boron is unusual because it has only three electrons to share and cannot acquire an octet of electrons by just sharing. Later, you'll learn more about boron's chemistry.

MiniLab 8.1

Trends in Atomic Radii

What are the periodic trends for the atomic radii of the main group elements? The reactivity of an atom depends on how easily the valence electrons can be removed, and that depends on their distance from the attractive force of the nucleus. In this MiniLab, you will study the periodic trends in the atomic radii of the first 36 main group elements from hydrogen through barium.

Procedure

1. Read and complete the lab safety form.
2. Obtain a **96-well microplate, straws** to fit the wells in the plate, **scissors,** and a **ruler.** The well plate should be oriented to correlate with the periodic table of the elements like so: Row 1 of the plate represents the first period, H1 is hydrogen, A1 is helium; Row 2 of the plate represents the second period, from H2 (lithium) to A2 (neon). Rows 3–7 correlate with periods 3–7; however, only the main group elements will be represented. Label the well plate *Atomic radius in pm* (picometers).
3. Use **Figures 8.1** and **8.3** to help you make your model. Look up the atomic radius of each of the elements in **Table D.5** on page 850.
4. Convert atomic radius in picometers to an enlarged scale in centimeters by multiplying the atomic radius in picometers by the conversion factor, 1 cm/40 pm. For example, the atomic radius of hydrogen in centimeters is calculated thus: 78 pm × 1 cm/40 pm = 1.95 cm, or 2.0 cm. To represent the atomic radius of hydrogen, cut a piece of straw 2.0 cm long. Cut a piece of straw to scale for each element, and insert each piece into the appropriate well of the plate.

Analysis

1. **Interpret** How do the atomic radii change as you go from left to right across a period? Explain your observation on the basis of the electron configurations of the elements.
2. **Interpret** How do atomic radii change as you move from top to bottom within a group? Explain your observation on the basis of the electron configurations of the elements.
3. **Explain** Why are the atomic radii of the elements described as a periodic property?

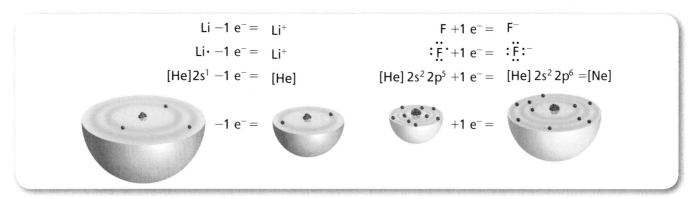

Carbon, nitrogen, oxygen, and fluorine are nonmetals. Carbon, with the configuration [He]$2s^2 2p^2$, and nitrogen, with the configuration [He]$2s^2 2p^3$, share electrons to attain the noble-gas configuration of neon, [He]$2s^2 2p^6$. Oxygen, with the configuration [He]$2s^2 2p^4$, gains two electrons to form the oxide ion, O^{2-}. Fluorine, with the configuration [He]$2s^2 2p^5$, gains one electron to become the fluoride ion, F^-. **Figure 8.7** illustrates an example of the loss and gain of electrons by two elements in period 2. The ionic compound formed by these elements results when each atom attains a noble-gas configuration.

The Main Group Metals and Nonmetals

The main group elements are those in groups 1, 2, and 13–18. These elements represent almost the complete spectrum of chemical and physical properties—elements that are highly reactive and elements that do not normally react at all; elements that are solids, liquids, and gases at room temperature; and elements that are metals, nonmetals, and metalloids. Because they represent such a breadth of properties, they are also sometimes called the representative elements. As you study the main group elements, remember that their valence electrons are arranged in s and p orbitals in the same way for each member of the group. As a result, group members share similar chemical properties.

The alkali metals The group 1 elements—lithium (Li), sodium (Na), potassium (K), rubidium (Rb), cesium (Cs), and francium (Fr)—are called the **alkali metals**. The alkali elements are soft, silvery-white metals, and good conductors of heat and electricity. Their chemistry is relatively uncomplicated; they lose their s valence electron and form a 1+ ion with the stable electron configuration of the noble gas in the preceding period.

Because all of the alkali metals react by losing their single s valence electrons, the most reactive alkali metal is the one that has the least attraction for this electron. Remember that the bigger the atom, the farther away the valence electron is from the nucleus and the less tightly it is held. In the alkali metal family, francium is the largest atom and probably the most reactive, but francium has not been widely investigated because it is scarce and radioactive. Cesium (Cs) is usually considered the most active alkali metal—in fact, the most active of all the metals. Lithium (Li), the smallest of the alkali metals, is the least reactive element in group 1. **Figure 8.8** illustrates some reactions and uses of the alkali metals.

■ **Figure 8.7** When an alkali metal such as lithium (Li) loses an electron, it adopts the electron configuration of the noble gas in the preceding period. When a nonmetal such as fluorine (F) gains an electron, it acquires the configuration of the noble gas at the end of its period. Ionic compounds such as LiF are combinations of a positive metal ion and a negative nonmetal ion, each having a noble-gas configuration.

FOLDABLES
Incorporate information from this section into your Foldable.

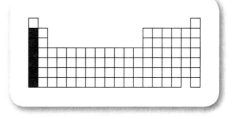

FIGURE 8.8

Visualizing Alkali Metals

Metal Compounds
Because of their chemical reactivities, the alkali metals do not exist as free elements in nature. Sodium, for example, is found mostly combined with chlorine in sodium chloride in salt. Metallic sodium is obtained from sodium chloride through a process called electrolysis, in which an electric current is passed through the molten salt.

Alkali Metals and Water
Oil protects pure or elemental sodium from spontaneous reaction with oxygen or moisture in the air (top). Sodium metal is soft enough to be cut with a knife, and inside you can see the shiny metallic surface (left). Sodium and the other group 1 elements are among the most reactive of all the metals. All group 1 metals react vigorously with water. When they do, they replace hydrogen and form a hydroxide, as shown in the following equation.

$$2Na^+ + 2H_2O \rightarrow H_2 + 2NaOH$$
Sodium   water   hydrogen   sodium hydroxide

Spontaneous Reactivity
So much heat is generated in the rapid reaction of potassium and water that the hydrogen gas produced in the reaction bursts into flames. The pink hue of the water is due to the presence of the indicator phenolphthalein, which turns pink when the solution is alkaline. The pink color of the flame is characteristic of potassium. Potassium hydroxide (KOH) formed in the reaction makes the solution alkaline. Hydroxides are common ingredients in household and industrial chemicals.

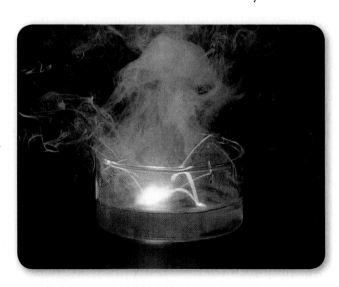

Industrial Uses
Among other industrial uses, sodium hydroxide (NaOH) is used in the digestion of pulp in the paper-making process (left). Other stages of the paper-making process require an alkaline environment. For these processes, sodium hydroxide is used to maintain the appropriately alkaline conditions.

Sodium hydroxide is also used to make soap, in petroleum refining, in the reclaiming of rubber, and in the manufacture of rayon (right). The rayon threads at right are being spun from a thick solution made in part by soaking cellulose in sodium hydroxide.

Biological and Household Uses
In your household chemical storehouse, you'll find sodium hydroxide (lye) in oven cleaners and in the granular material you use to unclog drains. It is sodium hydroxide's ability to convert fats to soap that makes it effective as a kitchen drain cleaner. Among the biologically important functions of the alkali metals, the ions of sodium (Na^+) and potassium (K^+) play key roles in transmitting nerve impulses. Potassium is also an essential nutrient for plants.

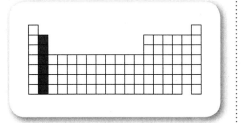

The alkaline earth metals The group 2 elements—beryllium (Be), magnesium (Mg), calcium (Ca), strontium (Sr), barium (Ba), and radium (Ra)—are called the **alkaline earth metals**. Their properties are similar to those of the group 1 elements. Like the alkali metals, they are too reactive to be found as free elements in nature. They lose both of their s valence electrons and form 2+ ions with the stable electron configuration of the noble gas in the preceding period. Because the group 2 elements must lose two electrons rather than one, these metals are less reactive than the group 1 elements. Each alkaline earth metal is denser and harder and has a higher melting point than the alkali metals of group 1.

The most reactive element in the alkaline earth group is the one with the largest atomic radius and the least amount of attraction to its two valence electrons. With this in mind, you can predict that radium, the largest atom in the group, is the most reactive.

The relationship between increasing reactivity and increasing size of atom for the alkaline earth metals is illustrated by the reaction of the elements with water, shown in **Figure 8.9.** Beryllium does not react with water. Magnesium reacts with hot water. But calcium reacts with water to form calcium hydroxide ($Ca(OH)_2$), as shown by this equation.

$$Ca^{2+} + 2H_2O \rightarrow H_2 + Ca(OH)_2$$

The alkaline earth metals play important roles in many natural and human-made systems. For example, magnesium is present at the center of chlorophyll molecules. Chlorophyll is the pigment in plants responsible for photosynthesis, a process on which much of life ultimately depends. **Figure 8.10** illustrates several additional applications of the alkaline earth metals.

■ **Figure 8.9** No reaction is visible when beryllium is placed in water, but the reaction of calcium with water of hydrogen produces bubbles. Strontium, barium, and radium react with water with increasing strength.

FIGURE 8.10

Visualizing Alkaline Earth Metals

Uses of the Alkaline Earth Metals
Beryllium has become a strategically important metal in the nuclear and weapons industries. Magnesium and beryllium are valued for their individual properties, but they are especially useful when alloyed with other metals. Alloys of magnesium are used when light weight and strength are essential, as in this jet engine. Magnesium resists corrosion because it reacts with oxygen in the air to form a coating of magnesium oxide (MgO). The coating of magnesium oxide protects the metal underneath from further reaction with oxygen.

Reactions of Magnesium and Calcium
Magnesium oxide is also formed when magnesium is heated in air. It burns vigorously, producing a brilliant white light and magnesium oxide. In the process, magnesium loses two electrons to form the Mg^{2+} ion, and oxygen gains two electrons to form the O^{2-} ion. Together, they form the ionic compound MgO. The following equation shows what happens.

$$2Mg^{2+} + O_2 \rightarrow 2MgO$$

Magnesium and calcium are essential elements for humans and plants. Plants need magnesium for photosynthesis because a magnesium atom is located at the center of every chlorophyll molecule.

Calcium ions are essential in your diet. They maintain heartbeat rate and help blood to clot. But the largest amount of dietary calcium ions is used to form and maintain bones and teeth. Bone is composed of protein fibers, water, and minerals, the most important of which is hydroxyapatite ($Ca_5(PO_4)_3OH$), a compound of calcium, phosphorus, oxygen, and hydrogen—all main group elements.

Strontium's Red Glare
Strontium is a less well-known element of group 2, but it is important nevertheless. Because of its chemical similarity to calcium, strontium can replace calcium in the hydroxyapatite of bones and form $Sr_5(PO_4)_3OH$. This could be a problem only if the strontium atoms are the radioactive isotope strontium-90, which is hazardous if it is incorporated into a person's bones. Strontium can be identified by a brilliant red color in a fireworks display. Strontium also produces a red color in laboratory flame tests.

Section 8.1 • Main Group Elements

CHEMLAB

SMALL SCALE

REACTIONS AND ION CHARGES OF THE ALKALINE EARTH ELEMENTS

Background

The alkaline earth elements, group 2 on the periodic table, include beryllium (Be), magnesium (Mg), calcium (Ca), strontium (Sr), barium (Ba), and radium (Ra). The positive ions of most of these elements react with negative oxalate ions ($C_2O_4^{2-}$) to form insoluble compounds. In this ChemLab, you will study the reactions of alkaline earth metals with oxalate ions and determine the formulas for the insoluble products.

Question

In what ratios do the positive ions of calcium, strontium, and barium react with negative oxalate ions to produce compounds?

Objectives

- **Observe** the reactions of the calcium, strontium, and barium ions with oxalate ions.
- **Determine** the formulas for the insoluble products and the charges on the ions of the alkaline earth elements.
- **Relate** the ion charges of the alkaline earth elements to their electron configurations.

Preparation

Materials

96-well microplates (3)
microtip pipettes (4)
black paper
toothpicks (3)
marking pen
0.1M calcium nitrate solution
0.1M strontium nitrate solution
0.1M barium nitrate solution
0.1M sodium oxalate solution

Safety Precautions

WARNING: *The solutions are toxic if ingested and may irritate the skin. Do not touch or ingest the solutions. Dispose of the products as instructed by your teacher.*

Procedure

1. Read and complete the lab safety form.
2. Obtain three 96-well microplates and four labeled micropipettes containing the four solutions: calcium nitrate solution, strontium nitrate solution, barium nitrate solution, and sodium oxalate solution. Label the microplates *calcium, strontium,* and *barium.*
3. Lay the calcium microplate near the edge of the lab bench with row H aligned with the edge of the bench. You will use only wells H1 through H9. You will see the product best if the microplate rests on a piece of black paper.
4. Add one drop of the calcium solution to well H1, two to well H2, three to well H3 and so on to well H9, which receives nine drops.

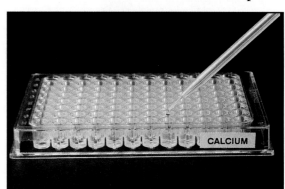

5. Add one drop of the sodium oxalate solution to well H9, two drops to well H8, and three drops to well H7. Continue in this way until you get to H1, which receives nine drops.
6. Use a toothpick to stir the mixtures.
7. It may take several minutes for the reactions to be complete. When the insoluble product has settled in the bottoms of the wells, stoop so that your eyes are level with the microplate.
8. Determine and record the identity of the well (or wells) that contains the greatest depth, and therefore the greatest amount, of the insoluble product. Record your observations in a data table like the one shown in Data and Observations.

9. Repeat procedures 3–8 with the strontium and oxalate solutions.
10. Repeat procedures 3–8 with the barium and oxalate solutions.
11. Dispose of the substances in all three well plates as directed by your teacher. Rinse the well plates with tap water and then with distilled water.

Analyze and Conclude

1. **Interpret** What ratio of drops of the reactant solutions produced the maximum amount of insoluble product for each of the three reactions you performed?
2. **Interpret** All the solutions you used were 0.1 M, which means that they contained equal numbers of ions per drop. For example, if the maximum amount of product was produced from two drops of calcium solution and eight drops of oxalate solution, the insoluble product contained one calcium ion for each four oxalate ions. The formula for such a compound would be $Ca(C_2O_4)_4$. Interpret your results to determine the formulas for calcium oxalate, strontium oxalate, and barium oxalate.
3. **Conclude** The oxalate ion has a charge of 2−, and it combines with the positive alkaline earth ions in such a way that neutral compounds result. What do you think are the ion charges of calcium, strontium, and barium ions? Explain how your experimental results support your conclusion.

Apply and Assess

1. **Apply** Write the electron configurations of calcium, strontium, and barium atoms. Relate the charges of calcium, strontium, and barium ions to the electron configurations of their respective atoms.
2. **Predict** Use the same reasoning that you used to answer question 1 to predict the ion charges of potassium in group 1 and gallium in group 13. Explain your reasoning.

INQUIRY EXTENSION
Formulate When lead nitrate and potassium iodide mix, a yellow precipitate of lead iodide forms. Develop a plan similar to the procedure in this ChemLab to assess lead's charge in this precipitate.

Data and Observations

Data table			
Sodium oxalate combined with	Number of well with maximum precipitate	Drops of oxalate solution	Drops of group 2 ion solution
Calcium nitrate			
Strontium nitrate			
Barium nitrate			

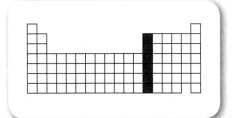

Group 13 The first element in group 13, boron (B), is the only metalloid in the group. The other group 13 elements—aluminum (Al), gallium (Ga), indium (In), and thallium (Tl)—are metals. None of the metals are as active as the metals in groups 1 and 2, but they are good conductors of heat and electricity. They are silver in appearance and fairly soft. Group 13 metals tend to share electrons in covalent compounds rather than form ionic compounds; in this respect, they resemble boron. Their valence electron configuration is s^2p^1, and they exhibit the 3+ oxidation number in most of their compounds.

Aluminum is the most abundant metallic element in Earth's crust and is becoming one of the world's most widely used metals. Aluminum's many uses result from its properties—low density, good electrical and thermal conductivity, malleability, ductility, and resistance to corrosion. Aluminum, like magnesium, is a self-protecting metal. With exposure to oxygen in the air, a protective layer of aluminum oxide (Al_2O_3) forms over the surface of the aluminum, preventing further reactions.

$$4Al + 3O_2 \rightarrow 2Al_2O_3$$

Aluminum is obtained from its ore through a process that consumes 4.5% of the electricity produced in the United States. Recycling aluminum reduces costs by lowering the need for power.

Figure 8.11 illustrates applications of the group 13 elements.

FIGURE 8.11

Visualizing Group 13

Uses of Boron
Boron is a metalloid found in boric acid (H_3BO_3) and borax ($Na_2B_4O_7 \cdot 10H_2O$). Boric acid is one of the active ingredients in eyewash or contact lens cleaning solution. Borax is the abrasive used in some tough cleansing powders.

The Importance of Aluminum
Think about how important aluminum is in your life. Aluminum foil and aluminum soft-drink cans are everywhere. The antacid in your medicine cabinet may contain aluminum hydroxide, $Al(OH)_3$. Your antiperspirant or deodorant might contain an aluminum zirconium hydroxide or aluminum chlorohydrate. Because aluminum is neither as hard nor as strong as steel, it is often alloyed with other metals to make structural materials. Aluminum alloys are used in automobile engines, airplanes, and truck bodies where high strength and light weight are important. At home, you may find bicycles, outdoor furniture, ladders, and pots and pans made from aluminum.

Group 14 The group 14 elements—carbon (C), silicon (Si), germanium (Ge), tin (Sn), and lead (Pb)—exhibit a variety of properties. Carbon is a nonmetal, silicon and germanium are metalloids, and tin and lead are metals. Because the valence electron configuration for these elements is s^2p^2, a gain or loss of four electrons results in a noble-gas configuration. However, it's unusual for any element to gain or lose four electrons. Instead of gaining electrons to attain a noble-gas configuration, carbon, silicon, and germanium react by sharing electrons. But tin and lead, like the metals in the preceding groups, react by losing electrons. In fact, larger elements at the bottom of the group lose electrons more easily than the smaller nonmetals at the top of the group. This is a result of their atomic radii and the reduced attraction the nucleus has for the outermost electrons. As a group, the most common oxidation number is 4+. **Figure 8.12** illustrates applications of the group 14 elements. The History Connection on page 270 discusses a significant historical connection to lead, the heaviest stable group 14 element.

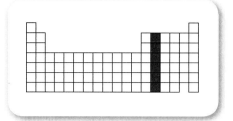

FIGURE 8.12

Visualizing Group 14

You'll find group 14 elements in many of the products in your household chemical stockroom—carbon in charcoal briquettes, lead pencils, diamond jewelry, and almost every food item in your home. You'll learn more about carbon and carbon compounds in Chapters 9 and 18.

From Sand to Many Uses
Silicon, like boron, is a metalloid. It is present in sand as silicon dioxide (SiO_2)—sometimes called silica. About 59 percent of Earth's crust is made up of silica. In its elemental form, silicon is a hard, gray solid with a relatively high melting point, 1410°C. Silicon is in window glass and in the chips that run computers. Compounds of silicon are found in lubricants, caulking, and sealants.

Tin Cans and Alloys
Tin (Sn) is best known for its use as a protective coating for steel cans used for food storage. The coating protects the steel from corrosion. Tin is also a principal component in the alloys bronze, solder, and pewter. Tin is a soft metal that can be rolled into thin sheets of foil.

The Lead-Acid Storage Battery
Lead (Pb) has been known and used since ancient times. It's obtained from the ore galena (PbS). The most important use of lead is in the lead-acid storage batteries used in automobiles.

History Connection

Lead Poisoning in Rome

Could lead poisoning be partially responsible for the fall of the Roman Empire? Some researchers think so; they found that lead poisoning occurred in early history.

How lead poisoning happens Lead gets into the body when materials containing lead compounds are ingested with food and drink, or when lead-containing dust in the air is inhaled or absorbed through the skin. As lead intake increases, the body's ability to get rid of it decreases. Over a period of time, the accumulation of lead in the liver, kidneys, bones, and other body tissues becomes critical. Symptoms of abdominal pains, anemia, lethargy, and nerve paralysis of hands and feet develop.

An old and versatile metal Lead was highly valued in ancient times. The Egyptians refined and used lead as early as 3000 B.C. Deposits of galena (PbS) located near Athens, Greece, were processed and used by the Greeks in the sixth century B.C. But it was the Romans in the first century B.C. who realized the full potential of lead. They processed it for a wide variety of purposes. Rome's famous system for supplying water to the populace, the aqueduct, was built with lead pipes. Beer and wine were stored in lead-glazed pottery, shown in **Figure 1,** and served in lead goblets. Cooking pots were made of lead.

Lead was everywhere The lead water pipes of the Roman plumbing system allowed lead to dissolve in the drinking water. Many types of food and drink were sweetened by a thick, syrup called sapa. Sapa was made by boiling wine in a lead pot until much of the water and alcohol had evaporated. What remained was a tasty but poisonous confection. A small portion of sapa was lead(II) acetate, also known as sugar of lead. It was impossible for anyone living in ancient Rome to avoid ingesting lead.

Some researchers think the Roman ruling class was slowly being poisoned to death by the lead used in almost every aspect of their lives.

Figure 1 Lead-glazed pottery that stored food and drink

Besides lead being used in pottery and water pipes, it was also used in paint pigments, as a food seasoning, in face powder and mascara, and as one of many ingredients in coins. Long-term lead poisoning can cause many health problems, including sterility, kidney damage, aggressive behavior, reduced IQ, irritability, and behavior or attention problems.

Lead emissions When Rome was at its peak in lead production, it produced 80,000 metric tons every year. In studies of changes in atmospheric composition throughout history, researchers measured lead residue from ancient Rome and Greece found in British peat bogs and in Swedish lake sediments. Lead emissions from Roman smelters at the height of Roman power from 500 B.C. to 300 A.D. were nearly as great as they were during the years of the Industrial Revolution in England from 1760 to 1840.

Connection to Chemistry

1. **Acquire Information** Find out how lead can get into water supplies today.
2. **Compare and Contrast** Use the library to find out about the problems the United States has with lead pollution and lead poisoning.
3. **Interpret** The formula for the acetate ion is $C_2H_3O_2^-$. What is the formula for lead(II) acetate?

Group 15 The trend in metallic properties is obvious as you move from the top of group 15 to the bottom—from nitrogen (N) to phosphorus (P) to arsenic (As) to antimony (Sb) and bismuth (Bi). Nitrogen and phosphorus are nonmetals. They form covalent bonds to complete a noble-gas configuration. Arsenic and antimony are metalloids and either gain or share electrons to complete their octets. Bismuth is more metallic and often loses electrons.

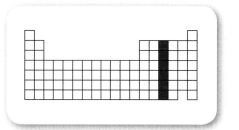

Group 15 elements have five valence electrons. Their valence-electron configuration is s^2p^3. They need only three electrons to attain the configuration of the noble gas at the end of their period. Nitrogen, phosphorus, and arsenic have an oxidation number of 3− in some of their compounds, but they can also have oxidation numbers of 3+ and 5+. Nitrogen is a component of the proteins deoxyribonucleic acid (DNA), shown in **Figure 8.13,** and ribonucleic acid (RNA). Phosphorus is equally important because the phosphate group (PO_4^{3-}) is a repeating link in the DNA chain. The DNA molecule carries the genetic code that controls the activities of cells for many living organisms. Another important biological molecule, adenosine triphosphate (ATP) also contains phosphate groups that store and release energy in living organisms.

Nitrogen, as the chemically unreactive molecule N_2, makes up 78 percent by volume of Earth's atmosphere. Plants and animals need nitrogen to survive, but they cannot use it in the molecular form found in the atmosphere. Lichens, bacteria in soil, and bacteria in the root nodules of beans, clover, and other similar plants convert nitrogen to ammonia and nitrate compounds. Lightning also converts atmospheric nitrogen to nitrogen monoxide (NO). Plants use these simple nitrogen compounds to make proteins. Complex nitrogen compounds eventually become part of the food chain. Natural sources of nitrogen in soil, plants, and the products of the decomposition of organic matter and waste, do not always supply enough nitrogen for optimum crop yields. Farmers apply fertilizers rich in nitrogen and phosphorous, group 15 elements, to produce a crop with a higher yield. **Figure 8.14** and the Everyday Chemistry feature on page 273 illustrate additional applications of the group 15 elements.

■ **Figure 8.13** The group 15 elements nitrogen and phosphorous are vital components of DNA, the molecule that contains the genetic code for every living organism. Phosphorous is the central atom in the phosphate groups that are part of the molecule's structure. Nitrogen is an important component of the bases that branch off of the structure.

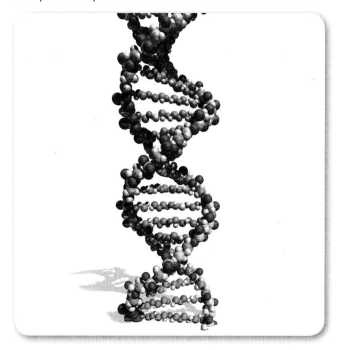

Section 8.1 • Main Group Elements **271**

FIGURE 8.14

Visualizing Group 15

A Common Element: Nitrogen
Commercially, elemental nitrogen (N_2) is obtained from liquid air by fractional distillation. Much of it is converted to ammonia (NH_3), the familiar ingredient in household cleaners.

Ammonia, the Essential Fertilizer
Ammonia is used as a liquid fertilizer applied directly to soil(top), or it can be converted to solid fertilizers such as ammonium nitrate, NH_4NO_3; ammonium sulfate, $(NH_4)_2SO_4$; or ammonium hydrogen phosphate, $(NH_4)_2HPO_4$. The bag of fertilizer shows the percentages of the essential main group elements—nitrogen, phosphorus, and potassium—that this fertilizer contains (left). Fertilizers are formulated in a variety of ways to provide the proper nutrients for different plant growth needs.

Two Allotropes of Phosphorus
White and red phosphorus are two common allotropes of phosphorus. Notice that the white phosphorus is photographed under a liquid because this form of phosphorus, which has the formula P_4, reacts spontaneously with oxygen in the air. Red phosphorus is used to make matches. You can read about it in the Everyday Chemistry feature in this chapter.

Gallium Arsenide Semiconductors
Arsenic is a metalloid found widely distributed in Earth's crust. An increasingly important use of the element is in the form of the binary compound gallium arsenide, GaAs. Because of better conductivity and performance, gallium arsenide is now replacing silicon in some of its semiconductor applications in electronic circuitry.

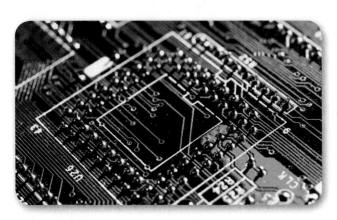

Everyday Chemistry

The Chemistry of Matches

The making and lighting of friction matches and safety matches involves a lot of chemistry.

Making friction matches Pinewood matchsticks, shown in **Figure 1**, are cut and dipped into a solution of borax ($Na_2B_4O_7 \cdot 10H_2O$, sodium tetraborate) or ammonium phosphate (($NH_4)_2PO_4$) to make the matches safer. Next, the match head end is dipped into paraffin and then into a mixture of glue, coloring, a combustible material, and an oxidizing agent. Sulfur or diantimony trisulfide (Sb_2S_3) is used as the combustible material. Potassium chlorate ($KClO_3$) and manganese dioxide (MnO_2) are common oxidizing agents. Adding the tip—a mixture of tetraphosphorus trisulfide (P_4S_3), powdered glass, and a binder—is the final step.

Figure 1 Friction matches light when struck on any surface.

Chemistry of friction matches The striking surface shown is powdered glass and glue. The P_4S_3 tip has a low kindling temperature. When the tip is rubbed on the striking surface, the heat from the friction causes it to ignite.

$P_4S_3(s) + 6O_2(g) \rightarrow P_4O_5(g) + 3SO_2(g) + heat$

The heat produced decomposes the $KClO_3$.

$2KClO_3(s) \rightarrow 2KCl(s) + 3O_2(g)$

The oxygen given off, combined with the heat from the first reaction, causes the sulfur to catch fire, which ignites the paraffin. The burning paraffin carries the flame from the head to the wooden stem.

Figure 2 Safety matches light when struck on packet.

How safety matches work The paperboard sticks of safety matches, shown in **Figure 2**, are treated similarly. Their heads contain diantimony trisulfide or sulfur, potassium chlorate or some other oxidizing agent, ground glass, and glue with paraffin underneath. These matches are called safety matches because they generally ignite only when they are rubbed across the striking surface on their box or packet. The striking surface serves the same function as the tip on the friction match; it ignites the head. The striking surface is a layer of red phosphorus, powdered glass, and glue. The friction of the match on the striking surface changes red phosphorus to white phosphorus.

$P(red) + heat \rightarrow P(white)$

White phosphorus forms and ignites spontaneously in air, giving off enough heat to ignite the match head.

$4P(white)(s) + 5O_2(g) \rightarrow P_4O_{10}(s) + heat$

Explore Further

1. **Compare and Contrast** The first step in lighting a safety match is the conversion of red phosphorus to white phosphorus. Compare the chemical reactivities of these allotropes.
2. **Apply** Devise a method for making a match that would produce a colored flame.

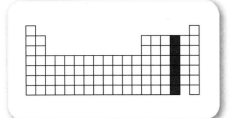

Table 8.1	Oxygen + Metals and Nonmetals
O_2 + Metal	O_2 + Nonmetal
$O_2 + 4Na \rightarrow 2Na_2O$	$O_2 + C \rightarrow CO_2$
$O_2 + 2Ca \rightarrow 2CaO$	$O_2 + S \rightarrow SO_2$

Group 16 Within group 16, oxygen (O), sulfur (S), selenium (Se), and tellurium (Te) are nonmetals, and polonium (Po) is a metalloid. Their valence electron configuration is s^2p^4. With rare exceptions, oxygen gains two electrons and forms an oxide ion (O^{2-}) with a neon configuration. Oxygen reacts with both metals and nonmetals and, among the nonmetals, is second only to fluorine in chemical reactivity.

Oxygen is the most abundant element on Earth. It makes up 21 percent by volume of Earth's atmosphere and nearly 50 percent by mass of Earth's crust. Oxygen is present in the compound water as well as oxides of other elements. Both metals and nonmetals react directly with molecular oxygen to form oxides, as shown in the equations in **Table 8.1**. Like nitrogen, oxygen gas (O_2) is obtained from fractional distillation of liquefied air.

There are two allotropes of oxygen—O_2, the most common, and O_3, called ozone. Ozone is a highly unstable and reactive gas that is considered a pollutant in the lower atmosphere. However, in the upper atmosphere, ozone protects Earth by absorbing harmful ultraviolet radiation from the Sun. Ozone is responsible for the acrid odor you may notice during thunder and lightning storms or while operating your computer and other electronic equipment.

Like oxygen, sulfur gains two electrons and forms the sulfide ion (S^{2-}) when it reacts with metals or with hydrogen. But in its reactions with nonmetals, sulfur can have other oxidation numbers. Much of the sulfur produced in the United States is taken from deposits of elemental sulfur by the Frasch process, shown in **Figure 8.15**. Several other important uses of the group 16 elements are illustrated in **Figure 8.16**.

■ **Figure 8.15** In the 1890s, the Frasch process (below) was invented by Herman Frasch. Hot water is pumped into an underground sulfur deposit where it melts the sulfur. The liquid sulfur is then brought to the surface by forcing air into the deposit. Liquid sulfur is shown solidifying after removal from a deposit (right).

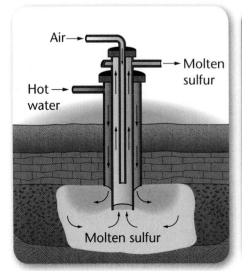

FIGURE 8.16

Visualizing Group 16

The Importance of Oxygen
The largest industrial use of oxygen is in the production of steel, which is described in Section 8.2. It's also used in the treatment of wastewater, as a part rocket fuel, and in medicine to assist in respiration.

Unstable Hydrogen Peroxide
In your household chemical stockroom, you'll find oxygen in solutions of hydrogen peroxide, shown in a brown bottle. Peroxides are unstable compounds that decompose to produce water and molecular oxygen. The brown container helps to slow the decomposition of hydrogen peroxide by excluding light. The reaction is described by this equation.

$$2H_2O_2 \rightarrow 2H_2O + O_2$$

Oxygen as an Antiseptic and Bleach
It's oxygen gas that produces the foam when you use hydrogen peroxide as an antiseptic to clean a cut or scrape. It's also oxygen that bleaches hair when a peroxide bleach is used. Some household cleansers use oxygen bleach rather than chlorine bleach.

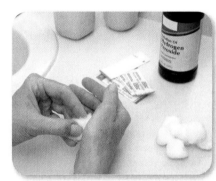

A Variety of Uses for Sulfuric Acid
Most elemental sulfur is converted to sulfuric acid (H_2SO_4), a key chemical in the production of a variety of products, such as fertilizers, automobile batteries, detergents, pigments, fibers, and synthetic rubber, as shown.

An Application of Selenium's Photosensitivity
The chemistry of selenium and tellurium is similar to that of sulfur. Selenium has the property of increased electrical conductivity when exposed to light. This property has applications in security devices and mechanical opening and closing devices, where the interruption of a beam of light triggers an electrical response. But the most important application of selenium is in xerography, a process employed in modern photocopiers, as shown here.

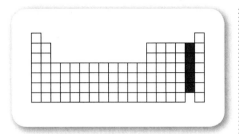

TRY AT HOME 🏠 LAB

See page 871 for **The Solubility of Iodine**

■ **Figure 8.17** Iodine is an essential dietary component. Most of the body's iodine concentrates in the thyroid, where it is used to make certain hormones. This map illustrates risk levels for iodine deficiency around the world.

The halogens The **halogens**—fluorine (F), chlorine (Cl), bromine (Br), iodine (I), and astatine (At)—are active nonmetals. Because of their chemical reactivity, they don't exist as free elements in nature. Their chemical behavior is characterized by a tendency to gain one electron to complete their s^2p^5 valence electron configuration and form a 1− ion with a noble-gas configuration. Chlorine, for example, has the configuration [Ne]$3s^23p^5$. When it gains an electron, the chloride ion Cl⁻ is formed with the argon configuration [Ne]$3s^23p^6$. The halogens can also achieve a noble-gas configuration by sharing electrons.

Because the halogens react by gaining an electron, the most reactive element in the group is the one with the strongest attraction for an electron. Fluorine is the smallest of the halogens so it has the greatest ability to attract and hold an electron. As the size of the atoms increases down the group, the ability of the nucleus to attract and hold outer-level electrons decreases. Iodine is the least active halogen because of its large atomic radius. Astatine, the largest of the halogens, is probably even less active than iodine, but it is scarce and radioactive.

The elemental halogens exist as diatomic molecules that are both highly reactive and toxic. Many household products contain chlorine compounds that can generate chlorine gas (Cl_2), if not handled properly. However, people all over the world safely consume many of these same halogens as ions. As you know, the chloride ion, Cl⁻, is part of common table salt. In proper concentrations, the iodide ion, I⁻, is essential for the health of the thyroid gland, a gland that regulates our metabolism. **Figure 8.17** shows how concentrations of dietary iodine vary around the world. **Figure 8.18** and the Biology Connection on page 278 illustrate several other important applications of the halogens.

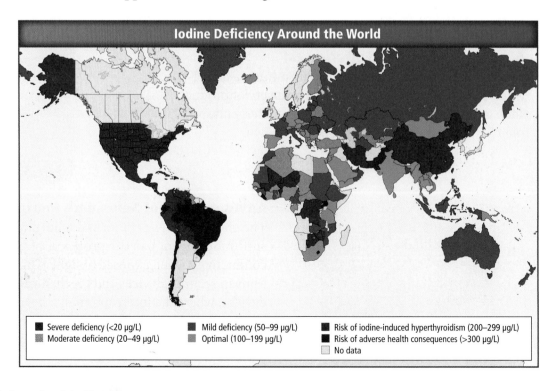

FIGURE 8.18

Visualizing Halogens

Everyday Halogen Use
Fluorine and chlorine are both abundant in nature, and both are present in biologically essential compounds. Table salt (NaCl), used to flavor foods, provides all the chloride ions needed for a healthy diet.

Fluorides Prevent Tooth Decay
Many towns and cities add fluorides to their water supply, and sodium fluoride (NaF) or tin(II) fluoride (SnF_2) is often added to toothpastes to prevent tooth decay.

Iodine as an Antibacterial
The halogens are important as antibacterial agents. Doctors use an iodine solution to sterilize the skin before surgery.

Chlorine Makes Water Safe
Chlorine is used in the water supply of most cities and towns and in swimming pools to kill bacteria. Chlorine added to swimming pools makes the water slightly acidic and can irritate your eyes and skin.

Silver Bromide Coats Photographic Film
The compounds of the halogens are more important than the free elements. Compounds of chlorine with carbon, such as carbon tetrachloride and chloroform, are useful solvents. Silver bromide (AgBr) is important in the light-sensitive coating on film.

Section 8.1 • Main Group Elements

Biology Connection

Fluorides and Tooth Decay

Do you worry that you might get a cavity? If so, it's not surprising because total prevention of cavities is not yet possible. However, adding fluorides to your oral hygiene can greatly reduce tooth decay.

Protection from tooth decay Drinking water with minute amounts of fluorides helps prevent tooth decay, according to the American Dental Association and the Centers for Disease Control and Prevention. They state that when children drink water containing 1 part per million (ppm) of sodium silicofluoride (Na_2SiF_4) they have 18–40 percent fewer cavities than those who drink non-fluoridated water.

Protection from tooth decay can also be obtained by receiving fluoride gel treatments from the dentist, using fluoride rinses at home, or using toothpaste containing fluorides. Fluoride gels use fluoride in the form of sodium fluoride (NaF), fluoride rinses use stannous fluoride (SnF_2), and fluorides in toothpaste are in the form of sodium fluoride (NaF), stannous fluoride (SnF_2), or sodium monofluorophosphate or MFP (Na_2FPO_3).

The decay process Tooth enamel shown in **Figure 1** is up to about 2.5 mm thick and composed of roughly 96 percent hydroxyapatite (crystalline calcium phosphate), $Ca_5(PO_4)_3OH$.

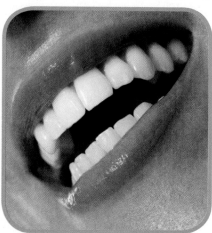

Figure 1 Tooth enamel dissolves in acidic saliva.

Although hydroxyapatite is essentially insoluble in water, tiny amounts dissolve in acidic saliva, produced by bacteria in your mouth, in a process called demineralization.

The reverse process, remineralization, is the body's defense against tooth enamel dissolving from bacterial acids. The body tries to replace the enamel dissolved by acidic saliva using minerals that come from the foods that you chew and carbon dioxide, CO_2, from your breathing. When the rates of demineralization and remineralization are equal, equilibrium is established and a cavity will not form.

Bacteria and cavities Oral bacteria feed on sugar in your mouth and produce lactic acid as a bi-product. The acid causes the pH of saliva, which is normally between 5.5-6.5, to drop below 5.0. When that happens, the rate of demineralization increases because hydroxyapatite dissolves quickly in low pH environments and tooth decay occurs.

Fluorides prevent cavities Fluoride compounds dissociate in water to form fluoride ions. The fluoride ions replace the hydroxide ions in some of the $Ca_5(PO_4)_3OH$ in tooth enamel and form fluorapatite, $Ca_5(PO_4)_3F$. Fluorapatite is about 100 times less soluble than hydroxyapatite; it is also harder and denser, so tooth enamel is stronger and more resistant to bacterial attack. Even though the American Dental Association and the Centers for Disease Control advocate fluoridating your teeth, opposition of the practice exists. However, the chemistry behind the practice supports fluoridation of the teeth.

Connection to Chemistry

1. **Acquire Information** Find out why not all of the cities in the United States fluoridate their water supply.

2. **Apply** Check the ingredients on five different toothpastes to see whether they contain a fluoride compound, and if so, which one.

■ **Figure 8.19** Because it does not burn helium, rather than hydrogen, is used to fill balloons.

Group 18 Helium (He), neon (Ne), argon (Ar), krypton (Kr), xenon (Xe), and radon (Rn), the noble gases, were originally called the inert gases because chemists could not make them react. Their lack of reactivity is understandable; all the noble gases have a full outer energy level and, therefore, no tendency to gain or lose electrons. **Figure 8.19** shows one use of helium that results from its lack of reactivity. Helium is lightweight and nonreactive, making it ideal for filling balloons. In recent years, however, chemists have succeeded in making fluorine compounds of the heavier noble gases, krypton and xenon, but no reactions have been achieved for the lighter members of the group.

Trends in the chemical properties for each group of the main group elements are directly related to changes in atomic radii. Within groups of elements that form compounds by losing electrons, the larger the atom, the more readily the atom gives up its electrons and reacts. Within groups of elements that form compounds by gaining electrons, the larger the element, the less attraction it has for electrons and the less reactive the atom is. You will see similar trends among the transition elements.

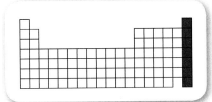

SUPPLEMENTAL PRACTICE

For more practice using trends on the periodic table, see Supplemental Practice, page 818.

Section 8.1 Assessment

Section Summary

▶ In a period of the periodic table, the number of valence electrons increases as atomic number increases.

▶ From left to right across a period, atomic radius decreases. Down a group, atomic radius increases.

▶ The metal with the biggest atom and smallest number of valence electrons is the most active metal.

▶ The nonmetal with the smallest atom and greatest number of valence electrons is the most active nonmetal.

1. **MAIN Idea Determine** Using only the periodic table, determine which is bigger: the atomic radius of potassium or the atomic radius of iodine. Which of these elements is a metal?

2. **Explain** why atomic radii decrease from left to right across a period and increase from top to bottom down a group.

3. **Compare** How does the size of a positive ion compare with the size of the neutral atom? How does the size of a negative ion compare with the size of the neutral atom? Explain your answer.

4. **Think Critically** Compare the way metals and nonmetals form ions and explain why they are different.

5. **Evaluate** Soap scum forms when soap is used with hard water. This is because of the presence of magnesium and calcium ions. Which of the two elements, magnesium or calcium, reacts more easily with water?

Section 8.1 • Main Group Elements

Section 8.2

Objectives
- **Relate** the chemical and physical properties of the transition elements to their electron configurations.
- **Predict** the chemical behavior of transition elements from their positions in the periodic table.

Review Vocabulary
halogen: an element from group 17 (F, Cl, Br, I, and At) that reacts with metals to form salts

Transition Elements

MAIN Idea The properties of the transition elements result from electrons filling d orbitals; for the inner transition elements, electrons fill f orbitals.

Real-World Reading Link Think for a moment about the many ways that people use metals. The might and the power of kings and queens have traditionally been displayed through ornaments, jewelry, and works of art made of precious metals. Iron is often considered the world's most important structural material and can be alloyed with other metals to lend hardness and resistance to corrosion. Copper is known for its electrical conductivity. You'll find gold, iron, copper, and many other important metals among the transition elements.

Trends in Properties

Each of the transition elements has its own properties that result from its atomic structure. For example, iron is strong. It is used for the structural framework of bridges and skyscrapers. But iron can also be reduced to a pile of reddish-brown rust if it is exposed to water and oxygen. Other transition metals might not be as strong as iron, but they do have important uses. Additionally, transition elements can be used together in alloys, as shown in **Figure 8.20**.

■ **Figure 8.20** When lightness, durability, and strength are needed for applications such as these athletes' prosthetic legs, a combination of transition elements alloyed with iron can provide the necessary properties.

280 Chapter 8 • Periodic Properties of the Elements

Melting points and boiling points With the exception of the group 12 elements (zinc, cadmium, and mercury), the transition metals have higher melting and boiling points than almost all of the main group elements. For example, in the fourth period (scandium–copper), the melting points range from 1083°C for copper (Cu) to 1890°C for vanadium (V). When you compare these melting temperatures to the melting temperatures of the main group metals, you find that only beryllium (Be) melts above 1000°C. Most of the other main group elements melt well below this temperature.

In any period, the melting and boiling points of the transition metals increase from group 3 to a maximum value in group 5 or 6. Then they decrease across the remainder of the period. Tungsten (W) in period 6, group 6 has a melting point of 3410°C, the highest of any metal. It is because of this high melting point that tungsten is used as the filament in lightbulbs, as you'll see in the How It Works feature. Mercury (Hg) in period 6, group 12 melts at −38°C, the lowest melting point of any metal. **Figure 8.21** compares the properties of tungsten and mercury.

Multiple oxidation states Transition elements are characterized by multiple oxidation states. Remember that iron gives up two electrons and forms the Fe^{2+} ion in its oxide, FeO. In another oxide, Fe_2O_3, iron gives up its two 4s electrons and one 3d electron to form the Fe^{3+} ion. Many of the transition elements can have multiple oxidation numbers ranging from 2+ to 7+. Variation in oxidation state results from the involvement of the d electrons in bonding. Recall that only some of the heavier main group elements such as tin, lead, and bismuth have multiple oxidation states. These elements also have d electrons that can be involved in bonding.

> **VOCABULARY**
> **WORD ORIGIN**
> **Tungsten**
> comes from the Swedish words *tung*, meaning *heavy*, and *sten*, meaning *stone*

> **FOLDABLES**
> Incorporate information from this section into your Foldable.

■ **Figure 8.21** Tungsten has the highest melting point of any metal. This property makes tungsten useful in situations that require resistance to heat. For example, in gas tungsten arc welding, the tip of the welder is a tungsten electrode (left). This electrode does not melt and is not consumed, despite the high welding temperatures. Mercury, on the other hand, has the lowest melting point of any metal—low enough, in fact, to make mercury a liquid at room temperature (right).

Tungsten arc welding

Mercury

HOW IT WORKS

Inert Gases in Lightbulbs

Tungsten (W) is used as a filament in incandescent lightbulbs as shown in **Figure 1** because it has a high melting point, 3422°C, and boiling point, 5555°C. But, nothing lasts forever. Because electricity continually passes through the filament, the metal eventually breaks as it vaporizes, and the bulb burns out as shown in **Figure 2**.

Figure 1 Lightbulb with tungsten (W) filament

❶ If a lightbulb were filled with air, the filament would react with oxygen, burn, break, and lose its ability to provide light. If there were no gas inside the bulb, the filament would quickly vaporize and no electricity would flow.

❸ To prevent the filament from burning and to slow its sublimation, lightbulbs traditionally have been filled with a mixture of nitrogen and argon. These gases are chemically inert, so the tungsten filament does not burn in them. Also, each argon atom or nitrogen molecule is massive enough to block an escaping tungsten atom that's trying to leave the filament and possibly cause the tungsten atom to rebound back onto the filament. The argon and nitrogen gases thus prolong the life of the filament; traditional bulbs can last 750–1000 hours.

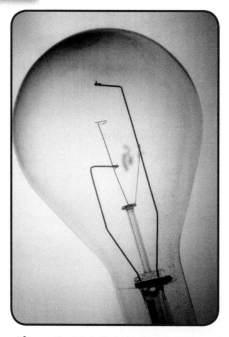

Figure 2 Lightbulb with a broken tungsten filament

❷ If you've ever looked closely at a burned-out bulb, you may have seen a black coating of condensed metal on the inside of the bulb from the sublimated filament.

❹ In the search for longer-lasting lightbulbs, manufacturers are experimenting with various inert gases to replace the nitrogen/argon mixture. The most promising gases are xenon and krypton, which produce an incandescent lightbulb that lasts 7500 to 10,000 hours—ten times longer than an ordinary incandescent lightbulb.

Think Critically

1. **Infer** How would you go about finding a substance other than tungsten to use as the filament of a lightbulb?
2. **Explain** What might be a disadvantage of using lightbulbs containing the relatively rare elements argon (Ar), krypton (Kr), and xenon (Xe)?

MiniLab 8.2

The Ion Charges of a Transition Element

How does the electron configuration of iron relate to the formation of iron compounds? Iron, the most common transition element, has the electron configuration $[Ar]4s^2 3d^6$. The two electrons in the highest energy level, $4s^2$, are the ones most likely to be involved in the chemical reactions of iron. However, iron is a transition element and, like other transition elements, it has a partially filled d sublevel. Electrons in the d sublevel may also be involved in reactions. In this MiniLab, you will investigate reactions that produce iron compounds and relate the results to the electron configuration of iron.

Procedure

1. Read and complete the lab safety form.
2. Place about 20 mL of aqueous **FeCl₃ solution** into a **125-mL Erlenmeyer flask** labeled **FeCl₃**. Add 2 drops of **1M NaOH solution**. Describe the results in your data table.
3. Place about 20 mL of aqueous **FeCl₂ solution** into a second **125-mL Erlenmeyer flask** labeled **FeCl₂**. Add 2 drops of **1M NaOH solution**. Describe the results in your data table.
4. **Stopper** the flask labeled **FeCl₂**. Swirl and shake the mixture in the flask. Every 30 seconds, stop shaking the flask and remove the stopper for a moment to incorporate more oxygen. Put the stopper back on the flask and resume shaking until a change occurs.
5. Record your observations in your data table.

Analysis

1. **Describe** the colors of the two precipitates formed in the reaction of $FeCl_3$ and $FeCl_2$ with NaOH.
2. **Infer** What are the charges on the iron ion in the two precipitates? Which electrons of the iron atom are probably lost to form the two ions?
3. **Explain** Examine your results from step 4. Explain what happened to the iron ions when oxygen entered the flask.
4. **Apply** Iron(II) ions (Fe^{2+}) are useful as nutrients, whereas iron(III) ions (Fe^{3+}) are not. Using the results of this MiniLab, suggest a reason why elemental iron (Fe) is often added to breakfast cereals as a dietary supplement rather than an iron compound containing Fe^{2+} ions.

Patterns in atomic size You learned in Section 8.1 that atomic radius decreases from left to right in a period within the main group elements because of increasing nuclear charge. A similar trend exists for the transition elements, but the changes in atomic radii for the transition elements are not as great as the changes for the main group elements. You also learned that as you move down a column in a main group, atomic radius increases. In the transition elements, this trend is a little more difficult to see. For example, there is little change in atomic radius as you move from the fifth period to the sixth period within the transition elements.

Because atomic size affects reactivity, you can expect the transition elements in periods 5 and 6 to have similar chemical properties, as shown in **Figure 8.22**. Atomic radius does increase as you move from the fourth period to the fifth period. Many of the transition metals form alloys—mixtures composed of more than one metal— with each other or with elements outside of the transition metal group. Due to differences in electron configuration, they do not always react the same way. For example, iron can exist in multiple oxidation states. **Figure 8.23** illustrates the importance of iron, the group 4 element that is the most common transition metal.

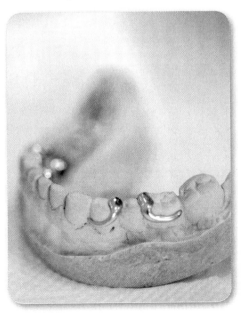

■ **Figure 8.22** Transition metals, including palladium, nickel, and chromium, are incorporated into alloys to construct dental crowns.

FIGURE 8.23

Visualizing Iron

Iron: First Among the Transition Elements
Iron, and many steel alloys that are made from it, have been known and used since ancient times. Iron is the fourth most abundant element in Earth's crust and the second most abundant metal after aluminum.

Iron Is Essential for Life
Besides its importance as a structural metal, iron is an essential element in biological systems. It is the iron ion at the center of the heme molecule (right) that forms bonds with oxygen. Iron in heme is the Fe^{2+} ion, and only the Fe^{2+} ion can bind oxygen. Heme molecules are a component of the proteins hemoglobin and myoglobin. Hemoglobin supports life by transporting oxygen through the blood from the lungs to every cell of the body. Myoglobin stores oxygen for use in certain muscles.

Separation of Iron from Its Ore
Iron is obtained from its ore (oxides) in a blast furnace. Iron oxide, Fe_2O_3, is mixed with carbon and limestone ($CaCO_3$) and fed continuously into the top of the furnace. Hot air is fed into the bottom of the furnace. Temperatures of 2000°C are reached as the mixture reacts while falling through the furnace. The process is complex, but the following equations are the important steps. First, carbon is ignited in the presence of the hot air and converted to carbon dioxide.

$$C(s) + O_2(g) \rightarrow CO_2(g)$$

Then, CO_2 reacts with more carbon to form carbon monoxide.

$$CO_2(g) + C(s) \rightarrow 2CO(g)$$

Carbon monoxide then converts iron ore (Fe_2O_3) to iron.

$$Fe_2O_3(s) + 3CO(g) \rightarrow 2Fe(l) + 3CO_2(g)$$

The molten iron, called pig iron, drops to the bottom of the furnace and is drawn off as a liquid. Impurities, called slag, form a layer on top of the iron and are drawn off separately. Pig iron, as it comes from the blast furnace, is a crude product containing impurities such as carbon, silicon, and manganese. The crude iron is refined, and then most of it is converted to steel.

Steelmaking
The first step in the production of steel is the removal of impurities from the pig iron. The second step is the addition of carbon, silicon, or any of a variety of transition metals in controlled amounts. Different elements give special properties to the final steel. Some steels are soft and pliable and are used to make things like fence wire (left). Others are harder and are used to make railroad tracks, girders, and beams.

The hardest steels are used in surgical instruments, drills, and ordinary razor blades. These steels are made from iron mixed with small amounts of carbon, molybdenum, and nickel. The chromium in surgical stainless steel contributes to the metals scratch resistance and molybdenum helps maintain a harder cutting edge.

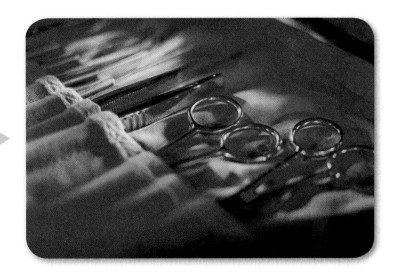

Heat-Treating
Heat-treating is a final step in the production of steel. When purified steel is heated to 500°C, the small amount of carbon contained in it combines with iron to form a carbide (Fe_3C), which dissolves in the steel. This makes the steel harder. Cooling the steel quickly in oil or water makes the hardness permanent.

CHEMISTRY AND TECHNOLOGY

Carbon and Alloy Steels

Steelmaking was already a well-advanced art in ancient times. The earliest known steel objects were made between 1500 and 1200 B.C. Around 300 B.C., wootz steel was developed in India by heating a mixture of iron ore and wood in a sealed container. Wootz steel later became known as Damascus steel, shown in **Figure 1,** when Europeans first encountered high-quality swords made from the steel in Damascus, Syria. Damascus steel became famous because swords made from this steel kept their sharpness after many battles, but were flexible enough not to break. Damascus steel also has a characteristic wavy surface pattern which forms during its production.

The knowledge of how to make Damascus steel was lost in the 1800s. New superplastic steel is similar in strength to the ancient Damascus steel and can be produced with a wavy surface pattern, but the internal structure of the superplastic steel produced today is not the same as the ancient steel.

■ **Figure 1** Damascus steel has a wave pattern that is formed as it is produced.

What is steel and how is it classified?

Steel is an iron alloy containing small amounts of carbon (0.02–2.1 percent) and sometimes other elements such as chromium, cobalt, manganese, nickel, tungsten, molybdenum, copper, silicon, and vanadium. Carbon steel contains only iron and carbon. Steel alloys contain iron and carbon, along with small amounts of the previously listed elements. Steel cannot contain more than 2.1 percent carbon without becoming brittle. Commercial iron, also called cast iron, contains 2-4 percent carbon and is very brittle. The most common carbon steels are the mild steels, which have less than 0.30 percent carbon content. Mild-carbon steels make up the largest portion of steel production. They are commonly used to manufacture steel food cans and in motors and electrical appliances.

Table 1	Carbon Steels		
Name of Steel	Composition	Characteristics	Uses
Mild	Fe, 0.02–0.29% C	malleable	steel food cans, automobile bodies
Medium	Fe, 0.30–0.59% C	less malleable	structural purposes—beams, bridge supports
High	Fe, 0.60–1.5% C	hard, brittle	cutting tools, masonry nails, drill bits, high-strength tension wires, truck springs
UHCS, Superplastic*	Fe, 1.5–2.1% C	corrosion- and wear-resistant, highly malleable	knives, axles, punches, earth movers, underside of tractors

*A few UHCSs are alloy steels.

■ **Figure 2** Automobile bodies made from mild-carbon steel

Ninety Percent of Steel Is Carbon Steel

The properties of carbon steel depend upon the percent of carbon present. They are classified as mild, medium, and high on this basis. Because mild-carbon steels are malleable, sheets can be cold-formed to mold fenders and other automobile body parts like those shown in **Figure 2**. Medium-carbon steels are stronger but are less malleable so they are used as structural materials, such as high tension wires on bridges shown in **Figure 3**. High-carbon steels are hard and brittle; they are used for wear-resistance purposes.

■ **Figure 3** High-strength tension wires made from medium-carbon steel

■ **Figure 4** Knives made from superplastic steel

Superplastic Steel Is New Again

A new steel has been developed, called superplastic or ultrahigh-carbon steel (UHCS), which can be formed into complex shapes such as the knives in **Figure 4**. This new steel is also sometimes called Damascus steel because of the strength of the steel and the wavy surface pattern that can be formed during production, even though the internal structure is different from the ancient Damascus steel.

Ultrahigh-carbon steels are important because of their durability and because they can be manufactured into their final shape in one step. For a metal to exhibit superplastic behavior, it must have fine grain sizes at temperatures of about 725 °C. At this high temperature, the metal is highly sensitive to stress and stain rates. This high sensitivity prevents the growth of mechanical instabilities during the manufacture of the superplastic steel. The lack of mechanical instabilities allows the superplastic steel to stretch up to a thousand percent of its length without breaking.

The limit for the stretching of most steels is 50–100 percent of their length. At high temperatures, ultrahigh-carbon steel pulls like taffy. The superplastic behavior of the steel allows materials to be formed in one step, almost eliminating waste and secondary processing from the manufacturing process. Therefore, materials and energy are conserved.

Table 2 Steel Alloys

Interactive Table Explore steel alloys at glencoe.com.

Name of Steel	Composition in %*	Characteristics	Uses
Alnico	Ni-20, Al-12, Co-5	strongly magnetic	loudspeakers, ammeters, microphones
Invar	Ni-36 to 50	low coefficient of thermal expansion	precision instruments, measuring tapes
Manganeses steel	Mn-12 to 14	holds hardness and strength	safes
18-8 Stainless	Cr-18, Ni-8	corrosion resistant	surgical instruments, cooking utensils, jewelry
Tungsten steel	W-5	stays hard when hot	high-speed cutting tools

*All alloys contain iron and 0.1–1.5% carbon

Steel Alloys Contain Carbon and Other Elements

In steel alloys, iron is mixed with carbon and varying amounts of other elements, mainly metals (Table 2). Added metals produce desired properties such as corrosion resistance (Cr), resistance to wear (Mn), toughness (Ni), and heat resistance (W and Mo). Stainless steel is a well-known, corrosion-resistant steel alloy, shown in **Figure 5.** It contains 10 to 30 percent chromium and sometimes nickel, silicon, and phosphorous. Because of its outstanding magnetic properties, Alnico steel is used to make permanent magnets. Alnico magnets are used in voltmeters to rotate the wire coil connected to the pointer and in microphones, as in **Figure 6.**

■ **Figure 6** Alnico steel is used in mircophones.

Working Steel

Two methods are used to shape steel into various shapes for specific purposes: hot working and cold working. In the hot-working process, steel is hammered, rolled, pressed, or extruded while it is above its recrystallization temperature before crystal structures have formed. Hot working manipulates the steel's shape and size and does not alter the steel's properties. Hammering and pressing are called forging. These processes were originally done by hand; in fact, hand forging continues as a craft in blacksmith shops. However, steam-powered hammers and hydraulic presses are used to forge most of the steel produced today.

■ **Figure 5** Stainless steel is used in cooking pots.

Extrusion involves forcing molten steel through a die that is cut to the desired shape, such as the steel beams shown in **Figure 7.** Rolling is the most widely used method for shaping steel. The metal is passed between two rollers that move in opposite directions. The shape of the finished product determines the type of rollers used. Railroad tracks and I-beams are shaped this way.

Cold working manipulates the steel's hardness and strength by deforming the steel through rolling, extrusion, or drawing while it is below its recrystallization temperature and crystal structures have already formed. Cold working increases the strength and the hardness of a metal by introducing defects into the metal's crystal structure, which reduces the grain size of the metal and prevents slippage of the grains. This process results in increased strength and hardness. Cold working steel is often used to decrease the thickness of sheet metal, while increasing the strength. Often cold-working processes follow hot working to produce a stronger product.

■ **Figure 8** Steel-reinforced concrete makes modern skylines possible.

Because steel is closely linked to infrastructure and economic development, the performance of the steel industry is often considered an indicator of economic progress worldwide and in individual countries. The growth of the economies in China and India has caused an increase in the demand for steel. China uses over one-third of the steel produced in the world. China, Japan, and the United States are the leading producers of steel.

■ **Figure 7** Steel beams formed by extrusion

The development of methods of producing and working steel led to a revolution in construction in the 1880s. Concrete, reinforced by steel, became an important structural material. Steel beams made skyscrapers possible and changed the shape of modern cities. Even though the Sears Tower, shown in **Figure 8,** is the tallest building in the United States, it has many neighboring buildings that also dominate the skyline of Chicago.

Discuss the Technology

1. **Apply** Using the Steel Alloys Table, what type of steel is used in airplane engines if the steel is composed of Fe, C, Cr, and Ni? What are the functions of Cr and Ni?

2. **Infer** Using the Steel Alloys Table, how would you classify most of the metals used? (Hint: Look at their locations on the periodic table.) Draw a conclusion about most metals used in alloy steels.

3. **Acquire Information** Find out how heat treatments, quenching, tempering, and annealing further alter the properties of different steels.

> **FACT of the Matter**
>
> The principal ore of zirconium, a group 4 transition element, is a colorless crystal called zircon. Often used in costume jewelry, zircon crystals are called cubic zirconium by jewelers.

Other Transition Elements: A Variety of Uses

You've learned that some transition elements are important in the production of steel because they impart particular properties to the steel. In addition, most of the transition elements, because of their individual properties, have a variety of uses in the production of infrastructure and consumer products of the modern world.

The iron triad, the platinum group, and the coinage metals Iron (Fe), cobalt (Co), and nickel (Ni) have nearly identical atomic radii, so it isn't surprising that these three elements have similar chemical properties. Like iron, both cobalt and nickel are naturally magnetic. Because of their similarities, the three elements are called the iron triad (group of three). Notice the relative positions of iron, cobalt, and nickel on the periodic table, as shown in **Figure 8.24**. They are in period 4, groups 8, 9, and 10. The elements below the iron triad in periods 5 and 6—ruthenium (Ru), rhodium (Rh), palladium (Pd), osmium (Os), iridium (Ir), and platinum (Pt)—all resemble platinum in their chemical behavior and are called the platinum group. The platinum group elements are often used as catalysts to speed up chemical reactions. Copper (Cu), silver (Ag), and gold (Au) in group 11 are the traditional metals used for coins because they are malleable, relatively unreactive, and in the case of silver and gold, rare. You might have predicted that these elements would have similar chemical properties because they are in the same group. Notice the positions of the platinum group and the coinage metals in **Figure 8.24**.

■ **Figure 8.24** The nearly identical atomic radii of the iron triad—iron, cobalt, and nickel—help to explain the similar chemistry of these three elements. The platinum group elements in periods 5 and 6 emphasize the fact that there is little difference between the atomic radii of these elements, whereas the coinage metals show an expected similarity among elements in the same group.

Group Number →	3	4	5	6	7	8	9	10	11	12
Period Number ↓										
4	Scandium Sc 21	Titanium Ti 22	Vanadium V 23	Chromium Cr 24	Manganese Mn 25	Iron Fe 26	Cobalt Co 27	Nickel Ni 28	Copper Cu 29	Zinc Zn 30
5	Yttrium Y 39	Zirconium Zr 40	Niobium Nb 41	Molybdenum Mo 42	Technetium Tc 43	Ruthenium Ru 44	Rhodium Rh 45	Palladium Pd 46	Silver Ag 47	Cadmium Cd 48
6	Lanthanum La 57	Hafnium Hf 72	Tantalum Ta 73	Tungsten W 74	Rhenium Re 75	Osmium Os 76	Iridium Ir 77	Platinum Pt 78	Gold Au 79	Mercury Hg 80

Iron triad Platinum group Coinage metals

Chromium When chromium is alloyed with iron, durable, hard steels or steels that are corrosion-resistant are formed. Chromium is also alloyed with other transition metals to produce structural alloys used in jet engines which must withstand high temperatures. A self-protective metal, chromium is often plated onto other materials to protect them from corrosion.

Chromium has the electron configuration $[Ar]4s^1 3d^5$ and exhibits oxidation numbers 2+, 3+, and 6+. When chromium loses two electrons, it forms the Cr^{2+} ion and has the configuration $[Ar]3d^4$. The Cr^{3+} ion results when chromium loses a second 3d electron.

Chromium can also lose six valence electrons and have an oxidation number of 6+. When it does, it loses all of its s and d electrons and assumes the electron configuration of argon. Potassium chromate (K_2CrO_4) and potassium dichromate ($K_2Cr_2O_7$) are two compounds in which chromium's oxidation number is 6+. **Figure 8.25** shows the brilliant colors that are typical of compounds of chromium. Chromium gets its name from the Greek word for color, *chroma*, and many of its compounds are brightly colored—yellow, orange, blue, green, and violet.

Zinc Like chromium, zinc is a corrosion-resistant metal. One of its principal uses is as a coating on iron and steel surfaces to prevent rusting. In the process called galvanizing, a surface coating of zinc is applied to iron by dipping the iron into molten zinc. Zinc is also important when alloyed with other metals. The most important of these alloys is the combination of zinc with copper in brass. Brass is used for making bright and useful objects like those shown in **Figure 8.26**.

■ **Figure 8.25** The brilliant colors of these chromium compounds are typical of many transition metal compounds. In both the yellow potassium chromate (K_2CrO_4) and the orange potassium dichromate ($K_2Cr_2O_7$), chromium has a 6+ oxidation number.
Determine *the oxidation numbers of potassium (K) and oxygen (O) in these two compounds.*

■ **Figure 8.26** Brass can be worked into smooth shapes and drawn into the long, thin-walled tubes needed for musical instruments. Look around and see how many brass items you find that are both decorative and useful.

Section 8.2 • Transition Elements

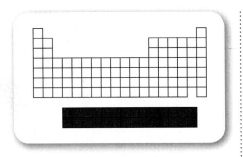

Lanthanides and Actinides: The Inner Transition Elements

The inner transition elements are found in the f block of the periodic table. In the lanthanides, electrons of highest energy are in the 4f sublevel. The lanthanides were once called rare earth elements because all of these elements occurred in Earth's crust as "earths", an older term for oxides, and seemed to be relatively rare. The highest-energy electrons in the actinides are in the 5f sublevel. You probably won't find these elements among your household chemicals. Their names are unfamiliar except for uranium and plutonium, which are the elements associated with nuclear reactors and weapons. Many of these elements, especially lanthanides, have important practical uses.

Cerium: the most abundant lanthanide Cerium is the principal metal in an alloy called misch metal. Misch metal is 50 percent cerium combined with lanthanum, neodymium, and a small amount of iron. Misch metal is used to make the flints for lighters. Cerium is often included in iron and magnesium alloys. A high-temperature alloy of 3 percent cerium with magnesium is used for jet engines. Some of cerium's compounds—for example, cerium(IV) oxide (CeO_2)—are used to polish lenses, mirrors, and television screens; in glass manufacturing to decolorize glass; and to make porcelain coatings opaque.

Other lanthanides Other lanthanides are used in the glass industry. Neodymium (Nd) is used not only to decolorize glass but also to add color to glass. When added to the glass used for welders' goggles, neodymium and praseodymium (Pr) absorb the eye-damaging radiation from welding, as shown in **Figure 8.27**. They also decrease reflected glare when used in the glass of a television screen. A combination of the oxides of yttrium (Y), a transition element, and europium (Eu) produce a phosphor that glows a brilliant red when struck by a beam of electrons, such as in a TV picture tube. This phosphor is used with blue and green phosphors to produce realistic-looking television pictures.

> **FACT of the Matter**
>
> More than half of the world's supply of the lanthanides comes from a single mine in California. Many of these elements are used to make ceramic superconductors.

■ **Figure 8.27** Intense light from a welding torch can harm eyes. The lanthanides neodymium and praseodymium, incorporated into the lenses of goggles, absorb the damaging wavelengths.

Because europium, gadolinium (Gd), and dysprosium (Dy) are good absorbers of neutrons, they are used in control rods in nuclear reactors. Promethium (Pm) is the only synthetic element in the lanthanide series. It is obtained in small quantities from nuclear reactors and is used in specialized miniature batteries. Samarium (Sm) and gadolinium are used in electronics. Terbium (Tb) is used in solid-state devices and lasers.

Radioactivity and the actinides Uranium (U) is a naturally occurring, radioactive element used as a source for nuclear fuel and other radioactive elements. Plutonium (Pu) is one of the elements obtained from the use of uranium as a nuclear fuel. The isotope Pu-238 emits radiation that is easily absorbed by shielding. Pu-238 is used as a power source in heart pacemakers and navigation buoys. Other isotopes of plutonium are used as nuclear fuel and in nuclear weapons. Plutonium is the starting material for the synthetic production of the element americium, which is used in smoke detectors, as shown in **Figure 8.28.**

Some actinides have medical applications; for example, radioactive californium-252 (Cf) is used in cancer therapy. Better results in killing cancer cells have been achieved using this isotope of californium than by using the more traditional X-ray radiation.

Connecting Ideas

If you can locate an element on the periodic table, you can predict its properties. Each element has unique characteristics because of its unique electron configuration. Together, the elements, their alloys, and compounds provide a wide variety of materials for countless applications. Compounds of the elements range from ionic to covalent, from polar to nonpolar. They have different sizes and shapes. In Chapter 9, you'll learn more about the formation of compounds and how to predict their shape and polarity.

■ **Figure 8.28** Inside one kind of smoke detector, small amounts of the radioactive actinide americium are used to generate an electric current. Smoke that enters the detector will disrupt this current, causing the alarm to sound.

Section 8.2 Assessment

Section Summary

▶ Transition elements form the d block of the periodic table.

▶ Multiple oxidation states are characteristic of transition elements.

▶ Transition elements with similar atomic radii often have similar chemical properties. The iron triad, platinum group, and coinage metals are examples.

▶ The inner transition elements, the lanthanides and actinides, form the f block of the periodic table.

6. **MAIN Idea Describe** What are transition elements? What are inner transition elements? Describe where all the transition elements are found on the periodic table.

7. **Describe** How do the electron configurations of the transition metals and inner transition elements differ from those of the main group metals?

8. **Explain** Iron, aluminum, and magnesium all have important uses as structural materials, yet iron corrodes (rusts), whereas aluminum and magnesium usually do not corrode. Explain what causes iron to rust.

9. **Think Critically** Why do the transition metals have multiple oxidation numbers whereas the alkali metals and the alkaline earth metals have only one oxidation number, 1+ and 2+, respectively?

10. **Explain** Name three transition metals that are added to iron and steel to improve their properties and help prevent corrosion. Explain how each additive works.

CHAPTER 8 Study Guide

BIG Idea Trends among elements in the periodic table include their sizes and their abilities to lose or gain electrons.

Section 8.1 Main Group Elements

MAIN Idea For the main group elements, both metallic character and atomic radius decrease from left to right across a period.

Vocabulary
- alkali metal (p. 261)
- alkaline earth metal (p. 264)
- halogen (p. 276)

Key Concepts
- In a period of the periodic table, the number of valence electrons increases as atomic number increases.
- From left to right across a period, atomic radius decreases. Down a group, atomic radius increases.
- The metal with the biggest atom and smallest number of valence electrons is the most active metal.
- The nonmetal with the smallest atom and greatest number of valence electrons is the most active nonmetal.

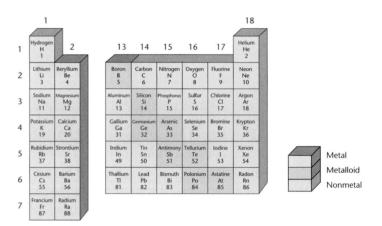

Section 8.2 Transition Elements

MAIN Idea The properties of the transition elements result from electrons filling d orbitals; for the inner transition elements, electrons fill f orbitals.

Key Concepts
- Transition elements form the d block of the periodic table.
- Multiple oxidation states are characteristic of transition elements.
- Transition elements with similar atomic radii often have similar chemical properties. The iron triad, the platinum group, and the coinage metals are examples.
- The inner transition elements, the lanthanides and the actinides, form the f block of the periodic table.

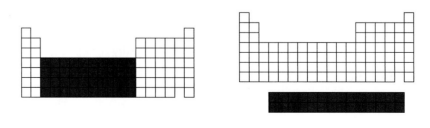

Chapter 8 Assessment

Understand Concepts

11. From each pair of atoms, select the one with the larger atomic radius.
    a) K, Ca
    b) F, Na
    c) Mg, Ca
    d) Rb, Cs
    e) Ca, Sr
    f) S, C

12. What is the effect of changes in atomic radii on the chemical reactivity of the halogens? Which is the most active halogen? Which is the least active?

13. Which atoms increase in size when they become ions? Explain your answer. Cs, I, Zn, O, Sr, Al

14. Why do elements in a group have similar chemical properties?

15. What are the products of a reaction between an alkali metal and water? An alkaline earth metal and water?

■ Figure 8.29

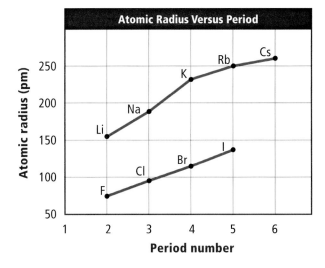

16. Use the graph of atomic radius versus period number in **Figure 8.29** to answer these questions. How does the size of a lithium atom compare with that of a cesium atom? How does the size of a fluorine atom compare with that of an iodine atom? Which has the larger radius, lithium or fluorine? Based on their atomic radii, which is the most active alkali metal? Which is the most active halogen? Explain.

Apply Concepts

How It Works

17. Why are noble gases used in lightbulbs? What advantages do they offer over a vacuum?

Chemistry and Technology

18. Explain the difference between carbon steel and a steel alloy. Which type makes up the majority of steel produced?

History Connection

19. Use the periodic table to determine the oxidation numbers of lead and sulfur in galena, PbS, the common ore of lead.

Biology Connection

20. Sodium silicofluoride, Na_2SiF_4, is added to water supplies to help prevent tooth decay. What is the oxidation number of silicon in Na_2SiF_4? Explain how you got your answer.

21. Nitrogen, oxygen, and phosphorus are essential elements for maintaining life. What essential biological molecules contain these elements?

22. Describe one way in which nitrogen (N_2) is converted into a form that plants can use.

23. Explain why magnesium and aluminum are considered corrosion-resistant metals.

Think Critically

Compare and Contrast

24. Compare the chemical behavior of oxygen when it combines with an active metal such as calcium to its behavior when it combines with a nonmetal such as sulfur.

Relate Concepts

25. Is carbon dioxide (CO_2) produced by sharing or transferring electrons? Draw the Lewis dot diagram for carbon dioxide and explain how each element achieves an octet of electrons.

Chapter 8 Assessment

Relate Concepts

26. **MiniLab 2** Write the electron configurations of the neutral iron atom and the ions Fe^{2+} and Fe^{3+}.

Make Predictions

27. Both sodium and cesium react with water. Predict the products of the reactions of sodium and cesium with water. Write balanced equations for both reactions.

Cumulative Review

28. Describe the physical properties of metals, metalloids, and nonmetals. (*Chapter 3*)

29. Write the electron configurations for each of the following atoms. Use the appropriate noble-gas inner core abbreviations. (*Chapter 7*)
    a) fluorine
    b) aluminum
    c) titanium
    d) argon

30. Explain why many compounds are a combination of a group 1 or group 2 element with a group 16 or group 17 element. Give the formula of a compound formed with an element from group 1 and an element from group 17. Give the formula of a compound formed from a group 1 and group 16 elements. (*Chapter 4*)

31. Write formulas for the following compounds. Are they ionic or covalent? (*Chapter 5*)
    a) aluminum sulfate
    b) dinitrogen monoxide
    c) sodium hydrogen carbonate
    d) lead(II) nitrate

32. The use of toothpaste containing fluoride results in the formation of fluorapatite, from hydroxyapatite, in the enamel of teeth.

    $Ca_5(PO_4)_3OH(s) + F^-(aq) \rightarrow$

    $Ca_5(PO_4)_3F(s) + OH^-(aq)$

    Which of the five types of chemical reactions is this? (*Chapter 6*)

Skill Review

Make and Use Graphs

33. Use the data in **Figure 8.30** below to graph atomic radius versus atomic number for the second and third period main group elements. Describe any patterns you observe.

■ **Figure 8.30**

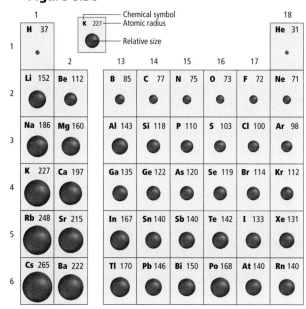

WRITING in Chemistry

34. Design a brochure explaining the properties of elements called metalloids. Use the concept of periodicity as the basis for your information. List the chemical symbols of the metalloids and describe some of the properties that relate metalloids to both metals and nonmetals.

Problem Solving

35. Sketch the outline of a periodic table. Use the data from the diagram in Question 26 of the Skill Review to help you draw arrows on the table showing trends in atomic radius. Write a sentence describing the relationship between reactivity and atomic radius.

Cumulative Standardized Test Practice

1. Across a period on the periodic table to the right, the atomic sizes of elements
   a) increase.
   b) decrease.
   c) vary.
   d) remain the same.

2. Down a group on the periodic table, the atomic sizes of elements
   a) increase.
   b) decrease.
   c) vary.
   d) remain the same.

3. With respect to atomic numbers, the atomic sizes of elements
   a) increase.
   b) decrease.
   c) vary.
   d) remain the same.

4. Why is the size of an aluminum atom larger than an atom of silicon?
   a) The atomic radii of metal atoms are larger than the atomic radii of nonmetal atoms.
   b) The positive charges in an aluminum atom's nucleus have a greater attraction on the atom's electron cloud.
   c) The positive charges in a silicon atom's nucleus have a greater attraction on the atom's electron cloud.
   d) The silicon atom contains more electrons.

5. Which statement about the Heisenberg uncertainty principle is true?
   a) The principle states that the location of an electron in the electron cloud cannot be determined.
   b) The principle states that the position and energy of an electron cannot be known at the same time.
   c) The principle states that the exact location of an electron around an atom cannot be measured.
   d) The principle states that the exact energy of an electron around an atom cannot be accurately measured.

6. Hydrocarbons are most commonly used as
   a) acids.
   b) bases.
   c) fuels.
   d) explosives.

7. What is the product of this synthesis reaction?
   $Cl_2(g) + 2NO(g) \rightarrow ?$
   a) NCl_2
   b) $2NOCl$
   c) N_2O_2
   d) $2ClO$

8. The chemical formula for baking soda $NaHCO_3$. Identify the elements and calculate the number of atoms in a baking soda molecule.
   a) 1 sodium atom; 1 hydrogen atom; 3 cobalt atoms
   b) 1 nitrogen atom; 1 helium atom; 3 cobalt atoms
   c) 1 sodium atom; 1 hydrogen atom; 1 carbon atom; 3 oxygen atoms
   d) 1 sodium atom; 1 helium atom; 1 carbon atom; 3 oxygen atoms

9. Which is true about transition metals?
   a) Transition metals tend to be the strongest metals.
   b) Transition metals tend to have higher melting points than the main group elements.
   c) Transition metals rarely have multiple oxidation states.
   d) Transition metals have some of the largest atomic masses.

10. Which is NOT true about nonmetals?
    a) They make up a large percent of Earth's crust.
    b) They are found in most compounds.
    c) They are good conductors of electricity.
    d) They have melting points lower than that of metals.

NEED EXTRA HELP?

If You Missed Question...	1	2	3	4	5	6	7	8	9	10
Review Section...	8.1	8.1	8.1	8.1	7.1	5.2	6.1	1.1	8.2	3.2

CHAPTER 9
Chemical Bonding

BIG Idea Atoms bond by sharing or transferring electrons.

9.1 Bonding of Atoms
MAIN Idea The difference between the electronegativities of two atoms determines the type of bond that forms.

9.2 Molecular Shape and Polarity
MAIN Idea The shape of a molecule and the polarity of its bonds determine whether the molecule as a whole is polar.

ChemFacts

- Tight networks of interlocking logs are used to construct log cabins.
- These dovetail joints provide a stable structure for the cabin.
- When atoms or ions form chemical bonds, the ratio and type of particles determine the structure and stability of the solid.

Start-Up Activities

LAUNCH Lab

Oil and Vinegar Dressing

When you mix different substances, do they always produce a homogeneous mixture?

Materials
- Beral-type pipette
- vinegar
- vegetable oil

Procedure
1. Read and complete the lab safety form.
2. Fill the bulb of a Beral-type pipette about 1/3 full of vinegar and 1/3 full of vegetable oil. Shake the pipette and its contents. Record your observations.
3. Allow the contents to sit for about five minutes. Record your observations.

Analysis
1. **Observe** Do oil and vinegar mix?
2. **Explain** What explanation can you give for your observations in this experiment?
3. **Infer** Why do the instructions on many types of salad dressings read "shake well before using"?

Inquiry Are there ways to make oil and vinegar or oil and water stay mixed? Research your answer.

Visit glencoe.com to:
- study the entire chapter online
- explore **concepts In Motion**
- take Self-Check Quizzes
- use Personal Tutors
- access Web Links for more information, projects, and activities
- find the Try at Home Lab, Breaking Covalent Bonds

FOLDABLES Study Organizer

Bond Character Make the following Foldable to help you organize your study of the three major types of bonding.

STEP 1 Collect three sheets of paper. Fold each sheet in half. Measure and draw a line about 3 cm from the left edge. Cut along the line to the fold. Repeat for each sheet of paper.

STEP 2 Label the left margin *Bond Character* and label each top flap with a type of bonding: *Ionic*, *Covalent*, and *Polar covalent*.

STEP 3 Staple the sheets together along the inside edge of the narrow flaps.

FOLDABLES Use this Foldable with Section 9.1. As you read this section, summarize what you learn about bond character and how it affects the properties of compounds.

Chapter 9 • Chemical Bonding **299**

Section 9.1

Objectives
- **Predict** the type of bond that forms between atoms by calculating electronegativity differences.
- **Compare and contrast** characteristics of ionic, covalent, and polar covalent bonds.
- **Interpret** the electron sea model of metallic bonding.

Review Vocabulary
alkali metal: any element from group 1: lithium, sodium, potassium, rubidium, cesium, or francium

New Vocabulary
electronegativity
shielding effect
polar covalent bond
malleable
ductile
conductivity
metallic bond

Concepts In Motion

Interactive Table Explore physical properties of ionic and covalent compounds at glencoe.com.

Bonding of Atoms

MAIN Idea The difference between the electronegativities of two atoms determines the type of bond that forms.

Real-World Reading Link You've probably used glue to repair a broken object or to create something crafty. The glue joins separate pieces together to form a stable product. Similarly, the glue that holds atoms together in a compound is the transfer and sharing of electrons between the atoms.

A Model of Bonding

Recall from Chapter 4 that atoms either transfer electrons to form ionic compounds or they share electrons to form covalent compounds. In each instance, the bond forms because of an increase in stability. By forming bonds, atoms acquire an octet of electrons and the stable electron configuration of a noble gas. Atoms are often more stable when they're bonded in compounds than when they exist as free atoms.

Atoms form two different kinds of bonds—ionic and covalent. If you are aware of the type of bonds in a compound, you can predict many of its physical properties. **Table 9.1** compares the physical properties of ionic compounds and covalent compounds. You can reverse this relationship. If you know the physical properties of an unknown compound, you can predict its bond type. But predictions might not always be correct because there is no clear-cut division between ionic compounds and covalent compounds. A compound may be partly covalent and partly ionic.

Degrees of sharing A realistic view of bonding is to consider that all chemical bonds involve a sharing of electrons. Electrons may be shared equally, shared only slightly—or almost not at all. The properties of any compound, particularly the compound's physical properties, are related to how equally the electrons are shared.

Table 9.1 Physical Properties of Ionic and Covalent Compounds

Property	Ionic Compound	Covalent Compound
State at room temperature	crystalline solid	liquid, gas, or solid
Melting point	high	low
Conductivity in liquid state	yes	no
Water solubility	high	low
Conductivity of aqueous solution	yes	no

In Chapter 4, you explored the two extremes in this range of electron sharing: the ionic bond and the covalent bond. A purely ionic bond results when the sharing is so unequal that it is best described as a complete transfer of electrons from one atom to another. A purely covalent bond results when electrons are shared equally. Most compounds fall somewhere in between these two extremes; they have characteristics of both ionic and covalent compounds.

Electronegativity—An Attraction for Electrons

You can think of bonding atoms as being in a tug-of-war over the shared valence electrons. To use this model of electron sharing, you need some way of determining how much tug each atom exerts on the shared electrons. An atom's tendency to tug is called electronegativity. **Electronegativity** is a measure of the ability of an atom in a bond to attract electrons. How each atom fares in a tug-of-war for shared electrons is determined by comparing the electronegativities of the two bonded atoms. **Figure 9.1** illustrates how bond character changes for several different compounds.

Assigning electronegativity values Where do the values used to determine electronegativity difference—and thus assess bond character—come from? There are several electronegativity scales, but the original and also most common scale was developed by American chemist Linus Pauling. Pauling assigned the most electronegative element, fluorine, a value of 4.0. The other elements' values were established relative to fluorine. Atoms with large electronegativity values attract shared valence electrons more strongly than atoms that have small electronegativity values.

VOCABULARY
WORD ORIGIN
Electronegative
comes from the Greek *elektron* meaning *amber* and from the Latin *negare* meaning *to deny*

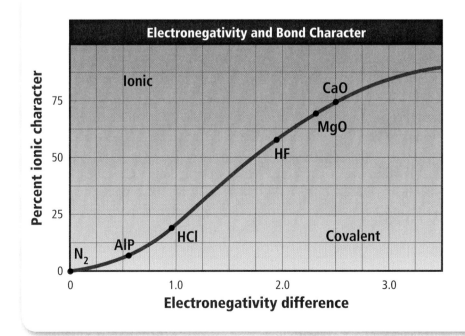

■ **Figure 9.1** The bonding between atoms in compounds can be viewed as a range of electron sharing measured by electronegativity difference, ΔEN. This range contains three main classes of bonds—ionic, polar covalent, and covalent. ΔEN is explained further on page 303.
Determine *the percent ionic character of calcium oxide.*

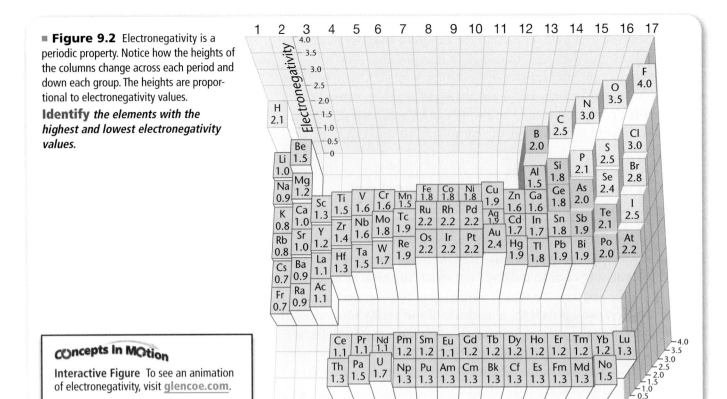

■ **Figure 9.2** Electronegativity is a periodic property. Notice how the heights of the columns change across each period and down each group. The heights are proportional to electronegativity values.
Identify the elements with the highest and lowest electronegativity values.

Concepts In Motion

Interactive Figure To see an animation of electronegativity, visit glencoe.com.

The electronegativities of most of the elements are illustrated in **Figure 9.2.** This figure might not always be at your fingertips, but you usually will have a periodic table. Electronegativity is a periodic property—it varies in a predictable pattern across a period and down a group on the periodic table. Electronegativity values tend to increase as you move from left to right across a period on the periodic table. Within any group, electronegativity values tend to decrease as you move down the group. Thus, the most electronegative elements are in the upper-right corner of the table and the least electronegative elements are in the lower-left corner. The noble gases are considered to have electronegativity values of zero and do not follow the periodic trends due to their stable electron configurations.

Electronegativity and shielding The decrease in electronegativity as you move down a column occurs because the number of energy levels increases, and the valence electrons are located farther away from the positively charged nucleus. Thus, there is less attraction between the nucleus and its valence electrons. Also, electrons in the inner energy levels tend to block the attraction between the nucleus and the valence electrons. This is known as the **shielding effect.** The shielding effect increases as you move down a column because the number of inner electrons increases. For example, magnesium has the electron configuration $1s^2 2s^2 2p^6 3s^2$ and an electronegativity of 1.2; calcium has the configuration $1s^2 2s^2 2p^6 3s^2 3p^6 4s^2$ and an electronegativity of 1.0. Although both have two valence electrons, calcium has eight more inner electrons than magnesium. These electrons shield the outer electrons from the attraction of the nucleus, as shown in **Figure 9.3.**

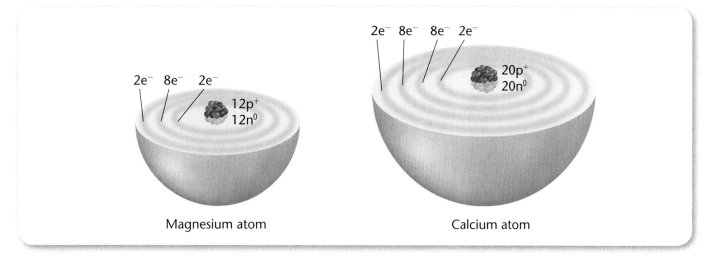

Figure 9.3 The electrons in the first and second energy levels shield the two valence electrons in the magnesium atom from attraction by 12 nuclear protons. Calcium's two valence electrons are shielded from the full attraction of the 20 nuclear protons by inner electrons in the first, second, and third energy levels. Because calcium has eight more (inner) electrons than magnesium, the valence electrons of calcium are held less tightly.

Electronegativity and nuclear charge As you look across a row from left to right on the periodic table, the number of protons in the nucleus increases. With an increase in nuclear charge, the attraction between the nucleus and the valence electrons increases, and therefore electronegativity tends to increase across a row. In period 4, as shown in **Figure 9.2,** potassium in group 1 has an electronegativity of 0.8, while bromine has a value of 2.8.

Because electronegativity varies in a periodic way, you can make predictions about electronegativity difference by looking at the distance between bonding atoms on the table. In general, the farther the bonding atoms are from each other on the periodic table, the greater their electronegativity difference. **Figure 9.4** illustrates this periodic trend in electronegativity.

Ionic Character

The greater the difference between the electronegativities of the bonding atoms, the more unequally the electrons are shared. The electronegativity difference between two bonding atoms is often represented by the symbol ΔEN, where EN is an abbreviation for electronegativity and Δ is the Greek letter *delta* meaning "difference." ΔEN is calculated by subtracting the smaller electronegativity from the larger, so ΔEN is always positive. For example, ΔEN for a bond between cesium and fluorine is $4.0 - 0.7 = 3.3$.

FOLDABLES
Incorporate information from this section into your Foldable.

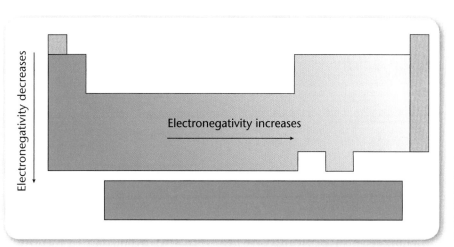

Figure 9.4 Electronegativity increases from left to right across a period and decreases from top to bottom down a group.
Infer *why noble gases don't have electronegativity values.*

Section 9.1 • Bonding of Atoms **303**

Highly unequal sharing Using the tug-of-war analogy, consider the situation in which a sumo wrestler is at one end of a rope and a small child is at the other end. There is no contest. The same is true in a chemical bond when the ΔEN between bonding atoms is 2.0 or greater.

When the electronegativity difference in a bond is 2.0 or greater, the sharing of electrons is so unequal that you can assume that the electron on the less electronegative atom is transferred to the more electronegative atom. This electron transfer results in the formation of one positive ion and one negative ion. The bond formed by the two oppositely charged ions is classified as an ionic bond. Many bonds are classified as ionic, but they have varying degrees of ionic character. The greater the difference in the electronegativities of the two atoms, the more ionic the bond. It can therefore be assumed that ionic bonds tend to form between atoms separated by a large distance on the periodic table.

Ionic bonding in sodium chloride Electrons are transferred when sodium chloride forms. The electronegativity of sodium is 0.9, and the electronegativity of chlorine is 3.0, one of the highest values on the table. The ΔEN for NaCl is:

$$\Delta EN = 3.0 - 0.9 = 2.1$$

The sharing of a pair of electrons between sodium and chlorine is so unequal that the electrons are both essentially bonded to the chlorine atom, creating a chloride ion (Cl^-). The sodium atom cannot compete with chlorine for a share in the electron pair and becomes a sodium ion (Na^+). Sodium ions and chloride ions combine to form sodium chloride (NaCl). Sodium chloride is best described as an ionic compound.

Figure 9.5 compares the formation of sodium chloride with the formation of lithium fluoride (LiF) and potassium bromide (KBr). For each of these salts, the ΔENs are equal to or greater than 2.0. Like sodium chloride, lithium fluoride and potassium bromide are classified as ionic compounds. Notice that the two atoms in each bond are separated from each other on the periodic table.

■ **Figure 9.5** The electronegativity differences in lithium fluoride (LiF), sodium chloride (NaCl), and potassium bromide (KBr) indicate that they are best represented as ionic compounds. Of these three compounds, lithium fluoride bonds have the highest degree of ionic character, and potassium bromide bonds have the lowest.

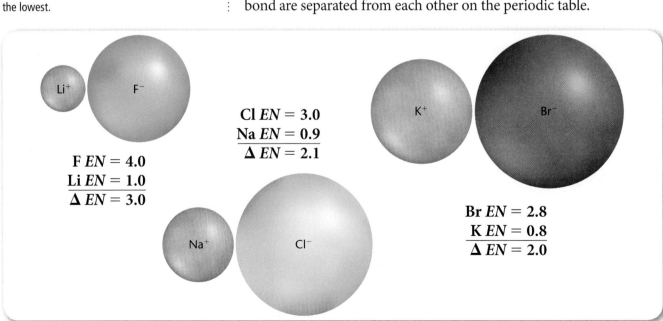

History Connection

Linus Pauling: An Advocate of Knowledge and Peace

Some proclaim him one of the 20 greatest scientists of all time. But Linus C. Pauling claimed merely to have been well-prepared and to have been in the right place at the right time. The time was the mid-1920s at the beginning of quantum physics.

Quantum theory of chemistry In 1925, Pauling was awarded a Ph.D. in chemistry from the California Institute of Technology, where he studied the crystal structure of materials. A year later, he was granted a Guggenheim Fellowship and traveled to Europe to study the quantum theory of the atom. Returning to CalTech, he merged his knowledge of material structures and quantum theory into the concept of the chemical bond, shown in **Figure 1**. Pauling's book *The Nature of the Chemical Bond* was influential in providing a framework for researchers to study and predict the structures and properties of inorganic, organic, and biochemical compounds. To acknowledge the importance of his work in understanding the chemical bond, Pauling was

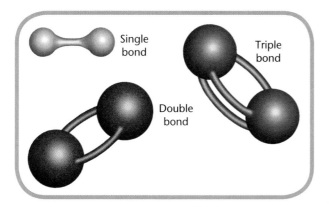

Figure 1 Pauling's chemical bonds

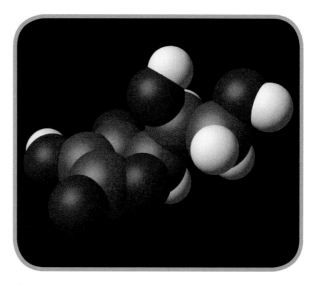

Figure 2 Vitamin C might provide many health benefits.

awarded the 1954 Nobel Prize for Chemistry.

Antinuclear armaments Pauling was an outspoken critic of the atmospheric testing of nuclear bombs. He was convinced that the radioactive fallout from such testing would be hazardous to humans for many generations. Petitioning scientists worldwide, Pauling pleaded for an international ban on testing nuclear weapons. The announcement that Pauling had been awarded the 1962 Nobel Prize for Peace was made on the day that the world's first partial nuclear test ban went into effect.

Vitamin C In the early 1970s, Pauling became an advocate of the health benefits of taking megadoses (large doses) of vitamin C, illustrated in **Figure 2**. His book *Vitamin C and the Common Cold* became a best-seller. Although his ideas are controversial, Pauling thought vitamin C might help eliminate minor ailments and be a possible cure for cancers.

Connection to Chemistry

1. **Apply** Why is understanding the nature of chemical bonding important?
2. **Acquire Information** Investigate the role Pauling played in the discovery of the structure of DNA.

FOLDABLES
Incorporate information from this section into your Foldable.

Covalent Character

You have seen that when there is a large electronegativity difference between two atoms (ΔEN = 2.0 or greater), the bond that forms between the two atoms is considered mostly ionic. What happens when there is no difference in electronegativity or if the electronegativity difference is small?

Equal sharing Imagine a different tug-of-war, this one between two teams exactly matched in terms of size, strength, and endurance. As you can see in **Figure 9.6,** the match will be even, with neither team gaining the upper hand. This situation is similar to when two of the same atoms form a bond—for example, when two fluorine atoms form the fluorine molecule (F_2). Here is the Lewis dot diagram for the fluorine molecule.

$$:\!\ddot{F}\!:\!\ddot{F}\!:$$

Because two atoms of the same element are forming the bond, the difference in electronegativities is zero. In the fluorine molecule, a pair of valence electrons are shared equally. This type of bond is a pure covalent bond. All other diatomic elements (Cl_2, Br_2, I_2, O_2, N_2, and H_2) have pure covalent bonds. In all these molecules, the electrons are shared equally.

■ **Figure 9.6** When the two teams are evenly matched in strength and size, neither team wins the match. Similarly, when atoms with the same electronegativity values form bonds, valence electrons are shared equally.

Sharing that's close to equal Electronegativities of bonding atoms can be close, but not necessarily the same. For example, carbon's electronegativity is 2.5 and hydrogen's is 2.1. A ΔEN of greater than zero always means unequal electron sharing. However, when the electronegativity difference is less than or equal to 0.5, the slightly unequal sharing of the electrons doesn't have a significant effect on the properties of the molecule. Thus, a bond in which the electronegativity difference is less than or equal to 0.5 is considered a covalent bond with electrons that are shared almost equally. All of the compounds formed between carbon and hydrogen are considered covalent.

Physical properties of covalent compounds Low boiling points and melting points are characteristic of pure covalent compounds. Most of the elemental diatomic molecules are gases at room temperature—Cl_2, F_2, O_2, N_2, and H_2. Carbon disulfide (CS_2), methane (CH_4), and nitrogen dioxide (NO_2), shown in **Figure 9.7,** are examples of covalent compounds in which the electron sharing is slightly unequal. These molecules are either gases or low-boiling point liquids at room temperature.

Polar Covalent Bonds

Now consider a tug-of-war in which a 300-pound sumo wrestler is on one end of the rope and a 400-pound sumo wrestler is on the other end. The unequal pull on the ends of the rope brings the middle of the rope closer to the 400-pound competitor. This is an analogy to a bond in which electrons are shared unequally. This type of bond falls between the two extremes of electron sharing: equal sharing (a covalent bond) and completely unequal sharing (an ionic bond). Bonds in which the pair of electrons is shared unequally have electronegativity differences between 0.5 and 2.0.

■ **Figure 9.7** Carbon disulfide is an important solvent for wax and grease. Methane is the principal component of natural gas. Nitrogen dioxide is used for making nitric acid and is also an atmospheric pollutant. All three compounds contain covalent bonds in which the sharing of electrons is more or less equal.

Identify *Which of these three covalent compounds has electrons that are shared most equally? Which has electrons that are shared least equally?*

The C—S bonds in carbon disulfide are pure covalent bonds. The $\Delta EN = 0$, even though the atoms are different

The $\Delta EN = 0.4$, in the bonds of methane is not sufficient to significantly affect the properties of the compound.

The N—O bonds in nitrogen dioxide have a greater degree of unequal electron sharing than the C—H bonds, but NO_2 is still considered a covalent compound.

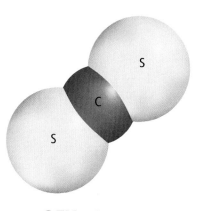

C EN = 2.5
S EN = 2.5
Δ EN = 0.0

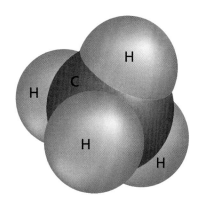

C EN = 2.5
H EN = 2.1
Δ EN = 0.4

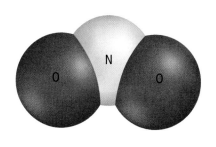

O EN = 3.5
N EN = 3.0
Δ EN = 0.5

Unequal sharing in covalent bonds When the electronegativity difference between bonding atoms is between 0.5 and 2.0, the electron sharing is not so unequal that a complete transfer of electrons takes place. Instead, there is a partial transfer of the shared electrons to the more electronegative atom. The less electronegative atom still retains some attraction for the shared electrons. The bond that forms when electrons are shared unequally is called a **polar covalent bond.** A polar covalent bond has a significant degree of ionic character.

Polar covalent bonds are called *polar* because the unequal electron sharing creates two poles across the bond. Just as a car battery or a flashlight battery has separate positive and negative poles, polar covalent bonds have poles, as shown in **Figure 9.8.** The negative pole is centered on the more electronegative atom in the bond. This atom has a share in an extra electron. The positive pole is centered on the less electronegative atom. This atom has lost a share in one of its electrons. Because there was not a complete transfer of an electron, the charges on the poles are not 1+ and 1−, but δ^+ and δ^-. These symbols, *delta* plus and *delta* minus, represent a partial positive charge and a partial negative charge. This separation of charge, resulting in positively and negatively charged ends of the bond, gives the polar covalent bond a degree of ionic character.

Physical properties of polar covalent compounds
Compounds with polar covalent bonds have different properties than compounds with pure covalent bonds. Purely covalent compounds tend to have low melting points and boiling points. Carbon disulfide, CS_2, as shown in **Figure 9.7,** is a triatomic molecule, with a ΔEN equal to zero. Carbon disulfide boils at 46°C. Water is also a triatomic molecule, but the bonding in water is polar covalent. Even though water is a much lighter molecule than carbon disulfide, its boiling point is 100°C.

Concepts In Motion

Interactive Figure To see an animation of bond types, visit glencoe.com.

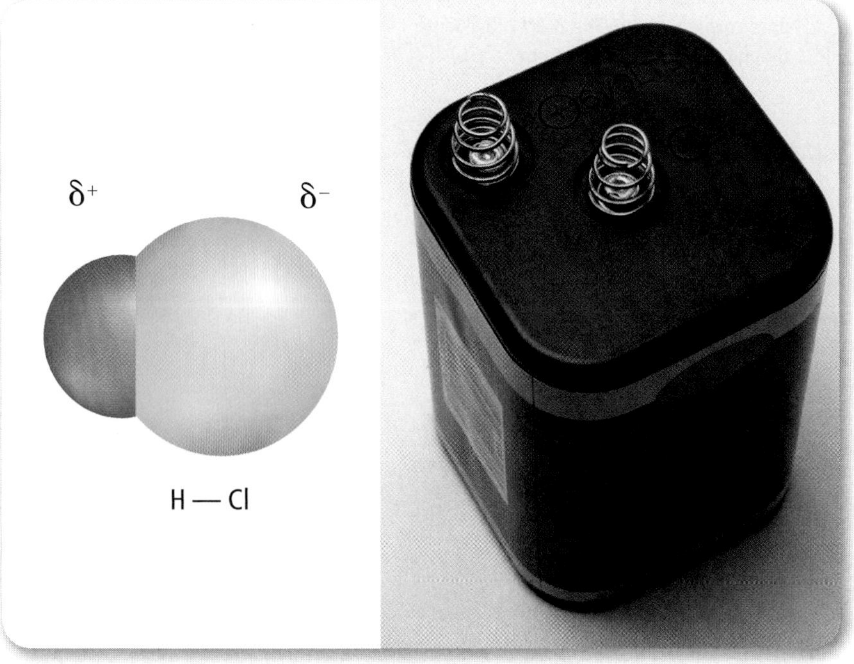

■ **Figure 9.8** When the sharing of electrons in a bond isn't equal, the bond is polar, as in the H—Cl bond. Just like this battery (right), the bond has two poles, one positive and one negative. The symbols δ^+ and δ^- (delta plus and delta minus) are used to show the distribution of partial charges in a polar covalent bond (left).

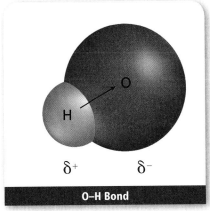 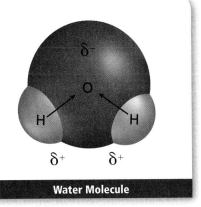

O–H Bond | **Water Molecule**

■ **Figure 9.9** Because oxygen is more electronegative than hydrogen, the electrons in an O—H bond spend more time near the oxygen atom than near the hydrogen atom. This distribution leads to a partial negative charge on oxygen and a partial positive charge on hydrogen. In a water molecule, there are two O—H bonds. In both of these bonds, electrons are attracted more to the oxygen atom.

The ΔEN for the O—H bond is 1.4, so water molecules have polar bonds. When hydrogen and oxygen atoms bond by sharing electrons, the shared pair of electrons is attracted toward the more electronegative oxygen. This unequal sharing causes an imbalance in the distribution of charge about the two atoms, as shown in **Figure 9.9**.

The effect of polar covalent bonds on boiling point is illustrated by comparing the boiling point of water with the hydrides of group 16 elements—sulfur, selenium, and tellurium. A hydride is a compound formed between any element and hydrogen. Water is the first hydride of group 16. ΔEN for the O—H bond is 1.4. But ΔEN for both the S—H and the Se—H bonds is only 0.4, and ΔEN for the Te—H bond is 0.3. These electronegativity differences tell you that H_2O contains polar covalent bonds, but the bonds in H_2S, H_2Se, and H_2Te are essentially covalent. **Figure 9.10** shows that even though H_2O is the lightest of the four hydrides, its boiling point is significantly higher than the hydrides of the other members of group 16.

TRY AT HOME 🏠 LAB

See page 872 for **Breaking Covalent Bonds.**

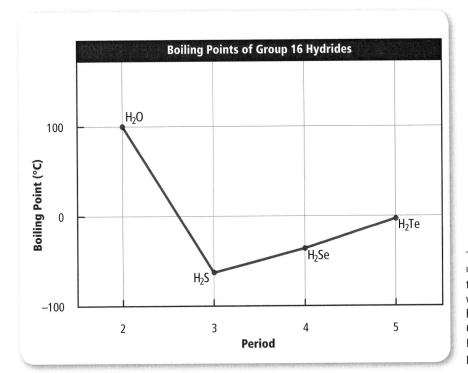

■ **Figure 9.10** On the basis of the mass of the molecules, you might predict that H_2O would have the lowest boiling point of the four hydrides. Another factor, the polarity of the O—H bond, makes that prediction incorrect. Later you'll learn how polar bonds often result in a polar molecule, as in the case of water.

Section 9.1 • Bonding of Atoms

PRACTICE Problems

Solutions to Problems Page 859

1. Calculate ΔEN for the pairs of atoms in the following bonds.
   a) Ca–S
   b) Ba–O
   c) C–Br
   d) Ca–F
   e) H–Br

2. Use ΔEN to classify the bonds in Question 1 as covalent, polar covalent, or ionic.

MiniLab 9.1

Coffee Filter Chromatography

Can you determine the composition of a substance using chromatography? Paper chromatography is a method of separating substances based upon their different attractions to the paper. The paper is called the stationary phase of the system, and the solvent is called the moving phase. In this investigation, the moving phase is water, which is a polar molecule.

Procedure

1. Read and complete the lab safety form.
2. Obtain a **set of markers** and **two circular coffee filters**.
3. Cover the bottom of a **plastic cup** with a small amount of **water**.
4. Lay the two coffee filters together on a dry surface. With the filters still together, fold the circles into eighths as shown.
5. Unfold the filters, but do not separate them. Find the center spot where all folds converge.
6. About 5 cm from the center, along a crease, make a dark mark with one of the pens. The mark should appear on both filters.
7. Continue for all eight creases, using a different color marker on each crease.
8. Separate the filters. Keep one filter unfolded as a control. Refold the other filter.
9. With a small amount of water in the bottom of the cup, gently place the folded coffee filter, tip down, into the water.
10. When water reaches the top of the paper, gently remove the paper.

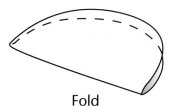
Fold

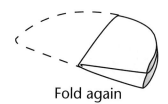
Fold again

Fold again

11. Gently open the paper. Compare each mark on the control filter with the corresponding mark on the chromatographed filter.

Analysis

1. **Compare** the marks on the control paper with the corresponding marks on the chromatographed filter. How are they different?
2. **Observe** Which colors on the chromatographed filter are different from those on the control filter?
3. **Infer** Polar substances tend to be attracted to other polar substances such as found in paper. Nonpolar substances move faster through the paper. Which of the inks contains the most polar substances?

Bonding in Metals

Bonding in metals does not result in the formation of compounds, but it does result in an interaction that holds metal atoms together and accounts for some of the common properties of metals and alloys. What are some of these properties?

Properties that reflect metallic bonding Metals and alloys are malleable and ductile, and they conduct electricity. When a metal can be pounded or rolled into thin sheets, it is said to be **malleable.** Gold is an example of a malleable metal. A chunk of gold can be flattened and shaped by hammering until it is a thin sheet. **Ductile** metals can be drawn into wires. For example, copper can be pulled into thin strands of wire and used in electric circuits, as illustrated in **Figure 9.11.** Electrical **conductivity** is a measure of how easily electrons can flow through a material to produce an electric current. Metals such as silver are excellent conductors because there is low resistance to the movement of electrons in the metal. These properties—malleability, ductility, and electrical conductivity—are the result of the way that metal atoms bond with each other.

Sea of valence electrons The valence electrons of metal atoms are loosely held by the positively charged nucleus. Sometimes, metal atoms form ionic bonds with nonmetals by losing one or more of their valence electrons and forming positive ions. However, in metallic bonding, metal atoms don't lose their valence electrons. As shown in **Figure 9.11,** metal atoms release their valence electrons into a sea of electrons shared by all of the metal atoms. The bond that results from this shared pool of valence electrons is called a **metallic bond.** This model of metallic bonding is referred to as the electron sea model.

> **FACT of the Matter**
>
> Among all of the metals, silver is the best conductor of electricity. Copper is the second best. Because silver is rare and more costly than copper, copper is the metal of choice in electric circuits.

■ **Figure 9.11** Copper exhibits the typical properties of metals, such as ductility and conductivity, because of the way in which metal atoms are held together.
Explain *Why are electrons in metals known as delocalized electrons?*

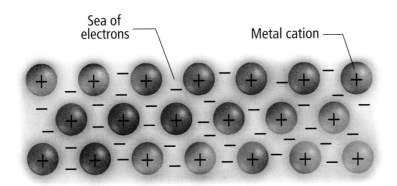

The valence electrons in metals (shown as a blue cloud of minus signs) are evenly distributed among the metallic cations (shown in red). Attractions between positive cations and the negative sea hold the metal atoms together in a lattice.

Copper is ductile and a good conductor of electricity. It is the most common metal used in electric circuits.

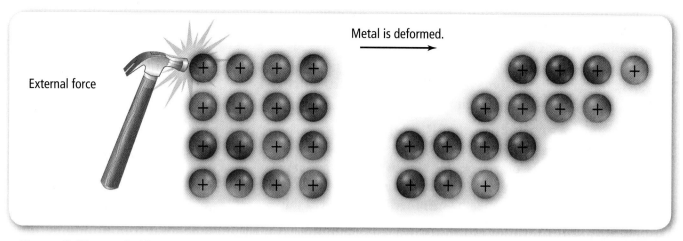

■ **Figure 9.12** An applied force causes metal ions to move through delocalized electrons, making metals malleable and ductile.

The bonding interaction at the submicroscopic level explains what you observe at the macroscopic level. Although the metal atoms are bonded together in a large network, they are not bonded to any single atom. Because atoms are not rigidly bonded to one another, metals often are malleable and ductile. As the metal in **Figure 9.12** is struck by the hammer, the atoms slide through the electron sea to new positions while continuing to maintain their connections to each other. The same ability to reorganize explains why metals can be pulled into long, thin wires.

The conductivity of metals can also be explained by the sea of electrons model of metallic bonding. Because the valence electrons of all the metal atoms are not attached to any one metal atom, they can move through the metal when an external force, such as that provided by a battery, is applied.

Section 9.1 Assessment

Section Summary

- Bond character varies from ionic to covalent. There is no clear-cut division between the types of bonds.
- Electronegativity—a measure of the attraction that an atom has for shared electrons—can be estimated from the periodic table.
- Electronegativity difference (ΔEN) is a measure of the degree of ionic character in a bond.
- A ΔEN = 2.0 or greater occurs when elements from ionic bonds; ΔEN = 0.5 – 2.0 reflects polar covalent bonds; and ΔEN < 0.5 reflects covalent bonds.
- Metal atoms bond by sharing in a sea of valence electrons.

3. **MAIN Idea Classify** Use ΔEN to classify the bonds in the following compounds as covalent, polar covalent, or ionic.
   a) H—S bond in H_2S
   b) S—O bond in SO_2
   c) Mg—Br bond in $MgBr_2$
   d) N—O bond in NO_2
   e) C—Cl bond in CCl_4

4. **Organize** Using only a periodic table, rank these atoms from the least to most electronegative: Na, Br, I, F, Hg.

5. **Classify** Rank these bonds from the least to the most polar.
   a) C—F
   b) O—F
   c) Al—Br
   d) Cl—F
   e) O—H

6. **Predict** The boiling points of the group 16 hydrides are shown in **Figure 9.10**. Based on the information in this figure, predict the order, from highest to lowest, of the boiling points of the first four group 17 hydrides (HF, HCl, HBr, and HI).

7. **Explain** What aspect of metallic bonding is responsible for the malleability, ductility, and conductivity of metals?

8. **Infer** Do you expect LiF or LiCl to have a higher melting point? Explain.

Section 9.2

Objectives

- **Diagram** Lewis dot diagrams for molecules.
- **Formulate** three-dimensional geometry of molecules from Lewis dot diagrams.
- **Predict** molecular polarity from three-dimensional geometry and bond polarity.

Review Vocabulary

electronegativity: a measure of the ability of an atom in a bond to attract electrons

New Vocabulary

double bond
triple bond
polar molecule

Molecular Shape and Polarity

MAIN Idea The shape of a molecule and the polarity of its bonds determine whether the molecule as a whole is polar.

Real-World Reading Link You've probably seen small-scale models of structures such as buildings or bridges. Architects and engineers find it helpful to work with small models before producing a detailed plan for a full-scale structure. Models can also be enlarged versions of objects that are so small that they can't be seen or handled.

The Shapes of Molecules

Models can help you visualize the three-dimensional structures of molecules. Consider the simplest molecule that exists—hydrogen, (H_2). Two hydrogen atoms share a pair of electrons in a nonpolar covalent bond, as shown in this Lewis dot diagram.

$$H:H$$

A model of this molecule might be made by connecting two gumdrops of the same color by a toothpick. The gumdrops represent the hydrogen atoms, and the toothpick represents the shared pair of electrons that makes up the covalent bond.

What does a gumdrop model tell about the shape of a hydrogen molecule? As **Figure 9.13** shows, there's only one way to put this model together. When the two gumdrops are connected with a toothpick, they lie in a straight line. Therefore, a hydrogen molecule has a linear structure. You could model other diatomic molecules, such as oxygen (O_2), nitrogen (N_2), chlorine (Cl_2), iodine (I_2), fluorine (F_2), and even hydrogen chloride (HCl). The model always predicts the same linear geometry.

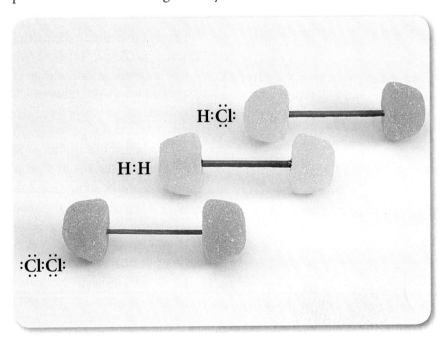

■ **Figure 9.13** Just as two gumdrops can be connected in only one way, all diatomic molecules are linear, whether they're composed of the same two atoms, as in H_2 and Cl_2, or different atoms, as in HCl.

■ **Figure 9.14** This gumdrop model of a water molecule shows that the three atoms form a bent structure.

Water A gumdrop model of a water molecule (H₂O) is shown in **Figure 9.14**. How would you build this model of a water molecule from scratch? First, you would draw the Lewis dot diagram. Remember that the dot diagram models the arrangement of valence electrons for a molecule in two dimensions. The Lewis dot diagram for a water molecule shows that each hydrogen shares a pair of electrons with the oxygen.

The eight valence electrons are distributed in such a way that the oxygen atom has an octet of electrons and attains the stable electron configuration of the noble gas neon. Each hydrogen has two valence electrons and the stable configuration of helium. Two pairs of valence electrons are involved in the bonding. These electrons are called bonding pairs. The other two pairs of valence electrons are not involved in the bonding. These are nonbonding pairs, also called lone pairs.

From Lewis dot diagram to model To make a model of the water molecule, you need two colors of gumdrops, such as red for oxygen and yellow for hydrogen. You also need two toothpicks to represent the two covalent bonds, and two additional toothpicks to represent the lone pairs of electrons. Even though the lone pairs of electrons are not part of any bonds, they play a major role in determining the shape of a molecule. They are present, and they take up space.

How should the hydrogen atoms be connected to the oxygen atom? Clearly, some rules are needed. Look at the Lewis dot diagram for water pictured above. Four pairs of electrons surround the oxygen. These electrons are all negatively charged, and because they all have the same charge, they repel each other. Therefore, they will form the three-dimensional arrangement around the oxygen atom that allows them to be as far from each other as possible. This arrangement is a geometric shape called a tetrahedron. In a tetrahedral arrangement, the repulsions between the electron pairs are minimized.

To create a model with tetrahedral geometry, place the four toothpicks in the red gumdrop in a three-dimensional arrangement with the largest possible angle between each adjacent toothpick. Two yellow gumdrops should then be placed at the ends of any two of the toothpicks. These represent the two O—H bonds. The other toothpicks represent the lone pairs. As you can see, the gumdrop model in **Figure 9.14** visualizes both the shape of the bonds and the tetrahedral arrangement of four electron pairs.

Can the gumdrop model of water be translated into other types of models? **Figure 9.15** shows two additional models of a water molecule. The first model uses balloons to show the areas occupied by the four different electron pairs. The second model is called a space-filling model. Space-filling models very clearly show each atom in a molecule and are one of the more common types of molecular models in use.

From model to water molecule The models in **Figure 9.14** and **Figure 9.15** suggest that the three atoms in the water molecule are arranged in a bent structure. The angles in a perfect tetrahedron each measure 109.5°. The angle between the two bonds in the water molecule is roughly 105°—a little less than the angle predicted by the gumdrop model. The difference between the predicted and experimental bond angle comes about because the lone pairs of electrons repel each other more than the shared pairs. In effect, the nonbonding electrons require more room, so they distort the tetrahedral arrangement by squeezing the bonding pairs closer together and decreasing the bond angle from 109.5° to 105°.

Four balloons, inflated equally and held together at a central point, arrange themselves in a tetrahedral shape. This is the most space-efficient arrangement for four things about a center point. In this model, the balloons represent the four electron pairs of the water molecule. The red balloons represent bonding pairs, while the white balloons represent lone pairs.

■ **Figure 9.15** Four balloons can be used to illustrate the bonding and lone pairs of electrons in a water molecule. Space-filling models show the electron clouds of each atom as spheres.

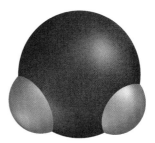

In a space-filling model, the electron cloud of each type of atom is represented with a different color. In this case, red indicates the oxygen atom and blue the hydrogen atoms. The clouds overlap when two atoms form a bond.

In the Field

Meet Dr. William Skawinski
Chemist

An ordinary copy of Mathematics for Physicists is about two inches thick. However, the Braille version on Dr. Skawinski's bookshelf requires more than almost a meter of space. The chemist, who began losing his sight in childhood, consults this reference as he creates three-dimensional models of molecules. In the following interview, Dr. Skawinski talks about his innovative work and his love of chemistry.

On the Job

Q Dr. Skawinski, can you tell us how you go about building molecular models?

A I start by putting information about molecular structures into a CAD (computer-aided design) program. The program uses the information to calculate a structure for the molecule. Then stereolithography is used to make a physical model from the calculations. This is basically how it works. A table covered by a polymer film sits at the top of a tank containing liquid plastic. A laser traces a cross section of the bottom of the molecule in the film, and a slice of the model hardens. Then the table moves down a little and the laser traces another cross section, and so on to the top of the molecular design. After 8 to 12 hours, I get one solid piece of plastic that replicates a molecule of a particular compound.

Q By what factor are these models bigger than the real thing?

A A carbon atom, for example, has a radius of about 0.2 nanometers. The model is about an inch in diameter—many billions of times larger.

Q Why are these three-dimensional models useful?

A Dealing with a physical model of a mathematical distribution can give anyone—blind or sighted—a better insight into what it represents.

Q Do you have a favorite of the molecular models you've built?

A Beta-cyclodextrin has doughnut-shaped molecules that are interesting. People have remarked that many of the molecular models look like pieces of artwork.

Early Influences

Q Do you remember specific incidents from your childhood that influenced your interest in science?

A I can remember being two years old and watching the flames of a gas water heater. The bright blue lights on a perfectly black background fascinated me. I also remember at about the age of five hammering red, yellow, and gray rocks in the backyard, breaking them up into powder, and mixing them with water to make colored suspensions.

Q What were things like for you in high school?

A Even though I was having serious problems with my vision by that time, I was involved in an amateur rocket group. These rockets weren't the little four-inch rockets you buy through the mail and send up a couple of hundred feet. One rocket for which I mixed the chemical propellant reached an altitude of 42,000 feet. I loved the idea of propelling a vehicle into space—and I still do. In fact, if anyone offered me a ticket to another planet, I'd only pause long enough to pack a toothbrush!

Q Did your high school chemistry class interest you?

A My chemistry teacher made chemistry a part of the real world. Once he brought in a round, green glob about the size of a walnut. It was a flawed emerald. He explained that just a slight difference in chemical makeup differentiated that worthless stone from a priceless one.

Personal Insights

Q Does the wonderment about science that you experienced as a child remain with you today?

A Of course. I think you really have to retain that trait to enter the field of science. Scientists are just a bunch of big kids who are fascinated with the world.

Q How have you dealt with the challenge of your blindness?

A I was able to see enough through college to learn a lot of critical information. But near the end of college, because my field of vision had narrowed so much, I could see just part of a word on a page at a time, and that was only with a strong magnifying glass. I managed to earn my master's degree and then my Ph.D. mostly through determination. My approach has always been to acknowledge obstacles and then concentrate on finding ways around them.

CAREER CONNECTION

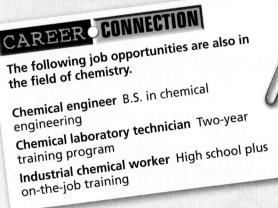

The following job opportunities are also in the field of chemistry.

Chemical engineer B.S. in chemical engineering

Chemical laboratory technician Two-year training program

Industrial chemical worker High school plus on-the-job training

Chapter 9 • In the Field   **317**

Everyday Chemistry

Jiggling Molecules

How often have you warmed a snack by microwaves—that is, in a microwave oven? Maybe you were too hungry to notice that the food was a lot hotter than the dish or container. A clue to the cause of this sometimes-overlooked observation is the steam you blew away from the food as you waited for it to cool.

Microwave heating Microwave radiation is a form of electromagnetic energy. The microwave radiation produced by a microwave oven has a wavelength of 11.8 cm and a frequency of 2.45 billion hertz. (A shorter way of writing a frequency of 2.45 billion hertz is 2.45 gigahertz, or 2.45 GHz.) The radiation travels through space and materials in the form of a moving electromagnetic field; that is, a field having an electric-field component and a magnetic-field component. The frequency of the wave is the number of oscillations or waves each second. The frequency also measures the energy of the waves.

Microwaves have little effect on most molecules. However, the microwave's oscillating electromagnetic field interacts with positively and negatively charged polar molecules. As a result, the charged molecules oscillate (move back and forth). The molecules move by stretching, translation (linear movement), and rotation, as shown in **Figure 1**. The increased motion means that the molecules have increased kinetic energy and generate heat. (Kinetic energy is energy of motion and is directly related to temperature.) As microwaves are absorbed by substances containing polar water molecules, the temperature rises rapidly. Heat is transferred by conduction from the water molecules to other parts of the substance.

Microwave decomposition If you have ever shaken a garden rake to dislodge a leaf, you know that violent motion can be used to separate things. Researchers are applying the same concept to decompose molecules of some toxic substances by using microwaves. Molecules of compounds such as dihydrogen sulfide, sulfur dioxide, and nitrogen dioxide are polar.

These molecules contribute to air pollution and can be decomposed into their nontoxic elements by the oscillations caused by microwave radiation. The oscillations become so violent that the attractions between the atoms in the molecule are no longer great enough to hold them together, and the molecule decomposes or falls apart. Here, the energy of the microwaves has overcome the energy of the chemical bond. Similar research is being done to initiate reactions in which toxic polar organic compounds are reacted with other substances to form nontoxic products. Applications of the research may produce cost-effective methods of using microwaves to control air pollution and clean up hazardous wastes.

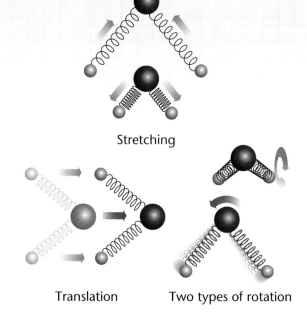

Figure 1 Types of molecular movement

Explore Further

1. **Apply** Why can microwave radiation be used to detoxify trichloromethane ($CHCl_3$) but not tetrachloromethane (CCl_4)?

2. **Infer** Why do the cooking instructions for most foods packaged for microwave preparation suggest that after microwaving, the food should stand for several minutes before being served?

Carbon dioxide Are all triatomic molecules bent like the water molecule? To find out, you can model carbon dioxide, CO_2, as you did water. Begin by drawing the Lewis dot diagrams for the two atoms. Carbon has four valence electrons, and each oxygen atom has six.

$$\cdot \overset{\cdot}{\underset{\cdot}{C}} \cdot \quad \cdot \overset{\cdot\cdot}{\underset{\cdot\cdot}{O}} :$$

To obtain a stable octet of electrons, the carbon atom needs four more electrons, and each oxygen atom needs two more electrons. Therefore, each oxygen atom must share two pairs of electrons with the carbon atom. A bond formed by sharing two pairs of electrons between two atoms is called a **double bond**, as illustrated by the Lewis dot diagram for carbon dioxide. If you count all the shared and unshared electrons around each of the three atoms, you'll see that each atom has an octet.

$$:\!\overset{\cdot\cdot}{O}::C::\overset{\cdot\cdot}{O}\!:$$

How can you determine the three-dimensional geometry of a carbon dioxide molecule? First, look at the arrangement of electrons around the central atom. Consider each double bond as one cloud of shared electrons. Two clouds surround the carbon. What geometry puts the two electron clouds as far apart as possible? Linear geometry, as illustrated in **Figure 9.16,** separates the clouds as much as possible. The model suggests that the three atoms in carbon dioxide are arranged in a straight line. This structure of CO_2 was validated by experimentation.

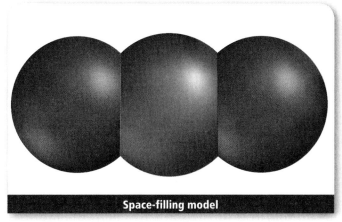
Gumdrop model

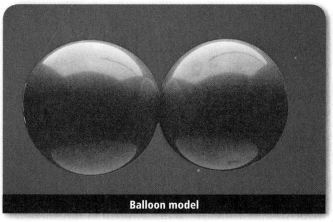
Balloon model

■ **Figure 9.16** A gumdrop model of carbon dioxide (CO_2) predicts a linear structure with the C—O bonds pointed in opposite directions. The space-filling model of CO_2 is a good representation of the molecule. The balloon model represents the clouds of bonding electrons on either side of the carbon atom.

Explain *What causes the molecule of carbon dioxide to be linear?*

Section 9.2 • Molecular Shape and Polarity

Balloon model

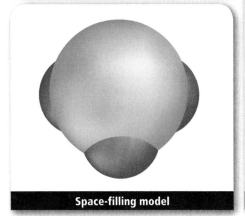
Space-filling model

Gumdrop model

■ **Figure 9.17** Ammonia has one lone electron pair and three bonding electron pairs. When these are arranged around the central atom in the gumdrop model, the arrangement is tetrahedral, but the geometry of the four atoms forms a triangular pyramid. The lone pair causes the H—N—H angles in ammonia to be 107°, slightly less than the predicted tetrahedral angle of 109.5°. The triangular pyramid shape is represented by clouds of bonding and lone electron pairs in the balloon and space-filling models above.

Ammonia Ammonia, NH_3, has bonds from a central nitrogen to three hydrogen atoms. Nitrogen has five valence electrons, and each hydrogen has one valence electron. Each hydrogen shares a pair of electrons with the nitrogen. Nitrogen's remaining two electrons form a nonbonding pair. This arrangement gives nitrogen a complete octet of electrons.

$$H:\ddot{N}:H$$
$$H$$

Count the number of electron pairs around the central nitrogen atom. There are four pairs—three bonding pairs and one lone pair. The four pairs avoid each other by organizing themselves in a tetrahedral arrangement, just as they did in the water molecule. But this time, three of the positions are N—H bonds, and the fourth position is the lone pair. The atoms in ammonia form a triangular pyramid structure. The three hydrogen atoms form the base of the pyramid with the nitrogen at the peak, as shown in **Figure 9.17**. Based on this tetrahedral arrangement, you would predict that the H—N—H bond angle is 109.5°. The experimentally determined structure is a triangular pyramid with a bond angle of 107°.

Methane The geometry of a methane molecule, CH_4, is shown in **Figure 9.18**. Methane is the simplest hydrocarbon compound. Hydrocarbons are organic compounds composed of only hydrogen and carbon. The Lewis dot diagram for methane consists of a central carbon atom with four C—H single bonds.

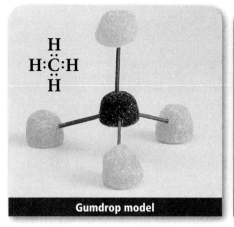

■ **Figure 9.18** Each of the four electron pairs around the carbon atom is a C—H bonding pair, so the structure is symmetrical with all bond angles of 109.5°. The balloon and space-filling models of methane display the symmetry of the molecule.

Concepts In Motion

Interactive Figure To see an animation of molecular shapes, visit glencoe.com.

Four pairs of electrons are positioned in a tetrahedral arrangement around the carbon atom, just as they are in the water and ammonia molecules. In methane, all four pairs of electrons are shared between the carbon and the four hydrogen atoms. Because there are no lone pairs requiring extra space, the structure of methane is a perfect tetrahedron with bond angles of 109.5°.

Ethane Ethane, C_2H_6, is the second member of the hydrocarbon series known as the alkanes. Methane is the first and simplest. Alkanes are hydrocarbons that contain carbon and hydrogen atoms with single bonds between the atoms. An ethane molecule has two carbon atoms that form a bond to each other and to three hydrogen atoms a piece. The Lewis dot diagram for ethane is shown in **Figure 9.19**. Each carbon atom has four single bonds. Just as in methane, discussed above, a tetrahedral arrangement of bonding electron pairs around each carbon atom exists.

■ **Figure 9.19** The geometric arrangement of atoms around each carbon atom in ethane is tetrahedral. All bond angles are 109.5° as predicted from the geometry. The gumdrop and space-filling models of ethane show two tetrahedral arrangements around the two carbon atoms.
Compare and contrast *the space-filling model of methane in* **Figure 9.18** *with that of ethane.*

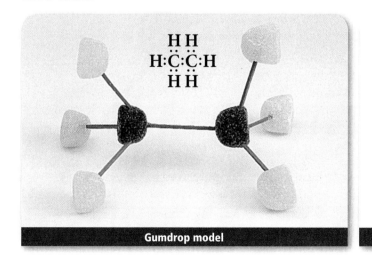
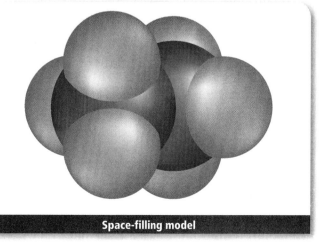

Section 9.2 • Molecular Shape and Polarity

Ethene Ethene (C_2H_4) is the first member of another hydrocarbon series called the alkenes. Ethene is related to ethane, but it has four hydrogens rather than six. The common name for ethene is ethylene. In order for the carbon atoms in ethene to acquire an octet of electrons, a double bond must exist between the carbons.

Each carbon atom has three bonds: two C—H bonds and one C=C double bond. The most space-efficient arrangement of three electron clouds about a central atom is a flat, triangular arrangement. Each carbon atom is at the center of three electron clouds and ethene is a flat molecule, as shown in **Figure 9.20**. The H—C—H and H—C—C bond angles are all approximately 120°.

You might have noticed as you built your model of ethane that the gumdrops were free to rotate about the single C—C bond. However, ethene's gumdrops are fixed about the double C=C bond—you can't rotate them at all. In addition to being true of the models, this is true of the molecules themselves. The geometry of ethene is rigid because of the double bond between the carbon atoms.

Ethene, the simplest of the alkenes, is a naturally occuring compound that is important in the development of many plants. For example, it plays a role in causing leaves to fall from some trees before the onset of winter. Ethene also causes fruit to ripen. As shown in **Figure 9.20,** you can take advantage of this property of ethene to ripen fruit. Ethene is also widely used within the petrochemical industry and is a primary ingredient in many plastics.

■ **Figure 9.20** Like the hydrocarbon ethane, the space-filling model of ethene has two geometric centers. The organization of hydrogen atoms around each carbon atom in ethene is a flat, triangular arrangement. All six atoms lie in the same plane, as illustrated in the gumdrop model below. You can take advantage of ethene's fruit-ripening properties at home. Placing unripe fruit in a bag will trap the ethene evolved by the fruit and hasten ripening.

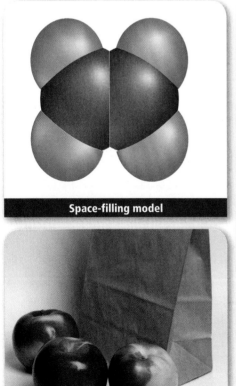

Space-filling model

Fruit ripening

Gumdrop model

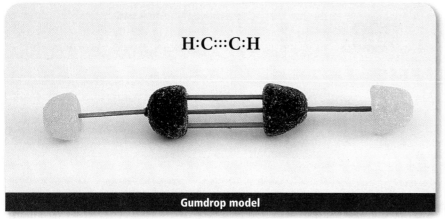
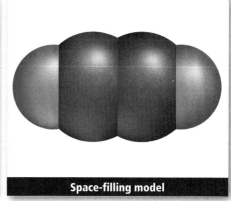

Ethyne Ethyne, shown in **Figure 9.21,** is the first member of a hydrocarbon series called the alkynes. Its formula is C_2H_2. Notice that ethyne has two carbons like ethane and ethene, but only two hydrogens. Ethyne is more commonly known as acetylene. It is the fuel used in torches for cutting steel.

You learned that in ethene, the two carbon atoms share two pairs of electrons in a double bond. In ethyne, the carbon atoms share three pairs of electrons to obtain a stable octet. A bond formed by sharing three pairs of electrons between two atoms is called a **triple bond**.

Each carbon has two bonds, a C—H single bond and a C≡C triple bond, so two electron clouds surround each carbon atom. Linear geometry places two electron clouds as far as possible from each other. In ethyne, the four atoms are arranged in a straight line, so ethyne is a linear molecule.

■ **Figure 9.21** The gumdrop model shows that the geometry of ethyne is linear. The triple bond makes the molecule rigid. The space-filling model of ethyne shows the arrangement of the electron clouds. Ethyne, when combined with excess oxygen, burns with a hot flame. It is used in welding torches.

MiniLab 9.2

Model Molecules

How can you model molecules? The ability to build and interpret models is an important skill in chemistry. This investigation will give you practice working with models.

Procedure

1. Read and complete the lab safety form.
2. Obtain a **model kit** from your teacher.
3. Draw the Lewis dot diagram for one of these molecules: H_2, HCl, H_2O, CO_2, NH_3, CH_4, C_2H_6, C_2H_4, C_2H_2.
4. Build a model of the molecule.
5. Draw a sketch of the geometric shape you predict for the molecule.
6. Repeat steps 3–5 for each of the molecules in the list.
7. Organize your results in a table that shows the formula, Lewis dot diagram, and a sketch of the predicted geometry for each molecule.

Analysis

1. **Explain** How did the Lewis dot diagram of each molecule help you predict its geometry?
2. **Summarize** Pick one of your assigned models. How many lone pairs of electrons does it have? How many bonding pairs of electrons are there?

CHEMISTRY AND TECHNOLOGY

Chromatography

Many of the interesting substances in the world are mixtures. Blood, dirt, air, pizza—to name a few—are mixtures. These substances are complex because they contain multiple components. Scientists have devised ways to analyze complex matter. One way is through chromatography.

Going with the Flow
The soggy poster in **Figure 1** is a good model of chromatography. Chromatography is a way of separating a mixture using differences in the abilities of the components to move through a material. All chromatography involves two phases—a stationary phase and a mobile phase. The movement of the mobile phase through the stationary phase allows separation to take place. Because the components of a mixture move at different rates, they eventually separate.

■ **Figure 1** Ink on the poster board becomes part of the mobile phase when it dissolves in water.

Paper Chromatography
One type of chromatography used to separate colored mixtures is paper chromatography, shown in **Figure 2**. A porous paper is used as the stationary phase. Water or some other solvent is used as the mobile phase. A spot or a line of the mixture to be separated is placed on the paper. The solvent moves upward along the paper because of capillary action. As it reaches the spot, the mixture dissolves in the solvent.

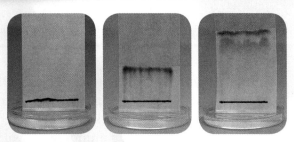

■ **Figure 2** Paper chromatography uses capillary action to separate mixtures.

Now, the components of the mixture begin to migrate upward on the paper along with the solvent. Those components that have little attraction to the paper move almost as quickly as the solvent. Components of the mixture that have greater attraction for the paper than for the solvent migrate at a slower rate. The differences in the migration rates result in differences in the distances the separated components travel.

Thin-Layer Chromatography
In thin-layer chromatography, shown in **Figure 3**, the stationary phase is a suspension made up of an adsorbent material such as silica gel or cellulose in a solvent. The suspension is applied as a thin coating on a glass or metal plate, which is then dried. The mixture being separated is applied to the bottom of the plate, and the plate is placed vertically in a solvent that acts as the mobile phase.

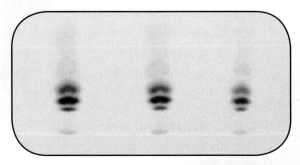

■ **Figure 3** Thin-layer chromatography uses adsorbent material to separate mixtures.

Gel Chromatography

Gel chromatography, shown in **Figure 4,** is often used to isolate the molecules of living, biological systems. The liquid phase carries the molecules through a stationary phase of a porous gel. The pore sizes can be manufactured precisely to affect the migration of molecules through the gel. Large molecules are blocked from moving through the gel. Because midsize molecules spend more time within the pores than smaller molecules, they take longer to migrate through the pores. As a result, gel chromatography can be used to isolate midsize molecules found in living materials.

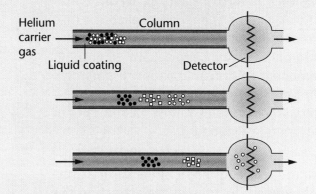

■ **Figure 5** Gas chromatography separates gas mixtures based on attraction of the components to the lining of the tube.

Portable gas chromatographs, used in conjunction with mass spectrometers, as shown in **Figure 6,** can analyze trace amounts of gases that contribute to air pollution. Among other things, gas chromatography is used by researchers to analyze complex mixtures of compounds that constitute aromas and flavors.

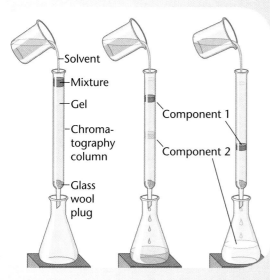

■ **Figure 4** In gel chromatography, the mobile phase travels through a gel stationary phase.

Gas Chromatography

In gas chromatography, shown in **Figure 5,** the mobile phase is a gas. The stationary phase is usually a liquid coating deposited on the interior of the tube through which the mixture, itself a mixture of gases, migrates. Helium is often used as the mobile phase. The extent to which the components of a gaseous mixture interact with the coating of the tube determines their migration rates. The separated components of the mixture arrive at the end of the tube at different times, where they are analyzed and identified by a light spectrometer or mass spectrometer.

■ **Figure 6** The gas chromatograph (left) separates mixtures and mass spectrometer (right) analyzes and identifies the components.

Discuss the Technology

1. **Analyze** Identify the stationary and mobile phases of a rain-soaked poster.

2. **Design** Design a laboratory method by which you could obtain pure samples of the components of a mixture that has been separated by using paper chromatography.

Chapter 9 • Chemistry and Technology

CHEMLAB

SEPARATING CANDY COLORS

Background
Yellow dye #5 is a food coloring approved by the FDA, but some people are allergic to this dye. Many candies contain Yellow #5 as part of a mixture to color the candies. Dye mixtures can be extracted from the candy and separated into their component colors using paper chromatography. The yellow food coloring that you buy in the grocery store contains Yellow #5 and can be used as a reference standard.

Paper chromatography is possible because different substances have different amounts of attraction for the paper. The greater the attraction the substance has for the paper, the slower it will move up the paper with the solvent.

Question
Are there any colored candies that a person with an allergy to Yellow #5 can safely eat?

Objectives
- **Observe** separation of colors in dye mixtures.
- **Interpret** data to determine which candies contain Yellow #5.

Preparation

Materials
10 cm × 10 cm piece of Whatman #1 filter paper
large beaker
colored candy
yellow food coloring
stirring rod
watch glass
water
salt
toothpicks
ruler
pencil

Safety Precautions

Procedure

1. Read and complete the lab safety form.
2. Make a data table like the one shown in Data and Observations.
3. Make a fine line with a pencil about 3 cm from one edge of the piece of filter paper.
4. Place a small amount of water on a watch glass.
5. Dip the tip of a toothpick into the water.
6. Dab the moistened tip of the toothpick onto a piece of colored candy to dissolve some of the colored coating.
7. Place the tip of the toothpick with dye onto the filter paper to form a spot along the pencil line, as shown in the photo below.

8. Remoisten the tip of the toothpick and dab the same piece of candy to dissolve additional coating. Place the tip of the toothpick onto the filter paper on the same spot made in step 7. Repeat this step until a concentrated spot is obtained.
9. Using a new toothpick and fresh water, repeat steps 4–8 with a different colored piece of candy. Make a new spot for each piece of candy, and keep a record in your data table.
10. Dip a fresh toothpick into a drop of the yellow food coloring to be used as a reference standard. Make a spot along the pencil line and mark the location of the reference spot.
11. Carefully roll the paper into a cylinder. The spots should be at one end of the cylinder. Staple the edges. Avoid touching the paper.
12. Add water to the beaker to a level of about 1.5 cm from the bottom. Sprinkle in a pinch of salt. Stir the solution with a stirring rod.

Analyze and Conclude

1. **Interpret** Do any of the candies contain Yellow #5? How can you tell?
2. **Compare and Contrast** Do any of the candies contain the same dyes? Explain.
3. **Infer** Which candies would be safe to eat if you were allergic to Yellow #5?

Apply and Assess

1. **Analyze** Where are the substances with the greater attraction for the paper prevalent? What conclusions can you draw about the molecular polarities of these dyes, given that the most polar component will have the greatest attraction for the paper?
2. **Infer** Why was it important to use a pencil instead of a pen to mark the paper?
3. **Summarize** Why was it important to do the investigation with a watch glass covering the mouth of the beaker?
4. **Explain** What makes the water move up the paper?
5. **Conclude** How did the rate of water movement up the paper change as the water got higher on the paper? Suggest reasons why it changed.

INQUIRY EXTENSION

Devise a better way to remove the dye from the candy and place a spot on the paper. Carry out your separation.

---

13. Place the filter-paper cylinder into the beaker so that the end with the spots is closest to the bottom. The water level must be at least 1 cm below the pencil line. Adjust the amount of water if necessary cover the beaker with a watch glass.
14. Allow the water to rise to about 1 cm from the top of the filter paper.
15. Carefully remove the filter paper, open it flat, and mark the solvent edge (the farthest point the water traveled) gently with a pencil. Lay the filter paper on a paper towel to dry.
16. For each piece of candy spotted, measure the distance from the original pencil line to the center of each separated spot. Record these data in your data table. Some candies may have more than one spot.
17. Measure and record the distance from the original pencil line to the marked solvent edge.
18. Record the distance from the original pencil line to the center of each spot separated from the reference spot of Yellow #5.

Data and Observations

Solvent distance : _____ (distance from first pencil mark to solvent edge)

Data Table			
Original spot	Distance (color 1)	Distance (color 2)	Distance (color 3)
Yellow #5 reference			
Candy 1			

The water on this leaf beads up because water molecules on the surface of the drop are attracted to other water molecules below them.

Similarly, a water drop suspended in air is shaped like a sphere because the molecules at the surface experience a net pull toward the center of the drop.

■ **Figure 9.22** Water molecules attract one another because they have positive and negative ends. These two photos illustrate the effect of this attraction on the shapes of water drops.

Polar and Nonpolar Molecules

Have you ever pulled clothing from the dryer and found that it was stuck together because of static cling? Static cling results from electrostatic attraction between positive charges and negative charges. Positive charges and negative charges arise in other ways besides the action of the clothes dryer, and some physical properties can be explained by them. For example, water molecules tend to bead up on smooth surfaces, and raindrops take a spherical shape, as shown in **Figure 9.22.** Why should water molecules stick together to form drops? The reason is that water molecules have positive ends and negative ends. Oppositely charged ends of the molecules attract one another, and the molecules stick together like the clothes in the dryer. Water is an example of how polar bonds and molecular geometry act together to affect the properties of compounds.

Water—a polar molecule You saw earlier that the water molecule has a bent structure. The ΔEN of the O—H bond is 1.4, so the two O—H bonds in a water molecule are polar bonds. The oxygen end of the bond has a partial negative charge, while the hydrogen end of the bond has a partial positive charge. Because of its bent shape, the water molecule as a whole has both a negative and a positive pole.

The water molecule is an example of how polar bonds, arranged in certain geometries, can produce a **polar molecule**, that is, a molecule that has a positive and a negative pole. A polar molecule is also called a dipole. Water is considered the universal solvent because it easily dissolves other polar molecules such as sucrose (sugar) and ammonia.

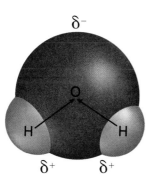

The O—H bonds in a water molecule are polar. Because of water's bent shape, the hydrogen side of the water molecule has a net positive charge, and the oxygen side has a net negative charge.

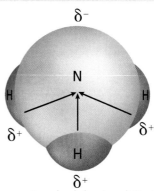

Like water, an ammonia molecule has two distinct sides. Because of polar bonds, the hydrogen side has a net positive charge and the nitrogen side has a net negative charge.

Ammonia—another polar molecule Ammonia, NH_3, is another example of a molecule with polar bonds. The N—H bond has a $\Delta EN = 0.9$. The geometry of an ammonia molecule is a triangular pyramid. When the three polar N—H bonds are arranged in this geometry, a polar molecule results. The net positive charge is located at the base of the pyramid, while a negative charge is associated with the nitrogen atom. **Figure 9.23** compares the charge distributions and shapes of water and ammonia molecules.

Carbon dioxide—a nonpolar molecule Carbon dioxide is another molecule with polar covalent bonds. The ΔEN for the C—O bond is 1.0, so the polarity of the C—O bond in carbon dioxide is similar to the polarity of the N—H bond in ammonia. But the geometry of CO_2 is different than the geometry of ammonia. The shape of a CO_2 molecule is linear. **Figure 9.24** shows that with the polar C=O bonds of a CO_2 molecule arranged in a straight line, the effects of the two polar bonds cancel each other. As a result, there is no separation of positive and negative charge in the molecule. So, although carbon dioxide has relatively strong polar covalent bonds, CO_2 is a nonpolar molecule.

■ **Figure 9.23** Water and ammonia are both polar molecules. In these models, the arrows indicate the directions in which electrons are pulled.

Explain *how the charge of an ammonia molecule is distributed due to the polar bonds.*

Chemistry Online

Personal Tutor For an online tutorial on polar and nonpolar molecules, visit glencoe.com.

■ **Figure 9.24** Carbon dioxide is a nonpolar molecule. Polar solvents such as water usually do not dissolve nonpolar substances. However, nonpolar CO_2 is slightly soluble, in water and under pressure, even more CO_2 dissolves.

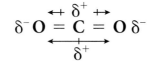

Both bonds in CO_2 have ΔEN equal to 1.0 and are polar. But, the polar bonds oppose each other and therefore cancel each other's effects.

Bottling soft drinks with CO_2 under pressure adds fizz. When you release the pressure by opening the bottle, the CO_2 comes out of solution—sometimes too fast, producing a gas-rich eruption of soft drink.

Section 9.2 • Molecular Shape and Polarity

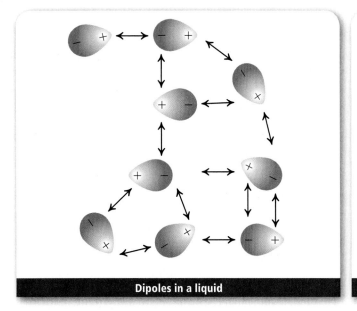
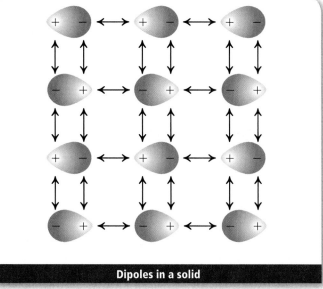

■ **Figure 9.25** The force between dipole molecules is an attraction of the positive end of one dipole for the negative end of another dipole. Shown here are representations of dipole-dipole attractions in liquids and solids.

Methane and water compared Polar molecules attract each other because they have positive and negative ends (dipoles). The diagram in **Figure 9.25** illustrates how these dipoles interact. Because of this attraction, the properties of polar molecules differ from those of nonpolar molecules. For example, melting points and boiling points of polar substances tend to be higher than those of nonpolar molecules of the same size.

When you compare the physical properties of the polar molecule water with the nonpolar molecule methane, you see major differences. Although water and methane are approximately the same size and both are covalently bonded, water is a liquid at room temperature, whereas methane is a gas. **Table 9.2** shows a comparison between the melting points and boiling points of methane and water. Notice that the boiling point of water is 264°C higher than the boiling point of methane. This is macroscopic evidence for the submicroscopic attractions at work among water molecules.

Ions, Polar Molecules, and Physical Properties

Recall from Chapter 4 that the submicroscopic interactions between the particles of a substance determine many of their macroscopic physical and chemical properties. In the case of ionic compounds, the strong attractive force that binds positive and negative ions into well-ordered crystals keeps the physical properties fairly similar.

Table 9.2	Comparison of Melting and Boiling Temperatures	
Compound	Melting Point	Boiling Point
Methane (nonpolar)	−183°C	−161°C
Water (polar)	0°C	100°C

Ionic compounds usually are solids at room temperature.

Covalent compounds can be solids, liquids, or gases at room temperature.

Several ionic solids are shown in **Figure 9.26,** along with several covalent compounds. The ionic compounds tend to have high melting points because of the strong attractions among their oppositely charged ions. Sugar is covalently bonded, but polar bonds create relatively strong interactions that hold the molecules together in a solid crystal structure at room temperature.

Water(H_2O) has covalent bonds and is a liquid at room temperature. As you learned from **Figure 9.10** on page 309, water has a much higher boiling point than many similar compounds because the O—H bond is polar. Propane (C_3H_8) is the least polar of the substances shown in **Figure 9.26.** Only very weak attractions form between propane molecules. As a result, propane is a gas at room temperature, but it can be compressed to a liquid for transportation and storage.

■ **Figure 9.26** Ionic compounds exhibit a narrower range of physical properties than covalent compounds. Ionic compounds tend to be brittle, solid substances with high melting points. Covalent substances may be solids, liquids, or gases at room temperature.

Connecting Ideas

Whether you use gumdrops, Lewis dot diagrams, or supercomputers, the ability to model bonding between atoms is useful. By determining the shape and polarity of a molecule, you can predict its behavior and properties. In Chapter 10, you'll learn more about the forces between particles and the effects they have on the physical states of different substances.

Section 9.2 Assessment

Section Summary

- Electron pairs around the central atom are either lone (nonbonding) pairs or bonding pairs.
- The polarity of the bonds and the shape of the molecule determine whether a molecule is polar or nonpolar.
- Interparticle forces determine many of the physical properties of substances.

9. **MAIN Idea Infer** Chloroform($CHCl_3$) is a molecule similar to methane in structure. Is chloroform a polar covalent or nonpolar molecule?

10. **Draw** the Lewis dot diagrams for each of the following molecules.
    a) PH_3 c) HBr
    b) CCl_4 d) OCl_2

11. **Describe** the shape of each molecule in question 10.

12. **Compare** a single bond, double bond, and triple bond?

13. **Apply** Sugar, water, and ammonia are all covalent compounds. How do these compounds demonstrate the variety of bond types in covalent compounds?

CHAPTER 9 Study Guide

BIG Idea Atoms bond by sharing or transferring electrons.

Section 9.1 Bonding of Atoms

MAIN Idea The difference between the electronegativities of two atoms determines the type of bond that forms.

Vocabulary
- conductivity (p. 311)
- ductile (p. 311)
- electronegativity (p. 301)
- malleable (p. 311)
- metallic bond (p. 311)
- polar covalent bond (p. 308)
- shielding effect (p. 302)

Key Concepts
- Bond character varies from ionic to covalent. There is no clear-cut division between the types of bonds.
- Electronegativity—a measure of the attraction that an atom has for shared electrons—can be estimated from the periodic table.
- Electronegativity difference (ΔEN) is a measure of the degree of ionic character in a bond.
- A ΔEN = 2.0 or greater occurs when elements form ionic bonds; ΔEN = 0.5 – 2.0 reflects polar colvalent bonds; and ΔEN <0.5 reflects covalent bonds.
- Metal atoms bond by sharing in a sea of valence electrons.

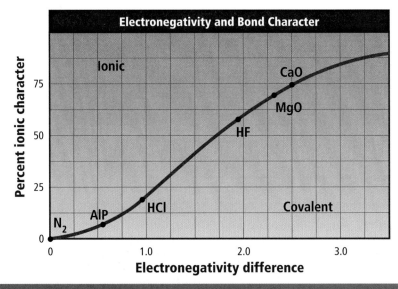

Section 9.2 Molecular Shape and Polarity

MAIN Idea The shape of a molecule and the polarity of its bonds determine whether the molecule as a whole is polar.

Vocabulary
- double bond (p. 319)
- polar molecule (p. 328)
- triple bond (p. 323)

Key Concepts
- Electron pairs around the central atom are either lone (nonbonding) pairs or bonding pairs.
- The polarity of the bonds and the shape of the molecule determine whether a molecule is polar or nonpolar.
- Interparticle forces determine many of the physical properties of substances.

Chapter 9 Assessment

Understand Concepts

14. Classify the following bonds as ionic, covalent, or polar covalent.
   a) Mg—O
   b) B—F
   c) S—Cl
   d) Ti—Cl

15. What is meant by a polar covalent bond? A nonpolar covalent bond?

16. Using only the periodic table, rank these bonds from the smallest to largest ΔEN: O—F, P—F, F—F, Al—F, Mg—F, N—F, K—F. Which do you think will be ionic bonds? Check your answer by calculating the actual ΔEN using the data in **Figure 9.2**.

17. Is carbon monoxide (CO) a polar or nonpolar molecule? Explain.

18. What is the shielding effect? Is shielding more significant in carbon or lead? Explain.

19. What experimental evidence supports the model of metallic bonding?

20. Compounds containing the ammonium ion are often used as fertilizers. What is the geometry of an ammonium ion (NH_4^+)?

■ **Figure 9.27**

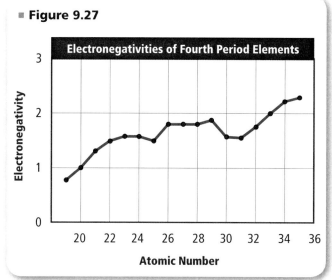

21. Electronegativity versus atomic number is graphed for the fourth period elements in **Figure 9.27** above. Describe the trend in electronegativity across period 4.

Apply Concepts

22. Sulfur dioxide (SO_2) is an atmosphereic pollutant resulting from the burning of sulfur-containing coal. What is the geometry of sulfur dioxide? Is sulfur dioxide a polar or nonpolar molecule?

Chemistry and Technology

23. Which chromatographic technique would be appropriate for separating a mixture of gaseous hydrocarbons?

Everyday Chemistry

24. Explain how microwaves used to cook food might be used to decompose some pollutants in the atmosphere. What other kinds of molecules could be decomposed in this way?

Think Critically

Make Predictions

25. Ethanol (an alcohol) and dimethyl ether both have the molecular formula C_2H_6O. Look up the structural formulas and draw Lewis dot diagrams for the two compounds. Use your diagrams to decide whether there is a difference in the polarity of the two molecules. Explain.

Design an Experiment

26. ChemLab A green dye could be produced by mixing a yellow dye with a blue dye. How could you determine whether a dye was a mixture of a yellow and blue dyes or whether it was a pure green dye?

Make Predictions

27. MiniLab 2 Below is the Lewis dot diagram for $AlBr_3$. What geometric shape do you predict for this molecule?

Chapter 9 Assessment

Cumulative Review

28. Suppose you have a cube of gold measuring 1 cm on each side. You hammer it into a square measuring 15 cm on each side as shown in **Figure 9.28**. What is the average thickness of the square? (*Chapter 3*)

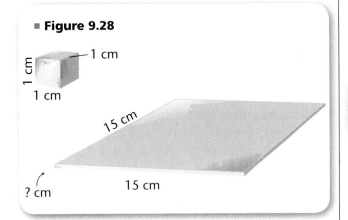

■ **Figure 9.28**

29. Carbon dioxide and carbon disulfide have identical Lewis dot diagrams. Why is this not surprising? (*Chapter 7*)

30. Does the mass number of an atom change when the atom forms a chemical bond? Explain. (*Chapter 2*)

Skill Review

31. **Organizing Information** Complete **Table 9.3**. Draw the Lewis dot structure for each compound listed. *Hint: Start with carbon as the central atom.* Predict the geometric arrangement of electron clouds around the central atom and use your prediction to determine the geometry of the molecule. From the geometry, decide whether the molecule is polar or nonpolar.

WRITING in Chemistry

32. Pick one period in Linus Pauling's career and research it. Locate a publication by Pauling from this period in his life and write a newspaper review about his accomplishments.

Problem Solving

33. There are three different structures for the hydrocarbon pentane (C_5H_{12}). These structures are called geometric isomers because each has a different shape. Draw Lewis dot diagrams for the three possibilities. Substitute a dash to represent electron pairs.

Table 9.3 Molecular Geometry and Polarity

Molecular Formula	Lewis Dot Diagram	Geometry of Electron Pairs About the Central Atom	Molecular Geometry	Polar Bonds	Nonpolar Bonds	Polar Molecule
H_2O	H:Ö:H	tetrahedral	bent	O—H	none	yes
CCl_4						
$CHCl_3$						
CH_2Cl_2						
CH_3Cl						
CH_4						

Cumulative Standardized Test Practice

Element	Electronegativity (EN)
Lithium	1.0
Iron	1.8
Sulfur	2.5
Nitrogen	3.0
Oxygen	3.5
Fluorine	4.0

Use the table above to answer Questions 1–4.

Halogen Electron Configuration	
Halogen	Electron Configuration
Fluorine	2,7
Chlorine	2,8,7
Bromine	2,8,18,7
Iodine	2,8,18,18,7
Astatine	2,8,18,32,18,7

Use the table to answer Questions 6–7.

1. What is the ΔEN of the bonds formed between oxygen and iron?
   a) 1.7
   b) 1.8
   c) 3.5
   d) 5.3

2. Which pair of elements will combine to form an ionic compound?
   a) lithium and iron
   b) oxygen and iron
   c) lithium and oxygen
   d) fluorine and oxygen

3. Which pair of elements will combine to form a covalent compound?
   a) lithium and iron
   b) fluorine and lithium
   c) iron and sulfur
   d) nitrogen and oxygen

4. Why is water a polar molecule?
   a) The bent structure of a water molecule forms a negative oxygen end and a positive hydrogen end.
   b) The bent structure of a water molecule forms a negative hydrogen end and a positive oxygen end.
   c) The hydrogen and oxygen atoms in a water molecule form a tetrahedral structure.
   d) The hydrogen and oxygen atoms in a water molecule form a linear end.

5. What is the shape of a carbon dioxide molecule?
   a) bent structure
   b) straight line
   c) tetrahedral
   d) right angle

6. Why do halogens have a −1 oxidation number?
   a) They have two electrons in their first orbital.
   b) They need one electron in the outer energy level.
   c) They lose one electron in one of their sublevels.
   d) They need one electron in their 2p orbitals.

7. What do all the halogens have in common?
   a) They have the same number of valence electrons.
   b) They have the same number of orbitals.
   c) They have the same number of sublevels.
   d) The have the same number of electrons in each orbital.

8. Lightning striking water to create hydrogen and oxygen gas is an example of
   a) synthesis.
   b) decomposition.
   c) single displacement.
   d) double displacement.

9. Yttrium, a metallic element with atomic number 39, will form
   a) positive ions.
   b) negative ions.
   c) both positive and negative ions.
   d) no ions at all.

NEED EXTRA HELP?

If You Missed Question . . .	1	2	3	4	5	6	7	8	9
Review Section . . .	9.1	9.1	9.1	9.2	9.2	8.1	8.1	6.2	5.1

CHAPTER 10: The Kinetic Theory of Matter

BIG Idea The kinetic theory of matter explains the properties of solids, liquids, and gases.

10.1 Physical Behavior of Matter
MAIN Idea The common states of matter are solid, liquid, and gas.

10.2 Energy and Changes of State
MAIN Idea Matter changes state when energy is added or removed.

ChemFacts

- The Japanese macaques (*Macaca fuscata*) are famous for their winter visits to the hot springs in central Japan.
- The three states of water are represented in this photo: solid — the snow on the rocks, liquid — the water in the springs, and gas — the invisible vapor in the air.
- All three states are able to exist in one place due to the cold air and the geothermal heat.

Start-Up Activities

LAUNCH Lab

Temperature and Mixing

The average speed of particles changes when heat is added or removed, resulting in a change in temperature. An increase in average speed can allow the mixing of two liquids to occur more rapidly. What is the relationship between temperature and rate of mixing?

Safety Precautions

Materials
- 250-mL beaker (3)
- 200 mL of ice water
- 200 mL of room temperature water
- 200 mL of hot water
- thermometer
- dropper
- food coloring

Procedure
1. Read and complete the lab safety form.
2. Set up the three beakers. Pour about 200 mL of ice water into the first beaker. Pour similar volumes of room temperature water and hot water into the second and third beakers respectively.
3. Use a dropper to add one drop of food coloring to each beaker. Observe the three beakers for 10 minutes.

Analysis
1. **Describe** How quickly did the food coloring disperse in the beaker of ice water? In room temperature water? In hot water?
2. **Explain** why things dissolve better in hot liquids than in cold liquids.

Inquiry Would you expect to observe differences if you used different colors of food dye? How might you test for these differences? Explain.

FOLDABLES Study Organizer

States of Matter Make the following Foldable to help you summarize information about the three common states of matter.

▶ **STEP 1** Fold a sheet of paper in half lengthwise. Make the back edge about 2 cm longer than the front edge.

▶ **STEP 2** Fold into thirds.

▶ **STEP 3** Unfold and cut along the folds of the top flap to make three tabs.

▶ **STEP 4** Label the tabs as follows: *Gases, Liquids,* and *Solids.* Label the back edge *States of Matter.*

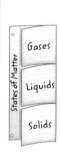

FOLDABLES Use this Foldable with Section 10.1. As you read the section, summarize information about the three common states of matter in your own words.

Visit glencoe.com to:
▶ study the entire chapter online
▶ explore Concepts In Motion
▶ take Self-Check Quizzes
▶ use Personal Tutors
▶ access Web Links for more information, projects, and activities
▶ find the Try at Home Lab, Estimating Metric Temperatures

Section 10.1

Objectives
- **Compare** characteristics of a solid, liquid, and gas.
- **Relate** the properties of a solid, liquid, and gas to the kinetic theory of matter.
- **Distinguish** among amorphous solids, liquid crystals, and plasmas.

Review Vocabulary
polar molecule: a molecule that has a positive and a negative pole

New Vocabulary
solid
liquid
gas
Brownian motion
kinetic theory of matter
ideal gas
pressure
crystal lattice
amorphous solid
liquid crystal
plasma

Physical Behavior of Matter

MAIN Idea The common states of matter are solid, liquid, and gas.

Real-World Reading Link You know matter exists as gases, liquids, and solids because you can smell the perfume of flowers, pour fruit punch into a glass, and stack logs of firewood. How would you describe the observable properties of a flower's aroma, fruit punch, and firewood?

States of Matter

Imagine trying to squeeze your textbook into a jelly jar. It can't be done. A **solid** is rigid with a definite shape, as shown in **Figure 10.1.** Solids are rigid because the atoms, ions, or molecules that make up a solid are fixed in place. These particles are tightly packed in solids. As a result, it is extremely difficult to press solids into a smaller volume. When solids are heated, their particles vibrate more, but generally move only slightly farther apart. It is important to realize that solids are not defined by rigidity or hardness. For example, at room temperature both concrete and candle wax are solids. However, they differ a great deal in terms of rigidity and hardness.

The characteristics of a liquid differ dramatically from those of a solid. A **liquid** is flowing matter with a definite volume but an indefinite shape. A liquid takes the shape of its container, as shown in **Figure 10.2.** If you have ever cleaned up a spill, you have witnessed how particles of a liquid move and easily glide over each other. You can feel how a liquid flows by standing under a running shower. Liquids are similar to solids in that their particles are very difficult to press into a smaller volume. Like solids, liquids generally do expand when heated, but only a relatively small amount.

■ **Figure 10.1** Whether a solid is natural, such as this crystal, or manufactured, such as laboratory glassware, a solid is rigid—it keeps its shape. Most elements are solids at room temperature.

■ **Figure 10.2** Unlike a solid, a liquid takes the shape of its container. When the water is in the pool, it takes the shape of the pool's bottom and sides. If sloshed over the edge, the water spreads out in a puddle.

Similar to liquids, gases also flow. For example, you can feel the flow of gases that make up the atmosphere on a windy day. The flow of a gas feels different than the flow of a liquid because the particles that make up a gas are farther apart than those of a liquid. If you blow up a balloon or a tire, you can observe some important properties of gases, as shown in **Figure 10.3.** From these observations, you see that a **gas** is flowing matter that has no definite volume or shape and is relatively easy to compress into a smaller volume. The particles that make up a gas are much farther apart than they are in solids and liquids, so they can be easily pushed together. Additionally, gases expand and contract in response to temperature changes much more readily than either solids or liquids.

FOLDABLES
Incorporate information from this section into your Foldable.

The Kinetic Theory of Matter

In 1827, Robert Brown, a Scottish botanist, was studying water samples with a microscope. He observed pollen grains suspended in the water moving continuously in irregular directions. Brown repeated his observations using dye particles in water, and noted that they too had random and erratic motions. This constant, random motion of tiny particles of matter suspended in a liquid or a gas is called **Brownian motion** in honor of Robert Brown.

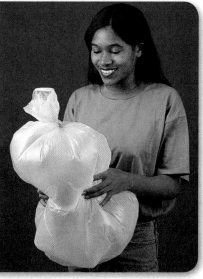

■ **Figure 10.3** The air completely fills the bag. Whatever the shape and volume of the bag, the air inside spreads out to fill it. If you squeeze the bag, you will observe that the air inside is compressible.

Do only molecules of water display random motion? Is all matter in motion? The **kinetic theory of matter** (sometimes called the kinetic-molecular theory) states that submicroscopic particles of matter are in constant, random motion. The energy of moving objects is called kinetic energy.

Kinetic model of gases In the kinetic theory, each particle of a gas moves like the air-hockey puck shown in **Figure 10.4.** The puck moves in a straight line until it strikes the side of the game board. As modeled by the kinetic theory, the particles in a sample of gas are in constant, random motion. Similar to the air-hockey puck, a gas particle can change direction only when it strikes the wall of its container or another gas particle.

Assumptions of the kinetic theory The kinetic theory makes several additional assumptions about the motion of gas particles. One of these assumptions is that the collisions made by gas particles are elastic collisions. Comparing the movement of gas particles with the air hockey puck may help you understand elastic collisions. This analogy oversimplifies the kinetic theory because in air hockey, after each collision with the wall of the game board, the puck loses speed because some of its energy is transferred to the wall. When the puck has lost all its kinetic energy, it will stop gliding. Unlike the puck, gas particles do not lose kinetic energy when they collide with the walls of their container or with other gas particles. Elastic collisions are defined as collisions in which no kinetic energy is lost. The kinetic theory assumes that collisions of particles in a gas are elastic collisions.

Other assumptions of the kinetic theory involve the size of gas particles and the distances separating them. Gas particles are very small and are separated from each other by relatively large volumes of empty space. As a result, kinetic theory assumes that the volume of the particles themselves can be ignored. Due to the distance between particles and the speeds at which they are moving, kinetic theory also assumes that gas particles experience no attractive or repulsive forces.

■ **Figure 10.4** The kinetic theory of matter states that submicroscopic particles of matter are in constant, random motion.

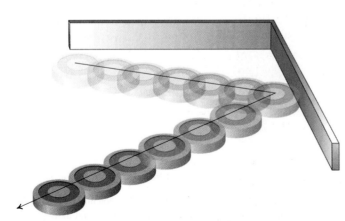

An air-hockey puck travels in a straight line until it strikes the side of the game board. Then, it rebounds in a straight line in a new direction.

Similarly, a gas particle moves through the space in its container in a straight line. The speed of the hockey puck is about 1 m per second, but the gas particle moves at a much faster rate of 10^2 to 10^3 m per second.

Ideal gases In general, the assumptions of the kinetic theory work very well for real gases. A gas that obeys the assumptions of the kinetic theory is called an **ideal gas**. Except at very low temperatures or very high pressures, most real gases behave like ideal gases. These limitations make sense when you think about them: at very low temperatures and high pressures, forces between particles and the size of the particles themselves begin to matter. Thus, the gases no longer follow the assumptions of the kinetic theory; in other words, the gases do not behave like ideal gases.

Gas particles and pressure The kinetic theory explains why gases fill their containers and why they exert pressure on the walls of the container. In the macroscopic world, blocking a volleyball is an example of how you can change the direction of motion. Remember the sting of the volleyball striking your hands? What you felt was pressure. **Pressure** is the force acting on a unit area of a surface; that is, for example, the force per square centimeter. Just as the volleyball exerted a force on a square centimeter of your skin, the particles in a gas exert a force on each square centimeter of the walls of the container when the walls deflect them.

MiniLab 10.1

Diffusion Rates

What inferences can you make about two molecular gases by observing the result of their movement through air?

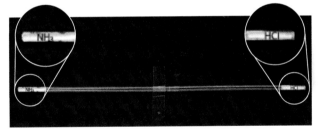

Procedure
1. Read and complete the lab safety form.
2. Make a tube by taping two **transparent straws** together end to end.
3. **Tape** the tube onto a black, horizontal surface. Label one end of the tube "NH₃" and the other "HCl."
4. Cut a **cotton swab** in half with **scissors** and wrap each cut end with **masking tape** thick enough to seal the tube.
5. Obtain containers of **concentrated solutions of ammonia (NH₃)** and **hydrochloric acid (HCl)** from your teacher. **WARNING:** *Both solutions can damage eyes, skin, and clothing. Handle them with care. If any skin contact or spillage occurs, notify your teacher immediately.*
6. Dip one swab into the ammonia and the other into the hydrochloric acid. The cotton should be saturated but not dripping.
7. Simultaneously insert the swabs into the appropriate ends of the tube, pushing them far enough so that wrapped tape handles seal the ends of the tube as shown above.
8. Do not jostle or move the straws. After a few seconds, look closely for the white ring of ammonium chloride (NH₄Cl), the reaction product.
9. Measure and record the distances from the respective cotton swabs to the ammonium chloride ring.

Analysis
1. **Describe** the movement of the two gases.
2. **Propose** an explanation for the difference in speed of each gas.

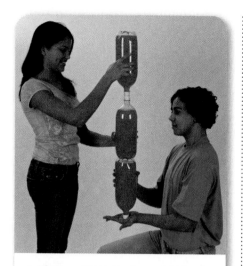

■ **Figure 10.5** The force of the bottle cap on the palm of the hand from three 2-L bottles is approximately equal to the force that the gases of the atmosphere exert on the same area of skin. Of course, the pressure from these three bottles on this person's hand is in addition to normal atmospheric pressure.

You can easily observe the effect of the pressure exerted by gas particles by blowing up a balloon. The outward pressure of the air inside a balloon is the force that keeps the balloon expanded. If the force is strong enough, the balloon will burst. You can feel this force by gently squeezing an inflated balloon or ball.

Earth's atmosphere and pressure Earth's atmosphere, which is a mixture of gases, exerts a pressure, too. Atmospheric pressure is the force exerted on you by molecules and atoms of air. Atmospheric pressure is related to the height of the column of air above you. This is why air pressure decreases as you gain elevation, but increases as you lose elevation. **Figure 10.5** illustrates an approximation of atmospheric pressure at sea level.

Humans and other forms of life on Earth have adapted to atmospheric pressure. We are sensitive only to changes in pressure. This explains why you don't really notice the atmospheric pressure you are immersed in throughout your life, but you would notice the roughly equivalent pressure of the soda bottles if you tried the simple investigation shown in **Figure 10.5.**

Kinetic model of liquids Just as a gliding air-hockey puck models a rebounding gas particle, **Figure 10.6** shows that magnetized spheres in a beaker model some behaviors of liquids. When liquids form puddles, interparticle forces maintain their volume but not their shape. The particles of a liquid can slide past each other, but they are so close together that they don't move as straight or as smoothly as an air-hockey puck. When you try to walk through a crowded hallway, you can't move quickly in a straight line, either.

■ **Figure 10.6** These magnetized spheres in a beaker model some behaviors of liquids.
Describe *how the magnetized marbles represent the behavior of liquids.*

Magnetized spheres spread out evenly to take the shape of their container. The volume they occupy cannot be reduced.

When the container is swirled, the spheres flow with a swirling motion.

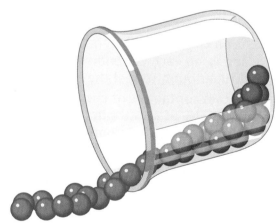

When the container is tipped, the magnetized spheres flow onto the table.

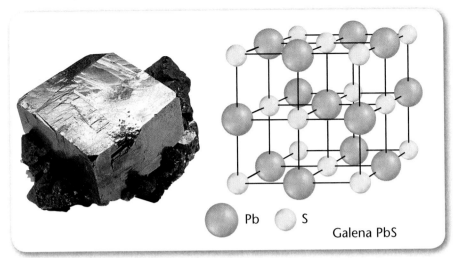

■ **Figure 10.7** Crystal lattices extend throughout solids. Shown here is the crystal lattice of lead(II) sulfide, or the mineral galena.

Compare *the arrangement of particles in a liquid and in a solid.*

Kinetic model of solids According to the kinetic theory, strong forces between particles explain the rigid structure of solids. Particles of a solid cannot move past each other, but they are in constant motion, bouncing between neighboring particles. In a solid, the particles occupy fixed positions in a well-defined, three-dimensional arrangement called a **crystal lattice**. The arrangement is repeated throughout the solid, as shown in **Figure 10.7**.

Other Forms of Matter

Some forms of matter cannot readily be described as solid, liquid, or gas. Maybe they look like solids or gases but sometimes behave like a liquid. These forms of matter are amorphous solids, liquid crystals, and plasmas.

Amorphous solids Is the peanut butter you spread on your bread a solid? What about the wax in candles? Although such materials have a definite shape and fixed volume, they are not classified as solids but rather as amorphous solids. An **amorphous solid** has a haphazard, disjointed, and incomplete crystal lattice. Candles and cotton candy are everyday examples of amorphous solids. Glass is another good example of an everyday amorphous solid. While there are different types of glass, some glass is made entirely of silicon dioxide (SiO_2), which has both crystalline and amorphous forms. **Figure 10.8** illustrates the difference between crystalline silicon dioxide and amorphous silicon dioxide.

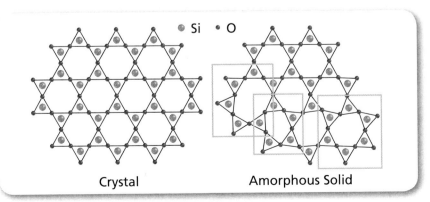

■ **Figure 10.8** The silicon dioxide crystal, SiO_2, has a regular honeycomb structure. If melted and then quickly cooled, silicon dioxide loses its regularity and becomes an amorphous solid.

Section 10.1 • Physical Behavior of Matter

Art Connection

Glass Sculptures

In 1976, an automobile accident cost Dale Chihuly the sight in his left eye. This accident could have ended the career of the artist who helped to found the Pilchuck Glass Center, a school for training glass artisans, in Stanwood, Washington in 1971. Although Chihuly lost his depth perception, which makes it difficult to blow glass safely, he continues to train and inspire others in the art of glass blowing.

Turning glass into art Glassblowing was developed in Syria in the first century B.C. Glass containers were produced for everyday and luxury use. Today, Chihuly's blown-glass sculptures, shown in **Figure 1,** are expensive pieces of art. Chihuly begins by drawing a design. One of his students or employees, shown in **Figure 2,** blows the glass to the specifications of the design. The glassblower sticks an iron blowpipe into the molten glass and blows a large glass bubble. When the bubble is the right size, the blower rolls it in tinted glass dust, which adheres to the surface. The different colored glass layers are fused by another firing in the furnace.

More blowing and careful shaping by pulling and pushing followed by further heating continues to transform the bubble. When cooled, it becomes a piece of art—an astonishing marvel of human creativity.

Figure 2 Chihuly and his students creating glass sculptures

Making glass Two types of glass are used for making glass sculptures: soda-lime glass and lead glass. Soda-lime glass is the most widely used glass in the world, being the same material comprising windows, bottles, jars, and lightbulbs. Soda-lime glass is composed of approximately 66 percent sand (SiO_2, silica), 15 percent soda (Na_2O or Na_2CO_3), and 10 percent lime (CaO or $CaCO_3$). It is inexpensive and easy to melt and shape. Therefore, it is suitable for many purposes, including art. Lead glass is made by replacing the lime and soda with lead oxide (PbO).

Lead glass is more expensive than soda-lime glass. It is easier to melt and form, however, and has a high refractive index, which gives the glass clarity and sparkle. It is also softer than most glass, which makes it easier to cut, engrave, and polish. Lead glass is used to make fine crystal and cut-glass objects, as well as art.

When any type of glass is molten, it looks just like the sugar syrup you might cook on your stove to make hard candy. The ingredients are heated in a furnace until they melt into a syrupy mass. When the mass cools, it hardens without forming crystals and becomes amorphous glass.

Figure 1 Chihuly glass sculptures

Connection to Chemistry

1. **Apply** What properties of glass make it ideal for repeated blowing and forming?
2. **Think Critically** Why is it scientifically incorrect to call a glass goblet a crystal goblet?

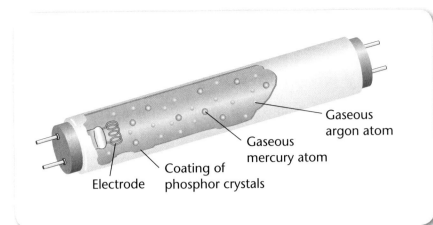

■ **Figure 10.9** When a small electrical current heats the electrode, some electrons in the electrode material acquire enough energy to leave the surface and collide with atoms of argon gas, which become ionized. As more electrons are freed, they also ionize some of the mercury atoms, forming a plasma. Electrons and mercury ions collide with mercury atoms and excite their electrons to higher energy levels. When the excited electrons return to lower energy levels, they radiate energy in the form of invisible ultraviolet light. The fluorescent tube is coated with phosphor crystals, which absorb the ultraviolet light and radiate visible light.

Liquid crystals When a solid melts, its crystal lattices disintegrate and its particles lose their three-dimensional pattern. However, when some materials called **liquid crystals** melt, they lose their rigid organization in only one or two dimensions. The interparticle forces in a liquid crystal are relatively weak. When weak interparticle forces in the lattice are broken, the crystal can flow like a liquid. Liquid crystal displays (LCDs) are used in watches, thermometers, calculators, and laptop computers because liquid crystals change with varying electric charge.

Plasmas The most common form of matter in the universe but the least common on Earth is plasma. The Sun and the other stars are made of plasma. Plasmas can also be found in fluorescent lights, as shown in **Figure 10.9**. A **plasma** is an ionized gas. It can conduct electric current but, like an ordinary conducting wire, it is electrically neutral because it contains equal numbers of free electrons and positive ions. Plasmas form at very high temperatures when matter absorbs energy and breaks apart into positive ions and electrons, or sometimes even into atomic nuclei and free electrons. In stars, the energy that ionizes the gases is produced by nuclear fusion reactions.

Section 10.1 Assessment

Section Summary

- The three common states of matter are solid, liquid, and gas.
- The kinetic theory of matter states that particles of matter are in constant motion.
- Interparticle forces determine whether a substance is a solid, liquid, or gas at a particular temperature and pressure.
- Plasma, the fourth state of matter, is the most common state of matter in the universe, but the least common on Earth.

1. **MAIN Idea Analyze and compare** the structures and shapes of liquids and gases in terms of particle spacing and particle motion.
2. **Describe** How do the particles of an ideal gas behave?
3. **Explain** Why is a plasma called a high-energy state of matter?
4. **Identify** how the kinetic energy of particles varies as a function of temperature.
5. **Explain** How could fluorescent light tubes be altered to supply indoor lighting in the blue and red region of the visible spectrum for raising plants?
6. **Define** a plasma and describe some of its properties. Where in the universe are plasmas commonly found? Where can plasmas be found on Earth?

Section 10.2

Objectives
- **Interpret** changes in temperature and changes of state of a substance in terms of the kinetic theory of matter.
- **Relate** Kelvin and Celsius temperature scales.
- **Analyze** the effects of temperature and pressure on changes of state.

Review Vocabulary
Brownian motion: constant, random motion of tiny particles of matter suspended in a liquid or a gas

New Vocabulary
temperature
absolute zero
Kelvin scale
kelvin (K)
diffusion
evaporation
sublimation
condensation
deposition
vapor pressure
boiling point
joule (J)
heat of vaporization
melting point
freezing point
heat of fusion

Energy and Changes of State

MAIN Idea Matter changes state when energy is added or removed.

Real-World Reading Link Have you ever forgotten to wash the dinner dishes before going to bed? The next morning your leftover dinner has dried and set hard on the plates. It's time to do the dishes. Will you use hot or cold water?

Temperature and Kinetic Energy

When you wash dishes with hot water, most of the water molecules are moving more rapidly than they do in cold water. They have more kinetic energy. Not all molecules of hot water in a sink have the same kinetic energy. They don't have the same speed. The same applies to a container of gas, such as an air-filled balloon. All the gas particles are moving randomly in the balloon at different speeds.

Temperature and particle motion Graph a in **Figure 10.10** shows the distribution of speeds of the particles in a container of gas. The most common speed of the particles is represented by the peak near the center of the graph. According to the kinetic theory of matter, **temperature** is a measure of the average kinetic energy of the particles in a material. Kinetic energy is related to both the mass and speed of the particles. Since the particles have the same mass, differences in kinetic energy will depend on differences in speed. As a gas is heated, the average kinetic energy of its particles increases. This increase in the average kinetic energy of the particles of the gas can be measured as an increase in the temperature of the gas, as shown in **Figure 10.10 graph b.** As a gas is cooled, the average kinetic energy and speed of its particles decreases.

■ **Figure 10.10** In a gas, some particles move slower or faster, but for most, the speed is near the average speed of the group.
Describe how the graph of particles versus speed would look for the same gas at a higher temperature.

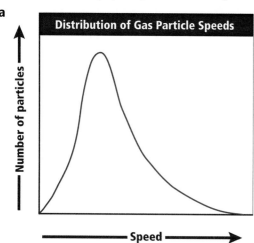

The peak of the graph represents the most common speed of the particles. The higher the temperature, the higher the average speed.

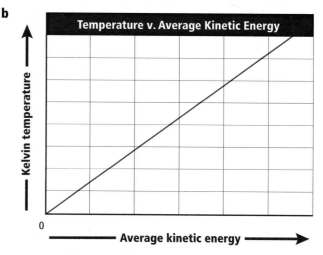

Because the graph is a straight line, the temperature of a gas is directly proportional to the average kinetic energy of its particles.

346 Chapter 10 • The Kinetic Theory of Matter

The Kelvin scale You know that, according to the kinetic theory of matter, particles of matter are in constant motion. According to the graphs in **Figure 10.10,** as the temperature of a gas increases, the average kinetic energy of its particles increases. As the temperature of a gas decreases, its kinetic energy decreases. In other words, as a substance is cooled, its particles lose kinetic energy.

Notice that **Figure 10.10 graph b** suggests that a substance would have zero kinetic energy based on the intersection of the graph's origin, (0, 0). This temperature is called **absolute zero.** In reality, there would be some particle motion even at this temperature, but it is the point at which all particles have the minimum amount of energy possible. The temperature of a substance can be lowered to values near absolute zero, but absolute zero can never be reached.

The scale used for temperature in **Figure 10.10** is the **Kelvin scale.** This scale is defined so that the temperature of a substance is directly proportional to the average kinetic energy of the particles. As a result, zero on the Kelvin scale corresponds to absolute zero. **Figure 10.11** relates the Kelvin scale to the Celsius scale, used throughout the world, and to the Fahrenheit scale, the scale used in the United States.

If the vertical temperature scale used in **Figure 10.10** had been the Celsius or Fahrenheit scale, the graph would be a straight line but the line would not pass through the origin. A zero reading on the Celsius or Fahrenheit scale does not correspond to absolute zero. Only on the Kelvin scale is the temperature reading directly proportional to the kinetic energy. The Kelvin scale makes calculations for solving problems about gases simple, as you will discover in Chapter 11.

The divisions of the Fahrenheit and Celsius scales are measured in degrees, but the divisions of the Kelvin scale are called **kelvins (K).** The kelvin is the SI unit of temperature. Note that the degree symbol is not used with temperatures expressed in kelvins. For example, absolute zero is written as 0 K. On the Celsius scale, the temperature reading for absolute zero is −273.15°C, usually rounded to −273°C. A Celsius degree and a kelvin are the same size. So, by shifting the Celsius scale up 273°, it will coincide with the Kelvin scale. Because the Kelvin scale measures temperatures above absolute zero and all readings are positive, the Kelvin scale is called the absolute temperature scale.

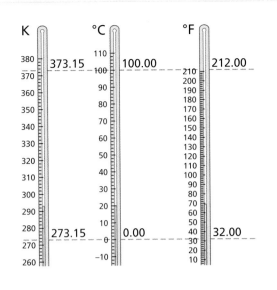

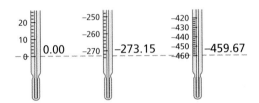

■ **Figure 10.11** The Celsius scale is defined so that the temperature interval from the freezing point of water to the boiling point of water measures 100 degrees. Water freezes at 0°C and boils at 100°C. On the Kelvin scale, water freezes at 273.15 K and boils at 373.15 K, an interval of 100. Thus, a Celcius degree and a kelvin are the same size.
Convert *Use Figure 10.11 to convert 50°C to kelvins.*

Section 10.2 • Energy and Changes of State **347**

> **TRY AT HOME 🏠 LAB**
>
> See page 872 for **Estimating Metric Temperatures.**

Temperature conversions Because the Celsius degree and the kelvin are the same size and kelvin readings are 273 degrees higher than Celsius readings, any Celsius reading can be easily expressed as a Kelvin reading. Simply add 273 to the Celsius reading.

$$T_K = (T_C + 273)\ K$$

For example, if the temperature of a room is 25°C, the kelvin reading can be found as follows.

$$T_K = (25 + 273)\ K = 298\ K$$

Similarly, a kelvin reading can be expressed as a Celsius reading by subtracting 273.

$$T_C = (T_K - 273)°C$$

For example, human body temperature, 310 K, can be expressed on the Celsius scale.

$$T_C = (310 - 273)\ °C = 37°C$$

Use **Figure 10.11** to verify that 37°C is correct.

Mass and Speed of Particles

When a bowling ball knocks over pins, you can see an effect of the kinetic energy of moving objects. You know that kinetic energy depends on speed, so you would expect that if the ball were rolling faster, it would have the kinetic energy to knock over more pins. You also know it's harder to move heavier objects than lightweight ones. In other words, it takes more kinetic energy to move heavier objects. **Figure 10.12** shows that the kinetic energy of a moving object, such as a train, depends on its mass and speed.

> **SUPPLEMENTAL PRACTICE**
>
> For more practice with temperature conversions, see Supplemental practice, page 820.

■ **Figure 10.12** Kinetic energy depends on both the mass and the speed of the moving body. The two trains in these photos may eventually move at the same speed, but their masses are vastly different. Thus, their kinetic energies are different. The train with the greater mass has the greater kinetic energy.

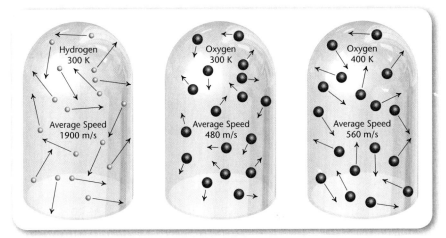

■ **Figure 10.13** Because the hydrogen gas in the left container and the oxygen gas in the center container are at the same temperature, they have the same average kinetic energy. Because the mass of an oxygen molecule is 32 u and the mass of a hydrogen molecule is only 2 u, the hydrogen molecules must have greater average speed. The two samples of oxygen—the center and right containers—are at different temperatures. Because both gases are oxygen, the particles in the two containers have the same mass. The molecules of oxygen gas in the container at the higher temperature have greater average kinetic energy because they are moving at a greater average speed.

Just as with the trains, the kinetic energy of gas particles depends on both the mass and the speed of the particles, as shown in **Figure 10.13.** You learned that temperature is a measure of the average kinetic energy of the particles of a substance, so samples of oxygen gas and hydrogen gas at the same temperature must have the same average kinetic energy. The particles in oxygen gas, however, are about 16 times more massive than the particles in hydrogen gas. Because the average kinetic energies of the samples must be equal, you can conclude that the hydrogen molecules at 300 K are moving faster than oxygen molecules at 300 K.

Diffusion The motions of particles of a gas cause them to spread out to fill the container uniformly. **Diffusion** is the process by which particles of matter fill a space because of random motion, as shown in **Figure 10.14.** If you have seen dye such as food coloring spreading through a liquid, you have watched diffusion. Your sense of smell depends on diffusion and air currents for you to detect molecules of a gas that waft by your nose. Diffusion is slow, but in your lungs, oxygen reaches your blood rapidly by diffusion. Oxygen diffuses across the walls of tiny blood vessels called capillaries from the air sacs of your lungs that fill with air each time you inhale. The rate of diffusion of a gas depends upon its kinetic energy, that is, on the mass and speed of its molecules.

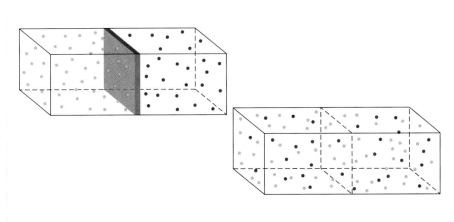

■ **Figure 10.14** Molecules of two different gases fill the left and right chambers. After the wall separating the two chambers has been removed, gas flows between chambers until the number of molecules flowing to the left chamber is the same as the number flowing to the right. At this point, the chambers have equal numbers of both kinds of molecules. The flow of gas from chamber to chamber continues, but there is no net change in the balance of molecules.

Explain *what property of gases makes diffusion possible.*

Section 10.2 • Energy and Changes of State

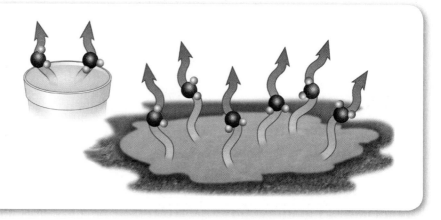

■ **Figure 10.15** The volume of water in the cup and puddle are the same, and they are at the same temperature. In the puddle, a larger surface area allows more molecules to escape.

Changes of State

You are very familiar with the changing states of water, from vapor to liquid water to ice. What environmental conditions are related to these changes of matter? When you remove ice from your freezer, it soon melts to water. When you boil vegetables in water, water vapor rises from the pot. From these observations, it is clear that temperature plays an important role in the changing of state of water, and indeed, of all matter.

Evaporation You have likely experienced evaporation after a tough workout on a hot, summer day. As your perspiration evaporates, your body cools as a result. You might have noticed that wet laundry hung on a clothesline dries faster on a hot day and slower on a cold day. **Evaporation** is the process by which particles of a liquid form a gas by escaping from the surface. The area of the surface, as well as the temperature and the humidity, affects the rate of evaporation, as shown in **Figure 10.15**.

Liquids that evaporate quickly, such as perfume and paint, are volatile liquids. Applying perfume with a spray bottle or splashing it on your skin increases its volatility and scent by increasing the size of the surface where evaporation takes place. Like the particles in a gas, the particles in a liquid have a distribution of kinetic energies. **Figure 10.16** applies the kinetic model to evaporation. As you can see, the kinetic energies of the particles are distrbuted over a fairly broad range, but only the particles at the high end of the distribution have the energy to escape the liquid. Because the molecules that escape have higher kinetic energy, the average kinetic energy of the remaining particles decreases. Therefore, as liquids evaporate, they cool.

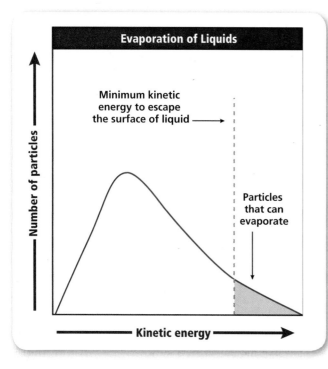

■ **Figure 10.16** Some particles in a liquid have enough kinetic energy to overcome interparticle forces. They leave the surface and become gas particles.
Explain *how the number of particles that can evaporate changes when the liquid is heated.*

350 Chapter 10 • The Kinetic Theory of Matter

Everyday Chemistry

Freeze Drying

Freeze drying, also known as lyophilization, is a dehydration process used to preserve a perishable material or to make a material weigh less, and therefore, making it easier to transport. It's likely some of the foods you have eaten have been freeze-dried. In this process, the material is frozen and placed in a vacuum where the surrounding pressure is lowered while enough heat is added to allow the frozen water in the material to sublimate directly from the solid phase to the gas phase. Then, the water vapor produced is removed from the chamber by a condenser.

Since the time of the Incas, people living in the Altiplano of Peru and Bolivia have eaten chuño, potatoes, shown in **Figure 1,** that have been preserved by drying in cold, dry air at 4000 m above sea level.

Advantages of freeze drying Materials that have been freeze-dried and sealed against the reabsorption of moisture can be stored safely at room temperature for many years before being used. Freeze drying is less destructive to materials than dehydration methods which require heat. Freeze drying also does not cause shrinkage of the material being dried, nor does it change the flavor or smells of food or decrease the vitamin content. Freeze-dried products can be rehydrated very quickly and used immediately.

Figure 1 Chuño potatoes

Figure 2 Freeze-dried ice cream

Freeze-dried foods, such as the freeze-dried ice cream shown in **Figure 2,** are lightweight and take up little room. Even foods such as chicken and potato salads have been freeze-dried, packed in airtight cans, and eaten as field rations on military maneuvers.

Uses of freeze drying The main two industries that use freeze drying are the pharmaceutical industry and the food industry. Pharmaceutical companies use freeze drying to increase the shelf life of vaccines and other injectable medications. The food industry uses freeze drying for a variety of products, including military and space rations and camping foods. Many foods can be freeze-dried successfully, such as meats, fish, fruits, vegetables, ice cream, and coffee. Freeze drying is also used for many other applications, such as preparing specimens for biological examination under a scanning electron microscope (SEM), restoring water-damaged books and documents, preservation of blood plasma and floral arrangements, and taxidermy, to name a few.

Explore Further

1. **Compare** Compare how chuño potatoes are preserved with the practice with freeze drying.
2. **Apply** Name some specific food products that are freeze-dried.

CHEMISTRY AND TECHNOLOGY

Fractionation of Air

In hospitals, at high altitudes, and on space walks, pure oxygen is used for life-support systems. Even greater supplies of oxygen are required by the steel and chemical manufacturing industries, where it is used in many reactions. Nitrogen, the most abundant gas in air, also has many industrial purposes. As a gas, it is used to insulate transformers and to exclude air from electric furnaces. Elemental nitrogen is used to manufacture fertilizers and explosives as well. Both oxygen and nitrogen are produced by the chemical industry through the fractional distillation of air.

Fractional Distillation

The components of dry air are nitrogen (78 percent), oxygen (21 percent), and smaller amounts of argon, carbon dioxide, neon, helium, krypton, hydrogen, xenon, and ozone (1 percent total). When a mixture such as air is fractionated, it is separated into its components. When the components are separated by differences in their boiling points, the method of separation is called fractional distillation. Follow the illustration in **Figure 1**.

1. Soot and dirt are removed from air by filters.
2. Water vapor is removed from air either by cooling and condensation or by absorbing the moisture with a silica gel sponge. Then carbon dioxide is removed as it reacts with lime.

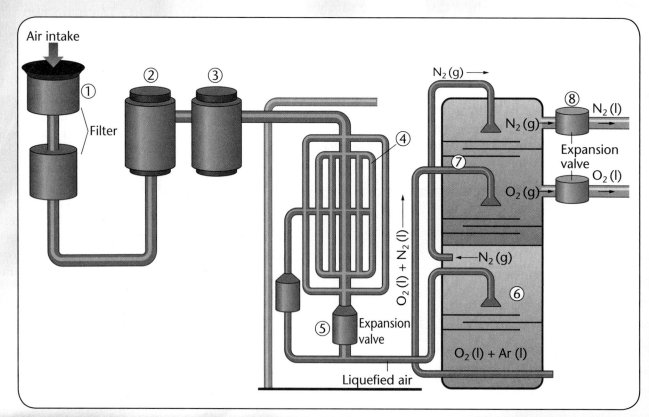

Figure 1 The fractional distillation process

3. The air is compressed to more than 100 times atmospheric pressure. Compression raises the temperature of the air.
4. In a heat exchanger, the air releases heat to the cooler surrounding fluid.
5. The cooled, compressed air passes through a nozzle into a chamber of larger diameter called an expansion valve. As the air passes through the valve, it expands and cools. (This cooling effect was first described by James Joule and William Thomson and is called the Joule-Thomson effect.) The temperature difference is so great that the air liquefies.
6. As the liquefied air flows over several heated trays, it is warmed to the boiling point of nitrogen (−195.8°C). Most of the nitrogen, with a trace of oxygen, vaporizes.
7. The liquid oxygen and remaining liquid nitrogen are collected and passed into an upper chamber for another distillation at a higher temperature. Here the oxygen and nitrogen vaporize and separate because of their different densities.
8. After passing through expansion valves, the separated gases are liquefied again and bottled as liquid nitrogen and oxygen.

Liquid oxygen and liquid nitrogen are shipped in special insulated semitrailer tanks, like that shown in **Figure 2,** at temperatures slightly lower than their boiling points. The boiling point for oxygen is −189.9 °C and the boiling point for nitrogen is −195.79 °C. Therefore, many of these tanks are insulated with a double walled-vacuum to withstand cryogenic temperatures, which are temperatures lower than −150 °C.

Figure 2 Semitrailer truck hauling liquid nitrogen

Discuss the Technology

1. **Analyze** Why is it necessary to liquefy air to separate its components?
2. **Infer** Why do you think liquid nitrogen is used to freeze food?

■ **Figure 10.17** When solid iodine is heated in the test tube, it forms a purplish gas, which is free molecules of I_2. The iodine rises to the top of the tube, where it cools. There, it condenses and forms solid iodine again.

Sublimation Some solid substances can change to the gaseous state directly, without melting first. The process by which particles of a solid escape from its surface and form a gas is called **sublimation.** For example, dry ice (solid carbon dioxide) does not melt but rather sublimes. Ice also sublimes. Some molecules in ice leave the surface and become water vapor. Vapor is used to describe the gaseous state of a substance that is a liquid at room temperature. Sublimation of ice is the reason that food stored for a long time in a freezer becomes freezer burned.

Condensation When dew forms, water vapor in the air condenses into its liquid state. Condensation is the reverse of evaporation. In **condensation,** gaseous particles come closer together—that is, they condense—and form a liquid. Sometimes, as in **Figure 10.17,** a solid forms directly from vapor. This process, the opposite of sublimation, is called **deposition.**

Vapor Pressure When a rain puddle dries, the water molecules leave the liquid state and become water vapor. However, if liquid water is in a closed container, only some of the water evaporates. **Figure 10.18** compares evaporation in an open container with evaporation in a closed container. In the closed container, equilibrium is established between the liquid and the vapor states. At that point, the rate at which particles enter the gas state from the liquid equals the rate at which particles enter the liquid state from the gas.

Chemistry Online

Personal Tutor For an online tutorial on vapor pressure, visit glencoe.com.

■ **Figure 10.18** Evaporation occurs in both open and closed containers. In open containers, water molecules that evaporate escape from the container. While some water molecules condense back into the liquid, the net flow of water molecules is into the gas phase. In a closed container, water vapor collects above the liquid. Eventually, an equilibrium is established: the movement of liquid molecules to the gas phase equals the movement of gas molecules back into the liquid.

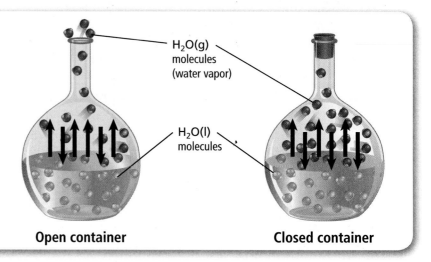

354 Chapter 10 • The Kinetic Theory of Matter

MiniLab 10.2

Vaporization Rates

What do you infer about liquids by observing their vaporizations? Not all particles in a liquid move at the same speed. If the liquid is in an open container, the fastest particles on the surface may have enough energy to escape from the liquid.

Procedure

1. Read and complete the lab safety form.
2. Half-fill a **250-mL beaker** with **water** and set it on a **hot plate**. Put a **thermometer** in the water and monitor its temperature until it reaches about 60°C. Remove the beaker from the hot plate with **beaker tongs** and set it on the lab table.
3. Label three **thin-stem pipettes** A, B, and C. Also, cut the bulbs of three other thin-stem pipets in half, creating three caps.
4. Obtain **small beakers of colored water, colored ethanol (C_2H_5OH), and hexane (C_6H_{14})** from your teacher.
5. Draw water into the bulb of pipette A until it is about one-third full. Then invert it so that its stem remains filled with liquid and cap it.
6. Grip the stem of the pipette with **forceps**, and immerse the bulb down in the hot water. Observe the liquid in the stem.
7. Measure and record the time it takes for the liquid to clear from the stem.
8. Repeat steps 4 to 6 for ethanol and hexane using pipettes B and C, respectively.

Analysis

1. **Relate** the rate of vaporization of these three liquids to the time they each take to clear from the pipette stem.
2. **Rank** the rate of vaporization of the three liquids from highest to lowest.
3. **Infer** Based on your observations, how do interparticle forces affect vaporization rates?

How is this equilibrium established? During evaporation of a liquid in a closed container, the amount of vapor and the pressure it exerts both increase. When equilibrium is reached, the pressure exerted by the vapor reaches its final, maximum value, and the volume of the liquid does not change. The pressure of a substance in equilibrium with its liquid is called its **vapor pressure.** The value of the vapor pressure of a substance indicates how easily the substance evaporates. When the pressure exerted by the vapor above the liquid equals the vapor pressure of the liquid, particles are evaporating and condensing at equal rates.

Volatile liquids and vapor pressure A volatile liquid such as ethanol has a high vapor pressure. The interparticle forces holding the ethanol molecules together as a liquid are weaker than those in water because ethanol molecules are less polar than water molecules. Therefore, it evaporates easily, and its vapor exerts more pressure in a closed cylinder. At the same temperature, a less volatile liquid such as water has a lower vapor pressure. Because the water molecule is polar, interparticle forces among water molecules are strong, and they tend to remain in the liquid state.

A liquid in a sealed container evaporates until its vapor pressure is high enough that rates of evaporation and condensation are equal. A liquid in an open container can never reach equilibrium with its vapor because the vapor is constantly escaping. When exposed to the atmosphere, liquids having high vapor pressures will evaporate more quickly than those with low vapor pressures.

■ **Figure 10.19** As you heat a beaker of water, water molecules gain kinetic energy. Vapor pressure increases (black arrows), but is initially less than atmospheric pressure (red arrows). However, when the water reaches its boiling point, the vapor pressure of the liquid equals the atmospheric pressure. At sea level, the boiling point of water is 100C°.

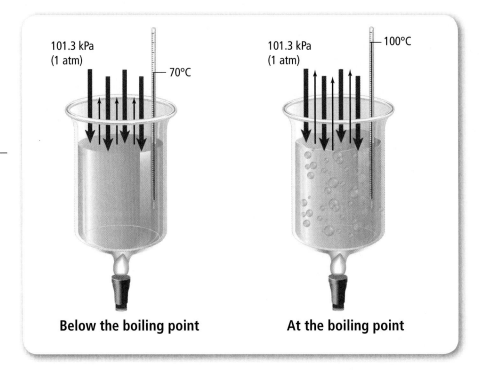

Below the boiling point   At the boiling point

Temperature, vapor pressure, and boiling Temperature and vapor pressure are related. The rate of evaporation is higher at higher temperatures. If the temperature of a liquid is raised high enough, not only will molecules of the liquid evaporate from its surface, but also, bubbles of the vapor will form below the surface.

You are probably familiar with bubbles of vapor forming below the surface of a liquid—it happens every time you boil water. **Boiling point** is the temperature at which a liquid's vapor pressure equals the pressure exerted on the surface of the liquid. **Figure 10.19** illustrates the boiling point of a substance. For a liquid in an open container, the pressure exerted on its surface is atmospheric pressure. Atmospheric pressure is the force per unit area that the gases in the atmosphere exert on the surface of Earth. Normal boiling point is the temperature at which a liquid boils in an open container at normal atmospheric pressure. For example, the normal boiling point of isopropyl alcohol is 82.3°C (355.5 K). However, the normal boiling point of ammonia is −33.35°C (239.80 K). The vapor pressure of liquid ammonia is so great that it boils well below room temperature.

Raising and lowering the boiling point When the pressure exerted on the surface of a liquid exceeds normal atmospheric pressure, vapor pressure must be higher than normal for the liquid to boil. To reach this higher vapor pressure, the liquid must be raised to a temperature higher than its normal boiling point. This is the principle behind the pressure cooker discussed in How It Works on page 357. Similarly, at pressures lower than normal atmospheric pressure, a liquid boils at a temperature below its normal boiling point. From these observations, you can conclude that the boiling point of a liquid increases when the pressure on the liquid increases and decreases when the pressure on the liquid decreases.

How It Works

Pressure Cookers

A pressure cooker is a heavy, covered pot designed to trap steam formed from boiling water. Because the pot is tightly sealed, the pressure of the steam builds up inside the pot, raising the boiling point of the water. Because both the water and steam are at higher temperatures, the food cooks faster.

A commercial pressure cooker reaches an internal temperature of up to 130°C (266°F) and cooks chicken in about 12 minutes—much shorter than the hour or so a conventional oven would require, but no faster than open-pan frying, which requires a much higher temperature and therefore has greater energy costs.

1 The sturdy construction of the pot allows pressure to build up safely inside the cooker.

2 The interlocking cover forms an airtight seal that keeps steam safely inside.

5 The food rack supports food or jars above the water level. As steam fills the inside of the cooker, the pressure above the boiling water increases and the boiling temperature increases, as does the temperature of the steam. Food cooks more quickly because both steam and water are at a higher temperature than in an open pot.

3 Some pressure cookers, such as those used for canning food, have a pressure gauge that indicates the internal pressure in pounds per square inch (psi) so that you can regulate the internal pressure by adjusting the external heat.

4 The rocker is a small weight that rests on top of a tiny valve built into the cover of the pressure cooker. This weight regulates the internal pressure and allows excess steam to escape safely. When the rocker vibrates gently, the cooker is heating at the proper rate.

Think Critically

1. **Explain** At higher altitudes, how does a pressure cooker offset the effects of the lower atmospheric pressure on the boiling of water?
2. **Describe** How does the rocker regulate the pressure inside?

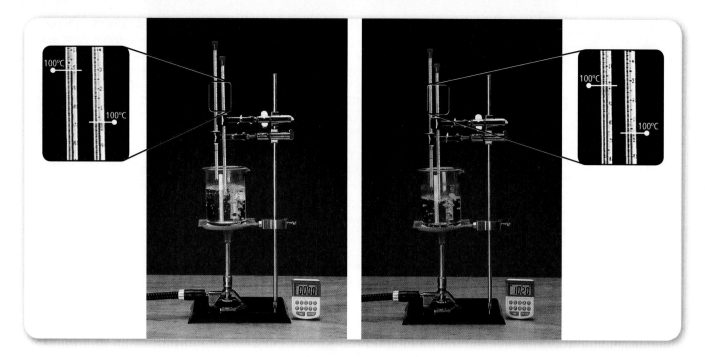

■ **Figure 10.20** The setup on the left at zero time shows water beginning to boil. After more than 10 minutes of boiling (right), the thermometers still have the same readings.

Compare *the kinetic energy of the molecules of steam to the kinetic energy of the molecules of liquid water.*

Heat of Vaporization

Placing a pan of water on a burner, applying heat, and bringing the water to its boiling point is probably a familiar process to you. You may be surprised to learn, however, what the temperatures of steam and water are, as shown in **Figure 10.20.** At the boiling point, the temperatures of the water and its vapor are equal. This will be true as long as there is any water left to boil.

You know that energy must be conserved. Therefore, the observation that adding energy to a system doesn't raise its temperature might seem strange or impossible. What happens to the energy supplied by the flame of the laboratory burner to the boiling water if it doesn't raise the temperature of the water? The rising bubbles of vapor within the boiling liquid are a clue.

The rising bubbles of vapor are less dense than the liquid water because the molecules in the bubbles are farther apart than those in the liquid. To separate the particles and form a vapor requires overcoming the interparticle forces holding them together in the liquid. The energy to overcome these forces comes from the heat of the flame. In other words, the energy supplied by the flame breaks the relatively strong forces of attraction between molecules of liquid water and causes the water to boil.

Imagine lifting a bag of 10 apples—approximately 1 kg of apples—from the ground to a counter top that is 1 m high. To overcome the force of gravity that keeps the apples on the ground requires energy. The energy expended to separate the bag of apples and the ground by 1 m is about 1 joule. The **joule** (J), is the SI unit of energy required to lift a 1-kg mass 1 m against the force of gravity. By comparison, the energy needed to move the molecules in 1 kg of water far enough apart that they form water vapor is about 2.26×10^6 J as shown in **Figure 10.21.** This amount of energy is expended regardless of whether the water is boiled or it evaporates.

Figure 10.21 About 2.26×10^6 J of energy are required to overcome the interparticle forces on the molecules in 1 kg of water. This number is the same regardless of whether the water evaporates or is boiled.

As with the apples, lifting 1 kg of water to a counter top 1 m high requires 1 J of energy. How much water could you lift a distance of 1 m with 2.26×10^6 J, the energy required to vaporize that kilogram of water? One kg of pure water has a volume of 1 L. Because raising a mass of 1 kg a distance of 1 m requires 1 J of energy, 2.26×10^6 kg of water—which is equal to 2.26×10^6 L of pure water—can be lifted a distance of 1 m. **Figure 10.22** compares a volume of 1 L with a volume of 2.26×10^6 L.

Energy and the heat of vaporization The energy absorbed when 1 kg of a liquid vaporizes at its normal boiling point is called its **heat of vaporization.** The heat of vaporization of water is 2.26×10^6 joules per kilogram, written as 2.26×10^6 J/kg, which is about 500 times larger than the energy required to raise the temperature of 1 kg of water by 1.0°C, 4180 J.

Because energy is always conserved, energy is released when vapor changes to a liquid. As the vapor condenses, the particles move closer together. For example, when 1 kg of steam condenses at water's normal boiling point, 100.0°C, it releases as heat the same energy it gained when it vaporized, 2.26×10^6 J. Perhaps you know that burns from steam are usually more severe than burns received from hot water. This occurs because the steam transfers a great deal of heat to the skin when it condenses.

Figure 10.22 On the left you see 1 L of water; 1 L of pure water weighs 1 kg. If you took the energy required to turn 1 kg of pure water into water vapor, which is roughly 2.26×10^6 J, you could lift a volume of pure water equal to the aquarium on the right 1 m off of the ground.

Section 10.2 • Energy and Changes of State

CHEMLAB

MOLECULES AND ENERGY

Background
Runners pick up speed in a race as muscles in their legs transfer energy from foods they ingest.

Question
How does energy transferred to or from a covalent compound affect the average kinetic energy of its molecules?

Objective
- **Observe** the temperature changes and changes of state when a covalent compound is heated and cooled.
- **Make and use** graphs to analyze temperature changes.
- **Interpret** temperature changes in terms of the changes in the average kinetic energy of a substance's molecules.

Preparation

Materials
timer
hot plate
20-mm × 150-mm test tube
400-mL beakers (2)
Celsius thermometers (2)
beaker tongs
test-tube holder
clamp and ring stand
stearic acid

Safety Precautions

WARNING: *Use beaker tongs when handling the beaker of hot water and a test-tube holder when handling the hot test tube.*

Procedure

1. Read and complete the lab safety form.
2. Prepare two tables like those shown on page 361.
3. Pour 300 mL of tap water into a 400-mL beaker and place the beaker on a hot plate.
4. Place a thermometer in the beaker of water. Turn on the hot plate and monitor the water temperature until it reaches 90°C. Keep the water temperature constant at 90°C by adjusting the heat or by adding cold water.
5. Half fill the test tube with stearic acid. Gently push the bulb of the second thermometer down into the substance. After the temperature of the thermometer has adjusted to the stearic acid, record this temperature in the first line of Data Table 1.
6. Attach the clamp to the test tube and immerse the tube in the beaker of hot water as shown. Read and record the temperature and the physical state or states of the stearic acid every 30 seconds until all of the material has melted and its temperature is about 80°C.

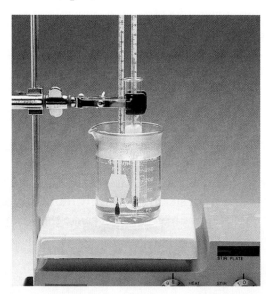

7. Pour 300 mL of cold tap water into the second 400-mL beaker.
8. Remove the test tube and contents from the first beaker and immerse it in the cold water in the second beaker. Read and record in Data Table 2 the temperature and physical state or states of the stearic acid every 30 seconds until the material has solidified.

360 Chapter 10 • The Kinetic Theory of Matter

Data and Observations

Data Table 1

Elapsed Time (s)	Temperature (°C)	Physical State
0		
30		
60		

Data Table 2

Elapsed Time (s)	Temperature (°C)	Physical State
0		
30		
60		

Analyze and Conclude

1. **Graph** the heating data by plotting temperature readings on the vertical axis and time on the horizontal axis. Connect the data points with straight lines or smooth curves. Label the appropriate segments of the graph solid, solid and liquid, or liquid. Graph the cooling data in the same way.
2. **Interpret** Divide each graph into three intervals by drawing two vertical lines at the points where the slope of the graph changes. Label the intervals of the first graph A, B, and C and those of the second graph D, E, and F.
3. **Conclude** According to your data, what is the approximate melting point of stearic acid?
4. **Describe** how the kinetic energy of the stearic acid molecules changed during each interval.

Apply and Assess

1. **Describe** how the molecular motion changed during each segment of the heating and cooling curves.
2. **Error Analysis** Identify a potential source of error in the procedure. How could you reduce the effect of this source of error?

INQUIRY EXTENSION

Suppose twice as much stearic acid were used. What would the graph look like? Make a sketch. Explain your graph.

FACT of the Matter

One definition of energy is "the ability to do work." The units of energy (joule, erg, foot-pound) are also the units of work.

Heat of Fusion

As with evaporation and condensation, the kinetic energies of the particles of a substance do not change during melting or freezing. If enough heat is applied to a solid, its crystal lattice breaks apart and it becomes a liquid. The **melting point** is the temperature of the solid when its crystal lattice begins to break apart. If more heat is applied after the solid has reached its melting point, the additional energy is used to overcome the interparticle forces until the crystal lattice collapses and becomes a liquid.

The temperature of a liquid when it begins to form a crystal lattice and becomes a solid is called its **freezing point.** If a liquid substance is cooled, the temperature falls until the substance reaches its freezing point. If more heat is transferred away from the substance, the temperature remains constant while the liquid becomes a solid. After all of the substance becomes a solid, the temperature can fall again.

The energy released as 1 kg of a substance solidifies at its freezing point is called its **heat of fusion.** The heat of fusion of water, for example, is 3.34×10^5 J/kg. The same energy is absorbed if 1 kg of a substance is heated until it melts.

Heating curves You can graph the temperature change of a substance as it is heated, as shown in **Figure 10.23**. This kind of graph is called a heating curve. On the heating curve for water, for example, the rising lines show the temperature change first as the solid ice is heated to the melting point, then as liquid water is heated to the boiling point, and finally as the temperature of the steam rises. The two plateaus represent periods when energy is being absorbed but the temperature is not changing. These periods occur at the melting point and at the boiling point. The general shape of the graph is characteristic of all solid substances being heated from a solid to a liquid and then to a gas.

■ **Figure 10.23** A 0.100-kg sample of ice at −20°C is heated to water vapor at 120°C. The rising line shows the rise in temperature of the ice, water, and water vapor. Notice how the graph is flat during changes of state. At these points, the added energy overcomes interparticle forces that hold the molecules in the solid or in the liquid state, and the temperature does not change. Once the change of state is completed, the temperature begins to rise again.

Infer Where would you place the photo of boiling water on the graph?

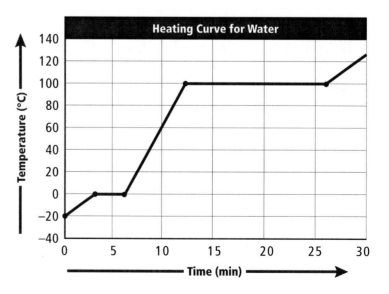

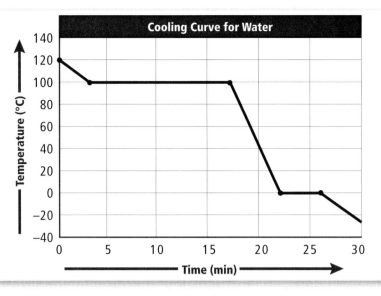

Cooling curves You can also create cooling curves for substances, as shown in **Figure 10.24.** Cooling curves are simply heating curves in reverse: the falling lines represent periods where heat is being released and the temperature is dropping; plateaus represent periods when energy is being released, but the temperature is not changing. Comparing **Figures 10.23** and **10.24,** you can see that the melting point of water—the plateau at 0°C on the heating curve—is the same as the freezing point: the plateau at 0°C on the cooling curve. The melting and freezing points of a substance are the same temperature, provided the pressure is the same.

■ **Figure 10.24** As a 0.100-kg sample of water vapor at 120°C is cooled, it liquefies to water, then freezes to ice, then cools to –20°C. The graph falls as the steam is cooled. It levels off while the steam is condensing, falls again after all the steam has condensed, levels off again while the liquid water freezes, and then falls after all the water has frozen.

Infer Where would you place the freezer full of frozen foods on the graph?

Connecting Ideas

The kinetic theory of matter explains the properties of solids, liquids, and gases, and explains changes of state in terms of interparticle forces and energy. It also quantitatively relates the pressure, volume, and temperature of gases. By studying how gases behave under different conditions, you will soon begin to understand how all matter behaves.

Section 10.2 Assessment

Section Summary

- Changes in the temperature of a substance indicate changes in the average kinetic energy of its particles.
- The Kelvin scale temperature of a substance is directly proportional to the average kinetic energy of its particles.
- At a substance's boiling point, the vapor pressure of the liquid equals the external pressure above the liquid.
- The temperature of a substance remains constant as it changes state at its melting and boiling points.

7. **MAIN Idea Explain** In terms of changes in total energy, how does the melting of 1 kg of water at 0°C differ from the freezing of 1 kg of water at 0°C?

8. **Rank** the following temperature readings in increasing order.

    32.0°F, 32.0°C, 32.0 K, 102.1°C, 102.1 K

9. **Infer** Why will water in a flask begin to boil at room temperature as air is pumped out of the flask?

10. **Analyze** Why are most elements solids at room temperature?

11. **Explain** why volatile liquids are often used as propellants in aerosol spray cans of such things as paints and deodorants.

CHAPTER 10 Study Guide

BIG Idea The kinetic theory of matter explains the properties of solids, liquids, and gases.

Section 10.1 Physical Behavior of Matter

MAIN Idea The common states of matter are solid, liquid, and gas.

Vocabulary
- amorphous solid (p. 343)
- Brownian motion (p. 339)
- crystal lattice (p. 343)
- gas (p. 339)
- ideal gas (p. 341)
- kinetic theory of matter (p. 340)
- liquid (p. 338)
- liquid crystal (p. 345)
- plasma (p. 345)
- pressure (p. 341)
- solid (p. 338)

Key Concepts
- The three common states of matter are solid, liquid, and gas.
- The kinetic theory of matter states that particles of matter are in constant motion.
- Interparticle forces determine whether a substance is a solid, liquid, or gas at a particular temperature and pressure.
- Plasma, the fourth state of matter, is the most common state of matter in the universe, but the least common on Earth.

Section 10.2 Energy and Changes of State

MAIN Idea Matter changes state when energy is added or removed.

Vocabulary
- absolute zero (p. 347)
- boiling point (p. 356)
- condensation (p. 354)
- deposition (p. 354)
- diffusion (p. 349)
- evaporation (p. 350)
- freezing point (p. 362)
- heat of fusion (p. 362)
- heat of vaporization (p. 359)
- joule (J) (p. 358)
- kelvin (K) (p. 347)
- Kelvin scale (p. 347)
- melting point (p. 362)
- sublimation (p. 354)
- temperature (p. 346)
- vapor pressure (p. 355)

Key Concepts
- Changes in the temperature of a substance indicate changes in the average kinetic energy of its particles.
- The Kelvin scale temperature of a substance is directly proportional to the average kinetic energy of its particles.
- At a substance's boiling point, the vapor pressure of the liquid equals the external pressure above the liquid.
- The temperature of a substance remains constant as it changes state at its melting and boiling points.

Chapter 10 Assessment

Understand Concepts

12. Compare and contrast evaporation and boiling.
13. What is the difference between an amorphous solid and a solid?
14. Why do most solid covalent compounds have lower melting points than ionic compounds?
15. Describe how vapor pressure is affected by increasing temperature.
16. Why are liquids less compressible than gases?
17. How is pressure exerted by a gas?
18. How does a real gas differ from an ideal gas?

Apply Concepts

19. What is the amount of energy absorbed when 5.00 g of water vaporize at 100°C?
20. Explain why a car's radiator is likely to boil over on a trip through the mountains, even though the air is colder.

Everyday Chemistry

21. Why is it necessary to place foods in low-pressure chambers during freeze drying?

Art Connection

22. How does glassblowing use the physical properties of gases?

Chemistry and Technology

23. Can fractional distillation be used to separate the elements in a compound? Explain.

Think Critically

Interpret Data

24. The boiling points of five liquids are provided in either Celsius or Kelvin temperatures. Calculate the missing information in **Table 10.1.** List the liquids in order from the one with the lowest boiling point to the one with the highest boiling point.

Table 10.1 Boiling Points

Liquid	Kelvin	Celsius
Acetone (C_3H_6O)	329 K	
Heptane (C_7H_{16})		98°C
Nitromethane (CH_3NO_2)	374 K	
Benzene (C_6H_6)		80°C
Sulfur trioxide (SO_3)	318 K	

Compare and Contrast

25. A sample of $Cl_2(g)$ and $N_2(g)$ are both at 25°C. Compare the average kinetic energy and the most probable speed of the molecules of the gases.

Make Predictions

26. Predict whether the boiling point of water is greater or less than 100°C at the shoreline of the Dead Sea, which is 400 m below sea level. Explain your prediction.

Infer

27. What shape would you expect an evaporating dish to have? Explain.

Compare and Contrast

28. Why does HCl have a higher boiling point than H_2?

Infer

29. Explain why an amorphous solid has a range of melting points rather than a fixed melting point.

Make Comparisons

30. The particles of which of the following gases have the highest average speed? The lowest average speed? Use the fact that NF_3 is the most massive and CH_4 is the least massive.
    a) carbon monoxide (CO) at 90°C
    b) nitrogen trifluoride (NF_3) at 30°C
    c) methane (CH_4) at 90°C
    d) carbon monoxide (CO) at 30°C

Chapter 10 Assessment

Use a Table

31. Examine **Table 10.2** and answer the following questions.
    a) Which of the substances are gases at 50°C? At 250°C?
    b) Which of the substances are liquids at 50°C? At 250°C?
    c) Which of the substances are solids at 50°C? At 250°C?
    d) Which substance has the smallest temperature range as a liquid?
    e) Would it have been easier to answer part (d) if the Kelvin temperature scale had been used? Explain.

Table 10.2 — Freezing and Boiling Points

Substance	Freezing Point (°C)	Boiling Point (°C)
Bromine	−7	58
Mercury	−39	357
Propane	−188	−42
Radon	−71	−62
Silver	961	2195

Interpret Graphs

32. **ChemLab** What can you say about the material with a heating curve such as the one shown in **Figure 10.25**.

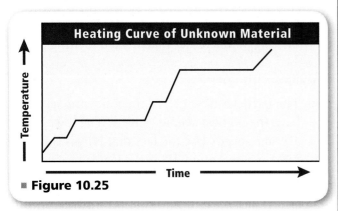
■ Figure 10.25

Make Predictions

33. **MiniLab 10.1** How would the result of the experiment change if the NH_3 were first heated?

Infer

34. **MiniLab 10.2** How is the evaporation rate of nonpolar covalent molecules affected by molecular mass?

Cumulative Review

35. What is the modern periodic law? How does it differ from the periodic law according to Mendeleev? (*Chapter 3*)

36. Explain how atomic radii influence the chemical reactivity of the alkaline earth metals. (*Chapter 8*)

Skill Review

37. Ethanol boils at 79°C and melts at −114°C. A few grams of ethanol are heated from −130°C to 130°C. Graph the heating curve for ethanol. Show time on the horizontal axis and temperature on the vertical axis.

WRITING in Chemistry

38. Compare the crystal structure of sodium chloride, diamond, and a metal such as copper in an essay. Illustrate each crystal with a drawing.

39. Research liquid crystals. Write a report about their most common characteristics and properties.

Problem Solving

40. Ethanol melts at −114°C. Express the melting point of ethanol on the Kelvin scale.

Cumulative Standardized Test Practice

Elemental Metals Boiling and Melting Points

Metal Element	Boiling Point (°C)	Boiling Point (K)	Melting Point (°C)	Melting Point (K)
Uranium	1132	?	3818	?
Gold	?	1337	?	3080
Copper	1083	1356	2567	?
Silver	962	1235	962	?
Lead	?	600	?	2013

Use the table above for Questions 1–3.

1. The boiling point of uranium is
   a) 859 K.
   b) 1132 K.
   c) 1405 K.
   d) 4091 K.

2. The melting point of lead is
   a) 1740 °C.
   b) 1740 K.
   c) 2286 °C.
   d) 2286 K.

3. Which element has the highest melting point in kelvins?
   a) lead
   b) copper
   c) gold
   d) uranium

4. Which state of matter has a definite volume but no definite shape?
   a) gas
   b) liquid
   c) solid
   d) plasma

5. Which CANNOT be used to speed up a reaction?
   a) increase temperature
   b) increase concentration
   c) add a catalyst
   d) decrease temperature

6. The d orbital holds a maximum of
   a) 2 electrons.
   b) 6 electrons.
   c) 10 electrons.
   d) 18 electrons.

7. Which is true of the alkaline earth metals?
   a) Alkaline earth metals have the same properties as the alkali metals.
   b) Alkaline earth metals are the most reactive metals.
   c) Alkaline earth metals have a greater density and hardness than Alkali metals.
   d) Alkaline earth metals tend to lose three valence electrons and form positively charged ions.

8. Which is true about the shielding effect?
   a) The shielding effect increases as you move up a column of elements.
   b) Inner energy levels block the attraction of the nucleus on valence electrons.
   c) Added energy levels block the attraction of other ions on the nucleus of an atom.
   d) The shielding effect on atoms creates an increase in the atom's electronegativity.

9. An atom of plutonium
   a) can be divided into smaller particles that retain all the properties of plutonium.
   b) cannot be divided into smaller particles that retain all the properties of plutonium.
   c) does not possess all the properties of a larger quantity of plutonium.
   d) cannot be imaged using current technology.

NEED EXTRA HELP?

If You Missed Question...	1	2	3	4	5	6	7	8	9
Review Section...	10.2	10.2	10.2	10.1	6.3	7.1	8.1	9.1	2.1

CHAPTER 11
Behavior of Gases

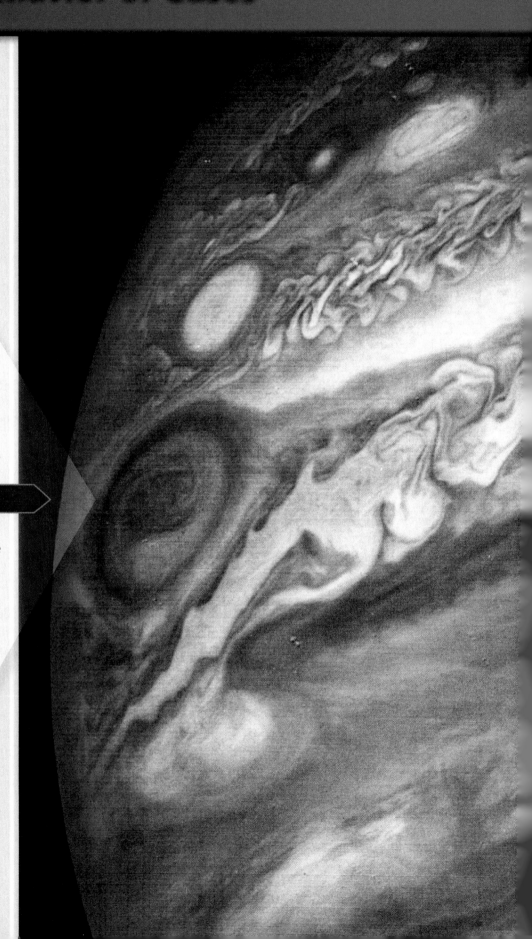

BIG Idea Gases respond in predictable ways to pressure, temperature, volume, and changes in number of particles.

11.1 Gas Pressure
MAIN Idea Gas pressure is related to the mass of the gas and to the motion of the gas particles.

11.2 The Gas Laws
MAIN Idea For a fixed amount of gas, a change in one variable—pressure, temperature, or volume—affects the other two.

ChemFacts

- The Great Red Spot on the planet Jupiter is a storm of gases within the turbulent atmosphere.
- Four planets in our solar system—Jupiter, Saturn, Uranus, and Neptune—lack solid surfaces and are referred to as gas giants.
- The interior temperatures of these planets range from 7,000 to more than 20,000 Kelvin.

Start-Up Activities

LAUNCH Lab

Volume and Temperature of a Gas

How does a temperature change affect the air in a balloon?

Materials
- 5-gal bucket
- round balloon
- ice
- string

Procedure
1. Read and complete the lab safety form.
2. Inflate a **round balloon** and tie it closed.
3. Half-fill the **bucket** with cold **water** and add **ice**.
4. Use a **string** to measure the circumference of the balloon.
5. Stir the water in the bucket to equalize the temperature. Submerge the balloon in the ice water for 15 minutes.
6. Remove the balloon from the water. Measure the circumference.

Analysis
1. **Infer** What happens to the size of the balloon when its temperature is lowered?

Inquiry What might you expect to happen to the balloon's size if the temperature is raised?

Chemistry Online

Visit glencoe.com to:
- study the entire chapter online
- explore Concepts in Motion
- take Self-Check Quizzes
- use Personal Tutors
- access Web Links for more information, projects, and activities
- find the Try at Home Lab, Crushing Cans

The Gas Laws Make the following Foldable to help you organize your study of the gas laws.

STEP 1 Stack two sheets of paper with the top edges about 2 cm apart vertically.

STEP 2 Fold up the bottom edges of the paper to form three equal tabs. Crease the fold to hold the tabs in place.

STEP 3 Staple along the fold. Label the tabs from top to bottom as follows: Boyle, Charles, and Combined.

FOLDABLES Use this Foldable with Section 11.2. As you read the section, summarize the gas laws in your own words.

Chapter 11 • Behavior of Gases **369**

Section 11.1

Objectives

- **Model** the effects of changing the number of particles, the mass, temperature, pressure, and volume on a gas using kinetic theory.
- **Evaluate** atmospheric pressure.
- **Demonstrate** the ability to use dimensional analysis to convert pressure units.

Review Vocabulary

diffusion: process by which particles of matter fill in a space because of random motion.

New Vocabulary

barometer
standard atmosphere (atm)
pascal (Pa)

Gas Pressure

MAIN Idea Gas pressure is related to the mass of the gas and to the motion of the gas particles.

Real-World Reading Link You might be surprised to see that an overturned tractor trailer can be uprighted by inflating several air bags placed beneath it. But, you're not at all surprised to see that an uprighted truck is supported by 18 tires inflated with air. Air can lift the tractor and support the truck because air is a mixture of gases, and gases exert pressure.

Defining Gas Pressure

Unless a ball has an obvious dent, you can't tell whether it is underinflated by looking at it. You have to squeeze it. When you squeeze the ball, you are checking the pressure of the air inside the ball. If it's soft, you know it needs to be pumped up with more air. **Figure 11.1** shows how the pressure of air inside a soccer ball rises as air is added.

How can changes in gas pressure be explained by the kinetic theory? Recall from Chapter 10 that the kinetic theory of matter states all submicroscopic particles of matter—atoms, ions, and molecules—are in constant, random motion. Particles in the air are in constant motion. The pressure they exert results from the force of the collisions when the particles strike the walls of the ball. The pressure exerted by particles of air inside a ball balances two other pressures—the pressure exerted by the atmosphere (the air outside the ball) and the pressure exerted by the wall of the ball itself.

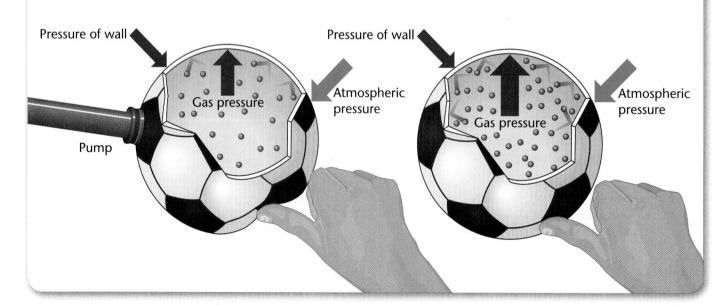

■ **Figure 11.1** Pumping more air into the deflated soccer ball (left) increases the number of gas particles inside. As a result, particles strike the inner wall of the ball more often and the pressure increases (right). This increase in pressure is counterbalanced by an increase in pressure from the tough wall. As a result, the ball becomes firm and bouncy.

■ **Figure 11.2** The mass of the ball on the left is greater than the mass of the one on the right. The ball on the left has greater mass because it has more air inside and, therefore, is at a higher pressure than the ball on the right.
Explain *why pressure increases when mass increases.*

Number of gas particles and pressure Recall from Chapter 10 that the pressure of a gas is the force per unit area that the particles in the gas exert on the walls of their container. As you would expect, the greater the number of air particles inside the ball, the greater the mass. In **Figure 11.2,** one basketball is not fully inflated. The other identical ball has been pumped up with more air. The ball that contains more air has a higher pressure and therefore greater mass.

From similar observations and measurements, scientists from as long ago as the eighteenth century learned that the pressure of a gas is directly proportional to its mass. According to the kinetic theory, all matter is composed of particles in constant motion, and pressure is caused by the force of gas particles striking the walls of their container. The more often gas particles collide with the walls of their container, the greater the pressure. Therefore, the pressure is directly proportional to the number of particles. For example, doubling the number of gas particles in a basketball doubles the pressure.

Alternately, if the number of gas particles in a basketball is cut in half, the pressure is also cut in half.

■ **Figure 11.3** The position of the piston relates to the pressure of the particles on each face of the piston.
Describe *the pressure on each side of the piston in terms of the number of collisions between the gas particles and the surface of the piston.*

The constant bombardment of molecules and atoms in air exerts a pressure on the outside face of the piston. This pressure balances the pressure caused by the confined gas bombarding the inside face of the piston (right).

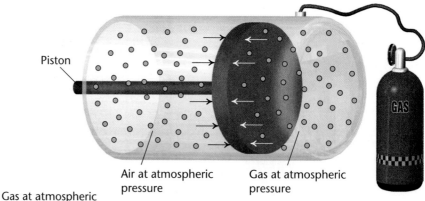

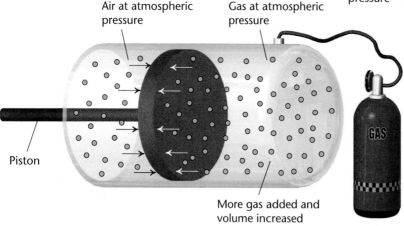

When gas is added to the cylinder, the piston is pushed out (left). When the pressure of the new volume inside the cylinder balances air pressure, the piston stops moving out.

Personal Tutor For an online tutorial on using the pressure of gas to do work, visit glencoe.com.

To demonstrate how a gas at constant temperature can be used to do work, examine the action of the piston shown in **Figure 11.3**. The piston inside the cylinder is like the cover of a jar because it makes an airtight seal, but it also acts like a movable wall. When gas is added, as shown in the top cylinder, the gas particles push out the piston until the pressure inside balances the atmospheric pressure outside.

If more gas is pumped into the cylinder, the number of gas particles increases, and the number of collisions on the walls of the container increases. Because the force on the inside face of the piston is now greater than the force on the outside face, the piston moves outward. As the gas spreads out into the larger volume, the number of collisions per unit area on the inside face decreases. Then, the pressure of the gas inside the container falls until it equals the pressure of the air outside and the piston comes to rest in its new position. The air pressure remains constant while the piston moves and the gas expands.

Temperature and pressure How does temperature affect the behavior of a gas? You know from Chapter 10 that at higher temperatures, the particles in a gas have greater kinetic energy. They move faster and collide with the walls of the container more often and with greater force resulting in an increase in pressure. If the volume of the container and the number of particles of gas do not change, the pressure of a gas increases in direct proportion to the Kelvin temperature.

What if the volume, such as in the piston in **Figure 11.3,** is not held constant? How would a change in temperature affect the volume and pressure of a gas? If the temperature of the gas inside the piston chamber increases, the pressure will increase momentarily. But, if the gas is permitted to expand as the temperature increases, the pressure remains equal to the atmospheric pressure and the volume increases. Thus, the volume of a gas at constant pressure is directly proportional to the temperature in kelvins.

When the cylinder shown in **Figure 11.3,** is in an automobile engine, the other end of the piston rod is attached to the crankshaft. When the mixture of gasoline and air in the cylinder ignites and burns, gases are produced. The gas inside the cylinder expands because the heat produced by combustion increases the temperature. The pressure of expanding gases drives the piston outward. As the piston rod moves out and then back in, it turns the crankshaft, delivering power to the wheels that propel the vehicle forward.

> **TRY AT HOME 🏠 LAB**
> See page 873 for **Crushing Cans.**

MiniLab 11.1

Mass and Volume of a Gas

Can you determine some properties of gases from sublimed dry ice? Recall from Chapter 4 that one property of carbon dioxide is that it changes directly from a solid (dry ice) to a gas. In other words, it sublimes.

Procedure
1. Read and complete the lab safety form.
2. Place a small, **self-sealing plastic bag** on the pan of a zeroed **balance.** Wearing gloves and using **tongs,** insert a 20- to 30-g piece of **dry ice** into the bag.
3. Measure and record the mass of the bag and its contents. Quickly press the air out of the bag and zip it closed.
4. Immediately put the bag into a larger, **clear plastic bag** so that you can observe the sublimation process. After the inner bag is filled with gas, unzip the closure through the outer bag. Immediately remove the bag with dry ice from the larger bag. Press out the gas and zip it closed.
5. Measure and record the mass of the bag and its contents.
6. Use tongs and gloves to remove and dispose of the dry ice from the bag.

7. Determine the volume of the inner bag by filling it with water and pouring the water into a graduated cylinder. Record its volume.
8. Repeat steps 2–7 with a smaller self-sealing bag and a smaller piece of dry ice.

Analysis
1. **Calculate** the mass of carbon dioxide that has sublimed in each trial.
2. **Calculate** the mass ratio by dividing the mass of sublimed CO_2 in Trial 1 by the mass that sublimed in Trial 2. Calculate the volume ratio of CO_2.
3. **Compare** How do the volume and mass ratios compare?
4. **Infer** What can you infer about the relationship between the volume and mass of a gas?

Devices to Measure Pressure

Although you can compare a property such as gas pressure by touching partially and fully inflated basketballs, this method does not give you an accurate measure of the two pressures. What is needed is a device to measure pressure.

The barometer One of the first instruments used to measure gas pressure was designed by the Italian scientist Evangelista Torricelli (1608–1647). He invented the **barometer,** an instrument that measures the pressure exerted by the atmosphere. His barometer was so sensitive that it showed the difference in atmospheric pressure between the top and bottom of a flight of stairs. **Figure 11.4** explains how Torricelli's barometer worked.

In the barometer, the height of the mercury column measures atmospheric pressure. We live at the bottom of an ocean of air. It is the pressure of the ocean of air particles pressing down on the mercury in the dish that support the column of mercury in the closed tube. If you've ever been diving in the ocean, you know that the deeper you go, the higher the pressure exerted by the water on your body. Similarly, the highest atmospheric pressures occur at the lowest altitudes. When you climb a mountain, atmospheric pressure decreases because the overall depth of air above you is less than at the base of the mountain.

One unit used to measure pressure is defined by using Torricelli's barometer. The **standard atmosphere** (atm) is defined as the pressure that supports a 760-mm column of mercury. This definition can be represented by the following equation.

$$1.00 \text{ atm} = 760 \text{ mm Hg}$$

Because atmospheric pressure is measured with a barometer, it is often called barometric pressure. Rising barometric pressure indicates increasing air pressure; falling barometric pressure indicates decreasing air pressure.

The pressure gauge Unfortunately, the barometer can measure only atmospheric pressure. It cannot measure the air pressure inside a bicycle tire or in an oxygen tank. You need a device that can be attached to the tire or tank. This pressure gauge must make some regular, observable response to pressure changes. If you have ever measured the pressure of an inflated bicycle tire, you're already familiar with such a device.

When you apply a pressure gauge to the valve stem of a tire, the pressurized air from the tire rushes into the gauge and is measured. You are measuring pressure above atmospheric pressure. The recommended tire inflation pressures listed by manufacturers are gauge pressures; that is, pressures read from a gauge. A barometer measures absolute pressure; that is, the total pressures exerted by all gases, including the atmosphere. To determine the absolute pressure of an inflated tire, you must add the barometric pressure to the gauge pressure.

VOCABULARY
WORD ORIGIN

Barometer
from the Greek word *baros,* meaning *heavy* and the Greek word *metron,* meaning to *measure*

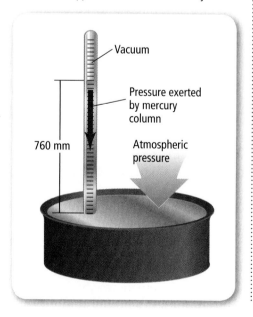

■ **Figure 11.4** Torricelli was the first to show that the atmosphere exerted pressure. A barometer consists of a tube of mercury that stands in a dish of mercury. Because the mercury stands in a column in the closed tube, you can conclude that the atmosphere exerts pressure on the open surface of the mercury in the dish. This pressure, transmitted through the liquid in the dish, supports the column of mercury.

How It Works

Tire-Pressure Gauge

A tire-pressure gauge, shown in **Figure 1,** is a device that measures the pressure of the air inside an inflated tire or a basketball. Because an uninflated tire contains some air at atmospheric pressure, a tire-pressure gauge records the amount that the tire pressure exceeds atmospheric pressure.

The most familiar tire gauge is about the size and shape of a ballpoint pen. It is a convenient way to check tire pressure for proper inflation regularly. Proper inflation ensures tire maintenance and safety.

Figure 1 Tire-pressure gauge

1 The pin in the head of the gauge pushes downward on the tire-valve inlet and allows air from the tire to flow into the gauge.

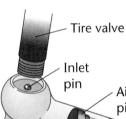

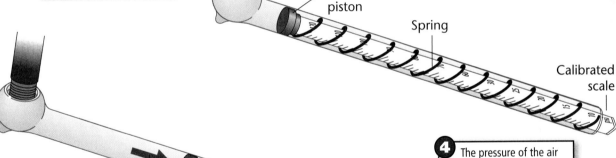

2 The air flowing into the gauge pushes against a movable piston that pushes a sliding calibrated scale.

3 The piston moves along the cylinder, compressing the spring until the force of air pressing on the surface of the piston is equal to the force of the compressed spring on the piston.

4 The pressure of the air in the cylinder, which is the same as the pressure of the air in the tire, is read from the scale.

Think Critically

1. **Describe** What does the calibration of the scale indicate about the relationship between the compression of the spring and pressure?
2. **Explain** whether a tire-pressure gauge should be recalibrated for use at locations where atmospheric pressure is less than that at sea level.

> **FACT of the Matter**
>
> The weight of a postage stamp exerts a pressure of about one pascal on the surface of an envelope.

Pressure Units

You have learned that atmospheric pressure is measured in mm Hg. Recall from Chapter 10, atmospheric pressure is the force per unit area that the gases in the atmosphere exert on the surface of Earth. **Figure 11.5** shows two additional units that are used to measure pressure of one standard atmosphere.

The SI unit for measuring pressure is the **pascal (Pa),** named after the French physicist Blaise Pascal (1623–1662). Because the pascal is a small pressure unit, it is more convenient to use the kilopascal. As you recall from Chapter 1, the prefix kilo- means 1000; so, 1 kilopascal (kPa) is equivalent to 1000 pascals. One standard atmosphere is equivalent to 101.3 kilopascals.

Table 11.1 on page 377 presents the standard atmosphere in equivalent units. Because there are so many different pressure units, the international community of scientists recommends that all pressure measurements be made using SI units, but pounds per square inch continues to be widely used in engineering and almost all nonscientific applications in the United States.

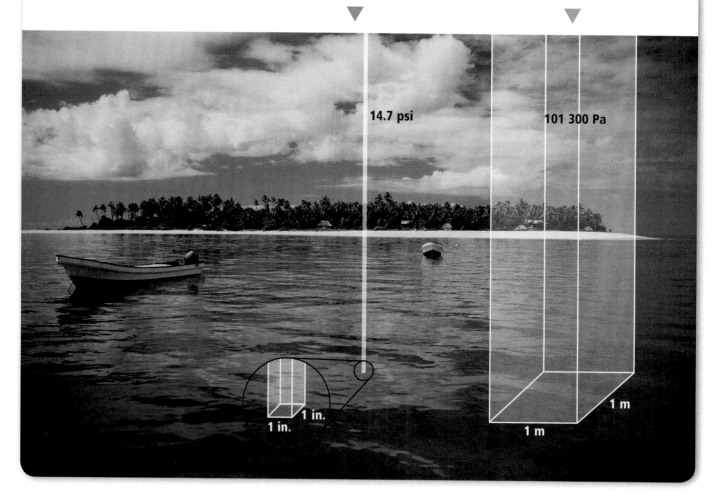

■ **Figure 11.5** The weight of the air in each column pushing on the area beneath it exerts a pressure of one standard atmosphere. Each column of air extends to the outer limits of the atmosphere.

If the unit of area is the square inch and the unit of force is the pound, then the unit of pressure is the pound per square inch (psi). Expressed in these units, one standard atmosphere is 14.7 pounds per square inch or 14.7 psi.

Expressed in SI units, one standard atmosphere is 101,300 pascals. Note that SI units are based on the square meter area and not the square inch.

Pressure conversions You can use **Table 11.1** to convert pressure measurements to other equivalent units. For example, you can now find the absolute pressure of the air in a bicycle tire. Suppose the gauge pressure is 44 psi. To find the absolute pressure, add the atmospheric pressure in pounds per square inch to the gauge pressure. Because the gauge pressure is given in pounds per square inch, use the value of the standard atmosphere that is expressed in psi. One standard atmosphere equals 14.7 psi.

$$44 \text{ psi} + 14.7 \text{ psi} = 59 \text{ psi}$$

Now, you can convert psi to kPa using the values from **Table 11.1**.

$$59 \text{ psi} \times \left(\frac{101.3 \text{ kPa}}{14.7 \text{ psi}}\right) = 410 \text{ kPa}$$

The following Example Problems show how to use the values in **Table 11.1** to express pressure in other units.

Table 11.1	Equivalent Pressures
	1.00 atm
	760 mm Hg
	14.7 psi
	101.3 kPa

EXAMPLE Problem 11.1

Converting Barometric Pressure Units In weather reports, barometric pressure is often expressed in inches of mercury. What is one standard atmosphere expressed in inches of mercury?

1 Analyze
You know that one standard atmosphere is equivalent to 760 mm of Hg. What is that height expressed in inches? A length of 1.00 inch measures 25.4 mm on a meterstick.

2 Set up
Select the appropriate equivalent values and units given in **Table 11.1**. Multiply 760 mm by the number of inches in each millimeter to express the measurement in inches.

$$760 \text{ mm} \times \left(\frac{1.00 \text{ in.}}{25.4 \text{ mm}}\right)$$

The factor on the right of the expression above is the conversion factor.

$$\cancel{760} \text{ mm} \times \left(\frac{1.00 \text{ in.}}{\cancel{25.4} \text{ mm}}\right)$$

$$= 29.9 \text{ in.}$$

3 Solve
Notice that the units are arranged so that the unit mm will cancel properly and the answer will be in inches.

4 Evaluate
Because 1 mm is much shorter than 1 in., the number of mm (760) should be much larger than the equivalent number of inches (29.9), which it is.

■ **Figure 11.6** Dimensional analysis can be used to calculate the number of pizzas that must be ordered for a party.

Apply *How many pizzas will you need if 32 people eat 3 slices per person and there are 8 slices in each pizza?*

SUPPLEMENTAL PRACTICE

For more practice with converting measurements, see Supplemental Practice, page 821.

The method used to convert units of measurement in Example Problem 1 to another unit is called dimensional analysis. This method is a powerful and systematic approach to problem solving. It will be useful to you both inside and outside of the classroom. **Figure 11.6** gives an example of how you might use the dimensional analysis in your everyday life.

Now, study the Example Problem again. To set this problem up, you need to find a conversion factor that helps you translate from millimeters of mercury to inches of mercury. The following equation gives the conversion factor because it contains both the given unit and the desired unit.

$$1.0 \text{ in.} = 25.4 \text{ mm}$$

You know that you can divide both sides of an equation by the same value and maintain equality. For example, you can divide both sides of the equation 1.00 in. = 25.4 mm by 25.4 mm, as shown below.

$$\frac{1.00 \text{ in.}}{25.4 \text{ mm}} = \frac{25.4 \text{ mm}}{25.4 \text{ mm}}$$

Now, simplify the right side. The conversion factor is on the left.

$$\frac{1.00 \text{ in.}}{25.4 \text{ mm}} = 1$$

In the Example Problem, the height of mercury in millimeters was being multiplied by this conversion factor. Because multiplying any quantity by 1 doesn't affect its value, the height of the mercury column isn't changed—only the units are. The dimensional analysis changes the units of the measurement without affecting the value. You will use the dimensional analysis to solve problems in this chapter and in several of the following chapters. To find more information on dimensional analysis, refer to pages 800-801 in the Chemistry Handbook.

EXAMPLE Problem 11.2

Converting Pressure Units The reading of a tire-pressure gauge is 35 psi. What is the equivalent pressure in kilopascals?

Problem-Solving Hint
In dimensional analysis, terms are arranged so that units will cancel out.

1 Analyze
- The given unit is pounds per square inch (psi), and the desired unit is kilopascals (kPa). According to Table 11.1, the relationship between these two units is 14.7 psi = 101.3 kPa.

2 Set up
- Write the conversion factor with kPa units as the numerator and psi units as the denominator. Note that the psi units will cancel and only the kPa units will appear in the final answer.

$$35 \text{ psi} \times \left(\frac{101.3 \text{ kPa}}{14.7 \text{ psi}}\right)$$

3 Solve
- Multiply and divide the values and units.

$$35 \text{ psi} \times \left(\frac{101.3 \text{ kPa}}{14.7 \text{ psi}}\right) = 240 \text{ kPa}$$

Notice that the given units (psi) will cancel properly and the quantity will be expressed in the desired unit (kPa) in the answer.

4 Check
- Because 1 psi is a much greater pressure than 1 kPa, the number of psi, 35, should be much smaller than the equivalent number of kilopascals.

PRACTICE Problems
Solutions to Problems Page 859

Your skill in converting units will help you to relate measurements of gas pressure in different units. Use **Table 11.1** and the equation 1.00 in. = 25.4 mm to convert the following measurements.

1. 59.8 in Hg to psi
2. 7.35 psi to mm Hg
3. 1140 mm Hg to kPa
4. 19.0 psi to kPa
5. 202 kPa to psi

Section 11.1 Assessment

Section Summary
- The pressure of a gas at constant temperature and volume is directly proportional to the number of gas particles.
- The volume of a gas at constant temperature and pressure is directly proportional to the number of gas particles.
- At sea level, the pressure exerted by gases of the atmosphere equals one standard atmosphere (1 atm).

6. **MAIN Idea Predict** A cylinder containing 32 g of oxygen gas is placed on a balance. The valve is opened and 16 g of gas escapes. How will the pressure change?

7. **Compare and contrast** how a barometer and a tire-pressure gauge measure gas pressure.

8. **Predict** At atmospheric pressure, a balloon contains 2.00 L of nitrogen gas. How would the volume change if the Kelvin temperature were only 75 percent of its original value?

9. **Interpret Data** Sketch graphs of pressure versus time, volume versus time, and number of particles versus time for a large, plastic garbage bag being inflated until it ruptures.

10. **Infer** Why do tire manufacturers recommend that tire pressure be increased if the recommended number of car passengers or load size is exceeded?

Section 11.2

Objectives

- **Analyze** data that relate temperature, pressure, and volume of a gas.
- **Predict** the effect of changes in pressure and temperature on the volume of a gas.
- **Model** Boyle's law and Charles's law using kinetic theory.
- **Explain** gas volume in terms of the kinetic theory of gases.

Review Vocabulary

pascal: SI unit for measuring pressure

New Vocabulary

Boyle's law
Charles's law
combined gas law
standard temperature and pressure, (STP)
law of combining gas volumes
Avogadro's principle

FOLDABLES Incorporate information from this section into your Foldable.

The Gas Laws

MAIN Idea For a fixed amount of gas, a change in one variable—pressure, temperature, or volume—affects the other two.

Real-World Reading Link An uninflated air mattress doesn't make a comfortable bed because comfort depends upon the pressure of the air inside it. When the mattress is filled and its air valve is closed, it is far more comfortable. As you sit on the mattress, you can feel the air inside press down before it supports your weight.

Boyle's Law: Pressure and Volume

As you know from squeezing a balloon, a confined gas can be compressed into a smaller volume. Robert Boyle (1627–1691), British Chemist, used a simple J-shaped tube apparatus to experimentally compress trapped air in the closed end of the tube. As he added mercury to the open end, the weight of the mercury compressed air trapped in the closed end.

Figure 11.7 shows a similar design, using a piston and a cylinder, to summarize Boyle's results. The number of air particles inside the cylinder does not change. As the piston is pressed down into the cylinder, the air particles can occupy less volume in the cylinder. This results in an increase in pressure. As you can see from the figure, each time the pressure doubles, the volume of air in the cylinders is cut in half.

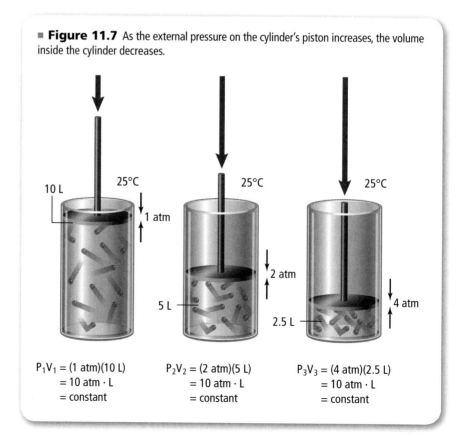

■ **Figure 11.7** As the external pressure on the cylinder's piston increases, the volume inside the cylinder decreases.

$P_1V_1 = (1\ \text{atm})(10\ \text{L})$
$= 10\ \text{atm} \cdot \text{L}$
$= \text{constant}$

$P_2V_2 = (2\ \text{atm})(5\ \text{L})$
$= 10\ \text{atm} \cdot \text{L}$
$= \text{constant}$

$P_3V_3 = (4\ \text{atm})(2.5\ \text{L})$
$= 10\ \text{atm} \cdot \text{L}$
$= \text{constant}$

Boyle's findings After performing many experiments with gases at constant temperatures, Boyle had four findings.
1) If the pressure of a gas increases, its volume decreases proportionately.
2) If the pressure of a gas decreases, its volume increases proportionately.
3) If the volume of a gas increases, its pressure decreases proportionately.
4) If the volume of a gas decreases, its pressure increases proportionately.

Because the changes in pressure and volume are always opposite and proportional, the relationship between pressure and volume is an inverse proportion. By using inverse proportions, all four findings can be included in one statement called Boyle's law. **Boyle's law** states that the pressure and volume of a gas at constant temperature are inversely proportional.

Graphing an inverse proportion Recall from Chapter 10 that the kinetic energy of an ideal gas is directly proportional to its temperature and that the graph of a direct proportion is a straight line. The graph of an inverse proportion is a curve like that shown in **Figure 11.8.** Just as a road runs both ways, you can think of a gas following the curve A-B-C or C-B-A. The path A-B-C represents the gas being compressed, which forces its pressure to rise. As the gas is compressed by half, from 10.0 L to 5.0 L, the pressure doubles from 1.0 atm to 2.0 atm. As it is compressed by half again, from 5.0 L to 2.5 L, the pressure again doubles from 2.0 atm to 4.0 atm. The reverse path C-B-A represents what happens when the pressure of the gas decreases and the volume increases accordingly.

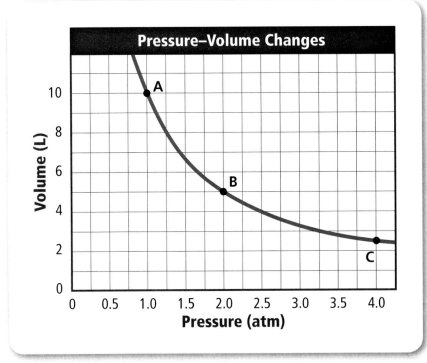

■ **Figure 11.8** This graph shows the inverse relationship between pressure and volume. As you follow the curve from left to right, the pressure increases and the volume decreases.

Apply *Use the graph to determine the volume if the pressure is 2.5 atm.*

■ **Figure 11.9** At an altitude of 15 km, atmospheric pressure is only 1/10 as great as the air pressure at sea level. When a balloon that was filled with helium at 1.00 atm reaches this altitude, its volume is about ten times as large as it was on the ground.

Applying Boyle's Law The weather balloon in **Figure 11.9** illustrates Boyle's law. When helium gas is pumped into the balloon, the balloon inflates, just as a soccer ball does, until the pressure of gas inside equals the pressure of the air outside. Because helium is less dense than air, the mass of helium gas is less than the mass of the same volume of air at the same temperature and pressure. As a result, the balloon rises. As it climbs to higher altitudes, atmospheric pressure decreases. According to Boyle's law, because the pressure on the helium decreases, its volume increases. The balloon continues to rise until the pressures inside and outside are equal. It hovers at this altitude and records weather data.

Boyle's law can also be used to explain the discomfort you may feel in your ears when ascending in an airplane or driving up into the mountains. When you fly or drive to a higher altitude, the air pressure around you and in your ears decreases, causing the volume of the air in your ears to increase. Air is released from your ears resulting in slight discomfort that is relieved as soon as the air bubbles in your ears pop.

Kinetic explanation of Boyle's Law You know that by compressing air in a tire pump, air pressure increases and volume of air decreases. According to the kinetic theory, if the temperature of a gas is constant and the gas is compressed, its pressure must increase. Boyle's law, based on volume and pressure measurements made on confined gases at constant temperature, quantified this relationship by stating that the volume and pressure are inversely proportional. **Figure 11.10** shows how the kinetic theory explains Boyle's observations of gas behavior.

■ **Figure 11.10** The kinetic theory relates pressure and the number of collisions per unit time for a gas.

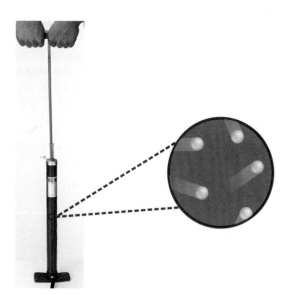

When the bicycle pump is pulled out as far as it will go, the pressure of the air inside the pump equals that of the atmosphere.

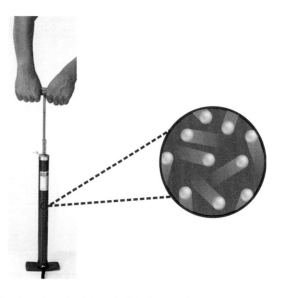

If the piston is pushed down half the length of the pump, the air particles are squeezed into a space half the original size. Pressure doubles because the frequency of collisions between the gas particles and the inner wall of the pump has doubled.

Earth Science Connection

Weather Balloons

If you watch the sky any day at 6 A.M. Eastern Standard Time (EST), you may see a weather balloon like the one in **Figure 1.** Every 12 hours, an instrument-packed helium or hydrogen balloon is launched by the National Weather Service at each of about 100 sites in and around the United States. The balloons provide the data used to forecast weather across the country.

Weather balloon system Since World War II, meteorologists have used weather balloons to provide profiles of temperature, pressure, relative humidity, and wind velocity in the upper atmosphere. There are about 900 launching sites around the world. The balloons are released at the same time in all countries—at 00:00 hours and 12:00 hours Coordinated Universal Time (UTC), which is 6 A.M. and 6 P.M. EST.

Data from weather stations all over the world reach the United States via a central computer near Washington D.C. From there, the data are sent to regional weather service stations throughout the country for release to newspapers and television and radio networks.

Sending the balloons aloft The synthetic rubber weather balloons are inflated with either helium or hydrogen to a diameter of about 2 m. An instrument package, called a radiosonde, is attached to each balloon. As the balloon rises, it expands to about 6 m. When it reaches a height of about 35 km, the volume of the expanded gas is so great that the balloon bursts. Then a parachute carries the radiosonde safely to the ground. At the highest altitudes, the radiosondes endure temperatures as cold as -95°C and wind speeds of almost 320 km/hr.

The flights last for about 2 hours and the balloons can drift to 200 km away. About 20 percent of the 75,000 balloons launched each year in the United States are found and returned to the National Weather Service. The National Weather Service then fixes and reuses the radiosondes.

Figure 1 Helium-filled weather balloon

Gathering data While aloft, the radiosonde transmits temperature, relative humidity, and barometric pressure data to the ground by radio transmission every 1–2 seconds. The system transmits the different kinds of data in rotation, for example, temperature first, followed by humidity, and then pressure.

Relative humidity readings are obtained from a device containing a polymer that swells when it becomes moist. The swelling causes an increase in the electrical resistance of a carbon layer in the polymer. Electrical measurements of the resistance of the material indirectly measure changes in relative humidity. Wind velocity and wind direction are determined by tracking the balloon by radar or by using a global positioning system (GPS).

Connection to Chemistry

1. **Think Critically** Some people thought that when weather satellites were deployed, weather balloons would become obsolete. Why do you think this did not happen?
2. **Infer** Why would a meteorologist have to understand the gas laws?

MiniLab 11.2

How Straws Work

Why do straws work? Each of us has used straws to sip soft drinks, milk, or another liquid from a container.

Procedure

1. Read and complete the safety form.
2. Obtain a clean, empty **food jar** with a **screw-on lid.** Fill the jar halfway with tap **water.**
3. Put the ends of two **soft-drink straws** into your mouth. Place the other end of one of the straws into the water in the jar and allow the end of the other straw to remain in the air. Try to draw water up the straw by sucking on both straws simultaneously. Record your observations.
4. Using a **hammer** and a **nail,** punch a hole of about the same diameter as that of a straw in the lid of the jar. Push the straw 2 to 3 cm into the hole and seal the straw in position with **wax** or **clay.**
5. Fill the jar to the brim with water, and carefully screw the lid on so that no air enters the jar. Try to draw the water up the straw. Record your observations.

Analysis

1. **Explain** your observations in step 3.
2. **Describe** your observations in step 5.
3. **Diagram** how a soft-drink straw works.

EXAMPLE Problem 11.3

Boyle's Law: Determining Volume The maximum volume of air that a weather balloon can hold without rupturing is 22,000 L. It is designed to reach an altitude of 30 km. At this altitude, the atmospheric pressure is 0.0125 atm. What maximum volume of helium gas should be used to inflate the balloon before it is launched?

1 Analyze
- At 30 km, the pressure exerted by the helium in the balloon equals the atmospheric pressure at that altitude, 0.0125 atm. What volume does this amount of helium occupy at 1 atm? The pressure is greater at high altitude by the factor 1.0 atm/0.0125 atm, which is greater than 1. According to Boyle's law, the volume at launch is less than at high altitude by the factor 0.0125 atm/1.0 atm, which is less than 1.

2 Set up
- Multiply the maximum volume by the factor given by Boyle's law.

$$V = 22{,}000 \text{ L} \times \left(\frac{0.0125 \text{ atm}}{1.0 \text{ atm}}\right)$$

3 Solve
- Multiply and divide the values and units.

$$V = 22{,}000 \text{ L} \times \left(\frac{0.0125 \text{ atm}}{1.0 \text{ atm}}\right) = \frac{22{,}000 \text{ L} \times 0.0125}{1.0} = 275 \text{ L}$$

4 Check
- As expected, the volume is less at sea level because the gas is compressed.

EXAMPLE Problem 11.4

Boyle's Law: Determining Pressure Two liters of air at atmospheric pressure are compressed into the 0.45-L canister of a warning horn. If its temperature remains constant, what is the pressure of the compressed air in the horn?

1 Analyze
- The initial pressure of the air that was forced into the canister was atmospheric pressure, 1.00 atm. Because the volume of air is reduced, its pressure increases. Multiply the pressure by the factor with a value greater than 1.

2 Set up

$$P = 1.00 \text{ atm} \times \left(\frac{2.0 \text{ L}}{0.45 \text{ L}}\right)$$

3 Solve

$$P = 1.00 \text{ atm} \times \left(\frac{2.0 \text{ L}}{0.45 \text{ L}}\right) = \frac{1.00 \text{ atm} \times 2.0}{0.45} = 4.4 \text{ atm}$$

4 Check
- Estimate and reason: Because the volume of the air was reduced from 2 L to about half a liter, the volume changed by the factor $\frac{0.5}{2}$ or about $\frac{1}{4}$. Thus, the pressure changed by a factor of about $\frac{4}{1}$. The final pressure, 4.4 atm, is about four times as large as the initial pressure, 1.00 atm.

PRACTICE Problems

Solutions to Problems Page 860

Assume that the temperature remains constant in the following problems.

11. Bacteria produce methane gas in sewage treatment plants. This gas is often captured or burned. If a bacterial culture produces 60.0 mL of methane gas at 700.0 mm Hg, what volume would be produced at 760.0 mm Hg?

12. At one sewage treatment plant, bacteria cultures produce 1000 L of methane gas per day at 1.0 atm pressure. What volume tank would be needed to store one day's production at 5.0 atm?

13. Hospitals buy 400-L cylinders of oxygen gas compressed at 150 atm. They administer oxygen to patients at 3.0 atm in a hyperbaric oxygen chamber. What volume of oxygen can a cylinder supply at this pressure?

14. If the valve in a tire pump with a volume of 0.78 L fails at a pressure of 9.00 atm, what would be the volume of air in the cylinder just before the valve fails?

15. The volume of a scuba tank is 10.0 L. It contains a mixture of nitrogen and oxygen at 290.0 atm. What volume of this mixture could the tank supply to a diver at 2.40 atm?

16. A 1.00-L balloon is filled with helium at 1.20 atm. If the balloon is squeezed into a 0.500-L beaker and doesn't burst, what is the pressure of the helium?

SUPPLEMENTAL PRACTICE

For more practice with Boyle's law, see Supplemental Practice, page 822.

CHEMLAB

BOYLE'S LAW

Background
The quantitative relationship between the volume of a gas and the pressure of the gas, at constant temperature, is known as Boyle's Law. By measuring quantities directly related to the pressure and volume of a tiny amount of trapped air, you can deduce Boyle's law.

Questions
What is the relationship between the volume and the pressure of a gas at constant temperature?

Objectives
- **Observe** the length of a column of trapped air at different pressures.
- **Examine** the mathematical relationship between gas volume and gas pressure.

Preparation

Materials
thin-stem pipette
double-post screw clamp
fine-tip marker
food coloring
matches
metric ruler
scissors
small beaker
water

Safety Precautions
WARNING: *Use care in lighting matches and melting the pipette stem.*

Procedure

1. Read and complete the lab safety form.
2. Cut off the stepped portion of the stem of the pipette with the scissors.
3. Place about 20 mL of water in a beaker, add a few drops of food coloring, and swirl to mix.
4. Draw the water into the pipette, completely filling the bulb and allowing the water to extend about 5 mm into the stem of the pipette.
5. Heat the tip of the pipette gently above a flame until it is soft. **WARNING:** *If the stem accidentally begins to burn, blow it out.* Use a metallic or glass object to flatten the tip against the countertop so that the water and air are sealed inside the pipette.
6. Holding the pipette by the bulb, tap any water droplets in the stem down into the liquid. You should observe a cylindrical column of air trapped in the stem of the pipette.
7. Center the bulb in a double-post screw clamp, and tighten the clamp until the bulb is just held firmly. Mark the knob of the clamp with a fine-tip marker, and tighten the clamp three or four turns so that the length of the air column is 50 to 55 mm.

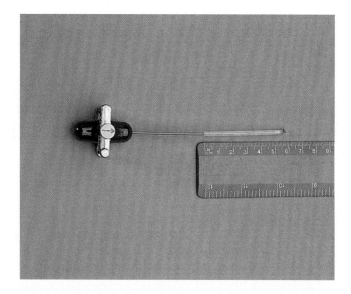

8. Record this number of turns T under Trial 1 in a data table like the one shown. Measure and record the length L of the air column in millimeters for Trial 1.
9. Turn the clamp knob one complete turn, and record the trial number and the total number of turns. Measure and record the length of the air column.
10. Repeat step 9 until the air column is reduced to a length of 25 to 30 mm.

Analyze and Conclude

1. **Observe** Explain whether the volume V of the air in the stem is directly proportional to the length L of the air column.
2. **Infer** What inferences can be made about the pressure, P, of the air in the column and the number of turns, T, of the clamp screw?
3. **Interpret Data** Calculate the product LT and the quotient L/T for each trial. Which calculations are more consistent? If L and T are directly related, L/T will yield nearly constant values for each trial. On the other hand, if L and T are inversely related, LT will yield almost constant values for each trial. Are L and T directly or inversely related?
4. **Draw Conclusions** Explain whether the data indicate that gas volume and gas pressure at constant temperature are directly related or inversely related.

Apply and Assess

1. **Interpret** For a mercury column barometer to measure atmospheric pressure as 760 mm Hg, the column containing the mercury must be completely evacuated. However, if the column is not completely evacuated, the barometer can still be used to correctly measure changes in barometric pressure. How is the second statement related to Boyle's law?
2. **Explain** Using the kinetic theory, explain how a decrease in the volume of a gas causes an increase in the pressure of the gas.

Data and Observations

Data Table				
Trial	Turns, T	Length of Air Column, L	Numerical Value, LT	Numerical Value, L/T

INQUIRY EXTENSION

Observe the length of a column of trapped air at different pressures. Examine the mathematical relationship between gas volume and gas pressure.

CHEMISTRY AND TECHNOLOGY

Health Under Pressure

You live, work and play in air that is generally about 1 atm in pressure and 21% oxygen. Have you ever wondered what might happen if the pressure and the oxygen content of the air were greater? Would you recover from illness or injury more quickly? These questions are at the heart of hyperbaric medicine.

Hyperbaric medicine The prefix *hyper-* means *above or excessive,* and a bar is a unit of pressure equal to 100 kPa, roughly normal atmospheric pressure. Thus, the term *hyperbaric* refers to pressure that is greater than normal. Patients receiving hyperbaric therapy are exposed to pressures greater than atmospheric pressure at sea level.

The oxygen connection Greater pressure is most often combined with an increase in the concentration of oxygen a patient receives. The phrase *hyperbaric oxygen therapy* (HBOT) refers to treatment with 100% oxygen. **Figure 1** shows a chamber that might be used for HBOT. Inside the hyperbaric chamber, pressures can reach five to six times normal atmospheric pressure. At hyperbaric therapy centers across the country, HBOT is used to treat a wide range of conditions, including burns, decompression sickness, slow-healing wounds, anemia, and some infections.

Figure 1 During HBOT, the patient lies in a hyperbaric chamber. A technician controls the pressure and oxygen levels.

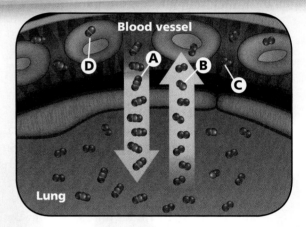

Figure 2 Gases are exchanged between the lungs and the circulatory system.

Carbon-monoxide poisoning Use **Figure 2** to help you understand how HBOT aids in the treatment of carbon-monoxide poisoning.
Normal gas exchange Oxygen (O_2) moves from the lungs to the blood and binds to the hemoglobin in red blood cells. Carbon dioxide (CO_2) is released, as shown by **A**.
Abnormal gas exchange If carbon monoxide (CO) enters the blood, as shown by **B**, it binds to the hemoglobin instead of oxygen. Cells in the body begin to die from oxygen deprivation.
Oxygen in blood plasma In addition to the oxygen carried by hemoglobin, oxygen is dissolved in the blood plasma, as shown by **C**. HBOT increases the concentration of dissolved oxygen to an amount that can sustain the body.
Eliminating carbon monoxide Pressurized oxygen also helps remove any carbon monoxide bound to hemoglobin, as shown by **D**.

Discuss the Technology

1. **Think Critically** What two factors cause the remarkable effects of HBOT? Explain how.
2. **Hypothesize** How might HBOT be effective for treating second- and third-degree burns over large portions of the body?

Charles's Law: Temperature and Volume

You may have observed the beautiful patterns and graceful gliding of a hot-air balloon, but what happens to a balloon when it is cold? **Figure 11.11** shows some dramatic effects of cooling a gas-filled balloon. In **Figure 11.11**, balloons are placed in a beaker of liquid nitrogen at a temperature of 77 K. As the liquid nitrogen rapidly cools the balloons, they shrink. When the balloons are removed from the beaker, warm air in the room causes the balloons to expand.

The French scientist Jacques Charles (1746–1823) didn't have liquid nitrogen, but he was a pioneer in hot-air ballooning. He investigated how changing the temperature of a fixed amount of gas at constant pressure affected its volume. The relationship Charles found is demonstrated in **Figure 11.12**.

■ **Figure 11.11** The volume of a gas is directly related to temperature. Balloons placed in a beaker of liquid nitrogen shrink instantly. When they are removed from the beaker and are exposed to warmer room temperatures, the balloons expand.

FOLDABLES
Incorporate information from this section into your Foldable.

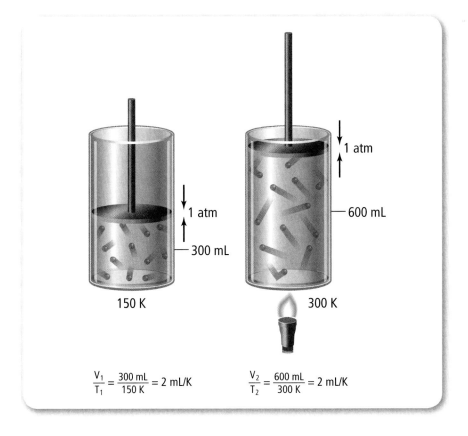

$$\frac{V_1}{T_1} = \frac{300 \text{ mL}}{150 \text{ K}} = 2 \text{ mL/K} \qquad \frac{V_2}{T_2} = \frac{600 \text{ mL}}{300 \text{ K}} = 2 \text{ mL/K}$$

■ **Figure 11.12** When the cylinder is heated, the kinetic energy of the gas particles increases, causing them to push the piston outward.

Section 11.2 • The Gas Laws

■ **Figure 11.13** The three straight lines show that the volume of each gas is directly proportional to its Kelvin temperature. The solid part of each line represents actual data. Part of each line is dashed because, as you recall from Chapter 10, when the temperature of a gas falls below its boiling point, a gas condenses to a liquid.

Compare the volumes of gases A and B at 350 K.

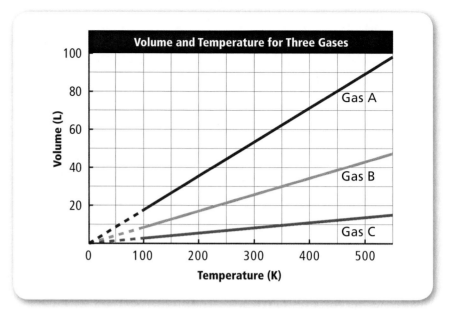

Concepts In Motion

Interactive Figure To see an animation of the gas laws, visit glencoe.com.

Charles's law states that at constant pressure, the volume of a gas is directly proportional to its Kelvin temperature, as shown in the graph in **Figure 11.13.** The straight lines for each gas indicate that volume and temperature are in direct proportions. For example, if the Kelvin temperature doubles, the volume doubles, and if the Kelvin temperature is halved, the volume is halved.

Kinetic explanation of Charles's law Why did the air in the balloons in **Figure 11.11** contract when cooled with liquid nitrogen? Study **Figure 11.14** to learn how the kinetic theory of matter explains the balloons and Charles's law. From the figure, you can see that decreasing the temperature decreases both the number of collisions and the force of the collisions that the particles of air make with the walls of the balloon. If the walls of the balloon were rigid (constant volume), the result would be a decrease in pressure. However, because the walls of the balloon flex at constant pressure, the result is a decrease in volume. Heating the air inside the balloon has exactly the opposite result.

■ **Figure 11.14** When the balloon is heated, the temperature of the air inside increases, and the average kinetic energy of the particles in the air also increases. They exert more force on the balloon, but the pressure inside does not rise above the original pressure because the balloon expands.

When the balloon cools, the temperature of the air inside falls and the average kinetic energy of the particles in the air decreases. The particles move slower and strike the balloon less often and with less force. The balloon contracts and the pressure of the air inside the balloon continues to balance the pressure of the atmosphere.

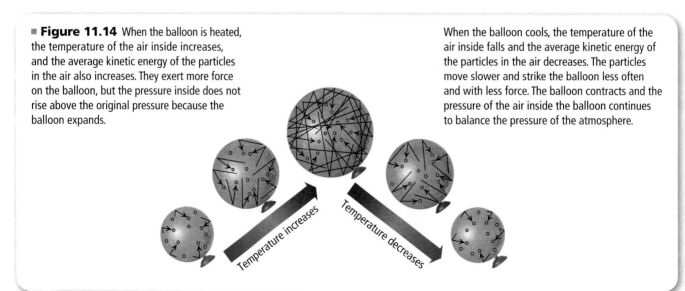

EXAMPLE Problem 11.5

Charles's Law A balloon is filled with 3.0 L of helium at 22°C and 760 mm Hg. It is then placed outdoors on a hot summer day when the temperature is 31°C. If the pressure remains constant, what will the volume of the balloon be?

1 Analyze

Because the volume of a gas is proportional to its Kelvin temperature, you must first express the temperatures in this problem in kelvins. As in Chapter 10, add 273 to the Celsius temperature to obtain the Kelvin temperature.

$T_K = T_C + 273$
$T_K = 22 + 273 = 295$ K
$T_K = 31 + 273 = 304$ K

Because the temperature of the helium increases from 295 K to 304 K, its volume increases in direct proportion. The temperature increases by the factor 304 K/295 K. Therefore, the volume increases by the same factor.

2 Set up

$V = 3.0 \text{ L} \times \left(\dfrac{304 \text{ K}}{295 \text{ K}}\right)$

3 Solve

$V = 3.0 \text{ L} \times \left(\dfrac{304 \text{ K}}{295 \text{ K}}\right) = \dfrac{3.0 \text{ L} \times 304}{295} = 3.1 \text{ L}$

4 Check

Does the answer have volume units? Does the volume increase as expected? Because the answer to both questions is yes, this solution is reasonable. Check your calculations to make sure your answer is correct.

> **Problem-Solving Hint**
> Remember that gas volumes and pressures are proportional to temperature only if the temperature is expressed in kelvins.

PRACTICE Problems

Solutions to Problems Page 860

Assume that the pressure remains constant in the following problems.

17. A balloon is filled with 3.0 L of helium at 310 K and 1 atm. The balloon is placed in an oven where the temperature reaches 340 K. What is the new volume of the balloon?

18. A 4.0-L sample of methane gas is collected at 30.0°C. Predict the volume of the sample at 0°C.

19. A 25-L sample of nitrogen is heated from 110°C to 260°C. What volume will the sample occupy at the higher temperature?

20. The volume of a 16-g sample of oxygen is 11.2 L at 273 K and 1.00 atm. Predict the volume of the sample at 409 K.

21. The volume of a sample of argon is 8.5 mL at 15°C and 101 kPa. What will its volume be at 0.00°C and 101 kPa?

SUPPLEMENTAL PRACTICE

For more practice with Charles's law, see Supplemental Practice, page 822.

FOLDABLES
Incorporate information from this section into your Foldable.

Combined Gas Law

You know that according to Boyle's law, if you double the volume of a gas while keeping the temperature constant, the pressure falls to half its initial value. You also know that according to Charles's law, if you double the Kelvin temperature of a gas while keeping the pressure constant, the volume doubles. One application of these gas laws can be seen at hot-air balloon events, as shown in **Figure 11.15**. What do you think would happen to the pressure if you doubled the volume and doubled the temperature?

Like Robert Boyle and Jacques Charles, you could investigate this question experimentally. You could measure temperature, pressure, and volume for a sample of gas; expand the gas to twice the volume; raise the temperature to twice as high; and then measure the pressure. Following accepted principles of scientific methods, you would make many such experiments covering a wide range of data. You could then look for relationships among your three variables.

As you performed your experiments, any one of the variables could be kept constant while varying another and measuring the effect on the third. Ultimately, you would find that doubling the volume first and the temperature second has exactly the same result as doubling the temperature first and the volume second. These results should not be surprising in light of the kinetic explanations of Boyle's and Charles's laws. Each of these laws is based on the behavior of the particles that make up a gas. Any time you change one of the three variables—pressure, temperature, or volume—you affect those particles in a predictable, consistent way.

Combining laws: step-by-step Because Boyle's law of gases and Charles's law of gases are equally valid, it is possible to determine one of the three variables, regardless of the order in which the other two are changed. If you doubled the volume and the temperature at the same time, you would get exactly the same result as if you had doubled one first, then the other. This means that you can approach problems where two of the three variables change in a simple, step-by-step fashion: first look at the effects of changing one variable, then the effects of changing the second. The order that you take the variables does not matter.

For example, suppose you had 3 L of gas at 200 K and 1 atm. If you double the volume, according to Boyle's law, the pressure falls to 0.5 atm. If you then double the temperature, according to Charles's law, the pressure is increased to twice as high again, or 1 atm. Taken in the other order, doubling the temperature send the pressure to 2 atm; doubling the volume brings it down to 1 atm, the same answer.

Note that doubling the volume and then doubling the temperature by the same factor brings a sample of gas back to its initial pressure. The effect on the pressure of doubling the volume and doubling the temperature offset each other because pressure is inversely proportional to volume but directly proportional to temperature.

■ **Figure 11.15** Hot-air ballooning is a popular sport all around the world. The height of the balloon over the ground is controlled by varying the temperature of the air within the balloon.

The combination of Boyle's law and Charles's law is called the **combined gas law**. The factors are the same as in the previous example problems, but you may have more than one factor in a problem because more than one quantity may vary. **Table 11.2** identifies factors that are held constant in each gas law. Some of the problems that follow will ask you to consider gases at **standard temperature and pressure**, or **STP**. STP is defined as a temperature of 0.00°C or 273 K and a pressure of 1 atm.

Table 11.2 The Gas Laws

Law	Boyle's	Charles's	Combined
What is constant?	amount of gas, temperature	amount of gas, pressure	amount of gas

EXAMPLE Problem 11.6

Determining Volumes at STP A 154-mL sample of carbon dioxide gas is generated by burning graphite in pure oxygen. If the pressure of the generated gas is 121 kPa and its temperature is 117°C, what volume would the gas occupy at standard temperature and pressure, STP?

1 Analyze

Reducing the pressure from 121 kPa to 101 kPa increases the volume of the carbon dioxide gas. Therefore, Boyle's law gives the factor 121 kPa/101 kPa. Because this factor is greater than 1, when it is multiplied by the volume, the volume increases. Cooling the gas from 117°C to 0.00°C reduces the volume of gas. To apply Charles's law, you must express both temperatures in kelvins.

$$T_K = T_C + 273 \qquad T_K = T_C + 273$$
$$= 117 + 273 \qquad\quad = 0.00 + 273$$
$$= 390 \text{ K} \qquad\qquad = 273 \text{ K}$$

Because the temperature decreases, the volume decreases. Charles's law gives the factor 273 K/390 K, which is less than 1. Therefore, multiplying the volume by this factor decreases the volume.

2 Set up

Multiply the volume by these two factors.

$$V = 154 \text{ mL} \times \left(\frac{121 \text{ kPa}}{101 \text{ kPa}}\right) \times \left(\frac{273 \text{ K}}{390 \text{ K}}\right)$$

3 Solve

The combined gas law equation is solved in the following steps.

$$V = 154 \text{ mL} \times \left(\frac{121 \text{ kPa}}{101 \text{ kPa}}\right) \times \left(\frac{273 \text{ K}}{390 \text{ K}}\right) = \frac{154 \text{ mL} \times 121 \times 273}{101 \times 390} = 129 \text{ mL}$$

4 Evaluate

Estimate to see whether the answer is reasonable. The reduction in pressure would expand the volume by a factor of about 12/10. Cooling the gas would contract it by a factor of about 7/10. Both factors together would alter the volume by a factor of about 84/100, which is less than 1. The final volume should be less than the initial volume. The final volume, 129 mL, is less than 154 mL, the initial volume.

SUPPLEMENTAL PRACTICE

For more practice with the combined gas law, see Supplemental Practice, page 822.

PRACTICE Problems

Solutions to Problems Page 861

22. A 2.7-L sample of nitrogen is collected at 121 kPa and 288 K. If the pressure increases to 202 kPa and the temperature rises to 303 K, what volume will the nitrogen occupy?

23. A chunk of subliming carbon dioxide (dry ice) generates a 0.80-L sample of gaseous CO_2 at 22°C and 720 mm Hg. What volume will the carbon dioxide gas have at STP?

The Law of Combining Gas Volumes

When water decomposes into its elements—hydrogen and oxygen gas—the volume of hydrogen produced is always larger than the volume of oxygen produced. Consider the formula H_2O; because matter is conserved, in the reverse synthesis reaction, the volume of hydrogen gas that reacts must always be twice the volume of oxygen gas.

Experiments with other gas reactions indicate that volumes of gases always react in ratios of small whole numbers. **Figure 11.16** presents the volume ratios for the decomposition of gaseous hydrogen chloride (HCl) into hydrogen gas and chlorine gas. These volume ratios are the same for the reverse reaction, the synthesis of HCl from hydrogen gas (H_2) and chlorine gas (Cl_2). The observation that at the same temperature and pressure, volumes of gases combine or decompose in ratios of small whole numbers is called the **law of combining gas volumes**.

FOLDABLES
Incorporate information from this section into your Foldable.

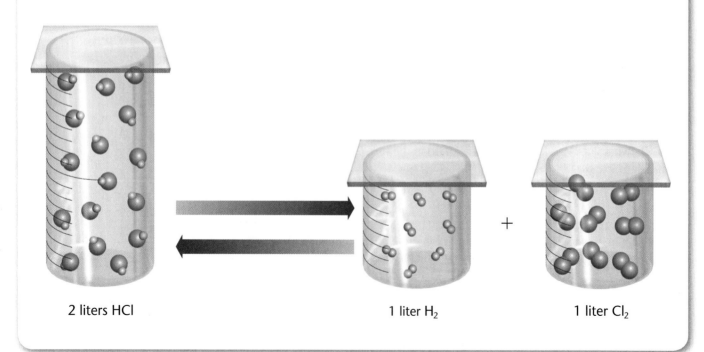

■ **Figure 11.16** When two liters of hydrogen chloride gas decompose to form hydrogen gas and chlorine gas, equal volumes of hydrogen gas and chlorine gas are formed—1 L of each. The ratio of hydrogen to chlorine is 1 to 1, and the ratio of volumes of hydrogen to hydrogen chloride is 1/2 to 1 or 1 to 2, the same as the ratio of chlorine to hydrogen chloride.

Infer *Consider the reverse reaction—the synthesis of hydrogen gas with chlorine gas to form hydrogen chloride gas. What is the ratio of the reactants and the ratio of the product to each reactant?*

2 liters HCl   1 liter H_2 1 liter Cl_2

Everyday Chemistry

Popping Corn

You recognize the smell, and that characteristic popping noise is a dead giveaway. Popcorn! What causes the kernels of popcorn to experience explosive change?

History of popcorn The oldest known ears of popcorn are 5600 years old. They were discovered in a cave in New Mexico. One-thousand-year-old popcorn has been discovered in a tomb in Peru. Native Americans popped popcorn in special jugs that they put in heated sand. These jugs allowed all the popcorn to be heated evenly and stopped the corn from going everywhere. Popcorn was also used as decoration for ceremonial headdresses, necklaces, and ornaments on statues of gods. Originally, popcorn only grew in North and South America and China, so Europeans did not find out about popcorn until the New World was discovered. Today, people in the United States eat about 66 liters of popcorn per year; more than any other people in the world.

Popcorn kernels Popcorn is the seed made by a popcorn plant. There are 25 different varieties of corn (maize) that are used for popcorn. A popcorn kernel is extremely small and hard. It has a hard, airtight shell, called a hull. The hull protects the embryo and its food supply. This food supply is starch located within the endosperm. Each kernel also contains a small amount of water.

Exploding the kernel We now know that popcorn pops because there is water inside each kernel. When heated to about 204°C, the water in the kernel turns to steam. The pressure from the expanding steam makes the hard shell pop with an explosive force. The shell explodes turning the kernel inside out, as shown in **Figure 1**. The white starchy center is now on the outside. This is why all popped popcorn is white, no matter what the color of the hull. When the shell explodes, the popcorn bursts open to 20 to 40 times its original size. The heat released by the steam bakes the starch into the fluffy product people like to eat.

Figure 1 Time-lapse photo of an exploding popcorn kernel

Water content The amount of water in the kernel is an important aspect of popping. Food chemists have found that the kernel must contain about 13.5 percent water by mass to pop properly. The kernels that do not pop properly do not have enough moisture in them. This can happen two ways: the hull is damaged, allowing the inside moisture to evaporate, or there was not enough moisture inside the hull when the seed formed. When popcorn is too dry, three things can happen: the kernels may pop smaller than normal, they may only split when heated, or they may not pop at all.

Explore Further

1. **Hypothesize** Suggest why too much water present in the kernel greatly increases the number of unpopped kernels.
2. **Apply** Why is popcorn better stored in the freezer or refrigerator rather than on the shelf at room temperature?
3. **Acquire Information** More than 1000 varieties of corn are grown in the world today. Write a report about as many uses for corn and corn products as you can.

■ **Figure 11.17** Avogadro determined that water is composed of particles.

Gas volume and the number of particles In the synthesis of water, two volumes of hydrogen react with one of oxygen, but only two volumes of water vapor are formed. You might expect three. In the reverse reaction, which is the decomposition of water, two volumes of water vapor yield one volume of oxygen gas and two volumes of hydrogen gas. That mysterious third volume appears again. How can we account for it?

Amedeo Avogadro (1776–1856), an Italian physicist, made the same observations and asked the same question. He first noticed that when water forms oxygen and hydrogen, two volumes of gas become three volumes of gas, and he hypothesized that water vapor consisted of particles, as shown in **Figure 11.17**. Each particle in the water vapor broke up into two hydrogen parts and one oxygen part. Two volumes of hydrogen and one volume of oxygen formed because there were twice as many hydrogen particles as oxygen particles. Today, we can show this by doing something that Avogadro could not do—write the chemical equation for the formation of water.

$$2H_2(g) + O_2(g) \rightarrow 2H_2O(g)$$

Avogadro's principle Avogadro was the first to interpret the law of combining volumes in terms of interacting particles. He reasoned that the volume of a gas at a given temperature and pressure must depend on the number of gas particles. **Avogadro's principle** states that equal volumes of gases at the same temperature and pressure contain equal numbers of particles.

Connecting Ideas

Knowing that equal volumes of gases at constant temperature and pressure have the same number of particles is the first step toward determining how many particles are present in a sample of gas. Knowing the number of particles in a reactant and the ratio in which the particles interact enables us to predict the amount of a product, which you will investigate in chapter 12.

FACT of the Matter

Because of his observations of reactions involving oxygen, nitrogen, and hydrogen gases, Avogadro was the first to suggest that these gases were made up of diatomic molecules.

Section 11.2 Assessment

Section Summary

▶ Boyle's law states that the pressure and volume of a confined gas are inversely proportional.

▶ Charles's law states that the volume of any sample of gas at constant pressure is directly proportional to its Kelvin temperature.

▶ Avogadro's principle states that equal volumes of gases at the same temperature and pressure contain equal numbers of particles.

24. **MAIN Idea** **Summarize** The gas laws relate the following variables—the number of gas particles, temperature, pressure, and volume. Which of these are held constant in Boyle's law, in Charles's law, and in the combined gas law?

25. **Explain** how an air mattress supports the weight of a person lying on it.

26. **Infer** Why is it important to determine the volume of a gas at STP?

27. **Analyze** If the volume of a helium-filled balloon increases by 20 percent at constant temperature, what will the percentage change in the pressure be?

28. **Explain** Use the kinetic theory to explain why pressurized cans carry the message "Do not incinerate."

29. **Explain** Which of the three laws that apply to equal amounts of gases are directly proportional? Which are inversely proportional?

CHAPTER 11 Study Guide

Download quizzes, key terms, and flash cards from glencoe.com.

BIG Idea Gases respond in predictable ways to pressure, temperature, volume, and changes in number of particles.

Section 11.1 Gas Pressure

MAIN Idea Gas pressure is related to the mass of the gas and to the motion of the gas particles.

Vocabulary
- barometer (p. 374)
- pascal (Pa) (p. 376)
- standard atmosphere (atm) (p. 374)

Key Concepts
- The pressure of a gas at constant temperature and volume is directly proportional to the number of gas particles.
- The volume of a gas at constant temperature and pressure is directly proportional to the number of gas particles.
- At sea level, the pressure exerted by gases of the atmosphere equals one standard atmosphere (1 atm).

Section 11.2 The Gas Laws

MAIN Idea For a fixed amount of gas, a change in one variable—pressure, temperature, or volume—affects the other two.

Vocabulary
- Avogadro's principle (p. 396)
- Boyle's law (p. 381)
- Charles's law (p. 390)
- combined gas law (p. 393)
- law of combining gas volumes (p. 394)
- standard temperature and pressure, STP (p. 393)

Key Concepts
- Boyle's law states that the pressure and volume of a confined gas are inversely proportional.
- Charles's law states that the volume of any sample of gas at constant pressure is directly proportional to its Kelvin temperature.
- Avogadro's principle states that equal volumes of gases at the same temperature and pressure contain equal numbers of particles.

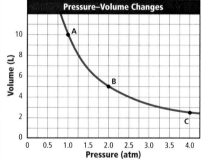

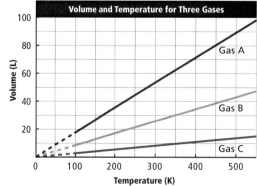

Chapter 11 Assessment

Understand Concepts

30. Name three ways to increase the pressure inside an oxygen tank.

31. What temperature in degrees Celsius corresponds to absolute zero?

32. In a chemical plant, xenon hexafluoride is stored at STP. Express these conditions using the Kelvin scale.

33. What physical conditions are specified by the term STP?

34. What factors affect gas pressure?

35. Convert the pressure measurement 547 mm Hg to all of the following units.
    a) psi   c) in of Hg
    b) kPa   d) atm

36. An oxygen storage tank contains 12.0 L at 25°C and 3 atm. After a 4-L canister of oxygen at 25°C and 3 atm is emptied into the storage tank, what do you expect the temperature and pressure of the tank to be?

Apply Concepts

37. The passenger cabin of an airplane is pressurized. Explain what this term means and why it is done.

38. Explain why balloons filled with helium rise in air but balloons filled with carbon dioxide sink.

39. Complete the concept map in **Figure 11.18** that shows how Boyle's law and Charles' law are derived from the combined gas law.

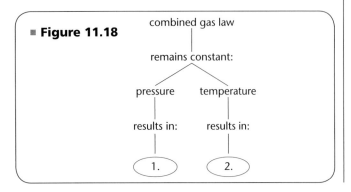

■ **Figure 11.18**

40. At 1250 mm Hg and 75.0°C, the volume of a sample of ammonia gas is 6.28 L. What volume would the ammonia occupy at STP?

41. The volume of a gas is 550 mL at 760 mm Hg. If the pressure is reduced to 380 mm Hg and the temperature is constant, what will be the new volume?

42. A 375-mL sample of air at STP is heated at constant volume until its pressure increases to 980 mm Hg. What is the new temperature of the sample?

43. A sample of neon gas has a volume of 822 mL at 160°C. The sample is cooled at constant pressure to a volume of 586 mL. What is its new temperature?

44. The volume of a gas is 1.5 L at 27°C and 1 atm. What volume will the gas occupy if the temperature is raised to 127°C at constant pressure?

45. A 50.0-mL sample of krypton gas at STP is cooled to -73°C at constant pressure. What will be the volume?

46. Use Boyle's law to calculate the missing values in **Table 11.3**.

Table 11.3	Boyle's Law		
V_1	P_1	V_2	P_2
1.4 L	?	3.0 L	1.2 atm
2.0 L	0.82 atm	1.0 L	?
2.5 L	0.75 atm	?	1.3 atm
?	1.2 atm	3.0 L	1 atm

47. How many liters of hydrogen will be needed to react completely with 10 L of oxygen in the composition reaction of hydrogen peroxide?

48. The volume of gas at 101 kPa is 400 mL. If the pressure changes at constant temperature to 404 kPa, what will the new volume be?

49. A 50-mL sample of ammonia gas at STP is cooled to 73°C at constant pressure. What will the volume be?

Chapter 11 Assessment

50. A sample of carbon dioxide is collected at STP. Its volume is 500.0 L. What volume will it have if the pressure is 380.0 mm Hg and the temperature is 100.0°C?

51. A 120-mL sample of nitrogen is collected at 900.0 mm Hg and 0.0°C. What volume will the nitrogen have when its temperature rises to 100.0°C and the pressure increases to 15 atm?

Chemistry and Technology

52. Compare the amount of oxygen you would receive in an HBOT chamber with the amount you would receive breathing ordinary air.

Everyday Chemistry

53. Explain why the density of a popped kernel of popcorn is less than that of an unpopped kernel.

Earth Science Connection

54. Would chlorine gas be suitable for weather balloons? Explain.

Think Critically

Compare and Contrast

55. **ChemLab** How are inferences made about the relationship between volume and pressure in this chapter's ChemLab similar to inferences made from observing the apparatus described in **Figure 11.7**?

Observe and Infer

56. **MiniLab 11.1** How might you account for the possible popping of the bag containing the subliming dry ice?

Apply Concepts

57. **MiniLab 11.2** Explain how air pressure helps you transfer a liquid with a pipette.

Make and Use Graphs

58. Automobile tires become underinflated as temperatures drop during the winter months if no additional air is added to the tires at the start of the cold season. For every 12.2°C drop in temperature, the air pressure in a car's tires goes down by about 1 psi (14.7 psi equals 1 atm). Complete **Table 11.4.** Then make a graph illustrating how the air pressure in a tire changes over the temperature ranges from 5°C to –25°C, assuming you start with a pressure of 30.0 psi at 5°C.

Table 11.4	Tire Pressure Based on Temperature
Temperature (°C)	Pressure (psi)
5	
–5	
–15	
–25	

Make Predictions

59. Use Boyle's law to predict the change in density of a sample of air if its pressure is increased.

Cumulative Review

60. What are allotropes? Name three elements that occur as allotropes. (*Chapter 5*)

61. Can an atom of an element have six electrons in the 2p sublevel? (*Chapter 7*)

62. Classify the following bonds as ionic, covalent, or polar covalent using electronegativity values. (*Chapter 9*)
    a) NH c) BH
    b) BaCl d) NaI

Chapter 11 Assessment

63. How does an increase in atmospheric pressure affect the following? (*Chapter 10*)
    a) the boiling point of ice
    b) the melting point of ice

Skill Review

64. **Dimensional Analysis** In a laboratory, 400 L of methane are stored at 600 K and 1.25 atm. If the methane is forced into a 200-L tank and cooled to 300 K, how does the pressure change?

WRITING in Chemistry

65. Modern manufacturing applies the properties of gases in many ways. List three examples of things you see and use every day that apply gas properties. Include at least one household product. Write a short explanation of the property used in each example.

66. Research decompression sickness (also called the bends). Explain the causes and symptoms of decompression sickeness. Prepare an instruction pamphlet for novice scuba divers on how to avoid this problem.

67. Apply your knowledge of gas pressure to explain how a plunger can be used to open a clogged drainpipe.

68. Many early balloonists dreamed of completing a trip around the world in a hot-air balloon, a goal not achieved until 1999. Write about what you imagine a trip in a balloon would be like, including a description of how manipulating air temperature would allow you to control your altitude.

69. Investigate and explain the function of the regulators on the air tanks used by scuba divers.

Problem Solving

70. A sample of argon gas occupies 2.00 L at −33.0°C and 1.50 atm. What is its volume at 207°C and 2.00 atm?

71. A welding torch requires 4500 L of acetylene gas at 2 atm. If the acetylene is supplied by a 150 L tank, what is the pressure of the acetylene?

72. How many liters of hydrogen will be needed to react completely with 10 L oxygen in the synthesis reaction of hydrogen peroxide?

73. Use Charles' law to determine the accuracy of the data plotted in **Figure 11.19**.

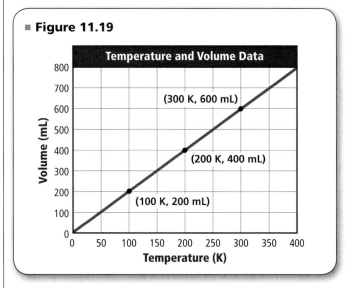

■ **Figure 11.19**

74. At a depth of 20 m, a 0.5-mL bubble of exhaled air is released from a scuba diver's mouthpiece. If the volume of the bubble just as it reaches the surface of the water is 1.5 mL, what was the pressure on the diver?

75. To produce 15.4 L of nitrogen dioxide at 300 K and 2 atm, how many liters of nitrogen gas and oxygen gas are required?

Cumulative Standardized Test Practice

1. Which of the following units is equivalent to 1.00 atm?
   a) 760 mm Hg
   b) 14.7 psi
   c) 101.3 kPa
   d) all of the above

2. Which of the following is a major component of organic compounds?
   a) sodium
   b) calcium
   c) carbon
   d) potassium

Use the graph to answer Questions 3–4.

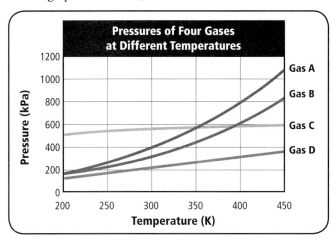

3. The graph shows that
   a) as temperature increases, pressure decreases.
   b) as pressure decreases, volume decreases.
   c) as temperature decreases, moles decrease.
   d) as pressure decreases, temperature decreases.

4. The predicted pressure of gas B at 310 K is
   a) 260 kPa.
   b) 620 kPa.
   c) 1000 kPa.
   d) 1200 kPa.

5. A sample of argon gas is compressed into a volume of 0.712 L by a piston exerting a pressure of 3.92 atm of pressure. The piston is slowly released until the pressure of the gas is 1.50 atm. The new volume of the gas is
   a) 0.272 L.
   b) 3.67 L.
   c) 1.8 L.
   d) 4.19 L.

6. How can a reaction be slowed down?
   a) add an inhibitor
   b) increase the activation energy
   c) add more reactants
   d) all of the above

7. Elements with d orbitals are called
   a) noble gases.
   b) transition elements.
   c) metals.
   d) nonmetals.

8. Which of the following is not a member of group 15?
   a) potassium
   b) arsenic
   c) bismuth
   d) antimony

9. Which compound will have a ΔEN equal to zero?
   a) CO
   b) H_2O
   c) O_2
   d) CH_4

10. Which statement explains the behavior of liquids?
    a) Liquid particles are attached to each other, and they can easily slide past each other.
    b) Liquid particles are attached and easily flow in straight lines when moved.
    c) Liquid particles are separated by spaces that allow them to move over each other.
    d) Liquid particles are held stiffly together until high temperatures loosen their bonds.

NEED EXTRA HELP?										
If You Missed Question . . .	1	2	3	4	5	6	7	8	9	10
Review Section . . .	11.1	5.2	11.1	11.1	11.2	6.3	7.2	8.1	9.1	10.1

CHAPTER 12 Chemical Quantities

BIG Idea The mole represents a large number of extremely small particles.

12.1 Counting Particles of Matter
MAIN Idea A mole always contains the same number of particles; however, moles of different substances have different masses.

12.2 Using Moles
MAIN Idea Balanced chemical equations relate moles of reactants to moles of products.

ChemFacts

- There are roughly 2,500 varieties of apples grown in the United States and 7,500 varieties grown worldwide.
- The mass of four small apples or three medium apples is about 500 g.
- Rather than counting or weighing apples, you can measure them by the bushel basket (about 35 L).

Start-Up Activities

LAUNCH Lab

How much is a mole?

Counting large numbers of items is easier when you use counting units like the dozen. Chemists use a counting unit called the mole.

Materials
- centimeter ruler
- paper clip

Procedure
1. Measure the length of a paper clip to the nearest 0.1 cm.
2. Measure the mass of a paper clip to the nearest 0.1 g.

Analysis
1. **Calculate** If a mole is 6.02×10^{23} items, how far will a mole of paper clips, placed end to end lengthwise, reach into space?
2. **Determine** What is the mass of a mole of paper clips? How does this compare to the mass of Earth (roughly 6.0×10^{24} kg)?

Inquiry How many light-years (ly) would the paper clips extend into space? (1 light-year = 9.46×10^{15} m). How does the distance you calculated compare with the following astronomical distances: nearest star (other than the Sun) = 4.3 ly; center of our galaxy = 30,000 ly; nearest galaxy = 2×10^6 ly?

Chemistry Online

Visit glencoe.com to:
- study the entire chapter online
- explore concepts in Motion
- take Self-Check Quizzes
- use Personal Tutors
- access Web Links for more information, projects, and activities
- find the Try at Home Lab, Measuring Moles of Sugar

Conversion Factors Make the following Foldable to organize information about conversion factors.

STEP 1 Fold two sheets of paper in half horizontally.

STEP 2 On the first sheet, make a 3-cm cut on the fold line on each side of the paper. On the second sheet, cut along the fold line to within 3 cm of each edge.

STEP 3 Slip the first sheet through the cut in the second sheet to make a eight-page book.

FOLDABLES Use this Foldable with Sections 12.1 and 12.2. As you read the sections, record information about conversion factors and summarize the steps involved in each conversion.

Section 12.1

Objectives

- **Define** the mole as a counting number.
- **Relate** counting particles to weighing samples of substances.
- **Solve** stoichiometric problems using molar mass.

Review Vocabulary

Avogadro's principle: equal volumes of gases at the same temperature and pressure contain equal numbers of particles

New Vocabulary

stoichiometry
mole
Avogadro's number
molar mass
molecular mass
formula mass

Counting Particles of Matter

MAIN Idea A mole always contains the same number of particles; however, moles of different substances have different masses.

Real-World Reading Link If you've ever stood on a ridge and looked out at the forest spread below your feet, you might have wondered how many leaves there are on all of those trees. Can you think of any ways to estimate such a number? Chemists need to handle even larger numbers.

Stoichiometry

Recall from Chapter 11 that volumes of gases always combine in definite ratios. This observation, called the law of combining volumes, is based on measurements of the gas volumes. When Avogadro suggested that gases combine in fixed ratios because equal volumes of gases at the same temperature and pressure contain equal numbers of particles, he might have been thinking of particles rearranging themselves. Individual gas particles are so small that their rearranging cannot be observed, but the volumes of gases can be measured directly. Avogadro's principle is one of the earliest attempts to relate the number of particles in a sample of a substance to a direct measurement made on the sample.

Today, by using the methods of stoichiometry, we can measure the amounts of substances involved in chemical reactions and relate them to one another. **Stoichiometry** is the study of quantitative relationships between reactants and products in a chemical reaction. For example, a sample's mass or volume can be converted to a count of the number of its particles, such as atoms, ions, or molecules.

However, expressing the number of individual particles in even a small sample of matter results in an extremely large number. Fortunately, there's an easier way to express large numbers of atoms, ions, and molecules. **Figure 12.1** shows how some everyday items are grouped for counting. Would you rather count 1,500 individual sheets of paper or three reams of paper? Chemists group the particles that make matter in a similar fashion.

■ **Figure 12.1** Different units are used to count different types of objects. A pair is two objects, a dozen is 12, a gross is 144, and a ream is 500.

List What other counting units are you familiar with?

404 Chapter 12 • Chemical Quantities

The mole As an example, imagine the formidable task of counting the number of pennies in a large barrel. Should you count each penny individually? Or, can you measure the pennies by grouping them in some conveniently large quantity and then counting the number of groups to find the total number of pennies in the drum?

Figure 12.2 illustrates one way to count large numbers of pennies. To count the pennies individually, you would have to handle every one—thousands and thousands of pennies. But you could also just make three measurements: the weight of the barrel of pennies, the weight of the empty barrel, and the weight of a group of 1000 pennies. **Figure 12.2** shows how a simple calculation based on these three measurements arrives at an accurate count of the pennies.

Avogadro's number For counting atoms, you will need a much, much larger group than the group of 1000 that you used for the pennies. Atoms are so tiny that an ordinary-sized sample of a substance contains so many of these submicroscopic particles that counting them by grouping them in thousands would be unmanageable. Even grouping them by millions would not help. The group or unit of measure used to count numbers of atoms, molecules, ions, or formula units of substances is the **mole** (abbreviated mol). The number of things in one mole is 6.02×10^{23}. This big number is called **Avogadro's number**. **Figure 12.3** shows one mole of three common substances—water, copper, and table salt.

All kinds of submicroscopic particles can be conveniently counted using Avogadro's number. There are 6.02×10^{23} carbon atoms in a mole of carbon, and there are 6.02×10^{23} carbon dioxide molecules in a mole of carbon dioxide. There are 6.02×10^{23} sodium ions in a mole of sodium ions. In a mole of eggs there are 6.02×10^{23} eggs, but eggs are so large compared with the particles that interact in chemistry that you would have no need to use Avogadro's number to estimate how many there are in a typical sample.

■ **Figure 12.2** If you wanted to determine the number of pennies in a barrel, you could use the mass of all the pennies (70,125 g) and the mass of 1000 pennies (2890.7 g). The ratio 1000 pennies/2890.7 g converts the mass of the pennies to the number of pennies.

$$70{,}125 \text{ g} \times \frac{1000 \text{ pennies}}{2890.7 \text{ g}} = 24{,}259 \text{ pennies}$$

The drum contains 24,259 pennies, worth $242.59.

■ **Figure 12.3** The amount of each substance shown is 6.02×10^{23}, or 1 mol of representative particles of water, copper, and salt.

■ **Figure 12.4** You can use a balance to measure 500 g of methanol. But how many molecules are in 500 g of methanol? And how much carbon dioxide and hydrogen gas do you need to synthesize 500 g of methanol? You need a way to relate macroscopic measurements, such as mass, to the molecules involved in chemical reactions.

FACT of the Matter
The most precise value of Avogadro's number is 6.02214179 × 10^{23}. For most purposes, rounding to 6.02 × 10^{23} is sufficient.

FOLDABLES
Incorporate information from this section into your Foldable.

Molar Mass

How does knowing the size of a mole help solve problems in chemistry? Unlike paper that is packaged in reams, chemicals in the lab don't come in convenient bundles of moles all set for the counting. But you can use a balance to measure the mass of a sample of matter.

For example, you might want to produce 500 g of methanol, shown in **Figure 12.4,** according to the balanced chemical equation below.

$$CO_2(g) + 3H_2(g) \rightarrow CH_3OH(l) + H_2O(g)$$

To produce 500 g of methanol, how many grams of CO_2 and H_2 would you need? How many grams of water would be produced? Those are stoichiometry questions about the masses of reactants and products. But the balanced chemical equation shows that three molecules of hydrogen gas react with one molecule of carbon dioxide gas. The equation relates the molecules of reactants and products, not their masses.

Like Avogadro, you need to relate the macroscopic measurements—the masses of carbon dioxide and hydrogen—to the number of molecules of methanol. To find the mass of carbon dioxide and the mass of hydrogen needed to produce 500 g of methanol, you first need to know how many molecules of methanol are in 500 g of methanol.

Remember the drum of pennies? By knowing the mass of one group of 1000 pennies and knowing the mass of all the pennies, you can find the number of groups of 1000 pennies. Then, finding the total number of pennies is an easy matter. You have a suitable unit of measure—the mole—and you know the mass of methanol you want to produce. But, you still need to know the mass of one mole of methanol molecules.

■ **Figure 12.5** A dozen limes has approximately twice the mass of a dozen eggs. The difference in mass is reasonable because limes are different from eggs in composition and size.

Molar mass of an element You probably would expect that a dozen limes and a dozen eggs have different masses, as shown in **Figure 12.5**. This makes sense—a single lime does not have the same mass as a single egg. Similarly, different elements have different masses. Recall from Chapter 2 that the average atomic masses of the elements are given on the periodic table. For example, the average mass of one iron atom is 55.8 u, where u means atomic mass units.

How does the mass of one atom relate to the mass of one mole of that atom? The atomic mass unit is defined so that the atomic mass of an atom of the most common carbon isotope is exactly 12 u, and the mass of one mole of the most common isotope of carbon atoms is exactly 12 g. The mass of one mole of a pure substance is called its **molar mass.** You can find the molar mass of the elements on the periodic table: the molar mass of an element is simply the average atomic mass of that element stated in grams rather than atomic mass units. So, the molar mass of iron is 55.847 g, and the molar mass of platinum is 195.08 g. **Figure 12.6** shows one mole of iron weighed on a balance.

If an element exists as a molecule, remember that the particles in one mole of that element are themselves composed of atoms. For example, the element oxygen exists as molecules composed of two oxygen atoms, so a mole of oxygen molecules contains two moles of oxygen atoms. Therefore, the molar mass of oxygen molecules is twice the molar mass of oxygen atoms: $2 \times 16.00 \text{ g} = 32.00 \text{ g}$.

VOCABULARY
WORD ORIGIN
Stoichiometry
comes from the Greek words *stoichen*, which means *element*, and *metreon*, which means *to measure*

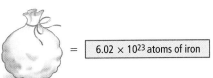

1 mol of iron = 6.02×10^{23} atoms of iron

■ **Figure 12.6** One mole of iron, represented by a bag of particles, contains Avogadro's number of atoms and has a mass equal to its atomic mass in grams.
Apply *What is the molar mass of copper?*

Concepts In MOtion
Interactive Figure To see an animation of molar mass, visit glencoe.com.

Section 12.1 • Counting Particles of Matter **407**

MiniLab 12.1

Counting by Mass

Can you determine how many buttons are in the bag without counting? Chemists and chemical engineers usually need to control the number of atoms, molecules, and ions in their reactions carefully. These particles are too small to be seen and too numerous to count, but their numbers can be determined by measuring their masses. Simulate this process by using mass to determine the approximate number of buttons (or other small, uniform objects) in a plastic bag.

Procedure

1. Read and complete the lab safety form.
2. Count out ten **buttons**. Measure and record the mass of the buttons.
3. Measure and record the mass of an empty, self-sealing **plastic storage bag**.
4. Completely fill the bag with buttons and seal it.
5. Measure and record the mass of the bag plus the buttons.

Analysis

1. **Design** a plan to determine the number of buttons in the sealed bag without opening it.
2. **Calculate** the number of buttons in the bag.
3. **Verify** your calculation by opening the bag and counting the buttons. Did your hand count match your calculation from question 2? Explain your answers.

TRY AT HOME LAB
See page 873 for **Measuring Moles of Sugar.**

Molar mass of a compound Recall from Chapter 4 that covalent compounds are composed of molecules and that ionic compounds are composed of formula units. The **molecular mass** of a covalent compound is the mass in atomic mass units of one molecule. Its molar mass is the mass in grams of one mole of its molecules. The **formula mass** of an ionic compound is the mass in atomic mass units of one formula unit. Its molar mass is the mass in grams of one mole of its formula units. How to calculate the molar masses for ethanol, a covalent compound, and for calcium chloride, an ionic compound, is shown below.

Ethanol (C_2H_5OH), a covalent compound

2 C atoms	2×12.0 u	=	24.0 u
6 H atoms	6×1.00 u	=	6.00 u
1 O atom	1×16.0 u	=	+16.0 u
molecular mass of C_2H_6O			46.0 u

mass of 1 mol C_2H_6O molecules	=	46.0 g
molar mass of ethanol		46.0 g/mol

Calcium chloride ($CaCl_2$), an ionic compound

1 Ca atom	1×40.1 u	=	40.1 u
2 Cl atoms	2×35.5 u	=	+71.0 u
formula mass of $CaCl_2$			111.1 u

mass of 1 mol $CaCl_2$ formula units	=	111.1 g
molar mass of calcium chloride		111.1 g/mol

EXAMPLE Problem 12.1

Number of Atoms in a Sample of an Element The mass of an iron bar is 16.8 g. How many iron (Fe) atoms are in the sample?

> **Problem-Solving Hint**
> Remember that the units of molar mass are grams per mole, which can be used as a conversion factor.

1 Analyze
Use the periodic table to find the molar mass of iron. The average mass of an iron atom is 55.8 u. Therefore, the mass of one mole of iron atoms is 55.8 g.

2 Set Up
To convert the mass of the iron bar to the number of moles of iron, use the mass of one mole of iron atoms as a conversion factor.

$$(16.8 \text{ g Fe})\left(\frac{1 \text{ mol Fe}}{55.8 \text{ g Fe}}\right)$$

Now, use the number of atoms in a mole to find the number of iron atoms in the bar.

$$(16.8 \text{ g Fe})\left(\frac{1 \text{ mol Fe}}{55.8 \text{ g Fe}}\right)\left(\frac{6.02 \times 10^{23} \text{ Fe atoms}}{1 \text{ mol Fe}}\right)$$

3 Solve

$$(16.8 \text{ g Fe})\left(\frac{1 \text{ mol Fe}}{55.8 \text{ g Fe}}\right)\left(\frac{6.02 \times 10^{23} \text{ Fe atoms}}{1 \text{ mol Fe}}\right) = \frac{16.8 \times 6.02 \times 10^{23} \text{ Fe atoms}}{55.8} = 1.81 \times 10^{23} \text{ Fe atoms}$$

4 Check
The units of measure cancel to leave only the number of iron atoms as the unit.

EXAMPLE Problem 12.2

Number of Formula Units in a Sample of a Compound The mass of a quantity of iron(III) oxide is 16.8 g. How many formula units are in the sample?

1 Analyze
Use the periodic table to calculate the mass of one formula unit of Fe_2O_3.

2 Fe atoms	2×55.8 u =	111.6 u
3 O atoms	3×16.0 u =	+ 48.0 u
formula mass of Fe_2O_3		159.6 u

Therefore, the molar mass of Fe_2O_3 is 159.6 g/mol.

2 Set Up
To convert the mass of the iron(III) oxide to the number of moles of iron(III) oxide, use the mass of one mole of iron(III) oxide molecules as a conversion factor.

$$(16.8 \text{ g Fe}_2\text{O}_3)\left(\frac{1 \text{ mol Fe}_2\text{O}_3}{159.6 \text{ g Fe}_2\text{O}_3}\right)$$

Multiply the moles of iron oxide by the number of formula units in a mole.

$$(16.8 \text{ g Fe}_2\text{O}_3)\left(\frac{1 \text{ mol Fe}_2\text{O}_3}{159.6 \text{ g Fe}_2\text{O}_3}\right)\left(\frac{6.02 \times 10^{23} \text{ Fe}_2\text{O}_3 \text{ formula units}}{1 \text{ mol Fe}_2\text{O}_3}\right)$$

3 Solve

$$(16.8 \text{ g Fe}_2\text{O}_3)\left(\frac{1 \text{ mol Fe}_2\text{O}_3}{159.6 \text{ g Fe}_2\text{O}_3}\right)\left(\frac{6.02 \times 10^{23} \text{ Fe}_2\text{O}_3 \text{ formula units}}{1 \text{ mol Fe}_2\text{O}_3}\right) =$$

$$\frac{16.8 \times 6.02 \times 10^{23} \text{ Fe}_2\text{O}_3 \text{ formula units}}{159.6} = 6.34 \times 10^{22} \text{ Fe}_2\text{O}_3 \text{ formula units}$$

4 Check
The units of measure show that the calculation is set up correctly. The multiplication also checks.

EXAMPLE Problem 12.3

Mass of a Number of Moles of a Compound What mass of water must be weighed to obtain 7.50 mol of H_2O?

1 Analyze
The molar mass of water is obtained by expressing its molecular mass in units of g/mol.

2 H atoms	2 × 1.00 u =	2.0 u
1 O atom	1 × 16.0 u =	+ 16.0 u
molecular mass of H_2O		18.0 u

The molar mass of water is 18.0 g/mol.

2 Set up
Use the molar mass to convert the number of moles to a mass measurement.

$$(7.50 \text{ mol } H_2O)\left(\frac{18.0 \text{ g } H_2O}{1 \text{ mol } H_2O}\right)$$

3 Solve

$$(7.50 \text{ mol } H_2O)\left(\frac{18.0 \text{ g } H_2O}{1 \text{ mol } H_2O}\right) = 7.50 \times 18.0 \text{ g } H_2O = 135 \text{ g } H_2O$$

4 Check
Using dimensional analysis, the units cancel to leave the final units of grams of H_2O, as expected. The number of grams, 135, makes sense given a starting point of 7.50 mol H_2O. The multiplication also checks.

SUPPLEMENTAL PRACTICE

For more practice converting between mass and moles, see Supplemental Practice, page 824.

PRACTICE Problems   Solutions to Problems Page 861

1. Without calculating, decide whether 50.0 g of sulfur or 50.0 g of tin represents the greater number of atoms. Explain how you arrived at your decision. Then verify your answer by calculating.

2. Determine the number of atoms in each sample below.
   a) 98.3 g mercury, Hg
   b) 45.6 g gold, Au
   c) 10.7 g lithium, Li
   d) 144.6 g tungsten, W

3. Determine the number of moles in each sample below.
   a) 6.84 g sucrose, $C_{12}H_{22}O_{11}$
   b) 16.0 g sulfur dioxide, SO_2
   c) 68.0 g ammonia, NH_3
   d) 17.5 g copper(II) oxide, CuO

4. Determine the mass of the following molar quantities.
   a) 3.52 mol Si
   b) 1.25 mol aspirin, $C_9H_8O_4$
   c) 0.550 mol F_2
   d) 2.35 mol barium iodide, BaI_2

Art Connection

Asante Brass Weights

Imagine waking up in the morning after a rain and finding that gold has popped from the ground. Does it sound like a fairy tale? In the 1700s and 1800s, this was a common experience for people living in Ghana, West Africa. Their land is an alluvial gold field. Besides picking up gold that has been bared by erosion and finding gold by panning streams, the Asante, one of about 10 major people groups in Ghana, recovered gold by digging into sediment where smaller streams joined larger ones and by crushing rock rich in gold and quartz, then separating the gold by washing. Gold dust was their most common national currency.

Asante standard weights As early as about A.D 1400, the Asante created artistic, standardized weights to weigh gold dust for commercial transactions. Their first weights were simple blocks, but later they made cubes, pyramids, triangular blocks, and rectangular blocks with carved designs. They also cast natural objects for their weights, such as seeds, beetles, shells, and crocodiles, as shown in **Figure 1**. In the early 1700s, while Europeans were colonizing America, the Asante were creating works of art for their weights, which included carving elaborate designs on flattened shapes and fashioning weights that replicated human and animal figures, as shown in **Figure 2**.

Figure 2 Brass statue representing a bird

The Asante people were skilled metalworkers. They cast their weights in brass by the lost wax method, which preserves fine detail in the finished product. In this process, shapes were carved in wax and then encased in clay and baked. The hot wax was poured out, leaving a ceramic mold. Molten metal was poured into the mold. When the metal cooled, the mold was broken to remove the newly formed brass weight. Errors in the casting process could not be detected until the mold was destroyed, and even then, few repairs were possible.

Today, Asante weights are collected as art forms—evidence of the advanced Asante culture of the past few centuries. Compared to the plain weights used in the United States, Asante weights add interest and beauty to everyday life.

Figure 1 Crocodiles with a shared stomach

Connection to Chemistry

1. **Hypothesize** Do you think standard weights will soon be obsolete in scientific laboratories? Why?
2. **Apply** Why does gold form few compounds?

■ **Figure 12.7** Each sample contains 6.02×10^{23} molecules of a covalent compound or 6.02×10^{23} formula units of an ionic compound. The samples have the same number of particles, but different masses.

Infer Which compound has the largest molar mass? The smallest? Explain.

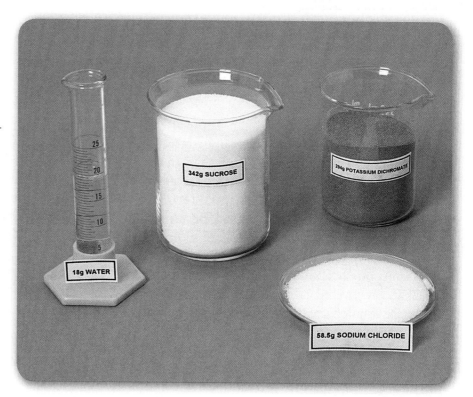

Molar mass and stoichiometry

The concept of molar mass makes it easy to determine the number of particles in a sample of a substance by simply measuring the mass of the sample. The concept is also useful in relating masses of reactants and products in chemical reactions. Recall the question that started the discussion of molar mass: How does knowing the size of a mole help you synthesize 500 g of methanol? By using stoichiometry, you can find the number of moles of methanol in 500 g; the masses of carbon dioxide and hydrogen needed to react to produce this amount of methanol; and the masses of carbon dioxide, hydrogen, and water produced as byproducts. In the next section, you will learn how to use moles to convert between moles of reactants and moles of products. **Figure 12.7** shows you what one mole of some familiar compounds looks like.

Section 12.1 Assessment

Section Summary

- Stoichiometry relates the amounts of products and reactants in a chemical equation to one another.
- The mole is a unit used to count particles of matter. One mole of a pure substance contains Avogadro's number of particles, 6.02×10^{23}.
- Molar mass can be used to convert mass to moles or moles to mass.

5. **MAIN Idea** **Compare and Contrast** How is counting a truckload of pennies like counting the atoms in 10.0 g of aluminum? How is it different?

6. **Assess** The average atomic mass of nitrogen is 14 times greater than the average atomic mass of hydrogen. Would you expect the molar mass of nitrogen gas, N_2, to be 14 times greater than the molar mass of hydrogen gas (H_2)? Why?

7. **Explain** Why is the following statement meaningless? An industrial process requires 2.15 mol of a sugar-salt mixture.

8. **Determine** the mass in grams of one average atomic mass unit.

9. **Infer** If the mole is truly central to quantitative work in chemistry, why do balances in chemistry labs measure mass and not moles?

Section 12.2

Objectives

- **Predict** quantities of reactants and products in chemical reactions.
- **Determine** mole ratios from formulas for compounds.
- **Identify** formulas of compounds by using mass ratios.

Review Vocabulary

mole: group or unit of measure used to count numbers of atoms, molecules, or formula units of a substance

New Vocabulary

molar volume
ideal gas law
theoretical yield
percent yield
empirical formula

FOLDABLES
Incorporate information from this section into your Foldable.

Using Moles

MAIN Idea Balanced chemical equations relate moles of reactants to moles of products.

Real-World Reading Link Have you ever baked cookies from scratch? If so, you know that the ingredients must be mixed in the proper proportions to make a certain amount of cookie dough. To double the amount of dough, you need to double amounts of the ingredients. Chemical reactions are similar—specific amounts of reactants result in specific amounts of products.

Using Molar Masses in Stoichiometric Problems

In Chapter 11, you balanced chemical equations to indicate the volume of gas required for a reaction and the volume of gas produced. Similarly, you can use balanced chemical equations and moles to predict the masses of reactants or products. **Figure 12.8** summarizes the steps for moving from the mass of one substance in a chemical reaction to the mass of a second substance.

It is very important to note that you can't move directly from the mass of one substance to the mass of the second—you must always first convert a given mass to moles. Recall from Chapter 6 that the coefficients in a balanced chemical equation tell you the number of particles of each chemical involved in the reaction. Therefore, they also tell you the number of moles of each chemical in the reaction. Once you know the number of moles of any reactant or product, use the coefficients in the equation to convert to moles of the other reactants and products. The example problem on the next page gives you practice in solving these types of problems.

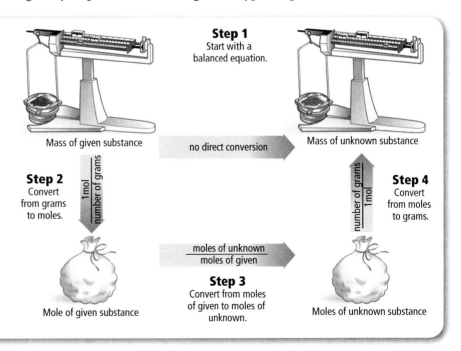

■ **Figure 12.8** Follow the steps from the balanced equation to the mass of the unknown. Note that there is no shortcut from the mass of the given substance to the mass of the unknown substance. The route goes through the mole. However, you can jump from step 1 to step 3 if the amount of the given substance is in moles.

EXAMPLE Problem 12.4

> **Problem-Solving Hint**
> The balanced chemical equation is the source of conversion factors relating moles of one substance to moles of another substance.

Predicting Mass of a Reactant and Product Ammonia gas is synthesized from nitrogen gas and hydrogen gas according to the balanced chemical equation below.

$$N_2(g) + 3H_2(g) \rightarrow 2NH_3(g)$$

How many grams of hydrogen gas are required for 3.75 g of nitrogen gas to react completely? What mass of ammonia is formed?

1 Analyze

Reactants and products are related in terms of moles, not mass. The amount of hydrogen needed depends on the moles of nitrogen present in 3.75 g and the ratio of moles of hydrogen gas to moles of nitrogen gas in the balanced chemical equation. Similarly, the amount of ammonia formed depends on the ratio of moles of ammonia to moles of nitrogen.

2 Set Up

Find the number of moles of N_2 molecules by using the molar mass of nitrogen.

$$(3.75 \text{ g } N_2)\left(\frac{1 \text{ mol } N_2}{28.0 \text{ g } N_2}\right)$$

To find the mass of hydrogen needed: The balanced chemical equation shows that 3 mol of H_2 react with 1 mol of N_2. Multiply the number of moles of N_2 by this ratio.

$$(3.75 \text{ g } N_2)\left(\frac{1 \text{ mol } N_2}{28.0 \text{ g } N_2}\right)\left(\frac{3 \text{ mol } H_2}{1 \text{ mol } N_2}\right)$$

To find the mass of hydrogen, multiply the moles of hydrogen by the mass of 1 mol of H_2.

$$(3.75 \text{ g } N_2)\left(\frac{1 \text{ mol } N_2}{28.0 \text{ g } N_2}\right)\left(\frac{3 \text{ mol } H_2}{1 \text{ mol } N_2}\right)\left(\frac{2.00 \text{ g } H_2}{1 \text{ mol } H_2}\right)$$

To find the mass of ammonia produced: Use the mole ratio of ammonia molecules to nitrogen molecules to find the moles of ammonia formed.

$$(3.75 \text{ g } N_2)\left(\frac{1 \text{ mol } N_2}{28.0 \text{ g } N_2}\right)\left(\frac{2 \text{ mol } NH_3}{1 \text{ mol } N_2}\right)$$

Use the molar mass of ammonia, 17.0 g, to find the mass of ammonia formed.

$$(3.75 \text{ g } N_2)\left(\frac{1 \text{ mol } N_2}{28.0 \text{ g } N_2}\right)\left(\frac{2 \text{ mol } NH_3}{1 \text{ mol } N_2}\right)\left(\frac{17.0 \text{ g } NH_3}{1 \text{ mol } NH_3}\right)$$

3 Solve

grams of H_2:

$$(3.75 \text{ g } N_2)\left(\frac{1 \text{ mol } N_2}{28.0 \text{ g } N_2}\right)\left(\frac{3 \text{ mol } H_2}{1 \text{ mol } N_2}\right)\left(\frac{2.00 \text{ g } H_2}{1 \text{ mol } H_2}\right) = \frac{3.75 \times 1 \times 3 \times 2.00 \text{ g } H_2}{28.0} = 0.804 \text{ g } H_2$$

grams of NH_3:

$$(3.75 \text{ g } N_2)\left(\frac{1 \text{ mol } N_2}{28.0 \text{ g } N_2}\right)\left(\frac{2 \text{ mol } NH_3}{1 \text{ mol } N_2}\right)\left(\frac{17.0 \text{ g } NH_3}{1 \text{ mol } NH_3}\right) = \frac{3.75 \times 1 \times 2 \times 17.0 \text{ g } NH_3}{28.0} = 4.55 \text{ g } NH_3$$

4 Check

The units cancel and leave grams, the desired unit. The mass of hydrogen in the reaction makes sense because the molar mass of hydrogen is much less than the molar mass of nitrogen. The mass of ammonia formed makes sense because the reacting amounts of hydrogen and nitrogen add up to the calculated mass of ammonia.

PRACTICE Problems

Solutions to Problems Page 825

10. The combustion of propane, C_3H_8, a fuel used in backyard grills and camp stoves, produces carbon dioxide and water vapor.

$$C_3H_8(g) + 5O_2(g) \rightarrow 3CO_2(g) + 4H_2O(g)$$

What mass of carbon dioxide forms when 95.6 g of propane burns?

11. Solid xenon hexafluoride is prepared by allowing xenon gas and fluorine gas to react.

$$Xe(g) + 3F_2(g) \rightarrow XeF_6(s)$$

How many grams of fluorine are required to produce 10.0 g of XeF_6?

12. Using the reaction in Practice Problem 11, how many grams of xenon are required to produce 10.0 g of XeF_6?

Using Molar Volumes in Stoichiometric Problems

In Chapter 11, you used the law of combining volumes. Avogadro inferred from that law that equal volumes of gases contain equal numbers of particles. In terms of moles, Avogadro's principle states that equal volumes of gases at the same temperature and pressure contain equal numbers of moles of gases.

The **molar volume** of a gas is the volume that a mole of a gas occupies at a pressure of one atmosphere (equal to 101 kPa) and a temperature of 0.00°C. Under these conditions of STP, the volume of one mole of any gas is 22.4 L, as shown in **Figure 12.9**. For example, at STP one mole of helium gas and one mole of chlorine gas both occupy 22.4 L.

Even though all gases occupy 22.4 L at STP, their masses are still different. Helium particles are very light, so one mole of helium particles will have less mass than one mole of the particles that make up air. As a result, 22.4 L of helium is less dense than 22.4 L of air, which is why helium balloons float in air. On the other hand, 22.4 L of chlorine gas weighs more than 22.4 L of air, so a balloon filled with chlorine would sink due to its greater density. Like molar mass, molar volume is used in stoichiometric calculations, but always make sure you are working with the volume of gas at STP—not the mass of the gas—before applying the molar volume.

Chemistry Online

Personal Tutor For an online tutorial on using conversion factors, visit glencoe.com.

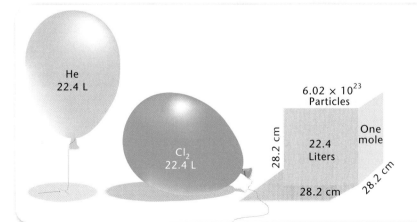

■ **Figure 12.9** One mole of any gas at STP occupies 22.4 L. How large is that? It is the volume of a cube that is 28.2 cm on each edge. Each such volume contains 6.02×10^{23} particles of gas, but the mass of 1 mol is different for each type of gas.

Explain Why does one mole of helium float?

EXAMPLE Problem 12.5

Using Molar Volume In the space shuttle, exhaled carbon dioxide gas is removed from the air by passing it through canisters of lithium hydroxide. The following reaction takes place.

$$CO_2(g) + 2LiOH(s) \rightarrow Li_2CO_3(s) + H_2O(g)$$

How many grams of lithium hydroxide are required to remove 500.0 L of carbon dioxide gas at 101 kPa pressure and 25.0°C?

1 Analyze
In this problem, you use the molar volume to find the number of moles.

2 Set Up
The volume of gas at 25.0°C must be converted to a volume at STP.

$$V = (500.0 \text{ L CO}_2)\left(\frac{273 \text{ K}}{298 \text{ K}}\right) = 458 \text{ L CO}_2$$

Now, find the number of moles of CO_2 gas as below.

$$(458 \text{ L CO}_2)\left(\frac{1 \text{ mol CO}_2}{22.4 \text{ L CO}_2}\right)$$

The chemical equation shows that the ratio of moles of LiOH to CO_2 is 2 to 1. Therefore, the number of moles of lithium hydroxide is given by the expression below.

$$(458 \text{ L CO}_2)\left(\frac{1 \text{ mol CO}_2}{22.4 \text{ L CO}_2}\right)\left(\frac{2 \text{ mol LiOH}}{1 \text{ mol CO}_2}\right)$$

To convert the number of moles of LiOH to mass, use its molar mass, 23.9 g/mol.

$$(458 \text{ L CO}_2)\left(\frac{1 \text{ mol CO}_2}{22.4 \text{ L CO}_2}\right)\left(\frac{2 \text{ mol LiOH}}{1 \text{ mol CO}_2}\right)\left(\frac{23.9 \text{ g LiOH}}{1 \text{ mol LiOH}}\right)$$

3 Solve

$$(458 \text{ L CO}_2)\left(\frac{1 \text{ mol CO}_2}{22.4 \text{ L CO}_2}\right)\left(\frac{2 \text{ mol LiOH}}{1 \text{ mol CO}_2}\right)\left(\frac{23.9 \text{ g LiOH}}{1 \text{ mol LiOH}}\right)$$

$$= \frac{458 \times 2 \times 23.9 \text{ g LiOH}}{22.4} = 977 \text{ g LiOH}$$

4 Check
A mass of 1000 g of lithium hydroxide is about 40 mol of the compound. According to the chemical equation, about half as much, or 20 mol of CO_2, will be removed from the air. At STP, 20 mol of any gas occupy about 450 L, which will expand to about 500 L at 25°C.

SUPPLEMENTAL PRACTICE
For more practice using molar volume, see Supplemental Practice, page 824.

PRACTICE Problems

Solutions to Problems Page 825

13. What mass of sulfur must burn to produce 3.42 L of SO_2 at 273°C and 101 kPa? The reaction is $S(s) + O_2(g) \rightarrow SO_2(g)$.

14. What volume of hydrogen gas can be produced by reacting 4.20 g of sodium in excess water at 50.0°C and 106 kPa? The reaction is
$2Na + 2H_2O \rightarrow 2NaOH + H_2$.

Everyday Chemistry

Air Bags

In 1990, on a Virginia hilltop, two cars collided in a head-on crash. Both drivers walked away with only minor injuries. A chemical reaction, along with safety belts, saved their lives. This was the first recorded head-on collision between two cars with air bags. Statistics show that air bags reduce the risk of dying in a head-on collision by 30 percent.

How front air bags function There are three steps to an air bag inflating, as shown in **Figure 1**, and slowing a driver's or passenger's forward motion.

1. When a car collides with a rigid barrier at a speed of 10 to 15 mph (16 to 24 km/h), a sensor on the front of the car sends an electric current to fire the control unit 0.010 s after impact.

2. Within 0.030 s, a chemical reaction in the stored air bag creates a gas that causes rapid inflation.

3. The bag begins to deflate in 0.045 s as the gas escapes through holes at the base of the bag.

4. The driver or passenger strikes the deflating bag 0.050 s after impact.

The chemical reactions Sodium azide (NaN_3) is a chemical that produces nitrogen gas to inflate the air bag. A pellet containing a mixture of NaN_3, potassium nitrate (KNO_3), and silicon dioxide (SiO_2) is contained in a gas generator, which is stored under a breakaway cover in the steering wheel or dash along with an inflation system and a tightly folded nylon air bag. The sensor sends an electrical impulse to ignite the gas-generator mixture providing the high temperature needed for NaN_3 to decompose into nitrogen gas and sodium.

The highly reactive sodium metal immediately reacts with KNO_3 to form potassium oxide (K_2O), sodium oxide (Na_2O), and more N_2 gas. The metal oxides (K_2O and Na_2O) react with silicon dioxide to form harmless and stable silicate glass.

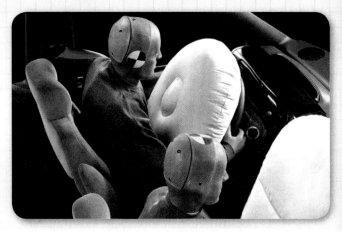

Figure 1 Air bags inflating

Reliability Front air bags have been required in all passenger vehicles since 1998. However, this increase in vehicles with air bags has not coincided with an increase in air bag defects. Why are they so dependable? There are no moving parts to wear out, all exposed components are tightly sealed, and the gold-plated electrical connectors corrode slowly.

Gas-volume relationships A typical driver's air bag requires 0.0650 m^3 of nitrogen to inflate—no more, no less. The passenger's air bag typically needs 0.1340 m^3. The pellet must have the exact amount of sodium azide needed to produce the correct amount of nitrogen. As with all expanding gases, pressure and temperature affect the amount of sodium azide needed.

Nitrogen behaves as an ideal gas. Therefore, the ideal gas law can be used to calculate the amount of nitrogen necessary to fill the airbag to a specific pressure. After the amount of nitrogen is calculated, the exact amount of sodium azide is carefully measured.

Explore Further

1. **Apply** If 130 g of sodium azide are needed for the driver's air bag, how much is needed for the passenger's bag? Explain.
2. **Infer** What effect does the heat from the sodium azide reaction have on the pressure and volume of the nitrogen gas formed?

MiniLab 12.2

Determine the Amount of Reactant

How much baking soda and vinegar need to react to fill a bag with carbon dioxide? Chemists often need to determine the most efficient and cost effective amounts of reactants and products. In this lab, you will determine the amount of baking soda that reacts with vinegar to yield just enough carbon dioxide to fill a one-quart, self-sealing plastic bag.

Procedure

1. Read and complete the lab safety form.
2. Write the balanced equation for the reaction of baking soda (sodium hydrogen carbonate) and vinegar (acetic acid) that produces sodium acetate, water, and carbon dioxide.
3. Find the volume of a **1-quart, self-sealing plastic bag** by filling it with **water** and then pouring the water into a **graduated cylinder** or **measuring cup**.
4. Calculate the mass of sodium hydrogen carbonate that will fill the bag with CO_2 gas when the compound reacts with excess acetic acid.
5. Weigh the calculated amount of **sodium hydrogen carbonate** and place it in a bottom corner of the bag. Use a **plastic-coated twist tie** to seal off this corner.
6. Pour about 60 mL of **1M acetic acid** into the other bottom corner of the bag. Be careful not to allow the reactants to mix. Squeeze the air from the bag and seal the zipper top.
7. Place the bag in a **trash can** or behind an **explosion shield.**
8. Release the twist tie, quickly mix the reactants, and allow the reaction to proceed.

Analysis

1. **Explain** the calculations that you used to determine the required mass of sodium hydrogen carbonate.
2. **Calculate** What mass of sodium hydrogen carbonate would be required to react with excess acetic acid to produce 20,000 L of carbon dioxide gas at STP for a water-treatment plant?
3. **Analyze** What would have happened in step 8 if the amount of acetic acid were insufficient to react with all of the baking soda?
4. **Infer** Suppose the pressure of the gas in the bag was measured to be 101.5 kPa. Is this pressure consistent with the ideal gas law? Assume $T = 20°C$ and $P = 101$ kPa.

■ **Figure 12.10** The pressure in the tire increases as the amount of air present increases.

Ideal Gas Law

In Chapter 11 you solved problems where the temperature, pressure, and volume of gases changed. All of these problems assumed that the number of particles of gas present was constant. But what happens if the number of gas particles changes? For example, adding air to a tire, as shown in **Figure 12.10,** changes the number of gas particles inside the tire. The **ideal gas law** describes the behavior of an ideal gas in terms of pressure, P, volume, V, temperature, T, and number of moles of gas, n.

$$PV = nRT$$

In the equation above, R represents a special constant called the ideal gas constant. The value of R can be determined using the definition of molar volume. At STP, one mole of gas occupies 22.4 L. Therefore, when $P = 101.3$ kPa, $V = 22.4$ L, $n = 1$ mol, and $T = 273.15$ K, the equation for the ideal gas law can be written as follows.

$$101.3 \text{ kPa} \times 22.4 \text{ L} = 1 \text{ mol} \times R \times 273.15 \text{ K}$$

Now, we can solve for R.

$$R = \frac{(101.3 \text{ kPa})(22.4 \text{ L})}{(1 \text{ mol})(273.15 \text{ K})} = 8.31 \frac{\text{kPa} \cdot \text{L}}{\text{mol} \cdot \text{K}}$$

Continue to use the gas laws as described in Chapter 11 if the number of gas particles does not change. However, if the number of gas particles present changes—as in **Figure 12.10** and in chemical reactions that evolve gases—then use the ideal gas law. You can also use the ideal gas law to find the moles in a sample of gas, as in the example problem below.

EXAMPLE Problem 12.6

Using the Ideal Gas Law How many moles are contained in a 2.44-L sample of gas at 25.0°C and 202 kPa?

1 Analyze

Solve the ideal gas law for n, the number of moles.

$$n = \frac{PV}{RT}$$

2 Set up

$$n = \frac{202 \text{ kPa} \times 2.44 \text{ L}}{\left(\frac{8.31 \text{ kPa} \cdot \text{L}}{\text{mol} \cdot \text{K}}\right) \times 298 \text{ K}}$$

3 Solve

$$n = \frac{202 \text{ kPa} \times 2.44 \text{ L}}{\left(\frac{8.31 \text{ kPa} \cdot \text{L}}{\text{mol} \cdot \text{K}}\right) \times 298 \text{ K}} = 0.199 \text{ mol}$$

4 Check

First, find the volume that 2.44 L of a gas would occupy at STP.

$$V = (2.44 \text{ L})\left(\frac{273 \text{ K}}{298 \text{ K}}\right)\left(\frac{202 \text{ kPa}}{101 \text{ kPa}}\right) = 4.47 \text{ L}$$

Then, find the number of moles in this volume.

$$(4.47 \text{ L})\left(\frac{1 \text{ mol}}{22.4 \text{ L}}\right) = 0.200 \text{ mol}$$

Solving the problem using the molar volume confirms the solution obtained with the ideal gas law. (The two solutions differ only in rounding.)

PRACTICE Problems

Solutions to Problems Page 863

15. How many moles of helium are contained in a 5.00-L canister at 101 kPa and 30.0°C?
16. What is the volume of 0.020 mol Ne at 0.505 kPa and 27.0°C?
17. How much zinc must react in order to form 15.5 L of hydrogen, $H_2(g)$, at 32.0°C and 115 kPa?

    $Zn(s) + H_2SO_4(aq) \rightarrow ZnSO_4(aq) + H_2(g)$

SUPPLEMENTAL PRACTICE

For more practice using the ideal gas law, see Supplemental Practice, page 825.

Theoretical Yield and Actual Yield

The amount of product of a chemical reaction predicted by stoichiometry is called the **theoretical yield**. As discussed earlier in this section, if 3.75 g of nitrogen completely react with hydrogen, a theoretical yield of 4.55 g of ammonia would be produced. The actual yield of a chemical reaction is usually less than predicted. The collection techniques and apparatus used, time, and the skills of the chemist might affect the actual yield.

When actual yield is less than theoretical yield, you express the efficiency of the reaction as percent yield. The **percent yield** of a reaction is the ratio of the actual yield to the theoretical yield expressed as a percent. For the nitrogen-hydrogen reaction, suppose the actual yield was 3.86 g of ammonia. Then, the percent yield could be calculated by using the following equation.

$$\text{percent yield} = \left(\frac{\text{actual yield}}{\text{theoretical yield}}\right) \times 100$$

$$\frac{3.86 \text{ g NH}_3}{4.55 \text{ g NH}_3} \times 100 = 84.8\%$$

This means that 84.8 percent of the possible yield of ammonia was obtained from the reaction. Calculating percentage yield is similar to calculating a baseball player's batting average, as shown in **Figure 12.11**.

A manufacturer is interested in producing chemicals as efficiently and inexpensively as possible. High yields make commercial manufacturing of substances possible. For example, taxol, a naturally occurring complex compound, is a strong agent against cancer. For ten years, chemists tried to synthesize this compound in the lab. In 1994, two independent academic research groups succeeded. However, the process is so complicated and time-consuming that the percent yield is probably not even one percent.

Chemistry Online

Personal Tutor For an online tutorial on finding and average, visit glencoe.com.

■ **Figure 12.11** You can calculate a player's batting average by dividing hits by at bats. For example, a player who gets 352 hits over the course of 1000 at bats has a batting average of 0.352.

$$\frac{\text{hits}}{\text{at bats}} = \frac{352}{1000} = 0.352$$

Just as batting averages measure a hitter's efficiency, percent yield measures a reaction's efficiency.

Determining Mass Percents

As you know, the chemical formula of a compound tells you the elements that comprise it. For example, the formula for geraniol (the main compound that gives a rose its scent) is $C_{10}H_{18}O$. The formula shows that geraniol is comprised of carbon, hydrogen, and oxygen. Because all these elements are nonmetals, geraniol is probably covalent and comprised of molecules.

The formula $C_{10}H_{18}O$ tells you that each molecule of geraniol contains ten carbon atoms, 18 hydrogen atoms, and one oxygen atom. In terms of numbers of atoms, hydrogen is the major element in geraniol. Which is the major element by mass? You can answer this question by determining the mass percents of each element in geraniol.

Suppose you have a mole of geraniol. Its molar mass is 154 g/mol. Of this mass, how many grams do the carbon atoms contribute? The formula shows that one molecule of geraniol includes ten atoms of carbon. Therefore, 1 mol of geraniol contains 10 mol of carbon. Multiply the mass of 1 mol of carbon by 10 to get the mass of carbon in 1 mol of geraniol.

$$(10 \text{ mol}) \left(\frac{12.0 \text{ g C}}{\text{mol}} \right) = 120 \text{ g C}$$

Personal Tutor For an online tutorial on percentages, visit glencoe.com.

Now, use this mass of carbon to find the mass percent of carbon in geraniol.

$$\text{Mass percent of C} = \frac{120 \text{ g C}}{154 \text{ g } C_{10}H_{18}O} \times 100 = 77.9\%$$

The mass percents of the other elements are calculated below in a similar way.

Mass of hydrogen in 1 mol geraniol:

$$(18 \text{ mol}) \times \left(\frac{1.00 \text{ g H}}{\text{mol}} \right) = 18.0 \text{ g H}$$

$$\text{Mass percent of H} = \frac{\text{mass of H}}{\text{mass of geraniol}} \times 100$$

$$= \frac{18.0 \text{ g H}}{154 \text{ g } C_{10}H_{18}O} \times 100 = 11.7\% \text{ H}$$

Mass of oxygen in 1 mol geraniol:

$$(1 \text{ mol}) \left(\frac{16.0 \text{ g O}}{\text{mol}} \right) = 16.0 \text{ g O}$$

$$\text{Mass percent of O} = \frac{\text{mass of O}}{\text{mass of geraniol}} \times 100$$

$$= \frac{16.0 \text{ g O}}{154 \text{ g } C_{10}H_{18}O} \times 100 = 10.4\% \text{ O}$$

The steps above will let you determine the composition, in terms of percent by mass, of any compound whose formula you know. **Figure 12.12** displays the composition of geraniol in a circle graph.

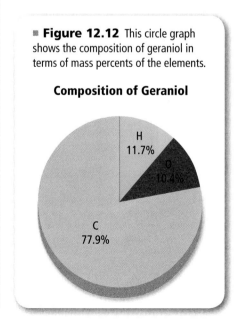

■ **Figure 12.12** This circle graph shows the composition of geraniol in terms of mass percents of the elements.

CHEMLAB

ANALYZE A MIXTURE

Background
Chemists often analyze mixtures to determine their compositions. Their analytical procedures may include gas chromatography, mass spectrometry, or infrared spectroscopy. In this ChemLab, you will use a double-displacement reaction between strontium chloride and sodium sulfate to analyze a mixture of sodium sulfate and sodium chloride.

Question
What is the mass percent of sodium sulfate in a mixture of sodium sulfate and sodium chloride?

Objectives
- **Observe** the double-displacement reaction between strontium chloride and sodium sulfate.
- **Quantify** the amount of strontium sulfate produced.
- **Compare** the mass of strontium sulfate produced with the mass of sodium sulfate that reacted.

Preparation

Materials
funnel
wash bottle with distilled water
filter paper
250-mL beakers (2)
50-mL graduated cylinder
Sodium sulfate-sodium chloride mixture
0.500M strontium chloride solution
stirring rod
ring stand
iron ring
spatula
weighing dish or paper
balance

Safety Precautions

Procedure

1. Read and complete the lab safety form.
2. Measure the mass of the weighing dish or weighing paper and record its mass in your data table where indicated.
3. Transfer a 0.500- to 0.600-g sample of the sodium sulfate-sodium chloride mixture to the dish or paper and measure its mass. Record the total mass of the sample and container in your data table.
4. Transfer all of the mixture to a 250-mL beaker, add approximately 50 mL of distilled water, and stir slowly until the sample is completely dissolved.
5. Use a graduated cylinder to measure 15 mL of 0.500M strontium chloride solution, then pour it into the solution in the beaker. Stir slowly for 30 s to completely precipitate the strontium sulfate.
6. Obtain a piece of filter paper, fold it in half, and tear off one corner. Fold the paper in half again. Measure the mass of the folded paper and record its mass in your data table where shown.
7. Clamp a funnel support or an iron ring to a ring stand, put the funnel in the support or ring, insert the folded filter paper in the funnel, and wet the paper with a small amount of distilled water from your wash bottle. Set an empty 250-mL beaker under the funnel to receive the filtrate.

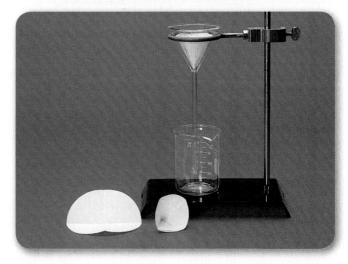

8. Carefully transfer the mixture, including all of the precipitate, from the beaker to the filter paper, using a stirring rod and a wash bottle as demonstrated by your teacher.

9. Rinse the precipitate by pouring about 10 mL of distilled water into the filter paper.

10. When liquid no longer drips from the funnel, carefully remove the filter paper and residue from the funnel. Unfold the paper and spread it on a paper towel to dry overnight. Write your name on the paper towel.

11. Dispose of the filtrate according to your teacher's instructions.

12. Measure the mass of the dry filter paper the following day, and record the mass of filter paper with residue in your data table.

Data and Observations

Data Table	
Mass of sample + dish	
Mass of dish	
Mass of sample	
Mass of $SrSO_4$ + filter paper	
Mass of filter paper	
Mass of $SrSO_4$	

Analyze and Conclude

1. **Interpret** What is the balanced equation for the reaction between sodium sulfate and strontium chloride in aqueous solution?

2. **Infer** After the reaction, does the sodium chloride produced and the unreacted sodium chloride from the original sample appear in the precipitate or the filtrate?

3. **Calculate** What mass of strontium sulfate is produced in the reaction? Subtract the mass of filter paper from the mass of strontium sulfate plus filter paper.

4. **Calculate** What is the mass and mass percent of sodium sulfate in the original sample?

Apply and Assess

1. **Predict** Could you precipitate the sulfate from the sodium sulfate in the sample by adding an excess of aluminum chloride solution? Explain.

2. **Error Analysis** If your strontium sulfate precipitate was not completely dry when you measured its mass, how would it affect your value of the mass percent of sodium sulfate in the sample?

INQUIRY EXTENSION

How could you have tested the precipitate to be sure it contained strontium?

CHEMISTRY AND TECHNOLOGY

Improving Percent Yield in Chemical Synthesis

The single most important industrial chemical in the world is probably sulfuric acid. World-wide sulfuric acid production is about 170 million tons annually, with the United States producing 40 million tons per year. The strength of the U.S. economy can be gauged by the amount of sulfuric acid produced because many jobs are associated with its production and its many uses. Over the years the manufacturing process of sulfuric acid has been improved to provide a higher and more economical yield.

Uses of Sulfuric Acid
Sixty percent of all manufactured sulfuric acid in the world is used to make fertilizers. It is also used in manufacturing detergents, photographic film, synthetic fibers, pigments, paints, pharmaceuticals, and other acids. It is the electrolyte in lead-acid automobile batteries, acts as a catalyst and a dehydrating agent, is a component in refining petroleum and metal ores, and is used in chemical synthesis to manufacture other chemicals.

The Lead-Chamber Process— A Low-Yield Process
The first industrial method used to manufacture sulfuric acid was the lead-chamber process. This process in no longer commonly used because its purity is low (62–72 percent sulfuric acid) and its percent yield is only 60–80 percent. However, it is much less expensive than the later and more productive contact process. For this reason, the lead-chamber process is still used for manufacturing some sulfuric acid for applications that do not demand high purity, such as the production of fertilizers. High purity sulfuric acid, about 95 percent, is needed for pharmaceuticals and pigments.

The Contact Process— The High-Yield Process
The contact process is the most widely used commercial method. It is more expensive than the lead-chamber process, but it is simple and it produces high-purity sulfuric acid at a high percent yield—about 98 percent. In addition, it creates no by-products that pollute the atmosphere. The contact process has four steps.

Maximizing Percent Yield
Reaction yields in the contact process were increased in several ways—by selecting an optimum temperature, using an efficient catalyst, removing a product from a reaction that does not go to completion, and by controlling the rate of reaction of SO_3 with water. Keeping operating pressures at the correct values also increases yield. As you read the following steps in the contact process, refer to **Figure 1**.

Step 1
Sulfur is burned in air to produce sulfur dioxide, a stable compound. This reaction takes place quickly and readily.

$$S(s) + O_2(g) \rightarrow SO_2(g)$$

Impurities produced in combustion are removed from the sulfur dioxide so they will not react with and impede the catalyst in the next step.

Step 2
Because sulfur dioxide reacts slowly with excess oxygen, a catalyst, either vanadium pentoxide (V_2O_5) or finely divided platinum at —produces sulfur trioxide.

$$2SO_2(g) + O_2(g) \xrightarrow{catalyst} 2SO_3(g)$$

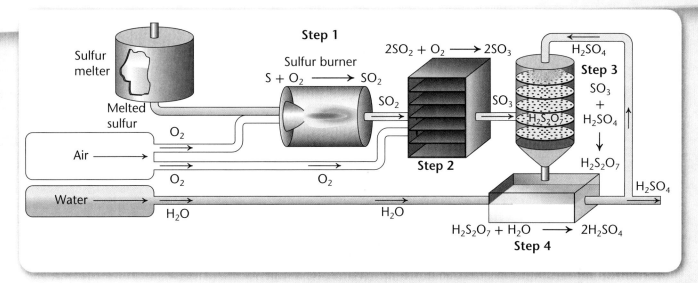

Figure 1 Contact process of manufacturing sulfuric acid

Sulfur trioxide is quickly removed from the contact chamber because it tends to form sulfur dioxide and oxygen again. Removing sulfur trioxide promotes the production of more sulfur trioxide.

Step 3
Because sulfur trioxide reacts violently with water, it is bubbled through 98 percent concentrated sulfuric acid and forms pyrosulfuric acid, $H_2S_2O_7$.

$$SO_3(g) + H_2SO_4(l) \rightarrow H_2S_2O_7(l)$$

Step 4
When water is added to the pyrosulfuric acid, high-quality, 98 percent concentrated sulfuric acid is formed.

$$H_2S_2O_7(l) + H_2O(l) \rightarrow 2H_2SO_4(l)$$

Using Scientific Methods to Attain the Best Yields
To develop higher yields that are closer to the theoretical values, chemists adjust temperatures, pressures, or other conditions in their industrial processes. They look for better catalysts or new ways to deal with undesirable side reactions. Unwanted by-products are serious concerns to the chemist. Because they may cause environmental damage or be expensive to dispose of, by-products may raise production costs.

Green Chemistry
The contact process is economically sound because the reactants are abundant and inexpensive and because it produces no unwanted by-products, which may be expensive to store or dispose of and which may pollute the environment. Ecologically sound chemical manufacturing is called green chemistry or green technology.

Discuss the Technology

1. **Hypothesize** Roasting iron sulfide (FeS_2), which is the mineral pyrite or marcasite, can replace burning sulfur in Step 1 to produce sulfur dioxide. Write an equation to show what you think the reaction with iron sulfide is.
2. **Acquire Information** Find out how sulfuric acid is used to manufacture hydrochloric acid and write the equation(s) of the reactions.

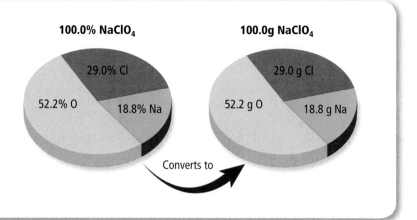

■ **Figure 12.13** Keep this figure in mind when doing problems using percent composition. You can always assume that you have a 100-g sample of the compound. In a 100-g sample, the percents of the elements simply become the masses of the elements present in the sample.

Determining Chemical Formulas

Suppose you analyzed an unknown compound and found that, by mass, it was 18.8 percent sodium, 29.0 percent chlorine, and 52.2 percent oxygen. Because this compound contains some metal and nonmetal elements, it may be an ionic compound. To determine its chemical formula, find the relative numbers of sodium, chlorine, and oxygen atoms in the formula unit of the compound.

Suppose you have 100.0 g of the unknown compound, as shown in **Figure 12.13.** Because you know the sample includes 18.8 g of sodium, 29.0 g of chlorine, and 52.2 g of oxygen, use the molar mass to find the number of moles of each element.

$$\left(18.8 \text{ g Na}\right)\left(\frac{1 \text{ mol Na}}{23.0 \text{ g Na}}\right) = 0.817 \text{ mol Na}$$

$$\left(29.0 \text{ g Cl}\right)\left(\frac{1 \text{ mol Cl}}{35.5 \text{ g Cl}}\right) = 0.817 \text{ mol Cl}$$

$$\left(52.2 \text{ g O}\right)\left(\frac{1 \text{ mol O}}{16.0 \text{ g O}}\right) = 3.26 \text{ mol Na}$$

You know that atoms combine in ratios of small whole numbers to form compounds. To find the whole-number ratios, divide the mole numbers by the smallest one.

$$\frac{0.817 \text{ mol Na}}{0.817} = 1.00 \text{ mol Na}$$

$$\frac{0.817 \text{ mol Cl}}{0.817} = 1.00 \text{ mol Cl}$$

$$\frac{3.26 \text{ mol O}}{0.817} = 3.99 \text{ mol O}$$

These values are whole numbers or very close to whole numbers. Therefore, the mole ratio for this compound is 1 mol Na : 1 mol Cl : 4 mol O. Similarly, the ratio of atoms in this compound is 1 Na : 1 Cl : 4 O.

The formula of a compound having the smallest whole-number ratio of atoms in the compound is called the **empirical formula.** The empirical formula of this unknown compound is $NaClO_4$.

VOCABULARY
WORD ORIGIN
empirical
comes from the Greek word *empeirikos*, which means *working from experience*

From the empirical formula to the chemical formula What is the chemical formula for this compound? You have learned that the formula for an ionic compound represents the simplest possible ratio of the ions present and is called a formula unit. Chemical formulas for most ionic compounds are the same as their empirical formulas. Because the unknown compound is ionic, the chemical formula for a formula unit of the compound is the same as its empirical formula, $NaClO_4$. The compound is called sodium perchlorate.

Chemical formula differs from emprical formula As another example, suppose the mass percents of a compound are 40.0 percent carbon, 6.70 percent hydrogen, and 53.3 percent oxygen. Because all its elements are nonmetals, the compound is covalent. Imagine you have 100 g of the compound. Then you have 40.0 g of carbon, 6.70 g of hydrogen, and 53.3 g of oxygen. Use the molar masses of these elements to find the number of moles of each element.

$$(40.0 \text{ g C})\left(\frac{1 \text{ mol C}}{12.0 \text{ g C}}\right) = 3.33 \text{ mol C}$$

$$(6.70 \text{ g H})\left(\frac{1 \text{ mol H}}{1.00 \text{ g H}}\right) = 6.70 \text{ mol H}$$

$$(53.3 \text{ g O})\left(\frac{1 \text{ mol O}}{16.0 \text{ g O}}\right) = 3.33 \text{ mol O}$$

Now, divide all the mole numbers by the smallest one.

$$\frac{3.33 \text{ mol C}}{3.33} = 1.00 \text{ mol C}$$

$$\frac{6.70 \text{ mol H}}{3.33} = 2.01 \text{ mol H}$$

$$\frac{3.33 \text{ mol O}}{3.33} = 1.00 \text{ mol O}$$

Because the ratio of moles is the same as the ratio of atoms, CH_2O is the empirical formula for this compound. But, the empirical formula is not always the chemical formula. Many different covalent compounds have the same empirical formula, as demonstrated in **Figure 12.14,** because atoms can share electrons in different ways.

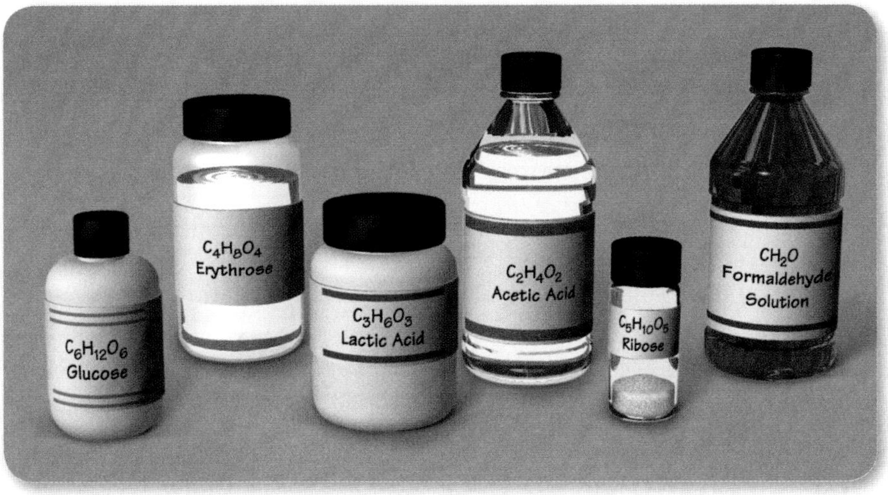

■ **Figure 12.14** For each of these compounds, the ratio of atoms is 1C : 2H : 1O. Because these compounds all have a different number of atoms in a molecule, each one is a different compound with its own molecular formula and properties.

Describe *one difference in the physical properties of acetic acid and ribose.*

To decide which multiple of the empirical formula is the correct molecular formula, you need the molar mass of the compound. Suppose a separate analysis shows that the molar mass of the compound is 90.0 g/mol. Therefore, the molecular mass of the compound is 90.0 u. The molecular mass of a CH_2O molecule is 30.0 u. By dividing the molecular mass of the compound by 30.0 u, you find the multiple.

$$\frac{90.0 \text{ u}}{30.0 \text{ u}} = 3$$

The molecular formula of the compound contains three empirical formula units. The molecular formula is $C_3H_6O_3$. The compound is lactic acid, the sour-tasting substance in spoiled milk. Lactic acid is also produced by active muscles. If lactic acid builds up in muscle cells at a faster rate than the blood can remove it, muscle fatigue results. An immediate burning sensation and soreness a few days later is an indication that lactic acid was produced in the muscles during exercise. **Figure 12.15** shows strenuous activity that can cause muscle fatigue and soreness.

Connecting Ideas

In this chapter, you learned to solve several kinds of stoichiometric problems. For each kind, you used the mole concept because when substances react, their particles interact. The number of particles at this submicroscopic level controls what happens macroscopically. In the next chapter, you will use the mole concept and the particle nature of matter to study the mixtures of substances called solutions.

■ **Figure 12.15** During strenuous activity, lactic acid is produced in muscle cells. Soreness in muscles a day or two after the activity is a sign of lactic acid formation.

Section 12.2 Assessment

Section Summary

▶ A balanced chemical equation provides mole ratios of the substances in the reaction.

▶ The ideal gas law is expressed in the equation $PV = nRT$.

▶ Percent yield measures the efficiency of a chemical reaction.

$$\text{Percent yield} = \frac{\text{actual yield}}{\text{theoretical yield}} \times 100$$

▶ Percent composition can be determined from the chemical formula of a compound; the empirical formula of a compound can be determined from its percent composition.

▶ The chemical formula of a compound can be determined if the molar mass and the empirical formula are known.

18. MAIN Idea List Octane (C_8H_{18}) is one of the many components of gasoline. Write the balanced chemical equation for the combustion of octane to form carbon dioxide gas and water vapor. Identify as many mole ratios among the substances in the combustion as you can.

19. Calculate If 25.0 g of octane burn as in Problem 1, how many grams of water will be produced? How many grams of carbon dioxide?

20. Evaluate A manufacturer advertises a new synthesis reaction for methane with a percent yield of 110 percent. Comment on this claim.

21. Predict The reaction of iron(III) oxide and aluminum is called the thermite reaction because of its intense heat. The iron produced is molten and was formerly used to weld railroad tracks in remote areas. The balanced chemical equation for the thermite reaction is shown below.

$$Fe_2O_3(s) + 2Al(s) \longrightarrow Al_2O_3(s) + 2Fe(l)$$

If 20.0 g of each reactant are used, which is the limiting reactant?

22. Determine Indigo, the dye used to color blue jeans, is prepared using sodium amide. Sodium amide contains the following mass percents of elements: hydrogen, 5.17 percent; nitrogen, 35.9 percent; and sodium, 58.9 percent. Find an empirical formula for sodium amide.

CHAPTER 12 Study Guide

Download quizzes, key terms, and flash cards from glencoe.com.

BIG Idea The mole represents a large number of extremely small particles.

Section 12.1 Counting Particles of Matter

MAIN Idea A mole always contains the same number of particles; however, moles of different substances have different masses.

Vocabulary
- Avogadro's number (p. 405)
- formula mass (p. 408)
- molar mass (p. 407)
- mole (p. 405)
- molecular mass (p. 408)
- stoichiometry (p. 404)

Key Concepts
- Stoichiometry relates the amounts of products and reactants in a chemical equation to one another.
- The mole is a unit used to count particles of matter. One mole of a pure substance contains Avogadro's number of particles, 6.02×10^{23}.
- Molar mass can be used to convert mass to moles or moles to mass.

Section 12.2 Using Moles

MAIN Idea Balanced chemical equations relate moles of reactants to moles of products.

Vocabulary
- empirical formula (p. 426)
- ideal gas law (p. 418)
- molar volume (p. 415)
- percent yield (p. 420)
- theoretical yield (p. 420)

Key Concepts
- A balanced chemical equation provides mole ratios of the substances in the reaction.
- The ideal gas law is expressed in the equation $PV = nRT$.
- Percent yield measures the efficiency of a chemical reaction.

$$\text{Percent yield} = \left(\frac{\text{actual yield}}{\text{theoretical yield}}\right) \times 100$$

- Percent composition can be determined from the chemical formula of a compound; the empirical formula of a compound can be determined from its percent composition.
- The chemical formula of a compound can be determined if the molar mass and the empirical formula are known.

Chapter 12 Assessment

Understand Concepts

23. A manufacturer must supply 20,000 connector units, each consisting of a bolt, two washers, and three nuts. How many of each part are needed?

24. Your uncle leaves you a barrel of nickels. You determine that the nickels weigh 1345 pounds and that 50 nickels weigh 232 g. How many dollars is the barrel of nickels worth?

25. Determine the molecular mass of the following substances.
    a) bromine
    b) argon
    c) lead(II) nitrate
    d) dinitrogen tetroxide

26. Explain how you would use weighing to count 40,000 washers.

27. Which has the largest mass?
    a) three atoms of magnesium
    b) one molecule of sucrose ($C_{12}H_{22}O_{11}$)
    c) ten atoms of helium

28. Which has the largest mass?
    a) 3 mol of magnesium
    b) 1 mol of sucrose ($C_{12}H_{22}O_{11}$)
    c) 10 mol of helium

29. A reaction requires 0.498 mol of Cu_2SO_4. How many grams must you weigh to obtain this number of formula units?

30. What is the molecular mass of UF_6? What is the molar mass of UF_6?

Table 12.1 Molar Masses of Some Compounds

Compound	Molar Mass
C_6H_5Br	?
$K_2Cr_2O_7$	?
$(NH_4)_3PO_4$	?
$Fe(NO_3)_3$	?

31. Determine the missing molar masses in Table 12.1.

32. What mass of copper contains the same number of atoms as 68.7 g of iron?

33. Explain why the mass of copper is not equal to the mass of iron in question 5.

34. Calculate the mass of 0.345 mol of sodium nitrite, $NaNO_2$.

35. Calculate each of the following:
    a) the mass in grams of 0.254 mol of calcium sulfate
    b) the number of atoms in 2.0 g of helium
    c) the number of moles in 198 g of glucose, $C_6H_{12}O_6$
    d) the total number of ions in 10.8 g of magnesium bromide

36. Calculate the mass of silver chloride (AgCl) and dihydrogen monosulfide (H_2S) formed when 85.6 g of silver sulfide (Ag_2S) reacts with excess hydrochloric acid, HCl.

37. Aluminum nitrite and ammonium chloride react to form aluminum chloride, nitrogen, and water. What mass of aluminum chloride is present after 43.0 g of aluminum nitrite and 43.0 g of ammonium chloride have reacted completely?

38. What are the molar masses of the following elements?
    a) nitrogen
    b) iodine
    c) oxygen
    d) nickel

39. Sodium nitrate decomposes upon heating to form sodium nitrite and oxygen gas. This reaction is sometimes used to produce small quantities of oxygen in the lab. How many grams of sodium nitrate must be heated to produce 128 g of oxygen?

40. Calculate the mass of silver chloride (AgCl) and dihydrogen monosulfide (H_2S), which are formed when 85.6 g of silver sulfide (Ag_2S) reacts with excess hydrochloric acid (HCl).

41. An oxide of nitrogen is 26 percent nitrogen by mass. The molar mass of the oxide is approximately 105 g/mol. What is the formula of the compound?

42. How many moles are in each of the following?
    a) 43.6 g NH_3
    b) 5.0 g aspirin ($C_9H_8O_4$)
    c) 15.0 g CuO

Chapter 12 Assessment

43. A 10.0-g sample of magnesium reacted with excess hydrochloric acid to form magnesium chloride by the reaction below. When the reaction was complete, 30.8 g of magnesium chloride were recovered. What was the percent yield?

    $Mg(s) + 2HCl(aq) \rightarrow MgCl_2(aq) + H_2(g)$

44. What is the molecular formula for each of the following compounds?
    a) empirical formula: CH_2; molar mass: 42 g/mol
    b) empirical formula: CH; molar mass: 78 g/mol
    c) empirical formula: NO_2; molar mass: 92 g/mol

Apply Concepts

45. Hydrogen fuels are rated with respect to their hydrogen content. Determine the percent hydrogen for the following fuels.
    a) ethane, C_2H_6
    b) methane, CH_4
    c) whale oil, $C_{32}H_{64}O_2$

46. Hydrogen peroxide and sodium nitrate are good oxidizing agents. Compare the mass percent of oxygen in each of these compounds.

Art Connection

47. How does the United States standardize measurements of mass and weight?

48. Serotonin is a compound that conducts nerve impulses in the brain. The molar mass is 176 g/mol. Use the circle graph in **Figure 12.16** to help determine its molecular formula.

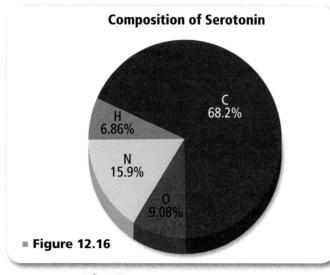

■ Figure 12.16

Everyday Chemistry

49. How many grams of sodium azide, NaN_3, are needed to fill a 0.0650-m^3 driver's air bag with nitrogen if the pressure required is 2.00 atmospheres and the gas temperature is 40.0°C? (Hint: 1 m^3 = 1000 L)

Chemistry and Technology

50. What are some uses of sulfuric acid?

Think Critically

Compare and Contrast

51. Distinguish between molar mass, formula mass, and molecular mass.

52. **ChemLab** Suggest a procedure to determine the mass percent of silver nitrate ($AgNO_3$) in a mixture of silver nitrate and sodium nitrate ($NaNO_3$).

Compare and Contrast

53. **MiniLab 1** Explain how your procedure in this activity is like using the molar mass to determine the number of particles in a sample of a known substance.

Relate Cause and Effect

54. **MiniLab 2** If you used calcium carbonate ($CaCO_3$) in a reaction with vinegar instead of baking soda ($NaHCO_3$), how would the amount of reactants needed change?

Cumulative Review

55. Distinguish between atomic number and mass number. How do these two numbers compare for two different isotopes of an element? (*Chapter 2*)

56. Why does the second period of the periodic table contain eight elements? (*Chapter 7*)

57. The volume of a gas at STP is 8.50 L. What is its volume if the pressure is 1250 mm Hg and the temperature is 75.0°C? (*Chapter 11*)

Chapter 12 Assessment

Table 12.2 Using Stoichiometry

Item	Formula	Molar Mass (g/mol)	Number of Particles	Rank
The formula units in 10.0 g of calcium fluoride				
The sodium ions in 10.0 g of sodium chloride				
The water molecules in 10.0 g of water				
The hydrogen atoms in 10.0 g of water				
The carbon dioxide molecules in 10.0 g of carbon dioxide				
The carbon monoxide molecules in 10.0 g of carbon monoxide				
The aspirin molecules, $C_9H_8O_4$, in 10.0 g of aspirin				
The carbon atoms in 10.0 g of aspirin				
The valence electrons in 10.0 g of aspirin				

58. Answer the following questions about sugar and table salt. (*Chapter 4*)
   a) What type of compound is each substance?
   b) Which compound is an electrolyte?
   c) Which compound is made up of molecules?

Skill Review

59. Using a Table Complete **Table 12.2** above. In the last column, rank the number of particles from smallest to largest. Did you use the number of moles or the number of individual particles for your ranking? Explain.

WRITING in Chemistry

60. A mole is such a large number that it is hard to envision just how big 6.02×10^{23} is. Write three of your own examples of the number of things in a mole. One example must use time, one must use distance, and the third is up to you.

Problem Solving

61. A student carries out the following sequence of chemical reactions on a 0.635-g sample of pure copper.

$$Cu(s) + 2HNO_3(aq) \rightarrow Cu(NO_3)_2(aq) + H_2(g)$$

$$Cu(NO_3)_2(aq) + 2NaOH(aq) \rightarrow Cu(OH)_2(s) + 2NaNO_3(aq)$$

$$Cu(OH)_2(s) \rightarrow CuO(s) + H_2O(l)$$

$$CuO(s) + H_2SO_4(aq) \rightarrow CuSO_4(aq) + H_2O(l)$$

$$CuSO_4(aq) + Mg(s) \rightarrow Cu(s) + MgSO_4(aq)$$

The student isolates and weighs the product at the end of each step before proceeding to the next. What is the theoretical yield of the copper product in each step? Based on the law of the conservation of matter, how should the yield of copper metal in the final step compare to the starting mass of copper metal?

Cumulative Standardized Test Practice

1. The most common form of matter in the universe is
   a) gas.
   b) liquid.
   c) solid.
   d) plasma.

2. Charles's Law states that the volume of a gas is
   a) inversely proportional to the temperature of the gas.
   b) directly proportional to the temperature of the gas.
   c) inversely proportional to the pressure of the gas.
   d) directly proportional to the pressure of the gas.

Interpreting Graphs: *Use the graph below to answer Questions 3–4.*

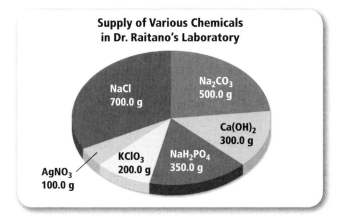

Supply of Various Chemicals in Dr. Raitano's Laboratory
- NaCl 700.0 g
- Na_2CO_3 500.0 g
- $Ca(OH)_2$ 300.0 g
- NaH_2PO_4 350.0 g
- $KClO_3$ 200.0 g
- $AgNO_3$ 100.0 g

3. Pure silver metal can be made using the reaction shown below:

   $Cu(s) + 2AgNO_3(aq) \rightarrow 2Ag(s) + Cu(NO_3)_2(aq)$

   How many grams of copper metal will be needed to use up all of the $AgNO_3$ in Dr. Raitano's laboratory?
   a) 18.70 g
   b) 37.3 g
   c) 74.7 g
   d) 100 g

$Na_2CO_3(aq) + Ca(OH)_2(aq) \rightarrow$
$2NaOH(aq) + CaCO_3(s)$

4. The LeBlanc process, shown above, is the traditional method of manufacturing sodium hydroxide. Using the amounts of chemicals available in Dr. Raitano's lab, the maximum number of moles of NaOH that can be produced is
   a) 4.05 mol.
   b) 4.72 mol.
   c) 8.097 mol.
   d) 9.43 mol.

5. Which chemical has the greatest molar mass?
   a) KCl
   b) CH_3OH
   c) $Ba(OH)_2$
   d) CH_3COOH

6. Moving down a group on the periodic table, which two atomic properties follow the same trend?
   a) atomic radius and ionization energy
   b) ionic radius and atomic radius
   c) ionization energy and ionic radius
   d) ionic radius and electronegativity

7. Which of the following metallic properties does a sea of valence electrons NOT explain?
   a) conductivity of electricity
   b) conductivity of heat
   c) malleability
   d) low boiling point

NEED EXTRA HELP?

If You Missed Question . . .	1	2	3	4	5	6	7
Review Section . . .	10.1	11.2	12.2	12.2	12.1	3.2	9.1

CHAPTER 13 Water and Its Solutions

BIG Idea Water plays a major role in many chemical reactions.

13.1 Uniquely Water
MAIN Idea The molecular shape of water gives it unusual properties.

13.2 Solutions and Their Properties
MAIN Idea Water dissolves a large number of ionic and covalent compounds.

ChemFacts

- Havasu Falls form where Havasu Creek plummets over a 37 m cliff.
- The distinct blue-green color of the creek's waters result from the creek's high mineral content, mainly calcium carbonate.
- The look and course of Havasu Creek can change a good deal from year to year due to minerals that crystallize on any debris that fall into the creek.
- When the carbonate minerals dissolved in the creek come out of solution, they form travertine, a type of sedimentary rock.

Start-Up Activities

LAUNCH Lab

Solution Formation

The interparticle forces among dissolving particles and the attractive forces between solute and solvent particles result in an overall energy change. Can this change be observed?

Materials
- balance
- 50-mL graduated cylinder
- 100-mL beakers (2)
- stirring rod
- ammonium chloride (NH_4Cl)
- calcium chloride ($CaCl_2$)
- water

Procedure

1. Read and complete the lab safety form.
2. Add 10 g of ammonium chloride (NH_4Cl) to a 100-mL beaker.
3. Pour 30 mL of water into the beaker of NH_4Cl and mix with a stirring rod.
4. Feel the bottom of the beaker and record your observations.
5. Repeat the procedure, replacing ammonium chloride with calcium chloride ($CaCl_2$).

Analysis

1. **Infer** Which dissolving process is exothermic? Which is endothermic?

Inquiry What are everyday applications for dissolving processes that are exothermic? Endothermic?

Chemistry Online

Visit glencoe.com to:
- study the entire chapter online
- explore Concepts in MOtion
- take Self-Check Quizzes
- use Personal Tutors
- access Web Links for more information, projects, and activities
- find the Try at Home Lab, Measuring Capillarity

FOLDABLES Study Organizer

Properties of Water Make the following Foldable to organize information about the water molecule.

STEP 1 Fold up the bottom of a horizontal sheet of paper about 5 cm as shown.

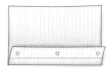

STEP 2 Fold the paper in half.

STEP 3 Unfold once and staple to make two pockets. Label the pockets *Water* and *Aqueous Solutions*.

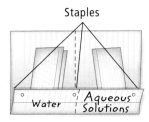

FOLDABLES Use this Foldable with Sections 13.1 and 13.2. As you read the sections, use index cards or quartersheets of paper to summarize what you learn about how the molecular structure of water contributes to the unique properties of water and its solutions. Insert these into the appropriate pockets of your Foldable.

Chapter 13 • Water and Its Solutions **435**

Section 13.1

Objectives

- **Describe** the uniqueness of water as a chemical substance.
- **Model** the three-dimensional geometry of a water molecule.
- **Relate** the physical properties of water to its molecular structure.

Review Vocabulary

interparticle forces: the forces between the particles that make up a substance

New Vocabulary

hydrogen bonding
surface tension
capillarity
specific heat

Uniquely Water

MAIN Idea The molecular shape of water gives it unusual properties.

Real-World Reading Link Water, which covers 71% of the surface of Earth, is one of the most common substances, but its properties are anything but common. Water vapor in the atmosphere contributes to changes in the weather, mountain streams provide fresh drinking water, and the oceans are a nutrient-rich ecosystem. Even the chemical reactions that sustain living things require water—as a reactant, as a place for the reaction to happen, or both. What properties of water make it so indispensable?

Physical Properties of Water

Because water is so much a part of life, its properties are easy to take for granted. If you step back and examine water scientifically, you will find that it is unusual among the compounds found on Earth. **Table 13.1** compares the physical properties of water against a group of molecules of similar mass. Look at the physical states at room temperature, melting points, and boiling points of these molecules. Water definitely behaves unlike other small molecules. It is a unique substance.

Water is most often thought of as a liquid. When you try to think of other common liquids, you are probably thinking of water-based, or aqueous, solutions. However, solid water, called ice, and gaseous water, called steam or water vapor, also exist in large quantities on Earth. Gaseous water is found around geysers and hot springs and in the atmosphere, where it plays a major role in determining weather. Ice occurs in glaciers and ice caps and acts as a huge reservoir of fresh water. Water is the only substance on Earth that exists in large quantities as all three common states of matter.

Water is also unique because most substances are more dense as solids than they are as liquids. Water is less dense as a solid than as a liquid. For this reason, ice floats in liquid water rather than sinks. If ice were like most solids and denser than the liquid state, lakes and ponds would freeze from the bottom up.

Table 13.1 The Uniqueness of Water

Concepts In Motion
Interactive Table Explore the uniqueness of water at glencoe.com.

Substance	Formula	Molecular Mass (u)	State at Room Temperature	Melting Point (°C)	Boiling Point (°C)
Methane	CH_4	16	gas	−183	−161
Ammonia	NH_3	17	gas	−78	−33
Water	H_2O	18	liquid	0	100
Nitrogen	N_2	28	gas	−210	−196
Oxygen	O_2	32	gas	−218	−183

Figure 13.1 The fact that ice is less dense than liquid water is the reason that ice collects at the top of the liquid rather than at the bottom. If it were not for this fact, ice-skating would not be possible.

The fact that the ice forms on the tops of lakes, ponds, and rivers allows aquatic life in the liquid water below to survive the harsh winter season. Also, if ice were like most solids, popular sports such as ice fishing, figure skating, and ice hockey, shown in **Figure 13.1**, would never be possible.

What factors explain the unique physical properties of water in its various states? **Figure 13.2** reviews what you have already learned about the composition of the water molecule, its electron distribution, and its three-dimensional structure. Because oxygen is so electronegative compared to hydrogen, the hydrogen ends of the bent water molecule are strongly positive. Additionally, the oxygen atom has two nonbonding electron pairs attached to it. Due to both the electronegativity and the nonbonding pairs, the oxygen end of the molecule is strongly negative.

Interparticle Forces in Water

Recall from Chapter 9 that molecules that are dipoles, such as water, have stronger interparticle attractive forces than nonpolar molecules. You can model the behavior of a group of water molecules by imagining that the molecules act like little bar magnets.

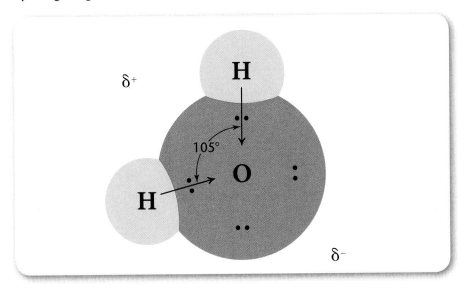

Figure 13.2 The arrangement of valence electrons around the oxygen atom of a water molecule relates to its three-dimensional geometry. There is a large electronegativity difference between the covalently bonded hydrogens and oxygen. Therefore, electron pairs are shared unequally. Because of the molecule's bent structure, the poles of positive and negative charge in the two bonds do not cancel, and the water molecule as a whole is polar.

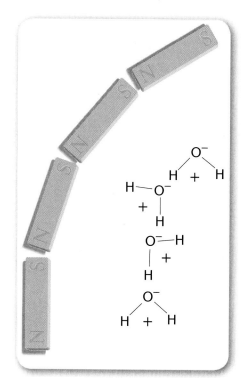

Figure 13.3 The attraction between water molecules is analogous to the attraction between magnets. However, water's interparticle forces result from opposite electric charges, not from magnetism. The effects of interparticle attractions are especially noticeable at low temperatures.

Infer *Why might interparticle forces have greater effects at low temperature?*

Figure 13.3 illustrates water molecules with positively charged ends attract negatively charged ends similar to the way north poles of magnets attract south poles. A group of magnets will tend to orient themselves with the opposite poles of the different magnets attracted to one another. Although dipoles aren't held by magnetic forces, a group of water molecules will orient in a similar way at the molecular scale because of electrical forces. Opposites attract and create order among molecules.

If you moved one of the magnets in **Figure 13.3,** the position of all the magnets would change. The same thing would happen if you could reach in a group of water molecules and move just one. This model demonstrates that attractive forces between objects do not create interactions between only two objects. The forces can combine to give organization to groups of objects, whether bar magnets or water molecules.

Hydrogen bonding Although the interparticle forces in water create order that involves large numbers of molecules, you can get a good picture of what's going on by modeling the interactions between just a few water molecules, as shown in **Figure 13.4**. Notice that the oxygen on one water molecule is attracted to the hydrogen atoms on other water molecules. These connections between the molecules are not covalent bonds, but they are still fairly strong.

The formation of such a connection between the hydrogen atoms on one molecule and a highly electronegative atom on another is called **hydrogen bonding.** Hydrogen bonds are strong dipole–dipole attractions between molecules that contain a hydrogen atom bonded to a small, highly electronegative element with at least one lone pair of electrons. Atoms that can form hydrogen bonds include oxygen, fluorine, and nitrogen. Notice that these three atoms meet all of the requirements for hydrogen bonding listed above: They are small, which means they can closely approach the hydrogen in a separate molecule. They are very electronegative, which means their chemical bonds with hydrogen have highly positive and negative regions. Finally, they have at least one lone pair of electrons. It is this lone pair that interacts with the positive hydrogen end of a separate molecule to form a hydrogen bond.

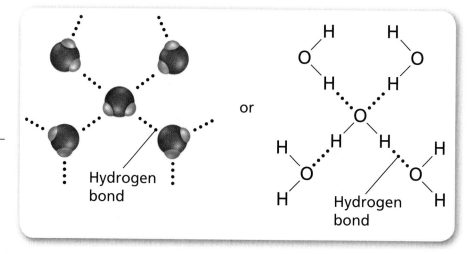

Figure 13.4 The hydrogen bonds between water molecules are stronger than typical dipole-dipole attractions because the bond between hydrogen and oxygen is highly polar.

Explain *Which forces are interparticle and which ones are intramolecular?*

438 Chapter 13 • Water and Its Solutions

Molecules that form hydrogen bonds Oxygen-hydrogen groups (O—H) in molecules tend to promote interparticle hydrogen bonding. Oxygen is sufficiently electronegative to attract hydrogen's single electron so strongly that the hydrogen atom almost becomes an exposed proton. In pure water, each water molecule may form hydrogen bonds with four other water molecules, as shown in **Figure 13.4.** Because the strong partial charges can interact closely with four other molecules, water is very good at forming networks of hydrogen bonds.

While water is particularly effective at forming hydrogen bonds, any molecule that contains O—H bonds has the potential for hydrogen bonding. Alcohols and organic compounds that contain O—H bonds also form hydrogen bonds, as shown in **Figure 13.5.** Hydrogen bonds contribute to some of the similarities in the physical properties of alcohols and water.

Because they have many O—H bonds, biological molecules such as proteins, nucleic acids, and carbohydrates also have the ability to form hydrogen-bond networks. These networks can be extensive, and their three-dimensional shape is important in determining their functions in living organisms.

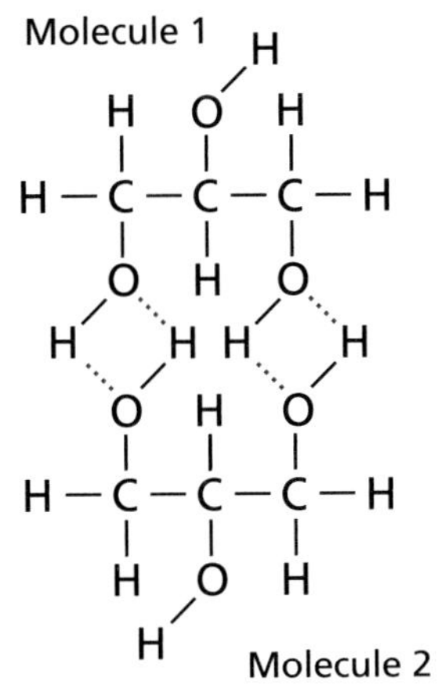

States of Water

Look at **Table 13.2,** which identifies the two molecules from **Table 13.1** that are closest to water in size. You can see again that compared to similar-sized molecules, water's physical properties are quite different. For example, water has very high melting and boiling points for such a small molecule. Many of these unique physical properties are a result of the hydrogen bonding. This is just another example of the relationship between submicroscopic structure and macroscopic properties, the directly observable behavior of substances.

■ **Figure 13.5** Alcohols, such as the glycerol ($C_3H_8O_3$) molecules above, contain O—H groups. As with water, these O—H groups can form networks of hydrogen bonds.
Determine *the possible number of hydrogen bonds a glycerol molecule can form with a second molecule.*

Table 13.2	Properties of Three Molecular Compounds		
Compound	Molecular Structure	Molar Mass (g)	Boiling Point (°C)
Methane (CH_4)		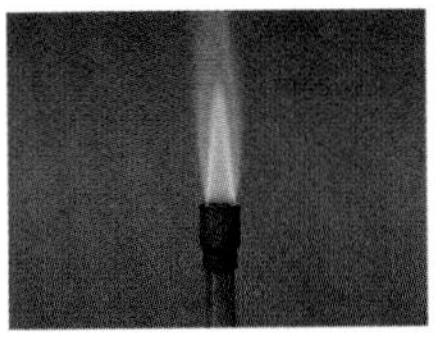 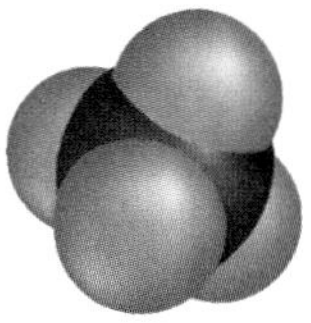 16	−161
Ammonia (NH_3)		 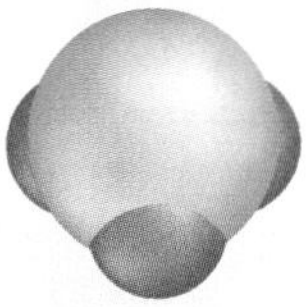 17	−33
Water (H_2O)		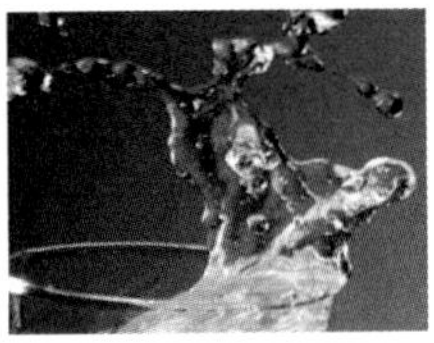 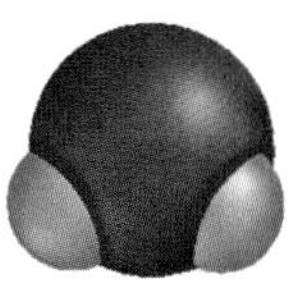 18	100

FOLDABLES Incorporate information from this section into your Foldable.

Solid and liquid water Water occurs primarily in the liquid and solid states on Earth, rather than as a gas. The hydrogen bonds hold water molecules together so strongly that they cannot easily evaporate into the gaseous state at ordinary temperatures. That is why water has such a high boiling point for such a small molecule (100°C). In order for water to boil, hydrogen bonds must be broken, which requires the addition of energy in the form of heat. When the temperature reaches 100°C, hydrogen bonds break and water molecules separate to form water vapor.

Density of ice If you drop an ice cube into a glass of water, the ice floats. This action reveals that the density of the solid water is less than that of liquid water. **Figure 13.6** shows how the density of water changes with respect to temperature. Like most substances, liquid water shrinks and becomes denser as it cools. As water cools from 60°C, its volume decreases and its density increases. The water molecules move less rapidly, and they are drawn closer together by interparticle forces. Because these molecules pull together, the volume of the water decreases. Meanwhile, the mass of water stays the same, so density increases.

Water's density turning point When water is cooled to about 4°C, something unusual happens. The volume stops decreasing; the density has reached a maximum. At this point, the water molecules are as close together as they can get. As the temperature falls below 4°C, the volume of water expands. The mass of water is still the same, so its density decreases.

How can you account for this change? Below 4°C, the water molecules are beginning to approach the solid state, which is highly organized. The water molecules begin to form an open arrangement. This arrangement results from hydrogen bonding and is the most stable structure for the molecules in or near the solid state.

The fact that liquid water expands when it freezes enables aquatic life to survive the winter cold (ice floats), but also causes frozen water pipes to break and sidewalks to crack during the winter cold. The forces involved in the expansion of freezing water are even strong enough to break huge boulders into small pieces. **Figure 13.7** illustrates the phase changes of water.

■ **Figure 13.6** Water achieves its minimum volume, and therefore its maximum density, at 3.98°C.

Infer *When you pour water at a temperature of 5 °C into water at a temperature of 15 °C, will the cold water sink to the bottom or remain on the surface before the temperature reaches equilibrium?*

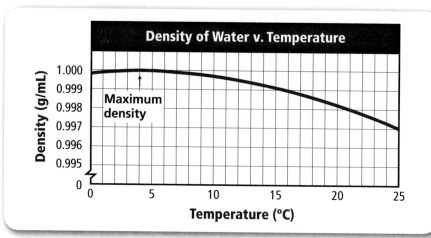

FIGURE 13.7
Phase Changes of Water

 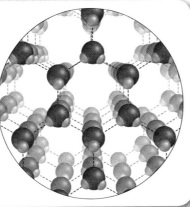

The six-pointed snowflakes reflect the open arrangement of bonds. In order to freeze, molecules of liquid water separate slightly from one another as a result of hydrogen bonding.

At 0°C, liquid water freezes to solid water, ice. The volume rises and the density drops dramatically, from 1.000 g/mL to 0.99984 g/mL, as the water molecules assume an open arrangement. The volume expansion that occurs when water freezes has some disadvantages, as you can see by looking at this burst jar. This action is also an important agent of erosion, with continued cycles of freezing and thawing capable of splitting large boulders.

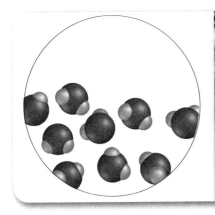 

Ice begins to melt when the temperature climbs above the freezing point of water, 0°C. As the ice melts to form liquid water, the molecules become less densely packed and are able to move about freely. Water bubbling at the surface of a pool illustrates the properties of liquid water and this inherent ability to flow.

When liquid water comes to a boil at 100°C, molecules are free to move at a fast and furious pace. As the temperature increases, density decreases. When liquid water evaporates to form a vapor, it becomes less dense than the surrounding air, at which point it rises. In the atmosphere, it cools and condenses to form clouds.

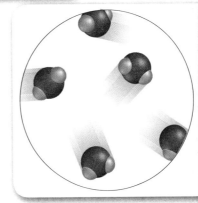 

Section 13.1 • Uniquely Water

More Evidence for Water's Interparticle Forces

The strong hydrogen bonds that hold water molecules together are evident in other properties of water. These observable properties provide more macroscopic validity to the scientific model of the submicroscopic structure of water molecules. Keep in mind that the source of all of these properties is the unique combination of the water molecule's geometry and the electron distribution that makes the O—H bond highly polar.

Surface tension Have you ever watched water drip from a faucet? Each drop is composed of an enormous number of water molecules—roughly 2×10^{21}. That this large number of molecules can hold together to form a single drop is more evidence for the presence of hydrogen bonding.

A water molecule forms a drop because of **surface tension**, which is the force needed to overcome interparticle attractions and break through the surface of a liquid to spread out the liquid. The higher the surface tension of a liquid, the more resistant the liquid is to having its surface broken. **Figure 13.8** shows a molecular model of the surface of water. The net downward force at the surface makes the surface of the drop contract and seem to toughen, behaving like a sort of skin.

Examples of surface tension Examples of water's surface tension are all around you. For instance, you might have observed that water drops on a freshly waxed car form tight beads. The water molecules have a strong attraction for each other, but little attraction to the car wax. As a result of surface tension, water forms small beads. Filling a glass with water until the surface of the water bulges above the level of the glass's rim is another example of surface tension. The spider in **Figure 13.8** illustrates a third good example.

FOLDABLES
Incorporate information from this section into your Foldable.

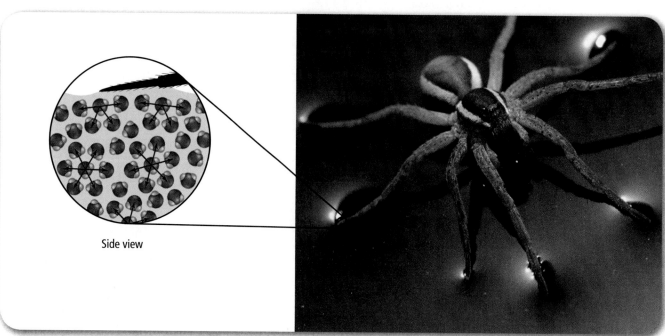

■ **Figure 13.8** At the surface of water, the particles are drawn toward the interior until attractive and repulsive forces are balanced.
Infer *How do you know that the spider is not floating on the water?*

Side view

MiniLab 13.1

Surface Tension

How many drops of water and how many drops of an aqueous detergent solution can a penny hold?

Procedure

1. Read and complete the lab safety form.
2. Lay a **penny** flat on your lab table.
3. Fill a **microtip pipette** with **tap water,** and count the number of drops you can deposit on the penny before water spills over the edge. Record the number of drops.
4. Repeat step 3 with a **detergent solution** prepared by your teacher.

2. **Infer** Which has the lower surface tension: the water or the aqueous detergent solution? What might account for this fact?

Analysis

1. **Explain** How is surface tension demonstrated in this investigation?

Liquids other than water also exhibit surface tension. Mercury is a liquid that has strong interparticle attractive forces and a high surface tension. In fact, mercury's interparticle forces are so strong that the beads it forms when placed on a smooth surface sometimes appear to be completely round in shape.

Capillarity Have you ever had a small sample of blood drawn from a finger during a routine blood test? After your finger is pricked, the blood is drawn by touching a thin glass tube, called a capillary tube, to the blood drop. Even without any suction applied, the blood rises into the tube. **Figure 13.9** shows this effect demonstrated with capillary tubes in water. Notice that the water level is higher in the narrower capillary tube than it is in the wider capillary tube.

VOCABULARY
WORD ORIGIN
Capillarity
comes from the Latin word *capillus*, which means *hair*

Adhesion   Cohesion

The force of attraction between the water molecules and the silicon dioxide in the glass cause the water molecules to creep up the glass.

Water molecules are attracted to each other (cohesion) and to the silicon dioxide molecules in the glass (adhesion).

■ **Figure 13.9** Water molecules have cohesive and adhesive properties.
Infer *why the water level is higher in the smaller diameter tube.*

Section 13.1 • Uniquely Water   **443**

■ **Figure 13.10** Water (left) forms a concave meniscus, whereas mercury (right) produces a convex meniscus. With mercury, there is essentially no attractive force between the mercury atoms and the silicon dioxide atoms of the glass tube to compete with the attractive forces between the mercury atoms themselves. The mercury forms a convex (high-centered) meniscus because the only force is the interparticle attractive force between mercury atoms, which produces surface tension.

TRY AT HOME 🏠 LAB
See page 874 for **Measuring Capillarity.**

These examples of the rising of liquids in narrow tubes are illustrations of **capillarity,** or, as it is sometimes called, capillary action. Capillarity results from interparticle attractive forces between the molecules of liquid and the attractive forces between the liquid and the tube that contains it.

In glass tubes, water molecules form hydrogen bonds with the oxygen atoms in silicon dioxide (SiO_2), a main component of glass. The attractive force between the water and the SiO_2 draws the water up the walls of the tube. Because the water molecules are also attracted to each other, they draw other water molecules upward, too. If the tube is very narrow, the liquid will be drawn high into the tube. Because nearly all the molecules are close to the walls that they are strongly attracted to, more water molecules form bonds with the glass. Thus, the water level is higher in the narrow tubes.

Meniscus formation If the glass tube is of larger diameter, like a graduated cylinder, you see the upward capillary effect near the tube walls. However, the larger volume of water in the cylinder has too much mass to be pulled upward as it is in a capillary tube. As a result, the liquid surface forms the meniscus shape shown in **Figure 13.10.** The photos also contrast the behavior of water in a glass tube with the behavior of mercury. The particles in mercury have very strong attractions to each other but almost no attraction to the walls of the glass tube. Therefore, the meniscus formed by mercury is the opposite of water.

Specific heat Have you ever jumped into a pool of cool water in the morning on a hot day in the summer? Despite the fact that the air temperature is high, the water temperature is lower because it changes more slowly. However, if you jump into that same pool in the evening when the air temperature has dropped, the pool water will be warmer than the outside air. The water has once again lagged behind its surroundings in changing temperature. Once the water is warm, it does not cool quickly. You might notice a similar effect when you put a pot of water on to boil and wait a long time for it to heat up.

Table 13.3 Specific Heats of Some Substances

Substance	Specific Heat (J/g°C)
Gold, Au	0.129
Copper, Cu	0.385
Iron, Fe	0.450
Glass	0.84
Cement	0.88
Wood	1.76
Ethanol, $C_2H_5OH(l)$	2.46
Water, $H_2O(l)$	4.18

What you observe in these examples is the high specific heat of water. **Specific heat** measures the amount of heat, in joules, needed to raise the temperature of 1 g of substance by 1°C. Because different substances have different compositions, each substance has its own specific heat. The specific heats of water and some other common substances are shown in **Table 13.3.**

The specific heat of water Notice that water has the highest specific heat of any of the substances listed in **Table 13.3.** This means that water must absorb or release more heat for its temperature to change by one degree Celsius than any of the other substances. In other words, its temperature will rise and fall more slowly than substances with lower specific heats, all other factors being equal.

The specific heat of water is 4.18 J/g • °C. (The unit is read "joules per gram per degree Celsius.") In order to raise the temperature of 1 g of water by 1°C, you must add 4.18 J of heat. You must also remove 4.18 J of heat from a 1-g sample of water to lower its temperature by 1°C. This large amount of heat transfer is why it takes time for the water in a swimming pool, like that in **Figure 13.11,** to warm up in the early part of the day. It also accounts for why the water, once warmed, cools off slowly when the outdoor temperature drops.

■ **Figure 13.11** The water in this swimming pool heats up during the day. Even though the outside temperature drops at night, the water tends to remain warm. Because of its high specific heat, the warm water has stored a great deal of energy.

Section 13.1 • Uniquely Water

■ **Figure 13.12** A desert's temperature changes greatly at twilight. Without water, heat is given off quickly.

Water serves as a great heat reservoir that moderates the temperature of Earth's surface. During the day, heat from the Sun is absorbed by the oceans. Despite all of the heat the water absorbs, the temperature of oceans does not increase detectably because of water's high specific heat and large volume. In the evening, the stored heat in Earth's oceans is released, helping to maintain the air temperature.

If there were no water on Earth's surface to serve as a temperature moderator, the daytime temperature would soar. Rocks have a much lower specific heat than water does. In the evening, Earth would turn frigid because rocks have so little capacity to store heat. You have noticed this effect if you have ever experienced the extreme temperature difference between day and night in a desert, such as the one shown in **Figure 13.12,** during clear weather.

Heat of vaporization Recall from Chapter 10 that vaporization is the change in state of liquid to gas, and the amount of heat required to vaporize a quantity of a liquid is called the heat of vaporization. During the vaporization of a liquid, interparticle forces must be overcome. Therefore, vaporization of a liquid is an endothermic (energy-absorbing) process. Condensation, the formation of a liquid from a gas, is an exothermic (energy-releasing) process. Water absorbs a great deal of heat in order to overcome hydrogen bonding and move the molecules farther apart. Thus, it has a high heat of vaporization. Water also loses a great deal of heat when it condenses. This quality explains the effectiveness of steam in heating a room or in causing burns.

Evaporation is vaporization from the surface of a liquid. Evaporation of water is one of the ways your body regulates its temperature. The water in the perspiration absorbs your body heat and evaporates, lowering your body temperature. The sweat glands that help regulate body temperature are located all over the skin, but are particularly concentrated in the palms of the hands, soles of the feet, and forehead, as shown in **Figure 13.13.**

■ **Figure 13.13** Perspiration is a mechanism by which your body passes water to the surface of your skin via sweat glands. The vaporization of water is an endothermic (energy-absorbing) process, and as the water evaporates, it removes heat from the surface of your skin.

Infer *Compared to other areas of the skin, why might sweat glands on the forehead be particularly effective in helping to regulate body temperature?*

Chemistry & Society

Water Treatment

With a twist of your wrist, you can make clean water flow from the faucet. Like most other people in industrialized countries, you may never stop to think how fortunate you are to have accessible, safe water. In many parts of the world, people are forced to carry water long distances. Then, after all their hard work, they have no assurance that the water is safe to drink. The water you use is probably safe because it is purified in a reliable water-treatment plant, a diagram of which is shown in **Figure 1,** through the following series of steps.

1. Water is often transported over long distances through pipes from a lake or other freshwater source to the treatment plant. When the water enters the intake basin, it passes through bar screens that remove large, suspended solids and trash.

2. After the screening, pumps lift the water 6 m or more so that it can flow through the rest of the tanks under the influence of gravity.

3. Some particles present in water are so fine that they can be trapped only by coagulation. To do this, chemicals such as alum, chlorine, and lime are added by chemical applicators.

The purpose of the alum—$Al_2(SO_4)_3 \cdot 18H_2O$—is to produce alum floc, a coagulant that causes fine, suspended solids to clump together. Chlorine kills bacteria. Lime causes precipitation of calcium carbonate and clears the water.

4. In the settling tank, bacteria, silt, and other impurities stick to the coagulants and settle to the bottom. The rest of the water moves on to a filtering tank.

5. The filtering tank allows the water to trickle through sand, gravel, and sometimes charcoal for a final cleansing.

6. When the water reaches the reservoir, it receives another treatment of chlorine. The purified water remains in the reservoir until it is pumped to homes, factories, and businesses.

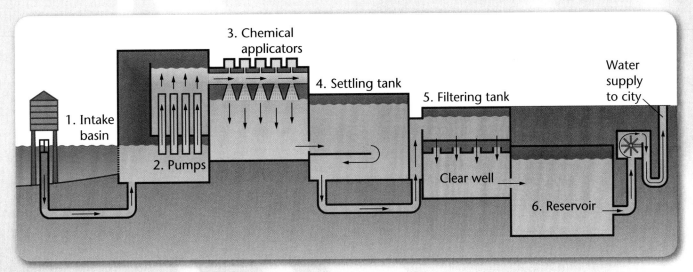

■ **Figure 1** Diagram of a water treatment plant

ANALYZE the Issue

1. **Acquire Information** Research and design a brochure on how biological treatment is used to clear water of organic waste from industries.

2. **Write** Prepare a public service announcement that explains community precedents that could be set to ensure that local drinking water does not become polluted.

In the Field

Meet Alice Arellano
Wastewater Operator

When alarm clocks buzz on weekday mornings, millions of people head for the bathroom to prepare for the day. Showers run, toilets flush, and teeth are brushed, generating a tremendous volume of wastewater that flows toward water treatment plants. Wastewater operators like Alice Arellano are on the job, making sure this water is properly treated and protecting the health of people and the environment.

On the Job

Q: Ms. Arellano, can you explain your job responsibilities to us?

A: I'm a wastewater control-room operator. Our plant is responsible for cleaning wastewater from this part of Austin, Texas.

Q: How does the cleaning process work?

A: First, the wastewater comes into our plant through a big pipe, about 54 inches in diameter. I've seen some pretty nasty stuff come through that pipe! The water goes through some bar screens, which remove the debris. The heavy solids, such as inorganic metals and sand, settle to the bottom of the chamber, where augers move them to a dump truck. The truck takes the inorganic waste to a landfill. The material that stays in suspension, the floatable stuff, goes into primary treatment. After primary treatment, then it goes to secondary treatment and some aeration tanks. Microorganisms in the tank break down the harmful bacteria. As I explain it in nontechnical language, the microorganisms start at the top of the tank and eat and party all the way down.

Q: After the water has been cleaned, where does it go?

A: Into the creeks and eventually into the Colorado River and Lake Powell. The water is filtered and used again. There's a finite amount of water on Earth, so we have to use it over and over.

Q: Does the weather have an effect on your work?

A: When it rains, my work can get really tough. Ordinarily, the plant treats about 40 million gallons of water a day. Capacity is about 120 million gallons. But an unusually heavy storm can overwhelm the system with 400 million gallons. Then, all we can do is add chlorine, open the valves, and let the water pass through.

Q: How do people react when you talk about your profession?

A: Some people are jealous because the job pays well and is steady work. Not everyone would like to do this job, though. The odor of ammonia can be strong. Once in a while, I get stomach viruses even though I wear gloves and glasses and use all my safety equipment. However, I find the work fascinating. I'm trying to learn everything that goes on at the plant, so I ask everyone questions—the mechanics, the instrumentation people, and the electricians.

Early Influences

Q: Was this position your first one after you left school?

A: I got my GED late because I became a mother quite young. I had a friend, a teammate on my softball team, who worked in the plant and encouraged me to get certified as an operator. I took classes in wastewater management because I really wanted to better myself.

Q: Do you find a use for the math and chemistry you took in school?

A: I have to know the formulas for the amount of chlorine to add, as well as for the sulfur dioxide. That chemical removes the chlorine so the water discharge won't harm fish in the creeks. More than my knowledge of chemistry, I use all my senses. It's often a matter of experience.

Personal Insights

Q: Your job isn't a typical one for a woman, is it?

A: Men are in the majority here, but that's changing. My daughter came to Take-Our-Daughters-to-Work Day and became interested in what I do.

Q: Do you have any advice for kids who might not like school?

A: Just to try hard, and stay in school. I encourage my own daughter to leave the adult behavior until the time she's ready for adult responsibilities. There are always about a dozen kids who make my home their gathering place these days. Because I had problems when I was young, they trust me to help them with their problems.

CAREER CONNECTION

Other jobs vital for keeping water clean include the following.

Environmental Health Inspector College degree plus licensing

Environmental Technician Two-year accreditation program

Industrial Machinery Mechanic High school plus apprenticeship program

Chapter 13 • In the Field **449**

■ **Figure 13.14** Everything you drink is an aqueous solution. Soft drinks, tea, coffee, spring water, and even tap water are all aqueous solutions. Even distilled water is an aqueous solution; it contains dissolved gases.

Water: The Universal Solvent

Most of the water on Earth is not pure but is present in solutions. Water is difficult to keep pure because it is an excellent solvent for a variety of solutes. Water is such a versatile solvent that it is sometimes called the universal solvent. Its ability to act as a solvent is one of water's most important physical properties. It is the attraction of water molecules for other molecules, as well as for one another, that accounts for these solvent properties.

As **Figure 13.14** shows, nearly all the water you consume is in aqueous solutions. Aqueous solutions provide efficient means of transporting nutrients in plants, as well as the nutrients in your blood. Almost all of the life-supporting chemical reactions that occur take place in an aqueous environment. In the absence of water, these reactions would not occur.

Section 13.1 Assessment

Section Summary

- The polarity of the water molecule is the source of many of water's unusual physical properties.
- Hydrogen bonds form between the hydrogen atom of one molecule and a highly electronegative atom of another.
- Specific heat is the amount of heat needed to raise the temperature of 1 g of a substance by 1°C.

1. **MAIN Idea** **List** five physical properties that distinguish water from most other compounds of similar molecular size.
2. **Model** hydrogen bonding.
3. **Describe** surface tension. Use the concept of surface tension to explain why a drop of water forms a sphere.
4. **Compare** Ethylene glycol is a molecule with the formula $C_2H_4(OH)_2$. Each molecule has two O—H bonds. Ethanol (C_2H_5OH) has one O—H bond. How would you expect the boiling points of ethanol and ethylene glycol to compare?
5. **Infer** Why is hot weather more uncomfortable if the humidity is high?
6. **Explain** why water forms a meniscus when it's in a graduated cylinder.

Section 13.2

Solutions and Their Properties

Objectives

- **Compare and contrast** the ability of water to dissolve ionic and covalent compounds.
- **Determine** the concentrations of solutions.
- **Compare and contrast** colligative properties of solutions.

Review Vocabulary

capillarity: the rising of a liquid in a narrow tube, sometimes called capillary action

New Vocabulary

dissociation
unsaturated solution
saturated solution
supersaturated solution
heat of solution
osmosis
colloid
Tyndall effect

MAIN Idea Water dissolves a large number of ionic and covalent compounds.

Real-World Reading Link When you make a cup of tea, you might place a tea bag into a cup of hot water and wait. After a few minutes, you remove the tea bag. You know that something from the tea must have dissolved into the water because the water is darker in color and tastes different. Clearly the bag and the remaining parts of the tea leaves didn't dissolve. Why do some things, but not others, dissolve in water?

The Dissolving Process

Just as with the other physical properties of water, you can relate water's solvent properties to its molecular structure. The submicroscopic interactions that occur between water molecules and various solute particles determine the extent to which water is able to dissolve a solute.

Water dissolves many ionic substances You have probably added table salt, sodium chloride, to water and noticed that the salt dissolved, as illustrated by **Figure 13.15**. If you made a careful measurement, you would discover that 26 g of sodium chloride dissolves in 100 mL of water at room temperature. Salt, like many ionic compounds, is soluble in water. The salt solution is also an excellent conductor of electricity. Electrical conductivity is always observed when ionic compounds dissolve to a significant extent in water.

Dissociation Why does a salt solution conduct electricity? Your model of water and its interactions explains why salt and many other ionic compounds dissolve in water and why the solutions conduct electricity.

■ **Figure 13.15** When water is added to table salt, the polar water molecules surround the sodium and chloride ions, and the ionic compound dissociates. Because water molecules are polar, they are attracted to both the positively charged sodium ions and the negatively charged chloride ions. Water molecules surround both types of ions.

Explain why water molecules are attracted to both negative ions and positive ions.

Concepts In Motion

Interactive Figure To see an animation of the dissolution of compounds, visit glencoe.com.

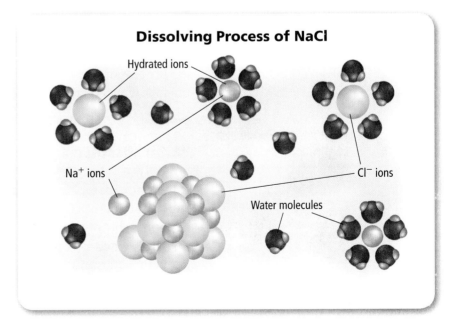

MiniLab 13.2

Hard and Soft Water

How well do different types of water produce suds when they are mixed with a soap solution? Water is considered to be hard if it has significant concentrations of certain ions, such as calcium ions (Ca^{2+}) and magnesium ions (Mg^{2+}). Water in which calcium and magnesium ions are present in low concentrations only or in which they have been substantially replaced with sodium ions (Na^+) is called soft water. Distilled water contains only dissolved gases and exceedingly few ions. Test hard water, soft water, and distilled water with a sodium oxalate ($Na_2C_2O_4$) solution. This solution will cause Ca^{2+} and Mg^{2+} to precipitate as calcium oxalate (CaC_2O_4) and magnesium oxalate (MgC_2O_4).

Procedure

1. Read and complete the lab safety form.
2. Add about 1 mL of **hard water** to a **small test tube**, about 1 mL of **soft water** to another small test tube, and about 1 mL of **distilled water** to a third small test tube.
3. Add to each of the three solutions 2 drops of **0.1M acetic acid** and 2 drops of **0.1M sodium oxalate**. Mix the contents of each of the tubes by tapping them with your finger. Observe the solutions against a black background and record your observations.
4. Add about 2 mL of hard water to a **large test tube**, about 2 mL of soft water to another large test tube, and about 2 mL of distilled water to a third large test tube.
5. Add to each of the three solutions about 1 mL of **soap solution.**
6. Shake each of the three solutions 10–15 times by holding the test tube in your hand with your thumb over the top. Measure and record the height of the suds in each tube.

Analysis

1. **Conclude** If the oxalate hardness test is negative (that is, if no precipitate is observed), is a complete absence of calcium and magnesium ions confirmed?
2. **Infer** How does the hardness of water affect the ability to make soap suds?

FOLDABLES
Incorporate information from this section into your Foldable.

Remember that ionic solids are composed of a three-dimensional network of positive and negative ions, which form strong ionic bonds. Take another look at **Figure 13.15** to see how a crystal of sodium chloride (NaCl) dissolves in water. Sodium chloride dissolves in water because of the favorable interactions between the ions of which it is composed and the polar water molecules.

The process by which the charged particles in an ionic solid separate from one another is called **dissociation**. You can represent the process of dissolving and dissociation in shorthand fashion by the following equation.

$$NaCl(s) \xrightarrow{H_2O} Na^+(aq) + Cl^-(aq)$$

Water dissolves many covalent substances Water is not only good at dissolving ionic substances. It also is a good solvent for many covalent compounds. Consider the covalent substance sucrose ($C_{12}H_{22}O_{11}$), commonly known as sugar. You have probably observed this substance dissolve in water. For instance, when you prepare lemonade by mixing lemonade powder with water, you dissolve a great amount of sucrose in water. In fact, sugar is highly soluble. It is possible to dissolve almost 200 g of sugar in 100 mL of water.

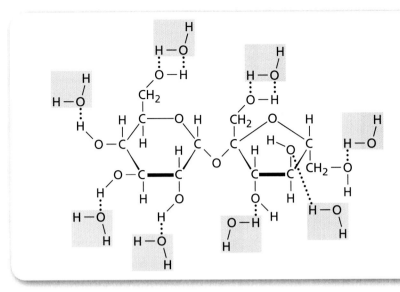

Figure 13.16 shows the structure of a sucrose molecule. Notice that the structure has a number of O—H bonds. As you learned earlier, if a molecule contains O—H bonds, it can form hydrogen bonds. One of the reasons that sugar is a solid at room temperature is that sugar molecules have the ability to form strong hydrogen bonds with each other. They can also form hydrogen bonds with water, which is why sugar dissolves so readily in water.

Molecules versus ions in solution For both sucrose and sodium chloride, interparticle attractive forces between the solvent and solute particles overcome attractive forces between solute particles. However, because sucrose is covalent, the sucrose molecules are simply separated from one another by water molecules. They do not dissociate into charged particles, but rather they remain neutral molecules. Neutral molecules cannot conduct electricity because they have no charge. Therefore, an aqueous solution of sucrose does not conduct electricity, as shown in **Figure 13.17**. The dissolving of sugar is represented by the following simple equation.

$$C_{12}H_{22}O_{11}(s) \xrightarrow{H_2O} C_{12}H_{22}O_{11}(aq)$$

■ **Figure 13.16** Sucrose molecules contain eight O–H bonds and are polar. Polar water molecules form hydrogen bonds with the O–H bonds, which draws the sucrose into solution.

■ **Figure 13.17** Sodium chloride (left) conducts electricity well because it is an electrolyte. Sucrose (right) does not conduct electricity because it is not an electrolyte. Sucrose molecules do not form ions when they dissolve.

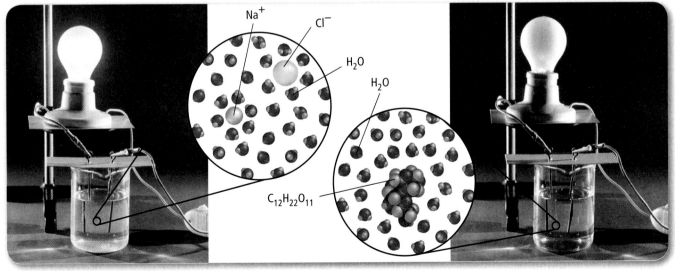

Like dissolves like Although water dissolves an enormous variety of substances, both ionic and covalent, it does not dissolve everything. The phrase that scientists often use when predicting solubility is "like dissolves like." The expression means that dissolving occurs when similarities exist between the solvent and the solute.

Consider the two examples of salt and sugar dissolving in water. In the case of the ionic salt, water is similar to a charged ionic compound in the sense that it is polar, meaning that it has partially charged ends. The interactions that occur between water molecules—primarily hydrogen bonds—are somewhat similar to the water-ion interactions that occur when salt dissolves. Thus, "like dissolves like." Water is polar and it tends to dissolve ionic substances.

In the case of sugar, water is like sucrose in that both substances are made of polar molecules, with partially positive ends and negative ends. The molecular interactions between sucrose molecules are similar to the molecular interactions between water molecules. Again, like dissolves like. Water is polar and has hydrogen bonding, and it tends to dissolve substances that are polar or that form hydrogen bonds.

Oil and water don't mix Oil and water are a classic example of two substances that do not mix; they do not form a solution. Oil is a mixture of nonpolar covalent compounds composed of primarily carbon and hydrogen. Given the composition of oil, you should not be surprised that oil does not dissolve in water. When you add oil to water, there is little interparticle attraction between solvent and solute and mixing is inhibited. As shown in **Figure 13.18,** even after vigorous shaking, oil and water will rapidly separate into layers. However, two different oils, both of which are nonpolar, are similar enough to remain mixed.

■ **Figure 13.18 a)** When a bottle containing a mixture of baby oil (colorless) and water (dyed blue) is shaken, the oil and water seem to mix at first. **b)** Upon sitting, the substances separate into two layers. **c, d)** In contrast, olive oil (greenish-gold) and safflower oil (nearly colorless) mix readily and remain mixed.

Everyday Chemistry

Soaps and Detergents

Has a slippery piece of pizza ever made a greasy stain on your favorite jeans? You pour water on the telltale spot, but the water just runs off. You are learning the lesson that oil and water don't mix, unless you add soap. Oil is normally insoluble in water, but soap molecules achieve the seemingly impossible by causing oil to mix with water.

Making soap Though most people use soap for bathing and other detergents for washing clothes, there are important similarities between the two.

People may first have learned how to make soap when ashes from a fire fell into a pot of boiling fat. The people soon found uses for the smooth white gel that floated to the top of the mixture. The ashes supplied lye, which is a strong base such as sodium hydroxide or potassium hydroxide. When fats or oils react chemically with lye, the end products are soap—often sodium stearate—and glycerin. Examine the chemical formula of sodium stearate.

$$CH_3-(CH_2)_{16}-COO-Na^+$$

A negative carboxyl group ion ($-COO^-$) is tied to a positive sodium ion (Na^+) and a long chain of $-CH_2-$ groups.

Like soaps, detergents have molecules with a polar end and a nonpolar end. Detergents typically contain sulfonates, which have $-SO_3$ groups attached to a carbon chain or ring. The detergents are typically sodium salts of these sulfonates, and their molecules have an $-SO^{3-}$ end rather than the $-COO^-$ end that is typical of soaps.

Soap forms a scumlike precipitate in the presence of the magnesium ions and calcium ions that are present in hard water. Detergents, on the other hand, form soluble sulfonate salts in the presence of magnesium ions and calcium ions. That makes detergents more effective than soaps in hard water.

Figure 1 Soap and detergents are used for cleaning all around your home.

How soaps and detergents work A soap molecule has two different parts: one end is hydrophilic, attracted to water, and the other end is hydrophobic, repelled by water. The hydrophobic part of the soap molecule consists of a long hydrocarbon chain that is structurally similar to oil and is therefore soluble in oil.

The hydrocarbon chains in soap are attracted to particles of oily grease and dirt. That part of the soap molecule forms a protective layer around the oily material. The hydrophilic end of the soap molecule (the $-COONa$ end) is attracted by polar water molecules. That causes the entire soap molecule, together with the oily material, to pull toward the wash water. The oil-soap complex is suspended in the water and rinsed away. Because of these attractions, the soaps and detergents shown in **Figure 1** are effective at cleaning.

Explore Further

1. **Apply** How might rubbing and scrubbing with soap help to remove a greasy stain from clothing?

2. **Infer** Some ancient peoples, such as the Egyptians and Romans, washed by rubbing themselves with oil, which was then scraped off. Compare the effectiveness of this method with the use of soap and water today.

CHEMLAB

SMALL SCALE

IDENTIFY SOLUTIONS

Background

In an aqueous solution, ionic compounds are completely dissociated into ions. For example, an aqueous solution of barium nitrate ($Ba(NO_3)_2$), contains Ba^{2+} ions and NO_3^- ions. If aqueous solutions of ionic compounds are mixed, some ions may interact to form an insoluble product called a precipitate. For example, if aqueous solutions of barium nitrate and sodium sulfate are mixed, insoluble barium sulfate ($BaSO_4$) will produce a white precipitate.

$Ba(NO_3)_2(aq) + Na_2SO_4(aq) \rightarrow$
$\qquad 2NaNO_3(aq) + BaSO_4(s)$

In this ChemLab, you will work with the aqueous solutions of six unknown ionic compounds. By observing the solutions and their interactions, you will be able to determine the identities of the compounds.

Question

What are the identities of six unknown aqueous solutions?

Objectives

- **Observe** the interactions of the aqueous solutions of six compounds.
- **Interpret** the results of the interactions and use the results to identify the solutions.

Preparation

Materials

96-well microplate
10 mL each of 0.1M solutions of sodium carbonate, sodium iodide, copper(II) sulfate, copper(II) nitrate, lead(II) nitrate, barium nitrate—identified only as A–F
6 microtip pipettes, labeled A–F
15 toothpicks
distilled water in a rinse bottle

Safety Precautions

WARNING: *Some of the solutions are toxic.*

Procedure

1. Read and complete the lab safety form.
2. Obtain a 96-well microplate. To produce all the combinations of solutions, you will be using only the 15 wells in the upper-right-hand corner.
3. Use the appropriately labeled microtip pipettes to place three drops of solution A in each of five wells in the first horizontal row, three drops of B in each of four wells in the second row, three drops of C in each of three wells in the third row, three drops of D in each of two wells in the fourth row, and three drops of E in one well in the fifth row. The first letter in each box in the table in Data and Observations shows the arrangement.
4. For your second additions, use the appropriately labeled microtip pipettes to add three drops of the solutions to the same wells to obtain the combinations shown in the Data and Observations table. Thus, you will place three drops of solution B in the single well in the first column, three drops of C in both of the wells in the second column, and so on up to the fifth column, to which you will add three drops of solution F. You have now created all the possible combinations of pairs of solutions in the wells.
5. Stir the solutions in each well, using a clean toothpick for each.
6. Place the microplate on a white surface and look down through each well to detect the presence of a precipitate. Cloudiness is evidence of a suspended precipitate. Repeat the procedure on a black surface. Record in your data table the color of any precipitates. If no precipitate is formed, write NR, meaning "no reaction."

456 Chapter 13 • Water and Its Solutions

Data and Observations

Data Table					
Combinations					
A + B	A + C	A + D	A + E	A + F
	B + C	B + D	B + E	B + F
		C + D	C + E	C + F
			D + E	D + F
				E + F

Analyze and Conclude

1. **Formulate Models** Write balanced equations for all 15 possible double-replacement reactions, regardless of whether they actually occurred. Show both possible products in each case. If the two possible products are the same as the two reactants, you can rule out the possibility of reaction, so write NR (no reaction) after the equation.

2. **Classify** Use a handbook of chemistry to find out the colors of each of the products and whether they are soluble or insoluble in water.

3. **Draw Conclusions** Identify the six solutions by relating your data and observations to the predicted colors and interactions. Assume that the color of an aqueous solution is generally the same as that of the corresponding solute compound, except that solutions of white compounds are colorless.

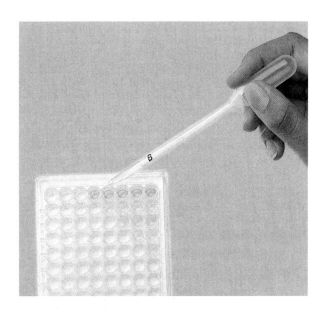

INQUIRY EXTENSION
Infer What are some other methods you might have used to identify the solutions?

Apply and Assess

1. **Explain** in general terms the reasoning you used in your identification process.

FACT of the Matter

The Dead Sea in Israel receives water from several streams but has no outflow. Instead, water evaporates from it, which concentrates salt. The Dead Sea has a salinity of 25 percent—as compared with only about 3.5 percent for ocean water—and is saturated with salt. The salt makes the water uninhabitable for fish but so buoyant that it is nearly impossible for swimmers to sink in it.

Solution Concentration

Suppose someone handed you a bottle to use to wash windows and said, "This is an aqueous ammonia solution." You'd know that it consists of ammonia dissolved in water, but you wouldn't know how much. In other words, you wouldn't know the concentration of the solution, which is the relative amount of solute and solvent.

Concentrated versus dilute If you were making tea, you might choose the approximate concentration you desire based upon personal taste. If you like strong tea, you make what a chemist might call a concentrated solution of tea: a relatively large amount of tea is dissolved in the water, so the concentration is high. On the other hand, if you like weak tea, a chemist might call the resulting mixture a dilute solution: relatively little tea is dissolved in the water, so the concentration is low. **Figure 13.19** illustrates this terminology. Chemists never apply the terms *strong* and *weak* to solution concentrations. As you'll see in the next chapter, these terms are used in chemistry to describe the chemical behavior of acids and bases. Instead, use the terms *concentrated* and *dilute*.

Unsaturated versus saturated Another way of providing information about solution composition is to express how much solute is present relative to the maximum amount the solution can hold. If the amount of solute dissolved is less than the maximum that can be dissolved, the solution is called an **unsaturated solution**. The oceans of Earth are examples of unsaturated saltwater solutions. They could hold a higher concentration of salt than they do now. At roughly 20°C, the maximum concentration of salt in water is approximately 36 g of salt dissolved per every 100 g of water, or 36 percent by mass. Such a solution, which holds the maximum amount of solute per amount of the solution under the given conditions, is called a **saturated solution**.

■ **Figure 13.19** When your tea is strong, it's concentrated. If you like weak tea, you prefer a dilute solution.

Supersaturated solutions An interesting third category of solutions is called a **supersaturated solution**. Such solutions contain more solute than the usual maximum amount and are unstable. They cannot permanently hold the excess solute in solution and may release it suddenly. Supersaturated solutions, as you might imagine, have to be prepared carefully. Generally, this is done by dissolving a solute in the solution at an elevated temperature, at which solubility is higher than at room temperature, and then slowly cooling the solution. The fudge-making process, shown in **Figure 13.20,** involves preparation of a supersaturated solution.

Effect of temperature on solubility As in the fudge-making process, the solubility of sugar increases as the temperature increases. More and more solute is able to dissolve at higher and higher temperatures. Temperature has a significant effect on solubility for most solutes. As a general rule, the solubility of most solid solutes increases with increasing temperature. As you will learn later in this section, however, the solubility of gases always decreases as temperature increases.

Figure 13.21 shows how the solubilities of six different solutes change with temperature. Notice that each solute behaves differently with temperature. The solubilities of some solutes, such as sodium nitrate and potassium nitrate, increase dramatically with increasing temperature. Notice how steeply the curves climb upward in the figure. Other solutes, like sodium chloride (NaCl) and potassium chloride (KCl), show only slight increases in solubility with increasing temperatures. A few solutes, such as cerium(III) sulfate, $Ce_2(SO_4)_3$, decrease in solubility as temperature increases.

■ **Figure 13.20** In making fudge, you heat a highly concentrated mixture of sugar, chocolate, and milk to a high enough temperature to make a sugar solution that is supersaturated. Next, you slowly cool the mixture with a great deal of stirring. If you've done everything right, your fudge will be soft and creamy because the sugar will form as very small crystals.

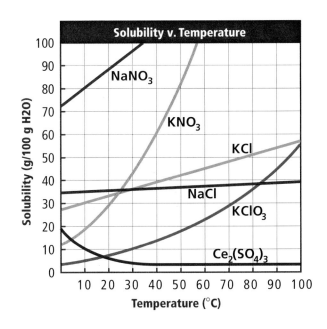

■ **Figure 13.21** The amount of solute required to achieve a saturated solution in water depends upon the temperature, as this graph shows. Most solutes increase in solubility as temperature increases.

Interpret What is the solubility of NaCl at 80°C?

Heat of solution As you can see, heat plays an important role in determining solubility. The heat absorbed or released in a dissolving process is called the **heat of solution**. For most solutes, the process of dissolving in a solvent is an endothermic process. For example, the dissolving of ammonium nitrate (NH_4NO_3) in a solvent is endothermic. You can write the equation for the process, incorporating the heat term, as follows.

$$NH_4NO_3(s) + \text{heat} \xrightarrow{H_2O} NH_4^+(aq) + NO_3^-(aq)$$

Notice that the heat in an endothermic process is written as if it were a reactant because it must be added to the substances that will form the products. To increase the solubility of NH_4NO_3, add more heat, which increases the temperature. This forces the process toward the production of more aqueous ions. When working with a solute that dissolves endothermically, you might notice that the mixture becomes cooler during the dissolving. The heat required for the process is taken from the solvent. The solution cools as this heat of solution is absorbed.

The dissolving of some solutes is exothermic. For these solutes, the dissolving process releases heat. Calcium chloride ($CaCl_2$) is a good example of such a solute. Note that in the equation for the dissolving process, heat shows up on the product side of the reaction.

$$CaCl_2(s) \xrightarrow{H_2O} Ca^{2+}(aq) + 2Cl^-(aq) + \text{heat}$$

Hot and cold packs for sports injuries, shown in **Figure 13.22,** use solutes whose dissolving is highly exothermic or endothermic. Therefore, the solutes have large heats of solution.

Molarity Suppose you're an employee in a hospital pharmacy lab and you must prepare a salt solution that matches the salt concentration of a patient's blood. You need to measure things carefully. Blood is a dilute salt solution, but the term *dilute* gives only qualitative information. In this case, you need a quantitative concentration unit.

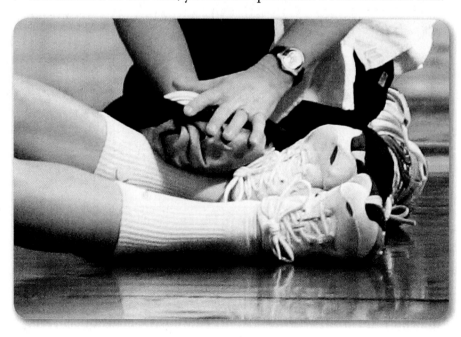

■ **Figure 13.22** When hot or cold packs for sports injuries are activated, a solute that dissolves exothermically in the case of a hot pack or endothermically in the case of a cold pack mixes with water. Hot packs generally use calcium chloride ($CaCl_2$) and cold packs generally use ammonium nitrate (NH_4NO_3).

Concentration units can vary greatly. They express a ratio that compares an amount of the solute with an amount of the solution or the solvent. For chemistry applications, the concentration term molarity is generally the most useful. Molarity is defined as the number of moles of solute per liter of solution.

$$\text{molarity} = \frac{\text{moles of solute}}{\text{liter of solution}}$$

Note that the volume is the total solution volume that results, not the volume of solvent alone. Suppose you need 1.0 L of the salt solution mentioned on the previous page. In order to be at the same concentration as the salt in the patient's blood, it needs to have a concentration of 0.15 moles of sodium chloride per liter of solution. In other words, it must have a molarity of 0.15. To save space, you refer to the solution as $0.15M$ NaCl, where the M stands for "moles/liter" and represents the word *molar*. Thus, you need 1.0 L of a 0.15-molar solution of NaCl. How will you prepare it?

Solution preparation To make an aqueous solution, you need to know three things when working quantitatively: the concentration, the amount of solute, and the total volume of solution needed. **Figure 13.23** shows the general solution-making steps in more detail using equipment that is available in labs where solutions are prepared quantitatively. To determine the amount of solute you need to prepare a solution or to determine the molarity of a solution you have prepared, examine the following example problems.

■ **Figure 13.23** Preparing solutions is important lab work. In step 1, measure the mass of the solute. In step 2, transfer the solute to a volumetric flask, which holds a known volume. In step 3, add enough water to dissolve the solute, and then add more water to bring the solution volume up to the calibration mark on the flask.
Infer *In preparing this solution, why couldn't you simply add water to the flask's calibration mark and then stir in the solute?*

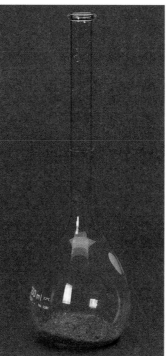

Problem-Solving Hint

Remember that a conversion factor is any relationship that compares two different units.

EXAMPLE Problem 13.1

Preparing a Different Volume of a Glucose Solution How would you prepare 5.0 L of a 1.5M solution of glucose ($C_6H_{12}O_6$)?

1 Analyze
- You need to determine the number of grams of glucose to add to a 5.0-L container. The 1.5M solution must contain 1.5 mol of glucose per liter of solution.

2 Set up
- Use the solution molarity as a conversion factor (1.5 mol glucose/L solution) to change from liters of solution to moles of glucose. Then, use the molar mass of glucose as another conversion factor to change from moles to grams of glucose. To find the molar mass of glucose, add the atomic masses of 6 C, 12 H, and 6 O, and apply the unit grams/mole to the sum. The result is 180 g/mol.

3 Solve
- The proper setup, showing the conversion factors, is as follows.

$$\left(\frac{5.0 \text{ L solution}}{}\right)\left(\frac{1.5 \text{ mol glucose}}{1 \text{ L solution}}\right)\left(\frac{180 \text{ g glucose}}{1 \text{ mol glucose}}\right)$$

Cancel units and carry out the calculation.

$$\left(\frac{\cancel{5.0 \text{ L solution}}}{}\right)\left(\frac{\cancel{1.5 \text{ mol glucose}}}{\cancel{1 \text{ L solution}}}\right)\left(\frac{180 \text{ g glucose}}{\cancel{1 \text{ mol glucose}}}\right) = 1400 \text{ g glucose}$$

To prepare 5.0 L of 1.5M glucose solution, weigh 1400 g of glucose and add it to a 5.0-L container. Then, add enough water to dissolve the glucose. Finally, fill with water to the 5.0-L mark.

4 Check
- Check to be sure that all units were correctly cancelled and that the calculation can be repeated with the same result.

PRACTICE Problems   Solutions to Problems Page 863

7. How would you prepare 1.00 L of a 0.400M solution of copper(II) sulfate, $CuSO_4$?

8. How would you prepare 2.50 L of a 0.800M solution of potassium nitrate, KNO_3?

9. What mass of sucrose, $C_{12}H_{22}O_{11}$, must be dissolved to make 460 mL of a 1.10M solution?

10. What mass of lithium chloride, LiCl, must be dissolved to make a 0.194M solution that has a volume of 1.00 L?

11. Complete **Table 13.4** by determining the mass of solute necessary to prepare the indicated solutions.

Table 13.4	Preparing Solutions	
Solution	Volume	Mass of solute
0.1M $CaCl_2$	1.0 L	
0.20M $CaCl_2$	500.0 mL	
3.0M NaOH	250 mL	

EXAMPLE Problem 13.2

Calculating Molarity You add 32.0 g of potassium chloride to a container and add enough water to bring the total solution volume to 955 mL. What is the molarity of this solution?

1 Analyze
- You have all the information you need about the preparation of the solution. You know the solute, the mass, and the total volume of the solution in milliliters. You need to know the molarity (moles KCl/L) of the solution.

2 Set up
- You are given that there are 32.0 g of solute per 955 mL of solution, so this relationship can be expressed in fraction form with the volume in the denominator. Therefore, the initial part of the setup is as follows.

$$\frac{32.0 \text{ g KCl}}{955 \text{ mL solution}}$$

Determine that the molar mass of KCl is 74.6 g/mol by adding the atomic masses of K and Cl and applying the units grams/mole to the sum. The conversion factor that must be used to convert from grams to moles of KCl is 1 mol KCl/74.6 g KCl.

$$\left(\frac{32.0 \text{ g KCl}}{955 \text{ mL solution}}\right)\left(\frac{1 \text{ mol KCl}}{74.6 \text{ g KCl}}\right)$$

Next, convert milliliters to liters. There are 1000 mL solution/L solution, so use that conversion factor in the setup.

$$\left(\frac{32.0 \text{ g KCl}}{955 \text{ mL solution}}\right)\left(\frac{1 \text{ mol KCl}}{74.6 \text{ g KCl}}\right)\left(\frac{1000 \text{ mL solution}}{1 \text{ L solution}}\right)$$

3 Solve
- Cancel units and carry out the calculation, using the setup just developed.

$$\left(\frac{32.0 \text{ g KCl}}{955 \text{ mL solution}}\right)\left(\frac{1 \text{ mol KCl}}{74.6 \text{ g KCl}}\right)\left(\frac{1000 \text{ mL solution}}{1 \text{ L solution}}\right)$$

$= 0.449$ mol KCl/L solution
$= 0.449M$ KCl

4 Check
- Check to be sure that all units were correctly handled and that the calculation can be repeated with the same result. All aspects of setup and calculation appear to be correct.

PRACTICE Problems
Solutions to Problems Page 864

12. What is the molarity of a solution that contains 14 g of sodium sulfate (Na_2SO_4) dissolved in 1.6 L of solution?

13. Calculate the molarity of a solution, given that its volume is 820 mL and that it contains 7.4 g of ammonium chloride (NH_4Cl).

Chemistry Online

Personal Tutor For an online tutorial on calculating molarity and molality, visit glencoe.com.

SUPPLEMENTAL PRACTICE

For more practice with molarity, see Supplemental Practice, page 827.

Properties and Applications of Solutions

Throwing salt onto an icy sidewalk, as shown in **Figure 13.24**, and adding coolant to a car radiator—what do these familiar activities have in common? They both relate to a set of interesting and useful solution properties that depend only on the concentration of solute particles. These properties are freezing-point depression and boiling-point elevation.

Freezing-point depression A solution always has a lower freezing point than the corresponding pure solvent. If you are interested only in aqueous solutions, this means that any aqueous solution will have a freezing point lower than 0°C. The amount that the freezing point is depressed relative to 0°C depends only upon the concentration of the solute. The identity of the solute particles—for example, whether they are sodium ions (Na^+), sulfate ions (SO_4^{2-}), or glucose molecules ($C_6H_{12}O_6$)—does not matter. All that matters is the number of those particles in the given volume of solvent.

You have observed freezing-point depression if you have de-iced walkways in the winter. Salt placed on an icy sidewalk causes the ice to melt by disrupting the highly organized crystals of ice. The salt dissolves in the water that makes the ice and forms a solution that has a lower freezing point than pure water, as illustrated in **Figure 13.25**. Making ice cream at home is a second application of freezing point depression. In order to reach the proper consistency, the ice cream needs to be mixed at temperatures lower than the freezing point of water. Packing a mixture of water, ice, and salt around the mixing canister achieves the required temperature.

■ **Figure 13.24** Solute particles interfere with the attractive forces among solvent particles, preventing the solvent from becoming a solid at its normal freezing point. Salting icy sidewalks and roads is one of the most common applications of this property, called freezing-point depression.

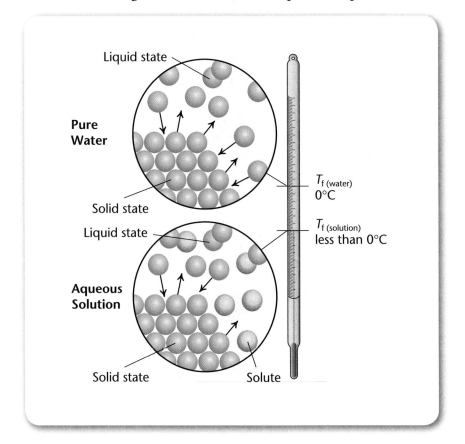

■ **Figure 13.25** A solute disrupts the high level of organization that is necessary if water is to be in the solid state. Thus, the freezing point of the solution is lower than the freezing point of the pure water.

Infer *How does the freezing point vary when the amount of solute particles varies?*

■ **Figure 13.26** Freezing-point depression has some practical uses, such as in deicing airplanes, a process that uses ethylene glycol as a solute.

Freezing point and the number of particles
One mole of an ionic solute produces a greater freezing point depression than one mole of a covalent one because it dissociates into ions. One mole of ionic NaCl thus produces two moles of solute particles. It produces twice the interference with the freezing process and twice the desired effect as one mole of sucrose. Likewise, one mole of $CaCl_2$ has three times the effect of one mole of sucrose because each formula unit of $CaCl_2$ dissociates into three ions: one Ca^{2+} and two Cl^-.
Figure 13.26 gives an additional application of freezing-point depression.

Boiling-point elevation You have just learned that the freezing point of a solution is lower than the freezing point of the pure solvent. It turns out that the boiling point of a solution is higher than the boiling point of the pure solvent. For aqueous solutions, this means that the boiling point of a solution will be greater than 100°C, assuming standard atmospheric pressure. The solute must also be nonvolatile; that is, not able to evaporate readily. For instance, if you add a sufficient amount of salt to water when you cook spaghetti, as shown in **Figure 13.27,** the temperature of the boiling water will be greater than 100°C and the spaghetti will cook faster.

Solute particles affect boiling point because they take up space on the surfaces where the liquid solvent and gaseous atmosphere meet, which interferes with the ability of the solvent particles to escape the liquid state. The solute particles lower the vapor pressure of the solvent, so it requires higher temperature to bring the vapor pressure up to atmospheric pressure and cause boiling. Similar to the trend for freezing point depression, the higher the concentration of solute particles, the greater the boiling-point elevation.

■ **Figure 13.27** When salt is added to water, the boiling point becomes elevated. This means that the temperature of the boiling water will be higher and the spaghetti will cook faster.

Section 13.2 • Solutions and Their Properties

Everyday Chemistry

Antifreeze

In 1885, Karl Benz of Germany invented and patented the first radiator for an automobile. This was a big change over simply cooling the engine by evaporation, which removed 4 L of water each hour and required constant replenishment. A radiator was able to recirculate the water used to cool the engine. After air-cooled water removed the heat created when the engine burned fuel, the water returned to the radiator to be recooled.

How chemical coolants work The first use of a chemical engine coolant, ethylene glycol, was tried in England in 1916 for high-performance military aircraft engines.

Ethylene glycol ($C_2H_4(OH)_2$) is still the major ingredient of most automobile antifreeze solutions used in the United States. When it is added to water, the solution that is formed has a higher boiling point. This keeps the water in the radiator from boiling off, as it would without the coolant. Recall that the boiling point of a substance is the temperature at which the vapor pressure of the liquid phase equals the atmospheric pressure. When a solute is present, it crowds out some of the water molecules at the surface of the solution. This reduces the number of water molecules that can escape as water vapor. As a result, the solution has a lower vapor pressure than pure water. If the vapor pressure of the solution in the radiator is lowered, additional kinetic energy is needed to raise it to the same level as the atmospheric pressure. This makes the boiling point of the solution higher than that of water.

Ethylene glycol as an antifreeze The process of adding a solute to water also lowers the temperature at which water freezes. The number of degrees the freezing point is lowered is proportional to the number of solute particles that are dissolved in the water. Any solute added to water decreases its freezing point.

A solution with a low freezing point, such as an ethylene glycol solution in a car radiator shown in **Figure 1,** is less likely to freeze in winter.

Properties of a good engine coolant Besides raising the boiling point and lowering the freezing point, a good coolant doesn't corrode the materials in the radiator and the pump. Over the years, material and design changes have caused problems of corrosion of metal parts in contact with coolants. Adding inhibitors to the coolant or changing the materials solves the problem.

A good coolant should also be easily disposed. Ethylene glycol is biodegradable. However, it is toxic to mammals if ingested. It has a sweet smell and taste, which appeals to animals and sometimes to children. Care must be taken in disposing of these chemicals until a safer product is found.

Figure 1 Green, blue, or red dye make identification of type of antifreeze easier.

Explore Further

1. **Acquire Information** Investigate how ice cream is made at home. Relate the information gathered to the use of an antifreeze in an automobile during winter.

2. **Apply** If a coolant becomes oily, murky, or rusty, what would this suggest?

Osmosis Have you ever seen water being sprayed onto vegetables in a supermarket? Some of the water is absorbed by the vegetables, which makes them plumper, crisper, and fresher looking. The water is able to move inside because cell membranes on the outside of the vegetables are selectively permeable—that is, they allow certain materials, such as water, to pass through them.

As you may recall from Chapter 10, gas molecules have a tendency to diffuse from an area of high concentration to an area of low concentration. Particles in a liquid have a similar tendency to move. In the case of the vegetables, the water that is naturally inside them contains solutes, such as sugars and salts. Because of the presence of these solutes, there is less water per unit volume than in pure water. If pure water is sprayed onto the outside of the vegetables, the water will tend to diffuse into the vegetables. It moves from the area of higher water concentration outside to the area of lower water concentration inside the vegetables. This flow of solvent molecules through a selectively permeable membrane, driven by concentration difference, is called **osmosis**.

Figure 13.28 shows a system that illustrates osmosis. Pure water is on the left side of a selectively permeable membrane, and an aqueous solution of sucrose is on the right side. The membrane allows water molecules to pass, but not sucrose molecules.

VOCABULARY
WORD ORIGIN
Osmosis
comes from the Greek word *osmos*, which means *impulse* and the Greek word *othein* which means *push*

■ **Figure 13.28** Note what takes place when pure water and a sucrose solution are separated by a selectively permeable membrane.

Concepts In Motion
Interactive Figure To see an animation of osmosis, visit **glencoe.com**.

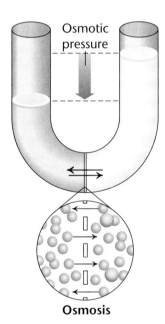
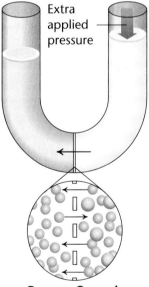

As osmosis begins, water molecules diffuse more rapidly from the water into the sucrose solution than they diffuse from the sucrose solution into the water. As a result, the sucrose solution gains water, becomes more dilute, and the volume rises.

The increasing height of the sucrose solution exerts a pressure that opposes the diffusion of water molecules from left to right. Eventually, the pressure becomes high enough that the rates of diffusion of water in both directions become the same.

Adding extra pressure to the side with the sucrose solution can cause the water diffusion to move in the opposite direction, forcing water out of the solution. This process, called reverse osmosis, can be used to purify water, as discussed in How It Works.

HOW IT WORKS

A Portable Reverse Osmosis Unit

Reverse osmosis (RO) is the process by which water from a solution is forced through a selectively permeable membrane by the application of pressure to the solution side of the membrane. Portable RO units are commercially available that can purify seawater of its salts and make it drinkable. Approximately 27 atm of pressure needs to be applied to seawater in order to counteract the flow of solvent into the seawater through a selectively permeable membrane. In order to get a usable amount of water through the membrane, you need to apply about twice that pressure. A portable RO unit, shown below, can apply these large pressures.

Portable reverse osmosis units are sold for personal use. They can be used by people who live in rural areas without clean water, by people who live in areas where the water source is polluted, and by people who are camping, fishing, or on long boating trips. Marine aquarium owners use portable reverse osmosis units to remove excessive chemicals from tap water. The units are also used by the military on the battlefield and in training.

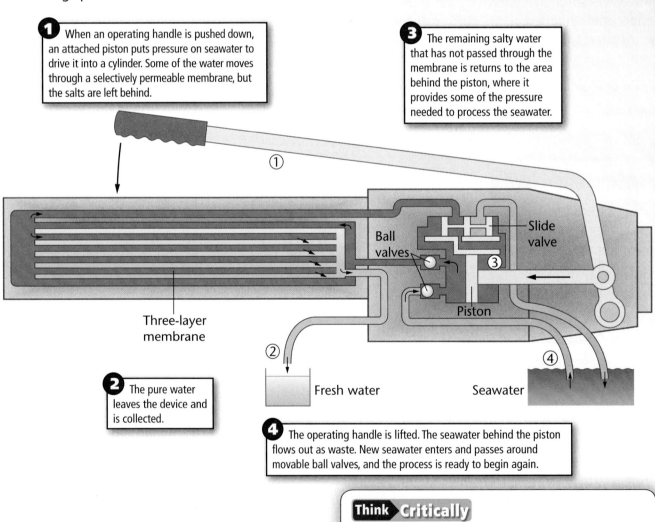

1 When an operating handle is pushed down, an attached piston puts pressure on seawater to drive it into a cylinder. Some of the water moves through a selectively permeable membrane, but the salts are left behind.

2 The pure water leaves the device and is collected.

3 The remaining salty water that has not passed through the membrane is returns to the area behind the piston, where it provides some of the pressure needed to process the seawater.

4 The operating handle is lifted. The seawater behind the piston flows out as waste. New seawater enters and passes around movable ball valves, and the process is ready to begin again.

Think Critically

1. **Explain** Why must the pressure required to operate a reverse osmosis unit be more than 27 atm?
2. **Infer** Why isn't reverse osmosis more widely used as a purification method for water?

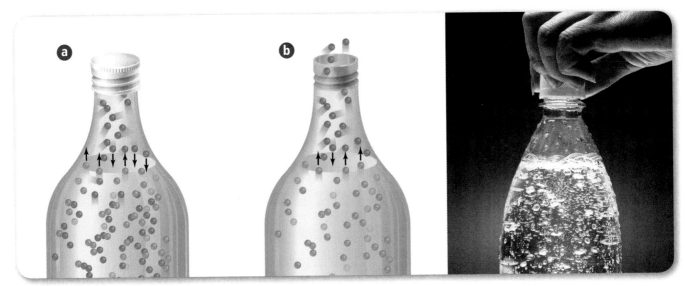

Solutions of Gases in Water

If you observe an unopened bottle of club soda, the liquid inside looks just like pure water. However, when you unscrew the cap, small bubbles of gas appear throughout the liquid and rise to the top. The solubility of a gas in a liquid depends on the pressure of the gas pushing down on the liquid. The higher the pressure, the more soluble is the gas. When you remove the cap from a bottle of soda, the pressure suddenly drops and bubbles of gas come out of solution. **Figure 13.29** illustrates this behavior.

You might have noticed that if the club soda is warm, the fizziness can be so intense that the liquid rises and spurts out the top. In addition to being dependent on pressure, the solubility of a gas in a liquid depends on temperature. For solutions of gases in liquids, gas solubility decreases as temperature increases. That's why club soda fizzes more vigorously when it is warm. Life in lakes and oceans depends on gases that are dissolved in water and can be greatly affected by changes in pressure and temperature, as shown in **Figure 13.30.**

■ **Figure 13.29** Club soda is a solution of carbon dioxide in water. **a)** In the unopened bottle, the gas is dissolved under pressure in the water. **b)** When the bottle is opened, the carbon dioxide gas trapped above the liquid escapes and the pressure drops. The solution is now supersaturated in carbon dioxide. Some of the dissolved carbon dioxide reenters the gaseous state and forms the bubbles that you see.

■ **Figure 13.30** Deep under water, pressures are high. The nitrogen gas from the air in the lungs of divers gets dissolved at higher-than-normal concentrations in their blood. As they ascend from the depths, the pressure decreases and so does the blood solubility of the nitrogen. If divers come up too rapidly, the released nitrogen can form dangerous and painful bubbles in the blood vessels in a condition called the bends. Fish depend on dissolved oxygen in the water. If water temperature climbs too high, the solubility of the oxygen may fall too low, and the fish may die.

CHEMISTRY AND TECHNOLOGY

Versatile Colloids

Colloids are mixtures composed of tiny particles of one substance that are dispersed or evenly distributed in another substance. The particles of a colloid are intermediate in size between those of suspensions and true solutions. They range somewhere between the size of large molecules and a size great enough to be seen through a microscope.

■ **Figure 2** Soot is a solid aerosol.

charged plates of this type of air cleaner attract colloidal soot particles and remove them from the air. This is how air pollution has been reduced in many industrial cities.

■ **Figure 1** Fog is a liquid aerosol of water vapor in air.

Liquid Aerosols
Fog shown in **Figure 1** is an example of a liquid aerosol, which forms when fine liquid droplets disperse in a gas. Fog appears when moist air near Earth's surface is cools to the point at which the water vapor begins to condense. Other liquid aerosols with which you are familiar include spray deodorants and hair spray.

Solid Aerosols
At times, solid particles disperse in a gas. One example of a solid aerosol is the polluting soot particles shown in **Figure 2** that may be released into the air from an industrial smokestack. Such a problem could be remedied by placing an electrostatic precipitator in the smokestack. The

■ **Figure 3** Salad dressings are examples of emulsions

Emulsions
Milk, mayonnaise, and the creamy salad dressing shown in **Figure 3** are examples of emulsions. Emulsions are dispersions of fine droplets of liquid—generally a fat—in another liquid. Many emulsions are able to maintain their stabilities with the help of materials such as gums. Gums and other stabilizers thicken the liquid phase, making it less likely for the dispersed droplets to come together.

■ **Figure 4** Paint is a sol.

Sols

A sol is a fluid colloidal system in which fine solid particles are dispersed in a liquid medium. Most household paints, such as the one shown in **Figure 4**, are sols with finely ground pigments mixed with acrylic resins dissolved in water. Paint is applied to a surface as a liquid. With time, the water dries and the resins harden, leaving a thin solid film on the surface.

■ **Figure 5** Jelly contains pectin, which gives it a gel structure.

Gels

The food industry uses many colloids that have the ability to thicken or gel liquid foods. Gels are dispersions of giant macromolecules in liquid. Many gels are natural gums found in seaweeds and land plants. The natural substance pectin, found in fruits, is responsible for the gel structure in jelly, as shown in **Figure 5**.

Pastes

The basis for a paste is a concentrated dispersion of solids in a limited amount of liquid. To make a beautiful porcelain vases like the ones in **Figure 6**, ground quartz and feldspar are mixed with a white clay called kaolin in a small portion of water. A paste is formed in which the water adheres to the surface of the clay, making the clay easy to work with.

■ **Figure 6** Porcelain vases are made of pastes.

Foam

You are familiar with the foam that forms when you beat egg whites. Foam is a dispersion of gas bubbles in a liquid. Yeasts provide another kind of foam, as can be seen in bread dough. They do this by fermenting carbohydrates and giving off carbon dioxide gas, which produces tiny holes in the dough and the bread.

Discuss the Technology

1. **Classify** Classify several health and beauty products according to the kinds of colloids described in this feature.

2. **Think Critically** Colloids do not pass through selectively permeable membranes. What can you conclude from this?

3. **Hypothesize** Destruction of a colloid through a clumping process called coagulation can usually be achieved by heating. How would heat accomplish this?

■ **Figure 13.31** Brilliant colors can be given to glass, an amorphous solid, by adding certain solids. One part of nickel oxide particles is added to 50,000 parts of the ingredients normally used in making milk glass, which is frosty white, in order to produce a glass with a yellow tint. One part cobalt oxide particles added to 10,000 parts of the ordinary ingredients in glass produces blue glass. Gold, copper, or selenium oxide particles are used to produce red glass.

VOCABULARY
WORD ORIGIN
Colloid
comes from the Greek word *kolla* meaning *glue* and the Greek word *eidos* meaning *form*

Colloids

Sometimes, mixtures are partway between true, homogeneous solutions and heterogeneous mixtures. Such mixtures, called **colloids**, contain particles that are evenly distributed through a solvent, and remain distributed over time rather than settling out. The major difference between a colloid and a solution is the size of the solute particles. Colloid particles are generally clumps that are 10–100 times larger than typical ions or molecules dissolved in solutions. Because of their relatively large particle sizes, colloids play important roles in a variety of processes. Sometimes solid particles are dispersed in another solid, like the stained-glass shown in **Figure 13.31**. Some biological molecules, such as proteins, are large enough that their behaviors are often best understood by classifying them as colloids.

Because colloid particles are evenly dispersed, it is sometimes difficult to distinguish colloids from homogeneous solutions. However, the larger particle size gives colloids some unique properties that help in identifying them. For example, a beam of light looks different when it passes through a solution than it does through a colloid. In the solution, the beam's path is hardly visible. As the light moves into the colloid, the light is partially scattered and reflected by the dispersed particles, and the beam becomes visible and broadens. You can observe the same phenomenon when you see the path of sunbeams through dusty air or the path of headlight beams on a foggy night. This light-scattering effect is called the **Tyndall effect**. It occurs because the dispersed colloid particles are about the same size as the wavelength of visible light (400–700 nanometers). The solute particles in a true solution are too small to produce this effect. **Figure 13.32** illustrates the Tyndall effect.

■ **Figure 13.32** Notice how the light beam becomes easily visible in the colloid because of light scattering. Headlight beams are visible in fog for the same reason.
Determine *which mixture is the colloid.*

Connecting Ideas

Now that you have learned about some of the properties of water, solutes, and colloids, you can understand why it is sometimes said that the chemistry of water is the chemistry of solutions. As you have discovered in this chapter, the physical properties of water make it a unique and important substance. Its ability to form solutions gives it an essential role in every aspect of life, including the reactions of acids and bases. As you will see, water is a subtle but important partner in these reactions.

Section 13.2 Assessment

Section Summary

- Like dissolves like.
- Interparticle forces between solvent and solute strongly influence solution formation.
- Ionic compounds dissociate when they dissolve in water.
- Solutions can be unsaturated, saturated, or supersaturated.
- Temperature affects solubility.
- Molarity is the number of moles of solute dissolved per liter of solution.
- Colligative properties of solutions, such as freezing-point depression and boiling-point elevation, depend only on the concentration of solute particles.

14. **MAIN Idea** **Explain** how a water molecule can be attracted to both a positive ion and a negative ion when dissolving an ionic compound.

15. **Apply** Write equations for the dissociation of the following ionic compounds when they dissolve in water.
    a) Na_2SO_4 b) NaOH   c) $CaCl_2$

16. **Characterize** each solution as unsaturated, saturated, or supersaturated.
    a) a solution that will produce a large amount of crystalline solid if only a small additional amount of solute is added
    b) a solution in which additional solute can be dissolved and remain in solution
    c) a solution with undissolved solute at the bottom of the container

17. **Compare** Determine which solution has the highest concentration. Then rank the solutions from the one with the lowest to the one with the highest freezing point, and explain your answer.
    a) 0.10 mol of KBr in 100.0 mL of solution
    b) 1.1 mol of NaOH in 1.00 L of solution
    c) 1.6 mol of $KMnO_4$ in 2.00 L of solution

18. **Explain** Oil and aqueous solutions normally do not mix. However, if oil and vinegar or lemon juice are beaten together with egg, a stable mixture is formed. Explain how this is possible.

CHAPTER 13 Study Guide

Download quizzes, key terms, and flash cards from glencoe.com.

BIG Idea Water plays a major role in many chemical reactions.

Section 13.1 Uniquely Water

MAIN Idea The molecular shape of water gives it unusual properties.

Vocabulary
- capillarity (p. 444)
- hydrogen bonding (p. 438)
- specific heat (p. 445)
- surface tension (p. 442)

Key Concepts
- The polarity of the water molecule is the source of many of water's unusual physical properties.
- Hydrogen bonds form between the hydrogen atom of one molecule and a highly electronegative atom of another.
- Specific heat is the amount of heat needed to raise the temperature of 1 g of a substance by 1°C.

Section 13.2 Solutions and Their Properties

MAIN Idea Water dissolves a large number of ionic and covalent compounds.

Vocabulary
- colloid (p. 472)
- dissociation (p. 452)
- heat of solution (p. 460)
- osmosis (p. 467)
- saturated solution (p. 458)
- supersaturated solution (p. 459)
- Tyndall effect (p. 472)
- unsaturated solution (p. 458)

Key Concepts
- Like dissolves like.
- Interparticle forces between solvent and solute strongly influence solution formation.
- Ionic compounds dissociate when they dissolve in water.
- Solutions can be unsaturated, saturated, or supersaturated.
- Temperature affects solubility.
- Molarity is the number of moles of solute dissolved per liter of solution.
- Colligative properties of solutions, such as freezing-point depression and boiling-point elevation, depend only on the concentration of solute particles.

Chapter 13 Assessment

Understand Concepts

19. List several ways in which water is an unusual substance.
20. Explain why water is polar.
21. Describe how the density of water relates to temperature.
22. What types of molecules form hydrogen bonds?
23. Explain what capillarity is and give two examples of it.
24. When mercury is placed in a glass cylinder, it forms a convex meniscus. Why does it do this?
25. In **Figure 13.33** below, which shows water molecules, identify the bonds as covalent or as hydrogen bonds.

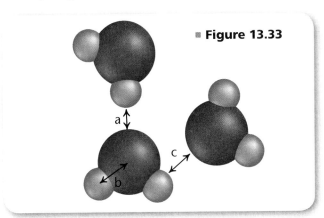

■ **Figure 13.33**

26. Why does sucrose dissolve so well in water?
27. Explain how you would prepare a supersaturated solution.
28. Define specific heat. Why is the high specific heat of water important to Earth?
29. Why does ice float? What is the significance of this fact for aquatic life?
30. How does your body use perspiration to stay cool?
31. What is the molarity of a solution prepared by dissolving 0.217 mol of ethanol, C_2H_5OH, in enough water to make 100.0 mL of solution?
32. How many grams of sodium carbonate are needed to make 1.30 L of 0.890M Na_2CO_3?

Apply Concepts

33. If equal masses of gold and iron sit in the Sun and absorb equal amounts of energy, which metal will experience the larger increase in temperature?
34. What mass of iron(III) chloride ($FeCl_3$) is needed to prepare 1.00 L of a 0.255M solution?
35. What is the molarity of a solution that contains 4.13 g of magnesium bromide, $MgBr_2$, in 0.845 L of solution?
36. What mass of potassium iodide, KI, is needed to prepare 5.60 L of a 1.13M solution?
37. Explain how adding antifreeze to your car's radiator protects it from freezing.
38. Calcium chloride is found in salt mixtures that are used to melt ice on roads in the winter. Dissolving $CaCl_2$ is an exothermic exothermic. Provide two reasons why $CaCl_2$ is a good choice for this application as compared to a salt such as NaCl or a salt that dissolves endothermically.
39. The specific heat of aluminum is 0.903 J/g°C; that of copper is 0.385 J/g°C. Suppose you have one cube each of Al and Cu; both cubes have a mass of 100 g and are at a temperature of 100°C. Which cube will release more heat when it cools to 20°C?
40. If you prepared a saturated aqueous solution of potassium chloride at 25°C, then heated the solution to 50°C, would you describe the resulting solution as saturated, unsaturated, or supersaturated? Explain.
41. 100.0 mL of an aqueous solution of 1.00M barium nitrate, $Ba(NO_3)_2$, is mixed with 100.0 mL of an aqueous solution of 1.00M sodium sulfate, Na_2SO_4. How many grams of barium sulfate will form?

$Ba(NO_3)_2(aq) + Na_2SO_4(aq) \rightarrow$
$\quad BaSO_4(s) + 2NaNO_3(aq)$

Everyday Chemistry

42. Explain why soap is effective in removing greasy dirt.

Chapter 13 Assessment

How It Works

43. Why would it be more difficult to use a reverse osmosis unit to purify water from the Dead Sea than water from the ocean?

Think Critically

Relate Cause and Effect

44. **MiniLab 1** If you carefully place a steel razor blade flat on the surface of water, the razor blade can be made to float. Explain this result, given that the density of steel is much greater than that of water.

Interpret Graphs

45. Use **Figure 13.21** to compare the solubilities in water of sodium chloride and potassium nitrate over the temperature range from 10°C to 40°C.

Draw Conclusions

46. **ChemLab** Suppose you have spilled some blue copper(II) sulfate solution on a shirt. When washing the shirt, why should you avoid using washing soda (sodium carbonate)?

Cumulative Review

47. The formula for a certain compound is $CHCl_3$. What does this formula tell you about the compound? (*Chapter 1*)

48. Explain how a potassium atom bonds to an atom of bromine. (*Chapter 4*)

49. What is the maximum number of electrons in thet second energy level? In the third energy level? Explain. (*Chapter 7*)

50. Why does your skin feel cool if you dab it with rubbing alcohol? (*Chapter 10*)

51. How many liters of nitrogen are needed to react completely with 28 L of oxygen in the synthesis of nitrogen dioxide? (*Chapter 11*)

Skill Review

52. **Written Summary of a Graph** Explain what is being graphed in **Figure 13.34**. Explain how the dashed line relates to trends on the periodic table and whether or not the boiling points of H_2O, HF, and NH_3 can be predicted from their groups' trends.

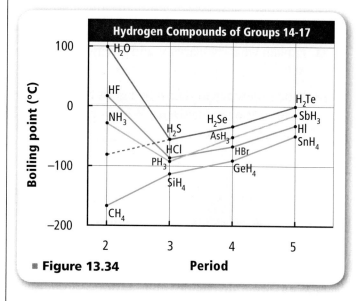

■ **Figure 13.34**

WRITING in Chemistry

53. Write an article giving a general overview of municipal sewage treatment and drinking-water treatment. Include a description of the route taken by water from your drain to the point where it is again safe for drinking.

Problem Solving

54. Suppose that you immersed a clean, dry capillary tube into a beaker of molten wax and withdrew the tube in such a way that the inside walls of the tube became coated with a thin film of solid wax. What would happen if you dipped this wax-coated tube into a beaker of water? How would the level of water inside the tube compare to a non-wax-coated capillary tube? Explain.

Cumulative
Standardized Test Practice

1. Which pair of elements will combine to form a polar covalent compound?
   a) calcium and chlorine
   b) lithium and oxygen
   c) fluorine and iron
   d) iron and sulfur

2. Snow turning directly into water vapor on a cold day is an example of
   a) sublimation.
   b) evaporation.
   c) vaporization.
   d) condensation.

3. While it is on the ground, a blimp is filled with 5.66×10^6 L of He gas. The pressure inside the grounded blimp, where the temperature is 25°C, is 1.10 atm. Modern blimps are non-rigid, which means their volume is changeable. If the pressure inside the blimp remains the same, what will be the volume of the blimp at a height of 2300 m, where the temperature is 12°C?
   a) 5.66×10^6 L
   b) 2.72×10^6 L
   c) 5.40×10^6 L
   d) 5.92×10^6 L

4. The variable P in the ideal gas law formula requires the unit
   a) atm.
   b) mm Hg.
   c) kPa.
   d) psi.

5. Why is water a unique substance?
   a) Water is a liquid at room temperature even though it has a low molecular mass.
   b) In its solid state, water has a lower density than in its liquid state.
   c) Water has a high boiling point for a substance with such as a low molecular mass.
   d) All off the above.

Interpreting Graphs: Use the graph to answer Questions 6–7.

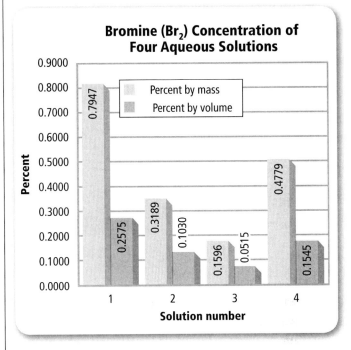

6. The volume of bromine (Br_2) in 7.000 L of Solution 1 is
   a) 55.63 mL.
   b) 8.808 mL.
   c) 18.03 mL.
   d) 27.18 mL.

7. How many grams of Br_2 are in 55.00 g of Solution 4?
   a) 3.560 g
   b) 0.084 g
   c) 1.151 g
   d) 0.2628 g

8. Which of the following is NOT true of an atom obeying the octet rule?
   a) obtains a full set of eight valence electrons
   b) acquires the valence configuration of a noble gas
   c) possess eight electrons total
   d) has a s^2p^6 valence configuration

9. What is the charge on the anion in $AlPO_4$?
   a) 2+
   b) 3+
   c) 2−
   d) 3−

NEED EXTRA HELP?									
If You Missed Question . . .	1	2	3	4	5	6	7	8	9
Review Section . . .	9.2	10.2	11.2	12.2	13.1	13.2	13.2	4.2	5.1

CHAPTER 14: Acids, Bases, and pH

BIG Idea Acids and bases can be defined in terms of hydrogen ions and hydroxide ions.

14.1 Acids and Bases
MAIN Idea Acids produce hydronium ions (H_3O^+) in solution; bases produce hydroxide ions (OH^-).

14.2 Strengths of Acids and Bases
MAIN Idea In solution, strong acids and bases ionize completely, but weak acids and bases ionize only partially.

ChemFacts

- There are over 280 species of biting ants, which are commonly known as fire ants.
- The average colony of fire ants contains 100,000 to 500,000 ants.
- The venom of a fire ant can cause allergic reactions.
- The bite of a fire ant can feel like being burned with a hot match and is caused by formic acid in the ant's venom.

Start-Up Activities

LAUNCH Lab

Testing Household Products

Can you separate household products into two groups?

Materials
- microplate
- labeled containers of household products, such as window cleaner, baking soda solution, vinegar, colorless carbonated soft drink, soap, distilled water, lemon juice, table salt solution, and drain cleaner
- red and blue litmus paper
- phenolphthalein

Procedure
1. Read and complete the lab safety form.
2. Place three or four drops of each of your household products into separate wells of a microplate. Draw a chart to show the position of each product.
3. Test each product with red and blue litmus paper. Place two drops of phenolphthalein in each sample. Record your observations.

WARNING: *Phenolphthalein is flammable. Keep away from flames.*

Analysis
1. **Classify** the products into two groups based on your observations.
2. **Summarize** how the groups differ.

Inquiry Choose one sample that reacted with the phenolphthalein. Can you reverse the reaction? Design an experiment to test your hypothesis.

Chemistry Online

Visit **glencoe.com** to:
- study the entire chapter online
- explore **Concepts In Motion**
- take Self-Check Quizzes
- use Personal Tutors
- access Web Links for more information, projects, and activities
- find the Try at Home Lab, Testing for Acid Rain

FOLDABLES Study Organizer

Acids and Bases Make the following Foldable to organize information about acids and bases.

STEP 1 Find the middle of a horizontal sheet of paper. Fold both edges to the middle and crease the folds.

STEP 2 Fold the paper in half, as shown.

STEP 3 Unfold and cut along the fold lines to make four tabs.

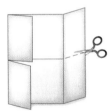

STEP 4 Label the tabs *strong acids, strong bases, weak acids,* and *weak bases.*

FOLDABLES Use this Foldable with Section 14.2. As you read the section, record information about the properties of strong and weak acids and strong and weak bases. Also, record examples of each type of acid and base.

Chapter 14 • Acids, Bases, and pH **479**

Section 14.1

Objectives
- **Distinguish** acids from bases by their properties.
- **Define** acids and bases in terms of their reactions in water.
- **Evaluate** the central role of water in the chemistry of acids and bases.

Review Vocabulary
osmosis: the flow of solvent molecules through a selectively permeable membrane, driven by a concentration difference

New Vocabulary
acid
hydronium ion
acidic hydrogen
ionization
base
acidic anhydride
basic anhydride

Acids and Bases

MAIN Idea Acids produce hydronium ions (H_3O^+) in solution; bases produce hydroxide ions (OH^-).

Real-World Reading Link Classifying substances into broad categories simplifies the study of chemistry. Though you might not realize it, two of the most common classifications that you encounter are acids and bases. You are familiar with acids and bases by the tart taste of some of your favorite beverages and by the slippery feel of soap.

Macroscopic Properties of Acids and Bases

Because they are present in so many everyday materials, acids and bases have been recognized as interesting substances since the time of alchemists. Simple, observable properties distinguish the two.

Taste and feel Although taste is not a safe way to classify acids and bases, you probably are familiar with the sour taste of acids. Lemon juice and vinegar, for example, are both aqueous solutions of acids. Bases, on the other hand, taste bitter.

Bases have a slippery feel. Like taste, feel is not a safe chemical test for bases, but you are familiar with the feel of soap, a base, on the skin. Bases, such as soap, react with protein in your skin, and skin cells are removed. This reaction is part of what gives soaps a slippery feel, as well as a cleansing action. This reaction also makes some bases excellent drain cleaners, as illustrated by **Figure 14.1.**

■ **Figure 14.1** Certain bases are excellent at dissolving hair, which is often the source of clogged drains. Hair is composed of protein.

Table 14.1 Common Industrial Acids and Bases

Concepts In Motion
Interactive Table Explore common industrial acids and bases at glencoe.com.

Substance	Acid/Base	Some Uses
Sulfuric acid (H_2SO_4)	Acid	Car batteries; manufacture of chemicals, fertilizer, and paper
Lime (CaO)	Base	Neutralization of acidic soil
Ammonia (NH_3)	Base	Fertilizer; cleaner; manufacture of rayon, nylon, and nitric acid
Sodium hydroxide (NaOH)	Base	Drain and oven cleaners; manufacture of soap and chemicals
Phosphoric acid (H_3PO_4)	Acid	Soft drinks; manufacture of detergents and fertilizers

Acids react with bases The reactions of acids and bases are central to the chemistry of living systems, the environment, and many important industrial processes. **Table 14.1** shows some of the top industrial acids and bases produced in the United States, along with some of their uses. You'll learn more about acid-base reactions in Chapter 15.

Litmus test and other color changes Acids and bases cause certain dyes to change color. Dyes such as these are called acid-base indicators because they are often used to indicate whether substances are acids or bases. The most common of these dyes is litmus. When mixed with an acid, litmus is red. When added to a base, litmus is blue. Therefore, litmus is a reliable indicator of whether a substance is an acid or a base. **Figure 14.2** shows how a naturally occurring substance changes color in the presence of an acid or a base.

■ **Figure 14.2** A chemical in red cabbage acts as an acid-base indicator. The colorful solutions shown below indicate a range of acids starting on the left, and bases on the right.

Section 14.1 • Acids and Bases

MiniLab 14.1

Reactions of Acids

How do acids react? Most acids tend to be reactive substances. Test the reactivities of three acids with several common substances, and develop an operational definition for acidic solutions.

Procedure

1. Read and complete the lab safety form.
2. Use a labeled **microtip pipette** to add 10 drops of **3M hydrochloric acid** (HCl) to wells D1–D6 of a clean, **24-well microplate**. In the same manner, add 10 drops of **3M sulfuric acid** (H_2SO_4) to wells C1–C6 and 10 drops of **3M acetic acid** (CH_3COOH) to wells B1–B6.
3. Dip **blue litmus paper** into wells D1, C1, and B1. Record your observations.
4. Add 2 drops of **bromthymol blue indicator solution** to wells D2, C2, and B2. This indicator turns from blue to yellow as the solutions become more acidic. Record your observations.
5. In a similar manner, add **marble chips** (calcium carbonate) to wells D3, C3, and B3; pieces of **zinc** to wells D4, C4, and B4; pieces of **aluminum** to wells D5, C5, and B5; and a small amount of **egg white** to wells D6, C6, and B6. Record your observations.
6. Dispose of all materials as directed by your teacher. Rinse the microplate with **tap water,** then **distilled water.**

Analysis

1. **Summarize** the reactions of the three acids with the substances you tested. This summary constitutes an operational definition of an acid.
2. **Identify** Which acid reacted less noticeably? Explain this behavior.

Metals' reactions with carbonates Another characteristic property of an acid is that it reacts with metals that are more active than hydrogen. Bases, on the other hand, do not commonly react with metals. **Figure 14.3** shows the reaction between iron metal (Fe) and hydrochloric acid (HCl) along with an activity series that compares hydrogen with some common metals. Notice that hydrogen is low on the series—there are only a handful of metals that are less active. This property explains why acids corrode most metals.

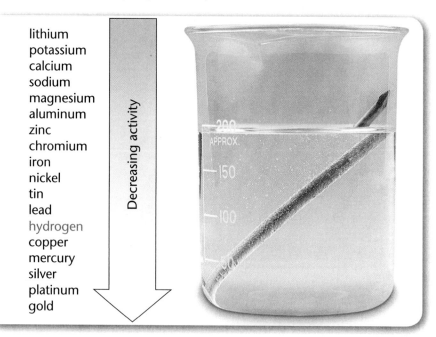

■ **Figure 14.3** Typical behavior in certain chemical reactions helps identify substances as acids. Acids react with metals that are more active than hydrogen to form both a compound of the metal and hydrogen gas.

$Fe(s) + 2HCl(aq) \rightarrow FeCl_2(aq) + H_2(g)$

Apply *Use the activity series to explain why HCl reacts with the iron in the nail in the photo.*

Vinegar, a solution of acetic acid, reacts with egg shell, which is primarily calcium carbonate, to produce carbon dioxide, calcium acetate, and water.

$$2HC_2H_3O_2(aq) + CaCO_3(s) \rightarrow CO_2(g) + Ca(C_2H_3O_2)_2(aq) + H_2O(l)$$

Calcium carbonate is the major component in limestone and marble. In the presence of acids in the environment, marble and limestone sculptures and buildings can be damaged or destroyed.

You might recognize the reaction between iron metal and hydrochloric acid as a single displacement reaction, which you read about in Chapter 6. Metal-acid reactions are one of the most common kinds of single displacement reactions, but you can use the activity series in **Figure 14.3** for any single displacement reaction involving these metals. For example, you can see that copper is more active than silver. Therefore, copper metal will react with silver nitrate to form silver metal and copper nitrate. However, silver metal will not react with copper nitrate.

Another simple test that distinguishes acids from bases is the reaction of acids with ionic compounds that contain the carbonate ion (CO_3^{2-}) to form carbon dioxide gas (CO_2), water, and another compound, as shown in **Figure 14.4**. A similar reaction, also shown in **Figure 14.4,** is the source of the destructive action of acid precipitation on marble and limestone sculptures. Bases do not react with carbonates.

Submicroscopic Behavior of Acids

The description of acids and bases in terms of their physical and chemical properties—macroscopic properties that are easy to observe—is useful for classification purposes. However, to understand these properties, you need to know about the behavior of acids and bases at the submicroscopic level.

Hydronium ion formation The submicroscopic behavior of acids when they dissolve in water can be described in several ways. The simplest definition is that an **acid** is a substance that produces hydronium ions when it dissolves in water. A **hydronium ion** (H_3O^+) consists of a hydrogen ion attached to a water molecule.

■ **Figure 14.4** Acids can be distinguished from bases by their reaction with the carbonate ion (CO_3^{2-}). One common occurrence of the carbonate ion is in calcium carbonate.

FACT of the Matter

Except for group 1 carbonates, carbonate-containing compounds are almost completely insoluble in water. This makes naturally occurring substances such as marble and limestone stable materials for sculpting and building.

VOCABULARY
WORD ORIGIN
Acid
comes from the Latin word *acidus*, meaning *sour*.

Section 14.1 • Acids and Bases

HC₂H₃O₂	HCl
Acetic acid	Hydrochloric acid
Monoprotic acids	
H₂SO₄	H₃C₆H₅O₇
Sulfuric acid	Citric acid
Diprotic acid	**Tri**protic acid

■ **Figure 14.5** If a hydrogen atom loses its electron, all that remains is a proton. Prefixes, used with the term *protic*, which refers to the remaining proton, indicate how many acidic hydrogens are present in an acid.

Infer Some acids, such as phosphoric acid, have three acidic hydrogens. What term would you use to tell someone that an acid has three acidic hydrogens?

For example, hydrochloric acid is produced by dissolving hydrogen chloride gas (HCl) in water. Remember from Chapter 13 that water is a polar molecule that is able to form strong hydrogen bonds with solutes that also form hydrogen bonds. When HCl dissolves in water, it produces hydronium ions by the reaction shown below. This reaction confirms that HCl is an acid.

$$HCl(g) + H_2O(l) \rightarrow H_3O^+(aq) + Cl^-(aq)$$

Acetic acid (CH₃COOH) undergoes a similar reaction when it dissolves in water to form a vinegar solution.

$$CH_3COOH(aq) + H_2O(l) \rightarrow H_3O^+(aq) + CH_3COO^-(aq)$$

Notice the similarities in these two reactions. In both cases, the solute reacts with water to form hydronium ions and a negatively charged ion.

Acidic hydrogen atoms How and why are hydronium ions formed? At the submicroscopic level, the reaction of an acid with water is a transfer of a hydrogen ion (H^+) from an acid to a water molecule. This transfer forms the positively charged hydronium ion (H_3O^+) and a negatively charged ion. In an acid, any hydrogen atom that can be transferred to water is called an **acidic hydrogen**. **Figure 14.5** shows that it is possible for acids to have more than one acidic hydrogen per molecule.

Which hydrogen is acidic? Take another look at the acetic acid example: acetic acid (CH₃COOH) has four hydrogen atoms. Which one can be transferred to a water molecule to form the hydronium ion? As shown in **Figure 14.6,** only the hydrogen atom bonded to oxygen, a highly electronegative element, is acidic. The hydrogen atoms bonded to carbon form nonpolar bonds and are not acidic.

While the electronegativity difference in the oxygen-hydrogen bond facilitates the transfer of the hydrogen atom to water, not all hydrogen atoms in polar bonds are acidic. For example, methanol (CH₃OH) contains an oxygen-hydrogen bond, but it is not an acid. However, you can be sure that hydrogen atoms in nonpolar bonds are not acidic. Thus, you can distinguish the acidic hydrogen in acetic acid. You can also infer that a compound such as benzene (C_6H_6) has no acidic hydrogens.

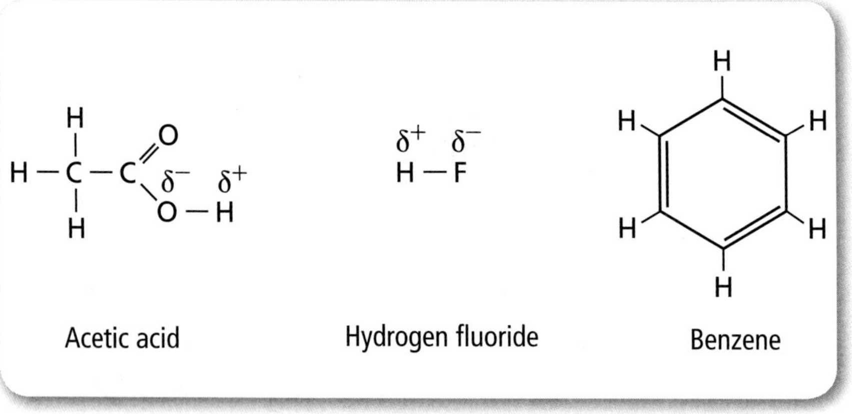

■ **Figure 14.6** Whether a hydrogen is ionizable depends in part on the polarity of its bond. Acetic acid and hydrogen fluoride both contain hydrogens attached to very electronegative elements—oxygen and fluorine, respectively. These hydrogens are acidic. The hydrogens in benzene, however, are not.

CHEMISTRY AND TECHNOLOGY

Manufacturing Sulfuric Acid

You might not expect that a simple acid would acquire worldwide status, but sulfuric acid has done just that. Most industrialized nations produce significant quantities of the chemical. The United States alone produces 40 million tons of sulfuric acid every year.

This product has many uses. Sixty percent of the sulfuric acid made in the United States is used in the production of liquid fertilizers and other inorganic chemicals. The rest is used in refining petroleum, in steel production, and in producing organic chemicals. Sulfuric acid is also useful in removing unwanted materials from ores and in lead-acid automobile batteries shown in **Figure 1**.

The Manufacturing Process

The production of sulfuric acid is fairly simple. It starts with burning sulfur to produce sulfur dioxide.

$$S(s) + O_2(g) \rightarrow SO_2(g)$$

■ **Figure 1** Lead-acid automobile battery

■ **Figure 2** Stored sulfuric acid

The next step in the process is called the contact method because the sulfur dioxide and oxygen molecules are in contact with a catalyst, usually vanadium pentoxide, V_2O_5. When the sulfur dioxide and oxygen gases pass through a heated tube that contains layers of the pellet-size catalyst, the sulfur dioxide is converted to sulfur trioxide. To make sure the reaction is complete, contact with the catalyst takes place twice.

$$2SO_2(g) + O_2(g) \xrightarrow{catalyst} 2SO_3(g)$$

Then the sulfur trioxide is bubbled through a solution of sulfuric acid to produce pyrosulfuric acid, $H_2S_2O_7$. Pyrosulfuric acid is then added to water to produce sulfuric acid, shown stored in **Figure 2**.

$$SO_3(g) + H_2SO_3(l) \rightarrow H_2S_2O_7(l)$$

$$H_2S_2O_7(l) + H_2O(l) \rightarrow 2H_2SO_4(l)$$

Discuss the Technology

1. **Hypothesize** Why do you suppose sulfur trioxide is bubbled through a solution of sulfuric acid instead of through water to produce sulfuric acid?
2. **Infer** Some people use the quantity of sulfuric acid produced by an industrialized nation as an economic indicator. Why is sulfuric acid production useful in this regard?

■ **Figure 14.7** Sulfuric acid, which is used to make steel, is an example of a diprotic acid.

Chemical reaction shorthand You know that you can write an equation for the ionization of a specific acid. However, it is sometimes handy to represent the formation of hydronium ions when acids dissolve in water by a general equation. In this general equation, any monoprotic acid is represented by the general formula HA. Compare this general equation to the specific equation for the ionization of HCl.

$$HCl(g) + H_2O(l) \rightarrow H_3O^+(aq) + Cl^-(aq)$$

$$HA + H_2O(l) \rightarrow H_3O^+(aq) + A^-(aq)$$

Although the form of the general equation written above is the most complete, it is more convenient to use a shorthand form of the reaction. In the shorthand form, water is not shown in the reaction, and the hydronium ion is represented as an aqueous hydrogen ion.

$$HA(aq) \rightarrow H^+(aq) + A^-(aq)$$

Similar equations apply to the transfer of hydrogen ions from polyprotic acids, such as sulfuric acid, which is used in processes such as steel making, shown in **Figure 14.7**. Equations depicting the transfer are shown in **Figure 14.8**.

When using this convenient shorthand style, keep in mind that the water molecule is almost always an active participant in the reaction, even though it is not written in the equation.

■ **Figure 14.8** Polyprotic acids lose their acidic hydrogens one at a time.

Describe the steps it takes for a diprotic acid and a triprotic acid to lose their acidic hydrogens.

	General:	Example:
Diprotic Acid	$H_2A(aq) \rightarrow H^+(aq) + HA^-(aq)$ $HA^-(aq) \rightarrow H^+(aq) + A^{2-}(aq)$	$H_2SO_4(aq) \rightarrow H^+(aq) + HSO_4^-(aq)$ $HSO_4^-(aq) \rightarrow H^+(aq) + SO_4^{2-}$
Triprotic Acid	$H_3A(aq) \rightarrow H^+(aq) + H_2A^-(aq)$ $H_2A^-(aq) \rightarrow H^+(aq) + HA^{2-}(aq)$ $HA^{2-}(aq) \rightarrow H^+(aq) + A^{3-}(aq)$	$H_3PO_4(aq) \rightarrow H^+(aq) + H_2PO_4^-(aq)$ $H_2PO_4^-(aq) \rightarrow H^+(aq) + HPO_4^{2-}(aq)$ $HPO_4^{2-}(aq) \rightarrow H^+(aq) + PO_4^{3-}(aq)$

Biology Connection

Measurement of Blood Gases

In blood, there are only four hydrogen ions for every 100,000,000 other ions and molecules. But enzyme reactions in the body are sensitive to small changes in the concentration of hydrogen ions, so it is crucial. Hydrogen ions affect acid-base relationships in body fluids.

Interpreting acid-base status in blood When a patient is ill, the doctor's role is to diagnose the patient's condition. Sometimes, this is difficult because different conditions may have similar symptoms. One helpful tool that the physician has is a blood test that provides information about the acid-base relationships in blood. **Figure 1** shows blood being drawn from a patient in preparation for a blood test. This particular blood test will provide data on acidity (pH), pressure caused by dissolved carbon dioxide (P_{CO_2}), pressure caused by dissolved oxygen (P_{O_2}), and hydrogen carbonate concentration (HCO_3^-). The normal ranges of these components are shown in the table below.

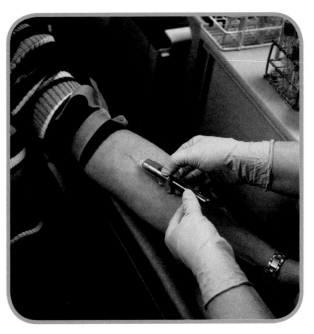

Figure 1 Blood drawn from a patient can be analyzed to determine the levels of acidity, carbon dioxide, oxygen, and carbonate ions.

Normal Ranges of Some Blood Components

Plasma Component	Normal Range
HCO_3^-	23–29 m Eq/liter*
P_{CO_2}	35–45 mm Hg
P_{O_2}	75–100 mm Hg
pH	7.35–7.45

*expressed in molar equivalents per liter

Case histories To see how a physician uses acid-base relationships, consider the following case history. After a certain pneumonia patient was put on a respirator, she failed to improve. Her blood test showed the following components.

HCO_3^-	18 m Eq/liter	P_{O_2}	75 mm Hg
P_{CO_2}	17 mm Hg	pH	7.65

The low carbon dioxide level and the high pH while on the respirator were unexpected. These levels led the physician to check the settings on the respirator. The volume adjustment on the respirator had slipped. The patient was receiving twice the recommended quantity of air. This caused respiratory alkalosis, a condition of decreased acidity of the blood and tissues. When the respirator was adjusted, the blood levels returned to normal as the acid-base balance was reestablished, and the patient began to recover.

Connection to Chemistry

1. **Hypothesize** In a heart attack, blood flow to some parts of the heart may be stopped or greatly reduced. How might this affect the acid-base relationship in blood in the heart muscle?
2. **Apply** Blood gases may fail to show major abnormalities while the patient is at rest. Suggest a way to overcome this problem, assuming the patient is not bedridden.

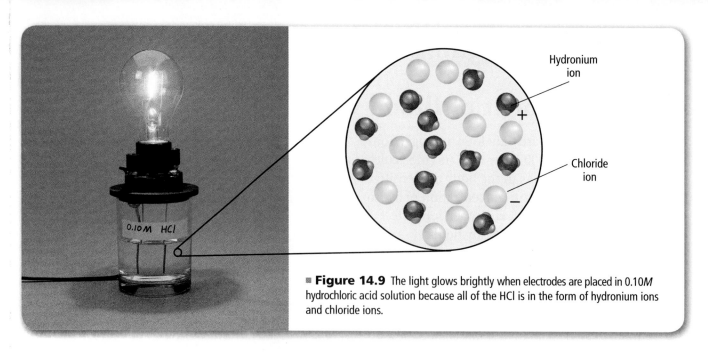

Figure 14.9 The light glows brightly when electrodes are placed in 0.10M hydrochloric acid solution because all of the HCl is in the form of hydronium ions and chloride ions.

Acids are electrolytes Recall from Chapter 4 that pure water does not conduct electricity. Some substances dissolve in pure water to form ions. These substances conduct electricity and are called electrolytes. The process of forming ions in solution is called **ionization**. Specifically, acids form ions in a process called acid ionization. Because acids ionize to form ions in water, acidic solutions conduct electricity. Acids are electrolytes.

As opposed to solutions of ionic compounds such as table salt, which are always excellent conductors of electricity, acidic solutions have electrical conductivities that range from strong to weak. **Figure 14.9** shows that a 0.10M solution of hydrochloric acid is an excellent conductor. Conversely, **Figure 14.10** shows that a solution of acetic acid at the identical 0.10M concentration does not conduct current nearly as well. The range of electrical conductivities demonstrates that acids vary in their abilities to produce ions.

Figure 14.10 When electrodes are placed in 0.10M acetic acid solution, the light is dim. Compare this illustration with **Figure 14.9**.
Explain *the difference in the brightness of the bulbs in terms of the concentration of ions in solution.*

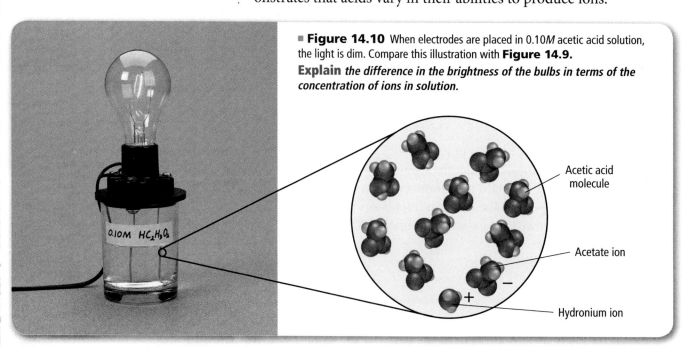

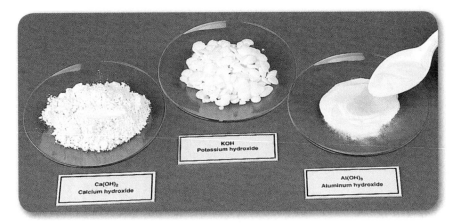

■ **Figure 14.11** All of these compounds are bases because they produce hydroxide ions when they dissolve in water. As with polyprotic acids, it is possible for a formula unit of a metal hydroxide to produce more than one hydroxide ion in water. Calcium hydroxide ($Ca(OH)_2$) and aluminum hydroxide ($Al(OH)_3$) are examples of such bases.

Submicroscopic Behavior of Bases

The behavior of bases is also described at the molecular level by the interaction of the base with water. A **base** is a substance that produces hydroxide ions (OH^-) when it dissolves in water. There are two mechanisms by which bases produce hydroxide ions when they dissolve in water.

Simple bases: metal hydroxides The simplest kind of base is a water-soluble ionic compound that contains the hydroxide ion as the negative ion. When sodium hydroxide dissolves in water, for example, it dissociates into sodium ions (Na^+) and hydroxide ions (OH^-).

$$NaOH(s) \rightarrow Na^+(aq) + OH^-(aq)$$

Any water-soluble metal hydroxide will be a base when added to water. Water plays a different role here than in the formation of hydronium ions when acids ionize in water. The hydroxide ion forms by simple ionic dissociation, and no transfer occurs between the base and the water molecules to form the hydroxide ions. **Figure 14.11** shows three common metal hydroxide bases.

Bases that accept H^+ A few bases are covalent compounds that produce hydroxide ions by an ionization process when dissolved in water. The ionization involves the transfer of a hydrogen ion from water to the base. The most common example of this type of base is ammonia (NH_3).

When ammonia gas dissolves in water, some of the aqueous ammonia molecules react with water molecules to form ammonium ions (NH_4^+) and hydroxide ions, as shown below.

$$NH_3(g) + H_2O(l) \leftrightarrow NH_4^+(aq) + OH^-(aq)$$

Ammonia produces hydroxide ions in water by a different mechanism than bases such as NaOH. In the reaction with ammonia, the water molecule is an active chemical reactant. Water molecules transfer hydrogen ions to ammonia molecules.

It is also helpful to have a general reaction for the ionization of a covalent base, which is represented by the letter B. Study the equation for the general reaction.

$$B + H_2O(l) \leftrightarrow BH^+(aq) + OH^-(aq)$$

In the Field

Meet Fe Tayag
Cosmetic Bench Chemist

It pays to be a careful reader of labels. Here's a hint from Ms. Tayag, who has formulated cosmetics for more than 20 years. Many cosmetic companies find it is a good selling point to add sunscreen to their products. But unless the container specifies an SPF (sun protection factor) as a number, there probably isn't enough sunscreen to do much good. In this interview, Ms. Tayag shares other cosmetics savvy that can make you a wiser shopper.

On the Job

Q: Ms. Tayag, what do you do on your job?

A: I'm in the Research and Development Department of a cosmetic laboratory. I formulate cosmetics, such as shampoos, lotions, and bubble baths, for cosmetic companies.

Q: Can you give us an idea of a typical formula for a shampoo?

A: A simple shampoo consists of a mixture of water, sodium lauryl sulfate, and an amide to make it foam. I heat the mixture and then I have to adjust the acidity. Most shampoos are neutral. If it's more basic than that, I adjust it with a citric acid solution. Then I cool it and check the viscosity, or how it flows. I don't want it to be either water-thin or molasses-thick, so I'll adjust the viscosity by using a 20 percent sodium chloride solution. Perfume and color are added to make it smell and look good.

Q: The label on a cosmetic usually lists water as the first ingredient. Does that mean it's mostly water?

A: Yes. Face creams have the lowest amount of water, about 60 percent. The amount of water used depends on the skin type.

Q: What colorings do you add to the cosmetics you make?

A: I use colors approved by the Food and Drug Administration. They come in basic colors, but I can combine them to produce other colors. Green is the most popular coloring for shampoo. That color seems to be associated with freshness and cleanliness.

Q: What trends do you see developing in cosmetics?

A: There are more cosmetics created especially for different ethnic groups. Sunscreen is being added to more and more cosmetics as people become increasingly aware of the damage that ultraviolet light can do to the skin. And, because a large part of the population is getting older, ingredients that are supposed to delay the aging effects, such as antioxidants, are becoming more popular. In addition, there's lots of interest in antiallergenic, natural cosmetics that are both unscented and uncolored.

Early Influences

Q: How did you get interested in becoming a chemist?

A: My father was a warehouse man for a cosmetics company in the Philippines, where I grew up. He was fascinated by the process of producing cosmetics. Because he thought the job of a chemist was a very dignified one, he encouraged me to study chemistry. Our family, which included eight children, was poor, but he somehow found a way to send me to college.

Q: Did you enjoy the study of chemistry in college?

A: Not at first. I had a tough time in college and kept asking myself, "Why did I take this course?" I cried at exam time, but I couldn't bear to disappoint my father. I prayed a lot, and I thought of my father struggling to find money to pay my tuition. That helped me find the courage to continue. My third year of college marked a turning point, and my studies became easier for me.

Q: Were there people besides your father who influenced you in your career?

A: A friend of my father helped me get a job in the cosmetics lab while I was still in school. I worked days and attended school at nights. That way, I got to see what chemistry was all about in the real world. I felt I was ahead of students who lacked practical experience. For instance, in my colloid chemistry class, I brought materials to school and demonstrated how to make a cleansing cream.

Personal Insights

Q: Do you consider chemistry to be a little like cooking?

A: Yes. When I cook, I rely heavily on my senses of sight and smell. In my opinion, a keen sense of smell and a good ability to make observations are very important for a chemist, too.

Q: Some people think that cosmetics are frivolous. Do you agree?

A: No. I think it's important for people's self-confidence to look nice. Because people want to remain attractive and young-looking, this is an industry that will never die out.

CAREER CONNECTION

These career opportunities are related to cosmetic chemistry.

Food and Drug Inspector College degree plus written examinations

Manufacturers' Sales Representative High school diploma

Cosmetologist State-administered exam

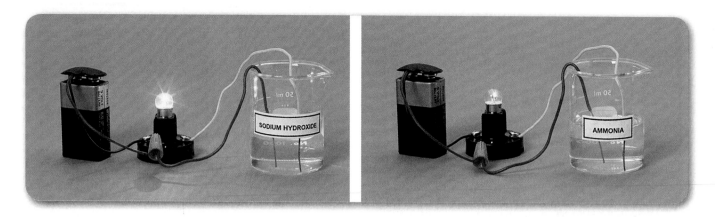

■ **Figure 14.12** The electrical conductivity of 1*M* NaOH is greater than that of 1*M* NH$_3$.
Explain Why are some bases good conductors of electric current in solution, while others conduct poorly?

Personal Tutor For an online tutorial on the transfer of hydrogen ions, visit glencoe.com.

■ **Figure 14.13** Ammonia is a polar molecule. The hydrogen bond that forms between the N end of NH$_3$ and the H end of H$_2$O is strong enough to pull an H$^+$ completely away from H$_2$O. The two electrons in the broken O—H bond remain as a lone pair on the O. The result is a stable hydroxide ion, OH$^-$. The H$^+$ bonds to the N in NH$_3$, using the lone pair of electrons on the N to form a fourth N—H bond and thus a stable ammonium ion, NH$_4^+$. Each atom in the Lewis dot structures has a stable number of valence electrons.

Bases are electrolytes Because a base in water produces ions, you can predict that aqueous solutions of bases will conduct electricity. **Figure 14.12** compares the conductivity of a 1*M* NaOH solution with that of a 1*M* NH$_3$ solution. As with acids, the ability of basic solutions to conduct electricity varies, depending upon the base. This variability demonstrates that differences exist in the ability of different bases to produce ions.

Why does water transfer H$^+$ to bases? Think back to why acids transfer hydrogen ions to water. The same model can be used to explain why water molecules lose hydrogen ions to covalent bases when they dissolve in water. Consider the example of ammonia. Ammonia is a polar molecule because it contains polar covalent N–H bonds. The nitrogen end of the molecule has a slight negative charge, and the hydrogen atoms each have a slight positive charge. A lone pair of electrons is also on the central nitrogen. Look at **Figure 14.13** to see what happens when ammonia molecules dissolve in water. As you review the figure, notice that the reaction is reversible. In this reaction, formation of NH$_3$ and H$_2$O is actually favored. The transfer of a proton from water to ammonia to form the ammonium ion and hydroxide ion only occurs to a small degree, with the great majority of molecules remaining un-ionized.

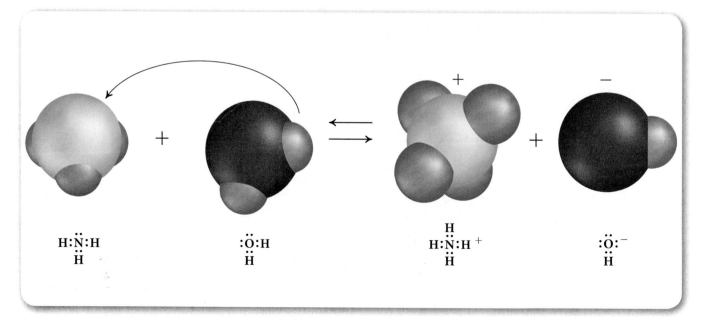

Other Acids and Bases: Anhydrides

Two related classes of compounds do not fit the previous models of acids and bases, but they also act as acids or bases. These compounds are both oxides, which are compounds containing oxygen bonded to just one other element. These oxides are called anhydrides, which means that they contain no water.

Anhydrides differ, depending upon whether the oxygen is bonded to a metal or a nonmetal. Nonmetal oxides that form acids when they react with water are called **acidic anhydrides.** On the other hand, metal oxides that react with water to form bases are called **basic anhydrides.** In both of these reactions, water is a reactant. Now, examine some examples of anhydrides.

Acidic anhydrides Probably the most familiar acidic anhydride is carbon dioxide (CO_2). Water that has had carbon dioxide bubbled through it turns blue litmus to red, indicating that carbon dioxide and water form an acid, carbonic acid (H_2CO_3). A solution of carbon dioxide also has a slightly sour taste, which is one of the reasons that carbonated water is such a refreshing beverage. **Figure 14.14** shows solutions of several common oxides, including carbon dioxide, and water.

Acid rain Carbon dioxide is a minor component in Earth's atmosphere and an important component in the carbon cycle. Because carbon dioxide is always in the atmosphere, it dissolves in rainwater, forming carbonic acid. The result is that rain is always slightly acidic.

If rain is always acidic, why is acid rain such an environmental concern? The acidity of normal rain does not damage the environment. However, other nonmetal oxides, such as sulfur oxides and nitrogen oxides, sometimes present in the atmosphere, can increase the acidity of normal rain to levels that can damage the environment.

Sources of sulfur oxides The major source of sulfur oxides in the atmosphere is the burning of sulfur-containing coal in power plants. As this type of coal burns in a furnace, sulfur dioxide gas (SO_2) is produced. The sulfur dioxide escapes into the atmosphere, where it reacts with more oxygen to form sulfur trioxide (SO_3).

> **TRY AT HOME LAB**
> See page 874 for **Testing for Acid Rain.**

■ **Figure 14.14** Each tube contains a solution of an oxide of the indicated element, water, and universal indicator. In the acidic range, universal indicator goes from red (most acidic) to light yellow. Thus, the tube labeled carbon shows that a solution of CO_2 is mildly acidic. In the basic range, universal indicator darkens from light brownish–green to deep purple.

Infer *Based on the photo, what can you infer about the nature of metal oxides in solution?*

■ **Figure 14.15** One of the major sources of nitrogen oxides in the atmosphere is automobiles.

Sources of nitrogen oxides At room temperature, the reaction between nitrogen (N_2) and oxygen (O_2) is slow and insignificant. At the high temperatures of an automobile engine, the reaction between the two gases goes quickly, and large amounts of nitrogen oxides are produced in exhaust, as shown in **Figure 14.15.**

When sulfur oxides, nitrogen oxides, and increased amounts of carbon dioxide dissolve in rain, they form acids and produce acid rain. Acid rain has been significantly reduced over the past decade as new mechanisms for trapping nonmetal oxides before they get into the atmosphere have been developed.

Basic anhydrides Unlike nonmetal oxides, which are covalent compounds, metal oxides are ionic compounds. When metal oxides react with water, they produce hydroxide ions. Refer back to **Figure 14.14** to see solutions of metal oxides in water.

Gardeners sometimes use calcium oxide, also called lime, to treat their soil. When calcium oxide is spread on soil, it reacts with water in the soil to form calcium hydroxide, ($Ca(OH)_2$). This compound then forms calcium (Ca^{2+}) and hydroxide (OH^-) ions.

$$CaO(s) + H_2O(l) \rightarrow Ca(OH)_2(aq)$$

$$Ca(OH)_2(aq) \rightarrow Ca^{2+}(aq) + 2OH^-(aq)$$

Soap making A similar reaction was used historically to produce soap, as shown in **Figure 14.16.** When wood burns, the metal atoms in the wood form solid metal oxides in the burning process. The metal oxides formed are mainly those of sodium, potassium, and calcium. These metal oxides are ionic so they are solids, even at the high temperature of a roaring fire. They are the major component of the ash that is left when the fire burns out. The ash was collected and turned into a solution called lye, which was mixed with animal fat to form soap. The reaction of water and sodium oxide (Na_2O), one of the metal oxides in wood ash, is similar to that shown for lime and water.

$$Na_2O(s) + H_2O(l) \rightarrow 2NaOH(aq)$$

$$NaOH(aq) \rightarrow Na^+(aq) + OH^-(aq)$$

■ **Figure 14.16** Lye was produced by collecting wood ash and soaking it in water. After several days, the highly basic solution was separated from the undissolved ash. This solution was then combined with animal fat and boiled, as shown in the photo. The lye reacted with the fat to make soap.

Chemistry & Society

Atmospheric Pollution

The air you breathe is literally a matter of life and death. Atmospheric oxygen is taken into your body and, during respiration, reacts with glucose to produce the energy required for all the life processes that keep you going.

Unfortunately, the same air, at times, might contain materials that cause respiratory diseases and bring about other harmful effects. Air is often polluted with chemicals produced by human activity. Sometimes, natural processes produce some of the same air pollutants. Examples include volcanic eruptions and forest fires.

Introducing the major air pollutants The major chemicals that pollute the air are carbon monoxide (CO); carbon dioxide (CO_2); sulfur dioxide (SO_2); nitrogen monoxide (NO); nitrogen dioxide (NO_2); hydrocarbons; and suspended particles.

In addition, pollutants form under the influence of sunlight when oxygen, nitrogen oxides, and hydrocarbons react. These reactions produce ozone (O_3), and aldehydes such as formaldehyde (CH_2O). Why are pollutants a problem?

■ **Figure 2** Pine trees killed by acid rain

Many industrial and power plants burn coal and oil. The smoke produced can contain sulfur oxides, suspended particles, and nitrogen oxides. Automobiles also contribute to the problem by emitting similar oxides. These chemicals react with water in the air to form acids, such as sulfuric acid. These acids reach the surface of Earth in precipitation. Acid rain can have a disastrous effect when it reaches bodies of water and waterways. But if a lake has a high limestone content, the limestone can somewhat neutralize the acid.

Smog Large cities with many automobiles may have another problem with airborne pollutants. It is called smog, which is a haze or fog that is made harmful by the chemical fumes and suspended particles it contains.

A type of smog known as photochemical smog frequently occurs when the pollutants from automobile exhaust enter the air and are exposed to sunlight. Photochemical smog is generally worse on hot days and between 11 A.M. and 4 P.M. when exhaust has accumulated in the air.

■ **Figure 1** Salmon killed by acidic water

Acid rain What do the salmon in **Figure 1** and the pine trees in **Figure 2** have in common? Both have succumbed to the acid environment in which they live. They are two of the many victims of acid rain.

ANALYZE the Issue

1. **Acquire Information** Research free trade in this country. How might the issue of free trade influence the problems of air pollution in this country?
2. **Debate** Older-model cars are responsible for the greatest amount of air pollutants being vented into the atmosphere. Hold a debate on whether these cars should be banished from the highways.

Table 14.2 Everyday Acids and Bases

Acids
- lemons
- tomatoes
- vinegar
- coffee
- tea

Bases
- oven cleaner
- soap
- bleach
- antacids

The Macroscopic-Submicroscopic Acid-Base Connection

Table 14.2 lists some everyday acids and bases. The properties of acids and bases are determined by the submicroscopic interactions between the acid or the base and water. For example, hydrochloric acid and acetic acid both form hydronium ions in water. Both solutions turn blue litmus red. However, the conductivity of $1M$ hydrochloric acid is much greater than that of $1M$ acetic acid.

A base interacts with water molecules to form hydroxide ions either by ionic dissociation or by the transfer of a hydrogen ion from a water molecule to the base. Consider a $1M$ sodium hydroxide solution and a $1M$ ammonia solution. Both solutions are basic. They both turn red litmus to blue. But the sodium hydroxide solution shows strong electrical conductivity, while the ammonia solution only weakly conducts.

SUPPLEMENTAL PRACTICE

For more practice with acids and bases, see Supplemental Practice, page 830.

Section 14.1 Assessment

Section Summary

- The concentrations of hydronium ions (H_3O^+) and hydroxide ions (OH^-) determine whether an aqueous solution is acidic, basic, or neutral.
- Acidic anhydrides are nonmetallic oxides that react with water to form acids. Basic anhydrides are metallic oxides that react with water to form bases.

1. **MAIN Idea Make a table** that compares and contrasts acids and bases.
2. **Demonstrate** Use a chemical equation to show how aqueous HNO_3 fits the definition of an acid.
3. **Identify** Consider the oxides MgO and CO_2. For each oxide, tell whether it is a basic anhydride or an acidic anhydride. Write an equation for each to show its acid-base chemistry.
4. **Explain** Chemists often call a hydrogen ion a proton. Explain why an acid is sometimes called a proton donor, and a base is sometimes called a proton acceptor.
5. **Apply** After using soap to wash dishes by hand, it is sometimes difficult to keep your hands from remaining slick. Explain why rinsing your hands in lemon juice would make them less slick.

Section 14.2

Objectives

- **Relate** different electrical conductivities of acidic and basic solutions to degree of dissociation or ionization.
- **Distinguish** strong and weak acids or bases by degree of dissociation or ionization.
- **Compare** and contrast the composition of strong and weak solutions of acids or bases.
- **Relate** pH to the strengths of acids and bases.

Review Vocabulary

ionization: the process in which ions form from a covalent compound.

New Vocabulary

strong base
strong acid
weak acid
weak base
pH

FOLDABLES Incorporate information from this section into your Foldable.

Strengths of Acids and Bases

MAIN Idea In solution, strong acids and bases ionize completely, but weak acids and bases ionize only partially.

Real-World Reading Link You wouldn't hesitate to use the acetic acid in vinegar on a salad, but you certainly wouldn't use hydrochloric acid on any type of food. On the other hand, hydrochloric acid is much better for cleaning bricks than vinegar is. This also applies to bases: ammonia solutions are great as mild cleaners, but not as drain openers. Sodium hydroxide is one of the bases to choose for opening drains.

Strong Acids and Bases

You know that when acids and bases are mixed with water, they form ions. Much of the behavior of acids and bases depends on how many ions are formed in water. The degree to which bases and acids produce ions depends on the nature of the acid or base.

Acids and bases are put into one of two categories depending upon their strengths, which is the degree to which they form ions. The strong category is reserved for those substances, such as sodium hydroxide (NaOH) and hydrochloric acid (HCl), that completely ionize when dissolved in water. All other acids and bases are classified as weak because they produce few ions when dissolved in water. **Figure 14.17** shows one use of the strong acid hydrochloric acid and one use of the strong base calcium hydroxide.

Strong bases A **strong base** is a base that completely dissociates into ions when dissolved in water. Calcium hydroxide ($Ca(OH)_2$) is a strong base because all $Ca(OH)_2$ formula units dissociate into separate calcium and hydroxide ions in solution. The dissociation of the base is complete.

■ **Figure 14.17** The strong acid hydrochloric acid (HCl), also called muriatic acid, is used to clean bricks and concrete. The strong base calcium hydroxide ($Ca(OH)_2$) is used to make whitewash.

Table 14.3 Common Strong Acids and Bases

Strong Acids	Strong Bases
Perchloric acid ($HClO_4$)	Lithium hydroxide (LiOH)
Sulfuric acid (H_2SO_4)	Sodium hydroxide (NaOH)
Hydriodic acid (HI)	Potassium hydroxide (KOH)
Hydrobromic acid (HBr)	Calcium hydroxide ($Ca(OH)_2$)
Hydrochloric acid (HCl)	Strontium hydroxide ($Sr(OH)_2$)
Nitric acid (HNO_3)	Barium hydroxide ($Ba(OH)_2$)
	Magnesium hydroxide ($Mg(OH)_2$)

The strength of a base is based on the percentage of formula units dissociated, not the overall solubility of the compound. For example, some bases, such as magnesium hydroxide ($Mg(OH)_2$) are not very soluble in water. However, they are still considered to be strong bases because all of the base that does dissolve completely dissociates into ions.

Table 14.3 lists some common strong acids and bases. The strong bases shown in **Table 14.3** are all ionic compounds that contain hydroxide ions. Sodium hydroxide and potassium hydroxide are the most common strong bases you will encounter. A $1M$ solution of NaOH and a $1M$ solution of KOH will each contain $1M$ OH^- because both compounds completely dissociate.

Strong acids A **strong acid** is an acid that completely ionizes in water. Hydrochloric acid, shown being used in **Figure 14.17,** is a strong acid because the hydrogen and the chlorine completely dissociate in a water solution. Because of the strong attraction between the water molecules and HCl, every unit of HCl ionizes. A $1M$ HCl solution contains $1M$ H_3O^+ and $1M$ Cl^-. Similarly, a $1M$ nitric acid (HNO_3) solution contains $1M$ H_3O^+ and $1M$ NO_3^-.

Because they are used so often, it is helpful to memorize the names and formulas of the strong acids and bases listed in **Table 14.3**. They all completely dissociate into ions when they dissolve in water. Most acids and bases not listed in this group are considered to be weak. However, the term weak is not absolute. Strengths of acids and bases cover a wide range from strong to extremely weak. Notice that the strong bases in **Table 14.3** are all hydroxides of group 1, the alkali metals, and group 2, the alkaline earth metals. In fact, *alkali* is a term frequently used to refer to materials that have noticeably basic properties.

Weak Acids and Bases

The weak category of acids and bases contains those with a wide range of strengths. This is the category into which most acids and bases fall. Instead of being completely ionized, weak acids and bases are only partially ionized.

Weak acids Acetic acid, CH_3COOH, is a good example of a weak acid. A **weak acid** is an acid that only partially ionizes in solution. In other words, most of the acid remains as CH_3COOH. For example, in a $1M$ CH_3COOH solution, less than 0.5 percent of the acetic acid molecules ionize. Roughly 99.5 percent of the acetic acid molecules remain un-ionized. Consider 1000 acetic acid molecules in a water solution. On the average, only five of the 1000 molecules transfer their single hydrogen ion to a water molecule. The molarity of hydronium ion produced in a $1M$ CH_3COOH solution is much less than $1M$ due to this partial ionization. Other common weak acids include phosphoric acid (H_3PO_4) and carbonic acid (H_3CO_3). Both are found in soft drinks.

The molecular structure of a weak acid determines the extent to which the acid ionizes in water. **Figure 14.18** shows the structures of some common weak acids.

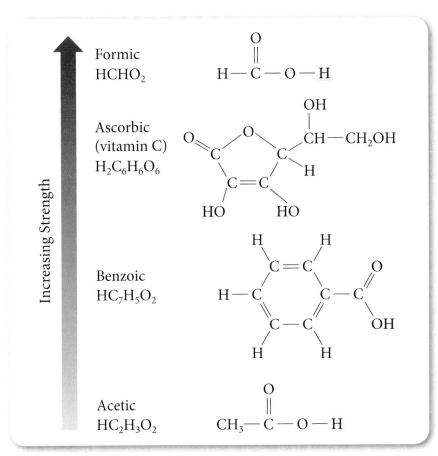

■ **Figure 14.18** Common weak acids vary in their structures. The acidic hydrogens are highlighted in blue.

Table 14.4 shows the dramatic difference in degree of ionization between a solution of a weak acid and one of a strong acid. A solution of weak acid contains a mixture of un-ionized acid molecules, hydronium ions, and the corresponding negative ions. The concentration of the un-ionized acid is always the greatest of the three concentrations for a weak acid. In fact, for the weakest acids, the concentration of un-ionized acid molecules is very nearly 100 percent, while the concentration of ionized molecules is very nearly zero percent.

Table 14.4	Strong and Weak Acids		
Type of acid	Concentration before ionization	Ionization	Equilibrium concentrations after ionization
Strong (such as HCl)	HA	Complete (100%)	H_3O^+ A^-
Weak (such as $HCHO_2$)	HA	Partial (percentage varies)	HA H_3O^+ A^-

Section 14.2 • Strengths of Acids and Bases

Weak bases Ammonia is an example of a weak base. Similar to a weak acid, a **weak base** ionizes only partially in solution. For example, in a 1M aqueous solution of ammonia, only about 0.5 percent of the ammonia molecules reacts with water to form ammonium and hydroxide ions. Most of the molecules do not react with water to form ions. About 99.5 percent of the ammonia molecules remain intact. The molarity of hydroxide ions in a 1M ammonia solution is thus much less than 1M. The major dissolved component in a solution of weak base is the un-ionized base. Other examples of weak bases include aluminum hydroxide ($Al(OH)_3$) and iron(III) hydroxide ($Fe(OH)_3$).

Weak is not insignificant Although most acids and bases are classified as weak, their behaviors are extremely significant. Most of the acid-base chemistry in living systems occurs through interactions between weak acids and bases. For example, amino acids, the small molecules that serve as the building blocks of proteins, have properties of both weak acids and weak bases. The amino portion of the molecule acts as a base when it comes into contact with a strong acid, and the acid part of the molecule acts as a weak acid when exposed to a base. Even the coiling of DNA into a double helix is due to the interactions between weak acids and bases. Weak does not mean "insignificant."

Strength is not concentration Although the terms *weak* and *strong* are used to compare the strengths of acids and bases, *dilute* and *concentrated* are terms used to describe the concentration of solutions. The combination of strength and concentration ultimately determines the behavior of the solution. For example, compare the solutions in **Figure 14.19**. Note on the labels that it is possible to have a concentrated solution of a weak acid or weak base or a dilute solution of a weak acid or weak base. Similarly, you can have a concentrated solution of a strong acid or strong base, as well as a dilute solution of a strong acid or strong base.

■ **Figure 14.19** Strength and concentration are separate properties of acids and bases. The number of ions an acid or a base contributes to solution depends on both strength and concentration.
Determine *which of the acids is more concentrated. Which is stronger? Which of the bases is more concentrated? Which is stronger?*

| 0.01M HCl | 6M NaOH | 6M $HC_2H_3O_2$ | 0.1M NH_3 (aq) |
| Strong Acid | Strong Base | Weak Acid | Weak Base |

The pH Scale

The range of possible concentrations of hydronium ions and hydroxide ions in solutions of acids or bases is huge. For example, a $6M$ solution of HCl has an H_3O^+ molarity of $6M$, but a $6M$ solution of $HC_2H_3O_2$ has an H_3O^+ molarity of $0.01M$. In most applications, the observed range of possible hydronium or hydroxide ion concentrations spans $10^{-14}M$ to $10^0 M$. To make this huge range of concentrations easier to work with, the pH scale was developed by S.P.L. Sørenson.

What is pH? pH is a mathematical scale in which the concentration of hydronium ions in a solution is expressed as a number from 0 to 14. A scale of 0 to 14 is much easier to work with than a range from 10^0 to 10^{-14}. The pH scale is a convenient way to describe the concentration of hydronium ions in acidic solutions, as well as the hydroxide ions in basic solutions. The pH value is the negative of the exponent of the hydronium ion concentration. For example, a solution with a hydronium ion concentration of $10^{-11}M$ has a pH of 11. A solution with a pH of 4 has a hydronium ion concentration of $10^{-4}M$.

Relating hydronium ions to hydroxide ions How do the numbers above relate to hydroxide ion concentrations? Experimental evidence shows that when the hydronium ion concentration and the hydroxide ion concentration in aqueous solution are multiplied together, the product is 10^{-14}. So, if the pH of a solution is 3, the hydronium ion concentration is $10^{-3}M$, so the hydroxide ion concentration is $10^{-14}/10^{-3}M$, which is $10^{-14-(-3)}M$, or $10^{-11}M$.

You can think of hydronium and hydroxide ions as being on a sliding scale: as the concentration of one increases, the concentration of the other decreases. Acidic solutions simply contain more hydronium ions than hydroxide ions, while basic solutions contain more hydroxide ions than hydronium ions. **Figure 14.20** will help you visualize the relationship between hydronium and hydroxide ion concentrations.

Neutral solutions In neutral solutions, hydronium and hydroxide ion concentrations are equal. Since the product of the concentrations must equal 10^{-14}, the concentrations of both hydronium and hydroxide ions in a neutral solution are 10^{-7}. This corresponds to a pH of 7.

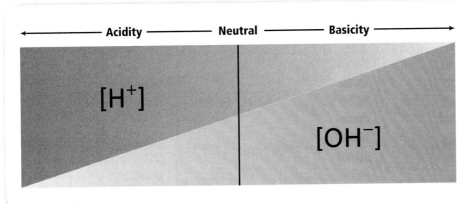

■ **Figure 14.20** Note how [H⁺] and [OH⁻] change simultaneously. As [H⁺] decreases to the right, [OH⁻] increases to the right.

Identify the point in the diagram at which the two ion concentrations are equal.

SUPPLEMENTAL PRACTICE

For more practice with pH and ion concentration, see Supplemental Practice, page 830.

PRACTICE Problems Solutions to Problems Page 864

Find the pH of each of the following solutions.

6. The hydronium ion concentration equals:
   a) $10^{-5} M$ b) $10^{-12} M$ c) $10^{-2} M$
7. The hydroxide ion concentration equals:
   a) $10^{-4} M$ b) $10^{-11} M$ c) $10^{-8} M$

Interpreting the pH scale You now know that the pH scale is divided into three areas. Solutions with a pH lower than 7 are acidic, and solutions with a pH greater than 7 are basic. If a solution has a pH of exactly 7, the solution is neutral. It is neither acidic nor basic. **Figure 14.21** shows two common ways to assess the pH of a solution. What does a change in pH indicate about the concentrations of hydronium and hydroxide ions in solution?

Decreasing pH As the pH decreases, the concentration of hydronium ions increases, and the concentration of hydroxide ions decreases. Every one unit decrease in the pH means a factor of 10 increase in the hydronium ion concentration. For example, a solution with a pH of 4 and a solution with a pH of 3 are both acidic because their pHs are less than 7. The solution with a pH of 3 has ten times the concentration of H_3O^+ compared to the solution with a pH of 4. Small changes in pH can mean big changes in hydronium ion concentration.

■ **Figure 14.21** Litmus paper can tell you whether a solution is acidic or basic—red for acids, blue for bases. If you want to know the pH of a solution, however, you will need pH paper or a pH meter.

The approximate pH of a solution can be obtained by wetting a piece of pH paper with the solution and comparing the color of the wet paper with a set of standard colors.

A portable pH meter, which is being used above to measure the pH of rainwater, provides a more accurate measurement in the form of a digital display of the pH.

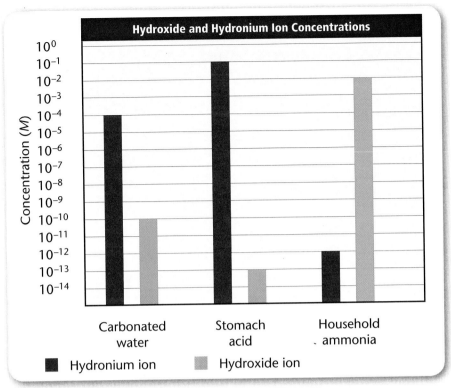

■ **Figure 14.22** Look at the hydronium ion and hydroxide ion concentrations for three common solutions.

Apply *For each solution, what is the product of the two concentrations?*

Increasing pH As the pH increases, the concentration of hydroxide ions increases, and the concentration of hydronium ions decreases. For example, consider a solution with a pH of 10 and another solution with a pH of 11. The solution with a pH of 11 has ten times more hydroxide ions than the solution with a pH of 10.

In a neutral solution, the concentration of hydroxide ions and the concentration of hydronium ions are equal. **Figure 14.22** compares hydroxide and hydronium ion concentrations for three solutions with different pHs.

pH of common substances **Figure 14.23** gives the pHs of some common materials. Notice the range of pHs from battery acid to oven cleaner. Pure water is exactly neutral.

Compare the pHs of vinegar and milk. Depending upon the brand and type of vinegar, the pH ranges from 2.4 to 3.4. If you have a bottle of vinegar with a pH of 3.4, the vinegar is definitely acidic. Milk, with a pH of 6.4, is also acidic, but much less so. The difference between 3.4 and 6.4 (3 pH units) may not seem like much, but remember that each unit of pH represents a power of 10. A difference of 3 pH units means that the hydronium ion concentration in the vinegar is 10^3, or 1000, times the hydronium concentration in the milk.

■ **Figure 14.23** Compare the pH values for these familiar substances.

Determine *whether seawater or detergent has a higher concentration of H^+ ions. How many times higher?*

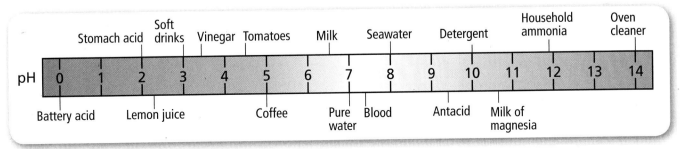

Section 14.2 • Strengths of Acids and Bases

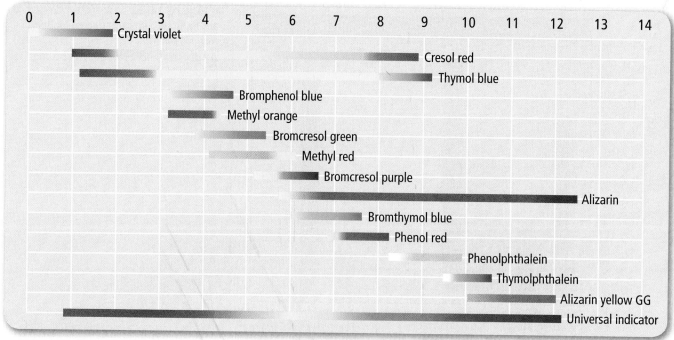

■ **Figure 14.24** Notice the colors of these different acid-base indicators across the pH scale. Some are better indicators at low pH, others at moderate pH, and still others at high pH.

Using Indicators to Measure pH

Perhaps you've seen someone using indicators to test the water of a swimming pool for pH. The pH tells the condition of the water and its suitability for swimming. The colored solutions that are used in this test are indicators that have different colors at different pHs. These dyes allow you to find the approximate pH by comparing the color to a standard chart. **Figure 14.24** shows the colors of several indicators at different pHs.

MiniLab 14.2

Antacids

How can you evaluate the effectiveness of antacids? Use your knowledge of acid chemistry to evaluate the effects of antacids that are commonly used to treat heartburn.

Procedure

1. Read and complete the lab safety form.
2. Obtain four snack-size **self-sealing bags**, and mark each with the name of an antacid to be tested.
3. To each of the four bags, add 5 mL of **white vinegar**, 10 mL of **water**, and enough **cabbage juice indicator** (probably 30 to 40 drops) to impart a distinct color.
4. Add the appropriate **antacid tablet** to each bag, squeeze out the excess air, and zip the bag closed.
5. Squeeze the antacid tablets to break them into small pieces. Record your observations. When the reactions have stopped, record the colors and approximate pH values of the solutions. Consult the pH-color chart from the ChemLab on page 506.

Analysis

1. **Describe** the different ways in which the antacids reacted with the vinegar.
2. **Infer** which of the antacids contain carbonates. Explain your answer.
3. **Conclude** Which of the antacids do you think was most effective? Explain.

504 Chapter 14 • Acids, Bases, and pH

Everyday Chemistry

Balancing pH in Cosmetics

With so many shampoos available, you might find it hard to know which type is best for your hair. Advertisements for each type tell you that their shampoo has more to offer than any other shampoo. How can you know which one to choose?

Cosmetic chemists have many tools for determining the effect of their products on different hair types. Using one technique, developed by NASA, they place a hair under a microscope connected to a TV screen that is hooked up to a computer. The computer evaluates the hair before and after treatment with the shampoo. Working with different types of hair allows these chemists to determine the best treatment for each type of hair.

Shampoos and pH balance The clear, outer layer of a strand of hair is the cuticle, which consists of the protein keratin. The cells of the cuticle are arranged like overlapping shingles. Shampoos that have a high pH make the entire hair shaft swell and push the cells of the cuticle away from the rest of the shaft. People with coarse, curly hair sometimes choose to use high-pH shampoos. These products expand and relax the hair, making it softer and less curly. However, the pH must be carefully balanced. As you have learned, some bases can dissolve hair. The harsh alkaline substances in solutions for permanents and hair coloring dissolve some of the cuticle, damaging the hair. Damaged hair is dull and dry.

In contrast, acidic substances in shampoos of low pH contract or tighten the hair shaft causing the cells of the cuticle to lie flat. The tightening of the hair shaft locks in moisture and smoothes the hair. Low-pH shampoos help restore damaged hair to its original condition and make it shine again. They also strengthen the keratin and increase the flexibility and elasticity of the hair.

Figure 1 pH balanced products

Why balance pH in skin products? The outer layer of skin has a keratin structure just as hair does. Products aimed at making the skin look brighter and clearer, such as some of the products in **Figure 1,** have a higher pH. Their purpose is to remove the top layer of keratin, which may consist of dead cells. The new cells underneath look fresh and vibrant. Occasional use of these products may be helpful, but regular use damages healthy skin by removing too many layers of cells.

Another problem with basic skin products is related to an acid mantle that bathes the top layer of the skin, the epidermis. This fluid—composed of oil, sweat, and other cell secretions—is a natural defense against bacterial infections. Strongly basic soaps can neutralize the protective acid mantle. People with acne or oily skin must be careful not to remove the acid mantle.

Explore Further

1. **Infer** If you live in an area where the water is hard, your hair might look dull. Why would rinsing with water to which lemon juice has been added help?

2. **Think Critically** Why would a person with acne use skin products that are pH-neutral or mildly acidic?

CHEMLAB

SMALL SCALE

HOUSEHOLD ACIDS AND BASES

Background

Indicators often are used to determine the approximate pH of solutions. In this ChemLab, you will make an indicator from red cabbage and use the indicator to determine the approximate pH values of various household liquids. The cabbage juice indicator contains a molecule called anthocyanin that accounts for the color changes.

Question

What are the approximate pH values of various household liquids?

Objectives

- **Measure and compare** the pH values for various household liquids.
- **Compare** the functions of the liquids to their chemical makeups.

Preparation

Materials

red cabbage
hot plate
beaker tongs
100-mL beakers (2)
distilled water
microtip pipettes (9)
96-well microplate
piece of white paper
100-mL graduated cylinder
toothpicks
solutions of:
  eyewash
  lemon juice
  white vinegar
  table salt
  soap
  baking soda
  borax
  drain cleaner

Safety Precautions

WARNING: *Use beaker tongs to handle hot beakers. Some of the solutions to be tested are caustic, especially the drain cleaner. Avoid all contact with skin and eyes. If contact occurs, immediately wash with large amounts of water and notify the teacher.*

Procedure

1. Read and complete the lab safety form.
2. Tear a red cabbage leaf into small pieces, and layer the pieces in a 100-mL beaker to a depth of about 2 cm. Add about 30 mL of distilled water.
3. Set the beaker on a hot plate, and heat until the water has boiled and become a deep purple color. Remove the beaker from the hot plate using beaker tongs, and allow it to cool. Pour off the cabbage juice indicator liquid into a clean beaker.
4. Set a clean microplate on a piece of white paper. Use the pipettes to add 5 drops of eyewash to well H1, lemon juice to H2, white vinegar to H3, and solutions of table salt to H4, soap to H5, baking soda to H6, borax to H7, and drain cleaner to H8. Use a clean pipette for each solution.
5. Draw the cabbage juice indicator solution into a clean pipette, and add 5 drops to each of the solutions in wells H1–H8. Stir the solution in each well with a clean toothpick.
6. Looking down through the wells, note and record the color of each solution in a data table such as the one shown. Using the color chart below, record in the data table the approximate pH of each of the solutions.

Indicator color	Relative pH
bright red	strong acid
red	medium acid
reddish purple	weak acid
purple	neutral
blue green	weak base
green	medium base
yellow	strong base

506 Chapter 14 • Acids, Bases, and pH

Analyze and Conclude

1. **Interpret** Are food items such as lemon juice or vinegar acidic or basic? These solutions are either tart or sour, so what ion probably accounts for this characteristic?
2. **Interpret** Were the cleaning solutions acidic or basic? What ion is probably involved in the cleaning process?
3. **Observe and Infer** How can you account for the great pH difference between lemon juice (citric acid solution) and eyewash (boric acid solution)?
4. **Use Variables, Constants, and Controls** Suppose that, in addition to the solutions, you tested a well containing pure distilled water. What purpose would this test serve?

Apply and Assess

1. **Explain** Would your indicator work well to determine the pH of ketchup? Explain.
2. **Apply** You might have noted that some shampoos are described as pH-balanced. What do the manufacturers mean by this phrase? Why would they do this to a soap or detergent?

INQUIRY EXTENSION
Predict how other solutions at home would react with the cabbage juice indicator. Explain your predictions.

Data and Observations

Data Table		
Solution	Color	Approximate pH
Eyewash		
Lemon juice		
White vinegar		
Table salt		
Soap		
Baking soda		
Borax		
Drain cleaner		

■ **Figure 14.25** As the concentration of carbon dioxide increases in the atmosphere, more of it dissolves in the oceans, making the oceans more acidic. Some of the ocean's organisms, such as corals, have difficulty forming skeletons and surviving in conditions of increased acidity.

Connecting Ideas

You may have noticed that this chapter, which introduces the properties of acids and bases, follows a chapter that focuses on the properties of water. Although acids and bases have different properties, the common link between the two is the role that water plays in making the chemistry of acids and bases happen. Many of the reactions that are most important for sustaining life take place in aqueous solutions, whether inside of cells or free in the ocean. **Figure 14.25** illustrates the importance of ocean pH to coral reefs.

Except for those cases where ionic hydroxides dissolve in water to form free hydroxide ions, the behavior and strength of acids and bases are due to their ability to cause hydrogen ions to move to water molecules from an acid or from water molecules to a base. This movement of hydrogen ions between particles in solution can be used to demonstrate different types of acid-base reactions.

Section 14.2 Assessment

Section Summary

- Most acids and bases are weak. Only a small percentage of their molecules dissociate to form ions.
- The pH scale is a convenient way to compare the acidity and basicity of solutions.
- Acidic solutions have a pH less than 7, basic solutions have a pH greater than 7, and neutral solutions have a pH of exactly 7.

8. **MAIN Idea** **Describe** How are the concentrations of hydronium ions and hydroxide ions in aqueous solution related? Describe what happens to one as the other increases or decreases.

9. **Classify** the following acids and bases as strong or weak: NH_3, KOH, HBr, $HCHO_2$, HNO_2, $Ca(OH)_2$.

10. **Identify** Other than water molecules, what particle has the highest concentration in an aqueous solution of ammonia? What has the lowest concentration? Why?

11. **Interpret Data** The pH of normal rain is about 5.5 due to dissolved CO_2. Consider a sample of rainfall with a pH of 3.5. How does the hydronium ion concentration in these two rain samples compare?

12. **Estimate** A solution of unknown pH gives a pink color with phenolphthalein indicator and a blue-gray color with universal indicator. Use **Figure 14.24** to estimate the pH of the unknown solution.

CHAPTER 14 Study Guide

BIG Idea Acids and bases can be defined in terms of hydrogen ions and hydroxide ions.

Section 14.1 Acids and Bases

MAIN Idea Acids produce hydronium ions (H_3O^+) in solution; bases produce hydroxide ions (OH^-).

Vocabulary
- acid (p. 483)
- acidic anhydride (p. 493)
- acidic hydrogen (p. 484)
- base (p. 489)
- basic anhydride (p. 493)
- hydronium ion (p. 483)
- ionization (p. 488)

Key Concepts
- The concentrations of hydrogen (H^+) ions and hydroxide ions (OH^-) determine whether an aqueous solution is acidic, basic, or neutral.
- Acidic anhydrides are nonmetallic oxides that react with water to form acids. Basic anhydrides are metallic oxides that react with water to form bases.

Section 14.2 Strengths of Acids and Bases

MAIN Idea In solution, strong acids and bases ionize completely, but weak acids and bases ionize only partially.

Vocabulary
- pH (p. 501)
- strong acid (p. 498)
- strong base (p. 497)
- weak acid (p. 498)
- weak base (p. 500)

Key Concepts
- Most acids and bases are weak. Only a small percentage of their molecules dissociate to form ions.
- The pH scale is a convenient way to compare the acidity and basicity of solutions.
- Acidic solutions have a pH less than 7, basic solutions have a pH greater than 7, and neutral solutions have a pH of exactly 7.

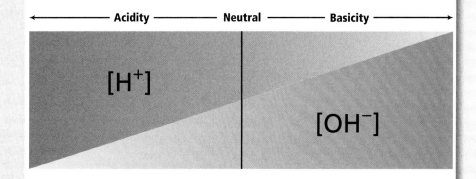

Chapter 14 Assessment

Understand Concepts

13. Show that magnesium hydroxide is a base when dissolved in water.

14. Predict if the following substances will cause an increase, decrease, or no change in pH when added to water.
    a) NaOH
    b) $H_2C_2O_4$
    c) NH_3
    d) CO_2
    e) CH_4
    f) Fe_2O_3

15. You test several solutions and find that they have pH values of 7.6, 9.8, 4.5, 2.3, 4.0, and 11.6. Which solution is the closest to being neutral? Assuming the concentrations are the same, which solution is the weakest base?

16. Write the chemical equation that shows the ionization of the acid HBr in water.

17. Write balanced equations that show what happens when NH_3 and KOH are placed separately in water. How do these reactions differ?

18. Explain the role of water in the ionization of an acid. Use the general equation to illustrate your explanation.

19. Explain the role of water in the ionization of a covalent base. Use the general equation for the reaction to illustrate your explanation.

20. Citric acid ($H_3C_6H_5O_7$) and vitamin C (ascorbic acid, $H_2C_6H_6O_6$), are polyprotic acids found in citrus fruits. Why are these acids classified as polyprotic acids? Write balanced equations that show the stepwise ionization of these acids in water.

21. Write the balanced equation for the reaction of nitric acid with sodium carbonate.

22. What product forms when sulfur trioxide reacts with water? Is the product an acid or is it a base?

23. Write the formula of each acid and identify each as a diprotic, a triprotic, or a monoprotic acid.
    a) sulfuric acid
    b) perchloric acid
    c) phosphoric acid
    d) hydrofluoric acid
    e) acetic acid

24. Give the formula for each of the following, and identify each as a strong base or a weak base.
    a) ammonia
    b) lithium hydroxide
    c) calcium hydroxide
    d) barium hydroxide

25. Dimethylamine is the strongest base of the three methyl derivatives of ammonia and trimethylamine is the weakest. How would the pH values of equal concentration solutions of these bases compare?

26. Why are soluble ionic hydroxides always strong bases?

27. What factor differentiates strong acids from weak acids?

28. Indicate whether each of the following pH values represents an acidic solution, a neutral solution, or a basic solution.
    a) 3.5
    b) 11.5
    c) 8.8
    d) 7.0

29. What are the hydronium ion concentrations of each of the solutions in the previous question?

Apply Concepts

30. Natural rainfall has a pH of approximately 5.5. Write the balanced equation that shows why natural rainfall is acidic.

31. How many times greater is the hydronium ion concentration in natural rainfall (pH = 5.5) compared to pure water (pH = 7.0)?

32. A $0.10M$ solution of dihydrogen sulfide (H_2S) has a pH of 4.0. What percent of the original H_2S molecules react with water to form hydronium ions?

33. Blood has a pH of 7.4. Milk of magnesia, a common antacid for upset stomachs, has a pH of 10.4. What is the difference in pH between the two bases? Compare the hydroxide ion concentrations of the two solutions.

34. An aqueous ammonia solution is sometimes called ammonium hydroxide. Why?

Chapter 14 Assessment

35. Why does acid rain dissolve marble statues? Write the reaction that occurs when rain containing sulfuric acid reacts with a statue made of marble ($CaCO_3$). Calcium sulfate is one of the products formed.

36. Finland, a country that has been hit hard by acid rain, has used lime (CaO) to try to return life to acidified lakes. Why was lime used?

Everyday Chemistry

37. Some people rinse their freshly shampooed hair in diluted lemon juice or vinegar. Why might doing this be beneficial to your hair?

Biology Connection

38. After exercise, muscles sometimes become sore because of a buildup of lactic acid. How might this buildup temporarily affect the acid-base balance in the blood?

Chemistry and Society

39. Why might city dwellers with respiratory diseases such as asthma or emphysema be advised to stay inside on hot days?

Chemistry and Technology

40. From what you know about reactions that acids will undergo, infer how sulfuric acid might be used to remove unwanted materials from ores.

Think Critically

Apply Concepts

41. What is the molarity of HCl, H_3O^+, and Cl^- in a 0.50M solution of HCl?

Relate Cause and Effect

42. Baking powder, when mixed with water, produces bubbles that make a cake rise. From what you know about the properties of acids, infer what types of compounds are contained in baking powder and what gas is contained in the bubbles released by the reaction.

Apply Concepts

43. Hard water deposits around sinks may be composed of calcium carbonate and magnesium carbonate. You can buy commercial cleaners to remove these insoluble compounds, or you could use something from your kitchen. What might you try?

Interpret Data

44. Identify the first compound in the following reactions as an acid or a base.
   a) $C_5H_5N + H_2O \rightarrow C_5H_5NH^+ + OH^-$
   b) $HClO_3 + H_2O \rightarrow H_3O^+ + ClO_3^-$
   c) $HCHO_2 + H_2O \rightarrow H_3O^+ + CHO_2^-$
   d) $C_6H_5SH + H_2O \rightarrow H_3O^+ + C_6H_5S^-$

45. One model of acid/base interactions focuses on the transfer of hydrogen ions in chemical reactions. In this model, acids are the H^+ donors and bases are the H^+ acceptors. Use the reactions in the previous question to classify the reactants as acids or bases.

Interpret Chemical Structures

46. What is an acidic hydrogen? Draw the dot structure for acetic acid, and use it to explain why only one of the four hydrogens is acidic.

Observe and Infer

47. **ChemLab** What would you expect the pH of an aqueous solution of the salt calcium chloride to be?

Form a Hypothesis

48. **MiniLab 1** Small pieces of zinc must be added to the acids instead of powdered zinc. Hypothesize why powdered zinc can't be used safely.

Observe and Infer

49. **MiniLab 2** Examine antacid ingredient lists. Some antacid tablets contain ingredients other than the antacid itself. What are some of these ingredients? Infer why the presence of these ingredients might affect the amount of antacid contained in an antacid tablet.

Chapter 14 Assessment

Cumulative Review

50. Write a balanced equation for each of the following reactions involving acids and bases, and classify each reaction as one of the five general reaction types. *(Chapter 6)*
    a) the reaction of water and lime to form calcium hydroxide
    b) the reaction of sulfuric acid with zinc
    c) the reaction of nitric acid with aqueous potassium hydroxide

51. What is the molar mass of each of the following substances? *(Chapter 12)*
    a) calcium carbonate   b) chlorine gas
    c) ammonium sulfate

52. How many grams of sodium hydroxide are needed to prepare 2.50 L of 2.00M NaOH? *(Chapter 13)*

Skill Review

53. **Make and Use Tables** Complete each row of the **Table 14.5** using the information provided. For some cells, there is more than one possible correct answer. Assume that all substances all have a molarity of 0.10M.

WRITING in Chemistry

54. Find out about how sulfuric acid is made and its many commercial uses. Write an advertisement for sulfuric acid, including any research information you found out. Include graphics that illustrate things such as uses of the acid or a graph that indicates the amount of production by different countries.

Problem Solving

55. A student buys some distilled water. She measures the pH of the water and finds it to be 6.0. She then boils the water, fills a container to the top with the hot water, and puts a lid on the container. When the water is at room temperature, the pH is 7.0. The student pours half the water into one container and stirs it for five minutes. She blows air through a straw into the other sample of water. She measures the pH of the samples of water and finds it to be 6.0 in both samples. Write an entry in your Science Journal that explains this series of pH measurements.

Table 14.5 Properties of Acids and Bases

Name of Substance	Formula of Substance	Acid, Base, or Neutral	Strong or Weak Electrolyte	pH: >7, <7, or 7	Litmus Color	Major Particle(s) in Solution
Acetic Acid						
	HCl					
		Acid	Strong			
Ammonia						
	LiOH					
		Base	Strong			
			Weak	<7		
		Neutral				Na^+, Cl^-
						H^+, NO_3^-
Phosphoric Acid						

Cumulative Standardized Test Practice

Use the table below to answer Questions 1–4.

Substances	H₃O⁺ concentration in M	pH
Eyewash	?	5
Apples	10^{-3}	?
Lemons	10^{-2}	?
Lye	10^{-14}	?
Household ammonia	?	12

1. What is the pH of apples?
   a) −3
   b) 3
   c) −10
   d) 10

2. The concentration of hydronium ions in eyewash is
   a) 3 times greater than the concentration of hydronium ions in lemons.
   b) 3 times less than the concentration of hydronium ions in lemons.
   c) 1,000 times greater than the concentration of hydronium ions in lemons.
   d) 1,000 times less than the concentration of hydronium ions in lemons.

3. The concentration of hydroxide ions in apples is
   a) $10^{-11} M$.
   b) $10^{11} M$.
   c) $10^{-3} M$.
   d) $10^{3} M$.

4. What is the H₃O⁺ ion concentration of household ammonia?
   a) $10^{-2} M$
   b) $10^{2} M$
   c) $10^{-12} M$
   d) $10^{12} M$

5. An increase in the temperature of a substance results in a(n)
   a) increase in its boiling point.
   b) increase in its rate of evaporation.
   c) decrease in its boiling point.
   d) decrease in its rate of evaporation.

6. When the temperature of the air inside a basketball is increased, the
   a) pressure of the air inside the ball increases.
   b) kinetic energy of the air particles decreases.
   c) the volume of the air increases.
   d) pressure of the air inside the ball decreases.

7. How many moles of helium occupy 26.4 L at a pressure of 103.0 kPa, and 10°C?
   a) 0.115 mol
   b) 0.865 mol
   c) 1.16 mol
   d) 32.7 mol

8. Why does water have a boiling point of 100°C?
   a) The heavier mass of water molecules prevents them from boiling at lower temperatures.
   b) The lighter mass of water molecules causes them to boil at lower temperatures.
   c) The hydrogen bonds between water molecules prevent them from boiling at lower temperatures.
   d) The hydrogen bonds between water molecules cause them to boil at lower temperatures.

9. Pure O₂ gas can be generated from the decomposition of potassium chlorate (KClO₃):

   $$2KClO_3(s) \rightarrow 2KCl(s) + 3O_2(g)$$

   If 100.0 g of KClO₃ are used and 12.8 g of oxygen gas are produced, the percent yield of the reaction is
   a) 12.8%.
   b) 32.7%.
   c) 65.6%
   d) 98.0%

NEED EXTRA HELP?

If You Missed Question ...	1	2	3	4	5	6	7	8	9
Review Section ...	14.2	14.2	14.2	14.2	10.2	11.1	12.2	13.1	12.2

CHAPTER 15
Acids and Bases React

BIG Idea The reaction of an acid with a base produces water and a salt.

15.1 Acid and Base Reactions
MAIN Idea The concentrations of the acid and the base determine the classification of an acid-base reaction.

15.2 Applications of Acid-Base Reactions
MAIN Idea Acid-base reactions play an important role in the human body.

ChemFacts

- Most of the human figures called moai date between 1250-1500 CE.
- The weathering process results in the breakdown of rock – and statues made of rock – due to contact with chemicals in Earth's atmosphere.
- Sulfur and nitrogen added to the atmosphere through human emissions results in acid precipitation.

Start-Up Activities

LAUNCH Lab

Buffers

What is the role of a buffer?

Materials
- plain aspirin and buffered aspirin
- mortar and pestle
- distilled water
- 250-mL beakers (2)
- magnetic stirring bars and stirrers (2)
- digital pH meter

Procedure
1. Read and complete the lab safety form.
2. Place a crushed aspirin tablet in 100 mL of distilled water in a 250-mL beaker.
3. Place a crushed, buffered aspirin tablet in 100 mL of distilled water in a 250-mL beaker.
4. Put magnetic stirring bars in both beakers and set them on magnetic stirrers.
5. With the stirrers set on low, place digital pH meter probes in both solutions.
6. Record the initial pH of each solution, the pH after 5 min, and the pH after 10 min.

Analysis
1. **Compare** the results you've obtained for the two types of aspirin.
2. **Infer** Aspirin is an acid, which means it will lower the pH as it dissolves in water. Based on your results, what do you think the role of the buffer in the buffered aspirin is?

Inquiry Which type of aspirin is best for a person who has ulcers or is prone to acid indigestion?

 **Acid-Base Reactions** Make the following Foldable to organize information about how acids and bases react.

STEP 1 Fold a horizontal sheet of paper into thirds.

STEP 2 Unfold the paper and fold it into thirds lengthwise. Unfold and draw lines along the folds.

STEP 3 Label the columns as follows: *Strong Base* and *Weak Base*. Label the rows *Strong Acid* and *Weak Acid*.

FOLDABLES Use this Foldable with Section 15.1. As you read the section, summarize information about each type of reaction in the appropriate box on the chart.

Chemistry Online

Visit glencoe.com to:
▶ study the entire chapter online
▶ explore **concepts in Motion**
▶ take Self-Check Quizzes
▶ use Personal Tutors
▶ access Web Links for more information, projects, and activities
▶ find the Try at Home Lab, Testing for Ammonia

Section 15.1

Objectives
- **Determine** the overall, ionic, and net ionic equations for an acid-base reaction.
- **Classify** acids and bases using the hydrogen transfer definition.
- **Predict and explain** the final results of an acid-base reaction.

Review Vocabulary
pH: mathematical scale in which the concentration of hydrogen ions in a solution is expressed as a number from 0 to 14

New Vocabulary
neutralization reaction
salt
ionic equation
spectator ion
net ionic equation
Brønsted-Lowry model

Acid and Base Reactions

MAIN Idea The concentrations of the acid and the base determine the classification of an acid-base reaction.

Real-World Reading Link When you watch a debate between two political candidates, you might end up agreeing with one candidate or the other. However, other times your opinion might be evenly split between the two. Acid-base reactions are similar: Sometimes the result favors the acid, sometimes the base; sometimes, it is completely neutral.

Types of Acid-Base Reactions

The reaction of an acid and a base is called a neutralization reaction because the properties of both the acid and base are diminished or neutralized when they react. A **neutralization reaction** is a reaction of an acid with a base in aqueous solution to produce water and a salt, as shown by the following equation:

$$\text{acid} + \text{base} \rightarrow \text{salt} + \text{water}$$

Salt is a general term used in chemistry to describe the ionic compound formed from the negative part of the acid and the positive part of the base. In the language of chemistry, sodium chloride, common table salt, is just one of a large number of ionic compounds that are called salts. Potassium chloride (KCl), ammonium nitrate (NH_4NO_3), and iron (II) phosphate ($Fe_3(PO_4)_2$) are other examples of salts.

Consider the following neutralization reaction. Hydrochloric acid, HCl, is a common household and laboratory acid. Muriatic acid is the common household name of hydrochloric acid. It is often sold in hardware stores to be used in masonry work to remove excess mortar from brick. Sodium hydroxide (NaOH) is a common household and laboratory base. The common name of sodium hydroxide is lye. It is the primary component of many drain cleaners. **Figure 15.1** shows litmus tests before and after mixing together sodium hydroxide and hydrochloric acid.

■ **Figure 15.1** A solution of hydrochloric acid (HCl) is added to exactly the same amount of a solution of basic sodium hydroxide (NaOH). The solutions react. Litmus papers show that the resulting salt solution is neither acidic nor basic.

$NaOH(aq) + HCl(aq) \rightarrow NaCl(aq) + H_2O(l)$

Table 15.1 **Types of Acid-Base Reactions**

Acid	Base
Strong	Strong
Strong	Weak
Weak	Strong
Weak	Weak

After the reaction, the mixture contains only the salt sodium chloride (NaCl), dissolved in water. The litmus test shows no acid or base present in the reaction products.

Because both acids and bases may be either strong or weak, four possible combinations of acid-base reactions occur. Those possibe combinations are summarized in **Table 15.1**. As you will find out later in this section, only three of the types are significant in everyday chemistry.

As long as one of the reactants is strong, the acid-base reaction goes to completion. As you learned in Chapter 6, a reaction goes to completion when the limiting reactant is completely consumed.

Although all of the reactions in **Table 15.1** are acid-base reactions, the submicroscopic interactions in each are different. Examine each possible type of acid-base reaction and see how they compare.

Strong Acid + Strong Base

A typical type of acid-base reaction is one in which both the acid and base are strong. The reaction of aqueous solutions of hydrochloric acid (HCl) and sodium hydroxide (NaOH) shown in **Figure 15.1** is a good example of this type of reaction.

A macroscopic view It is easy to write and balance equations for strong acid–strong base reactions. In **Figure 15.1,** HCl is the acid; NaOH is the base. The products are NaCl, which is a salt, and water. Now take a closer look at these reactants and their products.

The submicroscopic view: ionic equations Recall from Chapter 14 that HCl, when dissolved in water, completely ionizes into hydronium ions and chloride ions because HCl is a strong acid. The hydronium ion is more conveniently written in shorthand as H^+.

$$HCl(aq) \rightarrow H^+(aq) + Cl^-(aq)$$

You also know that sodium hydroxide in water completely dissociates into sodium ions and hydroxide ions because NaOH is a strong base.

$$NaOH(aq) \rightarrow Na^+(aq) + OH^-(aq)$$

FOLDABLES
Incorporate information from this section into your Foldable.

■ **Figure 15.2** When a conductivity apparatus is placed in solutions of each reactant and product of the reaction between a strong acid and a strong base, the lightbulb lights up because the acid, base, and salt exist as ions in solution.

An overall equation for the reaction between NaOH and HCl shows each substance involved in the reaction.

$$NaOH(aq) + HCl(aq) \rightarrow NaCl(aq) + H_2O(l)$$

An overall equation does not indicate whether these substances exist as ions in the solution. The best way for you to model the submicroscopic behavior of an acid-base reaction is to show reactants and products as they actually exist in solution. Instead of an overall equation, an **ionic equation**, in which substances that primarily exist as ions in solution are shown as ions, can be written.

$$H^+(aq) + Cl^-(aq) + Na^+(aq) + OH^-(aq) \rightarrow Na^+(aq) + Cl^-(aq) + H_2O(l)$$

Notice in the equation above that, in addition to showing the acid as completely ionized and the base as completely dissociated, the ionic compound NaCl is also dissociated. Water does not ionize much, so it is indicated as a molecule rather than H^+ and OH^- ions. **Figure 15.2** illustrates the ionic aspect of this equation.

MiniLab 15.1

Acidic, Basic, or Neutral?

How can you determine whether a solution is acidic, basic, or neutral? Test several aqueous salt solutions with bromthymol blue indicator to determine whether the solutions are acidic (yellow), basic (blue), or neutral (green).

Procedure

1. Read and complete the lab safety form.
2. Use labeled **microtip pipettes** to put six drops of **sodium acetate** solution in A1, **potassium nitrate** solution in A2, **ammonium chloride** solution in A3, **sodium carbonate** solution in A4, **sodium chloride** solution in A5, and **aluminum sulfate** solution in A6 of a **24-well microplate**.
3. Add two drops of **bromthymol blue indicator** solution to each of the salt solutions. Stir each with a separate **toothpick**.
4. Set the microplate on a piece of **white paper** and look down through each well to determine the color of the solution. Record your results.

Analysis

1. **Summarize** According to the color of each solution that has bromthymol blue indicator added, is each of the solutions acidic, basic, or neutral?
2. **Relate** the relative strengths of the parent acids and bases to the results of using the indicator with the salt solutions.
3. **Infer** What type of salt might be added to a product, such as shampoo, in order to make it slightly acidic so that it will not be harmful to skin and hair? Check the label of several brands of shampoo for the presence of such a salt.

How It Works

Taste

Although the total flavor of a food comes from the complex combination of taste, smell, touch, texture or consistency, and temperature sensations, taste is a major factor. Three of the four fundamental tastes are directly linked to acids and bases. Your tongue has four different types of taste buds—sweet, salty, bitter, and sour—that are located at different places on your tongue. Only certain molecules and ions can react with these specific buds to produce a signal that is sent to a certain region of your brain. When these signals are received, your brain processes them, and you sense taste.

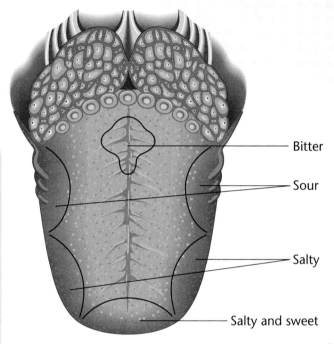

1 The taste buds that sense a bitter taste are located at the base of the tongue. Bases taste bitter. Many medications are basic, and pharmaceutical companies spend a lot of time and research trying to mask the bitter taste with other tastes.

3 The taste buds that detect the salty and sweet tastes are located at the tip of your tongue. A salt is the product of an acid-base reaction. The sweet taste seems to depend a great deal on the properties of both acids and bases that are combined on a single molecule, but this taste is not as clear-cut as the other tastes.

2 The taste buds that detect a sour taste are along the sides of the tongue. The sour taste comes from acids in your food. Sour-tasting foods include vinegar and citrus fruits.

Sweetness of Some Compounds

Compound	Relative Sweetness
Lactose	16
Glucose	74
Sucrose	100
Fructose	173
Aspartame	16,000
Saccharin	50,000

As the table shows, aspartame, the popular artificial sweetener in soft drinks, is 160 times sweeter than sucrose, common table sugar. The sweetening ability of aspartame comes from a molecular structure that creates an impact 160 times that of sucrose on your taste buds that detect the sweet taste.

Think Critically

1. How are three of the four fundamental tastes linked to the properties of acids and bases?
2. What types of companies do you think would support research on taste?
3. What benefits might artificial sweeteners, such as aspartame, have over natural sweeteners?

■ **Figure 15.3** The presence of spectators at a sporting event is important, but spectators do not actually participate in the game and do not determine the final outcome.

Spectator ions and the net ionic equations Note that the ionic equation gives more information about what happens in a strong acid-strong base reaction. When you examine the two sides of the ionic equation on page 518, you see that Na^+ and Cl^- are present both as reactants and as products. Although they are important components of an overall equation, they do not directly participate in the chemical reaction. They are called spectator ions. **Spectator ions** are present in the solution but do not participate in the reaction just as fans or spectators do not actually participate in a game, as shown in **Figure 15.3**.

Simplifying the overall reaction You might ask why include the spectator ions in an equation if they aren't really involved in the reaction? An ionic equation can be simplified to take care of that problem. Just as in a mathematical equation, items common to both sides of the equation can be subtracted. This process simplifies the equation so that the reactants and products that actually change can be seen more clearly.

$$H^+(aq) + \cancel{Cl^-(aq)} + \cancel{Na^+(aq)} + OH^-(aq) \rightarrow$$
$$\cancel{Na^+(aq)} + \cancel{Cl^-(aq)} + H_2O(l)$$

When ions common to both sides of the equation are removed from the equation, the result is called the **net ionic equation**. For the reaction of HCl with NaOH, it looks as follows:

$$H^+(aq) + OH^-(aq) \rightarrow H_2O(l)$$

The net ionic equation describes what is really happening at the submicroscopic level. Although solutions of HCl and NaOH are mixed, the net ionic equation is hydrogen ions reacting with hydroxide ions to form water.

Even though the strong acid and strong base in Example Problem 1 are different from those in the HCl reaction with NaOH, the net ionic equation is the same. Hydrogen ions from the acid react with hydroxide ions from the base to form water. This equation is always the net ionic equation for a strong acid-strong base reaction.

EXAMPLE Problem 15.1

Equations—Strong Acid, Strong Base
Write the overall, ionic, and net ionic equations for the reaction of sulfuric acid with potassium hydroxide.

> **Problem-Solving Hint**
> Coefficients should be in the smallest whole-number ratio possible.

1 Analyze
Decide whether the acid is a strong acid or a weak acid and whether the base is strong or weak. A list of strong acids and bases can be found in Chapter 14, Table 14.3. By looking at this table, you can see that sulfuric acid (H_2SO_4) is a strong acid. Potassium hydroxide (KOH) is a strong base.

2 Set up
Write an equation for the overall reaction. Because sulfuric acid is a diprotic acid, you need two moles of KOH for every one mole of H_2SO_4. Two moles of water and one mole of K_2SO_4 will be produced.

$$H_2SO_4(aq) + 2KOH(aq) \rightarrow K_2SO_4(aq) + 2H_2O(l)$$

Write the ionic equation by showing H_2SO_4, KOH, and K_2SO_4 as ions. You must keep track of the coefficients from the overall equation and the formulas of the substances when writing the coefficients of the ions.

$$2H^+(aq) + SO_4^{2-}(aq) + 2K^+(aq) + 2OH^-(aq) \rightarrow$$
$$2K^+(aq) + SO_4^{2-}(aq) + 2H_2O(l)$$

3 Solve
Look for spectator ions. In this reaction, K^+ and SO_4^{2-} are spectator ions. Subtract them from both sides of the equation to get the net ionic equation.

$$2H^+(aq) + \cancel{SO_4^{2-}(aq)} + \cancel{2K^+(aq)} + 2OH^-(aq) \rightarrow$$
$$\cancel{2K^+(aq)} + \cancel{SO_4^{2-}(aq)} + 2H_2O(l)$$

$$2H^+(aq) + 2OH^-(aq) \rightarrow 2H_2O(l)$$

Simplify the balanced net reaction by dividing coefficients on both sides of the equation by the common factor of 2.

$$H^+(aq) + OH^-(aq) \rightarrow H_2O(l)$$

4 Check
Take a final look at the net ionic equation to make sure no ions are common to both sides of the equation.

PRACTICE Problems
Solutions to Problems Page 864

Write overall, ionic, and net ionic equations for each of the following reactions.

1. hydroiodic acid (HI) and calcium hydroxide ($Ca(OH)_2$)
2. hydrobromic acid (HBr) and lithium hydroxide (LiOH)
3. sulfuric acid (H_2SO_4) and strontium hydroxide ($Sr(OH)_2$)
4. perchloric acid ($HClO_4$) and barium hydroxide ($Ba(OH)_2$)

SUPPLEMENTAL PRACTICE

For more practice writing net ionic equations, see Supplemental Practice, page 831.

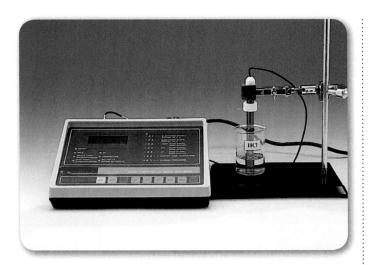

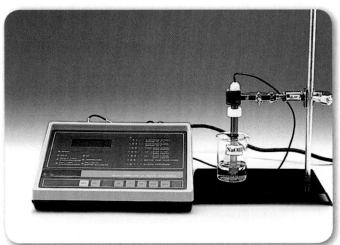

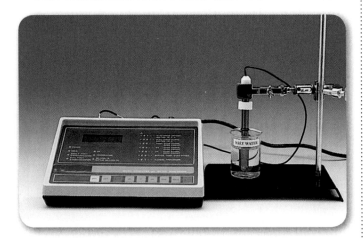

■ **Figure 15.4** The reaction of a strong acid and a strong base is definitely a neutralization. The pH of a 0.100M HCl solution (top) is 1. The pH of a 0.100M NaOH solution (middle) is close to 13.
Calculate *What is the pH of the resulting solution (bottom) when the reaction is complete?*

FOLDABLES
Incorporate information from this section into your Foldable.

Strong acid, strong base, and pH The net ionic equation also shows why the reaction of an acid and a base is called a neutralization reaction. The hydrogen ion from the acid reacts with the hydroxide ion from the base to form water, which has a neutral pH.

Figure 15.4 shows the reaction of 50.0 mL of 0.100M HCl with 50.0 mL of 0.100M NaOH, which is exactly the right amount of base to react with all of the acid. As the NaOH solution is added to the HCl solution, the pH of the solution increases. After all of the NaOH solution is added, the pH of the final solution is 7.

Strong Acid + Weak Base

Do the reactions change when the strength of an acid or base changes? Look at an example of what happens when a strong acid—hydrobromic acid (HBr)—and a weak base—aluminum hydroxide ($Al(OH)_3$)—are mixed together. Hydrobromic acid and aluminum hydroxide are the reactants and aluminum bromide and water are the products of the reaction. The overall equation is shown here.

$$3HBr(aq) + Al(OH)_3(s) \rightarrow AlBr_3(aq) + 3H_2O(l)$$

Hydrobromic acid, a strong acid, completely ionizes in water. All of the $Al(OH)_3$ that dissolves dissociates, so it is technically a strong base. However, because it is so insoluble, few OH^- ions are produced, and $Al(OH)_3$ acts as a weak base. Therefore, the ionic equation shows little dissociation of the base. The dissociated salt, $AlBr_3$, is also shown as ions.

$$3H^+(aq) + 3Br^-(aq) + Al(OH)_3(s) \rightarrow$$
$$Al^{3+}(aq) + 3Br^- + 3H_2O(l)$$

The spectator ions in this equation are bromide ions. They are removed from both sides of the equation to produce the following net ionic equation.

$$3H^+(aq) + Al(OH)_3(s) \rightarrow Al^{3+}(aq) + 3H_2O(l)$$

Compare this equation to the net ionic equation for a strong acid–strong base reaction. How are they the same and how are they different? When a strong acid reacts with a weak base, the base does not completely dissociate. Aluminum ions from the base dissolve in water, rather than water being the primary product of the reaction.

522 Chapter 15 • Acids and Bases React

A strong acid + ammonia Recall from Chapter 14 that the most common weak base does not contain the hydroxide ion. Consider an equation for the reaction between hydrochloric acid (HCl) and ammonia (NH_3).

$$HCl(aq) + NH_3(aq) \rightarrow NH_4Cl(aq)$$

Notice that although the product is a salt (NH_4Cl), no water is produced in this overall reaction. As before, use the net ionic equation to understand the submicroscopic processes for this reaction of a weak base with a strong acid.

The net ionic equation Recall from Chapter 14 that when ammonia dissolves in water, some of the ammonia molecules react with water to form ammonium ions (NH_4^+) and hydroxide ions (OH^-). However, most of the ammonia molecules (NH_3) remain as they are. Other than water, the major particle present in an aqueous solution of ammonia is ammonia molecules (NH_3).

A solution of ammonia is best represented by $NH_3(aq)$. A solution of HCl is best represented as $H^+(aq)$ and $Cl^-(aq)$. The ionic reaction is written by representing what is actually in the reactant and product solutions.

$$H^+(aq) + Cl^-(aq) + NH_3(aq) \rightarrow NH_4^+(aq) + Cl^-(aq)$$

The ionic salt NH_4Cl in aqueous solution is written as dissociated ions. **Figure 15.5** shows which of these particles are involved in the reaction.

For these examples of strong acid–weak base reactions, the net ionic equation differs from that for a strong acid and a strong base. The submicroscopic interactions in these strong acid–weak base reactions are between hydrogen ions and the bases.

FACT of the Matter
In medieval times, natural deposits of ammonium chloride, a compound derived from ammonia, were first mined in Egypt near a temple of the Egyptian god Ammon.

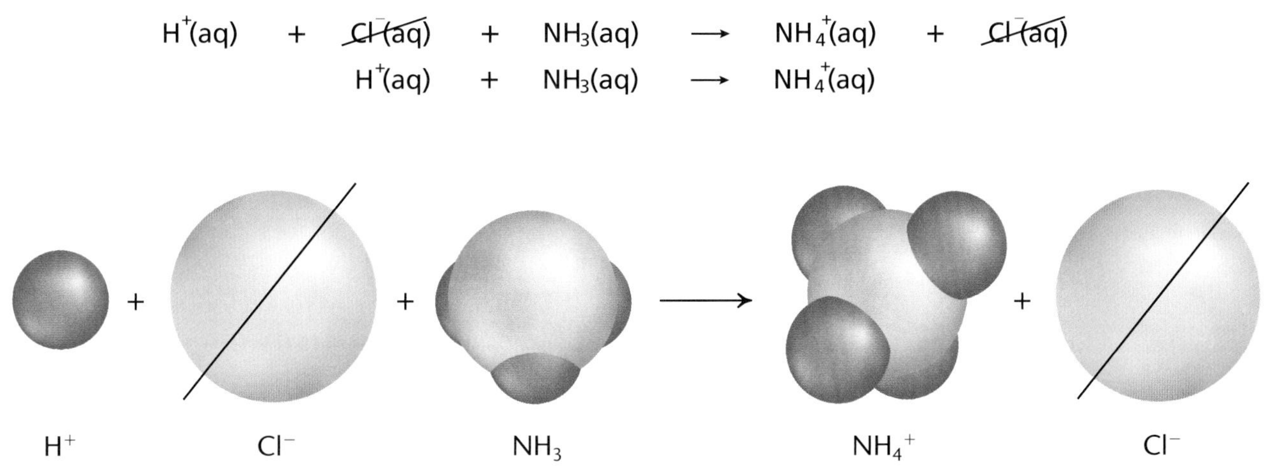

■ **Figure 15.5** A quick look at the ionic reaction shows that the chloride ion is a spectator ion because it appears on both sides of the reaction. You get the net ionic equation by subtracting the spectator Cl^- from both sides of the ionic equation.

$$H^+(aq) + \cancel{Cl^-(aq)} + NH_3(aq) \rightarrow NH_4^+(aq) + \cancel{Cl^-(aq)}$$
$$H^+(aq) + NH_3(aq) \rightarrow NH_4^+(aq)$$

Section 15.1 • Acid and Base Reactions

EXAMPLE Problem 15.2

Equations—Strong Acid–Weak Base Write the overall, ionic, and net ionic equations for the reaction of nitric acid (HNO_3) with ammonia (NH_3).

1 Analyze
Decide whether the acid is strong or weak and whether the base is strong or weak. Nitric acid (HNO_3) is a strong acid. Ammonia (NH_3) is a weak base.

2 Set up
Write an equation for the overall reaction.

$$HNO_3(aq) + NH_3(aq) \rightarrow NH_4NO_3(aq)$$

Write the ionic equation by showing HNO_3 and NH_4NO_3 as ions. Because HNO_3 is a strong acid, you write it as completely ionized. NH_3 is a weak base, so you write it as NH_3. You dissociate the salt, ammonium nitrate (NH_4NO_3), into its component ions because it is an ionic compound.

$$H^+(aq) + NO_3^-(aq) + NH_3(aq) \rightarrow NH_4^+(aq) + NO_3^-(aq)$$

3 Solve
Look for spectator ions. In this reaction, only the nitrate ion (NO_3^-) is a spectator ion. Subtract NO_3^- from both sides of the equation to get the net ionic equation.

$$H^+(aq) + \cancel{NO_3^-(aq)} + NH_3(aq) \rightarrow NH_4^+(aq) + \cancel{NO_3^-(aq)}$$

$$H^+(aq) + NH_3(aq) \rightarrow NH_4^+(aq)$$

Note that this is the same net equation as in the HCl and NH_3 example shown in **Figure 15.5**.

4 Check
Take a final look at the equation to make sure no ions are common to both sides of the equation.

PRACTICE Problems   Solutions to Problems Page 865

SUPPLEMENTAL PRACTICE

For more practice writing overall, ionic, and net ionic equations, see Supplemental Practice, page 831.

Write overall, ionic, and net ionic equations for each of the following reactions.

5. perchloric acid ($HClO_4$) and ammonia (NH_3)

6. hydrochloric acid (HCl) and aluminum hydroxide ($Al(OH)_3$)

7. sulfuric acid (H_2SO_4) and iron(III) hydroxide ($Fe(OH)_3$)

Earth Science Connection

Cave Formation

When you think of a cave, what images come to mind? Possibly what comes to mind is a massive, damp, cool, underground chamber housed within Earth or a fantasy underground world with stone icicles rising from the floor and dangling from the ceiling. Whatever pictures you imagine, caves are one of nature's wonders.

How caves are formed Caves form in layers of limestone rock throughout the world. Limestone is calcium carbonate, which is only slightly soluble in water. The caves that form within these rocks are called solution caves.

What causes natural water to be acidic? Most rainwater is slightly acidic because it contains carbon dioxide (CO_2) from the atmosphere. A small amount of the carbon dioxide dissolves in the water, but some of it reacts with the water to form carbonic acid (H_2CO_3).

$$CO_2(g) + H_2O(l) \rightarrow H_2CO_3(aq)$$

Carbonic acid forms a hydronium ion (H_3O^+) and a hydrogen carbonate ion (CHO_3^-).

$$H_2CO_3(aq) + H_2O(l) \rightarrow$$
$$H_3O^+(aq) + HCO_3^-(aq)$$

How does this acid form caves? The hydronium ions react with limestone to produce soluble calcium ions (Ca^{2+}) and hydrogen carbonate ions (HCO_3^-).

$$H_3O^+(aq) + CaCO_3(s) \rightarrow$$
$$Ca^{2+}(aq) + HCO_3^-(aq) + H_2O(l)$$

The acidic water dissolves the limestone rocks, producing open spaces that contain water.

Stalactites and stalagmites Stalactites and stalagmites, shown in **Figure 1**, are deposits of calcium carbonate that form by the drip of calcareous water. Stalactites resemble icicles and form from the ceiling or sides of a cave. Stalagmites resemble inverted stalactites and form on the floor of a cave.

Figure 1 Stalactites and Stalagmites are deposits of calcium carbonate.

During the second phase, clay, silt, sand, or gravel moves into the spaces. In the third phase, streams partially remove these materials and modify and enlarge the spaces. Stalactites and stalagmites now form by the reverse of the chemical and physical processes that formed the cave. The water containing dissolved carbon dioxide and carbonic acid is saturated with calcium hydrogen carbonate. As it seeps through the roof of the cave, the water in each droplet slowly evaporates.

Some of the carbonic acid changes back into carbon dioxide and water. The pH of the water increases, and the solubility of calcium hydrogen carbonate decreases. The calcium carbonate precipitates out slowly, forming stalactites over thousands of years. As the saturated water drops hit the floor, the same processes slowly form stalagmites. Sometimes, the two formations grow together, forming pillars.

Connection to Chemistry

1. **Apply** Identify two physical changes and two chemical changes that occur in cave formation.
2. **Think Critically** Write the equations for cave formation from the $MgCO_3$ part of dolomite ($CaCO_3 \cdot MgCO_3$).

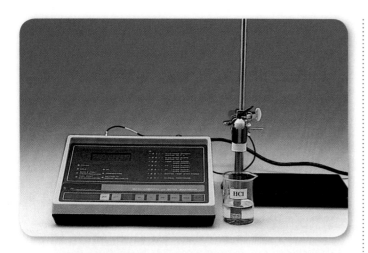
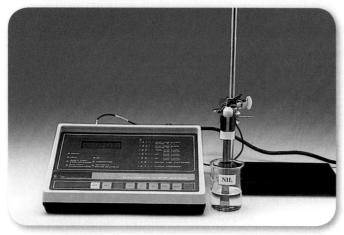
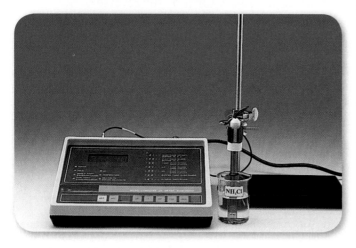

■ **Figure 15.6** The reaction of a strong acid and weak base is not quite a neutralization. The pH of a 0.100M HCl solution is 1 (top), and the pH of a 0.100M NH$_3$ solution is approximately 11 (middle).

Apply *What is the pH of the resulting solution (bottom) when equal volumes of the solutions are mixed and the reaction is complete?*

Strong acid, weak base, and pH

Figure 15.6 shows that a solution of 0.100M ammonia is definitely a base. It has a pH greater than 7. If you compare the pH of 0.100M NaOH in **Figure 15.4** with the pH of 0.100M NH$_3$, you see that ammonia is a weaker base because it has a lower pH.

Figure 15.6 also shows that when equal volumes of solutions of ammonia and hydrochloric acid of equal molarity are mixed, the pH of the final mixture is less than 7. A pH less than 7 means that the final reaction mixture is acidic. Therefore, more hydronium than hydroxide ions must be present in the final reaction mixture.

How can equal moles of base and acid react to produce a neutral solution in one reaction and an acidic solution in the second? The result must have something to do with the relative strengths of the acid and the base.

A Broader Definition of Acids and Bases

The reaction of a strong acid with a weak base demonstrates the need for a slightly broader definition of acids and bases. As you learned in the last chapter, much of the behavior of acids and bases in water can be explained by a model that focuses on the hydrogen ion transfer from the acid to the base. This model will also help explain why every acid-base reaction does not result in a neutral solution.

Hydrogen-ion donor or acceptor You can use the ability to exchange a hydrogen ion as the basis of a broader definition of an acid or a base. In this definition, called the **Brønsted-Lowry model** of acids and bases, an acid is defined as a substance that donates, or gives up, a hydrogen ion in a chemical reaction. A base, not surprisingly, is just the opposite. A base is a substance that accepts a hydrogen ion in a chemical reaction.

Take another look at the net ionic equation of hydrochloric acid (HCl) with ammonia (NH$_3$) on page 523. If you adhere strictly to the definition of a base as a hydroxide-ion producer in water, none of these equations define ammonia as a base. Remember that hydroxide ions are produced, but the amount is so small that it is not shown in the equations.

$$H^+(aq) + NH_3(aq) \rightarrow NH_4^+(aq)$$

A. In the transfer of H⁺ from HCl to a water molecule, HCl acts as an acid, and water acts as a base.

$$\overset{\frown}{HCl}(aq) + H_2O(l) \longrightarrow H_3O^+(aq) + Cl^-(aq)$$
$\quad\;\;$acid $\qquad\;\;$ base

B. In the transfer of H⁺ from a hydronium ion to an ammonia molecule, the hydronium ion acts as an acid, and the ammonia molecule acts as a base.

$$\overset{\frown}{H_3O^+}(aq) + NH_3(aq) \longrightarrow NH_4^+(aq) + H_2O(l)$$
$\quad\;\;$acid $\qquad\;\;\;$ base

C. Water also reacts with ammonia molecules. Water acts as an acid, and ammonia acts as a base.

$$\overset{\frown}{H_2O}(l) + NH_3(aq) \longrightarrow NH_4^+(aq) + OH^-(aq)$$
$\quad\;\;$acid $\qquad\;\;$ base

■ **Figure 15.7** In a reaction between aqueous HCl and aqueous NH₃, several H⁺ transfers occur.

TRY AT HOME 🏠 LAB
See page 875 for **Testing for Ammonia.**

Using the Brønsted-Lowry definition, you can definitely say that ammonia is acting as a base. The ammonia is accepting a hydrogen ion from the hydronium ion. **Figure 15.7B** shows this reaction written in its most complete form and clearly shows the hydrogen ion transfer. The hydronium ion (H_3O^+) is acting as the acid because it donates the hydrogen ion.

It takes two to transfer Notice from **Figure 15.7** that the definitions of an acid as a hydronium-ion producer and a base as a hydroxide-ion producer are included in this H^+ transfer definition. When HCl reacts with water, it acts as an H^+ donor, so it is an acid. Water acts as an H^+ acceptor, so it is a base (**Figure 15.7A**). When ammonia reacts with water, the ammonia molecule accepts H^+ from the water. Ammonia is the base and water is the acid (**Figure 15.7B**). Remember that it takes two substances to transfer, so for every acid (an H^+ donor), there must be a base (an H^+ acceptor).

Although you generally think of water as neutral, a unique property of water can now be observed. Water can act as either an acid or a base, depending on what else is in solution. In **Figure 15.7A,** the water acts as the base; in **Figure 15.7B,** the water acts as the acid.

Water not required Although most of the reactions that you will study occur in water, the H^+ transfer definition does not require water to be present. For example, the reaction between HCl(g) and NH₃(g) is shown in **Figure 15.8.** This reaction occurs in the gas phase. It involves the transfer of a hydrogen ion from a gaseous HCl molecule to a gaseous ammonia molecule to form a solid product. This gas reaction can now be classified as an acid-base reaction.

■ **Figure 15.8** HCl(g) and NH₃(g) react to form NH₄Cl(s). Gases from the concentrated aqueous solutions react to form ammonium chloride.

Weak Acid + Strong Base

Considering this new H+ transfer acid-base definition, take a look at the type of acid-base reaction in which the acid is weak and the base is strong. An example is the reaction of acetic acid ($HC_2H_3O_2$) the weak acid present in vinegar, with sodium hydroxide. The equation of the overall reaction is similar to that of a strong acid-strong base reaction.

$$HC_2H_3O_2(aq) + NaOH(aq) \rightarrow NaC_2H_3O_2(aq) + H_2O(l)$$

The ionic reaction: What's in solution?
As you know, acetic acid is a weak acid. In a solution of acetic acid, only a small fraction of the acetic acid molecules ionize. Other than water, the major particle present in an aqueous solution of acetic acid is $HC_2H_3O_2$. NaOH, as seen before, completely dissociates. The ionic equation shows what is present when the acid and base react. As in the previous cases, the salt is also written as dissociated ions (Na^+, $C_2H_3O_2^-$).

$$HC_2H_3O_2(aq) + Na^+(aq) + OH^-(aq) \rightarrow Na^+(aq) + C_2H_3O_2^-(aq) + H_2O(l)$$

Net ionic reaction: H+ transfer
The ionic equation contains the sodium ion as a spectator ion. Subtracting Na^+ from each side of the ionic equation produces the net ionic equation.

$$HC_2H_3O_2(aq) + \cancel{Na^+(aq)} + OH^-(aq) \rightarrow \cancel{Na^+(aq)} + C_2H_3O_2^-(aq) + H_2O(l)$$

$$HC_2H_3O_2(aq) + OH^-(aq) \rightarrow C_2H_3O_2^-(aq) + H_2O(l)$$

The net ionic equation shows that a weak acid and a strong base react by hydrogen ion transfer from the weak acid to the hydroxide ion. The acetic acid is the H^+ donor and serves as the acid. The hydroxide ion is the H^+ acceptor and serves as the base. Although this is an acid-base reaction, notice that H^+ is not involved as a reactant or product of the reaction. However, using the H^+ transfer definition, it is easy to include this reaction as an acid-base reaction.

Weak acid, strong base, and pH
The photos in **Figure 15.9** show what happens when $0.100M$ solutions of sodium hydroxide and acetic acid are mixed. Notice that the pH of the $0.100M$ $HC_2H_3O_2$ is greater than the pH of the $0.100M$ HCl solution in **Figure 15.6**.

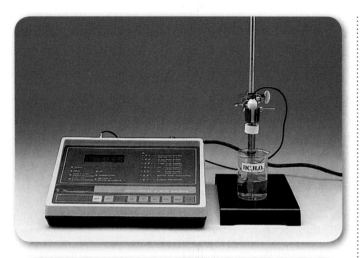
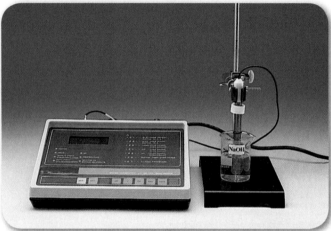
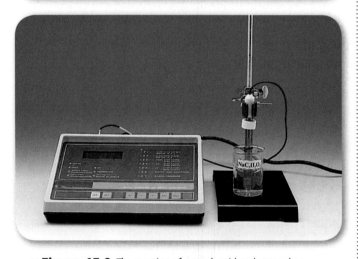

■ **Figure 15.9** The reaction of a weak acid and strong base does not result in a neutral solution. The pH of a $0.100M$ $HC_2H_3O_2$ solution (top) is approximately 3, and the pH of a $0.100M$ NaOH solution (middle) is approximately 13.

Calculate What is the pH of the resulting solution (bottom) when the reaction is complete?

FOLDABLES
Incorporate information from this section into your Foldable.

As the sodium hydroxide solution is added to the acetic acid solution, the pH increases as the hydrogen from acetic acid molecules react with the hydroxide ions. When equal volumes of the two are mixed, the final pH is greater than 7. The final reaction mixture is basic.

As in the case of the weak base–strong acid reaction, mixing equal moles of acid and base does not produce a neutral solution. Because the final pH in this case is basic, more hydroxide than hydronium ions must be present in the final reaction mixture.

VOCABULARY
WORD ORIGIN
Vinegar
comes from the French word *vinaigre* meaning *sour wine*

EXAMPLE Problem 15.3

Equations—Weak Acid, Strong Base Write the overall, ionic, and net ionic equations for the reaction of phosphoric acid with lithium hydroxide.

1 Analyze
Identify the acid and base as strong or weak. Phosphoric acid (H_3PO_4), is a weak acid. Lithium hydroxide (LiOH) is a strong base.

2 Set up
Write a balanced equation for the overall reaction. The salt is lithium phosphate, Li_3PO_4. Water is also a product. You'll need three moles of LiOH for each mole of H_3PO_4. Three moles of water will be produced.

$$H_3PO_4(aq) + 3LiOH(aq) \rightarrow Li_3PO_4(aq) + 3H_2O(l)$$

3 Solve
Now, write the ionic equation. Because H_3PO_4 is a weak acid, it is only partially ionized, and you write it in the ionic equation as H_3PO_4. LiOH is completely dissociated, as is the salt lithium phosphate. Be careful to keep the ionic equation balanced.

$$H_3PO_4(aq) + \cancel{3Li^+(aq)} + 3OH^-(aq) \rightarrow \cancel{3Li^+(aq)} + PO_4^{3-}(aq) + 3H_2O(l)$$

Check for spectator ions. Li^+ is a spectator ion. Subtract Li^+ from both sides of the equation to get the net ionic equation.

$$H_3PO_4(aq) + 3OH^-(aq) \rightarrow PO_4^{3-}(aq) + 3H_2O(l)$$

4 Check
Check to see that the reaction occurs through hydrogen ion transfer from phosphoric acid molecules to hydroxide ions. Every weak acid–strong base reaction occurs by this type of H^+ transfer.

FACT of the Matter

Vinegar is a dilute solution of acetic acid. Pure acetic acid is often called glacial acetic acid. It was first purified in 1700 by the distillation of vinegar. At room temperature, pure acetic acid is a liquid, but it freezes at 17°C. The term glacial means "icelike." In poorly heated chemistry labs, pure acetic acid freezes.

PRACTICE Problems   Solutions to Problems Page 865

Write overall, ionic, and net ionic equations for the reactions.

8. carbonic acid (H_2CO_3) and sodium hydroxide (NaOH)
9. boric acid (H_3BO_3) and potassium hydroxide (KOH)
10. acetic acid ($HC_2H_3O_2$) and calcium hydroxide ($Ca(OH)_2$)

SUPPLEMENTAL PRACTICE

For more practice with weak acid, strong base net ionic equations, see Supplemental Practice, page 831.

■ **Figure 15.10** The weak acetic acid does not react with the weak base aluminum hydroxide.

FOLDABLES Incorporate information from this section into your Foldable.

Weak and Weak: It's Uncertain

The strong-strong reaction plus the two types of weak-strong reactions are the predominant acid-base reactions. Looking at an acid-base reaction as occurring by H^+ transfer helps you to understand why the weak-weak reaction is not considered a predominant reaction.

Because neither a weak acid nor a weak base has a strong tendency to transfer a hydrogen ion, transfer between the two may occur, but it is uncommon. Reactions between a weak acid and a weak base generally do not play an important role in acid-base chemistry, as shown in **Figure 15.10**.

Section 15.1 Assessment

Section Summary

- Acid-base reactions are classified by the strengths of the acid and the base.

- Strong acid–strong base, weak acid–strong base, and weak base–strong acid are the predominant types of acid-base reactions.

- Acids and bases in reactions can be identified using a hydrogen-ion transfer definition. An acid is an H^+ donor; a base is an H^+ acceptor.

- When acids and bases react, the pH of the final solution is dependent on the nature of the reactants.

- Strong acid–strong base reactions yield neutral solutions.

- Weak acid–strong base reactions yield more basic solutions, whereas strong acid–weak base reactions produce acidic solutions.

11. **MAIN Idea** Write the overall, ionic, and net ionic equations for the following reactions.

    a) perchloric acid ($HClO_4$) and sodium hydroxide (NaOH)

    b) sulfuric acid (H_2SO_4) and ammonia (NH_3)

    c) citric acid ($H_3C_6H_5O_7$) and potassium hydroxide (KOH)

12. **Predict** whether the pH of the product solution is acidic, basic, or neutral for each of the reactions in question 1. Explain.

13. **Identify** the acid and the base in each of the following reactions.

    a) $HBr(aq) + H_2O(l) \rightarrow H_3O^+(aq) + Br^-(aq)$

    b) $NH_3(aq) + H_3PO_4(aq) \rightarrow NH_4^+(aq) + H_2PO_4^-(aq)$

    c) $HS^-(aq) + H_2O(l) \rightarrow H_2S(aq) + OH^-(aq)$

14. **Apply Concepts** Think about what acid and what base would react to form each of the following salts. When each of the salts is dissolved in water, will its solution be acidic, basic, or neutral? If not neutral, use an equation to explain why.

    a) NH_4Cl c) $LiC_2H_3O_2$

    b) NaCl   d) $NH_4C_2H_3O_2$

15. **Predict** Lactic acid is produced in muscles during exercise. It is also produced in sour milk due to the action of lactic acid bacteria. The formula for lactic acid is $HC_3H_5O_3$. Write the overall, ionic, and net ionic equations for the reaction of lactic acid with sodium hydroxide. Will the pH of the product solution be greater than 7, exactly 7, or less than 7?

Section 15.2

Objectives
- **Evaluate** the importance of a buffer in controlling pH.
- **Perform** acid-base titrations.
- **Calculate** results from titration data.

Review Vocabulary
spectator ions: ions that are present in a solution but do not participate in a reaction

New Vocabulary
buffer
titration
standard solution

Applications of Acid-Base Reactions

MAIN Idea Acid-base reactions play an important role in the human body.

Real-World Reading Link What do the chemistry of blood, rain and lakes, and your favorite shampoo have in common? In each of these cases, and in many others, an optimum acid-base balance must be maintained. Whether you become interested in the reactions that sustain life or simply in making a better shampoo, it is important to understand acid-base chemistry.

Buffers Regulate pH

Regulating pH is vital to life. For example, maintaining healthy blood chemistry depends on the acid-base balance in blood. The pH of your blood is slightly basic, about 7.4. If you're healthy, the pH of your blood does not vary by more than one-tenth of a pH unit.

If you stop and think about all of the materials that go into and out of your blood, it is amazing that the pH remains so constant. Many of those materials are acidic or basic, which is why the constancy of the pH of blood is fascinating. Blood is an example of an effective buffer.

The pH level isn't just vital to people, it's vital to many of the reactions that sustain living things. For example, the jellyfish in **Figure 15.11** don't have blood. But maintaining a narrow pH range of roughly 8.1 to 8.4 is important for the health of the jellies.

Buffers defined A **buffer** is a solution that resists changes in pH when moderate amounts of acids or bases are added. It contains ions or molecules that react with OH^- or H^+ if one of these ions is introduced into the solution.

Ammonia–ammonium chloride buffer Buffer solutions are prepared by using a weak acid with one of its salts or a weak base with one of its salts. For example, a buffer solution can be prepared by using the weak base ammonia (NH_3) and an ammonium salt, such as NH_4Cl. If an acid is added, NH_3 reacts with the H^+.

$$NH_3(aq) + H^+(aq) \rightarrow NH_4^+(aq)$$

If a base is added, the NH_4^+ ion from the salt reacts with the OH^-.

$$NH_4^+(aq) + OH^-(aq) \rightarrow NH_3(aq) + H_2O(l)$$

Acetic acid-sodium acetate buffer Look at another system that contains a weak acid, acetic acid ($HC_2H_3O_2$), and a salt, sodium acetate ($NaC_2H_3O_2$). If a strong base (OH^-) is added to the buffer system, the weak acid reacts to neutralize the solution.

$$HC_2H_3O_2(aq) + OH^-(aq) \rightarrow C_2H_3O_2^-(aq) + H_2O(l)$$

■ **Figure 15.11** To provide a healthy environment for these jellies, the pH of the aquarium at the Monterey Bay Aquarium must be adjusted to stay within the range of 8.1 to 8.4.
Predict what would happen if the pH were allowed to fall to 7.0.

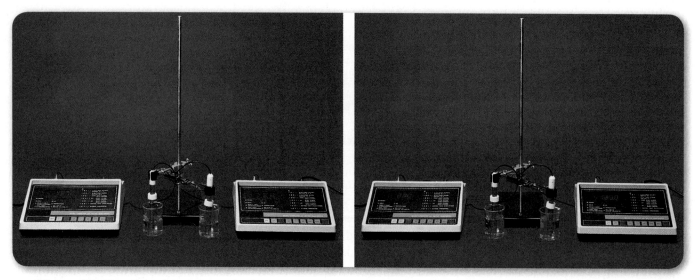

■ **Figure 15.12** Buffers influence the pH of a solution.

Compare what happens when acid and base are added to an acetic acid/sodium acetate buffer system at pH = 5 (left) to what happens when the same amount of acid and base are added to an unbuffered solution of pH = 5 (right).

This reaction neutralizes OH^-. If H^+ is added, the acetate ion from the $NaC_2H_3O_2$ is available to neutralize the H^+.

$$C_2H_3O_2^-(aq) + H^+(aq) \rightarrow HC_2H_3O_2(aq)$$

Notice in **Figure 15.12** that the pH does not remain constant in the buffer solution. It changes slightly in the direction of the pH of the added acid or base. These pH changes are insignificant when you compare them to the changes that occur in the unbuffered solution.

Blood buffer: dissolved CO_2 The NH_3–NH_4Cl buffer system and $HC_2H_3O_2$–$NaC_2H_3O_2$ buffer system are common ones used in many laboratories. How does the buffer system in your blood compare to these laboratory systems? Though different chemicals are involved, the operation of the system is essentially the same.

The ability of blood to maintain a constant pH of 7.4 is due to several buffer systems. Dissolved carbon dioxide (CO_2) makes up one of the systems. Remember that when carbon dioxide dissolves in water, it produces carbonic acid (H_2CO_3).

$$CO_2(g) + H_2O(l) \rightarrow H_2CO_3(aq)$$

The other part of the blood buffer is the hydrogen carbonate ion (HCO_3^-). If something happens to increase OH^- in your blood, H_2CO_3 reacts to lower the OH^- concentration and keep the pH from increasing.

$$H_2CO_3(aq) + OH^-(aq) \rightarrow HCO_3^-(aq) + H_2O(l)$$

If H^+ enters the blood, HCO_3^- reacts to keep the pH from decreasing.

$$HCO_3^-(aq) + H^+(aq) \rightarrow H_2CO_3(aq)$$

Blood pH and the lungs Keep in mind that the level of carbon dioxide in the blood, and therefore the level of carbonic acid, is ultimately controlled by the lungs. If the amount of H^+ in the blood increases, a large amount of carbonic acid is produced, which lowers the H^+ concentration.

In order to reduce the carbonic acid concentration produced, the lungs work to remove carbon dioxide through respiration. However, rapid and deep breathing can cause a deficiency of carbon dioxide in the blood. This problem is called hyperventilation. It often occurs when a person is nervous or frightened. In this case, the air in the lungs is exchanged so rapidly that too much carbon dioxide is released. Carbon dioxide blood levels drop, which causes the amount of carbonic acid in the blood to decrease. This causes the blood pH to increase and can be fatal if steps are not taken to stop this type of breathing.

When a person hyperventilates, he or she needs to become calm and breathe regularly. If the person breathes with a paper bag covering his or her nose and mouth, the concentration of carbon dioxide is increased in the air breathed. More carbon dioxide is forced into the blood through the lungs, and the pH of the blood drops to its normal level. **Figure 15.13** shows the narrow pH range of blood and summarizes the behavior of the buffer that controls the pH.

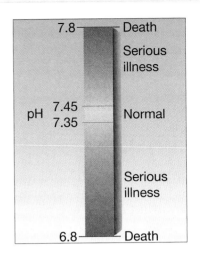

■ **Figure 15.13** The pH of human blood is maintained within a narrow range by a mixture of buffers. The H_2CO_3/HCO_3^- system is one of the important parts of the blood buffer.

MiniLab 15.2

Buffers

How can you change the pH of a solution?
Compare the amounts of acidic and basic solutions required to cause similar pH changes in two solutions: one a sodium chloride solution at a pH of 7 and the other a buffered solution with a pH of 7.

Procedure

1. Read and complete the lab safety form.
2. Obtain and number four **test tubes** 1–4. Place them in a **test tube rack**.
3. Use a **small graduated cylinder** to measure 5.0 mL of **NaCl solution** into tubes 1 and 3.
4. Use another small, graduated cylinder to measure 5.0 mL of **pH 7 buffer solution** into tubes 2 and 4.
5. Add two drops of **methyl orange indicator** to tubes 1 and 2. The methyl orange is yellow if the pH is greater than 4.4 and will change to orange-red as the pH is lowered from 4.4 to 3.2.
6. Add **0.100M HCl solution** drop by drop to tubes 1 and 2, stirring after each drop, just until the solution color changes to orange-red. Record the number of drops required for each solution.
7. Add two drops of **phenolphthalein indicator** to tubes 3 and 4. The phenolphthalein is colorless if the pH is lower than 8.2 and will change to pink or magenta as the pH is raised from 8.2 to 10.0.
8. Add **0.1M NaOH solution** drop by drop to tubes 3 and 4, stirring after each drop, just until the solution color changes to pink or magenta. Record the number of drops required for each solution.

Analysis

1. **Compare** the number of drops of acid required to lower the pH of the two solutions to produce an orange-red color change.
2. **Compare** the number of drops of base required to raise the pH of the two solutions to produce a magenta color change.
3. **Describe** the effect of the buffer on the pH of the solution.

Everyday Chemistry

Hiccups

Have you ever wondered why people hiccup? Hiccups often start for no apparent reason. They afflict almost everyone, but have no clear purpose.

What are hiccups? Most of the time, hiccups are a harmless annoyance lasting a few minutes to several hours. They are repeated, involuntary spasms of your diaphragm, which is the dome-shaped breathing muscle separating the chest area from the abdomen as shown in **Figure 1**. When hiccups start, the diaphragm jerks and the air coming in is stopped when a small flap (the epiglottis) suddenly closes the opening to the windpipe (the glottis) resulting in the familiar "hic." This process might start by abnormal stimulation of the phrenic nerve that controls the diaphragm and the glottis.

Possible onset of hiccups Hiccups sometimes occur when you eat too fast and swallow air along with your food, when your stomach is distended due to gas or overeating, or when you laugh, cough, or cry vigorously. Hiccups also sometimes occur when your body is exposed to sudden temperature changes, such as drinking a cold beverage with a hot meal, or when you are excited or stressed.

Possible cures A number of common cures might stop the rhythmic reflex. These methods include concentrating on breathing slowly and deeply, sipping ice water, eating a spoonful of sugar or honey, and biting on a lemon. Interruptions of normal breathing such as sneezing or coughing and sudden pain or fright may stop hiccups.

Most cures that seem to work are related to an increase of carbon dioxide in the blood, which slightly lowers the pH. The body has a buffer system to maintain a blood pH range of 7.35 to 7.45. Slight decreases in pH might stop the stimulation occurring to the phrenic nerve that causes hiccups.

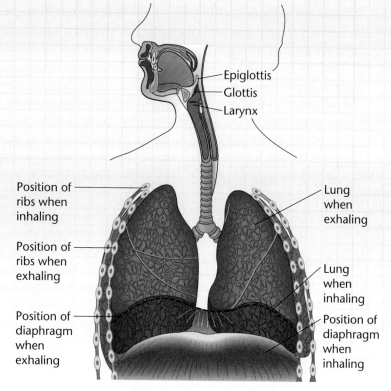

Figure 1 Positions of ribs and diaphragm when breathing

Normal breathing and hiccups The following steps describe what happens in normal breathing and hiccups.

A When you breathe in, the diaphragm contracts and flattens. The glottis and epiglottis open, and air rushes in.

B When you breathe out, the diaphragm relaxes and resumes its dome shape, pushing the air out.

C When you hiccup, your diaphragm twitches. Air is forced past the glottis and epiglottis, which snap shut and cause a hiccup.

Explore Further

1. **Compare and contrast** possible cures for hiccups.
2. **Hypothesize** as to why breathing in and out of a paper bag or holding your breath might stop hiccups.

Figure 15.14 If the rock and soil that compose and surround a lake bed are rich in limestone, the lake can neutralize acid rain by acid-base reactions. These lakes have a capacity to absorb acid rain without an appreciable change in pH. The aqueous solution of lakewater and $Ca(HCO_3)_2$ behaves as a buffer.

Acid rain versus acid lakes In Chapter 14, you learned about the sources of acid rain and about its impact on plant and animal life, as well as on human-made objects such as monuments and buildings. When acid rain falls on lakes and streams, it might be expected that the acidity of the water increases and the pH decreases.

This result is true in some cases. Many lakes in the northeastern United States, southern Canada, northern Europe, and the Scandinavian countries have a pH as low as 4.0. The pH of a healthy lake is approximately 6.5.

Other lakes, many of which reside in the midwestern United States, appear to receive the same amount of acid rain as these low-pH lakes, but they do not show a dramatic lowering of pH. The key to the ability of these lakes to resist the pH-lowering effects of acid rain is their local geology as shown in **Figure 15.14**.

Limestone is primarily calcium carbonate ($CaCO_3$). Calcium carbonate reacts with carbon dioxide (CO_2) and water to form calcium hydrogen carbonate ($Ca(HCO_3)_2$), a water-soluble compound.

$$CaCO_3(s) + CO_2(g) + H_2O(l) \rightarrow Ca(HCO_3)_2(aq)$$

Lakes in areas rich in limestone have significant concentrations of hydrogen carbonate ions (HCO_3^-). Just as in the regulation of pH in the blood, hydrogen carbonate ions produced from calcium hydrogen carbonate form a base that can neutralize acid in lakes.

$$HCO_3^-(aq) + H^+(aq) \rightarrow CO_2(g) + H_2O(l)$$

The Acid-Base Chemistry of Antacids

The acid-base chemistry of an upset stomach is big business. **Figure 15.15** shows a small selection commercial antacids available in pharmacies and grocery stores. Many people use these or similar products to calm their acid indigestion.

The stomach's acidic environment While stomach acid can cause painful problems, it is important to remember that you do need an acid stomach in order to be healthy. Remember that the pH of stomach acid, which is mostly hydrochloric acid, is about 2.5. The acidic environment in the stomach is extremely important for good digestion.

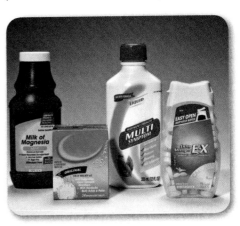

Figure 15.15 Antacids contain bases which act as buffers to combat stomach problems associated with acid indigestion.

Figure 15.16 A gastric ulcer forms when the membrane that protects the stomach lining breaks down and stomach acid attacks the stomach wall.

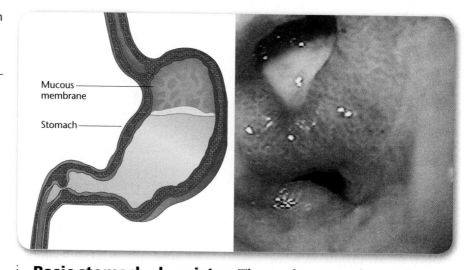

Basic stomach chemistry
The combination of an acidic environment and enzymes in the stomach works together to break down the complex molecules that you eat into smaller molecules that can be transported through your blood and delivered to every cell in your body. These smaller molecules are the source of structural material and energy for the cells.

Your stomach is made of protein that is not unlike the protein in a hamburger. The inner walls of your stomach are coated by a basic mucous membrane that protects the stomach from the digestive power of the acid and enzymes. If the stomach contents become too acidic, this basic membrane breaks down by acid-base neutralization reactions. At the point where the barrier is neutralized, the gastric juices can begin to digest the protein that makes up the stomach wall, as shown in **Figure 15.16**. This causes the discomfort of acid indigestion and may result in more serious problems.

In most individuals, this breakdown of the mucous membrane corrects itself in a short time, and there is no long-term damage to the stomach. However, in some cases, the damage may be more long-term, and if left untreated, a stomach ulcer may result. The correction process can be expedited by using an antacid.

Antacids = anti-acids = bases
Although a variety of over-the-counter and prescription antacids is available, they all share common acid-base chemistry. **Table 15.2** identifies the bases that are most widely used in antacids. The compounds can be divided into two categories: the hydroxide-containing bases and the carbonate-containing bases.

FACT of the Matter

An alternative approach to controlling acid indigestion is now available in over-the-counter medication. These drugs work by decreasing the secretion of acid in the stomach. These drugs were previously available only by prescription for people who suffer severe acid indigestion or who have a tendency for gastric ulcers.

Table 15.2 Compounds Used in Antacids

Insoluble Hydroxides	Carbonate-based
Aluminum hydroxide ($Al(OH)_3$)	Calcium carbonate ($CaCO_3$)
	Magnesium carbonate ($MgCO_3$)
Magnesium hydroxide ($Mg(OH)_2$)	Sodium hydrogen carbonate ($NaHCO_3$)
	Potassium hydrogen carbonate ($KHCO_3$)

Chemistry & Society

Artificial Blood

You have a serious accident and need a blood transfusion immediately. Would you rather receive someone else's blood or artificial blood? Under the described circumstances, artificial blood might be the better choice.

What is blood? Blood is made up of several ingredients which are important to the everyday processes that take place in the body. White blood cells fight disease, platelets allow the blood to clot when we are cut, and red blood cells, which make up the largest percentage of the blood's composition, are responsible for transporting oxygen and carbon dioxide through our bodies.

Why is artificial blood needed? One of the main uses of blood in medicine is to maintain the transport of oxygen to the vital organs during times of tremendous blood loss (over 40 percent of a person's blood volume). Transfusions have saved many lives over the past century. However, the process of blood transfusions has been questioned due to several issues: compatibility, effective storage, blood donor shortages, and risk of infection. The solution to these problems might be artificial blood.

Function of artificial blood The goal of artificial blood, shown in **Figure 1**, is to mimic the hemoglobin in red blood cells, which carries oxygen through the body. This type of artificial blood is called an oxygen-therapeutic blood substitute. There are two approaches used to produce an oxygen-therapeutic blood substitute. One approach is to utilize hemoglobin from cattle or pig blood, from outdated human blood from blood banks, or from genetically engineered bacteria.

Another approach to this problem is to use perfluorocarbon (PFC), a chemical compound that can carry and release oxygen. A PFC emulsion is made with antibiotics, water, vitamins, nutrients, and salts, and it performs many of the functions of biological blood. The PFC emulsion carries

■ **Figure 1** Artificial blood

oxygen so well that mice can survive by breathing the liquid mixture.

Artificial success? Hemoglobin-based blood substitutes are presently not approved by the FDA for human use, although some varieties have been approved for veterinarian use. It will likely not be long before oxygen-therapeutic blood substitutes are approved by the FDA for human use. Perfluorocarbon-based blood substitutes also are presently not approved by the FDA. However, many medical scientists believe this product has potential for future uses.

ANALYZE the Issue

1. **Acquire Information** Research the structure of hemoglobin and make a simplified model of hemoglobin.
2. **Infer** Analyze why scientists working to develop blood substitutes need an extensive knowledge of chemistry.
3. **Think Critically** What problems might arise with government control of blood substitute research? What advantages might exist?

Hydroxide antacids If you examine the ingredients on an antacid, you probably recognize any hydroxide-containing compounds as bases. The hydroxides used in antacids have low solubility in water. Because the pH of saliva is neutral to slightly basic, these insoluble hydroxides do not dissolve and react until they get past your mouth and upper digestive tract and into the highly acidic environment of the stomach.

Milk of magnesia, a suspension of magnesium hydroxide in water, is a good example of this type of antacid. If you have ever used unflavored milk of magnesia, you may have noticed that it has a bitter taste, which is typical of a base.

Carbonate antacids You are familiar with the acid-neutralizing abilities of the carbonates and hydrogen carbonates from the discussion about blood buffering and acid rain. Carbonate (CO_3^{-2}) and hydrogen carbonate (HCO_3^{-2}) antacids react with hydrochloric acid (HCl) to form carbonic acid (H_2CO_3), which decomposes into carbon dioxide gas and water.

$$CaCO_3(s) + 2HCl(aq) \rightarrow CaCl_2(aq) + H_2CO_3(aq)$$

$$NaHCO_3(aq) + HCl(aq) \rightarrow NaCl(aq) + H_2CO_3(aq)$$

$$H_2CO_3(aq) \rightarrow H_2O(l) + CO_2(g)$$

Many carbonates and hydrogen carbonates are insoluble in water and have great neutralizing power. They are the primary components of many over-the-counter antacid tablets. **Figure 15.17** shows an antacid of this type.

As with any over-the-counter medication, antacids are designed for occasional use. Anyone who needs to use them frequently for attacks of indigestion may have a more serious health problem and should consult a physician.

■ **Figure 15.17** Some antacids contain a carbonate or hydrogen carbonate and a weak acid, such as citric acid. When added to water, these compounds react, producing a basic salt and carbonic acid, which decomposes into water and carbon dioxide.

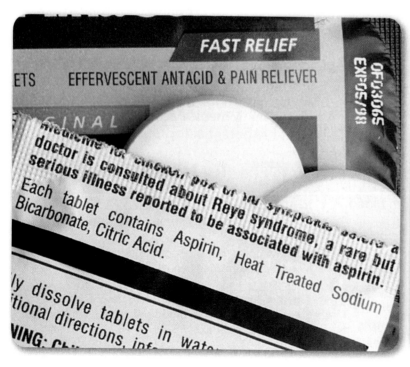

Acid-Base Titrations

Although acid-base reactions do not have to happen in aqueous solutions, many are water-based. In these reactions, it is important to know the concentrations of the solutions involved.

The general process of determining the molarity of an acid or a base through the use of an acid-base reaction is called an acid-base **titration**. In a titration, the molarity of one of the reactants, acid or base, is known, but the other is unknown. The known reactant molarity is used to find the unknown molarity of the other solution. Solutions of known molarity that are used in this fashion are called **standard solutions**.

Titration of an acid with a base Suppose you are in charge of cleaning up and doing inventory in the chemistry stockroom. Your first job is to catalog the variety of solutions that are on the shelf. One of the solutions you find is a tightly stoppered, well-labeled bottle of $0.100M$ NaOH.

On the next shelf, you find a tightly stoppered bottle that is not as well-labeled. The label has been damaged, and the only thing left of the label is M HCl. This label indicates that the solution contains hydrochloric acid, but it does not identify the molarity. For the inventory form, you need to know the molarity of the solution. What can you do?

The titration process You know that NaOH and HCl react completely in a neutralization reaction.

$$HCl(aq) + NaOH(aq) \rightarrow NaCl(aq) + H_2O(l)$$

You know the concentration of the NaOH solution from the stockroom, so it now becomes your standard solution. You can use the reaction, the volumes of acid and base used, and the molarity of the base to determine the molarity of the unlabeled HCl. Follow the first part of this process in **Figure 15.18.**

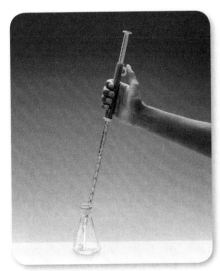
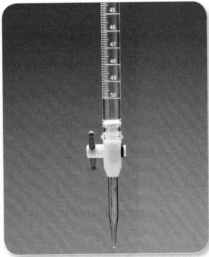

You need to carefully measure a volume of the unknown HCl solution into a flask. For careful volume measurements, you can use a pipette.

To add the NaOH to the HCl and to measure the amount of NaOH solution needed, a burette is used. A burette is a long, calibrated tube with a valve at the bottom.

■ **Figure 15.18** A titration requires mixing measured volumes of the standard solution and the unknown solution.

Using indicators in a titration When the solution is neutral, you know that you have added exactly enough base to react with the amount of acid present, and you are at what is known as the endpoint of the titration. But how do you know when the endpoint is reached? Probably the best way to indicate the endpoint of a titration is to use an acid-base indicator. Different indicators change color at different pH values.

Look at the graphs in **Figure 15.19**. For a titration of NaOH and HCl, the endpoint is reached when the solution achieves a pH of 7. Therefore, you need an indicator that changes color as close to pH=7 as possible. The best indicator for this titration is bromthymol blue, which changes from yellow to blue close to pH=7.

Collecting data Assume you have a burette containing an NaOH standard solution and a flask containing 20.0 mL of HCl of unknown concentration, as described in **Figure 15.18**. Use the bromthymol blue indicator in this titration. To actually perform the NaOH-HCl titration, follow the process outlined in **Figure 15.20,** and record the data collected.

At the start of the titration, you know the molarity of your standard solution—in this case, the NaOH—along with the volume of your unknown. The titration itself will give you the volume of standard solution needed to neutralize the unknown. These three quantities—the volume of the standard used, the molarity of the standard, and the volume of the unknown—will allow you to calculate the molarity of the unknown.

Acid-base indicators are useful when trying to determine the molarity of an unknown acidic or basic solution. **Figure 15.19** illustrates titration curves for the reactions between a strong acid and a strong base, a weak acid and a strong base, and a weak base and a strong acid, respectively. The graph represents titration results from a reaction between a strong acid and a strong base. A neutral solution results, and bromthymol blue is an effective indicator for such a reaction. When a weak acid and a strong base react, as illustrated in the middle graph, a slightly basic solution is produced and phenolphthalein is used as an indicator. The titration of a weak base with a strong acid produces a slightly acidic solution as shown by the methyl red endpoint in the bottom graph.

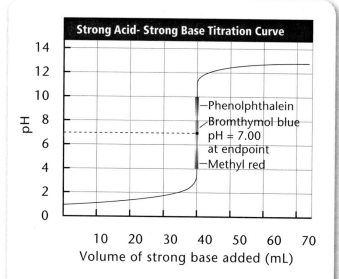

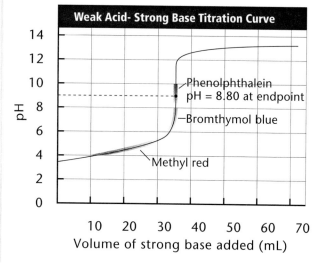

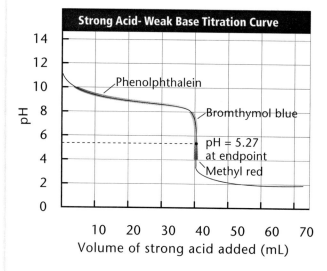

■ **Figure 15.19** The graphs show some of the more popular acid-base indicators used in the chemistry laboratory and the pH at which they change color.

Apply *At what pH do bromthymol blue, phenolphthalein, and methyl red change color?*

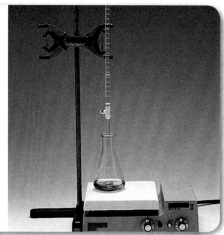

Determining concentration using stoichiometry

You now have all your experimental data. How do you use these experimental data to calculate the molarity of the HCl solution, which is the desired experimental result?

First, summarize what you know. You know that 20 mL of the HCl solution reacts with 19.9 mL of $0.100M$ NaOH solution to reach the endpoint. From the balanced equation for the reaction, you know that one mole of HCl reacts with one mole of NaOH. Therefore, the number of moles of HCl in 20.0 mL of the HCl solution equals the number of moles of NaOH in 19.9 mL of $0.100M$ NaOH solution.

Now, use dimensional analysis to solve this solution stoichiometry problem, just as you used it to solve other stoichiometry problems. Because you know the concentration of the NaOH solution, first find the number of moles of NaOH involved in the reaction.

$$19.9 \text{ mL soln} \left(\frac{1 \text{ L}}{10^3 \text{ mL}} \right) \left(\frac{0.100 \text{ mol NaOH}}{1.00 \text{ L soln}} \right) = 1.99 \times 10^{-3} \text{ mol NaOH}$$

Next, examine the balanced equation for the reaction and determine that, because their coefficients are the same, equal numbers of moles of NaOH and HCl react.

$$\text{NaOH(aq)} + \text{HCl(aq)} \rightarrow \text{NaCl(aq)} + \text{H}_2\text{O(l)}$$

Because 1.99×10^{-3} mol NaOH react, 1.99×10^{-3} mol HCl present in solution also react.

Finally, use the volume to find the molarity of the acid.

$$\left(\frac{19.9 \times 10^{-3} \text{ mol HCl}}{20.0 \text{ mL soln}} \right) \left(\frac{10^3 \text{ mL}}{1 \text{ L}} \right) = \left(\frac{0.0995 \text{ mol HCl}}{1 \text{ L soln}} \right) = 0.0995M \text{ HCl}$$

Based on your single titration, the molarity of the HCl solution is $0.0995M$. However, before you put this value on the label, you probably would repeat the titration several times in order to verify your analysis and be more confident of the value on the label. Study the following Example Problem for additional help solving solution stoichiometry problems.

■ **Figure 15.20** The following process shows how to perform and obtain data for an HCl-NaOH titration.
Left: Add a few drops of bromthymol blue indicator to the HCl solution in the flask. The solution turns yellow.
Middle: Titrate the HCl with the NaOH by slowly adding NaOH from the burette to the flask as the solution is constantly stirred.
Right: Continue to add the NaOH slowly until the color changes from yellow to blue.

Concepts In Motion

Interactive Figure To see an animation of titration, visit glencoe.com.

Personal Tutor For an online tutorial on acid-base indicators, visit glencoe.com.

EXAMPLE Problem 15.4

Finding Molarity A 15.0-mL sample of a solution of H_2SO_4 with an unknown molarity is titrated with 32.4 mL of 0.145M NaOH to the bromthymol blue endpoint. Based upon this titration, what is the molarity of the sulfuric acid solution?

1 Analyze
Because the molarity of the NaOH solution is known, the number of moles of NaOH involved in the titration can be calculated. The corresponding number of moles of H_2SO_4 can then be determined, and this figure can be used to calculate the molarity of the acid.

2 Set up
Write the balanced equation for the reaction. Remember that sulfuric acid is a diprotic acid.

$$H_2SO_4(aq) + 2NaOH(aq) \rightarrow 2H_2O(l) + Na_2SO_4(aq)$$

3 Solve
Because the concentration of the NaOH solution is known, find the number of moles of NaOH used in the titration.

$$32.1 \text{ mL soln} \left(\frac{1 \text{ L}}{10^3 \text{ mL}}\right)\left(\frac{0.145 \text{ mol NaOH}}{1.00 \text{ L soln}}\right) = 4.7 \times 10^{-3} \text{ mol NaOH}$$

Using the balanced equation, find the number of moles of H_2SO_4 that react with 4.70×10^{-3} mol NaOH.

$$4.7 \times 10^{-3} \text{ mol NaOH} \left(\frac{1 \text{ mol } H_2SO_4}{2 \text{ mol NaOH}}\right) = 2.35 \times 10^{-3} \text{ mol } H_2SO_4$$

Find the concentration of the H_2SO_4.

$$\left(\frac{2.35 \times 10^{-3} \text{ mol } H_2SO_4}{15.0 \text{ mL soln}}\right)\left(\frac{10^3 \text{ mL}}{1 \text{ L}}\right) = \left(\frac{0.157 \text{ mol } H_2SO_4}{1 \text{ L soln}}\right) = 0.157M \text{ } H_2SO_4$$

4 Check
Check to make sure that the final units are what they are supposed to be.

SUPPLEMENTAL PRACTICE

For more practice with finding molarity, see Supplemental Practice, page 832.

PRACTICE Problems
Solutions to Problems Page 865

16. A 0.100M LiOH solution was used to titrate an HBr solution of unknown concentration. At the endpoint, 21.0 mL of LiOH solution had neutralized 10.0 mL of HBr. What is the molarity of the HBr solution?

17. A 0.150M KOH solution fills a burette to the 0 mark. The solution was used to titrate 25.0 mL of an HNO_3 solution of unknown concentration. At the endpoint, the burette reading was 34.6 mL. What was the molarity of the HNO_3 solution?

18. A $Ca(OH)_2$ solution of unknown concentration was used to titrate 15.0 mL of a 0.125M H_3PO_4 solution. If 12.4 mL of $Ca(OH)_2$ are used to reach the endpoint, what is the concentration of the $Ca(OH)_2$ solution?

How It Works

pH Indicators

pH indicators are chemical dyes whose colors are affected by acidic and basic solutions. These indicators are weak acids and bases that have complicated structures and change color as they lose or gain hydrogen ions. Most pH indicators are weak acids with the ability to form positive hydrogen ions when the molecule ionizes in solution. For example, thymol blue, ($H_2C_{27}H_{28}O_5S$) is a weak acid that is used as a pH indicator.

1 At low pH (below 2), a solution containing thymol blue has two positive hydrogen ions attached to each molecule of thymol blue. In this form, the molecule is red.

3 At pH of 8.9, a solution containing thymol blue loses its second positive hydrogen ion per molecule. With no positive hydrogen ions, the molecule is blue.

2 When the pH of a solution rises above 2, the concentration of H^+ in the solution drops sufficiently so that each thymol blue molecule loses one of its positive hydrogen ions. This form of thymol blue is yellow.

However, a single indicator can give only a general indication of pH. For example, if a solution is blue with thymol blue, it does not reveal the actual pH. It indicates only that the pH is above 8.9. By combining indicators that span the range of pH from 1 to 14, it is possible to create color changes that allow more precise measurement of pH. pH paper is such a mixture. When the color of the paper is compared to the color chart, the pH can be determined to within about one pH unit. If more specific pH values are needed, a pH meter should be used.

Think Critically

1. What do you think is the relationship between the number of positive hydrogen ions that a pH indicator molecule loses and the number of color changes?

2. What might be some advantages of using pH indicator paper instead of a pH meter?

CHEMLAB

SMALL SCALE

TITRATION OF VINEGAR

Background
Vinegar is a solution of mostly acetic acid (CH_3COOH) in water. It varies in concentration from about three percent to five percent by volume. The acetic acid in vinegar may be neutralized by adding sodium hydroxide (NaOH).

$$CH_3COOH(aq) + NaOH(aq) \rightarrow H_2O(l) + NaCH_3COO(aq)$$

You will titrate several brands of commercial vinegar with sodium hydroxide solution of a known concentration, and use your data to calculate the molarities and volume percentages of acetic acid in the vinegars.

Question
What are the concentrations of acetic acid in several brands of vinegar?

Objective
Observe acid-base titrations of several vinegars with a standard sodium hydroxide solution.

Calculate the molarities and volume percentages of acetic acid in the vinegars.

Compare the acetic acid concentrations of various brands of vinegar.

Preparation

Materials
24-well microplate
several brands of vinegar
labeled microtip pipettes
standard NaOH solution
phenolphthalein solution
toothpicks
distilled water in a wash bottle

Safety Precautions
WARNING: *Sodium hydroxide is caustic and can damage skin and eyes. If you come into contact with any of this solution, rinse the affected area with a large volume of water and notify the teacher. Wash hands thoroughly when you complete the lab.*

Procedure

1. Read and complete the lab safety form.
2. Make a data table like the one shown.
3. Use a microtip pipette to add ten drops of the first type of vinegar to each of the wells—A1, B1, and C1—of the microplate. Record the brand of the first vinegar in your data table.

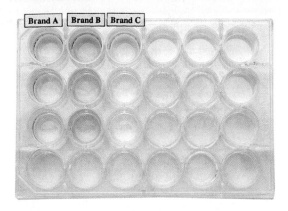

4. Use a clean microtip pipette to add one drop of phenolphthalein indicator solution to each of the three wells.
5. Set the microplate on a piece of white paper.

544 Chapter 15 • Acids and Bases React

6. Use a clean pipette to carefully add a drop of the standard NaOH solution to the solution in well A1, and stir with a toothpick. Pause for about 30 seconds and look down through the well for evidence of a persistent, light pink, phenolphthalein color that indicates the endpoint of the titration. Repeat this process with each drop until the endpoint is reached. Record the number of drops of sodium hydroxide solution required to titrate the vinegar to the endpoint.

7. Repeat procedure 6 with the second sample of this vinegar, which is in well B1. If the result differs by more than one drop from that of the first titration, repeat again with the sample in well C1.

8. Repeat procedures 3–7 with the other brands of vinegar, using other columns of wells of the microplate. Record your data after each titration.

Analyze and Conclude

1. **Interpret Data** From the two closest trials, find the average number of drops of NaOH required to titrate each vinegar. Use this average number of drops and the given molarity of the NaOH to calculate the molarity of acetic acid in each of the brands of vinegar. Assuming identical volumes for drops of vinegar and drops of NaOH solution, the ratio of reacting volumes in liters is the same as the ratio of reacting volumes in drops.

2. **Calculate** Use your results to calculate the volume percentage of acetic acid in each brand of vinegar according to the formula: M acetic acid $\times$ (1.00 percent acetic acid/ 0.175M acetic acid) = percent acetic acid by volume.

3. **Compare and Contrast** Which of the brands of vinegar you tested contained the highest volume percent of acetic acid?

Apply and Assess

1. If the cost and volume of each brand of vinegar are available, calculate the cost per percent of acetic acid per unit volume for each.

Data and Observations

Data Table

Type of Vinegar	Trial 1—Drops of NaOH	Trial 2—Drops of NaOH	Trial 3—Drops of NaOH

INQUIRY EXTENSION

Infer How could you have changed the experimental procedure in order to more accurately determine the concentrations of the vinegars?

■ **Figure 15.21** Silver tarnishes because it reacts with sulfur–containing compounds in air. To remove the tarnish, polishes containing aluminum or sodium bicarbonate are used.

Connecting Ideas

Acid-base reactions are usually double-displacement reactions. If you examine the oxidation number of each element involved in a double-displacement reaction, you can see that the oxidation numbers of the elements do not change. Are there types of reactions in which the oxidation numbers do change?

Many important reactions, such as the rusting of an old car or the burning of a fuel, involve chemical reactions in which oxidation numbers—the charge on an ion—do change. **Figure 15.21** shows the results of a reaction in which silver reacts with sulfur-rich substances in air. It's important to find out what causes these changes and investigate this important type of reaction.

Section 15.2 Assessment

Section Summary

▶ A buffer is a solution that maintains a relatively constant pH when H^+ ions or OH^- ions are added.

▶ Antacids are bases that react with stomach acid.

▶ An acid-base titration uses an acid-base reaction to determine the molarity of an unknown acid or base.

19. **MAIN Idea Apply** Use chemical equations to show how the buffer system in blood works.

20. **Explain** How does a solution that contains dissolved ammonia and ammonium chloride act as a buffer? Use net ionic equations to show how this buffer responds to added H^+ and OH^-.

21. **Sequence** the steps in an acid-base titration.

22. **Apply Concepts** From the following endpoint pHs, indicate whether the titration involves a weak acid–strong base, a weak base–strong acid, or a strong acid–strong base reaction. Use **Figure 15.19** to select the best indicator for the reaction. What color will you see at the endpoint?

    a) pH = 4.65   b) pH = 8.43   c) pH = 7.00

23. **Compare** You crush two antacid tablets, X and Y and add each to 100 mL of 1.00M HCl. After stirring, you titrate the leftover HCl in the resulting solution with 1.00M NaOH. The brand X tablet requires less NaOH than the brand Y tablet requires. Which antacid neutralizes more acid?

CHAPTER 15 Study Guide

Download quizzes, key terms, and flash cards from glencoe.com.

BIG Idea The reaction of an acid with a base produces water and a salt.

Section 15.1 Acid and Base Reactions

MAIN Idea The concentrations of the acid and the base determine the classification of an acid-base reaction.

Vocabulary
- Brønsted-Lowry model (p. 526)
- ionic equation (p. 518)
- net ionic equation (p. 520)
- neutralization reaction (p. 516)
- salt (p. 516)
- spectator ion (p. 520)

Key Concepts
- Acid-base reactions are classified by the strengths of the acid and base.
- Strong acid–strong base, weak acid–strong base, and weak base–strong acid are the predominant types of acid–base reactions.
- Acids and bases in reactions can be identified using a hydrogen-ion transfer definition. An acid is an H^+ donor; a base is an H^+ acceptor.
- When acids and bases react, the pH of the final solution is dependent upon the nature of the reactants.
- Strong acid–strong base reactions yield neutral solutions.
- Weak acid–strong base reactions yield more basic solutions, where-as strong acid–weak base reactions produce acidic solutions.

Section 15.2 Applications of Acid-Base Reactions

MAIN Idea Acid-base reactions play an important role in the human body.

Vocabulary
- buffer (p. 531)
- standard solution (p. 539)
- titration (p. 539)

Key Concepts
- A buffer is a solution that maintains a relatively constant pH when H^+ ions or OH^- ions are added.
- Antacids are bases that react with stomach acid.
- An acid-base titration uses an acid-base reaction to determine the molarity of an unknown acid or base.

Chapter 15 Assessment

Understand Concepts

24. Describe what it means for a reaction to go to completion.

25. Identify the three types of acid-base reactions that always go to completion? Give an example of each by writing an overall equation.

26. Write and balance the overall equation for each of the following reactions. Identify the type of acid-base reaction represented by the equation.
    a) potassium hydroxide (KOH) + phosphoric acid (H_3PO_4)
    b) formic acid ($HCHO_2$) + calcium hydroxide ($Ca(OH)_2$)
    c) barium hydroxide (BaOH) + sulfuric acid (H_2SO_4)

27. For each of the reactions in Question 26, write the ionic and net ionic equations.

28. For each of the reactions in Question 26, what ions are spectator ions? Will the final reaction mixture be acidic, basic, or neutral? Explain.

29. In words, not equations, explain the use and differences in the overall, ionic, and net ionic equations.

30. Define a buffer solution.

31. Write the pH control reactions for the carbon dioxide-based buffer in blood.

32. Why is it incorrect to define a buffer as a solution that maintains a constant pH?

33. If the H^+ concentration in blood increases, what happens to the concentrations of H_2CO_3, HCO_3^-, and H^+?

34. What role do the lungs play in regulating blood pH?

35. Write two reactions that explain why lakes in limestone areas are capable of resisting pH decreases due to acid rain.

36. Consider an antacid that contains aluminum hydroxide. Write the overall equation that shows how this antacid reduces the acidity of stomach acid.

37. A student found that 53.2 mL of a 0.232M solution of NaOH was required to titrate 25.0 mL of an acetic acid solution of unknown molarity to the endpoint. What is the molarity of the acetic acid solution?

38. A student neutralizes 30.0 mL of a sample of sodium hydroxide with 28.9 mL of 0.150M HCl. What is the molarity of the sodium hydroxide?

39. How does the endpoint pH of a strong acid-strong base titration compare with that of a weak acid-strong base titration?

40. How does the endpoint pH of a strong base-strong acid titration compare with that of a weak base-strong acid titration?

41. Use **Table 15.3** to determine the molarity of the ammonia solution.

Table 15.3	Volume and Molarity	
	NH_3	H_2SO_4
Volume (mL)	50.0	36.3
Molarity (M)		0.100

42. What is hyperventilation? How does it change the pH of blood?

Apply Concepts

43. Methylamine (CH_3NH_2) is a weak base, as ammonia is. When methylamine completely reacts with hydrochloric acid, the final solution has a pH less than 7. Why are the products of this "neutralization" reaction not neutral? Use the net ionic equation to help in your explanation.

44. How many milliliters of 0.200M HCl are required to react with 25.0 mL of 0.100M methylamine (CH_3NH_2)?

45. Why are magnesium hydroxide and aluminum hydroxide effective antacids, but sodium hydroxide is not?

46. How many milliliters of $0.100M$ NaOH are required to neutralize 25.0 mL of $0.150M$ HCl?

47. Complete and balance the following overall equations.
    a) KOH(aq) + HNO$_3$(aq)
    b) Ba(OH)$_2$(aq) + HCl(aq)
    c) NaOH(aq) + H$_3$PO$_4$(aq)
    d) Ca(OH)$_2$(aq) + H$_3$PO$_4$(aq)

48. Dihydrogen phosphate (H$_2$PO$_4^-$) and monohydrogen phosphate (HPO$_4^{2-}$) ions play an important role in maintaining the pH in intracellular fluid. Write equations that show how these ions maintain the pH.

49. The concentration of H$_2$CO$_3$ in blood is 1/20 of the concentration of HCO$_3^-$, yet the blood buffer is capable of buffering the pH against bases, as well as against acid. Explain.

50. Tartaric acid is often added to artificial fruit drinks to increase tartness. A sample of a certain beverage contains 1.00 g of tartaric acid (H$_2$C$_4$H$_4$O$_6$). The beverage is titrated with $0.100M$ NaOH. Assuming no other acids are present, how many milliliters of base are required to neutralize the tartaric acid?

51. How many grams of tartaric acid (H$_2$C$_4$H$_4$O$_6$) must be added to 150 mL of $0.245M$ NaOH to completely react?

Chemistry and Society

52. What function of blood is most important when developing artificial blood?

Earth Science Connection

53. What acid most likely causes groundwater to be acidic? How does groundwater become acidic?

Everyday Chemistry

54. How is the carbon dioxide concentration in the blood related to hiccups?

How It Works

55. Explain how molecules and ions are related to taste.

How It Works

56. Can an indicator provide an exact pH? Explain.

Think Critically

Observe and Infer

57. A sample of rainwater turns blue litmus red. Fresh portions of the rainwater turn thymol blue indicator yellow, bromphenol blue indicator green, and methyl red indicator red. Estimate the pH of the rainwater.

Interpret Data

58. When formic acid (HCHO$_2$) reacts completely with NaOH, the resulting solution has a pH greater than 7. Why are the products of this neutralization reaction not neutral? Use the net ionic equation to help in your explanation.

Applying Concepts

59. Write an overall equation for the acid-base reaction that would be required to produce each of the following salts.
    a) NaCl   d) (NH$_4$)$_2$SO$_4$
    b) CaSO$_4$ e) KBr
    c) MgCl$_2$

Observe and Infer

60. **ChemLab** Explain why different bottles of the same brand of vinegar might contain solutions that have different pHs.

Make Predictions

61. **MiniLab 1** Would a solution of iron(III) bromide (FeBr$_3$) be acidic, basic, or neutral?

Relate Cause and Effect

62. **MiniLab 2** Explain why phenolphthalein and methyl orange are used as indicators in MiniLab 2.

Chapter 15 Assessment

Cumulative Review

63. Give the name of the compound represented by the formula $Mn(NO_3)_2 \cdot 4H_2O$, and determine how many atoms of each element are present in three formula units of the compound. (*Chapter 5*)

64. Terephthalic acid is an organic compound used in the formation of polyesters. It contains 57.8 percent C, 3.64 percent H, and 38.5 percent O. The molar mass is known to be approximately 166 g/mol. What is the molecular formula of terephthalic acid? (*Chapter 12*)

65. Write equations for the dissociation of the following ionic compounds when they dissolve in water. (*Chapter 13*)
    a) $CuSO_4$
    b) $Ca(NO_3)_2$
    c) Na_2CO_3

66. What is a monoprotic acid? A triprotic acid? Give an example of each. (*Chapter 14*)

Skill Review

67. **Data Table** Solutions of five different monoprotic acids are all 0.100M. The pH of each solution is given in **Table 15.4**. Copy the Data Table below. Rank the acids in the Data Table from weakest to strongest. For each solution, use the indicator table in **Figure 14.22** to predict the color that each solution would produce with the given indicator.

Table 15.4	pH of Solution
Solution	pH
A	5.45
B	1.00
C	3.45
D	4.50
E	2.36

WRITING in Chemistry

68. Write an article about the effect of acid rain on a specific aspect of a local environment such as a lake or a forest. Give some history of the problem and indicate when local residents first realized a problem exists. What, if any, corrective measures have been taken to correct the problem? Is the environmental damage reversible?

Problem Solving

69. A 165-mg sample of an antacid tablet containing calcium carbonate is dissolved in 50.0 mL of 0.100M HCl. After complete reaction, the excess HCl is titrated with 15.8 mL of 0.150M NaOH. Sketch a flowchart that shows the steps in the analysis.

Data Table				
Solution (weakest to strongest acid)	pH	Color in Bromphenol Blue	Color in Methyl Red	Color in Thymol Blue

Cumulative
Standardized Test Practice

1. What is the electron configuration of xenon?
   a) 2, 8, 18, 18, 7, 1
   b) 2, 8, 18, 32, 18, 7, 1
   c) 2, 8, 18, 18, 8
   d) 2, 8, 18, 32, 18, 8

2. A heterogeneous mixture
   a) cannot be separated by physical means.
   b) is composed of distinct areas of composition.
   c) is also called a solution.
   d) has the same composition throughout.

3. Heat of vaporization and heat of fusion are measurements of
   a) temperature.
   b) mass.
   c) energy.
   d) volume.

4. Which of the following conditions is not STP?
   a) 0.00°C
   b) 760 mm Hg
   c) 273 K
   d) 22.4 L

5. Red mercury(II) oxide decomposes at high temperatures to form mercury metal and oxygen gas:

   $$2HgO(s) \rightarrow 2Hg(l) + O_2(g)$$

   If 3.55 moles of HgO decompose to form 1.54 moles of O_2 and 618 g of Hg, what is the percent yield of this reaction?
   a) 13.2%
   b) 42.5%
   c) 56.6%
   d) 86.8%

6. Water molecules form drops because of
   a) surface tension.
   b) capillarity.
   c) evaporation.
   d) sublimation.

7. Acids are defined as substances that
   a) produce hydronium ions in water.
   b) have hydrogen in their compounds.
   c) produce hydroxide ions in water.
   d) neutralize bases.

Use the table below to answer question 8.

Weak Acid Ionization Constants	
Weak Acid	Ionization Constant
Hydrofluoric acid	6.3×10^{-4}
Methanoic acid	1.8×10^{-4}
Ethanoic acid	1.8×10^{-5}
Hypochlorous acid	4.0×10^{-8}

8. The ionization constant of a weak acid is a calculation of the number of ionized molecules (products) in a dilute aqueous solution divided by the number of un-ionized molecules (reactants) in the solution. Which acid is the weakest?
   a) hydrofluoric acid
   b) methanoic acid
   c) ethanoic acid
   d) hypochlorous acid

9. How many milliliters of $0.225M$ HCl would be required to titrate 6.00 g of KOH?
   a) 0.0561 mL
   b) .0475 L
   c) 1.50 mL
   d) 475 mL

NEED EXTRA HELP?									
If You Missed Question...	1	2	3	4	5	6	7	8	9
Review Section...	7.2	1.1	10.2	11.2	12.2	13.1	14.1	15.1	15.2

CHAPTER 16 Oxidation-Reduction Reactions

BIG Idea Oxidation-reduction reactions involve the transfer of electrons.

16.1 The Nature of Oxidation-Reduction Reactions
MAIN Idea Oxidation and reduction are complementary—as an atom is oxidized, another atom is reduced.

16.2 Applications of Oxidation-Reduction Reactions
MAIN Idea Oxidation and reduction reactions are among the most common reactions in both nature and industry.

ChemFacts

- The coloration in these "red rocks" results from the oxidation of iron-rich minerals to form the mineral hematite, Fe_2O_3.
- When oxidation occurs, an atom or ion loses electrons. During the formation of hematite, iron changes from Fe^{2+} to Fe^{3+}.
- The opposite of oxidation is reduction, which describes when an atom or ion gains an electron. "Red rocks" lose their distinct coloration when the presence of water reduces iron to Fe^{2+}.

Start-Up Activities

LAUNCH Lab

Observe a Redox Reaction

The rust you see on objects such as cars and bridges is the result of a reaction between iron and oxygen. Can iron nails also react with substances other than oxygen?

Materials
- iron nail
- steel wool or sandpaper
- graduated cylinder
- test tube
- 1M copper(II) sulfate ($CuSO_4$)

Procedure
1. Read and complete the lab safety form.
2. Polish the end of a nail with steel wool or sandpaper.
3. Add about 3 mL 1.0M $CuSO_4$ to a test tube. Place the polished end of the nail into the $CuSO_4$ solution. Let stand and observe for about 10 minutes. Record your observations.

Analysis
1. **Describe** What happened to the color of the copper(II) sulfate solution? What is the substance found clinging to the nail?
2. **Summarize** Write the balanced chemical equation for the reaction you observed.

Inquiry Was the change to the nail chemical or physical? How could you determine this?

Chemistry Online

Visit **glencoe.com** to:
- study the entire chapter online
- explore **Concepts In Motion**
- take Self-Check Quizzes
- use Personal Tutors
- access Web Links for more information, projects, and activities
- find the Try at Home Lab, Testing the Oxidation Power of Bleach

FOLDABLES Study Organizer

Oxidation-Reduction Reactions
Make the following Foldable to help you organize information about oxidation-reduction reactions.

STEP 1 Fold up the bottom of a horizontal sheet of paper about 5 cm.

STEP 2 Fold the sheet in half.

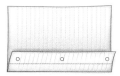

STEP 3 Open the paper and staple the bottom flap to make two compartments. Label as shown.

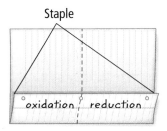

FOLDABLES Use this Foldable with Sections 16.1 and 16.2. As you read about oxidation-reduction reactions, define them and summarize information on index cards and store them in the appropriate compartments.

Section 16.1

Objectives

◗ **Analyze** the characteristics of an oxidation-reduction reaction.

◗ **Distinguish** between oxidation reactions and reduction reactions by definition.

◗ **Identify** the substances that are oxidized and those that are reduced in a redox reaction.

◗ **Differentiate** between oxidizing and reducing agents in redox reactions.

Review Vocabulary

buffer: a solution that resists changes in pH when moderate amounts of acids or bases are added

New Vocabulary

oxidation-reduction reaction
oxidation
reduction
oxidizing agent
reducing agent

The Nature of Oxidation-Reduction Reactions

MAIN Idea Oxidation and reduction are complementary—as an atom is oxidized, another atom is reduced.

Real-World Reading Link Keeping a car rust free can be a challenge, especially in areas that use a lot of road salt in the winter and in areas that are near the ocean. Why do metals rust? How can you prevent rusting? Oxidation-reduction reactions provide the key to answer these questions.

What is oxidation-reduction?

Oxygen is the most abundant element in Earth's crust. It is very reactive and can combine with almost every other element. An element that bonds to oxygen to form a compound called an oxide usually loses electrons because oxygen is more electronegative.

Recall from Chapter 9 that an electronegative element has a strong attraction for electrons. Because of this strong attraction, oxygen is able to pull electrons away from other atoms. The reactions in which elements combine with oxygen to form oxides were among the first to be studied by early chemists, who grouped them together and called them oxidation reactions.

Later, chemists realized that some other nonmetal elements can combine with substances in the same way as oxygen and that these reactions are similar to oxidation reactions. Modern chemists use the term oxidation to refer to any chemical reaction in which an element or compound loses electrons to another substance.

A common oxidation reaction occurs when iron metal loses electrons to oxygen. Each year in the United States, corrosion of metals—especially the iron in steel—costs billions of dollars as automobiles, ships, and bridges and other structures are slowly eaten away. **Figure 16.1** shows some of this damage and one way it can be prevented.

■ **Figure 16.1** One way to protect a steel structure from rust is to paint it. The coat of paint seals out the air and moisture that contribute to corrosion, but the paint deteriorates over time. Therefore, objects such as this bridge must be repainted often.

■ **Figure 16.2** Both buckets in the photo are steel. The bucket on the left was galvanized–protected from oxidation with a coating of a more active metal such as zinc. Zinc loses electrons to oxygen more readily than iron, so the zinc coating is oxidized preferentially, forming zinc oxide. The coating of zinc and zinc oxide prevents the formation of rust by keeping oxygen from reaching the iron in the steel.

Redox A second way to prevent rust is to galvanize steel. **Figure 16.2** compares a galvanized bucket with one that is not galvanized. What happens to the zinc (Zn) in galvanized steel? It reacts with oxygen to form zinc oxide (ZnO) in the following reaction.

$$2Zn(s) + O_2(g) \rightarrow 2ZnO(s)$$

You learned in Chapter 6 that this is classified as a synthesis reaction. The formation of zinc oxide falls into another, broader class of reactions characterized by the transfer of electrons from one atom or ion to another. A chemical reaction in which electrons are transferred from one atom or ion to another is called an **oxidation-reduction reaction**, commonly known as a redox reaction.

Many important chemical reactions are redox reactions. The formation of rust is one example; the combustion of fuels is another. In each redox reaction, one element loses electrons, and another element acquires them. However, remember that oxygen is not always involved in redox reactions. For example, the reaction of iron and chlorine to form iron(III) chloride is a redox reaction, even though oxygen is not involved.

Keeping track of electrons If you examine the equation above for the reaction between zinc and oxygen more closely, you can see which atoms are gaining electrons and which are losing them. You also can determine where the electrons go during a redox reaction by comparing the oxidation number of each type of atom or ion before and after the reaction takes place.

Recall from Chapter 5 that the oxidation number of an ion is equal to its charge. All atoms in their elemental form—whether uncombined or as diatomic molecules—have a charge of zero and are assigned an oxidation number of zero. Thus, both the zinc atom and the diatomic oxygen molecule that react to form zinc oxide have oxidation numbers of zero. In the ionic compound formed, each oxide ion has a 2– charge and an oxidation number of 2–. Because the compound must be neutral, the total positive charge must be 4+; thus, each zinc ion must have a charge and an oxidation number of 2+.

Section 16.1 • The Nature of Oxidation-Reduction Reactions

Oxidation The part of a redox reaction in which an element loses electrons is called the oxidation part of the reaction. **Oxidation** is the loss of electrons by the atoms or ions of a substance. The element that loses the electrons is oxidized and becomes more positively charged; that is, its oxidation number increases. Zinc is oxidized during the formation of zinc oxide because metallic zinc atoms each lose two electrons. The oxidation reaction can be written by itself to show how zinc changes during the redox reaction as follows.

$$Zn: \rightarrow [Zn]^{2+} + 2e^- \quad \text{(loss of electrons)}$$

Reduction What happens to the electrons that are lost by the zinc atom? Electrons do not wander around by themselves; they must be transferred to another atom or ion. Oxidation reactions are always paired with reduction reactions. **Reduction** is the gain of electrons by the atoms or ions of a substance. The element that acquires the electrons becomes more negatively charged. Its oxidation number decreases, and the element is said to be reduced. Because oxidation and reduction reactions occur together, each is referred to as a half-reaction.

In every redox reaction, at least one element undergoes reduction while another undergoes oxidation. Just as a successful pass in football requires a quarterback to throw the ball and a receiver to catch it, a redox reaction must have one element that gives up electrons and one that accepts them. The electronic structure of both reactants changes during a redox reaction.

Figure 16.3 illustrates the movement of electrons in the formation of zinc oxide. Oxygen accepts the electrons that zinc loses. Oxygen is reduced during the reaction between zinc and oxygen because each oxygen atom gains two electrons. Like the oxidation reaction, the reduction reaction can be written by itself as follows.

$$\cdot \ddot{O} : + 2e^- \rightarrow : \ddot{O} :^{2-} \quad \text{(gain of electrons)}$$

VOCABULARY

WORD ORIGIN

Reduction
comes from the Latin *re*, meaning *back*, and *ducere*, meaning *to lead*

■ **Figure 16.3** In the formation of zinc oxide, the zinc atom loses two electrons during the reaction, becoming a zinc ion. The oxygen atom gains the two electrons from zinc, becoming an oxide ion.
Apply How do the oxidation numbers of zinc and oxygen change during this reaction?

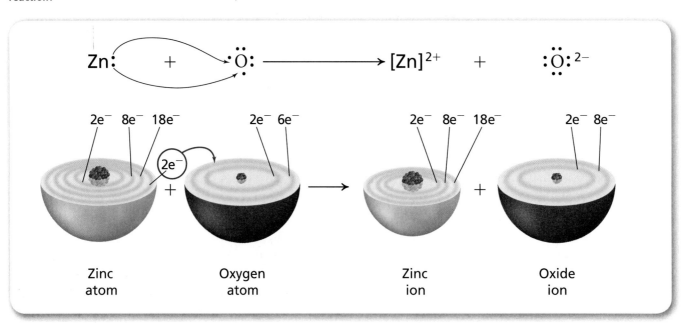

MiniLab 16.1

Corrosion of Iron

What factors affect the corrosion of a nail? Corrosion is the term generally used to describe the oxidation of a metal during its interaction with the environment. In this MiniLab, you will study the corrosion of a nail and determine the factors that affect this process.

Procedure
1. Read and complete the lab safety form.
2. Dissolve a **package of clear, unflavored gelatin** in about 200 mL of **warm water**. Stir in 2 mL of **phenolphthalein solution** and 2 mL of **potassium hexacyanoferrate(III) solution**. Pour the prepared solution into a **widemouth glass jar** or **petri dish** to a depth of about 1 cm.
3. In the liquid gelatin, place a plain **iron nail,** an **aluminum nail,** a **galvanized iron nail,** and a **painted iron nail**. Space the nails far apart.
4. Allow the gelatin to set for several hours or overnight.
5. Record your observations of the nails and the substances in the gelatin.

Analysis
1. **Identify** Which of the nails have reacted with the substances in the gelatin? What is the evidence of corrosion?
2. **Explain** Any blue color in the gelatin is due to a loss of electrons and the formation of iron(II) ions. Any pink or red color in the gelatin is due to the gaining of electrons by oxygen and water molecules. Which of these reactions is oxidation, and where does it occur on the reacting nail?
3. **Apply** Did any of the nails not corrode in the solution? What methods are commonly used to prevent or minimize corrosion?

Combining the half-reactions The equation for the reduction half-reaction shows one atom of oxygen reacting. However, oxygen is found in nature as a diatomic molecule (O_2). The reduction equation must be multiplied by 2 to reflect this. Thus, the balanced equation for the reduction reaction is written as follows.

$$O_2 + 4e^- \rightarrow 2\ddot{\underset{..}{O}}{:}^{2-}$$

Note that four electrons are gained by the oxygen molecule. Those four electrons come from two atoms of zinc that take part in the reaction. Therefore, the balanced equation for the oxidation reaction is written as follows.

$$2Zn{:} \rightarrow 2[Zn]^{2+} + 4e^-$$

The balanced overall equation for the reaction now can be written as shown in **Figure 16.4**.

Concepts In MOtion
Interactive Figure To see an animation of a redox reaction, visit glencoe.com.

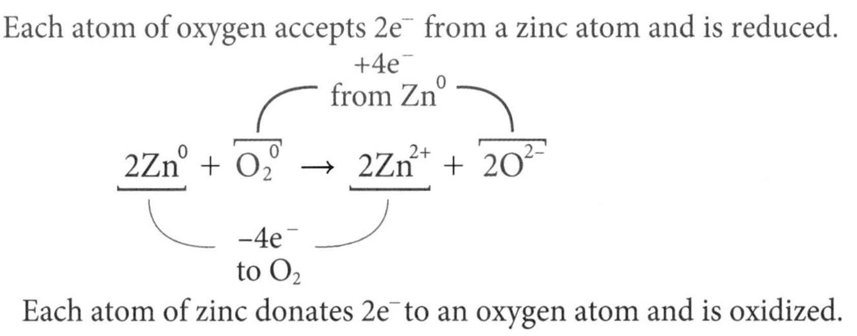

Each atom of oxygen accepts $2e^-$ from a zinc atom and is reduced.

$$\underline{2Zn^0} + \underline{O_2^{\,0}} \rightarrow \underline{2Zn^{2+}} + \underline{2O^{2-}}$$

+4e^- from Zn^0
−4e^- to O_2

Each atom of zinc donates $2e^-$ to an oxygen atom and is oxidized.

■ **Figure 16.4** Two zinc atoms combine with one diatomic oxygen molecule to form two formula units of zinc oxide.
Explain *why the equation is balanced in terms of electrons.*

The equation shown in **Figure 16.4** is the same as the equation for the formation of zinc oxide that you read at the beginning of the discussion on redox reactions. Now you know that it represents the net oxidation-reduction reaction and is the sum of an oxidation half-reaction and a reduction half-reaction. A memory aid might help you remember the distinction between oxidation and reduction. The phrase **L**oss of **E**lectrons is **O**xidation and **G**ain of **E**lectrons is **R**eduction is shortened to **LEO GER** (**LEO** the lion says **GER** or for short, **LEO GER**).

Origin of the term reduction Why is it called reduction when an element gains electrons? Electrons carry a negative charge. Thus, any atom that gains electrons undergoes a reduction in the charge or oxidation number. While this explanation makes intuitive sense, there is an older, historic reason for the use of the term *reduction*.

The term was first applied to processes in furnaces, illustrated in **Figure 16.5,** in which metals are isolated from their ores at high temperatures. Ordinarily, metals are found in ores which means they are combined with other elements. In a process called smelting, furnaces operate at high temperatures to separate the metal from these other elements. During this process of refining the ores to a more pure state, oxygen is removed from the ores in which it is combined with the metal. In other words, the ore is reduced to the free metal. The positively charged metal ions in the ore are reduced to the elemental state, while oxygen and other negatively charged elements are oxidized. This refining process results in a reduction in the amount of solid material and a considerable decrease in volume.

Identifying a Redox Reaction

The oxidation of zinc is a redox reaction in which oxygen is a reactant. You have learned that elements other than oxygen can accept electrons and become reduced during redox reactions.

■ **Figure 16.5** Copper smelting probably began over 7,000 years ago. The smelting of iron in furnaces goes back roughly 3,000 years.

A stone furnace once used for smelting iron

Iron smelting in a modern industrial blast furnace

■ **Figure 16.6** As a result of this redox reaction between zinc and copper sulfate solution, solid copper metal is deposited on the strip of metal.

You are already familiar with the explosive reaction in which sodium and chlorine combine to form table salt.

$$2Na(s) + Cl_2(g) \rightarrow 2NaCl(s)$$

Are electrons transferred during this reaction? Yes, because each sodium atom loses one electron to become a sodium ion with a charge of 1+. The oxidation number of sodium increases from 0 to 1+. Each chlorine atom gains one electron to form a chloride ion. The oxidation number of chlorine decreases from 0 to 1−. Therefore, this is another example of a redox reaction.

$$\overset{\overset{2e^-}{\frown}}{2Na^0 + Cl_2^0} \rightarrow 2Na^+ + 2Cl^-$$

Oxidizing and reducing agents Another redox reaction that doesn't involve oxygen occurs when a strip of zinc metal is placed in a solution of copper(II) sulfate, shown in **Figure 16.6.** The equations for this reaction are shown in **Figure 16.7.** The progress of this reaction can be followed easily because an observable change takes place. The blue copper(II) sulfate solution gradually becomes colorless if a strip of zinc metal is placed in it. The zinc gives up electrons, becoming oxidized to zinc ions. The colorless zinc ions that form go into solution. The copper ions pick up electrons from zinc and become reduced to copper metal atoms, which are deposited on the strip. As shown in **Figure 16.6,** copper metal begins to form on the zinc strip.

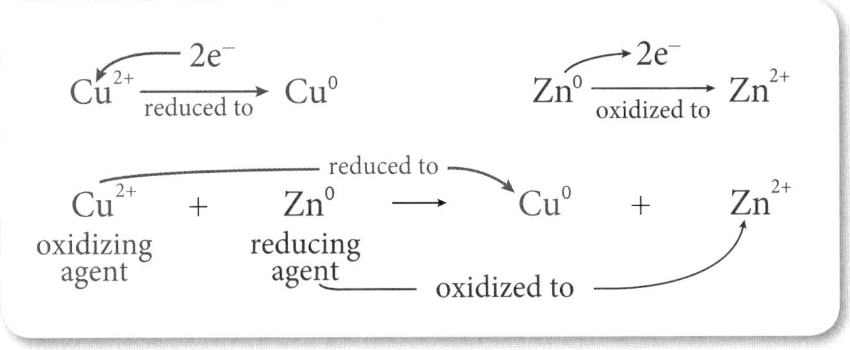

$$Cu^{2+} \xrightarrow[\text{reduced to}]{2e^-} Cu^0 \qquad Zn^0 \xrightarrow[\text{oxidized to}]{2e^-} Zn^{2+}$$

$$\underset{\substack{\text{oxidizing}\\\text{agent}}}{Cu^{2+}} + \underset{\substack{\text{reducing}\\\text{agent}}}{Zn^0} \longrightarrow Cu^0 + Zn^{2+}$$

■ **Figure 16.7** Copper (II) is the oxidizing agent that loses electrons and zinc is the reducing agent that accepts these electrons.

CHEMLAB

OXIDATION AND REDUCTION REACTIONS

Background
Copper atoms and ions often take part in reactions by losing or gaining electrons, which are called oxidation and reduction reactions, respectively. If copper atoms lose electrons to form positive ions, copper is oxidized. Other atoms or ions must gain the electrons that the copper atoms lose. These atoms or ions are reduced when they accept electrons and therefore acquire a negative charge. In this ChemLab, you will observe two reactions that involve the oxidation or reduction of copper.

Question
What are some common reactions that involve the oxidation or reduction of copper?

Objectives
- **Observe** reactions that involve the oxidation or reduction of copper.
- **Classify** the reactants as the substance that is oxidized, the reducing agent, the substance reduced, and the oxidizing agent.

Preparation

Materials
copper(II) oxide
powdered charcoal (carbon)
weighing paper
balance
large ovenproof glass test tubes (2)
1-hole rubber stopper fitted with glass tube with bend as shown
150-mL beakers (2)
graduated cylinder, small
glass stirring rod
Bunsen or Tirrill burner
ring stand
test-tube clamp
thermal glove
limewater (calcium hydroxide solution)

Safety Precautions
WARNING: *Use care in handling hot objects and when working around open flames. Do not breathe in the fumes that are produced during the teacher demonstration in step 1.*

Procedure

1. Read and complete the lab safety form.

2. **Teacher Demonstration** Your teacher will perform this reaction as a demonstration either in the fume hood or outside the building. **WARNING:** *Do not perform this procedure by yourself.* A 1-cm square of copper foil will be placed in a porcelain evaporating dish. First, 5 mL of water, then 5 mL of concentrated nitric acid (HNO_3) will be added. Note the color of the evolved gas and the color of the resulting solution. Record your observations in a table similar to the one under Data and Observations.

3. On a piece of weighing paper, mix approximately 1 g of copper(II) oxide with twice its volume of powdered charcoal thoroughly. Place the mixture in a clean, dry ovenproof glass test tube. Add about 10 mL of limewater to a second test tube, and stand it in a 150-mL beaker. Assemble the apparatus as shown below, with the copper-oxide test tube sloped slightly downward and the delivery tube extending into the limewater.

4. Heat the mixture in the test tube, gently at first and then strongly. As soon as you notice a change in the limewater, carefully remove the stopper and delivery tube from the reaction test tube. **WARNING:** *Do not stop heating as long as the tube is in the limewater.* Record your observations of the limewater in your table.

5. Continue heating the reaction test tube until a glow spreads throughout the reactant mixture. Turn off the burner.

6. After the reaction test tube has cooled to nearly room temperature, empty the contents into a beaker that is about half full of water. In a sink, slowly stir the mixture while running water into the beaker until all the unreacted charcoal has washed away. Observe the product that remains in the beaker, and record your observations.

Data and Observations

Data Table	Observations
Step 2: Gas	
Solution	
Step 4: Limewater	
Step 6: Product	

Analyze and Conclude

1. **Interpret Data** What evidence of chemical change did you observe in each reaction?

2. **Interpret Data** In the first reaction, a blue-colored solution indicates the presence of Cu^{2+} ions, the brown gas is NO_2, and the colorless gas is NO. In the second reaction, if limewater becomes cloudy and white, carbon dioxide gas has reacted with the calcium hydroxide to form insoluble calcium carbonate. Using this information, analyze your data and observations. Determine which reactants (Cu and HNO_3 for the first reaction, CuO and C for the second reaction) were oxidized and which were reduced in each reaction.

Apply and Assess

1. The mass of copper produced in the second reaction is less than the mass of the reacting copper(II) oxide. With this in mind, explain why the gain of electrons is known as reduction?

2. What mass of copper may be produced from the reduction of 1.0 metric ton of copper(II) oxide? Hint: *Determine the formula mass of copper(II) oxide.*

INQUIRY EXTENSION

Water-treatment plants utilize redox reactions that involve chlorine, which reacts with organic materials in the water. What is oxidized and what is reduced in these reactions? How do these reactions help make the water suitable for drinking?

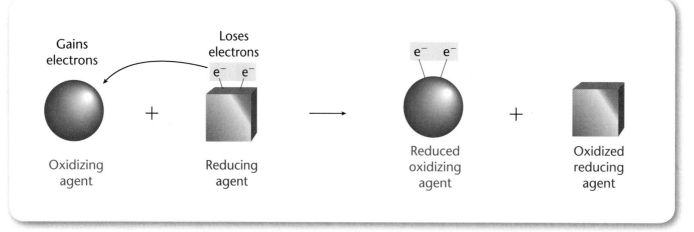

■ **Figure 16.8** In a redox reaction, the element that is reduced oxidizes another element by attracting electrons from it, so it is called an oxidizing agent. The element that is oxidized reduces the first element by transferring electrons to it, so it is called a reducing agent.

Oxidizing agents What role do the copper ions play in the redox reaction? Each copper ion is reduced to uncharged copper metal when it accepts electrons from the zinc metal. Because the copper ion is the agent that oxidizes zinc metal to form a zinc ion, Cu^{2+} is called an oxidizing agent. An **oxidizing agent** is the substance that gains electrons in a redox reaction. The oxidizing agent is the material that is reduced.

Reducing agents Because oxidation and reduction go hand in hand, a reducing agent must be present in the redox reaction. Zinc metal is the agent that supplies electrons and reduces the copper ion to copper metal; therefore, zinc is called the reducing agent. A **reducing agent** is the substance that loses electrons in a redox reaction. The reducing agent is the material that is oxidized. **Figure 16.8** summarizes the roles of oxidizing and reducing agents in redox reactions.

Section 16.1 Assessment

Section Summary

▶ Oxidation occurs when an atom or ion loses one or more electrons.

▶ Reduction takes place when an atom or ion gains electrons.

▶ Oxidation and reduction reactions always occur together in a net process called a redox reaction.

▶ An oxidizing agent gains electrons and is reduced during a redox reaction.

▶ A reducing agent loses electrons and is oxidized during a redox reaction.

1. **MAIN Idea** **Name and explain** the two half-reactions that make up a redox reaction.

2. **Identify** which reactant is reduced and which is oxidized in each of the following reactions.
   a) $2Al(s) + 3Cu^{2+}(aq) \rightarrow 2Al^{3+}(aq) + 3Cu(s)$
   b) $2Cr^{3+}(aq) + 3Zn(s) \rightarrow 2Cr(s) + 3Zn^{2+}(aq)$
   c) $2Au^{3+}(aq) + 3Cd(s) \rightarrow 2Au(s) + 3Cd^{2+}(aq)$

3. **Identify** What is the oxidizing agent when iron rusts? What is the reducing agent?

4. **Apply** The following equation represents the reaction between an acid and a base to form a salt and water. Determine the oxidation number for each element. Is this a redox reaction? Explain.

   $$2KOH(aq) + H_2SO_4(aq) \rightarrow K_2SO_4(aq) + 2H_2O(l)$$

5. **Explain** Compounds that are easily oxidized can act as antioxidants to prevent other compounds from being oxidized. Vitamins C and E protect living cells from oxidative damage by acting as antioxidants. Why does adding lemon juice to fruit salad prevent the fruit from browning?

Section 16.2

Objectives
- **Analyze** common redox processes to identify the oxidizing and reducing agents.
- **Identify** some redox reactions that take place in living cells.

Review Vocabulary
oxidation: reaction in which an element loses electrons

FOLDABLES
Incorporate information from this section into your Foldable.

Applications of Oxidation-Reduction Reactions

MAIN Idea Oxidation and reduction reactions are among the most common reactions in both nature and industry.

Real-World Reading Link Redox reactions are going on around you every day, everywhere. A few examples of common redox reactions include the combustion of gasoline in a car engine, the formation of rust on the surface of steel, photosynthesis in plants, and energy production by alkaline batteries. Redox reactions even occur in your body when you metabolize food.

Everyday Redox Reactions

Near the vents of explosive volcanoes, where lava or ash erupts from deep within Earth, enormous deposits of solid yellow sulfur can be found. **Figure 16.9** shows such a deposit. The element sulfur acts both as an oxidizing agent and as a reducing agent in the reaction that forms the sulfur deposits. Can you tell which sulfur compound serves each function in this reaction?

$$2H_2S(g) + SO_2(g) \rightarrow 3S(s) + 2H_2(g) + O_2(g)$$

Note that more than one element in a reaction can be oxidized or reduced. The sulfur in hydrogen sulfide (H_2S) and the oxygen in sulfur(SO_2) dioxide both are oxidized. Sulfur in sulfur dioxide and hydrogen in hydrogen sulfide both are reduced. Each reactant acts as both a reducing agent and an oxidizing agent.

Understanding natural redox reactions such as the one that occurs in gas-rich volcanoes has allowed chemists to develop many processes that make use of oxidation and reduction reactions. Without them, photographs and steel wouldn't exist, and stains would be much harder to remove from clothing.

■ **Figure 16.9** The brilliant yellow color of these deposits is characteristic of elemental sulfur. Hydrogen sulfide and sulfur dioxide gases are commonly released from volcanic vents. In addition to forming deposits such as these, they can produce acid rain and alkaline lakes.

■ **Figure 16.10** In a daguerreotype, a redox reaction between silver and iodine fumes produced a layer of light-sensitive silver iodide on the surface of the polished photographic plate. Exposure to light caused decomposition of the silver iodide into elemental silver, which was then treated with the fumes of heated mercury. The image of Paris shown here was made by Daguerre himself.

VOCABULARY
WORD ORIGIN
Photograph
comes from the Greek *photos*, meaning *light*, and *graphein*, meaning *to write*

Redox in photography Leonardo da Vinci described a primitive "camera" before 1519, in which someone had to trace images focused on a glass plate inside a box. However, it wasn't until 1838 that the French inventor L.J.M. Daguerre successfully fixed the images in a camera on highly polished, silver-plated copper to make the first photographs. These early photographs, like the one shown in **Figure 16.10,** were called daguerreotypes in his honor.

Modern photographic film is made of a plastic backing covered with a layer of gelatin, in which millions of grains of silver bromide are embedded. When light strikes a grain, silver and bromide ions are converted into their elemental forms through a redox reaction. The equation for this redox reaction is as follows.

$$2Ag^+ + 2Br^- \rightarrow 2Ag + Br_2$$

The reaction begins when the shutter on a camera is opened. Light from the scene being photographed passes through the camera's lens and shutter and strikes the light-sensitive silver bromide on the film. The light energy causes electrons to be ejected from a few of the bromide ions, oxidizing them to elemental bromine. The electrons are transferred to silver ions, reducing them to metallic silver atoms. These grains are now activated.

The developing chemicals continue the redox reaction. In areas where the light is brightest, more grains are activated, and after developing, they become the darker areas. No silver atoms form in areas of the film that are not struck by light, and that part of the film remains transparent. The exposed film is then developed into a negative, during which time the remaining AgBr and Br_2 is washed away. **Figure 16.11** describes the developing and printing processes.

FIGURE 16.11

Developing Film

Making photographic negatives by developing exposed film involves several steps. The process describes how black-and-white pictures are made. For color photos, light-sensitive dyes are combined with the silver bromide in layers on the film.

1. The exposed film is transferred to a canister, where it is developed using a solution of a reducing agent, called a developer. The organic compound hydroquinone is usually used for this purpose. The developer reduces all the silver ions to silver atoms in any grain of silver bromide that was hit by light, but it does not react with silver ions in grains that were not exposed to light. Because metallic silver is dark and silver bromide is light, an image having light and dark areas is produced.

2. After the film has been developed, a solution of a fixer containing thiosulfate ions ($S_2O_3^{2-}$) is added. Thiosulfate ions react with unreduced silver ions (Ag^+) to form a soluble complex $[Ag(S_2O_3)_2]^{3-}$, which is washed away. This prevents unreduced silver ions from becoming reduced and darkening slowly over time. The reaction follows.

$$AgBr(s) + 2S_2O_3^{2-}(aq) \rightarrow [Ag(S_2O_3)_2]^{3-}(aq) + Br^-(aq)$$

3. The fixed film is washed to remove any remaining developer or fixer solution. The photographic negative is the reverse of the image photographed; that is, light areas in the scene are dark on the film, and vice versa.

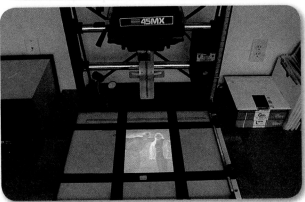

4. When light is shone through the negative onto light-sensitive photographic paper, a photographic print is made. The print is positive; light and dark areas are identical to those in the scene.

Section 16.2 • Applications of Oxidation-Reduction Reactions

Physics Connection

Rocket Booster Engines

If you have ever built and launched a model rocket, you probably noticed that the rocket engine was made of a solid, highly combustible material packed into a cardboard tube. After ignition, the expansion and expulsion of the gases produced enough downward force to launch the lightweight rocket quickly into the air. Space shuttles use a similar type of technology, but on a much larger scale.

Engine systems The space shuttle has two different engine systems. The three main engines attached directly to the shuttle operate on liquid hydrogen and liquid oxygen reservoirs carried in the large, centrally located disposable fuel tank. The two smaller, reusable, strap-on booster rockets on either side of the main fuel tank are loaded with a solid fuel, which undergoes a powerful, thrust-producing, oxidation-reduction reaction that helps boost the shuttle into orbit.

Solid rocket fuel The solid rocket fuel is a mixture containing 69.8 percent ammonium perchlorate, 16 percent aluminum powder, 12 percent polymer binder, 2 percent epoxy-curing agent, and 0.2 percent iron oxide powder as a catalyst. Once ignited, the engine cannot be extinguished. The extremely reactive ammonium perchlorate supplies oxygen to the easily oxidized aluminum powder, providing a tremendously exothermic and fast reaction. The purpose of the polymer binder is to hold the mixture together and to help it burn evenly. The overall redox reaction is shown here.

$$\overset{\overset{24e^-}{\frown}}{8Al + 3NH_4ClO_4} \rightarrow 4Al_2O_3 + 3NH_4Cl$$

Figure 1 Space shuttle solid rocket boosters firing

Shuttle forces Each solid rocket booster, shown firing in **Figure 1**, weighs 591,000 kg at liftoff, produces 11.5 million Newtons (N) of force, and operates for about two minutes into the flight. For comparison, a 1000-kg car accelerating from 0 to 26.8 m/s² (60 mph) in 7 seconds would require a force of only 3830 N. The release of energy and expansion of hot gases due to the oxidation-reduction reaction within the engine of the solid rocket booster produces the tremendous thrust needed to get the 2 million-kg shuttle from 0 to almost 700 m/s² (1500 mph) in just 132 seconds.

Connection to Chemistry

1. **Apply** Powdered aluminum is used in another highly exothermic reaction, the thermite reaction, which is used for welding metals. The reaction is as shown: $2Al + Fe_2O_3 \rightarrow Al_2O_3 + 2Fe$ What role does the powdered aluminum play in this reaction?

2. **Acquire Information** Investigate the lives and research of Robert Goddard and Werhner Von Braun, who both experimented with rockets in the 1930s and helped guide the United States into the space age. Write a short report about these men.

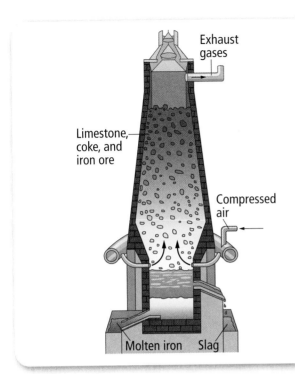

Redox in a blast furnace Iron is seldom found in the elemental form needed to make steel. Metallic iron must be separated and purified from iron ore—usually hematite (Fe_2O_3). This process takes place in a blast furnace through a series of redox reactions. The major reaction in which iron ore is reduced to iron metal uses carbon monoxide gas as a reducing agent.

First, a blast of hot air causes coke, a form of carbon, to burn, producing CO_2 and heat. Limestone ($CaCO_3$), which is mixed with the iron ore in the furnace, decomposes to form lime (CaO) and carbon dioxide, which then oxidizes the coke in a redox reaction to form carbon monoxide. The carbon monoxide is used to reduce the iron ore to iron. This process is illustrated in **Figure 16.12.**

$$CaCO_3(s) \rightarrow CaO(s) + CO_2(g)$$

$$CO_2(g) + C(s) \rightarrow 2CO(g)$$

$$2Fe^{3+}(s) + 3O^{2-}(s) + 3CO(g) \xrightarrow{6e^-} 2Fe^0(l) + 3CO_2(g)$$

■ **Figure 16.12** Iron ore (Fe_2O_3), coke (C), and limestone ($CaCO_3$) are added at the top of the furnace. Hot air at about 900°C burns the coke in an exothermic reaction. This reaction causes temperatures in a blast furnace to reach about 2000°C. Molten iron and slag are separately drawn off at the bottom of the furnace.

Redox in bleaching processes Bleaches can be used to remove stains from clothing. Where do the stains go? Bleach does not actually remove the chemicals in stains from the fabric; it reacts with them to form colorless compounds. In chlorine bleaches, an ionic chlorine compound in the bleach reacts with the compounds responsible for the stain. This ionic compound is sodium hypochlorite (NaOCl). The hypochlorite ions oxidize the molecules that cause dark stains.

NaOCl⁻(aq) + stain molecule(s) →
    (colored)   Cl⁻(aq) + oxidized stain molecule(s)
                    (colorless)

> **FACT of the Matter**
>
> The Stone, Bronze, and Iron Ages are historical periods that draw their names from the material most commonly used to make tools during each time. The Bronze Age came before the Iron Age because copper and tin, the elements that are alloyed to form bronze, are easier to smelt than iron. Iron requires higher temperatures to be reduced to its elemental form.

Figure 16.13 Hypochlorite is used in disinfectants to kill bacteria in swimming pools and in drinking water. In both cases, the hypochlorite ions act as oxidizing agents. Bacteria are killed when important compounds in them are destroyed by oxidation. In this photo, the amount of chlorine in the water is being monitored. Chlorine reacts with the water to form hypochlorite ions.

TRY AT HOME LAB
See page 875 for **Testing the Oxidation Power of Bleach.**

Bleaches containing hypochlorite should be used carefully because hypochlorite is a powerful oxidizing agent that can damage delicate fabrics. These bleaches usually have a warning label telling the user to test an inconspicuous part of the fabric before using the product.

In addition to acting as a bleaching agent, hypochlorite ions are also used as disinfectants to kill bacteria. Again, it is the role of hypochlorite as a powerful oxidizing agent that accounts for bleach's effectiveness as a disinfectant. Hypochlorite ions kill bacteria by destroying important compounds in the bacteria. Disinfectants containing hypochlorite are used to kill bacteria in swimming pools, as shown in **Figure 16.13,** in drinking water, and on surfaces such as kitchen counters.

MiniLab 16.2

Test for Alcohol

How can you identify alcohol in household products? Alcohols react with orange dichromate ions ($Cr_2O_7^{2-}$), producing blue-green chromium(III) ions (Cr^{3+}). This reaction is used in a breathalyser test to test for the presence of alcohol in a person's breath. In this MiniLab, you will use this reaction to test for the presence of alcohol in a number of household hygiene, cosmetic, and cleaning products.

Procedure
1. Read and complete the lab safety form.
2. Label five **small test tubes** with the names of the products to be tested.
3. Place approximately 1 mL of each product in the appropriate tube.
4. Add three drops of **dichromate reagent** to each tube, and stir to mix the solutions. **WARNING:** *Do not allow dichromate reagent to come into contact with skin. Wash with large volumes of water if it does.*
5. Observe and record any color changes that occur within one minute.

Analysis
1. **Determine** Which of the products that you tested contain alcohol? Was the presence of alcohol noted on the labels of these products?
2. **Identify** If the orange $Cr_2O_7^{2-}$ ion reacts with alcohol to produce the blue-green Cr^{3+} ion, what substance is the reducing agent in the reaction?

How It Works

Breathalyser Test

A breathalyser is a device for estimating blood alcohol content from a breath sample. The alcohol in beverages, hair spray, and mouthwash is ethanol. Ethanol is a volatile liquid that evaporates rapidly at room temperature. Because of its volatility, drinking an alcoholic beverage results in a level of gaseous ethanol in the breath that is proportional to the level of alcohol in the bloodstream. Almost 40 percent of all automobile accidents that result in a fatality are caused by intoxicated drivers. Law officers can determine quickly whether a person is legally intoxicated by using a breathalyser.

The first breathalysers depended on a solution of potassium dichromate oxidizing ethanol to form acetic acid, causing a color change in the process.

Modern breathalysers project an infrared beam of light through the captured breath in the sample chamber. The characteristic bond of alcohols is the O-H bond. Infrared light is absorbed by the O-H bond of alcohol. The more alcohol that is present, the more infrared light is absorbed, causing a decrease of infrared light reaching the detector on the other side. Less infrared light reaching the detector results in a higher alcohol reading.

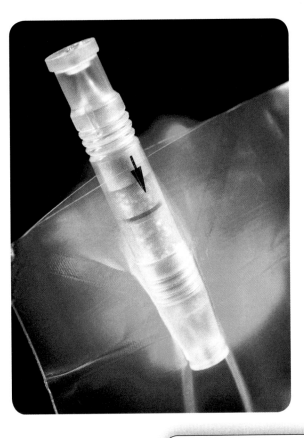

1 A simple breathalyser device has an inflatable plastic bag attached to a tube containing an orange solution of potassium dichromate and sulfuric acid.

2 During a breathalyser test, a person blows into the mouthpiece of the bag.

3 If alcohol vapors are present in the person's breath, ethanol undergoes a redox reaction with the dichromate. As ethanol is oxidized to acetic acid, the orange Cr^{6+} ions are reduced to blue-green Cr^{3+} ions.

4 The exact color produced depends on the amount of alcohol in the breath. The color change that is produced during the test is compared to standard color mixtures of the two chromium ions to get an estimate of the blood alcohol level.

Think Critically

1. Suppose a person used mouthwash shortly before taking a breathalyser test. What might be the result?
2. How would the color produced in a breathalyser test change as the ethanol content of the blood increases?

■ **Figure 16.14** The green color of the Statue of Liberty in New York Harbor is due to a layer of patina, or protective coating, that covers the copper sheets making up the statue.

Corrosion of metals Did you know that the Statue of Liberty is made of copper sheets attached to a steel skeleton? Why does it appear green rather than the reddish-brown color of copper? When copper is exposed to humid air that contains sulfur compounds, it undergoes a slow oxidation process. Under these conditions, the copper metal atoms each lose two electrons to produce Cu^{2+} ions, which form the compounds $CuSO_4$, $3Cu(OH)_2$ and $Cu_2(OH)_2CO_3$. These compounds are responsible for the green coat of patina found on the surface of copper objects that have been exposed to air for long periods of time, like the Statue of Liberty shown in **Figure 16.14**.

You have learned that iron is oxidized by oxygen in the air to form rust. Aluminum is a more active metal than iron. As a result of its greater activity, aluminum is oxidized more quickly than iron. If this is true, why does an aluminum can degrade much more slowly than a tin can, which is made of iron-containing steel that is coated with a thin layer of tin? The reason is that, like copper, aluminum is oxidized to form a compound that coats the metal and protects it from further corrosion, as illustrated in **Figure 16.15**. Aluminum reacts with oxygen to form aluminum oxide (Al_2O_3) in a redox reaction.

$$4Al(s) + 3O_2(g) \rightarrow 2Al_2O_3(s)$$

A coating of aluminum oxide is tough and does not flake off easily like iron oxide rust does. Because iron rust is porous and flaky, it does not form a good protective coating for itself. When rust flakes fall off a surface, additional metal is exposed to air and becomes corroded.

■ **Figure 16.15** A tin coating offers some protection to the iron. However, if a hole or crack develops in the thin tin coating, the underlying iron corrodes rapidly. A tin-coated steel will degrade completely in about 100 years. The aluminum oxide coating on an aluminum can is tough and densely packed. It protects the underlying aluminum from further corrosion. The can will take about 400 years to degrade.

Explain *Why are metals such as copper and aluminum sometimes called self-protecting metals?*

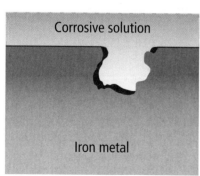

Steel can

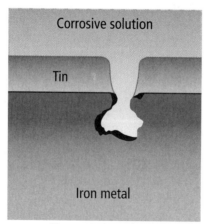

Tin-coated steel can

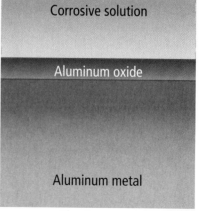

Aluminum can

Everyday Chemistry

Lightning-Produced Fertilizer

Did you know that one of the main nutrients plants need is nitrogen? Although the air surrounding Earth is 78 percent nitrogen, the nitrogen is in the form of N_2 molecules, a form that most plants and animals cannot use. Nitrogen from the air is converted to a form that plants can use by a process called nitrogen fixation. Plants can best use nitrogen when it is in the form of the ammonium ion (NH_4^+) where the nitrogen has an oxidation number of 3−, but they can also use the nitrate ion (NO_3^-) with nitrogen having an oxidation number of 5+.

Nitrogen fixation Nitrogen can be fixed for plants in three ways: by lightning, by nitrogen-fixing bacteria living in the roots of plants or in the soil, and by commercial synthesis reactions such as the Haber process, in which a reaction between nitrogen and hydrogen is forced to produce ammonia.

Nitrogen is a fairly inert gas because the triple bond of N_2 is strong and resists breaking. However, the exceptionally high temperatures of lightning, like that shown in **Figure 1**, which heat the surrounding air to 10,000 C°, can easily break bonds and allow for recombination of gases in the atmosphere.

Lightning-driven reactions In the process of lightning-driven nitrogen fixation, nitrogen and oxygen combine to form nitrogen monoxide. Nitrogen monoxide then combines with more oxygen to form nitrogen dioxide. This nitrogen dioxide mixes with water in the air to form nitric acid and more nitrogen monoxide, which is available to continue the cycle.

$$N_2 + O_2 \rightarrow 2NO$$

$$2NO + O_2 \rightarrow 2NO_2$$

$$3NO_2 + H_2O \rightarrow 2HNO_3 + NO$$

Figure 1 Lightning fixes nitrogen

Fertilizer production The pH of rainwater is naturally slightly acidic, and some of this acidity is due to the dissolved nitric acid (HNO_3) from nitrogen fixation. Bacteria in the soil convert the nitrate ions into ammonium ions when it rains.

How does nature manufacturing fixed nitrogen compare with commercial production of fixed nitrogen? Lightning may seem uncommon, but it is estimated that at any given moment, there are about 2000 thunderstorms in progress around the world. Lightning hits the earth an average of 100 times per second, or 8.6 million times a day. Due to the abundance of lightning strikes per day, approximately 36 billion kg of nitrogen are fixed yearly in the atmosphere. Biological agents such as bacteria fix about 122 billion kg of nitrogen yearly, and about 77 billion kg of nitrogen is fixed through the manufacture of fertilizer through the Haber process.

Explore Further

1. **Classify** Nitrogen fixation in the soil is accomplished by bacteria living in the roots of certain plants. Name some of these plants.
2. **Apply** In each of the three equations shown, what is oxidized and what is reduced?
3. **Acquire Information** The process by which nitrogen is put back into the air is called denitrification. Find out what conditions are necessary for this process and what reaction occurs.

VOCABULARY

WORD ORIGIN

Tarnish
comes from the old French *teme*, meaning *dull*

Silver tarnish: a redox reaction Imagine if, along with your usual chores of taking out the trash, washing dishes, feeding your pets, and taking care of your younger siblings, you also had to polish the silver—as people did back in your great-grandparents' days. How would you find time for any fun? Fortunately, other materials such as stainless steel have replaced most "silverware." Why do silver utensils have to be polished, but those made of stainless steel or aluminum do not? Silver becomes tarnished through a redox reaction that is a form of corrosion, similar to rusting. Tarnish is formed on the surface of a silver object when silver reacts with hydrogen sulfide (H_2S) in air. The product, black silver sulfide (Ag_2S), forms the coating of tarnish on the silver.

$$O_2(g) + 4Ag(s) + 2H_2S(g) \rightarrow 2Ag_2S(s) + 2H_2O(l)$$

Many commercial silver polishes contain abrasives that help to remove tarnish. Unfortunately, they also remove some of the silver. A gentle way to remove tarnish from the surface of a silver object involves another redox reaction. In this reaction, aluminum foil scraps act as a reducing agent.

$$2Al^0(s) + 6Ag^+(s) + 3S^{2-}(s) + 6H_2O(l) \rightarrow$$
$$6Ag^0(s) + 2Al^{3+}(aq) + 6OH^-(aq) + 3H_2S(g)$$

(6e⁻ transferred)

This reaction is essentially the reverse of the reaction that forms tarnish. Here, silver ions in the Ag_2S tarnish are reduced to silver atoms, while aluminum atoms in the foil are oxidized to aluminum ions. The tarnish-removing solution usually includes baking soda (sodium hydrogen carbonate) to help remove any aluminum oxide coating that forms and to make the cleaning solution more conductive. **Figure 16.16** describes how this method of silver cleaning is done.

■ **Figure 16.16** The tarnish on this silver pitcher results from an unwanted redox reaction, but another redox reaction can help remove it. A nest of crumpled aluminum foil scraps is made at the bottom of a large pot. The tarnished silver object is added, making sure the silver is in contact with the foil scraps. Baking soda is added, and the silver is covered with water. When the pot is heated on a stove, the silver sulfide tarnish is reduced to silver atoms, and the silver object becomes shiny and bright.

■ **Figure 16.17** Lightning is the light energy that is released through chemiluminescence.

Chemiluminescence Some redox reactions can release light energy at room temperature. The production of this kind of cool light by a chemical reaction is called chemiluminescence. The light from chemiluminescent reactions can be used in emergency light sticks that work without an external energy source. You may recall learning in Chapter 6 how these light sticks work. When you snap the glass capsule inside the plastic case, you start a chemical reaction that releases light energy as electrons are transferred. Now you know that the reaction that takes place when the two solutions in the light sticks are mixed involves an oxidation and a reduction. When the reactants are used up, light energy is no longer released and light is no longer produced.

Lightning Some chemiluminescent redox reactions occur naturally in the atmosphere as a result of lightning, as shown in **Figure 16.17**. When lightning is produced by an electrical discharge in the atmosphere, electrons in molecules of O_2 and N_2 gases are excited to higher energy levels. Energy from the electricity breaks the O_2 and N_2 molecules into atoms. When the atoms recombine to form other molecules and the electrons return to lower energy levels, light energy is released through chemiluminescence.

Forensic investigations Other chemiluminescent reactions involve luminol, an organic compound that emits cool blue-green light when it is oxidized. Luminol is a crystalline solid that is white to pale yellow in color, and soluble in water. Luminol reactions are utilized by forensic chemists to analyze evidence in crime investigations. They spray luminol onto a location where the presence of blood is suspected. If blood is present, the iron(II) ions in the hemoglobin of red blood cells oxidize the luminol to form a chemiluminescent compound that glows in the dark. The iron is reduced by the luminol and the luminol is oxidized. **Figure 16.18** shows the blue-green glow from the oxidized form of luminol.

■ **Figure 16.18** When luminol is oxidized and is observed in the dark, an eerie blue-green glow is produced through chemiluminescence.

CHEMISTRY AND TECHNOLOGY

Forensic Blood Detection

The gas station at the corner was robbed, and the cashier was shot. A bloody shoeprint and a bloody handprint, as shown in **Figure 1**, were found at the scene. On television, police announce that Suspect A has been taken into custody. They have confiscated a shoe, allegedly worn by the suspect. After preliminary examination by the police department, the shoe is sent to a forensic laboratory for scientific investigation including a test to determine whether or not there are blood stains on the shoe.

■ **Figure 1** A bloody handprint is visualized with luminol.

The Luminol Test

The laboratory technician may choose from several chemical tests for blood, all based on the fact that the hemoglobin in blood catalyzes the oxidation of a number of organic indicators to produce a colored product that emits light, or luminesces as shown in **Figure 1**.

The technician on this case chooses the luminol test. Luminol has an organic double-ring structure, shown in **Figure 2**. In 1928, German chemists first observed the blue-green luminescence when the compound was oxidized in alkaline solution. It was soon found that a number of oxidizing agents, such as hydrogen peroxide, cause luminescence. Later, workers noted that the luminescence was greatly enhanced by the presence of blood, which led to its current use in forensic investigations.

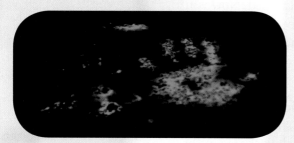

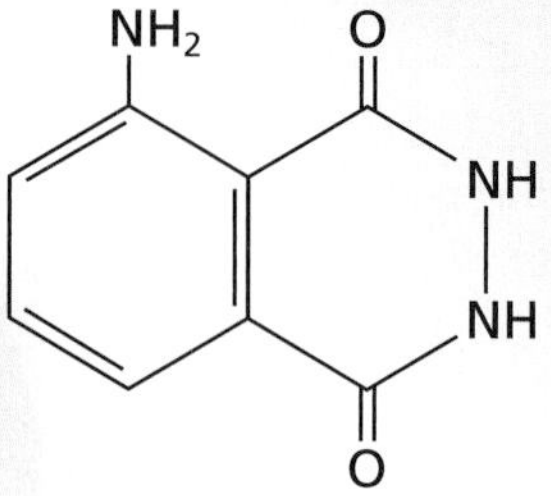

■ **Figure 2** Molecular structure of luminol

The technician carefully mixes an alkaline solution of luminol with aqueous sodium peroxide and in a darkened workplace sprays the solution onto the shoe. Bingo! An intense, blue-green chemiluminescence is emitted from several spots. The technician photographs the spots and their characteristic light.

Ruling Out with Luminol

You may wonder if this relatively simple procedure will serve to convict Suspect A. Certainly not. If the test had been negative and the bloody fingerprints did not match, Suspect A might have been cleared from suspicion. A negative result ensures that a stain is not blood. But, because this is not the case with the stains on the shoe, the luminol test is preliminary and will be used along with other tests.

The luminol test is especially useful because it works well with both fresh and dried blood. Luminol has one particularly useful feature. The same stains can be made luminescent over and over again if the spray is allowed to dry and the stains are resprayed.

A positive test should not be taken as absolute proof of blood because luminol reacts with copper and cobalt ions. However, it reacts much more strongly with hemoglobin. A large number of forensic authorities believe that the luminol test has value as a preliminary sorting technique.

Discuss the Technology

1. **Apply** If a luminol test yields a positive reaction, what is the next logical step?
2. **Infer** Why can it almost never be assumed that stains are uncontaminated, although stain evidence is important in a criminal investigation?

Biochemical redox processes How are bears able to stay warm enough to keep from freezing during their winter hibernation? How do marathon runners get the energy to finish a race without stopping to eat? In both cases, fats stored in the body are oxidized. Oxygen molecules from the air are reduced as they gain electrons to form water. In a series of redox reactions called respiration, energy is released. **Figure 16.19** shows one effect of this energy release in plants. Respiration will be discussed in Chapter 19.

Many other redox reactions take place in living things. Electrons are transferred between molecules in redox reactions during photosynthesis. You will study photosynthesis in Chapter 20.

Bioluminescence Some organisms can use the energy released during redox reactions to convert chemical energy into light energy, a process called bioluminescence. You are probably familiar with the flashing lights given off by fireflies during courtship, but do you know that many different organisms—including some fish, at least one type of mushroom, and a caterpillar known as a glowworm— are also bioluminescent? **Figure 16.20** shows bioluminescence in the firefly squid, *watasenia scintillans*.

Browning fruit Now that you have learned what redox reactions are and have read about some of the processes of which they are a part, you can reexamine the redox reaction that makes cut fruit turn brown. The color is due to brown pigments that are formed by the oxidation of colorless compounds normally present in the cells of the fruit. Oxygen in the air is the oxidizing agent that reacts with the colorless compounds to produce the brown pigments. The oxygen is reduced when it accepts electrons from the pigments, so the pigments function as reducing agents. As in all redox reactions, this combination of oxidation and reduction goes hand in hand because electrons that are lost by one element must be gained by another.

■ **Figure 16.19** Although it is common to think that only mammals keep warm, in truth, all plants and animals maintain a temperature at which their enzymes function best. Plants keep from freezing because heat is produced as a by-product of respiration and photosynthesis. One of the first plants to poke through the snow in early spring is the heat-producing skunk cabbage. The heat it releases allows it to get a head start on other plants and also contributes to the unpleasant odor that gives it its name.

■ **Figure 16.20** Squid use bioluminescence to attract prey and mates and to confuse predators. Light energy is released during an enzyme-catalyzed redox reaction. Luciferase is the name given to the enzyme that speeds up the reaction in which the organic molecule luciferin is oxidized.

Firefly Squid luminescening

Firefly Squid

Figure 16.21 Vitamin C owes its antioxidant properties to the fact that it reacts so readily with oxygen. When added to a food product, oxygen reacts preferentially with vitamin C, thereby sparing the food product from oxidation. Other antioxidant food additives include the synthetic compounds BHA and BHT and the natural antioxidant, vitamin E.

The skin of a fruit keeps oxygen out, which is why unbroken fruit does not turn brown. Coating cut fruit with an antioxidant can prevent browning and keep a fruit salad looking fresh longer. The vitamin C in lemons is a good antioxidant. If lemon juice is squirted onto cut banana or apple slices, they will not brown as quickly because the vitamin C reacts with oxygen more readily than do the fruit-browning compounds, as shown in **Figure 16.21**.

Connecting Ideas

Most reactions involve electron transfer and thus are redox reactions. You have learned to identify which element is reduced and which is oxidized when you are given the equation for a redox reaction. You might question why one element accepts electrons from another and whether you can accurately predict which element will be oxidized and which will be reduced. Learning to make those predictions is the next step in your study of electron-transfer processes in compounds. In Chapter 17, you will investigate electron transfer and relate this process to redox reactions in batteries.

Section 16.2 Assessment

Section Summary

- Redox reactions can reduce metals found in ores to their pure states.
- Bleach oxidizes the molecules that cause stains to form colorless molecules.
- Metals such as copper and aluminum react with oxygen to form protective coatings to prevent corrosion.
- Chemiluminescent reactions in emergency light sticks, lightning, and the luminol reaction convert the energy of chemical bonds into light energy.

6. **MAIN Idea** List two important industrial redox reactions and two important natural redox reactions.

7. **Describe** How is most of the iron that is used for making steel purified from iron ores?

8. **Explain** Why do aluminum cans degrade more slowly than cans made of iron?

9. **Apply** Oxygen is required for the production of light by fireflies. What role does the oxygen play in the reaction?

10. **Consider** Why can't rust stains be removed with bleach?

11. **Infer** What chemical process do hibernating animals use to stay warm?

12. **Summarize** Why is aluminum metal used to remove tarnish from silver?

13. **Recognize** Why is gold rather than copper used to coat electrical connections in expensive electronic equipment?

CHAPTER 16 Study Guide

STUDY TO GO — Download quizzes, key terms, and flash cards from glencoe.com.

BIG Idea Oxidation-reduction reactions involve the transfer of electrons.

Section 16.1 The Nature of Oxidation-Reduction Reactions

MAIN Idea Oxidation and reduction are complementary—as an atom is oxidized, another atom is reduced.

Vocabulary
- oxidation (p. 556)
- oxidizing agent (p. 562)
- oxidation-reduction reaction (p. 555)
- reducing agent (p. 562)
- reduction (p. 556)

- Oxidation occurs when an atom or ion loses one or more electrons.
- Reduction takes place when an atom or ion gains electrons.
- Oxidation and reduction reactions always occur together in a net process called a redox reaction.
- An oxidizing agent gains electrons and is reduced during a redox reaction.
- A reducing agent loses electrons and is oxidized during a redox reaction.

$$Cu^{2+} \xrightarrow[\text{reduced to}]{2e^-} Cu^0 \qquad Zn^0 \xrightarrow[\text{oxidized to}]{2e^-} Zn^{2+}$$

$$\underbrace{Cu^{2+}}_{\text{oxidizing agent}} + \underbrace{Zn^0}_{\text{reducing agent}} \longrightarrow Cu^0 + Zn^{2+}$$

(reduced to / oxidized to)

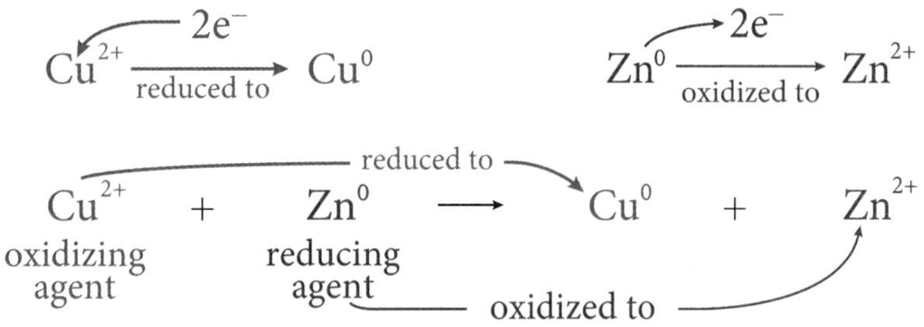

Section 16.2 Applications of Oxidation-Reduction Reactions

MAIN Idea Oxidation and reduction reactions are among the most common reactions in both nature and industry.

- Redox reactions can reduce metals found in ores to their pure states.
- Bleach oxidizes the molecules that cause stains to form colorless molecules.
- Metals such as copper and aluminum react with oxygen to form protective coatings to prevent corrosion.
- Chemiluminescent reactions in emergency light sticks, lightning, and the luminol reaction convert the energy of chemical bonds into light energy.

Luminol

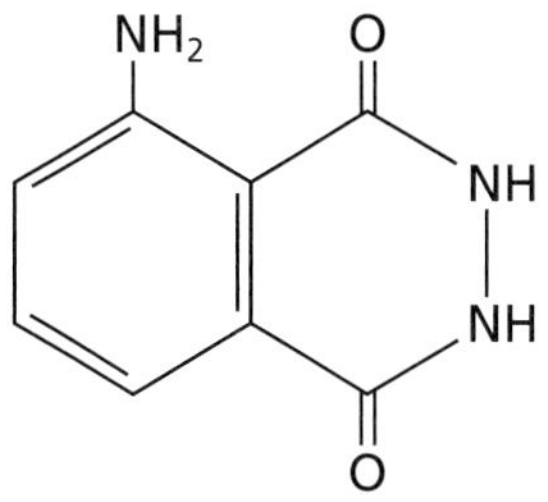

Chapter 16 Assessment

Understand Concepts

14. What is the difference between an oxidizing agent and a reducing agent?

15. Which of the changes indicated are oxidations and which are reductions?
    a) Cu becomes Cu^{2+}
    b) Sn^{4+} becomes Sn^{2+}
    c) Cr^{3+} becomes Cr^{6+}
    d) Ag becomes Ag^+

16. Identify the oxidizing agent in each of the following reactions.
    a) $Cu^{2+}(aq) + Mg(s) \rightarrow Cu(s) + Mg^{2+}(aq)$
    b) $Fe_2O_3(s) + 3CO(g) \rightarrow 2Fe(l) + 3CO_2(g)$

17. What is the oxidizing agent in household bleach?

18. Why does a photographic negative need to be fixed?

19. In which direction do electrons move during a redox reaction: from oxidizing agent to reducing agent or reducing agent to oxidizing agent?

20. Write the equation for the redox reaction that occurs when a piece of zinc metal is dipped in a solution of copper(II) sulfate.

21. Identify the the reaction in **Figure 16.22** as an oxidation reaction or a reduction reaction.

    $Fe^{2+} \rightarrow Fe^{3+} + e^-$

    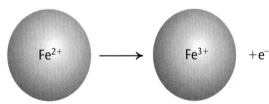

    ■Figure 16.22

22. Determine the oxidation number of the bold-face element in each of these compounds.
    a) H**N**O$_3$
    b) **Sb**$_2$O$_5$

Apply Concepts

23. If galvanized nails, which have been coated with zinc, are placed in a brown solution containing I_2, the solution slowly turns colorless. Adding a few drops of bleach to the colorless solution results in a return of the brown color. Explain what makes these changes occur.

24. List several ways in which a steel chain-link fence could be treated to prevent corrosion.

25. Explain why oxidation and reduction reactions must always occur together.

26. Write the equation for a reaction that is not a redox reaction. Are electrons transferred in this reaction?

27. Is oxygen a necessary reactant for an oxidation reaction? Explain.

Everyday Chemistry

28. Just before World War I, a German chemist named Fritz Haber developed a process for fixing atmospheric nitrogen into ammonia. The ammonia produced this way can be converted into ammonium nitrate, an important and explosive fertilizer.

    $$3H_2 + N_2 \rightarrow 2NH_3$$

    a) What element is oxidized during this reaction? What is reduced?
    b) What is the oxidizing agent? What is the reducing agent?

Physics Connection

29. By passing an electric current through water, the water can be separated into its component elements in the reverse of the reaction used to power the main stage of the space shuttle.

    $$2H_2O(l) + energy \rightarrow 2H_2(g) + O_2(g)$$

    a) Is this a redox reaction? If so, what element is oxidized?
    b) Where does the energy for this endothermic reaction come from?

Chapter 16 Assessment

How It Works

30. If ethanol were less volatile, how might the usefulness of a breathalyser test be affected? Explain.

Chemistry and Technology

31. Why should a positive result from the luminol test not be taken as proof of the presence of blood?

Think Critically

Use a Table

32. Table 16.1 lists some of the most common compounds that are used as oxidizing agents.

Table 16.1	Common Oxidizing Agents	
Formula	Name	Use
O_2		
H_2O_2		
$KMnO_4$		
Cl_2		
$K_2Cr_2O_7$		
HNO_3		
$NaClO$		
$KClO_3$		

a) Name each of the compounds in the table.
b) List at least one practical application, mentioned in this chapter or from a reference book, of each of these oxidizing agents.

Make Predictions

33. Hydrogen peroxide (H_2O_2) can be used to restore white areas of paintings that have darkened from the reaction of lead paint pigments with polluted air containing hydrogen sulfide gas.

$$PbS + 4H_2O_2 \rightarrow PbSO_4 + 4H_2O$$
(black) (white)

Could hydrogen peroxide be used to remove tarnish from silver objects? Would the reaction have any undesirable effects?

Interpret Data

34. ChemLab Write the balanced equation for the reaction that caused the limewater to become cloudy. Is this a redox reaction? Explain.

35. MiniLab 1 Why do you think corrosion seems to occur predominantly at the head and point of a nail?

Make Inferences

36. MiniLab 2 When a pile of orange ammonium dichromate is ignited, it decomposes in an exothermic reaction in which the green product and flames shoot upward like an erupting volcano. (**WARNING:** *Do NOT perform this reaction.*)

$$(NH_4)_2Cr_2O_7(s) \rightarrow Cr_2O_3(s) + N_2(g) + 4H_2O(g)$$

a) What is the reducing agent in this reaction? The oxidizing agent?
b) How is this reaction similar to the Breathalyser reaction?

Cumulative Review

37. Name each of the ionic compounds. (*Chapter 5*)
a) NaF   d) $Na_2Cr_2O_7$
b) CaS   e) KCN
c) $Al(OH)_3$ f) NH_4Cl

38. How many grams of nitrogen are needed to react completely with 346 g of hydrogen to form ammonia by the Haber process? (*Chapter 12*)

$$N_2 + 3H_2 \rightarrow 2NH_3$$

39. Draw Lewis dot diagrams for each of the following covalent molecules. (*Chapter 9*)
a) $CHCl_3$
b) CH_3CH_2OH
c) CH_3CH_3

40. A gaseous sample occupies 32.4 ml at −23°C and .075 atm. What will it occupy at STP? (*Chapter 11*)

Chapter 16 Assessment

41. Identify each as a pure substance or a mixture. (*Chapter 1*)
    a) petroleum
    b) fruit juice
    c) smog
    d) diamond
    e) milk
    f) iron ore

42. List some characteristic properties of metals. (*Chapter 3*)

Skill Review

43. **Design an Experiment** Do you think silver will tarnish more quickly in clean air or in polluted air? Design an experiment to test your hypothesis.

WRITING in Chemistry

44. Research the evidence that suggests that the antioxidant properties of vitamin C may help prevent cancer in people who take large doses of this vitamin. Write a summary of your findings in which you propose how to conduct more tests to determine whether vitamin C has anticarcinogenic properties.

45. A flask filled with acid-washed steel wool is fitted with a long, thin glass tube in a rubber stopper. When the flask is inverted so the tube opening is in a beaker of colored water, the water slowly begins to rise in the tube. Write a summary of this experiment, as if you had performed it. Explain what makes the water rise. Predict what portion of the flask will be filled with water at the end of the experiment.

46. The patina coating on the Statue of Liberty has preserved most of the copper metal in the statue. Some damage does occur wherever steel rivets are in contact with copper and exposed to water. Research the Statue of Liberty to determine why those sites are more susceptible to corrosion than the rest of the statue. Prepare a poster that includes a detailed description of how patina forms and a redox reaction to explain the process.

Problem Solving

47. Metallic lithium reacts vigorously with fluorine gas to form lithium fluoride.
    a) Write an equation for this process.
    b) Is this an oxidation-reduction reaction?
    c) If it is an oxidation-reduction, which element is oxidized? Which is reduced?
    d) If 2.0 g of lithium are reacted with 0.1 L fluorine at STP, which reactant is limiting?
    e) If 0.04 g of lithium fluoride is formed in reaction in part d., what is the percent yield?

48. Identify the oxidizing reagent in each of the following reactions.
    a) $C_2H_5OH(l) + 3O_2(g) \rightarrow 2CO_2(g) + 3H_2O(l)$
    b) $CuO(s) + H_2(g) \rightarrow Cu(s) + H_2O(l)$
    c) $2FeO(s) + C(s) \rightarrow 2Fe(s) + CO_2(g)$
    d) $2Fe^{2+}(aq) + Br_2(l) \rightarrow 2Fe^{3+}(aq) + 2Br^-(aq)$

49. When coal and other fossil fuels containing sulfur are burned, sulfur is converted to sulfur dioxide:
    $$S(s) + O_2(g) \rightarrow SO_2(g)$$
    a) Is this an oxidation-reduction reaction?
    b) If it is an oxidation-reduction, which element is oxidized? Which is reduced?
    c) If 7.0×10^3 kg of fuel containing 3.5 percent sulfur are burned in a city on a given day, how much SO_2 will be produced? Assume that the sulfur reacts completely.

50. Sodium nitrite is formed when sodium nitrate reacts with lead:
    $$NaNO_3(s) + Pb(s) \rightarrow NaNO_2(s) + PbO(s)$$
    a) What is the oxidizing reagent in this reaction? What is the reducing reagent?
    b) If 5.00 g of sodium nitrate reacts with an excess of lead, what mass of sodium nitrite will form if the percent yield is 100%?

Cumulative Standardized Test Practice

Data for Elements in the Redox Reaction $Zn + HNO_3 \rightarrow Zn(NO_3)_2 + NO_2 + H_2O$

Element	Oxidation Number	Complex ion of which element is a part
Zn	0	none
Zn in $Zn(NO_3)_2$	+2	none
H in HNO_3	+1	none
H in H_2O	?	none
N in HNO_3	?	NO_3^-
N in NO_2	+4	none
N in $Zn(NO_3)_2$	?	NO_3^-
O in HNO_3	−2	NO_3^-
O in NO_2	?	none
O in $Zn(NO_3)_2$	?	NO_3^-
O in H_2O	−2	none

Use the table above to answer Questions 1–3.

1. Which element forms a monatomic ion that is a spectator in the redox reaction?
   a) Zn   b) O   c) N   d) H

2. The oxidation number of N in $Zn(NO_3)_2$ is
   a) +3.   b) +5.   c) +1.   d) +6.

3. The element that is oxidized in this reaction is
   a) Zn.   b) O.   c) N.   d) H.

Use the chemical equation to answer Questions 4 and 5.

$$Mg(OH)_2(aq) + 2HCl \rightarrow MgCl_2(aq) + 2H_2O(l)$$

4. Which of the compounds in the equation is considered a base?
   a) $Mg(OH)_2$ c) $MgCl_2$
   b) HCl   d) H_2O

5. Which of the compounds in the equation is considered a salt?
   a) $Mg(OH)_2$ c) $MgCl_2$
   b) HCl   d) H_2O

6. A volume of gas has a pressure of 1.4 atm. What would the pressure be if you doubled the volume and reduced the temperature in kelvins by half?
   a) 0.35 atm   c) 2.8 atm
   b) 1.4 atm   d) 5.6 atm

7. How many grams of hydrogen make up 3.5 moles of glucose ($C_6H_6O_6$)?
   a) 1.7 g   c) 21.0 g
   b) 3.5 g   d) 101.5 g

8. Why do the ingredients in Italian salad dressing form a heterogeneous solution?
   a) The water is polar, and the oil is nonpolar.
   b) The oil is polar, and the water is nonpolar.
   c) Both water and oil are ionic substances.
   d) Both oil and water are covalent substances.

9. A triprotic acid is
   a) an acid with three hydrogen atoms.
   b) an acid with three acidic hydrogen atoms.
   c) an acid with three hydronium ions.
   d) an acid with three extra protons.

NEED EXTRA HELP?

If You Missed Question...	1	2	3	4	5	6	7	8	9
Review Section...	16.1	16.1	16.1	14.1	15.1	11.2	12.2	13.2	14.1

CHAPTER 17 Electrochemistry

BIG Idea Chemical energy can be converted to electric energy and electric energy to chemical energy.

17.1 Voltaic Cells: Electricity from Chemistry
MAIN Idea In voltaic cells, oxidation takes place at the anode, yielding electrons that flow to the cathode, where reduction occurs.

17.2 Electrolysis: Chemistry from Electricity
MAIN Idea In electrolysis, a power source causes nonspontaneous reactions to occur in electrochemical cells.

ChemFacts

- The temperature of a lightning bolt is hotter than the surface temperature of the Sun.
- The air around a lightning strike becomes so hot that it explodes—causing thunder.
- In a storm cloud, the upper part of the cloud has a positive charge, and the lower portion of the cloud has a negative charge.
- The path taken by lightning is not predictable, and it is not always the tallest object that is struck.

Start-Up Activities

LAUNCH Lab

A Lemon Battery

You can purchase a battery at any convenience store. You also can craft a working battery from a lemon. How do these power sources compare?

Materials
- lemon pieces
- zinc metal strip
- copper metal strip
- voltmeter with leads

Procedure
1. Read and complete the lab safety form.
2. Insert the zinc and copper strips into the lemon, about 2 cm apart from each other.
3. Attach the black lead from a voltmeter to the zinc and the red lead to the copper. Read and record the potential difference (voltage) from the voltmeter.
4. Remove one of the metals from the lemon and observe what happens to the potential difference on the voltmeter.

Analysis
1. **Infer** What is the purpose of the zinc and copper metals?
2. **Infer** What is the purpose of the lemon?
3. **Compare and contrast** a lemon battery and a 9 V battery.

Inquiry Could you obtain the same results from other fruits or vegetables? Why or why not?

Chemistry Online

Visit glencoe.com to:
- study the entire chapter online
- explore Concepts In Motion
- take Self-Check Quizzes
- use Personal Tutors
- access Web Links for more information, projects, and activities
- find the Try at Home Lab, Removing Electroplating

FOLDABLES Study Organizer

Electrochemical Cells Make this Foldable to help you compare voltaic cells to electrolytic cells.

STEP 1 Fold a sheet of paper in half.

STEP 2 Fold down about 3 cm at the top.

STEP 3 Open the paper and draw lines along the fold lines. Label the columns *Voltaic Cells* and *Electrolytic Cells*.

FOLDABLES Use this Foldable with Sections 17.1 and 17.2. As you read about electrochemical cells, summarize information about them in the appropriate columns.

Chapter 17 • Electrochemistry **583**

Section 17.1

Objectives

- **Relate** the construction of a voltaic cell to how it functions to produce a voltage and an electrical current.
- **Trace** the movement of electrons in a voltaic cell.
- **Relate** chemistry in a redox reaction to separate reactions occurring at electrodes in a voltaic cell.

Review Vocabulary

reduction: reaction in which an element gains one or more electrons

New Vocabulary

electric current
voltaic cell
anode
cathode
potential difference
voltage
cation
anion

Voltaic Cells: Electricity from Chemistry

MAIN Idea In voltaic cells, oxidation takes place at the anode, yielding electrons that flow to the cathode, where reduction occurs.

Real-World Reading Link What would you do with one half of a one-dollar bill? Without the other half, you couldn't spend it. Similarly, voltaic cells are made of two half cells. You need both halves to make a useful voltaic cell.

Redox and Electrochemical Cells

You have learned that oxidation and reduction always occur simultaneously. Think about the chemistry of corrosion you studied in Chapter 16. When iron metal reacts with oxygen, a redox reaction creates rust, iron oxide. Electrons are always transferred when a redox reaction occurs. In the rust reaction, electrons are transferred from the reducing agent, iron, to the oxidizing agent, oxygen.

Suppose you could separate the oxidation and reduction parts of a redox reaction and cause the electrons to flow through a wire. The flow of charged particles such as electrons in a particular direction is called an **electric current**. In other words, you are using a redox reaction to produce an electric current. This is what occurs in a battery—one form of an electrochemical cell in which chemical energy is converted to electrical energy.

Electrochemical cells Alessandro Volta is credited with inventing the first electrochemical cell in 1800. His cell consisted of layers of two different metals separated by cloth or cardboard soaked in an acidic solution. An image of cells based on Volta's early cells is shown in **Figure 17.1**. The layers served to separate the two halves of a spontaneous redox reaction, and thus were capable of producing electric current.

■ **Figure 17.1** Early batteries consisted of alternating layers of silver and zinc separated by pieces of paper soaked in salt water. These layers were repeated over and over to form a tall pile. Wires and an acid as the necessary electrolyte completed the apparatus, which was called a voltaic pile.

Electrochemical cells that use spontaneous redox reactions to create an electric current are named **voltaic cells** in honor of Volta. Modern batteries use different materials, but they still are simply one or more voltaic cells contained in a single package.

The lemon battery How does the lemon battery produce electrical energy? You made a lemon battery in the Launch Lab at the beginning of this chapter. The lemon itself is a container for a solution of electrolyte—the lemon juice. Recall from Chapter 4 that an electrolyte is a solution that conducts electricity. Lemon juice is sour; that is, it is acidic. The hydrogen ions from partially dissociated citric acid give it a sour taste and also provide the ions for conduction of charge through the lemon battery. The two dissimilar metal strips are the electrodes at which an oxidation reaction and a reduction reaction take place to provide the battery's power source. The **anode** is the electrode at which the oxidation reaction occurs. The **cathode** is the electrode at which the reduction reaction occurs.

These two electrodes are identified for the lemon battery in **Figure 17.2.** The electrode made of the metal that is more easily oxidized becomes the anode. The second electrode becomes the cathode. The substance in a lemon that is most easily reduced is the abundant hydrogen ion of the electrolyte. When these two reactions occur together in the same cell, they combine to produce a spontaneous redox reaction. This type of reaction is represented by the equation below, where M represents the metal that is oxidized and H^+ represents the hydrogen ion that is reduced.

$$M + 2H^+ \rightarrow M^{2+} + H_2$$

This spontaneous reaction generates the cell voltage of the lemon battery by producing a different electrical potential at each electrode.

Electric current Electrons in the metal electrodes of the lemon battery move through the external circuit as a current and can do useful work. For example, a portable radio needs power to work, and it gets this power from the electrons flowing through the wires from the anode to the cathode of the battery. The chemical reaction at the anode gives off electrons, which enter the metal and then flow through the external part of the circuit connecting the anode to the cathode. At the cathode, the electrons are used up in a reduction reaction.

FOLDABLES
Incorporate information from this section into your Foldable.

VOCABULARY
WORD ORIGIN
Spontaneous
comes from the Latin word *sponte* meaning *of free will*

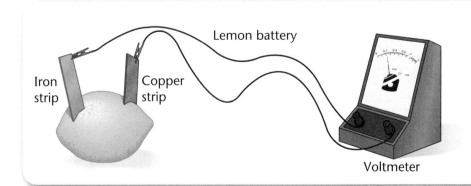

■ **Figure 17.2** A battery can be made by inserting iron and copper strips into a lemon and connecting them with a conducting wire in an external circuit. Electrons travel through the wire by metallic conduction and through the lemon by electrolytic conduction. The iron strip is the anode, where oxidation occurs. The copper strip is the cathode, where reduction occurs.

MiniLab 17.1

Lemon Voltage

What happens if you use different types of metals to make a lemon battery?

Procedure
1. Read and complete the lab safety form.
2. Gently knead a **lemon** without breaking the skin. Make two slits about 1 cm in depth on opposite sides of the lemon.
3. Insert a **strip of zinc** in one of the slits and a **strip of lead** in the other slit.
4. Connect an **alligator clip wire** to each of the metal strips, touching or connecting the other end of each wire to the poles of a **voltmeter.** If the voltmeter gives no reading, reverse the wires.
5. Read and record the voltage.
6. Repeat steps 2–5 with **strips of lead** and **aluminum,** making a new slit for the aluminum and lightly buffing the metal with **fine steel wool** immediately before inserting to remove the oxide coating. **WARNING:** *Discard the lemon. Do not use for food.*

Analysis
1. **Interpret** What causes the potential difference between the zinc and lead strips?
2. **Explain** Why is the potential difference greater when aluminum is substituted for zinc?
3. **Infer** If strips of zinc and magnesium, rather than zinc and lead, were used in the MiniLab, would the reaction that occurs at the zinc strip be the same? Explain.

■ **Figure 17.3** In this voltaic cell, the oxidation of zinc at the anode supplies electrons that flow to the cathode and reduce copper ions.
Infer *When does the spontaneous reaction stop?*

Potential difference Just as adding water to a container raises the level of the water, adding electrons builds up a negative potential at the anode. This electric potential is often described as a force, or a pressure of electrons. **Figure 17.3** illustrates the result of this pressure.

Why do the electrons travel from the anode to the cathode and not from the cathode to the anode? The electron pressure at the cathode is kept low by the reduction reaction, and the electrons flow from a region of high pressure (negative potential at the anode) to a region of low pressure (positive potential at the cathode). This difference in electron pressure at the cathode and at the anode is called **potential difference**. This potential difference between the electrodes in the lemon battery causes an electric current to flow. If there is no potential difference between the electrodes, no current will flow.

The size of the current depends upon the size of the potential difference. As the electrons move from a region of more negative potential to a region of more positive potential, they lose energy, so the discharge of a lemon battery is a spontaneous process. Potential energy stored in chemical bonds is released as electrical energy and, finally, as heat. An electrical potential difference is called **voltage** and is expressed in units of volts in honor of Alessandro Volta.

Table 17.1 Ease of Oxidation of Common Metals

Easily oxidized ↑

Li	$Li \rightarrow Li^+ + e^-$
K	$K \rightarrow K^+ + e^-$
Na	$Ca \rightarrow Ca^{2+} + 2e^-$
Na	$Na \rightarrow Na^+ + e^-$
Mg	$Mg \rightarrow Mg^{2+} + 2e^-$
Al	$Al \rightarrow Al^{3+} + 3e^-$
Mn	$Mn \rightarrow Mn^{2+} + 2e^-$
Zn	$Zn \rightarrow Zn^{2+} + 2e^-$
Cr	$Cr \rightarrow Cr^{3+} + 3e^-$
Fe	$Fe \rightarrow Fe^{2+} + 2e^-$
Ni	$Ni \rightarrow Ni^{2+} + 2e^-$
Sn	$Sn \rightarrow Sn^{2+} + 2e^-$
Pb	$Pb \rightarrow Pb^{2+} + 2e^-$
Cu	$Cu \rightarrow Cu^{2+} + 2e^-$
Ag	$Ag \rightarrow Ag^+ + e^-$
Hg	$Hg \rightarrow Hg^{2+} + 2e^-$
Pt	$Pt \rightarrow Pt^{2+} + 2e^-$
Au	$Au \rightarrow Au^{3+} + 3e^-$

Not easily oxidized

Ease of oxidation Iron is readily oxidized partly because the transfer of electrons from iron to an oxidizing agent releases a large amount of energy. You learned in Chapter 16 that some metals are oxidized in corrosion reactions. Yet, others form protective coating because they are not easily oxidized. Different substances release different amounts of energy when they become oxidized, and this fact may be used to construct a table such as **Table 17.1.** It may be used as a general guide to the ease with which a substance will lose electrons. By examining this table, you can see why silver, platinum, and gold are among the metals most often used in jewelry. All three are hard to oxidize and are thus resistant to corrosion.

Voltaic cells In the lemon battery, a redox reaction occurs spontaneously to produce a separation of charge at the two electrodes. The reaction begins as soon as the two electrodes are connected by a wire so that current can flow. As you have learned, an electrochemical cell in which an oxidation-reduction reaction occurs spontaneously to produce a potential difference is called a voltaic cell. In a voltaic cell, chemical energy is converted into electrical energy. *Voltaic cells* are sometimes called *galvanic cells*; both terms refer to the same device.

FACT of the Matter

Chemists at the University of California at Irvine have made the world's smallest voltaic cell. It is too small to be seen without an electron microscope and much smaller than most human cells. The voltaic cell consists of two mounds each of copper and silver attached to a graphite surface. Although it probably will never be used as a practical battery, it may allow scientists to study redox reactions at the atomic level.

The simple magnesium-copper voltaic cell shown in **Figure 17.4** illustrates the chemical change that takes place in a voltaic cell. Two beakers are used in order to separate the oxidation half reaction from the reduction half reaction. Because magnesium is more easily oxidized than copper, as you can see in **Table 17.1,** the magnesium (Mg) loses electrons and becomes oxidized, forming Mg^{2+} ions. The potential of the magnesium anode becomes more negative because of the increased electrical pressure from the released electrons. At the same time, the Cu^{2+} ions pick up electrons from the copper electrode and are reduced to copper metal (Cu). The potential of the copper electrode becomes more positive because electrical potential is lowered as electrons are removed from the cathode.

The complete voltaic cell In order for electric current to flow, two more things are necessary to complete the circuit between the half reactions. First, a wire connects the electrodes and allows electrons to flow from the magnesium electrode (anode) to the copper electrode (cathode). Second, a salt bridge—the tube shaped like an inverted U in **Figure 17.4** which contains a salt solution— connects the two beakers, completing the circuit.

The salt bridge What function does the salt bridge serve? A cell voltage registers because a potential difference exists between two electrodes. For the cell shown in **Figure 17.4,** Mg^{2+} ions are released into the solution at the anode, and Cu^{2+} ions are removed at the cathode. Positive ions, such as Mg^{2+} and Cu^{2+}, are called **cations**. Negative ions are called **anions**. Ions must be free to move between the electrodes to neutralize positive charge (Mg^{2+} cations) created at the anode and negative charge (anions) left over at the cathode. The solution of ions in the salt bridge allows ionic conduction to complete the electric circuit and prevent a buildup of excess charge at the electrodes.

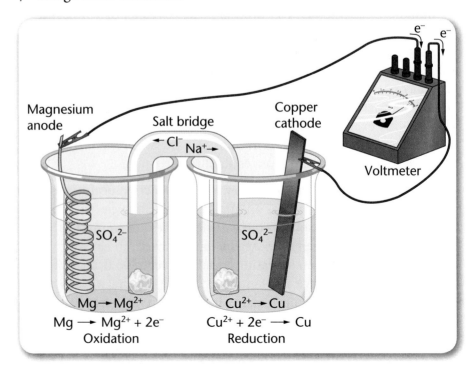

■ **Figure 17.4** A piece of magnesium metal is placed in a beaker containing a solution of magnesium sulfate, and a piece of copper metal is placed in a beaker containing a solution of copper(II) sulfate. An external circuit containing a voltmeter connects the two metal electrodes. As the blue copper ions are reduced, they move out of the solution and are deposited as copper metal on the copper strip. As a result, the blue copper solution becomes less and less blue-colored.

Infer What is the role of the salt bridge?

The Mg/Cu²⁺ reaction without an external circuit The same overall redox reaction occurs if the magnesium metal is placed directly into a solution of copper sulfate, illustrated in **Figure 17.5.** However, this is not a voltaic cell because the electrons do not flow through an external circuit. Instead, the electrons move directly from the magnesium metal to the copper ions, forming copper metal. The more easily oxidized magnesium forms colorless Mg^{2+} ions, which dissolve in the solution. The blue copper(II) ions are reduced to the red-brown copper metal that can be seen at the bottom of the beaker. This is a way to make copper metal from copper ions, but it is not a way to make electric power.

You can see that for every spontaneous redox reaction, you theoretically can construct a voltaic cell that can capture the energy released by the reaction. The amount of energy released depends on two properties of the cell: the amount of material that is present and the potential difference between the electrodes. The more material there is in the electrode, the more electrons it can produce during the course of the reaction. The potential difference depends on the nature of the reaction that takes place; that is, it corresponds to the relative positions of the two substances in a table such as **Table 17.1.** The farther apart the two substances are in the table, the greater the potential difference between the electrodes, and the greater the energy delivered by each electron that flows through the external wire.

How do you know which substance will be oxidized and which reduced in any cell? Look back at **Table 17.1.** Experimental chemists such as Humphry Davy and his student Michael Faraday did many experiments from which this type of table could be made. The table is used today to predict the outcome of new experiments. For example, in a Zn-Cu voltaic cell, zinc will be oxidized and copper reduced. Because zinc is more easily oxidized than copper, electrons will flow from zinc to copper.

The voltmeter When the circuit is complete, the voltmeter in the external circuit reads a voltage of 2.696 V. The energy released during discharge of the cell can be used to power a device such as a radio by connecting the wire from the electrodes through the radio. The overall reaction in the copper-magnesium cell is a redox reaction.

$$Mg(s) \rightarrow Mg^{2+}(aq) + 2e^-$$
Oxidation half-reaction

$$Cu^{2+}(aq) + 2e^- \rightarrow Cu(s)$$
Reduction half-reaction

$$Mg(s) + Cu^{2+}(aq) \rightarrow Mg^{2+}(aq) + Cu(s)$$
Net redox reaction

■ **Figure 17.5** When magnesium metal is added to a blue solution of $CuSO_4$, both the magnesium metal and the blue color disappear.

■ **Figure 17.6** When a simple voltaic cell does useful work, it is called a battery. If the external circuit is connected with a wire, electrons flow from the site of oxidation at the magnesium strip and through the LED to the surface of the copper strip, where reduction of Cu^{2+} ions takes place. The voltage pushes electrons through the LED, causing it to light up.

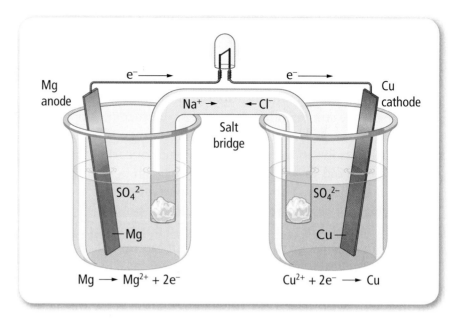

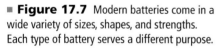

Interactive Figure To see an animation of a voltaic cell, visit glencoe.com.

The end of the reaction Useful work may be done if the voltmeter is replaced by wires connecting the voltaic cell to a lightbulb. In **Figure 17.6,** wires connect the cell to a light source called a light-emitting diode, or LED. If the circuit to the cell is complete, the LED lights up, showing that the cell is doing useful work. With time, the light intensity will fade. Why doesn't it stay lit indefinitely? Eventually, all of the magnesium in the anode becomes oxidized. The capacity of the battery has been exceeded, the magnesium is gone and, if there is no electrode, there can be no cell.

Batteries

Although the voltaic cell made from magnesium and copper in **Figure 17.4** can do useful work, it isn't something you'd want to bring along on a camping trip. The wet solutions could be sloppy, the glass could break easily, and the capacity is limited. Fortunately, scientists have developed much better batteries that are smaller, lighter, provide higher voltages, and last longer.

Figure 17.7 shows an assortment of commonly used batteries. Experimental batteries no thicker than a sheet of paper have already been developed. And although you might think batteries always have to be made of metal and acids, some batteries of the future may be made of microorganisms that use the energy in sugar to make electricity. A living fuel cell has been developed that someday could be used to power an automobile for up to 15 miles on two pounds of sugar.

How are batteries designed? The farther apart two metals are in **Table 17.1,** the larger the voltage of a battery that can be constructed from them. If you wanted to make a high-voltage battery to power your radio, you would choose metals that are far apart in the table. A copper penny with an iron nail will yield a larger voltage than a penny with a piece of nickel because copper is farther away from iron in the table than it is from nickel.

■ **Figure 17.7** Modern batteries come in a wide variety of sizes, shapes, and strengths. Each type of battery serves a different purpose.

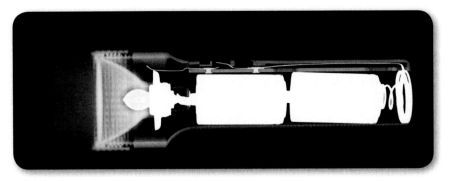

■ **Figure 17.8** This X-ray shows the completed circuit inside of a flashlight. When the flashlight was turned on, the metal contact at the top of the image slid forward and hit the metal cuff around the bulb. This completed the path for the electrons to flow from the negative terminal of the batteries, through the bulb, and back to the positive terminals.

Although the term *battery* usually refers to a series of voltaic cells connected together, some batteries have only one such cell. Other batteries may have a dozen or more cells. When you put a battery into a flashlight, radio, or CD player, you complete the electrical circuit of the voltaic cell(s), providing a path for the electrons to flow through as they move from the reducing agent (the site of oxidation) to the oxidizing agent (the site of reduction). **Figure 17.8** shows the battery contacts inside a flashlight. When the flashlight is turned on, the contacts complete the circuit, and the spontaneous redox reaction begins.

The most powerful batteries combine strong oxidizing agents and strong reducing agents to give the largest possible potential difference. But those agents aren't necessarily safe, convenient, or economical to use. To get a higher voltage from a cell type with a relatively small potential difference, several of the cells can be connected in series, as shown in **Figure 17.9.**

In **Figure 17.9,** the positive terminal of one cell—or lemon, in this case—is connected to the negative terminal of the next. The D batteries in the flashlight in **Figure 17.8** are also connected in series. For both the lemons and the flashlight batteries, the total voltage is the sum of the voltages of the individual cells.

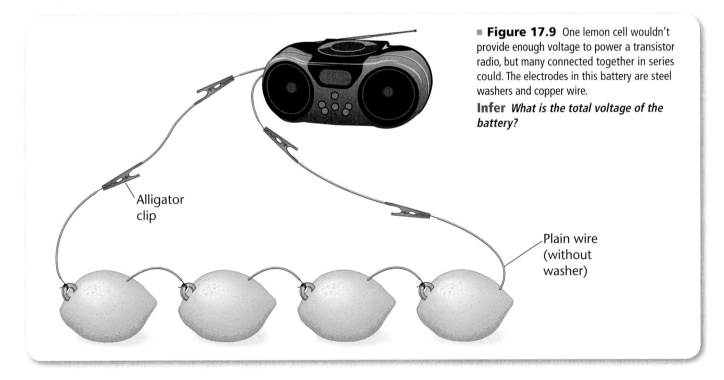

■ **Figure 17.9** One lemon cell wouldn't provide enough voltage to power a transistor radio, but many connected together in series could. The electrodes in this battery are steel washers and copper wire.

Infer *What is the total voltage of the battery?*

CHEMLAB

OXIDATION-REDUCTION AND ELECTROCHEMICAL CELLS

Background
Redox reactions involve the loss and gain of electrons. By separating the oxidation process from the reduction process and connecting them electrically through an external circuit, many spontaneous redox reactions can be utilized to produce an electric potential and an electric current. Devices that perform these functions are called electrochemical cells. In this ChemLab, you will investigate a redox reaction and use it to construct an electrochemical cell.

Question
How can a spontaneous redox reaction be used to construct an electrochemical cell?

Objectives
- **Observe** a simple oxidation-reduction reaction.
- **Relate** the reaction to the oxidation tendencies of the reactants.
- **Utilize** the reaction to construct an electrochemical cell that can operate electric devices.

Preparation

Materials
craft stick support with V-cut and slit cut
25-mm (flat diameter) dialysis tubing (15 cm in length)
magnesium ribbon (10 cm in length)
magnesium ribbon (1 cm in length)
copper foil (10 cm × 1 cm strip)
copper foil (1 cm × 2 mm piece)
metric ruler
250-mL beaker
10 × 100 mm test tubes (2)
wire leads with alligator clips (2)
DC voltmeter with a 2-V or 3-V scale
flashlight bulb for 2 AAA batteries
9-V transistor radio
0.5M sodium chloride (NaCl)
0.5M copper(II) chloride ($CuCl_2$)
0.1M magnesium chloride ($MgCl_2$)

Safety Precautions

Procedure

1. Read and complete the lab safety form.
2. Soak the dialysis tubing in tap water for about ten minutes while you complete steps 3 and 4. Tie two knots near one end of the tubing, and open the other end by sliding the material between your fingers.
3. Pour a small amount of the copper(II) chloride solution into a 10 × 100 mm test tube, and drop a 1-cm length of magnesium ribbon into the solution. Observe the system for about one minute, and record your observations in a data table similar to the table in Data and Observations. Pour the solution down the drain with a lot of water and discard the piece of magnesium in a wastebasket.
4. Repeat step 3 using magnesium chloride solution and the small piece of copper foil.
5. Pour copper(II) chloride solution into the open end of the tubing to a depth of 6 cm to 8 cm, and insert the strip of copper foil. Slide the top of the tubing and copper strip into the V-cut in the craft stick support shown below. Suspend the tubing in the beaker using the craft stick support.

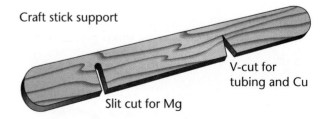

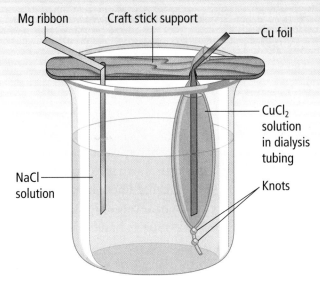

6. Slide the length of magnesium ribbon into the slit cut in the craft stick, as shown above.
7. Pour about 200 mL of sodium chloride solution into the beaker.
8. Connect leads to the pieces of copper and magnesium, and touch the leads to a DC voltmeter—the lead from the Cu electrode to the 1 terminal and the lead from the Mg electrode to the 2 terminal. Read and record the potential difference.
9. Cooperate with other lab groups in the following way to light the bulb and to operate the transistor radio. The flashlight bulb requires a voltage of about 3V, and the radio requires a voltage of about 9V. Connect your electrochemical cells in series (copper to magnesium) to provide the desired voltages. **WARNING:** *Be sure to connect your cells to the battery terminals of the radio in the correct polarity.* Connected in series, the voltages are additive; for example, five 2-V cells in series yield a voltage of 10 V. Such combinations of electrochemical cells are called batteries.
10. Disassemble your cell, observing the pieces of copper and magnesium and recording your observations. Rinse the pieces of copper and magnesium, and dispose of them according to the instructions of your teacher.

Data and Observations

Data Table

	Data and Observations
$Mg - Cu^{2+}$	
$Cu - Mg^{2+}$	
Voltage	
Pieces of Cu and Mg	

Analyze and Conclude

1. **Interpret** Write the balanced equation for the single-replacement reaction between magnesium and copper(II) chloride that occurred in step 3. Which metallic element, Cu or Mg, has the greater tendency (or oxidation potential) to lose electrons?
2. **Relate Concepts** In an electrochemical cell, oxidation occurs at the anode, and reduction occurs at the cathode. Which metal was the anode and which was the cathode? Write the equations for the half-reactions.

Apply and Assess

1. **Apply** When an electrochemical cell is used to operate an electrical device, in which direction do the electrons move in the external circuit?

INQUIRY EXTENSION

Infer Is it possible to construct an electrochemical cell in which lead is the anode and lithium is the cathode?

■ **Figure 17.10** A standard D battery is shown both whole and cut in half to reveal the structure of the zinc-carbon dry cell. Beneath the outside paper cover of the battery is a cylinder casing made of zinc. The zinc serves as the anode and will be oxidized in the redox reaction. The carbon rod in the center of the cylinder—surrounded by a moist, black paste of manganese(IV) oxide (MnO_2) and carbon black—acts as a cathode. Ammonium chloride (NH_4Cl) and zinc chloride ($ZnCl_2$) serve as electrolytes. Alkaline batteries contain potassium hydroxide (KOH) in place of the ammonium chloride electrolyte, and they maintain a high voltage for a longer period of time.

Zinc-carbon dry cell

Carbon-Zinc Dry Cell

When you put two or more common D batteries into a flashlight, you are connecting them in series. They must be in the correct orientation so that electrons flow through both cells. These batteries are carbon-zinc voltaic cells, and they come in several types, including standard, heavy-duty, and alkaline. This type of battery is often called a dry cell because there is no aqueous electrolyte solution; a semisolid paste serves that role. Examine the zinc-carbon battery in **Figure 17.10** and locate the parts of the voltaic cell it contains.

What is missing in this voltaic cell? Notice that the circuit is not complete, so the electrons that are produced at the zinc cylinder have no external conductor through which to travel to the carbon. This is by design and is not a defect in the battery. The circuit will be complete when the battery is placed in something designed to be powered by it, such as the flashlight in **Figure 17.8.** When the flashlight is turned on, the redox reaction starts. Electrons travel out of the zinc casing into a piece of metal built into the flashlight. There, they travel through a bulb, causing it to light. The electrons then reenter the battery at the top and move down through the carbon rod and into the black paste, where they take part in the reduction reaction.

The carbon-zinc dry cell reactions The flow of electrons from the zinc cylinder through the electrical circuits of an appliance and back into the battery provides the electricity needed to power a flashlight, radio, CD player, toy, clock, or other item. When electrons leave the casing, zinc metal (Zn) is oxidized.

$$Zn \rightarrow Zn^{2+} + 2e^-$$

The reactions in the carbon rod and the paste are much more complex, but one major reduction that takes place is that of manganese in manganese(IV) oxide (MnO_2). In this reaction, the oxidation number of manganese is reduced from 4+ to 3+.

$$2MnO_2 + H_2O + 2e^- \rightarrow Mn_2O_3 + 2OH^-$$

Adding the two half-reactions together gives the major redox reaction taking place in a carbon-zinc dry cell.

$$Zn + 2MnO_2 + H_2O \rightarrow Zn^{2+} + Mn_2O_3 + 2OH^-$$

HOW IT WORKS

The Pacemaker: Helping a Broken Heart

Your heart is made of cardiac muscle tissue that contracts and relaxes continuously. This beating results from electric impulses moving along pathways throughout your heart. A group of specialized cells in the upper wall of the heart's right atrium—upper chamber—generates electric impulses. If these cells fail to function or the electric impulse pathways are interrupted, the heart does not beat normally. A pacemaker is an electrical device that can monitor and correct an irregular heartbeat. How does it work?

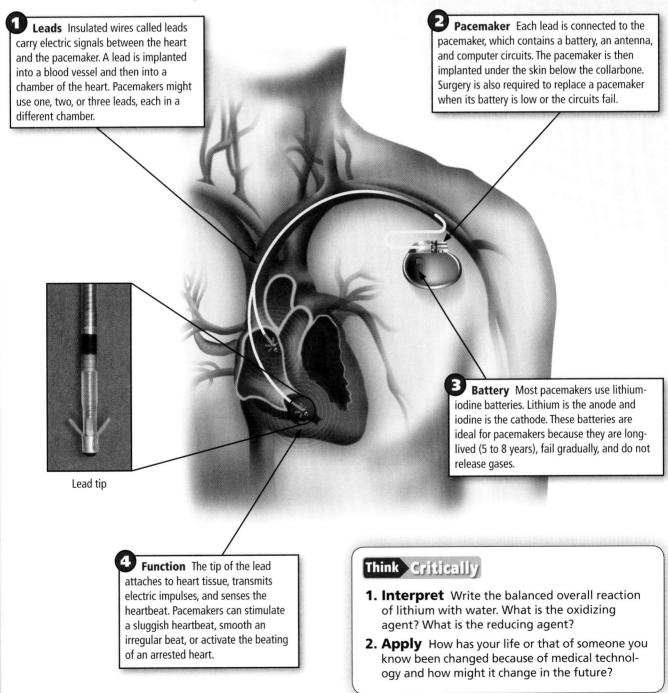

1 Leads Insulated wires called leads carry electric signals between the heart and the pacemaker. A lead is implanted into a blood vessel and then into a chamber of the heart. Pacemakers might use one, two, or three leads, each in a different chamber.

2 Pacemaker Each lead is connected to the pacemaker, which contains a battery, an antenna, and computer circuits. The pacemaker is then implanted under the skin below the collarbone. Surgery is also required to replace a pacemaker when its battery is low or the circuits fail.

3 Battery Most pacemakers use lithium-iodine batteries. Lithium is the anode and iodine is the cathode. These batteries are ideal for pacemakers because they are long-lived (5 to 8 years), fail gradually, and do not release gases.

4 Function The tip of the lead attaches to heart tissue, transmits electric impulses, and senses the heartbeat. Pacemakers can stimulate a sluggish heartbeat, smooth an irregular beat, or activate the beating of an arrested heart.

Lead tip

Think Critically

1. **Interpret** Write the balanced overall reaction of lithium with water. What is the oxidizing agent? What is the reducing agent?

2. **Apply** How has your life or that of someone you know been changed because of medical technology and how might it change in the future?

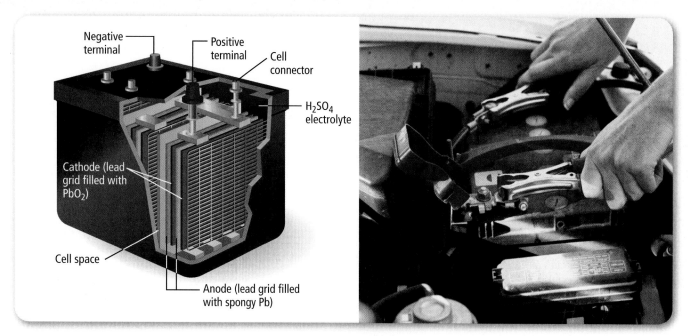

■ **Figure 17.11** Lead-acid batteries contain lead plates and lead(IV) oxide plates. The electrolyte is a solution of sulfuric acid. When the battery is in use, the sulfuric acid is depleted and the electrolyte becomes less dense. A car with a dead battery can still be started. Electricity from a second car is used to jump-start the car, bypassing the dead battery. When a car battery is dead, it should be disposed of properly.

Automobile Lead Storage Battery

The most common battery used in cars is a lead-acid, 12-volt storage battery. It contains six 2-volt cells connected in series. This type of battery is durable, supplies a large current, and can be recharged. When you turn your key in the ignition, the battery supplies electricity to start the car. It also provides energy for any demands not met by the car's alternator, such as using the radio or the lights when the engine is off. Leaving the lights or radio on for too long with the engine off can make the battery go dead because it is the engine that recharges the battery as the car runs.

Each cell in a lead-acid battery has two electrodes—one made of a lead(IV) oxide (PbO_2) plate and the other of spongy lead metal (Pb), as shown in **Figure 17.11**. In each cell, lead metal is oxidized as lead(IV) oxide is reduced. The lead metal is oxidized to Pb^{2+} ions, releasing two electrons at the anode. The Pb^{4+} ions in lead oxide gain two electrons, forming Pb^{2+} ions at the cathode. The Pb^{2+} ions combine with SO_4^{2-} ions from the sulfuric acid electrolyte to form lead(II) sulfate ($PbSO_4$) at each electrode. Thus, the net reaction that takes place when a lead-acid battery is discharged results in the formation of lead sulfate at both of the electrodes.

$$PbO_2 + Pb + 2H_2SO_4 \rightarrow 2PbSO_4 + 2H_2O$$

Recharging the battery The reaction that discharges a lead-acid battery is spontaneous and requires no energy. The reverse reaction, which recharges the battery, is not spontaneous and requires an input of energy from the car's alternator. This energy drives the conversion of lead sulfate and water back into lead(IV) oxide, lead metal, and sulfuric acid. Sulfuric acid is corrosive and it is important to be careful when working around a car battery.

$$2PbSO_4 + 2H_2O \rightarrow PbO_2 + Pb + 2H_2SO_4$$

How It Works

Nicad Rechargeable Batteries

The nickel-cadmium, or nicad, cell, shown in **Figure 1,** is a common storage battery that can usually be discharged and recharged more than 500 times. These batteries are used in calculators, cordless power tools such as the cordless drill in **Figure 2,** vacuum cleaners, and rechargeable electric toothbrushes and shavers. Once nicad batteries have been spent, disposal presents a problem because cadmium is toxic. Nicads can be recycled, but the process is expensive. Although rechargeable batteries containing less toxic metals are being developed, none have been found that can sustain a constant rate of discharge as well as the nicad.

■ **Figure 2** Cordless power drill

1 Newly purchased nicad batteries must be charged before use.

2 The nicad voltaic cell has cadmium anodes, hydrated nickel oxide cathodes, and KOH as the electrolyte. The electrodes are arranged in jelly-roll fashion.

3 In the redox reaction that takes place during discharge, nickel oxide is reduced at the cathode, and cadmium is oxidized at the anode.

$Cd + NiO \rightarrow CdO + Ni$

4 The electrolysis reaction that takes place when an external source of electricity is used to recharge the cell is the reverse of the discharge reaction.

$CdO + Ni \rightarrow Cd + NiO$

5 Nicad batteries are not suitable for devices that are left idle for long stretches—such as smoke detectors, cameras, and flashlights—because they will lose about one percent of their charges daily even when not being used.

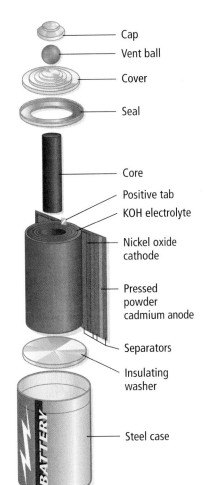

■ **Figure 1** Nicad battery

Think Critically

1. **Determine** What are the equations for the oxidation and reduction half-reactions that occur while a nicad battery is recharging?
2. **Explain** What might be an environmental advantage to using nicad batteries?

How It Works

Hydrogen-Oxygen Fuel Cell

Recall that the combustion of a fuel is a redox reaction in which the fuel molecules are oxidized and oxygen is reduced to form an oxide. For years, scientists have worked to find a way to separate the oxidation and reduction reactions to make them produce an electric current. The simplest fuel cell involves the oxidation of the fuel hydrogen gas to form water. Today, hydrogen-oxygen fuel cells are used to supply electricity to the space shuttle orbiters. The fuel cells have a weight advantage over storage batteries, and the water produced during their operation can be used for drinking.

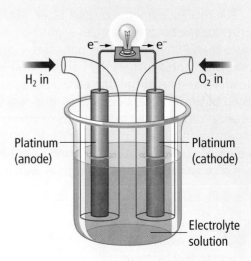

■ **Figure 2** Hydrogen-oxygen fuel cell with platinum electrodes

1 A simple hydrogen-oxygen fuel cell differs in two major ways from a voltaic cell: the electrodes are made of an inert material—such as carbon in **Figure 1** or platinum in **Figure 2**—that doesn't react during the process, and hydrogen and oxygen gas are fed in continuously.

2 Hydrogen is fed onto an electrode on one side of the fuel cell, and oxygen is fed onto an electrode on the other side.

3 Concentrated KOH serves as the electrolyte in the fuel cell.

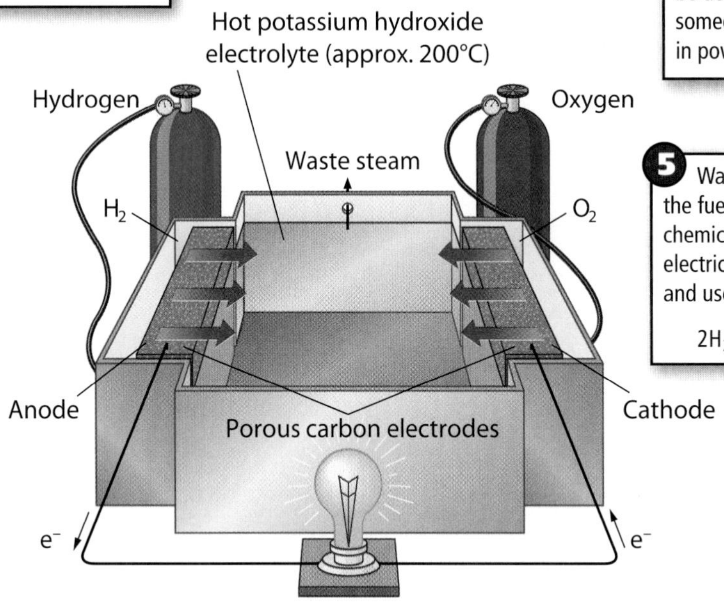

4 The electrons lost by hydrogen molecules, which are oxidized at the anode, flow out of the fuel cell, through a circuit, and then back into the fuel cell at the cathode, where oxygen is reduced.

5 Water vapor—steam—is produced in the fuel cell, as up to 75 percent of the chemical energy is converted into electricity. The steam can be condensed and used for drinking water.

$$2H_2(g) + O_2(g) \rightarrow 2H_2O(g) + \text{energy}$$

6 If more inexpensive and longer-lasting fuel cells can be developed, they may someday produce electricity in power plants.

■ **Figure 1** Hydrogen-oxygen fuel cell with carbon electrodes

Think Critically

1. **Describe** What causes electrons to flow from hydrogen to oxygen in a fuel cell?
2. **Determine** If fuel cells are about 75 percent efficient, what happens to the rest of the potential energy?

Electric Cars

At the end of the nineteenth century, most cars were powered by steam or by electric batteries; today most cars are powered by gasoline. A return to electric cars could help reduce our dependence on fossil fuels, cause less pollution, and be more economical in the long run. However, they have several disadvantages, such as high initial cost, limited driving range, low speed, and long recharge time. They also present a disposal problem because cadmium from the nicad batteries commonly used in electric cars is a toxic metal.

These disadvantages would disappear if a battery that is cheap enough, powerful enough, and safe enough for running an electric car could be developed. Two new experimental types of batteries for use in electric cars show early promise as candidates. One is a rechargeable, nickel-metal hydride, or NiMH, battery. This type of battery is less toxic and has a higher storage capacity than the batteries now used in electric cars.

Another experimental battery is a lithium battery with a water-based electrolyte. Lithium is more easily oxidized than any other metal but has a drawback that has limited its use in batteries: it explodes violently when it comes into contact with water. Lithium is used in some batteries to power camcorders, but they require an expensive, nonaqueous electrolyte. **Figure 17.12** shows the construction of the experimental aqueous lithium battery. This battery is less toxic and will probably be cheaper to manufacture than the nickel-cadmium batteries used in most electric cars in operation today.

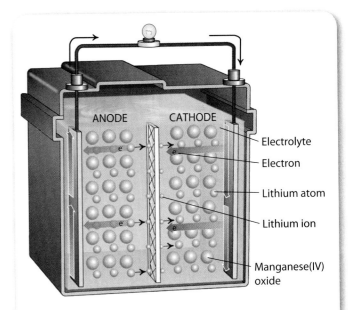

■ **Figure 17.12** How can a lithium battery have an aqueous electrolyte? Two facets of the construction of this new battery keep the lithium metal from reacting with water. First, the lithium is in the form of individual atoms embedded in a material such as manganese(IV) oxide, rather than as a solid metal. Second, the electrolyte is full of dissolved lithium salts, so the lithium ions that are produced travel to the site of reduction without reacting with water.

SUPPLEMENTAL PRACTICE

For more practice with voltaic cells, see Supplemental Practice, page 835.

Section 17.1 Assessment

Section Summary

- A potential difference between two substances is a measure of the tendency of electrons to flow from one to the other.
- A voltaic cell is a chemical system that produces an electric current through a spontaneous redox reaction.
- Batteries contain one or more voltaic cells.

1. **MAIN Idea** Describe the movement of electrons in a voltaic cell.
2. **Draw** a diagram of a simple voltaic cell.
3. **Compare** zinc-carbon and lead-acid batteries.
4. **Analyze** A piece of copper metal is placed in a 1M solution of silver nitrate ($AgNO_3$).
   a) Use **Table 17.1** to predict which metal will be reduced and which will be oxidized.
   b) Write an equation for the net redox reaction that occurs. HINT: Cu^{2+} is formed.
   c) Is this system a voltaic cell? Explain.
5. **Explain** why a dry cell cannot really be dry.

Section 17.2

Objectives

- **Explain** how a non-spontaneous redox reaction can be driven forward during electrolysis.
- **Relate** the movement of charge through an electrolytic cell to the chemical reactions that occur.
- **Apply** the principles of electrolysis to its applications such as chemical synthesis, refining, plating, and cleaning.

Review Vocabulary

voltage: electric potential difference, expressed in units of volts

New Vocabulary

electrolysis
electrolytic cell

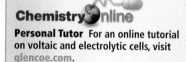

Personal Tutor For an online tutorial on voltaic and electrolytic cells, visit glencoe.com.

Electrolysis: Chemistry from Electricity

MAIN Idea In electrolysis, a power source causes nonspontaneous reactions to occur in electrochemical cells.

Real-World Reading Link When you ride a bicycle downhill, you don't have to do any work. You just coast. What is different when you ride a bicycle uphill? You have to provide a lot of energy by pedaling. Reversing redox reactions is similar to riding a bicycle uphill.

Reversing Redox Reactions

When a battery generates electric current, electrons given up at the anode flow through an external circuit to the cathode. Some kinds of batteries can be recharged by passing a current through them in the opposite direction. For example, plugging a cell phone into its charger and the charger into the wall draws a current into the phone that charges the battery. Charging a battery is the opposite process of discharging it.

Figure 17.13 compares these opposite processes. You are already familiar with one type of electrochemical cell, the voltaic cell. In it, a spontaneous redox reaction causes electrons to pass through a wire and do work. The second cell in **Figure 17.13** demonstrates electrolysis. **Electrolysis** is the process in which electric energy is used to drive a nonspontaneous chemical reaction. Electrolysis takes place in a second type of electrochemical cell called an **electrolytic cell**.

In the electrolytic cell shown in **Figure 17.13**, the block labeled *Voltage source* represents the source of energy needed to drive the nonspontaneous reaction. For example, this could be the outlet that you plug a cell phone into—or more precisely, the electric plant that supplies the outlet. Always remember that the reactions in electrolytic cells are not spontaneous, and they require an outside source of energy to make them go.

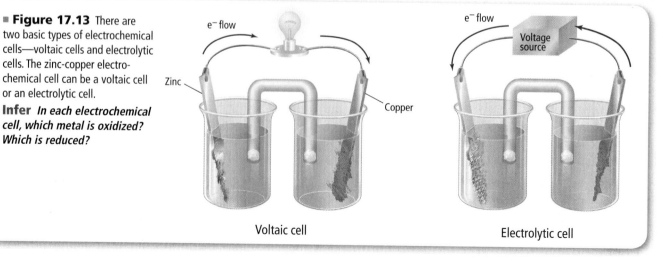

■ **Figure 17.13** There are two basic types of electrochemical cells—voltaic cells and electrolytic cells. The zinc-copper electrochemical cell can be a voltaic cell or an electrolytic cell.

Infer *In each electrochemical cell, which metal is oxidized? Which is reduced?*

600 Chapter 17 • Electrochemistry

Electrolysis

As you can see from **Figure 17.13**, an electrochemical cell is a device that uses a redox reaction to produce electric energy or that uses electric energy to cause a chemical reaction. Not too long after Volta's invention of the first electrochemical cell, the British chemist Humphry Davy built a cell of his own and used it to pass electricity through molten salts.

The anode and cathode in electrolysis

In Davy's electrolysis of molten sodium chloride (NaCl), sodium ions (Na^+) were reduced to metallic sodium (Na) at the cathode. The oxidation of chloride ions (Cl^-) to chlorine gas (Cl_2) occurred at the the anode. This fits the definition of the cathode and the anode from the discussion of voltaic cells: reduction takes place at the cathode and oxidation at the anode.

However, the electrodes in the two different types of cells are not identical. For example, in a voltaic cell, electrons build up at the anode and flow spontaneously to the cathode. Because electrons carry a negative charge, the anode of a voltaic cell is negative relative to the cathode, which is positive. The situation in an electrolytic cell is reversed: electrons are pumped away from the anode. The anode is positive relative to the cathode, which is negative.

The half-reactions in the electrolysis of molten sodium chloride are as follows.

$$2Na^+(l) + 2e^- \rightarrow 2Na(l) \text{ and}$$
$$2Cl^-(l) \rightarrow Cl_2(g) + 2e^-$$

When the equations for the two half-reactions are combined, the equation for the overall reaction can be written as

$$2Na^+(l) + 2Cl^-(l) \rightarrow 2Na(l) + Cl_2(g)$$

Figure 17.14 shows the modern commercial electrolysis of molten rock salt. Rock salt is sodium chloride (NaCl). In this process, pure sodium metal and chlorine gas are produced.

Electrolysis and new elements
Davy discovered several elements in this way, beginning in 1807. After releasing purified potassium metal from potassium hydroxide, it took him only a year to produce magnesium, strontium, barium, and calcium. Fewer than 30 elements had been isolated by 1800, but by 1850, more than 50 were known. Most of these new elements were isolated using electrolysis.

> **FOLDABLES**
> Incorporate information from this section into your Foldable.

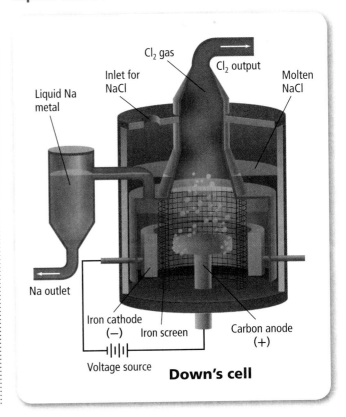

■ **Figure 17.14** In a Down's cell, electrons supplied by a generator are used to reduce sodium ions. As electrons are removed from the anode, chloride ions are oxidized to chlorine gas.
Explain What is the role of the iron screen?

Section 17.2 • Electrolysis: Chemistry from Electricity

VOCABULARY
WORD ORIGIN
Electrolysis
comes from the Latin word *electricus* meaning *involving electricity* and the Greek word *lytikos* meaning *to split*

The electrolysis process During the process of electrolysis, electrons are transferred between the metal electrodes and the ions or atoms at the electrode surfaces. In the liquid, the charge is conducted by ions, such as the Na^+ and Cl^- in the molten rock salt. Of course, ions contain electrons too, but these electrons are held tightly to individual ions. Ions can conduct current through a liquid only when they move through the liquid. This type of conduction is called electrolytic conduction. For example, in the electrolysis of molten NaCl, Na^+ ions move toward the cathode, which is negatively charged. The negative anions, Cl^- ions, move toward the positively charged anode.

At the cathode What happens when the moving ions reach an electrode surface? If the electrodes are inert, which means that they don't react chemically with the ions in the solution, then only electron transfer will take place at the electrodes. Electrons are being pumped from the battery toward the cathode, where reduction will occur. At the cathode, the ion that reacts is the one that most readily reacts with electrons. In molten NaCl, Na^+ and Cl^- are both present in the liquid near the surface of the cathode. Na^+ accepts electrons more readily, so each Na^+ cation gains one electron, being reduced to the metal.

At the anode At the anode, electrons are transferred from the ion that most easily gives them up to the anode. In this case, Cl^- holds onto its electrons more loosely, so each Cl^- anion loses an electron and is oxidized to a chlorine atom. Chlorine atoms then combine to form Cl_2 molecules. The electrons released by the chloride ions flow through the external circuit to the battery and are recycled to the cathode, where they continue the reduction reaction.

MiniLab 17.2

Electrolysis

How can the current from a 9-V battery produce chemical changes in a simple electrolytic cell?

Procedure
1. Read and complete the lab safety form.
2. Pour about 200 mL of **0.5M copper(II) sulfate** into a **250-mL beaker**.
3. Use **wires with alligator clips** to attach two **4-inch pencil leads** (actually graphite, not lead) to the terminals of a **9-V battery**.
4. Put the pencil leads into the copper(II) sulfate solution. Be sure to keep the pencil leads as far from each other as possible.
5. Observe the reactions that occur at the two pencil leads for five minutes. Record your observations, including the polarities (+ and −) of the leads.

Analysis
1. **Describe** the reaction that occurs at the positive electrode, the anode, of the electrolytic cell. Write the equation for the oxidation reaction.
2. **Describe** the reaction that occurs at the negative electrode, the cathode. Write the equation for the reduction reaction.
3. **Explain** how you might use an electrolytic cell to silver plate an iron spoon.

Because only as many electrons are available at the cathode as are removed at the anode, the reduction process at the cathode must always occur together with the oxidation process at the anode. Charge transfer at the two electrodes must exactly balance because, just as in the redox reaction, the liquid and its contents must always remain electrically neutral. Therefore, in an electrolytic cell, the overall result of the two electrolysis processes is to carry out a balanced redox reaction, even though the two half-reactions take place at different locations.

Producing chemicals by electrolysis In the electrolysis of molten sodium chloride, a redox reaction taking place in an electrolytic cell can be used to generate chemicals that are important commercially. Because electrolysis consumes large amounts of energy when it is carried out on a commercial scale, many companies that use this process have made their homes in locations where electric power is inexpensive. The abundant hydroelectric power available from Niagara Falls has made that area of New York State a prime location for companies that use electrolysis.

One important commercial electrolytic process is the electrolysis of rock salt solutions to produce chlorine, hydrogen, and sodium hydroxide. The overall change taking place in this process is both a redox reaction and a substitution reaction.

$$2NaCl(aq) + 2H_2O(l) \rightarrow Cl_2(g) + H_2(g) + 2NaOH(aq)$$

How is electrolysis of a sodium chloride solution different from electrolysis of molten sodium chloride? In molten NaCl, the only ions present are Na^+ and Cl^-. What ions are present in an aqueous solution of rock salt? Recall that water dissociates slightly to form H^+ and OH^- ions. Therefore, a rock salt solution contains Na^+, H^+, Cl^-, and OH^- ions. **Figure 17.15** shows the electrolysis of brine. Compare **Figure 17.15** with **Figure 17.14,** the electrolysis of molten NaCl.

At the anode At the anode, the chloride ions lose electrons more easily than the other ions present, just as they did in the electrolysis of molten rock salt. This oxidation forms chlorine gas that can be used for making PVC plastic and other consumer products.

■ **Figure 17.15** In the electrolysis of brine (aqueous NaCl), sodium is not a product because water is easier to reduce. The chlorine gas is used to make PVC plastics.
Name *the species that is oxidized in the electrolysis of brine.*

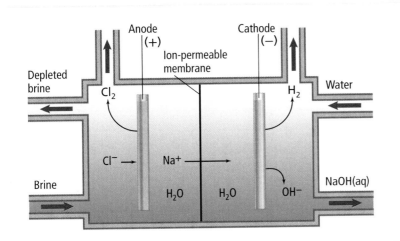

At the cathode At the cathode, the reaction that occurs in molten salt doesn't occur. Instead, H^+ ions are easier to reduce than Na^+, Cl^-, or OH^- ions, so hydrogen ions pick up electrons from the cathode and are reduced to form hydrogen gas. Hydrogen is used in industrial processes such as the catalytic hydrogenation of vegetable oils to form margarine. The Na^+ and OH^- ions are left dissolved in water after the electrolysis process has removed H^+ and Cl^- ions. This is a solution of the base sodium hydroxide (NaOH). Sodium hydroxide is an important industrial and household chemical.

Electrochemical transformation of a simple salt solution produces three valuable products. Each can be sold to pay for the electrical energy that must be invested to make them, with a little left over for a profit to the manufacturer.

Other Applications of Electrolysis

Electrolysis has numerous useful applications in addition to the generation of chemical substances. The process can also be used to purify metals from ores, coat surfaces with metal, and purify contaminated water. The applications range from the world of art to the world of heavy industry.

Refining ores Just as sodium can be produced from melted NaCl by electrolysis, many metals are separated from their ores using electrolysis—a process called refining ores. Today, the metal produced in largest quantity by electrolysis is aluminum from aluminum oxide. First, bauxite ore is heated, driving off the water and leaving aluminum oxide (Al_2O_3). Pure aluminum oxide melts at about 2000°C, so cryolite (Na_3AlF_6) is added to lower the melting point to about 1000°C. The molten aluminum salt solution is placed in a large electrolytic cell lined with carbon, which acts as the cathode during electrolysis. Large carbon anodes are dipped into the molten salt to complete the cell, as shown in **Figure 17.16**.

■ **Figure 17.16** Aluminum is refined in smelters similar to the diagram. Note the carbon (graphite) serves as both the anode and the cathode. Recycled aluminum is often fed into the cell with the new aluminum. Molten aluminum is drawn off from the bottom of the electrolytic cell, where it accumulates during the electrolysis process.

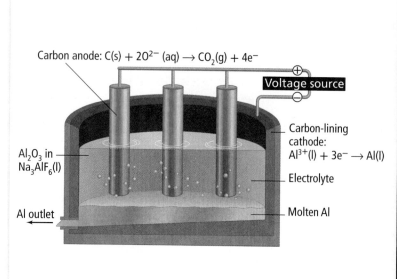

The electron transfers cause complex reactions during electrolysis of molten Al_2O_3 / Na_3AlF_6, but the net reactions are simple. At the cathode, electrons reduce aluminum ions to aluminum metal, which is molten at this temperature. At the anodes, oxide ions lose electrons to form oxygen. The oxygen then combines with the carbon anodes to produce carbon dioxide. The carbon anodes must be replaced periodically because they are gradually used up. The overall redox reaction follows.

$$2Al_2O_3(s) + 3C(s) \rightarrow 4Al(l) + 3CO_2(g)$$

This electrolytic method for producing aluminum is called the Hall-Héroult process because it was developed simultaneously by Charles Martin Hall of the United States and Paul Héroult of France in 1886. Aluminum was so rare before this process was developed that it was more expensive than silver or gold. **Figure 17.17** illustrates the historical value of aluminum. Today, about 10 million metric tons of aluminum are produced per year worldwide using this process.

Recycling aluminum The Hall-Héroult process is expensive and consumes large amounts of electrical energy. Recycling aluminum metal prevents some of the expense of producing new aluminum by electrolysis. A lot of waste aluminum is available from discarded aluminum containers, and the energy invested in the original refining of aluminum can be saved if metallic aluminum can be melted down. **Figure 17.18** shows recycled aluminum being melted. It takes only about seven percent as much energy to make new aluminum cans from old ones as it does to make new cans from aluminum ore.

■ **Figure 17.17** The cap of the Washington Monument is a 2.8-kg aluminum pyramid. The cap was placed in 1884, two years before the development of the Hall-Héroult process, when aluminum was still a rare and expensive metal.

■ **Figure 17.18** Recycling of aluminum provides a relatively cheap source of this metal. Currently, 60 percent of aluminum beverage cans are recycled in the United States. Once collected, scrap aluminum is melted in a furnace and then reformed into other products.

CHEMISTRY AND TECHNOLOGY

Copper Ore to Wire

It is hard to imagine life without many of the common metals that are used today. Copper, for example, is used in many of the pots and pans in your kitchen, the cooling coils in the air-conditioning system, and some of the pennies you carry in your pocket. More importantly, copper is the metal of choice for most of the electrical wiring in your appliances, homes, and cars because of its good conductivity and low cost. How does copper get from rocks to the finished wire?

Copper has a history of use that is at least 10,000 years old. In Roman times, much of the copper used was obtained from the island of Cyprus; hence, copper became known as cyprium. This term was then shortened to cuprum, from which the English word copper is derived. By 3000 B.C., many civilizations were adept at the processes of hammering, annealing, oxidation, and reduction, smelting, alloying, and removing impurities. Unfortunately, these processes could produce only small quantities of copper. Large-scale production was not mastered until modern furnaces and rolling techniques were developed.

1. Mining
Copper can be found occurring naturally as an uncompounded mineral in many parts of the world. However, most copper is obtained from the minerals chalcopyrite ($CuFeS_2$), chalcocite (Cu_2S), and cuprite (Cu_2O). The ores are typically surface mined, as shown in **Figure 1,** and ground into powders.

■ **Figure 1** Copper surface mine

2. Ore Enrichment
Because each copper ore only contains 0.40–12 percent copper, the ore must be concentrated by the flotation process, shown in **Figure 2.** A frothing agent such as pine oil is mixed with the powdered copper ore, and air is blown in to froth the mixture. Because copper ores are hydrophobic (not wetted by water), the copper and iron sulfides cling to the oil and float to the top, where they can be continuously removed, increasing the copper concentration to 20–40 percent.

■ **Figure 2** Copper and iron sulfides bubble to the top and are removed.

3. Roasting

The traditional preparation of copper for use involves roasting the ore, shown in **Figure 3**, with oxygen to convert the metallic sulfides to metallic oxides, which drives off some of the sulfur as sulfur dioxide. Usually, both iron and copper are present in the mix.

$$2Cu_2S(s) + 3O_2(g) \rightarrow 2Cu_2O(s) + 2SO_2(g)$$

$$2FeS(s) + 3O_2(g) \rightarrow 2FeO(s) + 2SO_2(g)$$

4. Smelting

The copper and iron oxides are smelted by mixing them with silica, air, and limestone and then heating the mixture to about 1300°C. This further concentrates the metal, called matte, to 45–70 percent copper. Melting the matte and blowing air through the liquid metal, as shown in **Figure 4**, drives off more iron and sulfur, which forms dense blister copper (98.5 percent copper), lightweight iron(II) calcium silicate slag, and gaseous SO_2. The blister copper can be drawn off the bottom and cast into large blocks, as shown in **Figure 5**.

■ **Figure 5** Blister copper is cast into large blocks.

5. Purification

The copper can be purified further by electrolysis, an oxidation-reduction process, to 99.95 percent purity. The large blocks of blister copper are used as anodes suspended in a solution of aqueous copper(II) sulfate, as shown in **Figure 6**. Pure copper is used as the cathode. During electrolysis, copper is oxidized at the anode, moves through the solution as Cu^{2+} ions, and is deposited on the cathode. Waste products left after the dissolution of the anode produce sludge on the bottom of the electrolysis vessel. The sludge, which is often rich in silver and gold, can be recovered.

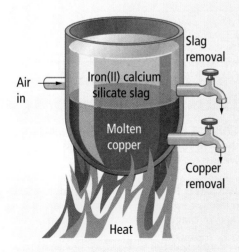

■ **Figure 4** Copper removal from smelting

■ **Figure 6** Blister copper blocks are used as anodes.

Chapter 17 • Chemistry and Technology

6. Wire Bar Production
The electrolytic copper is cast into wire bars, shown in **Figure 7,** that weigh about 113 kg. The wire bars are heated to 700–1200°C and rolled without further reheating into rods approximately 1 cm in diameter.

■ **Figure 7** Wire bars are heated to 700–1200°C.

7. Drawing Wire
As illustrated by **Figure 8,** the 1-cm rods of copper are drawn through successively smaller dies until the desired size of wire is reached. The dies must be made of exceptionally hard materials, such as tungsten carbide or diamond, because of the tremendous amount of wear from drawing the wire. The wire is also heavily lubricated to lessen wear on the dies.

■ **Figure 8** Dies for drawing wire

8. Coatings
The finished wire may be coated with various materials, such as plastic, polymers, ceramics, enamel, glass, or another metal to help protect it from moisture and oxidation, or for electrical insulation. **Figure 9** shows several of these coated wires.

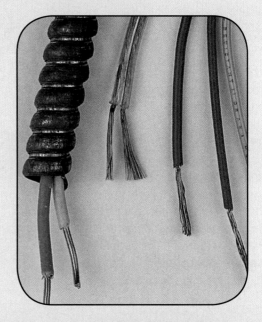

■ **Figure 9** Coatings for wire protect the wire from moisture and oxidation.

Discuss the Technology

1. **Think Critically** Why is it important that the slag formed during smelting be lightweight?

2. **Acquire Information** Copper can be made into many different forms and alloys. Research and list several of the forms and uses for each.

3. **Hypothesize** How might the gold and silver be removed from the sludge formed beneath the anode?

Electroplating You have learned that many metals can be protected from corrosion by plating them with other metals. Zinc coatings are often used to keep iron from rusting. Metal garbage cans are galvanized by dipping them into molten zinc. This process produces an uneven, lumpy coating both inside and outside of the can. That's acceptable for a garbage can, but the lumpy surface wouldn't look good under the bright red paint job on a new sports car. That's why automobile manufacturers electroplate zinc onto the steel used for car bodies. This process involves dragging a sheet of steel across the surface of an electrolyte in an electrolytic cell. The process produces a thin (cost-saving), uniform (smooth and clean) coating of zinc on only one side of the sheet of steel, saving half the coating cost. Because only one side of the car body is exposed to the corrosive effects of water and salt, it is not worth the cost of coating the inside. **Figure 17.19** shows other uses of electroplating.

In zinc electroplating, zinc ions are reduced to zinc atoms at the surface of the metal object to be coated, which becomes the cathode in an electrolytic cell. At the anode, which is made of zinc, atoms of zinc are oxidized to ions. The electrolyte solution contains dissolved zinc salt. The thickness of the zinc metal coating can be controlled exactly by controlling the total charge (number of electrons) used to plate the object. Only that portion or side of the object that is immersed in the cell electrolyte receives a zinc coat.

Because coatings adhere best to a chemically clean foundation, objects to be coated are usually degreased, cleaned with soap, and then treated with a corrosive fluid to remove any dirt on their surface before electroplating. Then, the object is immersed in an electrolyte containing the salt of the metal to be deposited. Because the object acts as the cathode of the electrolytic cell, it must be a conductor. Metal objects are electroplated most often because metals usually are excellent conductors.

TRY AT HOME LAB
See page 876 for **Removing Electroplating.**

■ **Figure 17.19** Chromium is often electroplated onto a softer metal to improve its hardness, stability, and appearance. Just as chromium can be plated onto another metal, copper can be plated onto zinc. In this process, Cu^{2+} ions in solution are reduced to Cu metal at the zinc cathode. In the early 1980s, because of high inflation of the U.S. currency and a worldwide shortage of copper, the cost of producing a copper penny became nearly equal to the value of the copper metal itself. The U.S. Mint began to produce pennies by electroplating zinc disks with a copper coat.

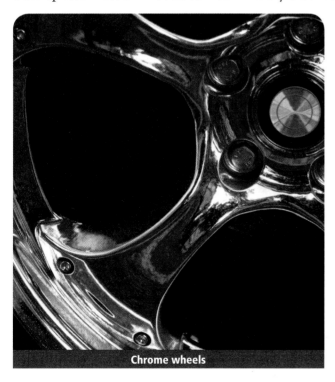
Chrome wheels

Copper-coated pennies

Section 17.2 • Electrolysis: Chemistry from Electricity

■ **Figure 17.20** Chemistry played a key role in restoring many items taken from the wreck of the Titanic. Electrolysis was used to clean and stabilize many metal artifacts. The study of objects from the ship may help scientists compile information for long-term storage and containment under seawater.

VOCABULARY
WORD ORIGIN
Corrosion
comes from the Latin words *com* meaning *thoroughly* and *rodere* meaning *to gnaw*

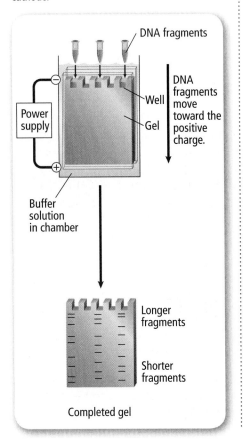

■ **Figure 17.21** Electrophoresis is a valuable laboratory tool used to separate and identify large charged particles such as DNA. In this process, a solution containing DNA is loaded on a gel medium. When the loaded gel is placed in an electrophoresis tank and the current is turned on, the DNA fragments separate. Negatively charged particles move toward the anode. Positively charged particles move toward the cathode.

The anode is made of the same metal that is being plated so that it will replenish the metal ions in the electrolyte that are removed as the plating proceeds. The net effect is that when an electric current is passed through the electroplating bath, metal is transferred from the anode and distributed over the cathode. Eventually, the entire object becomes coated with a thin film of the desired metal.

Electrolytic cleaning Electrolysis can be used to clean objects by pulling ionic dirt away from them. The process has been used to restore some of the many metal artifacts taken from the shipwrecked cruise ship *Titanic*, such as shown in **Figure 17.20,** which sank in the northern Atlantic Ocean in 1912. Coatings of salts containing chloride ions, which came from the seawater, were removed by electrolysis. The electrolysis cell for this cleaning process includes a cathode that is the object itself, a stainless steel anode, and an alkaline electrolyte. When an electric current is run through the cell, the chloride ions are drawn out. Hydrogen gas forms and bubbles out, helping to loosen corrosion. Among the objects that have been recovered are a porthole, a chandelier, and buttons from the uniforms of crew members.

Electrophoresis Electrophoresis is another electrochemical process that was used to restore some of the ceramic and organic artifacts from the Titanic. Electrophoresis involves placing an artifact in an electrolyte solution between positive and negative electrodes and applying a current. The current breaks up salts, dirt, and other particles as their charged components migrate to the electrodes. Electrophoresis is also used in biological laboratories to separate and identify large molecules. The process, illustrated in **Figure 17.21,** can separate DNA fragments based on its negative charge. The DNA is loaded on the negative end of the electrophoresis gel and when current is applied, the DNA fragments move towards the positive end of the gel. Smaller fragments will move faster and farther.

Everyday Chemistry

Manufacturing a Hit CD

The sound from your new compact disc (CD) is so clear and crisp that it seems as if the musicians are in the same room. The clarity of the sound is possible thanks to the chemical process of metallic depositing and electroplating used in the manufacturing of CDs.

Data pickup A CD is a collection of binary data in a long, continuous spiral that runs from the inside edge to the outside. Nothing touches the information stored on the CD except laser light, which is reflected from the CD and read by the CD player computer as a binary signal. This signal is then transformed by a tiny computer into audio signals, which are amplified to produce the sound heard from your speakers.

Master copy production Your CDs are stamped from a master copy called a stamper master. But how is this stamper made? The musical data are first transferred onto a glass disc using a high-powered laser that etches small pits in the glass as shown in **Figure 1**. This glass master copy then contains all of the musical information in binary data form.

The master is coated with a dilute solution of silver diammine complex—$[Ag(NH_3)_2]^+$—followed by a solution of formaldehyde that acts as a reducing agent for the silver. The result is a redox reaction that deposits a thin silver mirror that plates the etched disc. This mirrored master disc is produced in the following reaction.

$$2Ag^+ + HCHO + H_2O \rightarrow 2Ag + HCOOH + 2H^+$$

Electrodeposition The silver coating forms the surface onto which a thin layer of nickel is electrodeposited to make a nickel-coated disc called the mother disc.

$$Ni^{2+}(aq) + 2e^- \rightarrow Ni(s)$$

A second coating of nickel is then electrodeposited onto the mother disc. This nickel layer, the stamper master, is peeled off and used to stamp the data impression onto melted polycarbonate plastic discs.

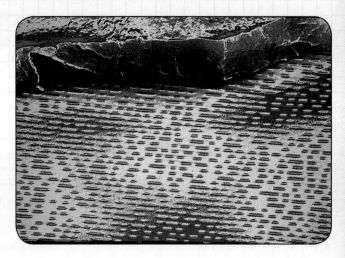

Figure 1 The etched pits on this glass master are magnified 1200×.

The polycarbonate disc now has all of the pits found on the original glass master disc etched by the laser. Because polycarbonate is clear, it is vacuum coated with a thin aluminum film to produce the reflective layer required by the laser. This delicate aluminum layer is covered with a protective layer of polycarbonate to prevent aluminum oxidation and marring of the data surface. The back of the CD can now be covered with information in the form of art and lettering.

Stamping characteristics Each nickel stamper master can make about 20,000 copies before it wears out. At this point, the nickel stamper can be recycled into an aqueous nickel solution and used to make more nickel stampers. A hit CD may go through as many as 50 nickel stampers plated onto the mother disc. CD-ROMs for your computer are made the same way.

Explore Further

1. **Hypothesize** Why do you think nickel is used for stamping the polycarbonate discs?

2. **Acquire Information** Polycarbonate is the base material used for CDs. Research its properties and some of its other major uses.

In the Field

Meet Harvey Morser
Metal Plater

A seasonal highlight in Reno, Nevada, is a celebration called Hot August Nights. On display are thousands of beautifully restored antique cars. "Muscle cars"—like a 1934 Ford sedan with a $12,000 custom paint job and immaculately plated chrome—draw admiring crowds. Harvey Morser watches with pride as the cars sporting his shop's work on grilles and door handles parade by.

On the Job

 Mr. Morser, will you tell us what you do on the job?

 Although I'm now the owner of Western Metal Finishing, I still go out on the shop floor and do plating. All iron metal needs some type of protective coating to keep it from rusting. There are different types of applications that you can use. Cadmium plating is probably the best, but it's very toxic. Hardchrome plating also releases fumes into the atmosphere. So, because of environmental concerns, electroless nickel is the favored process. It's a chemically applied nickel plating done without electricity. In my opinion, this process is twice as effective because it goes on easily and consistently, unlike hard plating, which you have to apply and then grind back down. Chemists have developed an electroless nickel that has the same Rockwell factors (the hardness factors) as hardchroming. I know that a major heavy-equipment manufacturer converted from hardchrome over to electroless nickel and saved something like $3 million the first year.

 What kinds of metal products do you plate?

 One of my major accounts is for the metal straps on hearing protectors that the Navy uses on aircraft carriers. At this plant, we also plate things like computer chassis and covers for stereo systems.

 Why does the electroless nickel process produce a more even coating?

 Visualize a flat, square plate. When you put a hook in one edge of it and hang it in the tank to hardplate it, the electrical current will reach the corners first, then travel down the side edges, and finally spread into the center of the plate. The plating will go on in the same way, building up probably twice as fast on the edges as in the center. So the edges might have eight ten-thousandths of an inch of plating, whereas the center might only have four ten-thousandths. With electroless nickel, a metal part will plate perfectly evenly because it's put on chemically. Electroless nickel has the highest corrosion resistance next to cadmium plating, so these plated parts resist corrosion as well as being uniform.

Early Influences

 What training did you have in metal plating?

 All my training came on the job. I started out polishing the metal prior to plating. That's a tough and dirty job, but an important one. Plating duplicates a surface, so it has to be polished like a mirror. Otherwise, plating will magnify even a tiny pit. In those early days, I had to work a side job in a bowling alley as a pin chaser, unsticking the balls and unjamming the pins. Along the way, I picked up carpentry, welding, and electrical skills, which have come in very handy in doing the maintenance at the metal plant.

 How did you work your way up to owning the plant?

 It all boiled down to learning quickly, not whining, and working hard. Plating is tough and heavy work. The heat on the lines is terrible, with the humidity and steam off the tanks. All the tanks are running about 160°, so in the summertime it will be 104° on the line. The joke here is that we don't charge extra for the steam bath. Nine years ago, I became the owner of the business. To me, that's the Great American Dream.

Personal Insights

 If someone came to you just out of high school, would you give him or her a chance on the job?

 In a heartbeat. I believe people need to get their schooling, but they also need to have common sense. Every employee here gets on-the-job training, just like I did. What I look for in a prospective employee is honesty and dependability, plus an ability and desire to learn.

 Is this a stressful business to be in?

 Absolutely! I carry a pager all the time and even take a cellular phone out on the lake when I go fishing. There's always the possibility of an industrial accident. Earthquakes don't announce that they are on their way. If I see the numbers 1–5 on my pager, I know it's all clear. Seeing five zeros is what I dread.

 What appeals to you about the plating business?

 That I could take something that looked terrible, like an old car part, and make it look gorgeous. I want even the most modest plating job to look good, even when it's on a part that probably won't be visible after it's installed.

CAREER CONNECTION

These jobs also involve working with metals.

Metallurgical Technician Two-year training program

Mining Engineer Bachelor's degree in engineering

Scrap Metal Processing Worker On-the-job training after high school

■ **Figure 17.22** Hybrid cars have both a gasoline engine and an electric motor, which is powered by rechargeable batteries. These batteries are recharged by regenerative braking. In this process, the electric motor serves as a generator. As the driver brakes, the generator transforms some of the car's energy of motion into potential energy stored in the batteries.

Explain how regenerative braking depends on fossil fuels and why it does not violate the law of conservation of energy.

SUPPLEMENTAL PRACTICE

For more practice with electrolytic cells, see Supplemental Practice, page 836.

Electrolysis of toxic wastes The plating baths used in the various applications of electrolysis often contain toxic materials or produce toxic by-products. After bath solutions have been used for a period of time, they must be changed and the toxic contents disposed of in a safe manner. Remarkably, electrolysis offers one of the safest and most thorough means of cleaning up toxic metal-containing wastes. When the bath solution is subjected to electrolysis, the toxic metal ions are reduced to free metal at the cathode. The metal can then be recycled or disposed of safely.

Connecting Ideas

While spontaneous redox reactions can provide electric energy—and electric energy can drive nonspontaneous redox reactions—most of the electricity you use comes not from redox reactions but from the fossil fuels petroleum, natural gas, and coal. Generating electricity in a coal-burning power plant and recharging the batteries in the hybrid car shown in **Figure 17.22** are both examples of electric power that ultimately depends on fossil fuels. Fossil fuels represent one type of a large group of chemicals, the carbon-containing organic compounds. In the next chapter, you'll learn that in addition to being important sources of fuel, organic chemicals also provide us with most medicines, dyes, plastics, and textiles.

Section 17.2 Assessment

Section Summary

▶ An electrolytic cell is a chemical system that uses an electric current to drive a nonspontaneous redox reaction.

▶ Electrolysis can be used to produce compounds, separate metals from ores, clean metal objects, and plate metal coatings onto objects.

6. **MAIN Idea** Draw and label the parts of an electrolytic cell.

7. **Explain** the value of electroplating. Why is the process used?

8. **Describe** how electrolysis is used for cleaning objects.

9. **Apply Concepts** What effect would electroplating steel jewelry with gold have on the rate of corrosion of the jewelry?

10. **Apply** Magnesium in seawater is found mostly as $Mg(OH)_2$, which can be converted to $MgCl_2$ by reacting it with HCl. Magnesium metal can then be purified by electrolysis of molten $MgCl_2$.

    a) What reaction takes place at the cathode during electrolysis?
    b) What reaction takes place at the anode during electrolysis?
    c) Write an equation for the net reaction that occurs.

CHAPTER 17 Study Guide

Download quizzes, key terms, and flash cards from glencoe.com.

BIG Idea Chemical energy can be converted to electric energy and electric energy to chemical energy.

Section 17.1 Voltaic Cells: Electricity from Chemistry

MAIN Idea In voltaic cells, oxidation takes place at the anode, yielding electrons that flow to the cathode, where reduction occurs.

Vocabulary
- anion (p. 588)
- anode (p. 585)
- cathode (p. 585)
- cation (p. 588)
- electric current (p. 584)
- potential difference (p. 586)
- voltage (p. 586)
- voltaic cell (p. 585)

Key Concepts
- A potential difference between two substances is a measure of the tendency of electrons to flow from one to the other.
- A voltaic cell is a chemical system that produces an electric current through a spontaneous redox reaction.
- Batteries contain one or more voltaic cells.

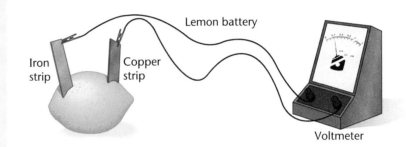

Section 17.2 Electrolysis: Chemistry from Electricity

MAIN Idea In electrolysis, a power source causes nonspontaneous reactions to occur in electrochemical cells.

Vocabulary
- electrolysis (p. 600)
- electrolytic cell (p. 600)

Key Concepts
- An electrolytic cell is a chemical system that uses an electric current to drive a nonspontaneous redox reaction.
- Electrolysis can be used to produce compounds, separate metals from ores, clean metal objects, and plate metal coatings onto objects.

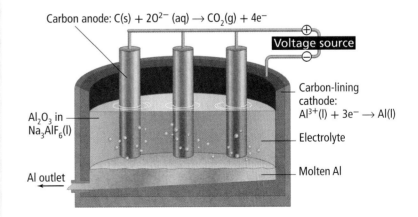

Chapter 17 Assessment

Understand Concepts

11. What is a voltaic cell?
12. What is the difference between an electrolytic cell and a voltaic cell?
13. What is the function of the salt bridge in a voltaic cell?
14. What happens to the case of a carbon-zinc dry cell as the cell is used to produce an electric current?
15. Why is the electrolyte necessary in both voltaic and electrolytic cells?
16. How can you make a non-spontaneous redox reaction take place in a cell?
17. If a strip of copper metal is placed in a solution of silver nitrate, will a redox reaction take place? Will a current be produced? Explain.
18. What products are formed from the electrolysis of an aqueous solution of rock salt?
19. What is the function of the acid in the lead-acid storage battery used in cars?
20. By what process can chlorine gas be prepared commercially?

Apply Concepts

21. What will happen if a rod made of aluminum is used to stir a solution of iron(II) nitrate?
22. Can a solution of copper(II) sulfate be stored in a container made of nickel metal? Explain.
23. How would gold-electroplated jewelry compare to jewelry made of solid gold in terms of price, appearance, and durability?
24. To plate a steel pendant with gold, the pendant can be placed in an electrolytic cell that has a gold electrode and an electrolyte solution containing gold ions. The pendant functions as the second electrode. A battery may function as the power source. Is the pendant the anode or cathode?
25. How could lead be removed from drinking water by electrolysis?

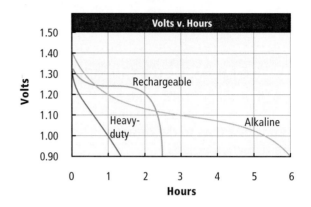

■ **Figure 17.23**

26. Tests conducted on different types of common commercial batteries involved measuring the voltage drop over time during simulated non-stop use of a motorized toy. Based on the data in **Figure 17.23**, which battery type would be best to use if you wanted to run the toy for a long period of time? Which battery type goes dead abruptly?

How It Works

27. What are the advantages of using lithium-iodine batteries in pacemakers?

How It Works

28. What are the advantages and disadvantages of using rechargeable batteries instead of conventional types?

How It Works

29. Write the equations for the two half-reactions that take place in a hydrogen-oxygen fuel cell.

Chemistry and Technology

30. Should the anode or the cathode be made of pure copper in an electrolytic cell designed for refining copper metal? Explain.

Everyday Chemistry

31. The information on a CD stamper master is the reversal of that on the original glass master cut by the recording laser. Explain why this reversal is necessary.

Chapter 17 Assessment

Think Critically

Draw Conclusions

32. MiniLab 1 Why should the metal pieces used as electrodes in the lemon battery be cleaned with steel wool?

Make Predictions

33. MiniLab 2 Would it be possible to plate a silver spoon or a gold spoon with copper?

Relate Cause and Effect

34. ChemLab How would your result in step 8 of the ChemLab have been different if
  a) a piece of zinc were used instead of the piece of magnesium?
  b) silver were used instead of copper?

Make Decisions

35. What factors must be considered in designing or selecting batteries for the following applications?
  a) flashlight
  b) hearing aid
  c) pacemaker
  d) toy car

Make Predictions

36. If a strong tendency to be oxidized were the only consideration, what metals other than lithium might be used to power a cardiac pacemaker?

Form a Hypothesis

37. Why was electrophoresis rather than electrolysis used to restore ceramic and organic artifacts from the Titanic?

Cumulative Review

38. Draw Lewis dot diagrams for the ions listed. *(Chapter 2)*
  a) Ca^{2+}
  b) Cl^-
  c) OH^-
  d) O^{2-}

39. Give the chemical formulas for the following compounds. *(Chapter 5)*
  a) manganese(IV) dioxide
  b) potassium iodide
  c) copper(II) sulfate
  d) aluminum chloride
  e) sulfuric acid
  f) iron(II) oxide

40. List the names and symbols of all of the noble gases. *(Chapter 8)*

41. Which requires more energy: boiling 100 g of water or melting 100 g of ice? Explain. *(Chapter 10)*

42. A 0.543-g piece of magnesium reacts with excess oxygen to form magnesium oxide in the reaction

$$2Mg(s) + O_2(g) \rightarrow 2MgO(s).$$

How much oxygen reacts? What is the mass of magnesium oxide (MgO) produced? *(Chapter 12)*

43. What are the mass percents of magnesium and oxygen in magnesium oxide in the problem above? *(Chapter 12)*

44. Compare the hydronium ion concentrations in two aqueous solutions that have pH values of 9 and 11. *(Chapter 14)*

45. You test several solutions and find that they have the following pH values: 8.6, 4.7, 10.4, 13.1, 2.6, 6.1. Determine the pH of each of the following. *(Chapter 14)*
  a) the strongest acid
  b) the strongest base
  c) the weakest acid
  d) the solution that is closest to being neutral

46. Which of the changes indicated are oxidations and which are reductions? *(Chapter 16)*
  a) Cu^+ becomes Cu^{2+}
  b) S becomes S^{2-}
  c) Na becomes Na^+
  d) Mn^{4+} becomes Mn^{2+}

Chapter 17 Assessment

47. Green plants make glucose and oxygen in photosynthesis. The balanced chemical equation for this redox process is shown below. *(Chapter 16)*

    $6CO_2 + 6H_2O + energy \rightarrow C_6H_{12}O_6 + 6O_2$

    a) Is the carbon in carbon dioxide reduced or oxidized in this process?

    b) What kind of energy do you think is used in this process?

Skill Review

48. **Interpreting Scientific Illustrations** A process called cathodic protection is sometimes used to protect a buried steel pipeline from corrosion. In this process, the pipeline is connected to a more active metal such as magnesium, which is corroded preferentially before the iron. **Figure 17.24** below illustrates how the two metals are connected and shows the reactions that take place.

    a) What acts as the cathode in this process? What acts as the anode?

    b) What is the oxidizing agent?

    c) Write a short summary describing how the magnesium is preferentially corroded.

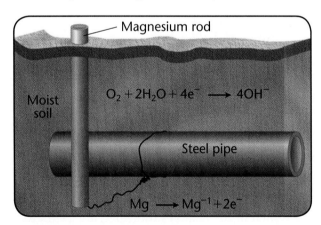

■ **Figure 17.24**

49. **Make Scientific Illustrations** Draw a diagram of a voltaic cell in which the reaction is $Ni(s) + 2Ag^+(aq) \rightarrow Ni^{2+}(aq) + 2Ag(s)$. Label the cathode and the anode. Show the ions present in both compartments, and indicate the direction of electron flow in the external circuit.

WRITING in Chemistry

50. Write an article about the development and uses for the Daniell cell, an early battery made in 1836 by John Frederic Daniell of Great Britain. Find out how it improved on the Volta cell and whether or not this type of battery is used much today.

Problem Solving

51. When a test tube containing copper(II) oxide and powdered charcoal (carbon) is heated, a gas and copper metal are produced in a redox reaction. If the gas is bubbled into a test tube containing limewater (aqueous calcium hydroxide solution), a milky white precipitate forms. Write a lab report of this experiment as if you had performed it. Answer the following questions in your report.

    a) What is the gas that forms?

    b) Write an equation for the redox reaction that takes place.

    c) What is the identity of the white precipitate?

52. In one type of fuel cell, methane gas (CH_4) is "burned" electrochemically to produce electricity:

    $CH_4(g) + H_2O(g) \rightarrow CO(g) + 3H_2(g) + energy$

    a) Is this a voltaic or an electrolytic cell?

    b) What acts as the oxidizing reagent?

    c) What acts as the reducing reagent?

    d) If 224 L of natural gas are burned in this fuel cell at STP, how many moles of carbon monoxide and hydrogen gases are produced?

53. What will happen to your gold ring if you leave it sitting in a solution of iron(II) chloride ($FeCl_2$) at room temperature?

54. What will happen to a copper bracelet that remains in contact with a solution of silver nitrate ($AgNO_3$) for several hours at room temperature?

Cumulative Standardized Test Practice

1. Which process is not an application of electrolysis?
   a) electroplating
   b) cleaning toxic wastes
   c) separating metals from ores
   d) creating synthetic diamonds

Use the table to answer Questions 2 and 3.

Standard Reduction Potentials	
Metal	Standard Reduction Potential (V)
Li	−3.0401
Al	−1.662
Cu	0.521
Ag	0.7996
Au	1.498

2. The table of standard reduction potential of substances lists common metals in order of increasing reduction potential and decreasing oxidation potential. Which metal will most easily corrode?
   a) Li
   b) Al
   c) Cu
   d) Au

3. What would happen if a lemon battery were constructed using a copper strip and a gold strip?
   a) Electrons would flow from the copper strip, where reduction would occur, to the gold strip, where oxidation would occur.
   b) Electrons would flow from the copper strip, where oxidation would occur, to the gold strip, where reduction would occur.
   c) Electrons would flow from the gold strip, where reduction would occur, to the copper strip, where reduction would occur.
   d) Electrons would flow from the gold strip, where oxidation would occur, to the copper strip, where reduction would occur.

4. A supersaturated solution is a
   a) solution without a solute.
   b) solution with less than the usual maximum dissolved solute.
   c) solution with the maximum dissolved solute.
   d) solution with more than the usual maximum dissolved solute.

5. Which of the following is the best description of the total conversion of the diprotic succinic acid ($H_2C_4H_4O_4$) into hydronium ions?
   a) $H_2C_4H_4O_4(aq) \rightarrow HC_4H_4O_4^-(aq) + H_3O(aq)$
   b) $H_2C_4H_4O_4(aq) \rightarrow H_2C_4O_4^{2-}(aq) + H_3O^+(aq)$
   c) $H_2C_4H_4O_4(aq) \rightarrow HC_4H_4O_4^-(aq) + H_3O(aq)$
      $\rightarrow C_4H_4O_4^{2-}(aq) + H_3O^+(aq)$
   d) $H_2C_4H_4O_4(aq) \rightarrow H_2C_4H_3O_4^-(aq)$
      $+ H_3O^+(aq) \rightarrow H_2C_4H_2O_4^{2-}(aq) + H_3O^+(aq)$

6. Which ions are excluded from a net ionic equation?
   a) weak acids or bases
   b) negative ions
   c) positive ions
   d) spectator ions

7. The term used to describe a chemical reaction in which a substance loses electrons to another substance is
   a) oxidation.
   b) redox.
   c) reduction.
   d) corrosion.

8. The following system is in equilibrium.

   $2S(s) + 5F_2(g) \rightleftharpoons SF_4(g) + SF_6(g)$

   Which of the following will cause the equilibrium to shift to the right?
   a) increased concentration of SF_4
   b) increased concentration of SF_6
   c) increased pressure on the system
   d) decreased pressure on the system

NEED EXTRA HELP?								
If You Missed Question . . .	1	2	3	4	5	6	7	8
Review Section . . .	17.2	17.1	17.1	13.2	14.1	15.1	16.1	6.3

CHAPTER 18 Organic Chemistry

BIG Idea Chains of carbon atoms are the backbone of organic chemistry.

18.1 Hydrocarbons
MAIN Idea Hydrocarbons contain only hydrogen and carbon, but differ in the lengths of their carbon chains and the presence of single, double, and triple bonds.

18.2 Substituted Hydrocarbons
MAIN Idea Functional groups can substitute for hydrogen atoms in hydrocarbons, resulting in diverse groups of organic compounds.

18.3 Plastics and Other Polymers
MAIN Idea Polymers are large organic molecules made up of many smaller repeating units.

ChemFacts

- Marshmallows are made of sugar, water, and gelatin.
- Roasted marshmallow is the result of oxidation reactions.
- The sugars contained in a marshmallow are largely made of carbon, which is what is left when a marshmallow is roasted to completion.

Start-Up Activities

LAUNCH Lab

Model Simple Hydrocarbons

Hydrocarbons are made of hydrogen and carbon atoms. Recall that carbon has four valence electrons and it can form four covalent bonds. How can you model simple hydrocarbons?

Materials
- molecular model kit

Procedure
1. Read and complete the lab safety form.
2. Use a molecular model kit to build a structure with two carbon atoms connected by a single bond.
3. Place hydrogen atoms in all of the unoccupied positions on your model so that each carbon atom has a total of four bonds.
4. Repeat steps 2–3 for models based on three, four, and five carbon atoms each. Be sure that each carbon atom is attached to a maximum of two other carbon atoms.

Analysis
1. **Make** a table listing the number of carbon and hydrogen atoms in each structure.
2. **Describe** the compostion of each structure with a molecular formula.
3. **Analyze** the pattern of the carbon-to-hydrogen ratio to develop a generic formula for hydrocarbons with single bonds.

Inquiry How do you think the molecular formula would be affected if the carbon atoms were attached by double and triple bonds?

Chemistry Online

Visit glencoe.com to:
- study the entire chapter online
- explore Concepts in Motion
- take Self-Check Quizzes
- use Personal Tutors
- access Web Links for more information, projects, and activities
- find the Try at Home Lab, Comparing Water and Alcohol

Organic Compounds Make the following Foldable to help you organize information about organic compounds.

STEP 1 Fold three sheets of notebook paper in half horizontally. Holding two sheets of paper together, make a 3-cm cut at the fold line on each side of the paper.

STEP 2 On the third sheet, cut along the fold line leaving a 3 cm portion uncut on each side of the paper.

STEP 3 Slip the first two sheets through the cut in the third sheet to make a 12-page book. Label your book *Organic Compounds*.

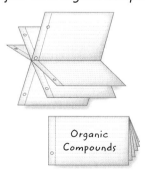

FOLDABLES Use this Foldable with Sections 18.1, 18.2, and 18.3. As you read these sections, use your book to record features of each type of organic compound, distinguishing characteristics, and real-world examples.

Section 18.1

Objectives
- **Write and interpret** structural formulas of linear, branched, and cyclic alkanes, alkenes, and alkynes.
- **Distinguish** the isomers of a given hydrocarbon.
- **Explain** the relationship between fossil fuels and organic chemicals.

Review Vocabulary
voltage: electric potential difference, expressed in units of volts

New Vocabulary
saturated hydrocarbon
alkane
isomer
unsaturated hydrocarbon
alkene
alkyne
aromatic hydrocarbon
fractional distillation
cracking
reforming

Hydrocarbons

MAIN Idea Hydrocarbons contain only hydrogen and carbon, but differ in the lengths of their carbon chains and the presence of single, double, and triple bonds.

Real-Word Reading Link After a rain shower, you might have noticed a spectrum of colors on the top of an otherwise drab puddle. What causes those bright colors to appear? They form as a result of pollution, namely small amounts of gasoline or oil that have leaked out of automobiles and formed a miniature spill on the puddle. The spill forms an extremely thin layer of hydrocarbon molecules on the water, which reflects sunlight.

Organic Compounds

Carbon is unique among elements in that it can bond to other carbon atoms to form chains containing as many as several thousand atoms. Because a carbon atom can bond to as many as four other atoms at once, these chains can have branches and can form closed-ring structures. These possibilities result in an extremely wide variety of compounds. In addition, carbon can bond strongly to elements such as oxygen and nitrogen, and it can form double and triple bonds. Thus, carbon forms an enormous number of compounds with chains and rings of various sizes, each with a variety of bond types and atoms of other elements bonded to them. **Figure 18.1** illustrates the wide variety of organic compounds in your everyday life.

It probably seems like an impossible task to memorize the structures and properties of millions of compounds. Fortunately, you don't need to study each of these compounds separately to understand organic chemistry. Despite their huge number, organic compounds can be classified into a relatively small number of groups that have similar structures and properties.

■ **Figure 18.1** From the molecules that make the bodies of living things to the modern high-tech polymers in many common manufactured items, organic compounds are all around you every day. In addition to the people in this photo, the food, the clothing, the furniture, and the plants all contain organic compounds.

Identify *two organic compounds that you have studied in a previous science course.*

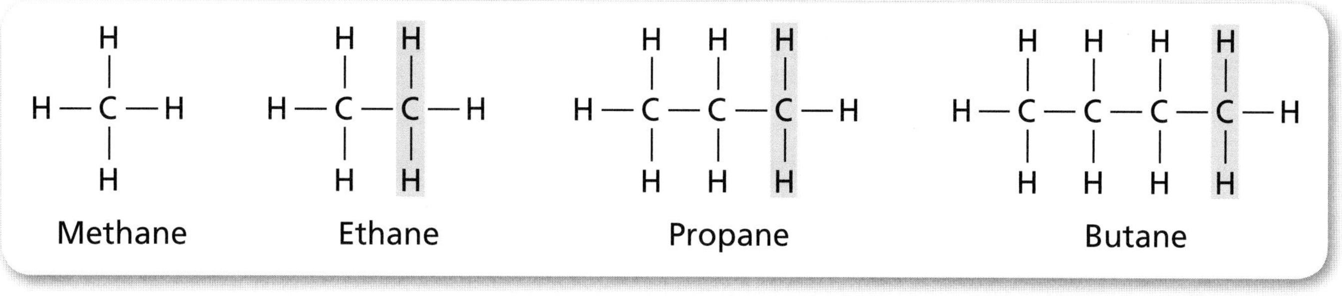

[Figure 18.2: Structural formulas of Methane, Ethane, Propane, Butane]

Saturated Hydrocarbons

Gasoline is a mixture of organic compounds that is derived from petroleum. Most of the compounds in gasoline are hydrocarbons. You might recall from Chapter 5 that hydrocarbons are organic compounds containing only hydrogen and carbon atoms. A hydrocarbon in which all the carbon atoms are connected to each other by single bonds is called a **saturated hydrocarbon.** Another name for a saturated hydrocarbon is an **alkane.** Although burning alkanes for fuel is their most common use, they are also used as solvents in paint removers, glues, and other products.

Alkanes Alkanes are the simplest hydrocarbons. The carbons in an alkane can be arranged in a chain or a ring, and both chains and rings can have branches of other carbon chains attached to them. Alkanes that have no branches are called straight-chain alkanes. Methane (CH_4), ethane (C_2H_6), propane (C_3H_8), and butane (C_4H_{10}) are all common fuels. Their structural formulas, as shown in **Figure 18.2,** differ from the next by an increment of —CH_2—.

Some alkanes have branched structures. In these compounds, a chain of one or more carbons is attached to a carbon in the longest continuous chain, which is called the parent chain. If a chain containing a single carbon is branched off the second carbon of a propane parent chain, a branched alkane with the following structure results.

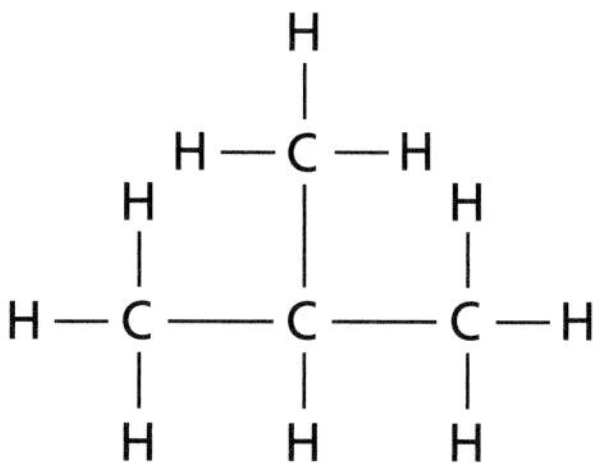

[Structure of 2-methylpropane]

The carbon atoms in alkanes can also link to form closed rings. The most common rings contain five or six carbons. The structures of these compounds can be drawn showing all carbon and hydrogen atoms.

[Structures of Cyclopentane and Cyclohexane]

■ **Figure 18.2** The structural formulas of the four shortest alkanes—methane, ethane, propane, and butane—are shown. Notice that each formula differs from its neighbor by one –CH_2– increment, as indicated by the blue highlights.

FOLDABLES
Incorporate information from this section into your Foldable.

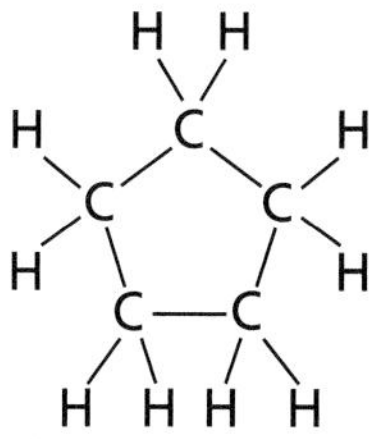

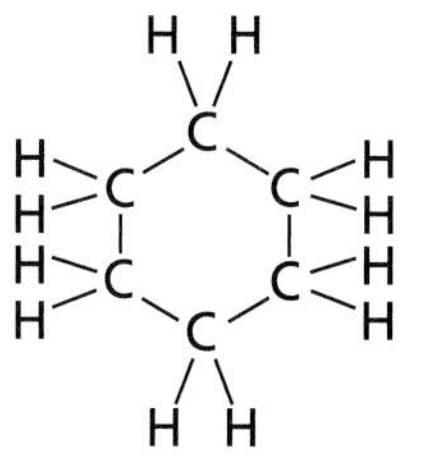

Table 18.1 Simple Alkanes

Concepts In Motion
Interactive Table Explore alkanes at glencoe.com.

Molecular Formula	Structural Formula	Ball-and-Stick Model	Space-Filling Model
Ethane (C_2H_6)	H–C(H)(H)–C(H)(H)–H		
Propane (C_3H_8)	H–C(H)(H)–C(H)(H)–C(H)(H)–H		
Butane (C_4H_{10})	H–C(H)(H)–C(H)(H)–C(H)(H)–C(H)(H)–H		

Alkanes in your gas tank The ball-and-stick and space-filling models of three simple straight-chain alkanes are shown in **Table 18.1**. The properties of a gasoline mixture—such as how well it burns in an engine—are determined by the relative amounts of various components present in the mixture illustrated in **Figure 18.3**. Hydrocarbons that contain rings and many branches burn at a more uniform rate than do straight-chain alkanes, which tend to explode prematurely, causing engine knock. The octane number on a gasoline pump is an index that indicates the relative amounts of branched and ring structures in that blend of gasoline.

■ **Figure 18.3** Gasolines are rated on a scale known as octane rating. The higher the octane rating, the greater the percentage of complex-structured hydrocarbons that are present in the mixture.

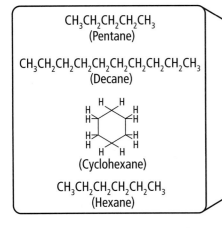

Simplifying structural diagrams The structural diagrams on the previous page can be written in a simplified form. For alkanes that are ring structures, structural diagrams can be simplified by using straight lines to represent the bonds between atoms in the rings. In these ring diagrams, each corner represents a carbon atom. For example, rings containing five and six carbons are drawn as a pentagon and a hexagon, respectively. Because carbon usually forms four bonds, it is understood that enough hydrogen atoms are bonded to each carbon to give it four bonds.

Cyclopentane   Cyclohexane

Structural diagrams of straight- and branched-chain hydrocarbons also can be written more simply by leaving out some of the bonds. For example, the formula for propane can be written as $CH_3—CH_2—CH_3$. In this type of shorthand structure, the bonds between C and H are understood. In an even simpler type of shorthand, the condensed structural formula for propane can be written as $CH_3CH_2CH_3$. Here, the bonds both between C and C and between C and H are understood.

Naming alkanes The names of the first ten straight-chain alkanes, shown in **Table 18.2,** are used as the basis for naming most organic compounds. To name a branched alkane, you must be able to answer three questions about its structure.

1. How many carbons are in the parent chain, the longest continuous chain of the molecule?
2. How many branches are on the parent chain and what are their sizes?
3. To which carbons in the parent chain are the branches attached?

Numbering carbon atoms For convenience, the carbon atoms in organic compounds are given position numbers. In straight-chain hydrocarbons, the numbering can begin at either end. It makes no difference. In branched hydrocarbons, the numbering begins at the end closest to the branch.

Examine the structure of this branched alkane.

$$\overset{}{\underset{}{\overset{CH_3}{\underset{|}{}}}}$$
$$\overset{1}{C}H_3\overset{2}{C}H\overset{3}{C}H_2\overset{4}{C}H_3$$

Four carbons are in the parent chain. **Table 18.2** shows that the name of the four-carbon alkane is *butane*. Butane is the parent chain and will be part of the compound's name. There is only one branch, and it contains one carbon. Instead of calling this a methane branch, change the -ane in methane to -yl. Thus, this is a methyl branch. Because the methyl branch is attached to the second carbon of the butane chain, this compound has the name 2-methylbutane.

Table 18.2	The First Ten Alkanes	
Formula	Name	Condensed Structural Formula
CH_4	methane	CH_4
C_2H_6	ethane	CH_3CH_3
C_3H_8	propane	$CH_3CH_2CH_3$
C_4H_{10}	butane	$CH_3(CH_2)_2CH_3$
C_5H_{12}	pentane	$CH_3(CH_2)_3CH_3$
C_6H_{14}	hexane	$CH_3(CH_2)_4CH_3$
C_7H_{16}	heptane	$CH_3(CH_2)_5CH_3$
C_8H_{18}	octane	$CH_3(CH_2)_6CH_3$
C_9H_{20}	nonane	$CH_3(CH_2)_7CH_3$
$C_{10}H_{22}$	decane	$CH_3(CH_2)_8CH_3$

Now, examine the structure of a different hydrocarbon.

Propane will be part of this compound's name because the parent chain has three carbons. Two methyl branches are present, both on the second carbon. To indicate the presence of more than one branch of the same kind, use the same Greek prefixes presented in Chapter 5 for naming hydrates and molecules. Thus, the name of this compound is 2,2-dimethylpropane. In general, follow the steps in Example Problem 18.1 to name branched-chain alkanes. For alkanes containing rings, place the prefix cyclo-before the name.

Chemistry Online
Personal Tutor For an online tutorial on naming hydrocarbons, visit glencoe.com.

EXAMPLE Problem 18.1

Naming Branched-Chain Alkanes Name the alkane shown.

1 Analyze
Follow the steps below for naming alkanes.

2 Solve
Step 1 Count the number of carbons in the parent chain. Because structural formulas can be written with chains oriented in various ways, you need to be careful in finding the longest continuous carbon chain. In this case, it is easy. The longest chain has eight carbon atoms, so the parent name is *octane.*

Step 2 Number each carbon in the parent chain. Number the chain in both directions. Numbering from the left puts the alkyl groups at carbons 4, 5, and 6. Numbering from the right puts alkyl groups at carbons 3, 4, and 5. Because 3, 4, and 5 are the lowest position numbers, they will be used in the name.

Step 3 Note the size and number of branches present. Identify and name the branches, making certain to change the –ane hydrocarbon suffix to –yl. There are one-carbon methyl groups attached to carbons 3 and 5, and a two-carbon ethyl group attached to carbon 4. Because there are two methyl groups, dimethyl will be part of the name. No prefix is needed for the one ethyl group.

One ethyl group: *no prefix*
Position and name: *4-ethyl*

Parent chain: *octane*

Step 4 Place the names of the alkyl branches in alphabetical order, ignoring prefixes such as di-. Use the number of the carbon to which each group is attached to identify its position. Use hyphens to separate numbers from words and commas to separate numbers from numbers. Thus, the branches appear in the name as 4-ethyl-3,5-dimethyl. To complete the name, add the name of the parent chain to the part of the name identifying the branches. The name in its final form should be written as 4-ethyl-3,5-dimethyloctane.

3 Check
The longest continuous carbon chain has been identified and numbered correctly. All branches have been designated with correct prefixes and names. Alphabetical order and punctuation are correct.

PRACTICE Problems

Solutions to Problems Page 866

1. Name the following structures.

   a. CH₃ CH₃
      | |
      CH₃CHCH₂CHCH₂CH₃

   b. CH₃ CH₃
      | |
      CH₃CCH₂CHCH₃
      |
      CH₃

   c. CH₃
      |
      CH₂   CH₃   CH₃
      | | |
      CH₃CHCH₂CH₂CHCH₂CHCH₃

2. Draw the structures of the following branched-chain alkanes.
   a. 2,3-dimethyl-5-propyldecane
   b. 3,4,5-triethyloctane

Isomers Butane and 2-methylpropane are two alkanes with different names and different structures. **Figure 18.4** shows familiar applications of both of these compounds. Is there any relationship between these two compounds? If you count the number of carbon and hydrogen atoms in each, you will find that they both have four carbons and ten hydrogens, which gives them the same molecular formula, C_4H_{10}. Compounds that have the same formula but different structures are called **isomers.** Butane and 2-methylpropane are known as structural isomers. Each has the molecular formula C_4H_{10}, but they have different structural formulas because the carbon chains have different shapes.

SUPPLEMENTAL PRACTICE

For more practice with naming alkanes, see Supplemental Practice, page 837.

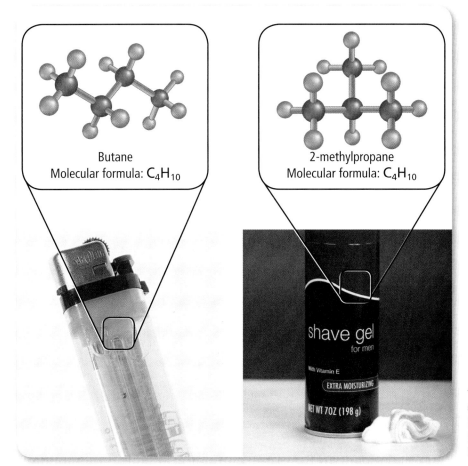

Butane
Molecular formula: C_4H_{10}

2-methylpropane
Molecular formula: C_4H_{10}

■ **Figure 18.4** Butane is a fuel used in lighters. 2-methylpropane, also called isobutane, is used as propellant in products such as shaving gel.

Section 18.1 • Hydrocarbons **627**

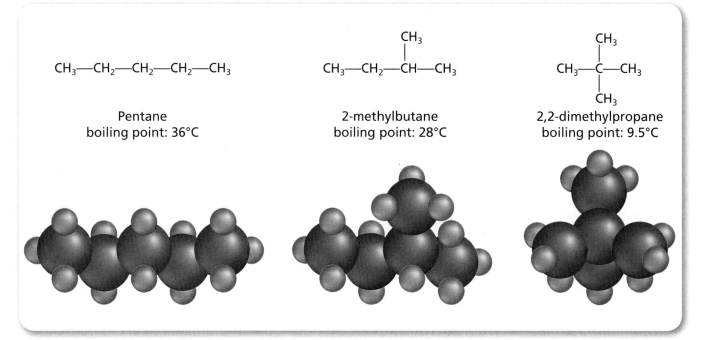

■ **Figure 18.5** Structure and properties are closely related, as you can see by examining the isomers of pentane. Although all three compounds have the formula C_5H_{12}, differences in the amount of branching affect their boiling points. Note the differences in the shapes of the molecules.

Describe *the effect of increased branching on boiling point for the pentane isomers. What does this indicate about the intermolecular attractive forces of each compound?*

Concepts In Motion

Interactive Figure To see an animation of the isomers of pentane, visit glencoe.com.

Despite their identical molecular formulas, isomers have different properties. The boiling and melting points of 2-methylpropane and butane are different, as are their densities and solubilities in water. In addition, their chemical reactivities are different. **Figure 18.5** shows boiling point differences in the isomers of pentane.

Although only two alkane isomers have four carbons, the number of possible isomers increases rapidly as carbon atoms are added to the parent chain. This is because longer chains provide more locations for branches to attach. A methyl branch on a six-carbon parent chain can be attached to either the second or the third carbon from the end of the chain, as shown in **Figure 18.6**. To be certain that these two compounds are really isomers, count the number of carbon and hydrogen atoms in the structures formed by placing the methyl group at those two positions, and write the molecular formulas. There are three isomers of pentane, five of hexane, and more than 4 billion isomers of the alkane with the formula $C_{30}H_{62}$.

■ **Figure 18.6** Moving a methyl group from carbon number 2 to carbon number 3 results in a different isomer. Both isomers have the formula C_7H_{16}.

Explain *why moving the methyl group from carbon 2 to carbon 5 does not result in a new isomer of C_7H_{16}.*

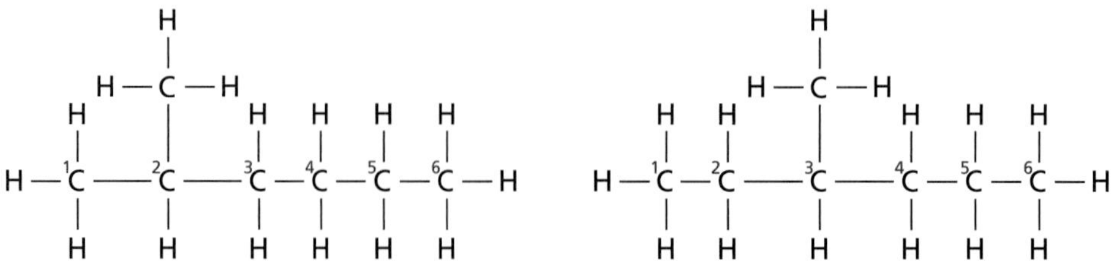

2-methylhexane
C_7H_{16}

3-methylhexane
C_7H_{16}

Properties of alkanes Properties are affected by the structure or arrangement of atoms present in a molecule. Another factor that affects properties of alkanes is chain length. In general, the more carbons present in a straight-chain alkane, the higher its melting and boiling points. At room temperature, straight-chain alkanes that have 1–4 carbon atoms are gases, those with 5–16 carbon atoms are liquids, and those with more than 16 carbon atoms are solids.

A property shared by all alkanes is their relative unreactivity. Recall from Chapter 9 that the carbon-carbon and carbon-hydrogen bonds found in alkanes are nonpolar. Because alkanes don't have any polar bonds, they undergo only a small number of reactions and will dissolve only those organic compounds that are nonpolar or that have low polarity, such as oils and waxes. The nonpolar and mostly unreactive nature of alkanes makes them good organic solvents. Paints, paint removers, and cleaning solutions often contain hexane or cyclohexane as solvents.

Alkenes

You learned in Chapter 9 that carbon atoms in organic compounds can be connected by single, double, or triple bonds. A hydrocarbon that has one or more double or triple bonds between carbons is called an **unsaturated hydrocarbon.** Molecules with single, double, and triple bonds are compared in **Table 18.3**. Note the similarities in the names of the three molecules presented in **Table 18.3**: ethane, ethene, and ethyne. Each one is a two-carbon molecule, but they differ in that there is a single, a double, and a triple bond between the carbons, respectively. From this observation, can you predict how the names of unsaturated hydrocarbons are made?

> **FACT of the Matter**
>
> The terms *saturated* and *unsaturated* originated before chemists understood the structures of organic substances. They knew that some hydrocarbons would take up hydrogen in the presence of a catalyst. Hydrocarbons that would take up additional hydrogen were said to be unsaturated, while those that would react with no more hydrogen were said to be saturated. Today, we know that unsaturated hydrocarbons contain double and triple bonds that will react with hydrogen to form single bonds.

Table 18.3 Comparing Single, Double, and Triple Bonds

	Ethane	Ethene	Ethyne
Structural formula	CH_3-CH_3	$CH_2=CH_2$	$CH\equiv CH$
Ball-and-stick model			
Lewis dot diagram	H:C:C:H (with H's above and below each C)	H:C::C:H (with H's above and below each C)	H:C:::C:H
Space-filling model			

MiniLab 18.1

Test for Unsaturated Oil

Can you determine the relative amount of unsaturation? Most animal fats are saturated hydrocarbons that are solids at room temperature, whereas most vegetable fats are unsaturated and are liquids at room temperature. Different amounts of unsaturation in fats can be compared by testing how quickly a red-brown iodine solution added to each fat decolorizes. Iodine adds to carbons that take part in multiple bonds, forming colorless organic halogen compounds when the double or the triple bond is broken.

Procedure

1. Read and complete the lab safety form.
2. Place 20 mL of **peanut oil** in one **small flask** and 20 mL of **canola oil** in another **flask.** Label the flasks.
3. Add five drops of **tincture of iodine** to each oil and swirl to mix well. Note the color of each solution.
4. Heat both flasks on a **hot plate** set on low.
5. Note which oil returns to its original color first. This oil is more unsaturated.
6. Now read the **food label** on each bottle of oil, and determine whether your test results agree with how much unsaturated fat the labels say are in each oil.

Analysis

1. **Describe** What happens to the red-brown iodine when it is added to the unsaturated oils?
2. **Conclude** Which oil is more unsaturated?
3. **Predict** Examine nutrition labels on bottles of the oils listed below. Predict which of each pair will decolorize faster.
   a) canola or corn oil
   b) coconut or sunflower oil

■ **Figure 18.7** Ethylene is the common name of ethene. This compound occurs naturally as a plant hormone that functions to speed up ripening of fruits and vegetables. Fruits and vegetables can be harvested unripe, which is efficient and extends shelf life. They can then be treated with ethylene so that they all ripen at the same time for purchase at the grocery store.

You have learned that gasoline is a mixture that contains several different alkanes. Gasoline also contains several hydrocarbons with double bonds. A hydrocarbon in which one or more double bonds link carbon atoms together is called an **alkene.**

Naming alkenes Alkenes are named using the root names of the alkanes, with the -ane ending changed to -ene. The simplest alkene is ethene (CH_2CH_2), which contains two carbons linked in a chain. Ethene, a gas at room temperature, is the most important organic compound used in the chemical industry. Almost half of the ethene used is converted into plastics. It is also used to make automobile antifreeze (ethylene glycol). **Figure 18.7** shows another use of ethene.

The alkene with three carbons in a chain is called propene, CH_2CHCH_3. When four or more carbons are present in a chain, the double bond can be located at more than one possible position. When naming these alkenes, a number must be added at the beginning of the name to indicate where the double bond is located. You should use the following steps in naming alkenes with long carbon chains.

1. Count the number of carbons in the longest continuous chain that contains the double bond, and assign the appropriate alkene name.
2. Number the carbons consecutively in the longest chain, starting at the end of the chain that will result in the lowest possible number for the first carbon to which the double bond is attached.
3. Write the number corresponding to the first carbon in the double bond, followed by a hyphen and then the alkene name.

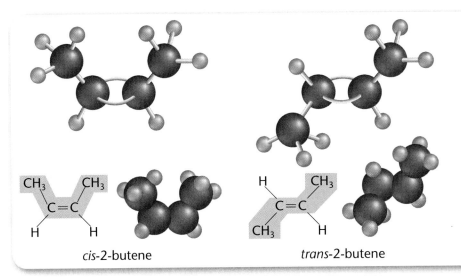

■ **Figure 18.8** *Cis*-2-butene and *trans*-2-butene are geometric isomers. Note their shapes in both ball-and-stick models and space-filling models. The properties of a pair of geometric isomers vary. *Trans*- isomers are more symmetrical than *cis*- isomers, and they can pack together more closely when in the solid state. This close packing makes the molecules harder to pull apart.

Infer what the close packing of *trans*-isomers means for their melting points.

What is the name of the compound that has the following structure?

$$CH_2=CHCH_2CH_3$$

This compound has four carbons in a chain with one double bond, so butene will be part of its name. Numbering the carbons starting on the left side of the compound gives the first carbon that is part of the double bond the position number of one. Thus, this compound is 1-butene.

$$\overset{1}{C}H_2=\overset{2}{C}H\overset{3}{C}H_2\overset{4}{C}H_3 \quad \text{1-butene}$$

Positional isomers An isomer of 1-butene is 2-butene ($CH_3CH=CHCH_3$). These two compounds have the same molecular formulas but different structures. They are called positional isomers because they differ only by the position of the double bond. Positional isomers have different properties, just as structural isomers do.

The formation of a double bond prevents the carbons on each side of the bond from rotating with respect to each other. If the two groups attached to either carbon are different, the alkene can have two different geometric structures. These structures are geometric isomers. Study the two geometric isomers of 2-butene that are modeled in **Figure 18.8.** In the isomer called *cis*-2-butene, the hydrogen atoms and —CH_3 groups are on the same side of the double bond. In *trans*-2-butene, the hydrogen atoms and —CH_3 groups are on opposite sides of the double bond.

Reactivity of alkenes Alkenes are more reactive than alkanes because the two extra electrons in the double bond are not held as tightly to the carbons as are the electrons in a single bond. Alkenes readily undergo synthesis reactions in which smaller molecules or ions bond to the atoms on either side of the double bond. An unsaturated alkene can be converted into a saturated alkane by adding hydrogen to the double bond. This reaction is called hydrogenation.

$$CH_2=CH_2 + H_2 \rightarrow CH_3CH_3$$
Ethene   Ethane
(unsaturated)   (saturated)

VOCABULARY

WORD ORIGIN

cis-
comes from the Latin word *cis* meaning *on this side*

trans-
comes from the Latin word *trans* meaning *across*

Biology Connection

Vision and Vitamin A

Vision is a process usually studied in biology class. That's where you may have learned that light rays pass through the eye to reach the retina, where the rods and cones are located. Rods and cones are nerve receptors that are excited by light. More than 120 million rods in each eye detect white light and provide sharpness of visual images. The 7 million cones in each eye detect color. The pigment molecules responsible for vision are attached to the ends of the rods and cones. One of these pigments is called rhodopsin. Rhodopsin has two parts—a protein called opsin and a small molecule called retinal.

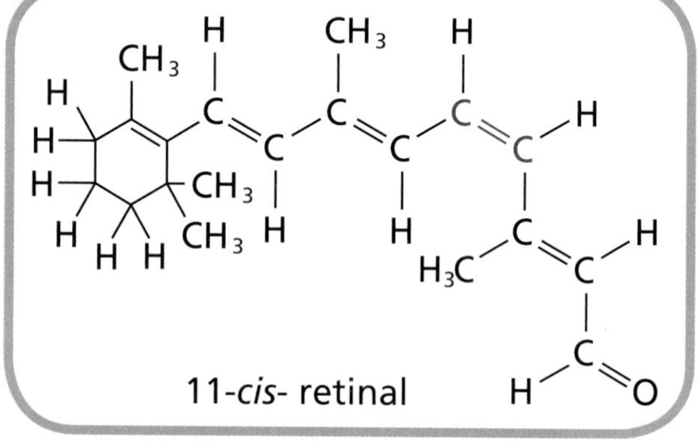

Figure 1 Structure of 11-*cis*-retinal

Chemistry of vision Beta-carotene, the natural orange pigment found in carrots, breaks down in your body to form vitamin A, which is then converted to 11-*cis*-retinal shown in **Figure 1**. In the retina of the eye, 11-*cis*-retinal attaches to the protein opsin to form rhodopsin. When light is absorbed by the 11-*cis* isomer portion of rhodopsin, the energy causes the pigment molecule to undergo a change in shape. The right end of the molecule rotates about a double bond to form all-*trans*-retinal, in which all the groups are in the *trans* position shown in **Figure 2**.

When the retinal portion of rhodopsin is isomerized to the all-*trans* form, it separates from the opsin. Thus, light energy causes the rhodopsin to break down into the substances from which it was formed.

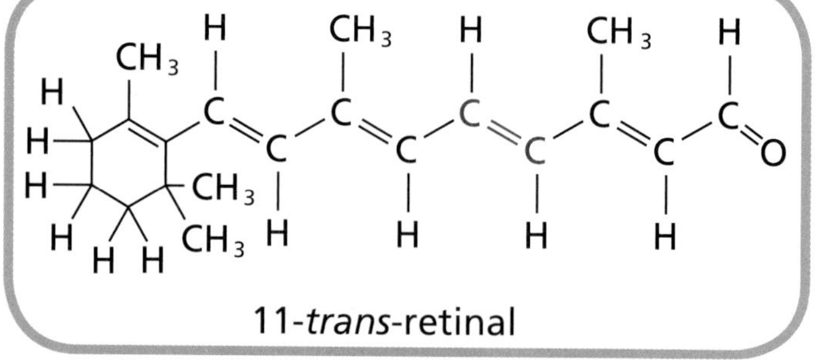

Figure 2 Structure of 11-*trans*-retinal

As the rhodopsin molecule splits, the rods become excited. This is probably due to ionic charges that develop on the splitting surfaces. These charges last only for a second, but they generate nerve signals that are transmitted to the optic nerve and then to the brain. After the rhodopsin is activated, the *trans*-retinal returns to the *cis* form and recombines with opsin. This process is relatively slow, which is why your eyes need time to adjust to dim light.

Night vision When large quantities of light energy strike the rods, large amounts of rhodopsin are broken down. As a consequence, the concentration of rhodopsin in the rods falls to a low level. If you leave a bright area and enter a darkened room, the quantity of rhodopsin in the rods at first is small. As a result, you experience temporary blindness. The concentration of rhodopsin gradually builds up until it becomes high enough for even a small amount of light to stimulate the rods. During dark adaptation, the sensitivity of the retina can increase as much as a thousandfold in only a few minutes and as much as 100,000 times in an hour or more.

Connection to Chemistry

1. **Hypothesize** Owls and bats avoid daylight and are active in the dark. Their eyes contain only rods. Hypothesize as to the relative amount of rhodopsin in their rods compared with that in human eyes, and explain.
2. **Analyze** What is the significance of the change from 11-*cis*-retinal to all-*trans*-retinal for the vision process?

■ **Figure 18.9** Because the triple bond is reactive, ethyne is very useful as a starting material in the manufacture of other compounds, such as the patio furniture shown here. Ethyne is a by-product of oil refining and can also be produced by reacting calcium carbide (CaC_2) and water (H_2O).

Write *the equation for the reaction of calcium carbide and water. (In addition to ethyne, calcium hydroxide is produced.)*

Alkynes

Another type of unsaturated hydrocarbon, called an **alkyne**, contains a triple bond between two carbon atoms. Alkynes are named using the alkane root name for a given carbon chain length and changing the -ane ending to -yne. Ethyne (C_2H_2), known more commonly as acetylene, is one of the most important commercial alkynes. Most ethyne produced in the United States is used to make vinyl and acrylic materials, like those shown in **Figure 18.9,** although about 10 percent is burned in oxyacetylene torches. These torches are used to cut and weld metals.

Physical and chemical properties of alkynes Few alkynes occur naturally because they are very reactive. However, they can be synthesized from other organic compounds. The names and structures of some small alkyne molecules are shown in **Table 18.4.**

Melting and boiling points of alkynes increase with increasing chain length, just as they did for alkanes and alkenes. Alkynes have physical and chemical properties similar to those of alkenes; their melting points are higher than those of alkanes, and they undergo synthesis reactions. For example, hydrogen molecules can be added to an alkyne in a stepwise fashion to form an alkene and then an alkane.

$$CH{\equiv}CH + H_2 \rightarrow CH_2{=}CH_2$$
$$CH_2{=}CH_2 + H_2 \rightarrow CH_3CH_3$$

Table 18.4 Simple Alkynes

Chemical Name	Common Name	Structure
ethyne	acetylene	HC≡CH
propyne	methylacetylene	HC≡CCH$_3$
1-butyne	—	HC≡CCH$_2$CH$_3$
2-butyne	—	CH$_3$C≡CCH$_3$

In the Field

Meet John Garcia
Pharmacist

"A treasure house going up in smoke"—that's how pharmacist John Garcia describes the fires set to clear land in the Amazon. He's concerned about the potential medicines that humans might be destroying as they make inroads into formerly wild areas. "However," he continues, "endangered sources for new medicines aren't only in jungles. Lincomycin, an antibiotic, was discovered in soil in Lincoln, Nebraska." In this interview, Mr. Garcia describes some of the changes he has seen in his four decades as a pharmacist.

On the Job

Q Mr. Garcia, can you tell us what you'll be doing first today in your pharmacy?

A This morning, I have to put together 100 suppositories for a patient with migraine headaches. At least that won't be as difficult as the order for a suppository a colleague of mine once had. That was for an elephant! The medication was made with an aluminum baseball bat mold. Just one dose cost $300.

Q Do you prepare prescriptions for other animals?

A I've made them for rabbits. I added raspberry flavor because rabbits enjoy the flavor.

Q Do your human patients get a choice of flavors, too?

A I've got about 40 flavors—piña colada, bubble gum, peppermint. Tutti-frutti is my personal favorite. These flavorings aren't just frills. For instance, I prepare chloroquine, an antimalarial drug for children, and put it in a chocolate base. Because the molecules of chocolate are fairly large, they block the taste buds and cover up the bitterness of the medicine.

Q You own and operate your own independent pharmacy. How does it differ from most pharmacies?

A This one is different because I specialize in compounding prescriptions. That means that I can prepare customized formulas in precise doses to meet patients' individual needs. In most pharmacies, the pharmacists do mostly what I call "count and pour." That is, they prepare standardized doses of drugs. A lot of times, I work in partnership with a patient's doctor. He or she will ask my advice about medication. Besides suggesting alternative drugs or doses, I can talk with the doctor about methods of administering a drug.

Q How has the practice of pharmacy changed in the more than 40 years you've been a pharmacist?

A New laws have given pharmacists the responsibility of explaining carefully to a patient the workings of a drug and its possible side effects. At the same time, changes in the insurance industry have added lots of paperwork and reduced our fees. So that means that pharmacists have to work more quickly while maintaining careful standards.

Q: Seriously ill patients are living longer than they used to. How has that change affected your practice?

A: I work with several hospices, which care for terminally ill patients. By the time patients are admitted to a hospice, they typically are in the last stages of an illness and have about 40 days to live. Our priority is to make the patient comfortable by using a variety of medications to relieve pain and discomfort.

Early Influences

Q: How did you come to be a pharmacist?

A: I got an after-school job in a small pharmacy when I was in high school. I started out making deliveries and eventually helped fill prescriptions. I liked working with the public, and the pharmacist, Nathan Fisher, encouraged me to apply to pharmacy school. I feel honored that he was my mentor.

Q: Were you, as a child, aware of pharmacies and medications?

A: Not really. I don't think I even saw a doctor until I needed a physical exam to play high school sports. My parents are from Mexico, and my mother knew many home remedies. She gave me things like charcoal or mint for an upset stomach.

Personal Insights

Q: What qualities do you think it takes to be a good pharmacist?

A: It's obvious that you have to like people and be accurate in mathematical and chemical calculations. I also like to say that pharmacy is a lot like cooking—a little of this, a little of that. Just like a cook, I want my preparations to contain the right proportion of ingredients, to taste good, and also to look elegant.

Q: What changes do you see coming in the field of pharmacy during the 21st century?

A: I think people will be able to insert a prescription in something like an automated teller machine. They'll also insert medical cards encoded with their personal medical history and their age, height, and weight. The machine will dispense their medication.

CAREER CONNECTION

These careers are all related to the field of pharmacy.

Pharmacologist Two to four years post-college study at a medical school or school of pharmacy

Pharmaceutical Technician Two years in a training program after high school

Pharmaceutical Production Worker High school diploma and on-the-job training

This structure of benzene, showing alternating single and double bonds, is now known to be incorrect because it does not account for benzene's inertness and bond lengths.

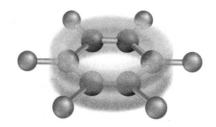
The flat benzene molecule is shown with clouds of shared electrons above and below the plane of the ring. These delocalized electrons help explain benzene's inertness.

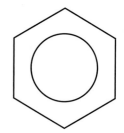
This hexagon diagram provides an accurate shorthand representation of benzene. Each corner represents a carbon atom. The circle in the middle of the structure represents the cloud of six electrons that are shared equally by the six carbon atoms in the molecule.

■ **Figure 18.10** The stability of benzene is greater than expected for an unsaturated hydrocarbon. In addition, research showed all six C–C bonds to be the same. A new representation resulted.

FACT of the Matter

One of the biggest puzzles of nineteenth-century chemistry involved the structure of benzene, which seemed to have too few hydrogen atoms to exist as a straight-chain molecule. The German chemist August Kekulé (1829–1896) solved the puzzle after having a dream in which a dancing snake formed a ring and bit its own tail. Kekulé's insight into the ring structure of benzene is considered to be among the most important work in nineteenth-century theoretical organic chemistry.

Aromatic Hydrocarbons

Another group of unsaturated hydrocarbons has distinctive, six-carbon ring structures. The simplest compound in this group is benzene, with the molecular formula C_6H_6. Benzene contains six carbons joined together in a flat ring. Its structure was originally thought to contain alternating single and double bonds between the carbons, as illustrated in **Figure 18.10,** but this structure is now known to be incorrect.

Although it contains double bonds, benzene does not share most of the properties of alkenes. It is unusually resistant to hydrogenation, whereas most alkenes readily hydrogenate. To account for this inertness, chemists have suggested that the extra electrons are shared equally by all six carbons in the ring rather than being located between specific carbon atoms. The currently accepted structure for the benzene molecule is shown in **Figure 18.10.**

This sharing of electrons among so many carbons gives benzene and similar compounds unique properties. The name **aromatic hydrocarbon** is used to describe a compound that has a benzene ring or the type of bonding exhibited by benzene. Aromatic hydrocarbons were originally named because most of them have distinctive aromas. Naphthalene, formerly used in mothballs to prevent moth damage to woolen clothes, consists of two benzene rings attached side-by-side. Aromatic hydrocarbons are unusually stable because the carbons are bonded tightly together by so many electrons.

Sources of Organic Compounds

You have examined structures and learned how to name basic organic compounds, but have you stopped to think about where these compounds come from? Most hydrocarbons come from fossil fuels, especially petroleum, but also natural gas and coal. **Figure 18.11** illustrates two sources of organic compounds. Other important sources include wood and the fermentation products of plant materials.

Soot is full of aromatic hydrocarbons produced by combustion of organic materials such as wood or coal. In the 1930s, it was shown that a number of large hydrocarbons with ring structures found in the coal tar in soot were carcinogenic, or cancer-causing. Combustion engines in vehicles and the burning of wood and coal for warmth and energy to do work are sources of soot.

Valuable deposits of petroleum and natural gas often are found by off-shore oil drilling. These deposits are common under the oceans because microscopic marine organisms—including algae, bacteria, and plankton—died and settled on the seafloor, where their remains were altered by high temperatures and pressure.

Natural gas Natural gas deposits contain large quantities of methane, along with smaller amounts of alkanes up to about five carbons in length. In addition to these organic components, natural gas deposits also contain carbon dioxide gas, nitrogen gas, and helium gas. Natural gas is formed the same way petroleum is and is most often found with petroleum deposits.

Natural gas passes easily through pipes and is used mostly as a fuel. Because it burns cleanly, methane is the most desireable component of natural gas for use as a fuel. Therefore, the raw natural gas from deposits is processed to remove components other than methane before being burnt. Natural gas is also used as a raw material for making many small organic compounds.

Petroleum Petroleum is a complex mixture, mostly of alkanes and cyclic alkanes. As shown in **Figure 18.11,** petroleum is often found by off-shore drilling. Products we get from petroleum include gasoline, jet fuel, kerosene, diesel fuel, fuel oil, asphalt, and lubricating oil. To use these organic products, they have to be separated from one another. What properties of hydrocarbons might allow them to be separated?

Fractional distillation One property that allows hydrocarbons to be separated is boiling point. As you learned in Chapter 5, distillation is a technique used to separate substances that have different boiling points. In the petroleum industry, huge towers are used to distill petroleum into its component liquids. Inside the tower, many plates provide multiple surfaces on which repeated vaporization-condensation cycles take place. Repeated cycles provide for more efficient separations and allow fractions containing only one or a few different compounds to be isolated. This method of separation is called **fractional distillation.**

■ **Figure 18.11** Most hydrocarbons come from fossil fuels such as petroleum, natural gas, and coal.

Infer *why petroleum deposits are also found under dry land.*

VOCABULARY
WORD ORIGIN
Petroleum
comes from the Latin words *petra*, meaning *a rock*, and *oleum*, meaning *oil*

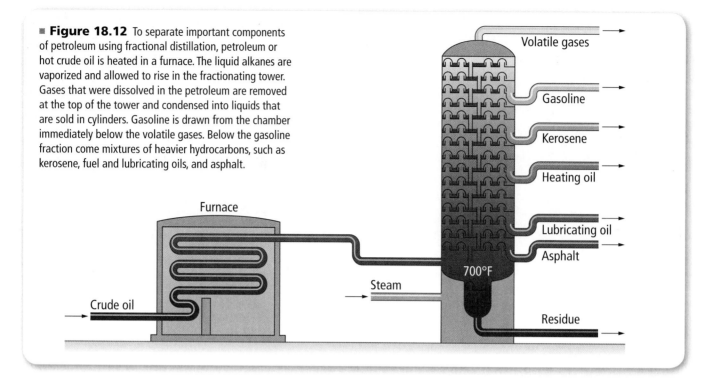

■ **Figure 18.12** To separate important components of petroleum using fractional distillation, petroleum or hot crude oil is heated in a furnace. The liquid alkanes are vaporized and allowed to rise in the fractionating tower. Gases that were dissolved in the petroleum are removed at the top of the tower and condensed into liquids that are sold in cylinders. Gasoline is drawn from the chamber immediately below the volatile gases. Below the gasoline fraction come mixtures of heavier hydrocarbons, such as kerosene, fuel and lubricating oils, and asphalt.

VOCABULARY

WORD ORIGIN

Distillation
comes from the Latin words *de*, meaning *down*, and *stillare*, meaning *to drop*

Figure 18.12 shows where different products are drawn off according to boiling points at different levels of the tower. The temperature of the tower is controlled so that it is hotter at the bottom and cooler at the top. Because boiling point increases along with molecular mass, the heavier, higher-boiling components condense near the bottom, and the lighter, lower-boiling molecules condense toward the top. Pipes connected at various points allow chemical engineers to draw off each fraction.

Cracking and reforming The process of fractional distillation does not produce quantities of hydrocarbons in the proportions needed in industry. For example, gasoline is usually the fraction of petroleum most in demand, but this fraction makes less than half of petroleum. How can the quantities of useful hydrocarbons be increased? The hydrocarbons in the fractions coming off the lower parts of the tower are mostly large alkanes. These can be converted into smaller, more useful alkanes and alkenes by a process called **cracking**. Cracking uses catalysts or high temperatures in the absence of air to break down or rearrange large hydrocarbons. Cracking is also used to increase the yield of natural gas by producing small alkenes from larger molecules. The cracking of propane (C_3H_8) produces some methane (CH_4) and ethene (C_2H_4), in addition to propene (C_3H_6) and hydrogen (H_2).

$$2CH_3CH_2CH_3 \xrightarrow{500-700°C} CH_4 + CH_2=CH_2 + CH_3CH=CH_2 + H_2$$

Another process that uses heat, pressure, and catalysts to convert large alkanes into other compounds is called **reforming**. It is used to form aromatic hydrocarbons.

The most inexpensive way to obtain coal that is not deeply buried is by surface-mining. Large areas of land, stripped of all vegetation, cause environmental problems. The exposed soil washes away after the coal is removed. Laws in the United States now require that these areas be restored.

Underground coal mining can be difficult, dirty, and dangerous work. The coal seams are often found deep underground, and tunnels must be dug so that miners can reach them. A mining machine breaks up the coal and delivers it to a conveyor belt, which carries the coal chunks out of the mine.

Coal Like petroleum and natural gas, coal also is a fossil fuel. It is formed from the remains of plants that became buried underwater and were subjected to increasing pressures as layers of mud built up. Coal is composed primarily of carbon but also contains many mineral impurities. It is used mostly as a fuel and as a source of aromatic hydrocarbons. Coal must be obtained from surface or underground mines, as shown in **Figure 18.13.** The environment in areas that are surface-mined must be restored to prevent soil erosion.

■ **Figure 18.13** Coal is our most abundant fossil fuel.

Section 18.1 Assessment

Section Summary

▶ Alkanes are saturated hydrocarbons that contain only single bonds between carbon atoms, whereas alkenes and alkynes are unsaturated hydrocarbons.

▶ Isomers are compounds that have the same molecular formula but different structures.

▶ Because properties depend upon structure, isomers often have different properties.

▶ Benzene is one of a group of unusually stable cyclic hydrocarbons. This stability is due to the sharing of six electrons by all the carbon atoms in the molecule.

▶ The many components in petroleum are separated by fractional distillation.

3. **MAIN Idea Name** the organic compounds shown.
   a) $CH_3CH=CHCH_2CH_2CH_3$
   b) CH_3
      $|$
      $CH_3CHCH_2CH_2CH_3$
   c) $CH_3CH_2CH_2CH_2CH_2CH_2CH_3$
   d) CH_3CH_3
      $||$
      $CH_3CH_2CHCH_2CHCH_3$

4. **Draw** structures that correspond to the names of the hydrocarbons given.
   a) hexane
   b) 3-ethyloctane
   c) trans-2-pentene
   d) ethyne
   e) 1,2-dimethylcyclopropane
   f) 1-butyne

5. **List** What are the major sources of hydrocarbons? Where are they found?

6. **Analyze** Halogen molecules, such as Br_2, can be added to double bonds in a reaction similar to hydrogenation. Draw the structure of the product that forms when Br_2 is added to propene.

7. **Infer** Would it be more important to store octane or pentane in a tightly sealed bottle at a low temperature? Why?

Section 18.2

Objectives

- **Compare and contrast** the structures of the major classes of substituted hydrocarbons.
- **Summarize** properties and uses of each class of substituted hydrocarbon.

Review Vocabulary

alkane: a saturated hydrocarbon that contains only single bonds between carbon atoms

New Vocabulary

substituted hydrocarbon
functional group

FOLDABLES Incorporate information from this section into your Foldable.

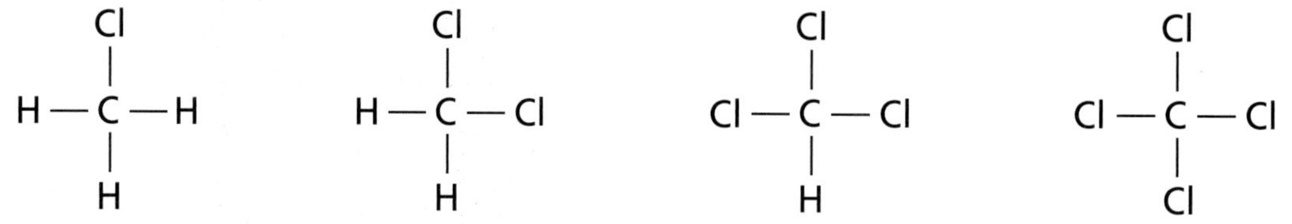

■ **Figure 18.14** Four products can be produced by reacting chlorine and methane gases—chloromethane, dichloromethane, trichloromethane, and tetrachloromethane. These four molecules have different properties and different applications.

Substituted Hydrocarbons

MAIN Idea Functional groups can substitute for hydrogen atoms in hydrocarbons, resulting in diverse groups of organic compounds.

Real-Word Reading Link Doesn't that apple pie fresh out of the oven look good? You can smell the apples, cinnamon, nutmeg, and vanilla. What causes these ingredients to have aromas? Apples, cinnamon, nutmeg, vanilla, and many other fruits and spices all contain molecules with a distinctive group of atoms located at the end of the molecule.

Functional Groups

When chlorine and methane gases are mixed in the presence of heat or ultraviolet light, an explosive reaction produces a mixture of the four products shown in **Figure 18.14**. Notice that the structures of these compounds are the same as hydrocarbons, except for the substitution of atoms of another element for hydrogen. Such compounds are called **substituted hydrocarbons.**

The part of a molecule having a specific arrangement of atoms that is largely responsible for the chemical behavior of the parent molecule is called a **functional group.** Functional groups can be atoms, groups of atoms, or bond arrangements. Notice that in the structures shown in **Figure 18.14,** one or more of the hydrogen atoms in methane is replaced by a chlorine atom.

Replacing part of a hydrocarbon with functional groups changes the compound's structure, properties, and uses. For example, chloromethane (CH_3Cl) is a gas used as a refrigerant; dichloromethane (CH_2Cl_2) is a liquid used as a solvent to decaffeinate coffee. Trichloromethane ($CHCl_3$), also known as chloroform, was an early anesthetic. Tetrachloromethane (CCl_4), commonly called carbon tetrachloride, is a solvent that was used in dry cleaning and in fire extinguishers.

Some functional groups are complex in structure and consist of a group of atoms rather than a single atom. These groups of atoms often contain oxygen or nitrogen, and some contain sulfur or phosphorus. Double and triple bonds also are considered to be functional groups. Many organic compounds contain more than one type of functional group. Simple organic compounds are grouped into categories depending on the functional group they contain. **Figure 18.15** will help you visualize the structures and functions of the most important functional groups.

Chloromethane
(methyl chloride)

Dichloromethane
(methylene chloride)

Trichloromethane
(chloroform)

Tetrachloromethane
(carbon tetrachloride)

FIGURE 18.15

Visualizing Functional Groups

Halogenated Compounds
Structure: R—X, where X=F, Cl, Br, or I
Functional group: halogen atoms
Properties: high density
Uses: refrigerants, solvents, pesticides, moth repellents, some plastics
Biological functions: thyroid hormones
Examples: chloroform, dichloromethane, thyroxin, chlorofluorocarbon refrigerant, DDT, PCBs, PVC

chlorofluorocarbon refrigerant

Many pesticides, which are used to protect crops from damage from pests such as insects and weeds, contain halogen functional groups. For example, atrazine, used around the world to control certain types of weeds, contains a chlorine atom attached to a carbon-nitrogen ring. As pesticides often work by poisoning pests, their use must be tightly regulated.

Spraying pesticides

Chlorofluorocarbons (CFCs) are substituted compounds containing chlorine and fluorine atoms bonded to carbon. The most common CFC, chlorofluorocarbon refrigerant, has the formula CCl_2F_2. CFCs were once widely used as propellants in aerosol cans; as solvents; as the foaming agent in the manufacture of plastic foam materials; and as refrigerants in air conditioners, refrigerators, and freezers. However, in 1987, the major industrial nations of the world agreed to a gradual reduction in the use of CFCs because they cause a depletion of the valuable ozone in Earth's upper atmosphere. CFCs are being replaced by other halogenated compounds that are not as damaging to the atmosphere.

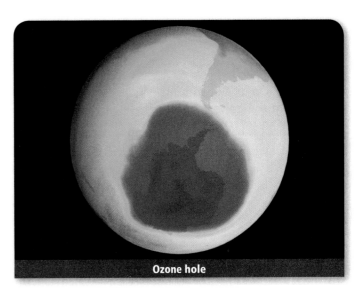
Ozone hole

Alcohols

Structure: R—O—H

Functional group: hydroxyl group (—OH); hydrogen atom bonded to an oxygen atom, which is bonded to the hydrocarbon part of the molecule

Properties: polar, so water molecules are attracted to it; high boiling point; alcohols with low molecular weights are water-soluble

Uses: solvents, disinfectants, mouthwash and hair-spray ingredient, antifreeze

Biological functions: reactive groups in carbohydrates, product of fermentation

Examples: methanol, ethanol, isopropanol (one type of rubbing alcohol), cholesterol, sugars

Rubbing alcohol

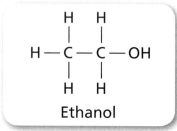
Ethanol

An organic compound that contains at least one hydroxyl group is called an alcohol. The name of an alcohol ends in -ol. Alcohols have many different uses. One of the most important is as a disinfectant for killing bacteria and other potentially harmful microorganisms. For this reason, ethanol is put into mouthwashes, and rubbing alcohol is used as a disinfectant. Antifreeze also is an alcohol.

Carboxylic Acids

Functional group: carboxyl group (—COOH); oxygen atom double-bonded to a carbon, which is also bonded to a hydroxyl group and the hydrocarbon part of the molecule

Properties: acidic, usually water-soluble, strong unpleasant odors, form metal salts in acid-base reactions

Uses: vinegar, tart flavoring, skin-care products, production of soaps and detergents

Biological functions: pheromones, ant-sting toxin; causes rancid-butter and smelly-feet odors

Examples: acetic acid (in vinegar), formic acid, citric acid (in lemons), salicylic acid

Fire ant

> **TRY AT HOME 🏠 LAB**
> See page 876 for **Comparing Water and Alcohol**

structure: $R-\overset{\overset{\displaystyle O}{\|}}{C}-O-H$

$H-\overset{\overset{\displaystyle O}{\|}}{C}-O-H$
Formic acid
(ant-sting toxin)

A compound containing a carboxyl group is known as a carboxylic acid, or organic acid. Many pheromones contain carboxyl functional groups. Pheromones are organic compounds used by animals to communicate with each other. When an ant finds food, it leaves behind a pheromone trail that other ants in its colony can follow to get to the food source.

Esters

Derived from carboxylic acids in which the —OH of the carboxyl group has been replaced with an —OR from an alcohol
Properties: strong aromas, volatile
Uses: artificial flavorings and fragrances, polyester fabric biological functions: fat storage in cells, DNA phosphate-sugar backbone, natural flavors and fragrances, beeswax
Examples: banana oil, oil of wintergreen, triglycerides (fats)

Wintergreen

Pineapple

structure: R—C(=O)—O—R'

A compound formed from the reaction of an organic acid and an alcohol is called an ester. Some esters are used in the production of polyester fabrics. Many esters are used as flavorings in food products. Natural flavors are often complex mixtures of esters and other compounds, whereas artificial flavors usually contain fewer compounds and may not taste exactly the same as the natural flavor.

Ethyl butyrate (pineapple flavoring)

Ether

Structure: R—O—R·; oxygen atom bonded to two hydrocarbon groups
Properties: mostly unreactive, insoluble in water, volatile
Uses: anesthetics, solvents for fats and waxes
Examples: diethyl ether

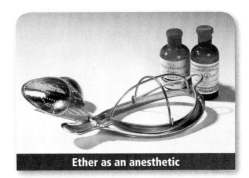
Ether as an anesthetic

An ether is an organic compound in which an oxygen atom is bonded to two hydrocarbon parts of the molecule. Diethyl ether, often simply called ether, is an effective anesthetic. Because it is insoluble in water, it passes readily through the membranes surrounding cells. Ether is rarely used as an anesthetic today because it is highly flammable and because it causes nausea.

Ethyl ether

Ketones and Aldehydes

Functional group: carbonyl group (—CO); carbon atom double bonded to an oxygen atom
Properties: very reactive, distinctive odors
Uses: solvents, flavorings, manufacture of plastics and adhesives, embalming agent
Examples: acetone; formaldehyde; cinnamon, vanilla, and almond flavorings

structures: ketones: $R-\overset{\overset{O}{\|}}{C}-R'$ aldehydes: $R-\overset{\overset{O}{\|}}{C}-H$

Aldehydes and ketones are compounds that contain a carbonyl group. If the carbonyl group is on the end of the carbon chain, the compound is called an aldehyde. If the carbonyl group is not at the end of the chain, the compound is a ketone. Acetone is a ketone solvent commonly used in nail-polish remover. Nail polish is not soluble in water—if it were, it would come off when you wash your hands. It is soluble in many organic compounds such as acetone, which is used to remove it.

Acetone

Amines and Amides

Structure: amine: R—NH$_2$
Functional group: amine: amino group (—NH$_2$); two hydrogen atoms bonded to a nitrogen atom, which is bonded to the hydrocarbon part of the molecule
Amide: amino group bonded to a carbonyl group (—CONH$_2$)
Properties: amines: basic, ammonia-like odor
    amides: neutral, most are solids
Uses: solvents, synthetic peptide hormones, fertilizer
Biological functions: in amino acids, peptide hormones, and proteins; distinctive odor of some cheeses
Examples: urea, putrescine, cadaverine, aspartame

amine: $R-NH_2$ amide: $R-\overset{\overset{O}{\|}}{C}-NH_2$

An organic compound containing an amino group is called an amine. Amines and amides are important biological molecules because they are part of proteins. When an organism dies, its proteins are broken down into many different compounds containing amino functional groups. These compounds have a distinctive, unpleasant smell that specially trained sniffer dogs can use to locate human remains and to help in forensic investigations. Cadaverine also contributes to bad breath.

Putrescine

MiniLab 18.2

A Synthetic Aroma

Can you synthesize an ester? When brought together under the proper conditions, an organic acid and an alcohol will react to form an ester. In this MiniLab, you will combine methanol and salicylic acid to produce the ester called methyl salicylate.

Procedure

1. Read and complete the lab safety form.
2. Set up a hot-water bath for the reaction by half-filling a **250- or 400-mL beaker** with **tap water** and setting it on a **hot plate**. Allow the water to warm but not boil.
3. Place 3 mL of **methyl alcohol** and 1 g of **salicylic acid** in a **large (20 mm × 150 mm) test tube**. Mix the reactants with a **glass stirring rod**.
4. Add about 0.5 mL (about ten drops from a **dropper**) of **concentrated sulfuric acid** to the reaction mixture and stir. Do not draw the acid into the rubber bulb of the dropper. **WARNING:** *Concentrated sulfuric acid is corrosive to skin, eyes, body tissues, and most other organic materials. If any contact occurs, notify your teacher immediately.*
5. Place the test tube and contents in the hot-water bath and allow it to warm for five or six minutes.
6. Pour the contents of the test tube into about 50 mL of **cold, distilled water** in a **small beaker**. Cover the beaker with a **watch glass** and allow it to stand for a minute or two.
7. Remove the watch glass and waft the methyl salicylate odor toward your nose. Record your observations.

Analysis

1. **Identify** What familiar odor is caused by methyl salicylate?
2. **Report** Look up the structural formulas for the reactants in a reference book, and write the equation for the reaction using structural formulas for all organic reactants and products.
3. **Infer** What organic reactants would be required to produce the ester ethyl butyrate?

The Sources of Functional Groups

Alkanes are not very reactive, so it can be difficult to react them directly to form substituted hydrocarbons. You read at the beginning of this section about one reaction used to substitute chlorine atoms for hydrogen atoms in methane. That reaction is not practical because a mixture of four different products is formed. Each must be separated and purified before it can be used.

Alcohols are a better choice than alkanes as the raw materials for synthesizing organic compounds. Alcohols can be converted into compounds containing almost every other type of functional group.

Where do all the alcohols needed by industry come from? Cracking of petroleum products produces alkenes, which can readily be converted into alcohols by reaction with water. This process is used to synthesize ethanol, the two-carbon compound that contains an alcohol functional group.

Ethene + Water ⟶ Ethanol

■ **Figure 18.16** Ethanol can be produced industrially by the fermentation of sugars and starches by yeast cells. These sugars and starches come from plant material such as the grapes grown in vineyards like this one.
Name *other sources of sugar and starches.*

SUPPLEMENTAL PRACTICE

For more practice with substituted hydrocarbons, see Supplemental Practice, page 838.

A second important natural source of ethanol provides a convenient supply of this reactive molecule. Ethanol is produced by yeast cells when they ferment sugars and starches found in plant materials, like the grapes in **Figure 18.16**. Fermentation produces much smaller quantities of ethanol than does the reaction between ethene and water; it is used in industry mainly to produce the ethanol in alcoholic beverages.

Section 18.2 Assessment

Section Summary

▶ Special bond organizations, atoms, or groups of atoms that give predictable characteristics to molecules are called functional groups.

▶ Atoms and groups of atoms are substituted in carbon rings and chains to form substituted hydrocarbons.

▶ Compounds containing a specific functional group share characteristic properties.

8. **MAIN Idea** **Explain** why organic compounds with similar structures often have similar uses.

9. **Distinguish** between an aldehyde and a ketone.

10. **Identify** Redraw the following structure of thyroxine, a thyroid hormone. Circle and name each type of functional group present.

    HO—⬡(I)(I)—O—⬡(I)(I)—CH₂CHCOOH
                                    |
                                    NH₂

11. **Explain** why alkenes and alkynes contain functional groups but are not substituted hydrocarbons.

12. **Infer** Ethylene glycol, used as an antifreeze in car radiators, has two hydroxyl groups, as shown below. What can you infer about the boiling point, freezing point, and solubility in water of this compound from its structure?

    ```
         OH  OH
          |   |
     H — C — C — H
          |   |
          H   H
    ```

Section 18.3

Objectives
▶ **Identify** the monomers that form specific polymers.
▶ **Draw** structural formulas for polymers made from given monomers.
▶ **Differentiate** between condensation and addition polymerization reactions.
▶ **Summarize** the relationship between structure and properties of polymers.

Review Vocabulary
functional group: the part of a molecule that is largely responsible for the chemical behavior of the molecule

New Vocabulary
polymer
monomer
addition reaction
condensation reaction
cross-linking
thermoplastic
thermosetting

FOLDABLES
Incorporate information from this section into your Foldable.

Plastics and Other Polymers

MAIN Idea Polymers are large organic molecules made up of many smaller repeating units.

Real-Word Reading Link On April 6, 1938, a young chemist named Roy Plunkett working on the preparation of a compound used in refrigeration, opened the valve on a tank of the reactant he planned to use in this process, tetrafluoroethene gas. When no gas came out after he opened the valve, he became curious and cut open the tank. He discovered a white, waxy solid in place of the gas he expected. Today, you are familiar with this solid as a nonstick coating on pots and pans. It is also used for dentures, artificial joints and heart valves, space suits, and fuel tanks on space vehicles.

Monomers and Polymers

How did a white, waxy solid form from the colorless gas tetrafluoroethene? One clue to what type of reaction occurred can be found in the properties of those two substances. Most nonpolar molecules that are relatively small tend to be gases at room temperature, whereas larger molecules usually are solids. When the structure of the waxy solid was determined, the molecule was found to consist of chains of carbons with fluorine atoms attached. Somehow, the tetrafluoroethylene molecules in the gas had reacted with each other to form these long-chain molecules.

A large molecule that is made up of many smaller, repeating units is called a **polymer.** A polymer forms when hundreds or thousands of these small individual units, which are called **monomers,** bond together in chains. The monomers that bond together to form a polymer may all be alike, or they may be different. When the waxy solid formed in the tank, many small tetrafluoroethylene monomers combined to form long polymers of polytetrafluoroethylene. The structures in **Figure 18.17** show a monomer and a polymer.

■ **Figure 18.17** The nonstick coating is useful on this skillet because food will not stick to it and it is unreactive. The stable molecule that makes up the nonstick coating is composed of tetrafluoroethylene molecules bonded together to form polytetrafluoroethylene.
Infer *Why is it important that a skillet coating is very stable and unreactive?*

Section 18.3 • Plastics and Other Polymers

■ **Figure 18.18** Polymers have a vast variety of properties and applications. Garden hoses, the lenses of many glasses, and styrofoam containers are just a few examples of the everyday polymers all around us.

The properties of a polymer are different from those of the monomers that formed it. For example, the polyethylene plastic in milk jugs is made when molecules of gaseous ethylene react to form long chains. The unique properties of a given polymer, such as tensile strength, water-repellency, or flexibility, are related to the polymer's enormous size and the way its monomers join together. **Figure 18.18** demonstrates the wide range of physical properties of various polymers.

Synthetic polymers As you can probably guess from viewing **Figure 18.18,** polymers are everywhere. Fabrics such as nylon and polyester, plastic wrap and bottles, rubber bands, and many more products you see every day are all made of polymers. How many polymers can you identify in **Figure 18.19**? What do the polymers that make up such different substances have in common? All are large molecules made of smaller, repeating units.

Chemists have been synthesizing polymers in laboratories for only about 100 years. Can you imagine what life was like before people had these synthetic polymers? You might have gotten wet on the way to school without a nylon raincoat, eaten a stale sandwich for lunch without plastic wrap or a container to put it in, and worn a heavier cotton uniform in your sports activity instead of a lightweight, synthetic fabric. Many polymers have added conveniences that we take for granted in our lives.

VOCABULARY
WORD ORIGIN
Polymer
comes from the Greek words *poly,* meaning *many,* and *meros,* meaning *part*

■ **Figure 18.19** Sports activities would be different today without synthetic polymers. Balls, uniforms, artificial turf, bandages used to wrap sprains, and protective pads are usually made of synthetic polymers. Modern artificial turf is generally about as safe to play on as natural grass.

CHEMLAB

IDENTIFY POLYMERS

Background
For centuries, polymers have been used for clothing in the form of cotton, wool, and silk. Chemists trying to improve these natural polymers to fit specific purposes have designed many synthetic polymers that also are used to make fabrics. Among these are nylon, rayon, acetate, and the polyesters.

Many synthetic fabrics are designed to look and feel like natural polymers but also have the superior properties for which they were designed, such as being wrinkle-resistant, water-repellent, or quick-drying. However, these imitators can't fool a chemist; simple tests that can be done in minutes will distinguish the real thing from a pretender. This is because differences in their structures lead to differences in properties. In this experiment, you will identify fabric samples by testing them for characteristic properties.

Question
What tests can be performed that will differentiate among the polymers used to make fabrics?

Objectives
- **Analyze** changes in fabric samples that are subjected to flame and chemical tests.
- **Classify** fabrics by polymer type based on test results.

Preparation

Materials
fabric samples A–G (7 types; six 0.5 × 0.5 cm squares of each type)
Bunsen burner
medium test tubes (4)
watch glass
red litmus paper
glass stirring rod
beaker containing water
forceps
test-tube holder
test-tube rack
calcium hydroxide ($Ca(OH)_2$)
$1M$ barium chloride ($BaCl_2$)
concentrated H_2SO_4
iodine solution
$0.05M$ copper(II) sulfate ($CuSO_4$)
100-mL beakers (2)
10-mL pipet
25-mL graduated cylinder
$3M$ sodium hydroxide (NaOH)
acetone

Safety Precautions
WARNING: *Use care when working with an open flame and handling concentrated acids and bases. Do not inhale odors from plastics.*

Procedure

Read and complete the lab safety form.

Flame Tests
1. Make a data table similar to Data Table 1 in Data and Observations on page 651.
2. Use forceps to hold a square of fabric A in a Bunsen burner flame for 2 seconds shown in the photos on page 650.
3. Remove the fabric from the flame, and blow out the flame if the fabric keeps burning.
4. Observe the odor by wafting smoke from the smoldering sample toward your nose. Be sure the fabric is no longer burning by immersing it in a beaker of water.
5. Record your observations about the way the fabric burns in the flame, odor observed, and characteristics of the residue left after burning.
6. Repeat steps 2–5 six times, using fabric samples B–G.
7. Use the following table on page 650 to make a preliminary identification of your samples.

Preliminary Identification Table			
Polymer Type	Type of Burning	Burning Odor	Residue Type
Silk or wool	Burns and chars	Hair	Crushable bead
Cotton	Burns and chars	Paper	Ash
Nylon, polyester, acetate, or acrylic	Burns and melts	Chemical	Plastic bead

Chemical Tests

Use your preliminary identifications to determine which tests are necessary to identify each fabric sample. You should not have to carry out every test on every sample. Make a data table similar to Data Table 2 on page 651, and record all your results.

1. **Nitrogen Test** Place a fabric square in a test tube and add 1 g of $Ca(OH)_2$. Using a test-tube holder, heat the tube gently in a Bunsen burner flame while holding a piece of red litmus paper with forceps over the mouth of the tube. If the litmus paper turns blue, nitrogen is present. Only silk, wool, nylon, and acrylic contain nitrogen.

2. **Sulfur Test** Add a fabric square to 10 mL of $3M$ NaOH in a test tube, and gently heat to boiling by holding the tube in a Bunsen burner flame. Be careful that the tube is not pointing toward anyone. Cool the solution, add 30 drops of $BaCl_2$ solution, and observe whether or not a precipitate forms. Only wool contains enough sulfur to give a barium sulfide precipitate.

3. **Cellulose Test** Place a fabric square in a beaker, cover with approximately 2 mL of concentrated H_2SO_4, and then carefully transfer the contents to another beaker containing ten drops of iodine solution in 25 mL of water. Rinse the empty beaker with large quantities of water. Cotton gives a dark blue color within 1–2 minutes, and acetate gives this color after 1–2 hours. **WARNING:** *Handle the beaker containing sulfuric acid with care.*

4. **Protein Test** Place a fabric square on a watch glass, and add ten drops of $0.05M$ $CuSO_4$. Wait 5 minutes, and then use forceps to dip the fabric into $3M$ NaOH in a test tube for 5 seconds. Silk and wool are protein polymers, and a dark violet color will appear on those fabrics after doing this test.

5. **Formic Acid Test** Place any sample to be subjected to the formic acid test in a test tube and take it to your teacher, who will perform this test in the fume hood. The teacher will add 1 mL of formic acid to the test tube and stir with a glass rod. Note whether or not the fabric dissolves in the solution. Silk, acetate, and nylon dissolve in formic acid.

Data and Observations

Data Table 1

Sample	Type of Burning	Burning Odor	Residue Type
A			

Data Table 2

Sample	Nitrogen	Sulfur	Cellulose	Protein	Formic Acid	Acetone
A						

6. **Acetone Test** Add a fabric square to 1 mL of acetone in a test tube, stir with a glass rod, and note whether or not the fabric dissolves. Only acetate dissolves in acetone. **WARNING: Be careful to do this test away from any open flames.**

7. Dispose of all products of your tests as directed by your teacher.

Analyze and Conclude

1. **Think Critically** Why do you think it was necessary to add NaOH and heat before adding $BaCl_2$ in the sulfur test?
2. **Compare and contrast** the burning of synthetic polymers with the burning of natural polymers.
3. **Classify** Use your data from the flame and chemical tests to classify each fabric sample you tested as silk, cotton, wool, nylon, acrylic, acetate, or polyester polymer.

Apply and Assess

1. **Explain** What does the plastic, beadlike residue left by some polymers after burning tell about the polymers' structures?
2. **Interpret** If burning silk and wool smell like burning hair, what does this tell about the structure of hair?
3. **Predict** What results of the flame and chemical tests would you expect to see if you were testing a fabric sample that was a polyester-cotton blend?

INQUIRY EXTENSION

Would you expect wood to also give a dark blue color in the cellulose test? Explain your hypothesis.

Natural spinneret

Industrial spinneret

■ **Figure 18.20** Long, fine threads are spun when polymer molecules are forced through tiny holes in spinnerets, both natural and industrial.

Natural polymers Laboratories are not the only place where polymers are synthesized. Living cells are efficient polymer factories. Proteins, DNA, the chitin exoskeletons of insects, wool, silky spiderwebs and moth cocoons, and the jellylike sacs that surround salamander eggs are polymers that are synthesized naturally. The strong cellulose fibers that give tree trunks enough strength and rigidity to grow hundreds of feet tall are formed from monomers of glucose, which is a sweet crystalline solid.

Many synthetic polymers were developed by scientists trying to improve on nature. For example, nylon was developed as a possible silk substitute. The idea for the process by which synthetic threads are formed in factories was borrowed from spiders. Observe in **Figure 18.20** the similarities between fibers from the spinneret of a spider and those from an industrial spinneret.

MiniLab 18.3

Absorbent Polymers

How much water can disposable diapers hold? Sometimes, the ability of certain polymers to repel water is useful—this property is what keeps you dry when you wear a raincoat. Other times, it is desirable for polymers to absorb water. Disposable diapers contain a superabsorbent polymer. In this MiniLab, you will determine how much water the polymers in two different brands of diapers can hold.

Procedure

1. Read and complete the lab safety form.
2. Obtain two **diapers** of approximately the same size but of different brands.
3. Fill a **100-mL graduated cylinder** with **tap water,** and slowly pour the water into the center of one of the diapers. Stop when the water begins leaking out of the diaper.
4. Record the volume of water the diaper holds.
5. Repeat steps 3 and 4 with the other diaper.

Analysis

1. **Compare** What could be different about the two diapers that makes one of them hold more water than the other?
2. **Analyze** How might placing a superabsorbent polymer around the roots of a houseplant help it to grow?

Cellulose   Glucose   Starch

Structure of polymers If you examine the structure of a polymer, you can identify the repeating monomers that formed it. Because polymer molecules are large, they are commonly represented by showing just a piece of the chain. The piece shown must include at least one complete repeating unit.

Look carefully at the structure of a segment of a cellulose molecule shown in **Figure 18.21**. Cellulose is a polymer found in the cell walls of plant cells, such as those of wood, cotton, and leaves. It is responsible for giving plants their structural strength. Can you find the portions of the cellulose structure that repeat? Notice that the ring parts of the molecule are all identical. These are the monomer units that combine to form the polymer. Glucose is the name of the monomer found in cellulose. In **Figures 18.21,** the glucose units are shown in simplified form without carbon and hydrogen atoms.

Another natural plant polymer that is formed from glucose monomers is starch. Examine the structure of starch in **Figure 18.21** to see if you can find the difference between starch and cellulose. Both cellulose and starch are made from only glucose monomers. The difference between them is the way that these monomers are bonded to each other. In starch, the oxygen atom joining each pair of monomers is pointed downward; in cellulose, the oxygen atom is pointed upward. This appears to be only a minor difference, but it changes the properties of the two polymers dramatically.

■ **Figure 18.21** Cellulose molecules range in length from several hundred to several thousand glucose units, depending on their source. Starch molecules are usually several hundred glucose units in length. In addition to forming linear chains, glucose molecules form a branched polymer.

■ **Figure 18.22** Cellulose is found in the cell walls of fruits and vegetables, and cannot be digested by humans. When starch is digested, it is broken down into glucose molecules that can be absorbed by the digestive tract.

Cellulose-containing foods

Starch-containing foods

Section 18.3 • Plastics and Other Polymers

$$\underset{\text{Ethylene monomers}}{\ce{H2C=CH2} + \ce{H2C=CH2}} \longrightarrow \underset{\text{Polyethylene}}{-\ce{CH2-CH2-CH2-CH2}-}$$

Monomers —Addition→ Polymer + More monomers —Addition→ Larger polymer

■ **Figure 18.23** Ethylene monomers undergo an addition reaction to form the polyethylene that is used in plastic bags, food wrap, and bottles. The extra pair of electrons from the double bond of each ethylene monomer is used to form a new bond to another monomer.

Polymerization

How do you think the many different types of polymers are formed? **Polymerization** is a type of chemical reaction in which monomers are linked together one after another to make large chains. The two main types of polymerization reactions are addition polymerization and condensation polymerization. The type of reaction that a monomer undergoes depends upon its structure.

Addition reactions The reaction by which polytetrafluoroethylene is made from its monomer, tetrafluoroethylene, is called an **addition reaction.** In this type of reaction, monomers that contain double bonds add onto each other, one after another, to form long chains. The product of an addition polymerization reaction contains all of the atoms of the starting monomers. In **Figure 18.23,** the ethylene monomer contains a double bond, whereas there are none in polyethylene. When monomers are added onto each other in addition polymerization, the double bonds are broken. Thus, all of the carbons in the main chain of an addition polymer are connected by single bonds. Other polymers made by addition polymerization are illustrated in **Figure 18.24.**

■ **Figure 18.24** These polymers are made by polymerization of monomers.

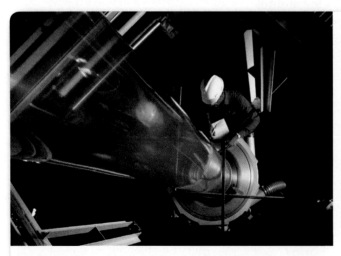

Molten low-density polyethylene can be formed into a film. This tough plastic forms a barrier to food odors, which makes it useful for wrapping and storing foods.

Polyvinyl acetate is another plastic made by addition polymerization. When mixed with sugar, flavoring, glycerol (for softening), and other ingredients, it becomes chewy candy.

Everyday Chemistry

Chemistry and Permanent Waves

A series of bad-hair days may drive you to the hair stylist in search of a permanent wave. Your hair endures the chemical changes caused by a perm.

What happens in a permanent wave?
Permanent-wave lotions are designed to penetrate the scales of the cuticle, the outer layer of the hair shaft. The lotions work because they affect the structure of the proteins that make up the hair. The amino acid cysteine, which contains an atom of sulfur, is found in human hair protein. Sulfur atoms in neighboring cysteine molecules within the hair protein form strong, covalent, disulfide (S—S) bonds shown in **Figure 1**. This cross-linking between cysteine molecules holds the strands of hair protein in place and affects its shape and strength.

When you have a permanent wave, the waving lotion that is applied in **Figure 2** breaks the disulfide bonds. Then, after the hair is set in a new style, it is treated with a neutralizing lotion, which creates new cross-links between cysteine molecules to hold the hair in the new style. Thus, two chemical reactions occur. In the first, the waving lotion reduces each disulfide bond to two —SH groups. In the second, the neutralizing lotion oxidizes the —SH groups to form new disulfide bonds.

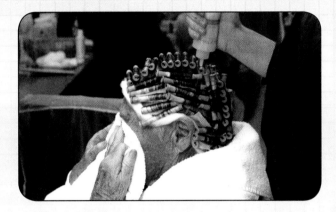

Figure 2 Applying permanent wave lotion reduces the number of disulfide bonds.

Kinds of permanent waves Two kinds of permanent-wave lotions are available. Alkaline lotions have as the reducing agent ammonium thioglycolic acid. The alkalinity causes the scales of the cuticle to swell and open, allowing the solution to penetrate rapidly. The advantages of the alkaline permanent are that it forms stronger, longer-lasting curls; takes a shorter time (usually 5 to 20 minutes); and occurs at room temperature. The alkaline permanent is used on hair capable of resisting damage.

Acid-balanced permanent-wave lotions contain monothioglycolic acid. Normally, acidic lotions penetrate the cuticle only slightly. That's why the process takes a longer time and requires heat in order for curls to develop. The advantages of the acid-balanced wave are that it forms softer waves, is more easily controlled, and can be used for delicate hair or hair that has been colored.

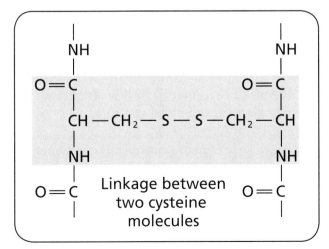

Figure 1 Cysteine molecules form disulfide bonds.

Explore Further

1. **Investigate** Find out more about the protein in hair. Where else is this protein found?
2. **Apply** Why might hair that has been colored or recently permed absorb lotion more rapidly than untreated hair?

$$\underset{\text{1,6-diaminohexane}}{H-\overset{\overset{H}{|}}{N}-(CH_2)_6-\overset{\overset{H}{|}}{N}-H} + \underset{\text{Adipic acid}}{H-O-\overset{\overset{O}{\|}}{C}-(CH_2)_4-\overset{\overset{O}{\|}}{C}-O-H} \rightarrow \underset{\text{Amide link}}{-\overset{\overset{H}{|}}{N}-\overset{\overset{O}{\|}}{C}-} + H_2O$$

■ **Figure 18.25** The condensation of two different monomers—1,6-diaminohexane and adipic acid—is used to make the most common type of nylon. Nylons are named according to the number of carbon atoms in each monomer unit. This type of nylon is called nylon 66.
State How many carbon atoms are there in each of these two monomers?

Condensation reactions In the second type of polymerization reactions, monomers add on one after another to form chains, as they do in an addition reaction. However, with every new bond formed, a small molecule—usually water—also forms from atoms of the monomers. In such a reaction, each monomer must have two functional groups so it can add on at each end to another unit of the chain. This type of polymerization is called a **condensation reaction** because a portion of the monomer is split out—usually as water—as the monomers combine.

In a condensation reaction, a hydrogen atom from one end of a monomer combines with the —OH group from the other end of another monomer to form water. The condensation reaction shown in **Figure 18.25** is used to make a type of nylon. Other plastics made by condensation polymerization include the hard, ovenproof plastic used for handles of toasters and cooking utensils; polyethylene terephthalate, which is used as fiber for clothing and carpets; and plastic wrap.

Cross-linking Another process that often occurs in combination with addition or condensation reactions is the linking together of many polymer chains and is illustrated in **Figure 18.26**. This is called **cross-linking,** and it gives additional strength to a polymer. In 1844, Charles Goodyear discovered that heating the latex from rubber trees with sulfur can cross-link the hydrocarbon chains in the liquid latex. The solid rubber that is formed can be used in tires and rubber balls. The process is called vulcanizing, named for Vulcan, the Roman god of fire and metalworking.

When World War II resulted in the cutting off of the Allies' supply of natural rubber, the polymer industry grew rapidly as chemists searched for rubber substitutes. Some of the most successful substitutes developed were gas- and oil-resistant neoprene, now used to make hoses for gas pumps, and styrene-butadiene rubber (SBR), which is now used along with natural rubber to make most automobile tires.

■ **Figure 18.26** When a piece of vulcanized, cross-linked rubber such as a rubber band is stretched and then released, the cross-links pull the polymer chains back to their original shape. Without vulcanization, the chains would slide past one another.

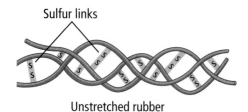
Unstretched rubber

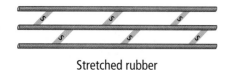
Stretched rubber

Chemistry & Society

Recycling Plastics

PETE Polyethylene terephthalate | **HDPE** High–density polyethylene | **V** Vinyl | **LDPE** Low–density polyethylene | **PP** Polypropylene | **PS** Polystyrene | **OTHER** All other plastics

The growing mass of throwaway materials has caused more and more landfills to reach their capacities. In response, many people sort their trash into categories: garbage, paper, cardboard, glass, and plastics. Because garbage, paper, and cardboard are biodegradable and glass can be reused, the focus is on plastics. Twelve percent of the volume of waste in the United States is composed of plastics. The process for recycling plastics is more complex than for most other materials. Five plastics are commonly found in landfills. They are polyethylene—both high- and low-density, polyethylene terephthalate, polystyrene, polyvinyl chloride, and polypropylene.

The Society of the Plastics Industry has designated code numbers and acronyms, shown above, to help people distinguish among plastics. This also helps to make communication uniform. The codes are useful in sorting plastics and in deciding what method will be used for recycling. In addition to having a distinct chemical composition, each kind of plastic has distinct physical properties that determine how it can be used.

How are plastics used? Polyethylene is the most widely used plastic. High-density polyethylene (HDPE) is used in rigid containers such as milk and water jugs and in household and motor-oil bottles. Low-density polyethylene (LDPE) is usually used for plastic film and bags. Polyethylene terephthalate (PETE) is found in rigid containers, particularly carbonated-beverage bottles. Polystyrene (PS) is best known as a foam in the form of plates, cups, and food containers, although in its rigid form, it is used for plastic knives, forks, and spoons.

Polyvinyl chloride (PVC) is a tough plastic that is used in plumbing and construction. It is also found as shampoo, oil, and household-product containers. Finally, polypropylene (PP) has a variety of uses, from snack-food packaging to battery cases to disposable-diaper linings.

Recycling plastics PETE beverage bottles and HDPE milk and water bottles receive the most attention because they are the easiest to collect and sort. PETE bottles require a drawn-out recycling process because they are made of several materials. Only the body of the bottle is PETE. The base is HDPE, the cap is another kind of plastic or aluminum, and the label has adhesives. The bottles are shredded and ground into chips for processing. The adhesives can be removed with a strong detergent. The lighter HDPE is separated from PETE in water because one sinks and the other floats. The aluminum is removed electrostatically. What is left are plastic chips that may be sold to manufacturers who can use the chips in making other plastics. However, the FDA prohibits the use of recycled plastics in food containers, making one major market out of bounds for recycled plastics.

ANALYZE the Issue

1. **Debate** Debate whether industries should be held responsible for recycling the packaging in which their products are sold.

2. **Write** Prepare a letter or article for a newspaper in which you support the use of less packaging for manufactured products.

■ **Figure 18.27** Molded plastics are used for making objects with strength and durability qualities that are superior to those of earlier materials. Whereas a wood deck eventually will decay, plastic is so difficult to break down that it can create disposal problems as plastic materials pile up in landfills. Many decks are made of recycled plastic.

Plastics

Although the terms plastic and polymer are often used synonymously, not all polymers are plastics. Plastics are polymers that can be molded into different shapes. What physical state has a fixed volume that you can pour into a mold and that will take the shape of its mold? Only liquids will. After a polymer has formed, it must be heated enough to become liquefied if it is to be poured into a mold. **Figure 18.27** shows an example of plastics that were allowed to cool into useful shapes.

Some plastics will soften and harden repeatedly as they are heated and cooled. This property is described as being **thermoplastic.** Thermoplastic materials are easy to recycle because each time they are heated, they can be poured into different molds to make new products. Polyethylene and polyvinyl chloride are examples of this type of polymer.

Other plastics harden permanently when molded. Because they are set permanently in the shape they first form, they are called **thermosetting** polymers. Thermosetting plastics usually are rigid because they have many cross-links. No matter how much they are reheated, they won't soften enough to be remolded; instead, they get harder when heated because the heat causes more cross-links to form. Resin bakeware is this type of molecule. Even though thermosetting polymers are more difficult to reuse than thermoplastics, they are durable. **Figure 18.28** shows the amounts of various plastics produced for packaging and the amounts of plastics recycled in 2006.

■ **Figure 18.28** The relative amounts of plastics produced for packaging for 2006 are shown here. Polyethylene terephthalate (PETE) and high-density polyethylene (HDPE) are the most recyclable forms of plastic because they are so easy to remelt and reprocess.

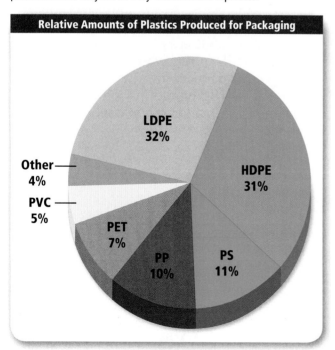

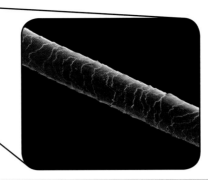

■ **Figure 18.29** Human hair is made up of fibrous structural protein called keratin.

Connecting Ideas

Although polymers are large, complex molecules, their structures are easier to understand when you realize they are made from chains of simpler building blocks. Living things consist mostly of organic compounds, many of them extremely complex polymer structures. For example, **Figure 18.29** shows an important structural protein, a type of biological polymer, with which you are very familiar. Keep in mind, however, that the chemical reactions through which living things get energy and building materials are exquisitely organized processes whereby a relatively small number of building blocks are used to make the many thousands of biochemicals that cells need. You will study these biomolecules in Chapter 19.

Section 18.3 Assessment

Section Summary

- Polymers are large molecules made from many smaller units called monomers that repeat over and over again.

- Many synthetic polymers can be made from smaller monomers. Living cells also make many polymers.

- Addition and condensation are the main reactions by which polymers are made. In addition reactions, all atoms of monomers add end-to-end to form chains. In condensation reactions, a molecule of water is released as each new bond forms between two monomer units.

- Thermoplastic polymers can be recycled for different purposes because they can be repeatedly softened by heating and hardened by cooling. Thermosetting plastics become hardened permanently.

13. **MAIN Idea Identify** the repeating unit that appears in each of the following polymers.
   a) polystyrene ~$CH_2CHCH_2CHCH_2CHCH_2CH$~ (each CH bonded to a benzene ring)

   b) rubber ~CH_2—C(CH$_3$)=CH—CH_2—CH_2—C(CH$_3$)=CH—CH_2~

14. **Draw** the structure of the polymer that will be formed from each of the monomers.
   a) $CH_2=CHCl$
   b) NH_2CH_2COOH
   c) $CH_2=CHOH$

15. **Predict** Can a polymer be made from alkane monomers by addition polymerization? Explain.

16. **Distinguish** Paints usually contain three components: a binder that hardens to form a continuous film, a colored pigment, and a volatile solvent that evaporates. In latex paints, one of these components is a polymer. Use what you have learned about polymer properties to decide which of the three most likely is a polymer. Explain.

CHAPTER 18 Study Guide

Download quizzes, key terms, and flash cards from glencoe.com.

BIG Idea Chains of carbon atoms are the backbone of organic chemistry.

Section 18.1 Hydrocarbons

MAIN Idea Hydrocarbons contain only hydrogen and carbon, but differ in the lengths of their carbon chains and the presence of single, double, and triple bonds.

Vocabulary
- alkane (p. 623)
- alkene (p. 630)
- alkyne (p. 633)
- aromatic hydrocarbon (p. 636)
- cracking (p. 638)
- fractional distillation (p. 638)
- isomer (p. 627)
- reforming (p. 638)
- saturated hydrocarbon (p. 623)
- unsaturated hydrocarbon (p. 629)

Key Concepts
- Alkanes are saturated hydrocarbons that contain only single bonds between carbon atoms, whereas alkenes and alkynes are unsaturated hydrocarbons.
- Isomers are compounds that have the same molecular formula but different structures.
- Because properties depend upon structure, isomers often have different properties.
- Benzene is one of a group of unusually stable cyclic hydrocarbons. This stability is due to the sharing of six electrons by all the carbon atoms in the molecule.
- The many components in petroleum are separated by fractional distillation.

Section 18.2 Substituted Hydrocarbons

MAIN Idea Functional groups can substitute for hydrogen atoms in hydrocarbons, resulting in diverse groups of organic compounds.

Vocabulary
- functional group (p. 640)
- substituted hydrocarbon (p. 640)

Key Concepts
- Special bond organizations, atoms, or groups of atoms that give predictable characteristics to molecules are called functional groups.
- Atoms and groups of atoms are substituted in carbon rings and chains to form substituted hydrocarbons.
- Compounds containing a specific functional group share characteristic properties.

Section 18.3 Plastics and Other Polymers

MAIN Idea Polymers are large organic molecules made up of many smaller repeating units.

Vocabulary
- addition reaction (p. 654)
- condensation reaction (p. 656)
- cross-linking (p. 656)
- monomer (p. 647)
- polymer (p. 647)
- thermoplastic (p. 658)
- thermosetting (p. 658)

Key Concepts
- Polymers are large molecules made from many smaller units called monomers that repeat over and over again.
- Many synthetic polymers can be made from smaller monomers. Living cells also make many polymers.
- Addition and condensation are the main reactions by which polymers are made. In addition reactions, all atoms of monomers add end-to-end to form chains. In condensation reactions, a molecule of water is released as each new bond forms between two monomer units.
- Thermoplastic polymers can be recycled for different purposes because they can be repeatedly softened by heating and hardened by cooling. Thermosetting plastics become hardened permanently.

Chapter 18 Assessment

Understand Concepts

17. Draw the structures of the following hydrocarbons.
  a) pentane
  b) propene
  c) 1-butyne
  d) nonane
  e) 2-methylpentane
  f) methylpropene

18. Name the following hydrocarbons.
  a) $CH_3CH_2CH_2CH_3$
  b) $CH_2=CHCH_2CH_2CH_2CH_3$
  c) $CH_3C\equiv CH$
  d) $CH_3CH_2CH_2CH_2CH_2CH=CH_2$

19. What molecule is usually released when monomers combine in a condensation reaction?

20. Name the branched alkanes shown.

  a)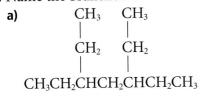

  b)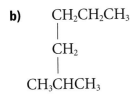

21. Describe how the boiling points of alkanes change as chain length increases.

22. Why is there no compound named 4-methylhexane? What is the correct name for this compound?

Apply Concepts

23. High-density polyethylene (HDPE) contains linear polymer chains, whereas low-density polyethylene (LDPE) contains highly branched chains. Predict which will be a harder polymer and explain why.

24. Glycerol and isopropanol are both alcohols. Each contains a three-carbon chain, but glycerol has three alcohol functional groups, whereas isopropanol has only one. Predict which alcohol will be more soluble in water.

25. Copy the following structure of penicillin G, an antibiotic. Circle all the functional groups in the molecule.

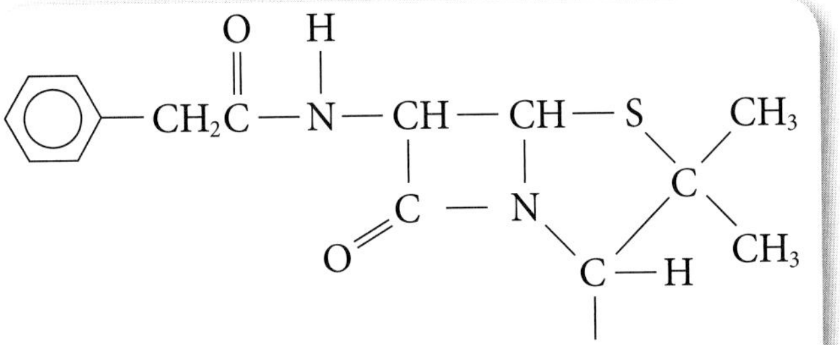

Penicillin G

Think Critically

Design an Experiment

26. ChemLab Design an experiment that shows how you could distinguish between two fabric samples if one is polyester and the other is a cotton-polyester blend.

Make Predictions

27. MiniLab 1 Predict whether iodine will be decolorized more quickly by melted butter or by melted margarine, assuming that both are at the same temperature.

Apply

28. MiniLab 2 What is the name of the ester formed by reacting ethanol and acetic acid?

Make Predictions

29. MiniLab 3 Disposable diapers contain two different polymers. Examine the structures of the polymers shown, and determine which would best function on the inside and which on the outside of a diaper.
  a) $\sim CH_2CH_2CH_2CH_2\sim$
  b)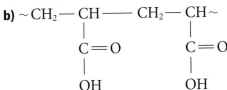

Chapter 18 Assessment

Cumulative Review

30. Explain why different compounds have different properties. (*Chapter 4*)

31. Write the formulas for the following compounds. (*Chapter 5*)
    a) acetic acid
    b) cesium chloride
    c) carbon disulfide

32. Compare the laws of Boyle and Charles. (*Chapter 11*)

Skill Review

33. **Interpret a Graph** Examine **Figure 18.30** and answer the questions.
    a) How does the number of carbon atoms in an alkene affect its boiling point?
    b) Is there a relationship between the amount of branching and the boiling point of an alkene? If so, what is the relationship?

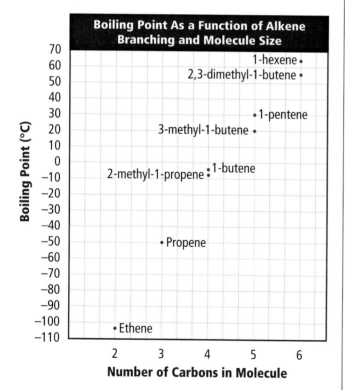

■ **Figure 18.30**

34. **Hypothesize** The gasoline blend sold in hot climates consists of hydrocarbons of larger molecular mass than the gasoline blend sold in cold climates. Write a report suggesting a reason why refiners might vary the blends this way.

35. **Design an Experiment** Imagine you are a chemist at a textile company and are in charge of a research project aimed at developing a synthetic cotton substitute. What kind of information would you have to collect or questions would you have to ask to solve the problem? Design an experiment for beginning this research.

WRITING in Chemistry

36. Research to find information about government regulations controlling levels of sulfur in coal-burning emissions. Write a report giving arguments for or against these regulations. If you can think of a better way to reduce sulfur emissions, give your ideas in your report.

Problem Solving

37. a) Write the balanced reaction for the synthesis of ethanol from ethene and water.
    b) If 448 L of ethene gas are reacted with excess water at STP, how many grams of ethanol will be produced?

38. a) Write a balanced equation for the complete combustion of hexane.
    b) Use the equation you wrote in part a. to determine how many moles of CO_2 will be produced from burning 4.25 moles of hexane, assuming 100 percent yield.

39. How might the greenhouse effect be affected by the development of new techniques that use atmospheric carbon dioxide as a carbon source for making organic compounds? Make a poster showing such possible effects.

Cumulative Standardized Test Practice

1. A hydrocarbon with the formula C_6H_6 will be a(n)
   a) carbon ring.
   b) alkane.
   c) straight chain of carbon atoms.
   d) bent chain of carbon atoms.

2. The compound pictured below is a(n)

$$CH_3CH_2CH_2-\overset{\overset{O}{\|}}{C}-H$$

   a) ether.
   b) aldehyde.
   c) ketone.
   d) alcohol.

3. Molarity is defined as
   a) the number of liters of solution per molecules of solute.
   b) the number of moles of solute per liter of solution.
   c) the number of liters of solute per molecules of solution.
   d) the number of moles of solution per liter of solute.

4. Bases are defined as substances that
   a) produce hydronium ions in water.
   b) have hydrogen in their compounds.
   c) produce hydroxide ions in water.
   d) neutralize acids.

5. Which of the following is true about a solution with a pH lower than 7?
   a) The solution is a strong acid.
   b) The quantity of hydronium ions is greater than the quantity of hydroxide ions.
   c) The quantity of hydroxide ions is greater than the quantity of hydronium ions.
   d) The solution is an equal mixture of moles of acid and moles of base.

6. The oxidation numbers of the elements in $CuSO_4$ are
   a) $Cu = +2, S = +6, O = -2$
   b) $Cu = +3, S = +5, O = -2$
   c) $Cu = +2, S = +2, O = -1$
   d) $Cu = +2, S = 0, O = -2$

7. Electrolysis is
   a) the process of using electric currents to speed up redox reactions.
   b) the process of using electric currents to speed up any reaction.
   c) the process of using electric currents to start reactions that do not occur.
   d) the process of using electric currents to stop reaction from occurring.

8. A plastic is a
   a) hardened polymer.
   b) moldable polymer.
   c) heated polymer.
   d) heat resistant polymer.

9. Which of the following is NOT a reason noble gases are unreactive?
   a) They have a stable electron configuration.
   b) They have filled energy levels.
   c) They have the maximum number of valence electrons.
   d) They have a very low melting point.

10. What is the amount of free energy associated with the conversion of liquid water into hydrogen and oxygen gas?
    a) -572 kJ/mol
    b) +572 kJ/mol
    c) -458 kJ/mol
    d) +458 kJ/mol

NEED EXTRA HELP?

If You Missed Question . . .	1	2	3	4	5	6	7	8	9	10
Review Section . . .	18.1	18.2	13.2	14.1	15.1	16.1	17.1	18.3	7.2	6.1

CHAPTER 19: The Chemistry of Life

BIG Idea Organisms are made of carbon-based molecules, many of which are involved in chemical reactions necessary to sustain life.

19.1 Molecules of Life
MAIN Idea Biological molecules—proteins, carbohydrates, lipids, and nucleic acids—interact to carry out activities necessary to living cells.

19.2 Reactions of Life
MAIN Idea Respiration is a metabolic process in which energy is obtained from biological molecules.

ChemFacts

- Bees have two stomachs, the regular stomach and the honeystomach which they use to store nectar.
- Nectar is 80 percent water with some complex sugars.
- Worker bees digest the nectar in their mouths and honey stomachs, breaking the complex sugars into simple sugars using enzymes.
- Water evaporates from the nectar when it is spread over the honeycomb. This produces honey, which is a much thicker syrup.

Start-Up Activities

LAUNCH Lab

Test for Simple Sugars

Your body constantly uses energy. Many different food sources can supply that energy, which is stored in the bonds of molecules called simple sugars. What foods contain simple sugars?

Materials
- 400-mL beaker
- hot plate
- 10-mL graduated cylinder
- Benedict's solution
- 10% glucose solution
- other food solutions such as 10% starch, 10% gelatin, or honey
- test tube
- tongs
- boiling chip
- stirring rod

Procedure
1. Read and complete the lab safety form.
2. Fill a 400-mL beaker one-third full of water. Place this beaker on a hot plate and begin to heat it to boiling.
3. Place 5.0 mL 10% glucose solution in a test tube.
4. Add 3.0 mL Benedict's solution to the test tube. Mix the two solutions using a stirring rod. Add a boiling chip to the test tube.
   Warning: *Benedict's solution is an eye and skin irritant*
5. Using tongs, place the test tube in the boiling water bath and heat for five minutes.
6. Record a color change from blue to yellow or orange as a positive test for a simple sugar.
7. Repeat the procedure using food samples such as a 10% starch solution, a 10% gelatin (protein) suspension, or a few drops of honey suspended in water.

Analysis
1. **Evaluate** Describe the color change that was observed.
2. **Interpret** Which foods tested positive for the presence of a simple sugar?

Inquiry Why might it be important to know if a food contains simple sugars?

FOLDABLES Study Organizer

Biomolecules Make the following Foldable to help you organize information about biomolecules.

▶ **STEP 1** Fold a sheet of notebook paper lengthwise, keeping the margin visible on the left side.

▶ **STEP 2** Cut the top flap into four tabs.

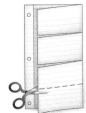

▶ **STEP 3** Label the margin flap *Biomolecules*, and the four flaps *Proteins, Carbohydrates, Lipids*, and *Nucleic Acids*.

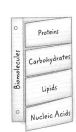

FOLDABLES Use this Foldable with **Section 19.1**. As you read, summarize the general structure and function of the biological molecules, and give examples of each.

Chemistry Online

Visit glencoe.com to:
▶ study the entire chapter online
▶ explore **Concepts In Motion**
▶ take Self-Check Quizzes
▶ use Personal Tutors
▶ access Web Links for more information, projects, and activities
▶ find the Try at Home Lab, Counting Nutrients

Chapter 19 • The Chemistry of Life **665**

Section 19.1

Objectives

- **Compare and contrast** the structures and functions of proteins, carbohydrates, lipids, and nucleic acids.
- **Analyze** the relationship between the three-dimensional shape of a protein and its function.

Review Vocabulary

cross-linking: the linking together of many polymer chains, giving the polymer increased strength

New Vocabulary

biochemistry
protein
amino acid
denaturation
substrate
active site
carbohydrate
lipid
fatty acid
steroid
nucleic acid
DNA
RNA
nucleotide
vitamin
coenzyme

Molecules of Life

MAIN Idea Biological molecules—proteins, carbohydrates, lipids, and nucleic acids—interact to carry out activities necessary to living cells.

Real World Reading Link Many of the most important molecules in your body are polymers. These molecules are part of you and all around you. For example, the food you eat contains proteins and carbohydrates, both of which are polymers. The complex reactions that sustain your body depend on these—and other—molecules of life.

Biochemistry

All living things, no matter how different they appear, are composed of a relatively few kinds of chemicals. Some of these chemicals are covalent and some are ionic. They are arranged into complex and highly organized materials that provide support, carry out energy transformations, or duplicate themselves. The study of the chemistry of living things is called **biochemistry**. This science explores the substances involved in life processes and the reactions they undergo. Other than water, which can account for 80 percent or more of the weight of an organism, most of the molecules of life are organic. These organic biological molecules are referred to as biomolecules.

The elemental composition of living things is different from the relative abundance of elements in Earth's crust. As shown in **Figure 19.1**, oxygen, silicon, aluminum, and iron are the most abundant atoms in Earth's crust. However, more than 95 percent of the atoms in your body are oxygen, carbon, hydrogen, and nitrogen. All four of these elements can form the strong covalent bonds found in organic molecules. Along with two other elements, sulfur and phosphorus, they are the only elements needed to make most of the proteins, carbohydrates, lipids, and nucleic acids found in every cell.

■ **Figure 19.1** The composition of Earth's crust and that of the human body are significantly different. The numbers represent percentages by mass in each. The elements oxygen and hydrogen are found in both, but carbon is concentrated in living things.

Element	Percent
oxygen	46.0
silicon	28.0
aluminum	8.0
iron	6.0
magnesium	4.0
calcium	2.4
potassium	2.3
sodium	2.1
hydrogen	0.9
other	0.3
	100.0%

Element	Percent
oxygen	65.0
carbon	18.5
hydrogen	9.5
nitrogen	3.0
calcium	1.5
phosphorus	1.0
potassium	0.4
sulfur	0.3
other	0.8
	100.0%

Biomolecules Are these biomolecules any different from the other organic chemicals that you have studied? Scientists once thought that these molecules possessed a vital force and did not follow the same chemical and physical principles that govern all other matter. Until the early 1800s, no biomolecule had been synthesized outside of a living cell, although many had been identified. But in 1828, a German physician and chemist named Friedrich Wöhler reported that he could "make urea without needing a kidney."

The organic molecule urea is normally made in your kidneys and excreted in urine to dispose of excess nitrogen. Wöhler had produced urea in the laboratory from ammonia and cyanic acid. He showed that it is possible to synthesize one of the molecules of life in the laboratory. Today, it is possible to synthesize artificially many thousands of complex biomolecules. However, living cells still are the most efficient laboratories, and it can take months or years for chemists to synthesize a large molecule that a cell can make in seconds or minutes.

Four types of macromolecules Some biomolecules are macromolecules. Macromolecules are large molecules formed by joining smaller molecules together. **Table 19.1** summarizes the four basic categories of biological macromolecules. Three of these four types of macromolecules—carbohydrates, proteins, and nucleic acids—are large polymers. As you will learn, lipids are made from smaller molecules, but they are usually not considered polymers.

◀ **FOLDABLES**
Incorporate information from this section into your Foldable.

Concepts In Motion
Interactive Table Explore biological macromolecules at glencoe.com.

Table 19.1 Biological Macromolecules

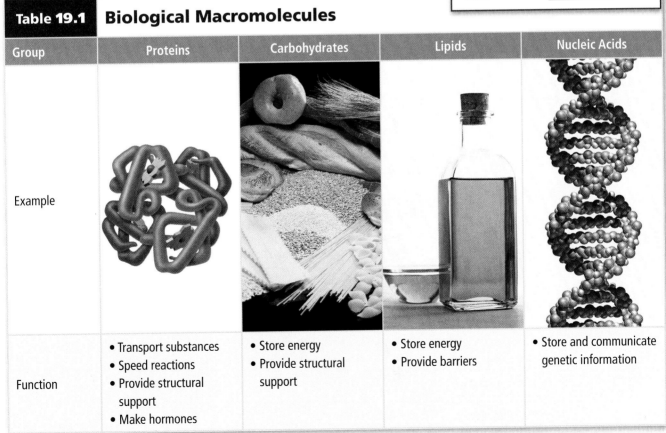

Group	Proteins	Carbohydrates	Lipids	Nucleic Acids
Example				
Function	• Transport substances • Speed reactions • Provide structural support • Make hormones	• Store energy • Provide structural support	• Store energy • Provide barriers	• Store and communicate genetic information

■ **Figure 19.2** Keratin, the protein in hair and nails, is also the protein in the soft fur and tough hooves of animals such as this moose.

> **FOLDABLES**
> Incorporate information from this section into your Foldable.

Proteins

What makes foods like chicken, beans, and fish so nutritious? They are composed largely of proteins. Proteins are one of the most important classes of biomolecules needed by living cells. A **protein** is a polymer formed from small monomers linked together by amide groups. When proteins were first discovered and named in the 19th century, their name was taken from the Greek word *proteios*, meaning "first" or "primary," because they were believed to be the chemical essence of life.

The role of proteins About half of the proteins in your body function as catalysts in cell reactions. Proteins play other important roles in living systems, as well. **Figure 19.2** shows some of these many functions. For example, structural proteins provide strength to your body and give it shape. Structural proteins include collagen and keratin. Collagen is found in ligaments, tendons, and cartilage, and it provides the structural glue that holds your cells and tissues together. Keratin is the protein in hair, and it is also found in fur, hooves, skin, and fingernails.

Other proteins make up your muscles and function to transport substances through your body in the blood. The hemoglobin that makes your blood red and carries oxygen from the lungs to the cells is a protein. Fibrin is another protein that is sometimes found in your blood. Fibrin is involved in clotting blood at the site of cuts and scrapes, as shown in **Figure 19.3**.

> **FACT of the Matter**
> Some arctic fishes make unique antifreeze proteins that prevent ice crystals from forming in their blood. Scientists have transferred the genes for these proteins to tomato plants to make the plants tolerant of cold weather.

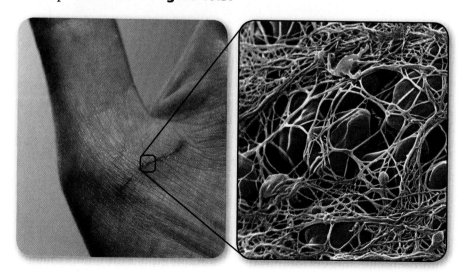

■ **Figure 19.3** A scab forms due to fibrin threads trapping blood cells and platelets.

Amino acids Proteins contain the elements carbon, hydrogen, oxygen, nitrogen, and sulfur. How are atoms of these elements arranged? The monomers that form proteins are organic compounds called **amino acids**. Although many different amino acids exist, only 20 are used by human cells to make proteins.

All amino acids have certain parts of their structure in common. Look at this general structure of an amino acid and see if you can recognize the two functional groups present. The name amino acid should give you a clue.

```
                       R   Variable side chain
                       |
Amino group   H₂N — C — C — OH   Carboxyl group
                    |   ||
Hydrogen atom       H   O
```

On the left side of the structure is an amino group, $-NH_2$, and on the right side is a carboxyl group, $-COOH$. Both of these groups are attached to a central carbon atom that also has a hydrogen atom and a side chain attached. The side chain, represented in the structure by the letter R, may consist of a number of different atoms or groups of atoms and is what makes each of the 20 amino acids unique with its own particular properties.

The structures of several different amino acids are shown in **Table 19.2**. Glycine is the simplest amino acid. Its side chain consists simply of a hydrogen atom. Alanine is the next simplest with a methyl-group side chain. Phenylalanine contains a benzene ring as part of its side chain, and cysteine contains a polar $-SH$ group.

Table 19.2 Amino Acid Structures

Glycine	Serine	Cysteine	Lysine
Alanine	Glutamine	Valine	Phenylalanine

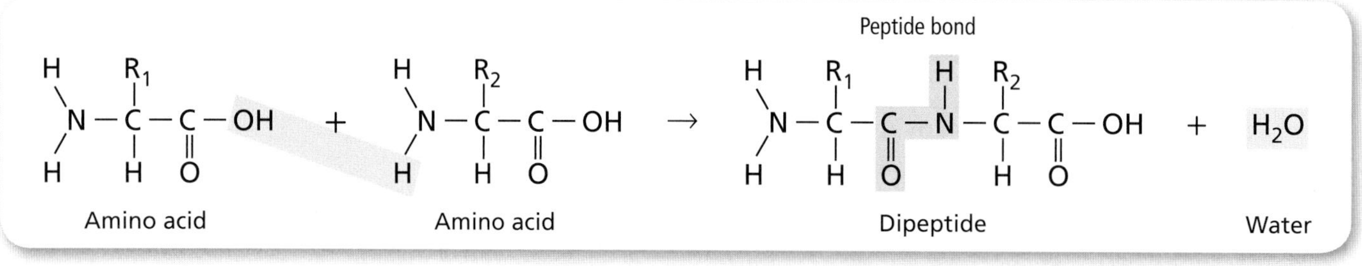

$$\underset{\text{Amino acid}}{\overset{H}{\underset{H}{N}}-\overset{R_1}{\underset{H}{C}}-\overset{}{\underset{O}{C}}-OH} + \underset{\text{Amino acid}}{\overset{H}{\underset{H}{N}}-\overset{R_2}{\underset{H}{C}}-\overset{}{\underset{O}{C}}-OH} \rightarrow \underset{\text{Dipeptide}}{\overset{H}{\underset{H}{N}}-\overset{R_1}{\underset{H}{C}}-\overset{\text{Peptide bond}}{\overset{}{\underset{O}{C}}-\overset{H}{\underset{}{N}}}-\overset{R_2}{\underset{H}{C}}-\overset{}{\underset{O}{C}}-OH} + \underset{\text{Water}}{H_2O}$$

■ **Figure 19.4** The amino group of one amino acid bonds to the carboxyl group of another amino acid to form a dipeptide and water. The organic functional group formed is an amide linkage called a peptide bond.
Describe *how an amide functional group forms.*

Making proteins: condensation reactions The polymerization reactions that link amino acids to form proteins are condensation reactions, similar to those that are used to make some of the plastics you studied in Chapter 18. When two amino acids bond together, a hydrogen (—H) from the amino group of one amino acid combines with the hydroxyl (—OH) part of the carboxyl group of the other amino acid to form a molecule of water (H_2O). When the water molecule is released, an amide group has formed, linking the two amino acids. Recall that an amide group has the following structure.

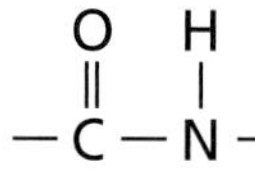

$$-\overset{O}{\underset{}{C}}-\overset{H}{\underset{}{N}}-$$

Biochemists call an amide group by the name *peptide bond* when it occurs in a protein. Although the names are different, the functional groups are identical. When two amino acids are linked by a peptide bond, the resulting chain of two amino acids is called a dipeptide, as shown in **Figure 19.4**.

Additional amino acids can be added to a dipeptide by the same reaction to form a long chain. The chain is called a polypeptide because it is a polymer of amino acids held together by peptide bonds. A protein can consist of a single polypeptide chain, although most proteins contain two or more different polypeptide chains.

Three-dimensional protein structure Proteins have a three-dimensional structure because the polypeptide chains of which they are composed fold up like those shown in **Figure 19.5**. The polypeptide chains are held together by hydrogen bonds, ionic bonds, and disulfide cross-links between the side chains of neighboring amino acids.

■ **Figure 19.5** Proteins can fold into either round, globular structures (left) or long, fibrous structures (right). The amino acid chains are held in place in three-dimensional structures by attractive forces between the side chains of different amino acids, which have been brought close together by the bending and folding of the polypeptide chains.

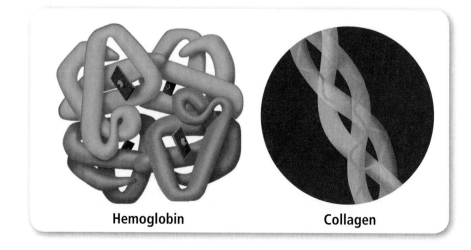

Hemoglobin   **Collagen**

The amino acids in a folded protein may be twisted into a helix or held tightly in sheets by hydrogen bonds that form between the N—H and C=O groups of neighboring amino acids. You may remember from Chapter 18 that disulfide bonds in keratin give hair its strength and shape. These bonds form between sulfur atoms in the side chains of two molecules of the amino acid cysteine.

Protein shape and properties A molecule's geometric shape is important in determining how chemical interactions take place. For example, the antibodies in your blood are folded into shapes so specific that each will bond to only one type of molecule specific to an invader. These antibodies protect your body from invasion by foreign organisms such as bacteria by bonding to molecules on the surface of the invader. Once bound, the invaders are destroyed by your immune system. Because bonding with invaders is so specific, your body must make a different kind of antibody for each kind of invader.

Denaturation What happens when the three-dimensional structure of a protein is disrupted? Think of the difference between the consistency of a raw egg white and that of a hard-boiled egg. When the forces holding a polypeptide chain in its three-dimensional shape are broken, the protein is unfolded in a process called **denaturation**. High temperatures can denature proteins, which is why cooking foods that contain proteins results in the proteins' denaturation. Denatured proteins form the solid white of a hard-boiled egg.

In addition to high temperatures, proteins can also be denatured by extremes in pH, mechanical agitation, and chemical treatments, as shown in **Figure 19.6.** Because the shape of a protein is essential to its function, denaturation results in a loss of function. For this reason, most organisms can live only in narrow temperature and pH ranges. However, some organisms do thrive at relatively high temperatures and low pHs. The proteins of these organisms have shapes held together by covalent bonds such as disulfide linkages. Disulfide linkages are stronger than the hydrogen bonds that hold together proteins of organisms living in more moderate conditions.

■ **Figure 19.6** Beating egg whites incorporates air and denatures some of the proteins, providing the firm structure of the meringue on a lemon meringue pie. The proteins of organisms that live in extremely cold or hot surroundings have strong bonding that prevents denaturation and loss of protein function. The proteins of bacteria that live in hot springs such as this one in Yellowstone National Park are resistant to heat denaturation.

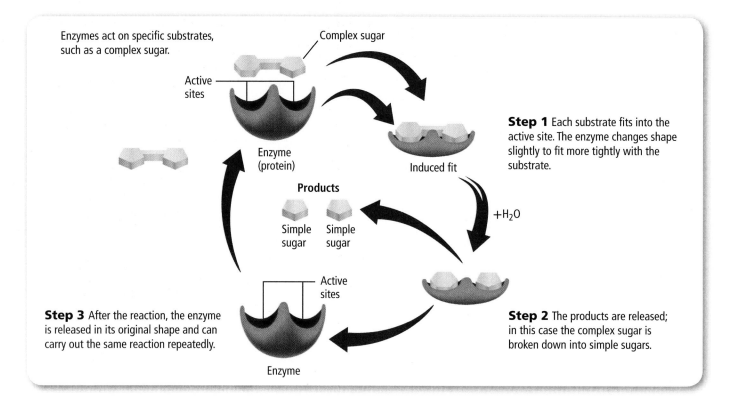

■ **Figure 19.7** Enzymes lower the activation energy needed for a reaction to occur. Enzymes change the speed at which chemical reactions occur without being altered themselves in the reaction.

Explain *in your own words how an enzyme works.*

The role of proteins as enzymes One of the most important roles of proteins is that of the biological catalysts called enzymes. Enzymes help speed up reactions without being changed themselves during a reaction. Spontaneous chemical reactions can occur so slowly that they seem to not occur at all. Enzymes greatly speed up a reaction that would occur eventually, but they cannot make a reaction go if it would never occur spontaneously.

Enzyme structure and function How are enzymes able to speed up reactions? **Figure 19.7** shows how enzymes work. Enzymes provide specific sites, usually a pocket or groove in their three-dimensional structure, where each reactant, called a **substrate**, can bind. The pocket that can bind a substrate taking part in the reaction is called the **active site**. After a specific substrate binds to its active site, the active site changes shape to fit the substrate. This process of recognition is called induced fit. The shape of the substrate matches the shape of the active site, much as the shape of a key matches the hole in the lock it opens. Only substrates that have a shape allowing them to bind to the active site can take part in a particular enzyme-catalyzed reaction.

Biological role of enzymes Almost all reactions that take place in the body are catalyzed by enzymes. During digestion, enzymes speed up the breakdown of foods into molecules small enough to be absorbed by cells. Enzymes also make possible the reactions required for cells to extract energy from these nutrients. Enzymes are used in medicine to treat disorders that result from their deficiencies. Lactase is an enzyme that breaks down milk sugar, called lactose, so that the two smaller sugar molecules can be absorbed.

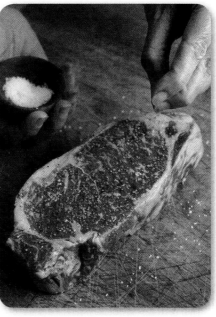

■ **Figure 19.8** A group of enzymes found in many common household products is the proteases. Proteases in contact lens-cleaning solutions help remove the grimy buildup on lenses that results from proteins secreted by cells around the eyes. These enzymes hydrolyze proteins into free amino acids. The proteases in meat tenderizer make steak more tender by breaking down the proteins.

People with lactose intolerance lack the ability to make the enzyme lactase. When this enzyme isn't present, eating dairy products causes lactose to accumulate in the intestines, resulting in bloating and diarrhea. If lactase pills are taken along with the dairy products, the enzyme breaks down the lactose just as the naturally produced enzyme would. **Figure 19.8** shows other applications of enzymes.

Carbohydrates

A **carbohydrate** is a biomolecule that contains the elements carbon, hydrogen, and oxygen in a ratio of about two hydrogen atoms and one oxygen atom for every carbon atom. Early chemists thought that carbohydrates were chains of carbon with water attached—hydrated carbon—which is how carbohydrates were named. Chemists now know that carbohydrates are not hydrated carbon chains, but the name persists.

The role of carbohydrates Foods such as pasta, bread, and fruit are rich in carbohydrates. When the carbohydrates in pasta and other foods are broken down in the body, the sugar glucose is formed. The oxidation of glucose in cells provides most of the energy needed for life. When excess glucose is produced, as when athletes eat large quantities of pasta, the energy that can't be used right away is stored in the cells. Animals store glucose in their livers and muscles as glycogen, a polymer of glucose. In plants, glucose is stored as starch, the polymer you studied in Chapter 18.

Carbohydrates also play a structural role in living things. You have learned that cellulose, also a glucose polymer, provides the strength and support needed in plants. Chitin, a natural polymer made from a sugar similar to glucose, forms the hard exteriors of insects and other arthropods, as shown in **Figure 19.9**. Carbohydrates are also found in cotton and rayon fabrics, wood, and paper.

FOLDABLES
Incorporate information from this section into your Foldable.

■ **Figure 19.9** Chitin is a structural carbohydrate that makes up the protective exoskeleton of arthropods such as beetles and lobsters. Chitin is one of the most abundant natural polymers on Earth due to the enormous number of insects on our planet.

Section 19.1 • Molecules of Life

CHEMLAB

CATALYTIC DECOMPOSITION

Background
You may have used hydrogen peroxide, H_2O_2, as an antiseptic when you cut or scratched your skin. If you have, you will have noticed that it seems to fizz and bubble when it contacts your skin. Hydrogen peroxide decomposes to form water and oxygen. The decomposition reaction is more rapid when the hydrogen peroxide comes into contact with your skin. Catalase, an enzyme found in the cells of your skin as well as in most other cells, is responsible for this rapid decomposition.

The activity of an enzyme is affected by environmental factors such as pH and temperature. Every enzyme has optimum conditions at which its reaction rate is fastest. In this ChemLab, you will study the decomposition of hydrogen peroxide as catalyzed by the catalase in carrot cells, and you will determine the optimum temperatures under which this enzyme works.

Question
How does temperature affect the decomposition of hydrogen peroxide by catalase from carrot cells?

Objectives
- **Observe** the action of catalase on the decomposition of hydrogen peroxide.
- **Compare** the rate of reaction at various temperatures.

Make and use graphs to interpret results.

Preparation

Materials
hot plate
small (13 mm × 100 mm), clean, unscratched test tubes (8)
test-tube rack
small beakers (4)
carrot-cell slurry
3% hydrogen peroxide solution
10-mL graduated cylinder
crushed ice
Celsius thermometer
glass stirring rod
timer or clock
metric ruler
marking pencil
thermal glove

Safety Precautions

WARNING: *Hydrogen peroxide can harm the eyes. Wear safety goggles when using it.*

Procedure

1. Read and complete the lab safety form.
2. Place carrot-cell slurry into each of four test tubes to a depth of about 2 cm.
3. Carefully measure 3.0 mL of 3% H_2O_2 solution into each of another four test tubes. If large quantities of bubbles are produced, obtain a new tube or thoroughly clean the original one.
4. Set one carrot tube and one hydrogen peroxide tube in each of four small beakers. Label the beakers *A, B, C,* and *D.* Add crushed ice and tap water to one beaker, hot tap water to a second beaker, and room-temperature tap water to the remaining two beakers.

5. Set the hot-water beaker and one of the room-temperature-water beakers on the hot plate, and heat until the water temperatures reach 60–65°C and 37–38°C, respectively.

6. Make a data table similar to the one shown under Data and Observations. Take the temperature of the water in each of the four beakers, and record the temperature and label of each beaker. Let the test tubes sit in the beakers of water at 0°C, room temperature (probably about 18–25°C), 37–38°C, and 60-65°C for five to ten minutes.

7. Pour the tubes of H_2O_2 into the tubes of carrot-cell slurry at the same temperature. Stir each tube quickly with a glass stirring rod, and start the timer at the same moment. Measure the height of the foam, from the top of the liquid mixture to the top of the foam, at one-minute intervals for four minutes or until the foam reaches the top of the test tube. Record your data in the table.

Analyze and Conclude

1. **Interpret Data** Use your data to construct and label a graph for each of your temperature trials. Plot foam height on the vertical axis and time on the horizontal axis.

2. **Interpret Data** Construct another graph, plotting foam height at three minutes on the vertical axis and temperature on the horizontal axis.

3. **Compare and Contrast** Which of the temperatures appears to be optimum for the catalyzed decomposition of hydrogen peroxide by catalase? How does this temperature compare with human body temperature?

Apply and Assess

1. **Summarize** a way in which the gas in the foam might have been tested to verify that it was oxygen.

INQUIRY EXTENSION
Explain Peroxide ions are produced in plants and animals as a result of cellular reactions. Given that these peroxides can oxidize and damage cell structures, why is the presence of catalase in cells beneficial?

Data and Observations

Data Table					
Beaker Number	Temperature of Reactants	Height of Foam After 1 Minute	Height of Foam After 2 Minutes	Height of Foam After 3 Minutes	Height of Foam After 4 Minutes
A					
B					
C					
D					

In the Field

Meet Dr. Lynda Jordan
Biochemist

The door to Dr. Lynda Jordan's office at the North Carolina Agricultural and Technical State University is nearly always open. Students and younger faculty members continuously stop by for help with lab problems and personal problems. Dr. Jordan comments, "I know what my life would have been like if there hadn't been people saying, 'You can do it.' That's why I decided to 'put my business on the street.' Students need to know that being a good, hard-working person does pay off."

On the Job

Q A Public Broadcasting Service TV special has honored you and your work with enzymes. Can you tell us about this project?

A I'm studying an enzyme involved in diseases such as asthma, arthritis, bronchial disorders, gastrointestinal disorders, preterm labor, and diabetes. I want to find out about this enzyme's structural and functional relationship and how it behaves in human cells. This enzyme is isolated from cells of the human placenta. This is not an easy task. My colleagues and I have classified it by its molecular weight, pH, and calcium dependencies.

Q Why is this research so important to you, personally?

A I first started working on the enzyme, which is called phospholipase A, as part of my post-doctoral research at the Institute Pasteur in Paris, France, in 1985. During that time, we discovered that phospholipase was present in human placentas, and we were one of the first laboratory groups to isolate this calcium-independent protein. It was a very exciting time for me.

I know many people that have the diseases associated with this enzyme. If I could understand how this enzyme functions, then perhaps I could make a contribution to understanding what happens in the diseased state. This information would help us in the scientific community to devise strategies that could circumvent these biomedical problems.

Q Which are the diseases that most interest you?

A Recent scientific evidence has shown a link between the enzyme phospholipase A_2 and diabetes. I am most interested in investigating the relationship between PLA_2 and glucose metabolism. Many of my family members, including myself, have diabetes. Understanding this disease on a molecular level would be rewarding for me on a personal as well as scientific level.

Early Influences

 What were you like as a high school student?

 I was a kid from an inner-city high school, with no hope for going to college. Although I was a good student, I used to skip classes. One day, when I was 15 years old, I was skipping class with a girlfriend in the rest room. The hall monitor came in, and I ran into the first room I came to, which happened to be an auditorium where an Upward Bound orientation was taking place. The speaker really got my attention. He asked me, 'What are you going to do the rest of your life—stand on the street corner and watch the world go by?' That's about all I was doing, and I was lucky not to end up a sad statistic. I signed up for the program right away.

 Can you describe the program?

It was a rigorous six-week, college-level program at Brandeis University. We lived in college dormitories and were not allowed to have contact with the inner-city environment. The objective was to change our environment, so we were scheduled from 6 A.M. to 11 P.M. every day. We were expected to give our very best, and we were pushed and pushed.

Personal Insights

 Mentors were crucial to your development. Are you yourself a mentor now?

 Of all the things I do, the main thing is developing people. It's the most taxing, especially when I'm dealing with people who are first-generation college students, as I was. But it's also the most rewarding. I try to help students and younger faculty members with both their scientific problems and their personal problems, so they'll have the confidence to survive in the big, bad world. Character has a lot to do with success. Honesty, trustworthiness, dependability, and acceptance of responsibility sometimes have to be learned.

 Can you draw a parallel between solving problems in the lab and solving them in everyday life?

 Both are step-by-step processes. First, you have to collect information. Then, you devise a strategy for reaching a goal. In the meantime, you need to think of backup strategies to utilize in case things don't work out as you anticipate. While you're progressing, you need to monitor how far you've come. If you're investigating careers, for example, and are considering being a doctor, the first thing to do is visit a hospital and maybe an emergency room. All along the way to solving any problem, you need to take little baby steps and monitor your progress.

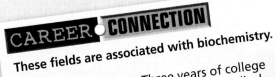

CAREER CONNECTION

These fields are associated with biochemistry.

Medical Technologist Three years of college followed by a one-year program in medical technology

Medical Laboratory Technician High school plus a two-year training program

Medical Records Technician High school plus a two-year training program

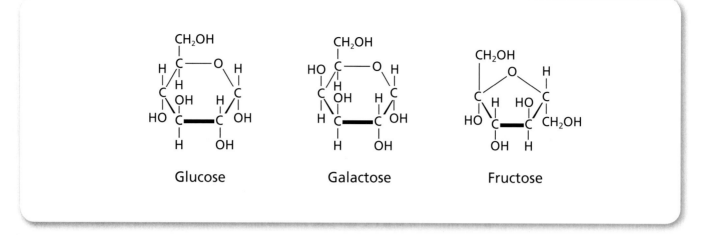

■ **Figure 19.10** Glucose, galactose, and fructose are monosaccharides.

The structures of carbohydrates Carbohydrates are a diverse group of molecules that include many simple sugar molecules, as well as larger molecules made by linking together different combinations of these sugars. Most simple sugars contain five, six, or seven carbon atoms arranged in a ring structure. An oxygen atom forms one corner of the ring, and hydroxyl groups are attached to each carbon. The most common simple sugars include glucose, galactose, and fructose. **Figure 19.10** shows the structures of these three common sugars.

Monosaccharides and disaccharides Simple sugar molecules are called monosaccharides. Two monosaccharides can link together in a condensation reaction to form a disaccharide, such as the most common carbohydrate in sweets and candies, sucrose. As shown in **Figure 19.11,** the structure of sucrose, which is also called table sugar, is made up of a glucose and a fructose monomer.

In order for a molecule to be classified as a nutrient, it must be able to enter cells. Disaccharides are too large to pass through cell membranes. However, enzymes in the digestive system catalyze the reaction, breaking sucrose into glucose and fructose molecules that are small enough to be absorbed and used by cells as nutrients. Other common disaccharides in foods are milk sugar (lactose) and malt sugar (maltose).

■ **Figure 19.11** When glucose and fructose bond, the disaccharide sucrose forms. Note that water is also a product of this condensation reaction. Remember that each ring structure is made of carbon atoms bonded to individual hydrogen atoms or to functional groups. The carbons in the rings and the individual hydrogen atoms are not shown for simplicity.

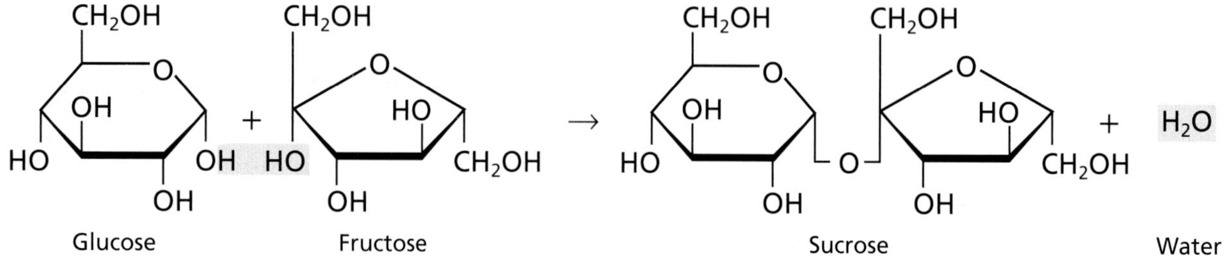

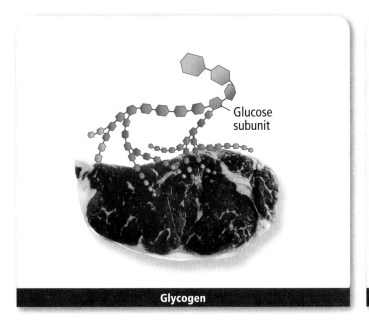
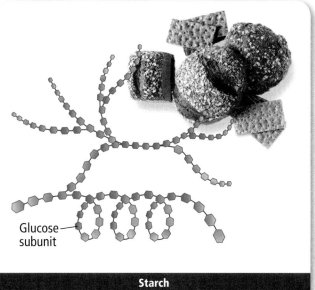

Polysaccharides What happens when you toast a slice of bread? Have you ever noticed that toast has a slightly sweet taste? The sweetness results from maltose, a disaccharide formed from two glucose molecules. Untoasted bread does not seem to taste sweet, and it contains no maltose. Where does the maltose come from when bread is toasted? It is the product of a reaction in which the large carbohydrates in bread—starch and cellulose—are broken down by heat.

As you recall, starch and cellulose are polymers that are formed from many glucose molecules linked together. Another common polymer consisting only of glucose is glycogen, also called animal starch. Starch, cellulose, and glycogen are examples of polysaccharides. Polysaccharides can contain hundreds or even thousands of monosaccharide molecules bonded together. A polysaccharide may contain only one type of monosaccharide, or it may contain more than one type.

The repeating glucose units are the same in the starch of a potato as in the cellulose in the wood of a pencil. So, how can the structures of these polymers be different? The major differences are the way the glucose molecules are bonded to each other, the number of glucose units in the polymer, and the amount of branching. These small differences in structure lead to large differences in properties and function. The structures of glycogen, starch and cellulose are compared in **Figure 19.12.**

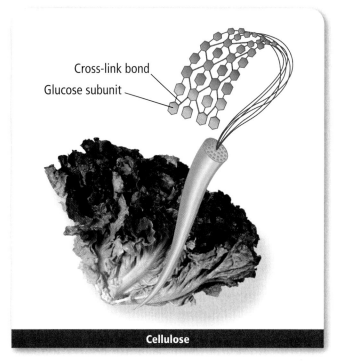

■ **Figure 19.12** The glycogen found in the muscle and liver of animals is a polysaccharide made of glucose. Glycogen is highly branched. Two other important polysaccharides are starch and cellulose. Starch molecules, which store energy in plants, can be branched or unbranched. Cellulose has a linear, unbranched structure that resembles a chain-link fence. Cellulose is an important structural molecule in plants.

Everyday Chemistry

Clues to Sweetness

What makes ripe strawberries taste so deliciously sweet? It has something to do with the sweetness message delivered to your brain when certain molecules from the strawberry fit into receptor sites in the taste buds on your tongue. The molecules lock onto the receptors in a specific way, determined by their structure. When this happens, chemical processes that produce the sweet response are stimulated. Even molecules of artificial sweeteners, many of which are not carbohydrates and are not metabolized by the body, are capable of matching the sweet receptor sites of the taste buds.

Breaking the code for sweetness Hydrogen bonding is the most common interaction between molecules. When a hydrogen atom is bonded to a highly electronegative atom, its electron is pulled toward the more electronegative atom, and the hydrogen develops a partial positive charge. Consequently, it is capable of attracting other atoms that have partial negative charges. Nitrogen, oxygen, fluorine, and other nonmetals are all highly electronegative atoms that are attracted to hydrogen.

Decades ago, chemists hypothesized that sweet-tasting molecules have two hydrogen-bonding sites—A and B—located close to each other as shown in **Figure 1.** They postulated that the sites are separated by 0.3 nm and that site A has a hydrogen atom covalently bonded to atom A. They called these two sites AH and B, and named the structure the AH, B system. The AH, B system interacts with a similar system, which consists of an NH group and a C=O group, located on the sweet receptors of the tongue.

Protein molecules form pairs of hydrogen bonds because they have both C=O and NH sites. The more electronegative oxygen in the C=O group seeks a hydrogen atom on another molecule. The hydrogen in the NH group seeks a more electronegative atom such as nitrogen or oxygen.

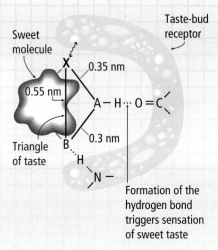

Figure 1 Triangle of sweetness

The C=O and NH sites are strategically placed about 0.3 nm apart. A molecule that has complementary sites, also 0.3 nm apart, can bind to the protein. The AH,B system of a sweet molecule, with sites AH and B, fits this description. If the sweet molecule bonds to the protein that forms part of the sweet taste receptors at the tip of the tongue, the brain receives a signal of sweetness.

Refining the sweetness model Chemists also noticed that sweet molecules shared a third common factor. Part of a sweet molecule is hydrophobic, which means it repels water. This part was called the X site. They found that X must be further from site B than from site AH. The whole molecular arrangement forms a triangle of sweetness.

Explore Further

1. **Analyze** What causes something to taste sweet?
2. **Acquire Information** Find a table comparing the relative sweetness of various sugars and artificial sweeteners. How do the following artificial sweeteners compare in sweetness with sucrose (table sugar): sucralose, aspartame, saccharin, and acesulfame-K?

■ **Figure 19.13 left** Lipbalm is a mixture of mostly oils and waxes. These lipids help keep the skin of the lips soft and moist. **right** The avocados in this guacamole contain large amounts of oils.

Lipids

A biomolecule that contains a large proportion of C—H bonds and less oxygen than in carbohydrates is called a **lipid**. Fats, oils, and waxes are all lipids. Lipids are insoluble in water and soluble in nonpolar organic solvents. In general, lipids that are derived from animals are called fats, and plant lipids are called oils. Lipid molecules are found in the fatty foods you eat such as butter, margarine, peanut butter, and the oil your french fries are cooked in. The waxes you use to polish the family car or to wax the bottom of a pair of skis also are lipids. Waxes are produced by both plants and animals. **Figure 19.13** shows some other common lipids.

The structure of lipids Most of the oils and fats in your diet consist of long-chain carboxylic acids called **fatty acids** that are bonded to a glycerol molecule. Glycerol is a short carbon chain with three hydroxyl functional groups. Three fatty acid molecules combine with a single glycerol in a condensation reaction to form three molecules of water and a lipid molecule with three ester functional groups. Each fatty acid contributes the hydroxyl part of its carboxyl (—COOH) group, and each hydroxyl group on the glycerol contributes the hydrogen atom to form the water molecules. The lipid formed is called a triglyceride. **Figure 19.14** illustrates the formation of triglycerides.

FOLDABLES
Incorporate information from this section into your Foldable.

■ **Figure 19.14** Triglycerides are formed when the hydroxyl groups of glycerol combine with the carboxyl groups of the fatty acids.

$$\begin{array}{c} CH_2OH \\ | \\ CHOH \\ | \\ CH_2OH \end{array} + \begin{array}{c} O \\ \| \\ HOC(CH_2)_{14}CH_3 \\ O \\ \| \\ HOC(CH_2)_{16}CH_3 \\ O \\ \| \\ HOC(CH_2)_{18}CH_3 \end{array} \rightarrow \begin{array}{c} CH_2-O-\overset{O}{\underset{\|}{C}}-(CH_2)_{14}-CH_3 \\ | \\ CH-O-\overset{O}{\underset{\|}{C}}-(CH_2)_{16}-CH_3 \\ | \\ CH_2-O-\overset{O}{\underset{\|}{C}}-(CH_2)_{18}-CH_3 \end{array} + 3H_2O$$

Glycerol   3 Fatty acids   Triglyceride   Water

Oleic acid

$$\underset{HO}{\overset{O}{\diagdown}}CCH_2CH_2CH_2CH_2CH_2CH_2CH_2CH=CHCH_2CH_2CH_2CH_2CH_2CH_2CH_3$$

Stearic acid

$$\underset{HO}{\overset{O}{\diagdown}}CCH_2CH_2CH_2CH_2CH_2CH_2CH_2CH_2CH_2CH_2CH_2CH_2CH_2CH_2CH_2CH_2CH_3$$

■ **Figure 19.15** Oleic acid is an unsaturated 18-carbon fatty acid that is found in olive oil. Stearic acid is a saturated 18-carbon fatty acid that is found in butter.

Identify *how the structure of the molecule is affected by the presence of a double bond.*

Saturated and unsaturated fatty acids The most common fatty acids are chains of 12 to 26 carbon atoms with a carboxylic acid group at one end. Saturated fatty acids, like saturated hydrocarbons, have only single bonds connecting the carbon atoms. Monounsaturated fatty acids have one double bond between two of the carbon atoms, as you can see in **Figure 19.15**. Fatty acids that are polyunsaturated have two or more double bonds.

Animal lipids tend to be solids at room temperature, while plant lipids tend to be liquids. Think of the bacon fat congealed in a cooled skillet compared to a bottle of vegetable oil. In general, animal lipids are more saturated than plant lipids. The bending of the molecules around the double bonds in plant lipids prevents them from packing closely. The closer packing of the molecules in animal lipids causes them to become solids at lower temperatures. That's why you need to pour off bacon grease before the skillet cools, but your bottle of vegetable oil is always easy to pour.

Steroids Another class of lipids is the steroids. A **steroid** is a lipid that has a specific four-ring structure: all steroids are built on the basic four-ring structure shown in **Figure 19.16**. Important steroids include cholesterol, some sex hormones, and vitamin D. Steroids that are hormones function to regulate metabolic processes. Cholesterol is an important structural component of cell walls.

■ **Figure 19.16** Many different steroids occur in plants, animals, and fungi, but all of them are based on the four-ring structure illustrated here.

Steroid ring structure

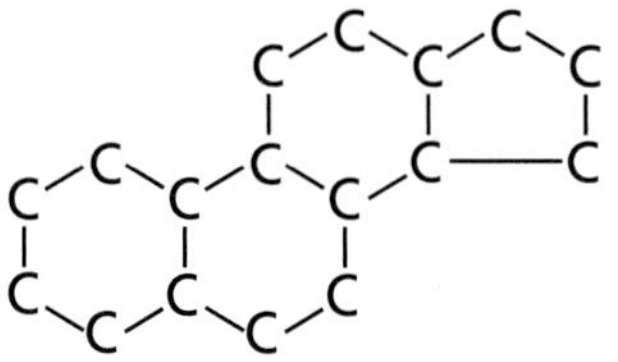

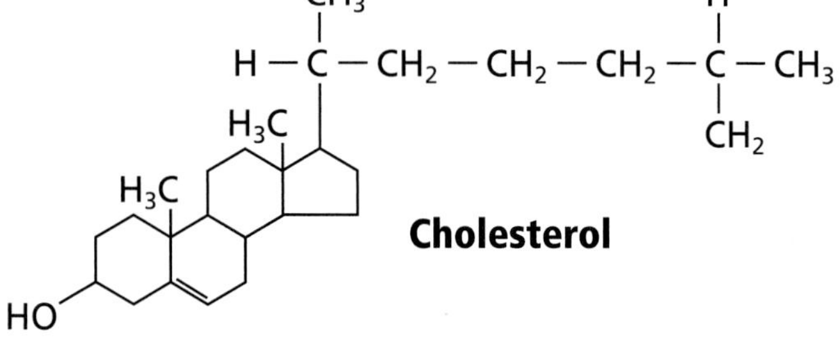

Cholesterol

682   Chapter 19 • The Chemistry of Life

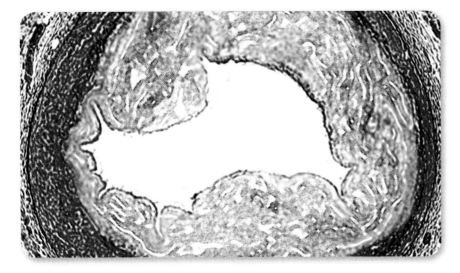

■ **Figure 19.17** Cholesterol forms part of the yellowish plaque material clogging this human artery. People with high levels of cholesterol in the blood have an increased risk of developing heart disease due to blocked arteries. Although cholesterol has a bad reputation, cells require it for making membranes, steroid hormones, and bile salts. The human body synthesizes about 1 g of cholesterol each day.

Lipids and diet Eating a diet high in saturated fats has been linked to the development of cardiovascular problems such as heart disease. The reason for this is not fully understood, but it may be due partly to the liver's ability to break down fatty acids into small pieces used to make cholesterol. High levels of cholesterol in the blood are associated with a stiffening and thickening of the artery walls shown in **Figure 19.17**. This condition can result in high blood pressure and heart disease. Reducing the amount of saturated fats and cholesterol in the diet—especially animal fats found in eggs, cheese, and red meats—is one way to lower blood cholesterol levels.

The functions of lipids Lipids have two major biochemical roles in the body. When an organism takes in and processes more food than it needs, excess energy is produced. The organism stores this excess energy for future use by using it to bond atoms together in lipid molecules. Later, when energy is needed, enzymes break these same bonds, releasing the energy used to form them.

Lipids also form the membranes that surround cells. These include cholesterol and phospholipids. Phospholipids are molecules in which a phosphate group and two fatty acids, rather than three fatty acids, are bonded to the three carbon atoms of glycerol. **Figure 19.18** shows how phospholipids align to form the double layer, called a phospholipid bilayer, that makes up the cell wall.

TRY AT HOME 🏠 LAB

See page 877 for **Counting Nutrients.**

FACT of the ▶ Matter

Many vegetable oils for cooking are promoted as being cholesterol-free. Because cholesterol occurs only in animal tissues, all plant products—fruits, vegetables, and all vegetable oils—are cholesterol-free.

■ **Figure 19.18** A phospholipid has a polar head and two nonpolar tails. The membranes of living cells are formed by a double layer of the lipids, called a bilayer.

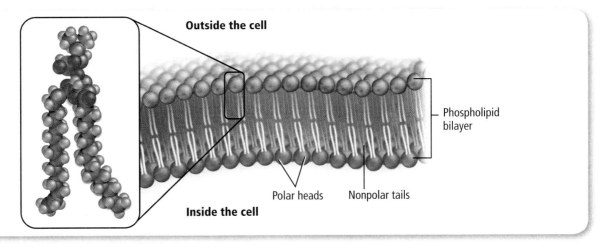

Section 19.1 • Molecules of Life

Everyday Chemistry

Fake Fats and Designer Fats

Fats have a reputation for causing health problems. At the same time, fats are an essential nutrient and are why many foods taste, look, and feel good as you eat them. Food scientists around the world are working on how to provide food that promotes good health, but also keeps the pleasing taste, appearance, and texture of fats.

Carbohydrates as fake fats Two kinds of carbohydrates—starches and cellulose—are being used to replace fats in foods. From the structures of these molecules in the chapter, you might wonder how starch and cellulose molecules can mimic fat's properties. When starch is mixed with water, it forms gels that have the texture and bulk of fat. The gels can replace the fat in some foods, but they can't be used for frying.

Usually, carbohydrates contribute only four Calories per gram as opposed to the nine Calories per gram that come from fats. But cellulose, a carbohydrate in the cell walls of plants, contributes no Calories at all because the body is not able to metabolize it. Microcrystalline cellulose is a form of cellulose that mixes with water to produce a texture similar to fat. It can replace fats in frozen desserts and bakery products.

Proteins for fats Other fat substitutes are based on proteins. In order for a protein to mimic a fat, it has to be cut into tiny particles that measure only from 0.1 to 3.0 μm in size. In one version of this process, egg white and milk protein are subjected to high heat and stress to form small, spherical particles. The small size of these protein particles causes them to be perceived by the mouth as smooth and creamy. Like carbohydrates, protein replacers provide only four Calories per gram. However, they are not suitable for frying, and they lack the fat flavor.

Chemically altered fats Searching for the perfect no-Calorie, health-promoting fat, food chemists have come up with chemically altered fats. They alter the size, shape, or structure of real fat molecules so that the human body digests and uses them to a lesser extent or not at all. One chemically altered fat is olestra, a sucrose polyester used in the chips in **Figure 1.** Food technologists claim that olestra will satisfy your taste buds without providing saturated fats and Calories. It can be substituted for butter and grease, and can even be used for frying. A fat consists of three fatty acids attached to glycerol. The altered fat olestra has six to eight fatty acids, all derived from vegetable oils, attached to sucrose. Being larger than a typical fat molecule, it is not absorbed by the cells in the digestive system. Enzymes fail to break the sucrose-fatty acid bond, so the molecule passes through the body undigested.

Figure 1 Chips made with olestra taste like chips made with cooking oil, but olestra is not absorbed by the body.

Explore Further

1. **Analyze** What are the dangers of eating only foods in which all the fats have been replaced?
2. **Think Critically** Products made with nondigestible fake fats could trap fat-soluble vitamins or medicines in the intestines. What effect would this have on the body? How could this problem be solved?

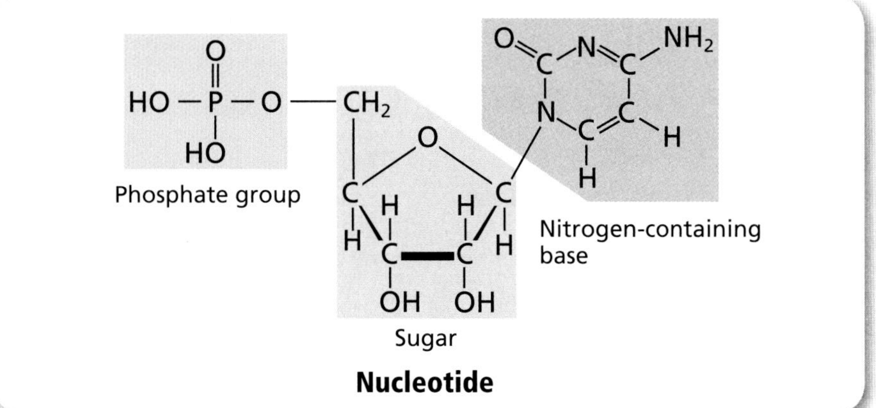

Figure 19.19 Each nucleotide contains three parts: a nitrogen-containing base, a five-carbon sugar, and a phosphate group.

Nucleic Acids

The fourth major class of biomolecules—nucleic acids—is not listed on food-product labels, although members of this group are found in every plant or animal cell that humans use for food. A **nucleic acid** is a large polymer containing carbon, hydrogen, oxygen, nitrogen, and phosphorus. Nucleic acids are present in cells only in tiny amounts and are not essential in the human diet because they can be made by the body from proteins and carbohydrates. Nucleic acids contain the coded genetic information that cells need to reproduce themselves, and they regulate the cell by controlling synthesis of the proteins that carry out so many functions in cells. The two kinds of nucleic acids found in cells are **DNA** (deoxyribonucleic acid) and **RNA** (ribonucleic acid). Nucleic acids are so named because they were first discovered in the nucleus of cells.

The structures of nucleic acids Nucleic acids are polymers made from building blocks that look complex at first glance. Understanding the structure of these building blocks, called **nucleotides**, is easier if you break them down into three parts, as shown in **Figure 19.19**.

Nucleotides can contain either of two similar sugars. RNA nucleotides contain a five-carbon sugar called ribose, whereas DNA nucleotides contain deoxyribose, which has the same general structure as ribose but with a single hydrogen atom in place of one of the hydroxyl groups. **Figure 19.20** shows the structures of ribose and deoxyribose. In both DNA and RNA nucleotides, the other two parts of the nucleotide—the phosphate group and the nitrogen-containing base—attach to the simple sugar.

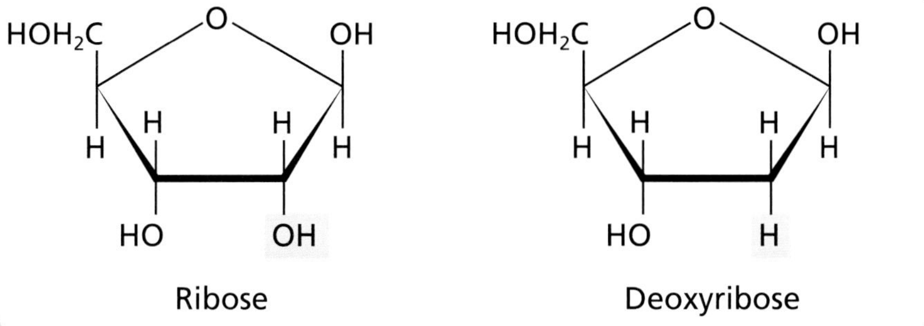

FOLDABLES Incorporate information from this section into your Foldable.

Figure 19.20 The five-carbon sugar ribose is the basis for RNA nucleotides, while the five-carbon sugar deoxyribose is the basis for DNA nucleotides. As their names suggest, the difference between the two is the omission of one atom of oxygen from deoxyribose.

Section 19.1 • Molecules of Life

■ **Figure 19.21** Nucleic acids are linear chains of alternating sugars and phosphates. Attached to every sugar is a nitrogen base. Because the nucleotides are offset, the chains resemble steps in a staircase.

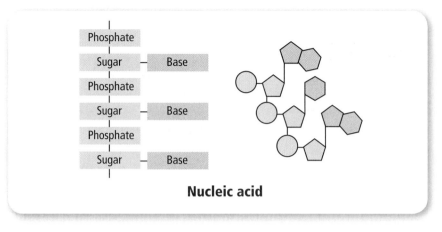

As shown in **Figure 19.21,** a nucleic acid polymer is made of chains in which the sugar of one nucleotide is linked to the phosphate of another. The chain of alternating sugars and phosphates is often called the backbone of the polymer. Five different nitrogen-containing bases are found in nucleotides. The names of these bases are often abbreviated by a single letter—A for adenine, C for cytosine, G for guanine, T for thymine, and U for uracil. DNA contains A, C, G, and T, but never U. In RNA, U replaces T. **Figure 19.22** shows the bases found in the nucleotides that make up DNA and RNA.

DNA In the three-dimensional structure of DNA, two sugar-phosphate backbone chains are held together by hydrogen bonds that form between the bases. Each base has a specific shape that allows it to bond to only one other base. In DNA, adenine bonds to thymine, and cytosine bonds to guanine. When the bases attached to the two DNA backbones bond together, the DNA forms a structure similar to a ladder. The DNA ladder twists into a spiral structure known as a double helix, as **Figure 19.22** shows.

The specific sequence of the bases in an organism's DNA forms its genetic code. This master code controls all the characteristics of the organism because it contains the instructions for the structure of every protein made by that organism. The code is passed from one generation to the next because offspring receive copies of their parents' DNA.

■ **Figure 19.22** Nucleotides are made of a phosphate, sugar, and a base. There are five different bases found in nucleotide subunits that make up DNA and RNA.
The structure of DNA is a double helix that resembles a twisted zipper. The two sugar-phosphate backbones form the outsides of the zipper.

Contrast What is a structural difference between guanine and cytosine?

Concepts In Motion

Interactive Figure To see an animation of the structure of DNA, visit glencoe.com.

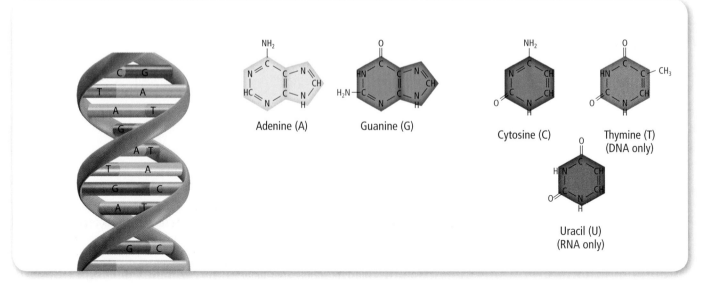

MiniLab 19.1

Extract DNA

How do you extract DNA from wheat germ cells? You have learned that DNA is a two-stranded molecule that consists of many polymerized nucleotides. DNA acts as the master blueprint for the cell's activity because it contains the coded instructions for every protein made by the cell. In addition, DNA enables a cell to pass on these instructions to the next generation because it is duplicated before the cell divides. Newly formed cells receive exact copies of the parent cell's DNA.

Procedure
1. Read and complete the lab safety form.
2. Use a clean **mortar and pestle** to grind approximately 5 g of **wheat germ** in 50 mL of **cell lysis solution** for one minute. Your teacher will supply the cell lysis solution. It contains chemicals that break open the wheat cells and remove unwanted cellular material.
3. Pour the mixture through a piece of **cheesecloth** or a **kitchen strainer** into a **250-mL or 400-mL beaker.** Discard the solids.
4. Add 100 mL of 91 percent **isopropanol** to the liquid, and stir briefly.
5. Carefully wind the strands of DNA onto a **glass rod.**
6. Spread a thin piece of the DNA on a **microscope slide,** and add two drops of **methylene blue dye.** Examine the DNA under a **microscope.**

Analysis
1. **Infer** Does the large amount of DNA that you extracted suggest that it is loosely arranged or compacted inside the cells?
2. **Explain** Does the physical appearance of the DNA enable you to describe its physical structure?

RNA Like DNA, RNA is also a polymer of nucleotides, but with some important differences. The sugars found in the two nucleic acids are different and the base uracil replaces the thymine found in DNA. Uracil binds to adenine in the same way that thymine does. The three-dimensional structure of RNA is also different. RNA has only a single chain of nucleotides, which is twisted into a helix. RNA carries the genetic information from DNA to the site of protein synthesis, where it directs the synthesis of the proteins. RNA is synthesized from a DNA template, as shown in **Figure 19.23.**

Vitamins

Cells require one more major group of organic molecules to carry out their functions. **Vitamins** are organic molecules that are required in small amounts in the diet.

SUPPLEMENTAL PRACTICE

For more practice with biomolecules, see Supplemental Practice, page 839.

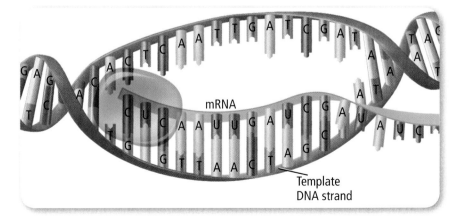

■ **Figure 19.23** One of the most central principles in biology is that DNA codes for RNA, and RNA then guides the synthesis of proteins. The process of making RNA from DNA, shown here, is called transcription.

■ **Figure 19.24** Vitamin A is needed to maintain healthy eyes, skin, and mucous membranes. It is stored in the body's fat cells, especially in the liver. Polar bears store huge quantities of vitamin A in their livers. In the nineteenth century, Arctic explorers died from vitamin A toxicity after eating large amounts of polar bear liver.

VOCABULARY

WORD ORIGIN

Vitamin
comes from the Latin word *vita*, which means *life*

Classes of vitamins Vitamins fall into two major classes: the water-soluble vitamins that dissolve in water and the fat-soluble vitamins that dissolve in nonpolar organic solvents. Vitamin D is a fat-soluble vitamin. When added to milk, it dissolves in the milk's fats.

Unlike the functions of proteins, carbohydrates, and lipids, vitamins are not used directly for energy or as building blocks for making structural materials in the body. Many serve as coenzymes in cell reactions. A **coenzyme** is an organic molecule that assists enzymes in catalyzing reactions. Vitamin C is a coenzyme for the reaction that modifies the protein collagen to make its structure stable enough to hold your tissues together.

Deficiencies of vitamins result in specific diseases. Recall the devastating effects of scurvy among British sailors caused by a deficiency of vitamin C. Too much of a fat-soluble vitamin also can result in disease because, unlike water-soluble vitamins, excess fat-soluble vitamins cannot be dissolved in urine and excreted. Fat-soluble vitamins include vitamins A, D, and E. **Figure 19.24** shows polar bears that store large amounts of vitamin A.

Section 19.1 Assessment

Section Summary

- There are four major classes of biomolecules.
- Many proteins function as enzymes that speed up reactions.
- Simple sugar units combine to form carbohydrate polymers.
- Lipids are insoluble in water and soluble in nonpolar solvents.
- Nucleic acids are polymers of units called nucleotides.

1. **MAIN Idea** **Summarize** the names, structures, and functions of the four major classes of biological macromolecules.
2. **Describe** the general structure of a triglyceride.
3. **Contrast** How do the structures of DNA and RNA differ?
4. **Apply** Why is it possible to form two different dipeptides from two given amino acids? Using the structural formulas for glycine and alanine in this section, draw the two possible dipeptides.
5. **Explain** Nutritionists believe that polyunsaturated fats in the diet are more healthful than saturated fats. What are polyunsaturated fats, and what is their source in the human diet?

Section 19.2

Objectives
- **Distinguish** between the reactions that cells use in the presence and in the absence of oxygen to extract energy from fuel molecules.
- **Explain** how a small number of biochemical building blocks can be used to make the extraordinary variety of molecules needed to perform life's chemical functions.

Review Vocabulary
substrate: the name given to a reactant in an enzyme-catalyzed reaction

New Vocabulary
metabolism
hormone
respiration
aerobic
ATP
electron transport chain
anaerobic
fermentation

Reactions of Life

MAIN Idea Respiration is a metabolic process in which energy is obtained from biological molecules.

Real-World Reading Link Bananas don't taste sweet until they ripen. As bananas ripen, starch breaks down into sugar. The reaction that breaks down starch is just one of many reactions that are constantly occurring in living cells.

Metabolism

What happens to the sugars and starch in a banana after you eat it? The sugars are small enough to be transported into the cells lining your digestive system. From there, they are moved into your bloodstream and sent to other cells. The starch molecules are too large to be transported into cells. As soon as you bite into the fruit, enzymes in your saliva begin to break down the starch into glucose. In the same way, the amazing variety of large and complex molecules in foods are broken down into smaller units before they can be absorbed into cells.

Enzymes catalyze the reactions in which proteins, carbohydrates, and lipids are broken down; this process is called digestion. Only small building blocks are able to enter cells to be used in the many reactions that are involved in metabolism. **Metabolism** is the sum of all the chemical reactions necessary for the life of an organism.

Chemical energy These numerous intertwined cellular reactions transform the chemical energy stored in the bonds of nutrients into other forms of energy and synthesize the biomolecules needed to provide structure and carry out the functions of living things. **Figure 19.25** summarizes the metabolic processes of breaking down and building up large biological molecules. What is the source of the chemical energy that is stored in nutrient molecules? This energy was converted from light energy by plants during photosynthesis. Ultimately, all the energy we need to live comes from sunlight. You will learn about photosynthesis and energy conversion in Chapter 20.

■ **Figure 19.25** A large number of different metabolic reactions take place in living cells. Some involve breaking down nutrients to extract energy, others involve using energy to build large biological molecules.
Describe Choose one food that you ate recently, and describe how it was metabolized.

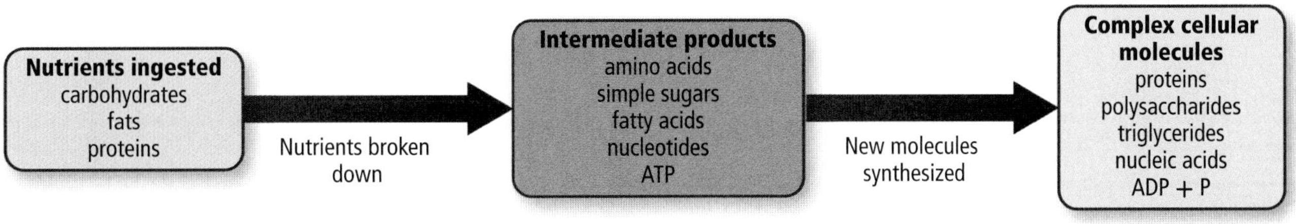

Nutrients ingested
carbohydrates
fats
proteins

→ Nutrients broken down →

Intermediate products
amino acids
simple sugars
fatty acids
nucleotides
ATP

→ New molecules synthesized →

Complex cellular molecules
proteins
polysaccharides
triglycerides
nucleic acids
ADP + P

SUPPLEMENTAL PRACTICE

For more practice with metabolism, see Supplemental Practice, page 840.

Controlling the reactions Cells perform only the reactions they need at any given time, which allows them to conserve both energy and material. How do cells switch the reactions off and on? They use an intricate series of control processes. Many of these can be triggered by signal molecules called **hormones** that are made in specific organs of the body and travel through the bloodstream to communicate with cells in other locations. Insulin is one example of a hormone. After a meal, insulin is released by the pancreas to signal to cells that glucose is available. One of the metabolic processes that insulin triggers in cells is the energy-releasing process called respiration.

Respiration

When cells need energy, they oxidize fuels such as carbohydrates and fats. This results in the formation of carbon dioxide and water. Energy is released in this reaction.

Oxidizing carbohydrates Most of the energy used by cells comes from the oxidation of carbohydrates. In oxidation, energy is needed initially to break bonds. In an automobile engine, hydrocarbons undergo combustion after the spark plugs heat the gasoline-oxygen mixture to the point where the molecules can react. Only then can the explosive reaction occur that drives the pistons in the engine.

How can oxidation of carbohydrates take place at normal body temperatures? The answer can be found in the catalytic power of enzymes. The complex series of enzyme-catalyzed reactions used to extract the chemical energy from glucose is called **respiration**. Respiration is an **aerobic** process, which means that it takes place only in the presence of oxygen. The athlete shown in **Figure 19.26** must breathe deeply to supply her body with oxygen.

To harness the energy from glucose, cells must break the bonds in which the energy is stored. The net exothermic reaction that takes place when the bonds in glucose are broken is similar to the combustion of hydrocarbons.

$$C_6H_{12}O_6 + 6O_2 \rightarrow 6CO_2 + 6H_2O + \text{energy}$$

VOCABULARY
WORD ORIGIN
Aerobic:
comes from the Greek word *aeros*, meaning *air*, and the Greek word *bios*, meaning *life*

Chemistry Online
Personal Tutor For an online tutorial on respiration, visit glencoe.com.

■ **Figure 19.26** Although many people might describe respiration as breathing, a biochemist's definition of respiration involves what happens to molecules in the cells when oxygen is available as a result of breathing. Glucose reacts with oxygen to form water and carbon dioxide, and energy is released.

■ **Figure 19.27** Burning a marshmallow releases the energy stored in the sugar, but not in a way that is useful to the cells of the body. Cellular respiration releases the energy in food in such a way that it is useful for sustaining life.

ATP and energy storage When gasoline burns in the engine of a car, large amounts of energy are released in a single explosive reaction. As illustrated by **Figure 19.27,** if the energy from the oxidation of sugars were to be released in the same way, the cells would not be able to use it all. Metabolic reactions require small amounts of energy. How can the energy in food be "packaged" so the cell can use it as needed?

When nutrients are broken down, the energy is transferred from the broken bonds to energy-storage molecules called adenosine diphosphate, or ADP for short. The structure of ADP is identical to one of the building-block nucleotides of nucleic acids, except that ADP has two phosphate groups attached to the ribose molecule, as shown in **Figure 19.28.**

Cells store energy by bonding a third phosphate group to ADP to form adenosine triphosphate—**ATP**. When the cell needs energy, the phosphate-phosphate bond in ATP is broken, producing ADP and a phosphate group and releasing stored energy. In cells, the conversion of ADP to ATP and vice versa occurs over and over as metabolic reactions release and use energy. Many of these reactions take place during respiration.

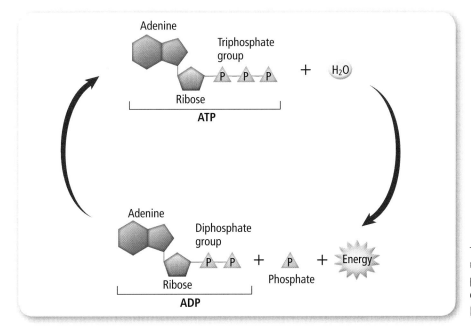

■ **Figure 19.28** The breakdown of ATP provides energy for cellular processes in living organisms.

Explain *where the energy is stored in ATP.*

Section 19.2 • Reactions of Life

Glycolysis—the first stage of respiration The process of respiration consists of many reactions that can be grouped into three stages. In the first stage of respiration, a series of nine reactions breaks down glucose into a pair of three-carbon compounds. Most of the energy from the broken bonds is transferred to ATP. A net yield of two molecules of ATP per molecule of glucose is produced during this stage. This series of reactions is called glycolysis, which means "splitting glucose."

Why do cells need ATP? Energy in the bonds of this molecule can be accessed much more quickly and in more manageable amounts than energy in triglycerides, starch, or glycogen.

The Krebs cycle—the second stage In the second stage of respiration, carbon dioxide is produced in a series of reactions. These reactions are called the Krebs cycle or the tricarboxylic acid (TCA) cycle because some of the molecules formed in the intermediate reactions have three carboxyl groups.

The electron transport chain—the third stage Most of the energy from glucose is released in the final stage of respiration, referred to as an **electron transport chain**. The electrons move step-by-step to lower energy levels, allowing for the controlled release of chemical potential energy. This is similar to the release of potential energy that happens at every bounce if a ball is dropped down a staircase. **Figure 19.29** shows another example of potential energy being released in a controlled fashion.

Energy is stored as ATP during these steps, as phosphate groups are transferred to ADP. A final redox reaction transfers the electrons to the oxygen you breathe, forming water. You can see that the electron transport chain can't operate without oxygen because oxygen is the final acceptor of the electrons that came from glucose. If oxygen is not present, the entire transport chain stops because the electrons have nowhere to go. **Figure 19.30** summarizes the three stages of respiration.

■ **Figure 19.29** If this dam were to break, all of the potential energy stored by holding water behind the walls of the dam would be released very rapidly. However, the functioning dam allows for the controlled release of the water's potential energy.

FIGURE 19.30

Respiration

In glycolysis, which is the first stage of respiration, a six-carbon glucose molecule is split into a pair of three-carbon molecules. Hydrogen ions and electrons also are produced in glycolysis. These combine with electron carrier ions called nicotinamide adenine dinucleotide (NAD^+) to form NADH. NADH is a coenzyme that is made from vitamin B_4, which is also called niacin or nicotinic acid. ATP and NADH serve as temporary storage sites for energy and electrons, respectively, during respiration. Two molecules of ATP are used, and four molecules are produced.

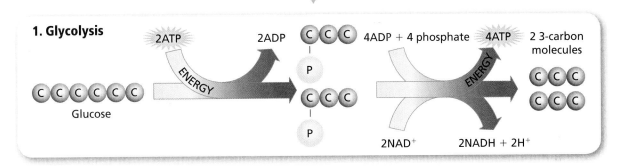

1. Glycolysis

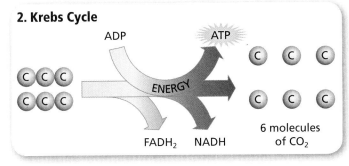

2. Krebs Cycle

In the second stage, the pair of three-carbon molecules that was made from glucose is converted into six molecules of carbon dioxide. More ATP and NADH are produced, along with another coenzyme molecule called flavin adenine dinucleotide ($FADH_2$).

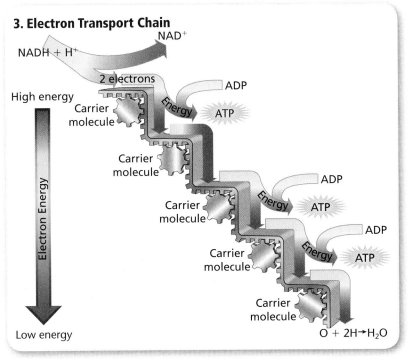

3. Electron Transport Chain

In the third stage, NADH and $FADH_2$ bring the electrons and hydrogen atoms from glucose to a series of carrier molecules called the electron transport chain. A chainlike series of redox reactions takes place. In the final reaction, the electrons and hydrogen are transferred to oxygen to produce water. At the top of the chain, the electrons have high energy. As they pass down the chain, the energy given off is captured in molecules of ATP. The energy available from the breakdown of each glucose molecule can be used to make as many as 36 ATP molecules, a net of 2 during glycolysis, and 2 during the Kreb's Cycle, and 32 from the electron transport chain.

Biology Connection

Function of Hemoglobin

In 1864, a British physiological chemist, Felix Hoppe-Seyler, discovered that hemoglobin, the pigment of the blood, binds and releases oxygen. As it does so, it changes color from red to bluish-red. Hemoglobin binds oxygen to its iron atoms in the alveoli of the lungs and transports it to the tissues of the body. The shape of the hemoglobin molecule changes as it moves from the lungs through the blood vessels to the capillaries near the cells. That change in shape is what causes it to drop off the oxygen where it is needed.

Structure and function of hemoglobin The physiological properties of hemoglobin can be explained by studying the structure of the heme group in **Figure 1**. The hemoglobin molecule (Hb) consists of two copies each of two slightly different polypeptide chains—alpha and beta. Each of the four polypeptide chains has a heme group near its center. Notice that a heme group consists of an iron atom associated with four nitrogen atoms, each of which is part of a ring structure. A heme group is the deep red iron component of hemoglobin that carries oxygen.

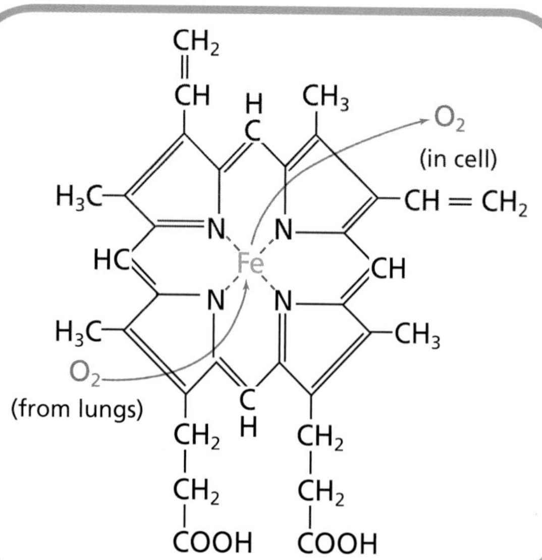

Figure 1 Heme group and oxygen

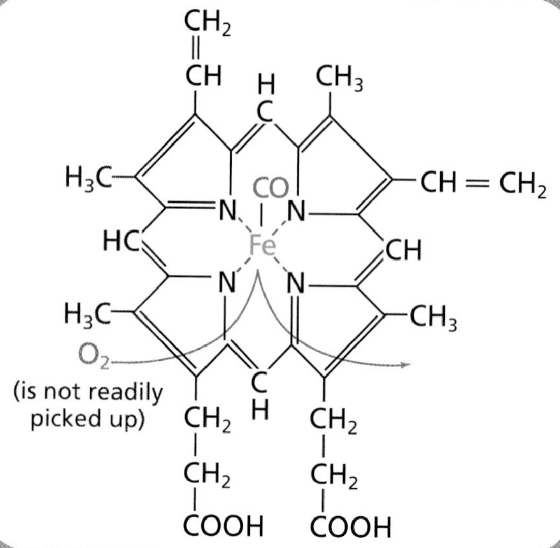

Figure 2 Heme group and carbon monoxide

Each iron atom in the hemoglobin molecule can bind with one molecule of diatomic oxygen, which contains two oxygen atoms. Thus, every hemoglobin molecule can hold eight oxygen atoms when saturated.

Carbon monoxide poisoning Because it is similar in size to oxygen, the poisonous gas carbon monoxide combines with hemoglobin in almost the same way that oxygen does (shown in **Figure 2**). Unfortunately, carbon monoxide (CO) is attracted to hemoglobin more than 200 times more strongly than oxygen. Both oxygen and carbon monoxide combine with hemoglobin at the same point on the molecule. Thus, they can't both be bound to hemoglobin at the same time. When the concentration of carbon monoxide rises to about 0.2 percent, the quantity of hemoglobin free to transport oxygen is too small to support life.

Connection to Chemistry

1. **Hypothesize** Anemia occurs when the number of red blood cells falls below normal. The patient feels weak and tired. Hypothesize the cause of these symptoms.
2. **Analyze** When a person is in danger of dying from carbon monoxide poisoning, pure oxygen is administered at six times the normal alveolar oxygen pressure. How might this help?

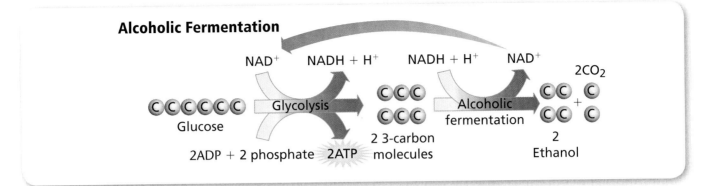

Fermentation

Metabolic processes that occur in the absence of oxygen are called **anaerobic** processes. Cells can generate energy from glucose anaerobically by the process of **fermentation**. Because the first stage of respiration can take place without oxygen, glucose is broken down through the reactions of glycolysis into the two smaller, three-carbon units before fermentation reactions proceed. There are two major types of fermentation. In one, ethanol and carbon dioxide are produced. In the other, the product is lactic acid.

Alcoholic fermentation Alcoholic fermentation occurs in some bacteria and in yeast cells. As shown in **Figure 19.31**, the process yields much less energy than aerobic respiration; much of the energy from glucose remains in the bonds of the ethanol. Although it is relatively inefficient, fermentation provides enough energy for cells to carry on their basic functions. Fermentation by yeast cells is used in the brewing, wine-making, and baking industries. The carbon dioxide produced during alcoholic fermentation causes bread dough to rise, making the dough light and spongy. The ethanol that is also produced evaporates during baking.

Lactic acid fermentation Rapidly contracting muscle cells sometimes use up oxygen faster than it can be supplied by the blood. When the cells run out of oxygen, a type of fermentation called lactic acid fermentation begins. **Figure 19.32** shows the process of lactic acid fermentation. If the anaerobic state continues for too long, lactic acid can accumulate in the muscles, causing stiffness.

■ **Figure 19.31** In alcoholic fermentation, each three-carbon molecule that is produced during glycolysis is split to form a two-carbon molecule—the alcohol ethanol—and a one-carbon molecule, carbon dioxide. Two molecules of ATP are produced during glycolysis.

■ **Figure 19.32** In an anaerobic state, glucose is converted into a pair of three-carbon molecules called lactic acid through glycolysis, followed by lactic acid fermentation. Two molecules of ATP are produced during glycolysis. Lactic acid fermentation occurs in some bacteria, in fungi, and in most animals including humans. The dairy industry uses bacteria to make yogurt and buttermilk by this process.

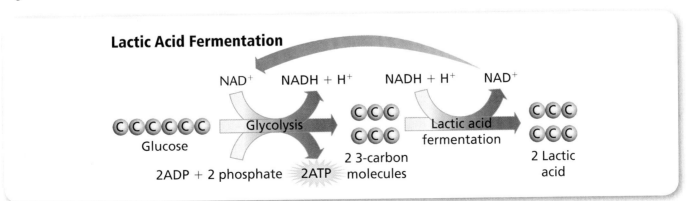

MiniLab 19.2

Fermentation by Yeast

Will yeast break down a disaccharide or a polysaccharide more quickly? You may have used yeast to make bread dough or pizza dough rise. Dry yeast is the dormant form of a single-celled fungus that, when given favorable living conditions and food in the form of a carbohydrate, begins to break down the carbohydrate. One of the products of respiration is carbon dioxide. In this MiniLab, you will mix yeast with a disaccharide, sucrose, and with a polysaccharide, the starch in flour, and compare the rates at which carbon dioxide is produced.

Procedure

1. Read and complete the lab safety form.
2. Obtain two sandwich-size, **self-sealing plastic bags,** and label them as sucrose or flour.
3. Fill a **water trough** or **plastic dishpan** about 2/3 full of **hot water,** and adjust the temperature to between 40°C and 50°C.
4. Put one package of **dry yeast** and one tablespoon of the appropriate **carbohydrate** into each of the bags. Mix well.
5. Working quickly, measure and add 50 mL (1/4 cup) of the warm water from the trough to each of the bags, and thoroughly mix the contents. Remove all the air you can and seal the bags. Start timing when both bags are sealed.
6. Put both bags into the trough of warm water. Record the time required for sufficient carbon dioxide to fill each bag completely. If either of the bags is not completely filled in 30 minutes, estimate the fraction of the bag that is filled.

Analysis

1. **Rank** Based upon your data, rank the rates at which yeast breaks down each type of carbohydrate.
2. **Explain** Why might the carbohydrates be broken down at different rates?

Connecting Ideas

Respiration involves the oxidation of fuel molecules. Animals must take in fuel molecules in the form of food, but how do plants get them? In Chapter 20, you will study how plants trap energy from sunlight and use it make food molecules. Energy in the form of heat and light is often involved in chemical processes. It is important to be able to understand the energy changes that accompany chemical reactions. Gaining those skills is the focus of the next chapter.

Section 19.2 Assessment

Section Summary

- Metabolism involves the capture and release of energy through chemical reactions.
- The process of cellular respiration results in the breakdown of glucose.
- In the absence of oxygen, cells can sustain respiration by fermentation.

6. **MAIN Idea Explain** how aerobic respiration provides cells with more energy per molecule of glucose than does fermentation.
7. **Contrast** How do alcoholic fermentation and lactic acid fermentation differ?
8. **Determine** How many moles of CO_2 are produced from one mole of glucose during respiration?
9. **Relate** Marathon runners go through a crisis called hitting the wall when they use up all their stored glycogen and start using stored lipids as fuel. What would be the advantage to carbohydrate loading, which means eating lots of complex carbohydrates the day before running a marathon?
10. **Explain** what aerobic exercise has in common with aerobic processes in cells.

CHAPTER 19 Study Guide

Download quizzes, key terms, and flash cards from glencoe.com.

BIG Idea Organisms are made of carbon-based molecules, many of which are involved in chemical reactions necessary to sustain life.

Section 19.1 Molecules of Life

MAIN Idea Biological molecules—proteins, carbohydrates, lipids, and nucleic acids—interact to carry out activities necessary to living cells.

Vocabulary
- active site (p. 672)
- amino acid (p. 669)
- biochemistry (p. 666)
- carbohydrate (p. 673)
- coenzyme (p. 688)
- denaturation (p. 671)
- DNA (p. 685)
- fatty acid (p. 681)
- lipid (p. 681)
- nucleic acid (p. 685)
- nucleotide (p. 685)
- protein (p. 668)
- RNA (p. 685)
- steroid (p. 682)
- substrate (p. 672)
- vitamin (p. 687)

Key Concepts
- There are four major classes of biomolecules.
- Many proteins function as enzymes that speed up reactions.
- Simple sugar units combine to form carbohydrate polymers.
- Lipids are insoluble in water and soluble in nonpolar solvents.
- Nucleic acids are polymers of repeating units called nucleotides.

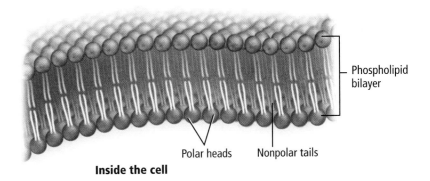

Section 19.2 Reactions of Life

MAIN Idea Respiration is a metabolic process in which energy is obtained from biological molecules.

Vocabulary
- aerobic (p. 690)
- anaerobic (p. 695)
- ATP (p. 691)
- electron transport chain (p. 692)
- fermentation (p. 695)
- hormone (p. 690)
- metabolism (p. 689)
- respiration (p. 690)

Key Concepts
- Metabolism involves the capture and release of energy through chemical reactions.
- The process of cellular respiration results in the breakdown of glucose.
- In the absence of oxygen, cells can sustain respiration by fermentation.

Chapter 19 Assessment

Understand Concepts

11. Name the two functional groups in an amino acid that become linked when peptide bonds form.

12. Describe the three-dimensional structure of the DNA double helix.

13. Why do unsaturated fats tend to be liquids rather than solids at room temperature?

14. Distinguish between a disaccharide and a polysaccharide.

15. What kind of functional group is formed when fatty acids combine with glycerol?

16. Compare the structures of a triglyceride and a phospholipid.

17. Describe the process by which a disaccharide forms from two simple sugar molecules.

18. What is the function of an enzyme?

19. What three structural units make up a nucleotide?

20. Compare the functions of DNA and RNA.

21. Where are hydrogen bonds found in the structure of DNA? What function(s) do they have?

22. Lipids are defined by their solubility properties. Why are lipids usually not soluble in water?

Apply Concepts

23. Enzymes in the mouth, stomach, and intestine break down protein and carbohydrate polymers into amino acids and simple sugar units, respectively. Humans often produce smaller quantities of these enzymes as we age. How does this affect the nutrients we get from food?

24. Plant seeds contain large amounts of starch. What do you think is the function of this starch?

25. What is a coenzyme and how does it function? Name a coenzyme that takes part in respiration.

26. List the types of bonds involved in holding a protein in its folded, three-dimensional shape.

27. Examine the structures of the molecules shown and decide whether each is a carbohydrate, a lipid, or an amino acid.

a)

b)

c)

28. Digestion of carbohydrates and proteins cannot take place in the absence of water. Explain.

29. Explain how changes in temperature and pH of a cell affect protein structure.

30. Which amino acid would you expect to be the most soluble in water: alanine, cysteine, or phenylalanine? Explain.

Everyday Chemistry

31. Why are protein molecules ideally suited for forming pairs of hydrogen bonds?

Chapter 19 Assessment

Everyday Chemistry

32. How is the structure of a fat molecule different from the structure of the chemically altered fat olestra?

33. Vitamins are classified by their solubility properties into two groups: water-soluble and fat-soluble. Use the structures of the vitamins shown to predict which group each falls into.

Vitamin A

Vitamin C

Vitamin D

Think Critically

Make Inferences

34. The odor of burning hair is due to large amounts of the amino acid cysteine in keratin, the major protein in hair. What element causes this smell?

Form a Hypothesis

35. When fresh pineapple is added to a gelatin solution, the gelatin will fail to gel, or solidify. When cooked or canned pineapple is used, the gelatin will gel. Explain.

Interpret Data

36. ChemLab Table 19.3 contains data obtained by measuring the activity of the enzyme salivary amylase in solutions buffered at different pH levels.
  a) At which pH is the enzyme most active?
  b) Can you identify any trend in activity as pH moves lower than the optimum pH? Higher?
  c) Why does pH affect enzyme activity?

Table 19.3	Enzyme Activity
pH	Time for Color to Disappear
4	10 min
5	8 min
6	1 min
7	20 s
8	40 s
9	4 min

Design an Experiment

37. MiniLab 19.1 Design an experiment that would show whether ethanol or isopropanol is more effective at precipitating DNA.

Relate Cause and Effect

38. MiniLab 19.2 Why was the experiment conducted in a 40–50°C water bath?

Cumulative Review

39. Draw Lewis dot diagrams for the following molecules. (*Chapter 3*)
  a) water
  b) carbon dioxide
  c) methane
  d) ethanol

 Chapter Test glencoe.com

Chapter 19 Assessment

40. Compare the boiling points of water, carbon dioxide, and ammonia, and explain their differences. (*Chapter 9*)

41. Explain what a buffer is and why buffers are found in body fluids. (*Chapter 15*)

42. Name the hydrocarbons shown. (*Chapter 18*)
    a) $CH_3CH_2CH_2CH_3$

    b)
    $$\begin{array}{c} CH_3 \\ | \\ CH_3CH_2CHCH_2CH \end{array}$$

    c)
    $$\begin{array}{c} CH_2 \\ / \quad \backslash \\ CH_2 \quad CH_2 \\ | \quad\quad | \\ CH_2 - CH_2 \end{array}$$

    d) $CH_3(CH_2)_5CH_3$

Skill Review

43. **Make and Use Graphs** Every enzyme has a pH at which it is most active. The following graph shows activity ranges for two digestive enzymes. Pepsin is found in the stomach, and trypsin is found in the intestine. Determine the optimum pH for each enzyme. Describe the pH where each enzyme is found. What would happen in the stomach if pepsin had an optimum pH of 8?

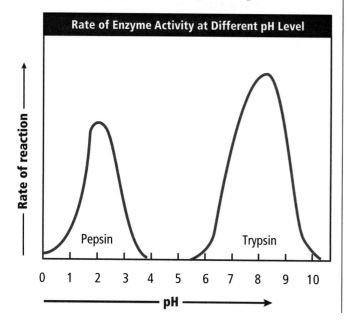

WRITING in Chemistry

44. Read *In Search of the Double Helix* by John Gribbin, and write a report discussing whether Watson and Crick or Rosalind Franklin really made the discovery of the DNA structure. Include a discussion of how serendipity played a part in the discovery of the structure.

45. Investigate and write a report about claims that ingesting large amounts of certain vitamins—especially C, A, and E—can help prevent cancer.

Problem Solving

46. Examine the structure of oxygenated hemoglobin on page 694 of this chapter and list the functional groups in this organic molecule.

47. One serving of cereal provides 26 g of carbohydrates, which is 9 percent of the recommended daily value for someone consuming 2000 Calories per day. Calculate how many total grams of carbohydrate that person should eat per day.

48. Write a balanced equation for the hydrogenation of oleic acid. How many moles of H_2 are needed to hydrogenate one mole of oleic acid? What is the name of the product?

49. Using what you have learned about the change in melting points of alkanes as carbon chain length increases, predict how an increase in the number of —CH_2— groups will affect the melting point of a saturated fatty acid. Give a short oral report to your class explaining your reasoning.

50. Why aren't the electrons from glucose transferred directly to oxygen during respiration? Why is the electron-transport chain needed? Write a short report in which you predict what would happen to a cell if the electrons were transferred directly to oxygen.

Cumulative
Standardized Test Practice

1. Electronegativity is
   a) a measure of an atom's ability in a bond to attract electrons.
   b) a measure of an atom's ability to form negative charges.
   c) a measure of an atom's ability to chemically combine with other elements.
   d) a measure of an atom's ability to physically combine with other elements.

2. Which is linked to high levels of cholesterol in the blood?
   a) severe bone malformations
   b) heart disease
   c) memory loss
   d) skin cancer

3. What volume of a $0.125M$ $NiCl_2$ solution contains 3.25 g $NiCl_2$?
   a) 406 mL
   b) 201 mL
   c) 38.5 mL
   d) 26.0 mL

4. Which is a strong acid?
   a) NaOH
   b) $HC_2H_3O_2$
   c) H_3O^+
   d) HCl

5. A Brønsted-Lowry acid is defined as a(n)
   a) acid that donates a hydrogen ion during a chemical reaction.
   b) acid that accepts a hydrogen ion during a chemical reaction.
   c) strong acid that donates the maximum number of hydrogen ions.
   d) weak acid that donates a limited number of hydrogen ions.

6. For the reaction $X + Y \rightarrow XY$, the element that will be reduced is the one that is
   a) more reactive.
   b) more massive.
   c) more electronegative.
   d) more radioactive.

7. The part of an electrochemical cell that carries electrons to a reacting ion is a(n)
   a) cathode.
   b) anode.
   c) electrode.
   d) electrolyte.

8. How many hydrogen atoms will an alkane with 11 carbon atoms have?
   a) 11
   b) 18
   c) 22
   d) 24

9. Which molecule releases energy during respiration?
   a) adenosine triphosphate
   b) adenosine diphosphate
   c) adenosine phosphate
   d) tricarboxylic acid

10. 18.7 psi equals
    a) 0.360 kPa
    b) 2.70 kPa
    c) 129 kPa
    d) 977 kPa

NEED EXTRA HELP?

If You Missed Question...	1	2	3	4	5	6	7	8	9	10
Review Section...	9.1	19.1	13.2	14.2	15.1	16.1	17.1	18.1	19.2	11.1

CHAPTER 20
Chemical Reactions and Energy

BIG Idea Energy is released when bonds form, but must be added to break bonds.

20.1 Energy Changes in Chemical Reactions
MAIN Idea Exothermic reactions release energy, and endothermic reactions absorb energy.

20.2 Measuring and Using Energy Changes
MAIN Idea Energy stored in chemical bonds can be converted to other forms and used to meet the needs of individuals and of societies.

20.3 Photosynthesis
MAIN Idea The process of photosynthesis converts energy from the Sun into chemical energy stored in the bonds of biological molecules.

ChemFacts

- The production of corn in the United States is twice that of any other crop grown.
- When burned, 1524 kg of shelled corn produces an equivalent amount of energy to one ton of hard coal.
- In one day, two and a half acres of corn produce enough oxygen through photosynthesis to meet the respiratory needs of approximately 325 people.

Start-Up Activities

LAUNCH Lab

Increasing the Rate of Reaction

Many chemical reactions occur so slowly that you don't even know they are happening. For example, hydrogen peroxide slowly decomposes upon sitting. Is it possible to alter the speed of a chemical reaction?

Materials
- hydrogen peroxide
- beaker or cup
- baker's yeast
- toothpicks

Procedure
1. Read and complete the lab safety form.
2. Create a "before and after" table to record your observations.
3. Pour about 10 mL of hydrogen peroxide into a small beaker or cup. Observe the hydrogen peroxide.
4. Add a pinch (roughly 1/8 tsp) of yeast to the hydrogen peroxide. Stir gently with a toothpick and observe the mixture again.

Analysis
1. **Identify** Into what two products does hydrogen peroxide decompose?
2. **Analyze** Why aren't bubbles produced in Step 3? What is the function of the yeast?

Inquiry Would substances other than yeast cause similar results? Develop a procedure to test your hypothesis.

Chemistry Online

Visit glencoe.com to:
- study the entire chapter online
- explore **Concepts In Motion**
- take Self-Check Quizzes
- use Personal Tutors
- access Web Links for more information, projects, and activities
- find the Try at Home Lab, Observing Entropy

FOLDABLES Study Organizer

Energy and Chemical Reactions Make the following Foldable to compare the changes that occur during chemical reactions.

▶ **STEP 1** Fold a sheet of paper in half. Fold in half again.

▶ **STEP 2** Unfold the sheet and cut along the fold line to make two tabs.

▶ **STEP 3** Label the tabs *Energy* and *Entropy*.

FOLDABLES Use this foldable with Section 20.1. As you read about chemical reactions and the forces that drive them, summarize information about energy and entropy and record it under the appropriate tabs.

Chapter 20 • Chemical Reactions and Energy

Section 20.1

Objectives

- **Compare and contrast** exothermic and endothermic chemical reactions.
- **Analyze** the energy diagrams for typical chemical reactions.
- **Illustrate** the meaning of entropy, and trace its role in various processes.

Review Vocabulary

spontaneous process: a physical or chemical change that once begun occurs without outside intervention

New Vocabulary

heat
law of conservation of energy
fossil fuel
entropy

Energy Changes in Chemical Reactions

MAIN Idea Exothermic reactions release energy, and endothermic reactions absorb energy.

Real-World Reading Link "Smile," says the photographer, as she pushes a button. The camera shutter opens and an electric current from a small lithium battery sparks across a gap in the flash unit. This spark ionizes xenon gas, creating a bright flash of light. The energy from the chemical reaction in the lithium battery has been successfully put to use.

Exothermic and Endothermic Reactions

Recall from Chapter 1 that chemical reactions can be exothermic or endothermic: an exothermic reaction releases heat, and an endothermic reaction absorbs heat. **Heat** is defined as the energy transferred from an object at high temperature to an object at lower temperature. **Figure 20.1** pictures an exothermic process as a reaction that's going downhill and an endothermic process as a reaction that's going uphill.

Exothermic reactions If you have ever started a campfire or built a fire in a fireplace, you know that burning wood is an example of an exothermic process. The reaction requires an input of energy from the burning match to get started. Once you have ignited the wood, the reaction generates enough heat to keep itself going. A net release of heat occurs, which is what makes the reaction exothermic.

■ **Figure 20.1** The exothermic reaction (left) gives off heat because the products are at a lower energy level than the reactants. The endothermic reaction (right) absorbs heat because the products are at a higher energy level than the reactants.

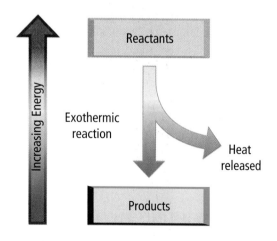 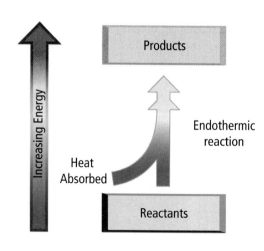

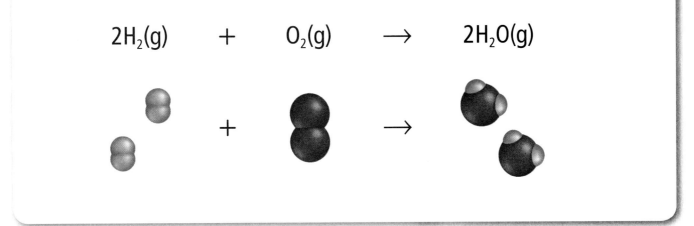

The reaction between hydrogen gas and oxygen gas to form water is another example of an exothermic reaction. Once a small amount of energy—often just a spark—is added to the mixture of gases, the reaction continues to completion, usually explosively. No additional input of energy from outside is needed to keep it going. Once energy has been supplied to break the covalent bonds in the first few molecules of hydrogen and oxygen, the atoms combine to form water and release enough energy to break the bonds in additional hydrogen and oxygen molecules. The reaction, shown in **Figure 20.2,** is exothermic because more energy is released in forming the bonds of water than is absorbed to break the bonds of hydrogen and oxygen gas molecules. The net energy is released as heat and light. In every case, energy must be supplied to break bonds, and energy is released as bonds form. In the case of an exothermic reaction, such as the formation of water from hydrogen and oxygen, more energy is released in forming the bonds of the products than is required to break the bonds of the reactants.

Figure 20.2 The reaction between hydrogen gas and oxygen gas is highly exothermic, requiring only a small spark to begin. The energy from this reaction has been used to power car and truck engines.

Infer For an endothermic reaction, how does the energy required to break the bonds of the reactants compare to the energy released by forming the bonds of the products?

Endothermic reactions Consider the reverse of the reaction just discussed. Just as water can be formed from hydrogen and oxygen, water can also be decomposed to reform hydrogen and oxygen. In the process of electrolysis, electrical energy is used to break the covalent bonds that unite the hydrogen atoms and the oxygen atoms in the water molecules. The hydrogen atoms pair up to form hydrogen molecules, and the oxygen atoms pair up to form oxygen molecules. The formation of the new bonds releases energy, but not as much as the amount required during the bond breaking. Additional energy must be added continuously during the electrolysis. The reaction absorbs energy and is, therefore, endothermic.

All endothermic reactions are characterized by a net absorption of energy. In the History Connection on page 56, you read about another example of an endothermic reaction: the decomposition of orange mercuric oxide into the elements mercury and oxygen. As long as heat is applied, the compound continues to decompose, but if the heat source is removed, the reaction stops. The net absorption of energy that is required is what makes the reaction endothermic.

HOW IT WORKS

Hot and Cold Packs

Instant hot and cold packs create aqueous solutions that form exothermically or endothermically and therefore release or absorb heat. A hot pack generates heat when a salt such as calcium chloride dissolves in water that is stored in the pack. The calcium chloride dissolves exothermically. A cold pack absorbs heat when a salt such as ammonium nitrate dissolves in water. The ammonium nitrate dissolves endothermically. In both cases, the salt and water are separated by a thin membrane. All you have to do is squeeze the pack to mix the components and you have instant heat or cold at your fingertips.

1 The outer casing is strong and flexible. It resists puncture and can be shaped to fit the area that you want to heat or cool.

2 Water is stored in an inner compartment separate from the solid salt.

3 The inner membrane breaks easily when you knead or squeeze the pack or strike it sharply.

4 Salt is stored in the outer compartment. When the inner membrane breaks, the salt and water mix. The salt dissolves in the water and releases (hot pack) or absorbs (cold pack) energy.

Outer casing
Water
Soluble salt
Membrane of water pack

Think Critically

1. Heat is defined as energy that flows from an object at higher temperature to an object at lower temperature. Diagram the flow of heat for a hot pack being used on a person's wrist. Then, diagram the flow of heat for a cold pack in use.

2. Another type of hand warmer contains fine iron powder and chemicals that cause the iron to rust. The rusting of iron lets the hand warmer maintain temperatures above 60°C for several hours. Explain how that is possible.

Heat and Chemical Reactions

The energy that is involved in exothermic and endothermic reactions is usually in the form of heat. As you have learned, heat is energy that is transferred from an object at high temperature to an object at lower temperature. Recall that energy is measured in joules; the symbol for joules is J. The symbol for a kilojoule, which is equal to 1000 J, is kJ.

Using symbols to show energy changes Energy changes are frequently included in the equation for a chemical reaction. The amount of heat absorbed or evolved during a reaction is a measure of the energy change that accompanies the reaction. When one mole (18.0 g) of liquid water is produced from hydrogen and oxygen gas, 286 kJ of energy are given off. This means that the energy of the uncombined hydrogen gas and oxygen gas is greater than the energy of the water. When one mole (18.0 g) of liquid water decomposes to form hydrogen gas and oxygen gas, 286 kJ of energy are absorbed. This also shows that the energy of the uncombined hydrogen and oxygen is greater than the energy of the water. The graphs in **Figure 20.3** illustrate this relationship.

Law of conservation of energy Scientists have observed that the energy released in the formation of a compound from its elements is always identical to the energy required to decompose that compound into its elements. This observation is an illustration of an important scientific principle known as the law of conservation of energy. The **law of conservation of energy** states that energy is neither created nor destroyed in a chemical change, but is simply changed from one form to another. In an exothermic reaction, the heat released comes from the change from reactants at higher energy to products at lower energy. In an endothermic reaction, the heat absorbed comes from the change from reactants at lower energy to products at higher energy.

◀ **FOLDABLES**
Incorporate information from this section into your Foldable.

VOCABULARY
WORD ORIGIN
Energy
comes from the Greek words *en*, meaning *in*, and *ergon*, meaning *work*

■ **Figure 20.3** As the graphs show, the energy produced when one mole of liquid water forms from hydrogen and oxygen gases is equal in magnitude to the energy absorbed when one mole of water decomposes to form hydrogen and oxygen gases.
Apply Which reaction is endothermic and which is exothermic?

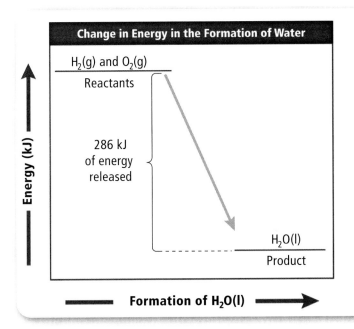
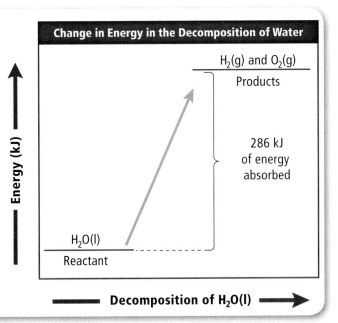

MiniLab 20.1

Heat In, Heat Out

Can you identify the energy change of a chemical reaction?

Procedure

1. Read and complete the lab safety form.
2. Pour approximately 40 mL of a **commercial chlorine bleach solution** into a **150-mL beaker.**
3. Put a **thermometer** into the solution and record the temperature. **WARNING:** *Be careful handling the bleach and avoid inhaling the fumes. Work in a well-ventilated room.*
4. Pour approximately 40 mL of a **0.5M sodium sulfite solution** into the beaker that contains the bleach, and stir gently for a few seconds.
5. Record the final temperature of the mixture.

Analysis

1. **Evaluate** Based on the temperature change, was the reaction that occurred exothermic or endothermic? Would the reverse reaction be exothermic or endothermic?
2. **Summarize** Write the balanced equation for the reaction. Identify the oxidizing agent and the reducing agent in the reaction.

Energy change and enthalpy The difference in energy between products and reactants in a chemical change is symbolized ΔH (delta H). The symbol Δ means a difference or change. The letter H doesn't actually represent energy; it represents a quantity called *enthalpy*. At constant pressure, the change in enthalpy of a reaction or system is the same as the heat gained or lost.

The energy absorbed or released in a reaction ($\Delta H_{reaction}$) is related to the energy of the products and the reactants by the following equation.

$$\Delta H_{reaction} = \Delta H_{products} - \Delta H_{reactants}$$

For exothermic reactions, ΔH is negative because the energy stored in the products is less than that in the reactants. For endothermic reactions, ΔH is positive because the energy of the products is greater than that of the reactants.

Writing ΔH in chemical equations The value of ΔH is often shown at the end of a chemical equation. For example, the exothermic formation of 2 mol of liquid water from hydrogen and oxygen gas would be written like this.

$$2H_2(g) + O_2(g) \rightarrow 2H_2O(l) \quad \Delta H = -572 \text{ kJ}$$

The equation for the endothermic decomposition of 2 mol of liquid water would be written like this.

$$2H_2O(l) \rightarrow 2H_2(g) + O_2(g) \quad \Delta H = +572 \text{ kJ}$$

The value 572 kJ in these equations is 2×286 kJ, which is the amount of energy released when 1 mol of liquid water forms. Note the use of the symbols (s), (l), and (g). When energy values are included with an equation for a reaction, it is especially important to show the states of reactants and products because the energy change in a reaction can depend greatly upon physical states.

Activation energy Fuels that are the remains of plants and other organisms that lived millions of years ago are called **fossil fuels.** Oil, natural gas, and coal are all examples of fossil fuels. Fossil fuels react with oxygen to produce carbon dioxide, water, and a great deal of energy in the form of heat. However, fossil fuels do not automatically burn on exposure to oxygen. Energy, usually in the form of heat or light, is needed to get the chemical reaction started. This input of energy is called activation energy. For example, the butane gas in a disposable lighter requires a spark to start the combustion of the gas. Once begun, the reaction proceeds spontaneously.

Activation energy is required in both exothermic and endothermic reactions. The fact that a fuel requires an input of energy—such as from a spark—to begin burning does not mean that the combustion reaction is endothermic. The reaction releases a net amount of heat, and so it is exothermic.

Activation energy and exothermic reactions Methane, the main component in natural gas, burns to yield carbon dioxide and water vapor. The equation for this reaction is as follows.

$$CH_4(g) + 2O_2(g) \rightarrow CO_2(g) + 2H_2O(g) \quad \Delta H = -802 \text{ kJ}$$

A graph of the energy change during the progress of this reaction is shown in **Figure 20.4.** The rise represents the activation energy, which is the energy difference between the reactants and the maximum energy stage in the reaction. The fall represents the energy liberated by the formation of new chemical compounds.

When one mole of methane burns, 802 kJ of heat is given off. That is, $\Delta H_{reaction} = -802$ kJ. The reaction is exothermic, as is shown by the negative sign of ΔH. The energy stored in the products is less than that stored in the reactants, so a net amount of energy is released. Some of the released energy provides the activation energy needed to keep the reaction going.

VOCABULARY
WORD ORIGIN
Combustion
comes from the Latin word *combustus*, meaning *burned*

■ **Figure 20.4** In order to occur, the combustion of methane, illustrated in the photo, requires an input of activation energy—in this case, provided by a match. Overall, the reaction releases 802 kJ of energy per mole of methane. Notice from the graph that the products are in a lower energy state than the reactants. The negative value of ΔH reflects this fact.

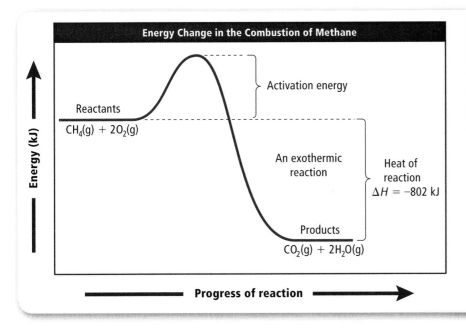

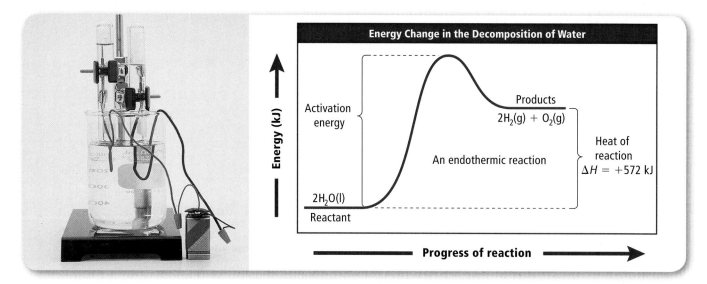

■ **Figure 20.5** The decomposition of water requires a continuous input of energy, such as electrical energy from a battery. Overall, the reaction absorbs 572 kJ of energy for every 2 mol of liquid water. The products are in a higher energy state than the reactant. The positive value of ΔH reflects this fact.

Activation energy and endothermic reactions Now, consider an example of an endothermic reaction—one that you have already read about—the decomposition of water. The equation for the reaction is as follows.

$$2H_2O(l) \rightarrow 2H_2(g) + O_2(g) \quad \Delta H_{reaction} = +572 \text{ kJ}$$

A graph of the energy changes during the progress of this reaction is shown in **Figure 20.5**. Notice how the energy curve rises above the level of the reactant, then falls only slightly when the products form. Therefore, the products have more stored energy than the reactants, and the graph clearly shows a net gain of energy. Because of this net gain, $\Delta H_{reaction}$ is positive. For this particular reaction, energy must be added continuously to keep the reaction going.

Activation energy and catalysts Recall from Chapter 6 that catalysts speed up a reaction. The catalyst provides a different reaction path—one for which the activation energy is lower. The catalyst thus creates a shortcut, or tunnel through the energy hill, between the reactants and products. More of the collisions between reactants will now produce products because less energy is required. **Figure 20.6** shows the effect of a catalyst on activation energy.

■ **Figure 20.6** Similar to a tunnel that shortens a trip in the mountains, a catalyst lowers a reaction's activation energy. As a result, the reaction goes faster than would be the case without the catalyst.

Compare *ΔH for the catalyzed pathway and for the uncatalyzed pathway.*

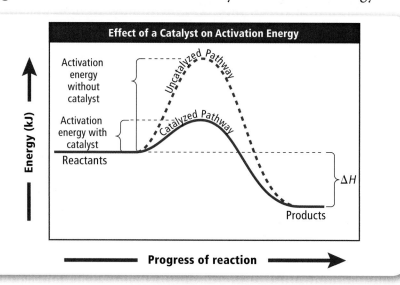

Everyday Chemistry

Catalytic Converters

Since 1975, every new car sold in the United States has a catalytic converter installed in the exhaust system. This device contains porous, heat-resistant material coated with catalysts. The purpose of the catalytic converter is to reduce air pollution by reducing the emissions that leave a car's exhaust system.

How a catalytic converter works A typical catalytic converter consists of a reduction catalyst and an oxidation catalyst deposited separately on a ceramic structure that is shaped like a honeycomb. The reduction catalyst uses platinum and rhodium to help reduce nitrogen monoxide (NO) and nitrogen dioxide (NO_2) emissions. When an NO or NO_2 molecule contacts the catalyst, the nitrogen atom is pulled out of the molecule and bound to the catalyst, freeing the oxygen and producing the nitrogen molecule (NO_2). The oxidation catalyst uses platinum and palladium to help reduce unburned hydrocarbons and carbon monoxide emissions. The pollutants are oxidized over platinum and palladium, producing carbon dioxide (CO_2) and water.

The honeycomb arrangement of catalytic materials in the converter shown in **Figure 1** provides a large surface area for the reactions to take place, increasing the rate at which the pollutants are removed. Catalytic converters have reduced the pollutants released into the air by vehicles by more than 90 percent.

Improving the catalytic converter The operating temperature of a catalytic converter is between 204° and 871°C. Below 204°C, the device does nothing. The pollutants emitted during the beginning of the driving period, when the converter is just heating up, slip through the catalytic converter without being changed.

In an effort to stop air pollution during the warm-up period, some catalytic converters heat up to 400°C within 5 seconds. The result is that almost as soon as the car is started, air pollutants in the exhaust are broken down.

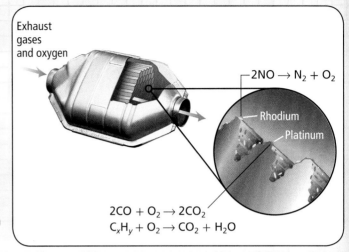

Figure 1 A porous, heat-resistant material coated with catalysts removes pollutants from car exhaust.

Heated catalytic converters are proving to be highly efficient. Someday, perhaps even those 5 seconds of dirty emissions before the converter is heated will be eliminated.

Engine performance Catalytic converters are one of the greatest emission control devices ever to be installed on vehicles. However, if the catalytic converter is not operating properly, engine performance and emission control may suffer.

Symptoms of an improperly operating catalytic converter include a drop in fuel economy, lack of high speed power, a rough idle or stalling of the vehicle, and elevated tailpipe emissions of hydrocarbons and carbon monoxide. Fortunately, catalytic converters are designed to operate efficiently for well beyond 100,000 miles or 161,000 km.

Explore Further

1. **Acquire Information** Find out why cars with catalytic converters cannot use leaded gasoline.

2. **Apply** Instead of the honeycomb structure of the catalytic converter described above, some earlier converters contained pellets coated with a platinum-palladium mixture. Why was the use of pellets effective?

Factors That Drive Chemical Reactions

When pure aluminum metal (Al) is exposed to chlorine gas (Cl_2), aluminum chloride ($AlCl_3$) is formed. This chemical change is spontaneous; that is, once it has begun it just happens on its own. The reaction is highly exothermic. The following equation summarizes the process.

$$2Al(s) + 3Cl_2(g) \rightarrow 2AlCl_3(s) \quad \Delta H_{reaction} = 1408 \text{ kJ}$$

If you reverse the equation for the reaction between aluminum and chlorine, you get the equation for the decomposition of aluminum chloride. This reaction is highly endothermic and it is not spontaneous. From this observation, it might be tempting to conlcude that only exothermic processes are spontaneous. But remember that ice melting at room temperature is a spontaneous, endothermic process. What determines whether a reaction or process is spontaneous? The answer has to do with the general forces that drive all chemical processes.

Entropy Scientists have observed two tendencies in nature that explain why chemical reactions occur. The first tendency is for systems to go from a state of high energy to a state of low energy. The state of low energy is more stable. Thus, exothermic reactions are more likely to occur than endothermic ones, all other things being equal. The second tendency is for the energy of a system to become more spread out, or dispersed.

To help you think of what the spreading out of energy means, imagine a deck of 52 cards, as shown in **Figure 20.7**. You shuffle the cards and deal out four hands of 13 cards each. It's possible that you will deal one hand of 13 spades, one hand of 13 hearts, one hand of 13 clubs, and one hand of 13 diamonds, each hand in order from 2 to ace. It's possible, but incredibly unlikely. From experience, you probably know that the suits and high and low cards will be spread out more evenly among the four hands. Why? It isn't that one hand is more likely than the other; in a random deal, all hands are equally possible. It's simply that there is only one way to deal out all the cards in order, but there are billions upon billions of ways to deal out hands with the cards spread out among them.

> **FOLDABLES**
> Incorporate information from this section into your Foldable.

■ **Figure 20.7** What are the chances that a shuffled deck of cards will deal out into four perfectly ordered rows, as shown in **a**? Your experience tells you the probablility of randomly getting this deal is negligible. A random deal similar to the one shown in **b** is far more likely.

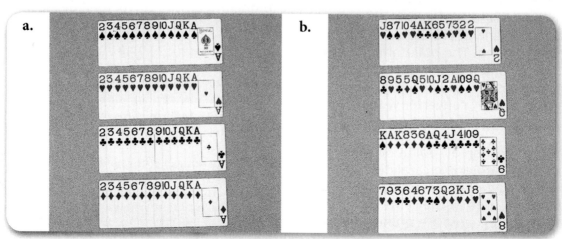

As with the cards, energy tends to spread out. Scientists use **entropy** as the term that describes and measures the spreading out of energy. Unlike energy, which is conserved in chemical changes and in the universe as a whole, entropy is not conserved. The natural tendency of entropy is to increase.

You might have heard increasing entropy equated with increasing disorder. Looking again at the deck of cards in **Figure 20.7,** this analogy makes intuitive sense. But remember that both of the hands shown are equally likely in a random deal. It's just that there are a whole lot more ways to deal a hand that "looks disordered" than there are to deal a hand that appears highly ordered. In dealing with chemical systems, the number of particles and energy states is far more staggering than the number of cards. Overwhelming probability favors spreading out, but this dispersion is not the same thing as disorder.

Some kinds of changes, such as melting and evaporation, are apt to increase entropy. These processes result in particles with greater freedom of movement. Entropy also increases during reactions and processes, such as dissolving sodium chloride in water to produce sodium and chloride ions, that result in an increase in the number of particles. When the number of particles increases, energy becomes more spread out. It is distributed among a larger number of particles.

The direction of a chemical reaction At room temperature, most exothermic reactions tend to proceed spontaneously forward. In other words, they favor the formation of products. In an exothermic reaction, the released energy, usually in the form of heat, raises the temperature of the products and many more atoms and molecules in the surroundings. The motion of a greater number of atoms and molecules increases. The energy is distributed among more molecules than it was before the reaction; in other words, the energy is more spread out. Therefore, entropy increases.

You have read that both energy and entropy play a role in determining spontaneity. In general, the direction of a chemical reaction is determined by the magnitude and direction of the energy and entropy changes. For example, a reaction will proceed in the forward direction, toward formation of products, if that direction results in both a release of heat and an increase in entropy. As an example, consider the combustion of butane (C_4H_{10}) shown in **Figure 20.8.**

$$2C_4H_{10}(g) + 13O_2(g) \rightarrow 8CO_2(g) + 10H_2O(g) + \text{heat}$$

This reaction occurs spontaneously in the forward direction because both energy and entropy changes are favorable. The energy decreases because the reaction is exothermic—a favorable change. The entropy increases partially because the total number of molecules increases, from 15 to 18—also a favorable change. The release of heat and increase in entropy combine to drive this reaction forward.

TRY AT HOME 🏠 LAB

See page 877 for **Observing Entropy.**

■ **Figure 20.8** Once a spark begins the combustion of butane, the reaction proceeds spontaneously forward until one of the reactants is used up. The reverse reaction isn't spontaneous. Even if a spark were added to a mixture of carbon dioxide and water, the butane wouldn't reform.

Table 20.1 Predicting Whether a Reaction Is Spontaneous

Interactive Table Explore reaction spontaneity at glencoe.com.

Energy Change	Entropy Change	Spontaneous?
Decrease (exothermic)	increase	yes
Decrease (exothermic)	decrease	yes at low temperature; no at high temperature
Increase (endothermic)	increase	no at low temperature; yes at high temperature
Increase (endothermic)	decrease	no

Look at another example of a reaction favored by both energy and entropy changes—the one between calcium carbonate ($CaCO_3$) and hydrochloric acid (HCl).

$$CaCO_3(s) + 2HCl(aq) \rightarrow CaCl_2(aq) + H_2O(l) + CO_2(g) + heat$$

This reaction favors formation of products because it produces gases and liquids. They are more disordered than the solid calcium carbonate, so entropy increases. Heat is given off, so the products are at a lower energy state than the reactants. That is also favorable. Once again, the decreased energy and increased entropy both drive the reaction in the forward direction.

What happens if only one of the changes—either energy or entropy—is favorable? If the favorable change is great enough to outweigh the unfavorable one, the reaction will still be spontaneous. Thus, endothermic reactions are spontaneous if entropy increases greatly. Also, reactions that decrease the entropy of the chemical system are spontaneous if they are exothermic enough.

Spontaneity depends on the balance between energy and entropy factors. **Table 20.1** summarizes these factors. Note that temperature plays a role in determining spontaneity when one factor is favorable and the other is not.

SUPPLEMENTAL PRACTICE For more practice with energy and entropy changes, see Supplemental Practice, page 841.

Section 20.1 Assessment

Section Summary

- Chemical reactions are either endothermic or exothermic.
- Energy can be converted from one form to another, but cannot be created or destroyed.
- Activation energy is the energy needed to get a reaction started.
- A spontaneous chemical reaction is a reaction that, once begun, occurs without any outside intervention.
- Entropy is a measure of disorder.

1. **MAIN Idea Classify** Suppose ΔH for a reaction is negative. Compare the energy of the products with the energy of the reactants. Is the reaction endothermic or exothermic?

2. **Describe** the energy changes that occur when a match lights.

3. **Summarize** Are reactions that occur spontaneously at room temperature generally exothermic or endothermic? Explain.

4. **Describe** each of the following processes as involving an increase or decrease in entropy.
   a) Water in an ice-cube tray freezes.
   b) You pick up scattered trash along a highway and pack it into a bag.
   c) Your campfire burns, leaving gray ashes.
   d) A cube of sugar dissolves in a cup of tea.

Section 20.2

Objectives

- **Sequence** the technique of calorimetry and illustrate its use.
- **Compare** the heat generated by some common fuels and by some foods.
- **Analyze** the efficiency of industrial processes and the need to conserve resources.

Review Vocabulary

entropy: term used to describe and measure the spreading out of energy

New Vocabulary

calorie
kilocalorie
Calorie

Measuring and Using Energy Changes

MAIN Idea Energy stored in chemical bonds can be converted to other forms and used to meet the needs of individuals and of societies.

Real-World Reading Link Have you ever read the nutrition facts on the label of one of your favorite foods? From this label, you can find out the size of a single serving and how many Calories are in that serving.

Calorimetry

The heat generated in chemical reactions can be measured by a technique called calorimetry and a device called a calorimeter. **Figure 20.9** shows a type of calorimeter called a bomb calorimeter, which is used by food chemists. Food burns exothermically in the calorimeter, raising the temperature of the water. In the case of endothermic reactions, the surrounding water in the calorimeter supplies the heat and decreases in temperature.

The heat lost or gained by the surrounding water is calculated by means of the following equation.

$$q_w = (m)(C_w)(\Delta T)$$

In this equation, the symbol q_w stands for heat absorbed by water, m is the mass of the water, ΔT is the temperature change of the water, and C_w is the specific heat of water, which equals 4.184 J/g · °C.

The symbol $q_{reaction}$ stands for the heat change of the reaction. By the law of conservation of energy, there is no net creation or destruction of energy. Heat loss by the reaction means heat gain for the water, and heat gain for the reaction means heat loss for the water. Therefore, the heat of reaction is equal to the negative of the heat change of the water.

$$q_{reaction} = -q_w$$

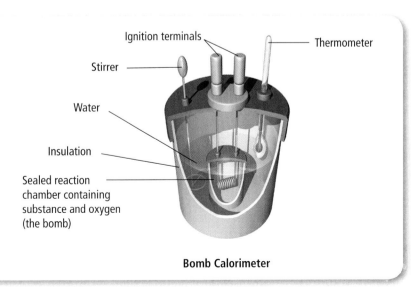

Bomb Calorimeter

■ **Figure 20.9** A sample is positioned in a steel inner chamber, called the bomb, which is filled with oxygen at high pressure. Surrounding the bomb is a measured mass of water stirred by a low-friction stirrer to ensure uniform temperature. The reaction is initiated by a spark, and the temperature is recorded until it reaches its maximum.

Infer Why is it important that the stirrer does not create friction?

Concepts In Motion

Interactive Figure To see an animation of calorimetry, visit glencoe.com.

EXAMPLE Problem 20.1

Calculating Heat of Reaction for Combustion Burning 1.60 g of methane in a bomb calorimeter raises the temperature of 1.52 kg of water from 20.0°C to 34.0°C. What is the heat of reaction for the combustion of one mole of methane?

1 Analyze

First, find the heat change for the water, q_w. This equals $-q_{reaction}$ for burning the 1.60 g of methane. Then, convert that value to find $q_{reaction}$ for one mole of methane. That will equal $\Delta H_{reaction}$.

2 Set Up

Use $\Delta T = T_{final} - T_{initial}$ to find the temperature change of the water. Then, use the equation $q_w = (m)(C_w)(\Delta T)$. The negative of the value of q_w will equal $q_{reaction}$ for the burning of 1.60 g of methane. Use the molar mass of methane, 16.0 g methane/mol, as a conversion factor to find the $q_{reaction}$ for one mole of methane.

3 Solve

- First, calculate ΔT.

$$\Delta T = T_{final} - T_{initial}$$
$$= 34.0°C - 20.0°C = 14.0°C$$

Then, calculate q_w.

$$q_w = (m)(\Delta T)(C_w)$$
$$= (1.52 \times 10^3 \text{ g})(14.0°C)\left(\frac{4.184 \text{ J}}{\text{g} \cdot °C}\right) = 8.90 \times 10^4 \text{ J} = 89.0 \text{ kJ}$$

Problem-Solving Hint
Be careful with signs in the calculation of ΔT. If the reaction is endothermic, the sign of ΔT will be negative.

- Now, find $q_{reaction}$ for the 1.60 g of methane.

$$q_{reaction} = -q_w = -89.0 \text{ kJ} = \text{heat released in burning 1.60 g methane}$$

- Now, use a molar-mass conversion factor to find $q_{reaction}$, or ΔH, for 1 mol of methane.

$$q_{reaction} = \left(\frac{-89.0 \text{ kJ}}{1.60 \text{ g methane}}\right)\left(\frac{16.0 \text{ g methane}}{\text{mol methane}}\right)$$
$$= -8.90 \times 10^2 \text{ kJ/mol methane}$$

$$q_{reaction} = \Delta H_{reaction} = -8.90 \times 10^2 \text{ kJ/mol methane}$$

4 Check

Check to be sure that all units were correctly handled, and review the calculation to be sure that it was sound. The result does check out.

SUPPLEMENTAL PRACTICE
For more practice with heat of reaction, see Supplemental Practice, page 841.

PRACTICE Problems

Solutions to Problems Page 866

5. How much heat is absorbed by a reaction that lowers the temperature of 500.0 g of water in a calorimeter by 1.10°C?

6. Aluminum reacts with iron(III) oxide to yield aluminum oxide and iron. Calculate the heat given off in the reaction if the temperature of the 1.00 kg of water in the calorimeter increases by 3.00°C.

7. Burning 1.00 g of a certain fuel in a calorimeter raises the temperature of 1.000 kg of water from 20.00°C to 28.05°C. Calculate the heat given off in this reaction. How much heat would one mole of the fuel give off, assuming a molar mass of 65.8 g/mol?

■ **Figure 20.10** Foods contain a variety of chemical compounds. They can be grouped into three general types—protein, carbohydrate, and fat—that have different energy contents. Your body processes transform some of the energy released when these foods are broken down into work and heat. However, when you take in more food than you need, your body stores the extra energy by producing fat for later use.

Energy Value of Food

Many years ago, chemists measured heat in calories instead of joules. A **calorie** is the heat required to raise the temperature of 1 g of liquid water by 1°C. A **kilocalorie** is a unit equal to 1000 calories. One calorie is equal to 4.184 J. One joule is equal to 0.239 calorie. The energy value of foods is measured in units called Calories. Note the capital C. One food **Calorie** is the same as 1 kilocalorie. One food Calorie is also equal to 4.184 kJ. Nutritionists and dietitians generally use Calories rather than kilojoules.

Chemical compounds in the food you eat provide you with energy. The compounds undergo slow combustion, combining with oxygen to produce the waste products carbon dioxide and water, along with compounds needed for body growth and development. Fats provide about 9 Calories per gram. In contrast, carbohydrates and proteins provide about 4 Calories per gram. **Figure 20.10** illustrates some types of nutrients. **Table 20.2** shows how many Calories you get from typical servings of some foods.

You can use calorimetry to measure the energy content of food. The food is burned rapidly in oxygen, rather than slowly, like it is in the body, but the amount of energy released is the same. The following example problem shows you how to calculate food Calories from calorimetry data.

Table 20.2	The Caloric Values of Some Foods		
Food	Quantity	Kilojoules	Calories
Butter	1 tbsp = 14 g	418	100
Peanut butter	1 tbsp = 16 g	418	100
Spaghetti	0.5 cup = 55 g	836	200
Apple	1	283	70
Chicken (broiled)	3 oz = 84 g	502	120
Beef (broiled)	3 oz = 84 g	1000	241

EXAMPLE Problem 20.2

Measuring Food Calories A 1.00-g sample of nuts reacts with excess oxygen in a calorimeter. The calorimeter contains 1.00 kg of water that has an initial temperature of 15.40°C and a final temperature of 20.20°C. Find the energy content of the nuts. Express your answer in kJ/g and Calories/g.

1 Analyze

You first need to find the heat change for the water, q_w. Because all the heat to raise the water's temperature came from burning 1.00 g of nuts, q_w is equal to the energy content of the 1.00 g of nuts.

2 Set Up

Find the temperature change of the water. Then, use the equation $q_w = (m)(C_w)(\Delta T)$ to find q_w and, therefore, the energy content of the nuts in joules. A conversion factor can then be used to convert to Calories.

3 Solve

- $\Delta T = T_{final} - T_{initial}$
  $= 20.20°C - 15.40°C = 4.80°C$

$$q_w = (m)(C_w)(\Delta T)$$
$$= (1.00 \times 10^3 \, g)(4.80°C)\left(\frac{4.184 \, J}{g \cdot °C}\right)$$
$$= 2.01 \times 10^4 \, J = 20.1 \, kJ$$

The nuts contain 20.1 kJ/g of energy.

- Now, apply a conversion factor to convert to Calories.

$$(20.1 \, kJ)\left(\frac{1 \, Calorie}{4.184 \, kJ}\right) = 4.80 \, Calories$$

The nuts contain 4.80 Calories/g of energy.

4 Check

Check to be sure that all units were correctly handled, and repeat the calculation to be sure that it was sound. The result does check out.

SUPPLEMENTAL PRACTICE

For more practice determining heats of reaction and the energy content of foods, see Supplemental Practice, page 841.

PRACTICE Problems Solutions to Problems Page 867

8. A group of students decides to measure the energy content of certain foods. They heat 50.0 g of water in an aluminum can by burning a sample of the food beneath the can. When they use 1.00 g of popcorn as their test food, the temperature of the water rises by 24°C. Calculate the heat released by the popcorn, and express your answer in both kilojoules and Calories per gram of popcorn.

9. Another student comes along and tells the group in problem 8 that she has read the label on a popcorn bag that states that 30 g of popcorn yields 110 Calories. What is that value in Calories/gram? How can you account for the difference?

10. A 3.00-g sample of a new snack food is burned in a calorimeter. The 2.00 kg of surrounding water change in temperature from 25.0°C to 32.4°C. What is the food value in Calories per gram?

Figure 20.11 The pie pans and other objects shown here are all made from aluminum.

Energy Economics

Have you ever wondered why recycling has become so important? Part of the answer has to do with energy. For many materials, the energy required for recycling used objects is less than the energy of simply throwing the used objects away and making new ones from fresh raw materials.

Aluminum production and recycling Aluminum is the metal of choice for pie pans, as shown in **Figure 20.11**. It is lightweight, has a low specific heat (0.902 J/g°C), and conducts heat rapidly. As a result of its low specific heat, the aluminum itself absorbs little heat, so the pie pan cools before the pie does. In Chapter 17, you learned that aluminum is produced by electrolysis of its principal ore, bauxite. The production of aluminum requires large amounts of energy, particularly electricity. Therefore, one reason to recycle aluminum is to conserve that energy and reduce the cost of producing aluminum products.

A comparison of the energy required to produce new aluminum cans and the energy required to produce cans from recycled aluminum reveals that 15–20 cans can be made from recycled aluminum for every one that can be made from aluminum ore. Thus, the cost of using recycled aluminum is much less than that of using new aluminum. In addition to saving money on energy costs, recycling aluminum conserves valuable raw materials and uses less fossil fuel, which is the ultimate source of most of the energy required to produce the aluminum. To understand how recycling conserves fossil fuels, it is important to look at the process used to generate electrical energy.

> **FACT of the Matter**
>
> Recycling aluminum consumes only about seven percent of the energy required to make the same amount of new aluminum from ore. Recycling 25 aluminum cans saves an amount of energy equivalent to that available from burning a gallon of gasoline.

CHEMLAB

ENERGY CONTENT OF FOOD

Background
Foods supply the energy and nutrients you require to build and sustain your body and maintain your levels of activity. The energy is released within cells during the process of respiration. In that process, oxygen combines with energy-storage substances from food, such as glucose, to produce carbon dioxide, water, and heat. The reaction is essentially a slow combustion process. In this ChemLab, you will compare the amounts of energy liberated from the combustion of three food items.

Question
How much energy is released during the combustion of some common food items?

Objective
- **Calculate** the energy released during the combustion of pecans, marshmallows, and a food item of your choice.
- **Compare** the energies obtained from the food items.
- **Infer** which types of foods contain the greatest amount of energy.

Preparation

Materials
oven mitt
bottle opener with can-piercing end
empty, clean soft-drink can with a tab closure
empty clean can, minus the top and bottom lids, of approximately the same diameter as the soft-drink can
ring stand
small iron ring
beaker tongs
Celsius thermometer
100-mL graduated cylinder
glass stirring rod
balance
large paper clip
matches

pecan half
2 small marshmallows
food sample of your choice

Safety Precautions
WARNING: *Be careful in using the match and in handling the cans after each experiment because they may be hot. Perform the experiment in a well-ventilated room because acrid fumes may be produced during the combustion process.*

Procedure

1. Read and complete the lab safety form.
2. Use a can opener to open holes near the bottom of the open-ended can. Bend a large paper clip to fashion the food support as shown in the figure.

3. Use the graduated cylinder to measure 100 mL of tap water, and pour the water into the soft-drink can. Support this can by the ring stand as shown in the figure. Put the thermometer through the opening in the top of the can and allow it to reach the temperature of the water. Record the initial water temperature to the nearest 0.1°C.

4. Measure the mass of a pecan half, record its mass, and impale it on the paper-clip food support. Hang the support on the edge of the open-ended can with the food sample inside, and rest the can on the ring stand.

5. Use a match to light the pecan. Quickly swing the water-filled soft-drink can assembly directly over the combustion can, allowing a small separation between the two cans.

6. Allow the pecan to burn as completely as possible. If the flame goes out during the process and the pecan is still mostly unburned, you must empty the water in the can and start over with fresh water and a new pecan half. After the pecan has burned out, allow the thermometer to reach the highest temperature. Record this temperature.

7. Carefully disassemble the apparatus, empty the water from the soft-drink can, and dispose of the burned food item.

8. Repeat steps 3–7 with two small marshmallows.

9. Repeat steps 3–7 with a small food item of your choice.

Analyze and Conclude

1. **Calculate** Use your data to calculate the number of kilojoules of energy liberated per gram of each food item. Assume that the water in the soft-drink can has a density of 1.00 g/mL and that the specific heat of the water is 4.19 J/g°C.

2. **Rank** the food items according to the amounts of energy liberated per gram.

3. **Infer** Based upon the chemical compositions of the foods you tested (protein, carbohydrate, fat, etc.), what type of food provides the greatest amount of energy per gram?

Apply and Assess

1. **Explain** Would it be desirable to eat mainly the type of food that provides the greatest amount of energy per unit mass? Explain.

2. **Error Analysis** What aspects of your procedure may have caused error in your calculated results? How would each of these sources of error affect your result?

INQUIRY EXTENSION
Develop a scale that relates the nutritional content of food to energy content. Choose an important nutrient and relate it to the energy content of the food. How do several foods that you commonly eat compare on your nutrition-energy scale?

Data and Observations

Data Table

Food Item	Mass (g)	Initial H$_2$O Temperature (°C)	Final H$_2$O Temperature (°C)
Pecan			
Marshmallow			
Other food			

MiniLab 20.2

Dissolving—Exothermic or Endothermic?

Can you classify dissolving as exothermic or endothermic? The dissolving of a solid in water, like most other processes, may liberate energy or absorb energy. If the dissolving process is exothermic, the liberated energy raises the temperature of the solution. If the process is endothermic, it absorbs energy from the solution, lowering the temperature. Examine the dissolution of several common solids in water.

Procedure

1. Read and complete the lab safety form.
2. Measure 100 mL of **water** into a **250-mL beaker.** Set a **Celsius thermometer** in the water and allow it to come to the water's temperature. The thermometer should not be touching the side or the bottom of the beaker. Record that temperature. Remove the thermometer from the water.
3. Add to the water approximately 1 tablespoon of the **solid to be tested,** and stir with a **stirring rod** for 20 seconds. Put the thermometer back into the solution. Record the temperature.
4. Pour the solution down the drain with large amounts of tap water.
5. Repeat steps 2–4 for each of the solids to be tested.

Analysis

1. **Classify** Which of the solids that you tested dissolve exothermically? Which dissolve endothermically?
2. **Apply** Which of the solids that you tested could be mixed with water inside a flexible plastic container to produce a cold pack?
3. **Explain** how the process of dissolving affects the entropy of a solid.

Converting chemical energy to electricity In the United States, most electricity must be produced by burning fossil fuels, typically coal. The chemical energy stored in the bonds of molecules that make up fossil fuels is released as these fuels are burned. Power plants convert this energy to electric energy. **Figure 20.12** illustrates the production of electricity in a coal-fired power plant. In addition to providing energy for industrial processes such as aluminum production, electricity provides light, heat, hot water, and the ability to run machines of all kinds.

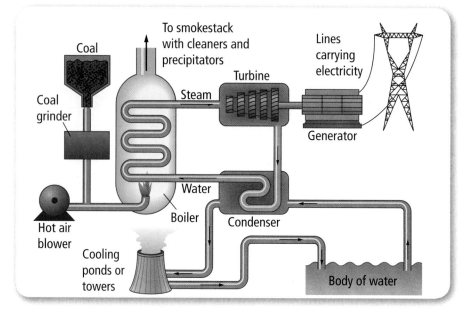

■ **Figure 20.12** In a typical power plant, coal burns, producing heat that boils liquid water into steam. The kinetic energy of the steam drives a turbine, which drives a generator that produces the electric energy.

Identify sources of energy other than fossil fuels that can be converted to electric energy.

Earth Science Connection

Bacterial Refining of Ores

Roughly 2000 years ago, Roman miners began to investigate a blue liquid that they often noticed around piles of discarded, low-grade copper ore. They associated the blue color with copper, probably based on their experience with blue minerals, such as turquoise, which were often found in the vicinity of that metal. When they heated those minerals in a charcoal fire, they obtained copper.

The blue color of the liquid made them suspect that copper was present in it. They heated the liquid to see whether it really did contain copper. Their efforts were successful. Pure copper was produced from copper salts in the liquid. What the Romans did not realize was that the copper in the liquid had been leached from the low-grade copper ore by bacteria.

Why bacteria are used to extract metals After crude ores are mined, they are crushed and then the metals are extracted by chemical treatment or by heating to high temperatures. Another way to extract metals from ore is to use bacteria to do part of the job. This method is less damaging to the environment, costs less in terms of energy input, and offers a way to use low-grade ores productively. The only disadvantage is that leaching ore with bacteria can take several years. Today, 25 percent of the world's copper production is the result of bioprocessing by bacteria.

One bacterium, *Acidithiobacillus ferrooxidans*, shown in **Figure 1,** obtains its energy by means of reactions involving various minerals. As a result, it produces acid and an oxidizing solution of iron(III) ions (Fe^{3+}), which can react with metals in crude ore.

The process is simple and requires little energy. Low-grade ore is first treated with sulfuric acid to stimulate the growth of bacteria. The microbes process the ore and release copper ions into solution. The metal is then extracted from the solution.

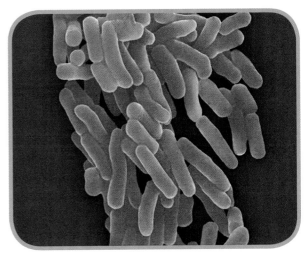

Figure 1 A colony of *Acidithiobacillus ferrooxidans*

Gold and bacteria Gold is another prospect for biorefining. As the sources of high-grade gold ore become depleted, miners are being forced to mine low-grade ores. The gold in these lesser ores is present in the form of sulfides. To burn off the sulfur, the ores are traditionally treated by roasting or pressure oxidation. Only then can the gold be extracted with cyanide ions.

In one bio-oxidation process, low-grade gold ore is mixed with a brew of bacteria in huge, stirred reactors. A newer process places open piles of gold ore on an impermeable base. Bacterial cultures and the fertilizers needed to nurture them are poured onto the gold ore. In both methods, *A. ferrooxidans* treats the gold sulfides at a lower cost and with an efficiency of recovery increased from 70–95 percent.

Connection to Chemistry

1. **Acquire Information** Thermophilic bacteria, which thrive at temperatures from 40°–70°C or higher, are being considered as possible candidates for biorefining. Find out what chemical advantage these bacteria have over other bacteria.
2. **Think Critically** What are the energy advantages of biorefining?

CHEMISTRY AND TECHNOLOGY

Alternative Energy Sources

Energy can be neither created nor destroyed. Does that mean that people can keep using the same energy sources at the present rate forever? Fossil fuels are a finite source of energy that should be conserved. Using alternative, renewable sources of energy can prolong the supply of fossil fuels.

Solar Energy
One alternative source of energy is solar energy. Earth receives more energy from the sun in just one hour than the world uses in a whole year. The demand for solar energy has increased 25 percent per year since 1991. The solar home in **Figure 1** was built to capture sunlight and convert it to heat that warms the rooms. Adobe walls, clay tile floors, triple windows, and heavily insulated walls and ceilings store the sun's warmth in winter. An overhang keeps the high-angle solar rays from entering the house in summer but does not block the lower-angle sun rays in winter. Solar energy can be used for heating water in solar water heaters, for cooking or drying food in solar cookers, for desalination, which is the removal of salt from seawater, and for electricity.

Figure 2 Photovoltaic cell

Photovoltaic cells, shown in **Figure 2**, use solar energy to produce electricity. These solar cells consist of layers of silicon with trace amounts of gallium or phosphorus and are capable of emitting electrons when struck by sunlight. Two main problems must be overcome before solar cells are a reliable source of electricity for everyday use. The first is the efficiency of conversion of solar energy to electrical energy. The second is finding efficient ways to store the electricity for use during the dark hours or during periods of cloudy weather.

Geothermal Energy
Magma, which is molten rock, can heat solid rock surrounding the magma chamber. When this occurs, water in the porous rock above the heated solid rock turns to steam. If there are cracks in the solid rock above, the steam escapes to the surface in the form of a geyser or hot spring. This is a natural source of geothermal energy. Many natural geothermal regions lie in the earthquake and volcano belts along Earth's crustal plate edges. Geysers in Yellowstone National Park are an indication of the heat that is stored beneath the ground. This energy can be tapped and used to run power plants to produce electricity. The largest geothermal power plant in the world is at The Geysers in California. It has the capacity to generate 1000 megawatts of electric power—enough to serve the needs of 1 million people.

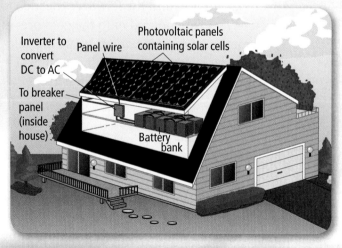

Figure 1 Solar home

Natural geothermal energy is only readily available for use in power plants in a few locations around the world. Although a geothermal power plant, such as the one shown in **Figure 3,** generates electricity less expensively than a nuclear power plant or a coal power plant, it is expensive to drill deep enough to access the heat generated from a magma chamber unless the magma chamber is close to the surface of Earth. Another consideration is that although carbon dioxide and nitrogen oxides are not produced from geothermal power plants, these power plants do produce air pollution in the form of hydrogen sulfide and ammonia.

Figure 4 Wind farm

Producing electricity from wind emits no air pollution. It requires no water for cooling. However, wind energy for large-scale energy production can be used only in areas that have reliable winds. When the winds die down, a backup system for producing electricity must also be available. Other considerations to producing electricity from wind energy are that the wind farms are unattractive and noisy and have a negative impact on wildlife. Some wind turbines kill birds of prey and bats, especially when the turbines are constructed in migratory flight paths.

Figure 3 Geothermal power plant

Wind Energy

In some parts of the country, wind energy is an almost unlimited source of electricity. Wind can turn turbines, shown in **Figure 4,** that produce mechanical energy. A generator converts the mechanical energy into electrical energy. Usually, hundreds of wind turbines are needed to run a power plant.

Discuss the Technology

1. **Evaluate** Which form of alternative energy might be feasible in your state? Explain.

2. **Acquire Information** Investigate and diagram how a solar furnace works.

3. **Think Critically** The three kinds of alternative energy sources discussed in this feature are not effective in all places. Despite that fact, how can they help solve the nation's energy problems?

Chapter 20 • Chemistry and Technology **725**

Saving energy with catalysts

You learned in Chapter 18 that polyethylene is a plastic used in many household products. It is made from ethylene, a simple hydrocarbon whose formula is C_2H_4. Ethylene is produced by removing hydrogen from ethane (C_2H_6), another hydrocarbon, according to the following equation.

$$C_2H_6(g) + \text{heat} \rightarrow C_2H_4(g) + H_2(g)$$

In theory, 1 mol of ethylene could be made with the addition of only 137 kJ of heat. In practice, the reaction above uses nearly four times that amount of energy to produce 1 mol of ethylene because of inefficiency in transferring energy. Over many years, chemists have introduced refinements into the production process. The development of better catalysts has lowered the energy requirements for this and other chemical processes. Research on energy-saving catalysts continues as energy becomes more expensive.

Entropy: one of the costs of using energy

Any time energy is produced or is converted from one form to another, as it is in power plants and industrial processes, some of the energy is lost. This may seem surprising because of the law of conservation of energy. However, in this case, the energy is not destroyed but is simply not usable to do work. What happens to it? It is wasted as heat. In many processes, the amount of energy lost as heat exceeds the amount of energy that can be harnessed to do useful work.

The waste heat generated in systems that use or produce energy cannot be reclaimed, reused, or recycled because it has become randomly dispersed to molecules in the environment. As a result of this dispersion, entropy increases, as shown in **Figure 20.13**. The environment cannot spontaneously reorganize itself to give back the energy that increased its entropy. This increase in entropy is part of the price of using energy.

> **FACT of the Matter**
>
> At one time, scientists thought that heat was a colorless, odorless, weightless fluid. This fluid, which they called caloric, was believed to cause other substances to expand when it was added to them.

■ **Figure 20.13** Every time energy is used to do useful work, some energy is lost to the environment, usually as heat. This heat increases the entropy of the environment and cannot be reclaimed.

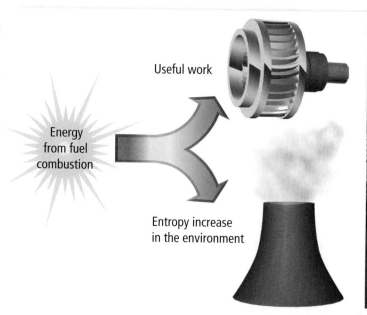

In a power plant, friction from moving parts, such as the turbine above, generates waste heat. This heat is dissipated to the environment through the plant's cooling towers.

This dissipated heat is useful in situations where an increase in temperature is desirable. For example, it could be used for heating greenhouses in cold climates.

Table 20.3 Efficiencies in Power Production

Maximum theoretical efficiency	63%
Efficiency of boiler	90%
Mechanical efficiency of turbine	75%
Efficiency of electrical generator	95%
Efficiency of power transmission	90%
Overall actual efficiency	36%

A process, such as recycling, that saves energy by conserving electricity does more than reduce the total consumption of electricity. It also saves the chemical energy of original fossil fuel that would otherwise be lost forever as waste heat and increased entropy.

Energy and efficiency Because of entropy increases, no power or industrial plant, no matter how well designed, can completely convert heat from chemical reactions into useful work. Waste heat is inevitable. This kind of loss is easiest to understand in terms of efficiency, which is the amount of work that can be obtained from a process compared to the amount that must go into it.

For example, a modern fossil-fuel-burning plant has a maximum theoretical efficiency of only 63 percent. This means that, at most, 63 percent of the chemical energy from the original fuel can be converted to useful work. The remainder is lost as waste heat.

Other inefficiencies further reduce the percentage of energy that is converted to useful work. **Table 20.3** illustrates some of these factors in a modern electrical power plant. The data in the table allow you to compute the plant's overall efficiency, which is simply the product of the efficiencies of all the individual steps multiplied by the maximum theoretical efficiency. Thus, the plant's overall efficiency can be calculated as follows.

$$(0.63)(0.90)(0.75)(0.95)(0.90)(100) = (0.36)(100) = 36\%$$

This low value is typical of the energy efficiency of power and industrial plants. When you use an electric appliance, such as the one shown in **Figure 20.14,** you are tapping a little over one-third of the energy stored in the original fossil fuel.

■ **Figure 20.14** In a coal-fired power plant, a maximum of 36 percent of the energy available from coal is converted to electric energy. The electric energy can be used to power electric appliances, such as this leaf blower.

■ **Figure 20.15** A cooling tower is a common way to get rid of waste heat. Inside the tower, hot water is sprayed into the air while large fans draw air through the droplets. Although this process only transfers the heat to the atmosphere, it is an improvement over pumping the hot water back into streams, an action that killed organisms in the stream.

Infer *What gas is released through cooling towers?*

Improving the efficiency of industrial processes

The efficiency of industrial processes can be increased, and valuable energy resources can be conserved if plants that use improved energy-saving devices are designed and built. Also, if uses are found for waste heat—such as heating buildings and growing food in cold climates—energy can be conserved. Cooling towers, **Figure 20.15**, are one way of transferring waste heat. The development of more energy-efficient refrigerators, air conditioners, and lightbulbs can also improve the efficiency of electricity use.

Finally, the energy conversion processes of organisms can be studied and perhaps eventually applied to industrial methods. Biological processes convert energy from one form to another and at the same time create highly ordered—that is, low-entropy—systems. These natural processes are far more efficient than most industrial processes.

Section 20.2 Assessment

Section Summary

▶ Chemical reactions are a source of energy because the energy of chemical bonds can be converted to heat, light, or electric energy.

▶ Heat changes can be measured by calorimetry.

▶ Whenever energy is converted from one form to another, some energy is lost as heat. The lost energy generally cannot be used again or converted back to a useful form of energy.

11. **MAIN Idea** **Explain** why it is impossible to convert 100 percent of the chemical energy in a fossil fuel to electrical energy.

12. **Infer** The specific heat of aluminum is 0.902 J/g°C. The specific heat of copper is 0.389 J/g°C. If the same amount of heat is applied to equal masses of the two metals, which metal will increase more in temperature? Explain.

13. **Calculate** How much heat, in kilojoules, is given off by a chemical reaction that raises the temperature of 700 g of water in a calorimeter by 1.40°C?

14. **Determine** Using Table 20.3 as a guide, determine the overall efficiency of a power plant that has a maximum theoretical efficiency of 50 percent, given that the other efficiencies are the same as those in the table.

15. **Compare** Using Table 20.2 as a guide, compare the Caloric content of 1 g of butter to 1 g of spaghetti. Which food gives you more Calories per gram?

Section 20.3

Objectives
- **Analyze** the process and importance of photosynthesis.
- **Compare** the energy efficiency of photosynthesis and processes that produce electricity.
- **Trace** how energy from the Sun passes through a food web.

Review Vocabulary
calorie: heat required to raise the temperature of 1 gram of water by 1°C

New Vocabulary
photosynthesis

Photosynthesis

MAIN Idea The process of photosynthesis converts energy from the Sun into chemical energy stored in the bonds of biological molecules.

Real-World Reading Link All organisms need energy to survive, and the main source of energy for Earth is the Sun. Somehow, the Sun's energy must be captured as chemical energy so that living things can use it.

The Basis of Photosynthesis

The process of converting the Sun's light energy to chemical energy for use by cells is called **photosynthesis.** Plants and other photosynthetic organisms have at least one obvious characteristic in common: they are green, the color of chlorophyll. The chlorophyll in cells of plants and algae is found in organelles called chloroplasts, as shown in **Figure 20.16**. The chlorophyll in photosynthetic bacteria is found on membranes spread throughout the cells.

Regardless of the exact location of the chlorophyll, the process works essentially the same way. Chlorophyll absorbs light and changes it to chemical energy, triggering a complex series of changes that together make up photosynthesis. Let's examine the chemical process in more detail. Note, as you read, that most of the changes absorb energy.

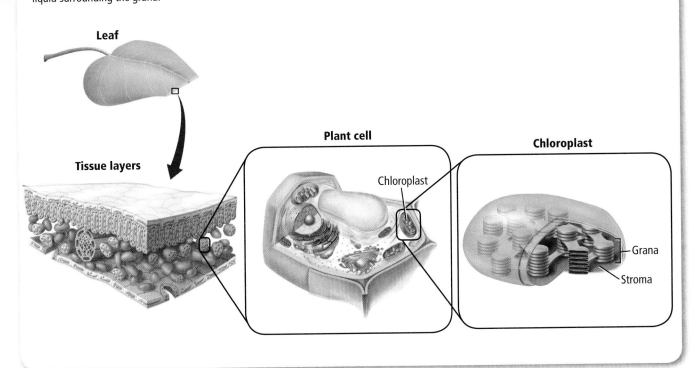

■ **Figure 20.16** Photosynthesis occurs inside pigmented organelles called chloroplasts. Chloroplasts contain the light-absorbing green pigment chlorophyll. The chlorophyll is located on stacks of membranes called grana. Sugar molecules are synthesized in the stroma, which is the liquid surrounding the grana.

VOCABULARY
WORD ORIGIN

Chlorophyll
comes from the Greek words *chloros*, meaning *green*, and *phyllon*, meaning *leaf*

Chemistry Online
Personal Tutor For an online tutorial on photosynthesis, visit glencoe.com.

■ **Figure 20.17** The light reactions
Infer *What is the energy source for photosynthesis?*

The Chemistry of Photosynthesis

The process of photosynthesis involves a series of reactions in which green plants and some other organisms manufacture carbohydrates from carbon dioxide and water using energy from sunlight. The net reaction that is carried out in photosynthesis is one that produces simple sugars and oxygen gas from carbon dioxide and water. The balanced equation is generally written as follows.

$$6CO_2 + 6H_2O \rightarrow C_6H_{12}O_6 + 6O_2$$

The formula $C_6H_{12}O_6$ represents glucose, a simple sugar. Photosynthesis is divided into two basic series of reactions—the light reactions and the Calvin cycle.

The light reactions When light reaches a chlorophyll molecule, the energy of the light is absorbed. The energy excites electrons in the molecule, as described in **Figure 20.17**. A high-energy electron is released from the chlorophyll and transfers some of its energy to a molecule of ADP, adenosine diphosphate. That causes the ADP to bond to a third phosphate to form ATP, adenosine triphosphate. As you may recall from Chapter 19, ATP is the main energy storehouse in cells and also plays a critical role in respiration.

As you can see by following the process shown in **Figure 20.17**, electrons lost by the chlorophyll in the first step are restored to it by a reaction in which water is split into elemental oxygen and hydrogen ions. The oxygen is released into the atmosphere and can later be used by organisms, including you, for respiration.

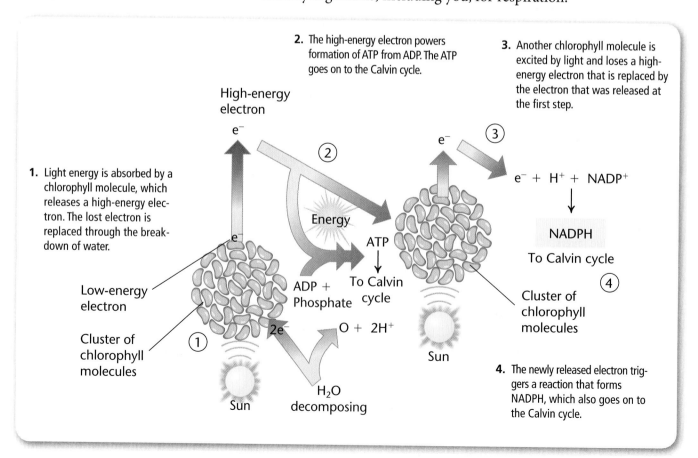

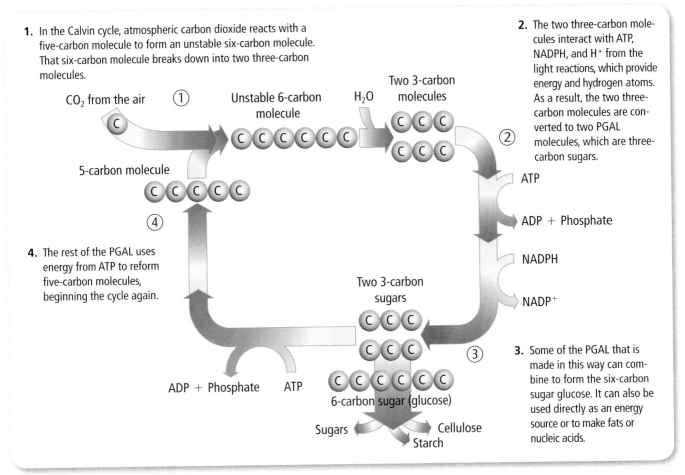

Figure 20.18 The Calvin cycle

The electron from the original chlorophyll has now lost its energy and moves on to join a different chlorophyll molecule—one that also absorbs light energy but is involved in a different reaction. The transferred electron replaces an electron that is energized by light absorption and released. This newly released electron is involved in another reaction, which forms an important coenzyme called NADPH. The NADPH is formed from NADP and from hydrogen ions that were freed in the earlier breakdown of water. Both the NADPH and the ATP formed as a result of the first absorption of energy continue to the second series of photosynthesis reactions, the Calvin cycle.

The Calvin cycle Once the light reactions have produced ATP and NADPH, the second series of photosynthesis reactions can occur. As shown in **Figure 20.18,** the Calvin cycle, which takes place in the stroma of chloroplasts, makes use of NADPH from the light reactions, as well as carbon dioxide taken in from the surroundings. Simple sugars, such as glucose, result.

In the Calvin cycle, ATP turns back into ADP and provides energy for a series of reactions. In the first of these, NADPH loses a hydrogen ion and becomes NADP. The hydrogen ion from the NADPH, together with other H^+ ions produced during the light reactions, provides hydrogen for the synthesis of the sugars. The carbon and oxygen for the sugars are provided mainly by the carbon dioxide.

Figure 20.19
Ultimately, the Sun is the source of the energy, and photosynthesis is the process that captures it and makes all life possible.

The oak tree absorbs energy from the Sun and stores it in the molecules it makes during photosynthesis. The stored energy allows it to carry out essential functions.

When plant-eating animals, such as caterpillars, consume the oak leaves, they can break down the energy-storage molecules in those leaves to release energy for their own life functions.

When meat-eating animals, such as some kinds of birds, consume the plant eaters, they make use of the energy that was stored in their prey.

Figure 20.20 This chemist spent fifteen days in a sealed chamber. His oxygen was supplied by wheat plants carrying out photosynthesis, and the carbon dioxide for photosynthesis came from his respiration. Experiments such as this are the starting point to determine whether humans could live in a similar but larger chamber on the moon or another planet.

Energy and the Role of Photosynthesis

As you read earlier, most of the reactions that occur during photosynthesis are endothermic. The absorbed energy is stored by using it to synthesize high-energy molecules. Only chlorophyll-containing organisms can carry out photosynthesis and make these high-energy molecules that they need in order to survive. The animals that eat those organisms get the energy they need from them, as shown in **Figure 20.19**. So, photosynthesis is the ultimate process for chemically storing the energy that all organisms in the food web need. You could thus describe photosynthesis as the entry point of energy for life on Earth, as **Figure 20.20** describes.

The organization and maintenance of your body, like those of every organism, depend upon energy that is used to power chemical changes. However, energy is not the only factor involved when systems undergo changes. There is also entropy. Natural systems have a tendency toward increased disorder. However, living things are highly ordered, and many of the processes that take place in them, such as building more complex molecules, involve decreases in entropy rather than increases.

How is it possible, then, for organisms to overcome the tendency to disorder and maintain their complex structures and life functions? The secret of their success is their ability to absorb energy from outside sources, such as light and food. These outside sources continue pumping in the energy needed to keep the otherwise-nonspontaneous processes going.

When the body converts stored chemical energy to other uses during respiration, heat is produced. The heat, if it built up, could lead to destructive entropy increases and a breakdown of structures and functions. Some of the heat produced is radiated to the environment. The heat given off raises the entropy of the surroundings rather than that of the body. The low entropy of living things is thus maintained partially at the expense of the entropy of the surroundings. The image in **Figure 20.21** shows the pattern of heat radiated from a person's hands and arms.

If you've ever felt cold and had to put on extra layers of clothes to stay warm, you might be thinking that radiating heat away isn't always the answer. If the core body temperature increases or decreases by more than a few degrees, there can be deadly results. Living things require a relatively narrow range of body temperatures to keep life going. In warm-blooded organisms, such as humans, some of the heat produced during respiration is retained by the body, which uses it to maintain a constant temperature. Striking the proper balance between warming and cooling allows various biochemical processes to take place at the proper rates.

Connecting Ideas

In this section, you learned that photosynthesis converts light energy from the Sun into chemical energy that sustains our lives. But where does the energy of the Sun itself come from? The Sun is absolutely huge—over 300,000 times the mass of Earth and containing 99.8% of the mass of the solar system. However, standard chemical processes, such as the combustion of wood or sugar, even on this massive scale could not come close to supplying the amount of energy for the length of time that the Sun has. In the next chapter, you'll study nuclear reactions, the ultimate source of energy in the Sun.

■ **Figure 20.21** Metabolic activities produce heat, some of which is radiated to the surrounding environment. In this thermogram image, the color white indicates the warmest areas, and the color blue indicates the coolest. The red color is intermediate in temperature.

SUPPLEMENTAL PRACTICE

For more practice with photosynthesis, see Supplemental Practice Problems, page 842.

Section 20.3 Assessment

Section Summary

▶ Photosynthesis is a process by which plants use light energy to synthesize simple carbohydrates.

▶ The reactions that occur during photosynthesis include the light reactions and the Calvin cycle. Most of the reactions that occur during this process are endothermic.

▶ The results of photosynthesis provide energy for life on Earth.

16. **MAIN Idea Summarize** Explain why a meat-eating animal such as an eagle depends on photosynthesis for its energy.

17. **Infer** Plants appear green because their leaves reflect green and yellow light. Explain what happens to the energy in the wavelengths of light that are not reflected.

18. **Explain** How does chlorophyll function in photosynthesis?

19. **Identify** the two simple compounds that plants need to take in from the environment for photosynthesis to occur.

20. **Predict** What would be the effect on Earth's atmosphere if all plants were temporarily unable to carry out the process of photosynthesis?

21. **Compare** the process of photosynthesis with the process of cellular respiration in Chapter 19. Explain why they are often considered opposite processes.

22. **Compare** How does the process by which plants obtain sugars differ from that by which most other organisms obtain sugars?

CHAPTER 20 Study Guide

Download quizzes, key terms, and flash cards from glencoe.com.

BIG Idea Energy is released when bonds form, but must be added to break bonds.

Section 20.1 Energy Changes in Chemical Reactions

MAIN Idea Exothermic reactions release energy, and endothermic reactions absorb energy.

Vocabulary
- entropy (p. 713)
- fossil fuel (p. 709)
- heat (p. 704)
- law of conservation of energy (p. 707)

Key Concepts
- Chemical reactions are either endothermic or exothermic.
- Energy can be converted from one form to another, but cannot be created or destroyed.
- Activation energy is the energy needed to get a reaction started.
- A spontaneous chemical reaction is a reaction that, once begun, occurs without any outside intervention.
- Entropy is a measure of disorder.

Section 20.2 Measuring and Using Energy Changes

MAIN Idea Energy stored in chemical bonds can be converted to other forms and used to meet the needs of individuals and of societies.

Vocabulary
- calorie (p. 717)
- Calorie (p. 717)
- kilocalorie (p. 717)

Key Concepts
- Chemical reactions are a source of energy because the energy of chemical bonds can be converted to heat, light, or electric energy.
- Heat changes can be measured by calorimetry.
- Whenever energy is converted from one form to another, some energy is lost as heat. The lost energy generally cannot be used again or converted back to a useful form of energy.

Section 20.3 Photosynthesis

MAIN Idea The process of photosynthesis converts energy from the Sun into chemical energy stored in the bonds of biological molecules.

Vocabulary
- photosynthesis (p. 729)

Key Concepts
- Photosynthesis is a process by which plants use light energy to synthesize simple carbohydrates.
- The reactions that occur during photosynthesis include the light reactions and the Calvin cycle. Most of the reactions that occur during this process are endothermic.
- The results of photosynthesis provide energy for life on Earth.

Chapter 20 Assessment

Understand Concepts

23. What is the difference between a positive ΔH and a negative ΔH in terms of energy and the kind of reaction involved?

24. Why is outside energy needed to start most exothermic reactions but not to sustain them?

25. Why is energy needed to sustain many endothermic reactions?

26. Nitrogen gas reacts with hydrogen gas to form ammonia gas (NH_3). The heat of reaction is -46 kJ/mol NH_3. Write a balanced chemical equation that includes the reaction heat. Is the reaction exothermic or endothermic?

27. What is meant by entropy? Explain why dissolving table salt (NaCl) in water increases entropy, but dissolving oxygen gas in water decreases entropy.

28. In terms of order and disorder, in what direction do most naturally occuring processes proceed?

29. Describe the general trends of spontaneous reactions in terms of energy and entropy change.

30. Substances like nitroglycerine are powerful explosives partially because they are chemically unstable. Use the following equation and concepts of energy, entropy, and spontaneity to explain why nitroglycerine is unstable.

    $4C_3H_5(NO_3)_3(l) \rightarrow$
    $\quad 6N_2(g) + O_2(g) + 12 CO_2(g) + 10H_2O(g)$

    $\Delta H = -5700$ kJ

31. Explain how your car converts energy from one form to another.

32. The efficiency of an automobile engine is improved by heating the interior of the car. Why is this the case?

33. Suppose that a coal-fired power plant converts energy from coal into electrical energy with 36 percent efficiency. How many kilojoules of energy from coal are needed to produce 1.0 kJ of electrical energy?

Apply Concepts

Chemistry and Technology

34. Explain why garbage could be an excellent fuel for the production of electricity.

How It Works

35. A reusable heat pack contains a supersaturated solution of a salt and a disc of metal. When the metal disc is bent, the solute begins to crystallize and releases heat. The pack can be reset by heating it in boiling water, which causes the salt to dissolve again. How can you account for the heat given off by this kind of pack?

Everyday Chemistry

36. Write balanced equations for the following catalytic converter reactions: **a)** the breakdown of nitrogen monoxide into oxygen and nitrogen, and **b)** the reaction of carbon monoxide with oxygen to produce carbon dioxide.

Earth Science Connection

37. When copper ore is roasted during refining, the solid compound $CuFeS_2$ is combined with oxygen gas to yield solid copper(II) sulfide, iron(II) oxide, and gaseous sulfur dioxide. Write a balanced equation for this reaction.

Think Critically

Interpret Data

38. **ChemLab** During an experiment, two different foods are burned in a calorimeter. The data from the experiment are shown in **Table 20.4**. Which food releases more heat per gram?

Table 20.4	Calorimetry Data	
	Mass (g)	Heat Released (Calories)
Sample 1	6.0	25
Sample 2	2.1	9.0

Chapter 20 Assessment

Relate Cause and Effect

39. MiniLab 1 Which factor—entropy, added heat, or both—promotes the dissolution in water of a solid that spontaneously dissolves endothermically? Which factor promotes dissolution in water of a solid that spontaneously dissolves exothermically?

Infer

40. MiniLab 2 A certain reaction proceeds in the forward direction with a ΔH of 216 kJ. Will the reverse reaction be endothermic or exothermic? What will be the value of ΔH for the reverse reaction?

Cumulative Review

41. Identify each of the following as either chemical or physical properties of the substance mentioned. *(Chapter 1)*
  a) The element mercury has a high density.
  b) At room temperatrure, solid carbon dioxide changes to gaseous carbon dioxide without passing through the liquid state.
  c) Iron rusts when exposed to moist air.
  d) Sucrose is a white crystalline solid.

42. Distinguish between atomic number and mass number. How do each of these two numbers compare for isotopes of an element? *(Chapter 2)*

43. Classify each of the following elements as a metal, a nonmetal, or a metalloid. *(Chapter 3)*
  a) molybdenum
  b) bromine
  c) arsenic
  d) neon

44. Why does the second period of the periodic table contain eight elements? *(Chapter 7)*

45. Explain why the metals zinc, aluminum, and magnesium are resistant to corrosion, whereas iron is not. *(Chapter 8)*

46. Compare and contrast the physical characteristics of the different states of matter. *(Chapter 10)*

Skill Review

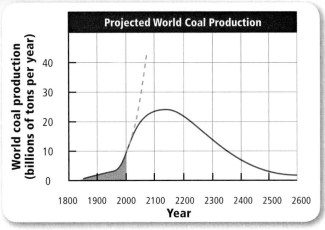

■ **Figure 20.22**

47. Use a Graph The graph in **Figure 20.22** projects the world production of coal (solid line) and also projects the demand for coal (dashed line).
  a) How much longer can the present growth rate in coal consumption continue?
  b) In what year will maximum coal production be reached, and how much coal will be produced that year?
  c) Compare the amount of coal produced in 1900 to the production in the peak year.

WRITING in Chemistry

48. Research the meaning of the term *thermal pollution*. Write a news article that explains the possible sources and effects of such pollution in the operation of a coal-fired power plant. Refer to **Figure 20.12** for assistance.

Problem Solving

49. An organic fuel is burned in a calorimeter. The reaction releases 194 kJ of heat. If the initial temperature of the water was 21.0°C, and the final temperature was 51.9°C, how much water was in the calorimeter?

Cumulative
Standardized Test Practice

Use the table below to answer Questions 1–3.

Chemical Reaction	Heat of Reaction (kJ/mol)
$4Fe(s) + 3O_2(g) \rightarrow 2FeO_3(s)$	−1625
$NH_4NO_3(s) \rightarrow NH_4^+(aq) + NO_3^-(aq)$	+27
$C_6H_{12}O_6(s) + 6O_2(g) \rightarrow 6CO_2(g) + 6H_2O(l)$	−2808
$H_2O(l) \rightarrow H_2O(g)$	+40.7

1. The rusting of an iron nail in which iron reacts with oxygen is a(n)
   a) decomposition reaction.
   b) displacement reaction.
   c) endothermic reaction.
   d) exothermic reaction.

2. How much energy is required to boil 0.500 L water into steam?
   a) 1.26 kJ
   b) 20.0 kJ
   c) 700 kJ
   d) 1260 kJ

3. What is the heat of reaction for the production of 360 g of glucose during the process of photosynthesis?
   a) 2800 kJ
   b) −2800 kJ
   c) 5600 kJ
   d) −5600 kJ

4. Which compound is a basic anhydride?
   a) CO_2
   b) SO_2
   c) CaO
   d) NO_2

5. A solution of 0.600M HCl is used to titrate 15.00 mL of KOH solution. The endpoint of the titration is reached after the addition of 27.13 mL of HCl. What is the concentration of the KOH solution?
   a) 9.000M
   b) 1.09M
   c) 0.332M
   d) 0.0163M

6. Why does electric current flow between the two electrodes of a lemon battery?
   a) The two electrodes have a potential difference maintaining a flow of electrons from an anode to a cathode.
   b) The two electrodes have a potential difference maintaining a flow of electrons from a cathode to an anode.
   c) A battery attached to the two electrodes creates a potential equivalence between the two different metals.
   d) A battery attached to the two electrodes creates a potential difference between the two different metals.

7. Isomers are
   a) compounds with the same chemical formula but different structures.
   b) compounds with the same structure but different chemical formula.
   c) compounds with the same number of carbon atoms but different types of bonds.
   d) compounds with a different number of carbon atoms but the same structure.

8. Which statement about enzymes is NOT true?
   a) Enzymes speed up chemical reactions during which their structure and chemical composition are altered.
   b) Enzymes provide pocket or grove sites in their three dimensional structures where substrates can bind.
   c) Active sites alter their initial shape to match the structural shape of a substrate after it binds to the site.
   d) Enzymes dramatically increase the rate of a chemical reaction, but they cannot cause a nonspontaneous reaction to happen.

NEED EXTRA HELP?

If You Missed Question...	1	2	3	4	5	6	7	8
Review Section...	20.1	20.1	20.1	14.1	15.2	17.1	18.1	19.1

CHAPTER 21: Nuclear Chemistry

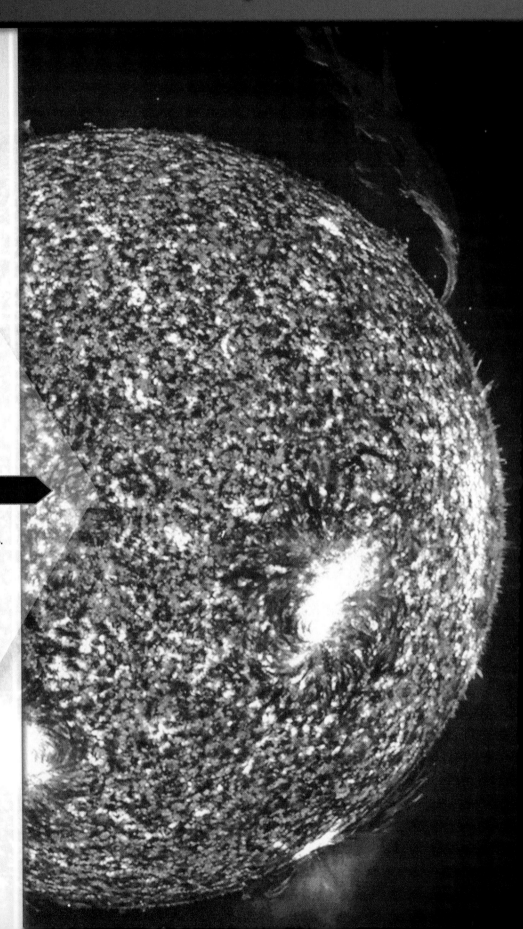

BIG Idea Nuclear chemistry has a vast range of applications, from the production of electricity to the diagnosis and treatment of disease.

21.1 Types of Radioactivity
MAIN Idea Alpha, beta, and gamma radiation are three types of radiation emitted as unstable nuclei decay into stable nuclei.

21.2 Nuclear Reactions and Energy
MAIN Idea Fission and fusion release tremendous amounts of energy.

21.3 Nuclear Tools
MAIN Idea Radiation has many useful applications, but it also has harmful biological effects.

ChemFacts

- By mass, the Sun is 70% hydrogen, 28% helium, and 2% trace elements.
- Temperature and pressure in the core of the Sun are so great that hydrogen nuclei combine to form helium nuclei in a thermonuclear reaction.
- Thermonuclear reactions produce heat and light energy that make life on Earth possible.

Start-Up Activities

LAUNCH Lab

The Penetrating Power of Radiation

Alpha, beta, and gamma radiation differ in their abilities to penetrate materials. What types of materials block these types of radiation?

Materials
- Geiger counter
- alpha, beta, and gamma sources
- shielding: sheet of paper, piece of aluminum foil, and a lead sheet

Procedure

WARNING: *Wash hands and arms thoroughly after handling radioactive sources. Do not handle radioactive sources if you have a break in the skin below the wrist. Immediately report any damage to any of the radioactive sources.*

1. Read and complete the lab safety form.
2. Obtain the alpha source. Slowly bring it near, but not touching, the Geiger counter. Record the Geiger counter's reading. Repeat the procedure three times, each time holding one type of shielding between the detector and the alpha source.
3. Repeat Step 2 for the beta and gamma sources.

Analysis
1. **Create** a table indicating what type of shielding was effective in blocking each type of radiation.
2. **Infer** Which type of radiation would be most useful for measuring the thickness of paper?

Inquiry Does distance also affect the amount of radiation reaching the Geiger counter? Develop a lab procedure that examines the effects of both the shielding material and the distance from the source.

FOLDABLES Study Organizer

Types of Radiation Make the following Foldable to help you organize information about the different types of radiation.

▶ **STEP 1** Collect two sheets of paper, and layer them about 2 cm apart vertically.

▶ **STEP 2** Fold up the bottom edges of the sheets to form three equal tabs. Crease the fold to hold the tabs in place.

▶ **STEP 3** Staple along the fold. Label the tabs *Gamma*, *Beta*, and *Alpha*. Label the front *Types of Radiation*.

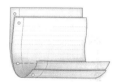

FOLDABLES Use this Foldable with Section 21.1. As you read, write about what happens to the nucleus during each type of decay. Include examples and sample equations.

Chemistry Online

Visit glencoe.com to:
▶ study the entire chapter online
▶ explore **concepts In Motion**
▶ take Self-Check Quizzes
▶ use Personal Tutors
▶ access Web Links for more information, projects, and activities
▶ find the Try at Home Lab, Modeling Fusion

Section 21.1

Objectives
- **Analyze** common sources of background radiation.
- **Compare and contrast** alpha, beta, and gamma radiation.
- **Apply** the concept of half-life of a radioactive element.

Review Vocabulary
photosynthesis: the process used by certain organisms to capture energy from the Sun

New Vocabulary
radioactivity
alpha particle
beta particle
gamma ray
half-life

Types of Radioactivity

MAIN Idea Alpha, beta, and gamma radiation are three types of radiation emitted as unstable nuclei decay into stable nuclei.

Real-World Reading Link If you wake up while it is still dark, the glowing numbers on your clock let you know what time it is. Many clocks use a type of radiation to make the numbers glow. The word *radiation* might cause you to think about nuclear power plants or dangerous, highly radioactive substances. However, less dangerous forms of radiation are often used in everyday objects such as clocks.

Discovery of Radioactivity

You are probably familiar with novelty items that glow in the dark. Maybe you even have one of those ceiling constellation maps with stars that give off an eerie, green glow long after the room lights are turned out. Have you ever wondered what causes the stars in such maps to glow? The material in them is a phosphorescent form of the compound zinc sulfide, which means that after it is exposed to light, it continues to give off the light it has absorbed, as shown in **Figure 21.1**.

A French scientist named Henri Becquerel was studying a phosphorescent uranium compound in 1896 when he made an accidental discovery that changed the course of history. Becquerel was testing the compound to see whether the phosphorescence was the cause of a recently discovered type of electromagnetic radiation called X-rays. After exposing the sample to sunlight, he placed paper-wrapped photographic film near it. X-rays would be able to pass through the paper and expose the film, producing an image of the sample on it, just as they make an image of your teeth on film when you have an X-ray taken. Becquerel found the image he expected. He believed that X-rays were the cause and that the uranium gave off these rays because of absorbed sunlight.

■ **Figure 21.1** When light shines on phosphorescent materials, such as these glow-in-the-dark icons on a watch face, the absorbed energy raises the electrons in the atoms to higher energy levels. When the electrons move back to lower energy levels, the extra energy is given off as light. The glow disappears gradually, as the electrons return to their more stable configurations.

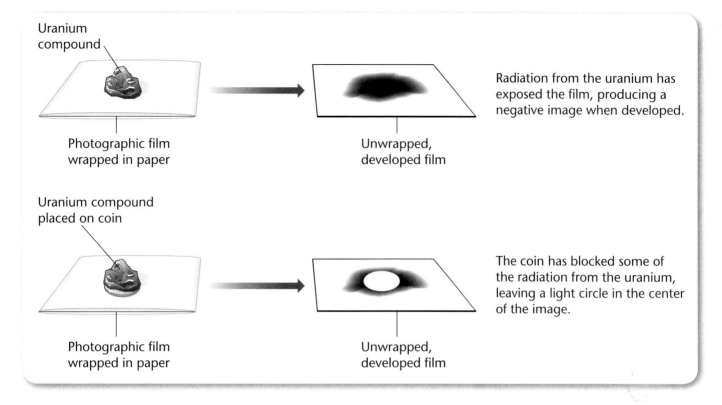

Figure 21.2 Becquerel discovered that uranium compounds spontaneously give off radiation that can pass through paper wrapping and produce an image on film. He also found that a coin prevents the radiation from reaching the film, producing a dark, unexposed circle in the image.

Radiation, not phosphorescence However, a subsequent finding contradicted this conclusion that X-rays resulting from absorbed light caused the image. Becquerel developed a piece of paper-wrapped film that had been in a dark drawer along with a crystal of the uranium. He did not expect to see an image on the film because the uranium sample had not been exposed to sunlight or any other source of energy. To his surprise, the film bore the image of the crystal. **Figure 21.2** summarizes Becquerel's uranium experiment.

What made the uranium give off radiation? It hadn't been exposed to light, so the radiation wasn't caused by phosphorescence. Was a chemical reaction taking place? No known chemical reactions had ever produced this effect. Becquerel suspected that the uranium was spontaneously emitting some type of radiation. Was the radiation the same as X-rays? No one could tell for sure. If so, it was odd that the rays could be produced without an input of energy, which was generally needed to produce X-rays.

The Curies Becquerel presented the problem to Marie and Pierre Curie for further study. Their conclusion was that a nuclear reaction was taking place within the uranium atoms. Marie Curie named this process radioactivity. **Radioactivity** is the spontaneous emission of radiation by an unstable atomic nucleus.

For the first time, the effects of changes in the nucleus of an atom had been observed and correctly interpreted. The Curies went on to study the properties of radioactivity and discovered elements other than uranium that also give off radiation. For example, **Figure 21.3** shows an image made by radioactive radium salts. Henri Becquerel and the Curies were awarded the 1903 Nobel Prize in Physics for their important work.

Figure 21.3 Radium salts are placed on a special emulsion on a photographic plate. After the plate is developed, the emulsion shows the dark tracks left by radiation emitted by the radium salts.

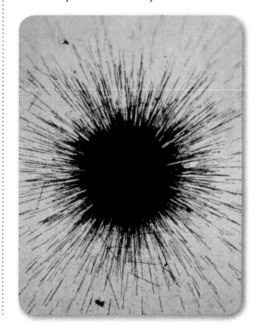

Table 21.1 Comparison of Chemical and Nuclear Reactions

Chemical Reactions	Nuclear Reactions
 • Occur when bonds are broken and formed • Involve only valence electrons • Associated with small energy changes • Atoms keep the same identity, although they gain, lose, or share electrons to form new substances. • Temperature, pressure, concentration, and catalysts affect reaction rates.	 • Occur when nuclei combine, split, and emit radiation • Can involve protons, neutrons, and electrons • Associated with large changes in energy • Atoms of one element are often converted into atoms of another element. • Temperature, pressure, and catalysts do not normally affect reaction rates.

Concepts In Motion

Interactive Table Explore chemical and nuclear reactions at glencoe.com.

Nuclear Notation

Nuclear reactions involve the protons and neutrons found in the nucleus. During nuclear reactions, a nucleus can lose or gain protons and neutrons. Adding or taking away protons changes the identity of an element. In nuclear reactions, atoms of one element are converted into atoms of another element. **Table 21.1** compares chemical reactions with nuclear reactions.

You can represent nuclear changes using equations, just as you do for chemical reactions. The reactants will be on the left and products on the right, following an arrow. However, now you will be considering the nuclei of elements as reactants and products.

For example, the element carbon has natural isotopes with three different atomic masses: carbon-12, carbon-13, and carbon-14. Two of these, carbon-12 and carbon-13, are stable isotopes and are not radioactive. However, carbon-14 is unstable and radioactive. To distinguish these, write the mass number as a superscript and the atomic number as a subscript to the left of the symbol for the element. The three isotopes of carbon are represented as $^{12}_{6}C$, $^{13}_{6}C$, and $^{14}_{6}C$. Each has six protons, but carbon-12 has a total of 12 protons and neutrons, carbon-13 has a total of 13, and carbon-14 has 14. **Figure 21.4** illustrates the isotopes of carbon and hydrogen.

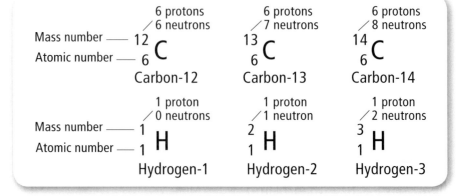

■ **Figure 21.4** Numerical superscripts and subscripts are used to show the mass numbers and atomic numbers of isotopes. The isotopes of carbon and of hydrogen are represented here.

Radioactive Decay

The release of radiation by radioactive isotopes—radioisotopes, for short—is called decay. The nuclei of such radioisotopes are unstable. However, not all unstable nuclei decay in the same way. Some give off more powerful radiation than others or different kinds of radiation. Between 1896 and 1903, scientists had discovered three types of nuclear radiation. Each type changes the nucleus in its own way. These three types were named after the first three letters of the Greek alphabet: alpha (α), beta (β), and gamma (γ).

Alpha decay Alpha radiation consists of streams of alpha particles. **Alpha particles** are helium nuclei, consisting of two protons and two neutrons. Alpha particles can be represented as ^4_2He or simply as α. Their relatively large size and charge, compared with other forms of radiation, cause them to collide frequently with atoms in air and objects they encounter. Alpha radiation does not deeply penetrate into matter and is easily stopped by a thin layer of material, such as paper and clothing, or even by air. Whenever alpha decay occurs, the decaying nucleus loses 2p and 2n, and turns into the nucleus of an element with an atomic number that is 2 less and a mass number that is 4 less than that of the original.

Radioactive radium-226 decays undergoes alpha decay, forming radon-222. When an alpha particle is emitted from a nucleus of radium-226, the radium atom loses two protons and two neutrons. The loss of protons makes the product radon, as shown in the following equation and in **Figure 21.5**. Note that radon-222 itself is radioactive and will decay into different products.

$$^{226}_{88}Ra \rightarrow {}^{222}_{86}Rn + {}^4_2He$$

Balancing alpha decay equations Is this equation balanced? Check to see whether the sum of the mass numbers on the right side of the equation is 226, which is the mass number of the radium isotope on the left side. It is, because the mass number of the radon atom is 222 and that of the alpha particle is 4. Also make sure that the sums of the atomic numbers on both sides of the equation are the same. Both sums equal 88, so the atomic number balances, and the nuclear equation as a whole is balanced. It is important to remember that a balanced nuclear equation differs from balanced chemical equations because the kinds of atoms involved need not remain the same.

FOLDABLES
Incorporate information from this section into your Foldable.

■ **Figure 21.5** A radium-226 nucleus undergoes alpha decay to form radon-222 and an alpha particle.
Evaluate *What are the numbers of protons and neutrons in radium-226 and radon-222?*

How It Works

Smoke Detectors

A smoke detector is a device that sounds an alarm in the presence of smoke particles from a smoldering or burning object. An ionizing smoke detector, like the one illustrated below, contains a sensor that detects smoke particles that interrupt an electric current. Ionizing detectors contain the unstable isotope americium-241 ($^{241}_{95}Am$), which produces the electric current needed in the detectors. Americium-241 decays to form neptunium-237 ($^{237}_{93}Np$) and helium-4 ($^{4}_{2}He$), two harmless isotopes.

Another type of smoke detector is the photoelectric smoke detector, which uses a light source and a light detector positioned at 90-degee angles. Normally, the light from the light source travels in a straight line and misses the detector. In the presence of smoke, the smoke particles scatter the light and some amount of light hits the sensor. Ionizing smoke detectors are most common because they are less expensive than photoelectric detectors and are better at detecting the small amounts of smoke.

1 A battery or other power source provides electricity.

2 Americium-241 provides alpha particles that ionize particles in air.

3 In the absence of smoke, the ions carry current between a positive and a negative electrode.

4 A microchip monitors the flow of current between the electrodes.

5 Under normal conditions, the microchip permits no current to flow through the alarm. The alarm does not sound.

6 If a fire occurs, smoke particles rise up into the detector and interfere with the flow of ions between the electrodes. The microchip senses the drop in current and allows electricity to flow through the alarm circuit, sounding the alarm.

Think Critically

1. **State** What kind of radioactive decay does americium-241 undergo?
2. **Explain** why the air that has been exposed to americium-241 is able to carry a current between charged electrodes.

Beta decay The second type of radiation from unstable atomic nuclei results in beta decay and produces beta radiation. Beta radiation is made up of particles that are much smaller and lighter than alpha particles, so they move much faster and have greater penetrating power. A **beta particle** is a high-energy electron with a 1− charge and is written as $_{-1}^{0}e$ or β^-. Beta particles pass through matter more easily than do alpha particles. They can be stopped only by thick materials such as stacked sheets of metal, blocks of wood, or heavy clothing.

The electron produced in beta decay is not one of the original electrons from outside the nucleus. It is produced by the change of a neutron into a proton and an electron. Whenever beta decay occurs, transmutation of elements occurs. The decaying nucleus turns into the nucleus of an element with an atomic number that is 1 greater than that of the original and a mass number that is the same.

Balancing beta decay equations Carbon-14 is an example of an isotope that decays by beta emission.

$$_{6}^{14}C \rightarrow {_{7}^{14}}N + {_{-1}^{0}}e$$

Notice that transmutation of elements has occurred. An atom of carbon-14 is converted into an atom of nitrogen-14. Because the beta particle has only the mass of an electron, the mass number remains 14. Because the beta decay changes a neutron to a proton, the atomic number of the nucleus increases by one unit, from 6 to 7. The atomic number on the left side of the equation, 6, is the same as the sum on the right side of the equation $(6 = 7 + (-1))$. The nuclear equation is balanced.

Gamma decay In the third type of decay, gamma radiation is produced. A **gamma ray** is a high-energy form of electromagnetic radiation without mass or charge. Because of its high energy and penetrating ability, gamma radiation can cause great harm to living cells. Gamma rays are written as γ. Such rays are much harder to stop than alpha or beta particles, as shown in **Figure 21.6.** They can pass easily through most types of material, and stopping them requires thick blocks of lead or even thicker blocks of concrete.

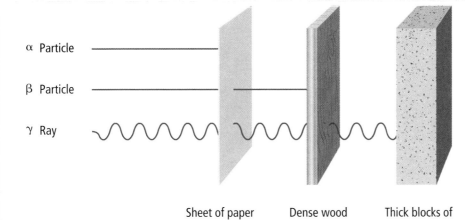

■ **Figure 21.6** The three types of radiation vary in their penetrating power. Alpha particles are stopped by even a thin sheet of paper, but beta particles require thicker shields, such as a dense piece of wood. The energetic gamma rays are stopped only by thick blocks of lead or concrete.

During gamma decay, only energy is released. Gamma radiation is often omitted from nuclear equations because it does not affect mass number or atomic number. Gamma decay generally does not occur alone, but rather accompanies other modes of decay. For example, the alpha decay of uranium to thorium involves simultaneous gamma decay.

$$^{238}_{92}U \rightarrow \,^{234}_{90}Th + \,^{4}_{2}He + \gamma$$

Inferring mode of decay Once you know what kind of nucleus is produced by a simple nuclear reaction, you can determine what type of decay has taken place. If both the atomic number and mass number decrease, alpha decay has occurred. If the mass number stays the same but the atomic number increases, beta decay has occurred. If neither atomic number nor mass number changes, only gamma radiation has been emitted.

In more complex reactions, more than one type of decay can take place at the same time. In some cases, one kind of radioactive nucleus decays to form another, which decays to form a third kind, and so on. The decay series ends with the production of a stable nucleus. For example, uranium-238 undergoes a 14-step process involving alpha, beta, and gamma decay, and eventually forms stable lead-206.

Detecting radioactivity Radioactivity cannot be seen, heard, or touched, and it has no taste or smell. Other means of detection must be used to tell whether radioactive materials are nearby. Photographic film can serve as a simple detection device. However, for film to be effective as a detector, the source of radioactivity has to be close. Instruments such as Geiger counters and scintillation counters are used to measure radiation.

Geiger counters The Geiger counter is one the most common and well known radioactivity detectors. The Geiger counter detects ionizing radiation. Ionizing radiation is radiation that is energetic enough to ionize matter with which it collides. **Figure 21.7** shows how a Geiger counter detects ionizing radiation.

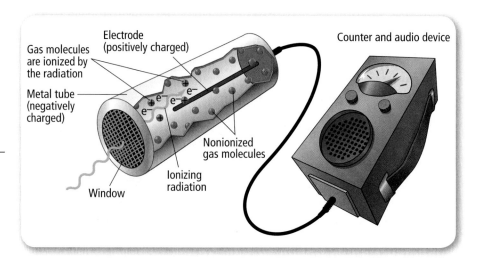

■ **Figure 21.7** A Geiger counter is used to detect and measure radiation levels. Ionizing radation produces an electric current in the counter. A measure of the current is displayed on a scaled meter, and a speaker produces audible sound.

As shown in **Figure 21.7,** ionizing radiation that enters a window at the end of a Geiger counter's metal tube ionizes gas particles contained inside the tube. When the gas ions and free electrons strike an electrode inside the tube, they cause an electric current. In turn, this current ultimately moves the display needle and causes the clicking you hear when a Geiger counter is detecting ionizing radiation.

Scintillation counters A scintillation counter registers the intensity of radiation by detecting light. Radioactive samples to be measured are mixed with compounds that emit a flash of light, called scintillation, when exposed to radiation. The level of radiation is measured by the number of flashes of light recorded by the device.

EXAMPLE Problem 21.1

Writing and Balancing a Nuclear Equation Potassium-40 can decay to form calcium-40. Write a balanced equation for this nuclear reaction and determine the type of decay.

Problem-Solving Hint
Do not be tempted to change mass number or atomic numbers of the knowns in solving this kind of problem.

1 Analyze
Write the symbols for the two isotopes involved. Each symbol must include the abbreviation of the element represented, the mass number as a superscript, and the atomic number as a subscript.

Potassium-40 is $^{40}_{19}K$ and calcium-40 is $^{40}_{20}Ca$.

2 Set up
Potassium is the reactant and goes on the left. The product, calcium, goes on the right.

$$^{40}_{19}K \rightarrow {}^{40}_{20}Ca + \,?$$

3 Solve
Because the mass numbers are already balanced (40 = 40), you have to balance only the atomic numbers. Potassium has an atomic number of 19, and the atomic number of calcium is 20. Because a beta particle ($^{0}_{-1}e$) has a negative charge and almost no mass, adding such a particle to the right side will balance the equation for atomic number.

$$^{40}_{19}K \rightarrow {}^{40}_{20}Ca + {}^{0}_{-1}e$$

4 Check
The mass number on the left equals the sum (40 + 0) on the right. The atomic number on the left is 19, which equals the sum (20 + (−1)) on the right. The equation is balanced. Because an electron (a beta particle) is released, beta decay has taken place.

PRACTICE Problems
Solutions to Problems Page 867

1. Write the balanced nuclear equation for the radioactive decay of radium-226 to give radon-222, and determine the type of decay.
2. Write a balanced equation for the nuclear reaction in which neon-23 decays to form sodium-23, and determine the type of decay.

SUPPLEMENTAL PRACTICE
For more practice with balancing nuclear equations, see Supplemental Practice, page 842.

CHEMLAB

MODEL RADIOACTIVE DECAY

Background
The nuclei of unstable atoms disintegrate or decay spontaneously, emitting alpha or beta particles and gamma radiation. Types of atoms that undergo this process are called radioactive isotopes. A decaying isotope is referred to as a parent atom, and the atom produced is a daughter atom. In this ChemLab, heads-up pennies represent individual parent atoms of the fictitious element pennium, and tails-up pennies represent the daughter atoms of the decay. You will study the decay characteristics of pennium and will determine its half-life, which is the time required for one-half of the atoms to decay.

Question
What is the half-life of the fictitious radioisotope pennium?

Objectives
- **Infer** the decay characteristics of pennium.
- **Analyze** your data to determine the half-life of pennium.
- **Make and use** graphs to interpret data.

Preparation

Materials
1 shoe box with lid
120 pennies
stopwatch or watch with a second hand

Procedure

1. Count 120 pennies and lay them heads-up in the bottom of the shoe box. In a table like the one in Data and Observations, record this as 0 daughter atoms and 120 parent pennium atoms at time 0.

2. Put the lid on the shoe box and, holding the lid on tightly, shake the box with moderate force up and down 20 times while your lab partner times the decay process to the nearest second. Assume that all the subsequent decay trial times are identical.

Data and Observations

Data Table		
Total Elapsed Time	Number of Tails-Up Daughter Atoms Removed	Number of Heads-Up Parent Atoms Remaining

3. Open the box and count the tails-up pennies, which represent the daughter atoms, and remove them from the box. Subtract this number from the original number of heads-up pennies to determine the number of remaining parent pennium atoms. Record the time taken for the shaking, the number of tails-up atoms removed, and the number of remaining parent atoms.

4. Repeat steps 3 and 4 four more times, recording your data after each trial. Simply add on the original shaking time repeatedly to arrive at the total elapsed time.

Analyze and Conclude

1. **Construct** a graph of your data, plotting the number of remaining parent pennies on the vertical axis (y-axis) and the total elapsed time on the horizontal axis (x-axis).
2. **Interpret** What is the half-life of pennium in your experiment? Explain how you arrived at this number.

Apply and Assess

1. **Infer** Does exactly the same fraction of pennium atoms decay during each half-life? What does this suggest about half-life? Why are such variations not likely to be obvious when actual atoms are involved?
2. **Calculate** If you had started with one mole of pennies (assuming you could actually take the time to count them), how many would remain after ten half-lives? Is this a large number?
3. **Evaluate** If you took a longer time to shake the box in this case, how would half-life be affected? Are the half-lives of atoms also controllable by a change in conditions?

INQUIRY EXTENSION

Redesign this procedure using common objects other than coins to represent radioisotopes. For your chosen object, specify what is meant by a half-life and how you determine which objects to remove with each half-life.

CHEMISTRY AND TECHNOLOGY

Archaeological Radiochemistry

What do a piece of wood from an Egyptian pyramid, a bone tool from an archaeological dig, and the Dead Sea Scrolls have in common? They all contain carbon from dead organisms. Thus, archaeologists can determine an object's age using a radioactive isotope, specifically carbon-14.

Development of Carbon-14 Dating
Archaeologists need accurate dating techniques for their studies of artifacts and remains of early humans, such as the skull in **Figure 1.** Before the advent of radiochemistry, which is the chemistry of radioactive substances, archaeological dating had to use indirect, time-consuming, and inaccurate methods. In 1949, Willard Libby, working at the University of Chicago, developed the carbon-14 method for dating objects that contain carbon. The first artifact to be dated using carbon-14 was a beam of cypress wood taken from a pharaoh's tomb. Since that time, carbon-14 dating has been widely used to date plant and animal materials.

■ **Figure 1** Ancestral human skulls are dated using carbon-14.

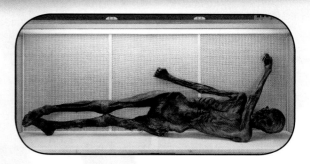

■ **Figure 2** Ötzi the Iceman lived 5300 years ago.

Carbon-14 forms in the upper atmosphere. When cosmic rays collide with atoms in the upper atmosphere, the atoms break up and release subatomic particles. Carbon-14 is made when a neutron collides with a nitrogen atom, causing it to lose a proton.

$$^{14}_{7}N + ^{1}_{0}n \rightarrow ^{14}_{6}C + ^{1}_{1}p$$

When the ratio of atoms of radioactive carbon-14 to atoms of radioactive carbon-12 in a once-living material is determined and compared with the ratio that is calibrated to have existed while the organism was living, the age of the material can be determined up to a maximum of 50,000–60,000 years old.

Ötzi the Iceman
In 1991, a hiker found a man who had been frozen in a glacier for many centuries in the Alps on the border between Austria and Italy. Using carbon-14 dating, scientists established that the man lived about 5300 years ago, making the human remains the oldest ever found intact. The ancient iceman shown in **Figure 2** had many secrets to reveal about himself and his way of life, for he was completely outfitted and carried various tools and devices with him. Scientists discovered that people at that time had developed wooden-handled copper axes. Also, the people cut their hair, had tattoos, and had knowledge of healthful properties of plants. Parts of the Copper Age history could now be rewritten, and what made it all possible was being able to find out the age of the remains.

Another secret that puzzled scientists was how the iceman, named Ötzi by scientists, had died. In 2001, an arrowhead was found in his back, but scientists were not sure that the arrowhead killed him because people can live with foreign materials imbedded in their bodies. Further study of where the arrowhead entered the body revealed arterial damage. This arterial damage would have caused massive blood loss. Scientists now conclude that Ötzi the Iceman bled to death shortly after being shot with the arrow. The position of the body also suggests that whoever shot Ötzi the Iceman turned him over to retrieve the arrow.

■ **Figure 4** Clay pots are dated using thermoluminescence.

Other Radiometric Dating Systems

Sometimes an artifact contains little or no carbon or is too old to be dated with carbon-14, so an alternative method has to be used. The method depends on what radioactive substances are present in the material. Two of the methods include rubidium-strontium dating and potassium-argon dating.

Archaeologists are also using a new method to date pottery, such as the clay pot shown in **Figure 4,** called the thermoluminescence (TL) method. As small amounts of radioactive materials such as uranium and thorium decay in clay, they excite other atoms in the clay to a higher energy state. When the clay is heated to at least 450°C, light called thermoluminescent glow is emitted as electrons fall back to their stable levels. Measuring this glow reveals roughly how long ago the artifact was made.

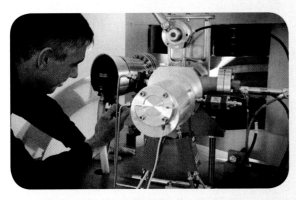

■ **Figure 3** Tandem accelerator mass spectrometer (TAMS) used in carbon-14 dating

TAMS: An Advanced Dating Method

In the standard lab method of carbon-14 dating, a piece of the object being tested has to be burned to produce carbon dioxide. The carbon-14 atoms in the carbon dioxide are then counted using a scintillator that counts the light flashes from beta radiation as the carbon-14 atoms decay. This method requires a fairly large sample because several grams of carbon are needed to accurately count the decay of carbon-14 atoms. The counting also takes days or weeks because of the slow decay of carbon-14. With the development of a device called a tandem accelerator mass spectrometer (TAMS) shown in **Figure 3,** carbon-14 dating can now be done quickly on only a few milligrams of sample in less than an hour.

Discuss the Technology

1. **Acquire Information** Carry out library research to obtain information on a radiometric dating technique not explained in this feature.

2. **Infer** Iron is not a once-living material. How, then, can iron tools be dated using the carbon-14 method?

3. **Apply** The thermoluminescence method is not highly accurate. However, it is useful in exposing new forgeries in pottery. Explain how this could be so.

Table 21.2	Half-Life Values for Some Radioactive Isotopes
Isotope	Half-Life
$^{3}_{1}H$	12.26 years
$^{14}_{6}C$	5730 years
$^{32}_{15}P$	14.282 days
$^{40}_{19}K$	1.25 billion years
$^{60}_{27}Co$	5.271 years
$^{85}_{36}Kr$	10.76 years
$^{93}_{36}Kr$	1.3 seconds
$^{87}_{37}Rb$	48 billion years
$^{99m}_{43}Tc$	6.0 hours*
$^{131}_{53}I$	8.07 days
$^{131}_{56}Ba$	12 days
$^{153}_{64}Gd$	242 days
$^{201}_{81}Tl$	73 hours
$^{226}_{88}Ra$	1600 years
$^{235}_{92}U$	710 million years
$^{238}_{92}U$	4.51 billion years
$^{241}_{95}Am$	432.7 years

*The m in the symbol tells you that this is a metastable element, which is one form of an unstable isotope. Technetium 99m gives off a gamma ray to become a more stable form of the same isotope, with no change in either atomic or mass number.

■ **Figure 21.8** During each half-life period, half of the radioactive nuclei in a sample decay. After one half-life, 50 percent of the sample remains. After two half-lives, 25 percent of the sample remains. After three half-lives, 12.5 percent remains, and so on.

Estimate *Use the graph to estimate how much material remains after 1.5 half-lives.*

Half-Life and Radioisotope Dating

Unlike chemical reaction rates, which are sensitive to factors such as temperature, pressure, and concentration, the rate of spontaneous nuclear decay cannot be changed. It is impossible to predict when a specific nucleus in a sample of a radioactive material will undergo decay. Because the overall rate of decay is constant, you can predict when a given fraction of the sample will have decayed.

Rate of decay The **half-life** is the time it takes for half of a given amount of a radioactive isotope to undergo decay. Half-life ($t_{1/2}$) is easy to measure and has been determined for many different radioisotopes. Some half-lives are only fractions of a second, whereas others are billions of years. Half-lives for some of the most commonly used radioisotopes are listed in **Table 21.2**. The concept of half-life is illustrated in the graph in **Figure 21.8**.

Radioactive decay has provided scientists with a technique for determining the age of fossils, geological formations, and human artifacts. Using a knowledge of the half-life of a given radioisotope, one can estimate the age of an object in which the isotope is found. Four different isotopes are commonly used for dating objects: carbon-14, uranium-238, rubidium-87, and potassium-40.

Carbon-14 dating All organisms take in carbon during their lifetime—plants from carbon dioxide gas and animals from the plants and animals they eat. Most carbon taken in by organisms is in the form of the stable isotopes $^{12}_{6}C$ and $^{13}_{6}C$, but about one carbon atom in every million is the radioisotope $^{14}_{6}C$. As long as an organism is alive, the $^{14}_{6}C$ in its cells remains relatively constant at about one in every million carbon atoms. After death, however, no new carbon—and no new $^{14}_{6}C$—is taken in from the surroundings. Because the organism no longer takes in any carbon at all, the proportion of $^{14}_{6}C$ slowly decreases as this radioisotope undergoes beta decay to stable nitrogen-14.

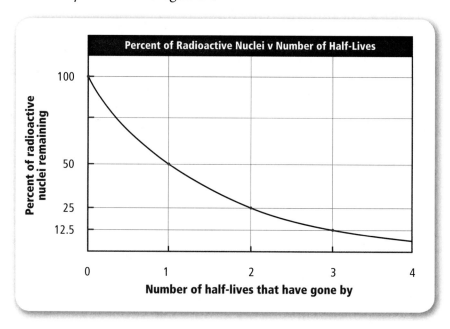

In carbon-14 dating, the fraction of $^{14}_{6}C$ that remains in material such as bone or skin is measured and compared with how much was in the material when it was alive. In this way, the age of the object can be estimated. For example, if half of the $^{14}_{6}C$ remains, one half-life period (5730 years) has passed, and the object is about 5730 years old. If only one-fourth of the $^{14}_{6}C$ remains, two half-life periods (11,460 years) have passed. **Figure 21.9** shows some examples in which carbon-14 dating proved useful.

Dating with other radioisotopes
The half-life of carbon-14, 5730 years, is short compared with the age of many fossils and geological formations. As a result, objects more than about 60,000 years old cannot be dated using this technique because the amount of carbon-14 left in them is too small to be measured accurately. To date objects that are more than 60,000 years old, other isotopes must be used. Rocks and minerals that are up to billions of years old can be dated using radioisotopes with long half-lives, such as potassium-40 ($t_{1/2}$ = 1.25 billion years), uranium-238 ($t_{1/2}$ = 4.5 billion years), and rubidium-87 ($t_{1/2}$ = 48 billion years). Objects dated by such techniques are illustrated in **Figure 21.10**.

■ **Figure 21.9** Carbon dating is used to determine the age of organisms and objects containing carbon. The mammoth skull in the photo to the left is 10,000–18,000 years old. The Inca artifacts in the photo to the right are from the fifteenth or sixteenth century.

■ **Figure 21.10** Objects, such as this 3.3-million-year-old skull and these 3.5-billion-year-old stromatolites, are too old to use carbon dating. Instead, dating is done with radioisotopes with long half-lives, such as potassium-40.

Art Forger, van Meegeren—Villain or Hero?

Hans van Meegeren, a skilled Dutch artist, painted forgeries in the style of the great seventeenth century Dutch painter Vermeer. In 1937, van Meegeren sold his best forged Vermeer, *Christ and His Disciples at Emmaus,* shown in **Figure 1,** to a Dutch museum for $280,520. He continued to forge and sell Vermeers and other Dutch masters until 1945.

Wartime forgery During the Nazi occupation of Holland in the early 1940s, van Meegeren sold his forged Vermeer paintings to members of the Nazi government. Part of the price Meegeren exacted for his paintings was the return of 200 works of Dutch art that had been looted earlier in the war by the Nazis. At the end of the war, in 1945, a forged Vermeer was found with the Nazis' treasures and traced to van Meegeren. When he was charged as a Nazi collaborator for selling Dutch national treasures, he confessed to the forgeries. He pointed out that by exchanging the forgeries, he had saved many Dutch art treasures from the Nazis. However, the authorities did not believe that he was a forger. He was sentenced to a year in prison, but died before serving it.

Radioactive dating resolves the problem It was not until 1968 that an American scientist named Bernard Keisch presented irrefutable evidence that van Meegeren was an art forger. Keisch's work made use of uranium-lead dating. White lead, a combination of $PbCO_3$ and $Pb(OH)_2$, is an important pigment in oil painting. The parent source of the stable lead-206 in white lead is uranium-238. Uranium-238 has a half-life of 4.5 billion years, decaying into a series of unstable isotopes, including radium-226 and polonium-210, until it finally forms stable isotope lead-206.

During the processing of white lead, most of the isotopes preceding polonium-210 in the uranium-decay series are removed. However, minute amounts of radium-226 remain. Because the radioactive elements that would be needed to make the polonium-210 are mostly removed, the amount of polonium-210 decreases as the pigment ages. The ratio between the amounts of the radium-226 and the polonium-210 is used to establish the age of paintings. A low ratio of radium to polonium indicates a recent painting.

Proof of a fake Keisch tested a white lead sample from the alleged fake "Emmaus" painting, shown in the image below, and found a low Ra:Po ratio. The ratio indicated that the painting was only about 30 years old, not 260, and therefore a forgery.

Figure 1 Emmaus painting that was exposed as a forgery

Connection to Chemistry

1. **Acquire Information** Belgian chemist Coremans also exposed the fraudulent Vermeers. Find out how.
2. **Interpret** You have two paintings. One is several hundred years old, the other one year old. Painting A has a Po concentration of 12 and a Ra concentration of 0.3. Painting B has a Po concentration of 5 and a Ra concentration of 6. Which painting is which?

EXAMPLE Problem 21.2

Determining the Age of a Fossil A fossilized tree killed by a volcano was studied. It had 6.25 percent of the amount of carbon-14 found in a sample of the same size from a tree that is alive today. When did the volcanic eruption take place?

Problem-Solving Hint
Repeatedly multiply 1/2 by itself until the number of factors produces the proper result.

1 Analyze
- The half-life of carbon-14 is 5730 years, and the fraction of carbon-14 remaining is 6.25 percent (0.0625).

2 Set up
- Calculate the number of half-lives that have passed for the carbon-14 in the sample. During each half-life of a radioactive isotope, one-half of the nuclei decay. The fossil sample has 6.25 percent (0.0625) of its original amount of carbon-14 left, so you need to find out how many times 1/2 (0.5) must be used as a factor to produce 0.0625 as a result. The answer is four times because $1/2 \times 1/2 \times 1/2 \times 1/2 = 0.5 \times 0.5 \times 0.5 \times 0.5 = 0.0625$ (or, carrying out the calculation another way, 100 percent $\times 1/2 \times 1/2 \times 1/2 \times 1/2 = 6.25$ percent). Therefore, four half-lives have gone by.

3 Solve
- Because four half-lives have gone by and each is 5730 years, multiply 5730 by 4.

$5730 \times 4 = 22{,}920$

The tree from which the sample was taken must have been killed by the volcano about 22,920 years ago.

4 Check
- Check to be sure that the calculation can be repeated with the same result. All aspects of setup and calculation appear to be correct.

PRACTICE Problems

Solutions to Problems Page 867

3. A rock was analyzed using potassium-40. The half-life of potassium-40 is 1.25 billion years. If the rock had only 25 percent of the potassium-40 that would be found in a similar rock formed today, calculate how long ago the rock was formed.

4. Ash from an early fire pit was found to have 12.5 percent as much carbon-14 as would be found in a similar sample of ash today. How long ago was the ash formed?

Section 21.1 Assessment

Section Summary
- Nuclear reactions involve protons and neutrons.
- Radioactivity can be detected and measured using film and instruments, including Geiger and scintillation counters.
- The half-life of a radioactive element is the time it takes for one-half of the nuclei in a sample to decay.

5. **MAIN Idea Explain** what is meant by radioactivity, and provide an example of a radioactive isotope.

6. **Describe** What is a Geiger counter used for? Describe its operation.

7. **Assess** If half the atoms in a sample decay in one year, why is it incorrect to conclude that the rest of them will decay in one more year?

8. **Infer** The oldest rocks dated so far on Earth are about 3.8 billion years old. Do you think that this fact was determined by means of carbon-14 dating? Explain.

9. **Evaluate** Suppose you were given an ancient wooden box. If you analyze the box for carbon-14 activity and find that it is 50 percent of that of a new piece of wood of the same size, how old is the wood in the box?

Section 21.2

Objectives
- **Compare and contrast** nuclear fission and nuclear fusion.
- **Demonstrate** equations that represent the changes that occur during radioactive decay.
- **Trace** the operation and structure of a fission reactor.

Review Vocabulary
half-life: the time it takes for half of a given radioactive isotope to decay into a different isotope or element

New Vocabulary
nuclear fission
nuclear reactor
nuclear fusion
deuterium
tritium

Nuclear Reactions and Energy

MAIN Idea Fission and fusion release tremendous amounts of energy.

Real-World Reading Link Both the United States and the former Soviet Union have sent nuclear reactors into space to power satellites. However, another giant nuclear reactor was there first—the Sun, our local star. Energy from the Sun sustains the plants that are the basis for all of our food, and warms our air and water. Is it possible to copy the reactions that take place in stars?

The Power of the Nucleus

Compared to chemical reactions, nuclear reactions involve enormous energy changes. The energy change is due in part to the conversion of small amounts of the mass of nuclear particles into energy. Albert Einstein was the first scientist to realize the enormous amount of potential energy available in matter. He realized that mass and energy are equivalent and related by the following equation.

$$E = mc^2$$

In this equation, E represents energy, m is mass, and c is the speed of light (3.00×10^8 m/s).

Because the value of the speed of light is so high, you can see that a small amount of mass can be converted into an enormous amount of energy. That explains why a nuclear weapon, such as an atomic bomb, is so much more powerful than a weapon that involves only chemical reactions, such as a bomb made with dynamite. Modern nuclear power plants, shown in **Figure 21.11**, operate on the same principle. However, in a power plant the reactions that convert mass to energy are carefully monitored and controlled.

■ **Figure 21.11** The main parts of a nuclear power plant are the reactor, which is located under the dome, and the cooling towers.

Nuclear Fission

Until 1938, all nuclear reactions observed involved the movement of radiation, such as alpha or beta particles, into or out of a decaying nucleus. Shortly before World War II, Otto Hahn, a German chemist, discovered that bombarding uranium with neutrons resulted in the formation of an isotope of barium that is only about 60 percent of the mass of the uranium isotope.

This result was so unusual that Hahn asked another physicist, Lise Meitner, shown in **Figure 21.12,** to help him interpret his findings. She determined that the bombardment of uranium with neutrons had resulted in the splitting of the nucleus into two pieces. She called this transformation nuclear fission. The word fission means "splitting." **Nuclear fission** is the splitting of an atomic nucleus into two or more smaller fragments, accompanied by a large release of energy.

The fission process Why did bombarding uranium atoms with neutrons result in fission? Remember that changing the ratio of neutrons to protons can make a nucleus less stable. Adding a neutron to a uranium-235 nucleus causes the nucleus to split. Atoms of two different elements are created along with more neutrons. Here is a typical reaction for uranium fission.

$$^{235}_{92}U + ^{1}_{0}n \rightarrow ^{140}_{56}Ba + ^{93}_{36}Kr + 3^{1}_{0}n$$

Notice that the sums of the mass numbers on the left and right sides are equal, and so are the sums of the atomic numbers. Therefore, the nuclear equation is balanced.

As World War II started, the race was on to find a way to sustain fission in a chain reaction. A chain reaction is a continuing series of reactions in which each produces a product that can react again. Each neutron produced during the fission of one atom of uranium has the potential to cause the fission of another nucleus. Notice that each fission reaction produces three neutrons—enough to keep the process going. The fission reactions can continue in a chain, as shown in **Figure 21.13**.

■ **Figure 21.12** Lise Meitner was a physicist who was born in Austria but had to flee after the Nazis gained control of her country. She first described nuclear fission. Fission can release amounts of energy that are far greater than those released during radioactive decay.

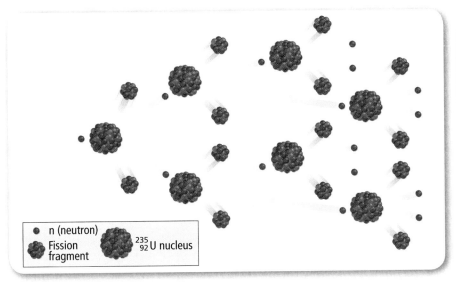

Concepts In Motion

Interactive Figure To see an animation of chain reactions, visit glencoe.com.

■ **Figure 21.13** During a chain reaction, each step produces a reactant for the next step. In the fission of uranium, a neutron enters a uranium-235 nucleus. As a result, the nucleus breaks into two smaller nuclei. At the same time, it releases more neutrons, which are absorbed by other uranium-235 nuclei, and the reaction continues.

Section 21.2 • Nuclear Reactions and Energy

Figure 21.14 Whether a nuclear reaction can be sustained depends on the amount of matter present. In a subcritical mass, the chain reaction does not start because neutrons escape before causing enough fission to sustain the chain reaction. In a supercritical mass, neutrons cause more and more fissions and the chain reaction accelerates.

Compare subcritical mass and critical mass.

Concepts In Motion
Interactive Figure To see an animation of critical mass, visit glencoe.com.

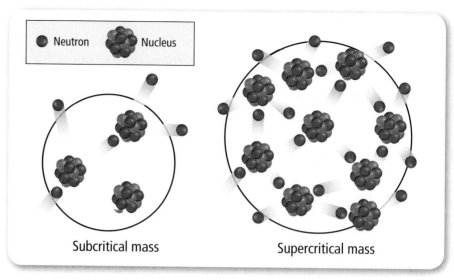

Critical mass For a chain reaction to take place, there must be enough fissionable material present for some of the neutrons produced to be able to collide with other nuclei, as illustrated in **Figure 21.14**. If there is not enough fissionable material present to sustain the chain reaction—in other words, if the sample isn't massive enough—the sample is said to have subcritical mass. A sample that has enough fissionable material to sustain a chain reaction has critical mass.

MiniLab 21.1

Model a Chain Reaction

Can you simulate a nuclear chain reaction with dominoes? Nuclear fission provides the energy in nuclear power plants. During the fission process, a heavy atomic nucleus generally absorbs a neutron, then splits into two smaller atomic nuclei and several more neutrons. If sufficient fissionable material is present, a chain reaction is sustained.

Procedure
1. Read and complete the lab safety form.
2. Obtain two sets of **domino tiles.**
3. Set the individual tiles on end in patterns such that if you first topple a single tile, it causes the other dominoes to fall in series.
4. Practice with possible arrangements until you decide on the best one; that is, one that will topple all the tiles in the quickest, most efficient way. Use a **stopwatch** to determine which arrangement caused the most tiles to fall in the shortest time. When you are ready, call the teacher to observe your domino chain reaction. You may perform a second trial if your first attempt is unsuccessful. The teacher will compare your best result with those of the class.

Analysis
1. **State** How many tiles were involved in your best chain reaction? How quickly did these tiles fall?
2. **Determine** If you were able to arrange one million dominoes so that each falling tile topples exactly two tiles, and assuming 1 second of time required for each of these double topples, how much time would be required to topple all the tiles?
3. **Explain** how this domino exercise simulates a nuclear fission chain reaction.

However, if there is a much greater amount of material present, called supercritical mass, the chain reaction quickly escalates. If the chain reaction takes place too quickly, an explosion will result and release an enormous amount of energy all at once. This is what happens when an atomic bomb explodes. If the rate of the fission chain can be controlled to release energy slowly, the energy can be used to heat materials such as water and then do useful work.

Fission reactors The first major step in developing usable, controllable nuclear power took place on December 2, 1942, in a squash court underneath an unused football field at the University of Chicago. It was there that Enrico Fermi, an Italian physicist, successfully carried out a sustained and controlled fission chain reaction.

Today, through a similar procedure, electrical energy is generated in nuclear power plants through the controlled fission of uranium. The device that is used to extract energy from a radioactive fuel is called a **nuclear reactor**. In the most common type of reactor used today, the fission of a sample of uranium enriched with the isotope uranium-235 releases energy that is used to generate electricity. It is, therefore, desirable to concentrate the uranium-235. Refer to the diagram of the nuclear reactor shown in **Figure 21.15** to learn how a fission reactor works.

VOCABULARY
WORD ORIGIN
Fission
comes from the Latin *findere* meaning *to split*

Concepts In Motion
Interactive Figure To see an animation of a nuclear power plant, visit glencoe.com.

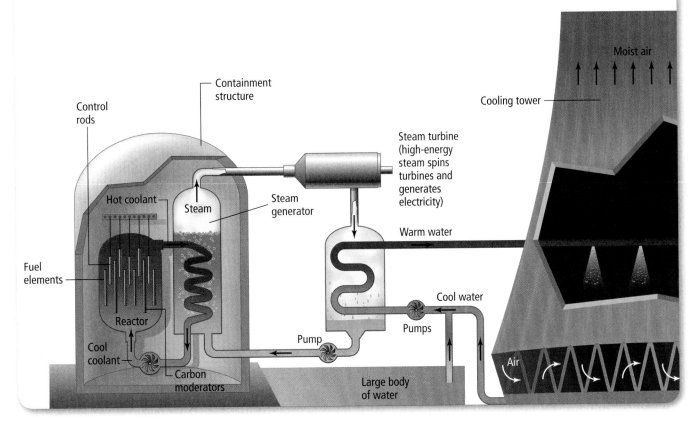

■ **Figure 21.15** In a nuclear reactor, uranium-235 in fuel rods produces fast-moving neutrons and heat in a fission chain reaction. The neutrons are slowed down by a moderator such as water or graphite (carbon) so that they are not moving too quickly to be absorbed by other uranium-235 nuclei. The rate of the reaction is regulated using control rods that absorb some of the neutrons. As illustrated, energy from this reaction produces steam that spins a turbine, producing electricity. The steam is eventually cooled and recycled.

Section 21.2 • Nuclear Reactions and Energy

■ **Figure 21.16** The interior of a reactor is filled with water. A crane is used to extract and replace fuel rods.

■ **Figure 21.17** In 1979, the Three Mile Island reactor became flooded with water contaminated with radioactive material, and radioactive gas was released into the atmosphere. In 1986 at the Chernobyl nuclear power plant, water used to cool the chamber decomposed into hydrogen and oxygen gas, which exploded. After the accident, the damaged reactor was encased in a concrete structure known as the sarcophagus.

More than 100 nuclear power plants are now in operation in the United States, providing almost 20 percent of all the electricity used. Thirty-three states have nuclear plants. Many more plants are operating in other countries. The country that gets the highest percentage of its electrical power from nuclear reactors is France. There, more than 70 percent of the electricity produced comes from nuclear fission reactors. **Figure 21.16** shows the core of a modern, functioning nuclear power plant.

Breeder reactors Plutonium can also be used to power a nuclear fission reactor. Plutonium-239 undergoes a chain reaction that eventually produces more plutonium-239 from uranium-238, as well as heat that is used to generate electricity. This type of fission reactor, which produces fissionable material as it operates, is called a breeder reactor. A breeder reactor can produce more fissionable material than it uses. The construction of this type of reactor has been discouraged in many countries because of the health hazard that plutonium presents, as well as the fact that the plutonium-239 generated can be used to make powerful fission bombs.

Nuclear reactors and pollution Unlike burning fossil fuels, nuclear reactions do not produce pollutants such as carbon dioxide and acidic sulfur and nitrogen compounds. However, the nuclear reactions do form highly radioactive waste that is hard to dispose of safely. Other serious problems include the potential release of radioactive materials into the environment, the limited supply of fissionable fuel, and the higher cost of producing electricity using nuclear fuels rather than fossil fuels. Nuclear reactors that have experienced serious accidents are shown in **Figure 21.17**.

Nuclear Fusion

Energy is also released in a type of nuclear reaction that is the opposite of fission. **Nuclear fusion** is the process of combining two or more nuclei to form a larger nucleus. Fusion is the nuclear process that produces energy in stars like our Sun. In the most typical fusion reaction in the Sun, hydrogen nuclei fuse to form helium.

Three Mile Island nuclear power plant

Chernobyl nuclear power plant

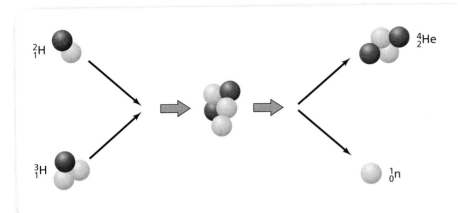

Figure 21.18 When a nucleus of deuterium (hydrogen-2) and a nucleus of tritium (hydrogen-3) undergo nuclear fusion, a nucleus of helium-4 and a neutron are produced.
Compare *the number of neutrons and protons before and after the reaction.*

The fusion process Some of the fusion processes in which hydrogen forms helium have been carried out and studied in laboratories. In one common fusion reaction, two different isotopes of hydrogen combine to form helium and a neutron, as shown in **Figure 21.18**.

$$_1^2H + _1^3H \rightarrow _2^4He + _0^1n$$

Notice that the nuclear equation is balanced. The isotope of hydrogen with a mass number of 2 that is one of the reactants in the equation above is called **deuterium** (D). The hydrogen isotope with a mass number of 3 is called **tritium** (T). Nuclei such as these, which have a small mass, can combine to produce heavier, more stable nuclei. The helium produced has a stable neutron:proton ratio of 1:1. Two deuterium nuclei can also fuse to form either tritium or helium, as shown in the following two equations.

$$_1^2H + _1^2H \rightarrow _1^3H + _1^1p$$

$$_1^2H + _1^2H \rightarrow _2^3He + _0^1n$$

In the first case, tritium and a proton are produced; in the second, $_2^3He$ and a neutron are produced. The fusion of hydrogen produces 20 times the energy produced by the fission of an equal mass of uranium. That large quantity of energy makes the idea of fusion reactors an appealing one. Also, deuterium, the required fuel in fusion, is abundant on Earth. Energy production through nuclear fusion has some other potential advantages over production through fission. No radioactive products are produced, so waste disposal would not be as difficult. A fusion reaction is also easier to control, which would help prevent fires and explosions.

However, there is one major problem with creating a fusion reactor. Compared to uranium nuclei, hydrogen nuclei have much less tendency to react. A large initial input of energy must be provided to start the fusion. In the Sun, enormous pressures and temperatures trigger fusion. To initiate a fusion reaction on Earth, a temperature in excess of 100 million kelvins would be required. Any material used to contain the reaction would melt at that temperature. The difficulty in initiating and containing nuclear fusion has prevented its use as a practical energy source.

VOCABULARY
WORD ORIGIN
Fusion
comes from the Latin *fundere* meaning *to pour*

TRY AT HOME 🏠 LAB
See page 878 for **Modeling Fusion.**

■ **Figure 21.19** In the JET (Joint European Torus) reaction chamber, fusion reactions occur in a plasma confined by strong magnets. For the reactions to occur, extremely large electric currents, beams of hydrogen isotopes, and radio waves must heat the plasma to temperatures in excess of 100 million kelvins. *Torus*—the *T* in *Jet*—refers to the doughnut shape of the chamber.

Tokamak reactors Scientists in a number of countries are researching ways to make commercially feasible fusion reactors. One promising reactor type is called a tokamak, which is shown in **Figure 21.19**. In this reactor, hydrogen nuclei are trapped by powerful magnetic fields produced by huge electromagnets. The nuclei are heated with radio waves to initiate fusion. Because a magnetic field confines the nuclei, no other container is needed. Such a reactor has been operated at the break-even point, where it produces as much energy as is required to run it, but because it produces no net energy, it is not useful. Better electromagnets and structural materials must be developed to correct the engineering problems encountered in the tokamak.

Section 21.2 Assessment

Section Summary

▶ Nuclear reactions convert a small mass into a large amount of energy, according to Einstein's equation $E = mc^2$.

▶ Critical mass is the minimum mass of a sample of fissionable material necessary to sustain a nuclear chain reaction.

▶ Uranium-235 undergoes a fission chain reaction in a nuclear reactor, which converts some of the released energy into electricity.

▶ Fusion in the Sun provides energy for life on Earth.

10. MAIN Idea Describe What is the difference between nuclear fission and nuclear fusion? List one current use for fission.

11. Explain How is fission controlled and maintained in a nuclear reactor?

12. Predict What advantages would a fusion reactor have over the fission reactors currently in use?

13. Compare and Contrast What are the advantages and disadvantages of using energy generated by nuclear fission reactors instead of energy from fossil fuels such as petroleum, natural gas, and coal?

14. Balance The Sun's energy source was a mystery to people for a long time. One nineteenth-century astronomer calculated that if the Sun were made entirely of coal, it would burn for only about 10,000 years. Today, it is known that the Sun's fuel is hydrogen, consumed in fusion reactions in which helium is formed. After the Sun has used up nearly all of its hydrogen, it is expected to begin a fusion reaction in which helium-4 nuclei combine to form carbon-12. Write the nuclear equation for this process, and verify that it is balanced.

Section 21.3

Objectives
- **Distinguish** the biological effects of radiation and the units used to measure levels of exposure.
- **Illustrate** medical and nonmedical uses for radioactivity.

Review Vocabulary
nuclear fusion: the process in which two or more nuclei combine to form a larger nucleus

New Vocabulary
gray
sievert

Nuclear Tools

MAIN Idea Radiation has many useful applications, but it also has harmful biological effects.

Real-World Reading Link Almost everyone gets cuts or scrapes from time to time. Usually, the first thing you do is clean the injury and cover it with a bandage. One of the many uses of radiation is to sterilize medical bandages.

Medical Uses of Radioisotopes

You are much more likely to be saved by some form of nuclear medicine than to be killed by the effects of radiation. For example, radioisotopes are used as tracers to find out where certain chemicals move in the body and to identify abnormalities. The use of radioisotopes in such applications is based on the fact that a radioisotope of any element undergoes the same chemical reactions as stable isotopes of that element. In other words, a molecule of glucose containing a radioactive atom will be metabolized by the body the same way as glucose containing no radioactive atoms. A detector can pick up the particular type of radiation coming from glucose molecules labeled with radiotracers. In this way, the path of the molecules can be followed through the body.

Radioisotopes are widely used in diagnosis to generate images of organs and glands, and in treatments for conditions such as cancer. **Figure 21.20** shows the use of iron-59 to produce an image of a patient's circulatory system. Other isotopes used in medical diagnosis and treatment include iodine-131 and technetium-99m. **Figure 21.21** summarizes several of the more common uses of radioisotopes in medicine.

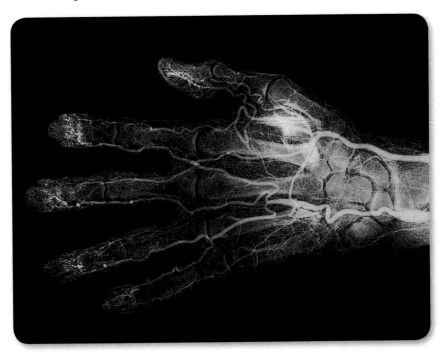

■ **Figure 21.20** Iron-59 was used to make this image of the circulatory system. Iron-59 emits gamma rays and beta particles. It has a half-life of 44.5 days.

FIGURE 21.21

Visualizing Medical Uses of Radioisotopes

Radioisotopes have become extremely useful tools in medical diagnosis and treatment. Using them allows doctors to detect many diseases early and to treat them more successfully than ever before.

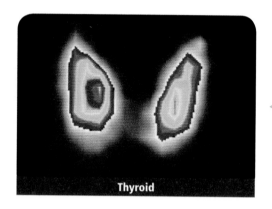
Thyroid

Most of the iodine that enters your body ends up in the thyroid gland, where it is incorporated into hormones that regulate growth and metabolism. If you ingest radioactive iodine-131, an image showing the size, shape, and activity of your thyroid gland can be made. The image is useful in diagnosing medical problems such as hyperthyroidism. The photo shows a normal thyroid gland.

Technetium-99m is a metastable isotope that gives off gamma rays. It is commonly used in medicine because it produces no alpha or beta particles that could cause unnescessary damage to cells and because it has a half life of only about six hours. Injected into the blood stream, technetium-99m attaches to red blood cells. Therefore, it is used to diagnose problems with the heart and circulation. Doctors can use images such as these to determine whether the heart is pumping properly.

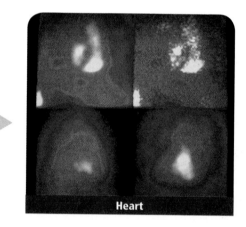
Heart

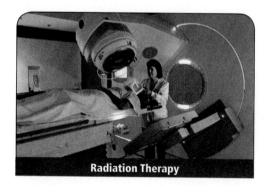
Radiation Therapy

Cobalt-60 is one of the isotopes commonly used in radiation therapy to treat cancer. To destroy cancerous cells while minimizing damage to normal cells, the cobalt-60 source moves around the outside of the patient's body in a circle, with the beam of rays sharply focused on the tumor at all times. This allows any given area of normal tissue to receive only a small dose of radiation while the tumor gets a large total dose.

Nonmedical Uses of Radioisotopes

Hospitals are not the only locations where you'll find radioisotopes in use. You have already read about several nonmedical uses for radioactivity: nuclear reactors, nuclear weapons, and radiodating techniques. As you will see, radioisotopes are also used in research and in the food industry.

Practical uses of tracers If a radioisotope is substituted for a nonradioactive isotope of the same element in a chemical reaction, all compounds formed from that element in a series of steps will also be radioactive. That makes it possible to follow the reaction pathway using instruments that can detect radiation. In this way, the series of steps involved in many important reactions has been studied.

Radioisotopes have many applications outside the chemistry lab as well. **Figure 21.22** shows the use of a radiotracer to study pesticide movement. If, for example, pesticides are detected in a stream, it can be very difficult to tell where they came from and how they got there. However, pesticides labeled with radioisotopes such as sulfur-35 can be used to trace passage of the pesticides from the field to the stream. Again, the principle is similar to the use of radioisotopes to image the body: radioisotopes go through all the same reactions and pathways as their nonradioactive counterparts. Tracers are also used to test structural weaknesses in mechanical equipment and to follow the pathways taken by pollutants.

■ **Figure 21.22** A pesticide that is sprayed onto a field may be transported to other areas by runoff into soil, streams, and groundwater or by movement of animals that have taken in the pesticide. If a pesticide is labeled with a radioisotope, such as sulfur-35, its movement can be traced. Samples of soil or water can be taken at a number of points, and the amount of radiolabeled pesticide at each point can be measured.

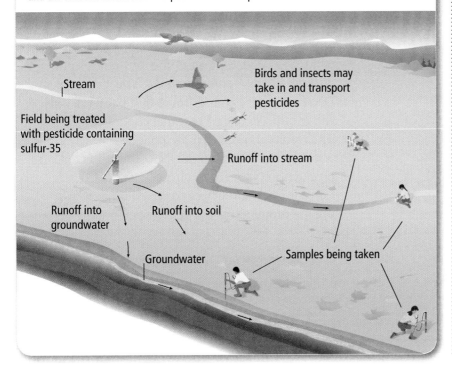

Biology Connection

A Biological Mystery Solved with Tracers

For many years, biologists searched for the identity of the genetic material of life. Some scientists thought that the genetic material was protein. Others believed it was nucleic acids. By 1944, strong evidence pointed to DNA. In 1952, Alfred Day Hershey and Martha Chase published the results of their experiments, confirming that DNA determines heredity.

Radioisotopes used as tracers solve scientific mysteries Radioactive and nonradioactive isotopes of the same element act the same way in a chemical reaction. When scientists want to put tags on a compound, they substitute a radioactive isotope for a nonradioactive isotope in the compound. Then they can use radiation detectors to track and locate the radioisotope tracer.

Hershey and Chase use tracers Bacteriophages (phages) are simple viruses that attack bacteria. They are composed of two parts: a protein coat and a DNA core. The protein is composed of carbon, hydrogen, oxygen, nitrogen, and sulfur. The DNA is composed of carbon, hydrogen, oxygen, nitrogen, and phosphorus. Because sulfur is found only in protein and phosphorus only in DNA, Hershey and Chase chose radioactive ^{35}S and ^{32}P as the tracers for these substances to compare the inheritance of protein and DNA.

Hershey and Chase asked: "When a phage infects a bacterial cell, does the phage inject its DNA core, its protein coat, or both?" They designed experiments using tracers to track the protein coat and DNA core separately. They grew bacteria in a culture containing ^{35}S. When they added phages to this ^{35}S-labeled culture, the new phages produced had protein coats containing ^{35}S. When they infected bacteria having no radioactive materials with the phages containing ^{35}S, they found the ^{35}S only in protein coats of phages outside the new bacteria. The phages made inside the new bacteria had no ^{35}S in their coats. This showed that the protein coat was not injected.

DNA is the genetic material They did a similar experiment shown in **Figure 1** with bacteria grown in a culture containing ^{32}P, to which they added phages. The new phages produced had DNA containing ^{32}P. This time, the DNA with ^{32}P entered unlabeled bacterial cells and produced new phages that contained ^{32}P. The results showed that the phage's DNA core was injected. The phage's DNA alone directed production of an entire phage, with coat. Therefore, Hershey and Chase concluded that DNA, not the protein, was the genetic material.

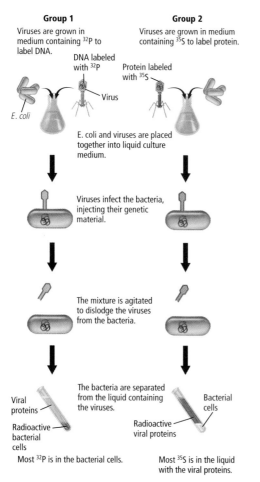

Figure 1 Bacteriophage experiment

Connection to Chemistry

1. **Compare** Compare the structure of a protein with that of DNA.
2. **Interpret** Explain how Hershey and Chase came to their conclusion.

■ **Figure 21.23** Exposure to gamma radiation can extend the shelf life of food by preventing spoilage. Which group of mushrooms do you think was irradiated before storing?

Food irradiation Gamma radiation disrupts metabolism in cells, sometimes enough to kill them or at least keep them from multiplying. This property makes it useful for sterilization of food and surgical instruments. Exposing food to gamma radiation produced by the decay of cobalt-60 nuclei can keep the food from spoiling. The radiation destroys microorganisms and larger organisms such as insects. The food itself does not become radioactive.

Irradiation can extend the shelf life of food so that it can be stored for long periods of time without refrigeration, as shown in **Figure 21.23.** However, people who oppose food irradiation are concerned about what are called unique radiolytic products—URPs, for short. These products result from chemical changes caused by the ionizing effects of radiation. Whether URPs present a hazard has not yet been determined for sure. The U.S. Food and Drug Administration (FDA) has approved the use of irradiation for most fruits and vegetables.

Sources of Radioisotopes

Using nuclear chemistry, scientists today can change one element into another and even produce elements artificially. How are elements made artificially? Some are produced as by-products in nuclear reactors. Most are made by bombarding nuclei with small particles that have been accelerated to high speeds. This is done in instruments like that shown in **Figure 21.24.**

■ **Figure 21.24** The large, expensive, and heavily shielded biomedical cyclotron is used to produce most radioisotopes used in PET scans. Hospitals use a different piece of equipment to make the short-lived technetium-99m isotope just before it is needed in diagnostic techniques.

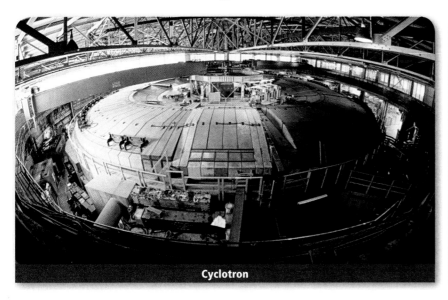

Cyclotron

Technitium-99m

Section 21.3 • Nuclear Tools **767**

Problems Associated with Radioactivity

The understanding of radioactivity has grown rapidly in the 100 years since its discovery. When the Curies worked with radioisotopes, they did not realize how harmful such materials could be. Marie Curie died of leukemia that was probably caused by her years of contact with radioisotopes. Many more radioactive isotopes exist than the few studied by the Curies. In fact, most of the roughly 2000 known isotopes of all elements are unstable and undergo nuclear decay. Fortunately, most of those do not occur naturally, but are produced synthetically. Your surroundings contain mostly stable isotopes of the common elements, so you are normally not exposed to enough radiation to do you much harm.

Background radiation You are constantly being bombarded with low levels of radiation called background radiation. Some of it is in the form of cosmic rays, which are particles that reach Earth from outer space. Small amounts of radioactive elements are found almost everywhere on Earth, as well—in wood and bricks used to make buildings, in the fabrics used in clothing, in the foods you eat, and even inside your body. Traces of uranium in rock layers beneath houses may also produce radioactive radon gas that enters the houses and can present health risks. Various sources of radiation are illustrated in **Figure 21.25.**

Ionizing radiation and health Exposure to radioactive elements can be hazardous to your health. That's because the radiation they give off is powerful enough to knock electrons loose from atoms and generate ions when it collides with neutral matter.

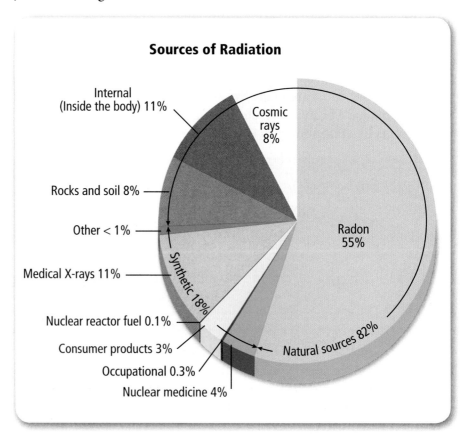

■ **Figure 21.25** As this circle graph shows, natural radioactivity accounts for about 82 percent of the radiation to which you are exposed. Most of the natural radiation is from radon produced by the decay of uranium present in rocks and soil. The remaining 18 percent of the radiation comes from sources produced by humans in the last century, including nuclear medical tools such as X-rays and nuclear fuel used in reactors.

Because nuclear radiation can result in the formation of ions, it is also known as ionizing radiation. In contrast, most electromagnetic radiation—ordinary visible light, for example—does not have enough energy to knock electrons out of neutral atoms and is, therefore, nonionizing. Only radiation such as cosmic rays and X-rays has sufficient energy to generate ions. It is mainly ionizing power that makes radioactive elements dangerous.

Because of the danger, elaborate and expensive precautions must be taken to protect people who work with radioisotopes. Highly radioactive waste products that can take many thousands of years to decay must be stored carefully. Choosing storage sites is difficult, and so is transport of the wastes to the storage sites. Many people are concerned about nuclear processes as energy sources. The threat of nuclear war and concern over radioactive fallout have also caused some people to be reluctant to use any form of radioactivity.

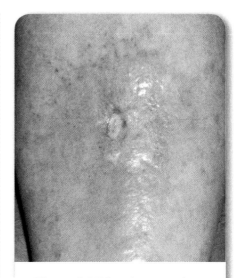

■ **Figure 21.26** Radiation can disrupt cell processes and damage skin.

Radiation exposure Radiation can do severe damage to parts of a single cell and can cause its death. **Figure 21.26** shows a skin lesion resulting from radiation damage to skin cells. Damage to DNA can be especially harmful because DNA is the blueprint from which genetic material for offspring is made. When a cell with damaged DNA divides, all cells made from it also have damaged DNA. When this damage occurs in egg or sperm cells, damaged or mutated DNA can be passed to offspring.

Measures of radiation exposure Several different units can be used to measure the energy in a given amount of radiation. A unit used to measure the received dose of radiation is called the gray. One **gray** is equivalent to the transfer of 1 joule of energy in the form of radiation to 1 kg of living cells. However, not all of the radiation that reaches living tissue gets absorbed by the tissue. The biological damage caused by radiation is indicated best in terms of how much is actually absorbed, which is measured by a unit called a sievert (Sv). One **sievert** is equal to 1 gray, multiplied by a factor that takes into account how much of the radiation hitting the tissue is absorbed by it. Likely harmful biological effects of single doses of radiation of various strengths are listed in **Table 21.3**.

Yearly radiation exposure Background ionizing radiation from natural sources results in a dose of about 0.003 Sv per year for the average person. The U.S. government recommends that your total exposure to sources other than background radiation should be limited to 0.005 Sv per year. Workers in nuclear plants are permitted exposure to 0.05 Sv per year. Both these doses are far below the 2- to 5-Sv single dose that can be lethal. Besides the natural background level, the only significant source of radiation for most people who do not work around radioactive materials is medical radiation, mainly X-rays. The dose equivalent of a chest X-ray is about 0.0005 Sv, and that of a dental X-ray is about 0.0002 Sv, presenting risks that most people consider slight when compared to the potential diagnostic benefits of the X-rays.

Table 21.3 Probable Physiological Effects of Radiation

Dose (Sv)	Effect
0–0.25	No immediate effect
0.25–0.50	Small temporary decrease in white blood cell count
0.50–1.0	Large decrease in white blood cell count, lesions
1.0–2.0	Nausea, hair loss
2.0–5.0	Hemorrhaging, possible death
>5.0	50 percent chance of death within 30 days of exposure

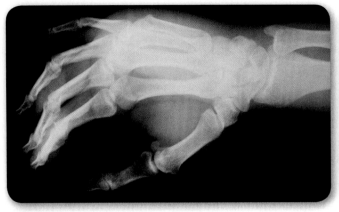 

■ **Figure 21.27** Medical and dental X-rays present only a slight risk due to radiation and are of great benefit in diagnosis. Cigarettes present more serious radiation hazards and provide no benefits.

Lifestyle and environment factors, represented in **Figure 21.27,** can result in more than the normal background exposure to radiation. Cigarette smoke contains significant amounts of radioactive material that can contribute to lung cancer. Smoking two packs of cigarettes a day results in exposure to 0.1 Sv per year. Living at a high altitude or taking frequent trips in airplanes increases your exposure to cosmic rays from outer space. That's because the higher up you are, the less atmosphere there is to block the incoming radiation.

MiniLab 21.2

Test for Radon

How can you test for the presence of radon gas? A high level of radon in a home is correlated to a greater risk of lung cancer for the occupants. Radon-222 is an eventual product of the decay of uranium-238, which occurs naturally in many rocks and soils.

Procedure

1. Read and complete the lab safety form.
2. Your teacher will supply the class with several **commercial radon-detection kits**—one for each group of students. The four-day exposure canisters will probably be the most practical to use; however, the 30-day units will usually give more accurate results.
3. Familiarize yourself with the types of homes, geological and geographic features, and factors such as industries in your area. As a class, decide where each group will place its detector.
4. Obtain permission to test for radon from those who own or live in the test site.
5. Follow the directions on the radon-testing kits, and expose the detectors in the selected home. Send the detectors to be evaluated according to the instructions.
6. Examine your data and collect and tabulate data from the rest of the class.
7. Look for any safety precautions in your radon kit. Be sure to follow them carefully.

Analysis

1. **Identify** Did any of the test results exceed the recommended maximum, which is 4 picocuries per liter? How many did so, and by how much?
2. **Analyze** the class data, looking for correlations between the radon levels and factors such as geological and geographic features, proximity to industrial sites, style of home, and materials used in home construction.
3. **Suggest** several ways in which radon levels in homes might be reduced.

Everyday Chemistry

Radon—An Invisible Killer

If you do not smoke or breathe in cigarette smoke, do you still have a chance of getting lung cancer? The answer is yes. Each year, as many as 15,000–20,000 deaths due to lung cancer are the result of radon pollution. Radon, element 86, is the densest noble gas. All of its isotopes are radioactive. Radon-222 has the longest half-life, 3.823 days. Radon is formed in uranium deposits in Earth's crust. Because it is a gas, it can seep through the rocks and soil to the surface.

How does radon cause lung cancer? Even though radon is radioactive, it is not highly dangerous. Most of it is inhaled and rapidly exhaled, like any other gas. However, radon has a short half-life and quickly changes into radioactive isotopes of polonium and lead.

$$^{222}_{86}Rn \rightarrow {}^{218}_{84}Po + {}^{4}_{2}He$$

$$^{218}_{84}Po \rightarrow {}^{214}_{82}Pb + {}^{4}_{2}He$$

These radioactive isotopes are solids that can collect on dust. When the dust is inhaled, the radioactive solids remain in the lungs. High-energy alpha particles from the polonium and lead damage the DNA of lung cells, sometimes causing cancer.

Can radon get into your home? Radon can enter homes and other buildings through cracks in concrete foundations and slab floors; porous cinderblock walls; unsealed, poorly ventilated crawl spaces; and small areas around water and sewer pipes. In the early 1970s, radon started to cause more extensive problems. People made their homes airtight to save energy and money. That helped to seal in the unwanted radon.

Is your home safe? Because radon is a dense gas, it accumulates near the bottoms of structures. It is simple to test for its presence using detection kits, like that shown in **Figure 1**.

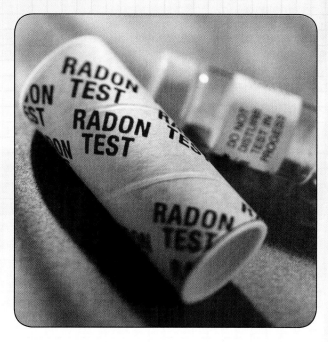

Figure 1 Radon detection tests measure the amount of radon in homes.

Radon detection kits are purchased at hardware stores. The Environmental Protection Agency (EPA) considers a result of more than 4 picocuries per liter (pCi/L) to be unsafe.

How can radon pollution be corrected? Fixing radon pollution problems usually requires only minor repairs or alterations. Installing an exhaust fan in the polluted area or within a foundation might solve the problem. Sealing large cracks or spaces around pipes can also help prevent the entrance of radon.

Explore Further

1. **Compare and Contrast** Compare the use, operation, and effectiveness of short- and long-term radon test devices.

2. **Think Critically** Some of the atoms of radon gas in the lungs can turn out to be dangerous. Explain.

■ **Figure 21.28** Most radioactive waste is low-level waste. That is, there is a fairly small amount of radioactive material in a large volume of waste. Additionally, the radioactive waste generated by hospitals often contains isotopes with short half-lives. Waste from isotopes with short half-lives is easier to deal with than waste containing isotopes with very long half-lives.

Explain why hospital waste often involves isotopes with short half-lives.

Waste disposal You probably associate radioactive waste with nuclear reactors, but more than 80 percent of all such waste is generated in hospitals. **Figure 21.28** shows the type of container used for typical radioactive medical waste. How can this hazardous waste be dealt with so that it cannot harm living things? Most of the radioactive waste material produced at hospitals contains isotopes with short half-lives. That kind of waste can simply be stored until the isotopes have decayed to a safe level.

Although waste from nuclear reactors may not be produced in quantities that are large compared to radioactive hospital waste, it is far more dangerous. For example, spent fuel rods from nuclear reactors contain both short-lived and long-lived radioisotopes produced by collisions of fast-moving particles with atoms in the nuclear fuel and in the walls of the reactor itself. These rods are usually stored at the reactor site for several decades until the isotopes with short half-lives have decayed.

What can be done after that point to isolate the radioisotopes with long half-lives? The plan that currently holds the most promise is to incorporate the unstable nuclei into stable material such as glass, which is then surrounded by canisters made of layers of steel and concrete. The canisters can then be buried deep underground in stable rock formations, like the one shown in **Figure 21.29.** The storage sites would be located in a dry, remote area.

■ **Figure 21.29** Waste from dangerous isotopes with long half lives is much more difficult to deal with. This waste is packaged as safely as possible and then buried deep underground. The Yucca Mountain facility is shown here.

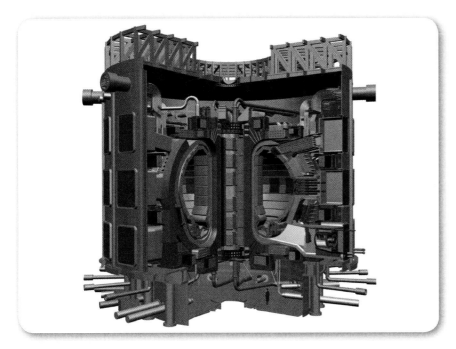

■ **Figure 21.30** Using fusion, rather than fission, to produce electricity would have significant advantages in terms of eliminating radioactive waste. The ITER project is an international project—including the United States, Japan, the European Union, China, India, the Russian Federation, and South Korea—designed to develop the potential for fusion reactors. While much progress has been made, there are still significant hurdles to overcome before human-controlled fusion reactions can be used to generate electricity.

Consider Are there currently any ways to harness the energy from fusion reactions for generating electricity?

An alternative proposal being tested for nuclear waste disposal is burial in clay sediments deep in the ocean. Recycling of uranium for reuse in reactors is also being considered. The uranium can be extracted from used rods that have been chemically dissolved. This uranium can then be made into new rods. However, this recycling process is currently too expensive to be viable. Whether or not recycling and disposal technologies can keep pace with the use of nuclear materials will influence the development of more technologies using these materials. **Figure 21.30** shows another technology that might one day eleminate the waste from fission reactors by using fusion.

Decisions regarding nuclear materials are not simply the business of government officials and scientists. You, too, will have to make decisions about such materials as you help decide how your community supplies energy, whether your country makes nuclear weapons, and, on a more immediate level, whether you undergo medical procedures that employ radioisotopes. How well you understand nuclear chemistry will influence your decisions.

SUPPLEMENTAL PRACTICE

For more practice with nuclear reactions, see Supplemental Practice page 843.

Section 21.3 Assessment

Section Summary

- Radioactive tracers have uses such as tracking compounds in an organism and charting the movement of pollutants.
- Radiation therapy involves selectively killing rapidly dividing cancer cells by targeting them with radiation.
- Gamma radiation can be used to irradiate food to keep it from spoiling.
- Most radiation exposure comes from natural sources.

15. **MAIN Idea Summarize** What is the effect of radiation on living cells?

16. **Identify** What disease is radiation therapy most often used to treat?

17. **Explain** what a PET scan does.

18. **Describe** How could a radioactive tracer be used to study the way glucose is metabolized in cells?

19. **List** What radioactive isotopes might be useful in scanning techniques to examine bones for fractures and other abnormalities? Include at least two elements not discussed in this chapter, and explain your answer.

20. **Explain** Why are foods irradiated? Does this make them radioactive?

CHAPTER 21 Study Guide

Download quizzes, key terms, and flash cards from glencoe.com.

BIG Idea Nuclear chemistry has a vast range of applications, from the production of electricity to the diagnosis and treatment of disease.

Section 21.1 Types of Radioactivity

MAIN Idea Alpha, beta, and gamma radiation are three types of radiation emitted as unstable nuclei decay into stable nuclei.

Vocabulary
- alpha particle (p. 743)
- beta particle (p. 745)
- gamma ray (p. 745)
- half-life (p. 752)
- radioactivity (p. 741)

Key Concepts
- Nuclear reactions involve protons and neutrons.
- Radioactivity can be detected and measured using film and instruments, including Geiger and scintillation counters.
- The half-life of a radioactive element is the time it takes for one-half of the nuclei in a sample to decay.

Section 21.2 Nuclear Reactions and Energy

MAIN Idea Fission and fusion release tremendous amounts of energy.

Vocabulary
- deuterium (p. 761)
- nuclear fission (p. 757)
- nuclear fusion (p. 760)
- nuclear reactor (p. 759)
- tritium (p. 761)

Key Concepts
- Nuclear reactions convert a small mass into a large amount of energy, according to Einstein's equation $E = mc^2$.
- Critical mass is the minimum mass of a sample of fissionable material necessary to sustain a nuclear chain reaction.
- Uranium-235 undergoes a fission chain reaction in a nuclear reactor, which converts some of the released energy into electricity.
- Fusion in the Sun provides energy for life on Earth.

Section 21.3 Nuclear Tools

MAIN Idea Radiation has many useful applications, but it also has harmful biological effects.

Vocabulary
- gray (p. 769)
- sievert (p. 769)

Key Concepts
- Radioactive tracers have uses such as tracking compounds in an organism and charting the movement of pollutants.
- Radiation therapy involves selectively killing rapidly dividing cancer cells by targeting them with radiation.
- Gamma radiation can be used to irradiate food to keep it from spoiling.
- Most radiation exposure comes from natural sources.

Chapter 21 Assessment

Understand Concepts

21. What part of the atom produces radioactivity?

22. What is ionizing radiation?

23. What is given off by a nucleus undergoing each of the following types of decay?
    a) alpha   b) beta   c) gamma

24. What types of material will block alpha particles? Beta particles? Gamma rays?

25. Balance the following equations, which represent nuclear reactions in the uranium-238 decay series.
    a) ? $\rightarrow ^{218}_{84}Po + ^{4}_{2}He$
    b) $^{234}_{90}Th \rightarrow ^{234}_{91}Pa + ?$
    c) ? $\rightarrow ^{214}_{82}Pb + ^{4}_{2}He$
    d) $^{226}_{88}Ra \rightarrow ^{4}_{2}He + ?$

26. Explain why carbon-14 dating cannot be used to determine how old a dinosaur fossil is.

27. What starts the chain reaction that occurs in nuclear fission reactors used today? What keeps it going? How is its rate controlled?

28. What type of nuclear reaction takes place in stars?

29. When each of the following is ejected from a nucleus, what happens to the atomic number and mass number of the atom?
    a) an alpha particle
    b) a beta particle
    c) gamma radiation

30. List three ways that radioisotopes are used in medicine.

31. How is cobalt-60 used to treat cancer?

Apply Concepts

32. Explain why radioisotopes with long half-lives are not administered internally in medical procedures. What are the half-lives of iodine-131, technetium-99m, and gadolinium-153, which are used in many medical procedures?

33. What new element is formed if magnesium-24 is bombarded with a neutron and then ejects a proton?

34. Some types of heart pacemakers contain plutonium-239 as a power source. Write a balanced nuclear equation for the alpha decay of plutonium-239.

Biology Connection

35. What results would Hershey and Chase have observed if they had
    a) radiolabeled only protein?
    b) radiolabeled only DNA?
    c) labeled both the protein and DNA with oxygen-18?

Chemistry and Technology

36. Carbon-14 formation in the atmosphere also results in formation of hydrogen. Explain.

Art Connection

37. If the white lead in a sample of paint taken from an alleged forgery of a famous old painting turns out to have a high ratio of radium-226 to polonium-210, what can be concluded?

Think Critically

Design an Experiment

38. **ChemLab** Design an experiment to determine how long the half-life of pennium is, given standardized shaking of the box.

Draw Conclusions

39. **MiniLab 21.1** Why do most chain reactions stop before all of the reactant elements have been used up?

Form a Hypothesis

40. **MiniLab 21.2** Would you expect to find higher levels of radon in basements of homes that have floors of dirt or floors of thick concrete? Explain.

Chapter 21 Assessment

Cumulative Review

41. List the change of state that takes place during each of the following. *(Chapter 10)*
   a) boiling
   b) melting
   c) sublimation
   d) condensation
   e) freezing
   f) deposition

42. What are four factors that influence the rate of a chemical reaction? *(Chapter 6)*

Skill Review

43. Use a Table Examine **Table 21.4,** which lists information about nuclear power use in selected countries in 2006. Then write a one-paragraph summary of the information, incorporating answers to the questions.

Table 21.4	Nuclear Power Use
Country	Percent Electricity from Nuclear Reactors
Argentina	6.9
Belgium	54
Brazil	3.1
France	78
Hungary	38
Netherlands	3.5
South Korea	39
Sweden	48
United Kingdom	18
United States	19

a) How many countries listed get more than 40 percent of their electricity from nuclear power?
b) Is the percentage of electricity from nuclear power in the United States greater or less than the average percentage for all countries listed?

WRITING in Chemistry

44. Write a history of the development of the tokamak fusion reactor. Research the progress being made in getting experimental tokamak fusion reactors to operate efficiently and economically. When, if ever, do you expect them to exceed the break-even point in terms of energy produced compared to energy input? Are reactors of different designs likely to replace the Tokamak reactor before it is ever used commercially?

45. *Fat Man and Little Boy* is a film about the people involved in the Manhattan Project, the operation that developed the first atomic bomb during World War II. Watch this movie and use library resources to read more about the Manhattan Project. Write a critical review of the film, including a discussion of how accurately you think the science and history are portrayed.

Problem Solving

46. Why is there more concern about radon levels now than there was 30 years ago?

47. What percentage of C-14 would you expect a piece of 34,000 year-old fossilized bone from a mastodon to have when compared to a similar piece of bone from a modern elephant?

48. How old is an Egyptian scroll made of papyrus that contains 75 percent of the amount of C-14 that would be found in a piece of paper today?

49. Mercury-190 has a half-life of 20 minutes. If you obtain a 36.0-mg sample, how much mercury-190 will remain after one hour?

50. The most common isotope of thorium, $^{232}_{90}$Th, is a radioactive alpha-particle emitter. What product results when thorium-232 decays by emitting an alpha particle? Write an equation for the process.

Cumulative
Standardized Test Practice

1. Which is true about beta decay?
   a) Beta particles can be stopped by heavy sheets of paper.
   b) During beta decay, an element will be transformed into a different element.
   c) A beta particle has a similar size and weight as an alpha particle.
   d) After beta decay, an atom will acquire a −1 charge.

2. A human skeleton is discovered by archeologists that is dated back to the year 9400 B.C. Approximately what percent of carbon-14 is still present in the skeleton compared with the amount of carbon-14 present in a living person?
   a) 12.5%
   b) 25%
   c) 45%
   d) 61%

3. The reaction between nickel and copper(II) chloride is:

   $Ni(s) + CuCl_2(aq) \rightarrow Cu(s) + NiCl_2(aq)$

   The half reactions for this redox reaction are
   a) $Ni \rightarrow Ni^{2+} + 2e^-$; $Cl_2 \rightarrow 2Cl^- + 2e^-$
   b) $Ni \rightarrow Ni^+ + e^-$; $Cu^+ + e^- \rightarrow Cu$
   c) $Ni \rightarrow Ni^{2+} + 2e^-$; $Cu^{2+} + 2e^- \rightarrow Cu$
   d) $Ni \rightarrow Ni^{2+} + 2e^-$; $2C^+ + 2e^- \rightarrow Cu$

4. An electrical potential difference is also called
   a) oxidation.
   b) reduction.
   c) corrosion.
   d) voltage.

5. Hydrocarbons containing a triple bond are called
   a) isomers.
   b) alkynes.
   c) alkenes.
   d) alkanes.

6. The general formula for a carbohydrate is
   a) $C_6H_{12}O_6c$)
   b) $C_{12}H_{24}O_{12}$
   c) CH_2O
   d) $C_2H_3O_2$

7. During a catalyzed reaction, the catalyst
   a) provides additional energy to increase the rate of the reaction.
   b) reduces the required activation energy to increase the rate of the reaction.
   c) transforms an endothermic reaction into an exothermic reaction.
   d) transforms an exothermic reaction into an endothermic reaction.

8. Albert Einstein's equation $E = mc^2$ indicates
   a) mass and energy are the two basic entities in the universe.
   b) matter and energy are connected and interchangeable at the speed of light.
   c) only small amounts of energy are needed to create large quantities of matter.
   d) small quantities of matter can be converted into enormous amounts of energy.

9. The immense energy released by the Sun is due to which of the following reactions occurring within its core?
   a) nuclear fission
   b) gamma decay
   c) nuclear fusion
   d) alpha decay

10. Which is NOT a chemical reaction?
    a) dissolution of sodium chloride in water
    b) combustion of gasoline
    c) fading of wallpaper by sunlight
    d) curdling of milk

NEED EXTRA HELP?

If You Missed Question...	1	2	3	4	5	6	7	8	9	10
Review Section...	21.1	21.1	16.1	17.1	18.1	19.1	20.1	21.2	21.2	6.1

Student Resources

For students and parents/guardians

The Chemistry Skill Handbook helps you review and sharpen your skills so you get the most out of understanding how to solve math problems involving chemistry. Reviewing the rules for mathematical operations such as scientific notation, fractions, and logarithms can also help you boost your test scores.

The Chemistry Data Handbook is another tool that will assist you. The practice problems and solutions are resources that will help increase your comprehension.

Table of Contents

Student Resources 778

Chemistry Handbook 783
- Measurement in Science 783
- Relating SI, Metric, and English Measurements 786
- Making and Interpreting Measurements 789
- Expressing the Accuracy of Measurements 790
- Expressing Quantities with Scientific Notation 792
- Computations with the Calculator 796
- Using Dimensional Analysis 798
- Organizing Information 800
- Ratios, Fractions, and Percents 805
- Operations Involving Fractions 806

Supplemental Practice 807

Safety Handbook 844
- Safety Guidelines in the Chemistry Laboratory 844
- First Aid in the Laboratory 844
- Safety Symbols 845

Reference Tables 846
- D.1 The Modern Periodic Table 846
- D.2 Color Key 848
- D.3 Symbols and Abbreviations 848
- D.4 Alphabetical Table of the Elements 849
- D.5 Properties of Elements 850
- D.6 Electron Configurations of the Elements 853
- D.7 Useful Physical Constants 855
- D.8 Names and Charges of Polyatomic Ions 855
- D.9 Solubility Guidelines 856
- D.10 Solubility Product Constants 856
- D.11 Acid-Base Indicators 856

Solutions to Problems 858

Try at Home Labs 868

Glossary/Glosario 879

Index 897

Make Comparisons

Why learn this skill?

Suppose you want to buy a portable mp3 music player, and you must choose between three models. You would probably compare the characteristics of the three models, such as price, amount of memory, sound quality, and size in order to determine which model is best for you. In the study of chemistry, you often make comparisons between the structures of elements and compounds. You will also compare scientific discoveries or events from one time period with those from another period.

Learn the Skill

When making comparisons, you examine two or more items, groups, situations, events, or theories. You must first decide what will be compared and which characteristics you will use to compare them. Then identify any similarities and differences.

For example, comparisons can be made between the two atoms in the illustration on this page. The structure of the hydrogen atom can be compared to the structure of the oxygen atom. By reading the labels, you can see that both atoms have protons, neutrons, and electrons. However, you can also see that the number of protons and electrons are different, as are the overall sizes of the atoms.

Practice the Skill

Create a table with the heading *Hydrogen and Oxygen Atoms*. Make three columns. Label the first column *Protons*. Label the second column *Neutrons*. Label the third column *Electrons*. Make two rows. Label the first row *Hydrogen*. Label the second row *Oxygen*. List the number of protons for each atom in the first column. Fill in the number of neutrons and electrons for each atom in the remaining columns. When you have finished the table, answer the following questions.

1. What is being compared? How are they being compared?
2. What do hydrogen and oxygen atoms have in common?
3. Describe how the differences between these two atoms affect the number of energy levels that each atom has.

Apply the Skill

Make Comparisons On page 600 you will find illustrations of a voltaic cell and an electrolytic cell. Compare these two illustrations carefully. Then, identify the similarities and the differences between the two cells.

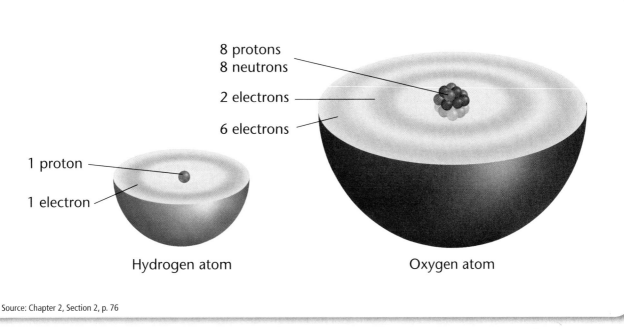

Hydrogen atom — 1 proton, 1 electron

Oxygen atom — 8 protons, 8 neutrons, 2 electrons, 6 electrons

Source: Chapter 2, Section 2, p. 76

Take Notes and Outline

Why learn this skill?

One of the best ways to remember something is to write it down. Taking notes—writing down information in a brief and orderly format—not only helps you remember, but also makes studying easier.

Learn the Skill

There are several styles of note taking, but all explain and put information in a logical order. As you read, identify and summarize the main ideas and details that support them and write them in your notes. Paraphrase—that is, state in your own words—the information rather then copying it directly from the text. Using note cards or developing a personal "shorthand"—using symbols to represent words—can help.

You might also find it helpful to create an outline when taking notes. When outlining material, first read the material to identify the main ideas. In textbooks, section headings provide clues to main topics. Identify the subheadings. Place supporting details under the appropriate heading. The basic pattern for outlines is as follows:

```
MAIN TOPIC
    I. First Idea or Item
        A. First Detail
            1. Subdetail
            2. Subdetail
        B. Second Detail
    II. Second Idea or Item
        A. First Detail
        B. Second Detail
            1. Subdetail
            2. Subdetail
    III. Third Idea or Item
```

Practice the Skill

Read the following excerpt from Chapter 19. Use the steps you just read about to take notes or create an outline. Then answer the questions that follow.

What happens to the sugars and starch in a banana after you eat it? The sugars are small enough to be transported into the cells lining your digestive system. From there, they are moved into your bloodstream and sent to other cells. As soon as you bite into the fruit, enzymes in your saliva begin to break down the starch into glucose. In the same way, the amazing variety of large and complex molecules in foods are broken down into smaller units before they can be absorbed into cells.

Enzymes catalyze the reaction in which proteins, carbohydrates, and lipids are broken down; this process is called digestion. Only small building blocks are able to enter cells to be used in the many reactions that are involved in metabolism. Metabolism is the sum of all of the chemical reactions necessary for the life of an organism.

These numerous intertwined cellular reactions transform the chemical energy stored in the bonds of nutrients into other forms of energy and synthesize the biomolecules needed to provide structure and carry out the functions of living things. What is the source of chemical energy that is stored in nutrient molecules? This energy was converted from light energy by plants during photosynthesis. Ultimately, all of the energy we need to live comes from sunlight.

1. What is the main topic of the excerpt?
2. What are the first, second, and third ideas?
3. Name one detail for each of the ideas.
4. Name one subdetail for each of the details.

Apply the Skill

Take Notes and Outline Go to Section 2.1 and take notes by paraphrasing and using shorthand or by creating an outline. Use the section title and headings to help you create your outline. Summarize the section using only your notes.

Analyze Media Sources

Why learn this skill?

To stay informed, people use a variety of media sources, including print media, broadcast media, and electronic media. The Internet has become an especially valuable research tool. It is convenient to use, and the information contained on the Internet is plentiful. Whichever media source you use to gather information, it is important to analyze the source to determine its accuracy and reliability.

Learn the Skill

There are a number of issues to consider when analyzing a media source. Most important is to check the accuracy of the source and content. The author and publisher or sponsors should be credible and clearly indicated. To analyze print media or broadcast media, ask yourself the following questions:

- Is the information current?
- Are the resources revealed?
- Is more than one resource used?
- Is the information biased?
- Does the information represent both sides of an issue?
- Is the information reported firsthand or secondhand?

For electronic media, ask yourself these questions in addition to the ones above.

- Is the author credible and clearly identified? Web site addresses that end in .edu and .gov tend to be credible and contain reliable information.
- Are the facts on the Web site documented?
- Are the links within the Web site appropriate and current?
- Does the Web site contain links to other useful resources?

Practice the Skill

To analyze print media, choose two articles, one from a newspaper and the other from a news magazine, on an issue on which public opinion is divided. Then, answer these questions.

1. What points are the articles trying to make? Were the articles successful? Can the facts be verified?
2. Did either article reflect a bias toward one viewpoint or another? List any unsupported statements.
3. Was the information reported firsthand or secondhand? Do the articles seem to represent both sides fairly?
4. How many resources can you identify in the articles? List them.

To analyze electronic media, visit glencoe.com and select Web Links. Choose one link from the list, read the information on that Web site, and then answer these questions.

1. Who is the author or sponsor of the Web site?
2. What links does the Web site contain? How are they appropriate to the topic?
3. What resources were used for the information on the Web site?

Apply the Skill

Analyze Sources of Information Think of an issue on which public opinion is divided. Use a variety of media resources to read about this issue. Which news source more fairly represents the issue? Which news source has the most reliable information? Can you identify any biases? Can you verify the credibility of the news source?

Debate Skills

New research leads to new scientific information. There are often opposing points of view on how this research is conducted, how it is interpreted, and how it is communicated. The *Chemistry and Society* features in your book offer a chance to debate a current controversial topic. Here is an overview on how to conduct a debate.

Choose a Position and Research

First, choose a scientific issue that has at least two opposing viewpoints. The issue can come from current events, your textbook, or your teacher. These topics could include human cloning or environmental issues. Topics are stated as affirmative declarations, such as "Cloning human beings is beneficial to society."

One speaker will argue the viewpoint that agrees with the statement, called the positive position, and another speaker will argue the viewpoint that disagrees with the statement, called the negative position. Either individually or with a group, choose the position for which you will argue. The viewpoint that you choose does not have to reflect your personal belief. The purpose of debate is to create a strong argument supported by scientific evidence.

After choosing your position, conduct research to support your viewpoint. Use resources in your media center or library to find articles, or use your textbook to gather evidence to support your argument. A strong argument is supported by scientific evidence, expert opinions, and your own analysis of the issue. Research the opposing position also. Becoming aware of what points the other side might argue will help you to strengthen the evidence for your position.

Hold the Debate

You will have a specific amount of time, determined by your teacher, in which to present your argument. Organize your speech to fit within the time limit: explain the viewpoint that you will be arguing, present an analysis of your evidence, and conclude by summing up your most important points. Try to vary the elements of your argument. Your speech should not be a list of facts, a reading of a newspaper article, or a statement of your personal opinion, but an analysis of your evidence in an organized manner. It is also important to remember that you must never make personal attacks against your opponent. Argue the issue. You will be evaluated on your overall presentation, organization and development of ideas, and strength of support for your argument.

Additional Roles There are other roles that you or your classmates can play in a debate. You can act as the timekeeper. The timekeeper times the length of the debaters' speeches and gives quiet signals to the speaker when time is almost up (usually a hand signal).

You can also act as a judge. There are important elements to look for when judging a speech: an introduction that tells the audience what position the speaker will be arguing, strong evidence that supports the speaker's position, and organization. The speaker also must speak clearly and loudly enough for everyone to hear. It is helpful to take notes during the debate to summarize the main points of each side's argument. Then, decide which debater presented the strongest argument for his or her position. You can have a class discussion about the strengths and weaknesses of the debate and other viewpoints on this issue that could be argued.

Chemistry Handbook

Measurement in Science

It's easier to determine if a runner wins a race than to determine if the runner broke a world's record for the race. The first determination requires that you sequence the runners passing the finish line—first, second, third, and so on. The second determination requires that you carefully measure and compare the amount of time that passed between the start and finish of the race for each contestant. Because time can be expressed as an amount made by measuring, it is called a quantity. One second, three minutes, and two hours are quantities of time. Other familiar quantities include length, volume, and mass.

The International System of Units In 1960, the metric system was standardized in the form of Le Système International d'Unités (SI), which is French for the "International System of Units." These SI units were accepted by the international scientific community as the system for measuring all quantities.

SI base units Seven independent qualities and their SI base units form the foundation of SI. These SI base units are listed in **Table 1**. All other units are derived from these seven units. The size of a unit is indicated by a prefix related to the difference between that unit and the base unit.

Table 1 SI Base Units

Quantity	Unit	Unit Symbol
Length	meter	m
Mass	kilogram	kg
Time	second	s
Temperature	kelvin	K
Amount of substance	mole	mol
Electric current	ampere	A
Luminous intensity	candela	cd

SI derived units You can see that quantities such as area and volume are missing from the table. The quantities are omitted because they are derived—that is, computed—from one or more of the SI base units. For example, the unit of area is computed from the product of two perpendicular length units. Because the SI base unit of length is the meter, the SI derived unit of area is the square meter, m^2. Similarly, the unit of volume is derived from three mutually perpendicular length units, each represented by the meter. Therefore, the SI derived unit of volume is the cubic meter, m^3. The SI derived units used in this text are listed in **Table 2**.

Chemistry Handbook

Table 2 — SI Derived Units

Quantity	Unit	Unit Symbol
Area	square meter	m^2
Volume	cubic meter	m^3
Mass density	kilogram per cubic meter	kg/m^3
Energy	joule	J
Heat of fusion	joule per kilogram	J/kg
Heat of vaporization	joule per kilogram	J/kg
Specific heat	joule per kilogram-kelvin	J/kg·K
Pressure	pascal	Pa
Electric potential	volt	V
Amount of radiation	gray	Gy
Absorbed dose of radiation	sievert	Sv

Metric units As previously noted, the metric system is a forerunner of SI. In the metric system, as in SI, units of the same quantity are related to each other by orders of magnitude. However, some derived quantities in the metric system have units that differ from those in SI. Because these units are familiar and equipment is often calibrated in these units, they are still used today. **Table 3** lists several metric units that you might use.

Table 3 — Metric Units

Quantity	Unit	Unit Symbol
Volume	liter (0.001 m^3)	L
Temperature	Celsius degree	°C
Specific heat	joule per kilogram-degree Celsius	J/kg·°C
Pressure	millimeter of mercury	mm Hg
Energy	calorie	cal

SI prefixes When you express a quantity, such as ten meters, you are comparing the distance to the length of one meter. Ten meters indicates that the distance is a length ten times as great as the length of one meter. Even though you can express any quantity in terms of the base unit, it might not be convenient. For example, the distance between two towns might be 25,000 m. Here, the meter seems too small to describe that distance. Just as you would use 16 miles, not 82,000 feet, to express that distance, you would use a larger unit of length, the kilometer (km). Because the kilometer represents a length of 1000 m, the distance between the towns is 25 km.

Chemistry Handbook

Table 4 — SI Prefixes

Prefix	Symbol	Meaning	Numerical Value	Expressed as scientific notation
Greater than 1				
giga–	G	billion	1,000,000,000	1×10^9
mega–	M	million	1,000,000	1×10^6
kilo–	k	thousand	1,000	1×10^3
Less than 1				
deci–	d	tenth	0.1	1×10^{-1}
centi–	c	hundredth	0.01	1×10^{-2}
milli–	m	thousandth	0.001	1×10^{-3}
micro–	µ	millionth	0.000001	1×10^{-6}
nano–	n	billionth	0.000000001	1×10^{-9}
pico–	p	trillionth	0.000000000001	1×10^{-12}

Table 4 lists the most commonly used SI prefixes. In SI, units that represent the same quantity are related to each other by some factor of ten, such as 10, 100, 1000, 1/10, 1/100, and 1/1000. In the example on page 786, the kilometer is related to the meter by a factor of 1000; namely, 1 km = 1000 m. As you see, 25,000 m and 25 km differ only in zeros and the units.

To change the size of the SI unit used to measure a quantity, you add a prefix to the base unit or derived unit of that quantity. For example, the prefix *centi-* designates one one-hundredth (0.01). Therefore, a centimeter (cm) is a unit one one-hundredth the length of a meter, and a centijoule (cJ) is a unit of energy one one-hundredth that of a joule. The exception to the rule is in the measurement of mass in which the base unit, kg, already has a prefix. To express a different size mass unit, you replace the prefix to the gram unit. Thus, a centigram, cg, represents a unit having one one-hundredth the mass of a gram.

PRACTICE Problems

Use Tables 1, 2, 3, and 4 to answer the following questions.

1. Name the quantities using SI prefixes. Then write the symbols for each.
   a) 0.1 m
   b) 1,000,000,000 J
   c) 10^{-12} m
   d) 0.000000001 m
   e) 10^{-3} g
   f) 10^6 J

2. Identify the quantity being expressed and rank the units in increasing order of size.
   a) cm, µm, dm
   b) Pa, MPa, kPa
   c) kV, cV, V
   d) pg, cg, mg
   e) mA, MA, µA
   f) dGy, mGy, nGy

Chemistry Handbook

Relating SI, Metric, and English Measurements

Any measurement tells you how much because it is a statement of quantity. However, how you express the measurement depends on the purpose for which you are going to use the quantity. For example, if you look at a room and say its dimensions are 9 ft by 12 ft, you are estimating these measurements from past experience. As you become more familiar with SI, you will be able to estimate the room size as 3 m by 4 m. Still, if you are going to buy carpeting for that room, you are going to make sure of its dimensions by measuring it with a tape measure no matter which system you are using.

Length, volume, and temperature Estimating and using any system of measurement requires familiarity with the names and sizes of the units and practice. As you can see in **Figure 1,** you are already familiar with some common units, which will help you become familiar with SI and metric units.

■ **Figure 1** Common units of measurement for length, volume, and temperature

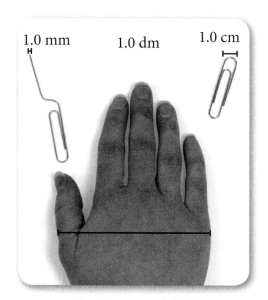

Length Your hand and a paper clip are useful approximations for SI units. For approximating, a meter and a yard are similar lengths. To convert length measurements from one system to another, you can use the relationships shown here.

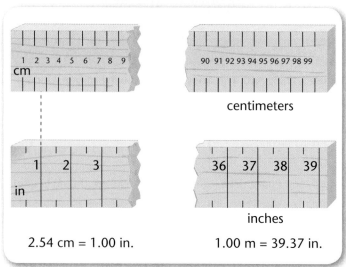

Chemistry Handbook

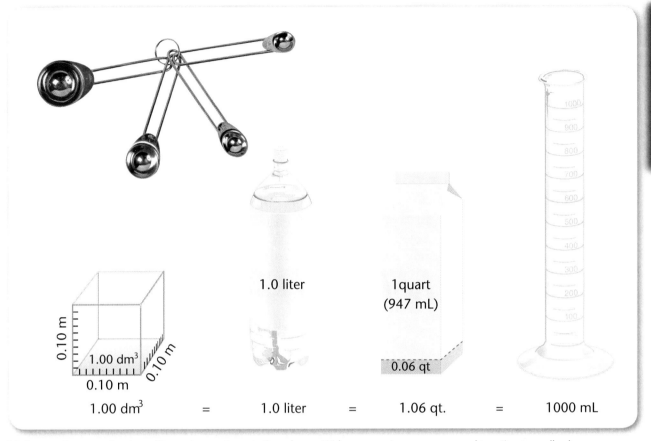

Volume For approximating, a liter and a quart are similar volumes. Kitchen measuring spoons are used to estimate small volumes.

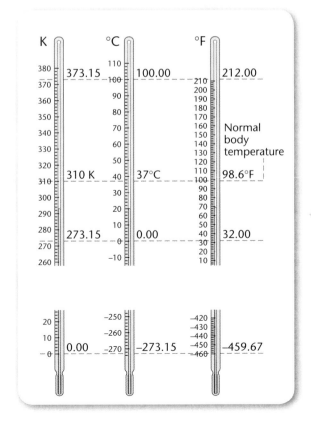

Temperature Having a body temperature of 37° or 310 sounds unhealthy if you don't include the proper units. In fact, 37°C and 310 K are normal body temperatures in the metric system and SI, respectively.

Chemistry Handbook

Mass and weight One of the most useful ways of describing an amount of stuff is to state its mass. You measure the mass of an object on a balance. Even though you are measuring mass, most people still refer to it as weighing. You might think that mass and weight are the same quantity. They are not. The mass of an object is a measure of its inertia; that is, its resistance to changes in motion. The inertia of an object is determined by its quantity of mass. A pint of sand has more matter than a pint of water; therefore, it has more mass.

The weight of an object is the amount of gravitational force acting on the mass of the object. On Earth, you sense the weight of an object by holding it and feeling the pull of gravity on it.

An important aspect of weight is that it is directly proportional to the mass of the object. This relationship means that the weight of a 2-kg object is twice the weight of a 1-kg object. Therefore, the pull of gravity on a 2-kg object is twice as great as the pull on a 1-kg object. Because it is easier to measure the effect of the pull of gravity rather than the resistance of an object to changes in its motion, mass can be determined by a weighing process.

One instrument used to determine mass is the triple-beam balance, illustrated in **Figure 2.** You can read how an electronic balance functions in *How It Works* in Chapter 12.

Chemistry Online
Personal Tutor For an online tutorial on mass and weight, visit glencoe.com.

■ **Figure 2** Mass and weight are not the same.

Mass In a triple-beam balance, weights of different sizes are placed at notched locations along each of two beams, and a third is slid along an arm to balance the object being weighed. To determine the mass of the object, you add the numbered positions of the three weights. Using a triple-beam balance is usually a much quicker way to determine the mass of an object than using a double-pan balance.

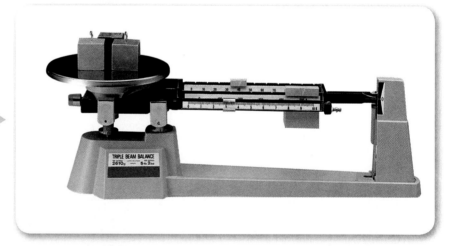

Weight Because weight and mass are so closely related, the contents of cans and boxes are labeled both in pounds and ounces, the British/American weight units, and grams or kilograms, the metric mass units.

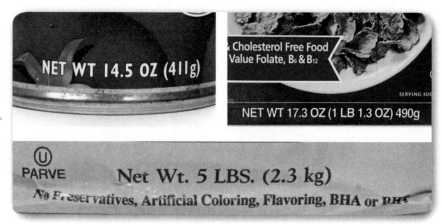

Chemistry Handbook

Making and Interpreting Measurements

Using measurements in science is different from manipulating numbers in math class. The important difference is that numbers in science are almost always measurements, which are made with instruments of varying accuracy.

As you will see, the degree of accuracy of measured quantities must always be taken into account when expressing, multiplying, dividing, adding, or subtracting them. In making or interpreting a measurement, you should consider two points. The first is how well the instrument measures the quantity you're interested in. This point is illustrated in **Figure 3.**

Because the top ruler in **Figure 3** is calibrated to smaller divisions than the bottom ruler, the upper ruler has more precision than the lower. Any measurement you make with it will be more precise than one made on the bottom ruler because it will contain a smaller estimated value.

The second point to consider in making or interpreting a measurement is how well the measurement represents the quantity you're interested in. Looking at **Figure 3,** you can see how well the tip of the pencil aligns with the value 27.65 cm. From sight, you know that 27.65 cm is a better representation of the pencil's length than 28 cm. Because 27.65 cm better represents the length of the pencil (the quantity you're interested in) than does 28 cm, it is a more accurate measurement of the strip's length.

■ **Figure 3** In the top ruler, you can see that the edge of the pencil lies between 27.6 cm and 27.7 cm. Because there are no finer calibrations between 27.6 cm and 27.7 cm, you estimate the length beyond the last calibration. In this case, you might estimate it as 0.05 cm. You would record this measurement as 27.65 cm. You and others should interpret the measurement as 27.65 ± 0.01 cm.

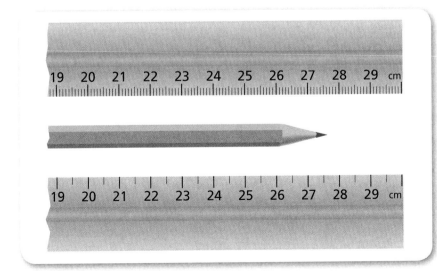

Using the calibrations on the bottom ruler, you know that the length of the pencil is between 27.5 cm and 28 cm. Because there are no finer calibrations between 27.5 cm and 28 cm, you have to estimate the length beyond the last calibration. An estimate would be 0.6 cm. Of course, someone might estimate it as 0.5 cm; others as 0.7 cm. Even though you record the measurement of the pencil's length as 27.6 cm, you and others reading the measurement should interpret the measurement as 27.65 ± 0.01 cm.

Chemistry Handbook

In **Figure 4,** you will see that a more precise measurement might not be the most accurate measurement of a quantity.

■ **Figure 4** Precision is different from accuracy.

Describing the width of the index card as 10.16 cm indicates that the ruler has 0.1-cm calibrations and the 0.06 cm is an estimation. Similarly, the uniform alignment of the edge of the index card with the ruler indicates that the measurement 10.16 cm is also an accurate measurement of width.

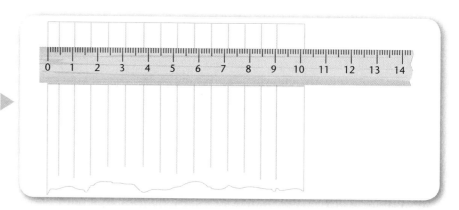

Describing the width of a brick as 10.16 cm indicates that this measurement is as precise as the measurement of the index card. However, 10.16 cm isn't a good representation of the width of the jagged-edged brick.

A better representation of the width of the brick is made using a ruler with less precision. As you can see, 10.2 cm is a better representation, and therefore a more accurate measurement, of the brick's width than 10.16 cm.

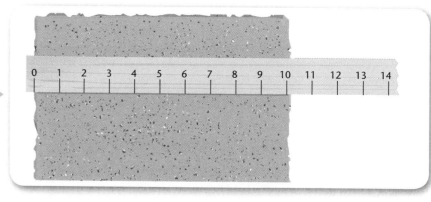

Expressing the Accuracy of Measurements

In measuring the length of the pencil as 27.65 cm and 27.6 cm in **Figure 3,** you were aware of the difference in the calibration of the two rulers. This difference appeared in the way each measurement was recorded. In one measurement, the digits 2, 7, 6, and 5 were meaningful. In the second, the digits 2, 7, and 6 were meaningful. In any measurement, meaningful digits are called *significant figures.* The significant figures in a measured quantity include all the digits you know for sure, plus the final estimated digit.

Significant figures The rules below can help you express or interpret significant figures. The fourth rule sometimes causes difficulty in expressing measurements. Because the zero is a placeholder in the measurement, it is not significant and 20 L has one significant figure. The measurement is interpreted as 20 L plus or minus the value of the least significant figure. A volume measurement of 20 L indicates 20 L ± 10 L, a range of 10-30 L. However, suppose you made the measurement with a device that is accurate to the nearest 1 L, making both the 2 and the 0 significant.

Adding a decimal point followed by a zero would indicate that the final zero past the decimal is significant (Rule 3), and the measurement would then be 20.0 ± 0.1 L. The measurement is expressed as 2.0×10^1 L. Now the zero is a significant figure because it is a final zero past the decimal (Rule 3).

Rules for Significant Figures

1. All nonzero digits of a measurement are significant.

Digital balance readout: 283.47 g

2	100.00 g	
8	10.00 g	
3	1.00 g	
4	0.10 g	
7 (±1)	0.01 g	
283.47 g		
5 significant figures		

Graduated cylinder

3	10.0 mL	
2	1.0 mL	
2 (±1)	0.1 mL	
32.2 mL		
3 significant figures		

2. Zeros occurring between significant figures are significant.

Digital balance: 56.06 g

5	10.00 g
6	1.00 g
0	0.10 g
6 (±1)	0.01 g
56.06 g	
4 significant figures	

1	10.0 mL
0	1.0 mL
7 (±1)	0.1 ml
10.7 mL	
3 significant figures	

3. All final zeros past the decimal point are significant.

Digital balance: 73.00 g

7	10.00 g
3	1.00 g
0	0.10 g
0 (±1)	0.01 g
73.00 g	
4 significant figures	

2	10.0 mL
0	1.0 mL
0 (±1)	0.1 mL
20.0 mL	
3 significant figures	

4. Zeros used as placeholders are not significant.

Digital balance: 0.09 g

0	1.00 g
0	0.10 g
9 (±1)	0.01 g
0.09 g	
1 significant figure	

0	1.0 mL
7 (±1)	0.01 mL
0.07 mL	
1 significant figure	

Chemistry Handbook

Expressing Quantities with Scientific Notation

The most common use of scientific notation is in expressing measurements of very large and very small dimensions. Using scientific notation is sometimes referred to as using *powers of ten* because it expresses quantities by using a small number between one and ten, which is then multiplied by ten to a power to give the quantity its proper magnitude. Suppose you went on a trip of 9000 km. You know that $10^3 = 1000$, so you could express the distance of your trip as 9×10^3 km. In this example, it may seem that scientific notation wouldn't be terribly useful. However, consider that an often-used quantity in chemistry is 602,000,000,000,000,000,000,000, the number of atoms or molecules in a mole of a substance. Recall that the mole is the SI unit of amount of a substance. Rather than writing out this huge number every time it is used, it's much easier to express it in scientific notation.

Determining powers of 10 To determine the exponent of ten, count as you move the decimal point left until it falls just after the first nonzero digit—in this case, 6. If you try this on the number above, you'll find that you've moved the decimal point 23 places. Therefore, the number expressed in scientific notation is 6.02×10^{23}.

Expressing small measurements in scientific notation is done in a similar way. The diameter of a carbon atom is 0.000000000000154 m. In this case, you move the decimal point right until it is just past the first nonzero digit—in this case, 1. The number of places you move the decimal point right is expressed as a *negative* exponent of ten. The diameter of a carbon atom is 1.54×10^{-13} m. You always move the decimal point until the coefficient of ten is between one and less than ten. Thus, scientific notation always has the form, $M \times 10^n$ where $1 \leq M < 10$.

Notice in **Figure 5** how the following examples are converted to scientific notation.

Personal Tutor For an online tutorial on scientific notation, visit glencoe.com.

■ **Figure 5** Scientific notation always has the form $M \times 10^n$ where $1 \leq M < 10$.

Quantities greater than 1			
17.16 g	17.16 →	1.716×10^1 g	Decimal point moved 1 place left.
152.6 L	152.6 →	1.526×10^2 L	Decimal point moved 2 places left.
73,621 kg	73,621. →	7.3621×10^4 kg	Decimal point moved 4 places left.

Quantities between 0 and 1			
0.29 mL	0.29 →	2.9×10^{-1} mL	Decimal point moved 1 place right.
0.0672 m	0.0672 →	6.72×10^{-2} m	Decimal point moved 2 places right.
0.0008 g	0.0008 →	8×10^{-4} g	Decimal point moved 4 places right.

Calculations with measurements You often must use measurements to calculate other quantities. Remember that the accuracy of measurement depends on the instrument used and that accuracy is expressed as a certain number of significant figures. Therefore, you must indicate which figures in the result of any mathematical operation with measurements are significant. The rule of thumb is that no result can be more accurate than the least accurate quantity used to calculate that result. Therefore, the quantity with the least number of significant figures determines the number of significant figures in the result. The method used to indicate significant figures depends on the mathematical operation.

Addition and Subtraction
The answer has only as many decimal places as the measurement having the least number of decimal places.

```
   190.2   g
    65.291 g
    12.38  g
   ───────
   267.871 g
```

The answer is rounded to the nearest tenth, which is the accuracy of the least accurate measurement.
267.9 g

Because the masses were measured on balances differing in accuracy, the least accurate measurement limits the number of digits past the decimal point.

Multiplication and Division
The answer has only as many significant digits as the measurement with the least number of significant digits.

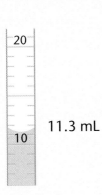

$$\text{density} = \frac{\text{mass}}{\text{volume}}$$

$$D = \frac{m}{V} =$$

$$\frac{13.78 \text{ g}}{11.3 \text{ mL}} = 1.219469 \text{ g/mL}$$

The answer is rounded to three significant digits, 1.22 g/mL, because the least accurate measurement, 11.3 mL, has three significant digits.

Multiplying or dividing measured quantities results in a derived quantity. For example, the mass of a substance divided by its volume is its density. But mass and volume are measured with different tools, which may have different accuracies. Therefore, the derived quantity can have no more significant digits than the least accurate measurement used to compute it.

Chemistry Handbook

PRACTICE Problems

3. Determine the number of significant figures in each measurement.
   - a. 64 mL
   - b. 0.650 g
   - c. 30 cg
   - d. 724.56 mm
   - e. 47,080 km
   - f. 0.072040 g
   - g. 1.03 mm
   - h. 0.001 mm

4. Write each measurement in scientific notation.
   - a. 76.0°C
   - b. 212 mm
   - c. 56.021 g
   - d. 0.78 L
   - e. 0.07612 m
   - f. 763.01 g
   - g. 10,301,980 nm
   - h. 0.001 mm

5. Write each measurement in scientific notation.
   - a. 73,000 ± 1 mL
   - b. 4000 ± 1000 kg
   - c. 100 ± 10 cm
   - d. 100,000 ± 1000 km

6. Solve the problems and express the answer in the correct number of significant digits.
   - a. $45.761 \text{ g} - 42.65 \text{ g}$
   - b. $1.6 \text{ km} + 0.62 \text{ km}$
   - c. $0.340 \text{ cg} + 1.20 \text{ cg} + 1.018 \text{ cg}$
   - d. $6000 \text{ µm} - 202 \text{ µm}$

7. Solve the problems and express the answer in the correct number of significant digits.
   - a. $5.761 \text{ cm} \times 6.20 \text{ cm}$
   - b. $\dfrac{23.5 \text{ kg}}{4.615 \text{ m}^3}$
   - c. $\dfrac{0.2 \text{ km}}{5.4 \text{ s}}$
   - d. $11.00 \text{ m} \times 12.10 \text{ m} \times 3.53 \text{ m}$
   - e. $\dfrac{4.500 \text{ kg}}{1.500 \text{ m}^2}$
   - f. $\dfrac{18.21 \text{ g}}{4.4 \text{ cm}^3}$

Adding and Subtracting Measurements in Scientific Notation Adding and subtracting measurements in scientific notation requires that the measurements must be expressed as the same power of ten. For example, the following three length measurements must be expressed in the same power of ten.

$$1.1012 \times 10^4 \text{ mm}$$
$$2.31 \times 10^3 \text{ mm}$$
$$+\ 4.573 \times 10^2 \text{ mm}$$

In adding and subtracting measurements in scientific notation, all measurements are expressed in the same order of magnitude as the measurement with the greatest power of ten. When converting a quantity, the decimal point is moved one place to the left for each increase in power of ten.

$$2.31 \times 10^3 \qquad 2.31 \times 10^3 \longrightarrow 0.231 \times 10^4$$
$$4.573 \times 10^2 \qquad 4.573 \times 10^2 \longrightarrow 0.4573 \times 10^3 \longrightarrow 0.04573 \times 10^4$$

Chemistry Handbook

$$1.1012 \times 10^4 \text{ mm}$$
$$0.231 \times 10^4 \text{ mm}$$
$$+ 0.04573 \times 10^4 \text{ mm}$$
$$\overline{1.37793 \times 10^4 \text{ mm} = 1.378 \times 10^4 \text{ mm (rounded)}}$$

Multiplying and dividing measurements in scientific notation Multiplying and dividing measurements in scientific notation requires that similar operations are done to the numerical values, the powers of ten, and the units of the measurements.

a) The numerical coefficients are multiplied or divided and the resulting value is expressed in the same number of significant figures as the measurement with the least number of significant figures.

b) The exponents of ten are algebraically added in multiplication and subtracted in division.

c) The units are multiplied or divided.

The following problems illustrate these procedures.

EXAMPLE Problems

Example Problem 1

$$(3.6 \times 10^3 \text{ m})(9.4 \times 10^3 \text{ m})(5.35 \times 10^{-1} \text{ m})$$
$$= (3.6 \times 9.4 \times 5.35) \times (10^3 \times 10^3 \times 10^{-1}) \text{ (m} \times \text{m} \times \text{m)}$$
$$= (3.6 \times 9.4 \times 5.35) \times 10^{(3+3+(-1))} \text{ (m} \times \text{m} \times \text{m)}$$
$$= (181.044) \times 10^5 \text{ m}^3$$
$$= 1.8 \times 10^2 \times 10^5 \text{ m}^3$$
$$= 1.8 \times 10^7 \text{ m}^3$$

Example Problem 2

$$\frac{6.762 \times 10^2 \text{ m}^3}{(1.231 \times 10^1 \text{ m})(2.80 \times 10^{-2} \text{ m})}$$
$$= \frac{6.762}{1.231 \times 2.80} \times \frac{10^2}{10^1 \times 10^{-2}} \times \frac{\text{m}^3}{\text{m} \times \text{m}}$$
$$= 1.961819659 \times 10^{(2-(+1-2))} \text{ m}^{(3-(+2))}$$
$$= 1.96 \times 10^{(2-(-1))} \text{ m}^{(1)}$$
$$= 1.96 \times 10^3 \text{ m}$$

PRACTICE Problems

8. Solve the addition and subtraction problems.
  a) 1.013×10^3 g $+ 8.62 \times 10^2$ g $+ 1.1 \times 10^1$ g
  b) 2.82×10^6 m $- 4.9 \times 10^4$ m

9. Solve the multiplication and division problems.
  a) 1.18×10^{-3} m $\times 4.00 \times 10^2$ m $\times 6.22 \times 10^2$ m
  b) 3.2×10^2 g $\div 1.04 \times 10^2$ cm^2 $\times 6.22 \times 10^{-1}$ cm

Chemistry Handbook

■ **Figure 6**
Subtracting Numbers in Scientific Notation with a Calculator
A quantity in scientific notation is entered by keying in the coefficient and then striking the [EXP] or [EE] key followed by the value of the exponent of ten. At the end of the calculation, the calculator readout must be corrected to the appropriate number of decimal places. The answer would be rounded to the second decimal place and expressed as 2.56×10^4 kg.

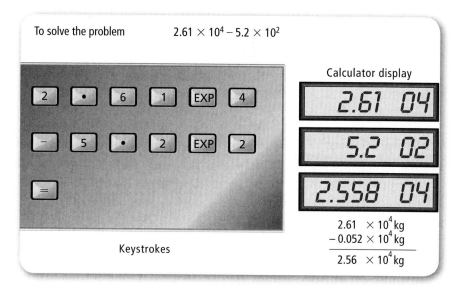

Computations with the Calculator

Working problems in chemistry will require you to have a good understanding of some of the advanced functions of your calculator. When using a calculator to solve a problem involving measured quantities, you should remember that the calculator does not take significant figures into account. It is up to you to round off the answer to the correct number of significant figures at the end of a calculation. In a multistep calculation, you should not round off after each step. Instead, you should complete the calculation and then round off. **Figure 6** shows how to use a calculator to solve a subtraction problem involving quantities in scientific notation.

Look at **Figure 7** to see how to solve multiplication and division problems involving scientific notation. The problems are the same as the two previous Example Problems.

■ **Figure 7**
Multiplying and Dividing Measurements Expressed in Scientific Notation
A negative power of ten is usually entered by striking the [EXP] or [EE], entering the positive value of the exponent, and then striking the [±] key.

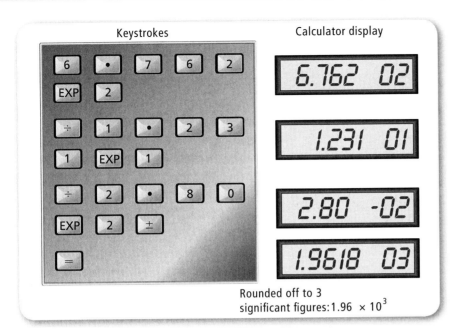

The numerical value of the answer can have no more significant figures than the measurement that has the least number of significant figures.

PRACTICE Problems

10. Solve the problems and express the answers in scientific notation with the proper number of significant figures.

a) 2.01×10^2 mL
3.1×10^1 mL
$+ 2.712 \times 10^3$ mL

b) 7.40×10^2 mm
-4.0×10^1 mm

c) 2.10×10^1 g
-1.6×10^{-1} g

d) 5.131×10^2 J
2.341×10^1 J
$+3.781 \times 10^3$ J

11. Solve the problems and express the answers in scientific notation with the proper number of significant figures.

a) $(2.00 \times 10^1 \text{ cm})(2.05 \times 10^1 \text{ cm})$

b) $\dfrac{5.6 \times 10^3 \text{ kg}}{1.20 \times 10^4 \text{ m}^3}$

c) $(2.51 \times 10^1 \text{ m})(3.52 \times 10^1 \text{ m})(1.2 \times 10^{-1} \text{ m})$

d) $\dfrac{1.692 \times 10^4 \text{ dm}^3}{(2.7 \times 10^{-2} \text{ dm})(4.201 \times 10^1 \text{ dm})}$

Chemistry Handbook

Using Dimensional Analysis

Dimensional analysis is used to express a physical quantity, such as the length of a pen, in any other unit that measures that quantity. For example, you can measure the pen with a metric ruler calibrated in centimeters and then express the length in meters.

If the length of a pen is measured as 14.90 cm, you can express that length in meters by using the numerical relationship between a centimeter and a meter. This relationship is given by the following equation.

$$100 \text{ cm} = 1 \text{ m}$$

If both sides of the equation are divided by 100 cm, the following relationship is obtained.

$$1 = \frac{1 \text{ m}}{100 \text{ cm}}$$

To express 14.90 cm as a measurement in meters, you multiply the quantity by the relationship, which eliminates the cm unit.

$$14.90 \text{ cm} \times \frac{1 \text{ m}}{100 \text{ cm}} = \frac{14.90 \text{ cm}}{1} \times \frac{1 \text{ m}}{100 \text{ cm}} = \frac{14.90}{100} \text{ m} = 0.1490 \text{ m}$$

Dimensional analysis doesn't change the value of the physical quantity because you are multiplying that value by a factor that equals 1. You choose the factor so that when the unit you want to eliminate is multiplied by the factor, that unit and the similar unit in the factor cancel. If the unit you want to eliminate is in the numerator, choose the factor which has that unit in the denominator. Conversely, if the unit you want to eliminate is in the denominator, choose the factor which has that unit in the numerator. For example, in a chemistry lab activity, a student measured the mass and volume of a piece of copper and calculated its density as 8.80 g/cm³. Knowing that 1000 g = 1 kg and 100 cm = 1 m, the student could then use dimensional analysis to express this value in the SI unit of density, kg/m³.

$$8.80 \, \frac{\text{g}}{\text{cm}^3} \times \frac{1 \text{ kg}}{1000 \text{ g}} \times \left(\frac{100 \text{ cm}}{1 \text{ m}}\right)^3 =$$

$$8.80 \, \frac{\text{g}}{\text{cm} \times \text{cm} \times \text{cm}} \times \frac{1 \text{ kg}}{1000 \text{ g}} \times \frac{100 \text{ cm}}{1 \text{ m}} \times \frac{100 \text{ cm}}{1 \text{ m}} \times \frac{100 \text{ cm}}{1 \text{ m}} =$$

$$\frac{8.80 \times (10^2 \times 10^2 \times 10^2)}{1000} \, \frac{\text{kg}}{\text{m}^3} = \frac{8.80 \times 10^{(2+2+2)}}{10^3} \, \frac{\text{kg}}{\text{m}^3} =$$

$$8.80 \times 10^{(6-3)} \text{ kg/m}^3 = 8.80 \times 10^3 \text{ kg/m}^3$$

Dimensional analysis can be extended to other types of calculations in chemistry. To use this method, you first examine the data that you have. Next, you determine the quantity you want to find and look at the units you will need. Finally, you apply a series of factors to the data in order to convert it to the units you need.

Chemistry Handbook

EXAMPLE Problems

Example Problem 1

The density of silver sulfide (Ag_2S) is 7.234 g/mL. What is the volume of a piece of silver sulfide that has a mass of 6.84 kg?

First, you must apply a factor that will convert kg of Ag_2S to g Ag_2S.

$$(6.84 \text{ kg } Ag_2S) \times \frac{(1000 \text{ g } Ag_2S)}{(1 \text{ kg } Ag_2S)}$$

Next, you use the density of Ag_2S to convert mass to volume.

$$(6.84 \text{ kg } Ag_2S) \times \frac{(1000 \text{ g } Ag_2S)}{(1 \text{ kg } Ag_2S)} \times \frac{(1 \text{ mL } Ag_2S)}{(7.234 \text{ g } Ag_2S)} = 946 \text{ mL } Ag_2S$$

Notice that the new factor must have grams in the denominator so that grams will cancel out, leaving mL.

Example Problem 2

What mass of lead can be obtained from 47.2 g of $Pb(NO_3)_2$?

Because mass is involved, you will need to know the molar mass of $Pb(NO_3)_2$.

$$Pb = 207.2 \text{ g}$$
$$2N = 28.014 \text{ g}$$
$$6O = 95.994 \text{ g}$$

Molar mass of $Pb(NO_3)_2 = 331.208$ g.
Rounding off according to the rules for significant figures, the molar mass of $Pb(NO_3)_2 = 331.2$ g.

You can see that the mass of lead in $Pb(NO_3)_2$ is 207.2/331.2 of the total mass of $Pb(NO_3)_2$.

Now you can set up a relationship to determine the mass of lead in the 47.2-g sample.

$$(47.2 \text{ g } Pb(NO_3)_2) \times \frac{(207.2 \text{ g Pb})}{(331.2 \text{ g } Pb(NO_3)_2)} = 29.52850242 \text{ g Pb}$$
$$= 29.5 \text{ g Pb rounded to 3 significant figures.}$$

Notice that the equation is arranged so that the unit g $Pb(NO_3)_2$ cancels out, leaving only g Pb, the quantity asked for in the problem.

PRACTICE Problems

12. Express each quantity in the unit listed to its right.
    a) 3.01 g   cg
    b) 6200 m   km
    c) 6.24×10^{-7} g μg
    d) 3.21 L   mL
    e) 0.2 L dm^3
    f) 0.13 cal/g   J/kg
    g) 5 ft, 1 in.   m
    h) 1.2 qt   L

Chemistry Handbook

Organizing Information

It is often necessary to compare and sequence observations and measurements. Two of the most useful ways are to organize the observations and measurements as tables and graphs. If you browse through your textbook, you'll see many tables and graphs. They arrange information in a way that makes it easier to understand.

Making and using tables Most tables have a title telling you what information is being presented. The table itself is divided into columns and rows. The column titles list items to be compared. The row headings list the specific characteristics being compared among those items. The information is recorded within the grid of the table. Any table you prepare to organize data taken in a laboratory activity should have these characteristics.

Consider, for example, that in a laboratory experiment, you are going to perform a flame test on various solutions. In the test, you place drops of the solution containing a metal ion in a flame, and the color of the flame is observed, as shown in **Figure 8.** Before doing the experiment, you might set up a data table like the one below.

While performing the experiment, you would record the name of the solution and then the observation of the flame color. If you weren't sure of the metal ion as you were doing the experiment, you could enter it into the table by checking oxidation numbers afterward. Not only does the table organize your observations, it could also be used as a reference to determine whether a solution of some unknown composition contains one of the metal ions listed in the table.

■ **Figure 8**
A drop of solution containing the potassium ion, K^+, causes the flame to burn with violet color.

Data Table	Flame Test Results	
Solution	Metal Ion	Color of Flame
KNO_3	K^+	violet-pink

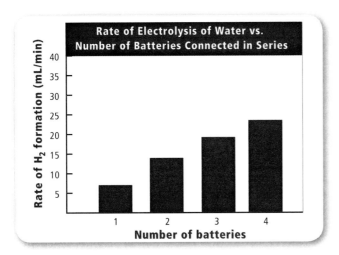

Figure 9
A sample bar graph

Making and using graphs After organizing data in tables, scientists usually want to display the data in a more visual way. Using graphs is a common way to accomplish that. There are three common types of graphs—bar graphs, pie graphs, and line graphs.

Bar graphs Bar graphs are useful when you want to compare or display data that do not continuously change. Suppose you measure the rate of electrolysis of water by determining the volume of hydrogen gas that forms. In addition, you decide to test how the number of batteries affects the rate of electrolysis. You could graph the results using a bar graph, as shown in **Figure 9**. Note that you could construct a line graph, but the bar graph is better because there is no way you could use 0.4 or 2.6 batteries.

Circle graphs Pie graphs are especially useful in comparing the parts of a whole. You could use a pie graph to display the percent composition of a compound such as sodium dihydrogen phosphate, NaH_2PO_4, as shown in **Figure 10**.

In constructing a pie graph, recall that a circle has 360°. Therefore, each fraction of the whole is that fraction of 360°. Suppose you did a census of your school and determined that 252 students out of a total of 845 were 17 years old. You would compute the angle of that section of the graph by multiplying $(252/845) \times 360° = 107°$.

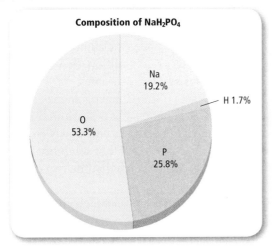

Figure 10
A sample pie graph

Chemistry Handbook

■ Figure 11
Constructing a Line Graph
1. Plot the independent variable on the x-axis (horizontal axis) and the dependent variable on the y-axis (vertical). The independent variable is the quantity changed or controlled by the experimenter. The temperature data in **Table 5** were controlled by the experimenter, who chose to measure the solubility at 10°C intervals.
2. Scale each axis so that the smallest and largest data values of each quantity can be plotted. Use divisions such as ones, fives, or tens or decimal values such as hundredths or thousandths.
3. Label each axis with the appropriate quantity and unit.
4. Plot each pair of data from the table as follows.
   - Place a straightedge vertically at the value of the independent variable on the x-axis.
   - Place a straightedge horizontally at the value of the dependent variable on the y-axis.
   - Mark the point at which the straightedges intersect.
5. Fit the best straight line or curved line through the data points.

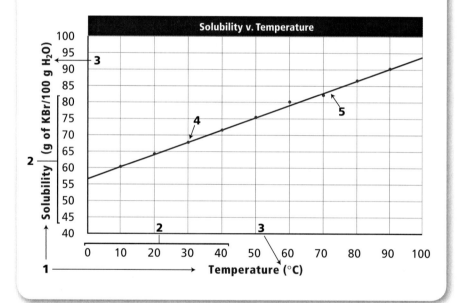

Table 5	Effect of Temperature on Solubility of KBr
Temperature (°C)	Solubility (g of KBr/100 g H$_2$O)
10.0	60.2
20.0	64.3
30.0	67.7
40.0	71.6
50.0	75.3
60.0	80.1
70.0	82.6
80.0	86.8
90.0	90.2

Line graphs Line graphs have the ability to show a trend in one variable as another one changes. In addition, they can suggest possible mathematical relationships between variables.

Table 5 shows the data collected during an experiment to determine whether temperature affects the mass of potassium bromide that dissolves in 100 g of water. If you read the data as you slowly run your fingers down both columns of the table, you will see that solubility increases as the temperature increases. This is the first clue that the two quantities may be related.

To see that the two quantities are related, construct a line graph as shown in **Figure 11**.

One use of a line graph is to predict values of the independent or dependent variables. For example, from **Figure 11** you can predict the solubility of KBr at a temperature of 65°C by the following method:

- Place a straightedge vertically at the approximate value of 65°C on the x-axis.
- Mark the point at which the straightedge intersects the line of the graph.
- Place a straightedge horizontally at this point and approximate the value of the dependent variable on the y-axis as 82 g/100 g H_2O.

To predict the temperature for a given solubility, you would reverse the above procedure.

PRACTICE Problems

Use Figure 11 to answer Questions 13 and 14.

13. Predict the solubility of KBr at each of the following temperatures.
    a) 25.0°C
    b) 52.0°C
    c) 6.0°C
    d) 96.0°C

14. Predict the temperature at which KBr has each of the following solubilities.
    a) 70.0 g/100 g H_2O
    b) 88.0 g/100 g H_2O

Graphs of Direct and Inverse Relationships Graphs can be used to determine quantitative relationships between the independent and dependent variables. Two of the most useful relationships are relationships in which the two quantities are directly proportional or inversely proportional.

When two quantities are directly proportional, an increase in one quantity produces a proportionate increase in the other. A graph of two quantities that are directly proportional is a straight line, as shown in **Figure 12**.

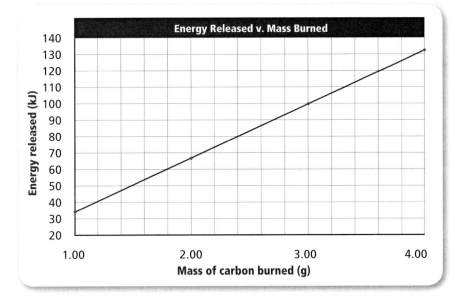

■ **Figure 12**
Graph of Quantities That Are Directly Proportional
As you can see, doubling the mass of the carbon burned from 2.00 g to 4.00 g doubles the amount of energy released from 66 kJ to 132 kJ. Such a relationship indicates that mass of carbon burned and the amount of energy released are directly proportional.

Chemistry Handbook

■ **Figure 13**
Graph of Quantities That Are Inversely Proportional
As you can see, doubling the pressure of the gas reduces the volume of the gas by one-half. Such a relationship indicates that the volume and pressure of a gas are inversely proportional.

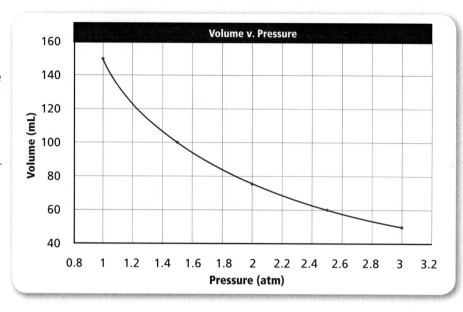

When two quantities are inversely proportional, an increase in one quantity produces a proportionate decrease in the other. A graph of two quantities that are inversely proportional is shown in **Figure 13**.

PRACTICE Problems

15. Plot the data in the table and determine whether the two quantities are directly proportional.

Effect of Temperature on Gas Pressure	
Temperature (K)	Pressure (kPa)
300.0	195
320.0	208
340.0	221
360.0	234
380.0	247
400.0	261

16. Plot the data in the table and determine whether the two quantities are inversely proportional.

Effect of Number of Mini-Lightbulbs on Electrical Current in a Circuit	
Number of mini-lightbulbs	Current (mA)
2	3.94
4	1.98
6	1.31
9	0.88

Ratios, Fractions, and Percents

When you analyze data, you may be asked to compare measured quantities. Or, you may be asked to determine the relative amounts of elements in a compound. For example, the relationship between molar masses can be expressed in three ways: a ratio, a fraction, or a percent.

Ratios You make comparisons by using ratios in your daily life. For example, the mass of one dozen limes, shown in **Figure 15,** is 12 times larger than the mass of one lime. In chemistry, the chemical formula for a compound compares the elements that make up that compound, as shown in **Figure 16**. A ratio is a comparison of two numbers by division. One way it can be expressed is with a colon (:). The comparison between the molar masses of oxygen and hydrogen can be expressed as follows.

$$\text{molar mass }(H_2):\text{molar mass }(O_2)$$
$$2.00\text{ g/mol}:32.00\text{ g/mol}$$
$$2.00:32.00$$
$$1:16$$

■ **Figure 15** The mass of one lime would be one-twelfth the mass of one dozen limes.

Notice that the ratio 1:16 is the smallest integer (whole number) ratio. It is obtained by dividing both numbers in the ratio by the smaller number, and then rounding the larger number to remove the digits after the decimal. The ratio of the molar masses is 1 to 16. In other words, the ratio indicates that the molar mass of diatomic hydrogen gas is 16 times smaller than the molar mass of diatomic oxygen gas.

Fractions Ratios are often expressed as fractions in simplest form. A fraction is a quotient of two numbers. To express the comparison of the molar masses as a fraction, place the molar mass of hydrogen over the molar mass of oxygen as follows.

$$\frac{\text{molar mass }H_2}{\text{molar mass }O_2} = \frac{2.0\text{ g/mol}}{32.00\text{ g/mol}} = \frac{2.00}{32.00} = \frac{1}{16}$$

In this case, the simplified fraction is calculated by dividing both the numerator (top of the fraction) and the denominator (bottom of the fraction) by 2.00. This fraction yields the same information as the ratio. That is, diatomic hydrogen gas has one-sixteenth the mass of diatomic oxygen gas.

Percents A percent is a ratio that compares a number to 100. The symbol for percent is %. The number of correct answers on an exam can be expressed as a percent. If you answered 90 out of 100 questions correctly, you would receive a grade of 90%.

The comparison between molar mass of hydrogen gas and the molar mass of oxygen gas described on the previous page can also be expressed as a percent by taking the fraction, converting it to decimal form, and multiplying by 100 as follows.

■ **Figure 16** In a crystal of table salt (sodium chloride), each sodium ion is surrounded by chloride ions, yet the ratio of sodium ions to chloride ions is 1:1. The formula for sodium chloride is NaCl.

$$\frac{\text{molar mass }H_2}{\text{molar mass }O_2} \times 100 = \frac{2.00\text{ g/mol}}{32.00\text{ g/mol}} \times 100 = 0.0625 \times 100 = 6.25\%$$

Diatomic hydrogen gas has 6.25% of the mass of diatomic oxygen gas.

Chemistry Handbook

$$\text{Quotient} = \frac{\overbrace{9 \times 10^8}^{\text{Dividend (numerator)}}}{\underbrace{3 \times 10^{-4}}_{\text{Divisor (denominator)}}}$$

■ **Figure 17** When two numbers are divided, the one on top is the numerator and the one on the bottom is the denominator. The result is called the quotient. When you perform calculations with fractions, the quotient can be expressed as a fraction or a decimal.

Operations Involving Fractions

Fractions are subject to the same type of operations as other numbers. Remember that the number on the top of a fraction is the numerator and the number on the bottom is the denominator. **Figure 17** shows an example of a fraction.

1. Addition and subtraction

Before two fractions can be added or subtracted, they must have a common denominator. Common denominators are found by finding the least common multiple of the two denominators. Finding the least common multiple is often as easy as multiplying the two denominators together. For example, the least common multiple of the denominators of the fractions $\frac{1}{2}$ and $\frac{1}{3}$ is 2×3, or 6. After you find the common denominator, simply add or subtract the fractions' numerators.

$$\frac{1}{2} + \frac{1}{3} = \left(\frac{3}{3} \times \frac{1}{2}\right) + \left(\frac{2}{2} \times \frac{1}{3}\right) = \frac{3}{6} + \frac{2}{6} = \frac{5}{6}$$

Sometimes, one of the denominators will divide into the other, which makes the larger of the two denominators the least common multiple. For example, the fractions $\frac{1}{2}$ and $\frac{1}{6}$ have 6 as the least common multiple denominator.

$$\frac{1}{2} + \frac{1}{6} = \left(\frac{3}{3} \times \frac{1}{2}\right) + \frac{1}{6} = \frac{3}{6} + \frac{1}{6} = \frac{4}{6}$$

In other situations, both denominators will divide into a number that is not the product of the two. For example, the fractions $\frac{1}{4}$ and $\frac{1}{6}$ have the number 12 as their least common multiple denominator, rather than 24, the product of the two denominators.

The least common denominator can be deduced as follows:

$$\frac{1}{6} + \frac{1}{4} = \left(\frac{4}{4} \times \frac{1}{6}\right) + \left(\frac{6}{6} \times \frac{1}{4}\right) = \frac{4}{24} + \frac{6}{24} = \frac{2}{12} + \frac{3}{12} = \frac{5}{12}$$

Because both fractions can be simplified by dividing numerator and denominator by 2, the least common multiple must be 12.

2. Multiplication and division

When multiplying fractions, the numerators and the denominators are multiplied together as follows:

$$\frac{1}{2} \times \frac{2}{3} = \frac{1 \times 2}{2 \times 3} = \frac{2}{6} = \frac{1}{3}$$

Note the final answer is simplified by dividing the numerator and the denominator by 2.

When dividing fractions, the divisor is inverted and multiplied by the dividend as follows:

$$\frac{2}{3} \div \frac{1}{2} = \frac{2}{3} \times \frac{2}{1} = \frac{2 \times 2}{3 \times 1} = \frac{4}{3}$$

Supplemental Practice

Chapter 1

Section 1.1

1. What is matter? Tell whether each choice is matter.
   a. microwaves
   b. helium gas inside a balloon
   c. heat from the Sun
   d. velocity
   e. a speck of dust
   f. the color blue

2. Identify each substance as an element, a compound, a homogeneous mixture, or a heterogeneous mixture.
   a. air
   b. blood
   c. antimony
   d. brass
   e. ammonia
   f. mustard
   g. water
   h. tin

3. What is an alloy? Identify each as either a pure metal or an alloy.
   a. zinc
   b. steel
   c. sterling silver
   d. copper
   e. bronze

4. Which compounds listed in **Table 1.3** contain the element nitrogen? Which compounds contain oxygen but do not contain carbon?

Section 1.2

5. The element rubidium melts at 39.5°C and boils at 697°C. What is the physical state of rubidium at room temperature?

6. Which substance has the highest melting point: ethanol, helium, or baking soda?

Table 1.3 Some Common Compounds

Compound Name	Formula	Compound Name	Formula
Acetaminophen	$C_8H_9NO_2$	Hydrochloric acid	HCl
Acetic acid	$C_2H_4O_2$	Magnesium hydroxide	$Mg(OH)_2$
Ammonia	NH_3	Methane	CH_4
Ascorbic acid	$C_6H_8O_6$	Phosphoric acid	H_3PO_4
Aspartame	$C_{14}H_{18}N_2O_5$	Potassium tartrate	$K_2C_4H_4O_6$
Aspirin	$C_9H_8O_4$	Propane	C_3H_8
Baking soda	$NaHCO_3$	Salt	NaCl
Butane	C_4H_{10}	Sodium carbonate	Na_2CO_3
Caffeine	$C_8H_{10}N_4O_2$	Sodium hydroxide	NaOH
Calcium carbonate	$CaCO_3$	Sucrose	$C_{12}H_{22}O_{11}$
Carbon dioxide	CO_2	Sulfuric acid	H_2SO_4
Ethanol	C_2H_6O	Water	H_2O
Ethylene glycol	$C_2H_6O_2$		

Supplemental Practice

7. Your friend says that a certain substance has a freezing point of −22°C and a melting point of −10°C. What do you think of your friend's statement?

8. Suppose you have a cube of an unknown metal with a mass of 102 g. You place the cube in a graduated cylinder filled with water to the 40-mL mark. The cube sinks to the bottom of the cylinder and the water level rises to the 54-mL mark. What is the density of the cube?

9. Identify each statement as either a chemical or a physical property of the substance.
    a. Krypton is an inert gas.
    b. Ethanol is a clear, colorless liquid.
    c. Hydrochloric acid reacts with many metals to form hydrogen gas.
    d. Barium sulfate is practically insoluble in water.
    e. The element tungsten has a very high density.

10. Identify each event as either a chemical or a physical change.
    a. An iron bar expands slightly when it is heated.
    b. A banana turns brown when it is left on the counter.
    c. Hydrogen burns in air to form water vapor.
    d. The surface of a pond freezes in winter.
    e. A balloon pops when it is pricked by a pin.

Chapter 2

Section 2.1

1. Describe the observations of the chemist Joseph Proust. What scientific law resulted from his studies?

2. Distinguish among a hypothesis, a theory, and a scientific law.

3. How did the discovery of isotopes help lead to a revised model of atomic structure?

4. What is the atomic number of vanadium? What does that number tell you about an atom of vanadium?

5. A radioactive isotope of iodine is used to diagnose and treat disorders of the thyroid gland. This isotope has 53 protons and 78 neutrons. What are the atomic number and mass number of this isotope? How many electrons does it have?

6. One atom has 45 neutrons and a mass number of 80. Another atom has 36 protons and 44 neutrons. What is the identity of each of these atoms? Which atom has a greater mass number?

7. The element boron has two naturally occurring isotopes. One has atomic mass 10.01, whereas the other has atomic mass 11.01. The average atomic mass of boron is 10.811. Which of the two isotopes is more abundant?

Section 2.2

8. A satellite can orbit Earth at almost any altitude, depending on the amount of energy that is used to launch it. How is this different from the way that an electron orbits the atomic nucleus?

9. Electromagnetic wave A has a wavelength of 10^4 m, whereas electromagnetic wave B has a wavelength of 10^{-2} m. Compare the frequencies and energies of the two waves.

10. For each element, tell how many electrons are in each energy level and draw the Lewis dot diagram for each atom.

   a. phosphorus, 15 electrons
   b. beryllium, four electrons
   c. carbon, six electrons
   d. helium, two electrons

Chapter 3

Section 3.1

1. Use the periodic table to separate these ten elements into five pairs of elements having similar properties.
   S, Ne, Li, O, Mg, Ag, Na, Sr, Kr, Cu

Section 3.2

2. For each element, write its group number and period number, its physical state at room temperature, and whether it is a metal, a nonmetal, or a metalloid.
   a. argon
   b. nickel
   c. antimony
   d. fluorine
   e. barium

3. Write the symbol and name of the element that fits each description.
   a. the second-lightest of the halogens
   b. the metalloid with the lowest period number
   c. the only group 16 element that is a gas at room temperature
   d. the heaviest of the noble gases
   e. the group 15 nonmetal that is a solid at room temperature

4. Element Q is in the third period. Its outer energy level has six electrons. How does the number of outer-level electrons of element Q compare with that of element Z, which is in the second period, group 14? Write the name and symbol of each element.

5. Why is the chemistry of the transition elements and the inner transition elements less predictable than the chemistry of the main group metals?

6. Write the symbol of the element that has valence electrons that fit each description.
   a. three electrons in the fourth energy level
   b. one electron in the second energy level
   c. eight electrons in the third energy level
   d. two electrons in the first energy level
   e. five electrons in the sixth energy level

7. Draw the Lewis dot diagram for each element. What is the group number of each element? Is the element a metal, a nonmetal, or a metalloid?
   a. As
   b. Ca
   c. S
   d. H
   e. Br
   f. Al
   g. Si
   h. Xe

8. The chemical formula for sodium oxide is Na_2O. Use the periodic table to predict the formulas of the following similar compounds.
   a. potassium oxide
   b. sodium sulfide
   c. potassium sulfide
   d. lithium oxide

Supplemental Practice

9. How can the electrical conductivity of a semiconductor such as silicon be increased?

10. Why are transistors, diodes, and other semiconductor devices useful?

Chapter 4

Section 4.1

1. Use carbon dioxide as an example to contrast the properties of a compound with those of the elements of which it is composed.

Section 4.2

2. How is a calcium ion different from a calcium atom? From a potassium ion?

3. Would you expect to find MgO_2 as a stable compound? Explain.

4. Describe the submicroscopic structure of a grain of salt. What is this structure called?

5. Aluminum metal reacts with fluorine to form an ionic compound. Use the periodic table to determine the number of valence electrons for each element. Draw Lewis dot diagrams to show how they would combine to form ions. What is the formula for the resulting compound?

6. Ethane is a compound with the formula C_2H_6. What kind of compound is ethane? Describe the formation of ethane from carbon and hydrogen atoms, and draw the Lewis dot diagram of ethane.

7. Classify each of the following compounds as ionic or covalent. Use the periodic table to determine whether each component element of the compound is a metal or nonmetal. Make a general statement about the type of elements that form covalent and ionic compounds.
   a. hydrogen iodide
   b. strontium oxide
   c. rubidium chloride
   d. calcium sulfide
   e. sulfur dioxide

8. You are given two clear, colorless solutions, and you are told that one solution consists of an ionic compound dissolved in water and the other consists of a covalent compound dissolved in water. How could you determine which is an ionic solution and which is a covalent solution?

9. You are given a white crystalline substance to identify. You find that it melts at about 90°C and is insoluble in water. Is this substance more likely to be ionic or covalent? Explain.

Sections 4.1 and 4.2

10. Sucrose, or table sugar, has the formula $C_{12}H_{22}O_{11}$. How are the properties of this compound different from those of its elements? Do you think sucrose is more likely to be an ionic or a covalent compound? Explain.

Chapter 5

Section 5.1

1. Write the symbols for two ions and one atom that have the same electron configuration as each ion.
   a. Br^-
   b. Ba^{2+}
   c. Na^+
   d. P^{3-}

2. Label each as a common name or a formal name.
   a. saltpeter
   b. sodium hydrogen sulfate
   c. slaked lime
   d. soda ash
   e. potassium nitrite

3. What is indicated by subscripts in chemical formulas?

4. Of the formulas $MgCl_2$ and Mg_2Cl_4, which is the correct formula for magnesium chloride? Explain your choice.

5. What is oxidation number? What determines an element's oxidation number? Write the most common oxidation numbers for each element.
   a. I
   b. P
   c. Cs
   d. Ba
   e. Se
   f. Al
   g. Cu
   h. Pb

6. Write the formula for each binary ionic compound.
   a. strontium chloride
   b. rubidium oxide
   c. sodium fluoride
   d. magnesium sulfide

Table 5.2 Common Polyatomic Ions

Name of Ion	Formula	Charge
Ammonium	NH_4^+	1+
Hydronium	H_3O^+	1+
Hydrogen carbonate	HCO_3^-	1−
Hydrogen sulfate	HSO_4^-	1−
Acetate	$C_2H_3O_2^-$	1−
Nitrite	NO_2^-	1−
Nitrate	NO_3^-	1−
Cyanide	CN^-	1−
Hydroxide	OH^-	1−
Dihydrogen phosphate	$H_2PO_4^-$	1−
Permanganate	MnO_4^-	1−
Carbonate	CO_3^{2-}	2−
Sulfate	SO_4^{2-}	2−
Sulfite	SO_3^{2-}	2−
Oxalate	$C_2O_4^{2-}$	2−
Monohydrogen phosphate	HPO_4^{2-}	2−
Dichromate	$Cr_2O_7^{2-}$	2−
Phosphate	PO_4^{3-}	3−

Supplemental Practice

Table 5.4	Names of Common Ions of Selected Transition Elements	
Element	**Ion**	**Chemical Name**
Chromium	Cr^{2+}	chromium(II)
	Cr^{3+}	chromium(III)
	Cr^{6+}	chromium(VI)
Cobalt	Co^{2+}	cobalt(II)
	Co^{3+}	cobalt(III)
Copper	Cu^{+}	copper(I)
	Cu^{2+}	copper(II)
Gold	Au^{+}	gold(I)
	Au^{3+}	gold(III)
Iron	Fe^{2+}	iron(II)
	Fe^{3+}	iron(III)
Manganese	Mn^{2+}	manganese(II)
	Mn^{3+}	manganese(III)
	Mn^{7+}	manganese(VII)
Mercury	Hg^{+}	mercury(I)
	Hg^{2+}	mercury(II)
Nickel	Ni^{2+}	nickel(II)
	Ni^{3+}	nickel(III)
	Ni^{4+}	nickel(IV)

7. Write the formula for the compound formed from each pair of elements.
   a. magnesium and bromine
   b. potassium and sulfur
   c. strontium and oxygen
   d. aluminum and phosphorus

8. Examine the list of common polyatomic ions in **Table 5.2** on page 811. Which element appears most often in these ions? Write the symbols of the ions that do not contain this element.

9. How many atoms of each element are present in four formula units of ammonium oxalate?

10. The metals in these compounds can have various oxidation numbers. Predict the charge on each metal ion, and write the name of each compound.
    a. Au_2S
    b. FeC_2O_4
    c. $Pb(C_2H_3O_2)_4$
    d. Hg_2SO_4

11. Write the formula for the compound made from each pair of ions.
  a. potassium and dichromate ions
  b. sodium and nitrite ions
  c. ammonium and hydroxide ions
  d. calcium and phosphate ions

12. Write the formula for each compound containing polyatomic ions.
  a. ammonium sulfate
  b. barium hydroxide
  c. sodium hydrogen sulfate
  d. calcium acetate

13. Write the names of the following compounds containing chromium.
  a. $CrBr_2$
  b. $Cr_2(SO_4)_3$
  c. CrO_3
  d. $CrPO_4$

14. Write the formula for the compound made from each pairs of ion.
  a. lead(II) and sulfite
  b. manganese(III) and fluoride
  c. nickel(II) and cyanide
  d. chromium(III) and acetate

15. Write the names of the following compounds that contain transition elements not listed in **Table 5.4** on page 812.
  a. $RhCl_3$
  b. WF_6
  c. Nb_2O_5
  d. OsF_8

16. Write the formulas for the compounds. What do the formulas have in common?
  a. hydrogen iodide
  b. calcium selenide
  c. cobalt(II) oxide
  d. gallium phosphide
  e. barium selenide

17. Compare the properties of hygroscopic substances and deliquescent substances.

18. Write the names of the hydrates.
  a. $MgSO_3 \cdot 6H_2O$
  b. $Hg(NO_3)_2 \cdot H_2O$
  c. $NaMnO_4 \cdot 3H_2O$
  d. $Ni_3(PO_4)_2 \cdot 7H_2O$

19. Write the formula for each hydrate.
  a. nickel(II) cyanide tetrahydrate
  b. lead(II) acetate trihydrate
  c. strontium oxalate monohydrate
  d. palladium(II) chloride dihydrate

20. What is a desiccant? Describe the properties of a compound that would make a good desiccant.

Supplemental Practice

Section 5.2

21. How do the macroscopic properties of molecular substances reflect their submicroscopic structure?

22. What is a molecular element? Name the seven nonmetals that exist naturally as diatomic molecular elements.

23. How are electrons shared differently in H_2, O_2, and N_2 molecules?

24. What is ozone? How is ozone harmful? How is it helpful?

25. Name the molecular compounds.
    a. SF_4
    b. CSe_2
    c. IF_5
    d. P_2O_5

26. Write the formulas for the molecular compounds.
    a. nitrogen trichloride
    b. diiodine pentoxide
    c. diphosphorus triselenide
    d. dichlorine heptoxide

27. Name the molecular compounds, all of which contain carbon.
    a. C_5H_{12}
    b. SiC
    c. CF_4
    d. C_9H_{20}

Sections 5.1 and 5.2

28. Write formulas for a nitrogen atom, a nitrogen ion, and a nitrogen molecule.

29. Which compounds are ionic and which are molecular?
    a. dihydrogen sulfide
    b. strontium oxide
    c. lithium carbonate
    d. decane
    e. rubidium hydroxide

30. How is $KC_2H_3O_2$ a substance with two different types of bonding?

Chapter 6

Section 6.1

1. If bread is stored at room temperature for a long period of time, it undergoes chemical changes that make it inedible. How can you tell that these chemical changes have occurred?

2. List one piece of evidence you expect to see that indicates that a chemical reaction is taking place in each situation.
    a. An apple that has been peeled begins to oxidize.
    b. A flashlight is turned on.
    c. An egg is fried.
    d. Acidic and basic substances in an antacid tablet react when the tablet is placed in water.

3. Use the equation to answer the questions.
    $3Zn(s) + 2FeCl_3(aq) \rightarrow 2Fe(s) + 3ZnCl_2(aq)$
    a. What is the physical state of iron(III) chloride? Of iron metal?
    b. What is the coefficient of zinc?
    c. What is the subscript of chlorine in zinc chloride?

Supplemental Practice

4. Balance the equations.
   a. Al(s) + HCl(aq) → AlCl$_3$(aq) + H$_2$(g)
   b. H$_2$O$_2$(aq) → H$_2$O(l) + O$_2$(g)
   c. HC$_2$H$_3$O$_2$(aq) + CaCO$_3$(s) → Ca(C$_2$H$_3$O$_2$)$_2$(aq) + CO$_2$(g) + H$_2$O(l)
   d. C$_2$H$_6$O(l) + O$_2$(g) → CO$_2$(g) + H$_2$O(g)

5. Write balanced chemical equations for the reactions described.
   a. sulfuric acid + sodium hydroxide → sodium sulfate solution + water
   b. pentane liquid + oxygen → carbon dioxide + water vapor + energy
   c. iron metal + copper(II) sulfate solution → copper metal + iron(II) sulfate solution

6. Magnesium metal will burn in air to form magnesium oxide. In the reaction, two magnesium atoms react with one oxygen molecule to form two formula units of magnesium oxide. If you react 80 billion atoms of magnesium with 30 billion molecules of oxygen, will all the reactants be used up? What do you think the results will be?

Section 6.2

7. Which of the five general types of reactions has only one product? Which of the five types always has oxygen as a reactant?

8. Classify each reaction as one of the five general types.
   a. Zn(s) + 2AgNO$_3$(aq) → Zn(NO$_3$)$_2$(aq) + 2Ag(s)
   b. Fe(s) + S(s) → FeS(s)
   c. 2KClO$_3$(s) → 2KCl(s) + 3O$_2$(g)
   d. CH$_4$(g) + 2O$_2$(g) → CO$_2$(g) + 2H$_2$O(g) + energy
   e. Na$_2$CO$_3$(aq) + MgSO$_4$(aq) → MgCO$_3$(s) + Na$_2$SO$_4$(aq)

9. Use word equations to describe the chemical equations given, and classify each reaction as one of the five major types.
   a. C$_9$H$_{20}$(l) + 14O$_2$(g) → 9CO$_2$(g) + 10H$_2$O(g)
   b. H$_2$SO$_4$(aq) + 2KOH(aq) → K$_2$SO$_4$(aq) + 2H$_2$O(l)
   c. 2KNO$_3$(s) + energy → 2KNO$_2$(s) + O$_2$(g)

10. Write a balanced equation for the combustion of ethene gas (C$_2$H$_4$). How many oxygen molecules will react with 15 trillion ethene molecules?

Sections 6.1 and 6.2

11. Nitrogen oxyfluoride gas (NOF) forms by a reaction between nitrogen monoxide and fluorine gases. Write a balanced chemical equation for this reaction, and classify the reaction as one of the five major types.

12. When an aqueous solution of copper(II) sulfate is combined with an aqueous solution of sodium hydroxide, a blue precipitate of copper(II) hydroxide forms.
    a. Write word and balanced chemical equations for this reaction.
    b. Classify this reaction as one of the five major types.

13. The unbalanced equation for the reaction between aqueous solutions of NaOH and MgSO$_4$ is as follows.

    NaOH(aq) + MgSO$_4$(aq) → Na$_2$SO$_4$(aq) + Mg(OH)$_2$(s)

    a. Balance the chemical equation.
    b. Write a word equation for the reaction.
    c. Classify this reaction as one of the five major types.

Supplemental Practice

Section 6.3

14. What is a reversible reaction? How can a chemical equation show that a reaction is reversible?

15. Distinguish between the terms *equilibrium* and *dynamic equilibrium*.

16. Describe Le Châtelier's principle. Why is this principle important to chemical engineers?

17. Will an endothermic reaction that is at equilibrium shift to the left or to the right to readjust after each of the following procedures is followed?
    a. More heat is added.
    b. Heat is removed.
    c. More products are added.
    d. More reactants are added.

18. In a closed container, the reversible reaction represented by the equation $N_2O_4(aq) \rightarrow 2NO_2(g)$ has come to equilibrium. The volume of the container is then decreased, causing the pressure in the container to increase. Will the reaction be shifted to the left or the right? Explain.

19. A certain exothermic reaction has a very high activation energy. Do you expect this reaction to take place spontaneously under normal conditions? Explain.

20. Why are many valuable historic documents stored in sealed cases with most of the air removed?

21. Road salt on a car speeds the process of rusting. If a car is coated with road salt, would it be more beneficial to wash the car before a period of cold weather or before a period of warm weather? Explain.

22. Hydrogen gas can be produced by reacting aluminum and sulfuric acid, as shown by the equation for the reaction.

    $2Al(s) + 3H_2SO_4(aq) \rightarrow Al_2(SO_4)_3(aq) + 3H_2(g)$

    In a particular reaction, 12 billion molecules of H_2SO_4 were mixed with 6 billion atoms of Al.
    a. Which reactant is limiting?
    b. How many molecules of H_2 are formed when the reaction is complete?

23. Fritz Haber carried out his process for the synthesis of ammonia at a temperature of about 600°C. High temperature increases the rate of reaction. Why didn't Haber use a higher temperature for his process?

24. If a catalyst is added to a reversible endothermic reaction that is at equilibrium, will the reaction shift to the left or to the right?

25. Why are catalysts often used in the form of a powder?

26. What are enzymes? Provide some examples of how your body uses enzymes, and list two common products that contain enzymes.

27. Examine the ingredients of some food products in your home, and list the inhibitors—usually called preservatives—that you find in these products.

28. The graph shown below represents the concentrations of three compounds, A, B, and C, as they take part in a reaction that reaches equilibrium.
    a. Which compound(s) represent the reactant(s) in this reaction? The product(s)?
    b. How long did it take for the reaction to reach equilibrium?
    c. Explain how the graph will change if more compound C is added one minute after the reaction reaches equilibrium.

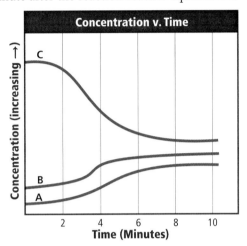

Sections 6.1 and 6.3

29. Sulfur dioxide gas and oxygen gas combine in a reversible reaction to form sulfur trioxide gas.
    a. Write a word equation for this reaction.
    b. Write a balanced chemical equation for this reaction.
    c. If more sulfur trioxide gas is added after the reaction reaches equilibrium, will the reaction be shifted to the left or to the right?

30. In one of a series of reactions used to produce nitric acid, ammonia gas and oxygen gas react to form nitrogen monoxide gas and water vapor. Write a balanced equation for this reaction. If 30 trillion ammonia molecules and 35 trillion oxygen molecules are available to react, which is the limiting reactant?

Chapter 7
Section 7.1

1. Given the colors yellow, red, and violet, which color has the lowest frequency? The shortest wavelength? The least energy?

2. How many f orbitals can there be in an energy level? What is the maximum number of f electrons in an energy level? What is the lowest energy level that can have f orbitals?

3. What is the Heisenberg uncertainty principle? This principle led to the development of what model to describe electrons in atoms?

4. What is the shape of an s orbital? How do s orbitals at different energy levels differ from one another?

Supplemental Practice

Section 7.2

5. Identify the elements that have the following electron configurations.
   a. $[Ar]3d^{10}4s^24p^6$
   b. $[Ne]3s^23p^1$
   c. $[Ar]4s^13d^5$
   d. $1s^22s^1$
   e. $[Xe]6s^24f^7$

6. Use the periodic table to help you write electron configurations for the following atoms. Use the appropriate noble gas, inner-core abbreviations.
   a. Br
   b. N
   c. Te
   d. Ni
   e. Ba

7. What are the valence electrons in the electron configuration of germanium, $[Ar]3d^{10}4s^24p^2$?

8. What is the highest occupied sublevel in the structure of each of the following elements?
   a. Ca
   b. B
   c. In
   d. He
   e. Bi

9. Refer to the periodic table and compare the similarities and differences in the electron configurations of the following pairs of elements: Ar and Kr, K and Ar, K and Sc.

10. Why do most of the transition elements have multiple oxidation numbers? What other category of elements exhibits multiple oxidation numbers?

Chapter 8

Section 8.1

1. From each pair of atoms, select the one with the larger atomic radius.
   a. Li, Be
   b. Ca, Ga
   c. P, As
   d. Br, O
   e. In, Ba

2. From each pair of ions, choose the one with the larger ionic radius and explain your choice.
   a. Se^{2-}, S^{2-}
   b. Te^{2-}, Cs^+
   c. Mg^{2+}, Al^{3+}
   d. B^{3+}, N^{3-}
   e. O^{2-}, P^{3-}

3. What is the electron configuration of the potassium ion? What noble gas has that configuration? Name two other ions with the same configuration.

4. Your friend says that the most active metal has a large atomic radius and a large number of valence electrons. Assess the accuracy of your friend's statement.

5. Compare and contrast the properties of the alkali metals and the alkaline earth metals.

6. Both potassium and strontium react with water. Predict the products of the reactions of potassium and strontium with water. Write balanced equations for both reactions.

7. Explain how and why the crystal structures of the ionic compounds CsCl and NaCl differ.

8. Boron is classified as a metalloid. What does this mean in terms of the way that boron reacts with other elements?

9. Describe some uses of the group 15 elements.

Sections 8.1 and 8.2

10. Give one reason why each metal is essential to human life.
    a. potassium
    b. magnesium
    c. calcium
    d. iron

Chapter 9

Section 9.1

1. In general, how do electronegativity values change as you move down a column of the periodic table? Across a row of the table? Explain these trends.

2. Using only a periodic table, rank these atoms from the least to the most electronegative.
   As, Ba, N, Mg, Cs, O

3. Classify the following bonds as ionic, covalent, or polar covalent.
   a. C—O
   b. Cu—Cl
   c. Ca—O
   d. Si—H
   e. Mn—O

4. Using only a periodic table, rank the bonds from the most ionic to the least ionic.
   a. Li—F
   b. Li—Br
   c. K—F
   d. Li—Cl

5. Do you expect KBr or KF to have a higher melting point? Explain.

6. Rank the bonds from the least to the most polar.
   a. S—O
   b. C—Cl
   c. Si—Cl
   d. H—Br

7. What factors determine whether a molecule is polar or nonpolar?

8. In metallic bonding, why are the valence electrons of the metal sometimes described as a "sea of electrons"?

9. Nitric oxide (NO) is a colorless gas used in the manufacture of nitric acid. Is nitric oxide a polar or nonpolar molecule? Explain.

Section 9.2

10. Draw the Lewis dot diagram for each molecule.
    a. CS_2
    b. HI
    c. CH_3Cl
    d. AsH_3

11. Describe the shape of each molecule in Question 10.

12. Dibromomethane, CH_2Br_2, is a molecule similar in structure to methane. Is dibromomethane a polar or nonpolar molecule? Explain.

13. Ethanol, carbon dioxide, and vitamin C are all covalent compounds. How do these compounds demonstrate the variety of types of covalent compounds?

14. Carbon tetrafluoride (CF_4) is used as a low temperature refrigerant and as a gaseous insulator. Are the bonds in CF_4 polar or nonpolar? Is CF_4 a polar molecule? Explain.

15. Compare the molecules phosphorus trichloride (PCl_3) and dichloromonoxide (Cl_2O). How many pairs of electrons surround the central atom? How many of these pairs are bonding? Nonbonding? What are the shapes of the molecules?

Supplemental Practice

16. Tetrachloroethylene, C_2Cl_4, is a derivative of ethene in which the hydrogen atoms have been replaced by chlorine atoms. Draw the Lewis dot diagram of C_2Cl_4. Is the molecule polar? Explain.

17. Formaldehyde, CH_2O, is a colorless gas that, when dissolved in water, is used in embalming fluids. Draw its Lewis dot diagram and describe its geometry.

18. Diethyl ether, $CH_3CH_2OCH_2CH_3$, can be viewed as a derivative of water, in which the hydrogens in water are replaced by CH_3CH_2 (ethyl) groups. Is diethyl ether a polar molecule? Explain.

19. In paper chromatography, what is the stationary phase? What is the mobile phase? Compare the distance migrated for a component with a strong attraction to the paper to the distance migrated for a component with a weak attraction to the paper.

20. Predict the relative boiling points of ammonia(NH_3), phosphine (PH_3), and arsine (AsH_3). Explain your prediction.

Chapter 10

Section 10.1

1. Why does a pollen grain suspended in a water droplet trace a random and erratic path?

2. Contrast the behavior of an air hockey puck when it collides with the wall of the game board to the behavior of an ideal gas particle when it collides with the wall of its container.

3. What is the numerical value of atmospheric pressure at sea level? Do humans notice the pressure of the atmosphere? Explain.

4. Compare solids and liquids in terms of particle spacing and particle motion.

Section 10.2

5. Write the Fahrenheit, Celsius, and Kelvin temperatures that correspond to
   a. the freezing point of water.
   b. the boiling point of water.
   c. absolute zero.

6. Rank the following temperature readings in increasing order.
   110°C, 212 K, 212°F, 273°C, 273 K

7. Complete the table.

Temperature	Celsius, °C	Kelvin, K
Boiling point of helium, He	−268.94	
Hot summer day		305
Melting point of zinc	419	
Boiling point of butane, C_4H_{10}		272.7

8. The boiling points of five gases are provided in either Celsius or Kelvin temperatures. List the gases in order from the one with the lowest boiling

point to the one with the highest boiling point.

Chlorine(Cl_2)	239 K
Krypton (Kr)	−153°C
Dimethyl ether (C_2H_6O)	−24°C
Dihydrogen sulfide (H_2S)	213 K
Sulfur dioxide (SO_2)	−10°C

9. Suppose that the outside temperature increases by 1.00 Celsius degree. What is the temperature increase in Fahrenheit degrees? In kelvins?

10. The particles of which of the following gases have the highest average speed? The lowest average speed?

   a. argon at 20°C
   b. nitrogen at 100°C
   c. nitrogen at 20°C
   d. helium at 100°C

11. Which gas particles in Question 10 have the highest average kinetic energy? Which have the lowest average kinetic energy?

12. Dihydrogen sulfide (H_2S) and sulfur dioxide (SO_2) are both colorless gases with unpleasant odors. Which of these gases has a higher rate of diffusion in air at the same temperature? Explain your answer.

13. Explain why liquids become cooler as they evaporate.

14. Compare and contrast sublimation and evaporation.

15. Can a liquid in an open container reach equilibrium with its vapor? Explain.

16. Describe how boiling point is affected by increasing pressure.

17. Predict whether the boiling point of water is greater or less than 100°C at your school.

18. Explain the relationship among the terms *freezing point*, *heat of fusion*, and *crystal lattice*.

19. What is the ratio of the amount of energy needed to boil 1 kg of water at 100°C to the amount of energy needed to melt 1 kg of ice at 0°C?

20. The compound cyclohexane (C_6H_{12}) boils at 80.7°C and melts at 6.5°C. A few grams of cyclohexane are cooled from +120°C to −20°C. Graph the cooling curve for cyclohexane. Show time on the horizontal axis and temperature on the vertical axis.

Chapter 11
Section 11.1

1. Using **Table 11.1** and the equation 1.00 in = 25.4 mm to convert the following pressure measurements.
   a. 147 mm Hg to psi
   b. 232 psi to kPa
   c. 67.2 kPa to mm Hg
   d. 214 in. Hg to atm

Table 11.1 Equivalent Pressures

1.00 atm	760 mm Hg	14.7 psi	101.3 kPa

Supplemental Practice

2. Convert the pressure measurement 3.50 atm to the following units.
   a. mm Hg
   b. psi
   c. kPa
   d. in Hg

3. What is the difference between gauge pressure and absolute pressure?

4. A tire gauge at a gas station indicates a pressure of 29.0 psi in your tires. What is the absolute pressure in your tires in inches of mercury?

5. Suppose that a certain planet's atmosphere consists primarily of CO_2 gas. Compare the atmospheric pressure on the planet's surface to the pressure that would be exerted if the atmosphere were the same thickness, but consisted primarily of methane gas.

6. Propose a reason why Torricelli used mercury in his barometer rather than some other liquid such as water.

Section 11.2

7. What factors affect gas volume?

8. A cylinder containing 48 g of nitrogen gas at room temperature is placed on a scale. The valve is opened and 12 g of gas are allowed to escape. Assuming that the temperature of the cylinder remains constant, how does the pressure change?

9. A cylinder contains 22 g of air. If 77 g of air are pumped into the cylinder at constant temperature, how does the pressure in the cylinder change?

10. A cylinder contains 36.5 g of argon gas at a pressure of 8.20 atm. The valve is opened and gas is allowed to escape until the pressure is reduced to 4.75 atm at constant temperature. How many grams of argon escaped?

11. If the gas pressure in an aerosol can is 182 kPa at 20.0°C, what is the pressure inside the can if it is heated to 251°C?

12. A 5.3-L sample of argon is at standard atmospheric pressure and 294°C. The sample is cooled in dry ice to −79°C at constant volume. What is its new pressure in mm Hg?

13. A tank for compressed gas can safely withstand a maximum pressure of 955 kPa. The pressure in the tank is 689 kPa at a temperature of 22°C. What is the highest temperature the tank can safely withstand?

14. You use a pressure gauge to measure the air pressure in your bicycle tires on a cold morning when the temperature is −5°C. The gauge reads 53 psi. The next afternoon, the temperature has warmed up to 11°C. If you measure the pressure in your tires again, what would you expect the reading on the pressure gauge to be? Assume the volume of air in the tires is constant.

15. A cylinder contains 98 g of air at 295 K. The pressure in the cylinder is 174 psi. If 32 g of air are allowed to escape and the cylinder is heated to 335 K, what is the new pressure?

16. A cylinder of compressed gas has a volume of 14.5 L and a pressure of 769 kPa. What volume would the gas occupy if it were allowed to escape into a balloon at a pressure of 117 kPa? Assume constant temperature.

17. At a sewage treatment plant, bacterial cultures produce 2400 L of methane gas per day at 1.0 atm pressure. If one day's production of methane is stored in a 310-L tank, what is the pressure in the tank?

Supplemental Practice

18. A 0.400-L balloon is filled with air at 1.10 atm. If the balloon is squeezed into a 250-mL beaker and doesn't burst, what is the pressure of the air?

19. A sample of argon gas is compressed, causing its pressure to increase by 25 percent at constant temperature. What is the percentage change in the volume of the sample?

20. Compare qualitatively the volume of a sample of air at STP to the volume of the same sample of air under normal conditions of temperature and pressure in your classroom.

21. A balloon is filled with 2.5 L of air at 23°C and 1 atm. It is then placed outdoors on a cold winter day when the temperature is −12°C. If the pressure is constant, what is the new volume of the balloon?

22. The volume of a sample of nitrogen is 75 mL at 25°C and 1 atm. What is its volume at STP?

23. At atmospheric pressure, a balloon contains 1.50 L of helium gas. How would the volume change if the Kelvin temperature were only 60 percent of its original value?

24. The volume of a sample of helium is 2.45 L at −225°C and 1 atm. Predict the volume of the sample at +225°C and 1 atm.

25. A cylinder contains 5.70 L of a gas at a temperature of 24°C. The cylinder is heated, and a piston moves in the cylinder so that constant pressure is maintained. If the final volume of the gas in the cylinder is 6.55 L, what is the final temperature?

26. A 3.50-L sample of nitrogen at 10°C is heated at constant pressure until it occupies a volume of 5.10 L. What is the new temperature of the sample?

27. A sample of argon gas has a volume of 425 mL at 463 K. The sample is cooled at constant pressure to a volume of 315 mL. What is the new temperature?

28. Use Charles's law to predict the change in density of a sample of air if its temperature is increased at constant pressure.

29. Suppose you have a 1-L sample of neon gas and a 1-L sample of nitrogen gas; both samples are at STP. Compare the numbers of gas particles in each sample. Compare the masses of the samples.

30. An 825-mL sample of oxygen is collected at 101 kPa and 315 K. If the pressure increases to 135 kPa and the temperature drops to 275 K, what volume will the oxygen occupy?

31. At 37.4 kPa and −35°C, the volume of a sample of krypton gas is 346 mL. What is the volume at STP?

32. A 44-g sample of CO_2 has a volume of 22.4 L at STP. The sample is heated to 71°C and compressed to a volume of 18.0 L. What is the resulting pressure?

33. A gas has a volume of 2940 mL at 58°C and 95.9 kPa. What pressure will cause the gas to have a volume of 3210 mL at 25°C?

Supplemental Practice

34. A balloon is filled with 4.50 L of helium at 15°C and 763 mm Hg. The balloon is released and it rises through the atmosphere. When it reaches an altitude of 2500 m, the temperature is 0°C and the pressure has dropped to 542 mm Hg. What is the volume of the balloon?

35. Compare the number of particles in 1.0 L of nitrogen gas at room temperature and 1.0 atm pressure to the number of particles in 2.0 L of oxygen gas at room temperature and 3.0 atm pressure.

36. How many liters oxygen will be needed to react completely with 15 L of carbon monoxide at the same temperature and pressure to form carbon dioxide?

37. How many liters of nitrogen gas and hydrogen gas must combine to produce 28.6 L of ammonia if all gases are at the same temperature and pressure?

38. A cylinder contains 14.2 g of nitrogen at 23°C and 5030 mm Hg. The cylinder is heated to 45°C and nitrogen gas is released until the pressure is lowered to 2250 mm Hg. How many grams of nitrogen are left in the cylinder?

39. A 75.0-mL sample of air is at standard pressure and −25°C. The air is compressed to a volume of 45.0 mL, and the temperature is adjusted until the pressure of the air doubles to 2.00 atm. What is the final temperature?

40. A gas has a volume of 146 mL at STP. What Celsius temperature will cause the gas to have a volume of 217 mL at a pressure of 615 mm Hg?

Chapter 12

Section 12.1

1. Your grandmother gives you a bag of pennies she has saved. If the pennies have a mass of 4.24 kg, and 50 pennies have a mass of 144 g, how many pennies did you receive?

2. What are the molar masses of these substances?
   a. palladium
   b. hydrogen
   c. calcium hydroxide
   d. diphosphorus pentoxide

3. Determine the number of atoms in each sample.
   a. 12.7 g silver (Ag)
   b. 56.1 g aluminum (Al)
   c. 162 g calcium (Ca)

4. Without calculating, decide whether 10.0 g of zinc or 10.0 g of silicon represents the greater number of atoms. Verify your answer by calculating.

5. Determine the number of moles in each sample below.
   a. 7.62 g cesium chloride (CsCl)
   b. 42.5 g propanol (C_3H_8O)
   c. 694 g ammonium dichromate (($NH_4)_2Cr_2O_7$)

6. Determine the mass of the following molar quantities.
   a. 0.172 mol ozone (O_3)
   b. 2.50 mol heptane (C_7H_{16})
   c. 0.661 mol iron(II) phosphate ($Fe_3(PO_4)_2$)

Supplemental Practice

7. Which has the largest mass?
   a. ten atoms of carbon (C)
   b. three molecules of chlorine gas (Cl_2)
   c. one molecule of fructose ($C_6H_{12}O_6$)

8. Which has the largest mass?
   a. 10.00 mol of carbon (C)
   b. 3.00 mol of chlorine gas (Cl_2)
   c. 1.000 mol of fructose ($C_6H_{12}O_6$)

9. Determine the number of molecules in 0.127 mol of formic acid (CH_2O_2). What is the mass of this quantity of formic acid?

10. Determine the number of molecules or formula units in each sample below. Identify each as a formula unit or a molecule.
    a. 85.3 g water (H_2O)
    b. 100.0 g chlorine (Cl_2)
    c. 0.453 g potassium chloride (KCl)
    d. 14.6 g acetaminophen ($C_8H_9NO_2$)

11. The average atomic mass of oxygen is 4.0 times greater than the average atomic mass of helium. Would you expect the molar mass of oxygen gas to be 4.0 times greater than the molar mass of helium gas? Explain.

12. Vitamin B2, also called riboflavin, has the chemical formula $C_{17}H_{20}N_4O_6$. What is the molecular mass of vitamin B_2? What is its molar mass?

13. The molecular formula of cholesterol is $C_{27}H_{46}O$. What is the molar mass of cholesterol? What is the mass in grams of a single molecule of cholesterol?

14. What mass of aluminum contains the same number of atoms as 125 g of silver?

Section 12.2

15. The combustion of methanol produces carbon dioxide gas and water vapor.
    $$2CH_3OH(l) + 3O_2(g) \rightarrow 2CO_2(g) + 4H_2O(g)$$
    What mass of water vapor forms when 52.4 g of methanol burn?

16. Disulfur dichloride is prepared by passing chlorine gas into molten sulfur.
    $$Cl_2(g) + 2S(l) \rightarrow S_2Cl_2(l)$$
    How many grams of chlorine will react to produce 20.0 g of S_2Cl_2?

17. Using the reaction in Question 16, how many grams of sulfur will react to produce 20.0 g of S_2Cl_2?

18. Calculate the mass of precipitate that forms when 250 mL of an aqueous solution containing 35.0 g of lead (II) nitrate reacts with excess sodium iodide solution by the following reaction.
    $$Pb(NO_3)_2(aq) + 2NaI(aq) \rightarrow 2NaNO_3(aq) + PbI_2(s)$$

19. What mass of carbon must burn to produce 4.56 L of CO_2 gas at STP? The reaction is
    $$C(s) + O_2(g) \rightarrow CO_2(g)$$

Supplemental Practice

20. What volume of hydrogen gas can be produced by reacting 5.40 g of zinc in excess hydrochloric acid at 25.0°C and 105 kPa? The reaction is
$$Zn(s) + 2HCl(aq) \rightarrow ZnCl_2(aq) + H_2(g).$$

21. Explain how the ideal gas law confirms the laws of Boyle and Charles.

22. How many moles of argon are contained in a 2.50-L canister at 125 kPa and 15.0°C?

23. What is the volume of 0.300 mol of oxygen at 46.5 kPa and 215°C?

24. What is the pressure of a 2.00-mol sample of nitrogen gas that occupies 10.5 L at −25°C?

25. Determine the value of the ideal gas constant R in units of atm•L/mol•K.

26. Using the value of the ideal gas constant determined in Question 25, calculate the number of moles in a 7.56-L sample of neon gas at −45°C and 4.12 atm. How many grams of neon does the sample contain?

27. Dodecane ($C_{12}H_{26}$) is one of the components of kerosene. Write the balanced chemical equation for the combustion of dodecane to form carbon dioxide gas and water vapor. If 3.00 moles of dodecane burn, how many moles of oxygen are consumed? How many moles of carbon dioxide and water vapor are produced?

28. If 60.0 g of dodecane burn as in Question 27, how many grams of water vapor are produced? How many grams of carbon dioxide?

29. Green plants produce oxygen by the reaction below. If a rosebush produces 50.5 L of oxygen at 27.0°C and 101.3 kPa, how many grams of glucose ($C_6H_{12}O_6$) did the plant also produce?
$$6CO_2(g) + 6H_2O(l) \xrightarrow{sunlight} C_6H_{12}O_6(aq) + 6O_2(g)$$

30. Calculate the mass of each product formed when 10.5 g of sodium hydrogen carbonate ($NaHCO_3$) react with excess hydrochloric acid.

31. When a solution of 25.0 g of silver nitrate in 100 g of water is mixed with a solution of 10.0 g of magnesium chloride in 100 g of water, a double displacement reaction occurs. The balanced chemical equation is shown below. Which is the limiting reactant?
$$2AgNO_3(aq) + MgCl_2(aq) \rightarrow Mg(NO_3)_2(aq) + 2AgCl(s)$$

32. What is the empirical formula of each compound?
    a. $K_2C_2O_4$
    b. $Na_2S_2O_3$
    c. $C_{20}H_{20}O_4$
    d. $Pb_3(PO_4)_2$
    e. $C_{15}H_{21}N_3O_{15}$
    f. $C_6H_{12}O_7$

33. What is the molecular formula of each compound?
    a. empirical formula: C_4H_4O;
       molar mass: 136 g/mol
    b. empirical formula: CH_2;
       molar mass: 154 g/mol
    c. empirical formula: As_2S_5;
       molar mass: 310 g/mol

34. Without calculating, which of the following compounds has the greater percentage of nitrogen: $Ca(NO_3)_2$ or $Ca(NO_2)_2$? How do you know?

35. An oxide of chromium is 68.4 percent chromium by mass. The molar mass of the oxide is 152 g/mol. What is the formula of the compound?

36. Determine the percent oxygen in each of the following oxides of nitrogen.
    a. nitrogen monoxide (NO)
    b. dinitrogen monoxide (N_2O)
    c. dinitrogen pentoxide (N_2O_5)

37. A mixture of sodium chloride and potassium chloride is analyzed and is found to contain 22 percent potassium. What percent of the mixture is sodium chloride?

38. Vanillin is a compound that is used as a flavoring agent in many food products. Vanillin is 63.2 percent C, 5.3 percent H, and 31.5 percent O. The molar mass is approximately 152 g/mol. What is the molecular formula of vanillin?

39. What is the formula for a hydrate that consists of 80.15 percent $ZnSO_3$ and 19.85 percent H_2O?

40. A 15.0-g sample of sodium carbonate reacts with excess sulfuric acid to form sodium sulfate by the reaction below. When the reaction is complete, 16.9 g of sodium sulfate are recovered. What is the percent yield?

$$Na_2CO_3(s) + H_2SO_4(aq) \rightarrow Na_2SO_4(aq) + CO_2(g) + H_2O(l)$$

Chapter 13

Section 13.1

1. Give four examples of molecular compounds, other than water, that will form hydrogen bonds.

2. Use **Figure 13.6** to answer the questions.
   a. What is the density of water at 10.0°C?
   b. At what temperature is the density of water 0.9998 g/mL?
   c. The density of water at 1.0°C is 0.9999 g/mL. At what other temperature does water have the same density?

3. The surface tension of water is about three times greater than that of ethanol. How would a drop of water placed on a flat tabletop behave differently from a drop of ethanol placed on the same surface?

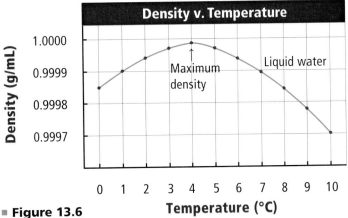

■ Figure 13.6

Supplemental Practice

4. Suppose you have a liquid in which both the interparticle attractive forces and the attractive forces between the liquid particles and silicon dioxide are moderately weak. You pour some of the liquid into a 25-mL glass graduated cylinder. How do you expect the liquid surface to appear?

5. Suppose you have 150 mL of water at 20°C in a 250-mL beaker, and 150 g of ethanol at 20°C in an identical beaker. You heat both liquids to 70°C. Which liquid absorbs more heat?

6. Suppose the specific heat of water were 2.0 J/g·°C. How would the climate on Earth be different?

Section 13.2

7. Write equations for the dissociation of the ionic compounds when they dissolve in water.
    a. $ZnCl_2$
    b. Rb_2CO_3
    c. $Mg(NO_3)_2$
    d. $(NH_4)_2SO_4$

8. When predicting solubility, scientists often use the phrase *like dissolves like*. Explain how water, a covalent compound, can be "like" an ionic compound.

9. Do you expect that the liquid compound hexane (C_6H_{14}) will mix with water to form a solution? Explain.

10. Compare and contrast the structures of detergent molecules and soap molecules. Which of the two is more effective in hard water? Explain.

11. Why don't chemists use the words strong and weak to describe solution concentrations? What terms do they use?

12. You dissolve 5.0 g of $KClO_3$ in 50.0 mL of water at 40°C, then you slowly cool the solution to 20°C. There is no visible change in the solution. Use **Figure 13.20** to determine whether the solution at 20°C is saturated, unsaturated, or supersaturated.

13. Use **Figure 13.20** to compare the solubilities in water of potassium chlorate and cerium(III) sulfate over the temperature range from 10°C to 30°C.

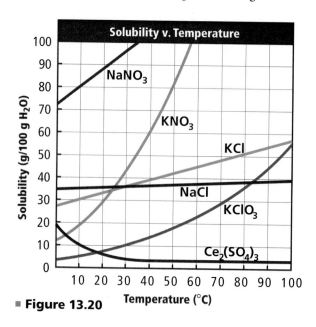

■ **Figure 13.20**

14. Is the process of dissolving in water exothermic or endothermic
    a. for most solid solutes?
    b. for a solute that is used in a hot pack?

15. How would you prepare 3.00 L of a 0.500M solution of calcium nitrate, $Ca(NO_3)_2$?

16. How would you prepare 1.50 L of a 1.75M solution of zinc chloride ($ZnCl_2$)?

17. What mass of ammonium sulfate (($NH_4)_2SO_4$) must be dissolved to make 720.0 mL of a 0.200M solution?

18. What mass of acetone, C_3H_6O, must be dissolved to make 2.50 L of a 1.15M solution?

19. What is the molarity of a solution that contains 2.50 g of potassium carbonate (K_2CO_3) in 1.42 L of solution?

20. Calculate the molarity of a solution that contains 185 g of methanol (CH_3OH) in 1150 mL of solution.

21. If you add 170.0 g of $NaNO_3$ to 1.00 L of water, have you created a 2.00M $NaNO_3$ solution? Explain.

22. For pure water, there is a 100°C difference between the boiling point and the freezing point. Is the corresponding temperature difference for a 1.5M sucrose solution greater or less than 100°C? Explain.

23. Rank these solutions from the one with the lowest to the one with the highest freezing point, and explain your answer.
    a. 1.2 mol of $ZnSO_4$ in 0.900 L of solution
    b. 0.40 mol of KCl in 550 mL of solution
    c. 3.4 mol of $NaNO_3$ in 1.80 L of solution

24. You have two aqueous solutions; one contains 151.6 g of KNO_3 dissolved in 1.00 L of solution, and the other contains 1026 g of sucrose dissolved in 2.00 L of solution. Calculate the molarity of each solution, and determine which solution has a higher boiling point.

25. Ethylenediamine is a molecule with the formula $C_2H_4(NH_2)_2$. Each molecule has two NH_2 groups. Ethylamine ($C_2H_5NH_2$) has one NH_2 group. How would you expect the boiling points of ethylamine and ethylenediamine to compare?

26. A selectively permeable membrane separates two aqueous solutions of sodium chloride. On the left side of the membrane is a solution composed of 74 g of NaCl dissolved in 420 g of water. On the right side of the membrane is a solution composed of 27 g of NaCl dissolved in 125 g of water. In which direction is the net solvent flow?

27. Describe two factors that affect the solubility of a gas in water.

28. Explain the cause of the dangerous condition known as the bends, which is sometimes experienced by divers.

29. What is a liquid aerosol? A foam? Give two examples of each.

30. Compare and contrast colloids and true solutions.

Supplemental Practice

Chapters 14

Section 14.1

1. Write the names and formulas of the acids and bases that were among the top ten industrial chemicals produced in the United States.

2. Write a chemical equation to show how aqueous perchloric acid fits the definition of an acid.

3. Write the balanced equation for the reaction of sulfuric acid with aluminum.

4. Write the formula for each acid and identify each as a monoprotic, a diprotic, or a triprotic acid.
   a. nitric acid
   b. hydrobromic acid
   c. citric acid
   d. carbonic acid
   e. benzoic acid

5. Identify the first compound in each reaction as an acid or a base.
   a. $HBrO + H_2O \rightarrow H_3O^+ + BrO^-$
   b. $N_2H_4 + H_2O \rightarrow N_2H_5^+ + OH^-$
   c. $C_{10}H_{14}N_2 + H_2O \rightarrow C_{10}H_{14}N_2H^+ + OH^-$
   d. $C_6H_5OH + H_2O \rightarrow H_3O^+ + C_6H_5O^-$

6. Classify each substance as an acid, a base, or neither when mixed with water.
   a. $HC_3H_5O_2$ c. HI   e. $Fe(OH)_3$
   b. Li_2O d. C_2H_6

7. Write the balanced equation for the reaction of calcium hydroxide with formic acid.

8. You want to prepare barium nitrate by an acid-base reaction. Write a reaction that will do this.

9. Consider the oxides SrO and SO_2. For each oxide, tell whether it is an acidic anhydride or a basic anhydride. Write an equation for each to demonstrate its acid-base chemistry.

Section 14.2

10. Give the formula for each hydroxide and identify each as a strong base or a weak base.
    a. potassium hydroxide
    b. aluminum hydroxide
    c. strontium hydroxide
    d. iron(III) hydroxide
    e. rubidium hydroxide

11. Hydrogen cyanide, HCN, is an extremely poisonous liquid that reacts very slightly with water to form relatively few hydronium ions and cyanide, CN^-, ions. Classify HCN as a strong acid, a weak acid, a strong base, or a weak base.

12. You test several solutions and find that they have pHs of 12.2, 3.5, 8.0, 5.7, 1.2, and 10.0. Which solution has the highest concentration of hydronium ions? Of hydroxide ions? Which solution is the closest to being neutral?

13. Find the pH values of the solutions with the following hydronium ion concentrations.

    a. $10^{-9} M$ b. $1 M$ c. $10^{-3} M$

14. Find the pH values of the solutions with the following hydroxide ion concentrations.

    a. $10^{-6} M$ b. $10^{-7} M$ c. $10^{-14} M$

15. A solution of sodium carbonate is tested and found to have a pH of 11. Compare the hydronium ion and hydroxide ion concentrations to those of a neutral solution.

16. In an aqueous solution of formic acid, compare the concentrations of hydroxide ions, hydronium ions, formate ions (CHO_2^-), and formic acid molecules.

17. Estimate the molarity of NH_3, NH_4^+, and OH^- in 0.25 M NH_3.

18. What are the molarities of HNO_3, H_3O^+, NO_3^-, and OH^- in a 0.010M solution of HNO_3? What is the pH of the solution?

19. What are the molarities of NaOH, Na^+, OH^-, and H_3O^+ in a 1.0M solution of NaOH? What is the pH of the solution?

Sections 14.1 and 14.2

20. Which solution is the best conductor of electricity? Which is the weakest conductor?

    a. 1.0M $HC_2H_3O_2$
    b. 0.1M $HC_2H_3O_2$
    c. 0.5M H_2SO_4

Chapter 15

Section 15.1 Write overall, ionic, and net ionic equations for each reaction in Questions 1–6.

1. nitric acid (HNO_3) and magnesium hydroxide ($Mg(OH)_2$)

2. formic acid ($HCHO_2$) and lithium hydroxide (LiOH)

3. sulfuric acid (H_2SO_4) and sodium hydroxide (NaOH)

4. hydrobromic acid (HBr) and ammonia (NH_3)

5. lactic acid ($HC_3H_5O_3$) and strontium hydroxide ($Sr(OH)_2$)

6. hydrochloric acid (HCl) and iron(III) hydroxide ($Fe(OH)_3$)

7. For each of the reactions in Questions 1–6, predict whether the pH of the product solution is acidic, basic, or neutral, and explain your prediction.

8. Identify the spectator ions in each reaction in Questions 1–6.

9. A reaction occurs in which perchloric acid neutralizes an unknown strong base. Write the net ionic equation for the reaction.

10. Benzoic acid ($HC_7H_5O_2$) is found in small amounts in many types of berries. Write the overall, ionic, and net ionic equations for the reaction of benzoic acid with lithium hydroxide. Will the pH of the product solution be greater than 7, exactly 7, or less than 7? Explain.

Supplemental Practice

11. When each salt is dissolved in water, will its solution be acidic, basic, or neutral? Explain.
    a. NaBr
    b. $NaC_2H_3O_2$
    c. NH_4NO_3

12. In terms of hydrogen ion transfer, how are acids and bases defined?

13. Identify the acid and the base in each reaction.
    a. $HC_2H_3O_2(aq) + NH_3(aq) \rightarrow NH_4^+(aq) + C_2H_3O_2^-(aq)$
    b. $HNO_2(aq) + H_2O(l) \rightarrow H_3O^+(aq) + NO_2^-(aq)$
    c. $H_2BO_3^-(aq) + H_2O(l) \rightarrow H_3BO_3(aq) + OH^-(aq)$

14. An amphoteric substance is one that may act as either an acid or a base. The hydrogen sulfite ion is amphoteric. Write reactions that demonstrate this property of HSO_3^-.

15. Provide an example of an acid-base reaction that does not require the presence of water. Explain how you know which reactant is the acid and which is the base.

16. Which type of acid-base reaction does not always go to completion? Explain why.

17. Complete and balance the overall equations for the acid-base reactions.
    a. $HNO_2(aq) + LiOH(aq) \rightarrow$
    b. $H_2SO_4(aq) + Al(OH)_3(s) \rightarrow$
    c. $H_3C_6H_5O_7(aq) + Mg(OH)_2(aq) \rightarrow$
    d. $HC_7H_5O_2(aq) + NH_3(aq) \rightarrow$

18. For each reaction in Question 17, identify the type of acid-base reaction represented by the equation.

19. Write an overall equation for the acid-base reaction that would be required to produce each salt.
    a. $Mg(NO_3)_2$
    b. NH_4I
    c. K_2SO_4
    d. $Sr(C_2H_3O_2)_2$

Section 15.2

20. If the OH^- concentration in blood increases, what then happens to the concentrations of H_2CO_3, HCO_3^-, and OH^-?

21. If a person's blood pH becomes too high, what can he or she do to cause the pH of the blood to drop to its normal level?

22. Consider an antacid that contains milk of magnesia. Write the overall equation that shows how this antacid reduces the acidity of stomach acid.

23. What is meant by the endpoint of a titration? At the endpoint, what happens to the pH of the solution that is being titrated?

24. How does the endpoint pH of a weak acid–strong base titration compare with that of a weak base–strong acid titration?

25. A 0.200M NaOH solution was used to titrate an HCl solution of unknown concentration. At the endpoint, 27.8 mL of NaOH solution had neutralized 10.0 mL of HCl. What is the molarity of of the HCl solution?

26. A student finds that 42.7 mL of 0.500M sodium hydroxide are required to neutralize 30.0 mL of a sulfuric acid solution. What is the molarity of the sulfuric acid?

Supplemental Practice

27. A student neutralizes 20.0 L of a potassium hydroxide solution with 11.6 mL of 0.500M HCl. What is the molarity of the potassium hydroxide?

28. An NaOH solution of unknown concentration was used to titrate 30.0 mL of a 0.100M solution of citric acid ($H_3C_6H_5O_7$). If 16.4 mL of NaOH are used to reach the endpoint, what is the concentration of the NaOH solution?

29. A 35.0-mL sample of a solution of formic acid ($HCHO_2$) is titrated to the endpoint with 68.3 mL of 0.100M $Ca(OH)_2$. What is the molarity of the formic acid?

30. A student finds that 31.5 mL of a 0.175M solution of LiOH was required to titrate 15.0 mL of a phosphoric acid solution to the endpoint. What is the molarity of the H_3PO_4?

31. How many milliliters of a 0.120M $Ca(OH)_2$ are needed to neutralize 40.0 mL of 0.185M $HClO_4$?

32. What volume of 0.100M HNO_3 is needed to neutralize 56.0 mL of 6.00M KOH?

33. A 14.5-mL sample of an unknown triprotic acid is titrated to the endpoint with 35.2 mL of 0.0800M $Sr(OH)_2$. What is the molarity of the acid solution?

34. The following are the endpoint pHs for three titrations. From each endpoint pH, indicate whether the titration involves a weak acid–strong base, a weak base–strong acid, or a strong acid–strong base reaction. Use **Figure 15.17** on page 834 to select the best indicator for each reaction.
    a. pH = 9.38   b. pH = 7.00   c. pH = 5.32

35. A 25.0-mL sample of an H_2SO_3 solution of unknown molarity is titrated to the endpoint with 76.2 mL of 0.150M KOH. What is the molarity of the H_2SO_3? Use **Figure 15.17** to select the best indicator for the reaction.

36. An antacid tablet containing $KHCO_3$ is titrated with 0.100M HCl. If 0.300 g of the tablet requires 26.5 mL of HCl to reach the endpoint, what is the mass percent of $KHCO_3$ in the tablet?

37. Vitamin C is also known as ascorbic acid ($HC_6H_7O_6$). A solution made by dissolving a tablet containing 0.500 g of vitamin C in 100 mL of distilled water is titrated to the endpoint with 0.125M NaOH. Assuming that vitamin C is the only acid present in the tablet, how many milliliters of the NaOH solution are needed to neutralize the vitamin C?

38. A student mixes 112 mL of a 0.150M HCl solution and 112 mL of an NaOH solution of unknown molarity. The final solution is found to be acidic. The student then titrates this solution to the endpoint with 16.7 mL of 0.100M NaOH. What was the molarity of the original NaOH solution?

Sections 15.1 and 15.2

39. Propanoic acid ($HC_3H_5O_2$) is a weak acid. When propanoic acid completely reacts with sodium hydroxide, the final solution has a pH slightly greater than 7. Why are the products of this neutralization reaction not neutral? Use the net ionic equation to help in your explanation.

Supplemental Practice

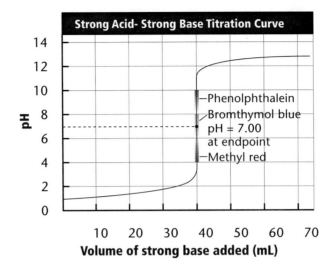

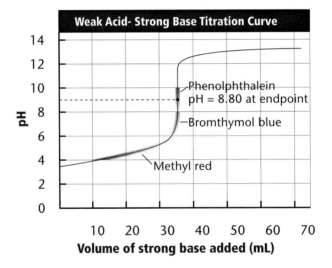

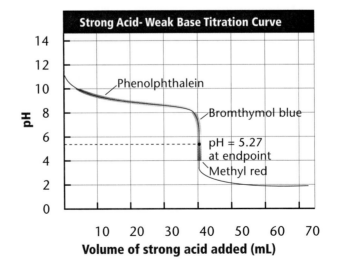

■ **Figure 15.17**

40. How does a solution that contains dissolved acetic acid and sodium acetate act as a buffer? Use net ionic equations to show how this buffer responds to added H⁺ and OH⁻.

Chapter 16

Section 16.1

1. What is the difference between an oxidation-reduction reaction and a reaction that is not oxidation-reduction?

2. Which changes are oxidations and which are reductions?
   a. Mg^{2+} becomes Mg
   b. K becomes K^+
   c. Fe^{2+} becomes Fe^{3+}
   d. Cl_2 becomes $2Cl^-$

3. Identify the reducing agent in each reaction.
   a. $Zn(s) + 2Ag^+(aq) \rightarrow Zn^{2+}(aq) + 2Ag(s)$
   b. $Cl_2(g) + 2Na(s) \rightarrow 2NaCl(s)$
   c. $2SO_2(g) + O_2(g) \rightarrow 2SO_3(g)$

4. Write the equation for the redox reaction that occurs when magnesium metal reacts with nitric acid. Name the oxidizing agent and the reducing agent.

5. The equation represents the neutralization reaction between hydrochloric acid and calcium hydroxide. Determine the oxidation number for each element. Is this a redox reaction? Explain.
   $2HCl(aq) + Ca(OH)_2(aq) \rightarrow CaCl_2(aq) + 2H_2O(l)$

6. For each reaction, determine whether or not it is a redox reaction. If it is, identify which reactant is reduced and which is oxidized.
   a. $Fe^{2+}(aq) + Mg(s) \rightarrow Fe(s) + Mg^{2+}(aq)$
   b. $2NaI(aq) + Pb(NO_3)_2(aq) \rightarrow PbI_2(s) + 2NaNO_3(aq)$
   c. $4Al(s) + 3O_2(g) \rightarrow 2Al_2O_3(s)$

7. When a mixture of iron filings and sulfur crystals is heated, the elements combine to form iron(II) sulfide.
   $Fe(s) + S(s) \rightarrow FeS(s)$
   Is this a redox reaction? If so, which element is oxidized and which is reduced?

8. Mercury(II) oxide is a red crystalline powder that decomposes into mercury and oxygen on exposure to light. Write a balanced equation for this reaction. Is it a redox reaction? If so, which element is oxidized and which is reduced?

Section 16.2

9. Name the compounds that form the green protective coating, or patina, on the Statue of Liberty. What is the oxidation number of copper in these compounds? What useful purpose does the patina serve?

10. What is chemiluminescence? Name two practical uses of chemiluminescence.

Supplemental Practice

Chapter 17

Section 17.1

1. A piece of aluminum metal is placed in a 0.50M solution of lead(II) nitrate ($Pb(NO_3)_2$).

    a. Use **Table 17.1** to predict which metal will be reduced and which will be oxidized.

    b. Write an equation for the net redox reaction that occurs.

    c. Is this system a voltaic cell? Explain.

Table 17.1 Ease of Oxidation of Common Metals

Li	$Li \rightarrow Li^+ + e^-$
K	$K \rightarrow K^+ + e^-$
Ca	$Ca \rightarrow Ca^{2+} + 2e^-$
Na	$Na \rightarrow Na^+ + e^-$
Mg	$Mg \rightarrow Mg^{2+} + 2e^-$
Al	$Al \rightarrow Al^{3+} + 3e^-$
Mn	$Mn \rightarrow Mn^{2+} + 2e^-$
Zn	$Zn \rightarrow Zn^{2+} + 2e^-$
Cr	$Cr \rightarrow Cr^{3+} + 3e^-$
Fe	$Fe \rightarrow Fe^{2+} + 2e^-$
Ni	$Ni \rightarrow Ni^{2+} + 2e^-$
Sn	$Sn \rightarrow Sn^{2+} + 2e^-$
Pb	$Pb \rightarrow Pb^{2+} + 2e^-$
Cu	$Cu \rightarrow Cu^{2+} + 2e^-$
Ag	$Ag \rightarrow Ag^+ + e^-$
Hg	$Hg \rightarrow Hg^{2+} + 2e^-$
Pt	$Pt \rightarrow Pt^{2+} + 2e^-$
Au	$Au \rightarrow Au^{3+} + 3e^-$

(Arrow from bottom "Not easily oxidized" to top "Easily oxidized")

2. Can a solution of magnesium chloride be stored in a container plate with chromium metal? Explain.

3. Which voltaic cell would you expect to have the highest voltage? The lowest voltage? Explain your answers.

    a. a zinc-copper cell

    b. an iron-copper cell

    c. a magnesium-copper cell

4. What change would occur if you placed

    a. a strip of zinc into a solution of silver nitrate?

    b. a strip of zinc into a solution of sodium sulfate?

5. Draw a diagram of a voltaic cell in which the reaction is

    $Zn(s) + Sn^{2+}(aq) \rightarrow Zn^{2+}(aq) + Sn(s)$.

    Label the cathode and the anode. Show the ions present in both compartments, and indicate the direction of electron flow in the external circuit.

6. What are some potential advantages of electric cars? Why are electric cars uncommon in the United States?

Supplemental Practice

Section 17.2   **7.** Answer the following questions pertaining to an electrolytic cell.
   a. What are positive ions called? What are negative ions called?
   b. Which electrode attracts positive ions? Which attracts negative ions?
   c. Where does oxidation occur? Where does reduction occur?

8. Describe the movement of electrons during electrolysis.

9. Name four uses of electrolysis.

10. What is produced by the Hall-Héroult process? Why was the development of this process so important? What are some drawbacks of the process, and what can people do to circumvent these drawbacks?

Chapter 18

Section 18.1   **1.** Name the hydrocarbons.
   a. CH_3CH_3
   b. $CH_3CH_2C \equiv CCH_3$
   c. $CH_3CH = CHCH_2CH_2CH_2CH_3$
   d.

   (cyclopentene structure with H atoms)

2. Name the branched alkanes shown.
   a. $CH_3CH_2CHCH_2CHCH_2CH_3$ with CH_3 and CH_2CH_3 branches
   b. $CH_3CHCH_2CHCH_2CH_2CHCH_3$ with CH_3, CH_2CH_3, and CH_3 branches
   c. $CH_3CHCH_2CH_3$ with CH_2CH_3 branch

3. Why is there no compound named 3,6-dimethylheptane? What is the correct name for this compound?

4. Compare the reactivities of alkanes and alkenes. Explain the difference.

5. The molecular formula of the cycloalkanes follow the pattern C_nH_{2n}, where n is the number of carbon atoms. What pattern is followed by the cycloalkenes with one double bond?

Supplemental Practice

6. Draw structures of each hydrocarbon, and identify each as an alkane, alkene, or alkyne.
   a. heptane
   b. *cis*-2-pentene
   c. propyne
   d. 1,2-dimethylcyclopentane
   e. 3-ethyl-2-methylhexane
   f. 2-methyl-2-butene

7. Write the names and structures of one positional isomer, one geometric isomer, and two structural isomers of *cis*-2-butene.

8. How many H_2 molecules would be required to fully hydrogenate one molecule of vitamin D_2, the structure shown here?

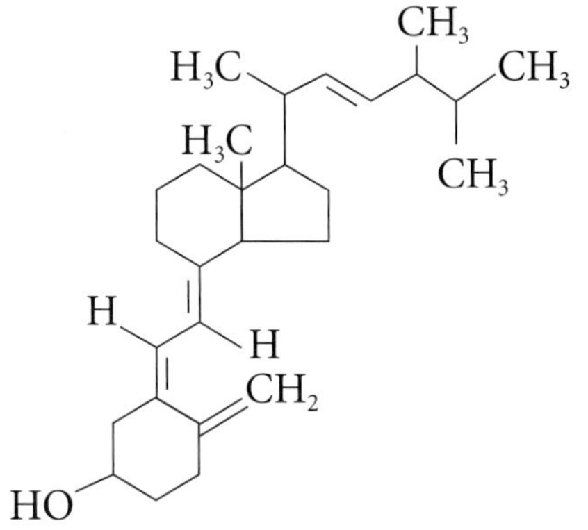

9. Halogen molecules such as Cl_2 can be added to double bonds in a reaction similar to hydrogenation. Draw the structure of the product that forms when Cl_2 is added to 2-butene.

10. Show the structure of the product of each of the following reactions.
    a. $CH_3CH = CHCH_3 + H_2 \rightarrow ?$
    b. $CH \equiv CH + 2H_2 \rightarrow ?$
    c. $CH_3C \equiv CH + H_2 \rightarrow ?$

11. How are the properties of benzene unlike those of most alkenes? How do chemists account for benzene's behavior?

12. What is petroleum? Where is it found? Name six useful products that are derived from petroleum.

13. What technique is commonly used to separate petroleum into its component liquids? What two processes are used to convert large alkanes into smaller hydrocarbons?

Section 18.2

14. Classify each of the following molecules as belonging to one of the categories of substituted hydrocarbons described in the chapter.

    a. $CH_3CH_2C(=O)OH$

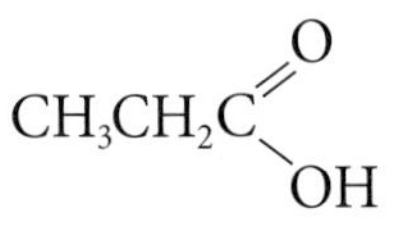

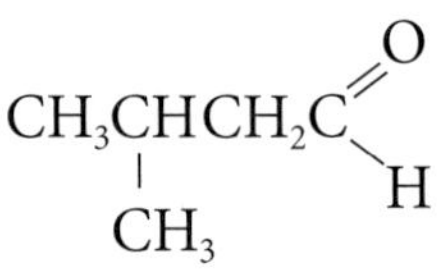

    b. $CH_3-O-CH(CH_3)CH_3$

    c. $CH_3CH(CH_3)CH_2C(=O)H$

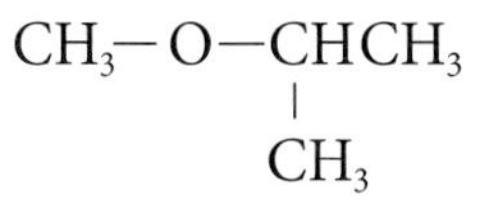

    d. $CH_3CH(Cl)CH(Cl)CH_3$

15. Distinguish between
    a. an amide and an amine.
    b. an ether and an ester.

16. Draw the structures.
    a. an ester, where R is $-CH_2CH_3$ and R' is $-CH_3$
    b. a ketone, where R is $-CHCH_3$
       $\quad\quad\quad\quad\quad\quad\quad\quad\quad\;\; |$
       $\quad\quad\quad\quad\quad\quad\quad\quad\quad\; CH_3$
       and R' is $-CH_2CH_3$

17. The structure of tyrosine, an amino acid, is shown. Name the functional groups in the molecule.

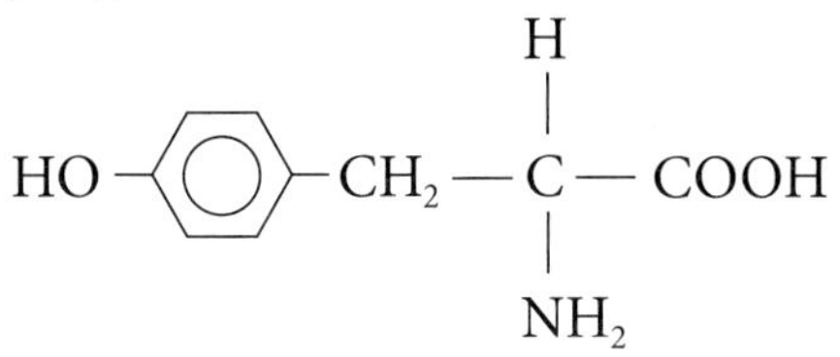

    $$HO-\langle\bigcirc\rangle-CH_2-\underset{\underset{NH_2}{|}}{\overset{\overset{H}{|}}{C}}-COOH$$

Sections 18.1 and 18.2

18. Name a common example of each class of organic compounds.
    a. ester
    b. ketone
    c. cycloalkane
    d. aromatic hydrocarbon

Section 18.3

19. Draw the structure of the polymer that will be formed from each of the monomers shown.
    a. $CH_2=CH_2$
    b. $CH_2=CHCH_3$
    c. $CH_2=CCH_3$
       $\quad\quad\quad\quad |$
       $\quad\quad\quad\; C=O$
       $\quad\quad\quad\quad |$
       $\quad\quad\quad\quad O$
       $\quad\quad\quad\quad |$
       $\quad\quad\quad\; CH_3$

20. Provide one example each of a thermoplastic polymer and a thermosetting polymer. Which type of polymer is easier to recycle? Which type is more durable? Explain.

Chapter 19

Section 19.1

1. Which elements form proteins? Name some of the roles that proteins play in living systems.

2. Describe briefly what enzymes are and how they work.

3. Name two common disaccharides found in foods. Are these molecules classified as nutrients? Explain.

4. Describe the functions of lipids and their solubility properties. Are animal lipids or plant lipids generally more saturated?

Supplemental Practice

5. Examine the structure of each of the following molecules shown and decide whether each is a carbohydrate, a lipid, or an amino acid.

   a.

   H–N(H)–C(H)(CH$_2$CH$_2$SCH$_3$)–C(=O)–OH

   b.

   CH$_3$—(CH$_2$)$_{14}$—C(=O)—OH

   c. (pentose sugar ring with HOH$_2$C, O, OH, H substituents)

   d.

   H$_2$N–C(H)(CH$_2$–C(=O)–OH)–C(=O)–OH

6. What monomers form each type of biomolecule?
   a. protein
   b. nucleic acid
   c. polysaccharide

7. What are nucleic acids? What functions do they perform in cells? Why are nucleic acids not listed on food-product labels? What are the two kinds of nucleic acids found in cells?

8. Name two diseases that can result from vitamin deficiencies in the diet. Can too much of a vitamin cause disease? Explain.

9. What is a coenzyme? Give an example in which a vitamin acts as a coenzyme.

Section 19.2 10. What are hormones? Give an example of a hormone and describe its function.

Chapter 20

Section 20.1

1. Nitric oxide gas (NO) can be prepared by passing air through an electric arc. The heat of reaction is +90.4 kJ/mol NO. Write a balanced chemical equation that includes the reaction heat. Is the reaction exothermic or endothermic?

2. Sketch a graph of the energy changes during the progress of an endothermic reaction with a high activation energy. On the same axes, sketch a graph of energy change versus progress of the reaction if a catalyst is used to lower the activation energy.

3. Describe each process as involving an increase or decrease in entropy. Explain your answers.
   a. Water boils, forming water vapor.
   b. Baking soda reacts with vinegar to form sodium acetate solution, water, and carbon dioxide gas.
   c. Two moles of nitrogen gas combine with six moles of hydrogen gas to form four moles of ammonia gas.

Section 20.2

4. How much heat is released by a reaction that raises the temperature of 756 g of water in a calorimeter from 23.2°C to 37.6°C?

5. When 5.00 g of a certain liquid organic compound are burned in a calorimeter, the temperature of the surrounding 2.00 kg of water increases from 24.5°C to 40.5°C. All products of the reaction are gases. Calculate the heat given off in this reaction. How much heat would 1.00 mol of the compound give off, assuming a molar mass of 46.1 g/mol?

6. Explain the relationship among a calorie, a Calorie, and a kilocalorie. If you eat a candy bar that contains 180 Calories, how many calories have you consumed? How many kilocalories?

7. Using **Table 20.2** as a guide, compare the Caloric contents of 1 g of peanut butter and 1 g of broiled chicken. How might you explain the difference?

Table 20.2	The Caloric Value of Some Foods		
Food	Quantity	Kilojoules	Calories
Butter	1 tbsp = 14 g	418	100
Peanut butter	1 tbsp = 16 g	418	100
Spaghetti	0.5 cup = 55 g	836	200
Apple	1	283	70
Chicken (broiled)	3 oz = 84 g	502	120
Beef (broiled)	3 oz = 84 g	1000	241

8. A 10.0-g sample of a certain brand of cereal is burned in a calorimeter. The 4.00 kg of surrounding water increase in temperature from 22.00°C to 30.20°C. What is the food value in Calories per gram?

Sections 20.1 and 20.2

9. List two factors that would make an exothermic reaction less likely to occur.

Supplemental Practice

Section 20.3   **10.** In photosynthesis, which three products of the light reactions take part in the reactions of the Calvin cycle?

Chapter 21

Section 21.1   **1.** Write the balanced nuclear equation for each nuclear reaction and determine the type of decay.

    a. Uranium-234 decays to form thorium-230.
    b. Bismuth-214 decays to form polonium-214.
    c. Thallium-210 decays to form lead-210.

2. A compound containing 200 mg of technetium-99m is injected into a patient to get a bone scan, but equipment problems make it impossible to scan the patient before 24 hours have passed. Will the nuclear medical technician be able to get a good scan after that period of time, or should more technetium-99m be injected? The half-life of technetium-99m is 6.0 hours.

3. A fossilized piece of wood is found to have 3.12 percent of the amount of carbon-14 that would be found in a new piece of similar wood. How old is the fossilized wood?

4. A scientist analyzing a rock sample finds that it contains 75 percent of the potassium-40 that would be found in a similar rock formed today. Use **Figure 21.7** and **Table 21.1** to estimate the age of the rock.

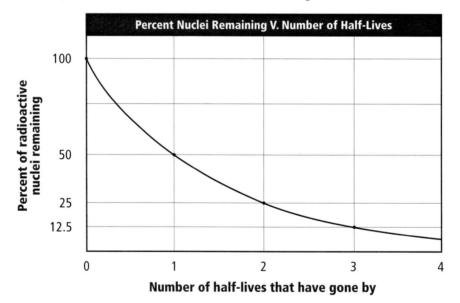

■ **Figure 21.7**

Supplemental Practice

Table 21.1	Half-Life Values for Commonly Used Radioactive Isotopes
Isotope	**Half-Life**
$^{3}_{1}H$	12.26 years
$^{14}_{6}C$	5730 years
$^{32}_{15}P$	14.282 days
$^{40}_{19}K$	1.25 billion years
$^{60}_{27}Co$	5.271 years
$^{85}_{36}Kr$	10.76 years
$^{93}_{36}Kr$	1.3 seconds
$^{87}_{37}Rb$	48 billion years
$^{99m}_{43}Tc$	6.0 hours*
$^{131}_{53}I$	8.07 days
$^{131}_{56}Ba$	12 days
$^{153}_{64}Gd$	242 days
$^{201}_{81}Tl$	73 hours
$^{226}_{88}Ra$	1600 years
$^{235}_{92}U$	710 million years
$^{238}_{92}U$	4.51 billion years
$^{241}_{95}Am$	432.7 years

*The m in the symbol tells you that this is a metastable element, which is one form of an unstable isotope. Technetium 99m releases a gamma ray to become a more stable form of the same isotope, with no change in either atomic or mass number.

Section 21.2

5. In the early 1900s, Albert Einstein developed a famous equation relating mass and energy in nuclear reactions. Use this equation to explain why nuclear reactions release very large amounts of energy.

6. Distinguish between deuterium and tritium. Why is deuterium potentially important as a source of energy?

Sections 21.1 and 21.2

7. How do nuclear reactions differ from chemical reactions?

Section 21.3

8. What is a radioactive tracer? Name three nonmedical uses for radioactive tracers.

9. The quantity of nuclear waste generated in hospitals is much greater than the quantity generated in nuclear reactors. Why, then, is nuclear waste from nuclear reactors more dangerous than nuclear waste from hospitals?

10. Using information on pages 768–770 of your text, compare the radiation exposure from a chest X-ray to the exposure from smoking two packs of cigarettes a day for a year.

Safety Handbook

Safety Guidelines in the Chemistry Laboratory

The chemistry laboratory is a safe place to work if you are aware of important safety rules and if you are careful. You must be responsible for your own safety and for the safety of others. The safety rules given here and the first aid instruction in **Table 1** will protect you and others from harm in the lab. While carrying out procedures in any of the **Launch Labs, MiniLabs, ChemLabs, or Try at Home Labs,** notice the safety symbols and warning statements. The safety symbols are explained in the chart on the next page.

1. Always obtain your teacher's permission to begin a lab.
2. Study the procedure. If you have questions, ask your teacher. Be sure you understand all safety symbols shown.
3. Use the safety equipment provided for you. Goggles and a safety apron should be worn when any lab calls for using chemicals.
4. When you are heating a test tube, always slant it so the mouth points away from you and others.
5. Never eat or drink in the lab. Never inhale chemicals. Do not taste any substance or draw any material into your mouth.
6. If you spill any chemical, wash it off immediately with water. Report the spill immediately to your teacher.
7. Know the location and proper use of the fire extinguisher, safety shower, fire blanket, first aid kit, and fire alarm.
8. Keep all materials away from open flames. Tie back long hair.
9. If a fire should break out in the classroom, or if your clothing should catch fire, smother it with the fire blanket or a coat, or get under a safety shower. **NEVER RUN.**
10. Report any accident or injury, no matter how small, to your teacher.

Follow these procedures as you clean up your work area.
1. Turn off the water and gas. Disconnect electrical devices.
2. Return materials to their places.
3. Dispose of chemicals and other materials as directed by your teacher. Place broken glass and solid substances in the proper containers. Never discard materials in the sink.
4. Clean your work area.
5. Wash your hands thoroughly after working in the laboratory.

Table 1	First Aid in the Laboratory
Injury	Safe Response
Burns	Apply cold water. Notify your teacher immediately.
Cuts and bruises	Stop any bleeding by applying direct pressure. Cover cuts with a clean dressing. Apply cold compresses to bruises. Notify your teacher immediately.
Fainting	Leave the person lying down. Loosen any tight clothing and keep crowds away. Notify your teacher immediately.
Foreign matter in eye	Flush with plenty of water. Use eyewash bottle or fountain.
Poisoning	Note the suspected poisoning agent and call your teacher immediately.
Any spills on skin	Flush with large amounts of water or use safety shower. Notify your teacher immediately.

Safety Handbook

These safety symbols are used in laboratory and investigations in this book to indicate possible hazards. Learn the meaning of each symbol and refer to this page often. *Remember to wash your hands thoroughly after completing lab procedures.*

SAFETY SYMBOLS	HAZARD	EXAMPLES	PRECAUTION	REMEDY
DISPOSAL	Special disposal procedures need to be followed.	certain chemicals, living organisms	Do not dispose of these materials in the sink or trash can.	Dispose of wastes as directed by your teacher.
BIOLOGICAL	Organisms or other biological materials that might be harmful to humans	bacteria, fungi, blood, unpreserved tissues, plant materials	Avoid skin contact with these materials. Wear mask or gloves.	Notify your teacher if you suspect contact with material. Wash hands thoroughly.
EXTREME TEMPERATURE	Objects that can burn skin by being too cold or too hot	boiling liquids, hot plates, dry ice, liquid nitrogen	Use proper protection when handling.	Go to your teacher for first aid.
SHARP OBJECT	Use of tools or glassware that can easily puncture or slice skin	razor blades, pins, scalpels, pointed tools, dissecting probes, broken glass	Practice common-sense behavior and follow guidelines for use of the tool.	Go to your teacher for first aid.
FUME	Possible danger to respiratory tract from fumes	ammonia, acetone, nail polish remover, heated sulfur, moth balls	Make sure there is good ventilation. Never smell fumes directly. Wear a mask.	Leave foul area and notify your teacher immediately.
ELECTRICAL	Possible danger from electrical shock or burn	improper grounding, liquid spills, short circuits, exposed wires	Double-check setup with teacher. Check condition of wires and apparatus.	Do not attempt to fix electrical problems. Notify your teacher immediately.
IRRITANT	Substances that can irritate the skin or mucous membranes of the respiratory tract	pollen, moth balls, steel wool, fiberglass, potassium permanganate	Wear dust mask and gloves. Practice extra care when handling these materials.	Go to your teacher for first aid.
CHEMICAL	Chemicals that can react with and destroy tissue and other materials	bleaches such as hydrogen peroxide; acids such as sulfuric acid, hydrochloric acid; bases such as ammonia, sodium hydroxide	Wear goggles, gloves, and an apron.	Immediately flush the affected area with water and notify your teacher.
TOXIC	Substance may be poisonous if touched, inhaled, or swallowed.	mercury, many metal compounds, iodine, poinsettia plant parts	Follow your teacher's instructions.	Always wash hands thoroughly after use. Go to your teacher for first aid.
FLAMMABLE	Open flame may ignite flammable chemicals, loose clothing, or hair.	alcohol, kerosene, potassium permanganate, hair, clothing	Avoid open flames and heat when using flammable chemicals.	Notify your teacher immediately. Use fire safety equipment if applicable.
OPEN FLAME	Open flame in use, may cause fire.	hair, clothing, paper, synthetic materials	Tie back hair and loose clothing. Follow teacher's instructions on lighting and extinguishing flames.	Always wash hands thoroughly after use. Go to your teacher for first aid.

Eye Safety Proper eye protection should be worn at all times by anyone performing or observing science activities.

Clothing Protection This symbol appears when substances could stain or burn clothing.

Animal Safety This symbol appears when safety of animals and students must be ensured.

Radioactivity This symbol appears when radioactive materials are used.

Handwashing After the lab, wash hands with soap and water before removing goggles

Table D.1 PERIODIC TABLE OF THE ELEMENTS

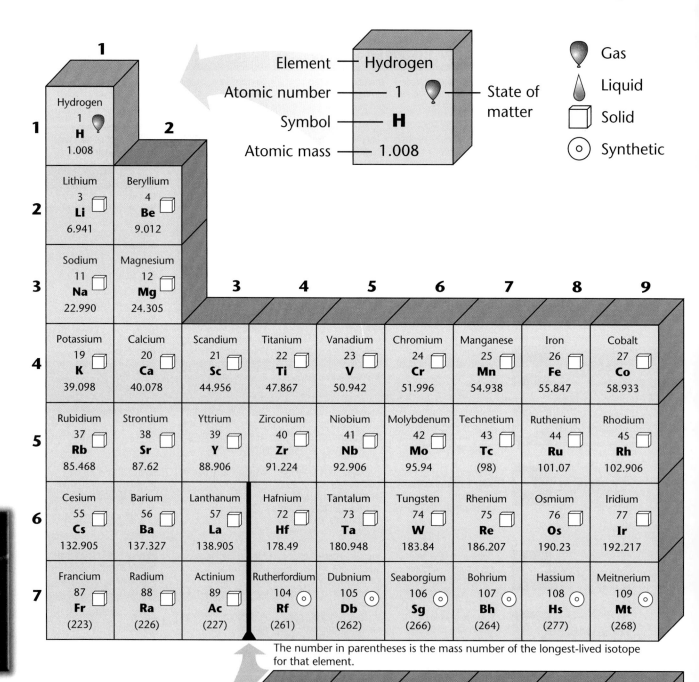

Reference Tables

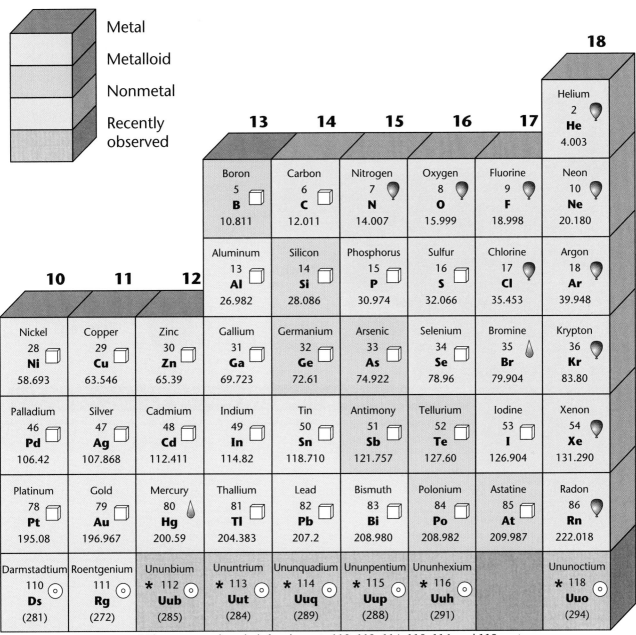

*The names and symbols for elements 112, 113, 114, 115, 116, and 118 are temporary. Final names will be selected when the elements' discoveries are verified.

Reference Tables

Table D.2 Color Key

Carbon	Bromine	Sodium/Other metals
Hydrogen	Iodine	Gold
Oxygen	Sulfur	Copper
Nitrogen	Phosphorus	Electron
Chlorine	Silicon	Proton
Fluorine	Helium	Neutron

Table D.3 Symbols and Abbreviations

α	=	particles from radioactive materials, helium nuclei	K_{eq}	=	equilibrium constant
β	=	particles from radioactive materials, electrons	K_{sp}	=	solubility product constant
γ	=	rays from radioactive materials, high-energy quanta	kg	=	kilogram
Δ	=	change in	M	=	molarity
λ	=	wavelength	m	=	mass, molality
ν	=	frequency	m	=	meter (*length*)
Π	=	osmotic pressure	mol	=	mole (*amount*)
A	=	ampere (*electric current*)	min	=	minute (*time*)
Bq	=	becquerel (*nuclear disintegration*)	N	=	newton (*force*)
°C	=	Celsius degree (*temperature*)	N_A	=	Avogadro's number
C	=	coulomb (*quantity of electricity*)	n	=	number of moles
c	=	speed of light	P	=	pressure, power
cd	=	candela (*luminous intensity*)	Pa	=	pascal (*pressure*)
C_p	=	specific heat	p	=	momentum
D	=	density	q	=	heat
E	=	energy, electromotive force	R	=	gas constant
F	=	force, Faraday	S	=	entropy
G	=	free energy	s	=	second (*time*)
g	=	gram (*mass*)	Sv	=	sievert (*absorbed radiation*)
Gy	=	gray (*radiation*)	T	=	temperature
H	=	enthalpy	U	=	internal energy
Hz	=	hertz (*frequency*)	u	=	atomic mass unit
h	=	Planck's constant	V	=	volume
h	=	hour (*time*)	V	=	volt (*electromotive force*)
J	=	joule (*energy*)	v	=	velocity
K	=	kelvin (*temperature*)	W	=	watt (*power*)
K_a	=	ionization constant (*acid*)	w	=	work
K_b	=	ionization constant (*base*)	x	=	mole fraction

Table D.4 Alphabetical Table of the Elements

Element	Symbol	Atomic number	Atomic mass	Element	Symbol	Atomic number	Atomic mass
Actinium	Ac	89	227.027 8*	Neodymium	Nd	60	144.24
Aluminum	Al	13	26.981 539	Neon	Ne	10	20.179 7
Americium	Am	95	243.061 4*	Neptunium	Np	93	237.048 2
Antimony	Sb	51	121.757	Nickel	Ni	28	58.6934
Argon	Ar	18	39.948	Niobium	Nb	41	92.906 38
Arsenic	As	33	74.921 59	Nitrogen	N	7	14.006 74
Astatine	At	85	209.987 1*	Nobelium	No	102	259.100 9*
Barium	Ba	56	137.327	Osmium	Os	76	190.2
Berkelium	Bk	97	247.070 3*	Oxygen	O	8	15.999 4
Beryllium	Be	4	9.012 182	Palladium	Pd	46	106.42
Bismuth	Bi	83	208.980 37	Phosphorus	P	15	30.973 762
Bohrium	Bh	107	262*	Platinum	Pt	78	195.08
Boron	B	5	10.811	Plutonium	Pu	94	244.064 2*
Bromine	Br	35	79.904	Polonium	Po	84	208.982 4*
Cadmium	Cd	48	112.411	Potassium	K	19	39.098 3
Calcium	Ca	20	40.078	Praseodymium	Pr	59	140.907 65
Californium	Cf	98	251.079 6*	Promethium	Pm	61	144.912 8*
Carbon	C	6	12.011	Protactinium	Pa	91	231.035 88
Cerium	Ce	58	140.115	Radium	Ra	88	226.025 4
Cesium	Cs	55	132.905 43	Radon	Rn	86	222.017 6*
Chlorine	Cl	17	35.452 7	Rhenium	Re	75	186.207
Chromium	Cr	24	51.996 1	Rhodium	Rh	45	102.905 50
Cobalt	Co	27	58.933 20	Roentgenium	Rg	111	272
Copper	Cu	29	63.546	Rubidium	Rb	37	85.467 8
Curium	Cm	96	247.070 3*	Ruthenium	Ru	44	101.07
Darmstadtium	Ds	110	271*	Rutherfordium	Rf	104	261*
Dubnium	Db	105	262*	Samarium	Sm	62	150.36
Dysprosium	Dy	66	162.50	Scandium	Sc	21	44.955 910
Einsteinium	Es	99	252.082 8*	Seaborgium	Sg	106	263*
Erbium	Er	68	167.26	Selenium	Se	34	78.96
Europium	Eu	63	151.965	Silicon	Si	14	28.085 5
Fermium	Fm	100	257.095 1*	Silver	Ag	47	107.868 2
Fluorine	F	9	18.998 403 2	Sodium	Na	11	22.989 768
Francium	Fr	87	223.019 7*	Strontium	Sr	38	87.62
Gadolinium	Gd	64	157.25	Sulfur	S	16	32.066
Gallium	Ga	31	69.723	Tantalum	Ta	73	180.947 9
Germanium	Ge	32	72.61	Technetium	Tc	43	97.907 2*
Gold	Au	79	196.966 54	Tellurium	Te	52	127.60
Hafnium	Hf	72	178.49	Terbium	Tb	65	158.925 34
Hassium	Hs	108	265*	Thallium	Tl	81	204.383 3
Helium	He	2	4.002 602	Thorium	Th	90	232.038 1
Holmium	Ho	67	164.930 32	Thulium	Tm	69	168.934 21
Hydrogen	H	1	1.007 94	Tin	Sn	50	118.710
Indium	In	49	114.82	Titanium	Ti	22	47.88
Iodine	I	53	126.904 47	Tungsten	W	74	183.85
Iridium	Ir	77	192.22	Ununbium	Uub	112	285*
Iron	Fe	26	55.847	Ununhexium	Uuh	116	291
Krypton	Kr	36	83.80	Ununoctium	Uuo	118	294
Lanthanum	La	57	138.905 5	Ununpentium	Uup	115	288
Lawrencium	Lr	103	260.105 4*	Ununquadium	Uuq	114	289*
Lead	Pb	82	207.2	Ununtrium	Uut	113	284
Lithium	Li	3	6.941	Unununium	Uuu	111	272*
Lutetium	Lu	71	174.967	Uranium	U	92	238.028 9
Magnesium	Mg	12	24.305 0	Vanadium	V	23	50.941 5
Manganese	Mn	25	54.938 05	Xenon	Xe	54	131.29
Meitnerium	Mt	109	266*	Ytterbium	Yb	70	173.04
Mendelevium	Md	101	258.098 6*	Yttrium	Y	39	88.905 85
Mercury	Hg	80	200.59	Zinc	Zn	30	65.39
Molybdenum	Mo	42	95.94	Zirconium	Zr	40	91.224

* The mass of the isotope with the longest known half-life.

Reference Tables

Table D.5 Properties of Elements

Element	Symbol	Atomic Number [Z]	Atomic Mass* [u]	Melting Point [°C]	Boiling Point [°C]	Density [g/cm³]	Atomic Radius [pm]	First Ionization Energy (kJ/mol)	Enthalpy of Fusion (kJ/mol)	Specific Heat (J/g·°C)	Enthalpy of Vaporization (kJ/mol)	Abundance in Earth's Crust (%)	Major Oxidation States
Actinium	Ac	89	[227.0278]	1050	3300	10.07	203	666	14.3	0.120	293	trace	3+
Aluminum	Al	13	26.981539	660.37	2517.6	2.699	143	577.5	10.71	0.9025	290.8	8.1	3+
Americium	Am	95	[243.0614]	994	2600	13.67	183	579	10	—	238.5	—	2+, 3+, 4+
Antimony	Sb	51	121.757	630.7	1635	6.697	161	834	19.5	0.2072	193	2 × 10⁻⁵	3+, 5+
Argon	Ar	18	39.948	−189.37	−185.86	0.001784	98	1521	1.18	0.52033	6.52	4 × 10⁻⁶	—
Arsenic	As	33	74.92159	816	615	5.778	121	947	27.7	0.3289	—	1.9 × 10⁻⁴	3+, 5+
Astatine	At	85	[209.98037]	300	(sublimes)	—	—	916	23.8	—	(sublimes)	trace	1−, 5+
Barium	Ba	56	137.327	726.9	1845	3.62	222	502.9	8.012	0.2044	90.3	0.039	2+
Berkelium	Bk	97	[247.0703]	986	—	14.78	170	601	—	—	140	—	3+, 4+
Beryllium	Be	4	9.012182	1287	2468	1.848	112	899.5	7.895	1.824	297.6	2 × 10⁻⁴	2+
Bismuth	Bi	83	208.98037	271.4	1564	9.808	151	703	10.9	0.1221	179	8 × 10⁻⁷	3+, 5+
Bohrium	Bh	107	[262]	—	—	—	—	—	—	—	—	—	—
Boron	B	5	10.811	2080	3865	2.46	85	800.6	50.2	1.026	504.5	9 × 10⁻⁴	3+
Bromine	Br	35	79.904	−7.25	59.35	3.1028	119	1139.9	10.571	0.47362	29.56	2.5 × 10⁻⁴	1−, 1+, 3+, 5+
Cadmium	Cd	48	112.411	320.8	770	8.65	151	867.7	6.19	0.2311	100	1.6 × 10⁻⁵	2+
Calcium	Ca	20	40.078	841.5	1500.5	1.55	197	589.8	8.54	0.6315	155	4.66	2+
Californium	Cf	98	[251.0796]	900	—	—	186	608	—	—	—	—	3+, 4+
Carbon	C	6	12.011	3620	4200	2.266	77	1086.5	104.6	0.7099	711	0.018	4−, 2+, 4+
Cerium	Ce	58	140.115	804	3470	6.773	181.8	541	5.2	0.1923	313	0.007	3+, 4+
Cesium	Cs	55	132.90543	28.4	674.8	1.9	262	375.7	2.087	0.2421	67	2.6 × 10⁻⁴	1+
Chlorine	Cl	17	35.4527	−101	−34	0.003214	91	1255.5	6.41	0.47820	20.41	0.013	1−, 1+, 3+, 5+
Chromium	Cr	24	51.9961	1860	2679	7.2	128	652.8	20.5	0.4491	339	0.01	2+, 3+, 6+
Cobalt	Co	27	58.9332	1495	2912	8.9	125	758.8	16.192	0.4210	382	0.0028	2+, 3+
Copper	Cu	29	63.546	1085	2570	8.92	128	745.5	13.38	0.38452	304	0.0058	1+, 2+
Curium	Cm	96	[247.0703]	1340	3540	13.51	174	581	—	—	—	—	3+, 4+
Darmstadtium	Ds	110	[271]	—	—	—	—	—	—	—	—	—	—
Dubnium	Db	105	[262]	—	—	—	—	—	—	—	—	—	—
Dysprosium	Dy	66	162.5	1407	2600	8.536	178.1	572	10.4	0.1733	250	6 × 10⁻⁴	2+, 3+
Einsteinium	Es	99	[252.0828]	860	—	—	186	619	—	—	—	—	3+
Erbium	Er	68	167.26	1497	2900	9.045	176.1	589	17.2	0.1681	293	3.5 × 10⁻⁴	2+, 3+
Europium	Eu	63	151.965	826	1439	5.245	208.4	547	10.5	0.1820	176	2.1 × 10⁻³	2+, 3+
Fermium	Fm	100	[257.0951]	—	—	—	—	627	—	—	—	—	2+, 3+
Fluorine	F	9	18.9984032	−219.7	−188.2	0.001696	69	1681	0.51	0.8238	6.54	0.0544	1−
Francium	Fr	87	[223.0197]	24	650	—	280	375	2	—	63.6	trace	1+
Gadolinium	Gd	64	157.25	1312	3000	7.886	180.4	592	15.5	0.2355	311.7	6.3 × 10⁻⁴	3+
Gallium	Ga	31	69.723	29.77	2203	5.904	134	578.8	5.59	0.3709	256	0.0018	1+, 3+
Germanium	Ge	32	72.61	945	2850	5.323	123	761.2	31.8	0.3215	334.3	1.5 × 10⁻⁴	2+, 4+
Gold	Au	79	196.96654	1064	2808	19.32	144	889.9	12.4	0.12905	324.4	3 × 10⁻⁷	1+, 3+

* [ ] indicates mass of longest-lived isotope.

Table D.5 Properties of Elements (continued)

Element	Symbol	Atomic Number [Z]	Atomic Mass* [u]	Melting Point [°C]	Boiling Point [°C]	Density [g/cm³]	Atomic Radius [pm]	First Ionization Energy (kJ/mol)	Enthalpy of Fusion (kJ/mol)	Specific Heat (J/g·°C)	Enthalpy of Vaporization (kJ/mol)	Abundance in Earth's Crust (%)	Major Oxidation States
Hafnium	Hf	72	178.49	2227	4691	13.28	159	654.4	29.288	0.1442	661	3 × 10⁻⁴	4+
Hassium	Hs	108	[265]	—	—	—	—	—	—	—	—	—	—
Helium	He	2	4.002602	−269.7 (2536 kPa)	−268.93	0.00017847	31	2372	0.02	5.1931	0.084	—	—
Holmium	Ho	67	164.9032	1461	2600	8.78	176.2	581	17.1	0.1646	251	1.5 × 10⁻⁴	3+
Hydrogen	H	1	1.00794	−259.19	−252.76	0.0000899	78	1312	0.117	14.298	0.904	—	1−, 1+
Indium	In	49	114.82	156.61	2080	7.29	167	558.2	3.26	0.2407	231.8	2 × 10⁻⁵	1+, 3+
Iodine	I	53	126.90447	113.6	184.5	4.93	138	1008.4	15.517	0.21448	41.95	4.6 × 10⁻⁵	1−, 1+, 5+, 7+
Iridium	Ir	77	192.22	2447	4550	22.65	135.5	880	26.4	0.1306	563.6	1 × 10⁻⁷	3+, 4+, 5+
Iron	Fe	26	55.847	1536	2860	7.874	126	759.4	13.807	0.4494	350	5.8	2+, 3+
Krypton	Kr	36	83.8	−157.2	−153.35	0.0037493	112	1351	1.64	0.2480	9.03	—	—
Lanthanum	La	57	138.9055	920	3420	6.17	187	538	8.5	0.1952	402	0.0035	3+
Lawrencium	Lr	103	[260.1054]	—	—	—	—	—	—	—	—	—	3+
Lead	Pb	82	207.2	327	1746	11.342	175	715.6	4.77	0.1276	178	0.0013	2+, 4+
Lithium	Li	3	6.941	180.5	1347	0.534	156	520.2	3	3.569	148	0.002	1+
Lutetium	Lu	71	174.967	1652	3327	9.84	173.8	524	11.9	0.1535	414	8 × 10⁻⁵	3+
Magnesium	Mg	12	24.305	650	1105	1.738	160	737.8	8.477	1.024	127.4	2.76	2+
Manganese	Mn	25	54.93805	1246	2061	7.43	127	717.5	12.058	0.4791	219.7	0.1	2+, 3+, 4+, 6+, 7+
Meitnerium	Mt	109	[266]	—	—	—	—	635	—	—	—	—	—
Mendelevium	Md	101	[258.0986]	—	—	—	—	1007	—	—	—	—	2+, 3+
Mercury	Hg	80	200.59	−38.9	357	13.534	151	685	2.2953	0.13950	59.1	2 × 10⁻⁶	1+, 2+
Molybdenum	Mo	42	95.94	2623	4679	10.28	139	530	36	0.2508	590	1.2 × 10⁻⁴	4+, 5+, 6+
Neodymium	Nd	60	144.24	1024	3111	7.003	181.4	2081	7.13	0.1903	283.7	0.004	2+, 3+
Neon	Ne	10	20.1797	−248.61	−246.05	0.0008999	71	597	0.34	1.0301	1.77	—	—
Neptunium	Np	93	237.0482	640	3900	20.45	155	736.7	9.46	—	336	—	2+, 3+, 4+, 5+, 6+
Nickel	Ni	28	58.6934	1455	2883	8.908	124	664.1	17.15	0.4442	375	0.0075	2+, 3+, 4+
Niobium	Nb	41	92.90638	2477	4858	8.57	146	1402	26.9	0.2648	690	0.002	4+, 5+
Nitrogen	N	7	14.00674	−210	−195.8	0.0012409	71	642	0.72	1.0397	5.58	0.002	3−, 2−, 1−, 1+, 2+, 3+, 4+, 5+
Nobelium	No	102	[259.1009]	—	—	—	—	840	31.7	—	—	—	2+, 3+
Osmium	Os	76	190.2	3045	5025	22.57	135	1313.9	0.44	0.130	627.6	2 × 10⁻⁷	4+, 6+, 8+
Oxygen	O	8	15.9994	−218.8	−183	0.001429	60	805	17.6	0.91738	6.82	45.5	2−, 1−
Palladium	Pd	46	106.42	1552	2940	11.99	137	1012	19.7	0.2441	362	3 × 10⁻⁷	2+, 4+
Phosphorus	P	15	30.973762	44.2	280.5	1.823	109	868	2.8	0.76968	49.8	0.11	3−, 3+, 5+
Platinum	Pt	78	195.08	1769	3824	21.41	138.5	585	19.86	0.1326	510.4	1 × 10⁻⁶	2+, 4+
Plutonium	Pu	94	[244.0642]	640	3230	19.86	162	585	2.8	0.138	343.5	—	3+, 4+, 5+, 6+
Polonium	Po	84	[208.9824]	254	962	9.4	164	813	3.81	0.125	103	—	2−, 2+, 4+, 6+
Potassium	K	19	39.0983	63.2	766.4	0.862	231	418.8	2.334	0.7566	76.9	1.84	1+

* [ ] indicates mass of longest-lived isotope.

Table D.5 Properties of Elements (continued)

Element	Symbol	Atomic Number [Z]	Atomic Mass * [u]	Melting Point [°C]	Boiling Point [°C]	Density [g/cm³]	Atomic Radius [pm]	First Ionization Energy (kJ/mol)	Enthalpy of Fusion (kJ/mol)	Specific Heat (J/g·°C)	Enthalpy of Vaporization (kJ/mol)	Abundance in Earth's Crust (%)	Major Oxidation States
Praseodymium	Pr	59	140.90765	935	3343	6.782	182.4	522	11.3	0.1930	332.6	9.1×10^{-4}	3+, 4+
Promethium	Pm	61	[144.9128]	1168	2460	7.2	183.4	536	8.17	—	293	—	3+
Protactinium	Pa	91	231.03588	1552	4227	15.37	163	568	14.6	—	481	trace	3+, 4+, 5+
Radium	Ra	88	226.0254	700	1630	5	228	509.1	8.36	—	136.8	—	2+
Radon	Rn	86	[222.0176]	−71	−62	0.00973	140	1037	16.4	—	16.4	—	—
Rhenium	Re	75	186.207	3180	5650	21.232	137	760	33.4	0.1368	707	1×10^{-7}	3+, 4+, 6+, 7+
Rhodium	Rh	45	102.9055	1960	3727	12.39	134	720	21.6	0.2427	494	1×10^{-7}	3+, 4+, 5+
Roentgenium	Rg	111	[272]	—	—	—	—	—	—	—	—	—	—
Rubidium	Rb	37	85.4678	39.5	697	1.532	248	403	2.19	0.36344	69.2	0.0078	1+
Ruthenium	Ru	44	101.07	2310	4119	12.41	134	711	25.5	0.2381	567.8	—	2+, 3+, 4+, 5+
Rutherfordium	Rf	104	[261]	—	—	—	—	—	—	—	—	—	—
Samarium	Sm	62	150.36	1072	1800	7.536	180.4	542	8.9	0.1965	191	7×10^{-4}	2+, 3+
Scandium	Sc	21	44.95591	1539	2831	3	162	631	15.77	0.5677	304.8	0.0022	3+
Seaborgium	Sg	106	[263]	—	—	—	—	—	—	—	—	—	—
Selenium	Se	34	78.96	221	685	4.79	117	940.7	5.43	0.3212	26.3	5×10^{-6}	2−, 2+, 4+, 6+
Silicon	Si	14	28.0855	1411	3231	2.336	118	786.5	50.2	0.7121	359	27.2	2+, 4+
Silver	Ag	47	107.8682	961	2195	10.49	144	730.8	11.65	0.23502	255	8×10^{-6}	1+
Sodium	Na	11	22.989768	97.83	897.4	0.968	186	495.9	2.602	1.228	97.4	2.27	1+
Strontium	Sr	38	87.62	776.9	1412	2.6	215	549.5	7.4308	0.301	137	0.0384	2+
Sulfur	S	16	32.066	115.2	444.7	2.08	103	999.6	1.7272	0.7060	9.62	0.03	2−, 4+, 6+
Tantalum	Ta	73	180.9479	2980	5505	16.65	146	760.8	36.57	0.1402	737	2×10^{-4}	4+, 5+
Technetium	Tc	43	97.9072	2200	4567	11.5	136	702	23.0	—	577	—	2+, 4+, 6+, 7+
Tellurium	Te	52	127.6	450	990	6.25	138	869	17.4	0.2016	50.6	2×10^{-7}	2−, 2+, 4+, 6+
Terbium	Tb	65	158.92534	1356	2800	8.272	177.3	564	10.3	0.1819	293	1×10^{-4}	3+, 4+
Thallium	Tl	81	204.3833	303.5	1457	11.85	170	589.1	4.27	0.1288	162	7×10^{-5}	1+, 3+
Thorium	Th	90	232.0381	1750	4787	11.78	179	587	16.11	0.1177	543.9	8.1×10^{-4}	4+
Thulium	Tm	69	168.93421	1545	1727	9.318	175.9	596	18.4	0.1600	213	5×10^{-5}	—
Tin	Sn	50	118.71	232	2623	7.265	141	708.4	7.07	0.2274	296	2.1×10^{-4}	2+, 4+
Titanium	Ti	22	47.88	1666	3358	4.5	147	658.1	14.146	0.5226	425	0.63	2+, 3+, 4+
Tungsten	W	74	183.85	3680	6000	19.3	139	770.4	35.4	0.1320	806	1.2×10^{-4}	4+, 5+, 6+
Uranium	U	92	238.0289	1130	3930	19.05	156	584	12.6	0.11618	423	2.3×10^{-4}	3+, 4+, 5+, 6+
Ununbium	Uub	112	[285]	—	—	—	—	—	—	—	—	—	—
Ununhexium	Uuh	116	[291]	—	—	—	—	—	—	—	—	—	—
Ununoctium	Uuo	118	[294]	—	—	—	—	—	—	—	—	—	—
Ununpentium	Uup	115	[288]	—	—	—	—	—	—	—	—	—	—
Ununquadium	Uuq	114	[289]	—	—	—	—	—	—	—	—	—	—
Ununtrium	Uut	113	[284]	—	—	—	—	—	—	—	—	—	—
Vanadium	V	23	50.9415	1917	3417	6.11	134	650.3	22.84	0.4886	459.7	0.0136	2+, 3+, 4+, 5+
Xenon	Xe	54	131.29	−111.8	−108.09	0.0058971	131	1170	2.29	0.15832	12.64	—	—
Ytterbium	Yb	70	173.04	824	1427	6.973	193.3	603	7.66	0.1545	155	—	2+, 3+
Yttrium	Y	39	88.90585	1530	3264	4.5	180	616	17.15	0.2984	393	3.4×10^{-4}	3+
Zinc	Zn	30	65.39	419.6	907	7.14	134	906.4	7.322	0.3884	115	0.0035	2+
Zirconium	Zr	40	91.224	1852	4400	6.51	160	659.7	20.92	0.2780	590.5	0.0162	4+

Table D.6 Electron Configurations of the Elements

	Elements	1s	2s	2p	3s	3p	3d	4s	4p	4d	4f	5s	5p	5d	5f	6s	6p	6d	7s	7p
1	Hydrogen	1																		
2	Helium	2																		
3	Lithium	2	1																	
4	Beryllium	2	2																	
5	Boron	2	2	1																
6	Carbon	2	2	2																
7	Nitrogen	2	2	3																
8	Oxygen	2	2	4																
9	Fluorine	2	2	5																
10	Neon	2	2	6																
11	Sodium	2	2	6	1															
12	Magnesium	2	2	6	2															
13	Aluminum	2	2	6	2	1														
14	Silicon	2	2	6	2	2														
15	Phosphorus	2	2	6	2	3														
16	Sulfur	2	2	6	2	4														
17	Chlorine	2	2	6	2	5														
18	Argon	2	2	6	2	6														
19	Potassium	2	2	6	2	6		1												
20	Calcium	2	2	6	2	6		2												
21	Scandium	2	2	6	2	6	1	2												
22	Titanium	2	2	6	2	6	2	2												
23	Vanadium	2	2	6	2	6	3	2												
24	Chromium	2	2	6	2	6	5	1												
25	Manganese	2	2	6	2	6	5	2												
26	Iron	2	2	6	2	6	6	2												
27	Cobalt	2	2	6	2	6	7	2												
28	Nickel	2	2	6	2	6	8	2												
29	Copper	2	2	6	2	6	10	1												
30	Zinc	2	2	6	2	6	10	2												
31	Gallium	2	2	6	2	6	10	2	1											
32	Germanium	2	2	6	2	6	10	2	2											
33	Arsenic	2	2	6	2	6	10	2	3											
34	Selenium	2	2	6	2	6	10	2	4											
35	Bromine	2	2	6	2	6	10	2	5											
36	Krypton	2	2	6	2	6	10	2	6											
37	Rubidium	2	2	6	2	6	10	2	6			1								
38	Strontium	2	2	6	2	6	10	2	6			2								
39	Yttrium	2	2	6	2	6	10	2	6	1		2								
40	Zirconium	2	2	6	2	6	10	2	6	2		2								
41	Niobium	2	2	6	2	6	10	2	6	4		1								
42	Molybdenum	2	2	6	2	6	10	2	6	5		1								
43	Technetium	2	2	6	2	6	10	2	6	5		2								
44	Ruthenium	2	2	6	2	6	10	2	6	7		1								
45	Rhodium	2	2	6	2	6	10	2	6	8		1								
46	Palladium	2	2	6	2	6	10	2	6	10										
47	Silver	2	2	6	2	6	10	2	6	10		1								
48	Cadmium	2	2	6	2	6	10	2	6	10		2								
49	Indium	2	2	6	2	6	10	2	6	10		2	1							
50	Tin	2	2	6	2	6	10	2	6	10		2	2							
51	Antimony	2	2	6	2	6	10	2	6	10		2	3							
52	Tellurium	2	2	6	2	6	10	2	6	10		2	4							
53	Iodine	2	2	6	2	6	10	2	6	10		2	5							
54	Xenon	2	2	6	2	6	10	2	6	10		2	6							
55	Cesium	2	2	6	2	6	10	2	6	10		2	6			1				
56	Barium	2	2	6	2	6	10	2	6	10		2	6			2				

Table D.6 Electron Configurations of the Elements

	Elements	1s	2s	2p	3s	3p	3d	4s	4p	4d	4f	5s	5p	5d	5f	6s	6p	6d	7s	7p
57	Lanthanum	2	2	6	2	6	10	2	6	10		2	6	1		2				
58	Cerium	2	2	6	2	6	10	2	6	10	2	2	6			2				
59	Praseodymium	2	2	6	2	6	10	2	6	10	3	2	6			2				
60	Neodymium	2	2	6	2	6	10	2	6	10	4	2	6			2				
61	Promethium	2	2	6	2	6	10	2	6	10	5	2	6			2				
62	Samarium	2	2	6	2	6	10	2	6	10	6	2	6			2				
63	Europium	2	2	6	2	6	10	2	6	10	7	2	6			2				
64	Gadolinium	2	2	6	2	6	10	2	6	10	7	2	6	1		2				
65	Terbium	2	2	6	2	6	10	2	6	10	9	2	6			2				
66	Dysprosium	2	2	6	2	6	10	2	6	10	10	2	6			2				
67	Holmium	2	2	6	2	6	10	2	6	10	11	2	6			2				
68	Erbium	2	2	6	2	6	10	2	6	10	12	2	6			2				
69	Thulium	2	2	6	2	6	10	2	6	10	13	2	6			2				
70	Ytterbium	2	2	6	2	6	10	2	6	10	14	2	6			2				
71	Lutetium	2	2	6	2	6	10	2	6	10	14	2	6	1		2				
72	Hafnium	2	2	6	2	6	10	2	6	10	14	2	6	2		2				
73	Tantalum	2	2	6	2	6	10	2	6	10	14	2	6	3		2				
74	Tungsten	2	2	6	2	6	10	2	6	10	14	2	6	4		2				
75	Rhenium	2	2	6	2	6	10	2	6	10	14	2	6	5		2				
76	Osmium	2	2	6	2	6	10	2	6	10	14	2	6	6		2				
77	Iridium	2	2	6	2	6	10	2	6	10	14	2	6	7		2				
78	Platinum	2	2	6	2	6	10	2	6	10	14	2	6	9		1				
79	Gold	2	2	6	2	6	10	2	6	10	14	2	6	10		1				
80	Mercury	2	2	6	2	6	10	2	6	10	14	2	6	10		2				
81	Thallium	2	2	6	2	6	10	2	6	10	14	2	6	10		2	1			
82	Lead	2	2	6	2	6	10	2	6	10	14	2	6	10		2	2			
83	Bismuth	2	2	6	2	6	10	2	6	10	14	2	6	10		2	3			
84	Polonium	2	2	6	2	6	10	2	6	10	14	2	6	10		2	4			
85	Astatine	2	2	6	2	6	10	2	6	10	14	2	6	10		2	5			
86	Radon	2	2	6	2	6	10	2	6	10	14	2	6	10		2	6			
87	Francium	2	2	6	2	6	10	2	6	10	14	2	6	10		2	6		1	
88	Radium	2	2	6	2	6	10	2	6	10	14	2	6	10		2	6		2	
89	Actinium	2	2	8	2	6	10	2	6	10	14	2	6	10		2	6	1	2	
90	Thorium	2	2	6	2	6	10	2	6	10	14	2	6	10		2	6	2	2	
91	Protactinium	2	2	6	2	6	10	2	6	10	14	2	6	10	2	2	6	1	2	
92	Uranium	2	2	6	2	6	10	2	6	10	14	2	6	10	3	2	6	1	2	
93	Neptunium	2	2	6	2	6	10	2	6	10	14	2	6	10	4	2	6	1	2	
94	Plutonium	2	2	6	2	6	10	2	6	10	14	2	6	10	6	2	6		2	
95	Americium	2	2	6	2	6	10	2	6	10	14	2	6	10	7	2	6		2	
96	Curium	2	2	6	2	6	10	2	6	10	14	2	6	10	7	2	6	1	2	
97	Berkelium	2	2	6	2	6	10	2	6	10	14	2	6	10	9	2	6		2	
98	Californium	2	2	6	2	6	10	2	6	10	14	2	6	10	10	2	6		2	
99	Einsteinium	2	2	6	2	6	10	2	6	10	14	2	6	10	11	2	6		2	
100	Fermium	2	2	6	2	6	10	2	6	10	14	2	6	10	12	2	6		2	
101	Mendelevium	2	2	6	2	6	10	2	6	10	14	2	6	10	13	2	6		2	
102	Nobelium	2	2	6	2	6	10	2	6	10	14	2	6	10	14	2	6		2	
103	Lawrencium	2	2	6	2	6	10	2	6	10	14	2	6	10	14	2	6	1	2	
104	Rutherfordium	2	2	6	2	6	10	2	6	10	14	2	6	10	14	2	6	2	2?	
105	Dubnium	2	2	6	2	6	10	2	6	10	14	2	6	10	14	2	6	3	2?	
106	Seaborgium	2	2	6	2	6	10	2	6	10	14	2	6	10	14	2	6	4	2?	
107	Bohrium	2	2	6	2	6	10	2	6	10	14	2	6	10	14	2	6	5	2?	
108	Hassium	2	2	6	2	6	10	2	6	10	14	2	6	10	14	2	6	6	2?	
109	Meitnerium	2	2	6	2	6	10	2	6	10	14	2	6	10	14	2	6	7	2?	
110	Darmstadtium	2	2	6	2	6	10	2	6	10	14	2	6	10	14	2	6	8	2?	
111	Roentgenium	2	2	6	2	6	10	2	6	10	14	2	6	10	14	2	6	9	2?	
112	Ununbium	2	2	6	2	6	10	2	6	10	14	2	6	10	14	2	6	10	2?	
113	Ununtrium	2	2	6	2	6	10	2	6	10	14	2	6	10	14	2	6	10	2?	
114	Ununquadium	2	2	6	2	6	10	2	6	10	14	2	6	10	14	2	6	10	2?	1?
115	Ununpentium	2	2	6	2	6	10	2	6	10	14	2	6	10	14	2	6	10	2?	2?
116	Ununhexium	2	2	6	2	6	10	2	6	10	14	2	6	10	14	2	6	10	2?	3?
118	Ununoctium	2	2	6	2	6	10	2	6	10	14	2	6	10	14	2	6	10	2?	6?

Table D.7 Useful Physical Constants

1 ampere is the constant current which, if maintained in two straight parallel conductors of infinite length, of negligible circular cross-section, and placed 1 meter apart in a vacuum, would produce a force of 2×10^{-7} newtons per meter of length between these conductors.

1 candela is the luminous intensity, in the perpendicular direction, of a surface of $1/600,000$ m_2 of a blackbody at the temperature of freezing platinum at a pressure of 101,325 pascals.

1 cubic decimeter is equal to 1 liter.

1 kelvin is $1/273.16$ of the thermodynamic temperature of the triple point of water.

1 kilogram is the mass of the international prototype kilogram.

1 meter is the distance light travels in $1/299,792,458$ of a second.

1 mole is the amount of substance containing as many elementary entities as there are atoms in 0.012 kilogram of carbon-12.

1 second is equal to 9,192,631,770 periods of the natural electromagnetic oscillation during that transition of ground state $^2S_{1/2}$ of cesium-133, which is designated $(F = 4, M = 0) \leftrightarrow (F = 3, M = 0)$.

Avogadro constant $= 6.0221367 \times 10^{23}$
1 electronvolt $= 1.60217733 \times 10^{-19}$ J
Faraday constant $= 96,485.309$ C/mole e^-
Ideal gas constant $= 8.314471$ J·mol·K $= 8.314471$ dm^3·kPa/mol·K
Molar gas volume at STP $= 22.41410$ dm^3
Planck's constant $= 6.626075 \times 10^{-34}$ J·s
Speed of light $= 2.99792458 \times 10^8$ m/s

Table D.8 Names and Charges of Polyatomic Ions

1−
Acetate, CH_3COO^-
Amide, NH^{2-}
Astatate, AtO_3^-
Azide, N_3^-
Benzoate, $C_6H_5COO^-$
Bismuthate, BiO_3^-
Bromate, BrO_3^-
Chlorate, ClO_3^-
Chlorite, ClO_2^-
Cyanide, CN^-
Formate, $HCOO^-$
Hydroxide, OH^-
Hypobromite, BrO^-
Hypochlorite, ClO^-
Hypophosphite, $H_2PO_2^-$
Iodate, IO_3^-
Nitrate, NO_3^-
Nitrite, NO_2^-
Perbromate, BrO_4^-
Perchlorate, ClO_4^-
Periodate, IO_4^-
Permanganate, MnO_4^-
Perrhenate, ReO_4^-
Thiocyanate, SCN^-
Vanadate, VO_3^-

2−
Carbonate, CO_3^{2-}
Chromate, CrO_4^{2-}
Dichromate, $Cr_2O_7^{2-}$
Hexachloroplatinate, $PtCl_6^{2-}$
Hexafluorosilicate, SiF_6^{2-}
Molybdate, MoO_4^{2-}
Oxalate, $C_2O_4^{2-}$
Peroxide, O_2^{2-}
Peroxydisulfate, $S_2O_8^{2-}$
Phosphite, HPO_3^{2-}
Ruthenate, RuO_4^{2-}
Selenate, SeO_4^{2-}
Selenite, SeO_3^{2-}
Silicate, SiO_3^{2-}
Sulfate, SO_4^{2-}
Sulfite, SO_3^{2-}
Tartrate, $C_4H_4O_6^{2-}$
Tellurate, TeO_4^{2-}
Tellurite, TeO_3^{2-}
Tetraborate, $B_4O_7^{2-}$
Thiosulfate, $S_2O_3^{2-}$
Tungstate, WO_4^{2-}

3−
Arsenate, AsO_4^{3-}
Arsenite, AsO_3^{3-}
Borate, BO_3^{3-}
Citrate, $C_6H_5O_7^{3-}$
Hexacyanoferrate(III), $Fe(CN)_6^{3-}$
Phosphate, PO_4^{3-}

1+
Ammonium, NH_4^+
Neptunyl(V), NpO_2^+
Plutonyl(V), PuO_2^+
Uranyl(V), UO_2^+
Vanadyl(V), VO_2^+

4−
Hexacyanoferrate(II), $Fe(CN)_6^{4-}$
Orthosilicate, SiO_4^{4-}
Diphosphate, $P_2O_7^{4-}$

2+
Mercury(I), Hg_2^{2+}
Neptunyl(VI), NpO_2^{2+}
Plutonyl(VI), PuO_2^{2+}
Uranyl(VI), UO_2^{2+}
Vanadyl(IV), VO^{2+}

Reference Tables

Table D.9 Solubility Guidelines

You will be working with water solutions, and it is helpful to have a few guidelines concerning what substances are soluble in water. A substance is considered soluble if more than 3 grams of the substance dissolve in 100 mL of water. The more common rules are listed below.

1. All common salts of the group 1 elements and ammonium ions are soluble.
2. All common acetates and nitrates are soluble.
3. All binary compounds of group 17 elements (other than F) with metals are soluble except those of silver, mercury(I), and lead.
4. All sulfates are soluble except those of barium, strontium, lead, calcium, silver, and mercury(I).
5. Except for those in Rule 1, carbonates, hydroxides, oxides, sulfides, and phosphates are insoluble.

Table D.10 Solubility Product Constants (at 25°C)

Substance	K_{sp}	Substance	K_{sp}	Substance	K_{sp}
AgBr	5.01×10^{-13}	$BaSO_4$	1.10×10^{-10}	Li_2CO_3	2.51×10^{-2}
$AgBrO_3$	5.25×10^{-5}	$CaCO_3$	2.88×10^{-9}	$MgCO_3$	3.47×10^{-8}
Ag_2CO_3	8.13×10^{-12}	$CaSO_4$	9.12×10^{-6}	$MnCO_3$	1.82×10^{-11}
AgCl	1.78×10^{-10}	CdS	7.94×10^{-27}	$NiCO_3$	6.61×10^{-9}
Ag_2CrO_4	1.12×10^{-12}	$Cu(IO_3)_2$	7.41×10^{-8}	$PbCl_2$	1.62×10^{-5}
$Ag_2Cr_2O_7$	2.00×10^{-7}	CuC_2O_4	2.29×10^{-8}	PbI_2	7.08×10^{-9}
AgI	8.32×10^{-17}	$Cu(OH)_2$	2.19×10^{-20}	$Pb(IO_3)_2$	3.24×10^{-13}
AgSCN	1.00×10^{-12}	CuS	6.31×10^{-36}	$SrCO_3$	1.10×10^{-10}
$Al(OH)_3$	1.26×10^{-33}	FeC_2O_4	3.16×10^{-7}	$SrSO_4$	3.24×10^{-7}
Al_2S_3	2.00×10^{-7}	$Fe(OH)_3$	3.98×10^{-38}	TlBr	3.39×10^{-6}
$BaCO_3$	5.13×10^{-9}	FeS	6.31×10^{-18}	$ZnCO_3$	1.45×10^{-11}
$BaCrO_4$	1.17×10^{-10}	Hg_2SO_4	7.41×10^{-7}	ZnS	1.58×10^{-24}

Table D.11 Acid-Base Indicators

Indicator	Lower Color	Range	Upper Color
Methyl violet	yellow-green	0.0–2.5	violet
Malachite green HCl	yellow	0.5–2.0	blue
Thymol blue	red	1.0–2.8	yellow
Naphthol yellow S	colorless	1.5–2.6	yellow
p-Phenylazoaniline	orange	2.1–2.8	yellow
Methyl orange	red	2.5–4.4	yellow
Bromphenol blue	orange-yellow	3.0–4.7	violet
Gallein	orange	3.5–6.3	red
2,5-Dinitrophenol	colorless	4.0–5.8	yellow
Ethyl orange	salmon	4.2–4.6	orange
Propyl red	pink	5.1–6.5	yellow
Bromcresol purple	green-yellow	5.4–6.8	violet
Bromxylenol blue	orange-yellow	6.0–7.6	blue
Phenol red	yellow	6.4–8.2	red-violet
Cresol red	yellow	7.1–8.8	violet
m-Cresol purple	yellow	7.5–9.0	violet
Thymol blue	yellow	8.1–9.5	blue
Phenolphthalein	colorless	8.3–10.0	dark pink
o-Cresolphthalein	colorless	8.6–9.8	pink
Thymolphthalein	colorless	9.5–10.4	blue
Alizarin yellow R	yellow	9.9–11.8	dark orange
Methyl blue	blue	10.6–13.4	pale violet
Acid fuchsin	red	11.1–12.8	colorless
2,4,6-Trinitrotoluene	colorless	11.7–12.8	orange

Solutions to Problems

Chapter 5

1. Write the formula for each compound.
   a. lithium oxide
      Li_2O
   b. calcium bromide
      $CaBr_2$
   c. sodium oxide
      Na_2O
   d. aluminum sulfide
      Al_2S_3

2. Write the formula for the compound formed from each pair of elements.
   a. barium and oxygen
      BaO
   b. strontium and iodine
      SrI_2
   c. lithium and chlorine
      $LiCl$
   d. radium and chlorine
      $RaCl_2$

3. Write the formula for the compound made from each set of ions.
   a. ammonium and sulfite ions
      $(NH_4)_2SO_3$
   b. calcium and monohydrogen phosphate ions
      $CaHPO_4$
   c. ammonium and dichromate ions
      $(NH_4)_2Cr_2O_7$
   d. barium and nitrate ions
      $Ba(NO_3)_2$

4. Write the formula for each compound.
   a. sodium phosphate
      Na_3PO_4
   b. magnesium hydroxide
      $Mg(OH)_2$
   c. ammonium phosphate
      $(NH_4)_3PO_4$
   d. potassium dichromate
      $K_2Cr_2O_7$

5. Write the formula for the compound made from each pair of ions.
   a. copper(I) and sulfite
      Cu_2SO_3
   b. tin(IV) and fluoride
      SnF_4
   c. gold(III) and cyanide
      $Au(CN)_3$
   d. lead(II) and sulfide
      PbS

6. Write the names of the compounds.
   a. $Pb(NO_3)_2$
      lead(II) nitrate
   b. Mn_2O_3
      manganese(III) oxide
   c. $Ni(C_2H_3O_2)_2$
      nickel(II) acetate
   d. HgF_2
      mercury(II) fluoride

13. Name the following molecular compounds.
    a. S_2Cl_2
       disulfur dichloride
    b. CS_2
       carbon disulfide
    c. SO_3
       sulfur trioxide
    d. P_4O_{10}
       tetraphosphorus decoxide

14. Write the formulas for the following covalent compounds.
    a. carbon tetrachloride
       CCl_4
    b. iodine heptafluoride
       IF_7
    c. dinitrogen monoxide
       N_2O
    d. sulfur dioxide
       SO_2

Chapter 6

Write word equations and chemical equations for reaction.

1. Magnesium metal and water combine to form solid magnesium hydroxide and hydrogen gas.
   magnesium + water →
              magnesium hydroxide + hydrogen
   $Mg(s) + 2H_2O(l) \rightarrow Mg(OH)_2(s) + H_2(g)$

2. An aqueous solution of hydrogen peroxide (dihydrogen dioxide) and solid lead(II) sulfide combine to form solid lead(II) sulfate and liquid water.
   hydrogen peroxide + lead(II) sulfide →
              lead(II) sulfate + water
   $4H_2O_2(aq) + PbS(s) \rightarrow PbSO_4(s) + 4H_2O(l)$

Solutions to Problems

3. When energy is added to solid manganese(II) sulfate heptahydrate crystals, they break down to form liquid water and solid manganese(II) sulfate monohydrate.
   manganese(II) sulfate heptahydrate + energy → water + manganese(II) sulfate monohydrate
   $MnSO_4 \cdot 7H_2O(s)$ + energy →
   $6H_2O(l) + MnSO_4 \cdot H_2O(s)$

4. Solid potassium reacts with liquid water to produce aqueous potassium hydroxide and hydrogen gas.
   potassium + water → potassium hydroxide + hydrogen gas
   $2K(s) + 2H_2O(l) \rightarrow 2\ KOH(aq) + H_2(g)$

Chapter 9

1. Calculate ΔEN for the pairs of atoms in the following bonds.
   a. Ca—S
   1.5
   b. Ba—O
   2.6
   c. C—Br
   0.3
   d. Ca—F
   3.0
   e. H—Br
   0.7

2. Use ΔEN to classify the bonds in Question 1 as covalent, polar covalent, or ionic.
   a. polar covalent
   b. ionic
   c. covalent
   d. ionic
   e. polar covalent

Chapter 11

1. 59.8 in Hg to psi

   $59.8 \text{ in Hg} \times \dfrac{25.4 \text{ mm Hg}}{1.00 \text{ in Hg}} \times \dfrac{14.7 \text{ psi}}{760 \text{ mm Hg}}$

   $= 59.8 \text{ in Hg} \times \dfrac{25.4 \text{ mm Hg}}{1.00 \text{ in Hg}} \times \dfrac{14.7 \text{ psi}}{760 \text{ mm Hg}}$

   $= \dfrac{59.8 \times 25.4 \times 14.7 \text{ psi}}{760} = 29.4 \text{ psi}$

2. 7.35 psi to mm Hg

   $7.35 \text{ psi} \times \dfrac{1 \text{ atm}}{14.7 \text{ psi}} \times \dfrac{760 \text{ mm Hg}}{1 \text{ atm}}$

   $= 7.35 \text{ psi} \times \dfrac{1 \text{ atm}}{14.7 \text{ psi}} \times \dfrac{760 \text{ mm Hg}}{1 \text{ atm}}$

   $= \dfrac{7.35 \times 760 \text{ mm Hg}}{14.7} = 3.80 \times 10^2 \text{ mm Hg}$

3. 1140 mm Hg to kPa

   $1140 \text{ mm Hg} \times \dfrac{1.00 \text{ atm}}{760 \text{ mm Hg}} \times \dfrac{101.3 \text{ kPa}}{1.00 \text{ atm}}$

   $= 1140 \text{ mm Hg} \times \dfrac{1.00 \text{ atm}}{760 \text{ mm Hg}} \times \dfrac{101.3 \text{ kPa}}{1.00 \text{ atm}}$

   $= \dfrac{1140 \times 101.3 \text{ kPa}}{760} = 152 \text{ kPa}$

4. 19.0 psi to kPa

   $19.0 \text{ psi} \times \dfrac{1.00 \text{ atm}}{14.7 \text{ psi}} \times \dfrac{101.3 \text{ kPa}}{1.00 \text{ atm}}$

   $= 19.0 \text{ psi} \times \dfrac{1.00 \text{ atm}}{14.7 \text{ psi}} \times \dfrac{101.3 \text{ kPa}}{1.00 \text{ atm}}$

   $= \dfrac{19.0 \times 101.3 \text{ kPa}}{14.7} = 131 \text{ kPa}$

5. 202 kPa to psi

   $202 \text{ kPa} \times \dfrac{1.00 \text{ atm}}{101.3 \text{ kPa}} \times \dfrac{14.7 \text{ psi}}{1.00 \text{ atm}}$

   $= 202 \text{ kPa} \times \dfrac{1.00 \text{ atm}}{101.3 \text{ kPa}} \times \dfrac{14.7 \text{ psi}}{1.00 \text{ atm}}$

   $= \dfrac{202 \times 14.7 \text{ psi}}{101.3} = 29.3 \text{ psi}$

Solutions to Problems

Assume that the temperature remains constant in the following problems.

11. Bacteria produce methane gas in sewage-treatment plants. This gas is often captured or burned. If a bacterial culture produces 60.0 mL of methane gas at 700.0 mm Hg, what volume would be produced at 760.0 mm Hg?

$$60.0 \text{ mL} \times \frac{700.0 \text{ mm Hg}}{760.0 \text{ mm Hg}}$$

$$= 60.0 \text{ mL} \times \frac{700.0 \text{ mm Hg}}{760.0 \text{ mm Hg}}$$

$$= \frac{60.0 \text{ mL} \times 700.0}{760.0} = 55.3 \text{ mL}$$

12. At one sewage-treatment plant, bacteria cultures produce 1000 mL of methane gas per day at 1.0 atm pressure. What volume tank would be needed to store one day's production at 5.0 atm?

$$1000 \text{ L} \times \frac{1 \text{ atm}}{5 \text{ atm}} = 1000 \text{ L} \times \frac{1 \text{ atm}}{5 \text{ atm}}$$

$$= \frac{1000 \text{ L}}{5} = 200 \text{ L}$$

13. Hospitals buy 400-L cylinders of oxygen gas compressed at 150 atm. They administer oxygen to patients at 3.0 atm in a hyperbaric oxygen chamber. What volume of oxygen can a cylinder supply at this pressure?

$$400 \text{ L} \times \frac{150 \text{ atm}}{3.0 \text{ atm}} = 400 \text{ L} \times \frac{150 \text{ atm}}{3.0 \text{ atm}}$$

$$= \frac{400 \text{ L} \times 150}{3} = 20{,}000 \text{ L}$$

14. If the valve in a tire pump with a volume of 0.78 L fails at a pressure of 9.0 atm, what would be the volume of air in the cylinder just before the valve fails?

$$0.78 \text{ L} \times \frac{9.0 \text{ atm}}{1.0 \text{ atm}} = 0.78 \text{ L} \times \frac{9.0 \text{ atm}}{1.0 \text{ atm}}$$

$$= \frac{0.78 \text{ L} \times 9.0}{1.0} = 7.0 \text{ L}$$

15. The volume of a scuba tank is 10.0 L. It contains a mixture of nitrogen and oxygen at 290.0 atm. What volume of this mixture could the tank supply to a diver at 2.40 atm?

$$10.0 \text{ L} \times \frac{290.0 \text{ atm}}{2.4 \text{ atm}} = 10.0 \text{ L} \times \frac{290.0 \text{ atm}}{2.4 \text{ atm}}$$

$$= \frac{10.0 \text{ L} \times 290.0}{2.40} = 1210 \text{ L}$$

16. A 1.00-L balloon is filled with helium at 1.20 atm. If the balloon is squeezed into a 0.500-L beaker and doesn't burst, what is the pressure of the helium?

$$1.20 \text{ atm} \times \frac{1.00 \text{ L}}{0.500 \text{ L}} = 1.20 \text{ atm} \times \frac{1.00 \text{ L}}{0.500 \text{ L}}$$

$$= \frac{1.20 \text{ atm}}{0.500} = 2.40 \text{ atm}$$

Assume that the pressure remains constant in problems 17 to 21.

17. A balloon is filled with 3.0 L of helium at 310 K and 1 atm. The balloon is placed in an oven where the temperature reaches 340 K. What is the new volume of the balloon?

$$3.0 \text{ L} \times \frac{340 \text{ K}}{310 \text{ K}} = 3.0 \text{ L} \times \frac{340 \text{ K}}{310 \text{ K}}$$

$$= \frac{3.0 \text{ L} \times 340}{310} = 3.3 \text{ L}$$

18. A 4.0-L sample of methane gas is collected at 30.0°C. Predict the volume of the sample at 0°C.

$T_1 = 30.0°C + 273 = 303 \text{ K}$
$T_2 = 0°C + 273 = 273 \text{ K}$

$$4.0 \text{ L} \times \frac{273 \text{ K}}{303 \text{ K}} = 4.0 \text{ L} \times \frac{273 \text{ K}}{303 \text{ K}}$$

$$= \frac{4.0 \text{ L} \times 273}{303} = 3.6 \text{ L}$$

19. A 25-L sample of nitrogen is heated from 110°C to 260°C. What volume will the sample occupy at the higher temperature?

$T_1 = 110°C + 273 = 383 \text{ K}$
$T_2 = 260°C + 273 = 533 \text{ K}$

$$25 \text{ L} \times \frac{533 \text{ K}}{383 \text{ K}} = 25 \text{ L} \times \frac{533 \text{ K}}{383 \text{ K}}$$

$$= \frac{25 \text{ L} \times 533}{383} = 35 \text{ L}$$

20. The volume of a 16-g sample of oxygen is 11.2 L at 273 K and 1.00 atm. Predict the volume of the sample at 409 K.

$$11.2 \text{ L} \times \frac{409 \text{ K}}{273 \text{ K}} = 11.2 \text{ L} \times \frac{409 \text{ K}}{273 \text{ K}}$$

$$= \frac{11.2 \text{ L} \times 409}{273} = 16.8 \text{ L}$$

21. The volume of a sample of argon is 8.5 mL at 15°C and 101 kPa. What will its volume be at 0.00°C and 101 kPa?

$T_1 = 15°C + 273 = 288$ K
$T_2 = 0°C + 273 = 273$ K

$$8.5 \text{ mL} \times \frac{273 \text{ K}}{288 \text{ K}} = 8.5 \text{ mL} \times \frac{273 \text{ K}}{288 \text{ K}}$$

$$= \frac{8.5 \text{ mL} \times 273}{288} = 8.1 \text{ mL}$$

22. A 2.7-L sample of nitrogen is collected at 121 kPa and 288 K. If the pressure increases to 202 kPa and the temperature rises to 303 K, what volume will the nitrogen occupy?

$$2.7 \text{ L} \times \frac{121 \text{ kPa}}{202 \text{ kPa}} \times \frac{303 \text{ K}}{288 \text{ K}}$$

$$= 2.7 \text{ L} \times \frac{121 \text{ kPa}}{202 \text{ kPa}} \times \frac{303 \text{ K}}{288 \text{ K}}$$

$$= \frac{2.7 \text{ L} \times 121 \times 303}{202 \times 288} = 1.7 \text{ L}$$

23. A chunk of subliming carbon dioxide (dry ice) generates a 0.80-L sample of gaseous CO_2 at 22°C and 720 mm Hg. What volume will the carbon dioxide gas have at STP?

$T_1 = 22°C + 273 = 295$ K
$T_2 = 0°C + 273 = 273$ K

$$0.80 \text{ L} \times \frac{720 \text{ mm Hg}}{760 \text{ mm Hg}} \times \frac{273 \text{ K}}{295 \text{ K}}$$

$$= 0.80 \text{ L} \times \frac{720 \text{ mm Hg}}{760 \text{ mm Hg}} \times \frac{273 \text{ K}}{295 \text{ K}}$$

$$= \frac{0.80 \text{ L} \times 720 \times 273}{760 \times 295} = 0.70 \text{ L}$$

Chapter 12

1. Without calculating, decide whether 50.0 g of sulfur or 50.0 g of tin represents the greater number of atoms. Verify your answer by calculating.

Because sulfur has a smaller molar mass, there will be a greater number of atoms in 50.0 g of sulfur than in 50.0 g of tin.

$$50.0 \text{ g S} \times \frac{1 \text{ mol S}}{32.1 \text{ g S}} \times \frac{6.02 \times 10^{23} \text{ S atoms}}{1 \text{ mol S}}$$

$$= 50.0 \text{ g S} \times \frac{1 \text{ mol S}}{32.1 \text{ g S}} \times \frac{6.02 \times 10^{23} \text{ S atoms}}{1 \text{ mol S}}$$

$$= \frac{50.0 \times 6.02 \times 10^{23} \text{ S atoms}}{32.1}$$

$$= 9.38 \times 10^{23} \text{ S atoms}$$

$$50.0 \text{ g Sn} \times \frac{1 \text{ mol Sn}}{119 \text{ g Sn}} \times \frac{6.02 \times 10^{23} \text{ Sn atoms}}{1 \text{ mol Sn}}$$

$$= 50.0 \text{ g Sn} \times \frac{1 \text{ mol Sn}}{119 \text{ g Sn}} \times \frac{6.02 \times 10^{23} \text{ Sn atoms}}{1 \text{ mol Sn}}$$

$$= \frac{50.0 \times 6.02 \times 10^{23} \text{ Sn atoms}}{119}$$

$$= 2.53 \times 10^{23} \text{ Sn atoms}$$

2. Determine the number of atoms in each sample below.

   a. 98.3 g mercury (Hg)

   $$98.3 \text{ g Hg} \times \frac{1 \text{ mol Hg}}{200.59 \text{ g Hg}} \times \frac{6.02 \times 10^{23} \text{ atoms Hg}}{1 \text{ mol Hg}}$$

   $$= 2.95 \times 10^{23} \text{ Hg atoms}$$

   b. 45.6 g gold (Au)

   $$45.6 \text{ g Au} \times \frac{1 \text{ mol Au}}{196.967 \text{ g Au}} \times \frac{6.02 \times 10^{23} \text{ Au atoms}}{1 \text{ mol Au}}$$

   $$= 1.39 \times 10^{23} \text{ Au atoms}$$

   c. 10.7 g lithium (Li)

   $$10.7 \text{ g Li} \times \frac{1 \text{ mol Li}}{6.941 \text{ g Li}} \times \frac{6.02 \times 10^{23} \text{ Li atoms}}{1 \text{ mol Li}}$$

   $$= 9.28 \times 10^{23} \text{ Li atoms}$$

   d. 144.6 g tungsten (W)

   $$144.6 \text{ g W} \times \frac{1 \text{ mol W}}{183.85 \text{ g W}} \times \frac{6.02 \times 10^{23} \text{ W atoms}}{1 \text{ mol W}}$$

   $$= 4.73 \times 10^{23} \text{ W atoms}$$

Solutions to Problems

3. Determine the number of moles in each sample below.

a. 6.84 g sucrose ($C_{12}H_{22}O_{11}$)

12 C atoms	12 × 12.0 u =	144.0 u
22 H atoms	22 × 1.0 u =	22.0 u
11 O atoms	11 × 16.0 u =	+ 176 u
molecular mass $C_{12}H_{22}O_{11}$		342 u
molar mass $C_{12}H_{22}O_{11}$		342 g/mol

$$6.84 \text{ g } C_{12}H_{22}O_{11} \times \frac{1 \text{ mol } C_{12}H_{22}O_{11}}{342 \text{ g } C_{12}H_{22}O_{11}}$$

$$= 2.00 \times 10^{-2} \text{ mol } C_{12}H_{22}O_{11}$$

b. 16.0 g sulfur dioxide (SO_2)

1 S atom	1 × 32.1 u =	32.1 u
2 O atoms	2 × 16.0 u =	+32.0 u
molecular mass SO_2		64.1 u
molar mass SO_2		64.1 g/mol

$$16.0 \text{ g } SO_2 \times \frac{1 \text{ mol } SO_2}{64.1 \text{ g } SO_2}$$

$$= 0.250 \text{ mol } SO_2$$

c. 68.0 g ammonia (NH_3)

1 N atom	1 × 14.0 u =	14.0 u
3 H atoms	3 × 1.0 u =	+3.0 u
molecular mass NH_3		17.0 u
molar mass NH_3		17.0 g/mol

$$68.0 \text{ g } NH_3 \times \frac{1 \text{ mol } NH_3}{17.0 \text{ g } NH_3}$$

$$= 4.00 \text{ mol } NH_3$$

d. 17.5 g copper (II) oxide (CuO)

1 Cu atom	1 × 63.5 u =	63.5 u
1 O atom	1 × 16.0 u =	+16.0 u
molecular mass of CuO		79.5 u
molar mass of CuO		79.5 g/mol

$$17.5 \text{ g CuO} \times \frac{1 \text{ mol CuO}}{79.5 \text{ g CuO}}$$

$$= 0.220 \text{ mol CuO}$$

4. Determine the mass of the following molar quantities.

a. 3.52 mol Si

$$3.52 \text{ mol Si} \times \frac{28.1 \text{ g Si}}{1 \text{ mol Si}}$$

$$= 98.9 \text{ g Si}$$

b. 1.25 mol aspirin ($C_9H_8O_4$)

9 C atoms	9 × 12.0 u =	108 u
8 H atoms	8 × 1.0 u =	8 u
4 O atoms	4 × 16.0 u =	+ 64 u
molecular mass $C_9H_8O_4$		180.0 u
molar mass $C_9H_8O_4$		180.0 g/mol

$$1.25 \text{ mol } C_9H_8O_4 \times \frac{180.0 \text{ g } C_9H_8O_4}{1 \text{ mol } C_9H_8O_4}$$

$$= 225 \text{ g } C_9H_8O_4$$

c. 0.550 mol F_2

2 F atoms	2 × 19.0 u =	38.0 u
molecular mass F_2		38.0 u
molar mass F_2		38.0 g/mol

$$0.550 \text{ mol } F_2 \times \frac{38.0 \text{ g } F_2}{1 \text{ mol } F_2}$$

$$= 20.9 \text{ g } F_2$$

d. 2.35 mol barium iodide (BaI_2)

1 Ba atom	1 × 137 u =	137 u
2 I atoms	2 × 127 u =	+ 254 u
molecular mass BaI_2		391 u
molar mass BaI_2		391 g/mol

$$2.35 \text{ mol } BaI_2 \times \frac{391 \text{ g } BaI_2}{1 \text{ mol } BaI_2}$$

$$= 919 \text{ g } BaI_2$$

10. The combustion of propane (C_3H_8) a fuel used in backyard grills and camp stoves, produces carbon dioxide and water vapor.

$$C_3H_8(g) + 5O_2(g) \rightarrow 3CO_2(g) + 4H_2O(g)$$

What mass of carbon dioxide forms when 95.6 g of propane burns?

molar mass C_3H_8 = 44.0 g/mol
molar mass CO_2 = 44.0 g/mol

$$95.6 \text{ g } C_3H_8 \times \frac{1 \text{ mol } C_3H_8}{44.0 \text{ g } C_3H_8} \times \frac{3 \text{ mol } CO_2}{1 \text{ mol } C_3H_8} \times \frac{44.0 \text{ g } CO_2}{1 \text{ mol } CO_2}$$

$$= 287 \text{ g } CO_2$$

11. Solid xenon hexafluoride is prepared by allowing xenon gas and fluorine to react.

$$Xe(g) + 3F_2(g) \rightarrow XeF_6(s)$$

How many grams of fluorine are required to produce 10.0 g XeF_6?

molar mass F_2 = 38.0 g/mol
molar mass XeF_6 = 245.3 g/mol

$$10.0 \text{ g } XeF_6 \times \frac{1 \text{ mol } XeF_6}{245.3 \text{ g } XeF_6} \times \frac{3 \text{ mol } F_2}{1 \text{ mol } XeF_6} \times \frac{38.0 \text{ g } F_2}{1 \text{ mol } F_2}$$

= 465 g F_2

12. Using the reaction in Practice Problem 11, how many grams of xenon are required to produce 10.0 g XeF_6?

molar mass XeF_6 = 245.3 g/mol
molar mass Xe = 131.3 g/mol

$$10.0 \text{ g } XeF_6 \times \frac{1 \text{ mol } XeF_6}{245.3 \text{ g } XeF_6} \times \frac{1 \text{ mol } Xe}{1 \text{ mol } XeF_6} \times \frac{131.3 \text{ g } Xe}{1 \text{ mol } Xe}$$

= 5.35 g Xe

13. What mass of sulfur must burn to produce 3.42 L of SO2 at 273°C and 101 kPa? The reaction is

$$S(s) + O_2(g) \rightarrow SO_2(g)$$

Note that the volume must be corrected to standard temperature (0°C) by applying the factor $\frac{273 \text{ K}}{(273 + 273)\text{K}} = \frac{273 \text{ K}}{546 \text{ K}}$.

$$\frac{3.42 \text{ L}}{1} \times \frac{273 \text{ K}}{546 \text{ K}} \times \frac{1 \text{ mol } SO_2}{22.4 \text{ L } SO_2} \times \frac{1 \text{ mol } S}{1 \text{ mol } SO_2} \times \frac{32.1 \text{ g } S}{1 \text{ mol } S}$$

$$= 3.42 \times 273 \times \frac{32.1 \text{ g } S}{546 \times 22.4} = 2.45 \text{ g } S$$

14. What volume of hydrogen gas can be produced by reacting 4.20 g sodium in excess water at 50°C and 106 kPa?

$$2Na + 2H_2O \rightarrow 2NaOH + H_2$$

$$4.20 \text{ g } Na \times \frac{1 \text{ mol } Na}{23.0 \text{ g } Na} \times \frac{1 \text{ mol } H_2}{2 \text{ mol } Na} \times \frac{22.4 \text{ L } H_2}{1 \text{ mol } H_2}$$

$$\times \frac{323 \text{ K}}{273 \text{ K}} \times \frac{101 \text{ kPa}}{106 \text{ kPa}}$$

$$\frac{4.20 \times 22.4 \text{ L } H_2 \times 323 \times 101}{23.0 \times 2 \times 273 \times 106} = 2.31 \text{ L } H_2$$

15. How many moles of helium are contained in a 5.00-L canister at 101 kPa and 30°C?

$$n = VP/RT = \frac{(5.00 \text{ L})(101 \text{ kPa})}{\left(\frac{8.31 \text{ kPa} \cdot \text{L}}{1 \text{ mol} \cdot \text{K}}\right)303 \text{ K}}$$

$$= \frac{5.00 \text{ L} \times 101 \text{ kPa} \times \text{mol} \times \text{K}}{8.31 \text{ kPa} \cdot \text{L} \times 303 \text{ K}}$$

$$= \frac{5.00 \times 101 \times 1 \text{ mol}}{8.31 \times 303} = 0.201 \text{ mol}$$

16. What is the volume of 0.020 mol Ne at 0.505 kPa and 27.0°C?

$$V = \frac{nRT}{P} = \frac{(0.020 \text{ mol Ne})\left(\frac{8.31 \text{ kPa} \cdot \text{L}}{1 \text{ mol} \cdot \text{K}}\right)(300 \text{ K})}{0.505 \text{ kPa}} = 99 \text{ L}$$

17. How much zinc must react in order to form 15.5 L of hydrogen, H_2 (g), at 32.0 °C and 115 kPa?

$$Zn(s) + H_2SO_4(aq) \rightarrow ZnSO_4(aq) + H_2(g)$$

First, use the ideal gas law to determine moles of H_2.

$$PV = nRT$$

$$n = \frac{PV}{RT}$$

$$n = \frac{(115 \text{ kPa})(15.5 \text{ L})}{\left(\frac{8.31 \text{ kPa} \cdot \text{L}}{1 \text{ mol} \cdot \text{K}}\right)(305 \text{ K})}$$

$$n = 0.703 \text{ mol } H_2$$

Now, determine the mass of zinc.

$$0.703 \text{ mol } H_2 \times \frac{1 \text{ mol } Zn}{1 \text{ mol } H_2} \times \frac{65.39 \text{ g } Zn}{1 \text{ mol } Zn}$$

= 46.0 g Zn

Chapter 13

7. How would you prepare 1.00 L of a 0.400M solution of copper(II) sulfate ($CuSO_4$)?

Dissolve 63.8 g of $CuSO_4$ in 1.00 L of solution;
Molar mass $CuSO_4$ = 159.61 g/mol

$$\left(\frac{1.00 \text{ L soln}}{1}\right)\left(\frac{0.400 \text{ mol } CuSO_4}{1 \text{ L soln}}\right)\left(\frac{159.61 \text{ g } CuSO_4}{1 \text{ mol } CuSO_4}\right)$$

= 63.8 g $CuSO_4$

Solutions to Problems

8. How would you prepare 2.50 L of a 0.800M solution of potassium nitrate (KNO_3)?

Dissolve 202 g of KNO_3 in 2.50 L of solution;
Molar mass KNO_3 = 101.10 g/mol

$$\left(\frac{2.50 \text{ L soln}}{1}\right)\left(\frac{0.800 \text{ mol KNO}_3}{1 \text{ L soln}}\right)\left(\frac{101.10 \text{ g KNO}_3}{1 \text{ mol KNO}_3}\right)$$

$= 202$ g KNO_3

9. What mass of sucrose ($C_{12}H_{22}O_{11}$) must be dissolved to make 460 mL of a 1.10M solution?

170 g of sucrose;
Molar mass $C_{12}H_{22}O_{11}$ = 342.30 g/mol

$$\left(\frac{460 \text{ mL soln}}{1}\right)\left(\frac{1 \text{ L}}{10_3 \text{ mL}}\right)\left(\frac{1.10 \text{ mol } C_{12}H_{22}O_{11}}{1 \text{ L soln}}\right)\cdots$$

$$\left(\frac{342.30 \text{ g } C_{12}H_{22}O_{11}}{1 \text{ mol } C_{12}H_{22}O_{11}}\right) = 170 \text{ g } C_{12}H_{22}O_{11}$$

10. What mass of lithium chloride (LiCl) must be dissolved to make a 0.194M solution that has a volume of 1.00 L?

8.23 g of LiCl;
Molar mass LiCl = 42.40 g/mol

$$\left(\frac{1.00 \text{ L soln}}{1}\right)\left(\frac{0.194 \text{ mol LiCl}}{1 \text{ L soln}}\right)\left(\frac{42.40 \text{ g/LiCl}}{1 \text{ mol LiCl}}\right)$$

$= 8.23$ g LiCl

11. Complete **Table 13.4** by determining the mass of solute necessary to prepare the indicated solutions.

Table 13.4 Preparing Solutions

Solution	Volume	Mass of solute
0.10M $CaCl_2$	1.0 L	11 g $CaCl_2$
0.20M $CaCl_2$	500.0 mL	2.2 g $CaCl_2$
3.0M NaOH	250 mL	30 g NaOH

Molar mass $CaCl_2$ = 110.98 g/mol
Molar mass NaOH = 40.00 g/mol

0.10M $CaCl_2$:

$$\left(\frac{0.10 \text{ mol CaCl}_2}{1.0 \text{ L}}\right)\left(\frac{110.98 \text{ g CaCl}_2}{1 \text{ mol CaCl}_2}\right)(1.0 \text{ L}) = 11 \text{ g CaCl}_2$$

0.10M $CaCl_2$:

$$\left(\frac{0.20 \text{ mol CaCl}_2}{1.0 \text{ L}}\right)\left(\frac{110.98 \text{ g CaCl}_2}{1 \text{ mol CaCl}_2}\right)\left(\frac{1 \text{ L}}{1000 \text{ mL}}\right)(500 \text{ mL})$$

$= 11$ g $CaCl_2$

3.0 M NaOH:

$$\left(\frac{3.0 \text{ mol NaOH}}{1.0 \text{ L}}\right)\left(\frac{40.00 \text{ g NaOH}}{1 \text{ mol NaOH}}\right)\left(\frac{1 \text{ L}}{1000 \text{ mL}}\right)(250 \text{ mL})$$

$= 30$ g NaOH

12. What is the molarity of a solution that contains 14 g of sodium sulfate (Na_2SO_4) dissolved in 1.6 L of solution?

0.062M Na_2SO_4;
Molar mass Na_2SO_4 = 142.06 g/mol

$$\left(\frac{14 \text{ g Na}_2SO_4}{1.6 \text{ L soln}}\right)\left(\frac{1 \text{ mol Na}_2SO_4}{142.06 \text{ g Na}_2SO_4}\right)$$

$= 0.062$ mol Na_2SO_4 /L soln
$= 0.062M$ Na_2SO_4

6. Calculate the molarity of a solution, given that its volume is 820 mL and that it contains 7.4 g of ammonium chloride (NH_4Cl).

0.17M NH_4Cl;
Molar mass NH_4Cl = 53.50 g/mol

$$\left(\frac{7.4 \text{ g NH}_4Cl}{820 \text{ mL}}\right)\left(\frac{1 \text{ mol}}{53.50 \text{ g}}\right)\left(\frac{10^3 \text{ mL}}{1 \text{ L}}\right)$$

$= 0.17$ mol NH_4Cl/L soln
$= 0.17M$ NH_4Cl

Chapter 14

Find the pH of each solution.

6. The hydronium ion concentration equals:
  a. $10^{-5}M$
  5
  b. $10^{-12}M$
  12
  c. $10^{-2}M$
  2

7. The hydroxide ion concentration equals:
  a. $10^{-4}M$
  10
  b. $10^{-11}M$
  3
  c. $10^{-8}M$
  6

Chapter 15

Write overall, ionic, and net ionic equations for each reaction.

1. hydroiodic acid (HI) and calcium hydroxide ($Ca(OH)_2$)

$2HI(aq) + Ca(OH)_2(aq) \rightarrow CaI_2(aq) + 2H_2O(l)$
$2H^+(aq) + 2I^-(aq) + Ca^{2+}(aq) + 2OH^-(aq) \rightarrow$
$\qquad Ca^{2+}(aq) + 2I^-(aq) + 2H_2O(l)$
$H^+(aq) + OH^-(aq) \rightarrow H_2O(l)$

Solutions to Problems

2. hydrobromic acid (HBr) and lithium hydroxide (LiOH)

   $HBr(aq) + LiOH(aq) \rightarrow LiBr(aq) + H_2O(l)$
   $H^+(aq) + Br^-(aq) + Li^+(aq) + OH^-(aq) \rightarrow$
   $\qquad Li^+(aq) + Br^-(aq) + H_2O(l)$
   $H^+(aq) + OH^-(aq) \rightarrow H_2O(l)$

3. sulfuric acid (H_2SO_4) and strontium hydroxide ($Sr(OH)_2$)

   $H_2SO_4(aq) + Sr(OH)_2(aq) \rightarrow SrSO_4(aq) + 2H_2O(l)$
   $2H^+(aq) + SO_4^{2-}(aq) + Sr^{2+}(aq) + 2OH^-(aq) \rightarrow$
   $\qquad Sr^{2+}(aq) + SO_4^{2-}(aq) + 2H_2O(l)$
   $H^+(aq) + OH^-(aq) \rightarrow H_2O(l)$

4. perchloric acid ($HClO_4$) and barium hydroxide ($Ba(OH)_2$)

   $2HClO_4(aq) + Ba(OH)_2(aq) \rightarrow$
   $\qquad 2H_2O(l) + Ba(ClO_4)_2(aq)$
   $2H^+(aq) + 2ClO_4^-(aq) + Ba^{2+}(aq) + 2OH^-(aq) \rightarrow$
   $\qquad 2H_2O(l) + Ba^{2+}(aq) + 2ClO_4^-(aq)$
   $H^+(aq) + OH^-(aq) \rightarrow H_2O(l)$

5. perchloric acid ($HClO_4$) and ammonia (NH_3)

   $HClO_4(aq) + NH_3(aq) \rightarrow NH_4ClO_4(aq)$
   $H^+(aq) + ClO_4^-(aq) + NH_3(aq) \rightarrow$
   $\qquad NH_4^+(aq) + ClO_4^-(aq)$
   $H^+(aq) + NH_3(aq) \rightarrow NH_4^+(aq)$

6. hydrochloric acid (HCl) and aluminum hydroxide ($Al(OH)_3$)

   $3HCl(aq) + Al(OH)_3(aq) \rightarrow AlCl_3(aq) + 3H_2O(l)$
   $3H^+(aq) + 3Cl^-(aq) + Al(OH)_3(aq) \rightarrow$
   $\qquad Al^{3+}(aq) + 3Cl^-(aq) + 3H_2O(l)$
   $3H^+(aq) + Al(OH)_3(aq) \rightarrow Al^{3+}(aq) + 3H_2O(l)$

7. sulfuric acid (H_2SO_4) and iron(III) hydroxide ($Fe(OH)_3$)

   $3H_2SO_4(aq) + 2Fe(OH)_3(aq) \rightarrow$
   $\qquad Fe_2(SO_4)_3(aq) + 6H_2O(l)$
   $6H^+(aq) + 3SO_4^{2-}(aq) + 2Fe(OH)_3(aq) \rightarrow$
   $\qquad 2Fe^{3+}(aq) + 3SO_4^{2-}(aq) + 6H_2O(l)$
   $3H^+(aq) + Fe(OH)_3(aq) \rightarrow Fe^{3+}(aq) + 3H_2O(l)$

8. carbonic acid (H_2CO_3) and sodium hydroxide (NaOH)

   $H_2CO_3(aq) + 2NaOH(aq) \rightarrow$
   $\qquad Na_2CO_3(aq) + 2H_2O(l)$
   $H_2CO_3(aq) + 2Na^+(aq) + 2OH^-(aq) \rightarrow$
   $\qquad 2Na^+(aq) + CO_3^{2-}(aq) + 2H_2O(l)$
   $H_2CO_3(aq) + 2OH^-(aq) \rightarrow CO_3^{2-}(aq) + 2H_2O(l)$

9. boric acid (H_3BO_3) and potassium hydroxide (KOH)

   $H_3BO_3(aq) + 3KOH(aq) \rightarrow K_3BO_3(aq) + 3H_2O(l)$
   $H_3BO_3(aq) + 3K^+(aq) + 3OH^-(aq) \rightarrow$
   $\qquad 3K^+(aq) + BO_3^{3-}(aq) + 3H_2O(l)$
   $H_3BO_3(aq) + 3OH^-(aq) \rightarrow BO_3^{3-}(aq) + 3H_2O(l)$

10. acetic acid, $HC_2H_3O_2$, and calcium hydroxide $Ca(OH)_2$

    $2HC_2H_3O_2(aq) + Ca(OH)_2(aq) \rightarrow$
    $\qquad Ca(C_2H_3O_2)_2(aq) + 2H_2O(l)$
    $2HC_2H_3O_2(aq) + Ca^{2+}(aq) + 2OH^-(aq) \rightarrow$
    $\qquad Ca^{2+}(aq) + 2C_2H_3O_2^-(aq) + 2H_2O(l)$
    $HC_2H_3O_2(aq) + OH^-(aq) \rightarrow$
    $\qquad C_2H_3O_2^-(aq) + H_2O(l)$

16. A 0.100M LiOH solution was used to titrate an HBr solution of unknown concentration. At the endpoint, 21.0 mL of LiOH solution had neutralized 10.0 mL of HBr. What is the molarity of the HBr solution?

    0.210M;

    $\left(\dfrac{21.0 \text{ mL LiOH}}{1}\right) \times \left(\dfrac{1 \text{ L}}{10^3 \text{ mL LiOH}}\right) \times$
    $\left(\dfrac{0.100 \text{ mol LiOH}}{1 \text{ L}}\right) \times \left(\dfrac{1 \text{ mol HBr}}{1 \text{ mol LiOH}}\right) \times$
    $\left(\dfrac{1}{10.0 \text{ mL HBr}}\right) \times \left(\dfrac{10^3 \text{ mL HBr}}{1 \text{ L}}\right)$
    $= 0.210M$ HBr

12. A 0.150M KOH solution fills a burette to the 0 mark. The solution was used to titrate 25.0 mL of an HNO_3 solution of unknown concentration. At the endpoint, the burette reading was 34.6 mL. What was the molarity of the HNO_3 solution?

    0.208M;

    $\left(\dfrac{34.6 \text{ mL KOH}}{1}\right) \times \left(\dfrac{1 \text{ L}}{10^3 \text{ mL KOH}}\right) \times$
    $\left(\dfrac{0.150 \text{ mol KOH}}{1 \text{ L}}\right) \times \left(\dfrac{1 \text{ mol HNO}_3}{1 \text{ mol KOH}}\right) \times$
    $\left(\dfrac{1}{25.0 \text{ mL HNO}_3}\right) \times \left(\dfrac{10^3 \text{ mL HNO}_3}{1 \text{ L}}\right)$
    $= 0.208M$ HNO_3

Solutions to Problems

18. A Ca(OH)$_2$ solution of unknown concentration was used to titrate 15.0 mL of a 0.125M H$_3$PO$_4$ solution. If 12.4 mL of Ca(OH)$_2$ are used to reach the endpoint, what is the concentration of the Ca(OH)$_2$ solution?

0.227M;

$$\left(\frac{15.0 \text{ mL H}_3\text{PO}_4}{1}\right) \times \left(\frac{1 \text{ L}}{10^3 \text{ mL H}_3\text{PO}_4}\right) \times \left(\frac{0.125 \text{ mol H}_3\text{PO}_4}{1 \text{ L}}\right) \times$$

$$\left(\frac{3 \text{ mol Ca(OH)}_2}{2 \text{ mol H}_3\text{PO}_4}\right) \times \left(\frac{1}{12.4 \text{ mL}}\right) \times \left(\frac{10^3 \text{ mL}}{1 \text{ L}}\right)$$

$= 0.227M$ Ca(OH)$_2$

Chapter 18

1. Name the following structures.

a. CH$_3$ CH$_3$
    | |
    CH$_3$CHCH$_2$CHCH$_2$CH$_3$

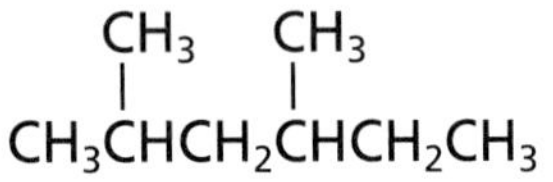

2, 4-dimethylhexane

b.   CH$_3$ CH$_3$
     | |
     CH$_3$CCH$_2$CHCH$_3$
     |
     CH$_3$

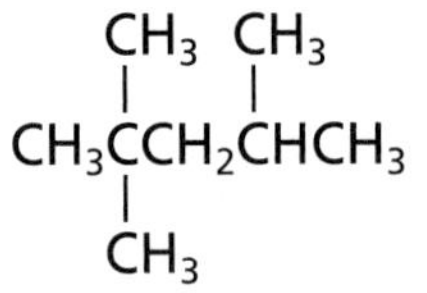

2, 2, 4-trimethylpentane

c.   CH$_3$
     |
     CH$_2$ CH$_3$ CH$_3$
     | | |
     CH$_3$CHCH$_2$CH$_2$CHCH$_2$CHCH$_3$

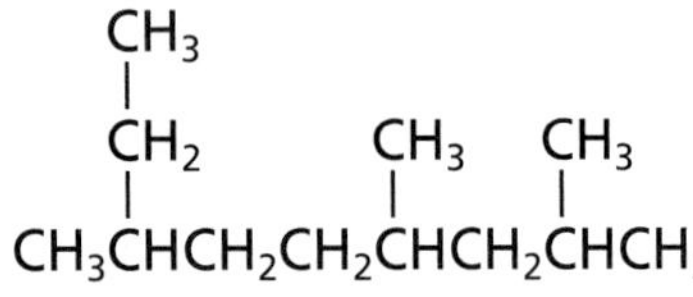

2, 4, 7-trimethylnonane

2. Draw the structure of the following branched chain alkanes.

a. 2, 3-dimethyl-5-propyldecane

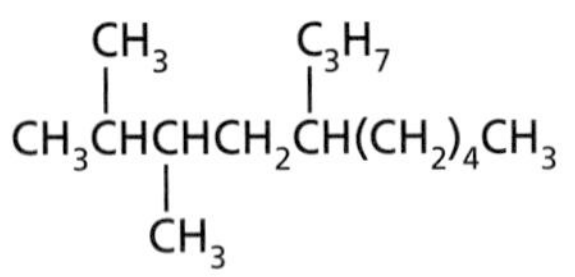

   CH$_3$ C$_3$H$_7$
   | |
   CH$_3$CHCHCH$_2$CH(CH$_2$)$_4$CH$_3$
        |
        CH$_3$

b. 3, 4, 5-triethyloctane

   C$_2$H$_5$ C$_2$H$_5$
   | |
   CH$_3$CH$_2$CHCHCHCH$_2$CH$_2$CH$_3$
             |
             C$_2$H$_5$

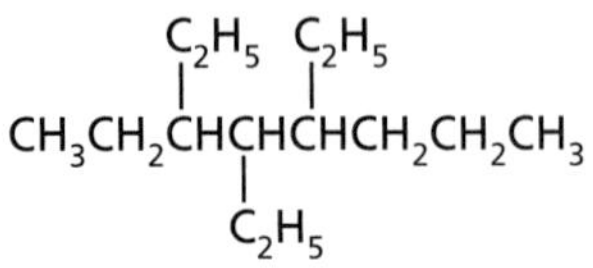

Chapter 20

5. How much heat is absorbed by a reaction that lowers the temperature of 500.0 g of water in a calorimeter by 1.10°C?

$q_w = (m)(\Delta T)(C_w)$
$= (500.0 \text{ g})(1.10°C)(4.184 \text{ J/g·°C})$
$= 2300 \text{ J}$
$= 2.30 \text{ kJ}$

6. Aluminum reacts with iron(III) oxide to yield aluminum oxide and iron. Calculate the heat given off in the reaction if the temperature of the 1.00 kg of water in the calorimeter increases by 3.00°C.

$q_w = (m)(\Delta T)(C_w)$
$= (1.00 \times 10^3 \text{ g})(3.00°C)(4.184 \text{ J/g·°C})$
$= 12,600 \text{ J}$
$= 12.6 \text{ kJ}$

7. Burning 1.00 g of a certain fuel in a calorimeter raises the temperature of 1.000 kg of water from 20.00°C to 28.05°C. Calculate the heat given off in this reaction. How much heat would one mol of the fuel give off, assuming a molar mass of 65.8 g/mol?

$\Delta T = 28.05°C - 20.00°C = 8.05°C$

For 1 g:

$q_w = (m)(\Delta T)(C_w)$
$= (1.000 \times 10^3 \text{ g})(8.05°C)(4.184/\text{g·°C})$
$= 33,700 \text{ J}$
$= 33.7 \text{ kJ}$

For 1 mol:

$q_w = \left(\frac{33.7 \text{ kJ}}{1 \text{ g}}\right)\left(\frac{65.8 \text{ g}}{1 \text{ mol}}\right)$
$= 2220 \text{ kJ/mol}$

8. A group of students decides to measure the energy content of certain foods. They heat 50.0 g of water in an aluminum can by burning a sample of the food beneath the can. When they use 1.00 g of popcorn as their test food, the temperature of the water rises by 24°C. Calculate the heat released by the popcorn, and express your answer in both kilojoules and Calories per gram of popcorn.

$$q_w = (m)(\Delta T)(C_w)$$
$$= (50.0 \text{ g})(24°C)(4.184 \text{ J/g}\cdot°C) \text{ for 1 g popcorn}$$
$$= 5000 \text{ J/g popcorn}$$
$$= 5.0 \text{ kJ/g popcorn}$$

$$= \left(\frac{5.0 \text{ kJ}}{1 \text{ g popcorn}}\right)\left(\frac{1 \text{ Cal}}{4.184 \text{ kJ}}\right) = 1.2 \text{ Cal/g popcorn}$$

9. Another student comes along and tells the group in problem 4 that she has read the label on a popcorn bag that states that 30 g of popcorn yields 110 Calories. What is that value in Calories/gram? How can you account for the difference?

$$\frac{110 \text{ Cal}}{30.0 \text{ g popcorn}} = 3.7 \text{ Cal/g popcorn}$$

1.2 Cal/g, the experimental value, is much lower than the value given on the popcorn bag because of loss of heat to the air in the student experiment. Also, the combustion of popcorn in the student experiment may be incomplete.

10. A 3.00-g sample of a new snack food is burned in a calorimeter. The 2.00 kg of surrounding water change in temperature from 25.0°C to 32.4°C. What is the food value in Calories per gram?

$$\Delta T = 32.4°C - 25.0°C = 7.4°C$$
$$q_w = (m)(\Delta T)(C_w)$$
$$= (2.00 \times 10^3 \text{ g})(7.4°C)(4.184 \text{ J/g}\cdot°C)$$
$$= 62,000 \text{ J} = 62 \text{ kJ}$$

$$\text{Calories per gram} = \left(\frac{62 \text{ kJ}}{3.0 \text{ g}}\right)\left(\frac{1 \text{ Cal}}{4.184 \text{ kJ}}\right) = 4.9 \text{ Cal/g}$$

Chapter 21

1. Write the balanced nuclear equation for the radioactive decay of radium-226 to give radon-222, and determine the type of decay.

$$^{226}_{88}\text{Ra} \rightarrow {}^{222}_{86}\text{Rn} + {}^{4}_{2}\text{He}; \text{ alpha decay}$$

2. Write a balanced equation for the nuclear reaction in which neon-23 decays to form sodium-23, and determine the type of decay.

$$^{23}_{10}\text{Ne} \rightarrow {}^{23}_{11}\text{Na} + {}^{0}_{-1}\text{e}; \text{ beta decay}$$

3. A rock was analyzed using potassium-40. The half-life of potassium-40 is 1.25 billion years. If the rock had only 25 percent of the potassium-40 that would be found in a similar rock formed today, calculate how long ago the rock was formed.

Because (0.5)(0.5) = 0.25, two half-lives have gone by.
2 half-lives × 1.25 billion years/half-life = 2.50 billion years

4. Ash from an early fire pit was found to have 12.5 percent as much carbon-14 as would be found in a similar sample of ash today. How long ago was the ash formed?

Because (0.5)(0.5)(0.5) = 0.125, three half-lives have gone by.
3 half-lives × 5730 years/half-life = 17,200 years

Try at Home Labs

TRY AT HOME LAB Comparing Frozen Liquids

Real-World Question How do different kitchen liquids react when placed in a freezer?

Possible Materials
five identical, narrow-necked plastic bottles or photographic film canisters
large cutting board or cookie sheet

- water
- orange juice
- vinegar
- soft drink
- cooking oil
- freezer

Procedure
1. Obtain permission to use the freezer before beginning this activity.
2. Fill one of the containers with water. The water should come to the top brim of the container.
3. Fill the other four containers in the same manner with the other four liquids.
4. Place the cutting board or cookie sheet in a freezer so that it is level and place the five containers on the board or sheet.
5. Leave the containers in the freezer overnight and observe the effect of the freezer's temperature on each liquid the following day.

Conclude and Apply
1. Describe the effect of the colder temperature on each liquid.
2. Infer why the water behaved as it did.
3. Infer why some liquids froze but others did not.

TRY AT HOME LAB Comparing Atom Sizes

Real-World Question How do the sizes of different atoms and subatomic particles compare?

Possible Materials
- metric ruler
- meterstick
- white sheet of paper
- fine-tipped black marker
- masking tape or transparent tape
- three plastic milk containers

Procedure
1. Using a black marker, draw a 0.1-mm-wide dot on one end of a white sheet of paper. This dot represents the diameter of an electron.
2. Measure a distance of 10 cm from the dot and draw a second dot. The distance between the two dots represents the diameter of a proton or a neutron.
3. Securely tape the paper to the top of a plastic milk container.
4. Measure a distance of 6.2 m from the milk container and place a second milk container. This distance represents the diameter of the smallest atom, a helium atom.
5. Measure a distance of 59.6 m from the first milk container and place a third milk container. This distance represents the diameter of a larger atom, a cesium atom.

Conclude and Apply
1. Considering the comparative sizes of protons and neutrons with the sizes of atoms, and infer what makes up most of an atom.

Try at Home Labs

TRY AT HOME LAB: Element Hunt

Real-World Question How many elements can be found in your home?

Possible Materials
periodic table of the elements
chemistry books or resources

Procedure
1. Create a chart to record the elements that are metals, metalloids, and nonmetals.
2. Use your chemistry textbook and resources to research common uses for elements.
3. Search your home for items made entirely or primarily of metal elements. List the items in your chart.
4. Search your home for items made primarily or entirely of metalloids and nonmetals. List the items in your data chart.

Conclude and Apply
1. List the two most common nonmetals found in your home and identify their locations.
2. List the most common metals found in your home.
3. Infer what group of metals is most commonly found in a home.

TRY AT HOME LAB: Mixing Ionic and Covalent Liquids

Real-World Question Do common ionic and covalent liquids found in a kitchen mix together?

Possible Materials
clear-glass container
water
rubbing alcohol
soft drink
vegetable oil
spoon or stirring rod
measuring cup with SI units
dish soap

Procedure
1. Create a data table to record your observations.
2. Measure 100 mL of water and pour it into the container. Measure 100 mL of rubbing alcohol, pour it into the container, and stir the mixture. Observe and record what happens to the two liquids.
3. Thoroughly wash and rinse out your container using dish soap.
4. Repeat steps 2 and 3 using a soft drink and vegetable oil.
5. Measure 100 mL of vegetable oil and pour it into the container. Measure 100 mL of rubbing alcohol, pour it into the container, and stir the mixture. Observe and record what happens to the two liquids.
6. Thoroughly wash and rinse out your container using dish soap.
7. Repeat steps 5 and 6 using a soft drink and rubbing alcohol.

Conclude and Apply
1. Research what type of compound water is.
2. Infer why the liquids mixed or did not mix the way they did.

Try at Home Labs

TRY AT HOME LAB Iron Ink

Real-World Question How can common household items be used to make nineteenth century ink?

Possible Materials
ceramic mug
tea bag
iron sulfate tablets
coffee filter
measuring cup
paintbrush
glass
water
funnel
oven mitt
white paper
microwave oven

Procedure
1. Pour 30 mL of water into the mug.
2. Lay a tea bag in the water and drop five iron sulfate tablets onto the bag.
3. Microwave the mug for 2 min. Remove the tea bag from the mug.
4. Place a coffee filter into a funnel, insert the funnel into a glass, and pour the mixture from the mug into the filter.
5. Place the glass of filtered solution in the refrigerator for 15 min.
6. Remove the funnel and use a paintbrush to write letters with the ink on a white sheet of paper. Set the paper aside to dry.

Conclude and Apply
1. Identify the molecular formula of the compound in the iron tablets that creates the dark color of the ink.
2. Infer the type of compounds involved in this chemical reaction.
3. Research the precipitate that forms and gives the ink its black color.

TRY AT HOME LAB Preventing a Chemical Reaction

Real-World Question How can the chemical reaction that turns apples brown be prevented?

Possible Materials
Seven identical glasses
SI measuring cup
100-mg vitamin C tablets
rolling pin
apple
masking tape
bottled water
wax paper
paper towels
kitchen knife
permanent black marker

Procedure
1. Pour 200 mL of water into each glass.
2. Label glass 1 *no vitamin C*, glass 2 *100 mg*, glass 3 *200 mg*, glass 4 *500 mg*, glass 5 *1000 mg*, glass 6 *2000 mg*, and glass 7 *3000 mg*.
3. Place a 100-mg vitamin C tablet between two sheets of wax paper and use a rolling pin to grind the tablet into powder.
4. Place the powder into glass 2 and stir the mixture vigorously.
5. Repeat steps 3 and 4 for glasses 3–7 using the appropriate mass of vitamin C.
6. Cut seven equal-sized wedges of apple and immediately place one into each glass.
7. After 5 min, lay the wedges on paper towels in front of the glasses in which they were soaking. Observe the apples every 5 min for 45 min.

Conclude and Apply
1. Describe the results of your experiment.
2. Infer why vitamin C prevents apples from turning brown.

Try at Home Labs

TRY AT HOME 🏠 LAB Comparing Orbital Sizes

Real-World Question How do the sizes of different electron orbitals compare?

Possible Materials
- metric ruler or meterstick
- 3-m-long piece of white paper (or tape several smaller pieces together)
- markers
- masking tape

Procedure
1. Lay the white paper out on a flat surface.
2. Use a marker to draw a thin, 3-cm-long line at one end of the paper.
3. Measure a 1-cm distance and draw a second line to represent the radius of an atom's nucleus.
4. From the first line, measure a distance of 37 cm to represent the radius of the first electron orbital, 156 cm to represent the radius of the second orbital, 186 cm for the third orbital, 227 cm for the fourth orbital, 248 cm to represent the radius of the fifth orbital, and 265 cm for the radius of the sixth orbital. Mark each length with a new line.
5. On your paper, label the radii with the appropriate orbital number.

Conclude and Apply
1. The average radius of an atom's nucleus is 1×10^{-12} m. Calculate the scale you used for your drawing.
2. Infer from your drawing what comprises most of an atom.

TRY AT HOME 🏠 LAB The Solubility of Iodine

Real-World Question How does the solubility of iodine differ in different liquids?

Possible Materials
- glass jar with lid (baby food jars work well)
- dropper
- mineral oil
- tincture of iodine
- water
- measuring cup
- tablespoon

Procedure
1. Half-fill a glass jar with water.
2. Add 4 drops of iodine to the water.
3. Tighten the lid and shake the jar vigorously until the solution turns a light brown color.
4. Measure 15 mL of mineral oil and pour it into the jar.
5. Shake the jar vigorously again and allow the mixture to sit for 5 min. Observe the mixture.

Conclude and Apply
1. Describe the mixture in the jar after you added the mineral oil and allowed the jar's contents to sit for 5 min.
2. Infer why the iodine behaved the way it did when the mineral oil was added.
3. Infer other elements or compounds that might behave similarly to the iodine. Explain your answer.

Try at Home Labs

TRY AT HOME LAB Breaking Covalent Bonds

Real-World Question What liquids will break the covalent bonds of polystyrene?

Possible Materials
- polystyrene packing peanuts or polystyrene cups
- large glass
- measuring cup
- nail polish remover
- shallow dish
- rubbing alcohol
- water
- cooking oil

Procedure
1. Pour 200 mL of water into a glass.
2. Drop a polystyrene packing peanut into the water and observe how the polystyrene and water react.
3. Thoroughly wash out the glass and repeat steps 1 and 2 using cooking oil and nail polish remover.
4. Drop several peanuts into any of the liquids that cause a chemical reaction with the polystyrene peanuts and observe what happens to them.

Conclude and Apply
1. Describe the reaction between the polystyrene peanuts and each of the four liquids.
2. Infer why the polystyrene reacted as it did with each of the liquids.

TRY AT HOME LAB Estimating Metric Temperatures

Real-World Question How can Americans estimate daily metric temperatures?

Possible Materials
- thermometer with only Fahrenheit temperature scale units (try not to use a mercury-filled thermometer)
- calculator

Procedure
1. Create a data table to record your temperature observations and calculations.
2. Place a Fahrenheit scale thermometer outside your home in a location where you can easily view it from a window or door.
3. Observe the air temperature in degrees Fahrenheit each day for two weeks. Observe the temperature at different times of the day.
4. After reading the temperature in degrees Fahrenheit each day, estimate the temperature in degrees Celsius and Kelvins.
5. Calculate the actual temperature in degrees Celsius.
6. Calculate the actual temperature in Kelvins.

Conclude and Apply
1. Compare how well you were able to estimate metric temperatures at the start of the activity with your ability at the end of the lab.
2. Infer why many Americans cannot estimate temperatures in degrees Celsius.

Try at Home Labs

TRY AT HOME 🏠 LAB Crushing Cans

Real-World Question How can an aluminum soft-drink can be used to demonstrate the gas laws?

Possible Materials
- aluminum soft-drink can (355 mL)
- stovetop or hot plate
- pan
- oven mitt
- large tongs
- water
- large, plastic container
- large bowl
- measuring cup with SI units
- watch with second hand or stopwatch
- ice

Procedure

1. Fill a pan with water, place the pan on a stovetop, and bring the water to a boil.
2. Fill a large, plastic container with cold water and ice.
3. Measure 25 mL of water and pour it into an aluminum soft-drink can.
4. Using an oven mitt, hold the can right side up in the boiling water for one minute.
5. Quickly invert the aluminum can and submerge it beneath the cold water in the container. Observe the result.

Conclude and Apply

1. Describe what happened to the can when you submerged it beneath the cold water.
2. Explain why the can behaved the way it did using both Charles's Law and Boyle's Law in your explanation.
3. Compare the amount of kinetic energy in the air molecules while the can was held in the boiling water to when the can was submerged in the cold water.

TRY AT HOME 🏠 LAB Measuring Moles of Sugar

Real-World Question How can you measure and calculate the number of moles in a sample of sugar?

Possible Materials
- sugar
- kitchen scale
- measuring cup
- small bowl
- calculator

Procedure

1. Measure the mass of a small bowl.
2. Measure approximately 100 mL of sugar and pour the sugar into the bowl.
3. Measure the mass of the sugar and bowl and calculate the mass of the sugar sample.

Conclude and Apply

1. Research the chemical formula of sucrose (table sugar).
2. Calculate the molar mass of sucrose.
3. Calculate the number of moles in your sugar sample.

Try at Home Labs

TRY AT HOME LAB Measuring Capillarity

Real-World Question How does the height of a meniscus differ in containers of different widths and shapes?

Possible Materials
- 30-mL medicine cup
- 750-mL bottle
- narrow-necked bottle
- 1-mL, 2-mL, and 5-mL droppers
- glass
- bowl
- water
- metric ruler

Procedure
1. Create a data table to record your data.
2. Measure the diameter of all the containers. Measure the width of the bottle's neck.
3. Suction up water into a 5-mL dropper until it is half-full.
4. Hold the dropper level and measure the height of the meniscus formed by the water.
5. Repeat steps 3 and 4 using 2-mL and 1-mL droppers.
6. Measure the heights of the meniscuses formed by water in a 30-mL medicine cup and the neck of a 750-mL bottle.

Conclude and Apply
1. Explain the relationship between container width and meniscus height.
2. Infer other factors, aside from a container's width, that will determine the height of a meniscus.
3. Infer the maximum width of a container in which water will still form a measurable meniscus.

TRY AT HOME LAB Testing for Acid Rain

Real-World Question Does your area have acid rain?

Possible Materials
- aquarium pH test kit
- small test tubes (from aquarium kit)
- clean plastic containers

Procedure
1. Create a data table to record your data.
2. Place a widemouth container outside in a location where it will capture rain. Set the container away from trees or other objects that might drop debris into the container.
3. Use an aquarium pH test kit to test the pH of your rainwater sample.
4. Collect water samples from streams or rivers in your area and test the pH.

Conclude and Apply
1. Research the pH value of normal rainwater.
2. Research the pH value of acid rain.
3. Test the pH of bottled water and compare the value to the pH of your water samples.
4. Infer whether or not your region has an acid rain problem.

Try at Home Labs

TRY AT HOME LAB: Testing for Ammonia

Real-World Question What substances elevate ammonia levels in natural waterways?

Possible Materials

- four glass jars with lids
- measuring cup
- raw chicken (6 ounces)
- water
- ammonia test kit
- kitchen scale
- kitchen knife
- masking tape
- black marker

Procedure

1. Use a kitchen scale to measure 28-g (1 ounce), 56-g (2-ounce), and 84-g (3-ounce) pieces of raw chicken.
2. Fill four jars with 500 mL of water.
3. Add nothing to the first jar. Place 28 g of raw chicken into the second jar, 56 g into the third jar, and 84 g into the fourth jar. Label your jars.
4. Create a data table to record your data.
5. Measure the amount of ammonia in each water sample each day for five days. Also observe the clarity of each sample.

Conclude and Apply

1. Identify possible procedural errors in your experiment.
2. Summarize your experiment results.
3. Infer the common cause for elevated ammonia levels in natural waterways.

TRY AT HOME LAB: Testing the Oxidation Power of Bleach

Real-World Question How well does bleach oxidize the compounds in different stains?

Possible Materials

- cotton swabs
- bleach
- mustard
- ketchup
- grass
- grape juice
- cranberry juice
- dropper or straw
- grape jam
- access to clothes dryer
- white cotton cloth
- rubber gloves
- bowl
- barbeque brush
- permanent black marker
- watch with second hand or stopwatch

Procedure

1. Take a peanut-sized dab of ketchup and place it on one end of a white cloth. Brush the ketchup into the cloth to form a stain that is about the size of a quarter and label the stain with a permanent marker.
2. Thoroughly wash out the brush and repeat step 1 using mustard and grape jam. Place the stains side by side.
3. Use a dropper to make quarter-sized stains of grape and cranberry juices next to the other three stains. Label each stain.
4. Place the cloth in a dryer until the stains dry.
5. Wearing rubber gloves, carefully pour bleach into a bowl.
6. Dip a cotton swab into the bleach and press the swab tip on the ketchup stain for 15 s.
7. Repeat step 6 for the other stains.

Conclude and Apply

1. Explain how bleach removes stains.
2. Describe the effectiveness of the bleach at oxidizing the compounds in the various stains you tested.
3. Infer why the bleach oxidized some stains better than others.

Try at Home Labs

TRY AT HOME LAB Removing Electroplating

Real-World Question How can the electroplating on aluminum pans be removed?

Possible Materials
- aluminum pan turned dark on the bottom
- water
- lemon
- kitchen knife
- measuring cup
- stovetop burner or hotplate

Procedure
1. Measure 500 mL of water and pour it into the pan.
2. Place the pan on a stovetop burner and bring the water to a boil.
3. Slice a lemon into several wedges and drop the wedges into the pan of boiling water.
4. Observe the chemical reaction that occurs at the bottom of the pan.

Conclude and Apply
1. Describe the results of your experiment.
2. Infer why the bottom of the pan changed in appearance.

TRY AT HOME LAB Comparing Water and Alcohol

Real-World Question How do the properties of water and an alcohol compare?

Possible Materials
- water
- stirring rod
- rubbing alcohol
- spoon
- vegetable oil
- kitchen scale
- three measuring cups with SI units
- stopwatch or watch with second hand
- three clear glasses
- two ice cubes

Procedure
1. Create a data table for comparing the physical properties of water and rubbing alcohol.
2. Compare the color and odor of each liquid.
3. Measure the masses of equal volumes of each liquid and calculate the density of each. Drop an ice cube into each glass and observe any differences in densities of the liquids.
4. Mix 20 mL of water with oil and stir the mixture. Do the same with the alcohol and oil.

Conclude and Apply
1. Describe what happened when an ice cube was dropped into each liquid.
2. Infer why the ice cube behaved as it did in each liquid.
3. Infer what would happen if water and the alcohol were mixed together.
4. Summarize your comparisons of the two liquids.

Try at Home Labs

TRY AT HOME LAB — Counting Nutrients

Real-World Question What percentage of the daily allowance of fats, carbohydrates, and proteins do you consume each day?

Possible Materials
- nutrition facts chart
- packages of foods and drinks consumed during a week
- nutrition guide book
- kitchen scale

Procedure
1. Create a data table to record the mass and percentage of the daily allowance of fats, carbohydrates, and proteins that you consume each day for a week.
2. After each meal or snack, check the nutrition facts charts on your food packaging or a nutrition guide to count up the number of grams of fat, carbohydrates, and proteins you consumed. Be certain you consider how many servings of each food or drink you consumed.
3. Count up the percentage of the daily allowance of fats, carbohydrates, and proteins that you consume each day.
4. Count the grams of fat, carbohydrates, and proteins you consume each day for a week and calculate a daily average for each nutrient.
5. Calculate a daily average for the percentage of the daily allowance of fats, carbohydrates, and proteins you consume.

Conclude and Apply
1. Compare your fat, carbohydrate, and protein intake with the amounts recommended on the bottom table of a nutrition facts chart.
2. Infer if this lab can determine whether or not a person has a healthy diet.
3. Infer possible dietary changes you may want to consider based upon the results of this lab.

TRY AT HOME LAB — Observing Entropy

Real-World Question How quickly do common household liquids enter into a state of entropy?

Possible Materials
- seven identical glass containers
- stopwatch or watch with second hand
- water
- corn syrup
- rubbing alcohol
- measuring cup
- clear soft drink
- vinegar
- milk
- cooking oil
- food coloring

Procedure
1. Fill identical glass containers with equal volumes of the seven different liquids. Label each container.
2. Create a data table to record your observations and measurements.
3. Quickly place one drop of food coloring into each container while a partner simultaneously starts a stopwatch.
4. Observe how the food coloring behaves in each liquid. Time how quickly the coloring and each liquid reach total entropy.

Conclude and Apply
1. Infer the relationship between the rate at which a liquid and the dye achieves a state of total entropy and the time measurement from your data table.
2. Infer why the entropy rates for the different liquids varied.

Try at Home Labs

TRY AT HOME LAB Modeling Fusion

Real-World Question How can fusion reactions be modeled?

Possible Materials
- 14 red gumdrops
- 12 green gumdrops
- toothpicks
- white paper (9 sheets)
- black marker

Procedure

1. Draw a large black arrow on 3 separate sheets of white paper. Draw a large plus sign on 6 separate sheets of paper.
2. While making your models in the steps below, use red gumdrops to represent protons and green gumdrops to represent neutrons.
3. Construct a deuterium atom, tritium atom, and helium atom. Arrange your three models and a single neutron into a common fusion reaction.
4. Construct two deuterium atoms and a tritium atom. Arrange your models and a single proton into a second fusion reaction.
5. Construct two deuterium atoms and a helium atom. Arrange your three models and a single neutron into a third fusion reaction.

Conclude and Apply

1. Describe what all three fusion reactions have in common.
2. Research the natural abundance of H-1 atoms and deuterium.
3. Infer from your models why nuclear fusion reactions create no radioactive waste.

Glossary/Glosario

A multilingual science glossary at glencoe.com includes Arabic, Bengali, chinese, English, Haitian Creole, Hmong, Korean, Portuguese, Russian, Tagalog, Urdu, and Vietnamese.

Pronunciation Key

Use the following key to help you sound out words in the glossary.

a	back (BAK)	ew	food (FEWD)
ay	day (DAY)	yoo	pure (PYOOR)
ah	father (FAH thur)	yew	few (FYEW)
ow	flower (FLOW ur)	uh	comma (CAHM uh)
ar	car (CAR)	u (+con)	rub (RUB)
e	less (LES)	sh	shelf (SHELF)
ee	leaf (LEEF)	ch	nature (NAY chur)
ih	trip (TRIHP)	g	gift (GIHFT)
i (i+con+e)	idea, life (i DEE uh, life)	j	gem (JEM)
oh	go (GOH)	ing	sing (SING)
aw	soft (SAWFT)	zh	vision (VIHZH un)
or	orbit (OR but)	k	cake (KAYK)
oy	coin (COYN)	s	seed, cent (SEED, SENT)
oo	foot (FOOT)	z	zone, raise (ZOHN, RAYZ)

Como usar el glosario en espanol:
1. Busca el termino en ingles que desees encontrar.
2. El termino en espanol, junto con la definicion, se encuentran en la columna de la derecha.

A

English	Español
absolute zero: zero on the Kelvin scale, which represents the lowest possible theoretical temperature; atoms are all in the lowest possible energy state. (Chap. 10, p. 347)	**absolute zero / cero absoluto:** equívale a cero grados en la escala de Kelvin y representa la temperatura teórica más fría posible; a esta temperatura todos los átomos se encuentran en el menor estado energético posible. (Cap. 10, pág. 347)
acid: a substance that produces hydronium ions when dissolved in water. (Chap. 14, p. 483)	**acid / ácido:** sustancia que produce iones de hidronio cuando se disuelve en agua. (Cap. 14, pág. 483)
acidic anhydride: a nonmetal oxide that reacts with water to form an acid. (Chap. 14, p. 493)	**acidic anhydride / anhídrido acídico:** óxido no metálico que reacciona con agua para formar un ácido. (Cap. 14, pág. 493)
acidic hydrogen: in an acid, any hydrogen that can be transferred to water. (Chap. 14, p. 484)	**acidic hydrogen / hidrógeno acídico:** en un ácido, es cualquier hidrógeno que puede transferirse al agua. (Cap. 14, pág. 484)
actinide: any of the second series of inner transition elements with atomic numbers from 90 to 103; all are radioactive. (Chap. 3, p. 102)	**actinide / actínido:** cualquiera de la segunda serie de elementos de transición interna con números atómicos de 90 a 103; todos son radioactivos. (Cap. 3, pág. 102)
activation energy: the amount of energy the particles in a reaction must have when they collide for the reaction to occur. (Chap. 6, p. 216)	**activation energy / energía de activación:** cantidad de energía que deben tener las partículas cuando chocan en una reacción para que ésta ocurra. (Cap. 6, pág. 216)
active site: on an enzyme, the pocket or groove that can bind a substrate taking part in a reaction. (Chap. 19, p. 672)	**active site / sitio activo:** en una enzima, es el abolsamiento o la ranura que puede unirse a un sustrato que participa en una reacción. (Cap. 19, pág. 672)

Glossary/Glosario

addition reaction: a reaction where monomers that contain double bonds add onto each other to form long chains; the product contains all the atoms of the starting monomers. (Chap. 18, p. 654)

aerobic: a metabolic process that takes place only in the presence of oxygen. (Chap. 19, p. 690)

alkali metal: any element from Group 1: lithium, sodium, potassium, rubidium, cesium, francium. (Chap. 8, p. 261)

alkaline earth metal: any element from Group 2: beryllium, magnesium, calcium, strontium, barium. (Chap. 8, p. 264)

alkane: a saturated hydrocarbon that consists of only carbon and hydrogen atoms with single bonds between all the atoms. (Chap. 18, p. 623)

alkene: a hydrocarbon in which one or more double bonds link carbon atoms. (Chap. 18, p. 630)

alkyne: an unsaturated hydrocarbon that contains a triple bond between two carbon atoms. (Chap. 18, p. 633)

allotrope: any of two or more molecules of a single element that have different crystalline or molecular structures. (Chap. 5, p. 173)

alloy: a solid solution containing different metals, and sometimes nonmetallic substances. (Chap. 1, p. 23)

alpha particle: a helium nucleus consisting of two protons and two neutrons. (Chap. 21, p. 743)

amino acid: an organic compound; a monomer that forms proteins. (Chap. 19, p. 669)

amorphous solid: a substance with a haphazard, disjointed, and incomplete crystal lattice. (Chap. 10, p. 343)

anaerobic: a metabolic process that takes place in the absence of oxygen. (Chap. 19, p. 695)

anhydrous: a compound in which all water has been removed, usually by heating. (Chap. 5, p. 165)

anion: a negative ion. (Chap. 17, p. 588)

anode: the electrode that takes electrons away from the reacting ions or atoms in solution. (Chap. 17, p. 585)

aqueous solution: a solution in which the solvent is water. (Chap. 1, p. 23)

aromatic hydrocarbon: a compound that has a benzene ring or the type of bonding exhibited by benzene; most have distinct odors. (Chap. 18, p. 636)

atom: the smallest particle of a given type of matter. (Chap. 2, p. 51)

atomic mass unit: one-twelfth the mass of a carbon-12 atom. (Chap. 2, p. 65)

addition reaction / reacción de adición: reacción donde monómeros que contienen dobles enlaces se unen unos con otros para formar cadenas largas; el producto contiene todos los átomos de los monómeros iniciadores. (Cap. 18, pág. 654)

aerobic / aeróbico: proceso metabólico que sólo se lleva a cabo en la presencia de oxígeno. (Cap. 19, pág. 690)

alkali metal / metal alcalino: cualquier elemento del Grupo 1: litio, sodio, potasio, rubidio, cesio, francio. (Cap. 8, pág. 261)

alkaline earth metal / metal alcalinotérreo: cualquier elemento del Grupo 2: berilio, magnesio, calcio, estroncio, bario. (Cap. 8, pág. 264)

alkane / alcano: hidrocarburo saturado formado únicamente por átomos de carbono e hidrógeno con enlaces sencillos entre todos los átomos. (Cap. 18, pág. 623)

alkene / alqueno: hidrocarburo en que uno o más enlaces dobles unen átomos de carbono. (Cap. 18, pág. 630)

alkyne / alquino: hidrocarburo insaturado que contiene un enlace triple entre dos átomos de carbono. (Cap. 18, pág. 633)

allotrope / alótropo: cualquiera de dos o más moléculas de un solo elemento que tienen estructuras cristalinas o moleculares diferentes. (Cap. 5, pág. 173)

alloy / aleación: solución sólida que contiene metales diferentes y, algunas veces, sustancias no metálicas. (Cap. 1, pág. 23)

alpha particle / partícula alfa: núcleo de helio que consiste en dos protones y dos neutrones. (Cap. 21, pág. 743)

amino acid / aminoácido: compuesto orgánico; monómero que forma proteínas. (Cap. 19, pág. 669)

amorphous solid / sólido amorfo: sustancia con una estructura cristalina aleatoria, desorganizada e incompleta. (Cap. 10, pág. 343)

anaerobic / anaeróbico: proceso metabólico que se lleva a cabo en ausencia de oxígeno. (Cap. 19, pág. 695)

anhydrous / anhidro: compuesto al que se le ha extraido toda el agua, generalmente por calentamiento. (Cap. 5, pág. 165)

anion / anión: ion negativo. (Cap. 17, pág. 588)

anode / ánodo: electrodo que quita electrones de iones o átomos reactivos en solución. (Cap. 17, pág. 585)

aqueous solution / solución acuosa: solución en la cual el disolvente es agua. (Cap. 1, pág. 23)

aromatic hydrocarbon / hidrocarburo aromático: compuesto que tiene un anillo de benceno o el tipo de enlace exhibido por el benceno; la mayoría posee olores distintivos. (Cap. 18, pág. 636)

atom / átomo: la partícula más pequeña de un cierto tipo de materia. (Cap. 2, pág. 51)

atomic mass unit/ unidad de masa atómica: la doceava parte de la masa de un atómica de carbono 12. (Cap. 2, pág. 65)

atomic number: the number of protons in the nucleus of an atom of an element. (Chap. 2, p. 64)

atomic theory: the idea that matter is made up of fundamental particles called atoms. (Chap. 2, p. 51)

ATP: adenosine triphosphate, the energy storage molecule in cells. (Chap. 19, p. 691)

aufbau principle: states that each electron occupies the lowest energy orbital available. (Chap. 7, p. 233)

Avogadro number: the number of things in one mole of a substance, specifically 6.02×10^{23}. (Chap. 12, p. 405)

Avogadro's principle: statement that at the same temperature and pressure, equal volumes of gases contain equal numbers of particles. (Chap. 11, p. 396)

B

barometer: an instrument that measures the pressure exerted by the atmosphere. (Chap. 11, p. 374)

base: a substance that produces hydroxide ions when it dissolves in water. (Chap. 14, p. 489)

basic anhydride: a metal oxide that reacts with water to form a base. (Chap. 14, p. 493)

beta particle: a high-energy electron with a 12 charge. (Chap. 21, p. 745)

binary compound: a compound that contains only two elements. (Chap. 5, p. 153)

biochemistry: the study of the chemistry of living things. (Chap. 19, p. 666)

boiling point: the temperature of a liquid where its vapor pressure equals the pressure exerted on its surface. (Chap. 10, p. 356)

Boyle's law: at a constant temperature, the volume and pressure of a gas are inversely proportional. (Chap. 11, p. 381)

Brønsted-Lowry model: a model of acids and bases in which an acid is a hydrogen-ion donor and a base is a hydrogen-ion acceptor. (Chap. 15, p.526)

Brownian motion: the constant, random motion of tiny chunks of matter. (Chap. 10, p. 339)

buffer: a solution that resists changes in pH when moderate amounts of acids or bases are added to it. (Chap. 15, p. 531)

atomic number / número atómico: número de protones en el núcleo del átomo de un elemento. (Cap. 2, pág. 64)

atomic theory / teoría atómica: la idea de que la materia está compuesta de partículas fundamentales llamadas átomos. (Cap. 2, pág. 51)

ATP / ATP: trifosfato de adenosina, la molécula de almacenamiento de energía en las células. (Cap. 19, pág. 691)

aufbau principle / principio de aufbau: establece que cada electrón occupa el orbital de energiá más bajo disponible. (Cap .7, pág. 233)

Avogadro number / número de Avogadro: el número de cosas en un mol de una cierta sustancia, específicamente 6.02×10^{23}. (Cap. 12, pág. 405)

Avogadro's principle / principio de Avogadro: afirmación de que a la misma temperatura y presión, volúmenes iguales de gases contienen números iguales de partículas. (Cap. 11, pág. 396)

barometer / barómetro: instrumento que mide la presión ejercida por la atmósfera. (Cap. 11, pág. 374)

base / base: sustancia que produce iones hidróxido cuando se disuelve en agua. (Cap. 14, pág. 489)

basic anhydride / anhídrido básico: óxido metálico que reacciona con agua para formar una base. (Cap. 14, pág. 493)

beta particle / partícula beta: electrón de alta energía con una carga de 12. (Cap. 21, pág. 745)

binary compound / compuesto binario: compuesto que contiene sólo dos elementos. (Cap. 5, pág. 153)

biochemistry / bioquímica: el estudio de la química de los seres vivos. (Cap. 19, pág. 666)

boiling point / punto de ebullición: temperatura de un líquido cuando su presión de vapor iguala a la presión ejercida sobre su superficie. (Cap. 10, pág. 356)

Boyle's law / ley de Boyle: a temperatura constante, el volumen y la presión de un gas son inversamente proporcionales. (Cap. 11, pág. 381)

Brønsted-Lowry model / modelo de Brønsted-Lowry: modelo de ácidos y bases en el que un ácido es un donante de iones hidrógeno y una base es un receptor de iones hidrógeno. (Cap. 15, pág. 526)

Brownian motion / movimiento browniano: movimiento aleatorio constante de trozos diminutos de materia. (Cap. 10, pág. 339)

buffer / amortiguador: solución que resiste cambios de pH cuando se le añaden cantidades moderadas de ácidos o bases. (Cap. 15, pág. 531)

C

calorie: the heat required to raise the temperature of 1 gram of liquid water by 1°C. (Chap. 20, p. 717)

Calorie: a food Calorie, equal to 1 kilocalorie, used to measure the energy value of foods. (Chap. 20, p. 717)

capillarity: the rising of a liquid in a narrow tube, sometimes called capillary action. (Chap. 13, p. 444)

carbohydrate: an organic molecule that contains the elements carbon, hydrogen, and oxygen in a ratio of about two hydrogen atoms and one oxygen atom for each carbon atom. (Chap. 19, p. 673)

catalyst: a substance that speeds up the rate of a reaction without being used up itself or permanently changed. (Chap. 6, p. 220)

cathode: the electrode that brings electrons to the reacting ions or atoms in solution. (Chap. 17, p. 585)

cation: a positive ion. (Chap. 17, p. 588)

Charles's law: at constant pressure, the volume of a gas is directly proportional to its Kelvin temperature. (Chap. 11, p. 390)

chemical change: the change of one or more substances into other substances. (Chap. 1, p. 38)

chemical property: a property that can be observed only when there is a change in the composition of a substance. (Chap. 1, p. 38)

chemical reaction: another term for chemical change. (Chap. 1, p. 38)

chemistry: the science that investigates and explains the structure and properties of matter. (Chap. 1, p. 4)

coefficient: a number placed in front of the parts of a chemical equation to indicate how many are involved; always a positive whole number. (Chap. 6, p. 197)

coenzyme: an organic molecule that assists an enzyme in catalyzing a reaction. (Chap. 19, p. 688)

colloid: a mixture that contains particles that are evenly distributed through a dispersing medium and do not settle out over time. (Chap. 13, p. 472)

combined gas law: the combination of Boyle's law and Charles's law. (Chap. 11, p. 393)

combustion: term for a reaction in which a substance rapidly combines with oxygen to form one or more oxides. (Chap. 6, p. 206)

compound: a chemical combination of two or more different elements joined together in a fixed proportion. (Chap. 1, p. 28)

concentration: the amount of a substance present in a unit volume. (Chap. 6, p. 217)

calorie / caloría: el calor requerido para incrementar la temperatura de 1 gramo de agua líquida en un 1°C. (Cap. 20, pág. 717)

Calorie / Caloría: una Caloría alimenticia (igual a 1 kilocaloría) se usa para medir el valor energético de los alimentos. (Cap. 20, pág. 717)

capillarity / capilaridad: elevación de un líquido por un tubo estrecho, llamado en ocasiones acción capilar. (Cap. 13, pág. 444)

carbohydrate / carbohidrato: una molécula orgánica que contiene los elementos carbono, hidrógeno y oxígeno en una proporción de aproximadamente dos átomos de hidrógeno y uno de oxígeno por cada átomo de carbono. (Cap. 19, pág. 673)

catalyst / catalizador: sustancia que acelera la velocidad de reacción sin que se la consuma o se cambie permanentemente. (Cap. 6, pág. 220)

cathode / cátodo: electrodo que lleva electrones a los iones o átomos reactivos en solución. (Cap. 17, pág. 585)

cation / catión: ion positivo. (Cap. 17, pág. 588)

Charles's law / ley de Charles: a presión constante, el volumen de un gas es directamente proporcional a su temperatura en Kelvin. (Cap. 11, pág. 390)

chemical change / cambio químico: cambio de una o más sustancias en otras sustancias. (Cap. 1, pág. 38)

chemical property / propiedad química: propiedad que puede observarse únicamente cuando hay un cambio en la composición de una sustancia. (Cap. 1, pág. 38)

chemical reaction / reacción química: otra forma de nombrar un cambio químico. (Cap. 1, pág. 38)

chemistry / química: ciencia que investiga y explica la estructura y las propiedades de la materia. (Cap. 1, pág. 4)

coefficient / coeficiente: número que se coloca antes de las partes de una ecuación química para indicar cuántas de ellas están involucradas; siempre es un número entero positivo. (Cap. 6, pág. 197)

coenzyme / coenzima: molécula orgánica que ayuda a una enzima a catalizar una reacción. (Cap. 19, pág. 688)

colloid / coloide: mezcla que contiene partículas uniformemente distribuidas a través de un medio dispersante y que no se estabiliza con el tiempo. (Cap. 13, pág. 472)

combined gas law / ley combinada de los gases: combinación de las leyes de Boyle y Charles. (Cap. 11, pág. 393)

combustion / combustión: término para una reacción en la cual una sustancia se combina rápidamente con oxígeno para formar uno o más óxidos. (Cap. 6, pág. 206)

compound compuesto: combinación química de dos o más elementos diferentes unidos en una proporción fija. (Cap. 1, pág. 28)

concentration / concentración: cantidad de sustancia presente en una unidad de volumen. (Cap. 6, pág. 217)

condensation: the process where gaseous particles come together, that is, condense, to form a liquid or sometimes a solid. (Chap. 10, p. 354)

condensation reaction: reaction to form a polymer where a small molecule, usually water, is given off as each new bond is formed. (Chap. 18, p. 656)

conductivity: a measure of how easily electrons can flow through a material to produce an electrical current. (Chap. 9, p. 311)

covalent bond: the attraction of two atoms for a shared pair of electrons. (Chap. 4, p. 138)

covalent compound: a compound whose atoms are held together by covalent bonds. (Chap. 4, p. 138)

cracking: the use of a catalyst or high temperatures in the absence of air to break down or rearrange large hydrocarbons. (Chap. 18, p. 638)

cross-linking: the linking together of many polymer chains, giving the polymer increased strength. (Chap. 18, p. 656)

crystal: a regular, repeating arrangement of atoms, ions, or molecules in three dimensions. (Chap. 4, p. 132)

crystal lattice: the three-dimensional arrangement repeated throughout a solid. (Chap. 10, p. 343)

condensation / condensación: proceso en que las partículas gaseosas se unen o se condensan formando un líquido o en algunas ocasiones un sólido. (Cap. 10, pág. 354)

condensation reaction / reacción de condensación: reacción para formar un polímero donde una molécula pequeña, generalmente agua, es eliminada al formarse un nuevo enlace. (Cap. 18, pág. 656)

conductivity / conductividad: medida de la facilidad con que pueden fluir los electrones a través de un material para producir una corriente eléctrica. (Cap. 9, pág. 311)

covalent bond / enlace covalente: atracción de dos átomos por un par compartido de electrones. (Cap. 4, pág. 138)

covalent compound / compueso covalente: compuesto cuyos átomos se mantienen unidos por enlaces covalentes. (Cap. 4, pág. 138)

cracking / cracking: uso de un catalizador o de altas temperaturas en ausencia de aire para romper o rearreglar hidrocarburos grandes. (Cap. 18, pág. 638)

cross-linking / entrecruzamiento: unión de varias cadenas de polímeros, produciendo un polímero con fuerza incrementada. (Cap. 18, pág. 656)

crystal / cristal: arreglo tridimensional, repetitivo y regular de átomos, iones, o moléculas. (Cap. 4, pág. 132)

crystal lattice / red cristalina: arreglo tridimensional repetido a través de un sólido. (Cap. 10, pág. 343)

D

decomposition: the name applied to a reaction where a compound breaks down into two or more simpler substances. (Chap. 6, p. 202)

deliquescent: a substance that takes up enough water from the air that it dissolves completely to a liquid solution. (Chap. 5, p. 165)

denaturation: the name given to the process of unfolding of a protein when the forces holding the polypeptide chain in shape are broken. (Chap. 19, p. 671)

density: the amount of matter (mass) in a given unit volume. (Chap. 1, p. 34)

deposition: the energy-releasing process by which a substance changes from a gas or vapor to a solid without first becoming a liquid. (Chap. 10, p. 354)

deuterium: the hydrogen isotope with a mass number of 2. (Chap. 21, p. 761)

diffusion: the process by which a gas enters a container and fills it, or when the particles of two gases or liquids mix together. (Chap. 10, p. 349)

dissociation: the process by which the charged particles in an ionic solid separate from one another, primarily when going into solution. (Chap. 13, p. 452)

decomposition / descomposición: nombre que se aplica a una reacción donde un compuesto se descompone en dos o más sustancias más simples. (Cap. 6, pág. 202)

deliquescent / delicuescente: sustancia que absorbe suficiente agua del aire como para disolverse completamente en una solución líquida. (Cap. 5, pág. 165)

denaturation / desnaturalización: nombre que se le da al proceso de desdoblamiento de una proteína que ocurre cuando se rompen las fuerzas que mantienen formada la cadena del polipéptido. (Cap. 19, pág. 671)

density / densidad: cantidad de materia (masa) en una unidad dada de volumen. (Cap. 1, pág. 34)

deposition / depositacíon: proceso de liberacíon de energia por el cual una sustancia cambia de gas o vapor a sólido sin antes convertirse en un líquido. (Cap. 10, pág. 354)

deuterium / deuterio: isótopo de hidrógeno con un número de masa de 2. (Cap. 21, pág. 761)

diffusion / difusión: proceso por el cual un gas entra a un contenedor y lo llena, o cuando se entremezclan las partículas de dos gases o líquidos. (Cap. 10, pág. 349)

dissociation / disociación: proceso por el cual las partículas cargadas en un sólido iónico se separan una de la otra, principalmente cuando entran en solución. (Cap. 13, pág. 452)

Glossary/Glosario

distillation: the method of separating substances in a mixture by evaporation of a liquid and subsequent condensation of its vapor. (Chap. 5, p. 171)

DNA: deoxyribonucleic acid. (Chap. 19, p. 685)

double bond: a bond formed by the sharing of two pairs of electrons between two atoms. (Chap. 9, p. 321)

double displacement: a type of reaction where the positive and negative portions of two ionic compounds are interchanged; at least one product must be water or a precipitate. (Chap. 6, p. 206)

ductile: property of a metal that means it can easily be drawn into a wire. (Chap. 9, p. 313)

dynamic equilibrium: term describing a system in which opposite reactions are taking place at the same rate. (Chap. 6, p. 209)

E

electric current: the flow of electrons in a particular direction. (Chap. 17, p. 584)

electrolysis: the process in which electrical energy causes a non-spontaneous chemical reaction to occur. (Chap. 17, p. 600)

electrolyte: any compound that conducts electricity when melted or dissolved in water. (Chap. 4, p. 142)

electrolytic cell: the electrochemical cell in which electrolysis takes place. (Chap. 17, p. 600)

electromagnetic spectrum: the whole range of electromagnetic radiation. (Chap. 2, p. 69)

electron: negatively-charged particle. (Chap. 2, p. 61)

electron cloud: the space around the nucleus of an atom where the atom's electrons are found. (Chap. 2, p. 75)

electron configuration: the most stable arrangement of electrons in sublevels and orbitals. (Chap. 7, p. 240)

electron transport chain: the controlled release of energy from glucose by the step-by-step movement of electrons to lower energy levels. (Chap. 19, p. 692)

electronegativity: the measure of the ability of an atom in a bond to attract electrons. (Chap. 9, p. 301)

element: a substance that cannot be broken down into simpler substances. (Chap. 1, p. 24)

---

distillation / destilación: método de separación de sustancias en una mezcla por evaporación de un líquido y la subsecuente condensación de su vapor. (Cap. 5, pág. 171)

DNA / DNA: ácido desoxirribonucleico. (Cap. 19, pág. 685)

double bond / enlace doble: enlace formado al compartir dos pares de electrones entre dos átomos. (Cap. 9, pág. 321)

double displacement / desplazamiento doble: tipo de reacción en la cual se intercambian las partes positivas y negativas de dos compuestos iónicos. Por lo menos, uno de los productos debe ser agua o un precipitado. (Cap. 6, pág. 206)

ductile / dúctil: propiedad de un metal que permite que se pueda estirar fácilmente formando un alambre. (Cap. 9, pág. 313)

dynamic equilibrium / equilibrio dinámico: término que describe un sistema en que las reacciones opuestas se llevan a cabo a la misma velocidad. (Cap. 6, pág. 209)

electric current / corriente eléctrica: flujo de electrones en cierta dirección. (Cap. 17, pág. 584)

electrolysis / electrólisis: proceso por el cual la energía eléctrica hace que ocurra una reacción química no espontánea. (Cap. 17, pág. 600)

electrolyte / electrolito: cualquier compuesto que conduce electricidad cuando se funde o disuelve en agua. (Cap. 4, pág. 142)

electrolytic cell / celda electrolítica: celda electroquímica en la cual se lleva a cabo la electrólisis. (Cap. 17, pág. 600)

electromagnetic spectrum / espectro electromagnético: rango completo de la radiación electromagnética. (Cap. 2, pág. 69)

electron / electrón: partícula cargada negativamente. (Cap. 2, pág. 61)

electron cloud / nube electrónica: espacio alrededor del núcleo de un átomo donde se encuentran los electrones del átomo. (Cap. 2, pág. 75)

electron configuration / configuración electrónica: arreglo de electrones más estable en subniveles y orbitales. (Cap. 7, pág. 240)

electron transport chain / cadena de transporte de electrones: liberación controlada de energía a partir de la glucosa por el movimiento gradual de electrones para disminuir los niveles energéticos. (Cap. 19, pág. 692)

electronegativity / electronegatividad: medida de la capacidad de un átomo en un enlace para atraer electrones. (Cap. 9, pág. 301)

element / elemento: sustancia que no puede separarse en sustancias más simples. (Cap. 1, pág. 24)

emission spectrum: the spectrum of light released from excited atoms of an element. (Chap. 2, p. 72)

empirical formula: the formula of a compound having the smallest whole-number ratio of atoms in the compound. (Chap. 12, p. 426)

endothermic: chemical reaction that absorbs energy. (Chap. 1, p. 41)

energy: the capacity to do work. (Chap. 1, p. 40)

energy level: the regions of space in which electrons can move about the nucleus of an atom. (Chap. 2, p. 73)

entropy: term used to describe and measure the degree of disorder in a process. (Chap. 20, p. 713)

enzyme: a biological catalyst. (Chap. 6, p. 220)

equilibrium: term for a system where no net change occurs in the amount of reactants or products. (Chap. 6, p. 209)

evaporation: the process by which particles of a liquid form a gas by escaping from the liquid surface. (Chap. 10, p. 350)

exothermic: chemical reaction that gives off energy. (Chap. 1, p. 40)

experiment: a set of controlled observations that test a hypothesis (Chap. 2, p. 57)

F

factor label method: The problem-solving method in chemistry that uses mathematical relationships to convert one quantity to another. (Chap. 11, p. 380)

family: see group. (Chap. 3, p. 96)

fatty acid: a long-chain carboxylic acid. (Chap. 19, p. 681)

fermentation: the anaerobic process of generating energy from glucose. (Chap. 19, p. 695)

formula: a combination of chemical symbols that show what elements make up a compound and the number of atoms of each element. (Chap. 1, p. 31)

formula mass: the mass in atomic mass units of one formula unit of an ionic compound. (Chap. 12, p. 408)

formula unit: the simplest ratio of ions in a compound. (Chap. 5, p. 154)

fossil fuel: a fuel such as oil or natural gas, comprised of hydrocarbons that are the remains of plants and other organisms that lived millions of years ago. (Chap. 20, p. 709)

fractional distillation: the distillation of a mixture by the use of repeated vaporization-condensation cycles to increase the efficiency of separation. (Chap. 18, p. 637)

emission spectrum / espectro de emisión: espectro de luz liberada por átomos excitados de un elemento. (Cap. 2, pág. 72)

empirical formula / fórmula empírica: fórmula de un compuesto que tiene la proporción de átomos en números enteros menor en el compuesto. (Cap. 12, pág. 426)

endothermic / endotérmica: reacción química que absorbe energía. (Cap. 1, pág. 41)

energy / energía: capacidad de hacer trabajo. (Cap. 1, pág. 40)

energy level / nivel energético: regiones del espacio en las cuales los electrones se pueden mover alrededor del núcleo de un átomo. (Cap. 2, pág. 73)

entropy / entropía: término para describir y medir el grado de degradación en un proceso. (Cap. 20, pág. 713)

enzyme / enzima: catalizador biológico. (Cap. 6, pág. 220)

equilibrium / equilibrio: término que describe un sistema donde no ocurre un cambio neto en la cantidad de reactivos ni de productos. (Cap. 6, pág. 209)

evaporation / evaporación: proceso por el cual las partículas de un líquido forman un gas al escaparse de la superficie del líquido. (Cap. 10, pág. 350)

exothermic / exotérmica: reacción química que libera energía. (Cap. 1, pág. 40)

experiment / experimento: conjunto de observaciones controladas que se realizan para probar una hipótesis. (Cap. 2, pág. 57)

factor label method / método del factor: método de solución de problemas en química que utiliza relaciones matemáticas para convertir una cantidad a otra. (Cap. 11, pág. 380)

family / familia: véase grupo. (Cap. 3, pág. 96)

fatty acid / ácido graso: ácido carboxílico de cadena larga. (Cap. 19, pág. 681)

fermentation / fermentación: proceso anaerobio de generación de energía a partir de glucosa. (Cap. 19, pág. 695)

formula / fórmula: combinación de símbolos químicos que muestra qué elementos forman un compuesto y el número de átomos de cada elemento. (Cap. 1, pág. 31)

formula mass / fórmula-masa: la masa en unidades de masa atómicas de una fórmula unitaria en un compuesto iónico. (Cap. 12, pág. 408)

formula unit / fórmula unitaria: la proporción más sencilla de iones en un compuesto. (Cap. 5, pág. 154)

fossil fuel / combustible fósil: combustible como el petróleo o el gas natural, compuesto de hidrocarburos que son restos de plantas y otros organismos que vivieron hace millones de años. (Cap. 20, pág. 709)

fractional distillation / destilación fraccionaria: destilación de una mezcla por el uso de ciclos repetidos de condensación y evaporación para aumentar la eficiencia de la separación. (Cap. 18, pág. 637)

freezing point: the temperature of a liquid when it becomes a crystal lattice. (Chap. 10, p. 362)

functional group: the part of a molecule that is largely responsible for the chemical behavior of the molecule. (Chap. 18, p. 640)

G

gamma ray: a high-energy form of electromagnetic radiation with no charge and no mass. (Chap. 21, p. 745)

gas: a flowing, compressible substance with no definite volume or shape. (Chap. 10, p. 339)

gray: the unit used to measure a received dose of radiation. (Chap. 21, p. 769)

group: the elements in a vertical column of the periodic table. (Chap. 3, p. 94)

H

half-life: the time it takes for half of a given radioactive isotope to decay (into a different isotope or element). (Chap. 21, p. 752)

halogen: any element from Group 17: fluorine, chlorine, bromine, iodine, astatine. (Chap. 8, p. 276)

heat: the energy transferred from an object at high temperature to an object at lower temperature. (Chap. 20, p. 704)

heat of fusion: the energy released as one kilogram of a substance solidifies at its freezing point. (Chap. 10, p. 362)

heat of solution: the heat taken in or released in the dissolving process. (Chap. 13, p. 460)

heat of vaporization: the energy absorbed when one kilogram of a liquid vaporizes at its normal boiling point. (Chap. 10, p. 359)

Heisenberg uncertainty principle: the principle that it is impossible to accurately measure both the position and energy of an electron at the same time. (Chap. 7, p. 238)

hormone: a signal molecule that tells cells whether to start or stop a reaction. (Chap. 19, p. 690)

hydrate: a compound in which there is a specific ratio of water to ionic compound. (Chap. 5, p. 164)

hydrocarbon: an organic compound that consists of only hydrogen and carbon. (Chap. 5, p. 180)

freezing point / punto de congelación: temperatura de un líquido cuando se convierte en una red cristalina. (Cap. 10, pág. 362)

functional group / grupo funcional: la parte de una molécula que es en gran parte responsable por el comportamiento químico de la molécula. (Cap. 18, pág. 640)

G

gamma ray / rayo gama: forma de radiación electromagnética de alta energía sin carga ni masa algunas. (Cap. 21, pág. 745)

gas / gas: sustancia comprensible y fluida sin volumen o forma definidos. (Cap. 10, pág. 339)

gray / gray: unidad que se usa para medir una dosis de radiación recibida. (Cap. 21, pág. 769)

group / grupo: elementos en una columna vertical de la tabla periódica. (Cap. 3, pág. 94)

H

half-life / media vida: tiempo que tarda la mitad de cierto isótopo radioactivo en descomponerse (en un isótopo o elemento diferentes). (Cap. 21, pág. 752)

halogen / halógeno: cualquier elemento del Grupo 17: flúor, cloro, bromo, yodo, astato. (Cap. 8, pág. 276)

heat / calor: tenergía transferida de un cuerpo a temperatura elevada hacia un cuerpo a una temperatura más baja. (Cap. 20, pág. 704)

heat of fusion / calor de fusión: energía liberada al solidificarse un kilogramo de sustancia en su punto de congelación. (Cap. 10, pág. 362)

heat of solution / calor de solución: calor consumido o liberado en el proceso de disolución. (Cap. 13, pág. 460)

heat of vaporization / calor de vaporización: energía absorbida cuándo un kilogramo de un líquido se vaporiza en su punto de ebullición normal. (Cap. 10, pág. 359)

Heisenberg uncertainty principle / principio de incertidumbre de Heisenberg: el principio que enuncia que resulta imposible medir exactamente la posición y energía de un electrón al mismo tiempo. (Cap. 7, pág. 238)

hormone / hormona: molécula mensajera que les indica a las células cuando empezar o parar una reacción. (Cap. 19, pág. 690)

hydrate / hidrato: compuesto en que hay una proporción específica de agua y compuesto iónico. (Cap. 5, pág. 164)

hydrocarbon / hidrocarburo: compuesto orgánico cuyos únicos componentes son hidrógeno y carbono. (Cap. 5, pág. 180)

hydrogen bonding: a connection between the hydrogen atoms on one molecule and a highly electronegative atom on another molecule, but not a full covalent bond. (Chap. 13, p. 438)

hydronium ion: a hydrogen ion attached to a water molecule. (Chap. 14, p. 483)

hygroscopic: a substance that absorbs water molecules from the air to become a hydrate. (Chap. 5, p. 165)

hypothesis: a prediction that can be tested to explain observations. (Chap. 2, p. 57)

I

ideal gas: a gas in which the particles undergo elastic collisions. (Chap. 10, p. 341)

ideal gas law: the equation that expresses exactly how pressure P, volume V, temperature T, and the number of particles n of a gas are related. $PV = nRT$. (Chap. 12, p. 418)

inhibitor: a substance that slows down a reaction. (Chap. 6, p. 221)

inner transition element: one of the elements in the two rows of elements below the main body of the periodic table; the lanthanides and the actinides. (Chap. 7, p. 248)

inorganic compound: a compound that does not contain carbon. (Chap. 5, p. 178)

insoluble: term describing a compound that does not dissolve in a liquid. (Chap. 6, p. 213)

interparticle forces: the forces between the particles that make up a substance. (Chap. 4, p. 142)

ion: an atom or group of combined atoms that has a charge because of the loss or gain of electrons. (Chap. 4, p. 132)

ionic bond: the strong attractive force between ions of opposite charge. (Chap. 4, p. 132)

ionic compound: a compound comprised of ions. (Chap. 4, p. 132)

ionic equation: an equation in which substances that primarily exist as ions in solution are shown as ions. (Chap. 15, p. 518)

ionization: the process where ions form from a covalent compound. (Chap. 14, p. 488)

isomer: a compound with a structure different from another compound with the same formula. (Chap. 18, p. 627)

isotope: any of two or more atoms of an element that are chemically alike but have different masses. (Chap. 2, p. 60)

hydrogen bonding / puente de hidrógeno: conexión entre los átomos de hidrógeno en una molécula y un átomo altamente electronegativo en otra molécula, pero no es un enlace covalente completo. (Cap. 13, pág. 438)

hydronium ion / ion hidronio: ion de hidrógeno unido a una molécula de agua. (Cap. 14, pág. 483)

hygroscopic / higroscópico: sustancia que absorbe moléculas de agua del aire para convertirse en un hidrato. (Cap. 5, pág. 165)

hypothesis / hipótesis: predicción que puede comprobarse para explicar observaciones. (Cap. 2, pág. 57)

ideal gas / gas ideal: gas cuyas partículas experimentan choques elásticos. (Cap. 10, pág. 341)

ideal gas law / ley del gas ideal: ecuación que expresa la relación exacta entre la presión P, el volumen V, la temperatura T y el número de partículas n de un gas. $PV = nRT$. (Cap. 12, pág. 418)

inhibitor / inhibidor: sustancia que decelera una reacción. (Cap. 6, pág. 221)

inner transition element / elemento de transición interna: uno de los elementos en dos filas de elementos debajo del cuerpo principal de la tabla periódica; los lantánidos y los actínidos. (Cap. 7, pág. 248)

inorganic compound / compuesto inorgánico: compuesto que no contiene carbono. (Cap. 5, pág. 178)

insoluble / insoluble: término que describe un compuesto que no se disuelve en un líquido. (Cap. 6, pág. 213)

interparticle forces / fuerzas interparticulares: fuerzas entre las partículas que componen una sustancia. (Cap. 4, pág. 142)

ion / ion: átomo o grupo de átomos combinados que tiene(n) una carga debido a la pérdida o la ganancia de electrones. (Cap. 4, pág. 132)

ionic bond / enlace iónico: fuerza de atracción intensa entre iones de carga opuesta. (Cap. 4, pág. 132)

ionic compound / compuesto iónico: compuesto formado por iones. (Cap. 4, pág. 132)

ionic equation / ecuación iónica: ecuación en la cual sustancias que existen principalmente como iones en solución se muestran como iones. (Cap. 15, pág. 518)

ionization / ionización: proceso en que los iones forman un compuesto covalente. (Cap. 14, pág. 488)

isomer / isómero: compuesto con una estructura diferente de otro compuesto con la misma fórmula. (Cap. 18, pág. 627)

isotope / isótopo: cualquiera de dos o más átomos de un elemento que son químicamente semejantes pero que tienen masas diferentes. (Cap. 2, pág. 60)

J

joule: the SI unit of energy; the energy required to lift a one-newton weight one meter against the force of gravity. (Chap. 10, p. 358)

K

kelvin (K): a division on the Kelvin scale; the SI unit of temperature. (Chap. 10, p. 347)

Kelvin scale: the temperature scale defined so that temperature of a substance is directly proportional to the average kinetic energy of the particles and so that zero on the scale corresponds to zero kinetic energy. (Chap. 10, p. 347)

kilocalorie: a unit equal to 1000 calories. (Chap. 20, p. 717)

kilopascal (kPa): 1000 pascals. (Chap. 11, p. 378)

kinetic theory: the theory that states that submicroscopic particles of all matter are in constant, random motion. (Chap. 10, p. 340)

L

lanthanide: one of the first series of inner transition elements with atomic numbers 58 to 71. (Chap. 3, p. 104)

law of combining gas volumes: the observation that at the same temperature and pressure, volumes of gases combine or decompose in ratios of small whole numbers. (Chap. 11, p. 394)

law of conservation of energy: statement that energy is neither created nor destroyed in a chemical change, but is simply changed from one form to another. (Chap. 20, p. 707)

law of conservation of mass: in a chemical change, matter is neither created nor destroyed. (Chap. 1, p. 40)

law of definite proportions: the principle that the elements that comprise a compound are always in a certain proportion by mass. (Chap. 2, p. 52)

Le Châtelier's principle: states that if a stress is applied to a system at equilibrium, the system shifts in the direction that relieves the stress. (Chap. 6, p. 212)

Lewis dot diagram: a diagram where dots or other small symbols are placed around the chemical symbol of an element to illustrate the valence electrons. (Chap. 2, p. 77)

---

joule / julio: unidad de energía del SI; energía requerida para levantar un metro el peso de un newton contra la fuerza de gravedad. (Cap. 10, pág. 358)

kelvin (K) / kelvin (K): una división en la escala Kelvin; la unidad de temperatura del SI. (Cap. 10, pág. 347)

Kelvin scale / escala Kelvin: escala de temperatura definida de tal forma que la temperatura de una sustancia es directamente proporcional a la energía cinética media de las partículas, de modo que cero en la escala corresponda a la energía cinética cero. (Cap. 10, pág. 347)

kilocalorie / kilocaloría: una unidad igual a 1000 calorías. (Cap. 20, pág. 717)

kilopascal (kPa) / kilopascal (kPa): 1000 pascales. (Cap. 11, pág. 378)

kinetic theory / teoría cinética: teoría que indica que las partículas submicroscópicas de toda la materia están en movimiento aleatorio constante. (Cap. 10, pág. 340)

lanthanide / lantánido: una de las primeras series de elementos de transición interna con números atómicos 58 al 71. (Cap. 3, pág. 104)

law of combining gas volumes / ley de combinación de volúmenes de gases: la observación de que a la misma temperatura y presión, los volúmenes de gases se combinan o se descomponen en proporciones de números enteros pequeños. (Cap. 11, pág. 394)

law of conservation of energy / ley de conservación de la energía: afima que la energía ni se crea ni se destruye durante un cambio químico, simplemente cambia de una forma a otra. (Cap. 20, pág. 707)

law of conservation of mass / ley de conservación de la masa: durante un cambio químico, la materia ni se crea ni se destruye. (Cap. 1, pág. 40)

law of definite proportions / ley de proporciones definidas: principio que dice que los elementos que forman un compuesto están siempre en una proporción de masa dada. (Cap. 2, pág. 52)

Le Châtelier's principle / principio de Le Châtelier: establece que si se aplica una perturbación a un sistema en equilibrio, el sistema cambia en la direción que reduce la perturbación. (Cap. 6, pág. 212)

Lewis dot diagram / diagrama de punto de Lewis: esquema en que se colocan puntos u otros símbolos pequeños alrededor del símbolo químico de un elemento para ilustrar los electrones de valencia. (Cap. 2, pág. 77)

limiting reactant: the reactant of which there is not enough; when it is used up, the reaction stops and no new product is formed. (Chap. 6, p. 218)

lipid: a biological compound that contains a large proportion of C—H bonds and less oxygen than in a carbohydrate; commonly called fats and oils. (Chap. 19, p. 681)

liquid: a flowing substance with a definite volume but an indefinite shape. (Chap. 10, p. 338)

liquid crystal: a material that loses its rigid organization in only one or two dimensions when it melts. (Chap. 10, p. 345)

M

malleable: property of a metal meaning it can be pounded or rolled into thin sheets. (Chap. 9, p. 311)

mass: the measure of the amount of matter an object contains. (Chap. 1, p. 4)

mass number: the sum of the neutrons and protons in the nucleus of an atom. (Chap. 2, p. 64)

matter: anything that takes up space and has mass. (Chap. 1, p. 4)

melting point: the temperature of a solid when its crystal lattice begins to break apart. (Chap. 10, p. 362)

metabolism: name given to the sum of all the chemical reactions necessary for the life of an organism. (Chap. 19, p. 689)

metal: an element that has luster, conducts heat and electricity, and usually bends without breaking. (Chap. 3, p. 101)

metallic bond: the bond that results when metal atoms release their valence electrons to a pool of electrons shared by all the metal atoms. (Chap. 9, p. 311)

metalloid: an element with some physical and chemical properties of metals and other properties of nonmetals. (Chap. 3, p. 103)

mixture: a combination of two or more substances in which the basic identity of each substance is not changed. (Chap. 1, p. 16)

molar mass: the mass of one mole of a pure substance. (Chap. 12, p. 407)

molar volume: the volume that a mole of gas occupies at a pressure of one atmosphere and a temperature of 0.00°C. (Chap. 12, p. 415)

mole: the unit used to count numbers of atoms, molecules, or formula units of substances. (Chap. 12, p. 405)

limiting reactant / reactivo limitante: el reactivo del que no hay suficiente; cuando se consume completamente, la reacción se detiene y no se forma ningún producto nuevo. (Cap. 6, pág. 218)

lipid / lípido: compuesto biológico que contiene una proporción considerable de enlaces C—H y menor cantidad de oxígeno que en un carbohidrato; llamados comúnmente grasas y aceites. (Cap. 19, pág. 681)

liquid / líquido: sustancia fluida con un volumen definido pero sin forma definida. (Cap. 10, pág. 338)

liquid crystal / cristal líquido: material que cuando se funde pierde su organización rígida en una o dos dimensiones únicamente. (Cap. 10, pág. 345)

malleable / maleable: propiedad de un metal que significa que éste puede golpearse o enrollarse en hojas delgadas. (Cap. 9, pág. 311)

mass / masa: medida de la cantidad de materia que contiene un cuerpo. (Cap. 1, pág. 4)

mass number / número de masa: suma de los neutrones y protones en el núcleo de un átomo. (Cap. 2, pág. 64)

matter / materia: todo aquello que ocupa espacio y tiene masa. (Cap. 1, pág. 4)

melting point / punto de fusión: temperatura a la cual se desintegr la estructura cristalina de un sólidoa. (Cap. 10, pág. 362)

metabolism / metabilismo: nombre dado a la suma de todas las reacciones químicas necesarias para la vida de un organismo. (Cap. 19, pág. 689)

metal / metal: elemento que posee lustre, conduce calor y electricidad y que generalmente se dobla sin romperse. (Cap. 3, pág. 101)

metallic bond / enlace metálico: enlace que resulta cuando átomos metálicos liberan sus electrones de valencia a un grupo de electrones compartidos por todos ellos. (Cap. 9, pág. 311)

metalloid / metaloide: elemento con algunas propiedades físicas y químicas de metales y otras propiedades de no metales. (Cap. 3, pág. 103)

mixture / mezcla: combinación de dos o más sustancias en la cual la identidad básica de cada sustancia no se altera. (Cap. 1, pág. 16)

molar mass / masa molar: la masa de un mol de una sustancia pura. (Cap. 12, pág. 407)

molar volume / volumen molar: volumen que ocupa un mol de gas a una presión de una atmósfera y una temperatura de 0.00°C. (Cap. 12, pág. 415)

mole / mole: unidad que se usa para contar números de átomos, moléculas, o fórmulas unitarias de sustancias. (Cap. 12, pág. 405)

Glossary/Glosario

molecular element: a molecule formed when atoms of the same element bond together. (Chap. 5, p. 172)

molecular mass: the mass in atomic mass units of one molecule of a covalent compound. (Chap. 12, p. 408)

molecular substance: a substance that has atoms held together by covalent rather than ionic bonds. (Chap. 5, p. 170)

molecule: an uncharged group of two or more atoms held together by covalent bonds. (Chap. 4, p. 138)

monomer: the individual, small units that make up a polymer. (Chap. 18, p. 647)

N

net ionic equation: the equation that results when ions common to both sides of the equation are removed, usually from an ionic equation. (Chap. 15, p. 520)

neutralization reaction: the reaction of an acid with a base, so called because the properties of both the acid and base are diminished or neutralized. (Chap. 15, p. 516)

neutron: a subatomic particle with a mass equal to a proton but with no electrical charge. (Chap. 2, p. 60)

noble gas: an element from Group 18 that has a full compliment of valence electrons and as such is unreactive. (Chap. 3, p. 98)

noble gas configuration: the state of an atom achieved by having the same valence electron configuration as a noble gas atom; the most stable configuration. (Chap. 4, p. 132)

nonmetal: an element that in general does not conduct electricity, is a poor conductor of heat, and is brittle when solid. Many are gases at room temperature. (Chap. 3, p. 103)

nuclear fission: the process in which an atomic nucleus splits into two or more large fragments. (Chap. 21, p. 757)

nuclear fusion: the process in which two or more nuclei combine to form a larger nucleus. (Chap. 21, p. 760)

nuclear reactor: the device used to extract energy from a radioactive fuel. (Chap. 21, p. 759)

nucleic acid: a large polymer containing carbon, hydrogen, and oxygen, as well as nitrogen and phosphorus; found in all plant and animal cells. (Chap. 19, p. 685)

nucleotide: the building blocks of nucleic acids; each consists of a simple sugar, a phosphate group, and a nitrogen-containing base. (Chap. 19, p. 685)

---

molecular element / elemento molecular: molécula que se forma cuando se unen átomos del mismo elemento. (Cap. 5, pág. 172)

molecular mass / masa molecular: masa en unidades de masa atómica de una molécula de un compuesto covalente. (Cap. 12, pág. 408)

molecular substance / sustancia molecular: sustancia que tiene átomos unidos por enlaces covalentes en lugar de unirse por enlaces iónicos. (Cap. 5, pág. 170)

molecule / molécula: grupo sin carga de dos o más átomos unidos por un enlace covalente. (Cap. 4, pág. 138)

monomer / monómero: unidades pequeñas e individuales que forman un polímero. (Cap. 18, pág. 647)

net ionic equation / ecuación iónica neta: ecuación que resulta cuando iones comunes a ambos lados de la ecuación se eliminan, generalmente de una ecuación iónica. (Cap. 15, pág. 520)

neutralization reaction / reacción de neutralización: reacción de un ácido con una base, llamada así porque las propiedades tanto del ácido como de la base disminuyen o se neutralizan. (Cap. 15, pág. 516)

neutron / neutrón: partícula subatómica con una masa igual a la de un protón pero sin carga eléctrica. (Cap. 2, pág. 60)

noble gas / gas noble: elemento del Grupo 18 que tiene completos todos sus electrones de valencia y por lo tanto no es reactivo. (Cap. 3, pág. 98)

noble gas configuration / configuración de gas noble: estado que logra un átomo al tener la misma configuración de electrones de valencia que un átomo de gas noble; la configuración más estable. (Cap. 4, pág. 132)

nonmetal / no metal: elemento que por lo general no conduce electricidad, es un mal conductor de calor y es quebradizo en estado sólido. Muchos son gases a temperatura ambiente. (Cap. 3, pág. 103)

nuclear fission / fisión nuclear: proceso en el cual un núcleo atómico se separa en dos o más fragmentos grandes. (Cap. 21, pág. 757)

nuclear fusion / fusión nuclear: proceso en el cual dos o más núcleos se combinan para formar un núcleo más grande (Cap. 21, pág. 760)

nuclear reactor / reactor nuclear: dispositivo que se emplea para extraer energía de un combustible radiactivo. (Cap. 21, pág. 759)

nucleic acid / ácido nucleico: polímero grande que contiene carbono, hidrógeno y oxígeno, así como nitrógeno y fósforo; se halla en toda célula vegetal y animal. (Cap. 19, pág. 685)

nucleotide / nucleótido: constituyentes de ácidos nucleicos; cada uno consiste en un azúcar sencillo, un grupo fosfato y una base nitrogenada. (Cap. 19, pág. 685)

nucleus: the small, dense, positively charged central core of an atom. (Chap. 2, p. 63)

O

octet rule: the model of chemical stability that states that atoms become stable by having eight electrons in their outer energy level except for some of the smallest atoms, which have only two electrons. (Chap. 4, p. 130)

orbital: the space in which there is a high probability of finding an electron. (Chap. 7, p. 239)

organic compound: a compound that contains carbon; a few exceptions exist. (Chap. 5, p. 178)

osmosis: the flow of molecules through a selectively permeable membrane driven by concentration difference. (Chap. 13, p. 467)

oxidation: a reaction in which an element loses electrons. (Chap. 16, p. 556)

oxidation number: the charge on an ion or an element; can be positive or negative. (Chap. 5, p. 155)

oxidation-reduction reaction: a reaction characterized by the transfer of electrons from one atom or ion to another. Also known as a redox reaction. (Chap. 16, p. 555)

oxidizing agent: the substance that gains electrons in a redox reaction. It is the substance that is reduced. (Chap. 16, p. 562)

P

pascal (Pa): the SI unit for measuring pressure. (Chap. 11, p. 376)

percent yield: the ratio of actual yield (from an experiment) to theoretical yield (from stoichiometric calculations) expressed as a percent. (Chap. 12, p. 420)

period: a horizontal row in the periodic table. (Chap. 3, p. 94)

periodic law: the statement that the physical and chemical properties of the elements repeat in a regular pattern when they are arranged in order of increasing atomic number. (Chap. 3, p. 92)

periodicity: the tendency to recur at regular intervals. (Chap. 3, p. 88)

pH: a mathematical scale in which the concentration of hydronium ions in a solution is expressed as a number from 0 to 14. (Chap. 14, p. 501)

photosynthesis: the process used by certain organisms to capture energy from the sun. (Chap. 20, p. 729)

nucleus / núcleo: la parte central y que está cargada positivamente, densa y pequeña de un átomo. (Cap. 2, pág. 63)

octet rule / regla del octeto: modelo de estabilidad química que establece que los átomos llegan a ser estables al tener ocho electrones en su nivel de energía externo, a excepción de los átomos más pequeños, los cuales tienen sólo dos electrones. (Cap. 4, pág. 130)

orbital / orbital: espacio en el cual hay una alta probabilidad de encontrar un electrón. (Cap. 7, pág. 239)

organic compound / compuesto orgánico: compuesto que contiene carbono; existen unas cuantas excepciones. (Cap. 5, pág. 178)

osmosis / ósmosis: flujo de moléculas a través de una membrana permeable selectiva debido a una diferencia de la concentración. (Cap. 13, pág. 467)

oxidation / oxidación: reacción en la cual un elemento pierde electrones. (Cap. 16, pág. 556)

oxidation number / número de oxidación: carga en un ion o elemento; puede ser positivo o negativo. (Cap. 5, pág. 155)

oxidation-reduction reaction / reacción de óxido-reducción: reacción caracterizada por la transferencia de electrones de un átomo o ion a otro. También conocida como reacción redox. (Cap. 16, pág. 555)

oxidizing agent / agente oxidante: sustancia que gana electrones en una reacción de redox. Es la sustancia que se reduce. (Cap. 16, pág. 562)

pascal (Pa) / pascal (Pa): unidad del SI para medir presión. (Cap. 11, pág. 376)

percent yield / porcentaje de rendimiento: razón del rendimiento real (de un experimento) al rendimiento teórico (de cálculos estequiométricos) expresado como porcentaje. (Cap. 12, pág. 420)

period / período: una fila horizontal en la tabla periódica. (Cap. 3, pág. 94)

periodic law / ley periódica: establece que las propiedades físicas y químicas de los elementos se repiten en un patrón regular cuando éstos se organizan en orden creciente de número atómico. (Cap. 3, pág. 92)

periodicity / periodicidad: tendencia de volver a ocurrir a intervalos regulares. (Cap. 3, pág. 88)

pH / pH: escala matemática en la cual la concentración de iones hidronio en una solución se expresa como un número entre 0 y 14. (Cap. 14, pág. 501)

photosynthesis / fotosíntesis: proceso que usan ciertos organismos para capturar energía solar. (Cap. 20, pág. 729)

physical change: a change in matter where its identity does not change. (Chap. 1, p. 20)

physical property: a characteristic of matter that is exhibited without a change of identity. (Chap. 1, p. 20)

plasma: an ionized gas. (Chap. 10, p. 345)

polar covalent bond: a bond where the electrons are shared unequally; there is some degree of ionic character to this type of bond. (Chap. 9, p. 308)

polar molecule: a molecule that has a positive pole and a negative pole because of the arrangement of the polar bonds; also called a dipole. (Chap. 9, p. 329)

polyatomic ion: an ion that consists of two or more different elements. (Chap. 5, p. 156)

polymer: a large molecule that is made up of many smaller repeating units. (Chap. 18, p. 647)

potential difference: the difference in electron pressure at the cathode (low) and at the anode (high) in an electrochemical cell. (Chap. 17, p. 586)

pressure: the force acting on a unit area of a surface. (Chap. 10, p. 341)

product: a new substance formed when reactants undergo chemical change. (Chap. 6, p. 190)

property: the characteristics of matter; how it behaves. (Chap. 1, p. 5)

protein: a polymer formed from small monomer molecules linked together by amide groups. (Chap. 19, p. 668)

proton: a positively charged subatomic particle. (Chap. 2, p. 60)

Q

qualitative: an observation made without measurement. (Chap. 1, p. 14)

quantitative: an observation made with measurement. (Chap. 1, p. 14)

R

radioactivity: the spontaneous emission of radiation by an unstable atomic nucleus. (Chap. 21, p. 741)

reactant: a substance that undergoes a reaction. (Chap. 6, p. 190)

reducing agent: the substance that loses electrons in a redox reaction. It is the substance that is oxidized. (Chap. 16, p. 562)

reduction: a reaction in which an element gains one or more electrons. (Chap. 16, p. 556)

physical change / cambio físico: cambio en la materia donde su identidad no cambia. (Cap. 1, pág. 20)

physical property / propiedad física: característica de la materia que se presenta sin un cambio de identidad. (Cap. 1, pág. 20)

plasma / plasma: gas ionizado. (Cap. 10, pág. 345)

polar covalent bond / enlace covalente polar: enlace en el cual los electrones no se comparten de igual manera; hay algún grado de carácter iónico en este tipo de enlace. (Cap. 9, pág. 308)

polar molecule / molécula polar: molécula con un polo positivo y uno negativo a causa del arreglo de los enlaces polares; también llamada dipolo. (Cap. 9, pág. 329)

polyatomic ion / ion poliatómico: ion que consta de dos o más elementos diferentes. (Cap. 5, pág. 156)

polymer / polímero: molécula grande compuesta de muchas unidades más pequeñas que se repiten. (Cap. 18, pág. 647)

potential difference / diferencia de potencial: diferencia en la presión electrónica en el cátodo (baja) y en el ánodo (alta) en una celda electroquímica. (Cap. 17, pág. 586)

pressure / presión: fuerza que actúa sobre una unidad de área de una superficie. (Cap. 10, pág. 341)

product / producto: sustancia nueva que se forma cuando los reactantes experimentan un cambio químico. (Cap. 6, pág. 190)

property / propiedad: características de la materia; la manera como se comporta. (Cap. 1, pág. 5)

protein / proteína: polímero formado a partir de pequeñas moléculas monoméricas unidas por grupos amida. (Cap. 19, pág. 668)

proton / protón: partícula subatómica cargada positivamente. (Cap. 2, pág. 60)

qualitative / cualitativa: observación hecha sin ninguna medición. (Cap. 1, pág. 14)

quantitative / cuantitativa: observación hecha con cierta medición. (Cap. 1, pág. 14)

radioactivity / radiactividad: emisión espontánea de radiación por un núcleo atómico inestable. (Cap. 21, pág. 741)

reactant / reactivo: sustancia que experimenta una reacción. (Cap. 6, pág. 190)

reducing agent / agente reductor: sustancia que pierde electrones en una reacción redox. Es la sustancia que se oxida. (Cap. 16, pág. 562)

reduction / reducción: reacción en la cual un elemento gana uno o más electrones. (Cap. 16, pág. 556)

reforming: the use of heat, pressure, and catalysts to convert large alkanes into other compounds, often aromatic hydrocarbons. (Chap. 18, p. 638)

respiration: the complex series of enzyme-catalyzed reactions that are used to extract chemical energy from glucose. (Chap. 19, p. 690)

RNA: ribonucleic acid. (Chap. 19, p. 685)

S

salt: the general term used in chemistry to describe the ionic compound formed from the negative part of an acid and the positive part of a base. (Chap. 15, p. 516)

saturated hydrocarbon: a hydrocarbon with all the carbon atoms connected to each other by single bonds. (Chap. 18, p. 623)

saturated solution: a solution that holds the maximum amount of solute under the given conditions. (Chap. 13, p. 458)

scientific law: a fact of nature that is observed so often that it is accepted as the truth. (Chap. 2, p. 57)

scientific method: a systematic approach used in scientific study; an organized process used by scientists to do research and to verify the work of others. (Chap. 2, p. 57)

scientific model: a thinking device, built on experimentation, that helps us understand and explain macroscopic observations. (Chap. 1, p. 9)

semiconductor: an element that does not conduct electricity as well as a metal, but that does conduct slightly better than a nonmetal. (Chap. 3, p. 103)

shielding effect: the tendency for the electrons in the inner energy levels to block the attraction of the nucleus for the valence electrons. (Chap. 9, p. 302)

sievert: the unit of radiation equal to one gray multiplied by a factor that assesses how much of the radiation striking tissue is actually absorbed by the tissue and so is a measure of how much biological damage is caused. (Chap. 21, p. 769)

single displacement: a type of reaction where one element takes the place of another in a compound. (Chap. 6, p. 203)

solid: a substance in which the particles occupy fixed positions in a well-defined, three-dimensional arrangement. (Chap. 10, p. 338)

soluble: term describing a substance that dissolves in a liquid. (Chap. 6, p. 215)

reforming / reformado: uso de calor, presión y catalizadores para convertir alcanos largos en otros compuestos, a menudo, hidrocarburos aromáticos. (Cap. 18, pág. 638)

respiration / respiración: serie compleja de reacciones catalizadas por enzimas que se usan para extraer energía química de la glucosa. (Cap. 19, pág. 690)

RNA / RNA: ácido ribonucleico. (Cap. 19, pág. 685)

salt / sal: término general que se usa en química para describir el compuesto iónico formado por la parte negativa de un ácido y la parte positiva de una base. (Cap. 15, pág. 516)

saturated hydrocarbon / hidrocarburo saturado: hidrocarburo con todos los átomos de carbono unidos unos a los otros por enlaces sencillos. (Cap. 18, pág. 623)

saturated solution / solución saturada: solución que tiene la cantidad máxima de soluto bajo las condiciones dadas. (Cap. 13, pág. 458)

scientific law / ley científica: un hecho de la naturaleza que se observa con tanta frecuencia que se acepta como cierto. (Cap. 2, pág. 57)

scientific method / métodos científicos: enfoque sistemático que se usa en los estudios científicos; proceso organizado que siguen los científicos para realizar sus investigaciones y verificar el trabajorealizado por otros científicos. (Cap. 2, pág. 57)

scientific model / modelo científico: instrumento de pensamiento, desarrollado sobre la experimentación, que nos ayuda a entender y explicar observaciones macroscópicas. (Cap. 1, pág. 9)

semiconductor / semiconductor: elemento que no conduce la electricidad tan bien como un metal, pero que la conduce ligeramente mejor que un no metal. (Cap. 3, pág. 103)

shielding effect / efecto protector: tendencia de los electrones en los niveles energéticos internos de bloquear la atracción del núcleo hacia los electrones de valencia. (Cap. 9, pág. 302)

sievert / sievert: unidad de radiación igual a un gray multiplicado por un factor que mide la cantidad de radiación que recibe un tejido y es absorbida por el mismo; por lo tanto es una medida del daño biológico causado. (Cap. 21, pág. 769)

single displacement / desplazamiento simple: tipo de reacción en el cual un elemento ocupa el lugar de otro en un compuesto. (Cap. 6, pág. 203)

solid / sólido: sustancia cuyas partículas ocupan posiciones fijas en un arreglo tridimensional bien definido. (Cap. 10, pág. 338)

soluble / soluble: término que describe una sustancia que se disuelve en un líquido. (Cap. 6, pág. 215)

solute: the substance that is being dissolved when making a solution. (Chap. 1, p. 23)

solution: a mixture that is the same throughout, or homogeneous. (Chap. 1, p. 23)

solvent: the substance that dissolves the solute when making a solution. (Chap. 1, p. 23)

specific heat: a measure of the amount of heat needed to raise the temperature of 1 gram of a substance 1°C. (Chap. 13, p. 445)

spectator ion: an ion that is present in solution but does not participate in the reaction. (Chap. 15, p. 520)

standard atmosphere (atm): the pressure that supports a column of mercury 760 millimeters in height. (Chap. 11, p. 374)

standard solution: a solution of known molarity used in a titration. (Chap. 15, p. 539)

standard temperature and pressure, STP: the set of conditions 0.00°C and 1 atmosphere. (Chap. 11, p. 393)

steroid: a lipid with a distinctive four-ring structure. (Chap. 19, p. 682)

stoichiometry: the study of relationships between measurable quantities, such as mass and volume, and the number of atoms in chemical reactions. (Chap. 12, p. 404)

strong acid: an acid that is completely ionized in water; no molecules exist in the water solution. (Chap. 14, p. 498)

strong base: a base that is completely dissociated into separate ions when dissolved in water. (Chap. 14, p. 497)

sublevel: the small energy divisions in a given energy level. (Chap. 7, p. 233)

sublimation: the process by which particles of a solid escape from its surface and form a gas. (Chap. 10, p. 354)

substance: matter with the same fixed composition and properties. (Chap. 1 p. 15)

substituted hydrocarbon: a compound that has the same structure as a hydrocarbon, except that other atoms are substituted for part of the hydrocarbon. (Chap. 18, p. 640)

substrate: the name given to a reactant in an enzyme-catalyzed reaction. (Chap. 19, p. 672)

supersaturated solution: a solution containing more solute than the usual maximum; they are unstable. (Chap. 13, p. 459)

surface tension: the force needed to overcome intermolecular attractions and break through the surface of a liquid or spread the liquid out. (Chap. 13, p. 442)

solute / soluto: la sustancia que se disuelve cuando se hace una solución. (Cap. 1, pág. 23)

solution / solución: mezcla uniforme u homogénea. (Cap. 1, pág. 23)

solvent / disolvente: sustancia que disuelve el soluto cuando se hace una solución. (Cap. 1, pág. 23)

specific heat / calor específico: medida de la cantidad de calor requerido para aumentar en 1°C la temperatura de 1 gramo de una sustancia. (Cap. 13, pág. 445)

spectator ion / ion espectador: ion que está presente en la solución pero que no participa en ella. (Cap. 15, pág. 520)

standard atmosphere (atm) / atmósfera estándar (atm): presión que sostiene una columna de mercurio de 760 milímetros de altura. (Cap. 11, pág. 374)

standard solution / solución estándar: solución de molaridad conocida que se usa en una titulación. (Cap. 15, pág. 539)

standard temperature and pressure, STP / temperatura y presión estándares, STP: las condiciones de 0.00°C y 1 atmósfera. (Cap. 11, pág. 393)

steroid / esteroide: lípido con una estructura distintiva de cuatro anillos. (Cap. 19, pág. 682)

stoichiometry / estequiometría: estudio de las relaciones entre cantidades mensurables, tal como masa y volumen y el número de átomos en reacciones químicas. (Cap. 12, pág. 404)

strong acid / ácido fuerte: ácido que se ioniza completamente en agua; no existen moléculas en solución acuosa. (Cap. 14, pág. 498)

strong base / base fuerte: base que se disocia completamente en iones separados cuando se disuelve en agua. (Cap. 14, pág. 497)

sublevel / subnivel: pequeñas divisiones energéticas en un nivel dado de energía. (Cap. 7, pág. 233)

sublimation / sublimación: proceso por cual las partículas de un sólido se escapan de su superficie y forman un gas. (Cap. 10, pág. 354)

substance / sustancia: materia con la misma composición fija y propiedades. (Cap. 1 p. 15)

substituted hydrocarbon / hidrocarburo sustituido: compuesto que tiene la misma estructura de un hidrocarburo, excepto que otros átomos reemplazan parte del hidrocarburo. (Cap. 18, pág. 640)

substrate / sustrato: nombre dado a un reactante en una reacción catalizada por enzimas. (Cap. 19, pág. 672)

supersaturated solution / solución supersaturada: solución que contiene más soluto que el máximo normal; son inestables. (Cap. 13, pág. 459)

surface tension / tensión superficial: fuerza necesaria para vencer las atracciones intermoleculares y penetrar la superficie de un líquido o para esparcirlo. (Cap. 13, pág. 442)

synthesis / síntesis

synthesis: the name applied to a reaction in which two or more substances combine to form a single product. (Chap. 6, p. 202)

vitamin / vitamina

synthesis / síntesis: reacción en que dos o más sustancias se combinan para formar un solo producto. (Cap. 6, pág. 202)

T

temperature: the measure of the average kinetic energy of the particles that make up a material. (Chap. 10, p. 346)

theoretical yield: in a chemical reaction, the maximum amount of product that can be produced from a given amount of reactant. (Chap. 12, p. 420)

theory: an explanation based on many observations and supported by the results of many experiments. (Chap. 2, p. 57)

thermoplastic: a plastic that will soften and harden repeatedly when heated and cooled. (Chap. 18, p. 658)

thermosetting: a plastic that hardens permanently when first formed. (Chap. 18, p. 660)

titration: the process of determining the molarity of an acid or base by using an acid-base reaction where one reactant is of known molarity. (Chap. 15, p. 539)

transition element: any of the elements in Groups 3 through 12 of the periodic table, all of which are metals. (Chap. 3, p. 101)

triple bond: a bond formed by sharing three pairs of electrons between two atoms. (Chap. 9, p. 323)

tritium: the hydrogen isotope with a mass number of 3. (Chap. 21, p. 761)

Tyndall effect: the scattering effect caused when light passes through a colloid. (Chap. 13, p. 472)

temperature / temperatura: medida de la energía cinética promedio de las partículas que componen un material. (Cap. 10, pág. 346)

theoretical yield / rendimiento teórico: la cantidad maxime de producto que se puede producir a partir de una cantidad dada de reactivo, durante una reacción química. (Cap. 12, pág. 420)

theory / teoría: explicación basada en muchas observaciones y sostenida por los resultados de muchos experimentos. (Cap. 2, pág. 57)

thermoplastic / termoplástico: plástico que se ablandará y se endurecerá repetidamente cuando se calienta y enfría. (Cap. 18, pág. 658)

thermosetting / fraguado: plástico que se endurece permanentemente cuando se le da forma. (Cap. 18, pág. 660)

titration / titulación: proceso para determinar la molaridad de un ácido o base usando una reacción ácido-base donde un reactivo es de molaridad conocida. (Cap. 15, pág. 539)

transition element / elemento de transición: cualquiera de los elementos en los Grupos 3 a 12 de la tabla periódica, todos los cuales son metales. (Cap. 3, pág. 101)

triple bond / enlace triple: enlace que se forma cuando dos átomos comparten tres pares de electrones. (Cap. 9, pág. 323)

tritium / tritio: isótopo de hidrógeno con un número de masa de 3. (Cap. 21, pág. 761)

Tyndall effect / efecto Tyndall: efecto de dispersión que ocurre cuando la luz pasa a través de un coloide. (Cap. 13, pág. 472)

U

unsaturated hydrocarbon: a hydrocarbon that has one or more double or triple bonds between carbon atoms. (Chap. 18, p. 629)

unsaturated solution: a solution in which the amount of solute dissolved is less than the maximum that could be dissolved. (Chap. 13, p. 458)

unsaturated hydrocarbon / hidrocarburo insaturado: hidrocarburo que tiene uno o más enlaces dobles o triples entre átomos de carbono. (Cap. 18, pág. 629)

unsaturated solution / solución insaturada: solución en que la cantidad de soluto disuelta es menor que el máximo que se puede disolver. (Cap. 13, pág. 458)

V

valence electron: an electron in the outermost energy level of an atom. (Chap. 2, p. 76)

vapor pressure: the pressure of a substance in equilibrium with its liquid. (Chap. 10, p. 355)

vitamin: an organic molecule required in small amounts; there are fat-soluble and water-soluble types. (Chap. 19, p. 690)

valence electron / electrón de valencia: electrón en el nivel energético más externo de un átomo. (Cap. 2, pág. 76)

vapor pressure / presión de vapor: presión de una sustancia en equilibrio con su líquido. (Cap. 10, pág. 355)

vitamin / vitamina: molécula orgánica requerida en cantidades pequeñas; son de tipo liposoluble e hidrosoluble. (Cap. 19, pág. 690)

Glossary/Glosario

volatile: description of a substance that easily changes to a gas at room temperature. (Chap. 1, p. 35)

voltage: an electrical potential difference, expressed in units of volts. (Chap. 17, p. 601)

voltaic cell: a type of electrochemical cell that converts chemical energy into electrical energy by a spontaneous redox reaction. (Chap. 17, p. 585)

W

weak acid: an acid in which almost all the molecules remain as molecules when placed into a water solution. (Chap. 14, p. 499)

weak base: a base in which most of the molecules do not react with water to form ions. (Chap. 14, p. 500)

volatile / volátil: se dice de una sustancia que se convierte fácilmente en un gas a temperatura ambiente. (Cap. 1, pág. 35)

voltage / voltaje: diferencia de potencial eléctrico, expresado en unidades de voltios. (Cap. 17, pág. 601)

voltaic cell / pila voltaica: tipa de celda electroquímica que convierte la energía química en energía elétrica mediante una reacción redox espontánea. (Cap. 17, pág. 585)

weak acid / ácido débil: ácido en que casi todas sus moléculas permanecen como tales cuando se introduce en una solución acuosa. (Cap. 14, pág. 499)

weak base / base débil: base en que la mayor parte de las moléculas no reaccionan con agua para formar iones. (Cap. 14, pág. 500)

Index

Index Key

Italic numbers = illustration/photo   **Bold numbers** = vocabulary term

A

Absolute zero, 347
Accelerators. *See* Particle accelerators
Accuracy, 799
Acetaminophen, 30 *table*, 815 *table*
Acetic acid (vinegar): buffer solution with sodium acetate, 531–532, 532 *illus.*; carboxyl group on, 642 *illus.*; dissolution of in water, 484; formula, 30 *table*, 815 *table*; glacial, 529; oil and vinegar dressing and, 299 *lab*; pH of, 503; production of hydronium ions by, 484; reaction of bones with, 169 *lab*; reaction with baking soda, 39 *illus.*, 190–191, 193, 418 *lab*; reaction with carbonates, 483 *illus.*; reaction with eggshells, 483 *illus.*; reaction with strong bases, 528–529, 529 *illus.*; titration of, 544–545 *lab*; uses of, 30 *table*; as weak acid, 498, 499 *illus.*
Acetic acid-sodium acetate buffer, 531–532
Acetone, 644 *illus.*
Acetylene, 323, 633
Acetylene welding, 633
Acetylsalicyclic acid, 29
Acid-base indicators, 481, 504; bromothymol blue, 518 *lab*; cabbage juice indicator, 506–507 *lab*; list of, 858 *table*; pH of household products and, 479 *lab*; pH range of different, 504 *illus.*; requirements for, 543; titration and, 540
Acid-base reactions, 481, 516–518; antacids and, 535–536, 538; blood gas measurements and, 487; buffers and, 531–533, 535; cave formation and, 525; gas-phase, 527, 527 *illus.*; ionic equations for, 517–518, 520, 521 prob.; salt produced by, 516, 518 *lab*; strong acid plus strong base, 517–518, 520, 521 prob., 522; strong acid plus weak base, 522–523, 524 prob., 526; types of, 517 *table*; weak acid plus strong base, 528–529, 529 prob.; weak acid plus weak bases, 530
Acid-base titration, 539–541; of an acid with a base, 539; concentration from, 541, 542 prob., 544–545 *lab*; how to perform, 540; indicators and, 540; standard solutions and, 539
Acidic anhydrides, **493**–494

Acidic hydrogen, **484**
Acid indigestion, 535–536
Acid ionization, **488**
Acid lakes, 535
Acid rain, 483, 493–494, 495, 535
Acids, 480–482, **483**, 484, 486, 488. *See also* Acid-base reactions; Bases; acid-base indicators and, 481, 506–507 *lab*; acidic anhydrides, 493–494; Bronsted-Lowry definition, 526–527; as electrolytes, 488; hyrdronium ion production, 483, 484, 486; importance of, 481, 481 *table*; ionization, 488, 498–499; mono- vs. polyprotic, 484 *illus.*, 486 *illus.*; names of common, 180 *table*; pH scale and, 501–504; reaction with carbonates, 482 *lab*, 483; reaction with metals, 234–235 *lab*, 482 *lab*, 482–483; strong, 498, 498 *table*, 499 *table*; taste and feel of, 480, 519; titration of, *see* Titration; weak, 498–499, 499 *illus.*, 499 *table*; Word Origin, 483
Actinides, 90–91 *illus.*, 92, **102**, 105 *illus.*, 248, 293
Activated charcoal, 175
Activation energy, **216**, 220, 709–710
Active site, **672**
Activities. *See* ChemLabs; Launch Labs; MiniLabs
Actual yield, 420
Addition reactions, 654
Adenine (A), 686
Adenosine diphosphate (ADP), 691, 693 *illus.*, 730
Adenosine triphosphate (ATP), 271, **691**, 693 *illus.*, 730
Adipic acid, polymerization, 656 *illus.*
ADP. *See* Adenosine diphosphate (ADP)
Aerobic, **690**; respiration, *see* Respiration; Word Origin, 690
Aerosols, 470
AFM. *See* Atomic force microscope (AFM)
Air: composition of, 352; density of, 34 *table*; exhaled vs. inhaled, 122 *table*; fractionation, 352–353; as homogeneous mixture, 23; space shuttle, filtering of on, 201, 416 prob.
Air bags, 417
Air pollution: acid rain, 483, 494, 495, 535; catalytic converters and, 711

Alanine, 669, 669 *table*
Alchemy, 25 *lab*
Alcoholic fermentation, 695, 695 *illus.*, 696 *lab*
Alcohols, 642 *illus.*; hydrogen bonding in, 439; production of compounds containing, 645–646; testing for presence of, 568 *lab*, 569; uses and properties of, 642 *illus.*; volatility of, 33
Aldehydes, 644 *illus.*
Alkali materials, 498
Alkali metals, 84, 97, **261**, 262–263 *illus.*; electron configuration, 261; ionic charge, 155 *table*; physical properties, 261; reactivity, 261, 262 *illus.*; valence electron configuration, 241
Alkaline earth metals, 97, 155 *table*, **264**; ion charges, 266–267 *lab*; properties, 264; reactivity, 96 *lab*, 264, 265 *illus.*, 266–267 *lab*; valence electron configuration, 241
Alkanes (saturated hydrocarbons), 321, **623**–629; first ten, 625 *table*; isomers, 627–628; naming, 625–626, 626–627 prob.; production of compounds containing, 645; properties, 629; structural diagrams, 623–625, 624 *table*, 626–627 prob.
Alkenes, 322, 629, **630**–631; isomers, 631; naming, 630–631; properties, 631, 636
Alkynes, 323, **633**, 633 *table*
Allotropes, **173**–177; of carbon, 174–176; of oxygen, 177, 274; of phosphorus, 173, 272 *illus.*
Alloys, **23**, 101 *illus.*, 104 *illus.*, 280; common, 23 *table*; copper and zinc, 25 *lab*; crystal structure, 106; misch metal, 292; shape-memory, 106–107; steel, *see* Steel
Alloy steels, 288, 288 *table*
Almond flavor, 644 *illus.*
Alnico steel, 288, 288 *table*
Alpha decay, 743
Alpha particles, 62, 228, **743**
Alpha radiation, 739 *lab*, 743
Alternative energy sources, 724–725; geothermal energy, 724–725; solar energy, 724–725; wind energy, 725
Alumina, 155
Aluminum, 94 *illus.*, 268; corrosion, 570 *illus.*; density, 34 *table*; in Earth's crust, 666 *illus.*; electron configuration, 243 *table*; Hall-Heroult method

Index **897**

of producing, 214–215, 604 *illus.*, 604–605, 719; importance of, 268; ionic compounds formed by, 133 *lab;* oxidation-reduction (redox) reactions, 570; reaction capacity, 234–235 *lab;* reaction with chlorine, 712; recycling, 268, 605, 719; uses of, 268

Aluminum chloride, 712

Aluminum hydroxide, 268, 489 *illus.;* reaction with strong acids, 522; as weak base, 500

Aluminum oxide, 155, 246, 268, 570

Aluminum zirconium, 268

Amethyst, 246

Amide groups, 644 *illus.*, 670

Amides, 644 *illus.*

Amines, 644 *illus.*

Amino acids, 500, 655, **669**

Amino group, 644 *illus.*, 669

Ammonia, 815 *table;* boiling point, 356; buffer solution with ammonium salt, 531; fertilizers from, 272 *illus.;* formula and uses, 30 *table;* Haber process for synthesizing, 214–215; molecular shape of, 320; polarity of, 329; reaction with strong acids, 523, 527, 528 *illus.;* water vs., 436 *table;* as weak base, 489, 492, 500

Ammonia-ammonium chloride buffer, 531

Ammonium chloride, 193, 523

Ammonium hydrogen phosphate, 272 *illus.*

Ammonium nitrate, 202 *illus.*, 272 *illus.*, 460 *illus.*, 706

Ammonium perchlorate, 566

Ammonium phosphate, 273

Ammonium sulfate, 167, 272 *illus.*

Ammonium thiocyanate, 41 *illus.*

Amorphous solids, **343**

Anaerobic, 695

Anhydrides, 494; acidic, 493–494; basic, 493, 494

Anhydrous compounds, **165**, 165 *illus.*

Animal fats, 630 *lab*, 682

Animal starch, 679, 679 *illus.*

Anions, **588**, 602

Anodes, **585**, 602

Antacids, 504 *lab*, 535–536, 538; carbonates, 538; compounds used in, 536 *table;* hydroxides, 538

Anthocyanin, 506–507 *lab*

Antibacterials, halogens as, 277 *illus.*

Antibodies, 671

Antifreeze, 466, 642 *illus.*

Antifreeze proteins, 668

Antimony, 27 *table*, 109, 271

Antioxidants, 576

Aqueous solutions, **23**, 450, 451–454. *See also* Colloids; Solutions; boiling-point elevation, 465; concentrated versus dilute, 458; of covalent substances, 452–453; endothermic/exothermic reactions in, 708 *lab;* freeze-point depression, 464–465; of gases in water, 469; guidelines for, 857 *table;* identification of, 456–457 *lab;* of ionic substances, 451–452, 452 *lab;* molarity and, 460–461; observe formation of, 435 *lab;* preparation of different volume of, 462 *prob.;* supersaturated, 459; temperature and solubility, 459–460; unsaturated vs. saturated, 458–459

Arcaheological radiometry, 750–751, 752–753

Argon, 130 *illus.*, 131, 243 *table*, 244 *table*, 245, 279

Aromas, synthetic, 645 *lab*

Aromatic hydrocarbons, 636, 638

Arsenic, 109, 271, 272 *illus.*

Art Connection: Asante brass weights, 411; Chinese porcelain, 161; glass sculptures, 344; radioactive dating of forged art, 754

Art forgeries, 754

Artificial blood, 537

Asante brass weights, 411

Ascorbic acid, 30 *table*, 499 *illus.*, 815 *table*

Aspartame, 5 *illus.*, 30 *table*, 519

Aspirin, 5 *illus.*, 8, 30 *table*, 815 *table;* buffered, 515 *lab;* derivation of from plants, 144; modeling, 9 *illus.;* structure, 8 *illus.;* synthesis, 29

Astatine, 276

Atherosclerosis, 683

Atmosphere: carbon dioxide in, 493; composition of, 352; nitrogen in, 271; oxygen in, 274; pollution of, 493–494, 495; pressure of, 342

Atmospheric pressure, 342, 374, 376; boiling and, 356; equivalent units, 377 *table*

Atomic bombs, 230, 756, 759

Atomic collisions, 128, 136–137

Atomic force microscope (AFM), 237

Atomic mass, 65, 66 *illus.*, 84, 90–91 *illus.*

Atomic mass unit, **65**

Atomic models, 8

Atomic number, **64**, 90–91 *illus.*

Atomic size (radii): atom size-ion size relationship, 258 *illus.;* main group element patterns, 257, 257 *illus.*, 260 *lab;* transition element patterns, 283

Atomic structure: Bohr's model of, 67–68, 228; discovery of, 59 63, 228–229; electron cloud model of, 75–77, 229, 238–239; nuclear model of, 63; periodic table and, 93–94, 96 *lab*, 96–97, 241–245; present-day model, 229

Atomic theory of matter, **51**, 228–229; conservation of matter and, 51, 53, 54–55 *lab;* Dalton's atomic theory and, 52–53, 228; early Greek ideas about, 50–51; Lavoisier's contributions to, 51; Proust's contributions to, 52

Atoms, 7, **51**; chemical changes and, 40–41; conservation of in chemical reactions, 40, 51, 53, 196–197; number in a sample, 409 *prob.;* particles of, 59–60, 64, 65 *table;* size of, 7, 63; Word Origin, 51

ATP. *See* Adenosine triphosphate (ATP)

Aurora borealis, 71

Austenite phase, 106

Automobile air bags, 417

Automobile batteries: lead-acid, 269 *illus.*, 596; rechargable for electric cars, 599

Avicel, 684

Avogadro, Amedeo, 396

Avogadro's number, 403 *lab*, **405**, 406

Avogadro's principle, **396**, 404

B

Bacteria: cavities and, 278; refining of ores by, 723

Baking soda, 141 *illus.;* endothermic reactions of, 41; formula for, 30 *table*, 815 *table;* reaction with vinegar, 190–191, 193, 418 *lab;* uses of, 30 *table*

Balances: double-pan, 796; triple-beam, 796

Balancing: of chemical equations, 196–197, 198–199 *prob.;* of nuclear equations, 742, 746 *prob.*

Balsa wood, density of, 34 *table*

Bar graphs, 811

Barium, 264, 266–267 *lab*

Barium hydroxide, 498 *table*

Barium hydroxide octahydrate, 41 *illus.*, 193

Barium nitrate, 456 *lab*

Barium sulfate, 456 *lab*

Barometers, 374

Barometric pressure units, converting, 377 *prob.*

Bartlett, Neil, 193

Bases, **489**, 492. *See also* Acid-base reactions; Acids; acid-base indicators and,

Index

BASF

481, 506–507 *lab;* antacids and, 504 *lab,* 535–536, 538; basic anhydrides, 493, 494; Bronsted-Lowry definition of, 526–527; as electrolytes, 492; hydroxide ions, production of by, 489; importance of, 481, 481 *table;* ionization of, 489; names of common, 180 *table;* pH scale and, 501–504; strong, 497–498, 498 *table;* taste and feel of, 480, 519; titration of, *see* Titration; weak, 500

BASF, 215
Basic anhydrides, 493, 494
Batteries, 590–591, 594. *See also* Electrochemical cells; carbon-zinc dry cell, 594; lead storage, 596; lemon, 583 *lab,* 585–586, 586 *lab,* 591; lithium, 595, 599; nicad, 597; rechargable, 596, 597, 598
Bauxite, 604
Becquerel, Henri, 740–741
Behavior of matter, 5, 5 *illus.*
Benz, Karl, 466
Benzene, 636, 636 *illus.*
Benzoic acid, 499 *illus.*
Beryllium, 264; atomic size, 257; electron configuration, 242, 243 *table;* melting point, 281; reactivity, 260, 264; uses, 265 *illus.*
Beta-carotene, 632
Beta decay, 745
Beta particles, 745
Beta radiation, 739 *lab,* 745
BHA, 576 *illus.*
BHT, 576 *illus.*
Binary compounds, 153–155; binary inorganic compounds, 179, 179 prob.; binary ionic compounds, 153–155; Word Origin (binary), 153
Biochemist, 676–677
Biochemistry, 666. *See also* Biomolecules
Biological specimens, freeze drying of, 351
Biology Connection: blood gases, measurement of, 487; fluoridation, 278; Hershey and Chase's experiments on DNA with tracers, 766; regulation of air quality in space vehicles, 201; vision and vitamin A, 632
Bioluminescence, 575
Biomolecules, 667. *See also* Metabolism; carbohydrates, 667, 673, 678–679; lipids, 667, 681–683, 684; nucleic acids, 667, 685–687, 687 *lab;* proteins, 667, 668–673, 674–675 *lab;* vitamins, 687–688
Biorefining, 723

Bismuth, 271
Bitter taste, 519
Black phosphorus, 173
Blast furnaces, 558 *illus.,* 567
Bleach, 192
Bleaching reactions, 192, 567–568
Blood: artificial, 537; buffers and, 531, 532–533; as complex mixture, 16 *illus.;* forensic detection of, 573, 574; hemoglobin in, 284 *illus.,* 668, 694; pH of, 531, 532–533
Blood gases, measurement of, 487
Blood sugar. *See* Glucose
Bohr, Neils, 67–68, 72, 228, 230
Bohr's atomic model, 67–68, 228, 228 *illus.,* 230
Boiling, 354, 356
Boiling point, 33, **356**; bond type and, 309; elevation of in solution, 465; periodic table trends, 86; polar vs. nonpolar molecules and, 330
Bonding electron pairs, 314
Bonds: covalent, *see* Covalent bonds; electronegativity and, 301–303; electronegativity difference between bonding atoms (Δ *EN*), 303–304, 307–309, 310 prob.; electron sharing model of, 301; hydrogen, *see* Hydrogen bonding; ionic, *see* Ionic bonds; metallic, 311–312; peptide, 670
Bone: osteoporosis detection, 764 *illus.;* reaction of with vinegar, 169 *lab;* scintigraphic scanning, 764 *illus.*
Borax, 268, 273
Boric acid, 268
Boron, 268; electron configuration, 243 *table;* reactivity of, 260; uses of, 268
Boyle, Robert, 380–381
Boyle's law, 381, 384–385 prob.; deduction of, 380–381, 386–387 *lab;* kinetic explanation of, 382; straws and, 384 *lab;* weather balloons and, 382, 383
Branched hydrocarbons, 623
Brass, 23 *table,* 291
Breakfast cereal, fortified, 28 *lab*
Breathalyser tests, 568 *lab,* 569
Breeder reactors, 760
Bromine, 84; chemical properties, 39 *illus.;* as diatomic element, 172; in halogen lamps, 105 *illus.;* physical state and color, 85; reactivity, 96 *lab,* 277 *illus.*
Bronsted-Lowry model of acids and bases, 526–527
Bronze, 23 *table,* 104 *illus.*
Bronze Age, 567
Brown, Robert, 339

Capillarity

Brownian motion, 339
Buckminsterfullerene, 176
Buffered lakes, 535
Buffers, 515 *lab,* **531**–533, 533 *lab;* acetic acid-sodium acetate, 530–531; ammonia-ammonium chloride, 530; blood, 531, 532–533; buffered lakes, 535
Butane, 142 *illus.,* 180 *table,* 623; formula, 30 *table,* 815 *table;* isomers, 627–628; structural diagrams, 624 *table;* uses of, 30 *table*
1-Butene, 631
***cis*-2-Butene,** 631 *illus.*
***trans*-2-Butene,** 631 *illus.*
Butter, 143 *illus.*

C

Cabbage juice acid-base indicator, 506–507 *lab*
Cadaverine, 644 *illus.*
Caffeine, 30 *table,* 815 *table*
Calcium: in Earth's crust, 666 *illus.;* electron configuration, 245; as essential element, 126; in human body, 666 *illus.;* importance of, 264; ionic compounds formed by, 133 *lab;* oxidation number, 155 *illus.;* reaction of bones with vinegar, 169 *lab;* reactivity of, 96 *lab,* 264, 266–267 *lab;* shielding effect in, 303 *illus.;* uses of, 265 *illus.*
Calcium-40, 746 prob.
Calcium carbonate. *See* Limestone (calcium carbonate)
Calcium chloride: in hot packs, 460 *illus.,* 706; reaction with potassium hydroxide, 213 *illus.*
Calcium fluoride, 154 *illus.*
Calcium hydroxide, 264, 489 *illus.,* 494, 497 *illus.,* 498 *table*
Calcium oxide, 155 *illus.,* 494
Calcium sulfate dihydrate, 165, 166
Calculators, computations with, 805–807
Californium-252, 293
Calorie (food), 717, 717 *table*
Calorie (heat), 717
Calorimeter, 715 *illus.*
Calorimetry: of food, 717, 717 *table,* 720–721 *lab;* heat of reaction and, 715, 716 prob., 718 prob.
Calvin cycle, 731
Cancer radiation therapy, 293, 764 *illus.*
Candle, burning of, 10–11 *lab*
Candy, chromatography of dyes in, 326–327 *lab*
Capillarity, 443–444; paper chromatog-

Capillary action

raphy and, 22 *lab*; Word Origin, 443
Capillary action, 22 *lab*, 443–444
Capillary tube, 443–444
Carbohydrates, 667, **673**, 678–679; disaccharides, 678; as fake fats, 684; hydrogen bonding in, 439; monosaccharides, 678; oxidation in respiration, 690; polysaccharides, 679; role of in living things, 673
Carbon, 105 *illus.*, 269; allotropes, 173, 174–176; electron configuration, 242, 243 *table*; as essential element, 126; in human body, 666 *illus.*; isotopes of, 742; properties, 123; reactivity, 261
Carbon-14, 742, 745
Carbon-14 dating, 750–751, 752–753, 755 prob.
Carbonate antacids, 538
Carbonated beverages, 493
Carbonates, reactions with acids, 483
Carbon blacks, 174
Carbon dioxide, 122–123, 815 *table*; as acidic anhydride, 493; acid rain and, 493–494, 495; air quality in space vehicles and, 201; blood buffering and, 532–533; disolved in sodas, 329 *illus.*; electron sharing in, 138–139; formation of, 197 *illus.*; formula and uses, 30 *table*; molecular modeling of, 319; nonpolarity of, 329; as pollutant, 495; reaction with sodium hydroxide, 197; respiration and, 691
Carbon disulfide, 178 *illus.*, 307 *illus.*, 308
Carbonic acid, 493, 532–533
Carbon monoxide, 495
Carbon monoxide poisoning, 388, 694
Carbon steels, 286 *table*, 286–289
Carbon tetrachloride, 277 *illus.*, 640
Carbonyl group, 643 *illus.*
Carbon-zinc dry cell batteries, 594
Carboxyl group, 642 *illus.*, 644 *illus.*
Carboxylic acids, 642 *illus.*
Career Connection. *See also* In the Field; chemical engineer, 317; chemical laboratory technician, 317; cosmetologist, 491; crime lab technologist, 13; environmental health inspector, 449; environmental technician, 449; fingerprint classifier, 13; food and drug inspector, 491; horticulturalist, 211; industrial chemical worker, 317; industrial machinery mechanic, 449; landscape architect, 211; manufacturer's sales representative, 491; medical laboratory technician, 677; medical records technician, 677; medical technologist, 677; metallurgical tech-

nician, 613; mining engineer, 613; pharmaceutical production worker, 635; pharmaceutical technician, 635; pharmacologist, 635; private investigator, 13; scrap metal processing worker, 613; soil conservation technician, 211
Cars: air bags, 417; batteries, *see* Automobile batteries; electric, 599
Catalase, 674–675 *lab*
Catalysts, **220**; activation energy and, 710; commercial energy production and, 726; Haber process (synthesis of ammonia) and, 215; platinum group, 290
Catalytic converters, 711
Catalytic decomposition, 674–675 *lab*
Cathode rays, 59, 60
Cathode-ray tubes, 59, 60
Cathodes, **585**, 602
Cations, **588**, 602
Caves, formation of, 525
CDs (compact discs), 611
Cell membranes, 683
Celluloid, 660
Cellulose, 652, 653, 673, 679; as fake fats, 684; structure of, 653 *illus.*
Celsius scale, 347–348, 795
Cement, 166, 445 *table*
Center of gravity of a mass, 26
Cereal, fortified, 28 *lab*
Cerium, 292
Cerium sulfate, 459
Cesium, 258 *illus.*, 261
Cesium chloride, 258 *illus.*
***Challenger* explosion**, 140
Changes of state, 33, 358–359, 362–363; condensation, 354; evaporation, 350; heat of fusion and, 362–363; heat of vaporization and, 358–359; kinetic energy and, 360–361 *lab*; pressure cookers and, 357; sublimation, 354; vapor pressure and boiling, 355 *lab*, 356
Charcoal, 123, 138, 175
Charles, Jacques, 140, 389
Charles's law, 369 *lab*, 389–**390**, 391 prob.
Chase, Martha, 766
Chemical bonds. *See* Bonds
Chemical change, **38**–41; atoms and, 40, 196; energy and, 40–41, 189 *illus.*, 193–194, 194 *lab*
Chemical coolants, 466
Chemical engineer, 317
Chemical equations: balancing, 196–197; converting word equations to, 191; energy and, 193–194; physi-

Chemistry Data Handbook

cal state and, 193; writing, 190–191, 193–194, 198–199 prob.
Chemical formulas. *See* Formulas
Chemical laboratory technician, 317
Chemical properties, **38**–41, 39 *illus.*; of kitchen chemicals, 18–19 lab
Chemical reaction rates. *See* Reaction rates
Chemical reactions, **38**–41. *See also* specific types; activation energy and, 709–710; atoms and, 40; classification of, 200, 202–203, 204–205 *lab*, 206, 207 *table*; direction of, 712–714, 714 *table*; endothermic, 41, 704, 705, 708 *lab*, 722 *lab*; energy changes and, 40–41, 193–194, 213, 704–705, 707–708; entropy and, 712–713; equations for, *see* Chemical equations; equilibrium and, 209, 212–213; evidence of, 117 *lab*; exothermic, 40–41, 704–705, 708 *lab*, 722 *lab*; heat of reaction of from calorimetry, 715, 716 prob., 718 prob.; nuclear reactions vs., 742 *table*; oxidation-reduction (redox), *see* Oxidation-reduction reactions; rate of, *see* Reaction rates; signs of, 187 *lab*, 188, 189 *illus.*, 204–205 *lab*; spontaneity of, 585, 712–714, 714 *table*; synthesis, 202; theoretical yield and actual yield, 420, 424–425; word equations for, 191
Chemical symbols: historic origins of, 27 *table*; periodic table, 90–91 *illus.*
Chemical worker, industrial, 317
Chemiluminescence, 195, 573, 573 *illus.*
Chemistry, 4
Chemistry and Society: artificial blood, 537; medicines from rain forests, 144; recycling of glass, 58; water-treatment plants, 447
Chemistry and Technology: alternative energy sources, 724–725; archaeological radiochemistry, 750–751; forensic blood detection, 574; fractionation of air, 352–353; Haber process, 214–215; hyperbaric oxygen (HBO) chambers, 388; microscopes, 236–237; refining of copper ore, 606–608; shape-memory metals, 106–107
Chemistry Data Handbook, 847–858; acid-base indicators, 858 *table*; alphabetical *table* of elements, 850 *table*; aqueous solutions, guidelines for, 857 *table*; electron configuration of each element, 854–855 *table*; periodic table of elements, 848–849; physical constants, 856 *table*; polyatomic ions, names and charges of, 856 *table*; solu-

bility product constants, 857 *table*; symbols and abbreviations, 847

Chemistry Skill Handbook, 791–814; calculators, 805–807; factor label method, 807–809; graphs, making and using, 811–813; measurements, making and intrepreting, 797–805; SI units, 791–793; *table*s, making and using, 810

Chemists, 316–317

ChemLabs. *See also* MiniLabs; acid-base indicators, 506–507 *lab*; alkaline earth metals, reactions and ion charges, 266–267 *lab*; aqueous solutions, identification of, 456–457 *lab*; Boyle's law, 386–387 *lab*; burning of a candle, 10–11 *lab*; catalytic decomposition, 674–675 *lab*; change of state and kinetic energy, 360–361 *lab*; chemical reactions, types of, 204–205 *lab*; classification of compounds as ionic or covalent, 170–171 *lab*; composition of a penny, 36–37 *lab*; conservation of matter, 54–55 *lab*; energy content of common foods, 720–721 *lab*; formation and decomposition of zinc iodide, 134–135 *lab*; kitchen chemicals, 18–19 *lab*; mass percents of compounds in a mixture, 422–423 *lab*; oxidation-reduction (redox) reactions in electrochemical cells, 592–593 *lab*; oxidation-reduction (redox) reactions involving copper, 560–561 *lab*; paper chromatography, 326–327 *lab*; periodic table trends, 98–99 *lab*; polymers, differentiating between, 649–651 *lab*; radioactive decay, 748–749 *lab*; reaction capacities of metals and valence electrons, 234–235 *lab*; titration of vinegar, 544–545 *lab*

Chernobyl nuclear power plan, 760 *illus.*
Chinese porcelain, 161
Chips, 111
Chitin, 673, 673 *illus.*
Chloride ion, 258 *illus.*
Chlorine, 84, 97; average atomic mass, 66 *illus.*; compounds of copper and chloring, 162 *table*; copper compounds containing, 162 *table*; as diatomic element, 172; electron configuration, 243 *table*; as essential element, 126; physical state and color of, 85 *illus.*; production of by electrolysis, 601 *illus.*; properties, 121; reaction with aluminum, 712; reaction with iron, 127; reaction with sodium, 131, 132 *table*; reactivity, 96 *lab*, 276; uses of, 277 *illus.*

Chlorofluorocarbons (CFCs), 641 *illus.*
Chloroform, 277 *illus.*, 640
Chloromethane, 640
Chlorophyll, 264, 729, 730–731; Word Origin, 730
Chloroplasts, 729
Cholesterol, 642 *illus.*, 682, 683, 683 *illus.*
Chromatography, 324–325; gas, 325; gel, 325; paper, 22 *lab*, 310 *lab*, 324, 326–327 *lab*; thin-layer, 324
Chrome plating, 104 *illus.*
Chromium, 291; colored compounds of, 101; electron configuration, 245 *illus.*, 291; electroplating, 609 *illus.*; as essential element, 126; glass and gem color and, 246, 247 *table*; properties of, 291
Cigarettes, radiation exposure from, 770 *illus.*
Cinnamon flavor, 644 *illus.*
cis **configuration**, 631; Word Origin (cis), 631
Citric acid, 642 *illus.*
Citrine, 246
Clock reactions, 218 *lab*
Clot captor, 107
Clouds, 125 *illus.*
Club soda, 469 *illus.*
Coal, 105 *illus.*, 197 *illus.*, 636, 639
Coal-fired power plants, 722 *illus.*
Cobalt, 290; electron configuration, 245 *illus.*; as essential element, 126; glass and gem color and, 246, 247 *table*; properties of, 290
Cobalt(II) chloride, 164 *lab*
Cobalt fluorides, 248
Codeine, 144
Coefficients, 197
Coenzymes, 688
Coffee filters, chromatography with, 310 *lab*
Coinage metals, 84, 108, 290
Cold packs, 460 *illus.*, 706
Cold-working of steel, 288, 289
Collagen, 668, 670 *illus.*
Colloids, 470–471, **472**; emulsions, 470; foams, 471; gels, 471; liquid aerosols, 470; pastes, 471; solid aerosols, 470; sols, 471; Tyndall effect and, 472, 473 *illus.*; Word Origin, 472
Color change, chemical reactions and, 188 *illus.*, 189 *illus.*
Color televisions, 105 *illus.*
Combined gas law, 392–**393**, 393–394 prob.
Combustion reactions, **206**, 207 *table*, 709; classification of reactions as,

204–205 *lab*; heat of reaction calculation, 716 prob.; reactants and products of, 10–11 *lab*; Word Origin, 206, 709

Common names, inorganic compounds, 179–180, 180 *table*
Compact discs (CDs), production of, 611
Composition of matter, 5, 5 *illus.*, 6; classification of matter by, 14
Compounds, **28**, 30–31. *See also* Covalent compounds; Inorganic compounds; Organic compounds; classification of, 170–171 *lab*; common, 30 *table*; decomposition of, 134–135 *lab*; elements vs., 151 *lab*; formula of unknown, 426–428; formulas for, 31, 136; formula units in a sample of, 408, 409 prob.; mixtures vs., 151 *lab*; molar mass of, 408, 410 prob.; natural vs. synthetic, 29, 30
Computer chips, 111
Concentrated solutions, 458
Concentration, **217**, 458–460; concentrated versus dilute solutions, 458; molarity and, 460–461; preparation of different solution volume, 462 prob.; rate of reaction and, 217; strength of acids and bases and, 500; from titration, 541, 542 prob., 544–545 *lab*
Condensation, 33, **354**, 446
Condensation reactions, **656**, 670
Conductivity, **311**, 312, 488 *illus.*
Conservation of energy, **707**
Conservation of mass, **40**, 196–197
Conservation of matter, 51 *illus.*, 53, 54–55 *lab*
Contact-lens cleaning solutions, 673 *illus.*
Contact process, production of sulfuric acid by, 424, 425, 485
Cooling curves, 363
Cooling towers, 728
Cool light, 195
Copper, 290; alloy of copper and zinc, 25 *lab*; biorefining of, 723; chemical symbol for, 27 *table*; chlorine-containing compounds, 162 *table*; as coinage metal, 84, 108, 290; conductivity of, 311; corrosion, 570 *illus.*; ductility, 311; electron configuration, 245 *illus.*; as essential element, 126; glass and gem color and, 247 *table*; magnesium-copper voltaic cell, 587–589; melting point, 281; oxidation-reduction (redox) reactions involving, 559, 560–561 *lab*, 562, 570; properties of, 104 *illus.*; refining of by electrolysis,

Index

Copper(II) nitrate

606–608; specific heat, 445 *table*; versatility of, 83 *lab*
Copper(II) nitrate, 203 *lab*
Copper(II) sulfate, 141 *illus.*, 559
Copper(II) sulfate pentahydrate, 165 *illus.*
Copper sulfate: hydrated, 165 *illus.*; reaction of aqueous with iron, 203 *illus.*
Coral reefs, 508 *illus.*
Cork, density of, 34 *table*
Corrosion, 554, 570, 572; ease of different metals, 587 *table*; of iron, 557 *lab*, 570; Word Origin, 610
Cosmetic bench chemist, Fe Tayag, 490–491
Cosmetics, pH of, 505
Cosmetologist, 491
Cottrell precipitators, 470
Covalent bonds, 138–139, 300–301; electronegativity difference between bonding atoms (Δ EN), 307–309; equal electron sharing, 306; nearly equal electron sharing, 307; polar, 308–309
Covalent compounds, 138–139, 168–180, 178 *table*; classification of compounds as, 170–171 *lab*; electronegativity difference between bonding atoms (Δ EN), 307–309; electron sharing in, 136–139; molecular elements, 172–177; naming of, 177–180; properties of, 142–143, 145, 168–169, 300 *table*, 330–331; separating from ionic substances, 168–169, 169 *lab*; solubility of, 452–453
Cracking, 638
Crayons, 168 *illus.*
Crime analysis. *See* Forensics
Crime lab technologists, 13
Cross-Curricular Connections. *See* Art Connection; Biology Connection; Earth Science Connection; Health Connection; History Connection; Literature Connection; Physics Connection
Cross-linking, 655, **656**
Crude iron, 284 *illus.*
Crystal lattices, 343
Crystals, **132**; color of, 246; liquid, 345
Curie, Marie, 741, 768
Curie, Pierre, 741
Cyclohexane, 623, 624, 626
Cyclononane, 626
Cyclopentane, 623, 624 *illus.*
Cyclotrons, 767 *illus.*
Cysteine, 669
Cytochrome *c*, 172 *illus.*

Cytosine (C), 686

D

Daguerre, L. J. M., 564
Daguerreotypes, 564
Dalton, John, 52–53, 59, 228
Dalton's atomic theory of matter, 52–53, 59, 228, 228 *illus.*
Damascus steel, 286, 287
da Vinci, Leonardo, 564
Davy, Humphrey, 11 *lab*, 601
DDT, 641 *illus.*
Dead Sea, 458
Decane, 180 *table*
Decomposition reactions, **202**, 204–205 *lab*, 207 *table*
Definite proportions, law of, 52
Deforestation, 144
Degrees, 347
Deicing, 465 *illus.*
Deliquescent substances, 165
Demineralization, 278
Democritus, 51
Denaturation, **671**, 674–675 *lab*
Density, **34**; of common materials, 34 *table*; composition of a penny and, 36–37 *lab*; determination of, 35 *illus.*; periodic table trends, 86; of water, 34 *table*, 436–437, 440
Dentrification, 571
Deoxyribonucleic acid. *See* DNA (deoxyribonucleic acid)
Deoxyribose, 685
Dependent variable, 812 *illus.*
Deposition, 354
Desalination, 468
Desiccants, 165
Designer fats, 684
Dessicants, 165
Detergents, 455
Deuterium (D), **761**
1,6-Diaminohexane, 656 *illus.*
Diamonds, 24 *illus.*, 105 *illus.*, 173, 175
Diapers, comparing polymers in, 652 *lab*
Diatomic elements, 172, 396; covalent bonds between, 306; gumdrop models of, 313
Dichloromethane, 641 *illus.*
Diethyl ether, 643 *illus.*
Diffusion, **349**; of gases, 341 *lab*, 349; in liquids, 349
Digestion, enzymes and, 672
Dilute solutions, 458
Dimensional analysis, 377 prob., 378, 379 prob., 807–809
2,2-Dimethylpropane, 628 *illus.*

Egg whites

Dinitrogen trioxide, 179 *illus.*
Diodes, 110–111
Dipeptides, 670
Dipole, 328
Diprotic acids, 484 *illus.*, 486 *illus.*
Direct relationships, graphs and, 813
Disaccharides, 678
Disorder, entropy and, 712–713
Displacement reactions, 203, 204–205 *lab*, 206
Dissociation, 452
Distillation, **168**, 169 *illus.*, 638; fractional, *see* Fractional distillation; Word Origin, 638
DNA (deoxyribonucleic acid), 269 *illus.*, **685**, 686; acid-base interactions in, 500; damage to by radiation, 769; double helix structure, 500, 686; electrophoresis of, 610; extraction of, 687 *lab*; Hershey and Chase's experiments on with tracers, 766; identification of as genetic material, 766; nucleotides in, 685–686
Dobereiner, J. W., 85–86
Dobereiner's triads, 85–86
***d* orbitals, periodic table and**, 241, 248
Double bonds, 319, 629 *illus.*
Double-displacement reactions, 204–205 *lab*, **206**, 207 *table*, 546
Double helix, DNA, 686
Double-pan balances, 796
Downs cell, 601 *illus.*
Drinking water, fluoridation of, 277 *illus.*, 278
Drugs, development of new, 29
Dry cells, carbon-zinc, 594
Dry ice, 122 *illus.*; sublimation of, 373 *lab*
Drying agents. *See* Dessicants
***d* sublevels**, 233, 238, 238 *table*
Ductile metals, 311
Dynamic equilibrium, **209**
Dysprosium, 293

E

Earth: elements in crust of, 124 *illus.*, 284 *illus.*, 666 *illus.*; surface temperature of and water, 446
Earth Science Connection: cave formation, 525; refining of ores by bacteria, 723; weather balloons, 383
Efficiency of industrial power production, 727 *table*, 727–728
Efficiency of reaction, 420
Eggshells: calcium in, 169 *lab*; reaction with vinegar, 483 *illus.*
Egg whites, protein denaturation in, 671 *illus.*

Einstein, Alfred, 756
Eka-aluminum, 89
Eka-silicon, 89
Elastic collisions, kinetic model of gases and, 340
Electrical conductivity. *See* Conductivity
Electric cars, batteries for, 599
Electric charge, observe, 227 *lab*
Electricity, 722
Electrochemical cells, 584–591; lemon batteries, 585–586, 586 *lab*; oxidation-reduction (redox) reactions in, 587–588, 592–593 *lab*; potential difference and, 586; voltaic, 584; voltaic cells, 587–590
Electrodeposition, 611
Electrolysis, 600–605, 609–610; chemical production and, 603–604; cleaning by, 610; electroplating and, 609–610; formation and decomposition of zinc iodide by, 134–135 *lab*; formation of metallic sodium from, 262 *illus.*; process of, 602 *lab*, 602–604; refining of ores by, 604 *illus.*, 604–605, 606–608; of toxic wastes, 614; Word Origin, 602
Electrolytes, 142; acids as, 488; bases as, 492
Electrolytic cells, 600, 602 *lab*
Electrolytic conduction, 602
Electromagnetic radiation, 68–70, 231
Electromagnetic spectrum, 69–70
Electron cloud model, 75–77, 229, 238–239
Electron configurations, 240, 242–245; of each element, 854–855 *table*; first period elements, 242; fourth period elements, 245; inner transition elements, 248; noble gases, 244; periodic table trends, 241–245, 248; prob. ability of distribution in 1s orbital, 244 *lab*; second period elements, 242; third period elements, 243–244; transition elements, 245, 248
Electron dot diagram: convert to molecular models, 314–315
Electronegativity, 301–303; covalent compounds and, 307–309; differences in, *see* Electronegativity difference (Δ *EN*); hydrogen bonding and, 438; ionic compounds and, 303–304, 306; periodic table trends, 302–303; Word Origin, 301
Electronegativity difference (Δ *EN*), 303–304, 310 prob.; covalent bonding and, 307–309; ionic bonding and, 303–304; polar covalent bonds and, 308–309

Electrons, **59**, 64, 65 *table*. *See also* Valence electrons; Bohr's model of atom and, 67–68; bonding pairs, 314; discovery of, 59–60, 228; electron cloud model of, 75–77, 238–239; energy levels and, *see* Energy levels; nonbonding, 314; orbits, 67–68, 68 *illus.*; sharing, 136–139; sublevels of, 233
Electron transport chain, **692**, 693 *illus.*
Electrophoresis, 610
Electroplating, 609–610, 611
Elements, 24–25, 27; alphabetical listing of, 850 *table*; cereal fortified with, 28 *lab*; compounds of, *see* Compounds; compounds vs., 151 *lab*; diatomic, 172; in Earth's crust, 124 *illus.*, 666 *illus.*; electron configuration of each, 854–855 *table*; emission spectrum of, 72, 74, 75 *lab*; essential for good health, 126; in human body, 666 *illus.*; ionic charges of some, 155 *table*; mass percent in compounds, 421; mixtures vs., 151 *lab*; molar mass of, 407; number of atoms in a sample of, 407, 409 prob.; Periodic table of, *see* Periodic table of elements; predicting properties of unknown, 87 *lab*; properties of each, 851–853 *table*; relative masses of, 407; symbols for, 27, 27 *table*, 90–91 *illus.*; synthetic, 100
Emergency light sticks, 195
Emission spectra, 72, 74, 75 *lab*, 231–232, 232 *illus.*, 232 *lab*
Empirical formula, **426**–428; determining correct molecular formula from, 428; determining mass percents from, 426–428; Word Origin, 426
Emulsions, 470
Endothermic reactions, **41**, 193, 213, 704 *illus.*, 705, 708 *lab*, 722 *lab*; activation energy and, 710; cold packs and, 706; decomposition of orange mercuric oxide, 56, 705; direction of, 714, 714 *table*
Energy, **40**, 362; alternative sources of, 724–725; changes in in chemical reactions, *see* Energy changes, chemical reactions and; conservation of, 707; direction of reaction and, 213; in food, 717, 717 *table*, 720–721 *lab*; in natural systems, 732–733; nuclear reactions and, 759; relationship with mass (E=mc²), 756; storage of in ATP and ADP, 691; Word Origin, 707
Energy changes, chemical reactions and, 40–41, 189 *illus.*, 194 *lab*, 704–705; activation energy and, 709–710;

chemical equations and, 193–194; direction of reaction and, 213; dissolution of solids in water and, 708 *lab*; endothermic reactions, 193–194, 704, 704 *illus.*, 705; exothermic reactions, 194, 194 *lab*, 704–705; heat of reaction, 708; hot and cold packs and, 706; symbols depicting, 707
Energy costs: alternative energy sources and, 724–725; catalysts and, 726; efficiency and, 727–728; electricity generation and, 719; increased entropy, 726–727; production and recycling of aluminum and, 719
Energy levels, **73**, 76–77, 233, 238–240; electron distribution in, 232 *illus.*, 238 *table*, 238–240; stable, 130–133
Engine coolants, 466
Enthalpy, 708
Entropy, **712**–713; energy production and, 726–727; living systems and, 732–733
Enviromental health inspector, 449
Environmental technician, 449
Enzymes, **220**, 672–673; active site model, 672 *illus.*; denaturation of, 671, 674–675 *lab*; metabolism and, 689
Equations: chemical, *see* Chemical equations; ionic, 517–518, 520, 521 prob.; nuclear, 742, 746 prob.
Equilibrium, **209**, 212–213; adding of reactants or energy and, 213; Haber process (synthesis of ammonia) and, 214–215; Le Chatelier's principle, 212–213; removing products and, 212–213; vapor-liquid, 354–355, 356
Essential elements, 126
Esters, 643 *illus.*, 645 *lab*
Estimated Safe and Adequate Dietary Intake (ESAI), 126
Ethane, 180, 180 *table*, 623; dehydrogenation, 726; molecular shape of, 321; structural diagrams, 624 *table*
Ethanol, 138 *illus.*, 642 *illus.*, 815 *table*; Breathalyzer tests for, 569; formula and uses, 30 *table*; production of by fermentation, 695; specific heat, 445 *table*; synthesis of, 645–646; vaporization of, 355 *lab*; vapor pressure of, 355
Ethene. *See* Ethylene
Ether (diethyl ether), 643 *illus.*
Ethers, 643 *illus.*
Ethnobotany, 144
Ethyl alcohol. *See* Ethanol
Ethyl butyrate, 643 *illus.*
Ethylene (Ethene), 630, 630 *illus.*; molecular modeling of, 322; polymer-

Ethylene glycol

ization reactions, 654 *illus.*; synthesis of, 726
Ethylene glycol, 30 *table*, 465 *illus.*, 466, 630, 642 *illus.*, 815 *table*
Ethyl ether, 643 *illus.*
Ethyne, 323, 633
Europium, 105 *illus.*, 292–293
Evaporation, **350**, 446
Everyday Chemistry: air bags, 417; bleaching reactions, 192; catalytic converters, 711; chemicals in food, 17; chemicals in the body, 17; coinage metals, 108; compact disc (CD) production, 611; essential elements, 126; fake fats and designer fats, 684; fireworks, 74; flameless ration heaters, 219; freeze drying, 351; gemstones and glass, color of, 246–247; hard water, 158; hiccups, 534; lightning-produced fertilizer, 571; matches, chemistry of, 273; microwave heating and decomposition, 318; permanent waves, 655; popping of popcorn, 395; soaps and detergents, 455; sweetness of foods, 680
Example Problems: alkanes, formulas and names of, 626 prob.; atoms, number in a sample of an element, 409 prob.; Boyle's law, 384–385 prob.; Charles's law, 391 prob.; combined gas law, 393; converting pressure units, 377 prob., 379 prob.; food calories, 717 *table*; formula for a binary ionic compound, 156 prob.; formula for a polyatomic ion, 159 prob.; heat of reaction for combustion, 716 prob.; ideal gas law and, 419 prob.; ionic and net ionic equations of strong acid/strong base reactions, 521 prob.; ionic and net ionic equations of strong acid/weak base reactions, 524 prob.; ionic and net ionic equations of weak acid/strong base reactions, 529 prob.; mass of a number of moles of a compound, 410 prob.; molarity from solution volume, 463 prob.; molarity from titration, 542 prob.; nuclear equations, writing and balancing, 746 prob.; number of formula units in a sample of a compound, 409 prob.; preparation of different solution volume, 462 prob.; radioactive dating of fossils, 755 prob.; writing chemical equations, 198 prob.
Exothermic reactions, **40**, 194, 194 *lab*, 704–705, 708 *lab*, 722 *lab*; activation energy and, 216, 709; direction of, 712–714, 714 *table*; hot packs and, 706

Extrusion, 288–289
Eye, vision and vitamin A, 632

F

Fabric, testing for synthetic fibers, 649–651 *lab*
FADH$_2$, 693 *illus.*
Fahrenheit scale, 347–348, 795
Fake fats, 684
Families. *See* Groups
Faraday, Michael, 10 *lab*
Fats, **681**. *See also* Lipids; fake, 684; fatty acids in, 681–682
Fat-soluble vitamins, 688
Fatty acids, **681**–682; monounsaturated, 682; polyunsaturated, 682; saturated, 682
Fermentation, **695**, 699; alcoholic, 646, 695, 695 *illus.*, 696 *lab*; lactic acid, 695, 699
Fermi, Enrico, 760
Fertilizers, 271, 272 *illus.*, 424, 571
Fillings, 594
Fingerprint classifier, 13
Fingerprints, development of, 12
Fire extinguishers, 123 *illus.*
Fireflies, 575
Firefly squid, 575
Fireworks, 74
First aid, 845
Fish, antifreeze proteins in, 668
Fission, nuclear. *See* Nuclear fission
Fission nuclear power plant, 759–760
Fixer, 565 *illus.*
Flameless ration heaters, 219
Flame tests, 232 *lab*
Flavin adenine dinucleotide (FADH2), 693 *illus.*
Flavorings, 643 *illus.*
Flexinol, 107
Fluids. *See* Gases; Liquids
Fluorapatite, 278
Fluorescent lights, 345
Fluoridation, 277 *illus.*, 278
Fluorides, 277 *illus.*, 278
Fluorine, 276; as diatomic element, 172; electron configuration, 243 *table*; electronegativity, 301; as essential element, 126; fluoridation and, 277 *illus.*, 278; hydrogen bonding by, 438; reaction with noble gases, 129; reactivity of, 261, 276
Fluorite, 154 *illus.*
Foam, 471
Fog, 470
Food: chemicals found in, 17; chromatography of dyes in, 326–327 *lab*;

Functional groups

energy content of, 717, 717 *table*, 720–721 *lab*; freeze drying of, 351; irradiation of, 767; spoilage of, 189 *illus.*, 221
Food and drug inspector, 491
Food irradiation, 767
Food webs, 732 *illus.*
Fool's gold, 52 *illus.*, 163 *illus.*
f orbitals, 248
Forensic Analytical Specialties, 12–13
Forensics: detection of blood stains with luminol, 573, 574; forensic scientists, 12–13
Forensic scientists, 12–13
Forging, 288
Formaldehyde, 644 *illus.*
Formic acid, 499 *illus.*, 642 *illus.*
Formula mass, **408**, 409 prob.
Formulas, **31**, 136; for binary ionic compounds, 153–155, 156 prob.; for covalent compounds, 178–179; determining unknown with mass percents, 426–428; empirical, 426–428; for hydrates, 165; multiple formula units in, 167; for polyatomic ions, 157, 159 prob., 160 prob.; for transition element compounds, 163, 163 *illus.*
Formula unit, **154**; formula mass and, 408, 409 prob.; representing multiple, 167
Fortified cereals, 28 *lab*
Fossil fuels, 636–639, **709**, 722; coal, 636, 639; natural gas, 636, 637; petroleum, 636, 637–639
Fossils, radioactive dating of, 752–753, 755 prob.
Fourth period elements: electron configuration, 245; properties, 280–281
Fractional distillation, 272 *illus.*, 352–353, **637**
Francium, 261
Frasch, Herman, 274 *illus.*
Frasch process, 274
Freeze drying, 351
Freeze-point depression, 464–465, 466
Freezing point, 33, **362**
Freon, 641 *illus.*
Frequency, 69 *illus.*, 231
Fructose, 678
Fruit: oxidation of cut surfaces, 575–576, 576 *illus.*; treatment with ethylene, 630 *illus.*
f sublevel, 233, 238, 238 *table*
Fudge, 459 *illus.*
Fuel cells, 598
Fullerenes, 173, 176
Functional groups, **640**–646; amino group, 644 *illus.*; carbonyl group,

644 *illus.*; carboxyl group, 642 *illus.*; halogens, 641 *illus.*; hydroxyl group, 642 *illus.*; production of compounds containing, 645–646
Furnaces, 558
Fusion, heat of, 362–363
Fusion, nuclear, 345, **760**–762; Word Origin, 761
Fusion reactors, 762

G

Gadolinium, 293
Gadolinium-153, 764 *illus.*
Galena, 269 *illus.*
Gallium, 89 *illus.*, 268
Gallium arsenide, 272 *illus.*
Galvani, Luigi, 584
Galvanizing, 291, 555, 555 *illus.*
Gamma decay, 745–746
Gamma radiation, 739 *lab*, 745–746
Gamma rays, 70 *illus.*, **745**–746; food irradiation with, 767
Garcia, John (pharmacologist), 634–635
Gardening, basic anhydrides and, 494
Gas chromatography, 325
Gases, 33, **339**; Avogadro's principle, 396; condensation of, 354; diffusion of, 341 *lab*, 349; gas laws governing behavior of, *see* Gas laws; ideal, 341; kinetic model of, 340–342, 341 *lab*, 390 *illus.*; mass of, 3 *lab*; mass-pressure relationship, 371 *illus.*, 371–372; mass-volume relationship, 373 *lab*; molar volume, 415, 416 prob.; particle speed distribution, 346 *illus.*; pressure of, *see* Gas pressure; properties of, 339 *illus.*; release of by chemical reactions, 189 *illus.*; solubility in liquids, 469; temperature-pressure relationship, 372–373; temperature-volume relationship, 389–390, 391 prob.
Gas laws, 380–382, 389–394; Boyle's law, 380–382, 384 *lab*, 384–385 prob., 386–387 *lab*; Charles's law, 389–**390**, 391 prob.; combined gas law, 392–393, 393–394 prob.; ideal gas law, 418 *lab*, 418–419, 419 prob.; law of combining gas volumes, 394, 396
Gasoline, 23, 143 *illus.*, 622, 623, 624 *illus.*; alkenes in, 630; combustion, 189 *illus.*, 206 *illus.*; octane rating of, 624; production, 638 *illus.*; volatility of, 33
Gas pressure, 341–342, 370–374, 376–378; devices that measure, 374, 375; number of particles and, 371–372; pressure-volume relationship (Boyle's law), 380–382, 384 *lab*, 384–385 prob., 386–387 *lab*; temperature and, 372–373, 389–390, 391 prob.; units used for, 376–378, 377 prob., 379 prob.
Gastric ulcers, 536, 536 *illus.*
Geiger counters, 739 *lab*, 747
Gel chromatography, 325
Gels, 38 *lab*, 471
Gemstones, colors of, 246
Genetic code, 686. *See also* DNA (Deoxyribonucleic acid)
Geochemical cycles, conservation of matter and, 53
Geometric isomers, 631
Geothermal energy, 724–725
Geraniol, 421
Germanium, 89 *table*, 269
Geysers, 725
Glacial acetic acid, 529
Glass: blowing of soda-lead for sculptures, 344; colored, 247, 292; lanthanides in, 292; recycling of, 58; specific heat, 445 *table*
Glass cullet, 58
Glazes, 161
Glucose, 673, 678, 679; metabolism of, 690, 692, 693 *illus.*; production of by photosynthesis, 730; structure, 653 *illus.*
Glutamine, 669 *table*
Glycerin, 455
Glycerol, 439 *illus.*, 681
Glycine, 669
Glycogen, 673, 679
Glycolysis, 692, 693 *illus.*
Gold. *See also* Fool's gold; Asante weights and, 411; attempts to convert common metals into, 25 *lab*; biorefining of, 723; chemical symbols for, 27 *table*; as coinage metal, 84, 108, 290; density of, 34 *table*; 18-karat, 23 *table*; 14-karat, 23 *table*; malleability of, 311; nuggets of, 24 *illus.*; specific heat, 445 *table*
Gold foil experiement, Rutherford's, 62, 228
Goodyear, Charles, 656
Granite, 21 *illus.*
Graphite, 105 *illus.*, 173, 174
Graphs, 811–813; bar graphs, 811; of direct relationships, 813; of inverse relationships, 814; line graphs, 811–812; pie graphs, 811
Gray (radiation unit), 769
Greeks: early ideas on matter, 50–51; use of scientific models by, 9
Green chemistry, 425
Green technology, 425
Group 1 elements. *See* Alkali metals
Group 2 elements. *See* Alkaline earth metals
Group 3 elements, 281
Group 5 elements, 281
Group 6 elements, 281
Group 13 elements, 268; ionic charge, 155 *table*; properties, 268; uses of, 268; valence electron configuration, 241, 242, 268
Group 14 elements, 269; properties, 269; uses of, 269, 269 *illus.*; valence electron configuration, 241, 242, 269
Group 15 elements, 271, 272 *illus.*; ionic charge of, 155 *table*; valence electron configuration, 241, 242, 271
Group 16 elements, 274, 275 *illus.*; ionic charge of, 155 *table*; properties, 274; uses of, 275 *illus.*; valence electron configuration, 229, 242, 274
Group 17 elements, 276, 277 *illus.*; ionic charge of, 155 *table*; reactivity, 96 *lab*, 276; uses of, 277 *illus.*; valence electron configuration, 242, 276
Group 18 elements (noble gases), **94**, 97, 279; electron configuration, 243–244, 244 *table*, 281; stability of, 129–130, 279; valence electron configuration, 242
Groups, **94**; atomic structure of elements in, 96–97; reactivity within, 96 *lab*
Guanine (G), 686
Gumdrop models, 313–315; of ammonia, 320; of carbon dioxide, 319; of diatomic elements, 313; of ethane, 321, 322; of ethyne, 323; of methane, 320–321; of water, 314–315

H

Haber, Fritz, 214–215
Haber process, 214–215, 571
Hafnium, 87 *illus.*
Hair: chemistry of perms, 655; dissolving of by bases, 480 *illus.*
Half-life, **752**; of commonly used isotopes, 752 *table*; modeling, 748–749 *lab*
Half reactions: electrolysis, 601; redox, 556, 557–558
Hall, Charles Martin, 605
Hall-Heroult process, 214–215, 604 *illus.*, 604–605, 719
Halogenated compounds, 641 *illus.*
Halogen lamp, 105 *illus.*
Halogens, 84, 97, **276**, 277 *illus.*; halo-

genated hydrocarbons, 641 illus.; properties of, 85, 85 table; reactivity, 276; uses, 276 illus., 277 illus.; valence electron configuration, 276

Halogen triad, 85 table, 85–86
Hard water, 158, 189 illus., 452 lab, 455
HBO therapy. See Hyperbaric oxygen (HBO) chambers
HDPE, 657
Health Connection: homeoglobin, 694
Heartburn, antacids and, 504 lab
Heart disease, 683
Heart imaging, 764 illus.
Heat, 704, 707–710; absorbtion of by chemical reactions, 704, 705. see also Endothermic reactions; reducing release of waste heat, 727–728; release of by chemical reactions, 704–705. see also Exothermic reactions; units of (joules), 707
Heating curves, 362
Heat of fusion, 362–363
Heat of reaction, 708; activation energy and, 709–710; measuring by calorimetry, 715, 718 prob.
Heat of solution, 460
Heat of vaporization, 358–359, 446
Heisenberg, Werner, 238
Heisenberg uncertainty principle, 238–239
Helium, 93, 94, 279; in airships, 140; density, 34 table; electron configuration, 130 illus., 242, 243, 244, 244 table; molar volume, 415
Heme, 284 illus., 694
Hemoglobin, 284 illus., 537, 668, 670 illus., 694
Heptane, 180 table
Heroult, Paul, 605
Hershey, Alfred Day, 766
Hertz (Hz), 69 illus.
Heterogeneous mixtures, 21–22; Word Origin (heterogeneous), 21
Hexane, 180 table; isomers, 628; vaporization of, 355 lab
Hiccups, 534
High-density polyethylene (HDOE), 657
Hindenburg, 140
History Connection: the *Hindenburg*, 140; Lavoisier, politics and chemistry, 56; lead poisoning in Rome, 270
Holes, semiconductors, 110
Homogeneous, Word Origin, 23
Homogeneous mixtures. See Solutions
Hormones, 690
Horticulturalist, 211
Hot-air balloons, 383, 389, 392 illus.
Hot packs, 460 illus., 706

Hot-working of steel, 288–289
Household chemicals, chemical and physical properties of, 18–19 lab
Household products: pH of, 479 lab, 506–507 lab
Housepaints, as colloids, 471
How It Works: breathalyser test, 569; cement, 166; emergency light sticks, 195; hot and cold packs, 706; hydrogen-oxygen fuel cell, 598; indicators, 543; lightbulbs, 282; lithium batteries in pacemakers, 595; nickle-cadmium (nicad) batteries, 597; pacemakers, 595; pressure cookers, 357; reverse osmosis units, 468; smoke detectors, 744; taste, 519; tire-pressure gauge, 375
Human body, composition of, 17, 666 illus.
Hydrates, 164–165; cement, 166; chemical weather predictors, 164 lab; formulas, 165; naming, 165
Hydrides, 309
Hydrobromic acid, 498 table, 522–523
Hydrocarbons, 180, 180 table, 495. See also Alkanes; Alkenes; Alkynes; air pollution and, 495; aromatic, 636; branching of, 623; cracking and, 638; fossil fuels, 636–639, 709; functional groups on substituted, 640–646; modeling of, 321–323, 323 lab, 621 lab; reforming and, 638; saturated, see Alkanes (saturated hydrocarbons); unsaturated, 629, 629–631, 630 lab, 633
Hydrochloric acid, 815 table; formula and uses, 30 table; polar covalent bonding, 308 illus.; production of hydronium when dissolved in water, 484, 486; reaction with calcium carbonate, 713; reaction with metals, 482, 482 lab; reaction with strong bases, 516 illus., 517–518; reaction with weak bases, 527, 528 illus.; as strong acid, 497 illus., 498, 498 table
Hydrogen, 93, 94; as diatomic element, 172; in Earth's crust, 666 illus.; electron cloud model of, 239 illus.; electron configuration, 242; emission spectrum, 72 illus., 231, 232 illus.; energy levels, 73 illus., 76 illus.; as essential element, 126; as fuel source, 140; in human body, 666 illus.; properties, 125; reaction with oxygen, 125 illus., 198 prob., 216, 705
Hydrogenation, 631
Hydrogen bonding, 438–439, 680. See also Water

Hydrogen carbonate ion, buffers and, 532–533, 535
Hydrogen chloride. See Hydrochloric acid
Hydrogen-oxygen fuel cell, 598
Hydrogen peroxide: catalytic decomposition of, 674–675 lab; chemical properties of, 39 illus.; decomposition of, 213 illus.; inhibitors and, 221; instability, 38, 275 illus.; reaction with magnesium dioxide, 39 illus.
Hydroiodic acid, 498 table
Hydronium ions, 483, 484, 486, 517; pH and, 501, 502 prob., 503
Hydroxide antacids, 538
Hydroxide ions, 156 illus., 489, 517; pH and, 501, 502 prob., 503
Hydroxides, 262 illus.
Hydroxyapatite, 278
Hydroxyl group, 642 illus.
Hygroscopic substances, 165
Hyperbaric oxygen (HBO) chambers, 388
Hyperventilation, 533
Hypochlorite bleaches, 567–568
Hypothesis, 57

I

Ice, 125 illus., 436; density, 34 table, 436–437, 440; freeze-melting, 440, 441 illus.; sublimation, 354, 373 lab
Icebergs, 26, 125 illus.
Ice man, dating of, 750–751
Ideal gases, 341
Ideal gas law, 418 lab, 418–419, 419 prob.
Incandescent lights, 129 illus.
Independent variable, 812 illus.
Indicators, acid-base. See Acid-base indicators
Indium, 268
Industrial chemical worker, 317
Industrial machinery mechanic, 449
Industrial processes. See also specific processes; aluminum production and recycling, 719; catalysts in, 726; efficiency of, 727 table, 727–728; energy production and, 722; entropy from, 726–727
Inert gases, 129–130, 281, 282. See also Noble gases
Infrared waves, 70 illus.
Inhibitors, 221
Inks, chromatography of, 22 lab
Inner transition elements, 102, 248, 292–293; electron configuration, 248, 292; radioactivity, 293; uses of, 292–293

Index

Inorganic compounds, 178; naming, 179–180
Insoluble compounds, 213
Insulin, 690
Integrated circuits, 111
International System of Units, 791–792
International System of Units (SI), 4
Interparticle forces, 142–143, 145
In the Field: biochemist, 676–677; chemist, 316–317; cosmetic bench chemist, 490–491; forensic scientist, 12–13; metal plater, 612–613; pharmacist, 634–635; plant-care specialist, 210–211; wastewater operator, 448–449
Invar, 288 *table*
Inverse relationships, graphs and, 814
Iodine, 84, 105 *illus.*, 276; as antibacterial, 277 *illus.*; biological importance of, 276, 276 *illus.*; as diatomic element, 172; as essential element, 126; halogen triad, 85, 85 *table*; ionic compounds formed by, 133 *lab*; properties of, 85 *table*; reaction with starch, 189 *illus.*, 218 *lab*; reactivity, 96 *lab*; sublimation, 354 *illus.*
Iodine-131, 764 *illus.*
Ion exchangers, 158
Ionic attraction, 133
Ionic bonds, 132, 133, 300–301; electronegativity difference between bonding atoms (Δ EN) and, 303–304
Ionic charges: predicting, 154–155; of representative elements, 155 *table*
Ionic compounds, 132; binary, 152–155; classification of compounds as, 170–171 *lab*; difference between electronegativities (Δ EN) of bonding atoms and, 303–304; dissolving of by water, 451–452; formation of by electron transfer, 132–133, 133 *lab*, 134–135 *lab*; formulas and names of, 153–155, 162–163, 167; hydrates, 164–165; identification of aqueous solutions and, 456–457 *lab*; interparticle forces in, 142 *illus.*; ion charge prediction, 154–155; polyatomic, 156–157; properties, 141–142, 300 *table*, 330–331; separating from covalent substances, 168–169, 169 *lab*; transition elements, 160, 162 *table*, 162–163
Ionic crystals, 132–133
Ionic equations, 518; for strong acid + strong base, 517–518, 520, 521 prob.; for strong acid + weak base, 522–523, 524 prob.; for weak acid + strong base, 528–529, 529 prob.
Ionic size (radii), 258; atom size-ion size relationship, 258 *illus.*; patterns in main group elements, 258–259
Ionic solids, 331
Ionization, 488; of acids, 488, 498–499; of bases, 489
Ionizing radiation, 769
Ionizing smoke detectors, 744
Ions, 132, 240; charge prediction, 154–155; spectator, 520
Iridium, 290
Iron, 284–285 *illus.*, 290; biological importance, 284 *illus.*; chemical symbol, 27 *table*; in Earth's crust, 666 *illus.*; electron configuration, 245 *illus.*, 283 *lab*; as essential element, 126; glass and gem color and, 246, 247 *table*; oxidation states, 281; properties, 6 *illus.*, 38, 39 *illus.*, 104 *illus.*, 280, 290; reaction of with chlorine, 127; reaction with copper sulfate, 203 *illus.*; reaction with HCL, 482; rusting, 39 *illus.*, 120 *lab*, 202 *illus.*, 248, 555, 557 *lab*, 570; separation of from ore, 284 *illus.*; smelting of in blast furnaces, 558 *illus.*, 567; specific heat, 445 *table*
Iron(II) oxide, 246
Iron(II) sulfate, 160 *illus.*
Iron(III) sulfate, 160 *illus.*
Iron Age, 567
Iron chloride, 138 *illus.*
Iron pyrite, 52 *illus.*, 163 *illus.*
Iron smelting, 558 *illus.*
Iron triad, 290
Irradiation, food, 767
Isomers, 627–628; geometric, 631; positional, 631; structural, 628
Isopropanol, 642 *illus.*
Isopropyl alcohol, 356, 642 *illus.*
Isotopes, 60. *See also* Radioisotopes; atomic mass and, 65, 66 *illus.*; half life of common, 752 *table*; nuclear notation and, 742; representing with pennies, 61 *lab*

J

Jewelry alloys, 23 *table*
Jordan, Lynda, 676–677
Joule (J), 358, 707

K

Kekule, August, 636
Kelvins (K), 347
Kelvin scale, 347–348, 795
Keratin, 505, 668
Kerosene, 638 *illus.*
Ketones, 644 *illus.*
Kilocalories, 717
Kilopascal (kPa), 376
Kinetic energy: diffusion and, 349; Kelvin scale and, 347–348; mass and speed and particles and, 348–349; temperature and, 346–348, 360–361 *lab*
Kinetic theory, 340–343; Boyle's law and, 382; Charles's law and, 390; of gases, 340–342, 341 *lab*; of liquids, 342; of solids, 343
Kitchen chemicals, 18–19 *lab*
Krypton, 129, 130 *illus.*, 244 *table*, 279

L

Lab(s). *See* Launch Lab(s)
Labs: ChemLab, *see* ChemLab; Launch Labs, *see* Launch Labs; MiniLab, *see* MiniLab; safety guidelines, 845; safety symbols, 846
Lactase, 673
Lactic acid, 428
Lactic acid fermentation, 695, 699
Lactose, 678
Lactose intolerance, 220, 673
Landscape architect, 211
Language Arts. *See* Literature Connection
Lanthanides, 90–91 *illus.*, 92, **102,** 105 *illus.*, 248, 292–293
Lanthanum, 292
Launch Labs: buffers, 515 *lab*; differences in mass, 3 *lab*; electric charge, 227 *lab*; elements, compounds, and mixtures, 151 *lab*; evidence of chemical reaction, 117 *lab*; hydrocarbons, 621 *lab*; lemon batteries, 583 *lab*; mole, 403 *lab*; observe a chemical reaction, 187 *lab*; oil and vinegar dressing, 299 *lab*; periodic table trends in melting point, 255 *lab*; pH of household products, 479 *lab*; radiation, penetration of, 739 *lab*; rate of reaction, 703 *lab*; redox reactions, 553 *lab*; solution formation, 435 *lab*; temperature and mixing, 337 *lab*; testing for simple sugars, 665 *lab*; versatile metals, 83 *lab*; volume and temperature of gas, 369 *lab*; What's inside?, 49 *lab*
Lavoisier, Antoine, 51, 56, 196
Law of combining gas volumes, 394, 396, 404
Law of conservation of energy, 707
Law of conservation of matter (mass),

40, 54–55 *lab;* discovery of, 51, 196; recycling and, 53
Law of definite proportions, 52
LCDs (liquid crystal displays), 345
LDPE, 657
Lead, 269; density, 34 *table;* symbol for, 27 *table;* uses, 269 *illus.*
Lead(II) iodide, 206 *illus.*
Lead(II) nitrate, 206 *illus.*
Lead-acid storage battery, 269 *illus.,* 596
Lead-chamber process, production of sulfuric acid by, 424
Lead poisoning, 270
Le Châtelier, Henri Louis, 212
Le Châtelier's principle, 212–213
LEDs, 590
Lemon batteries, 583 *lab,* 585–586, 586 *lab,* 591
Length, measuring, 794, 798
Levi, Primo, 95
Lewis dot diagrams, 77; for main groups of elements, 96 *illus.*
Libby, Willard, 750
Life, molecules of. *See* Biomolecules
Light, electromagnetic spectrum and, 231
Lightbulbs, 282
Light-emitting diodes (LEDs), 590
Lighter fuel, 627 *illus.*
Lightning, 571, 573
Light reactions, photosynthesis and, 730–731
Light sticks, 195, 573
Lime, 155 *illus.;* formation of, 209, 212; as soil treatment, 494
Limestone (calcium carbonate): acid dissolution, 483 *illus.,* 525; cave formation and, 525; decomposition, 209, 212; formula, 30 *table,* 815 *table;* hard water and, 158; insolubility in water, 483; neutralization of lakes by, 535; reaction with acids, 482 *lab,* 483; reaction with hydrochloric acid, 713
Limiting reactant, 218
Linear acetylenic carbon, 176
Linear molecular geometry, 313 *illus.,* 319 *illus.,* 323 *illus.*
Line graphs, 811–812
Lipid membranes, 683
Lipids, 667, **681**–683; fake fats and, 684; fatty acids and, 681–682; functions, 683; steroids, 682; structure, 681, 682
Liquid aerosols, 470
Liquid crystal displays (LCDs), 345
Liquid crystals, 345
Liquid hydrogen, 140
Liquids, 33, **338;** boiling, 356; diffusion in, 349; dissolved gases in, 469 *illus.;* evaporation, 350; freezing, 362–363; kinetic model of, 342; mass of, 3 *lab;* properties, 339 *illus.;* surface tension and, 442–443; volatile, 351, 355
Literature Connection: Jules Verne, icebergs, 26; Primo Levi, language of a chemist, 95
Lithium, 84, 95, 97, 260, 261; atomic size, 257; electron configuration, 242, 243, 243 *table;* ionic compounds formed by, 133 *lab;* ionic radii and, 259; reactivity, 259, 261
Lithium batteries: for electric cars, 599; in pacemakers, 595
Lithium fluoride, 261 *illus.,* 304
Lithium hydroxide, 201, 416 prob., 498 *table*
Litmus paper, 481
Lone electron pairs, 314
Low-density polyethylene (LDPE), 654 *illus.,* 657
Luciferase, 575 *illus.*
Luminol, 573, 574
Lung cancer, radon exposure and, 770, 771
Lye, 263 *illus.,* 455, 494

M

Macroscopic world, 7
Magma, 724
Magnesium, 264; in Earth's crust, 666 *illus.;* electron configuration, 243 *table;* as essential element, 126; ionic compounds formed by, 133 *lab;* magnesium-copper redox reaction, 588 *illus.;* reactivity, 96 *lab,* 264; shielding effect, 302, 303 *illus.;* uses of, 265 *illus.*
Magnesium alloys, 265 *illus.*
Magnesium chloride, reaction with silver nitrate, 198–199 prob.
Magnesium-copper voltaic cell, 587–589
Magnesium dioxide, 39 *illus.*
Magnesium hydroxide, 30 *table,* 498 *table,* 815 *table*
Magnesium oxide, 152, 265 *illus.*
Main group elements, 257–279; alkali metals, *see* Alkali metals; alkaline earth metals, *see* Alkaline earth metals; atomic radii trends, 257, 260 *lab;* halogens, *see* Halogens; ionic radii trends, 258–259; metallic property trends, 256 *illus.;* metal-metalloid-nonmetal-noble gas pattern, 256, 256 *illus.;* period 2 chemical reactivity, 259–261
Malachite, 246, 246 *illus.*
Malleability, 311
Maltose, 678
Manganese: electron configuration, 245 *illus.;* as essential element, 126; glass and gem color and, 247 *table*
Manganese(IV) oxide, reduction in carbon-zinc dry cells, 594 *illus.*
Manganese steel, 288 *table*
Manhattan project, 230
Manufacturers' sales representative, 491
Marble, acid dissolution of, 483, 483 *illus.*
Martensite phase, 106
Mass, 4, 4; atomic, *see* Atomic mass; comparing, 4 *illus.,* 5; conservation of, 40, 196–197; determine number of items by, 408 *lab;* measurement of, 796–797; molar, *see* Molar mass; molecular, 408; relationship with energy ($E=mc^2$), 756; relationship with gas volume, 373 *lab;* speed of particles and kinetic energy, 348–349; state of matter and, 3 *lab*
Mass formula, 408
Mass number, 64
Mass percents: determination of, 421, 422–423 *lab;* determining chemical formulas from, 426–428
Mass spectrometers, 325
Matches, chemistry of, 273
Matter, 4. *See also* Atoms; Compounds; Elements; States of matter; atomic model of, 9; change in state of, 350, 354–356, 358–359, 360–361 *lab,* 362–363; classification of, 14–16, 18, 20–24, 24 *illus.;* composition and behavior of, 5, 5 *illus.;* conservation of, 51 *illus.,* 53, 54–55 *lab;* density of, 34; early Greek ideas about, 50–51; kinetic theory of, 339–343; macroscopic view of, 7; mixtures of, 15–16, 20–23, 20–24; properties of, 5, 5 *illus.,* 5–6, 32–41; states of, 33, 338–345; submicroscopic view of, 7–8; substances, 14–15
Measurements: of length, 795, 798; of mass and weight, 796–797; precision and accuracy of, 799; as a quantitative observation, 14; scientific notation and, 801–805; significant digits and, 799–801; SI system for, 792–793; of temperature, 795; of volume, 795
Medical laboratory technician, 677
Medical records technician, 677
Medical technologist, 677
Medicines from rain forests, 144
Meitner, Lise, 757
Melting point, 33, **362;** periodic table

Mendeleev, Dmitri

Mendeleev, Dmitri (*continued*): trends, 86; polar vs. nonpolar molecules and, 330
Mendeleev, Dmitri, 86–89, 92
Meniscus, 444
Mercury: boiling point, 356; emission spectrum, 232 *illus.*; melting point, 281; meniscus formation, 444; physical properties, 104 *illus.*; surface tension, 443
Metabolism, **689**. *See also* Biomolecules; chemical energy and, 689; fermentation, 695, 696 *lab*; hormonal control of, 690; photosynthesis, 729–733; respiration, 690–692, 693 *illus.*, 698
Metal alloys. *See* Alloys
Metal hydroxides, 489
Metallic bonds, **311**
Metalloids, **103**; atomic structure, 103; periodic table trends, 98–99 *lab*; position of on periodic table, 100; properties, 105 *illus.*; valence electron configuration, 241
Metallurgical technician, 613
Metal plater (Harvey Morser), 612–613
Metals, 100, **101**–102; atomic structure, 103; bonding in, 311–312; conductivity of, 311, 312; corrosion, 557 *lab*, 570, 572; oxidation, ease of, 587 *table*; oxidation-reduction (redox) reactions involving, 570; periodic table trends, 98–99 *lab*; properties, 101, 103 *table*, 104–105 *illus.*, 311; reaction with acids, 482–483; rusting of, *see* Rust; shape-memory, 106–107; valence electron configuration, 234–235 *lab*, 241, 311–312; versatility of, 83 *lab*
Methane, 180, 180 *table*, 181 *illus.*, 307 *illus.*, 623 *illus.*, 815 *table*; combustion, 194, 709 *illus.*; deep-sea deposits of, 180; formula and uses, 30 *table*; molecular shape, 320–321; nonpolarity, 330; water vs., 330, 330 *table*, 436 *table*
Methanol, 406, 642 *illus.*
2-Methylbutane, 628 *illus.*
Methyl chloride, 640
2-Methylhexane, 628 *illus.*
3-Methylhexane, 628 *illus.*
2-Methylpropane, 623 *illus.*, 627–628
Methyl red, 540 *illus.*
Methyl salicylate, 645 *lab*
Metric system, 4, 792
Microscopes, 236–237; atomic force microscope (AFM), 237; scanning prob.e microscope (SPM), 236; scanning tunneling microscope (STM), 8, 237
Microwave decomposition, 318

Microwave heating, 318
Microwave ovens, 318
Microwave radiation, 69, 70 *illus.*, 318
Milk, 503
Milk of Magnesia, 538
Minerals, nutrition and, 126
MiniLabs. *See also* ChemLabs; alcohol, testing for presence of by redox, 568 *lab*; alloy of copper and zinc, 25 *lab*; antacids and acid-base chemistry, 504 *lab*; atomic radii, periodic table patterns, 260 *lab*; buffers, 533 *lab*; calcium (bones), reaction with vinegar, 169 *lab*; cereal fortified with elements, 28 *lab*; chemical weather predictors, 164 *lab*; chromatography, 22 *lab*, 310 *lab*; corrosion of iron and oxidation, 557 *lab*; determine number of items by weight, 408 *lab*; diffusion of gases in air, 341 *lab*; DNA extraction, 687 *lab*; electrolysis, 602 *lab*; electron configuration of iron, 283 *lab*; elements, predicting properties of unknown, 87 *lab*; emission spectrum, 75 *lab*, 232 *lab*; esters, aroma of, 645 *lab*; exothermic and endothermic reactions, 194 *lab*, 708 *lab*, 722 *lab*; gel, properties of, 38 *lab*; groups, reactivity within, 96 *lab*; isotopes, representing with pennies, 61 *lab*; lemon batteries, 586 *lab*; line emission spectra of elements, 75 *lab*; mass and volume of a gas, 373 *lab*; mixture of water and alcohol, 21 *lab*; molar volume of a gas, 418 *lab*; molecular shapes, modeling, 323 *lab*; neutralization reactions, 518 *lab*; nuclear fission chain reaction, 758 *lab*; paper chromatography, 22 *lab*, 310 *lab*; polymers, comparing, 652 *lab*; radon levels, 770 *lab*; rusting of iron, 120 *lab*; single-displacement reactions, 203 *lab*; 1s orbital, prob. ability of electron distribution in, 244 *lab*; starch-iodine clock reaction, 218 *lab*; straws and Boyle's law, 384 *lab*; vaporization rates, 355 *lab*; yeast fermentation, 696 *lab*
Mining engineer, 613
Misch metal, 292
Mixtures, 15–**16**, 20–24; analysis of, 422–423 *lab*; composition of, determining, 18–19 *lab*; compounds vs., 151 *lab*; elements vs., 151 *lab*; examples of everyday, 16 *illus.*; heterogeneous, 21; homogeneous (solutions), 22–23; separation of, 20, 324–325; of water and alcohol, 21 *lab*
Models, 8–9

Names

Molarity, 460–461; calculation of, 463 prob.; solution preparation and, 462 prob.; from titration, 541, 542 prob., 544–545 *lab*
Molar mass, 406–**407**, 412; of a compound, 408, 410 prob.; of an element, 407, 409 prob.; mass of a number of moles of a compound and, 410 prob.; number of atoms in a sample of an element from, 409 prob.; number of formula units in a sample of a compound and, 409 prob.; predicting the mass of a reactant and, 414–415 prob.
Molar volume, **415**, 416 prob., 418 *lab*
Molded plastics, 658
Mole, **405**; number of things in (Avogadro's constant), 403 *lab*, 405
Molecular compounds. *See* Covalent compounds
Molecular elements, **172**–177. *See also* Covalent compounds; allotropes, 173–177; diatomic elements, 172
Molecular formula, from empirical formula, 428
Molecular mass, **408**, 410 prob.
Molecular modeling, 313–315, 319–323, 323 *lab*; ammonia, 320; carbon dioxide, 319; diatomic elements, 313; ethane, 321, 322; gumdrop models, 313–315; methane, 320–321; water, 314–315
Molecular shape, 313–315. *See* Polarity
Molecular substances. *See also* Covalent compounds
Molecules, **138**–139
Monomers, 647
Monoprotic acids, **484** *illus.*
Monosaccharides, 678
Monounsaturated fatty acids, 682
Morser, Harvey (metal plater), 612–613
Mothballs, 33, 636
Moving phase, paper chromatography, 310 *lab*
MREs (Meal, Ready-to-Eat), 219
Muriatic acid, 497 *illus.*
Muscles, 695
Myoglobin, 284 *illus.*

N

NAD⁺, 693 *illus.*
NADH, 693 *illus.*
Nagaoka, Hantaro, 61
Nail polish remover, 643 *illus.*
Names: alkanes, 625–626, 626–627 prob.; alkenes, 630–631; alkynes, 633; binary inorganic compounds, 179 prob., 179–180; binary ionic com-

Naphthalene

pounds, 153; organic compounds, 180; polyatomic ions, 157; transition element compounds, 162, 162 *table*, 163

Naphthalene, 33, 636
Natural chemicals, synthetic, 29
Natural gas, 636, 637
Natural polymers, 652; comparing with synthetic, 652 *lab*; differentiating from synthetic polymers, 649–651 *lab*
Natural radiation, 768, 769
Natural systems, energy in, 732–733
Neodymium, 105 *illus.*, 292
Neon, 279; electron configuration, 242, 243, 243 *table*, 244 *table*; emission spectrum, 232 *illus.*; isotopes, 60 *illus.*, 64
Neon lights, 92 *illus.*, 129 *illus.*
Neptunium, 100
Net ionic equations, **520**, 521 prob.
Neutralization reaction, **516**–517, 518 *lab. See also* Acid-base reactions
Neutral solutions, 501
Neutrons, 60, 64; discovery of, 60; properties, 65 *table*
Nicad rechargeable batteries, 597
Nickel, 290; as coinage metal, 108; electron configuration, 245 *illus.*; as essential element, 126; properties, 290
Nickel-cadmium (Nicad) batteries, 597
Nickel-metal hydride batteries, 599
Nicotinimide adenine dinucleotide (NAD+), 693 *illus.*
Night vision, 632
NiMH batteries, 599
Nitinol, 106–107
Nitric acid, 498 *table*
Nitrogen, 271; biological importance of, 271; cycling, 53 *illus.*, 271; as diatomic element, 172, 173 *illus.*; electron configuration, 243 *table*; as essential element, 126; fertilizers from, 272 *illus.*; from fractional distillation, 272 *illus.*, 352–353; in human body, 666 *illus.*; hydrogen bonding by, 438; ionic compounds formed by, 133 *lab*; liquid, 105 *illus.*; reactivity, 261; water vs., 436 *table*
Nitrogen cycle, 53 *illus.*, 271
Nitrogen dioxide, 307 *illus.*
Nitrogen fixation, 53 *illus.*, 271, 571
Nitrogen-fixing bacteria, 214, 271
Nitrogen monoxide, 271, 495
Nitrogen oxides, 494 *illus.*, 495
Noble gas compounds, 193, 281
Noble gas configuration, **130**
Noble gases, **94**, 97, 279; applications, 129 *illus.*; electron configuration,

130 *illus.*, 243–244, 244 *table*, 281; periodic table and, 92; stability of, 129–130, 279
Nonane, 180 *table*
Nonbonding electron pairs, 314
Nonmetals, 100, **102**; atomic structure, 103; electron configuration, 241; periodic table and, 98–99 *lab*, 100; properties, 102, 103 *table*, 105 *illus.*
Nonstick coatings, 647–648
Normal boiling point, 356
npn-junctions, 110
n-type semiconductors, 110
Nuclear accidents, 760
Nuclear equations, 742; radioactive decay and, 743, 745–747; writing and balancing, 742, 746 prob.
Nuclear fission, 230, **757**–760; fission chain reaction, 757, 757 *illus.*, 758 *lab*; fission reactors, 759–760; Word Origin (fission), 759
Nuclear fusion, 345, **760**–762
Nuclear fusion reactors, 762 *illus.*
Nuclear medicine, 763, 764 *illus.*
Nuclear model of the atom, 63. *See also* Atomic structure
Nuclear reactions: chemical reactions vs., 742 *table*; energy potential of, 756; equations for, 742, 746 prob.; fission, 757 *illus.*, 757–760; fusion, 760–762; notation for, 742; radioactive decay, 743
Nuclear reactors, 759–760, 762 *illus.*
Nuclear waste disposal, 772–773
Nucleic acids, 667, **685**–687; deoxyribonucleic acid (DNA), 686, 687 *lab*; hydrogen bonding between, 439; nucleotides in, 685–686; ribonucleic acid (RNA), 687
Nucleotides, 685–686
Nucleus, 63, 228, 229 *illus.*
Nutrient cycles, conservation of matter and, 53
Nutrition, 126
Nylon, 648, 652

O

Oaktree automatic arm, 107
Observations, 14, 57; qualitative, **14**; quantitative, **14**
Octane, 180 *table*, 206 *illus.*
Octane ratings, 624
Octet rule, **130**
Odor: changes in and chemical reactions, 189 *illus.*; diffusion of gases in air and, 341 *lab*
Oils, 681. *See also* Lipids; Petroleum;

Oxidation-reduction reactions

action of soaps and detergents on, 455; fatty acids in, 681–682; insolubility in water, 454
Oleic acid, 682 *illus.*
Olestra, 684
Opsin, 632
Orange juice, 15 *illus.*
Orbitals, **238**–239; number of electrons in each, 238, 238 *table*; overlapping, 240; periodic table trends, 241–245, 248; size of, 248, 249 *illus.*; Word Origin, 238
Orbiter **space shuttle**, 201
Ores: refining of by bateria, 723; separation of iron from, 284 *illus.*, 558 *illus.*
Organic chemistry. *See* Biochemistry; Organic compounds
Organic compounds, **178**, 622. *See also* Saturated hydrocarbons; Unsaturated hydrocarbons; alkanes, 623–629, 626–627 prob.; alkenes, 629–631; alkynes, 633, 633 *table*; aromatic hydrocarbons, 636; functional groups, 640–646; naming, 180; plastics, 658; polymerization reactions, 654, 656; polymers, 647–648, 649–651 *lab*, 652 *lab*, 652–653; sources, 636–639
Osmium, 290
Osmosis, **467**; portable reverse osmosis units, 468; Word Origin, 467
Osteoporosis, detection of with radioisotopes, 764 *illus.*
Oxalate ions: as polyatomic ion, 156 *illus.*; reactivity of alkaline earth elements with, 266–267 *lab*
Oxidation number, **155**; multiple in transition elements, 248
Oxidation reactions, 554, **556**. *See also* Oxidation-reduction reactions; combining with reduction half-reactions, 557–558; metals, ease of, 587 *table*; oxidizing and reducing agents and, 559, 562
Oxidation-reduction reactions, 554, **555**–559, 560–561 *lab*, 562; alcohol, testing for presence of by, 568 *lab*; biochemical processes involving, 575; blast furnaces and, 567; bleaching processes and, 567–568; chemiluminescent, 573; copper atoms and ions and, 559, 560–561 *lab*, 562; corrosion of iron and oxidation, 557 *lab*; in electrochemical cells, 587–588, 592–593 *lab*; half-reactions, combining, 557–558; identification of, 558–559; lightning-produced fertilizer and, 571; observing, 553 *lab*; oxidizing and reducing agents and, 559,

Index

Oxides

Oxides 562; photography and, 564, 565 *illus.*; reversing through electrolysis, 600, 601; silver tarnishing, 572; tracking of electrons in, 555–558
Oxides, 493
Oxidizing agents, 559, **562**
Oxyacetylene torches, 633
Oxygen, 274. *See also* Oxidation reactions; abundance of, 274; allotropes, 177, 274; as diatomic element, 172, 173 *illus.*; dissolved in water, 469 *illus.*; in Earth's crust, 666 *illus.*; electron configuration, 97 *illus.*, 243 *table*; energy levels, 76 *illus.*; as essential element, 126; from fractional distillation, 274, 352–353; in human body, 666 *illus.*; hydrogen bonding by, 438; industrial uses of, 275 *illus.*; ionic compounds formed by, 133 *lab*; properties, 124; reactions with metals and nonmetals, 274, 274 *table*; reaction with hydrogen, 125 *illus.*, 198 prob., 216, 705; reactivity, 261, 274; water vs., 436 *table*
Ozone, 177, 274, 495

P

Pacemakers, lithium batteries in, 595
Pacific yew tree, 29, 30
Paints, 471
Palladium, 290
Paper chromatography, 22 *lab*, 310 *lab*, 324, 326–327 *lab*
Paper-making, 263 *illus.*
Parent acids, 518 *lab*
Parent bases, 518 *lab*
Particle accelerators, 100, 767 *illus.*
Particle generators, 767 *illus.*
Pascal, Blaise, 376
Pascal (Pa), **376**
Pastes, 471
Pauling, Linus, 301, 305
PCBs, 641 *illus.*
Pectin, 471
Pennies: composition of, 36–37 *lab*; creation of copper and zinc alloy with, 25 *lab*; electroplating of, 609 *illus.*; modeling radioactive decay with, 748–749 *lab*; representing isotopes with, 61 *lab*; surface tension and, 443 *lab*
Pentane, 180 *table*, 628
People in Chemistry. *See also* Career Connection
Peptide bonds, 670
Percent yield, **420**; improving in chemical synthesis of sulfuic acid, 424–425

Perchloric acid, 498 *table*
Period 2, chemical reactivity in, 259–261
Periodic, Word Origin, 94
Periodicity, 88, 92
Periodic law, **92**
Periodic table of elements, 27, 90–91 *illus.*, 92, 848–849. *See also* specific groups; atomic numbers, 64; atomic radius (size) patterns, 257, 260 *lab*; atomic structure and, 93–94, 96 *lab*, 96–97, 241–245; classes of elements on, 100–103; development of, 84–89, 87 *lab*, 92; electron configuration trends, 241–245, 243 *table*, 248; electronegativity trends, 302–303; information in each block of, 65 *illus.*; ionic charge trends, 154–155; ionic size patterns, 258–259; long version of, 102 *illus.*; melting point trends, 255 *lab*; metallic and nonmetallic characteristics and, 98–99 *lab*; modern version of, 92; orbital size pattern, 248, 249 *illus.*; physical states of elements and, 100; reactivity of period 2 and, 259–261; synthetic elements and, 100; valence electrons and, 229
Periods, **94**; atomic structure of elements in, 96–97
Permanent waves, 655
Permeable membranes, 468
Peroxides, 275 *illus.*
Perspiration, 446 *illus.*
Pesticides, tracking with tracers, 765
PET, 657
Petroleum, 143 *illus.*, 636, 637–639; cracking and, 638–639; fractional distillation of, 637–638; reforming and, 638; Word Origin, 637
pH, **501**–504, 502 prob.; of blood, 531, 532–533; buffers and, 531–533, 535; of common materials, 503, 503 *illus.*; of cosmetics, 505; development of pH scale, 505; of household products, 479 *lab*; interpretation of pH scale, 502–503; measuring with acid-base indicators, 504, 506–507 *lab*; protein denaturation and, 671; strong acid plus strong base reactions and, 522; strong acid plus weak base reactions and, 526; weak acid plus strong base reactions and, 528–529
Pharmaceutical production worker, 635
Pharmaceutical technician, 635
Pharmacist, John Garcia, 634–635
Pharmacologist, 635
Pharmacology, 634–635
Phenolphthalein, 540 *illus.*

Platinum group

Phenylalanine, 669, 669 *table*
Pheromones, 642 *illus.*
pH indicators. *See* Acid-base indicators
Phosphate group, 271, 686
Phospholipid bilayer, 683
Phosphorescence, 740 *illus.*
Phosphoric acid, 30 *table*, 498, 815 *table*
Phosphors, 292
Phosphorus, 271; allotropes, 173, 272 *illus.*; biological importance, 271; electron configuration, 243 *table*; as essential element, 126; in human body, 666 *illus.*
Phosphorus pentachloride, 212
Photochemical smog, 495
Photography: developing and printing pictures, 565 *illus.*; oxidation-reduction (redox) reactions and, 564; silver bromide in film, 277 *illus.*, 564; Word Origin (photograph), 564
Photosynthesis, 41 *illus.*, 122, 689, **729**–733; Calvin cycle, 731; chlorophyll in, 264; endothermic reactions in, 41, 732; light reactions, 730–731; net equation for, 730; redox reactions in, 575
Photovoltaic cells, 724
pH scale, 501–504; development of, 505; interpreting, 502–503
Physical changes, **20**, 34. *See also* Changes of state; separation of mixtures by, 20
Physical processes, 20
Physical properties, **20**, 32–34; composition of a penny and, 36–37 *lab*; kitchen chemicals, 18–19 *lab*; melting and freezing points, 33; polar vs. nonpolar molecules, 330
Physical state: in chemical equations, 193
Physics Connection: Aurora Borealis, 71; Bohr, Neils, 230; solid rocket booster engines, 566
Pie graphs, 811
Pig iron, 284 *illus.*
Pistons, 372 *illus.*, 373
Plant-care specialists, 210–211
Plants: cellulose in, 673; medicines from, 144; nitrogen fixation by, 271, 571; photosynthesis and, *see* Photosynthesis
Plasmas, 33, **345**
Plaster of paris, 165, 166
Plastics, 658; recycling of, 657; thermoplastic, 658; thermosetting, 658
Platen, 108
Platinum, 290, 711
Platinum group, 290

Plutonium, 100, 293, 760
Plutonium-238, 293
pnp-junctions, 110
Polar covalent bonds, 308–309; paper chromatography and, 310 *lab;* properties of compounds having, 309; water and, 309
Polarity, 328–329, 330; of ammonia, 329; chromatography and, *see* Chromatography; lack of in carbon dioxide, 329; of water, 309, 328, 330
Polar molecules, 328–330; ammonia, 329; attraction to other polar molecules, 330; properties, 330–331; water, 309, 328
Pollution: air, *see* Air pollution; nuclear reactors and, 760; tracking with tracers, 765
Polonium, 274
Polyatomic ions, 156–157; charge of, 157, 856 *table;* common, 157 *table,* 815 *table;* formulas for, 157, 159 prob., 160 prob.; naming, 157
Polyester, 643 *illus.,* 648
Polyethylene, 654, 657, 726
Polyethylene terphthalate (PET), 657
Polymerization reactions, 654, 656; addition reactions, 654; condensation reactions, 656; cross-linking and rubber, 656; plastics and, 658
Polymers, 647–648, 652–653, 656, 658–689. *See also* Plastics; comparing natural and synthetic, 652 *lab;* natural, 649–651 *lab,* 652; plastics, 658; polymerization reactions forming, 654, 656; structure, 653; synthetic, 648, 649–651 *lab;* Word Origin, 648
Polypeptide chains, 670. *See also* Proteins
Polypropylene (PP), 657
Polyprotic acids, 484 *illus.,* 486 *illus.*
Polysaccharides, 679
Polyunsaturated fatty acids, 682
Polyvinylacetate, 654 *illus.*
Polyvinyl alcohol, 38 *lab*
Polyvinyl chloride (PVC), 657
Popcorn, popping of, 395
***p* orbitals**, 239, 239 *illus.,* 240; periodic table and, 241–245
Porcelain, 161
Portable reverse osmosis units, 468
Positional isomers, 631
Positron emission tomography (PET), 764 *illus.*
Potassium, 84, 261; biological uses of, 263 *illus.;* decay of, 64 *illus.,* 746 prob.; in Earth's crust, 666 *illus.;* electron configuration, 245; as essential element, 126; symbol, 27 *table*
Potassium bromide, 304, 518 *lab*
Potassium chloride, 152, 459
Potassium chromate, 291
Potassium dichromate, 291
Potassium hydroxide, 262 *illus.;* reaction with calcium chloride, 213 *illus.;* as strong base, 498, 498 *table*
Potassium iodide, 152
Potassium perchlorate, 74
Potassium tartrate, 30 *table,* 815 *table*
Potential difference, 586, 586 *lab*
Power plants: coal-fired, 722 *illus.;* efficiency of, 727 *table;* nuclear, 759–760
Powers of ten. *See* Scientific notation
PP (polypropylene), 657
Practice Problems: alkanes, formulas and names of, 627 prob.; answers to, 859–868; Boyle's law, 385 prob.; Charles's law, 391 prob.; chemical equations, 199 prob.; converting pressure units, 379 prob.; electronegativity difference between bonding atoms (Δ EN), 310 prob.; food calories, 717 *table;* formulas for binary ionic compounds, 156 prob.; formulas for polyatomic ions, 160 prob.; half-life and rate of decay, 755 prob.; heat of reaction (Δ H), 718 prob.; hydronium ion and hydroxide ion concentrations, 502 prob.; ideal gas law and, 419 prob.; ionic and net ionic equations, 521 prob., 524 prob., 529 prob.; molarity, 542 prob.; molarity calculations, 463 prob.; molar mass, 409 prob., 410 prob.; molar volume, 416 prob.; molecular compounds, naming, 179 prob.; nuclear equations, writing and balancing, 746 prob.; predicting the mass of reactants and products, 415 prob.; solution preparation and, 462 prob.; supplemental prob.lems (Appendix B), 815–844; tritration, 542 prob.; word equations, 199 prob.
Praseodymium, 292
Precipitation, chemical reactions and, 189 *illus.,* 213
Precision, 799
Pressure, 341–342; converting units of, 377 prob., 377–378, 379 prob.; of gases, 341–342; gas pressure and volume and, *see* Boyle's law; Haber process and, 214; ideal gas law and, 418–419; measurement of, 374, 375, 376; standard atmosphere, 374; STP (standard temperature and pressure), 393, 393–394 prob.; units of, 376–378; vapor pressure, 355

Pressure cookers, 357
Pressure gauges, 374, 375
Pressure units, 376–378; converting between, 377 prob., 377–378, 379 prob.; equivalent, 377 *table*
Princeton Review. *See* Standardized Test Practice
Private investigator, 13
Products, 190; predicting mass of, 414–415 prob.; Word Origin, 190
Professional opportunities. *See* Career Connection; People in Chemistry
Promethium, 293
Propane, 30 *table,* 143 *illus.,* 180 *table,* 181 *illus.,* 331, 623, 624 *table,* 638, 815 *table*
Propene, 630
Properties of matter, 5, 5 *illus.,* 5–6
Prostratin, 144
Proteases, 220, 673 *illus.*
Proteins, 667, **668**–673; condensation reactions forming, 670; denaturation, 671; enzymes, 672–673, 674–675 *lab;* as fake fats, 684; hydrogen bonding and, 439; monomers of (amino acids), 669; role of in human body, 668; structure of, 669, 670–671
Protic, 484 *illus.*
Protons, 60, 64, 228; discovery of, 60; properties, 65 *table*
Proust, Joseph, 52
PS EPS, 657
***p* sublevels**, 233, 238
***p*-type semiconductors**, 110
Pure substances, 15
Putrescine, 644 *illus.*
PVC, 641 *illus.,* 657
Pyrite, 163 *illus.*

Q

Qualitative expression of composition, 32
Qualitative observations, 14
Quantitative expression of composition, 32
Quantitative observations, 14
Quantities, 791
Quantum theory of chemistry, 305
Quartz, 246
Quinine, 144

R

R (ideal gas law constant), 418–419
Radiation. *See* Electromagnetic radiation

Index

Radiation detectors, 747, 770 *lab*
Radiation exposure, 769
Radiation therapy, 764 *illus.*
Radient heat, 69
Radioactive dating, 752–753; Carbon-14 dating, 750–751, 752–753, 755 prob.; of forged art, 754
Radioactive decay, 743, 745–747; alpha particles, 743; beta particles, 745; gamma rays, 745–746; modeling, 748–749 *lab*; nuclear equations for, 743, 745–746, 746 prob.
Radioactive isotopes: dating methods using, 750–751, 752–753, 755 prob.; half life of, 752, 752 *table*; modeling decay of, 748–749 *lab*; nuclear equations for, 742, 746 prob.; radiation released from, 743, 745–746
Radioactive wastes, 772–773
Radioactivity, **741**; detection of, 747, 770 *lab*; discovery of, 740–742; hazards of exposure to, 768–770, 769 *table*, 769–770; nuclear equations and, 742; penetrating power of, 739 *lab*; radioactive dating and, 750–751, 752–753, 755 prob.; radioactive decay and, 743, 745–747, 748–749 *lab*, 752–753, 755 prob.; radon and, 770 *lab*, 771; waste disposal and, 772–773
Radioisotopes: Hershey and Chase's experiments on DNA with, 766; medical applications of, 763, 764 *illus.*; nonmedical applications of, 765, 767; release of radiation by, 743, 745–747; sources of, 767
Radiosonde, 383
Radio waves, 69, 70 *illus.*
Radium, 264
Radon, 130, 244 *table*, 279, 770 *lab*, 771
Radon-testing kits, 770 *lab*
Rain forests, medicines from, 144
Rare earth metals. *See* Lanthanides
Ration heaters, flameless, 219
Reactants, **190**; limiting, 218; predicting mass of, 414–415 prob., 418 *lab*
Reaction capacity, metals and valence electrons, 234–235 *lab*
Reaction rates, 216–218, 220–221; activation energy and, 216; catalysts and, 220; concentration and, 217; factors affecting speed of, 216–218, 220–221, 703 *lab*; inhibitors and, 221; starch-iodine clock reaction, 218 *lab*; temperature and, 217
Recommended Dietary Allowance (RDA), 126
Recycling, 53, 727; aluminum, 605, 719; conservation of matter and, 53; glass, 58; plastics, 657
Redox reactions. *See* Oxidation-reduction reactions
Red phosphorus, 173, 272 *illus.*
Reducing agent, 559, **562**
Reduction reactions, **556**. *See also* Oxidation-reduction reactions; combining with oxidation half-reactions, 557–558; historic use of term *reduction*, 558; oxidizing and reducing agents and, 559, 562; Word Origin (reduction), 556
Reforming, 638
Respiration, 122, 575, **690**–692, 693 *illus.*; ATP and energy storage and, 691; electron transport chain, 692; glycolysis, 692; photosynthesis and, 733; tricarboxylic acid cycle, 692
Retinal, 632
Reverse osmosis, 467, 468
Reverse osmosis units, 468
Reversible reactions, 208–209, 212–213
Rhodium, 290, 711
Rhodochrosite, 246
Rhodopsin, 632
Ribonucleic acid. *See* RNA (ribonucleic acid)
Ribose, 678, 685
RNA (ribonucleic acid), 271, **685**, 687; nucleotides in, 685–686
Robotic arms, 107
Rocket fuel, 566
Rockets, solid rocket booster engines, 566
Rolling, 289
Roman Empire, lead poisoning and, 270
Roman numerals, in names of transition-element compounds, 162 *table*, 162–163
Rose quartz, 246
Rubber: cross-linking in, 656; vulcanized, 656
Rubbing alcohol (isopropyl alcohol), 138 *illus.*, 356, 642 *illus.*
Rubidium, 261
Rubies, 246
Rust, 39 *illus.*, 120 *lab*, 202 *illus.*, 248, 555, 557, 557 *lab*, 570
Ruthenium, 290
Rutherford, Ernest, 62, 63, 228

Saccharin, 5 *illus.*
Safety guidelines, 845
Safety Handbook, 845–846
Safety symbols, 846
Salicylic acid, 29, 642 *illus.*
Salt (table salt). *See* Sodium chloride
Salt bridge, 588
Salts, **516**
Salty rainfall, 154
Samarium, 293
Sapa, 270
Sapphire, 246
Satellites, 67–68
Saturated fats, 683
Saturated fatty acids, 682
Saturated hydrocarbons (alkanes), **623**–629; isomers, 627–628; naming, 625–626, 626–627 prob.; properties, 629; structural diagrams, 623–625; structural diagrams and names, 626–627 prob.
Saturated solutions, **458**
Scandium, 245
Scanning probe microscope (SPM), 236
Scanning tunneling microscope (STM), 8, 237
Scientific laws, 57
Scientific method, 57
Scientific models, 9
Scientific notation, 801–804, 805–806
Scintigraphy, 764 *illus.*
Scintillation counters, 747
Scrap metal processing worker, 613
Scurvy, 688
Seawater: floatation of icebergs on, 26; as mixture, 16 *illus.*
Second period elements, 242, 243 *table*
Selectively permeable membranes, 467
Selenium, 274, 275 *illus.*; electron configuration, 97 *illus.*; as essential element, 126
Semiconductors, **103**, 109–111; diodes, 110–111; electrical conduction by, 109–110; gallium arsenide in, 272 *illus.*; n-type and p-type, 110; properties of at room temperature, 109; silicon in, 269 *illus.*
Sex hormones, 682
Shampoos, pH of, 505
Shape-memory metals, 106–107
Shielding effect, **302**, 303 *illus.*
Sievert (Sv), **769**
Significant digits, 799–800
Silica, 269 *illus.*
Silicon, 103, 105 *illus.*, 269; doping, 109, 110 *illus.*; in Earth's crust, 666 *illus.*; electron configuration, 243 *table*; as essential element, 126; glass from, 269 *illus.*; uses of, 269 *illus.*; valence electrons, 109 *illus.*
Silicon chips, 111 *illus.*
Silicon dioxide, 246, 269 *illus.*, 343 *illus.*
Silk, 652

Silver, 23, 290; as coinage metal, 84, 108, 290; conductivity, 311; symbol for, 27 *table*; tarnishing, 572
Silver bromide, 28 *illus.*, 277 *illus.*, 564, 565 *illus.*
Silver nitrate, 198–199 prob.
Silver sulfide, 572
Simple sugars. *See* Sugars
Single bonds, 629 *illus.*
Single-displacement reactions, 203, 203 *lab*, 204–205 *lab*, 207 *table*
SI units, 4, 791–793; base units, 791; derived units, 791, 792 *table*; English units and, 794; metric units and, 792, 794; prefixes for, 792–793, 793 *table*
Skawinski, William (chemist), 316–317
Skin care products, pH of, 505
Slag, 284 *illus.*, 567 *illus.*
Slime, creation of, 38 *lab*
Smog, 495
Smoke detectors, 744
Snowflakes, 441 *illus.*
Soap, 455, 494
Soap making, 494
Soap scum, 158, 189 *illus.*
Soda-lead glass, 344
Sodium, 97, 261; as coinage metal, 84; in Earth's crust, 666 *illus.*; from electrolysis, 601, 601 *illus.*, 603–604; electron configuration, 243, 243 *table*; as essential element, 126; properties, 121; reaction with chlorine, 131, 132 *table*, 262 *illus.*; reaction with water, 262 *illus.*; symbol for, 27 *table*
Sodium acetate, 531–532, 532 *illus.*
Sodium azide, air bags and, 417
Sodium bicarbonate, 141 *illus.*
Sodium bromide, 203 *illus.*
Sodium carbonate, 30 *table*, 165, 197, 815 *table*
Sodium chloride, 5 *illus.*, 118–121, 153; crystal structure of, 132–133; electrical conductivity, 119; electrolysis of, 601, 603–604; formation, 118–120, 131, 132 *table*, 516 *illus.*; formula, 30 *table*; ionic bonding, 304; mining, 118 *illus.*; properties, 119–120; solubility of, 451, 459; uses of, 30 *table*, 119 *illus.*
Sodium fluoride, 152, 277 *illus.*
Sodium hydrogen carbonate, 141 *illus.*
Sodium hydroxide, 815 *table*; as deliquescent substance, 164 *illus.*; formula, 30 *table*; production by electrolysis, 603–604; reaction with carbon dioxide, 197; reaction with strong acids, 516 *illus.*, 517–518, 522; reaction with weak acid, 528–529, 529 *illus.*; as strong base, 489, 498, 498 *table*; uses, 30 *table*, 263 *illus.*
Sodium hypochlorite, 192, 567–568
Sodium nitrate, 203 *lab*, 459
Sodium stearate, 158, 455
Sodium sulfate, 456 *lab*
Soft water, 452 *lab*
Soil, 16 *illus.*
Soil conservation technician, 211
Solar energy, 724
Solar stills, 168
Solar wind, 71
Solder, 16 *illus.*, 23 *table*
Solid aerosols, 470
Solid rocket booster engines, 566
Solid rocket fuel, 566
Solids, 33, **338**; kinetic model, 343; mass of, 3 *lab*; melting, 362–363; properties, 338 *illus.*; sublimation, 354, 373 *lab*
Sols, 471
Solubility: of a gas in a liquid, 469; of ionic compounds, 451; "like dissolves like", 454; temperature and, 459–460
Solubility product constants, 857 *table*
Soluble, 213
Soluble compounds, 213
Solutes, 23
Solution, heat of, 460
Solutions, 23. *See also* Aqueous solutions; Colloids; alloys, 23, 23 *table*; boiling-point elevation, 465; concentration, 458–459, 460–461; freezing-point depression, 464–465; of gases in liquids, 469; guidelines for, 857 *table*; heat of, 460; identification of, 456–457 *lab*; "like dissolves like", 454; molarity and, 460–461; observe formation of, 435 *lab*; preparation of different volume of, 462 prob.; saturated, 458–459; standard, 539; unsaturated, 458–459; volume when mixed, 21 *lab*
Solvents, 23
Soot, 470, 637 *illus.*
s orbitals, 238 *table*, 239; predicting distribution in, 244 *lab*
Sørenson, S.P.L., 501
Sour taste, 519
Space shuttle: regulation of air quality in, 201, 416 prob.; solid rocket booster engines, 566
Specific heat, 444, **445**, 445 *table*, 446
Spectator ions, 520, 522
Spectrum. *See also* Electromagnetic spectrum; Emission spectra; Word Origin, 231
Speed of reaction, 216–218
Spices, 16 *illus.*

Spinnerets, 652
SPM. *See* Scanning probe microscope (SPM)
Spontaneous reactions: entropy and, 712–714; predicting, 714 *table*; Word Origin (spontaneous), 585
Squid, 575
s sublevels, 233, 238
Stained-glass windows, 472 *illus.*
Stainless steel, 23 *table*
Stalactites, 525
Stalagmites, 525
Standard atmosphere (atm), 374
Standardized Test Practice, 47, 81, 115, 149, 185, 225, 253, 297, 335, 367, 401, 433, 477, 513, 551, 581, 619, 663, 701, 737, 777
Standard solutions, 539
Standard temperature and pressure (STP), **393**, 393–394 prob.
Starch, 653, 673, 679; as fat replacement, 684; reaction with iodine, 218 *lab*; structure, 653 *illus.*
Starch-iodine clock reaction, 218 *lab*
States of matter, 33, 338–345. *See also* Gases; Liquids; Plasmas; Solids; changes in, *see* Changes of state; periodic table trends, 90–91 *illus.*
Stationary phase, paper chromatography, 310 *lab*
Statue of Liberty, corrosion of, 570
Steam, 436
Stearic acid, 682 *illus.*
Steel, 23, 104 *illus.*, 280 *illus.*, 286–289; alloy steels, 288; carbon steels, 286 *table*, 286–289; classification, 286; earliest known objects made of, 286; galvanized, 291, 555, 555 *illus.*; production, 285 *illus.*; shaping (working), 288–289
Steelmaking, 285 *illus.*, 286–289; alloy steels, 288, 288 *table*; carbon steels, 286 *table*, 286–289; shaping of steel, 288–289
Sterling silver, 23 *table*
Steroids, 682
Stierle, Andrea, 29
Stierle, Donald, 29
STM. *See* Scanning tunneling microscope (STM)
Stoichiometry, 404–405; atoms in a sample of an element, 409 prob.; Avogadro's number and, 405; chemical formulas from, 426–428; formula units in a sample of a compound, 409 prob.; ideal gas law and, 418–419, 419 prob.; mass of a number of moles of a compound, 410 prob.; mass of a reac-

tant, predicting, 414–415 prob.; mass percents, 421, 422–423 lab; molarity from titration, 541, 542 prob., 544–545 lab; molar mass of a compound, 408, 410 prob.; molar mass of an element, 407, 409 prob.; molar volumes and, 415, 416 prob., 418 lab; percent yield and, 420; theoretical yield and actual yield, 420; Word Origin, 407

Stomach, acidity of and antacids, 535–536, 538
Stone Age, 567
Stoney, George, 60
STP. *See* Standard temperature and pressure (STP)
Straight-chain hydrocarbons, 623
Straws, gas laws and, 384 lab
Strong acids, 498, 498 table, 499 table; reactions with strong bases, 516 illus., 517–518, 520, 521 prob., 522; reactions with weak bases, 522–523, 524 prob.; titration curves and, 540 illus.
Strong bases, 497–498, 498 table; reactions with strong acids, 516 illus., 517–518, 520, 521 prob., 522; reaction with weak acid, 528–529, 529 prob.; titration curves and, 540 illus.
Strontium, 264; reactivity, 266–267 lab; uses of, 265 illus.
Strontium hydroxide, 498 table
Structural isomers, 628
Structural proteins, 668
Structure of matter, 7 illus.
Styrene-butadiene rubber, 656
Sublevels, 233
Sublimation, 354, 373 lab
Submicroscopic matter, 7–8
Subscripts: balancing chemical equations and, 197; formulas for binary ionic compounds, 153–154
Substances, 15, 24–25, 27–28, 30–31; compounds, *see* Compounds; elements, *see* Elements
Substituted hydrocarbons, 640–646. *See also specific groups*; production of, 645–646
Substrates, 672
Sucrose, 5 illus., 30 table, 143 illus., 678, 815 table; models of, 8; solubility, 453, 454; structure, 8 illus.; sweetness of, 519
Sugar of lead, 270
Sugars, 6 illus., 34 table, 665 lab, 678–679. *See also* Carbohydrates
Sulfate ion, 156 illus.
Sulfonates, 455
Sulfur, 274; electron configuration, 97 illus., 229 illus., 243 table; in human body, 666 illus.; mining of by Frasch process, 274, 274 illus.; oxidation-reduction (redox) reactions, 563; uses, 275 illus.
Sulfur dioxide, 495
Sulfuric acid, 275 illus., 815 table; in lead-acid batteries, 269 illus., 596; manufacturing, 424–425, 485; steelmaking and, 486 illus.; as strong acid, 498 table; uses, 30 table, 424, 485
Sulfur oxides, acid rain and, 493–494, 495
Sulfur volcanoes, 563
Superplastic steel, 286, 286 table, 287
Supersaturated solutions, 459
Supplemental Practice Problems, 815–844
Surface tension, 442–443, 443 lab
Sutliff, Caroline (plant-care specialist), 210–211
Sweetness, 519, 680
Symbols, element, 27 table, 90–91 illus.
Synthesis reactions, 202, 207 table; recognizing, 204–205 lab
Synthetic aromas, 645 lab
Synthetic chemicals, 29
Synthetic elements, 100
Synthetic polymers, 648; comparing with natural, 652 lab; differentiating between synthetic and natural, 649–651 lab
Synthetic rubber, 656, 656 illus.

T

Table salt. *See* Sodium chloride
Tables, making and using, 810
Table sugar. *See* Sucrose
Tandem accelerator mass spectrometer (TAMS), 751
Tandem Cascade Accelerator (TCA), 767 illus.
Tarnish, 572; Word Origin, 572
Taste, 519, 680
Taste buds, 519
Taxol, 29, 30, 420
Tayag, Fe (cosmetic bench chemist), 490–491
Technetium, 100
Technetium-99m, 764 illus., 767 illus.
Tellurium, 89, 92, 274, 275 illus.
Temperature, 346–348, 795; chemical reaction rate and, 217; gas pressure and, 356, 372–373; gas volume and, *see* Charles's law; Haber process (synthesis of ammonia) and, 214; ideal gas law and, 418–419; kinetic energy and, 346–348, 360–361 lab; measurement of, 795; protein denaturation and, 671, 674–675 lab; rate of mixing and, 337 lab; solubility and, 459–460; standard temperature and pressure (STP), 393
Temperature scales, 347–348, 795; converting between, 348
Terbium, 293
Tetrachloromethane, 640
Tetrafluoroethene, 647–648
Tetrahedral molecular geometry, 315 illus., 320 illus., 321 illus.
Textiles, polymers in, 649–651 lab
Thallium, 268
Theoretical yield, 420
Theories, 57
Thermoluminescence (TL) dating, 751
Thermoplastics, 658
Thermosetting, 658
Thin-layer chromatography, 324. *See also* Chromatography
Thiobacillus ferroxidans, 723
Third period elements, electron configuration, 243 table, 243–244
Thomson, J. J., 59–60, 228
Thomson's atomic model, 61 illus., 228, 228 illus.
Thornton, John, 12–13
Three Mile Island, 760 illus.
Thymine (T), 686, 687
Thymol blue, 543
Thyroid gland, iodine and, 276 illus., 764 illus.
Thyroxin, 641 illus.
Tin, 27 table, 269, 570 illus.
Tin(II) fluoride, 277 illus.
Tire-pressure gauges, 375
Titanic, **cleaning of by electrolysis**, 610
Titanium, 245 illus.; glass and gem color and, 246, 247 table
Titration, 539–541; concentration from, 541, 542 prob., 544–545 lab; how to perform, 540; indicators and, 540; standard solutions and, 539
Tokamak Fusion Reactor, 762 illus.
Tooth decay, prevention of by fluoridation, 277 illus., 278
Toricelli barometer, 374
Torricelli, Evangelista, 374
Toxic wastes: disposal of radioactive, 772–773; electrolysis of, 614
Tracers, radioisotope: Hershey and Chase's experiments on DNA with, 766; medical applications of, 763, 764 illus.; tracking of pollutants by, 765
trans **isomer configuration, 631**; Word Origin (trans), 631
Transistors, 110–111

Index

Transition elements, 92, **101**–102, 104 *illus.*. *See also* Actinides; Lanthanides; atomic size trends, 283; as coinage metals, 290; color of gemstones and, 246–247; electron configuration, 245, 248, 283 *lab*; formulas for compounds containing, 162–163, 163 *illus.*; ionic compounds of, 160, 162–163; multiple oxidation states, 281; naming of compounds having, 162, 162 *table*; oxidation number of, 160, 163; properties, 280–281, 284–285 *illus.*; uses, 284–285 *illus.*, 290–291
Transmutation, 745
Triads, Dobereiner's, 85–86
Triangular pyramid molecular geometry, 320 *illus.*
Tricarboxylic acid cycle, 692, 693 *illus.*
Trichloromethane, 640
Triglycerides, 681
Triple-beam balances, 796
Triple bonds, **323**, 629 *illus.*, 633
Triprotic acids, 484 *illus.*
Tritium (T), **761**
Tungsten: chemical symbols, 27 *table*; as light bulb filament, 282; melting point, 281, 282; Word Origin, 281
Tungsten steel, 288 *table*
Twenty Thousand Leagues Under the Sea, Jules Verne's, 26
Tyndall effect, **472**, 473 *illus.*

U

Ultrahigh-carbon steel (UHCS), 286 *table*, 287
Ultraviolet waves, 70
Unbranched hydrocarbons, 623
Uncertainty principle, Heisenberg's, 238
UNILAC particle accelerator, 100 *illus.*
Units of measurement, 791–797
Unsaturated fatty acids, 682 *illus.*
Unsaturated hydrocarbons, **629**; alkenes, 629–631; alkynes, 633, 633 *table*; aromatic hydrocarbons, 636; food oils and, 630 *lab*
Unsaturated solutions, 458
Uracil (U), 686, 687
Uranium, 293; alpha decay, 746; Becquerel's radioactivity experiment, 740–741; glass and gem color and, 247 *table*
Uranium experiment, Becquerel's, 740–741
Uranium-lead dating, 754
Urea, 667

V

Vacuum tubes, discovery of electrons and, 59
Valence electrons, **76**–77, 94; atomic collisions and, 128; atomic size and, 257; electron configurations and, 241–245; metal-acid reactions and, 234–235 *lab*; in metalloids, 103, 104 *illus.*; in metals, 103, 104 *illus.*; in nonmetals, 103, 104 *illus.*; periodic table trends, 229, 241–245, 248; semiconductors and, 109 *illus.*; sharing of, 301 *illus.*; shielding effect, 302, 303 *illus.*; sublevels and, 233
Valine, 669 *table*
Vanadium: electron configuration, 245 *illus.*; melting point, 281
Vanilla flavor, 644 *illus.*
van Meegeren, Hans, 754
Vaporization, 354–356; of different liquids, 355 *lab*; as endothermic process, 446; heat of, 358–359
Vapor pressure, 355–356
Vegetable fats, 630 *lab*
Vegetable oils, 682, 683
Verne, Jules, 26
Vinegar (acetic acid), 642 *illus.*; buffer solution with sodium acetate, 531–532, 532 *illus.*; carboxyl group on, 642 *illus.*; dissolution of in water, 484; formula, 30 *table*, 815 *table*; glacial, 529; oil and vinegar dressing and, 299 *lab*; pH of, 503; production of hydronium ions by, 484; production of hydronium when dissolved in water, 484; reaction of bones with, 169 *lab*; reaction with baking soda, 39 *illus.*, 190–191, 193, 418 *lab*; reaction with carbonates, 483 *illus.*; reaction with egg shells, 483 *illus.*; reaction with eggshells, 483 *illus.*; reaction with strong bases, 528–529, 529 *illus.*; titration of, 544–545 *lab*; tritration of, 544–545 *lab*; uses of, 30 *table*; as weak acid, 498, 499 *illus.*; Word Origin, 529
Visible light, 69 *illus.*, 70
Vision, vitamin A and, 632
Vitamin A, 632, 688
Vitamin C, 305, 499 *illus.*, 576 *illus.*, 688
Vitamin D, 688
Vitamin E, 576 *illus.*
Vitamins, **687**–688; coenzymes, 688; water- vs. fat-soluble, 688; Word Origin, 688
Volatile substances, 33, 351, 355
Volcanoes, sulfur-producing, 563

Volta, Alessandro, 584, 586
Voltage, **586**
Voltage source, 600
Voltaic cells, **584**, 587–590; carbon-zinc dry cell batteries, 594; lead storage batteries, 596; nicad rechargeable batteries, 597
Voltmeters, 589
Volume: combined gas law and, 392–393, 393–394 prob.; gas pressure and, *see* Boyle's law; gas pressure and temperature and, *see* Charles's law; ideal gas law and, 418–419; law of combining gas volumes and, 394, 396; measurement of, 795; relationship between volume and mass of a gas, 373 *lab*
Vulcanizing, 656

W

Wastewater operator (Alice Arellano), 448–449
Water, 5 *illus.*, 124–125, 436–450. *See also* Aqueous solutions; Ice; Steam; bond angle distortion, 315; changes in state of, 33, 440, 441 *illus.*; condensation, 354, 446; density, 34 *table*, 436–437, 440; dissolution of into hydrogen and oxygen, 40–41, 707, 708, 708 *lab*; distilled, 452 *lab*; electrolysis, 39 *illus.*; electron sharing in, 136–137; evaporation, 446; expansion of in freezing, 441 *illus.*; formation of, 125, 216, 705, 707 *illus.*; formula, 30 *table*, 815 *table*; freeze-melting, 33 *illus.*, 441 *illus.*; hard, 158, 452 *lab*; heat of fusion, 362–363; heat of vaporization, 358–359; hydrates, 164 *lab*, 164–165; hydrogen bonding in, 438–439, 440, 444, 454; intermolecular forces in, 437–438; meniscus formation, 444; methane vs., 330, 330 *table*; modeling shape of, 314–315; osmosis and, 467; polar nature of, 309, 328; properties, 39 *illus.*, 124–125, 330, 330 *table*, 331, 436 *table*, 436–437; regulation of Earth's temperature by, 446; snowflakes, 441 *illus.*; soft, 452 *lab*; as solvent, 23, 449, 451–454; specific heat of, 444–446; states of, 33, 124, 125 *illus.*, 436, 439–440; surface tension, 442, 443 *lab*; as universal solvent, 450; vapor pressure of, 355
Water softeners, 158
Water-soluble vitamins, 688
Water-treatment plants, 447, 467
Wavelength, 69 *illus.*, 231

Waves, 68–70

Waxes, 681. *See also* Lipids

Weak acids, **498**–499, 499 *illus*.; ionization of, 499 *table*; reaction with strong bases, 528–529, 529 *prob*.; reaction with weak bases, 530; titration curves and, 540 *illus*.

Weak bases, **500**; reaction with strong acids, 522–523, 524 *prob*., 526; reaction with weak acids, 530; titration curves and, 540 *illus*.

Weather balloons, 382, 383

Weather predictors, 164 *lab*

Weight, 796–797

White light, spectrum of, 69 *illus*.

White phosphorus, 173, 272 *illus*.

Willow bark, salicylic acid from, 29

Wind energy, 725

Wire, manufacture of copper, 606–608

Wohler, Frederich, 667

Wood, 445 *table*, 636

Wootz steel, 286

Word equations, 191, 198–199 *prob*.

Word Origin: acid, 483; aerobic, 690; allotrope, 173; atom, 51; aurora, 71; barometer, 374; binary, 153; capillarity, 443; chlorophyll, 730; cis-, 631; colloid, 472; combustion, 206, 709; corrosion, 610; distillation, 638; electronegativity, 301; empirical, 426; energy, 707; fission, 759; fusion, 761; heterogeneous, 21; homogeneous, 23; orbitals, 238; osmosis, 467; periodic, 94; petroleum, 637; photograph, 564; polymer, 648; product, 190; reduction, 556; spectrum, 231; spontaneous, 585; stoichiometry, 407; tarnish, 572; trans-, 631; tungsten, 281; vinegar, 529; vitamin, 688

Work, 40

Xenon, 129, 130 *illus*., 279

Xenon hexafluoride, 193, 415 *prob*.

Xerography, 275 *illus*.

X-rays, 70 *illus*., 740, 769, 770 *illus*.

Yeasts, fermentation by, 695, 696 *lab*

Yellow dye #5, chromatography of, 326–327 *lab*

Ytterbium, 105 *illus*.

Yttrium, 292

Z

Zinc, 291; alloy of copper and zinc, 25 *lab;* in carbon-zinc dry cells, 594; electron configuration, 245; as essential element, 126; oxidation-reduction (redox) reactions and, 555–558, 559

Zinc-carbon dry cells, 594

Zinc-copper voltaic cell, 589

Zinc electroplating, 609–610

Zinc iodide, electrolysis and, 134–135 *lab*

Zinc oxide formation, 555–558

Zircon, 290

Zirconium, 290

Credits

Photo Credits

Cover (tl) Steve Allen/Jupiterimages, (tr) Scott Speakes/Corbis, (cl) David Taylor/Science Photo Library, (cr) Victor Habbick Visions/Science Photo Library, (bl) Michael Melford/National Geographic Image Collection, (br) Shinichi Eguchi/Jupiterimages **2** Brand X Pictures/PunchStock **4** (l) Phil Degginger/Alamy (r) VisionsofAmerica/Joe Sohm/Getty Images **5** Matt Meadows **6** (tl) Jack Sullivan/Alamy (tr) Digital Vision/Getty Images (cl) Danilo Calilung/Corbis (cr) Matt Meadows (b) Chip Clark **8** (l) David Scharf/Photo Researchers, Inc., (r) Hewlett-Packard Laboratories/Photo Researchers, Inc. **9** J Richards/Alamy **12 13** Mark Tuschman **14** Michael Keller/Corbis **15** Nikreates/Alamy **16** (t) USDA, (cl) Brand X Pictures/PunchStock, (cr) liquidlibrary/PictureQuest, (b) XenGate/Alamy **17** Robert Farber/Corbis **20** Matt Meadows **21** The McGraw-Hill Companies Inc. **22** Matt Meadows **24** (l) Digital Vision/Getty Images, (r) Steve Hamblin/Alamy **26** Edward Joshberger, U.S. Geological Survey **27** Steve Raymer/Corbis **28** (l) John Cancalosi/Stock Boston, (c) Charles D. Winters/Photo Researchers, Inc., (r) Matt Meadows **29** Plantography/Alamy **32** Matt Meadows **33** Comstock/Corbis **34** Geoff Butler **35** Matt Meadows **36** (l) Mark Steinmetz, (r) Matt Meadows **37** Editorial Image, LLC/Alamy **39** (t br) Charles D. Winters/Photo Researchers, Inc., (cl) Richard Megna/Fundamental Photographs, (cr b) Matt Meadows **40** Charles D. Winters/Science Photo Library **41** (t) Matt Meadows, (b) Ariel Skelley/Corbis **42** Brian Elliott/Alamy **48** Equinox Graphics/Science Photo Library **50** (cl) PhotoLink/Getty Images, (t) Andre Jenny/Alamy, (cr) Digital Vision/PunchStock, (b) Sean Daveys/Australian Picture Library/Corbis **51** Richard Megna/Fundamental Photographs, NYC, **52** (l) The McGraw-Hill Companies, Inc. (r) Dirk Wiersma/Photo Researchers, Inc. **54** Matt Meadows **56** Popperfoto/Alamy **58** (l) Arthur S. Aubry/Getty Images, (r) James L. Amos/Photo Researchers, Inc. **59** Skip Comer **67** Index Stock/Alamy **69** (t) Dr. Parvinder Sethi, (b) Willard Clay/Getty Images **71** Sean White/Design Pics/Corbis **74** Akira Kaede/Getty Images **82** Mike Kemp/Getty Images **84** (tl) Lowell Georgia/Corbis, (tc) Visuals Unlimited/Corbis, (tr) Penny Tweedie/Corbis, (bl) PhotoLink/Getty Images, (bc) Bettmann/Corbis, (br) Phil Schermeister/Corbis **85** Andrew Lambert Photography/Science Photo Library **86** Stamp from the collection of Prof. C.M. Lang, photo by Gary Shulfer, University of Wisconsin, Stevens Point **87** Tetra Images/Alamy **89** The McGraw-Hill Companies, Inc. **92 93** Comstock Images/Jupiterimages **94** Wim Wiskerke/Alamy **95** Basso Cannarsa/AFP/Getty Images **99** Matt Meadows **100** AP Photo/Michael Probst **101** (l) David S. Holloway/Getty Images, (r) Laurie Chamberlain/Corbis **102** Kevin Schafer/Alamy **103** Andrew Brookes/Corbis **104** (t) Uripos/eStock Photo, (c) Jan Suttle/Alamy, (bl) Image Source/Corbis, (br) Martyn Chillmaid/Oxford Scientific **105** (t) Ed Wheeler, (tcl) Robert Nickelsberg/Getty Images, (tcr) Hugh Turvey/Photo Researchers, Inc., (bl) courtesy Texas Instruments Inc., (br) Kari Marttila/Alamy **106** NASA/Science Source **107** Gabe Palmer/Corbis **108** (l) Comstock Images/Jupiterimages, (r) Kirk Yeager/Photographer's Direct **110** Jon-Erik Lido/Photographer's Direct **111** (l) Dinodia Photo Library/Brand X/Corbis, (r) Dynamic Graphics/Jupiterimages **116** Peter Barritt/Alamy **118** (l) Enzo Signorelli/CuboImages srl/Alamy, (r) Luca Trovato/Getty Images **119** (tl) Rita Maas/The Image Bank/Getty Images, (tr) Don Tonge/Alamy, (b) Matt Meadows **120** Tom Grill/Corbis Premium RF/Alamy **121** (t) Lester V. Bergman/Corbis, (c) Charles D. Winters/Photo Researchers, Inc., (b) Andrew Lambert Photography/Photo Researchers, Inc. **122** (l) Richard Megna, Fundamental Photographs, NYC, (r) Charles D. Winters/Photo Researchers, Inc. **123** (t) Cordelia Molloy/Photo Researchers, Inc., (b) Tetra Images/Alamy **125** (t) Image Plan/Corbis, (b) Peter Scoones/Photo Researchers, Inc., **127** Charles D. Winters **128** Masterfile **129** (l) Digital Vision/Getty Images, (r) The McGraw-Hill Companies **132** Manfred Kage/Peter Arnold, Inc. **133** mediacolor's/Alamy **135** Matt Meadows **136** (t) Brand X Pictures/PunchStock, (b) Lawrence Manning/Corbis **136** Rachel Epstein/PhotoEdit **138** Matt Meadows **140** (t) NASA/Peter Arnold, Inc., (b) Bettmann/Corbis **141** (l) Charles D. Winters/Photo Researchers, Inc., (r) Wally Eberhart/Visuals Unlimited **142** Fundamental Photographs, NYC **143** (tl) Steven Mark Needham/FoodPix/Jupiterimages, (tr) Jeffrey Coolidge/Digital Vision/Getty Images, (bl) Paul Katz/Photodisc/Getty Images, (br) David R. Frazier/Photo Researchers, Inc. **144** Alison Wright/Corbis **150** Digital Vision/Getty Images **153** Charles D. Winters/Photo Researchers, Inc. **154** The McGraw-Hill Companies, Inc. **155** Lofman/Pix Inc./Time Life Pictures/Getty Images **158** Martyn F. Chillmaid/Photo Researchers, Inc. **160** (t) Andrew Lambert Photography/Photo Researchers, Inc., (bl) Matt Meadows, (br) Jonathan A. Meyers/Photo Researchers, Inc. **161** (t) Rèunion des Musèes Nationaux/Art Resource, NY, (b) Chinese School/The Bridgeman Art Library/Getty Images **163** Charles D. Winters/Photo Researchers, Inc. **164 165** Matt Meadows **166** (t) Pixtal/SuperStock, (b) Spencer Grant/PhotoEdit **167** Peter Maslej **168** Aaron Haupt **171** Matt Meadows **172** (l) Richard Feldmann/Phototake Inc./Alamy, (r) DLILLC/Corbis **173** Tim Courlas **174** (l) Ragnar Schmuck/Getty Images, (r) Andrew Holbrooke/Corbis **175** (l) Gerhard Gscheidle/Peter Arnold Inc., (r) Peter Maslej **176** Fridmar Damm/zefa/Corbis **177** Robert Essel NYC/Corbis **178** E. R. Degginger/Photo Researchers, Inc. **181** (l) John Kaprielian/Photo Researchers, Inc., (r) Lon C. Diehl/PhotoEdit **186** Lionel, Tim & Alistair/Science Photo Library **188** Deco Images/Alamy **189** (tl) Matt Meadows, (tr) The McGraw-Hill Companies, Inc., (c) Tim O'Hara/Corbis, (bl) Patti McConville/Getty Images, (br) Kayte M. Deioma/PhotoEdit **190** (l) Design Pics/PunchStock, (r) Tony Freeman/PhotoEdit **191** Matt Meadows **192** Masterfile **193** Matt Meadows **194** SuperStock **195** Matt Meadows **196** Richard Megna, Fundamental Photographs, NYC, 1994 **197** Brand X Pictures/PunchStock **200** (l) Brand X Pictures/PunchStock, (r) Jeremy Woodhouse/Getty Images **201** NASA/Roger Ressmeyer/Corbis **202** (t) The McGraw-Hill Companies, Inc., (b) Image Source/Corbis **203 205** Matt Meadows **206** (t) Matt Meadows, (b) VisionsofAmerica/Joe Sohm/Getty Images **208** David McNew/Getty Images **209** Steve Mason/Getty Images **210 211** Ted Corwin **212** Matt Meadows **213** (t) Gregg Otto/Visuals Unlimited, (b) Matt Meadows **214** (t) Robert W. Kelley/Time Life Pictures/Getty Images, (b) Terry Brandt/Grant Heilman Photography **215** James Shaffer/PhotoEdit **216** (t) Motoring Picture Library/Alamy, (b) Comstock Images/Jupiterimages, (inset top) Pixtal/SuperStock, (inset bottom) The Car Photo Library **217** (t) Tina Manley/Business/Alamy, (b) Phil Schermeister/Corbis **218** Aaron Haupt **219** Joe Raedle/Getty Images **220** Matt Meadows **221** The McGraw-Hill Companies **226** Ferrell McCollough/SuperStock **230** Wheeler Collection/AIP/Photo Researchers, Inc., **236** courtesy IBM Corporation, Research Division, Almaden Research Center **237** (t) IBM Research/Peter Arnold, Inc., (b) IBM Research **245** (l) Erich Schrempp/Photo Researcher, (r) Barry Mason/Alamy **246** Siede Preis/Getty Images **247** Peter Yates/Corbis **248** Asia Alan King/Alamy **254** Ben Plewes Conceptual Photography/Alamy **262** (t) Yoav Levy/PhotoTake NYC, (c) Chip Clark, (b) Stephen Frisch/Stock Boston **263** (t) Tony Freeman/PhotoEdit, (c) Fritz Goro/Time Life Pictures/Getty Images, (b) Dominic Burke/Alamy **264** Matt Meadows **265** (t) JG Photography/Alamy, (c) The McGraw-Hill Companies, Inc., (b) Getty Images **266** Matt Meadows **268** Corbis **269** (t) Comstock/PunchStock, (b) Custom Medical Stock Photo **270** C.M.Dixon/Ancient Art & Architecture Collection Ltd **271** Digital Art/Corbis **272** (t) Thomas Hovland/Alamy, (cl) Holt Studios International Ltd/Alamy, (cr) Tim Courlas, (b) Stockbyte/Getty Images **273** (l) Redfx/Alamy, (r) Danilo Calilung/Corbis **274** Farrell Grehan/Photo Researchers, Inc. **275** (tl) Colin Young-Wolff/PhotoEdit, (tr) Tom Pantages/Phototake, (bl) Roncen Patrick/Corbis KIPA, (br) John Maher **277** (t) Brand X Pictures/Punchstock, (cl) Michael Newman/PhotoEdit, (cr) Mark Karrass/Corbis, (b) Colin Hugill/Alamy **278** Adrianna Williams/zefa/Corbis **279** Jodi Jones/ZUMA/Corbis **280** Brendan Smialowski/Getty Images **281** (l) Leslie Garland Picture Library/Alamy, (r) AAA Photostock/Alamy **282** (t) Jupiterimages/Brand X/Alamy, (b) David Wasserman/Brand X Pictures/Jupiterimages **283** Imagemore Co., Ltd./Getty Images **284** (t) SPL/Photo Researchers, Inc., (c) Digital Vision/PunchStock, (b) FogStock/Index Stock **285** (t) Paul Johnson/Index Stock Imagery, (c) Lawrence Lawry/Getty Images, (b) Mediacolor's/Alamy **286** The Robert Elgood Collection, London **287** (tl) Edward Rozzo/Corbis, (tr) Hugh Threlfall/Alamy, (b) Photodisc/Getty Images **288** (t) Comstock Images/Alamy, (b) Image Source/PunchStock **289** (t) Scenics of America/PhotoLink/Getty Images, (b) Design Pics Inc./Alamy **291** (t) Matt Meadows, (b) David Muir/Masterfile **292** PhotoLink/Getty Images **293** S. Wanke/PhotoLink/Getty Images **298** INSADCO Photography/Alamy **305** Kenneth Eward/Photo Researchers, Inc. **306** Creatas/SuperStock **308** Peter Maslej **311** Jupiterimages/Comstock Images/Alamy **313–315** Matt Meadows **316 317** Richard Hutchings/Digital Light Source **319–321** Matt Meadows **322** (l) Peter Maslej, (r) Matt Meadows **323** Matt Meadows **324** (t) Matt Meadows, (bl) Peter Maslej, (br) Sinclair Stammers/Science Photo Library/Photo Researchers, Inc. **325** Dr. Jurgen Scriba/Photo Researchers, Inc. **326 327** Matt Meadows **328** (l) G Brad Lewis/Getty Images, (r) Comstock/PunchStock **329** Kristen Brochmann/Fundamental Photographs **331** Peter Maslej **336** Jeremy Woodhouse/Digital Vision/Getty Images **338** (l) Dr. Parvinder Sethi, (r) TEK Image/Science Photo Library **339** (t) Kerstin Hamburg/zefa/Corbis, (b) Matt Meadows **341 342** Matt Meadows **343** Doug Martin **344** (t) Peter Yates/Corbis, (b) age fotostock/SuperStock **348** (l) David Young-Wolff/PhotoEdit, (r) Charles Smith/Corbis **351** (t) Hugh Threlfall/Alamy, (b) Eitan Abramovich/Stringer/Getty Images **353** Daniel Dempster Photography/Alamy **354** (r) The McGraw-Hill Companies, Inc. **358** Matt Meadows **359** (l) Jupiterimages/Brand X/Alamy, (r) Ken Lucas/Visuals Unlimited **360** Matt Meadows **362** Jody Dole/The Image Bank/Getty Images **363** bobo/Alamy **368** Imagestate Pictor/Imagestate **371 373** Matt Meadows **375** David R. Frazier Photolibrary **376** Scott Gibson/Corbis **378** David Papazian/Beateworks/Corbis **382** (t) Staffan Widstrand/Corbis, (b) Matt Meadows **383** Staffan Widstrand/Corbis **388** Jason Cohn/Corbis **389** Matt Meadows **392** Open Door/Alamy **395** Bruce Dale/Contributor/Getty Images **402** Janusz Wrobel/Alamy **404–407** Matt Meadows **411** (t) Ariadne Van Zandbergen/Alamy, (b) Heini Schneebeli/The Bridgeman Art Library **412** Matt Meadows **417** Don Johnston/Getty Images **418** BananaStock/PunchStock **420** Randy Faris/Corbis **422 427** Matt Meadows **428** Ashley Cooper/Alamy **434** Brand X Pictures/PunchStock **437** LWA-Dann Tardif/Corbis **439** (t c) The McGraw-Hill Companies (b) Brand X Pictures/Punchstock **441** (t) Jim Zuckerman/Alamy, (cl) Organics Image Library/Alamy, (c) Matt Meadows, (cr) Susan Rayfield/Photo Researchers, Inc., (b) Brand X Pictures/Punchstock **442** Pier Munstermanu/Foto Nature/Minden Pictures **443** (t) Geoff Butler, (b) Richard Megna, Fundamental Photographs, NYC **444** Richard Megna/Fundamental Photographs **445** Roberto Soncin Gerometta/Alamy **446** (t) Ableimages/David Harrigan/Getty Images, (b) Neal Preston/Corbis **448 449** Daniel Schaefer **450** Matt Meadows **453** (b) Fundamental Photographs, NYC, (t) Richard Megna/Fundamental Photographs **454** Matt Meadows **455** The McGraw-Hill Companies **457** Matt Meadows **458** Mark Steinmetz/Amanita Pitures **459** Bob Pardue/Alamy **460** Jeff Brass/Getty Images **461** Matt Meadows **464** AFP/Getty Images **465** (t) Kelly-Mooney Photography/Corbis, (b) Matt Meadows **466** Kim Fennema/Visuals Unlimited **469** (t) Charles D. Winters/Photo Researchers, Inc., (bl) image100/PunchStock, (br) Bob Sacha/Corbis **470** (t) Robert Brook/Photo Researchers, Inc., (c) Goodshoot/PunchStock, (b) Masterfile **471** (t) Dana Hoff/Beateworks/Corbis, (c) Craft Alan King/Alamy, (b) Steve Lupton/Corbis **472** Brand X Pictures/PunchStock **473** Kip and Pat Peticolas/Fundamental Photographs **478** Patrick Lynch/Photographer's Direct **480** Gregg Otto/Visuals Unlimited **481** Andrew Lambert Photography/Photo Researchers, Inc. **482** Matt Meadows **483** (l) Matt Meadows, (r) Adam Hart-Davis/Science Photo Library **485** (t) Alex L. Fradkin/Getty Images, (b) IMS Communications Ltd/Capstone Design. All Rights Reserved. **486** Digital Vision/PunchStock **487** liquidlibrary/Jupiterimages **488 489** Matt Meadows **490 491** Mark Tuschman **492** Matt Meadows **493** Andrew Lambert Photography/Science Photo Library **494** (t) Ed Pritchard/Stone/Getty Images, (b) Terry Smith Images' Arkansas Picture Library/Alamy **495** (t) Dr. Parvinder Sethi, (b) Alex L. Fradkin/Getty Images **496** Matt Meadows **497** (l) Tom Pantages, (r) Michael Nicholson/Corbis **500** Matt Meadows **502** (l) Matt Meadows, (r) Andrew Lambert Photography/Science Photo Library/Photo Researchers, Inc. **505** The McGraw-Hill Companies **507** Matt Meadows **508** (l) Stockbyte/PunchStock, (r) Georgette Douwma/Science Photo Library **514** Bob Krist/Corbis **516 518** Matt Meadows **520** Tim O'Hara/Corbis **522** Matt Meadows **525** Christophe Boisvieux/Corbis **526** Matt Meadows **527** Charles D. Winters/Photo Researchers, Inc. **528 530** Matt Meadows **531** Sisse Brimberg/Getty Images **532** Matt Meadows **535** (t) Panoramic Images/Getty Images, (b) Matt Meadows **536** David M. Martin/Science Photo Library **537** Richard T. Nowitz/Phototake **538** Mark Steinmetz **539 541 543 544** Matt Meadows **546** Photo Agency Eye/Amana Images/Getty Images **552** Image Ideas/PictureQuest **554** Roger Ressmeyer/Corbis **555** Mark Burnett/Photo Researchers, Inc. **558** (l) Gary W. Carter/Corbis, (r) Photofusion Picture Library/Alamy **559–561** Matt Meadows **563** Photo Researchers, Inc. **564** Gernsheim Collection, Harry Ransom Humanities Research Center, University of TX, Austin **565** (cw from top) Mark Steinmetz, (2 3 4) Geoff Butler **566** NASA Images/Alamy **567** Patrick Pleul/dpa/Corbis **568** Phototake Inc./Alamy **569** Digital Vision/Alamy **570** PhotoLink/Getty Images **571** Brand X Pictures/Jupiterimages **572** Tim Courlas **573** (t) Warren Faidley/Corbis, (b) FogStock/Alamy **574** Mikael Karlsson/Alamy **575** (t) Angelina Lax/Science Photo Library/Photo Researchers, Inc., (b) Dante Fenolio/Photo Researchers, Inc. **576** Matt Meadows **582** Comstock Images/PunchStock **584** The Print Collector/Alamy **589 590** Matt Meadows **591** Jim Wehtje/Photodisc/Getty Images **594** StudiOhio **595** Tom Pantages **596** Stockbyte Platinum/Alamy **597** Judith Collins/Alamy **603** Tetra Images/Alamy **604** James L. Amos/Corbis **605** (t) Bettmann/Corbis, (bl) Jeff Greenberg/PhotoEdit, (br) M.E. Warren/Photo Researchers, Inc. **606** (t) Visions of America/Joe Sohm/Getty Images, (b) Erik De Castro/Reuters/

Corbis **607** (t) Science VU/AMAX/Visuals Unlimited, (b) Tom Hollyman/Photo Researchers, Inc. **608** (l) Charles E. Rotkin/Corbis, (r) Mark Steinmetz **609** (l) Jupiterimages, (r) Purestock **610** Ralph White/Corbis **611** Dr. Jeremy Burgess/Photo Researchers, Inc. **612 613** Courtesy of Harvey Morser **614** Car Culture/Corbis **620** Masterfile Corporation **622** Blend Images/SuperStock **624** JG Photography/Alamy **627** (l) Matt Meadows, (r) Janet Horton Photography **630** Frank Tschakert/Alamy **633** Tim Street-Porter/Beateworks/Corbis **634 635** Mark Tuschman **637** (l) John Nakata/Corbis, (r) Clark Dunbar/Corbis **639** (l) Photograph by H.E. Malde, USGS Photo Library, Denver, CO, (r) Bryan & Cherry Alexander Photography/Alamy **641** (l) Stockbyte/Digital Vision/PunchStock, (r) SVS/TOMS/NASA **642** (t) Charles D. Winters/Photo Researchers, Inc., (b) Jupiterimages/Creatas/Alamy **643** (t) Andreas Rentz/Getty Images, (c) John A. Rizzo/Getty Images, (b) CC Studio/Photo Researchers, Inc. **644** (t) Clayton Sharrard/PhotoEdit, (b) Berit Myrekrok/Digital Vision/Getty Images **646** D. Falconer/PhotoLink/Getty Images **647** Darren Greenwood/Design Pics/Corbis **648** (tl) Brand X Pictures/PunchStock, (tr) Phil Degginger/Alamy, (c) Lawrence Manning/Corbis, (bl) AP Photo/Al Behrman, (br) AP Photo/Amy E. Powers **650** Matt Meadows **652** (l) Scott Camazine/Alamy, (r) American Fiber Manufacturers Association **653** (l) Purestock/Getty Images, (r) DEA/G. Losito/Getty Images **654** (r) Eddie Gerald/Alamy **655** Lon C. Diehl/PhotoEdit **658** David R. Frazier Photolibrary, Inc./Alamy **659** (l) JGI/Blend Images/Corbis, (r) Medical-on-Line/Alamy **664** Gerry Ellis/Getty Images **666** (l) Corbis, (r) David Young-Wolff/PhotoEdit **667** (l) Charles D. Winters/Photo Researchers, Inc., (c) Jupiterimages/Brand X/Alamy, (r) Digital Art/Corbis **668** (tl) Eric Chen/Alamy, (tr) Jeff Vanuga/Corbis, (bl) Lauren Shear/Photo Researchers, Inc., (br) Dr. David M. Phillips/Visuals Unlimited **671** (l) Michael Lamotte/Cole Group/Getty Images, (r) Dr. Parvinder Sethi **673** (tl) Photodisc/Alamy, (tr) Evans Caglage/Dallas Morning News/Corbis, (b) IT Stock Free/Alamy **674** Mark Steinmetz **675** Matt Meadows **676 677** Photography by Otis Hairston, Jr. **679** (tl) FoodCollection/IndexStock, (tr) FoodCollection/Alamy, (b) Brand X Pictures/Alamy **681** (l) William Nicklin/Alamy **681** (r) Jupiterimages/Creatas/Alamy **682** (t) Jupiterimages/Brand X/Alamy, (b) D. Hurst/Alamy **683** Phototake Inc./Alamy **684** Eric Nathan/Alamy **688** Louise Murray/Alamy **690** Ezra Shaw/Getty Images **691** sumos/Alamy **692** David R. Frazier Photolibrary, Inc./Alamy **702** Digital Vision/PunchStock **706** Phil Degginger/Alamy **709** Matt Meadows **710** (t) Matt Meadows, (b) Dennis Frates/Alamy **712** Matt Meadows **713** Martyn F. Chillmaid/Science Photo Library **717** Tetra Images/Alamy **719** David Young-Wolff/PhotoEdit **723** Dr. Dennis Kunkel/Visuals Unlimited **724** U.S. Department of Energy/Photo Researchers, Inc. **725** (l) Robert Francis/Robert Harding World Imagery/Corbis, (r) Creatas/PunchStock **726** Arctic-Images/Corbis **727** (l) Larry Lee Photography/Corbis Premium RF/Alamy, (r) Jerry Arcieri/Corbis **728** Travel Ink/Getty Images **732** NASA **733** Scientifica/Visuals Unlimited **738** Brand X Pictures/PunchStock **740** Klaus Guldbrandsen/Science Photo Library **741** C. Powell, P. Fowler & D. Perkins/Photo Researchers, Inc. **742** (l) alwaysstock, LLC/Alamy, (r) Lee C. Coombs/Phototake **744** D. Hurst/Alamy **748** Geoff Butler **750** (t) Copper Age/The Bridgeman Art Library/Getty Images, (b) The Natural History Museum/Alamy **751** (l) James King-Holmes/Photo Researchers, Inc., (r) Richard A. Cooke/Corbis **753** (tl) Gyula Konrad/AFP/Getty Images, (tr) Ira Block/Getty Images, (bl) Photograph by Zeresenay Alemseged, (c)2006 Authority for Research and Conservation of Cultural Heritage ARCCH, (br) Georgette Douwma/Photo Researchers, Inc. **754** Museum Boijmans-van Beuningen, Rotterdam **756** vario images GmbH & Co.KG/Alamy **757** Corbis **760** (t) Yann Arthus-Bertrand/Corbis, (bl) Bettmann/Corbis, (br) Chuck Nacke/Time Life Pictures/Getty Images **762** EFDA-JET/Photo Researchers, Inc. **763** Vince Michaels/Stone/Getty Images **764** (t) Centre Jean Perrin/Photo Researchers, Inc., (c) Peter Berndt,M.D.,P.A./Custom Medical Stock Photo, (b) Luca Medical/Alamy **767** (t) Peticolas/Megna/Fundamental Photographs, (bl) Ed Young/Corbis, (br) 2004, Bristol-Myers Squibb Medical Imaging, Inc. All rights reserved. **769** Mediscan **770** (l) Wernher Krutein/Corbis, (r) Brand X Pictures/Jupiterimages **771** Steve Cole/Getty Images **772** (t) Kristopher Grunert/VEER, (b) Frederic Pitchal/Sygma/Corbis **773** published with permission of ITER **778** Tom Grill/Corbis **780** The McGraw-Hill Companies **781 782** Tim Fuller **786 787** The McGraw-Hill Companies **788** (t) Matt Meadows, (c b) The McGraw-Hill Companies **800** Doug Martin **805** Matt Meadows

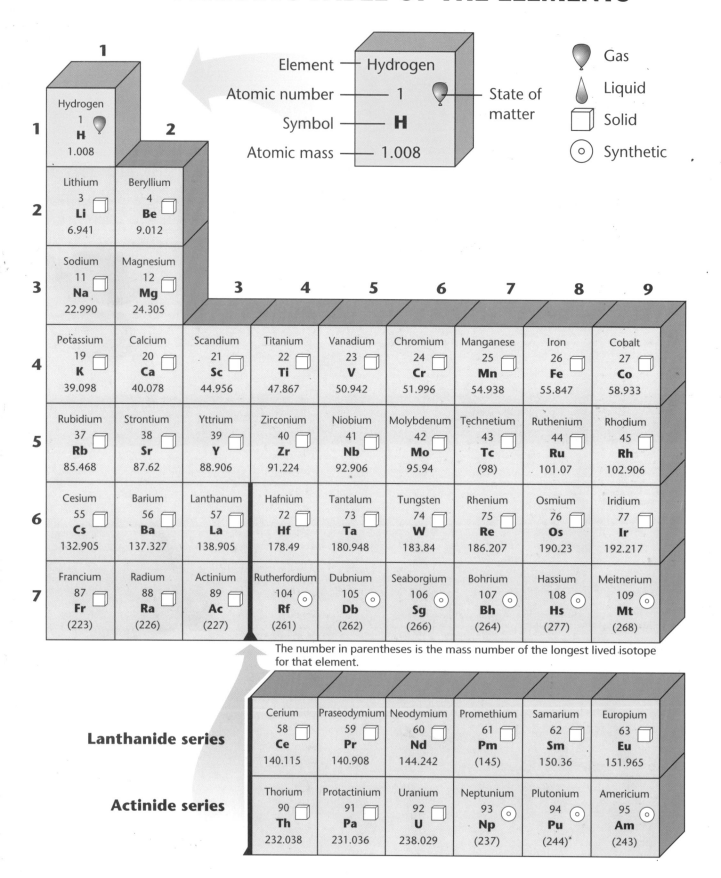